DESIGN AIDS

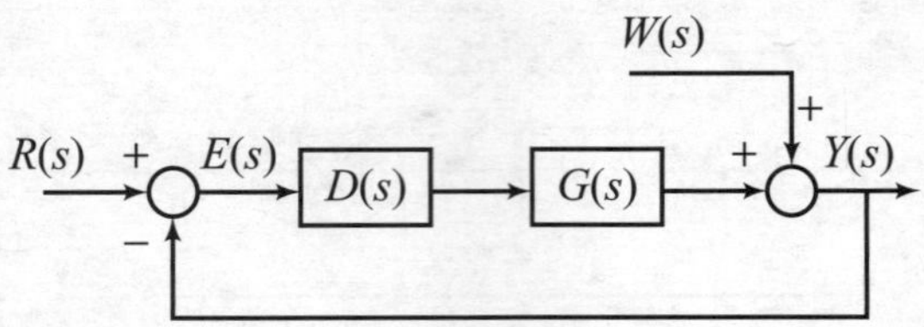

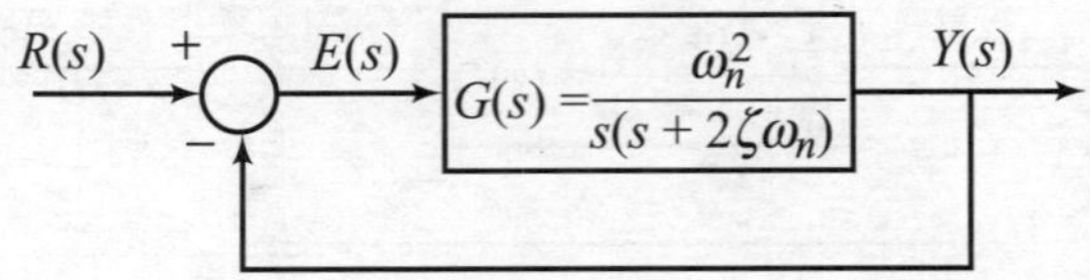

$M(s) = Y(s)/R(s)$; $M_w(s) = Y(s)/W(s)$;

$R(s)$: $1/s$ (unit step); $1/s^2$ (unit ramp); $1/s^3$(unit parabolic)

Final value theorem

$$\lim_{t\to\infty} y(t) = \lim_{s\to 0} sY(s)$$

when $sY(s)$ is analytic on the boundary and right half s-plane

Sensitivity

$$S_G^M = \frac{\partial M}{\partial G}\left(\frac{G}{M}\right) = \frac{1}{1 + D(s)G(s)}$$

$$S_K^M = \frac{\partial M(K,s)}{\partial K}\left[\frac{K}{M(K,s)}\right]$$

Steady-state Response

$$e_{ss} = \lim_{s\to 0} \frac{sR(s)}{1 + D(s)G(s)}$$

$$K_p = \lim_{s\to 0} D(s)G(s);\ e_{ss} = \frac{1}{1+K_p}$$

$$K_v = \lim_{s\to 0} sD(s)G(s);\ e_{ss} = \frac{1}{K_v}$$

$$K_a = \lim_{s\to 0} s^2 D(s)G(s);\ e_{ss} = \frac{1}{K_a}$$

Steady-state error due to $W(s)$

$$= \lim_{s\to 0} sM_w(s)W(s)$$

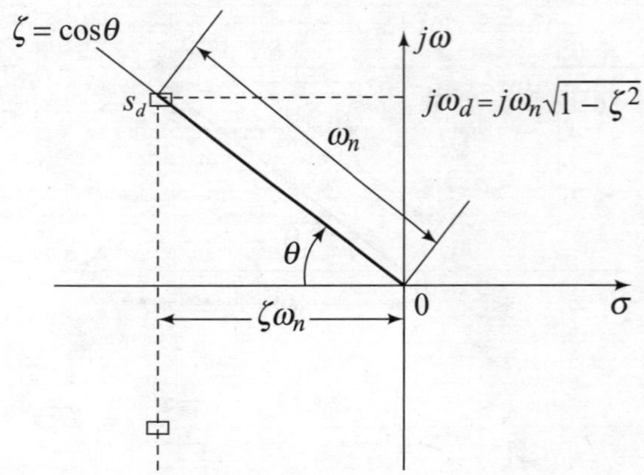

Transient Response

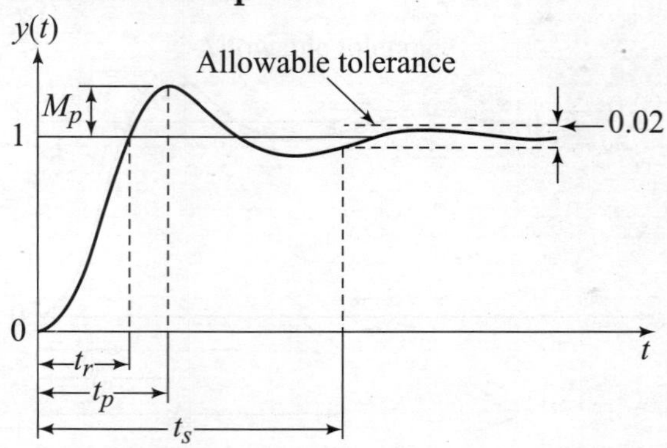

$$t_r = \frac{\pi - \cos^{-1}\zeta}{\omega_n\sqrt{1-\zeta^2}};\ t_p = \frac{\pi}{\omega_n\sqrt{1-\zeta^2}};\ t_s = \frac{4}{\zeta\omega_n}$$

$$M_p = e^{-\pi\zeta/\sqrt{1-\zeta^2}};\ \zeta^2 = \frac{\left[\ln(M_p)\right]^2}{\left[\ln(M_p)\right]^2 + \pi^2}$$

Controllers

PID: $D(s) = K_P + \frac{K_I}{s} + K_D s$

$$= K_c\left(1 + \frac{1}{T_I s} + T_D s\right)$$

Lead: $D(s) = \frac{s + 1/\tau}{s + 1/\alpha\tau}$; $\alpha < 1$

Lag: $D(s) = \frac{s + 1/\tau}{s + 1/\beta\tau}$; $\beta > 1$

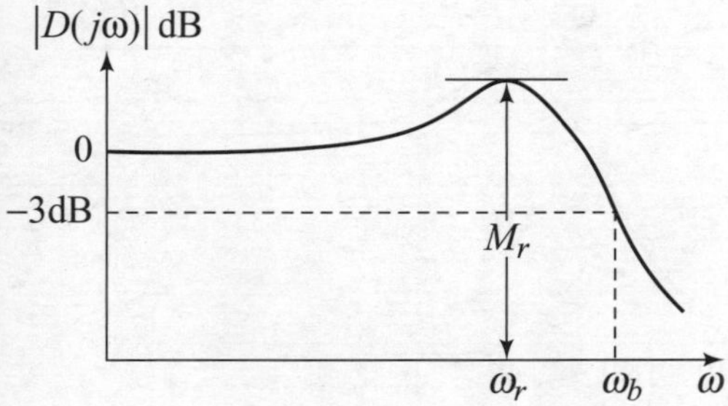

$$M_r = \frac{1}{2\zeta\sqrt{1-\zeta^2}} \; ; \; \omega_r = \omega_n\sqrt{1-2\zeta^2}$$

$$\omega_b = \omega_n\left[\left(1-2\zeta^2\right) + \sqrt{2-4\zeta^2+4\zeta^4}\right]^{1/2}$$

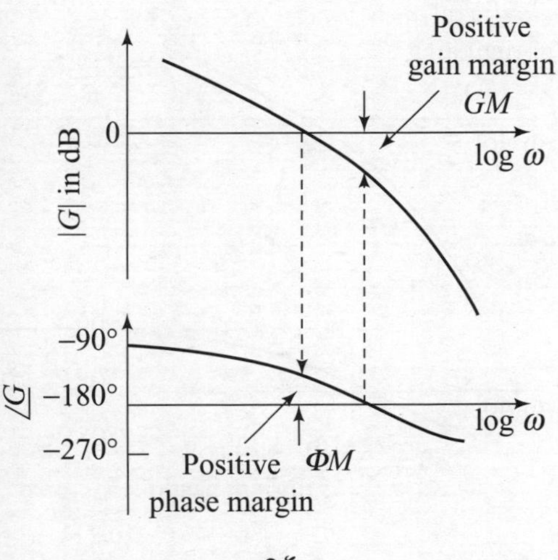

$$\Phi M = \tan^{-1}\frac{2\zeta}{\left\{\sqrt{1+4\zeta^4} - 2\zeta^2\right\}^{1/2}}$$

$$\zeta \cong \frac{\Phi M}{100} \; ; \; \omega_g = \omega_n\left[\sqrt{1+4\zeta^4} - 2\zeta^2\right]^{1/2}$$

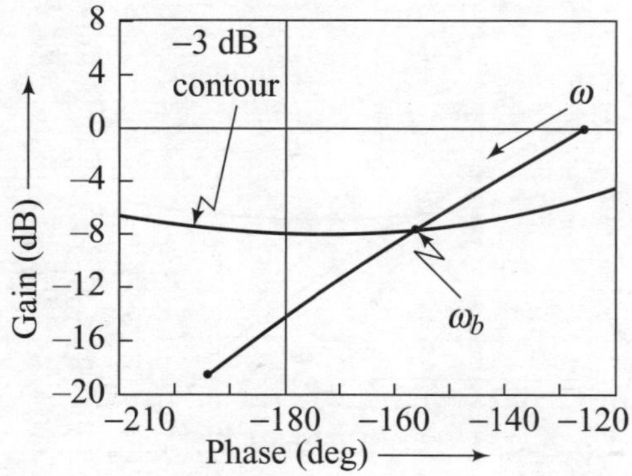

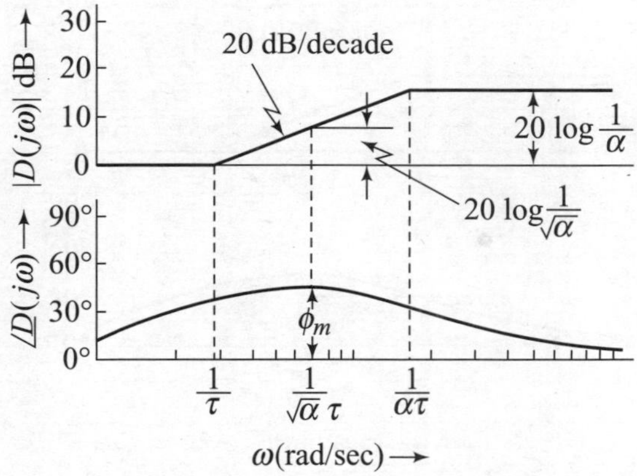

$$\alpha = \frac{1 - \sin\phi_m}{1 + \sin\phi_m}$$

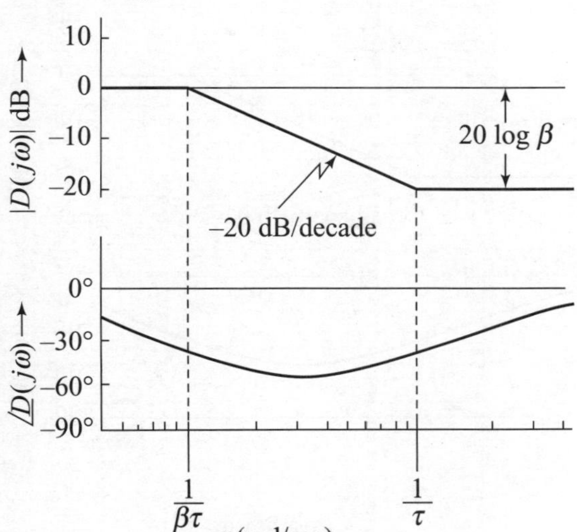

Control Systems

Principles and Design

Fourth Edition

ABOUT THE AUTHOR

M. Gopal, a Professor in Electrical Engineering at Indian Institute of Technology Delhi, is at present the Director of School of Engineering, Shiv Nadar University, Gautam Budh Nagar (U.P.) India. His teaching and research stints span three decades at the IITs.

Dr. Gopal is a globally known academician with excellent credentials as author, teacher, researcher, and administrator. He is the author/co-author of six books on *Control Engineering*. His books are used worldwide, and some of them have been translated into Chinese and Spanish as well. McGraw-Hill, Singapore has published his books for the Asia Pacific market and McGraw-Hill, USA for the US market. In India, his books have been serving as the major source of learning for more than three decades.

As a teacher, his potential is being used globally through a video course (http://www.youtube.com/iit), which is one of the most popular courses on You Tube by the IIT faculty across India. He has been conducting Faculty Development Programs on active learning and effective teaching.

A recognized researcher in the area of Machine Learning, Dr. Gopal has been a key-note speaker in many international conferences. He periodically runs executive programs as tutorial sessions in conferences, and short-term workshops, to empower the participants with the state-of-the-art techniques in pattern recognition and machine learning.

He is the author/co-author of over 135 research papers and his key contributions have been published in high impact factor journals. He has supervised 16 doctoral research projects. His current research interests are in the areas of Machine Learning, Soft-Computing Technologies, Pattern Recognition, and Intelligent Control.

In administrative capacity, he contributed to the growth of a large department at IIT Delhi with electrical (power), electronics, communications, computer technology, and information technology as the areas of activities. The department has acquired a respectable international standing in both teaching and research.

M. Gopal holds B.Tech (Electrical), M.Tech (Control Systems), and Ph.D. degrees from BITS, Pilani.

Control Systems

Principles and Design

Fourth Edition

M Gopal

Professor
Department of Electrical Engineering
Indian Institute of Technology Delhi
New Delhi

McGraw Hill Education (India) Private Limited

Published by McGraw Hill Education (India) Private Limited,
Registered Office: Anjana Complex No: 5/90 A, Butt Road, St. Thomas Mount, Chennai – 600016, Tamil Nadu, India.

Control Systems: Principles and Design, 4/e

22nd reprint 2025

ISBN -13: 978-0-07-133326-9
ISBN -10: 0-07-133326-6

23 24 25 26 27 IBH 29 28 27 26 25

Typeset at Mukesh Technologies Pvt. Ltd., Puducherry, India 605008, and
Printed and bound in India at IBH Print & Bind Pvt. Ltd., Plot No. 81, Ecotech -12, Greater Noida, U.P.

Visit us at: www.mheducation.co.in
Toll free in India: 1800 103 5875
Write to us at: support.india@mheducation.com
CIN: U80302TN2010PTC111532

To
my
Lakshmi

Contents

16 NONLINEAR SYSTEMS ANALYSIS

Preface

Introduction

This text provides an integrated treatment of all those aspects of control systems engineering that prepare the student for early productivity upon entering industrial practice. Important aspects of the subject that help realize this objective are:

- Knowledge of the basic control configurations that have been devised, and the characteristic performance features of each. This helps one in initially selecting one or more alternative configurations that have a potential for success. The vast majority of practical control systems are designed in this way rather than by some mathematical synthesis procedure. This text attempts to provide detailed description of fundamental control modes: proportional, integral, derivative, and specialized control schemes: cascade control, minor-loop feedback control, and state variable feedback.
- Knowledge of the tools of 'control system design toolbox' comprising time-tested design procedures; a particular application needs intelligent selection of tools from this box.

  Major emphasis in this book is on frequency-domain design methods based on transfer function representation of systems. Today, these are useful design tools in the hands of practicing control engineers.
- Familiarity with available hardware so that commercially available components to implement the design can be selected.

  In this book, innumerable realistic (but simple) application examples are included to develop familiarity with available hardware.
- Competence in modeling of physical systems using judicious assumptions.

  With the desire to use linear control design methods (general methods for nonlinear systems do not exist), we neglect some of the nonlinearities. Also many uncertainties are involved in the development of models for designing linear control systems. A robust design performs well for substantial variations in the dynamics of the plant from that used in the design. In this book, we exploit the range and power of feedback to achieve robustness. The industry, by and large, continue to rely on this approach.
- A modern development that is significant to practical design is the wide availability and low cost of Computer-Aided-Design (CAD) software.

  This book attempts to integrate learning of control system analysis and design with learning of how to compute the answers with MATLAB—a widely used CAD software package.

Salient Features of the Book

- Strong emphasis on development of models for practical control systems design; knowledge of approximations made in modeling is crucial in investigation of robustness of the design.
- Thorough introduction to PID Control, the basic building block of industrial controllers.
- Frequency response models introduced early in the text to mirror their importance in industrial practice.

- Abundance of solved examples, review examples, review questions, multiple-choice questions, and problems with final answers, to give the reader enough and graded practice.
- MATLAB/Simulink-based problem solving integrated with pen-and-paper practice through 16 chapter-wise MATLAB Modules given in web supplements of the book.
- Web supplements of the book include downloadable version of solutions manual for instructors only, and files of figures from the book to be used in presentation materials.

Chapter Organization

The contents of the book are organized into 16 chapters and 1 appendix. The appendix is given in the web supplement of the book. It consists of sixteen MATLAB modules, one for each chapter.

In **Chapter 1**, introduction to the control problem is given. A rich variety of practical problems are placed as examples which serve the dual purpose of (i) showing the full power of control theory, and (ii) defining basic structures of feedback, feedforward, and multivariable control systems.

The studies of control systems rely to a great extent on applied mathematics. For the study of classical control theory (frequency-domain design methods based on transfer function representation of systems), which represents a good portion of this text (Chapters 1–12), the required mathematical background includes Laplace transforms and Fourier transforms.

Chapter 2 provides a review of these topics, emphasizing those elements most needed in subsequent chapters. With this background, the concepts of transfer function models and frequency response models are introduced; and then related tools: block diagram and signal flow graph models are described. The chapter also covers dynamic response and steady-state response as used in control.

Chapter 3 deals with the mathematical models of different devices currently in use in industrial control practice. Models of typical motion control (control of speed or position of an inertia load) and process control (control of temperature or pressure of a process fluid) systems are also included.

Chapter 4 initiates a discussion on robustness and performance issues in control systems with feedback. Real systems are taken as examples. Effects of feedback on performance of uncertain systems are investigated quantitatively. Sensitivity studies in frequency domain included in this chapter build the foundation for design.

Chapter 5 deals exclusively with the stability problem of feedback control systems. It introduces the concepts of BIBO (bounded-input bounded-output) and asymptotic stability, and presents Routh stability criterion for investigation of stability. The chapter highlights the fact that the most important use of the Routh's method is in *design* and not in *analysis*.

The first step in formulating the design problem is taken in Chapter 5: finding ranges of the controller parameters for which the feedback system stays stable.

The second and the last step in formulating the design problem is taken up in **Chapter 6**: specifying the desired system performance. A number of performance measures for the speed of response, relative stability, and steady-state accuracy are introduced; and robustness issues examined.

Chapter 7 gives a thorough introduction to PID Control, the basic building block of industrial controllers. It introduces the characteristics of each of the terms P, I, and D; and presents closed-loop analysis for different combinations of these terms. It also includes controller tuning methods—an 'experimental' design procedure; and implementation aspects of the PID controller using op amp based networks.

The knowledge-base created through Chapters 1–7 provides a solid foundation for initiating the design process. Classical methods of design rely heavily on graphical tools. One of these tools, the root locus is the subject of **Chapter 8**. This chapter presents a good discussion of the fundamental definition and creation of the root locus. The root locus empowers us with qualitative and quantitative information about the stability and transient response of feedback control systems.

Chapter 9 extends the concept of root locus to the design of compensators. Cascade compensation (lead/lag) and minor-loop feedback compensation schemes are used. These compensation schemes are, in fact, PID control schemes with some additional features.

Frequency response methods are an alternative to the root locus for analyzing and designing feedback control systems. Chapters 10–12 deal with these methods. **Chapter 10** presents the fundamental definition and creation of Nyquist and Bode plots. Nyquist stability criterion and robust stability measures—the gain margin and the phase margin—are introduced in this chapter. **Chapter 11** gives control system performance specifications in the frequency domain. Correlation between time- and frequency-response is also established. **Chapter 12** uses frequency-response plots—Bode plot and Nichols chart—for compensator design. Lead/lag and PID compensation are covered in detail. It is demonstrated that the frequency-response and root-locus methods complement each other. Examples showing how the root locus and the Bode plot change with compensation for the same system, are given.

After considering the classical control theory for continuous-time systems, the book makes a smooth transition in **Chapter 13** to the digital control systems by introducing the concept of sampler and hold, and *z*-transforms. It is shown that the methods of analysis and design for digital control systems run parallel to those for analog control.

Chapter 14 deals with state variable methods for analyzing continuous-time systems. After a quick introduction of basic matrix theory, this chapter goes to state variable modeling. Then the gamut of concepts that the state-space approach brings into the analysis are explained in detail: the invariance property, the companion forms, importance of Eigen values, solution of state equations, controllability, observability, etc. A short description of these concepts is then given for discrete-time systems; the state space approach being quite similar in the two cases.

Chapter 15 treats the design of both analog and digital control systems in state space. It begins with the pole-placement problems, followed by the design of state observers. Finally, quadratic optimal control problem is introduced.

The book concludes with a chapter on nonlinear systems. **Chapter 16**, the final chapter, introduces some commonly used analysis techniques (describing function, phase plane method, and Lyapunov stability) which take account of system nonlinearities.

Web Supplements

The text is supported with online supplements for instructors and students. Instructor resources are available through *https://connect.mheducation.com* while student resources are available through *https://www.mheducation.co.in* (see the back cover for more details).

For Students

- 16 MATLAB Modules, one for each chapter, given in the Appendix.

For Instructors

- The Solutions Manual for the exercise problems given at the end of all chapters in the book.
- Also included are downloadable files of figures from the book to be used in presentation materials.

Acknowledgements

I would like to acknowledge the contributions of faculty, students and practicing engineers across the country, whose suggestions through previous editions have made a positive impact on this new edition. In particular, the considerable help and education I have received from my students and colleagues at Indian Institute of Technology Delhi, deserves sincere appreciation.

I also express my appreciation to the reviewers who offered valuable suggestions for this fourth and previous editions. The reviews have had a great impact on the project.

Finally, I would like to thank McGraw Hill, and its executives for providing professional support for this project through all phases of its development.

Generous participation of instructors, students, and practicing engineers to eliminate errors in the text (if any), and to refine the presentation will be gratefully acknowledged.

M. Gopal
mgopal@ee.iitd.ac.in

Publisher's Note

The response received from the readers in the last three editions has been overwhelming. In keeping with the standards and the quality, it will encourage us to receive your comments/ compliments/ ideas on this present edition. Please write to us at *support.india@mheducation.com* by mentioning the title and the author's name along with your feedback. Report of any piracy related problems will be highly appreciated, both by the author and the publisher.

INTRODUCTION TO THE CONTROL PROBLEM

1

Preview

In this chapter, our aim is to seek answers to questions like, why is control important? what is the significance of feedback? In addition, our objective is to learn the terms commonly used in control system engineering and the basic features and configurations of control systems. A rich variety of practical problems are placed as examples which serve the dual purpose of (a) showing the full power of control theory, and (b) defining the basic structures of control systems.

1.1 CONTROL SYSTEMS: TERMINOLOGY AND BASIC STRUCTURE

In recent years, control systems have assumed an increasingly important role in the development and advancement of modern civilization and technology. Practically, every aspect of our day-to-day activities is affected by some type of control system. A home-heating system, an automatic washing machine, a refrigerator, an air conditioner, and an automobile are all examples of control systems. Control systems are indispensable in modern industrial processes. We find control systems in all sectors of the industry, such as quality control of manufactured products, automatic assembly line, machine-tool control, space technology and weapon systems, transportation systems, power systems, robotics, and many others. Even such problems as inventory control, and socio-economic systems control may be approached from the theory of feedback control. In this book, we will deal with the control of *engineering systems* that are governed by the laws of physics and are therefore called *physical systems*.

In control parlance, the system to be controlled is given various names: *process*, *plant*, and *controlled system* being perhaps the most common. In the so-called process industries (chemicals, petroleum, steam power, fuel, etc.), one repeatedly encounters the need to control temperature, flow rate, liquid-level in vessels, pressure, humidity, chemical composition, and the like. Such applications are generally considered as *process control applications*. Historically, the wide practical application of control first took place in the process area. Most of the basic concepts were developed and brought to successful realization by the intuitive and experimentally oriented engineering methods typical of the 1900 to 1940 era.

Around the time of the Second World War, the technical needs of military systems like automatic airplane pilots, gun-positioning systems, radar antenna control systems, and the like, led to more scientific approaches in the control engineering field. A comprehensive mathematical theory; the theory of *servomechanism*[1]—aimed mainly at applications wherein the need to control mechanical motions (position, velocity, or acceleration), was developed around 1940 to 1950. Since then, the design methods of the process-control area, and the servomechanism area have gradually converged.

---

[1] The word *servomechanism* originated from the words 'servant' (or slave) and 'mechanism'.

The terminology of the two areas is also converging. The word 'process' is now in use for all types of controlled systems. Also, irrespective of the area of application, the word 'servomechanism' (or *servo system*) is used for a *command following system* wherein the controlled attribute of the system is required to follow a given command.

Figure 1.1 shows the input–output configuration of a process (or plant). Process *outputs* are the response variables which we require to behave in some specified fashion. Process *inputs* are flows of energy and/or material that cause the process to react or respond. The inputs are classified into *manipulated inputs* (subject to our control) and *disturbance inputs* (undesirable and unavoidable effects beyond our control, generated from outside the process environment, and from within). The presence of the disturbance is one of the main reasons of using the control. Judicious management of manipulated variables so as to counteract the effects of disturbances is the primary role of the *controller*. Figure 1.2 shows the input–output configuration of a *control system*.

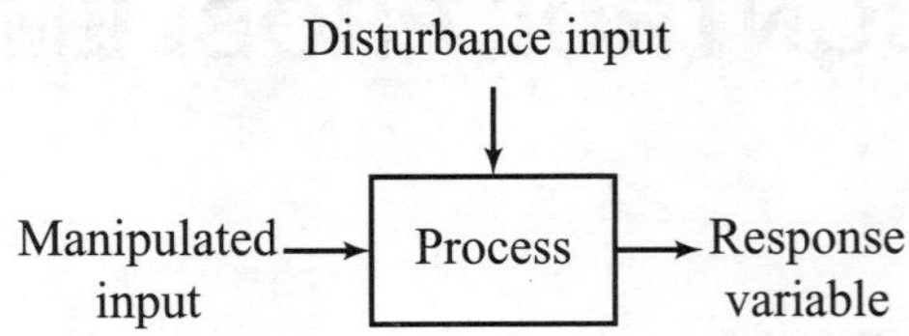

Fig. 1.1 *Input–output configuration of a process*

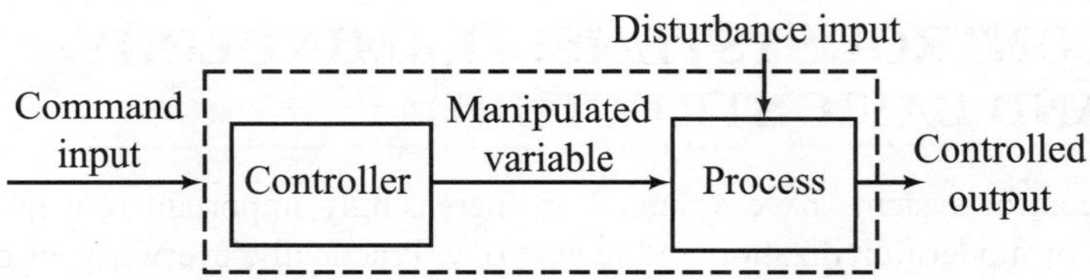

Fig. 1.2 *Input–output configuration of a control system*

When the desired value of the controlled output is more or less fixed and the main problem is to reject disturbance effects, the control system is sometimes called a *regulator*. The command input for a regulator becomes a constant and is called a *setpoint*, which corresponds to the desired value of the controlled output. The set-point may, however, be changed in time from one constant value to another, and the need for the control of manipulated variables arises from the requirements of both the setpoint changes and the disturbance rejection. The control problem is then called *resetting control problem*.

In the *follow-up* or *tracking system*, the controlled output is required to follow or track a time-varying command input. For such systems, the need for the control of manipulated variables arises from the requirements of command following as well as disturbance rejection.

In the configuration of Fig. 1.3, the controller receives 'information' about the desired value of the controlled output and uses this information as a means of control of the manipulated variables. In contrast to this configuration, the configuration of Fig. 1.4 utilizes measurement of the controlled output in order to compare the actual output with the desired output response. The controller then uses this

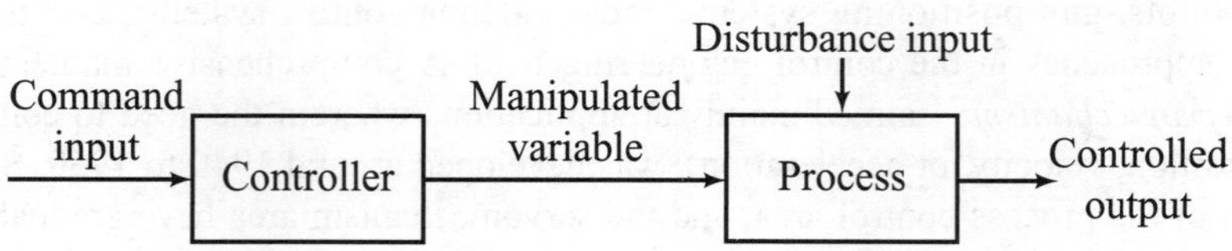

Fig. 1.3 *Input–output configuration of an open-loop control system*

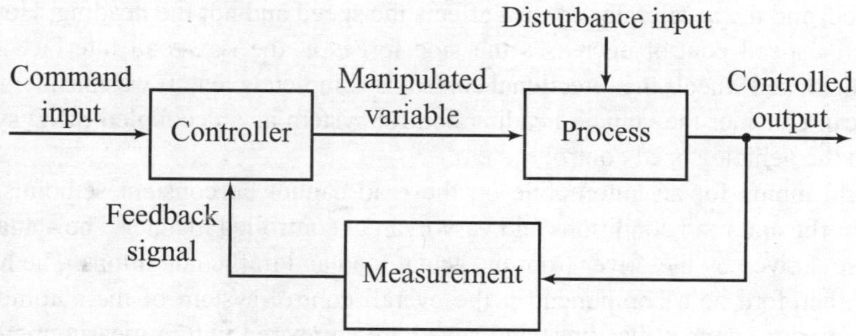

Fig. 1.4 *Input–output configuration of a closed-loop control system*

difference as a means of control of manipulated variables. The configuration of Fig. 1.3 is that of an *open-loop control system* while that of Fig. 1.4 is that of a *closed-loop control system* or *feedback control system*.

To make these definitions more concrete, let us consider some familiar examples of control systems.

Example 1.1 Automobile driving system

The system to be controlled has two inputs (steering and acceleration/braking) and two controlled outputs (heading and speed). The two command inputs are the direction of the highway and speed limits with traffic signals. A block diagram of this two-input, two-output system is shown in Fig. 1.5.

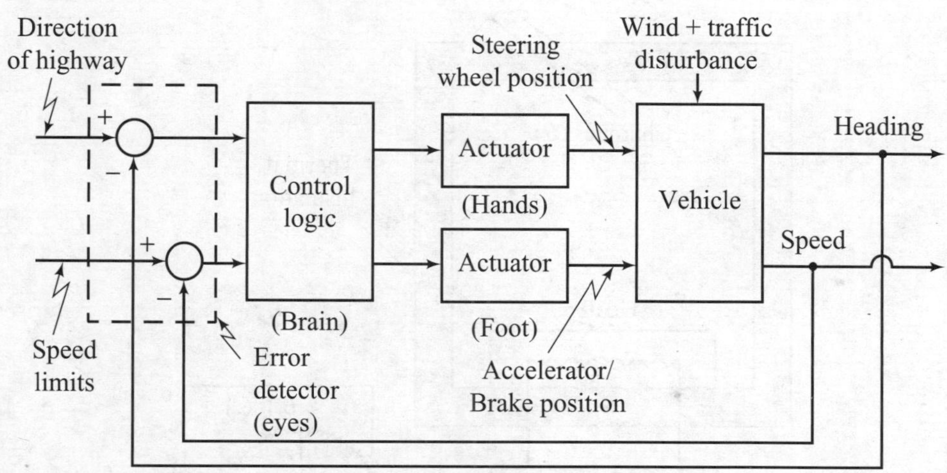

Fig. 1.5 *Automobile driving control system*

Systems with more than one controlled output and command input are called *multivariable* or *multi-input, multi-output* (MIMO) systems. On the other hand, in *single-input, single-output* (SISO) systems, a single output is controlled by a single input. In multivariable systems, an input that is meant to control a particular output also affects the other controlled outputs. This coupling is called *interaction*.

The automobile-driving system is a MIMO system. We can *decouple* this system into two SISO systems for the purpose of design because the interaction is negligible. Steering control affects the heading and not the speed, and the accelerator control affects the speed and not the heading. However, braking of the vehicle for speed control decreases the side forces at the tyre-road interface for directional control, and with locked wheels the directional control is completely lost. If we consider this interaction negligible, we can consider the vehicle-heading control system as a decoupled SISO system, with no interaction with the vehicle-speed control system.

The command inputs for an automobile on the road cannot be constant setpoints. These inputs depend on the traffic and road conditions and vary in an uncontrolled manner. The actual signal inputs to the system are derived by the driver from the actual road and traffic conditions. The human operator subsystem will therefore be a component in the overall control system of the automobile as shown in Fig. 1.5. The desired course (heading) and speed are compared with a measurement of the actual course and speed in order to generate a measurement of the error. The human operator then controls the manipulated variables (steering wheel position, accelerator/brake position) in a manner (control logic) which reduces the absolute error.

Example 1.2 A heating system

To this point, we have discussed in very general terms about the purpose of control and the open-loop and closed-loop control strategies. To further understand these concepts, let us consider the example of a heating system for a chamber, used for testing manufactured products for their thermal sensitivity. In this system, the chamber temperature, $\theta(°C)$ is the response variable of interest (controlled output), and the manipulated variable is the input power (kW) to heat the chamber. In the scheme of Fig. 1.6a, the desired

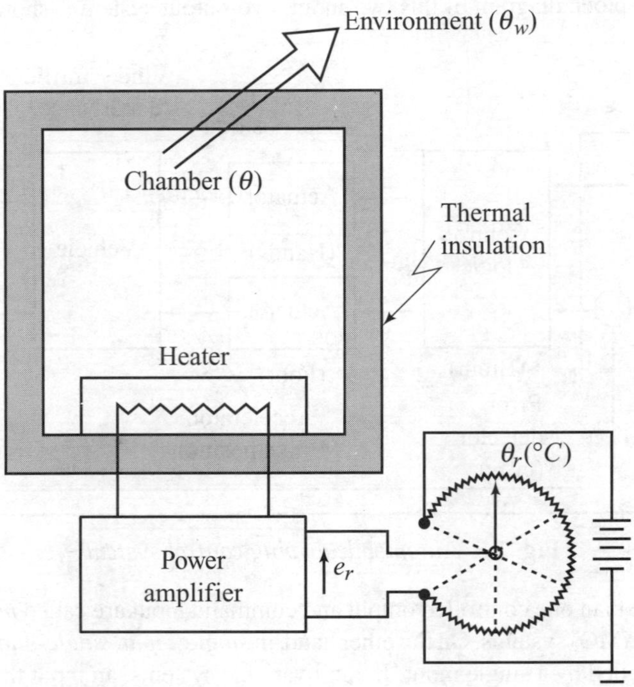

(a)

Fig. 1.6 *(Continued)*

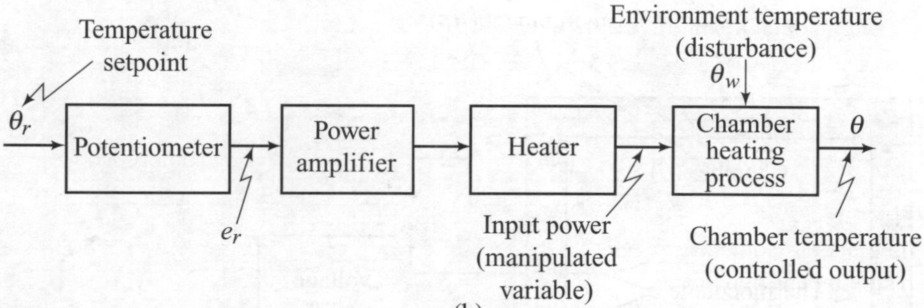

(b)

Fig. 1.6 *Open-loop temperature control system*

temperature, θ_r, which is the command input, is set on a calibrated dial. This positions the potentiometer wiper that applies a voltage signal, e_r (volts) to the power amplifier, which supplies the current to drive the heater element inside the chamber. The chamber temperature is thus, controlled by the command-input setting. It is also affected by the disturbance inputs; the main disturbance is the environment temperature, θ_w, which is external to the chamber. The chamber walls are not made of perfect insulator, and therefore some heat is lost through the walls. The amount of heat flow through the walls depends on the temperature differential that exists across the walls and the thermal properties of the wall material. Figure 1.6b shows the functional block diagram of the open-loop temperature control system.

The dial is calibrated when the environment temperature has certain value: each setting corresponds to the power required to regulate the chamber temperature as per command θ_r, in the face of *fixed* heat loss to the environment. If the environment temperature remains constant at the value it had at the time of calibration, then the open-loop temperature control scheme of Fig. 1.6 would give satisfactory performance. However, the environment temperature does fluctuate, and variations in this temperature result in a *fluctuating* heat loss which, for a fixed dial setting, attempts to change the chamber temperature. When the environment temperature changes significantly, the controlled temperature will deviate from the desired value by a large error, and hence, precise control will not be realized. The open-loop control scheme of Fig. 1.6 is used when performance requirements are not very stringent. ***In the open-loop scheme, there is no automatic adjustment of any errors in output temperature that may occur. We have to live with the consequences or redesign/recalibrate whenever major errors appear.***

The way to realize precise control is to replace the open-loop control scheme of Fig. 1.6 by a *closed-loop control scheme*, that can act on the information received on-line about what is going wrong. By measuring the process output (chamber temperature), we can find out how the process is behaving. Variations in the measured temperature show up the effects of command changes and disturbances. This information can be passed electronically to a comparator, which generates a measurement of error. The error signal then alters the input voltage to the power amplifier, and hence the heat input to the chamber. By doing this, we are *feeding back* information about the current situation; we are *closing the loop*.

Figure 1.7 gives one such feedback control scheme. For temperature measurements, we have selected a sensor which has, in addition to the basic device such as a thermocouple, a voltage amplifier to amplify weak signals from thermocouple. The output voltage, e_b, of the sensor is, say, $K\theta$ (i.e., $e_b = K\theta$), where θ is the chamber temperature and K is a constant. The desired temperature θ_r is converted into a reference voltage signal e_r; to make it compatible with e_b for comparison, $e_r = K\theta_r$. There is a comparator to generate the difference between the reference signal e_r and the feedback signal (sensor output) e_b.

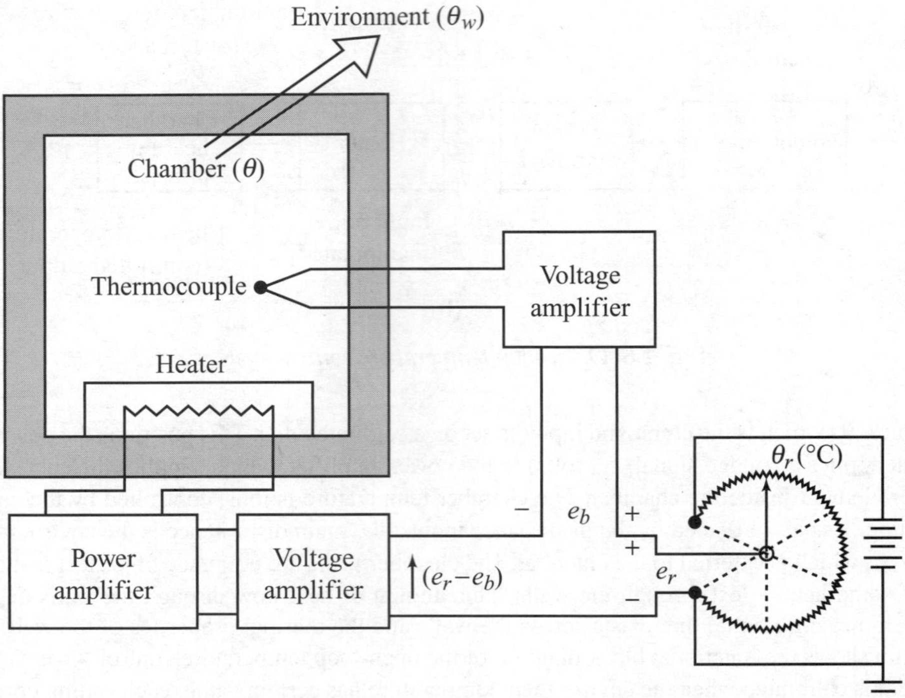

Fig. 1.7 *Closed-loop temperature control system*

The difference voltage from the comparator is amplified and applied to a power amplifier which supplies the current to drive the heater element inside the chamber.

The input heat causes the temperature θ as well as the feedback voltage e_b to increase (e_r is fixed for a given command input), and the difference signal $(e_r - e_b)$ to decrease (as a consequence of negative feedback). As the difference signal approaches the zero value, the drive to power amplifier is cut off and the supply of heat to the chamber is also cut off. As a result, the temperature inside the chamber corresponds to the reference signal e_r (i.e., $\theta = \theta_r$).

The thermal equilibrium within the chamber just established does not persist for long. The reason is the heat loss to the environment. It is a matter of time, then, before sufficient heat is lost through the walls, causing the temperature within the chamber to drop. When this happens, e_b will decrease and the increased difference signal will act on the power amplifier, which will, in turn, supply current to the heater element. The heater element will continue to supply heat to the chamber until its original temperature is restored, thus causing the error signal $(e_r - e_b)$ to become zero again and the heater to be cut off.

Thus, ***through the use of feedback, the system monitors the output signals of the process (which show up the effects of the disturbances) and takes corrective actions. The feedback control system provides automatic control of the chamber temperature***—it maintains the chamber temperature at the desired value, or at least close to the desired value. Figure 1.8 shows the functional block diagram of a closed-loop temperature control system.

Comparison of Figs 1.5 and 1.8 gives many common features of the two closed-loop systems. However, the two systems differ clearly in one aspect. In Fig. 1.5, both the feedback of the controlled outputs (actual course, and speed of travel) for comparison with command inputs (desired course,

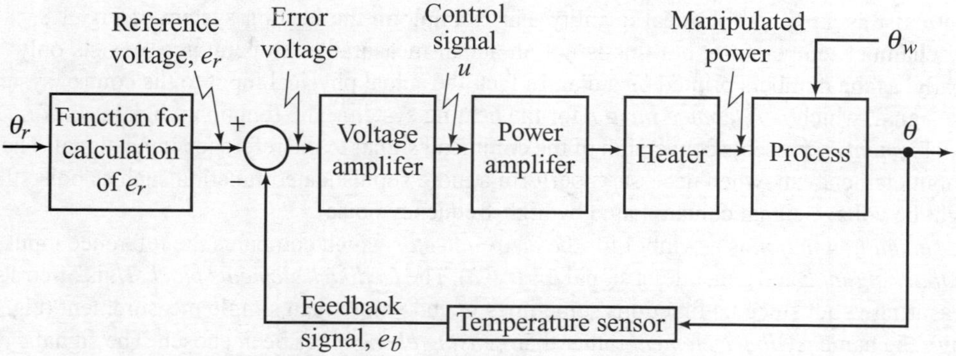

Fig. 1.8 *Functional block diagram of the system of Fig. 1.7*

and speed of travel), and the control actions (rotation of steering wheel, and acceleration/braking) take place through the human operator. In Fig. 1.8, the human operator has been replaced by automatic error-detecting and correcting devices; both feedback of the controlled output (actual chamber temperature), and the correcting action (control of the input power) take place without the need of a human operator. Such systems, therefore, are *automatic closed-loop control systems*. The systems with human operators included into the closed loop are *man-machine control systems*.

1.1.1 Basic Structure of a Feedback Control System

The familiar examples of feedback control cited earlier, fit into the basic structure shown in Fig. 1.9.

To understand Fig. 1.9, we begin with the *command signal* y_r—the desired value of the *controlled variable* y. This signal has the same units as the controlled variable (a physical quantity), but may or

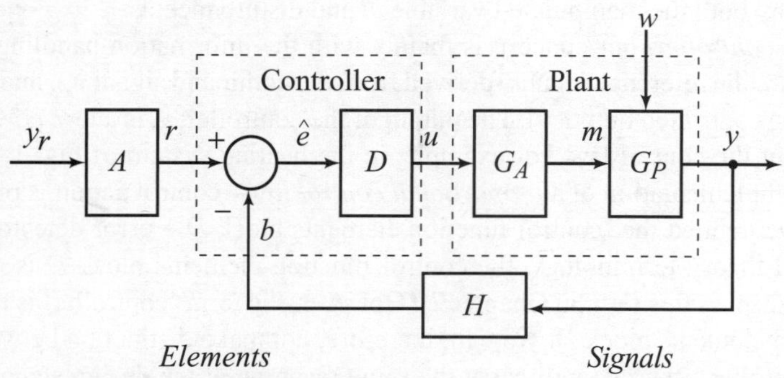

Elements		*Signals*	
A	: Reference function elements	y_r	: Command input
D	: Control function elements	r	: Reference input
G_A	: Actuator elements	$\hat{e}$	: Actuating error signal
G_P	: Controlled system elements	u	: Control signal
H	: Feedback elements	m	: Manipulated variable
		w	: Disturbance input
		y	: Controlled variable (output)
		b	: Feedback signal

Fig. 1.9 *Basic structure of a feedback control system*

may not exist as an actual physical quantity. For example in the heating system of Fig. 1.8, y_r is the desired chamber temperature, but this is not an actual measurable temperature; it exists only 'mathematically' as the numbers printed on a dial. In fact, the actual physical input to the control system is a voltage signal, which is *reference input r* for the heating system. The functional block, A, of *reference function elements* represents conversion of the command signal to the reference input signal. The reference input element can, when necessary, perform a more sophisticated function such as noise filtering (y_r might be voltage signal contaminated by high-frequency noise).

The *summing junction* is a symbol for the *error detector* which compares the reference input r with the *feedback signal b* and generates a signal $\hat{e} = (r - b)$. The *feedback elements block, H*, is often a sensor for measuring y, but since its functions sometimes include more than simple measurement (e.g., noise filtering), the name *feedback element* rather than *sensing element* has been chosen. The signal $\hat{e}$ defines the *actuating error signal*, and not the *system error signal*, which is logically defined as $e = (y_r - y)$. Note that r, b and $\hat{e}$ always have exactly the same dimensions.

The *control function elements block, D*, produces the *control signal u* which has the knowledge about the desired control action. The *controller* is generally thought of as a system that compares the reference input with the feedback signal and is also responsible for a suitable control action. In general, the control signal is a function of the error signal, which in turn is a function of the command input y_r and controlled output y; the functional relationship may therefore be expressed as $u = f(y_r, y)$. The primary objective set for this book is to study this functional relationship called the *control law*.

The *actuator elements block, G_A*, (final control elements) consists of devices that develop enough torque, pressure, heat, etc., (*manipulated variable m*) to influence the system variables under control. Thus, the actuator elements are the 'muscle' of a control system while the control function elements are the 'brain'.

Disturbance inputs w are the variables over which the designer has no control, and perhaps little information is available on their magnitude, functional form, or time of occurrence. The controlled variable y is influenced by both the manipulated variable m and disturbance w.

When we speak of a *controller*, our concern is mainly with the information handling devices; the sources of information are the reference input r (derived from the command signal y_r), and the feedback signal b (produced by the controlled output y). The output of the controller, u, is a low-power signal that carries information about the control law. For example, in the heating system of Fig. 1.8, the control signal u is the result of implementation of a *proportional control law*—control action is proportional to the error signal. We have defined the 'control function elements block D + error detector' as controller (Fig. 1.9). In control theory terminology, the control function elements block D is referred to as 'compensator' (we will discuss this term in Chapter 7). However, the term 'controller' is also in use for only the control function elements block D. We will, therefore, not make distinction between the terms 'compensator' and 'controller'; the two will carry the same meaning at the design stage. Appropriate additional functions will be added at the implementation stage.

The actuator, G_A, and the controlled process itself, G_P, are power handling devices. In a large number of practical situations, the controlled process and the actuator together, are treated as a single subsystem (plant). This is because of *loading effects*—we will discuss this aspect of control-system structure in the next chapter.

The structure shown in Fig. 1.9 defines the basic types of signals and components necessary for the description of any feedback control system. However, it must be adapted to the needs of each specific design. Alternative feedback structures will be introduced in Section 1.3.

1.2 THE GENESIS AND ESSENCE OF THE FEEDBACK CONTROL THEORY

From the simple and familiar examples of automatic control systems presented in Section 1.1, conclusions can be derived regarding the characteristics common to these systems, notwithstanding their manifestable heterogeneity. Automatic control systems share the following general features:

1. The principle of operation of the system uses the feedback idea, i.e., there exists a closed loop: error → manipulated variable → controlled variable → error.
2. The primary objective of the system is error self-nulling.

In all practical control systems, the objective of error self-nulling is required to be realized with a prescribed accuracy (steady-state errors, speed of response, etc.). A demand on accuracy, as we shall shortly see, may result in the loss of *stability*.[2] ***The reconciliation of the requirements of satisfactory accuracy and adequate stability margin is the principal accomplishment of the feedback control theory***. Detailed quantitative study of the effects of feedback on control system performance will be taken up in Chapter 4. The objective of the brief qualitative study in this section is to introduce the basic principles of the feedback control theory.

A brief consideration of the origins of the feedback control theory will be in order here. Examples of feedback control systems have been identified dating back to the third century BC. Ktesibios, a Greek living in Alexandria, is credited with building a self-regulating flow device for use in water clocks. Several examples of feedback control systems appear in windmill designs used by the Dutch in the fifteenth century.

The arrival of the machine age was accompanied by a large increase in the number of control system designs. The development of the steam engine led to the requirement for a speed control device to maintain constant speed in the presence of changes in load torque or steam pressure. In 1788, James Watt of Glasgow developed his now famous flyball governor for this purpose. The principle of the flyball governor is still used for speed control applications.

Analytical tools for control problems were first developed by J C Maxwell in 1868. Maxwell utilized a linear differential equation approach and derived stability conditions for a third-order system. In 1876, Edward John Routh was able to determine the stability conditions for higher-order systems. A solution to the stability problem was also obtained by Adolf Hurwitz and reported in 1895. Routh and Hurwitz were unaware of each other's work. In 1911, equivalence of the Routh and Hurwitz criteria was established.

From 1900 to 1940, significant developments occurred in large-scale power generation, aeronautics, chemical industries, and electronics. The development of the vacuum-tube amplifier, in particular, resulted in many analytical techniques for the design of feedback systems.

One of the key problems in electronics, especially since its application to long-distance telephony, was to develop amplifiers with good gain stability. The difficulty stemmed from the considerable scattering of characteristics in the mass-produced electronic tubes and from changes in characteristics taking place in time (e.g., due to emission loss). In 1927, Harold S Black discovered the negative feedback amplifier as a solution to this problem. Figure 1.10a shows the idea of operation.

[2] Stability is a notion that describes whether the system will be able to follow the input command (Chapter 5). In a nonrigorous sense, a system is said to be unstable if its output is out of control or increases without bound when the input is within bounds.

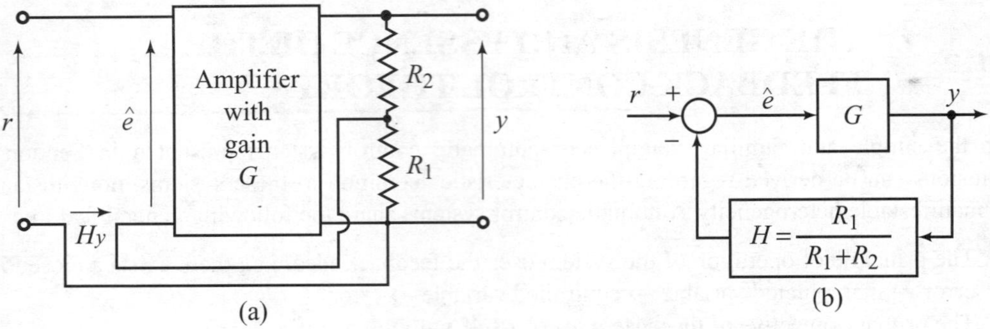

Fig. 1.10 *Amplifier with feedback*

A potential-dividing resistor delivers a part, Hy, of the amplifier's output voltage y to the amplifier's input, where it is subtracted from input voltage r. Figure 1.10b gives the equivalent block diagram of the system. It follows from this diagram that

$$y = G(r - Hy)$$

The closed-loop gain: $$M = \frac{y}{r} = \frac{1}{\frac{1}{G} + H}$$

This equation, specifying the relationship between the closed-loop gain M, the open-loop gain G, and the feedback gain H, is crucial for the feedback theory which was developed in electronics (the studies of Harry Nyquist and Hendrik W Bode in the 1930s) and which relied heavily on the then widely fostered frequency methods of analysis. The most important observation from this equation which determines feedback performance is that

$$M \cong \frac{1}{H} = \frac{R_1 + R_2}{R_1} \quad \text{for} \quad \frac{1}{G} << H \quad \text{or} \quad GH >> 1$$

where GH denotes the feedback-loop gain. It follows that under the conditions of strong feedback ($GH >> 1$), the closed-loop gain M does not depend upon the open-loop gain G; it depends only on the potential-dividing resistor. The gain M is practically constant even if gain G varies to a considerable extent, as long as it remains sufficiently high. Thus, by making use of feedback, it has become possible to convert an amplifier with very high but unsteady gain into an amplifier with a lower but steady gain. Furthermore, higher the gain G, more steady is M. On the other hand, an excessive increase of G results in the loss of stability (saturation of the output voltage sets in rapidly; the amplifier changes into a periodic oscillator). The reconciliation of the requirements of satisfactory performance and adequate stability margins was first the central problem and later the principal accomplishment of the feedback control theory.

During World War II, electronics engineers noticed an analogy between the feedback amplifier and the servomechanism of a radar command post, both in structure and principle of operation, and also with respect to the goal and the basic problem, i.e., reconciliation of the requirements of accuracy and stability. The feedback theory was then successfully applied to the design of servomechanisms, becoming more advanced in the process, and, in this way the *theory* of *servomechanisms* came into being.

The application of the theory of servomechanisms to the process industry led to the *feedback control theory*. At about the same time, Norbert Wiener founded *cybernetics* as the new field of study which transplanted elements of the feedback theory to economics, biology and other sciences.

Frequency-response methods have dominated the control systems area since the late 1940s as analysis and design tools. The Nyquist and Bode diagrams became indispensable for assessing the closed-loop stability of feedback control systems. Evans' root-locus method brought the complex variable-based approach to a fully developed state by 1948. A mature *classical* control systems engineering area thus evolved.

Further advances in control systems engineering resulted from the space programme and associated challenges—launching, manoeuvring, guidance, and tracking of space vehicles. The emergence of a *modern* control theory based on state-space ideas is generally traced back to the early 1960s.

After this brief consideration of the origins of the feedback control theory, let us focus our attention on the requirement of reconciling accuracy and stability, which makes the existence of a uniform feedback control theory fully justifiable. We have already considered one example—the electronic amplifier, wherein strong feedback results in insensitivity to variations in parameters of the electronic tube. There is, however, an upper limit on the improvement of performance beyond which stability suffers. In the following pages, we consider more examples.

Example 1.3 Servomechanism for steering of antenna

One of the earliest applications of radar tracking was for anti-aircraft fire control; first with guns and later with missiles. Today, many civilian applications exist as well, such as satellite-tracking radars, navigation-aiding radars, etc.

The radar scene includes the radar itself, a target, and the transmitted waveform that travels to the target and back. Information about the target's spatial position is first obtained by measuring the changes in the back-scattered waveform relative to the transmitted waveform. The time shift provides information about the target's range, the frequency shift provides information about the target's radial velocity, and the received voltage magnitude and phase provide information about the target's angle[3] [1, 2].

In a typical radar application, it is necessary to point the radar antenna towards the target and follow its movements. The radar sensor detects the error between the antenna axis and the target, and directs the antenna to follow the target. The servomechanism for steering the antenna in response to commands from radar sensor is considered here. The antenna is designed for two independent angular motions, one about the vertical axis, in which the azimuth angle is varied, and the other about the horizontal axis in which the elevation angle is varied (Fig. 1.11). The servomechanism for steering the antenna is described by two controlled variables—the azimuth angle β, and the elevation angle α. The desired values or commands are the azimuth angle β_r, and the elevation angle α_r, of the target. The feedback control problem involves error self-nulling, under conditions of disturbances beyond our control (such as wind power).

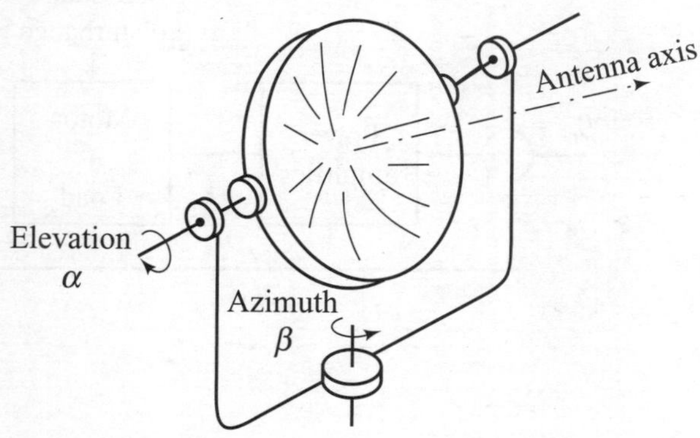

Fig. 1.11 *Antenna configuration*

The control system for steering of the antenna can be treated as two independent systems—the azimuth-angle servomechanism, and the elevation-angle servomechanism. This is because the

[3] The bracketed numbers are keyed to the list of references given at the end of the book.

interaction effects are usually small. The operational diagram of the azimuth-angle servomechanism is shown in Fig. 1.12. The steering command from the radar sensor which corresponds to the target azimuth angle is compared with the azimuth angle of the antenna axis. The occurrence of the azimuth-angle error causes an error signal to pass through the amplifier, which increases the angular velocity of the servomotor in a direction towards an error reduction. In the scheme of Fig. 1.12, the measurement and processing of signals (calculation of control signal) is digital in nature and is based on proportional control law. Figure 1.13a gives the functional block diagram of the control system. A simple model of the load (antenna) on the motor is shown in Fig. 1.13b. The moment of inertia J and the viscous friction

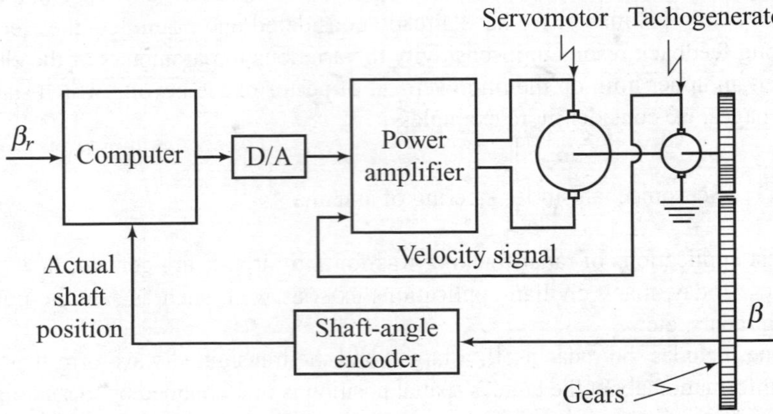

Fig. 1.12 *Azimuthal servomechanism for steering of antenna*

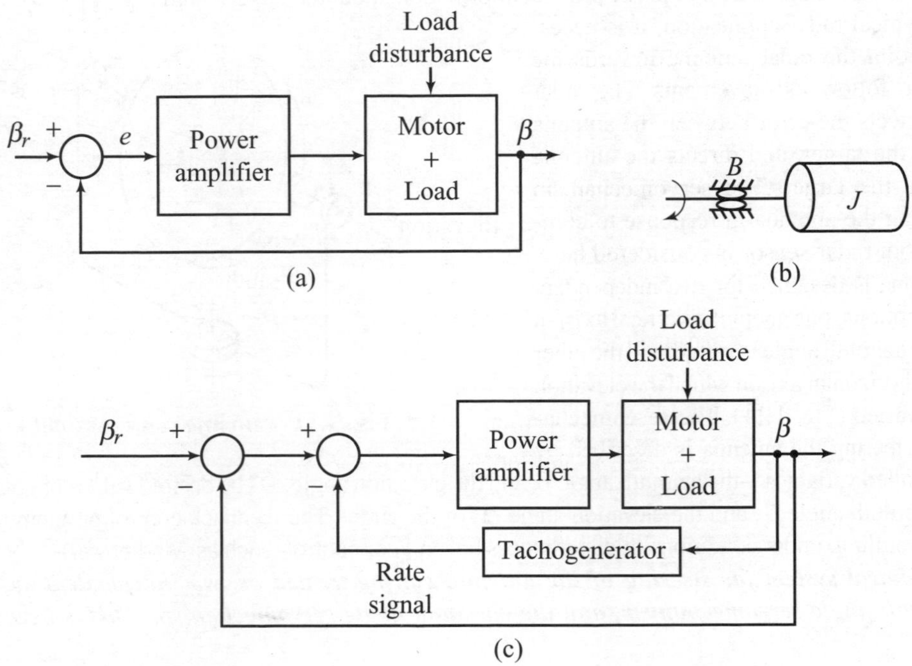

Fig. 1.13 *Functional block diagram of azimuthal servomechanism*

coefficient B are the parameters of the assumed model. The nominal load is included in the plant model for the control design. ***The main disturbance input is the deviation of the load from the nominal estimated value as a result of uncertainty in our estimate, effect of wind power, etc.***

In the tracking system of Fig. 1.13a, the occurrence of error causes the motor to rotate in a direction favoring the dissolution of error. Note that the components of our system cannot respond instantaneously because any real-world system cannot go from one energy level to another in zero time. Thus, in any real-world system, there is some kind of dynamic lagging behavior between the input and the output. In the servosystem of Fig. 1.13a, the control action on the occurrence of the deviation of the controlled output from the desired value (the occurrence of error), will be delayed by the cumulative dynamic lags of the shaft-angle encoder, digital computer and digital-to-analog converter, power amplifier, and the servomotor with load. Eventually, however, the trend of the controlled variable deviation from the desired value will be reversed by the action of the amplifier output on the rotation of the motor, returning the controlled variable towards the desired value. Now, if a strong correction (high amplifier gain) is applied (which is desirable from the point of view of the control system performance, e.g., strong correction improves the speed of response (Chapter 4)), the controlled variable overshoots the desired value (the 'run-out' of the motor towards an error with the opposite rotation), causing a reversal in the algebraic sign of the system error. Unfortunately, because of the system dynamic lags, a reversal of correction does not occur immediately and the amplifier output (acting on 'old' information) is now actually driving the controlled variable in the same direction as it is already going, rather than opposing its excursions, leading to a larger deviation. Eventually, the reversed error does cause a reversed correction, but the controlled variable again overshoots the desired value in the opposite direction and the correction is again in the wrong direction. The controlled variable is thus driven alternatively in opposite directions before it settles to an equilibrium condition. This oscillatory state is unacceptable as the behavior of the antenna-steering servomechanism. The considerable amplifier gain, which is necessary if high accuracies are to be obtained, aggravates the described unfavorable phenomenon.

The oscillatory effects can be controlled by generating the control signal with two components: one component proportional to the error signal, and the other proportional to the derivative of the error signal. Alternatively, a signal proportional to motor's angular velocity (called the rate signal) is subtracted from the error signal (Fig. 1.13c)—this loop within a loop (velocity feedback system embedded within a position feedback system) configuration is a control scheme, called *minor-loop feedback*[4].

Example 1.4 Variable speed dc drive

Many industrial applications require variable speed drives. For example, variable speed drives are used for pumping duty to vary flow rate or pumping pressure, rolling mills, harbor cranes, rail traction, etc. [3–5].

The variable speed dc drive is the most versatile drive available. The silicon controlled rectifiers (SCR) are almost universally used to control the speed of dc motors because of the considerable benefits that accrue from the compact static controllers supplied directly from the ac mains.

Basically, all the dc systems involving the SCR controllers are similar but with different configurations of the devices, different characteristics may be obtained from the controller. Figure 1.14 shows a dc motor driven by a full-wave rectified supply. Armature current of the dc motor is controlled by an SCR which is in turn controlled by the pulses applied by the SCR trigger control circuit.

[4] This, and other control schemes will be discussed in later chapters.

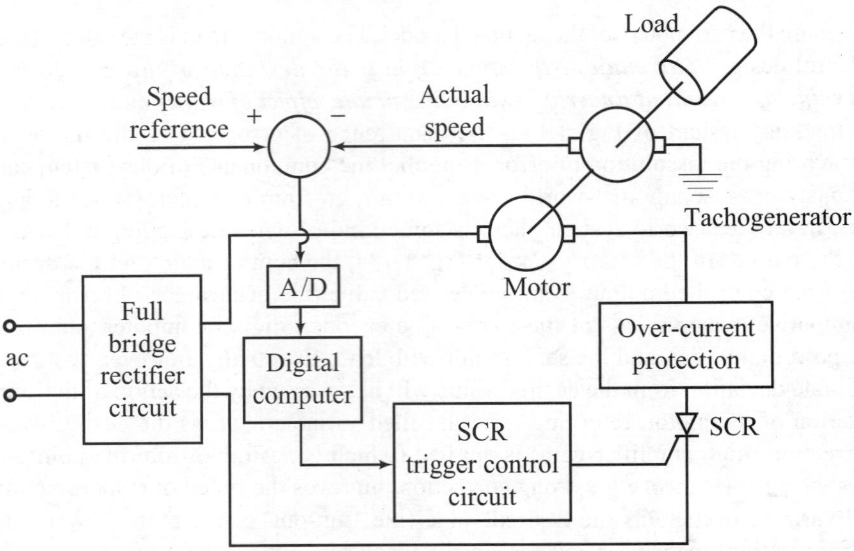

Fig. 1.14 *Variable speed dc drive*

Firing angle of the SCR controls the average armature current which in turn controls the speed of the dc motor. The average armature current (speed) increases as the trigger circuit reduces the delay angle of firing of the SCR, and the average armature current (speed) reduces as the delay angle of firing of the SCR is increased.

In the scheme of Fig. 1.14, the reference voltage which corresponds to the desired speed of the dc motor, is compared with the output voltage of the tachogenerator corresponding to the actual speed of the motor. The occurrence of the error in speed causes an error signal to pass through the trigger circuit which controls the firing angle of the SCR in a direction towards an error reduction. When the processing of the error signal (calculation of the control signal) is based on the proportional control law, a steady-state error exists between the actual speed and the desired speed (Chapter 4). The occurrence of the steady-state error can be eliminated by generating the control signal with two components: one component proportional to the error signal, and the other proportional to the integral of the error signal. This proportional-integral control law will be examined in detail in later chapters.

A point which needs a special mention here is that ***the presence of disturbances is the main reason for using feedback control*. *Without disturbances, there is probably no need for feedback control*.** Commonly occurring disturbances in the control system examples considered earlier are as follows.

1. Environment temperature in heating system (Example 1.2);
2. Wind power in servomechanism for steering of antenna (Example 1.3); and
3. Variations in load on dc motor in variable speed dc drive (Example 1.4).

Other disturbances which affect the performance of control systems are as follows.

1. Uncertainty in our estimate of the hardware parameters of the process and other components of the control system (the estimates of the parameters are used for the design of controller).
2. Changes in the parameters taking place in time (due to wear, ageing, environmental effects, etc.).
3. In systems controlling mechanical motions, mechanical vibrations may lead to high-frequency components (*noise*) in the output signals of position and speed sensors. Continuous stirring of fluid in tank reactors, gives rise to noise in output signals of temperature sensors.

As we shall see later in this book, ***feedback control systems***

(i) ***greatly reduce the effect on the controlled variable of all external disturbances except those associated with the sensor (measurement system), and***
(ii) ***are tolerant of variations in hardware parameters other than those of the sensors.***

The feedback signal coming from the sensor of controlled variable contains useful information related to disturbances (external disturbances and variations in hardware parameters), which is usually of a relatively low frequency. It may often include high frequency 'noise' introduced by the measurement sensors. Such noise signals are too fast for the control system to correct; *low-pass filtering* is often needed to allow good control performance.

1.3 FEEDFORWARD–FEEDBACK CONTROL STRUCTURE

This section presents the principle of one of the most profitable control schemes: *feedforward control*. In earlier sections, we studied the principle of feedback control; a very simple technique that compensates for any disturbance affecting the controlled variable. ***When a disturbance enters the process, the controlled variable deviates from its desired value (setpoint) and, on sensing the error, the feedback controller manipulates the process input in a way favoring the dissolution of error. The main limitation of a feedback control system is that in order for it to compensate for disturbances, the controlled variable must first deviate from its desired value. Feedback control acts upon an error between the setpoint and the controlled variable.*** This means that once a disturbance enters a process, it must propagate through the process and force the controlled variable to deviate from the setpoint before corrective action can be taken.

Feedforward control compensates for disturbances before they affect the controlled variable. In this scheme, the disturbances are measured before they enter the process and required value of the manipulated variable to maintain the controlled variable at its desired value is calculated; implementation of such a control results in undisturbed controlled variable. The design of this type of control scheme does not generally require any specialized control theory; basic system dynamics are sufficient in most cases. Success of *disturbance feedforward* control schemes depends on our ability to

(i) measure the disturbance, and
(ii) estimate the effect of the disturbance on the controlled variable, so that we can compensate for it.

Let us consider an example of disturbance feedforward.

Example 1.5 A tubular heat exchanger

The schematic diagram of a tubular heat exchanger is shown in Fig. 1.15 [13]. The process fluid flows inside the tubes of the heat exchanger and is heated by the steam condensing on the outside of the tubes (Fig. 1.15a). The objective is to control the outlet temperature θ of the process fluid in the presence of variations[5] in the process fluid flow Q and its inlet temperature θ_i. The objective of controlling the outlet temperature in the presence of disturbances can be accomplished by setting up a feedback control loop as shown in Fig. 1.15b. The flow controller consists of an electronic amplifier and an electropneumatic

[5] In a typical application, the heat exchanger is a part of the overall system; disturbances in Q and θ_i are caused by other processing operations taking place in the system (in the next section, we will study a distillation system wherein the heat exchanger is a subsystem).

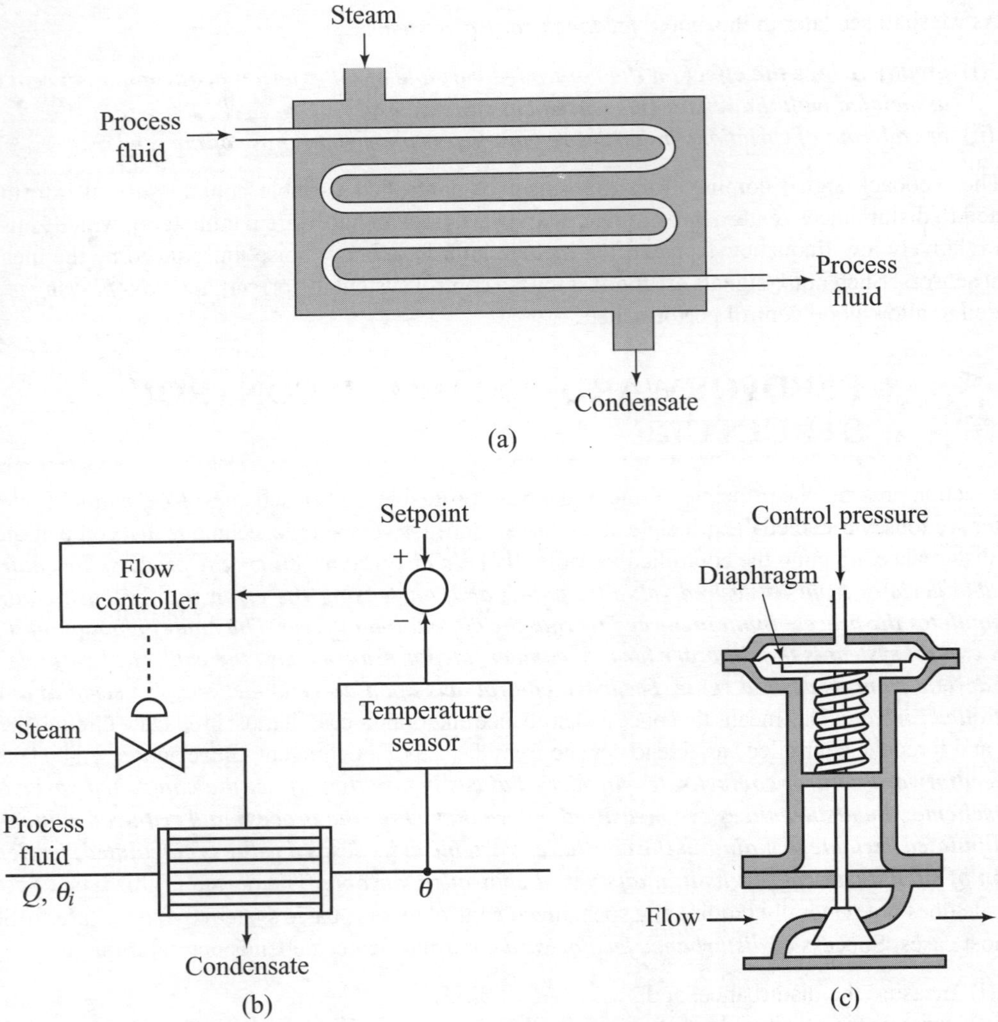

Fig. 1.15 *Feedback control loop for heat exchanger*

transducer, which converts the voltage corresponding to the error between the controlled temperature and its setpoint value to a pneumatic pressure variable (based on proportional control law). The error voltage is derived from the output voltage of the temperature sensor (corresponding to the output temperature, θ) and the setpoint voltage corresponding to the desired temperature. The pneumatic pressure actuates a pneumatic valve[6], whose essential features are shown in Fig. 1.15c. The actuator section of the valve comprises of a diaphragm whose position is set by the balance between the force exerted by a spring and the controller output pressure on the diaphragm.

---

[6] By far, the most common form of valve used in process control is the pneumatic actuating valve which can provide large power output. Pneumatic systems use a readily available working medium—compressed air supplies, and are commonly used in process industry because of their explosion-proof characteristics, simplicity and ease of maintenance.

In the control scheme of Fig. 1.15b, the feedback controller, which manipulates the heat input to the heat exchanger, will not act until an error has developed. If the system involves large time lags, it will take some time before any corrective action takes place.

If feedforward control is provided in such a system, then as soon as a change in Q and/or θ_i occurs, a corrective measure will be taken simultaneously by manipulating the heat input to the heat exchanger. Figure 1.16 shows the disturbance-feedforward control scheme.

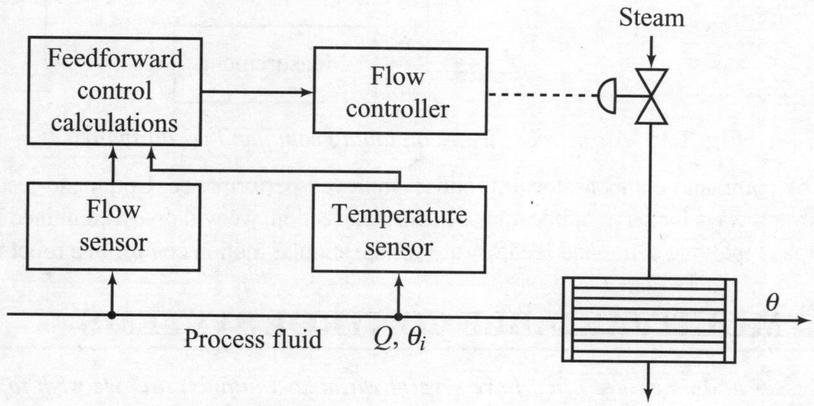

Fig. 1.16 *Disturbance compensated open-loop system*

Feedforward control can minimize the transient error. However, since it is an open-loop control, there are limitations to its functional accuracy. Feedforward control will not cancel the effects of unmeasurable disturbances. It is, therefore, necessary that a feedforward control system include a feedback loop as shown in Fig. 1.17. Essentially, the feedforward control minimizes the transient error caused by measurable disturbances, while the feedback loop compensates for any imperfections in the functioning of the feedforward control and provides corrections for unmeasurable disturbances.

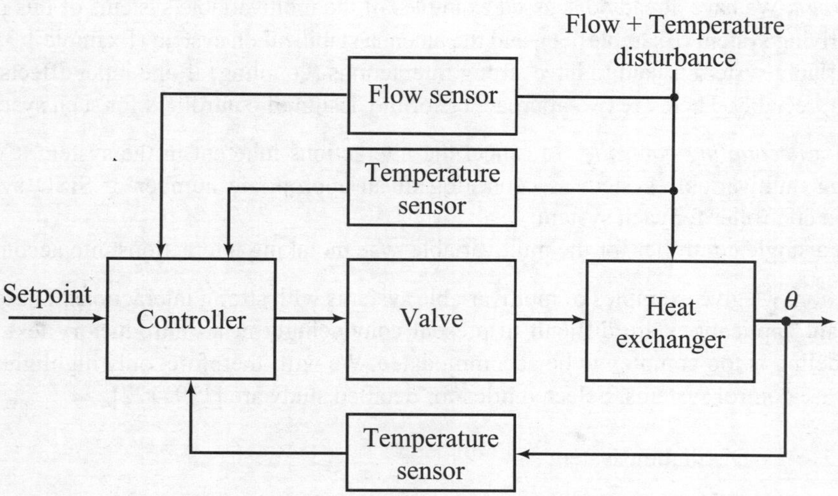

Fig. 1.17 *Feedforward–feedback control structure for the heat exchanger*

Another control scheme using the feedforward concept is the *command-feedforward* in combination with feedback loop (Fig. 1.18). Here, based on the knowledge of process characteristics, the command

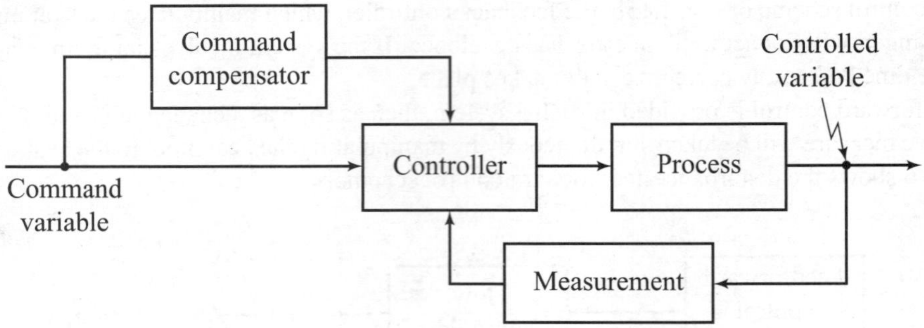

Fig. 1.18 *A control scheme employing command feedforward*

is augmented by command-compensator to produce improved performance. Command-feedforward has been used in several ways for servo applications. In the next section, we will give a combined feedforward-feedback scheme employing command feedforward for the angular motion control of a robot manipulator.

1.4 MULTIVARIABLE CONTROL SYSTEMS

Complex processes and machines often have several variables (outputs) that we wish to control, and several manipulated inputs available to provide this control. Sometimes, the control situation is simple; one input affects primarily one output and has only weak effect on the other outputs. In such situations, it is possible to ignore weak interactions (coupling) and design controllers under the assumption that one input affects only one output. Input–output pairing to minimize the effect of interactions and application of SISO control schemes to obtain separate controllers for each input–output pair, results in an acceptable performance. This, in fact, amounts to considering the multivariable system as consisting of an appropriate number of separate SISO systems. Coupling effects are considered as disturbances to the separate control systems and may not cause significant degradation in their performance if the coupling is weak. We have already discussed examples of the multivariable systems of this nature—the automobile driving system (Example 1.1), and the antenna stabilization system (Example 1.3).

A multivariable system is said to have strong interactions (coupling) if one input affects more than one output appreciably. There are two approaches for the design of controllers for such systems.

1. Design a *decoupling controller* to cancel the interactions inherent in the system. Consider the resulting multivariable system as consisting of an appropriate number of SISO systems, and design a controller for each system.
2. Design a single controller for the multivariable system, taking interactions into account.

In this section, we give examples of multivariable systems with strong interactions. These and many other important applications are difficult to present convincingly in an introductory text since their dynamic modeling is too complex to be accommodated. We will 'therefore' only highlight the salient features of these control systems. Selected titles for detailed study are [119–122].

Example 1.6 Distillation system

The purification of materials is a prominent operation of the chemical industry. There is scarcely any chemical process which does not require preliminary purification of raw materials or final separation of products from by-products. The towers of a modern petroleum refinery are the evidence of the important role played by these separations in a processing plant.

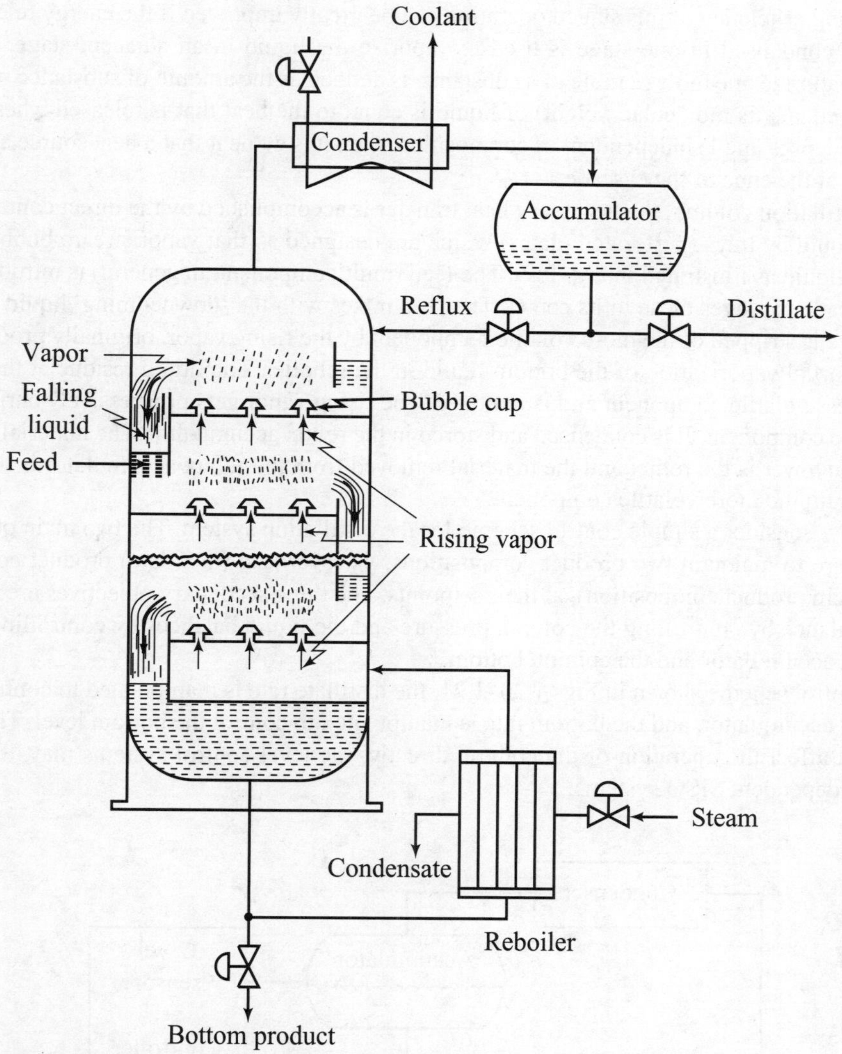

Fig. 1.19 *A distillation column with five primary controlled variables: two compositions, two levels, and pressure*

Figure 1.19 is a schematic diagram of a bubble-cup tray tower used for distillation [13].

Distillation is a method of separating the components of a liquid mixture by the repeated systematic application of vaporization and condensation. For example, if a liquid solution of ammonia and water is partially vaporized by heating, it is found that the newly created vapor phase and residual liquid both contain ammonia and water, but in proportions which at equilibrium are different for the two phases and different from those in the original solution. The vapor phase is richer in the more volatile component (ammonia) and the residual liquid is richer in the less volatile component (water), compared to the original solution. If the vapor and the liquid are separated from each other and the vapor is condensed, two solutions, one richer in ammonia and the other richer in water, are obtained. If the process of vaporization and condensation is continued over a number of stages, it eventually produces a product with a very high degree of purity in the more volatile component.

The thermal efficiency of this staged operation can be greatly improved if the energy released when the vapor is condensed in one stage is used to vaporize the liquid in an adjacent stage. If the heat required to vaporize one mole (a mole of a substance is defined as the amount of substance whose mass numerically equals its molecular weight) of liquid is equal to the heat that is released when one mole of vapor condenses and is independent of composition, then it is evident that a heat source and sink are needed only at the ends of the cascade.

In the distillation column, the necessary heat transfer is accomplished by the direct contact between vapor and liquid by trays (perforated plates) which are designed so that vapor stream bubbles directly through the liquid as illustrated in Fig. 1.19. The feed (multicomponent, in general) is introduced into a vertical cascade of stages close to its center. The feed mixes with the 'downcoming' liquid. The liquid at every stage is stripped of the more volatile component by the rising vapor, originally produced at the bottom by partial vaporization of the bottom liquid in the reboiler. The liquid residue at the bottom is rich in the less volatile component and is removed. The rising vapor gets progressively enriched in the more volatile component. It is condensed and stored in the reflux accumulator. The material returned at the top of the tower is the reflux and the material removed from the reflux accumulator is the distillate which is rich in the more volatile component.

Let us now consider a simple control scheme for the distillation system. The two main objectives of the control are to maintain two product compositions, x_D (overhead (distillate) product composition) and x_B (bottom product composition), at their setpoints. The three secondary objectives are to maintain the vapor balance by controlling the column pressure and the liquid balances by controlling the levels in the reflux accumulator and the column bottom.

In the control scheme shown in Figs 1.20–1.21, the distillate rate is manipulated to control the level of the reflux accumulator, and the bottom rate is manipulated to control the bottom level. The two variables do not affect the operation of the column directly; the level control systems may, therefore, be treated as independent SISO systems.

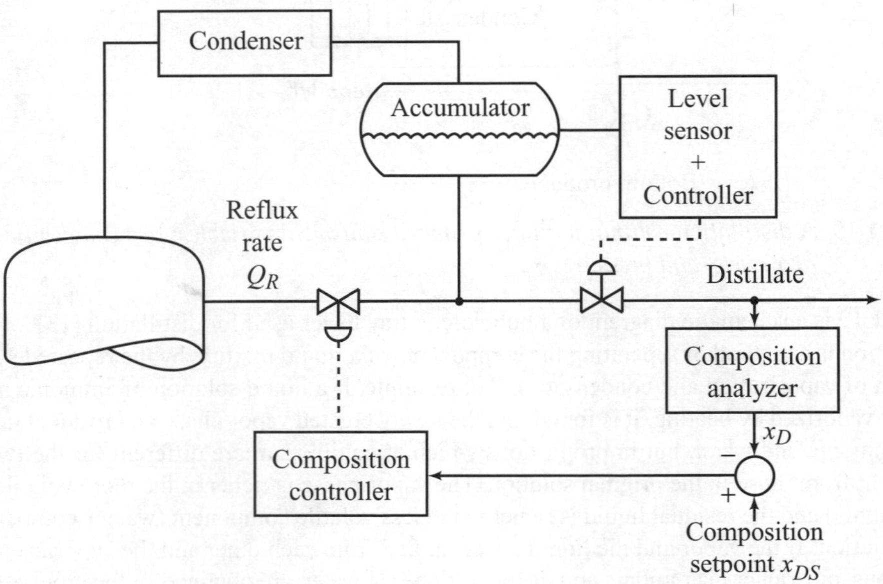

Fig. 1.20 *Distillation-column reflux*

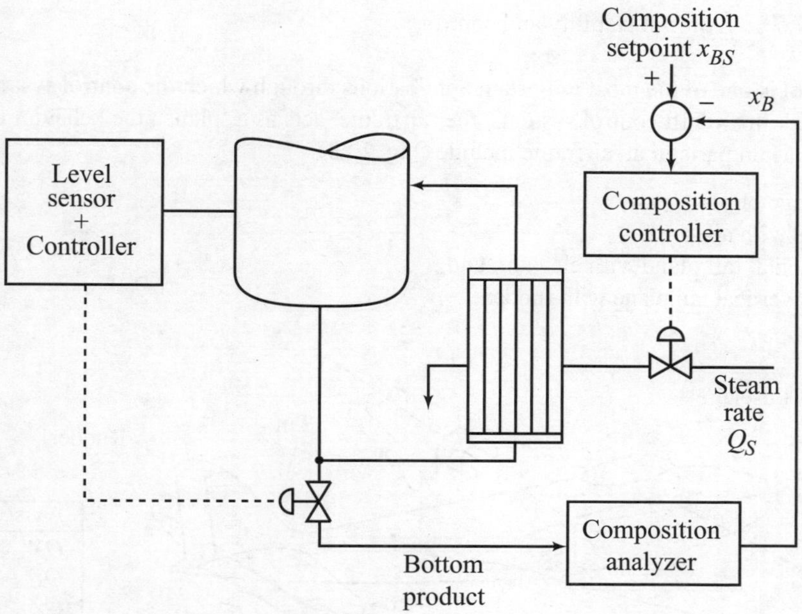

Fig. 1.21 *Distillation-column reboiler*

In a column, the pressure is determined solely by the heat balance. If heat is supplied in the reboiler at a higher rate than it is removed from the condenser, then the pressure in the column rises with time. Column pressure can be controlled by manipulating the reboiler heating rate or the condenser cooling rate. We will assume that the pressure controller manipulates the condenser cooling rate and results in a non-interacting control loop (not shown in the figures).

This reduces the problem to a two-by-two multivariable control problem: the manipulated variables are steam rate Q_S and reflux rate Q_R, and the controlled variables are the bottom product composition x_B and the distillate composition x_D. Figure 1.22 shows a block diagram representation of the resulting two-input, two-output system. Changes in the input variable Q_R cause responses in both the outputs, x_D and x_B. Similarly, input variable Q_S affects x_B as well as x_D.

In this book, we will consider many examples of liquid level, and temperature control systems. SISO design techniques will provide control law in a large number of cases.

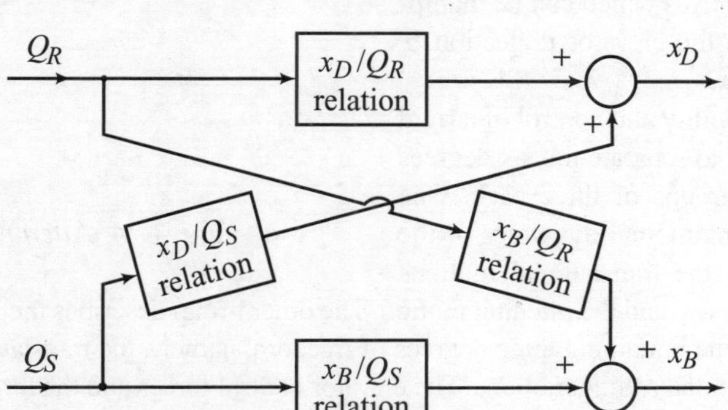

Fig. 1.22 *Block diagram representation of the multivariable control of a distillation column*

Example 1.7 Aircraft stability and control

Aircraft control is one of the most important applications through which the control system techniques have developed. In aircraft control systems, the 'airframe' acts as a 'plant'; the behavior of which is to be controlled. Main parts of an airframe include (Fig. 1.23):

1. Fuselage or mainbody,
2. Wings with ailerons,
3. Horizontal tail plane with elevator, and
4. Fin or vertical tail plane with rudder.

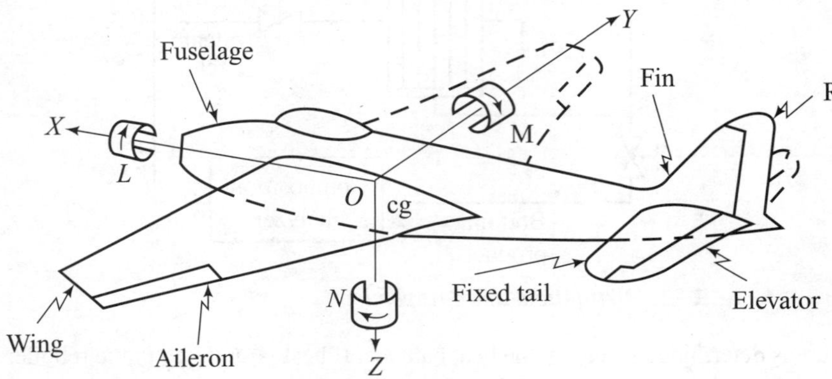

Axis	Moment
OX	*Rolling L*
OY	*Pitching M*
OZ	*Yawing N*

Fig. 1.23 *Control surfaces of an airframe*

Ailerons, elevator and rudder are the control surfaces. A control surface is a movable (hinged) flap which forms a portion of the fixed aerodynamic surface (wing, horizontal tail, vertical tail) and which is employed to vary the lift coefficient of the surface. Figure 1.24 shows aileron deflection; the aileron angle δ_a is defined as the mean value of the angular displacements of the two ailerons. The other two controls which can be manipulated by the pilot are the elevator deflection δ_e and rudder deflection δ_r.

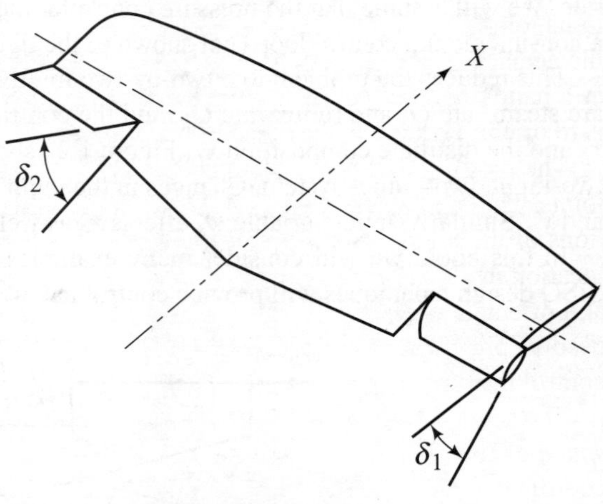

Fig. 1.24 *Aileron deflection*

In examining the stability and control of a rigid aircraft, it is customary to separate the six degrees of freedom into two groups of three each. One group describes the longitudinal dynamics of the aircraft and comprises the translational motions along the *OX*- and *OZ*-axes, and the pitching motion. The other group describes the lateral dynamics of the aircraft and comprises the remaining three degrees of freedom, namely, the translational motion along the *OY*-axis, and the yawing and rolling motions. The elevator control forces and the throttle affect the longitudinal motion, whereas the aileron and rudder primarily affect the lateral motion. It has been found that for the preliminary design work, the coupling of the lateral motion into longitudinal motion can be ignored.

In general, a control functions by causing a change in the pressure distribution on a surface, which results in a change in the lift coefficient of the surface. This change in the lift coefficient causes a change in the moment balance of the aeroplane and results in the angular moment about one or more of the axes. The principal control actions of the three control surfaces are:

1. The rudder controls yawing, i.e., motion about the *OZ*-axis,
2. The ailerons control rolling, i.e., motion about the *OX*-axis, and
3. The elevator controls pitching, i.e., motion about the *OY*-axis.

The actual positioning of the control surfaces is obtained by electric or hydraulic motors which are controlled by the pilot.

Later in this book, we shall analyze aircraft motions (SISO systems) under assumptions which neglect many details but preserve the essential nature of the problem. Mathematical modeling of aircraft dynamics is beyond the scope of this book. References [14–16] give a good treatment of the subject.

Example 1.8 Robotic control system

Machines that automatically load and unload, cut, weld or cast, are used by the industry in order to obtain accuracy, safety, economy, and productivity. Programable computers integrated with machines that often substitute for human labor in specific repeated tasks, are the modern *robots*.[7] Such devices even have anthropomorphic mechanisms, including what we might recognize as mechanical arms, wrists, and hands.

A *robot manipulator* is made of several links connected usually in series by the joints to form an arm. A link is revolute or prismatic depending upon the type of motion caused by the actuator attached to its joint. When the actuator of a joint causes rotational motion, the link is called *revolute*, and when the actuator produces translational motion, the link is called *prismatic*. A gripper, which is referred to as a hand or an end effector, is attached to the arm by means of wrist joints. The function of the wrist is to orient the end effector properly.

The motion of a manipulator end effector is caused by the movements of the actuators driving the joints. The joint actuators are electric or hydraulic motors, or pneumatically driven devices. The positions of the joints determine the *configuration of the arm*, which places the end effector at a specific location in the environment. The motion of the joints produced by the actuators determines the position and orientation of the end effector at any time. Transducers such as encoders and tachometers can be used to provide information for determining the position and orientation of the end effector, and to control the manipulator motion.

The set of all points that can be reached by the end effector of a manipulator arm forms the *workspace* of the manipulator. A particular position of the end effector is specified by three independent coordinates, which represent three degrees-of-freedom. Similarly, a specific orientation of the end-effector is determined by three independent variables. Thus, six independent variables are needed to describe the position and orientation of the end effector. Commercially available robots have typically six joints, giving six degrees-of-freedom. However, robots with less than six joints, designed for specific tasks, are also available. Figure 1.25 shows a picture of the PUMA (Programmable Universal Manipulator for Assembly), a revolute manipulator.

[7] The word 'robot' seems to originate from the Czech word *robota* meaning compulsory service.

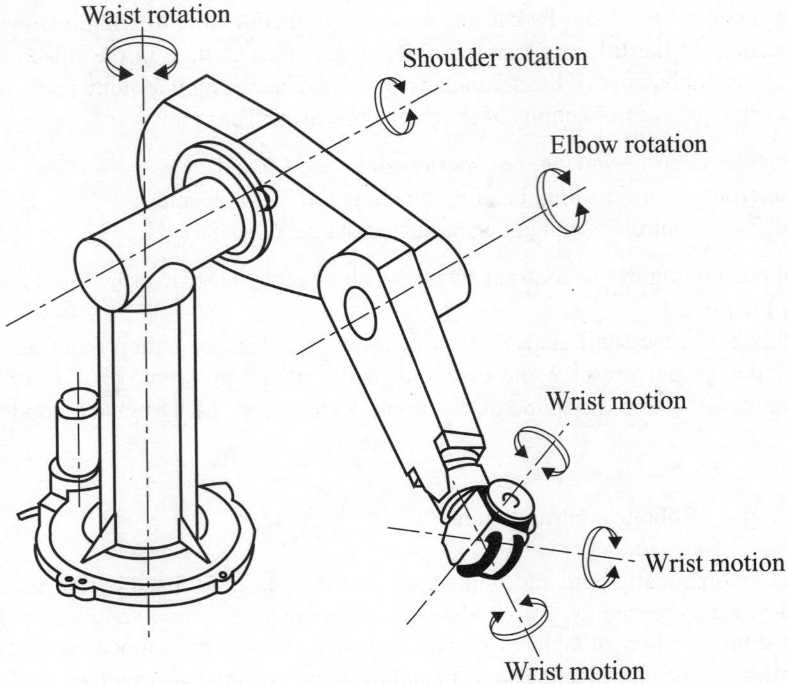

Fig. 1.25 *An industrial robot: PUMA*

The objective in providing the position control for a robot manipulator is to design an appropriate controller for the robot motors, so that the position and the orientation of the end effector follow the desired trajectory with no errors even in the presence of disturbances. A multilink manipulator is a highly complex system—each link of the robot arm has varying inertia and exerts varying torques on all the other links as the configuration of the robot arm changes. The equations of motion for a multilink manipulator describe a MIMO system; the equations are coupled representing the effects of interactions of the joints (***This MIMO system cannot be considered as consisting of an appropriate number of separate SISO systems***). For a six-joint robot, joint positions θ_i; $i = 1, \ldots, 6$ (represented by the position vector $\boldsymbol{\theta}$ in Fig. 1.26) are the controlled outputs, and actuator torques u_i; $i = 1, \ldots, 6$ (represented by the input vector $\mathbf{u}$ in Fig. 1.26) are the manipulated variables. $\boldsymbol{\theta}_r$ is the position vector corresponding to the desired trajectory.

Perfect tracking can be achieved by a scheme employing command feedforward. Let $\boldsymbol{\theta}_r(t)$ be the input to a feedforward controller with dynamics which are inverse of the dynamics of the manipulator. When the manipulator is driven by the output of such a controller, we get $\boldsymbol{\theta}(t) = \boldsymbol{\theta}_r(t)$.

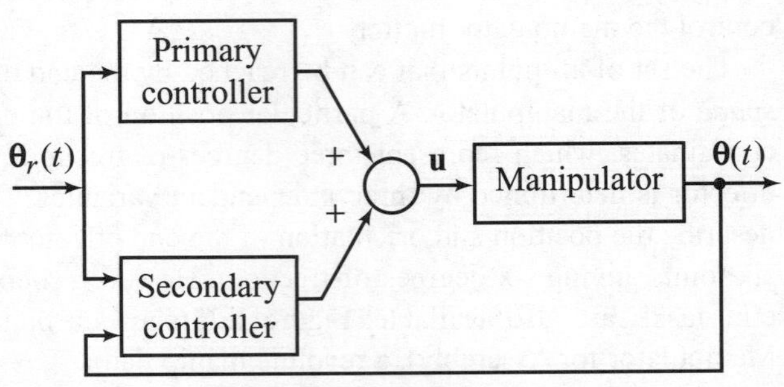

Fig. 1.26 *Manipulator system with primary and secondary controllers*

The command-tracking based on this *inverse dynamics* concept can be perfect only if we can exactly determine the dynamic model of the manipulator.

The robotic system is too complex to allow exact identification of the dynamic model. Also, the system is subject to unmeasurable disturbances. Undesirable deviations of the actual motion of the manipulator from the desired trajectory can be corrected by means of feedback control. A block diagram showing the total control system (primary controller (feedforward) and secondary controller (feedback)) is displayed in Fig. 1.26.

The problem of control of the multilink manipulator is too complex to be included in this introductory text. To bring robot application into our range, we will consider the problem of control of a single-link manipulator (SISO system). The problem of control of a two-link manipulator has been examined in the companion book [155]. Selected titles for further reading are [17–19].

CONCLUDING REMARKS

In this chapter, we discussed open- and closed-loop control systems. Basic understanding of control system engineering was given through some illustrative examples. Our description has been qualitative in nature. In Chapters 2 and 3, we will build the platform for quantitative description. Use of this platform to deliberate on issues related to feedback control will follow in later chapters.

What we have learnt?

- In industrial practice, a system is subjected to different commands (input) at different times; the system response (output) is required to follow these commands. We require a control mechanism to force the system to operate in command-following mode.
- A good control system that reacts to the uncertainties (disturbances generated from within the system and from outside) is necessary. One way to react is to measure the disturbances before they affect the system, and to manipulate the system so as to cancel their effects. Such a cancelation is, however, not possible because disturbances are usually unmeasurable. It is, therefore, necessary to let the disturbances affect the system, measure the resulting error in system output, *inform* the controller on-line what is going wrong, and let the controller react to it. By doing this, we are *feeding back* information about the current situation, to the system, i.e., we are creating a feedback control system. If there were no uncertainties in the system and its environment, there would be no need of a feedback control system; open-loop control would be feasible—even better.
- A feedback control system consists of interconnected subsystems to achieve a desired purpose. We have learnt these issues through functional block diagrams of two important classes of industrial control systems: (i) *motion control systems* with speed/position as response variable; and (ii) *process control systems* with temperature/pressure (liquid level in a tank) as response variable.

Review Questions

1.1 Define the following terms commonly used in control systems:
Plant, open-loop control system, closed-loop control system (feedback control system), automatic control system, regulator system, setpoint control, servomechanism (servo system), tracking system, SISO system, MIMO system, feedforward–feedback control system.

1.2 The introduction of feedback is mainly motivated by the incompleteness of the knowledge of the system to be controlled and the effects of external disturbances. More strictly, the use of feedback can be regarded as unreasonable if there is no uncertainty in the system. Describe this concept with the help of suitable control system examples.

1.3 The objective of a control system is to maintain the output of a process close to the desired value inspite of the disturbances—undesirable and unavoidable effects beyond our control, generated from outside the process environment and/or from within the feedback loop.
Explain qualitatively with the help of examples, the disturbances generated from outside and from within the feedback loop.

1.4 Draw a functional block diagram showing elements of a control system wherein the controlled variable is position θ of a shaft. Identify in feedback loop the sensor, the actuator, how appropriate action is determined (without and with minor-loop feedback); and some likely disturbances.

SYSTEM MODELS AND RESPONSE: TRANSFER FUNCTION, BLOCK DIAGRAM, SIGNAL FLOW GRAPH

2

Preview

The first step in an orderly sequence for the design of feedback control systems is to obtain a quantitative description of all the subsystems in the feedback loop with known functional block diagram of the type shown in Fig. 1.9. This amounts to developing a mathematical model for each subsystem and representing interconnections of the subsystems. For an individual subsystem, two forms of mathematical models are commonly used: (i) transfer function, and (ii) state equations. Interconnections of multiple subsystems are commonly represented as a block diagram model or a signal flow graph model. The objective set for this chapter is to learn the basics of the models: transfer function, block diagram, and signal flow graph; including methods for obtaining dynamic response of these models to command inputs and external disturbances. Development of models of real-life industrial control systems will appear in the next chapter. The use of state equations for analysis and design of feedback control systems will be discussed in Chapter 14 and onwards.

Transfer function models obtained using Laplace transformation of signals help in our understanding of the characteristics of signals and systems; they help in system analysis, and as we shall see in later chapters, understanding in transformed domain leads to better design. However, the design of feedback control systems in industry is probably accomplished using frequency-domain methods more often than other methods. It is the frequency-selective characteristics of signals in control systems that provide the motivation for transformation to frequency domain. Sinusoidal transfer function models (frequency response models) are used for design in frequency domain. The two methods do not compete with each other; they, in fact, complement each other—a practical design in industry is carried out using both of these tools; Laplace-transform-based technique is helpful in seeing what the frequency-domain design is accomplishing. We initiate the process of learning this practical design strategy by introducing in this chapter, the sinusoidal transfer function models and their relationship with the Laplace-transform-based transfer function models.

2.1 INTRODUCTION

The common form of representation in control systems is to use *block diagrams*—the system is represented by a block and the *signals* are represented by directed lines. Signals on the left-hand side of the block diagram of Fig. 2.1 are the *inputs* (excitations, causes) labeled $r_1(t)$, $r_2(t)$,...,$r_p(t)$. The signals emanating from the right-hand side of the block are the *outputs* (responses, effects) labeled $y_1(t)$, $y_2(t)$,...., $y_q(t)$. In general, the inputs and outputs vary with time, which is represented by the independent variable t; sometimes t will not be shown explicitly in order to simplify notation. The system of Fig. 2.1 is a *multi-input multi-output* (MIMO) system. In this book, we will be mostly concerned with situations where there is a single input and a single output. Figure 2.2 is the block diagram representation of *single-input single-output* (SISO) system.

The goal of mathematical modeling is to provide a mathematical description of the interrelationships between the system variables, as well as the relations between these variables and external inputs.

The family tree shown in Fig. 2.3 illustrates the major system classifications. The dashed lines in the figure indicate the existence of subdivisions similar to the others shown at the same level.

The significant variables in a system are distributed in space and they vary with spatial coordinates and time. The resulting dynamic models, called *distributed-parameter models*, consist of partial differential equations with time and space coordinates as independent variables. In *lumped-parameter models*, the matter is assumed to be lumped at some discrete points of the space, or the space is subdivided into cells and the matter is assumed to be lumped at these cells. The resulting dynamic models are ordinary differential equations with time as the only independent variable. For example, the assumption that a solid is a rigid body permits lumping of all its mass at its mass centre, resulting in lumped-parameter models of dynamics. Considerable simplification is achieved in mathematical models when lumped-parameter models are used. In this book, we employ lumped-parameter models for control system design.

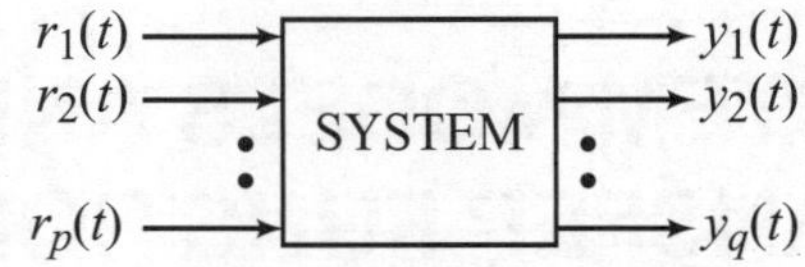

Fig. 2.1 *Multi-input multi-output (MIMO) system*

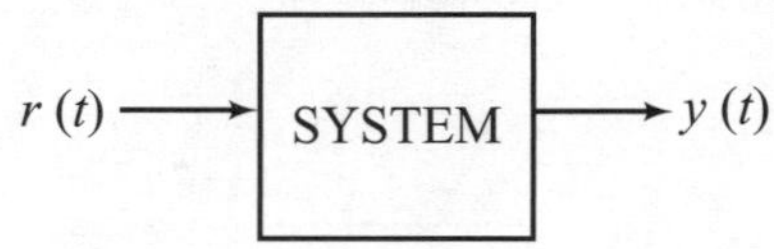

Fig. 2.2 *Single-input single-output (SISO) system*

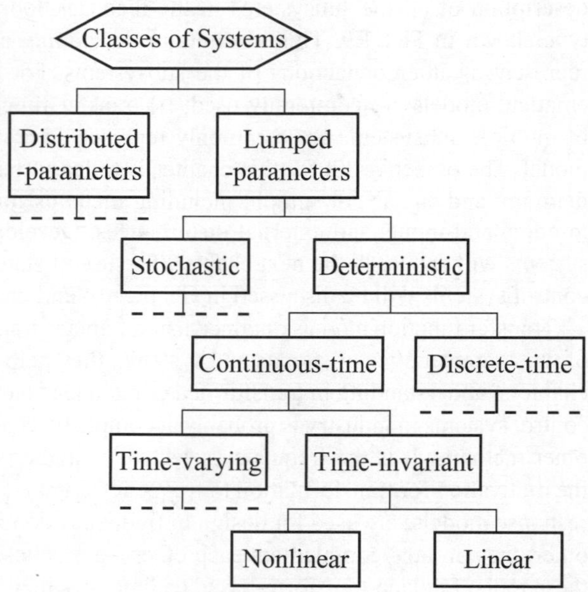

Fig. 2.3 *Major classes of system equations*

In physical systems, every signal (inputs, external disturbances, measured outputs) has associated with it certain amount of randomness (usually referred to by the generic term 'noise'). These uncertainties are inconsequential in many practical cases, and we proceed as though all signals are known functions of time. This assumption gives us a *deterministic model* of the system. When signals must be treated as random functions of time, *stochastic models* are used to characterize the system behavior. In this book, we will make the assumption of deterministic signals.

If all the elemental equations are defined for all time, then the system is a *continuous-time system*. If, as in sampled-data systems, some elemental equations are defined or used only at discrete points in time, a *discrete-time system* is the result. Continuous-time systems are governed by differential equations and discrete-time systems by difference equations. The primary focus in this book is on continuous-time systems. Chapters 13 to 15 provide an introduction to discrete-time systems.

A classification between large-scale and small-scale systems could be made in terms of *multi-input multi-output* (MIMO) and *single-input single-output* (SISO) systems. Our emphasis in this book will be on SISO models with scalar input $r(t)$ and scalar output $y(t)$.

If we set a system going at some time t_0 and apply the input $r(t)$; $t \in (t_0, \infty)$, monitoring the corresponding output $y(t)$; $t \in (t_0, \infty)$, the output, in general, is not uniquely determinable and one way of associating a unique y with each r consists in including the information on total energy stored within the system at time t_0 before application of the external input. This information may be given in terms of characterizing variables whose value at time t_0 gives the stored-energy status of the system at that time. Therefore, a complete mathematical model of a system includes *initial conditions* at the time of application of external input. The initial conditions of the model at t_0 give the energy status of the system at that time.

For instance, Newton's law of motion states that if a force of magnitude $F(t)$ is applied to a mass of M units, the displacement $x(t)$ of the mass is related to the force by the second-order differential equation, $M(d^2x(t)/dt^2) = F(t)$. A solution $x(t)$; $t \ge t_0$, can be obtained for a given input $F(t)$; $t \ge t_0$, if *initial conditions* for $x(t)$ and $(dx(t)/dt)$ are specified, i.e., if displacement $x(t)$ at $t = t_0$, and velocity $v(t) = (dx(t)/dt)$ at $t = t_0$ are known.

The output of a system at time $t_1 > t_0$ is uniquely determined by its initial conditions at t_0 and the input $r(t)$; $t \in [t_0, t_1]$. The input after t_1 has no effect on the system output at t_1; the system cannot foresee the future and responds only to inputs it has already received. This condition is called the *causality condition; causality is an intrinsic property of every physical system.*

A system is said to be relaxed at time t_0 if the energy stored at time t_0 is zero. This implies that initial conditions of the model at t_0 are zero and the output $y(t)$; $t \in (t_0, \infty)$, is uniquely determined by the input $r(t)$; $t \in (t_0, \infty)$.

If one or more parameters, or the very form of an elemental equation, vary in a known fashion with time, the system is said to be time varying. For example, in the guidance and control of a rocket, the mass of the rocket changes with time due to the depletion of fuel, and also the aerodynamic damping can change with time as the air density changes with altitude. The complexity of the control system design increases considerably when the parameters are time varying. In this book, ***we shall concern ourselves with those applications where the time-invariance assumption can safely be made, and therefore, the models we use will have constant parameters.***

Assume that a time-invariant system is relaxed and the input is applied at $t = t_0$. As long as relaxedness assumption is satisfied, a delay in the application of input by time τ will have no effect on the waveform of the output except for a corresponding delay by time τ. In general, as long as we start in the same energy state, any delay in application of input shifts the response correspondingly without affecting the magnitude. This suggests that ***in the study of time-invariant systems, the time origin t_0 is not important. For convenience, we can shift the origin to $t = 0$.***

Consider a system to have input r and output y. Suppose further that with the system at rest (relaxed system) at $t = 0$, we apply the input $r_1(t)$; $t \ge 0$, and observe the output $y_1(t)$; $t > 0$. After restoring the system to rest, we apply a second input $r_2(t)$; $t \ge 0$ and again observe the output, which we will call $y_2(t)$; $t > 0$. Then we form a composite input $r(t) = \alpha_1 r_1(t) + \alpha_2 r_2(t)$; $t \ge 0$, where α_1 and α_2 are real numbers. If superposition applies, then the response of the relaxed system to input $r(t)$ will be $y(t) = \alpha_1 y_1(t) + \alpha_2 y_2(t)$; $t > 0$. ***Superposition will apply, if and only if, the system is linear. Linear systems have the decomposition property which says that the response of a system to the input r starting in an energy state given by a set of initial conditions at $t = 0$ is identical with the zero-input response ($r = 0$) of the system starting in the given energy state plus the zero-state response (system at rest at $t = 0$) of the system to input r. A sufficient condition for linearity is that both the components of the response satisfy superposition property.***

Powerful mathematical tools that provide general methods of analysis and design are available for linear systems. For *nonlinear* systems, general methods are not available and each system must be studied as one of a kind. Practically, none of the systems is completely linear, but a linear model is often appropriate over certain ranges of operation.

Many real-world nonlinearities involve a 'smooth' curvilinear relation between an independent variable x and a dependent variable y:

$$y = f(x) \tag{2.1a}$$

A linear approximation to the curve, accurate in the neighborhood of selected operating point x_0, is the tangent line to the curve at that point. This approximation is given conveniently by the first two terms of the Taylor series expansion of $f(x)$:

$$y = f(x_0) + \left[\frac{df}{dx}\bigg|_{x=x_0} \right](x - x_0) + \left[\frac{d^2 f}{dx^2}\bigg|_{x=x_0} \right]\frac{(x - x_0)^2}{2!} + \cdots$$

The linear approximation is

$$y \cong f(x_0) + \left[\frac{df}{dx}\bigg|_{x=x_0} \right](x - x_0) \tag{2.1b}$$

Linearization over desired ranges of operation is usually possible and we will mostly employ linear models for systems with nonlinearities. For situations where this approximation becomes too crude to be employed and/or nonlinearities are deliberately introduced, mathematical tools such as Lyapunov stability analysis, describing function method and the phase-plane method may be employed (refer Chapter 16).

In this book, we will be mainly concerned with linear, time-invariant, lumped systems. Application of physical laws to such a system, as we shall see in this chapter, results in a set of ordinary differential equations which directly provide us with the rates of change of significant variables characterizing the dynamics of the system. Laplace transformation of these differential equations with zero initial conditions[1] gives us *transfer function* representation.

Through the examples given in this chapter, it will become clear that to obtain a transfer function model, one first writes the differential equations of the system which 'naturally' arise from the application of physical laws. Differential equations modeling is thus a 'natural' step required for transfer function modeling.

These differential equations, when rearranged in the form of a set of first-order differential equations, are called *state equations*. State variable analysis and design methods use the matrix theory and are, to some extent, different from those based on transfer function models. We have separated state variable methods from those based on transfer functions. State variable methods are explained in Chapter 14 and onwards, while the rest of the book uses the transfer function approach.

The transfer function blocks obtained for various subsystems of a feedback control system, can be organized into block diagrams or signal flow graphs to graphically depict the interconnections. Block diagram models are very convenient for designing and analyzing control systems. Block diagram algebra (used for manipulations of block diagrams) given in this chapter, will be used throughout the book as an important tool for the design process.

[1] The case of zero initial conditions is common in control system design because the signals are usually defined as perturbations from some steady state. When this is done, the initial value of the perturbation is, by definition zero.

Block diagrams are adequate for the representation of interrelationships in feedback control systems studied in this book. However, for a system with reasonably complex interrelationships, the block diagram algebra is cumbersome. Alternative representation is the signal flow graph model. Its advantage is the availability of a flow-graph gain formula which provides the relations between system variables directly, without any manipulation of the flow graph. We will introduce this alternative tool also in this chapter; this will prepare students for handling complex situations in industrial practice.

Since Laplace transformation of differential equations obtained by application of physical laws to a given system, gives us the transfer function model, and the interconnection of these models gives us the block diagram/signal flow graph model, a natural starting point for this chapter is the review of Laplace transform method. Our focus will be on Laplace transforms of signals and Laplace transform relationships useful for control system analysis, and the method for obtaining dynamic response of a transfer function model for an individual subsystem. The method will also be applicable for obtaining dynamic response of multiple subsystems interconnected by a block diagram/signal flow graph, after reducing the block diagram/signal flow graph model to a single transfer function.

2.2 THE LAPLACE TRANSFORM

The Laplace transform[2] is a technique of expressing a signal $f(t)$ as a continuous sum of exponential functions of the form e^{st} with $s = \sigma + j\omega;\ j \triangleq \sqrt{-1}$. The mapping[3]

$$f(t) \rightarrow F(s);\ f(t) = 0 \text{ for } t < 0$$

where

$$F(s) = \int_0^\infty f(t)e^{-st}dt;\ t > 0 \tag{2.2}$$

defines the Laplace transform of $f(t)$, denoted by $\mathscr{L}[f(t)]$ or $F(s)$, provided the integral exists [64].

The Laplace transform has proven to be a natural and powerful means of analyzing linear time-invariant systems. Most of the physical systems are nonlinear and time varying in nature; however many of these nonlinear systems have regions of operation in which their behavior is 'reasonably' linear. Also, their characteristics change very slowly in relation to variations in the inputs. A linear time-invariant model is often appropriate for many physical systems, making them amenable to analysis by techniques such as the Laplace transform. This is fortunate because the analysis techniques available for linear time-invariant systems are much more powerful and general than those that apply to nonlinear/time-varying systems. Thus, being able to make the assumption that a system is linear and time invariant is crucial.

The class of linear time-invariant systems, as we shall see later in this chapter, can be represented by ordinary differential equations with constant, real coefficients. Application of the Laplace transform turns a differential equation into an algebraic equation that is much easier to manipulate and interpret. The resulting Laplace transforms are ratios of polynomials in s with constant, real coefficients. The benefits of this representation will become clear in subsequent chapters where it leads to graphical techniques for designing compensation (controllers) for systems with feedback.

The condition of existence of integral in Eqn. (2.2) does not turn out to be a matter of great concern because the class of functions that arise in the analysis of linear time-invariant control systems are well

---

[2] We assume that the reader has had a previous introduction to the theory of Laplace transforms. Sections 2.2–2.5 provide a brief review of the Laplace transform techniques, adequate for the purposes of this text.

[3] It is conventional to use capital letters for s-functions and lower case for t-functions. Our notation mostly follows this convention.

behaved and the transform analysis exists. The brief introduction to the theory of Laplace transforms given in this chapter provides the working tools adequate for the purposes of this text.

Before proceeding, it is worth making sure that the nature of the mapping (2.2) is fully understood. The Laplace transform variable s is a complex variable:

$$\mathrm{Re}(s) + j\,\mathrm{Im}(s) = s = \sigma + j\omega;\ j \triangleq \sqrt{-1}$$

For a fixed value of σ, the locus of s is a vertical line in the complex s-plane (Fig. 2.4a). The vertical line corresponding to $\sigma = 0$ in the complex s-plane (Fig. 2.4b) will be of specific interest to us. This line is the *imaginary axis* of the s-plane.

Very often, Laplace transforms given by Eqn. (2.2) work out as the ratio of two constant-coefficient polynomials of the form:

$$F(s) = \frac{b_0 s^m + b_1 s^{m-1} + \cdots + b_m}{s^n + a_1 s^{n-1} + \cdots + a_n};\ m \le n \qquad (2.3a)$$

where all the a_i, and b_j are real.

The Laplace transform *zeros* are the values of s for which the transform is zero. The zeros are found by setting the numerator polynomial to 0 and solving the resulting equation. The Laplace transform *poles* are the values of s for which the transform is infinite. The poles are found by setting the denominator polynomial to 0 and solving the resulting equation. The *degree* of denominator polynomial is given by integer n, and that of the numerator is given by integer m.

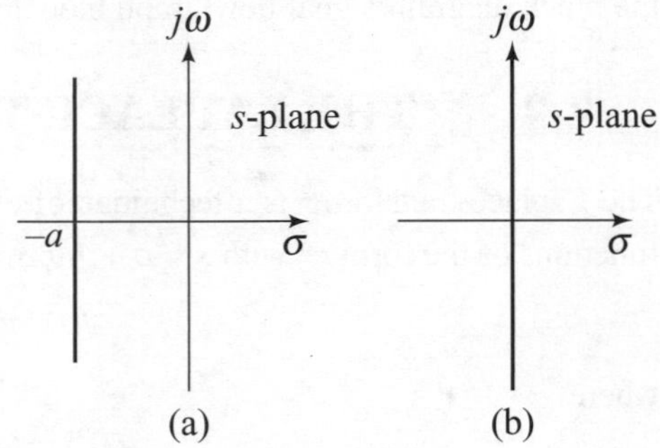

Fig. 2.4 *The complex s-plane*

The fact that the coefficients a_i, and b_j are real guarantees that the roots of the denominator and numerator polynomials of $F(s)$ will be either real or complex-conjugate pairs. Under these conditions, we can write

$$F(s) = \frac{K\prod_{i=1}^{m}(s+z_i)}{\prod_{i=1}^{n}(s+p_i)};\ m \le n \qquad (2.3b)$$

where zeros $s = -z_i$ and poles $s = -p_i$ are either real or complex-conjugate pairs; K is a constant.

To see the nature of mapping (2.2), we consider the function

$$f(t) = \begin{cases} e^{-at} & \text{for } t \ge 0 \\ 0 & \text{for } t < 0 \end{cases} \qquad (2.4)$$

For this function,

$$F(s) = \int_0^\infty e^{-at}e^{-st}\,dt = \int_0^\infty e^{-(s+a)}\,dt = \frac{-1}{s+a}\Big[e^{-(s+a)t}\Big]_{t=0}^{t=\infty} = \frac{1}{s+a}\Big[1 - \lim_{t\to\infty} e^{-(s+a)t}\Big] \qquad (2.5a)$$

$$\lim_{t\to\infty} e^{-(s+a)t} = \lim_{t\to\infty} e^{-(\sigma + j\omega + a)t} = \lim_{t\to\infty} e^{-(\sigma+a)t}e^{-j\omega t} \qquad (2.5b)$$

The term

$$e^{-j\omega t} = \cos \omega t - j \sin \omega t$$

will not approach a limit as $t \to \infty$. Therefore, for the limit in Eqn. (2.5b) to exit, the term $e^{-(\sigma+a)t}$ must approach the limit zero as $t \to \infty$. This will be true only if

$$-(\sigma + a) < 0$$

or equivalently

$$\sigma > -a \tag{2.6a}$$

Thus for $\sigma > -a$, Eqn. (2.5a) gives

$$F(s) = \frac{1}{s+a} \tag{2.6b}$$

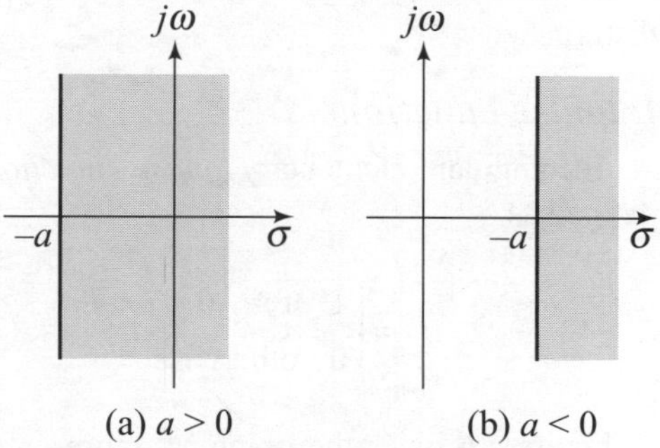

Fig. 2.5 *Region of convergence*

The regions of convergence for both $a > 0$ and $a < 0$ are shown in Fig. 2.5. The region is an infinite half plane that lies strictly to the right of the vertical line through $s = -a$. The same result holds if a is complex. For $a = \alpha + j\beta$, the region of convergence is an infinite half plane to the right of the vertical line through $s = -\alpha$.

2.3 LAPLACE TRANSFORM OF SIGNALS USEFUL FOR CONTROL SYSTEM ANALYSIS

To motivate the discussion in a control systems context, we revisit the generalized functional block diagram of Fig. 1.9. There are two inputs (external excitations) applied to a control system; a command input $y_r(t)$ (or reference input $r(t)$), and a disturbance input $w(t)$. We will first talk of disturbances.

2.3.1 Simple Models of Disturbances

The presence of disturbances is one of the main reasons for using feedback control. Without disturbances, there is probably no need for feedback control.

The disturbances may be classified as load disturbances, measurement errors, and parameter variations. ***Load disturbances are external disturbances from the process environment which influence the process variables. Measurement errors enter through the measuring instruments (sensors). Disturbances also enter in the form of variations in the parameters due to wear and tear, aging, environmental effects, etc., or due to unmodeled dynamics***. In Chapter 4, we will study the effects of these disturbances on control system performance. Here, our concern is with external disturbance signals and their Laplace transforms.

External disturbances, $w(t)$, are typically the uncontrolled variations in the load on a control system. In systems controlling mechanical motions, load disturbances may represent forces/torques, e.g., wind gusts on a stabilized antenna (refer to Example 1.3), variation in payload of a robot (refer to Example 1.8), etc. In voltage regulating systems, variations in electrical load are a major source of disturbances. In thermal systems, the load disturbances may be caused by variations in surrounding temperature (refer to Example 1.2). In fluidic systems, the load disturbances may result from variations in feed flow or variations in demanded flow (refer to Example 1.5).

Disturbances which are erratic, uncertain signals having no distinguishing waveform properties, are commonly referred to as *noise*. The accommodation of noise in a control problem is best handled by treating the noise as a *stochastic* or *random process*. The general theory of stochastic processes is quite complex, and is beyond the scope of this book.

In this book, we consider load disturbances which typically vary slowly. Such disturbances are normally well behaved uncertain signals having distinguishing waveform properties. As we shall see in Chapter 4, the errors due to load disturbances tend to be corrected by proper design of feedback control.

In the following, we consider simple, yet effective, models that characterize commonly occurring disturbances.

Impulse Function

A rectangular elementary *pulse function* is defined as

$$p_T(t) = \begin{cases} 1 & \text{for } 0 < t < T \\ 0 & \text{otherwise} \end{cases} \quad (2.7a)$$

Figure 2.6 gives the graphical representation of pulse functions.

$Ap_T(t - t_0)$ is a pulse of height A and duration T occurring at $t = t_0$:

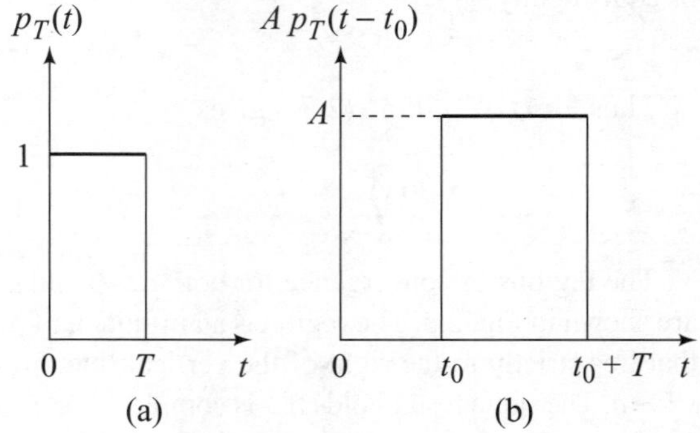

Fig. 2.6 *Pulse functions*

$$A\,p_T(t - t_0) = \begin{cases} A & \text{for } 0 < (t - t_0) < T \\ 0 & \text{otherwise} \end{cases} \quad (2.7b)$$

Consider the *unit-pulse function* $\delta_\Delta(t) = \dfrac{1}{\Delta} p_\Delta(t)$ (2.8)

It is a pulse of height $\dfrac{1}{\Delta}$ and duration Δ (i.e., of unit 'area') occurring at $t = 0$.

The properties of $\delta_\Delta(t)$ are such that its height increases whenever its duration decreases, but in such a way that its 'area' thereby remains constant and is equal to unity.

As $\Delta \to 0$, we obtain the limiting 'function'

$$\delta(t) \triangleq \lim_{\Delta \to 0} \delta_\Delta(t) \quad (2.9)$$

which is called the *unit-impulse function* or *δ-function*.

Properties of the δ-function are:

$$\delta(t) = \lim_{\Delta \to 0} \delta_\Delta(t) = \begin{cases} 0 & \text{for } t \neq 0 \\ \infty & \text{for } t = 0 \end{cases} \quad (2.10a)$$

$$\int_{-\infty}^{\infty} \delta(t)dt = \int_{-\infty}^{\infty} \lim_{\Delta \to 0} \delta_\Delta(t)dt = 1 \quad (2.10b)$$

From these properties it follows that $\delta(t)$ is not an ordinary function as we cannot assign a value to the dependent variable for each value of the independent variable. The quantitative characteristic of $\delta(t)$ does not lie in its amplitudes but is given solely by its 'area'. Consequently, $\delta(t)$ is not defined in the ordinary sense but $\int_{-\infty}^{\infty} \delta(t)dt$ is defined; $\delta(t)f(t)$ is not defined but $\int_{-\infty}^{\infty} \delta(t)f(t)dt$ is defined:

$$\int_{-\infty}^{\infty} \delta(t) f(t) dt = \int_{-\infty}^{\infty} \lim_{\Delta \to 0} \delta_\Delta(t) f(t) dt = \lim_{\Delta \to 0} \int_{-\infty}^{\infty} \frac{1}{\Delta} p_\Delta(t) f(t) dt = \lim_{\Delta \to 0} \frac{1}{\Delta} \int_0^{\Delta} f(t) dt$$

$$= f(0), \text{ assuming } f(t) \text{ is continuous at } t = 0 \tag{2.11}$$

In other words, the δ-function sifts out (draws out) the value of the function at the time it (the δ-function) occurs.

Figure 2.7 gives the graphical representation of impulse functions. The definition of impulse $\delta(t - t_0)$ occurring at $t = t_0$ is based on the following properties:

$$\left.\begin{aligned} &1. \int_{-\infty}^{\infty} \delta(t - t_0) dt = 1 \\ &2. \int_{-\infty}^{\infty} f(t)\delta(t - t_0) dt = f(t_0), \\ &\text{for any } f(t) \text{ that is continuous at } t = t_0 \end{aligned}\right\} \tag{2.12}$$

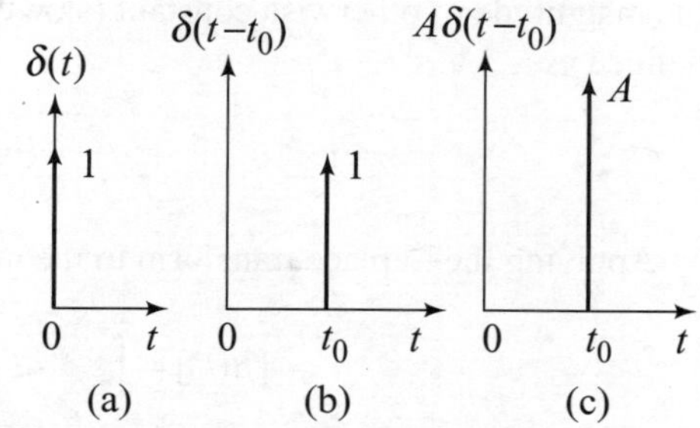

Fig. 2.7 *Impulse functions*

$A\delta(t - t_0)$ is an impulse of 'area' A (strength A) occurring at $t = t_0$.

Unit-impulse function has been listed as Fig. 2.8a in simple models of disturbances given in Fig. 2.8.

Applying the definition of the Laplace transform yields (refer to Eqns (2.7) – (2.9)).

$$\mathscr{L}[\delta(t)] = \int_0^{\infty} \delta(t) e^{-st} dt = \lim_{\Delta \to 0} \int_0^{\Delta} \frac{1}{\Delta} e^{-st} dt = \lim_{\Delta \to 0} \frac{-1}{\Delta s} [e^{-st}] \Big|_{t=0}^{t=\Delta} = \lim_{\Delta \to 0} \frac{1 - e^{-\Delta s}}{\Delta s}$$

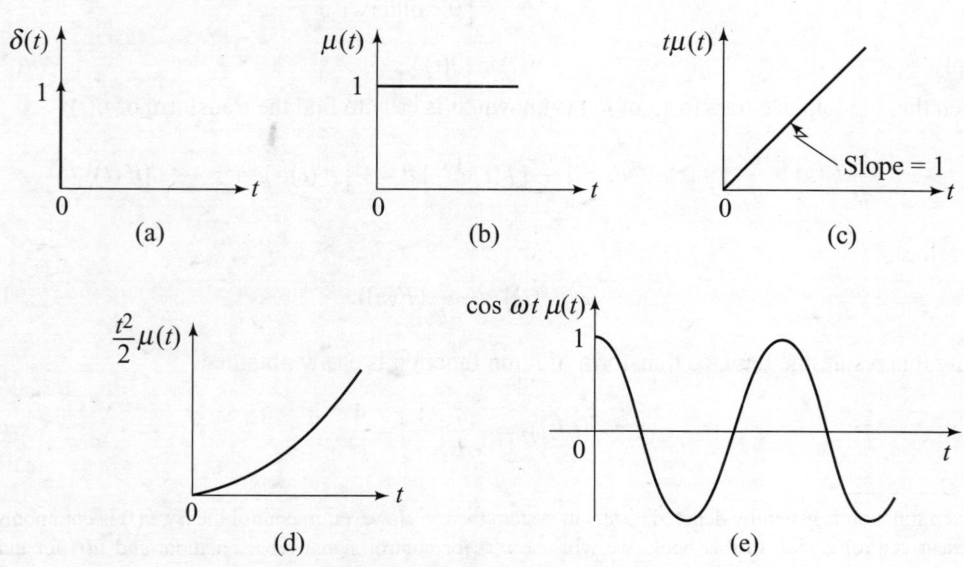

Fig. 2.8 *Simple models of disturbances*

This expression can be evaluated by treating numerator and denominator as functions of Δ and applying L Hospital's rule to obtain:

$$\lim_{\Delta\to 0}\frac{1-e^{-\Delta s}}{\Delta s}=\lim_{\Delta\to 0}\frac{\dfrac{d}{d\Delta}(1-e^{-\Delta s})}{\dfrac{d}{d\Delta}(\Delta s)}=\lim_{\Delta\to 0}\frac{se^{-\Delta s}}{s}=1$$

Therefore,

$$\mathscr{L}[\delta(t)]=1 \tag{2.13}$$

Step Function

A step signal is another prototype for a disturbance. It is typically used to represent a sudden jump in the magnitude of otherwise constant (slowly varying) disturbances. A *unit-step function*[4] (Fig. 2.8b) is defined as

$$\mu(t)=\begin{bmatrix}1 & \text{for } t>0\\ 0 & \text{otherwise}\end{bmatrix} \tag{2.14a}$$

Applying the Laplace transform to the unit step yields

$$\mathscr{L}[\mu(t)]=\int_0^\infty e^{-st}dt=\frac{-1}{s}[e^{-st}]\Big|_{t=0}^{t=\infty}=\frac{1}{s}\left[1-\lim_{t\to\infty}e^{-st}\right]=\frac{1}{s} \tag{2.14b}$$

where $\lim_{t\to\infty} e^{-st}=0$ if $\mathrm{Re}(s)=\sigma>0$. Thus, the region of convergence is the open-half plane to the right of the $j\omega$ axis (refer to Fig. 2.5).

Ramp Function

A unit-ramp function (Fig. 2.8c) is defined as

$$w(t)=\begin{bmatrix}t & \text{for } t>0\\ 0 & \text{otherwise}\end{bmatrix} \tag{2.15a}$$

or simply

$$w(t)=t\mu(t) \tag{2.15b}$$

Given that the Laplace transform of $f(t)$ is known, it is easy to find the transform of $tf(t)$:

$$\frac{d}{ds}F(s)=\frac{d}{ds}\int_0^\infty f(t)e^{-st}dt=\int_0^\infty\frac{d}{ds}[f(t)e^{-st}]dt=-\int_0^\infty tf(t)e^{-st}dt=-\mathscr{L}[tf(t)]$$

Therefore,

$$\mathscr{L}[tf(t)]=-\frac{d}{ds}(F(s)) \tag{2.16}$$

Using this result, the Laplace transform of ramp function is easily obtained.

$$\mathscr{L}[t\mu(t)]=-\frac{d}{ds}\left(\frac{1}{s}\right)=\frac{1}{s^2} \tag{2.17}$$

[4] Unit-step function is generally denoted as $u(t)$ in system theory. However, in control theory, $u(t)$ is commonly used to represent control signal. In this book, we will use $u(t)$ for control signal representation, and $\mu(t)$ for unit-step function.

The region of convergence is given by Re(s) > 0.

A ramp function is used to represent disturbances that suddenly start to drift away. In practice, the disturbances are often bounded; however, the ramp is a useful idealization within the usable range.

Parabolic Function

A parabolic function is one degree faster than the ramp function (in practice, we seldom find it necessary to use a signal faster than a parabolic function).

A *unit-parabolic function* (Fig. 2.8d) is defined as

$$w(t) = \begin{bmatrix} \dfrac{t^2}{2} & \text{for } t > 0 \\ 0 & \text{othewise} \end{bmatrix} \tag{2.18a}$$

or simply

$$w(t) = \frac{t^2}{2}\mu(t) \tag{2.18b}$$

Using the result in Eqn. (2.16), we obtain

$$\mathscr{L}\left[\frac{t^2}{2}\mu(t)\right] = \frac{1}{2}\mathscr{L}[t\{t\mu(t)\}] = -\frac{1}{2}\frac{d}{ds}\left(\frac{1}{s^2}\right) = \frac{1}{s^3} \tag{2.18c}$$

The region of convergence is given by Re(s) > 0.

Note that in deriving this result, we have used the following property of Laplace transforms:

$$\mathscr{L}[kf(t)] = k\mathscr{L}[f(t)]\text{; } k \text{ is a constant} \tag{2.19}$$

Sinusoidal Function

Sinusoidally time-varying signals (*sinusoids*, for short) are represented by sine or cosine functions. A sine function is defined as

$$w(t) = \begin{bmatrix} W_m \sin(\omega t + \phi) & \text{for } t \ge 0 \\ 0 & \text{otherwise} \end{bmatrix} \tag{2.20a}$$

or simply

$$w(t) = W_m \sin(\omega t + \phi)\mu(t) \tag{2.20b}$$

where W_m is the *amplitude* of the sinusoid, ω is the *angular frequency* in radians/sec, and ϕ is the *phase* in radians. The sinusoid repeats itself every T seconds; T is the period of the sinusoid ($\omega T = 2\pi$; $T = 2\pi/\omega$ seconds/cycle). Reciprocal of T is the number of cycles per second; $f = 1/T$ cycles/sec or hertz (Hz) is the *cyclic frequency*[5] of the sinusoid ($\omega = 2\pi f$). The argument ($\omega t + \phi$) of the sine function may be expressed in radians or in degrees.

Let us examine two sinusoids: $w_1 = W_{m1}\sin(\omega t)\mu(t)$, and $w_2 = W_{m2}\sin(\omega t + \phi)\mu(t)$, shown in Fig. 2.9. The first positive peak of w_2 occurs earlier than the first positive peak of w_1; we say that w_2 *leads* w_1 by ϕ or that w_1 lags w_2 by ϕ.

[5] Multiple use of the symbols cannot be avoided; earlier, f had been used for a function, and now we have used it for cyclic frequency. The reader is expected to be careful on interpretation.

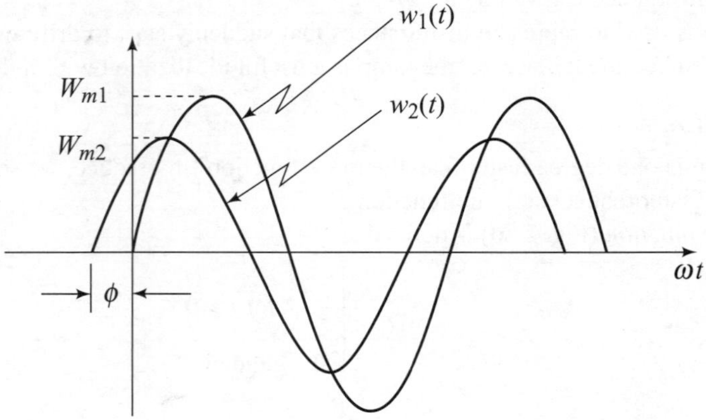

Fig. 2.9 *Leading and lagging sinusoids*

Sinusoids are easily expressed in terms of phasors, which are more convenient to work with than sine and cosine functions. A *phasor* is a complex number that represents the amplitude W_m and phase ϕ of the sinusoid. There are three representations for complex numbers: (i) polar, (ii) rectangular (or Cartesian), and (iii) complex exponential.

The sine function given in Eqn. (2.20b) can be expressed in terms of phasor W as

$$W = W_m \angle \phi, W_m \geq 0, -\pi \leq \phi \leq \pi \tag{2.21}$$

This is a *polar form* representation, shown in Fig. 2.10 as a vector of length W_m making an angle ϕ with horizontal axis. From this figure, we obtain the following rectangular form representation for phasor W:

$$W = W_m \cos \phi + jW_m \sin \phi \tag{2.22}$$

The phasor representation in *complex exponential form* is based on the trigonometric identity:

$$e^{j\theta} = \cos \theta + j \sin \theta \tag{2.23a}$$

or

$$\cos \theta = \frac{e^{j\theta} + e^{-j\theta}}{2}; \sin \theta = \frac{e^{j\theta} - e^{-j\theta}}{2j} \tag{2.23b}$$

From Eqns (2.22) and (2.23), we obtain

$$W = W_m (\cos \phi + j \sin \phi) = W_m \, e^{j\phi} \tag{2.24}$$

This is complex exponential form representation of a phasor.

A sinusoid can be expressed in either sine or cosine form. When comparing two sinusoids, it is expedient to express both as either sine or cosine with positive amplitude. This is achieved by using the following trigonometric identities:

$$\sin (A \pm B) = \sin A \cos B \pm \cos A \sin B \tag{2.25a}$$

$$\cos (A \pm B) = \cos A \cos B \mp \sin A \sin B \tag{2.25b}$$

Using these relationships, we can transform a sinusoid from sine form to cosine form or vice versa.

$$\sin(\omega t \pm 90°) = \pm \cos \omega t; \cos(\omega t \pm 90°) = \mp \sin \omega t \tag{2.25c}$$

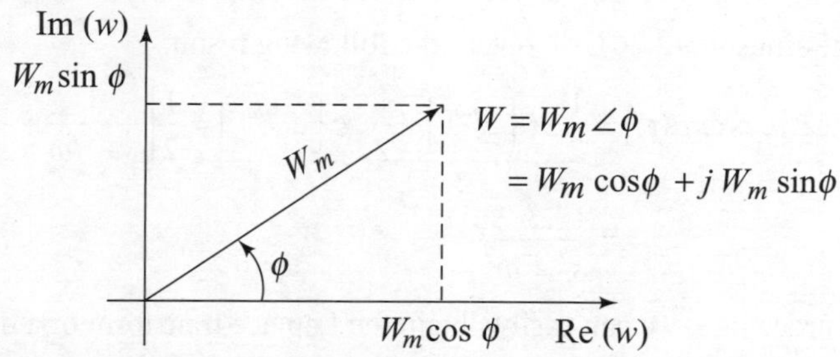

Fig. 2.10 *Phasor representation of a complex number*

We are interested in sinusoids primarily for two reasons:

1. Sinusoid is a prototype for a periodic disturbance—for example, waves in ship control systems, excitations due to mass eccentricity and misalignments in rotational systems, tooth meshing in gears, etc.
2. ***Through Fourier transformation, any time-varying signal may be resolved into a sum of sinusoidal functions $A_i \sin(\omega_i t + \phi_i)$ of different frequencies ω_i (established later in this chapter). Therefore, analysis of a linear system excited by any input function can be carried out indirectly in terms of analysis with sinusoidal excitations. This 'indirect' route of control system analysis has led to extremely important results for control system design—the frequency response methods.*** Chapters 10–12 are fully devoted to this indirect approach of analysis and design.

In the following, through an example, we illustrate the procedure of developing Laplace transforms of sinusoidal functions. Consider a sinusoid of unity amplitude and frequency ω (Fig. 2.8e):

$$w(t) = \cos\omega t\, \mu(t) \tag{2.26}$$

The transform of sinusoids can be found by expanding the function into complex exponential components (refer to Eqn. (2.23b)).

$$w(t) = \left[\frac{e^{j\omega t} + e^{-j\omega t}}{2}\right]\mu(t) \tag{2.27}$$

To obtain Laplace transform of this function, we will use linearity property of Laplace transforms:

$$\mathscr{L}[k_1\, f_1(t) \pm k_2\, f_2(t)] = k_1 \mathscr{L}[\, f_1(t)] \pm k_2 \mathscr{L}\,[\, f_2(t)] \tag{2.28}$$

where k_1 and k_2 are constants.

We also require Laplace transform of an exponential function (refer to Eqns (2.6)):

$$\mathscr{L}\left[e^{-at}\mu(t)\right] = \frac{1}{s+a}; \mathrm{Re}(s) = \sigma > -a \tag{2.29}$$

The same result holds if a is complex. For example, when $a = j\omega$,

$$\mathscr{L}\left[e^{-j\omega t}\mu(t)\right] = \frac{1}{s + j\omega}; \;\; \mathrm{Re}(s) = \sigma > 0 \tag{2.30}$$

Returning to the sinusoid (2.26), we obtain the following result:

$$\mathscr{L}[\cos\omega t\mu(t)] = \frac{1}{2}\mathscr{L}\left[e^{j\omega t}\right] + \frac{1}{2}\mathscr{L}\left[e^{-j\omega t}\right] = \frac{1}{2}\left[\frac{1}{s-j\omega} + \frac{1}{s+j\omega}\right]$$

$$= \frac{s}{s^2+\omega^2} \qquad (2.31)$$

Following this procedure, we can easily obtain the Laplace transform of a decaying sinusoid:

$$\mathscr{L}\left[e^{-at}\cos\omega t\mu(t)\right] = \frac{s+a}{(s+a)^2+\omega^2}; a > 0 \qquad (2.32)$$

Comparing the two transforms in Eqns (2.31) and (2.32), we observe that we substitute $(s + a)$ for s in the Laplace transform of $\cos\omega t\ \mu(t)$ to obtain Laplace transform of $e^{-at}\cos\omega t\ \mu(t)$. In general,

$$\mathscr{L}\left[e^{-at}f(t)\right] = F(s+a);\ a > 0 \qquad (2.33)$$

2.3.2 Standard Test Signals

In addition to the disturbance input, the other excitation signal to a control system is the command input (Fig. 1.9). The command (or reference) signals of some control systems are known to the designer. For example, the command signal of a heating system (Example 1.2) is a step function, whose magnitude is the desired temperature.

In many control systems, however, the command signals are not known fully ahead of time. For instance, in a radar tracking system for military applications (refer to Example 1.3), the position and speed of the target to be tracked vary in an unpredictable manner; it is thus difficult to express the actual input signals mathematically by simple functions. ***The characteristics of actual signals which severely strain a control system are***

(i) ***a sudden shock (modeled by impulse function),***
(ii) ***a sudden change (modeled by step function),***
(iii) ***a linear change with time (modeled by ramp function), and***
(iv) ***faster changes with time (a parabolic function is one degree faster than a ramp function).***

System dynamic behavior can be adequately judged and compared under the application of standard test signals—an impulse input, a step input, a ramp input, and a parabolic input.

Another standard test signal of great importance is the sinusoidal signal. When the response of a linear time-invariant system is analyzed in frequency domain, a sinusoidal function with variable frequency is used as input function. Input frequency is swept from zero to beyond the significant range of system characteristics; curves in terms of amplitude ratio and phase between output and input are drawn as functions of frequency. It is possible to predict the time-domain behavior of the system from the frequency-domain characteristics. We will take up this aspect of control system analysis in Chapters 10 and 11.

We summarize the Laplace transforms we have derived to this point plus some additional transforms in Table 2.1. All the transforms listed in the table can easily be derived from the first principles. It may be noted that in this transform table, regions of convergence have not been specified. In our applications of systems analysis, which involve transformation from time domain to s-domain and inverse transformation, the variable s acts as a dummy operator. If transform pairs for functions of interest to us are available, we are not concerned with the region of convergence.

Table 2.1 *Abbreviated table of Laplace transforms*

$f(t); t \geq 0$	$F(s)$	$f(t); t \geq 0$	$F(s)$
$\delta(t)$: unit impulse	1	$1 - e^{-at}$	$\frac{1}{s(s+a)}$
$\mu(t)$: unit step	$\frac{1}{s}$	$\sin \omega t$	$\frac{\omega}{s^2 + \omega^2}$
t	$\frac{1}{s^2}$	$\cos \omega t$	$\frac{s}{s^2 + \omega^2}$
$\frac{t^2}{2}$	$\frac{1}{s^3}$	$e^{-at} \sin \omega t$	$\frac{\omega}{(s+a)^2 + \omega^2}$
e^{-at}	$\frac{1}{s+a}$	$e^{-at} \cos \omega t$	$\frac{s+a}{(s+a)^2 + \omega^2}$
te^{-at}	$\frac{1}{(s+a)^2}$		

2.4 LAPLACE TRANSFORM INVERSION

A convenient method for obtaining the *inverse Laplace transform of* $Y(s)$ is to use the method of "partial fraction expansion". The idea of this method is to expand $Y(s)$ as a sum of partial fractions, and then find $y(t)$ as the sum of the inverse transforms of the individual members of the partial fraction expansion. The reason this works well is that the Laplace transforms that frequently arise in the study of control systems are ratios of polynomials in s with real coefficients (refer to Eqns (2.3)). The fact that the coefficients are real is crucial because it guarantees that the roots of the numerator and denominator polynomials will be either real or complex-conjugate pairs.

This, in turn, means that the individual terms in the partial fraction expansion of the transform will be simple in form. The transform pairs encountered using the partial fraction expansion technique will usually be those found in Table 2.1 (those not found in the table can easily be derived by using basic properties of Laplace transformation); thus inverse transform of each term in the expansion is known. As a consequence, with some practice, it is possible to do the inverse transformation by inspection.

Later in this chapter, we will illustrate the partial fraction expansion method by examples.

2.4.1 Convolution Integral

One important process in linear system theory is known as *convolution*. This, in fact, is the basis of the theory—the output of a system is the convolution of a stimulus and the system's impulse response. This aspect of linear system theory will appear in Section 2.7. Our objective here, is to define the convolution operation in time domain, and map this operation to transform domain using Laplace transformation.

The convolution of one signal $f_1(t)$; $f_1(t) = 0$ for $t < 0$, with another signal $f_2(t)$; $f_2(t) = 0$ for $t < 0$, is given by the integral

$$\int_0^t f_1(t-\tau) f_2(\tau) d\tau$$

This *convolution integral* is often written as $f_1(t) * f_2(t)$. Also note that if we substitute $t - \tau = \theta$, then

$$\int_0^t f_1(t-\tau)f_2(\tau)\,d\tau = -\int_t^0 f_1(\theta)f_2(t-\theta)\,d\theta = \int_0^t f_1(\tau)f_2(t-\tau)\,d\tau$$

Hence,

$$f(t) = f_1(t) * f_2(t) = \int_0^t f_1(t-\tau)f_2(\tau)d\tau \tag{2.34a}$$

$$= \int_0^t f_1(\tau)f_2(t-\tau)d\tau = f_2(t) * f_1(t) \tag{2.34b}$$

The application of Laplace transform to Eqn. (2.34a) yields

$$F(s) = \int_0^\infty f(t)e^{-st}dt = \int_0^\infty\left[\int_0^t f_1(t-\tau)\, f_2(\tau)d\tau\right]e^{-st}dt \tag{2.35}$$

Since $f_1(t-\tau) = 0$ for $\tau > t$ as implied by the condition: $f_1(t) = 0$; $t < 0$, the upper limit to t in Eqn. (2.35) can be set at ∞:

$$F(s) = \int_0^\infty\left[\int_0^\infty f_1(t-\tau)f_2(\tau)d\tau\right]e^{-st}dt$$

Changing the order of integrations, gives

$$F(s) = \int_0^\infty\left[\int_0^\infty f_1(t-\tau)e^{-st}dt\right]f_2(\tau)d\tau = \int_0^\infty\left[\int_0^\infty f_1(t-\tau)e^{-s(t-\tau)}dt\right]e^{-s\tau}f_2(\tau)d\tau$$

which, if we substitute $\theta = t - \tau$, can be expressed as

$$F(s) = \int_0^\infty\left[\int_{-\tau}^\infty f_1(\theta)e^{-s\theta}d\theta\right]e^{-s\tau}f_2(\tau)d\tau$$

Using the fact that $f_1(\theta) = 0$ for $\theta < 0$, we obtain

$$F(s) = \int_0^\infty\left[\int_0^\infty f_1(\theta)e^{-s\theta}d\theta\right]e^{-s\tau}f_2(\tau)d\tau = \left[\int_0^\infty f_1(\theta)e^{-s\theta}d\theta\right]\left[\int_0^\infty f_2(\tau)e^{-s\tau}d\tau\right] = F_1(s) \tag{2.36}$$

where

$$F_2(s) = \int_0^\infty f_2(t)e^{-st}dt \text{ and } F_1(s) = \int_0^\infty f_1(t)e^{-st}dt$$

We see that by applying the Laplace transform, a convolution integral is transformed into an algebraic equation.

Equation (2.36) shows that the multiplication of two transformed functions in the s-domain is equivalent to the convolution of two corresponding real functions of time in t-domain. An important fact to remember is that the inverse Laplace transform of the product of two functions in the s-domain is *not* equal to the product of the two corresponding real functions in the t-domain, i.e., in general

$$\mathscr{L}^{-1}\left[F_1(s)F_2(s)\right] \neq f_1(t)f_2(t) \tag{2.37}$$

2.5 LAPLACE TRANSFORM RELATIONSHIPS

When we begin to use Laplace transforms to analyze systems and their controllers, we will need to manipulate the transforms of signals and their time derivatives. We are required to know only a handful of useful properties and formulae, and some practice will soon ensure a good familiarity with their use.

Differentiator

We can find the Laplace transform of the derivative of a signal $y(t)$ by using the relationship,

$$\mathscr{L}\left[\frac{dy}{dt}\right] = sY(s) - y(0) \tag{2.38a}$$

where $y(0)$ is the initial value of $y(t)$, evaluated at $t = 0$.

We can prove this relationship using integration by parts as follows:

$$\int u dv = uv - \int v du$$

Setting $u = e^{-st}$ and $dv = \dfrac{dy}{dt}$, we get

$$\int_0^\infty e^{-st}\left(\frac{dy}{dt}\right)dt = [e^{-st}y(t)]\Big|_{t=0}^{t=\infty} - \int_0^\infty y(t)(-se^{-st})dt = -y(0) + s\int_0^\infty y(t)e^{-st}dt = sY(s) - y(0)$$

If the signal $y(t)$ has zero initial condition, then the result becomes

$$\mathscr{L}\left[\frac{dy}{dt}\right] = sY(s); \tag{2.38b}$$

a derivative operating on a signal can be represented as multiplying its Laplace transform by s. We see this in the block diagram for differentiator in the Fig. 2.11a.

Having found the Laplace transform of the first derivative of a signal, it is an easy matter to extend this result to higher derivatives.

$$\mathscr{L}\left[\frac{d^2y}{dt^2}\right] = s[sY(s) - y(0)] - \frac{dy}{dt}\bigg|_{t=0} = s^2Y(s) - sy(0) - \dot{y}(0) \tag{2.38c}$$

where the initial conditions are $y(0)$ and $\dot{y}(0)$. If the initial conditions are defined as zero,

$$\mathscr{L}\left[\frac{d^2y}{dt^2}\right] = s^2Y(s) \tag{2.38d}$$

Repeated application of this result will yield the general formula:

$$\mathscr{L}\left[\frac{d^n y(t)}{dt^n}\right] = s^nY(s) - \sum_{i=0}^{n-1} s^{n-1-i}\frac{d^i y(t)}{dt^i}\bigg|_{t=0} \quad ;n > 0, \frac{d^0 y(t)}{dt^0} \triangleq y(t) \tag{2.39}$$

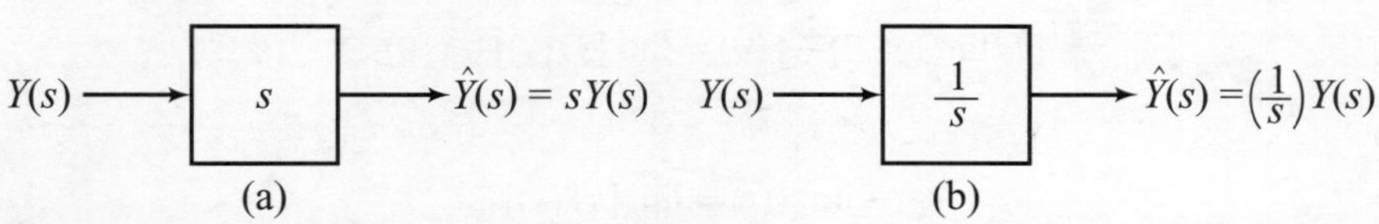

Fig. 2.11 *Block diagram representation of (a) differentiator (b) integrator in transform domain*

Integrator

The Laplace transform of the integral of a signal $y(t)$ results in the relation,

$$\mathscr{L}\left[\int_0^t y(\tau)d\tau\right] = \left(\frac{1}{s}\right)Y(s) \tag{2.40}$$

We can prove this relationship using integration by parts $\left(\int u dv = uv - \int v du\right)$. Setting $u = \int_0^t y(\tau)d\tau$ and $dv = e^{-st}$, we obtain

$$\int_0^\infty \left(\int_0^t y(\tau)d\tau\right)e^{-st}dt = \left(\int_0^t y(\tau)d\tau\right)\left(-\frac{1}{s}e^{-st}\right)\Bigg|_{t=0}^{t=\infty} - \int_0^\infty\left(-\frac{1}{s}e^{-st}\right)y(t)dt = 0 + \frac{1}{s}\int_0^\infty y(t)e^{-st}dt = \left(\frac{1}{s}\right)Y(s)$$

Thus an integrator operating on a signal can be represented as multiplying its Laplace transform by $(1/s)$. We see this in the block diagram for integrator in Fig. 2.11b.

Final Value Theorem

The final value theorem is frequently used in the study of control systems to quickly assess the final, steady-state value of the time response of a linear system. We will illustrate the use of this theorem in the next section. In the following, we state and prove the theorem.

The final value theorem is concerned with the evaluation of $y(t)$ as $t \to \infty$ assuming, of course, that $y(t)$ does approach a limit. Using partial fraction expansion for inverting Laplace transforms, it is simple to show that $y(t)$ ***approaches a limit as*** $t \to \infty$ ***if all the poles of*** $Y(s)$ ***lie in the left half of the complex plane*** ($Re(s) < 0$ ***in Fig. 2.4***). ***The imaginary-axis boundary is however excluded except for a single pole at*** $s = 0$. This is because purely sinusoidal signals, whose transforms will have poles on the imaginary axis, do not settle to a constant value as $t \to \infty$. Multiple poles at $s = 0$ are also excluded because, as we have already seen in the table of Laplace transforms, these correspond to unbounded signals like ramps. A more compact way of phrasing these conditions is to say that $sY(s)$ ***must be analytic on the boundary and the right half of the complex s-plane.*** The final value theorem states that when this condition on $sY(s)$ is satisfied, then

$$\lim_{t\to\infty} y(t) = \lim_{s\to 0} sY(s) \tag{2.41}$$

The proof is as follows:

$$\mathscr{L}\left[\frac{dy}{dt}\right] = \int_0^\infty \frac{dy}{dt}e^{-st}dt = sY(s) - y(0)$$

However,

$$\lim_{s\to 0}\int_0^\infty \frac{dy}{dt}e^{-st}dt = \int_0^\infty \frac{dy}{dt}dt = \lim_{t\to\infty}[y(t)] - y(0)$$

These two equations yield

$$\lim_{t\to\infty}[y(t)] - y(0) = \lim_{s\to 0}[sY(s)] - y(0)$$

or

$$\lim_{t\to\infty} y(t) = \lim_{s\to 0}[sY(s)]$$

The final value theorem can also be used to find the DC gain of a system. The ***DC gain is the ratio of the steady-state output of a system to its input*** (***presumed constant***). To find the DC gain, we assume that input is a unit step ($R(s) = 1/s$), and we use the final value theorem to compute the steady-state value of the output. Therefore, for a system with transfer function $G(s)$,

$$\text{DC gain} = \lim_{s \to 0} sG(s)\frac{1}{s} = \lim_{s \to 0} G(s) \tag{2.42}$$

Dead-time/ Transportation Lag

One modeling feature which we would include in our collection of simple models in this book is the feature known as *dead time* or *transportation lag*. Figure 2.12 shows modeling of the dead-time process. The Laplace transform of the dead-time model is given by

$$\mathscr{L}[y(t-\tau_D)] = e^{-s\tau_D}Y(s) \tag{2.43}$$

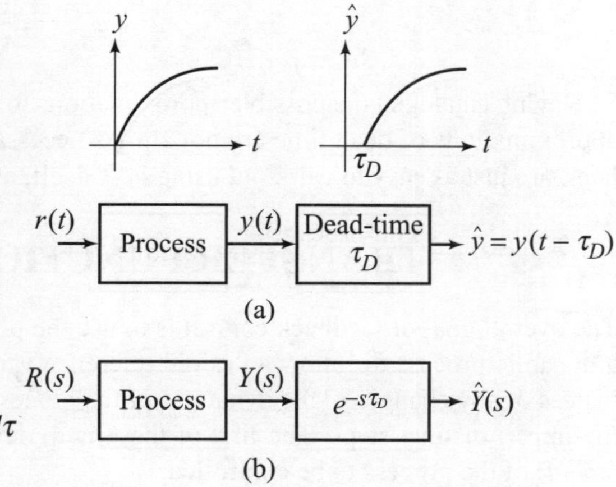

Fig 2.12 *Behavior of a dead-time element*

We can prove this relationship as follows.

$$\mathscr{L}[y(t)] = Y(s) = \int_0^\infty y(\tau)e^{-s\tau}d\tau = \int_0^\infty [y(\tau)\mu(\tau)]\, e^{-s\tau}d\tau$$

Changing the variable of integration by setting $\tau = t - \tau_D$ yields

$$Y(s) = \int_{\tau_D}^\infty y(t-\tau_D)\mu(t-\tau_D)e^{-s(t-\tau_D)}dt = e^{s\tau_D}\int_{\tau_D}^\infty y(t-\tau_D)\mu(t-\tau_D)e^{-st}dt$$

Since the function

$$\mu(t-\tau_D) = \begin{bmatrix} 1 & \text{for } t > \tau_D \\ 0 & \text{otherwise} \end{bmatrix}$$

we can write

$$Y(s) = e^{s\tau_D}\int_0^\infty y(t-\tau_D)\, e^{-st}dt = e^{-s\tau_D}\mathscr{L}[y(t-\tau_D)]$$

It is worth noting that Eqn. (2.43) results in a Laplace transform that is not a simple ratio of polynomials in s because variable s occurs in the argument of the exponential function. In the analysis of control systems using linear polynomial representations, it is sometimes inconvenient to use this exponential representation for the dead time.

There are several approximations to the dead time. The simplest is the Taylor series approximation. That is (refer to Eqn. (2.1))

$$\begin{aligned} e^{-s\tau_D} &= \left[e^{-s\tau_D}\right]_{s=0} + \left[\frac{d}{ds}e^{-s\tau_D}\right]_{s=0} s + \text{higher-order terms} \\ &= 1 - \tau_D\, s + \text{higher-order terms} \\ &\simeq 1 - \tau_D s \end{aligned} \tag{2.44}$$

Another is

$$e^{-s\tau_D} \simeq \frac{1}{1+\tau_D s} = 1-\tau_D s + \text{higher-order terms} \tag{2.45}$$

The Pade approximants [73] provide a family of approximations of increasing accuracy and complexity. These approximations are based on frequency-response comparisons; the simplest one being the first-order approximation:

$$e^{-s\tau_D} \simeq \frac{1-\tau_D s/2}{1+\tau_D s/2} \tag{2.46}$$

Having cataloged the possible approximations for the dead time, we would like to point out here that approximations of dead time are not always necessary; some of the design techniques, as we will see later, are just as easy to work out using $e^{-s\tau_D}$ itself, as it is using any of these approximations.

2.6 TRANSFER FUNCTION MODELS

The overall goal of feedback control is to use the principle of feedback to cause the output variable of a dynamic process to follow a desired reference variable accurately, regardless of any external disturbances or any changes in the dynamics of the process. This complex goal is met as a result of a number of simple, distinct steps. The first of these is to develop mathematical description (called a *dynamic model*) of the process to be controlled.

The term *model*, as it is used and understood by control engineers, means a set of differential equations that describe the dynamic behavior of the process. The differential equations developed in modeling are often nonlinear. Because nonlinear systems are significantly more challenging to solve than linear ones, and because linear models are usually adequate, the emphasis in the early chapters is primarily on linear models (Discussion on linearization and analysis of nonlinear systems is given in Chapter 16).

As we shall see, the application of physical laws to linear time-invariant systems results in a set of ordinary linear differential equations with constant coefficients (In the next chapter, we shall use the basic laws of physics to develop models of electromechanical, thermal and hydraulic systems that are commonly encountered in industrial practice). These differential equations when rearranged in the form of a set of first-order differential equations, result in a *state variable model* (we will defer discussion on state-space approach until Chapter 14). Laplace transformation of these differential equations with zero initial conditions gives the *transfer function model*. The Laplace transform variable s in a transfer function, as will be seen later in this chapter, can be associated with frequency. Hence, the transfer function description of a system is often referred to as in *frequency domain*, whereas the state variable description is said to be in *time domain*.

The transfer function model is highly suited to the frequency-domain approach of analysis and design of control systems. Since major emphasis in this book is on frequency-domain methods, the use of transfer functions will appear more frequently.

Consider the general form of differential equations, given below.

$$\frac{d^n y}{dt^n} + a_1 \frac{d^{n-1} y}{dt^{n-1}} + \cdots + a_{n-1}\frac{dy}{dt} + a_n y = b_0 \frac{d^m r}{dt^m} + b_1 \frac{d^{m-1} r}{dt^{m-1}} + \cdots + b_{m-1}\frac{dr}{dt} + b_m r \tag{2.47}$$

The coefficients a_i and b_j are real constants; m and n are integers with $m \le n$; $r = r(t)$ (the input) is the known time function, and $y = y(t)$ (the output) is the unknown solution of the equation; and n is called the *order* of the differential equation.

The coefficient of the highest derivative of y is assumed to be one. If it is not, it can be made one by dividing both sides of the equation by the original value of the coefficient.

To completely specify the problem so that a *unique solution* $y(t)$ can be obtained, a set of n *initial conditions* for $y(t)$ and its first $n-1$ derivatives must be specified. The set of initial conditions is

$$\left\{ y(0), \left.\frac{dy}{dt}\right|_{t=0}, \ldots, \left.\frac{d^{n-1}y}{dt^{n-1}}\right|_{t=0} \right\}$$

If a system is *relaxed* (*i.e.*, at rest) when the input $r(t)$ is switched on at $t = 0^+$, then the initial conditions for the differential equation model of the system are all zero. The implication of the relaxedness assumption is that the response depends only on the input $r(t)$. The input is applied at $t = 0^+$; $r(t)$ and its derivatives are zero at $t = 0$.

Applying the Laplace transform to Eqn. (2.47) under the assumption of zero initial conditions, yields (refer to Eqn. (2.39))

$$s^nY(s) + a_1s^{n-1}Y(s) + \cdots + a_{n-1}sY(s) + a_nY(s) = b_0s^mR(s) + b_1s^{m-1}R(s) + \cdots + b_{m-1}sR(s) + b_mR(s)$$

Solving for $Y(s)$:

$$Y(s) = \frac{b_0s^m + b_1s^{m-1} + \cdots + b_{m-1}s + b_m}{s^n + a_1s^{n-1} + \cdots + a_{n-1}s + a_n}R(s)$$

The ratio of polynomials in complex variable s is the transfer function $G(s)$ of the system.

$$G(s) = \frac{b_0s^m + b_1s^{m-1} + \cdots + b_{m-1}s + b_m}{s^n + a_1s^{n-1} + \cdots + a_{n-1}s + a_n} = \frac{Y(s)}{R(s)};\ m \le n \tag{2.48}$$

Thus, the ***transfer function of a linear time-invariant system is the ratio of the Laplace transforms of its output and input variables, assuming zero initial conditions.***

$$G(s) = \left.\frac{\mathscr{L}[y(t)]}{\mathscr{L}[r(t)]}\right|_{\substack{\text{system}\\ \text{relaxed}\\ \text{at } t_0=0}} = \left.\frac{Y(s)}{R(s)}\right|_{\substack{\text{system}\\ \text{relaxed}\\ \text{at } t_0=0}} \tag{2.49}$$

where $Y(s)$ = Laplace transform of the output variable $y(t)$,
$R(s)$ = Laplace transform of the input variable $r(t)$, and
$G(s)$ = transfer function of the system.

Figure 2.13 gives a block diagram of a SISO system shown in Fig. 2.2 in transform domain.

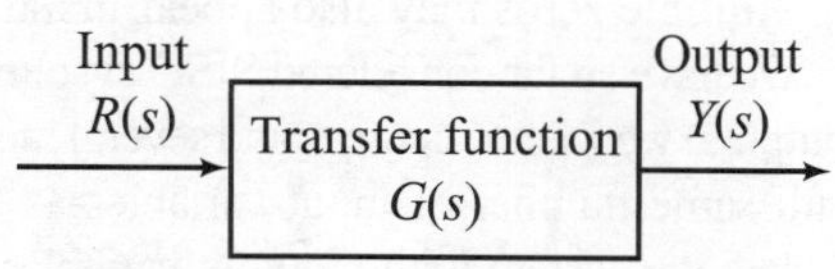

Fig. 2.13 *Description of a SISO system in transform domain*

Note that whenever a transfer function is used to describe a system, the system is always implicitly assumed to be linear, time invariant, and relaxed at $t_0 = 0$.

Since the transfer function given by Eqn. (2.48) is a rational function of a complex variable, we use the terminology of the theory of complex variables.

$G(s)$ is a rational function of s, i.e., a quotient of two polynomials of s. For $m \le n$, $G(s)$ is a *proper rational function*, and for $m < n$, $G(s)$ is a *strictly proper rational function*.

We will represent the numerator polynomial of $G(s)$ by $N(s)$, and the denominator polynomial by $\Delta(s)$:

$$G(s) = \frac{N(s)}{\Delta(s)} \tag{2.50}$$

where,

$$N(s) = b_0 s^m + b_1 s^{m-1} + \cdots + b_{m-1}s + b_m \quad (2.51a)$$

$$\Delta(s) = s^n + a_1 s^{n-1} + \cdots + a_{n-1}s + a_n \quad (2.51b)$$

Note that there is no loss in generality to assume that the coefficient of s^n in the polynomial $\Delta(s)$ is unity (i.e., the denominator polynomial $\Delta(s)$ is assumed to be *monic*[6]).

The highest power of the complex variable s in the denominator polynomial $\Delta(s)$ of the transfer function G(s) determines the order of the transfer function model. The denominator polynomial $\Delta(s)$ is called the characteristic polynomial.

The roots of the equation

$$\Delta(s) = 0 \quad (2.52a)$$

are called the poles of the transfer function G(s), and roots of the equation

$$N(s) = 0 \quad (2.52b)$$

are called the zeros.

Equation (2.52a) is called the characteristic equation; the poles are thus the characteristic roots.

The transfer function $G(s)$ given by Eqn. (2.48) can be expressed in the *pole-zero form* as

$$G(s) = \frac{K(s+z_1)(s+z_2)\cdots(s+z_m)}{(s+p_1)(s+p_2)\cdots(s+p_n)};\ m \le n \quad (2.53a)$$

where K, p_i and z_j are constants; $K = b_0$ is the *gain constant* of the transfer function, $-p_1, -p_2, \ldots, -p_n$ are the n poles (real or complex) of the transfer function (n roots of the characteristic equation (2.52a)), and $-z_1, -z_2, \ldots, -z_m$ are the m finite zeros (real or complex) of the transfer function (m roots of Eqn. (2.52b)).

In the representation of $G(s)$ in pole-zero form given by Eqn. (2.53a), it has been assumed that the poles and zeros are *distinct* (*simple*). If a pole, say at $s = -p_1$, is *multiple* (repeated) with multiplicity v, the corresponding transfer function in pole-zero form becomes

$$G(s) = \frac{K(s+z_1)(s+z_2)\cdots(s+z_m)}{(s+p_1)^v(s+p_{v+1})\cdots(s+p_n)};\ m \le n \quad (2.53b)$$

Multiple zeros may also appear in a transfer function model.

We have so far considered SISO systems. Complex processes and machines often have several variables (outputs) we wish to control, and several manipulated inputs to provide this control (refer to Fig. 2.1). Systems with same number of input variables as the number of output variables are quite common in industry. Consider a system with p output variables $y_1, y_2, \ldots, y_p$ and p input variables $r_1, r_2, \ldots, r_p$. A mathematical model of this system is, therefore, $p \times p$ *transfer function matrix* defined below:

$$\begin{bmatrix} Y_1(s) \\ Y_2(s) \\ \vdots \\ Y_p(s) \end{bmatrix} = \begin{bmatrix} G_{11}(s) & G_{12}(s) & \cdots & G_{1p}(s) \\ G_{21}(s) & G_{22}(s) & \cdots & G_{2p}(s) \\ \vdots & \vdots & & \vdots \\ G_{p1}(s) & G_{p2}(s) & \cdots & G_{pp}(s) \end{bmatrix} \begin{bmatrix} R_1(s) \\ R_2(s) \\ \vdots \\ R_p(s) \end{bmatrix} \quad (2.54a)$$

[6] A monic polynomial has a leading coefficient equal to unity.

An important general property of such systems (and a major source of design difficulties) is *interaction* (or *coupling*) between inputs r_i and outputs y_j in the sense that any input r_i; $1 \le i \le p$, will have a dynamic effect on all the outputs y_j; $1 \le j \le p$. If each input r_i has a dynamic effect only on y_i, the system is said to be *non-interacting* (or *decoupled*). For a non-interacting system, the transfer function matrix in Eqn. (2.54a) becomes a diagonal matrix:

$$\begin{bmatrix} Y_1(s) \\ Y_2(s) \\ \vdots \\ Y_p(s) \end{bmatrix} = \begin{bmatrix} G_1(s) & 0 & \cdots & 0 \\ 0 & G_2(s) & \cdots & 0 \\ \vdots & \vdots & & \vdots \\ 0 & 0 & \cdots & G_p(s) \end{bmatrix} \begin{bmatrix} R_1(s) \\ R_2(s) \\ \vdots \\ R_p(s) \end{bmatrix} \tag{2.54b}$$

Thus, our 'multivariable' system consists of p independent SISO systems. It is self-evident that each of these systems can be individually controlled by SISO feedback loops.

In simple control situations, it is possible to ignore weak interactions and design controllers under the assumptions that one input affects only one output. Interaction effects are considered as disturbances to the separate control systems and may not cause significant degradation in their performance if coupling is weak. Our focus in this book is on SISO systems; multivariable systems with weak coupling will also be considered.

There are two different ways in which transfer function models are usually obtained:

1. For lumped linear time-invariant systems, mathematical model building based on physical laws normally results in a set of (first order, and second order) differential equations. Applying Laplace transform to the differential equations results in a transfer function model of the system.

   We will be mostly concerned with this approach. A number of examples of mechanical and electrical systems will be taken up in this chapter to illustrate the modeling procedure.
2. The transfer function model of a system can be identified from experimentally obtained input-output data. Generally, the identification methods can be considered to be a type of curve-fitting, where the assumed transfer function is fitted to the available data in some optimal manner. The methods may be based on frequency response (sinusoidal input), step response (step input), impulse response (pulse input; pulse of small width approximating an impulse), or the response to more general inputs.

   Identification methods have not been covered in this book. Only simple methods based on step response and frequency response are given in later chapters. Refer to [155] for more details on identification methods.

2.6.1 Transfer Functions of Mass-Spring-Damper Systems

In the following section, we will formally show how mass-spring-damper systems can be modeled by transfer function, $G(s)$, that algebraically relates the Laplace transform of the output to the Laplace transform of the input.

The three idealized elements—the *mass*, the *spring* and the *damper*, represent three essential phenomena which occur in various ways in *mechanical translational* systems.

The ideal mass element, M, represents a particle of *mass* which is the lumped approximation of the mass of a body concentrated at the center of the body. The concept of the elastic deformation of a body is represented by the ideal element, K, symbolized as a helical *spring*. Spring is also used as a physical device in mechanical systems.

Friction exists in physical systems whenever mechanical surfaces are operated in sliding contact. Friction encountered in physical systems may be of many types.

1. *Coulomb friction force*: It is the force of sliding friction between dry surfaces. This force is substantially constant.
2. *Viscous friction force*: It is the force of friction between moving surfaces separated by viscous fluid, or the force between a solid body and a fluid medium. This force is approximately linearly proportional to velocity over a certain limited velocity range.
3. *Stiction*: It is the force required to initiate motion between two contacting surfaces (which is obviously more than the force required to maintain them in relative motion).

In many situations of interest, viscous friction predominates. Our models of mechanical systems are based on this assumption. Refer to Chapter 16 for situations where this approximation is not valid.

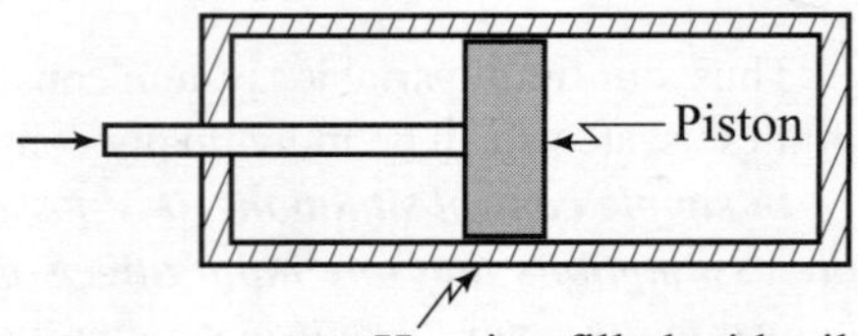

Fig. 2.14 *Dashpot*

Sometimes it may even be necessary to introduce viscous friction intentionally to improve the dynamic response of a system. A *dashpot* is a device that provides viscous friction or damping. It consists of a piston and an oil-filled cylinder with a narrow annular passage between the piston and the cylinder (Fig. 2.14). Any relative motion between the piston and the cylinder is resisted by oil.

The viscous friction between moving surfaces is represented by the ideal damping element B, symbolized as a stylized dashpot. Mechanical translational systems consisting of the three elements M, K, and B, are components of many control systems (Chapter 3 includes examples).

Consider the mass-spring-damper system shown in Fig. 2.15a. Let us obtain the transfer function model of this system. In order to obtain a linear model, we will assume that the friction force of the dashpot is proportional to velocity v (true for small velocities). Also the spring is considered to be a linear device; the spring force is proportional to displacement y (true for small displacements). Metric system of units for mechanical translational systems is as follows:

Force, F	: newtons or kg-m/sec^2	
Displacement, y	: meters (m)	
Velocity, v	: m/sec	
Acceleration	: m/sec^2	(2.55)
Mass, M	: kg	
Spring constant, K	: newtons/m	
Damping coefficient, B	: newtons/(m/sec)	

A systematic way of setting up dynamical equations for mass-spring-damper systems is to draw a *free-body diagram*. In a free-body diagram, each mass is isolated from the rest of the system; forces acting on each free-body are due to the rest of the system including the external forces. The free-body mass moves under the action of the resultant force. By applying Newtons law of motion to each free-body, we get the force equations describing the dynamics of the system.

Figure 2.15b shows the free-body diagram of the system of Fig. 2.15a. Applying Newton's law of motion, the force equation can be written as

$$F(t) - Bv(t) - Ky(t) = M\frac{dv(t)}{dt}$$

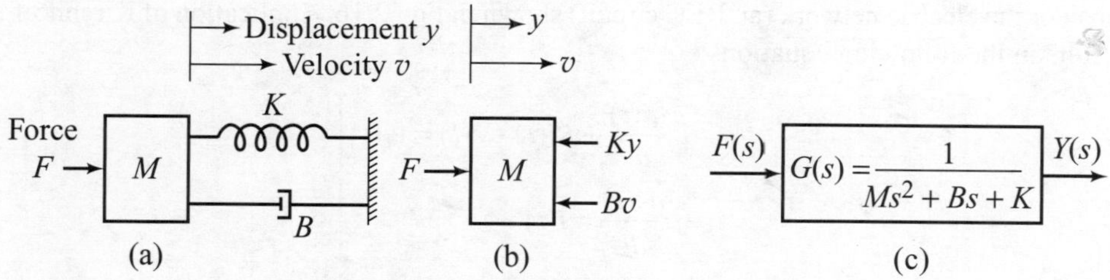

Fig. 2.15 *A mass-spring-damper system*

Since $v(t) \triangleq \dfrac{dy(t)}{dt}$, we have

$$M\frac{d^2y(t)}{dt^2} + B\frac{dy(t)}{dt} + Ky(t) = F(t) \tag{2.56}$$

This is a linear, constant-coefficient differential equation of second order.

Taking the Laplace transform of each term of this equation (assuming zero initial conditions), we obtain

$$Ms^2Y(s) + BsY(s) + KY(s) = F(s)$$

where

$$Y(s) \triangleq \mathscr{L}[y(t)];\ F(s) \triangleq \mathscr{L}[F(t)].$$

Therefore,

$$Y(s) = \frac{1}{Ms^2 + Bs + K}F(s) \tag{2.57a}$$

Figure 2.15c gives the block diagram representation of the system. The transfer function

$$G(s) = \frac{1}{Ms^2 + Bs + K} \tag{2.57b}$$

Since the highest power of s in the denominator polynomial of $G(s)$ is two, the transfer function model of the mass-spring-damper system is of second order.

2.6.2 Transfer Functions of Electrical Networks

We have shown that mass-spring-damper systems can be modeled by transfer function, $G(s)$, that algebraically relates the Laplace transform of the output to the Laplace transform of the input. Now we do the same for electric circuits—passive networks and operational amplifier circuits.

The three passive (passive means that there is no internal source of energy) linear components—the *resistor* (R), the *inductor* (L), and the *capacitor* (C)—are the basic elements of electric networks/circuits. The units of variables and parameters in electric circuits are as follows.

Voltage, e : volts
Current, i : amps
Resistance, R : ohms (2.58)
Inductance, L : henrys
Capacitance, C : farads

Consider the electric network (an RLC circuit) shown in Fig. 2.16. Application of Kirchhoff's voltage law results in the following equations:

$$\left.\begin{aligned} L\frac{di(t)}{dt} + Ri(t) + e(t) = e_i(t) \\ C\frac{de(t)}{dt} = i(t) \end{aligned}\right\} \tag{2.59}$$

$e(t)$ is the voltage across the capacitor, $i(t)$ is the current through the inductor, $e_i(t)$ is the voltage source, and R, L and C are parameters of the system.

Taking Laplace transform of all the terms of these equations, under the assumption of zero initial conditions, we obtain

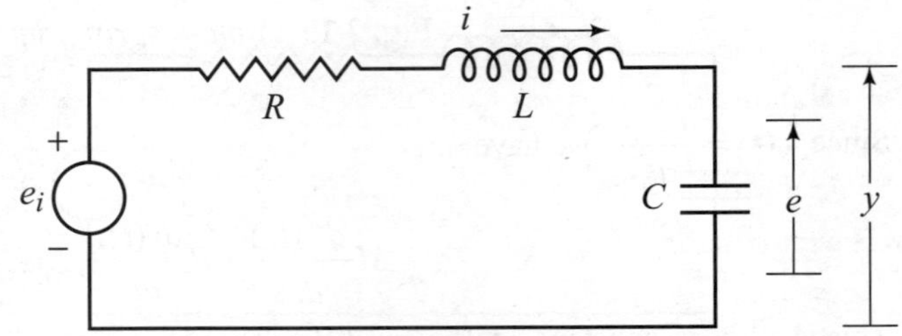

Fig. 2.16 *An RLC-series circuit*

$$C[sE(s)] = I(s);\; L[sI(s)] + RI(s) + E(s) = E_i(s)$$

where

$$E_i(s) \triangleq \mathscr{L}[e_i(t)]; E(s) \triangleq \mathscr{L}[e(t)]; I(s) \triangleq \mathscr{L}[i(t)]$$

Eliminating $I(s)$ from these equations, we get

$$LCs^2E(s) + RCsE(s) + E(s) = E_i(s)$$

The output $Y(s) = E(s)$.
The transfer function $G(s)$ of the RLC-series network is

$$G(s) = \frac{Y(s)}{E_i(s)} = \frac{1/LC}{s^2 + (R/L)s + 1/LC} \tag{2.60}$$

Since the highest power of s in the denominator polynomial of $G(s)$ is two, the transfer function model of the RLC-series circuit is of second order. Note that the differential equation model of this network is also of second order, obtained by combining the two first-order differential equations (2.59).

Impedance approach is useful for developing transfer functions of electrical circuits. Impedance $Z(s)$ of a passive circuit is the ratio of the Laplace transform of the voltage across the circuit to the Laplace transform of the current through the circuit under the assumption of zero initial conditions. For impedances in series, the equivalent impedance is equal to the sum of the individual impedances. For impedances in parallel, the reciprocal of the equivalent impedance is equal to the sum of the reciprocals of the individual impedances.

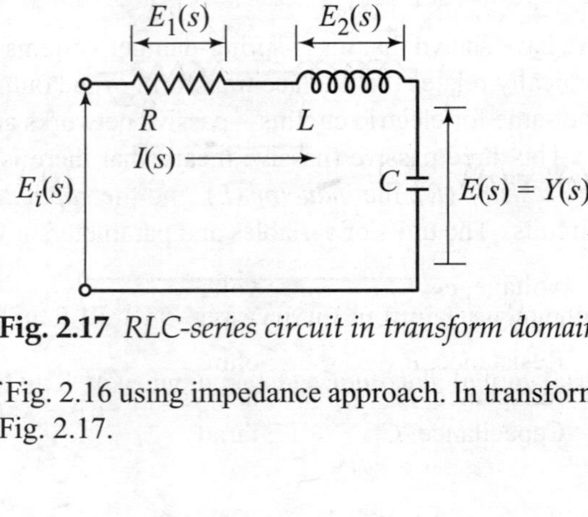

Fig. 2.17 *RLC-series circuit in transform domain*

Let us derive the transfer function of the circuit of Fig. 2.16 using impedance approach. In transform domain, the RLC-series circuit appears as shown in Fig. 2.17.

For the resistor,

$$e_1(t) = Ri(t);\ Z_1(s) = \frac{E_1(s)}{I(s)} = R$$

For the inductor, $$e_2(t) = L\frac{di(t)}{dt};\ Z_2(s) = \frac{E_2(s)}{I(s)} = sL$$

For the capacitor, $$i(t) = C\frac{de(t)}{dt};\ Z_3(s) = \frac{E(s)}{I(s)} = sC$$

Total impedance of the circuit

$$Z(s) = Z_1(s) + Z_2(s) + Z_3(s) = R + sL + \frac{1}{sC}$$

The circuit obeys the relation

$$E_i(s) = Z(s)\,I(s) = \left(R + sL + \frac{1}{sC}\right)I(s) \tag{2.61}$$

The voltage across the capacitor

$$E(s) = \frac{1}{sC}I(s)$$

Substituting $I(s)$ in Eqn. (2.61) and simplifying yields the same result as in Eqn. (2.60).

Operational Amplifiers

Passive electrical circuits was the topic of discussion up to this point. We will now discuss a class of active circuits used in control systems to implement controller transfer functions (Chapter 7 will give the implementation schemes). These are circuits built around an operational amplifier.

An *operational amplifier* or *op amp* for short, is an electronic device consisting of a complex arrangement of resistors, transistors, capacitors, and diodes. For our purposes, it will suffice to treat the op amp as a circuit building block and simply study what takes place at its terminals.

The circuit symbol for an operational amplifier (op amp) is shown in Fig. 2.18a. Two input terminals (pins) marked '+' and '–' are shown. These polarity signs refer to the phase relationship between the individual input signals and the output signal; the output signal is in phase with the signal applied to the input pin marked '+', known as the *non-inverting input*, and is in antiphase to the signal applied to the pin marked '–', known as the *inverting input*. The power supply connections are labeled V^+ for the positive voltage and V^- for the negative voltage. The op amp is usually shown as in Fig. 2.18b, without the power supply connections.

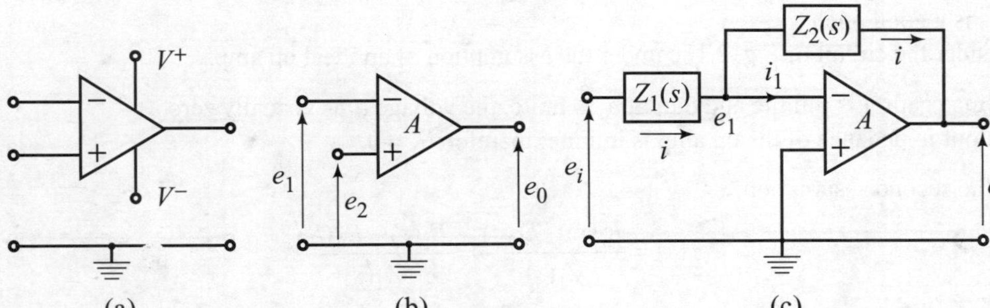

Fig. 2.18 *(a) Circuit symbol for an op amp (b) Simplified symbol (c) Op amp circuit building block*

The op amp senses the difference between the two inputs e_1 (voltage between the inverting terminal and ground) and e_2 (voltage between the non-inverting terminal and ground), multiplies it by gain A, and causes the resulting voltage e_0 to appear at the output:

$$e_0 = A(e_2 - e_1)$$

If e_2 is grounded, the amplifier is called an inverting amplifier, and the output becomes

$$e_0 = -Ae_1 \tag{2.62}$$

The gain A customarily has high values—of the order of 100,000. It is, therefore, necessary to restrict e_1 to those values that will keep e_0 atleast below the magnitude of the supply voltage; that is, e_1 must be limited to a *threshold value*. One of the design characteristics of op amp is that *saturation* of the output voltage sets in rapidly once the input signal reaches the threshold value; the saturation value is less than the supply voltage.

If two impedances are connected to the inverting operational amplifier as shown in Fig. 2.18c, we can derive an interesting result by assuming that the op amp is *ideal*. Characteristics of an ideal amplifier are listed below.

1. The impedance presented to a source applied at input terminal of the op amp is for all practical purposes infinite. The actual value may be of the order of several mega ohms but this is tantamount to infinity. This point is important because it means that the presence of the op amp in the circuit does not change the behavior of the signal source. In other words, there is no *loading effect* on the signal source connected to the input terminal of the op amp.
2. Op amp is considered to have a direct gain A from e_1 to e_0 (refer to Eqn. (2.62)) of infinity. Although we know this gain to be finite and in the vicinity of 100,000, this figure is high enough so that the practical consequences are the same as if it were treated to be of infinite value.
3. Output impedance of the op amp is zero. Output impedance is the impedance that appears across the output terminals looking back into the amplifier. The usual value is under 100 Ω, but for the level of load impedance values generally used across the output terminals of the op amp in applications, this value can be considered negligible. Keeping the output impedance small compared to load impedance, is important if most of the voltage generated by the op amp in response to input signals is to be available at the load, where it does useful work.
4. Bandwidth[7] of the op amp is infinite; it can amplify any signal frequency. The gain of an actual op amp is high (of the order of 100,000) typically at frequencies between zero (dc) and 10 Hz. Beyond this value of frequencies, the gain reduces considerably. In general, the op amps approach the ideal characteristics for dc and low-frequency signals.

Although assuming an ideal amplifier provides only an approximate analysis, most modern op amps have such large gains, large input impedances and small output impedances, that the approximate analysis is a good one.

Consider the circuit of Fig. 2.18c under the assumption of an ideal op amp.

1. Since gain A is infinite and output e_0 is finite, the voltage e_1 is virtually zero.
2. Input impedance of the op amp is infinite; therefore $i_1 = 0$.

Under the second assumption.

$$I(s) = \frac{E_i(s) - E_1(s)}{Z_1(s)} = \frac{E_1(s) - E_0(S)}{Z_2(s)}$$

[7] Precise definition of bandwidth with appear in Section 2.10.

Since $E_1(s) = 0$ as per first assumption, we get

$$I(s) = \frac{E_i(s)}{Z_1(s)} = -\frac{E_0(s)}{Z_2(s)}$$

or,

$$\frac{E_0(s)}{E_i(s)} = -\frac{Z_2(s)}{Z_1(s)} \tag{2.63}$$

where

$Z_1(s)$ = impedance connected to the input of the amplifier, and
$Z_2(s)$ = impedance connected between output and input of the amplifier (The output is fed back to the inverting terminal of the op amp, thus achieving *negative feedback*).

Note that the *closed-loop gain* (the ratio of the output voltage $E_0(s)$ to the input voltage $E_i(s)$) is almost *insensitive* to the open-loop gain A.

By choosing proper values of $Z_1(s)$ and $Z_2(s)$, operational amplifiers can be used for various purposes. We will use this feature in Chapter 7 to implement various control schemes.

Example 2.1 Consider the circuit of Fig. 2.19a. Under the assumption of an ideal op amp,

$$\frac{e_0}{e_i} = \frac{-R_F}{R_1} \tag{2.64a}$$

where
R_1 = resistance connected to the input of the inverting amplifier; and
R_F = feedback resistance

By choosing proper values of R_1 and R_F, we can design an op amp for the desired closed-loop gain. If an inverting circuit of unity gain is cascaded to the circuit of Fig. 2.19a, we get a non-inverting amplifier shown in Fig. 2.19b, with the gain

$$\frac{e_0}{e_i} = \frac{R_F}{R_1} \tag{2.64b}$$

If the circuit of Fig. 2.19a is modified to permit multiple inputs, the operational amplifier can perform addition. The output of circuit of Fig. 2.19c is a weighted sum of the input voltages with a sign change. The circuit is called a *summing amplifier*. It obeys the following equations:

$$\frac{e_{i1}}{R_1} + \frac{e_{i2}}{R_2} = -\frac{e_0}{R_F}$$

or

$$e_0 = -\left[\frac{R_F}{R_1}e_{i1} + \frac{R_F}{R_2}e_{i2}\right] \tag{2.64c}$$

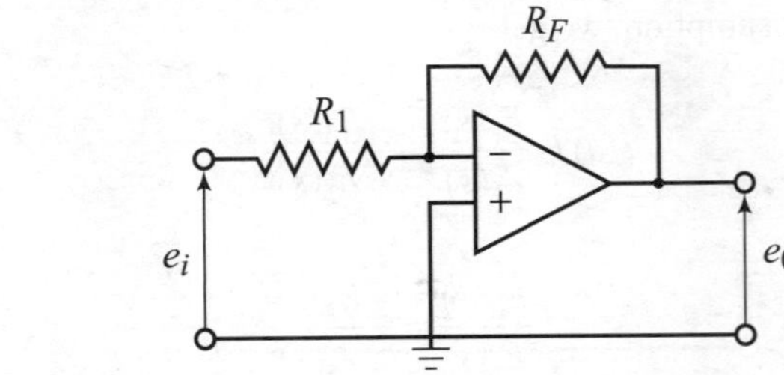

(a) Inverting amplifier

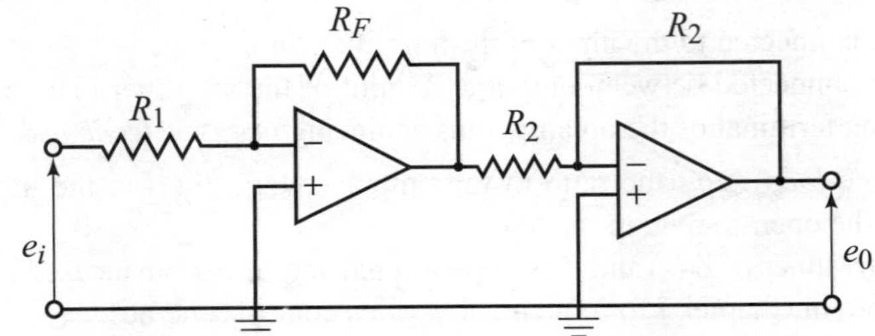

(b) Non-inverting amplifier

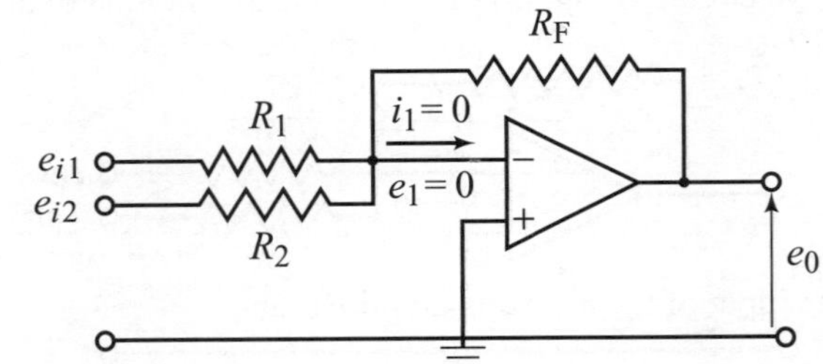

(c) Summing amplifier

Fig. 2.19

2.6.3 Analogous Systems

If two systems are described by dynamical equations of identical form, these are said to be *analogs* of each other. The concept of analogous systems is a useful technique for the study of various physical systems. Generally, it is convenient to study a non-electrical system: mechanical, thermal, hydraulic, etc., in terms of its electrical analog as electrical systems are more easily amenable to experimental study.

The only non-electrical systems we have worked with so far are mass-spring-damper systems. We will shortly see that the variables of RLC circuits behave exactly as the analogous variables of the mass-spring-damper systems (The concept will later be used for other non-electrical systems in Chapter 3).

The *principle of analogy* is the basic principle behind the use of *analog computers* for investigating system behavior—an analog to the physical system can be developed from the components of an analog computer. An analog computer provides an excellent means for developing control system prototypes.

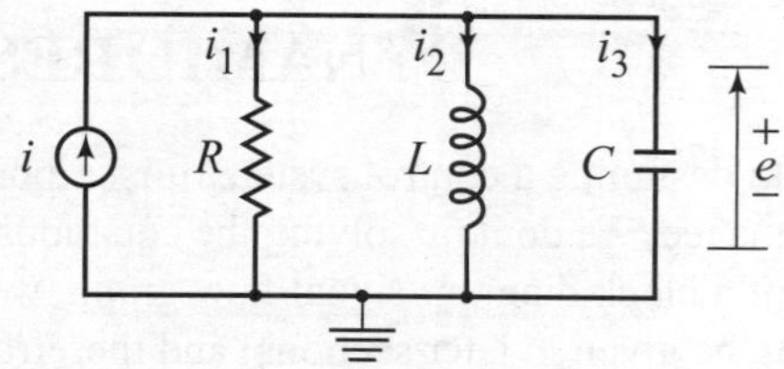

Fig. 2.20 *An RLC-parallel circuit*

We now consider a simple example. An electric analog of the mass-spring-damper system of Fig. 2.15 is the RLC-parallel circuit excited by a current source, shown in Fig. 2.20. This can be established as follows.

The mechanical system of Fig. 2.15 is governed by the differential equation (refer to Eqn. (2.56))

$$M\frac{d^2y(t)}{dt^2}+B\frac{dy(t)}{dt}+Ky(t)=F(t)$$

Rewriting this equation in terms of velocity variable $v(t)=\dot{y}(t)$ results in the following integro-differential equation:

$$M\frac{dv(t)}{dt}+Bv(t)+K\int_{-\infty}^{t}v(\tau)\,d\tau=F(t) \tag{2.65a}$$

The electric circuit of Fig. 2.20 obeys the relation,

$$I(s)=\frac{E(s)}{Z(s)}$$

where

$I(s)\triangleq\mathscr{L}[i(t)]; E(s)\triangleq\mathscr{L}[e(t)]$; and impedance $Z(s)$ of the RLC-parallel circuit is given by

$$\frac{1}{Z(s)}=\frac{1}{R}+\frac{1}{sL}+\frac{1}{1/sC}$$

Inverse Laplace transform of the equation

$$I(s)=\frac{E(s)}{R}+\frac{E(s)}{sL}+sCE(s)$$

results in the following integro-differential equation:

$$C\frac{de(t)}{dt}+\frac{1}{R}e(t)+\frac{1}{L}\int_{-\infty}^{t}e(\tau)\,d\tau=i(t) \tag{2.65b}$$

Comparing this equation with Eqn. (2.65a), we at once observe their mathematical similarity with respect to the following analogous pairs:

$$F(t)\leftrightarrow i(t);\ v(t)\leftrightarrow e(t) \tag{2.66}$$

$$M\leftrightarrow C;\ K\leftrightarrow 1/L;\ B\leftrightarrow 1/R$$

The commonly used characterizing variables in an electrical system—the voltage and the current—are respectively called *across* and *through* variables of the system. By analogy, velocity and force are, respectively, the across and through variables of mechanical systems. The analogous variables and element pairs in Eqn. (2.66) are appropriately called the *force-current analogy*.[8]

---

[8] Another possible analogy between mechanical and electrical systems is the *force-voltage analogy*. The corresponding analogous electrical system for the mechanical system of Fig. 2.15 will be an RLC-series circuit excited by a voltage source.

2.7 DYNAMIC RESPONSE

In designing a control system, it is important to see how well a trial design matches the desired performance. We do it by solving the equations of the system model. If system model is available in the form of a block diagram/signal flow graph, we reduce this model into a single transfer function (techniques to be given in later sections) and therefrom obtain the dynamic response.

The problem in hand for this section is: given the transfer function model $G(s)$ of a SISO system (Fig. 2.13), and the input $R(s)$, solve for the time response $y(t)$ of the system.

2.7.1 Response by Partial-Fraction Expansion

The easiest way to find response $y(t)$ of the system of Fig. 2.13 from its Laplace transform $Y(s) = G(s)R(s)$, if $Y(s)$ is rational, is to expand $Y(s)$ as a sum of simpler terms that can be found in the tables. The basic tool for performing this operation is called *partial fraction expansion*.

Basically four steps are involved in the computation of $y(t)$ from $Y(s) = G(s)R(s)$:

1. Given the input $r(t)$, find the Laplace transform $R(s)$ using results of Section 2.3 and/or transform pairs given in Table 2.1.
2. Compute the poles of $G(s)R(s)$.
3. Expand $Y(s)$ into partial fractions.
4. Obtain $y(t)$ by taking inverse Laplace transform of $Y(s)$ using transform pairs listed in Table 2.1 and the properties of the Laplace transforms.

In the following section, we will give examples to illustrate the procedure.

Example 2.2 Given the transfer function

$$G(s) = \frac{Y(s)}{R(s)} = \frac{1}{s^2 + 3s + 2};$$

find the response $y(t)$ to the input

(i) $r(t) = 5\,\mu(t)$; (ii) $r(t) = 5t\,\mu(t)$

Solution: (i) For the given step input, $R(s) = \frac{5}{s}$.

Therefore,

$$Y(s) = \frac{1}{s^2 + 3s + 2} R(s) = \frac{1}{(s+1)(s+2)}\left(\frac{5}{s}\right)$$

Expanding $Y(s)$ into partial fractions gives

$$Y(s) = \frac{5}{s(s+1)(s+2)} = \frac{A_1}{s} + \frac{A_2}{s+1} + \frac{A_3}{s+2}$$

$$A_1 = \lim_{s \to 0}\left[(s)\frac{5}{s(s+1)(s+2)}\right] = \frac{5}{2}$$

$$A_2 = \lim_{s\to -1}\left[(s+1)\frac{5}{s(s+1)(s+2)}\right] = -5$$

$$A_3 = \lim_{s\to -2}\left[(s+2)\frac{5}{s(s+1)(s+2)}\right] = \frac{5}{2}$$

or

$$Y(s) = \underbrace{\frac{5/2}{s}}_{\text{Excitation pole}} \underbrace{-\frac{5}{s+1}+\frac{5/2}{s+2}}_{\text{System poles}}$$

Inverting $Y(s)$, we obtain the time response:

$$y(t) = \left[\underbrace{\frac{5}{2}}_{\text{Steady-state response}} \underbrace{-5e^{-t}+\frac{5}{2}e^{-2t}}_{\text{Transient response}}\right]\mu(t) \tag{2.67}$$

The transient response terms correspond to system poles excited by the input. These terms vanish as $t \to \infty$.

The second response term arises due to the excitation pole, and has the same nature as the input itself except for a modification in magnitude caused by the system's behavior to the specified input. Since the input exists as $t \to \infty$, the second response term does not vanish and is called the *steady-state response* of the system. Figure 2.21a shows the response to the given step input.

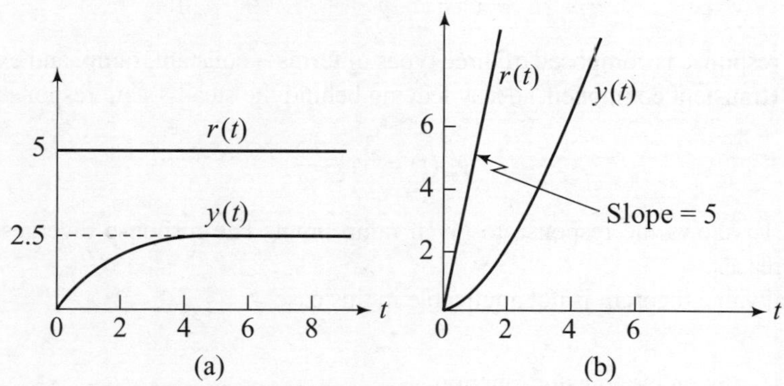

Fig. 2.21 *(a) Step response; (b) Ramp response*

The steady-state response can be quickly obtained without doing the complete inverse Laplace transform, by use of the *final value theorem* (Eqn. (2.41)), since $sY(s)$ has no poles on the imaginary axis and in the right half of the s-plane.

(ii) For the given ramp input, $R(s) = \dfrac{5}{s^2}$.

Therefore,

$$Y(s) = \frac{5}{s^2(s+1)(s+2)}$$

Expanding $Y(s)$ in partial fractions gives

$$Y(s) = \frac{A_1}{s^2} + \frac{A_2}{s} + \frac{A_3}{s+1} + \frac{A_4}{s+2}$$

$$A_1 = \lim_{s\to 0}\left[(s^2)\frac{5}{s^2(s+1)(s+2)}\right] = \frac{5}{2}$$

$$A_2 = \lim_{s\to 0}\frac{d}{ds}\left[(s^2)\frac{5}{s^2(s+1)(s+2)}\right] = \lim_{s\to 0}\left[\frac{-5(2s+3)}{(s^2+3s+2)^2}\right] = \frac{-15}{4}$$

$$A_3 = \lim_{s\to -1}\left[(s+1)\frac{5}{s^2(s+1)(s+2)}\right] = 5$$

$$A_4 = \lim_{s\to -2}\left[(s+2)\frac{5}{s^2(s+1)(s+2)}\right] = \frac{-5}{4}$$

or

$$Y(s) = \frac{5/2}{s^2} - \frac{15/4}{s} + \frac{5}{s+1} - \frac{5/4}{s+2}$$

Inverting $Y(s)$, we obtain the time response

$$y(t) = \left(\frac{-15}{4} + \frac{5}{2}t + 5e^{-t} - \frac{5}{4}e^{-2t}\right)\mu(t) \tag{2.68a}$$

The ramp response is composed of three types of terms—constant, ramp, and exponentials. The exponentials (transient component) decay leaving behind the steady-state response

$$y_{ss}(t) = \frac{5}{2}t - \frac{15}{4} \tag{2.68b}$$

Figure 2.21b shows the response to given ramp input. The response diverges from the ramp input as t increases.
Note that final value theorem is not applicable in this case.

Example 2.3 Given the transfer function

$$G(s) = \frac{Y(s)}{R(s)} = \frac{3}{s^2+2s+5};$$

find the response $y(t)$ to the unit-step input.

Solution: For a unit-step input, $R(s) = 1/s$.

Therefore,

$$Y(s) = \frac{3}{s(s^2+2s+5)}$$

This function has a pair of complex-conjugate poles. For purposes of finding the inverse Laplace transform of a complex function using Table 2.1, we may manipulate the function so that the complex poles fit the form of last two entries of the table. The given function $Y(s)$ can be expanded in the following form:

$$Y(s) = \frac{3/5}{s} - \frac{3}{5}\frac{s+2}{s^2+2s+5}$$

The second term needs to be manipulated and broken in two parts so that they match the last two entries in Table 2.1.

$$Y(s) = \frac{3/5}{s} - \frac{3}{5}\left[\frac{s+1}{(s+1)^2+(2)^2} + \frac{\frac{1}{2}(2)}{(s+1)^2+(2)^2}\right]$$

Thus, the response $y(t)$ is

$$y(t) = \frac{3}{5} - \frac{3}{5}e^{-t}\left(\cos 2t + \frac{1}{2}\sin 2t\right)$$

In order to visualize the solution, an alternate form of $y(t)$ obtained by trigonometric identities (2.25), is preferable. We may rewrite the preceding equation using the trigonometric identity:

$$\cos 2t \cos\theta + \sin 2t \sin\theta = \cos(2t - \theta)$$

Manipulation of $y(t)$ to fit this identity, gives the following result:

$$y(t) = \frac{3}{5} - \frac{3}{5}e^{-t}\left[\cos 2t \frac{(1)}{\sqrt{(1)^2+(1/2)^2}} + \sin 2t \frac{(1/2)}{\sqrt{(1)^2+(1/2)^2}}\right]\left[\sqrt{1+(1/2)^2}\right]$$

$$= 0.6 - 0.671e^{-t}(\cos 2t \cos\theta + \sin 2t \sin\theta);\ \theta = \tan^{-1} 1/2$$

This gives

$$y(t) = 0.6 - 0.671e^{-t}\cos(2t - 26.57°).$$

2.7.2 Response by Convolution

The fact that in s-domain, the output $Y(s)$ of the system of Fig. 2.13, can be represented as the multiplication of the transfer function $G(s)$ and the input function $R(s)$:

$$Y(s) = G(s)\,R(s) \tag{2.69a}$$

simply means that in t-domain, the output $y(t)(= \mathscr{L}^{-1}[Y(s)])$ is given by the convolution of the signals $g(t)$ $(=\mathscr{L}^{-1}[G(s)])$, and $r(t)$ $(= \mathscr{L}^{-1}[R(s)])$. This follows from Eqns ((2.34)–(2.36)):

$$y(t) = \int_0^t g(t-\tau)r(\tau)\,d\tau \tag{2.69b}$$

Let us interpret this result from the perspectives of system's theory. Let $r(t) = \delta(t)$, an impulse function. Then the response

$$Y(s)\big|_{\text{impulse input}} = G(s)\{\mathscr{L}[\delta(t)]\} = G(s)$$

and

$$y(t)\big|_{\text{impulse input}} = g(t) = \mathscr{L}^{-1}[G(s)]; t \ge 0 \tag{2.70}$$

Thus, $g(t)$ represents the *impulse response* of the system of Fig. 2.13; and the ***output*** $y(t)$ ***to any input*** $r(t)$; $t \geq 0$, ***is the convolution of the stimulus*** $r(t)$ ***and the system's impulse response*** $g(t)$. By inverting the time axis of the impulse response to give $g(t-\tau)$, we obtain a *memory function*. The convolution process then sums the effect of a stimulus multiplied by the memory function; the current output of the system is the cumulative response to the stimulus.

A very important property of the *impulse response model* is that it can be obtained from the input and output terminals of a system even if the system's internal structure is not known. If the system is known to be linear, time invariant, and relaxed at $t = 0$, then by applying $\delta_\Delta(t)$ function (refer to Eqn. (2.8)) at the input terminal, the response at the output terminal gives us immediately, the impulse response (approximate one; goodness of approximation is dependent on Δ being small) of the system. If the impulse response is known, the output of the system due to any input can be computed from Eqn. (2.69b).

2.7.3 Computer Simulation

The Laplace transform approach of investigating the dynamics of a linear system results in *closed form response* (refer to Examples 2.2–2.3)—the response variable y is given by a mathematical expression in terms of independent variable t. The method works well if the system model is relatively of low order, with relatively simple input functions.[9] If the model does not have these characteristics, we have to resort to numerical solution methods (*digital simulation*). The disadvantage of this type of solution is that generality is lost; the numerical values of the input functions, coefficients of the model, etc., must be specified [151–154]. However, often there is no alternative.[10]

Over the years, quite sophisticated, powerful and user-friendly software packages have been produced that also allow simulation operations. The usefulness and importance of these software packages is undeniable, because they greatly facilitate the analysis and design of control systems. They provide a tremendous tool in the hands of control engineers. A very nice feature of these packages is that they visually represent the simulation process by using *simulation block diagrams*. These block diagrams are essentially the same we use to describe control system structures and signal flow. The response at the output of any subsystem, in addition to the response of the total system consisting of interconnected multiple subsystems, can easily be displayed.

One such software package, MATLAB, has been extensively used in the book (refer to Appendix).

2.8 CHARACTERISTIC PARAMETERS OF FIRST- AND SECOND-ORDER MODELS

In this book, we will frequently come across first- and second-order transfer function models. Special emphasis on the characteristics of these models is justified.

---

[9] Most of the models used in the following chapters are of first-, second-, and third-order with simple input functions. Computer simulation is, therefore, not essential. However CAD facility, if available, can provide a better feel of how control system design is done in practice.

[10] The analog computer simulation provides another way of investigating system dynamics. This method is rarely used now because of the advantages of digital simulation. However, it still provides an excellent means of developing control system prototypes (refer to Section 2.6).

2.8.1 First-Order Models

We consider here a first-order model which can be represented by the first-order differential equation:

$$\frac{dy(t)}{dt} + ay(t) = br(t) \tag{2.71}$$

where $y(t)$ represents output of the system; $r(t)$ represents input; a and b represent model parameters (they depend on physical parameters of the system).

The transfer function of this first-order model becomes

$$\frac{Y(s)}{R(s)} = G(s) = \frac{b}{s+a}$$

A more common notation for first-order transfer function is

$$\frac{Y(s)}{R(s)} = G(s) = \frac{K}{\tau s+1} \tag{2.72}$$

since the physical meaning can be given to both K and τ.

In the following, we obtain the time response from Eqn. (2.72) to a unit-step signal. From the characteristics of the time response, the physical meaning of the parameters K and τ will become clear.

For a unit-step input, $R(s) = 1/s$. Therefore, (refer to Table 2.1)

$$Y(s) = \frac{K}{s(\tau s+1)} = K\left[\frac{1}{s} - \frac{\tau}{\tau s+1}\right] \tag{2.73a}$$

$$y(t) = K(1-e^{-t/\tau}); t \ge 0 = \underbrace{K}_{\text{Steady-state response}} - \underbrace{Ke^{-t/\tau}}_{\text{Transient response}} \tag{2.73b}$$

This solution is now investigated in detail. The step response as given by Eqns (2.73) is plotted in Fig. 2.22; the two components of the response are plotted separately along with the complete response. Note that the exponentially decaying transient term has an initial slope of K/τ, i.e.,

$$\left.\frac{d}{dt}(-Ke^{-t/\tau})\right|_{t=0} = \left.\frac{K}{\tau}e^{-t/\tau}\right|_{t=0} = \frac{K}{\tau}$$

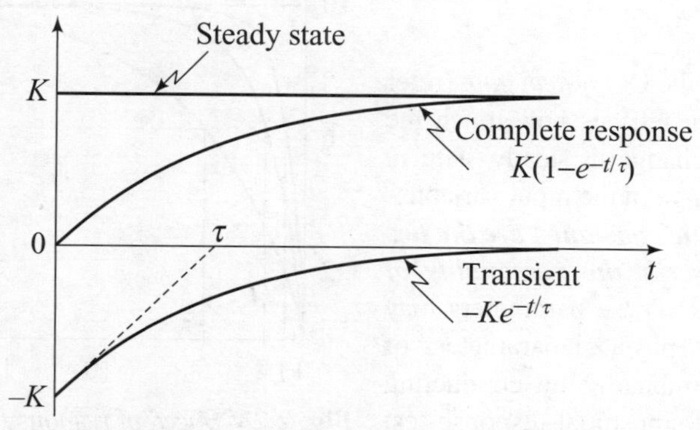

Fig. 2.22 *Step response of a first-order system*

Mathematically, the transient term does not decay to zero in a finite length of time. However, if the term continued to decay at its initial rate, the term would reach a value of zero in time $t = \tau$. The parameter τ is called the system *time constant*. It has units of time (seconds, minutes, etc.). This can be verified from the units of parameters of the physical system represented by the model (2.72).

The decay of the transient term is illustrated in Table 2.2 as a function of time constant τ. We observe that the transient term has decayed to less than five per cent of its initial value in three time constants, and to less than two per cent of its initial value in four time constants. In a practical sense, we consider the transient term to have decayed to zero in three to four time constants. Therefore, the time constant is related to the speed of response of the system. ***The slower the system responds to input, the larger the value of*** τ.

Table 2.2

t	$e^{-t/\tau}$
τ	$e^{-1} = 0.3679$
2τ	$e^{-2} = 0.1353$
3τ	$e^{-3} = 0.0498$
4τ	$e^{-4} = 0.0183$
5τ	$e^{-5} = 0.0067$

Figure 2.23 shows unit-step responses $y(t)$ of the following three systems:

$$\text{(i) } G_1(s) = \frac{10}{s+1}; \qquad \text{(ii) } G_2(s) = \frac{10}{2s+1}; \qquad \text{(iii) } G_3(s) = \frac{10}{5s+1}$$

The time constants of the system transfer functions are given by $\tau_1 = 1$, $\tau_2 = 2$, $\tau_3 = 5$.

The time constant represents the time taken for the output to rise to 63.2% of the change in output. In the plots of Fig. 2.23, the output changes from zero to a steady-state value $y_{ss} = 10$. We insert a line at $0.632y_{ss}$ on the graph so that the time constants are identified. As τ increases, the response becomes slower, i.e., it takes longer to reach 63.2% of the final output.

Another important parameter encountered in Eqns (2.73) is K. Note that

$$\lim_{t\to\infty} y(t) = K$$

The parameter K is the DC *system gain* (refer to Eqn. (2.42)), which tells us how much the output variable will change at steady state in response to a unit change in the input variable.

The gain K and time constant τ are the two parameters which describe the 'personality' of the first-order system. These parameters may be obtained from the physical parameters of the system or experimentally by conducting the step-response test/sinusoidal-response test (described later in Chapters 7 and 10).

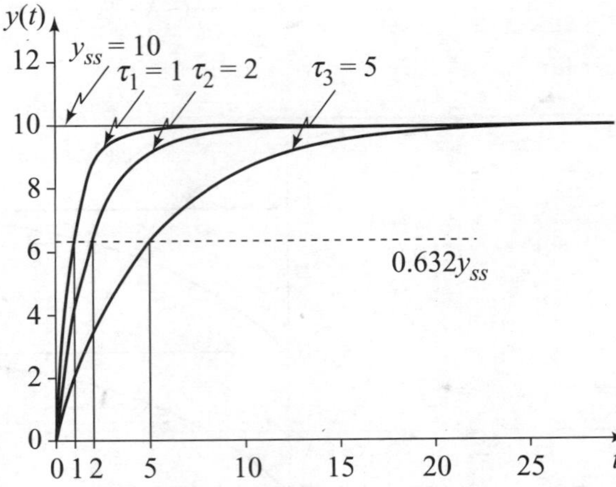

Fig. 2.23 *Speed of response for different time constants*

The transfer function

$$G(s) = \frac{K}{\tau s + 1} \tag{2.74}$$

is called the *time-constant form* for first-order transfer functions and will be encountered in all types of systems—electrical, mechanical, thermal, hydraulic, etc. A process described by this form of transfer functions is called a *first-order lag* or a *simple lag*.

2.8.2 Second-Order Models

We consider here a second-order model which can be represented by the second-order differential equation:

$$\frac{d^2 y(t)}{dt^2} + a_1 \frac{dy(t)}{dt} + a_2 y(t) = br(t) \tag{2.75}$$

where $y(t)$ represents the output of the system; $r(t)$ represents input; a_1, a_2 and b represent model parameters (they depend on the physical parameters of the system).

The transfer function of this second-order model becomes

$$\frac{Y(s)}{R(s)} = G(s) = \frac{b}{s^2 + a_1 s + a_2} \tag{2.76}$$

A more common notation for second-order transfer functions is

$$\frac{Y(s)}{R(s)} = G(s) = \frac{K}{\dfrac{1}{\omega_n^2}s^2 + \dfrac{2\zeta}{\omega_n}s + 1} = \frac{K\omega_n^2}{s^2 + 2\zeta\omega_n s + \omega_n^2} \tag{2.77}$$

since physical meaning can be given to the three parameters K, ζ and ω_n.

In the following, we obtain the time response from Eqn. (2.77) to a unit-step signal ($R(s) = 1/s$). From the characteristics of the time response, the physical meaning of the parameters K, ζ, and ω_n will become clear.

The response of the second-order transfer function (2.77) is dependent on the two poles of the transfer function, which are the roots of the characteristic equation:

$$s^2 + 2\zeta\omega_n s + \omega_n^2 = (s + \zeta\omega_n + j\omega_n\sqrt{1-\zeta^2})(s + \zeta\omega_n - j\omega_n\sqrt{1-\zeta^2}) = 0 \tag{2.78}$$

The unit-step response of the transfer function (2.77) with the parameter $\zeta = 0$, is given by (refer to Table 2.1).

$$Y(s) = \frac{K\omega_n^2}{s\left(s^2 + \omega_n^2\right)} = K\left[\frac{1}{s} - \frac{s}{s^2 + \omega_n^2}\right]$$

$$y(t) = K(1 - \cos \omega_n t);\ t \geq 0 \tag{2.79}$$

The output oscillates indefinitely about the steady-state value K; there is no decay in the magnitude of the oscillations. We will shortly see that the outputs of transfer functions with the parameter $\zeta > 0$ exhibit *damped oscillations*. This provides the physical significance of the parameter ζ, called the *damping ratio*.

It is an indicator of the type of transient behavior expected in a system's response. The damping ratio is a dimensionless parameter. This can be verified from the units of parameters of the physical system represented by the model (2.77).

All physical systems have some damping ($\zeta > 0$). If there were no damping ($\zeta = 0$) in the system, the output response would contain oscillations that did not decay. The frequency ω_n of these oscillations would be the *natural frequency* (sometimes referred to as the *undamped natural frequency*) of the system. The parameter ω_n has the dimension of reciprocal time, and its unit is rad/sec. This can be verified from the units of parameters of the physical system represented by the model (2.77).

We can classify second-order systems depending on the value of its damping ratio parameter ζ:

If $0 < \zeta < 1$, then the system is *underdamped.*

If $\zeta = 1$, then the system is *critically damped.*

If $\zeta > 1$, then the system is *overdamped.*

Underdamped Systems $(0 < \zeta < 1)$

Underdamped transfer function models have complex conjugate poles in the left half of the *s*-plane (refer to Eqn. (2.78)). The unit-step response is given by

$$Y(s) = \frac{K\omega_n^2}{s(s^2 + 2\zeta\omega_n s + \omega_n^2)} \tag{2.80a}$$

$$= K\left[\frac{1}{s} - \frac{s + 2\zeta\omega_n}{s^2 + 2\zeta\omega_n s + \omega_n^2}\right] \tag{2.80b}$$

From the last two entries in Table 2.1, we observe that the second term in Eqn. (2.80b) is the sum of the Laplace transforms of exponentially damped sine and cosine functions. For purposes of finding the time response from Table 2.1, we manipulate $Y(s)$ so that the resulting terms fit the form of entries of the table.

$$Y(s) = K\left[\frac{1}{s} - \frac{s + \zeta\omega_n}{(s + \zeta\omega_n)^2 + \omega_d^2} - \frac{\dfrac{\zeta}{\sqrt{1-\zeta^2}}\,\omega_d}{(s + \zeta\omega_n)^2 + \omega_d^2}\right] \tag{2.81}$$

where

$$\omega_d = \omega_n\sqrt{1-\zeta^2} \tag{2.82}$$

Taking the inverse Laplace transform produces

$$y(t) = K\left[1 - e^{-\zeta\omega_n t}\left(\cos\omega_d t + \frac{\zeta}{\sqrt{1-\zeta^2}}\sin\omega_d t\right)\right] \tag{2.83}$$

$$= K\left[1 - \frac{e^{-\zeta\omega_n t}}{\sqrt{1-\zeta^2}}\left(\sqrt{1-\zeta^2}\cos\omega_d t + \zeta\sin\omega_d t\right)\right]$$

We may rewrite the preceding equation in a more compact form using the trigonometric identity (refer to Eqns (2.25)):

$$\sin\theta\cos\omega_d t + \cos\theta\sin\omega_d t = \sin(\omega_d t + \theta), \text{ with } \tan\theta = \frac{\sqrt{1-\zeta^2}}{\zeta}$$

This results in the expression:

$$y(t) = K\left[1 - \frac{e^{-\zeta\omega_n t}}{\sqrt{1-\zeta^2}}\sin\left(\omega_d t + \tan^{-1}\frac{\sqrt{1-\zeta^2}}{\zeta}\right)\right] \; ; \; t \ge 0 \tag{2.84}$$

The step response as given by Eqn. (2.84) is plotted in Fig. 2.24 for various values of ζ, and fixed ω_n. For $0 < \zeta < 1$, the system response exhibits decaying oscillations of frequency $\omega_d = \omega_n\sqrt{1-\zeta^2}$. ω_d is called the *damped natural frequency*. With ζ approaching zero, ω_d approaches ω_n, and the system response exhibits continuous oscillations (refer to Eqn. (2.79)). With ζ approaching one, the system response approaches the final value without overshooting it and consequently there are no oscillations (this will be proved shortly). Systems with $0 < \zeta < 1$ are called underdamped systems.

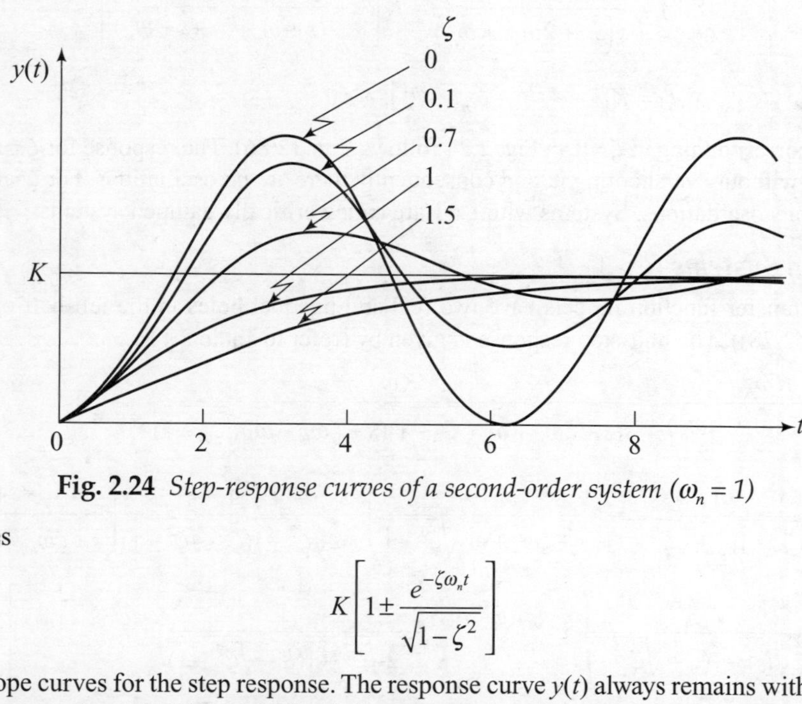

Fig. 2.24 *Step-response curves of a second-order system (ω_n = 1)*

The curves

$$K\left[1 \pm \frac{e^{-\zeta\omega_n t}}{\sqrt{1-\zeta^2}}\right] \tag{2.85a}$$

are the envelope curves for the step response. The response curve $y(t)$ always remains within this pair of envelope curves as shown in Fig. 2.25. The time constant of these envelope curves is $1/\zeta\omega_n$. The speed of decay of the transient response depends on the value of the time constant

$$\tau = \frac{1}{\zeta\omega_n} \tag{2.85b}$$

Critically Damped Systems ($\zeta = 1$)

Critically damped transfer function models have two real and equal poles in the left half of the s-plane (refer to Eqn. (2.78)). The unit-step response is given by (refer to Table 2.1).

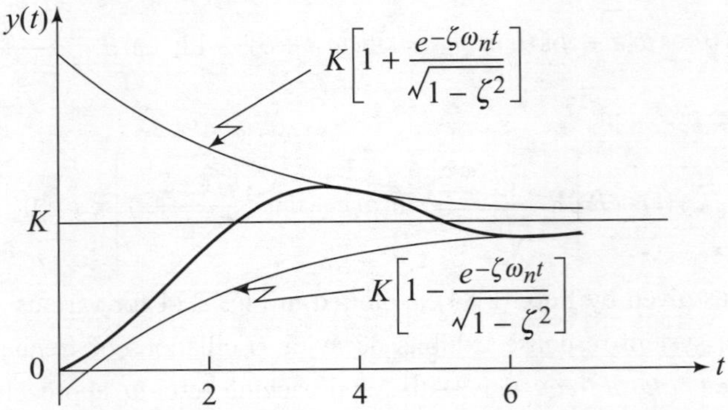

Fig. 2.25 *Envelope curves for the step response of a second-order system ($\omega_n = 1$, $\zeta = 0.7$)*

$$Y(s)=\frac{K\omega_n^2}{s\left(s^2+2\omega_n s+\omega_n^2\right)}=K\left[\frac{1}{s}-\frac{\omega_n}{(s+\omega_n)^2}-\frac{1}{(s+\omega_n)}\right]$$

$$y(t)=K[1-e^{-\omega_n t}-\omega_n t e^{-\omega_n t}];\ t\geq 0 \tag{2.86}$$

The curve corresponding to $\zeta = 1$ in Fig. 2.24 follows Eqn. (2.86). The response for $\zeta=1$ approaches the final value without overshooting it, and consequently there are no oscillations. For $\zeta<1$, the system response exhibits oscillations. Systems with $\zeta=1$ are called critically damped systems.

Overdamped Systems ($\zeta > 1$)

Overdamped transfer function models have two real and unequal poles in the left half of the *s*-plane (refer to Eqn. (2.78)). The unit-step response is given by (refer to Table 2.1).

$$Y(s)=\frac{K\omega_n^2}{s(s^2+2\zeta\omega_n s+\omega_n^2)}=\frac{K\omega_n^2}{s(s+\zeta\omega_n+\omega_n\sqrt{\zeta^2-1})(s+\zeta\omega_n-\omega_n\sqrt{\zeta^2-1})}$$

$$=K\left[\frac{1}{s}+\frac{1}{2\sqrt{\zeta^2-1}(\zeta+\sqrt{\zeta^2+1})}\left(\frac{1}{s+\zeta\omega_n+\omega_n\sqrt{\zeta^2-1}}\right)-\frac{1}{2\sqrt{\zeta^2-1}(\zeta-\sqrt{\zeta^2-1})}\left(\frac{1}{s+\zeta\omega_n-\omega_n\sqrt{\zeta^2-1}}\right)\right]$$

$$y(t)=K\left[1+\frac{1}{2\sqrt{\zeta^2-1}(\zeta+\sqrt{\zeta^2-1})}e^{-(\zeta+\sqrt{\zeta^2-1})\omega_n t}-\frac{1}{2\sqrt{\zeta^2-1}(\zeta-\sqrt{\zeta^2-1})}e^{-(\zeta-\sqrt{\zeta^2-1})\omega_n t}\right];\ t\geq 0 \tag{2.87}$$

Representative response curve for $\zeta>1$ is shown in Fig. 2.24. The effect of increasing damping beyond its critical value is to make the response sluggish. Systems with $\zeta>1$ are called overdamped systems.

The parameters K, ζ and ω_n describe the 'personality' of a second-order system. These parameters may be obtained from the physical parameters of a system or experimentally by conducting the sinusoidal response test (described later in Chapter 10).

The transfer function

$$G(s)=\frac{K}{\frac{1}{\omega_n^2}s^2+\frac{2\zeta}{\omega_n}s+1} \tag{2.88}$$

is the standard form for second-order transfer functions, and will be encountered in all types of systems—electrical, mechanical, thermal, hydraulic, etc. It is called a *second-order lag* or a *quadratic lag*. For $\zeta > 1$, the system no longer has a quadratic lag; it is then a cascade of two simple lags.

2.9 THE FOURIER TRANSFORM AND SINUSOIDAL TRANSFER FUNCTIONS

Consider the transfer function (refer to Eqn. (2.53a))

$$G(s) = \frac{Y(s)}{R(s)} = \frac{K\prod_{i=1}^{m}(s+z_i)}{\prod_{i=1}^{n}(s+p_i)} \tag{2.89}$$

where m and n are integers with $m \le n$; $s = -z_i$ are the zeros of the transfer function, and $s = -p_i$ are the poles. The poles are strictly in the left half of the *s*-plane. We will assume, without any loss of generality, that the poles are real and distinct.

Let the input $r(t) = R_0 \sin \omega t\, \mu(t)$; R_0 is a constant and ω is angular frequency measured in rad/sec. The Laplace transform of $r(t)$ is

$$R(s) = \frac{R_0\omega}{s^2+\omega^2} \tag{2.90}$$

For this sinusoidal input, the Laplace transform of the output can be written as

$$Y(s) = \frac{A}{s-j\omega} + \frac{A^*}{s+j\omega} + F(s);\ j \triangleq \sqrt{-1} \tag{2.91}$$

where the residue A^* is the conjugate of the residue A, and

$$F(s) = \sum_{i=1}^{n} \frac{A_i}{s+p_i}$$

with

$$A_i = [(s+p_i)Y(s)]|_{s=-p_i} = [(s+p_i)G(s)R(s)]|_{s=-p_i}$$

The inverse Laplace transform of $F(s)$ is (Table 2.1)

$$f(t) = \sum_{i=1}^{n} A_i e^{-p_i t}$$

If we are interested in the steady-state response, we need not solve for the residues A_i; when all the $p_i > 0$, the time function in $f(t)$ will decay to zero as $t \to \infty$. We need the residue A because it is a part of the steady-state response.

$$A = [(s-j\omega)Y(s)]|_{s=j\omega} = \left[(s-j\omega)\frac{R_0\omega}{s^2+\omega^2}G(s)\right]_{s=j\omega} = \frac{R_0}{2j}G(j\omega) = \frac{R_0}{2j}|G(j\omega)|\,e^{j\phi};\ \phi = \angle G(j\omega) \tag{2.92}$$

The output $y(t)$ is the inverse Laplace transform of the right-hand side of Eqn. (2.91):

$$y(t) = \mathscr{L}^{-1}[Y(s)] = \mathscr{L}^{-1}\left[\frac{A}{s-j\omega} + \frac{A^*}{s+j\omega} + F(s)\right]$$

$$= R_0 \,|\, G(j\omega) \,|\left[\mathscr{L}^{-1}\left\{\frac{e^{j\phi}}{2j(s - j\omega)}\right\} + \mathscr{L}^{-1}\left\{-\frac{e^{-j\phi}}{2j(s + j\omega)}\right\}\right] + \mathscr{L}^{-1}\left\{\sum_{i=1}^{n} \frac{A_i}{s + p_i}\right\}$$

$$= R_0 \,|\, G(j\omega) \,|\left\{\frac{e^{j(\omega t+\phi)} - e^{-j(\omega t+\phi)}}{2j}\right\} + \sum_{i=1}^{n} A_i e^{-p_i t} = R_0 \,|\, G(j\omega) \,|\, \sin(\omega t + \phi) + \sum_{i=1}^{n} A_i e^{-p_i t}$$

The manipulation in the preceding equation follows from trigonometric identities (2.23).
The steady-state response is

$$\lim_{t\to\infty} y(t) \triangleq y_{ss}(t) = R_0 \,|\, G(j\omega) \,|\, \sin(\omega t + \phi) \tag{2.93}$$

The steady-state response remains the same when we relax the assumption of all poles being real and distinct, and allow all types of poles but restricted to the left half of the s-plane (this will be established quantitatively in Chapter 5).

Thus, if a sinusoidal input is applied to a system $G(s)$, all of whose poles have a negative real part, the steady-state response is a scaled, phase-shifted version of the input. The scaling factor is $|G(j\omega)|$, and the phase shift ϕ is the phase of $G(j\omega)$.

The sinusoidal steady state that results after the transient component dies out, is by definition, the *frequency-response* of the system. For an input $r(t) = R_0 \sin \omega t$, the output $y(t)$ is of the form $Y_0 \sin (\omega t + \phi) = R_0|G(j\omega)|\sin(\omega t + \phi)$; both the amplitude ratio $Y_0/R_0 = |G(j\omega)|$ and the phase angle $\phi = \angle G(j\omega)$ change when we use different frequency ω, and our desired results are graphs of these two quantities against frequency, the so-called *frequency-response curves*.

For obtaining frequency-response, we need not resort to a complete solution for a given sinusoidal input. *Sinusoidal transfer function* provides a much better method. It will be shortly shown that if we replace s by $j\omega (j \triangleq \sqrt{-1}; \omega \triangleq$ angular frequency, radian/sec) in $G(s)$ to obtain $G(j\omega)$, we get a complex number $|\,G(j\omega)\,| \angle\phi(j\omega)$ where $|\,G(j\omega)\,|$ will be the amplitude ratio Y_0/R_0, and $\phi = \angle G(j\omega)$ will be the angle by which the output sine wave leads the input sine wave (negative ϕ means the output sine wave lags the input sine wave; refer to Fig. 2.9). $G(j\omega)$ is called the sinusoidal transfer function.

We call the plot of $|G(j\omega)|$ vs. ω the *magnitude frequency response*, and the plot of $\phi(j\omega)$ vs ω the *phase frequency response*. The combination of the magnitude and phase responses gives the *frequency response model* of the system.

2.9.1 Dynamic Response using Sinusoidal Transfer Function Model/Frequency Response Model

When we have the frequency-response curves for any system described by the transfer function of the form (2.89), evaluation of the steady-state response to a sinusoidal input of any frequency ω is straightforward. What is not obvious, but is extremely important, is that ***the frequency response model is really a complete description of the system's dynamic behavior and allows us to compute the response to any input not just sine waves.***

Consider a dynamic system with impulse response $g(t)$, input $r(t)$ and output $y(t)$. If the impulse response $g(t)$ of the system is known, we can compute the output $y(t)$ for any input $r(t)$ using convolution integral (refer to Eqn. (2.69b)):

$$y(t) = \int_0^t g(t - \tau) r(\tau) d\tau \tag{2.94}$$

Laplace transformation substitutes multiplication for integration (refer to Eqn. (2.69a)):

$$Y(s) = G(s)R(s) \tag{2.95}$$

where $Y(s)$,$G(s)$, and $R(s)$ are, respectively, Laplace transforms of $y(t)$, $g(t)$, and $r(t)$.

We now replace Laplace transformation by Fourier transformation and observe the consequences. The equation which defines the Fourier transform of a signal $r(t)$, denoted $\mathscr{F}[r(t)] \triangleq R(j\omega)$, is given by the complex integral [64]

$$\mathscr{F}[r(t)] = R(j\omega) = \int_{-\infty}^{\infty} r(t)e^{-j\omega t}\,dt \tag{2.96a}$$

provided the integral exists. A sufficient condition that the integral exists is that

$$\int_{-\infty}^{\infty} |r(t)|\,dt < \infty \tag{2.96b}$$

The quantity $R(j\omega)$ is a complex function of ω. Graphically, it can be plotted in two parts: (i) $|R(j\omega)|$ vs ω, and (ii) $\angle R(j\omega)$ vs ω. The first one is the *amplitude plot* and the second, the *phase plot*. Together they constitute the *frequency spectrum* of the signal $r(t)$.

The equation which defines the inverse Fourier Transform, is given as [64]

$$\mathscr{F}^{-1}[R(j\omega)] = r(t) = \frac{1}{2\pi}\int_{-\infty}^{\infty} R(j\omega)e^{j\omega t}\,d\omega \tag{2.97}$$

It is interpreted as *decomposition* of $r(t)$ in terms of continuum of basis functions $e^{j\omega t}$. Note that it is necessary to consider both the positive and negative frequencies—a physically existing signal can be realized from $e^{j\omega t}$ if we consider both the positive and negative values of ω. For example.

$$\cos \omega t = \frac{e^{j\omega t} + e^{-j\omega t}}{2}$$

The signal $r(t)$ can thus be resolved into a sum of sinusoidal functions of different frequencies; $R(j\omega)$ tells us what frequencies make the signal $r(t)$, and it dictates the magnitude and phase of the corresponding sinusoidal functions. The range of frequencies is symmetrical around zero frequency: ***negative frequencies is a necessary mathematical abstraction***.

The Fourier transform of the convolution integral in Eqn. (2.94), results in the following equation (This can easily be established by the following steps similar to Laplace transformation, given in Eqns ((2.35)–(2.36))) :

$$Y(j\omega) = G(j\omega)\,R(j\omega) \tag{2.98}$$

where $Y(j\omega) = \mathscr{F}[y(t)]$; $G(j\omega) = \mathscr{F}[g(t)]$; $R(j\omega) = \mathscr{F}[(r(t))]$

Once $Y(j\omega)$ is available as a result of multiplication of $G(j\omega)$ and $R(j\omega)$, we can obtain $y(t)$ by inverse Fourier transformation.

Thus $G(j\omega)$ is a complete description of the system's dynamic behavior, and allows us to compute the response to any input.

It is worthwhile to take some moments to compare the Laplace and Fourier transforms. For a given $g(t)$ that is nonzero for positive time only (i.e., $g(t) = 0$, $t < 0$), and $\int_0^{\infty} |g(t)|\,dt < \infty$, the two transforms are related by

$$G(j\omega) = G(s)\big|_{s=j\omega} \tag{2.99}$$

***If all the poles of** G(s) **lie in the left-half of** s**-plane, the condition** $\int_0^\infty |g(t)|\, dt < \infty$**, is satisfied (Chapter 5). Then, we can obtain the Fourier transform** $G(j\omega)$ **from the corresponding Laplace transform** $G(s)$ **by merely replacing** s **with** $j\omega$.* $G(j\omega)$ is called the *sinusoidal transfer function*—a frequently used *frequency response model.*

Note that the frequency-response model gives the equivalent representation of the system's transfer function $G(s)$; thus frequency-response model is a complete description of the system's dynamic behavior.

2.9.2 Why Use Frequency-Response Models?

One question would still arise — when the dynamic response $y(t)$ can easily be obtained from the relation $Y(s) = G(s)R(s)$ using inverse Laplace transformation, why use computationally complex Fourier transform? A related question is—when the characteristics of a dynamic system can be easily described in terms of complex frequency variable 's' ($s = \sigma + j\omega$), and the location of poles and zeros on the s-plane (i.e., in terms of the transfer function $G(s)$), why represent characteristics of the system in terms of frequency variable 'ω' and the frequency-response model (i.e., in terms of the sinusoidal transfer function $G(j\omega)$)?

The Fourier transform has given us a way of decomposition of a signal into sinusoidal components of various frequencies. A time-varying signal** $r(t)$ **may be resolved into a sum of sinusoidal functions** $A_i \sin(\omega_i t + \phi_i)$ **of different frequencies** ω_i**. We can manipulate the response** $y(t)$ **of a dynamic system by manipulating its frequency response at various frequencies that make input** $r(t)$**. This is helpful because inputs to control systems have frequency-selective characteristics: we can visualize clearly the low-frequency and high-frequency features, as will be seen in later chapters.

We will, in fact, integrate the use of Laplace transform and Fourier transform for analysis and design of feedback control systems. To test the performance of the design carried out in frequency domain resulting in a stable closed-loop system, we will replace $j\omega$ temporarily with s and perform analysis in s-domain. This iterative process of analysis and design will cash on the benefits of both the Fourier transform and the Laplace transform.

Another important factor in favor of frequency-response models is that for a dynamic system, the frequency-response model can be obtained directly from experimental measurements (Chapter 10 will give the procedure). The model obtained from application of physical laws is based on many approximating assumptions (This will be evident when we take up development of models in the next chapter). The modeling errors introduce uncertainties, and affect the robustness of the performance of a control system.

2.10 CHARACTERISTIC PARAMETERS OF FREQUENCY-RESPONSE MODELS

In Section 2.8, we learnt that ***time response of a first-order system is characterized by time constant** τ**; and that of a second-order system is characterized by damping ratio** ζ **and undamped natural frequency** ω_n*. Let us now examine the characteristic parameters of frequency response. In this book, we will frequently come across frequency response models with nearly unity magnitudes for some range of frequencies and nearly zero magnitudes for others. The corresponding systems, therefore, *pass* input signals with some range of frequencies and *reject* or attenuate others. The systems which give rise to such frequency-response models are called *filters*. ***As a frequency-selective system, a filter limits the frequency spectrum of an input signal to some specified band of frequencies.*** A *low-pass filter*, for example, is a frequency-selective system which passes without appreciable change only 'low' frequencies, while severely diminishing the output amplitude at 'high' frequencies. Similarly, one can speak of *high pass* and *band pass* filters. Since ***feedback control systems are designed to behave typically as low-pass filters*** (established in later chapters), our focus in this book will be on low-pass filters.

Ideally, a low-pass filter should totally pass all frequencies in the pass band (low frequencies), and totally reject all frequencies outside its pass band (high frequencies). Therefore, an *ideal filter* would have a clearly defined *bandwidth* ω_{BW}. An ideal filter passes the input signal with no waveform distortion, though a scale change and time delay would be acceptable in many practical situations. Such a filter should have a time-domain behavior given by

$$y(t) = K\, r(t - \tau_D) \tag{2.100}$$

where $r(t)$ is the input signal, K represents scale change and τ_D is the acceptable time delay. In frequency domain (through Fourier transformation; refer to Eqn. (2.43))

$$Y(j\omega) = Ke^{-jwt_D} R(j\omega)$$

or

$$\frac{Y(j\omega)}{R(j\omega)} = G(j\omega) = |G(j\omega)| \angle G(j\omega) \tag{2.101a}$$

$$|G(j\omega)| = K; \text{ and } \angle G(j\omega) = \angle -\omega\tau_D \tag{2.101b}$$

The amplitude and phase characteristics of such a distortionless filter $G(j\omega)$ with $K = 1$, are shown in Fig. 2.26. Sad to say, ideal filters are physically unrealizable, though they may be closely approximated [155]. A *practical filter* causes distortion in amplitude ($|G(j\omega)|$ is not a constant) as well as in phase ($\angle G(j\omega)$ is not a linear function of ω).

Consider, for example, a filter with transfer function model

$$G(s) = \frac{\omega_n^2}{s^2 + 2\zeta\omega_n s + \omega_n^2} \tag{2.102}$$

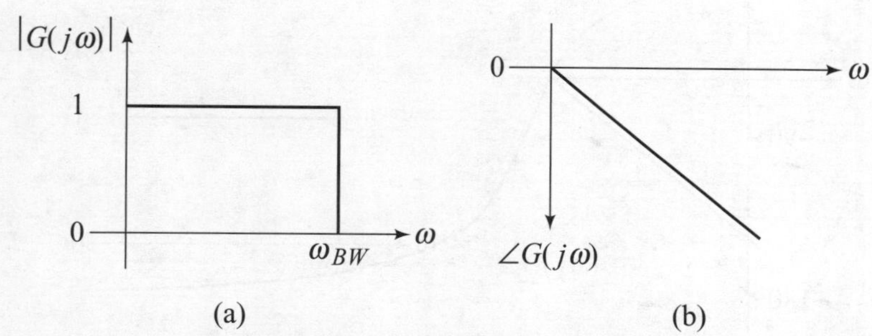

Fig. 2.26 *Ideal low-pass filter*

The sinusoidal transfer function of this filter is

$$G(j\omega) = \frac{\omega_n^2}{(j\omega)^2 + 2\zeta\omega_n(j\omega) + \omega_n^2} = \frac{\omega_n^2}{\omega_n^2 - \omega^2 + j2\zeta\omega_n\omega} \quad (2.103)$$

This gives

$$|G(j\omega)| = \frac{\omega_n^2}{\sqrt{(\omega_n^2 - \omega^2) + (2\zeta\omega_n\omega)^2}} \quad (2.104a)$$

$$\angle G(j\omega) = -\tan^{-1}\left(\frac{2\zeta\omega_n\omega}{\omega_n^2 - \omega^2}\right) \quad (2.104b)$$

Figure 2.27 shows a typical frequency response plot of the filter (2.104). We find that this practical low-pass filter deviates substantially from the *ideal* one shown in Fig. 2.26. This deviation is measured in terms of *height* and *width* of the practical low-pass filter.

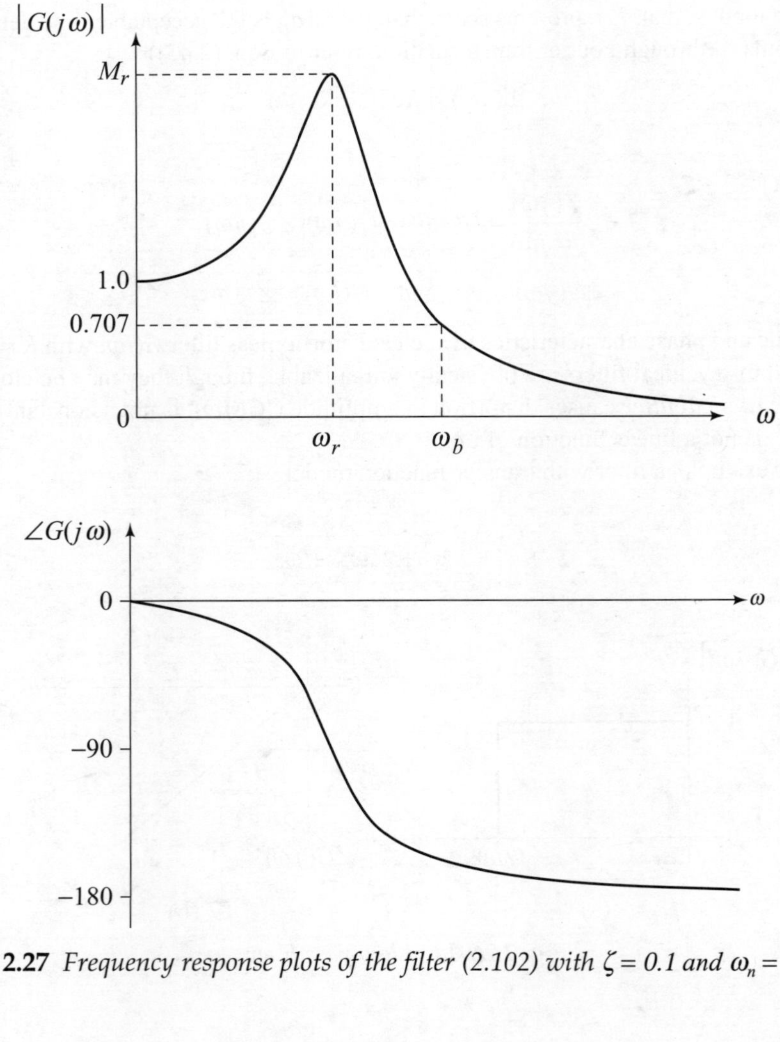

Fig. 2.27 *Frequency response plots of the filter (2.102) with $\zeta = 0.1$ and $\omega_n = 2000$*

The frequency response of the practical filter shown in Fig. 2.27 exhibits a sharp peak[11] M_r (called the *resonance peak*) in its amplitude characteristic at frequency ω_r (called the *resonance frequency*). It determines the height of the response curve. Also the practical filter has a gradual transition between pass band and rejection, unlike the ideal filter which has a sharp cut-off at ω_{BW}. Therefore, defining the bandwidth of a practical filter becomes somewhat arbitrary. A commonly used convention is to take ***bandwidth, ω_b, as that range of positive frequencies over which the amplitude ratio drops not lower than*** $1/\sqrt{2} = 0.707$ ***times the value at zero frequency***. Figure 2.27 shows the bandwidth of the filter (2.104).

The characteristic parameters that describe a system as a low-pass filter are (refer to Fig. 2.27):

(i) ***resonance peak, M_r;***
(ii) ***resonance frequency, ω_r; and***
(iii) ***bandwidth ω_b.***

For systems that exhibit no peak, bandwidth is the only parameter that represents filtering characteristics. Consider, for example, the filter

$$G(s) = \frac{1}{\tau s + 1}; \ \tau = \text{ time constant} \tag{2.105}$$

The sinusoidal transfer function of this filter is

$$G(j\omega) = \frac{1}{j\omega\tau + 1} \tag{2.106a}$$

This gives

$$|G(j\omega)| = \frac{1}{\sqrt{1+\omega^2\tau^2}}; \angle G(j\omega) = -\tan^{-1}\omega\tau \tag{2.106b}$$

Figure 2.28 shows a typical frequency response plot of this filter.

Example 2.4 Sinusoidal excitation is often used in testing of equipment and components. Consider the vibration-test system shown in Fig. 2.29a.

This system represents a *Shaker Table* which can be set to vibrate and then used to calibrate measuring instruments or to test the resilience of manufactured products [63].

Consider that an object, for example a camera, is placed on a shaker table and the input force $F(t)$ applied through a solenoid valve:

$$F(t) = K_s i(t) \tag{2.107}$$

where K_s (newtons/amp) is gain parameter of the solenoid. The input force is resisted by the spring (K) and the damping element (B). The net effect of these forces is the vertical displacement, $y(t)$, of the combined shaker table and camera.

The displacement $y(t)$ is considered with respect to position of the shaker table under the loaded condition. The gravitational forces, therefore, do not appear in the model being formulated.

[11] The resonance peak has a value greater than one for $0 < \zeta < 1/\sqrt{2}$; this will be established in Chapter 11.

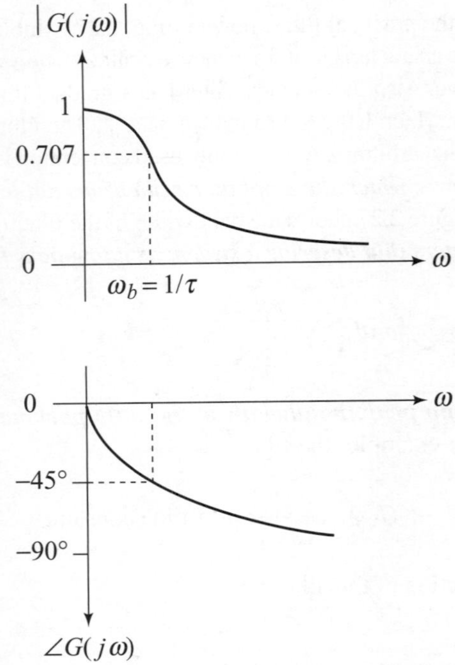

Fig. 2.28 *Frequency response of a first-order filter*

Figure 2.29b shows the free-body diagram for the system. From this figure, we have the following differential equation describing the dynamics of the system:

$$(M+m)\,\ddot{y}(t) = F(t) - K\,y(t) - B\,\dot{y}(t) \qquad (2.108)$$

Taking the Laplace transform of this equation, assuming zero initial conditions, we obtain

$$(M+m)s^2Y(s) = F(s) - KY(s) - BsY(s) \qquad (2.109)$$

Suppose that a transfer function is described between $I(s)$ and $Y(s)$; i.e., between the current input to the solenoid and the resulting displacement of the shaker table. From Eqns (2.107) and (2.109), we obtain

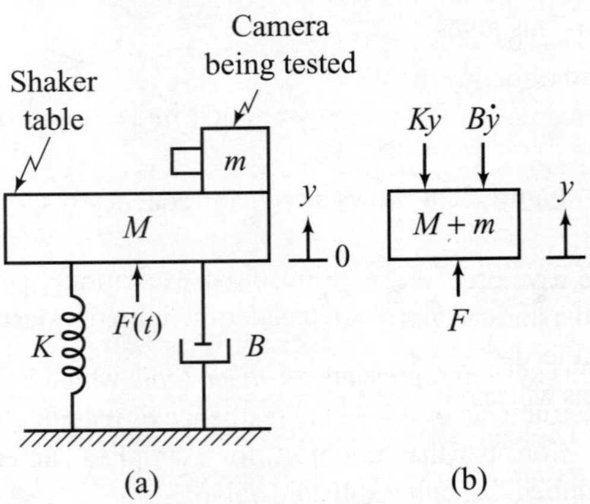

Fig. 2. 29 *Shaker table with camera being tested*

$$G(s) = \frac{Y(s)}{I(s)} = \frac{K_s/K}{\dfrac{M+m}{K}s^2 + \dfrac{B}{K}s + 1} \qquad (2.110)$$

With the parameters

$$M = 2.5 \text{ kg};\ m = 250 \text{ g};\ K = 1.1\times 10^7 \text{ newtons/m};$$

$$B = 1100 \text{ newtons /(m/sec)};\ K_s = 275 \text{ newtons/amp};$$

the transfer function becomes

$$G(s) = \frac{Y(s)}{I(s)} = \frac{2.5\times10^{-5}}{2.5\times10^{-7}s^2 + 1.0\times10^{-4}s + 1} \tag{2.111}$$

$$= \frac{2.5\times10^{-5}\omega_n^2}{s^2 + 2\zeta\,\omega_n s + \omega_n^2} = 2.5\times10^{-5}G_1(j\omega)$$

$$\text{where, } G_1(s) = \frac{\omega_n^2}{s^2 + 2\zeta\omega_n s + \omega_n^2} \tag{2.112}$$

with $\omega_n = 2000$ rad/sec, and $\zeta = 0.1$.

If we inject a sinusoidal input signal $i(t) = \dfrac{A}{2.5\times10^{-5}}$ sin $\omega t\,\mu(t)$; ω is the forcing frequency, the steady-state response

$$y_{ss}(t) = A|G_1(j\omega)|\sin(\omega t + \phi); \phi = \angle G_1(j\omega) \tag{2.113}$$

where (refer to Eqn. (2.103))

$$G_1(j\omega) = \frac{\omega_n^2}{\omega_n^2 - \omega^2 + j2\zeta\omega_n\omega} \tag{2.114}$$

The frequency-response plot of $G_1(j\omega)$ for $\zeta = 0.1$ and $\omega_n = 2000$, will be of the form shown in Fig. 2.27. The shaker table, thus, vibrates with the amplitude $A|G_1(j\omega)|$; the vibrations become violent as the forcing frequency ω approaches ω_r—the resonance frequency of the system, resulting in high dynamic stresses on the table.

2.11 BLOCK DIAGRAM MODELS

We have been working with individual systems represented as a *block* with an input *R*(*s*), an output *Y*(*s*), and a transfer function *G*(*s*) (Fig. 2.13). Control systems are composed of multiple subsystems with cascade and feedback interconnections based on cause-and-effect relationships. Examples of interconnected subsystems will appear in the next chapter. Here, we introduce a model for representation of such systems.

The cause-and-effect relationships can be represented by diagrammatic means using a *block diagram model*. Figure 2.30a shows an example of cascaded systems. Figure 2.30b shows the interconnections and the intermediate signal values in transform domain; and Fig. 2.30c gives the equivalent transfer function:

$$\frac{Y(s)}{R(s)} = \frac{Y(s)}{Y_2(s)}\times\frac{Y_2(s)}{Y_1(s)}\times\frac{Y_1(s)}{R(s)} = G_3(s)G_2(s)G_1(s) \tag{2.115}$$

which is the product of the subsystems' transfer functions.

Equation (2.115) has been derived under the assumption that interconnected subsystems do not *load* adjacent subsystems. That is, a subsystem's output remains the same whether or not subsequent subsystem is connected. If there is a change in the output, the subsequent subsystem loads the previous one, and the equivalent transfer function is *not* the product of the individual transfer functions.

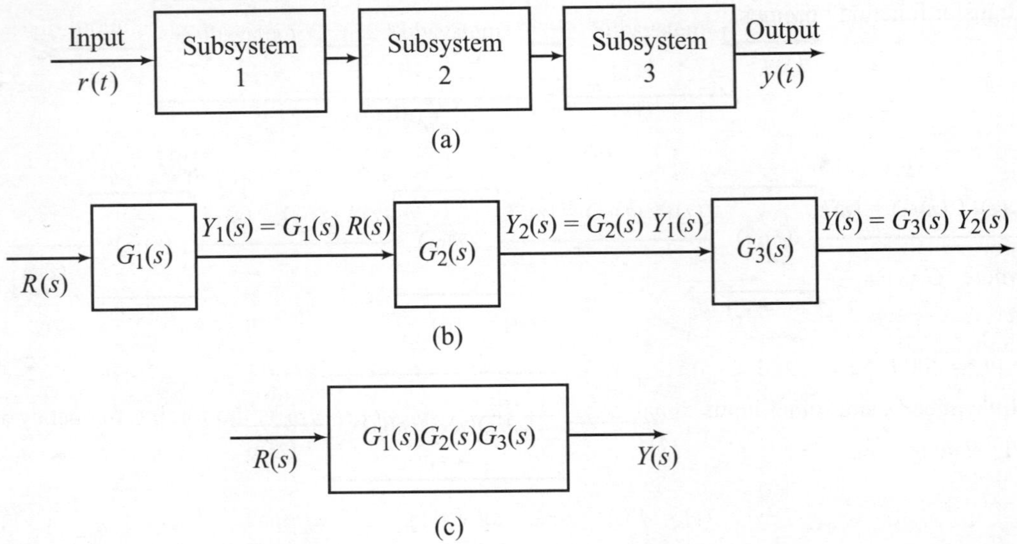

Fig. 2.30 *Cascaded nonloading subsystems*

Consider, for example, the servomechanism for steering of antenna (Fig. 1.13). A motor (actuator) is a subsystem that drives another subsystem—the antenna (load). Since antenna appreciably loads the motor (as will be seen in the next chapter), the two subsystems cannot be cascaded as per Fig. 2.30. The two together are treated as a single subsystem for which the transfer function model is derived from differential equations governing the dynamics of the (motor + antenna) subsystem.

In general, loading effects occur because when analyzing an isolated subsystem, we assume that no power is being withdrawn at its output location. When we later decide to attach another subsystem to the output of the first, this second subsystem does withdraw some power, contradicting our earlier assumption and thereby invalidating the analysis based on this assumption. Consider again the servomechanism shown in Fig. 1.13. The power amplifier has its own internal power supply (op amp (Fig. 2.18) is one example) and can thus supply power to a 'downstream' subsystem without drawing power from the 'upstream' subsystem. That is, such amplifiers have high input impedance and low output impedance. In other words, there is no loading effect on the signal source (error detector) connected to the input terminals of the power amplifier, and the voltage generated by the power amplifier in response to input signals is available at the (motor + load) subsystem, where it does useful work.

When we model chains of subsystems by simple multiplication of their individual transfer functions, the assumption will always be made that the interconnection of various subsystems is nonloading; loading effects are either not present, have been proven negligible or have been made negligible by interposing buffer amplifiers between loading subsystems. *Buffer amplifiers* may be electrical, hydraulic, pneumatic, or the like, to suit the physical nature of the connected subsystems; and have their own internal power supplies. Commonly used electrical buffer amplifiers are op amps.

2.11.1 Generalized Block Diagram of a Feedback System

We have already used block diagrams to represent the input–output behavior of a linear system; the signal into the *block* represents the input, the signal out of the block represents the output, while the block itself stands for the transfer function of the linear system (Fig. 2.31a). The flow of information (signal) is unidirectional from the input to the output, with the output being equal to the input multiplied by the transfer function of the block. A complex system comprising of several nonloading elements is represented by the interconnection of the blocks for the individual elements. The blocks are connected by lines and arrows indicating the unidirectional flow of information from the output of one block to the input of the other (Fig. 2.31b). In addition to this, a *summing junction* (Fig. 2.31c) in a block diagram represents the dynamic summation of signals, and a *branch point* (Fig. 2.31d) is a take-off point at which the information branches out and goes concurrently to other blocks or summing junctions.

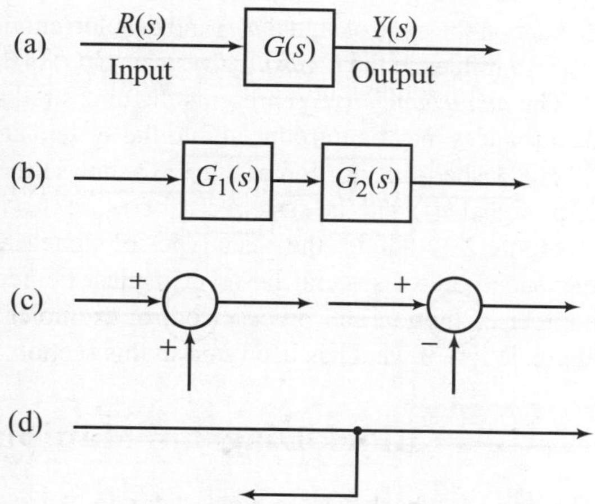

Fig. 2.31 *Elements of a block diagram*

Various feedback control configurations were described in Chapter 1 using block diagrams of general functional nature. With the transfer function concept introduced earlier in this chapter, we are now equipped to consider *operational block diagrams* necessary for design and analysis. We begin with the block diagram shown in Fig. 2.32. It represents the basic structure of feedback control systems (refer to Fig. 1.9). Not all systems can be forced into this format, yet it serves as a reference for discussion.

In Fig. 2.32, the variable $y(t)$ is the *controlled variable* of the system. The desired value of the controlled variable is $y_r(t)$; the *command input* $y_r(t)$ and $y(t)$ have the same units. In the next chapter, we will discuss many motion control systems with speed or position of an inertia load as controlled variable. Process control examples with temperature, or pressure of a process fluid as controlled variable will also appear.

The *feedback elements* with transfer function $H(s)$ are system components that act on the controlled variable $y(t)$ to produce the *feedback signal* $b(t)$. $H(s)$ typically represents the sensor action to convert the controlled variable $y(t)$ to an electrical sensor output signal $b(t)$.

The *reference function elements* with transfer function $A(s)$ convert the command signal $y_r(t)$ into a form compatible with the feedback signal $b(t)$. The transformed command signal is the actual physical input to the system. This actual signal input is defined as the *reference input*.

The comparison device (*error detector*) of the system compares the reference input $r(t)$ with the feedback signal $b(t)$ and generates the *actuating error signal* $\hat{e}(t)$. The signals $r(t)$, $b(t)$ and $\hat{e}(t)$ have the same units. The *controller* with transfer function $D(s)$ acts on the actuating error signal to produce the *control signal* $u(t)$.

The control signal $u(t)$ has the knowledge about the desired control action. The power level of this signal is relatively low. The *actuator elements* with transfer function $G_A(s)$ are the system components

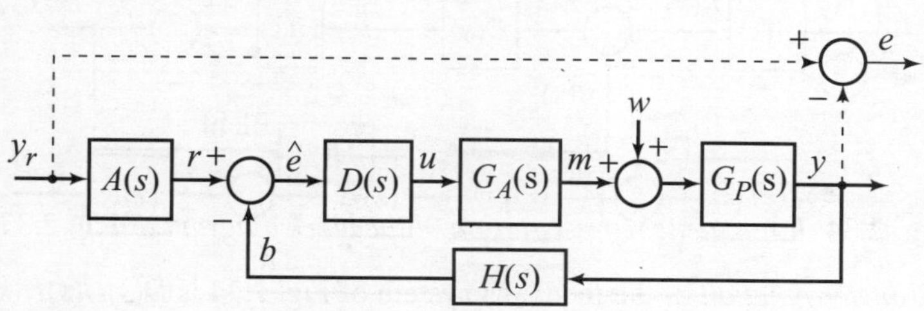

Fig. 2.32 *Generalized operational block diagram of a feedback system*

that act on the control signal $u(t)$ and develop enough torque, pressure, heat, etc., (*manipulated variable* $m(t)$) to influence the *controlled system*. $G_P(s)$ is the transfer function of the controlled system.

The *disturbance* $w(t)$ represents the undesired signals that tend to affect the controlled system. The disturbances may be introduced into the system at more than one location.

The dashed-line portion of Fig. 2.32 shows the *system error* $e(t) = y_r(t) - y(t)$. Note that the actuating error signal $\hat{e}(t)$ and the system error $e(t)$ are two different variables.

Figure 2.32 defines the basic types of signals and components necessary for the description of any feedback control system; however it must be adopted to the needs of each specific design. A large number of motion and process control examples given in the next chapter will surely help in better appreciation of what has been said in this section.

2.11.2 Block Diagram Manipulations

The basic feedback system block diagram of Fig. 2.32 is shown in an abridged form in Fig. 2.33. The output $Y(s)$ is influenced by the control signal $U(s)$ and the disturbance signal $W(s)$ as per the following relation:

$$Y(s) = G_P(s)G_A(s)U(s) + G_P(s)W(s) \tag{2.116a}$$

$$= G(s)U(s) + N(s)W(s) \tag{2.116b}$$

where $G(s)$ is the transfer function from the control signal $U(s)$ to the output $Y(s)$, and $N(s)$ is the transfer function from the disturbance input $W(s)$ to the output $Y(s)$. Using Eqns (2.116a) and (2.116b), we can modify the block diagram of Fig. 2.33 to the form shown in Fig. 2.34. Note that in the block diagram model of Fig. 2.34, the plant includes the actuator elements.

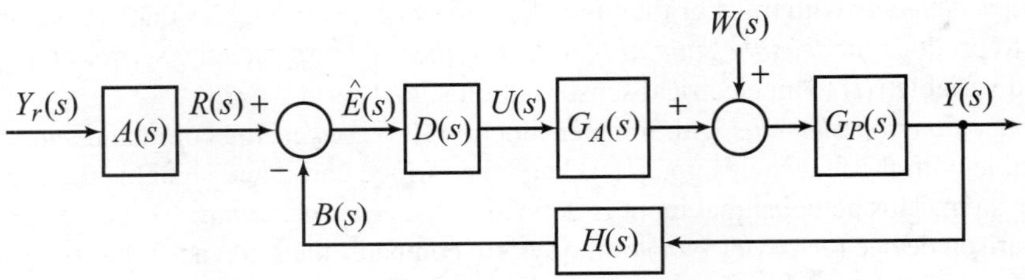

Fig. 2.33 *A general linear feedback system*

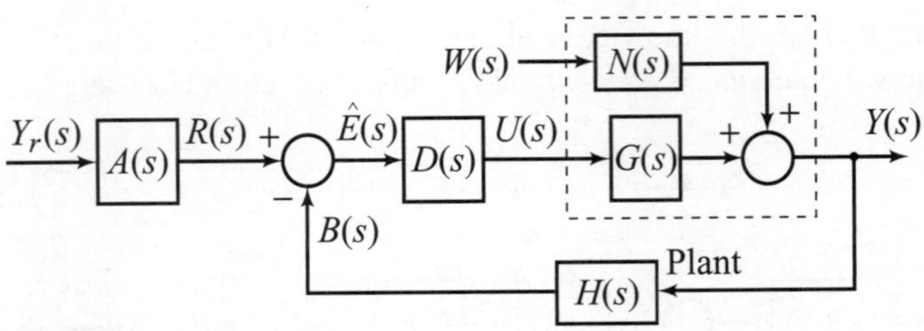

Fig. 2.34 *Equivalent representation of the block diagram of Fig. 2.33*

The *open-loop transfer function* of the feedback system of Fig. 2.34 is $D(s)G(s)H(s)$ (it is the transfer function between $R(s)$ and $B(s)$ if the feedback loop is broken at the summing point).

$$\text{Open-loop transfer function} \triangleq D(s)G(s)H(s) \tag{2.116c}$$

Note that for the closed-loop system, $D(s)G(s)H(s)$ is the transfer function product around the entire feedback loop. We call it *loop gain*, and denote it by $L(s)$:

$$L(s) \triangleq D(s)G(s)H(s) \tag{2.116d}$$

The actuating error signal

$$\hat{E}(s) = R(s) - B(s) = A(s)Y_r(s) - H(s)Y(s)$$

The control signal

$$U(s) = D(s)A(s)Y_r(s) - D(s)H(s)Y(s) \tag{2.117a}$$

$$= D(s)H(s)\left[\frac{A(s)}{H(s)}Y_r(s) - Y(s)\right] \tag{2.117b}$$

Using Eqns (2.117), we can simplify Fig. 2.34 to obtain the structure shown in Fig. 2.35.

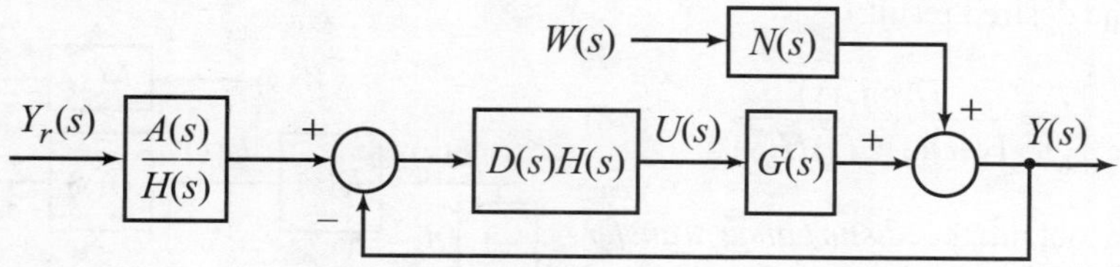

Fig. 2.35 *Simplification of the block diagram of Fig. 2.34*

A further simplification of Fig. 2.35 is possible if $H = A$. In this case, which is quite common, we can model the system as the *unity feedback system* as shown in Fig. 2.36, and take advantage of the fact that now the actuating signal is the system error $e(t)$.

The block diagrams in Figs 2.34–2.36, as we shall see later, are very useful for the purpose of system design. However, it should be clear that these block diagrams have lost the physical significance. For example, the block in Fig. 2.35 with transfer function $A(s)/H(s)$, does not refer to any physical portion of the original system. Rather, it represents the result of manipulating Eqn. (2.117a) into the form given by Eqn. (2.117b).

Thus, the reader is advised to ***think in terms of the equations that the block diagrams represent, rather than to attach any special significance to block diagrams themselves. For example, a unity-feedback block diagram does not convey that the sensor gain is necessarily unity. The only role played by a block diagram is that it is a convenient means of representing the various system equations, rather than writing them out explicitly. Block diagram manipulation is nothing more than the manipulation of a set of algebraic transform equations.***

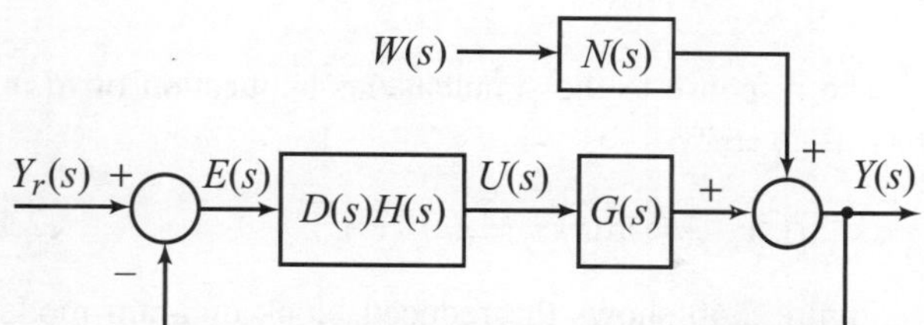

Fig. 2.36 *Unity feedback system*

For the analysis of a feedback system, we require the transfer function between the input—either reference or disturbance—and the output. We can use block diagram manipulations to eliminate all the signals except the input and the output. The reduced block diagram leads to the desired result.

Consider the block diagram of Fig. 2.37. This feedback system has two inputs. We shall use superposition to treat each input separately.

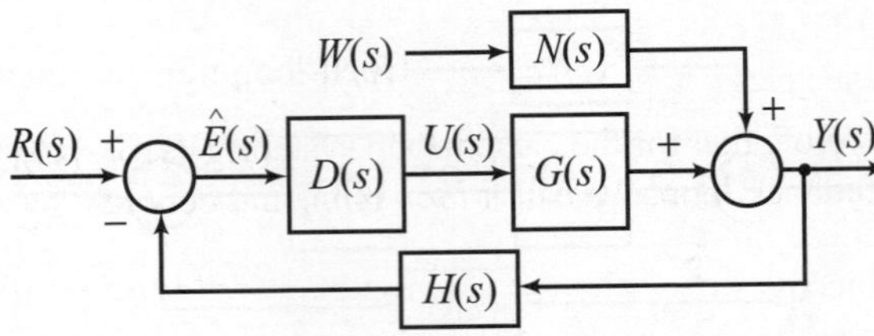

Fig. 2.37 *A typical feedback system with two inputs*

When the disturbance input is set to zero, the single-input system of Fig. 2.38 results. The transfer function between the input $R(s)$ and the output $Y(s)$ is referred to as the *reference transfer function* and will be denoted by $M(s)$. To solve for $M(s)$, we write the pair of transform equations,

$$\hat{E}(s) = R(s) - H(s)Y(s);$$
$$Y(s) = G(s)U(s) = G(s)D(s)\hat{E}(s)$$

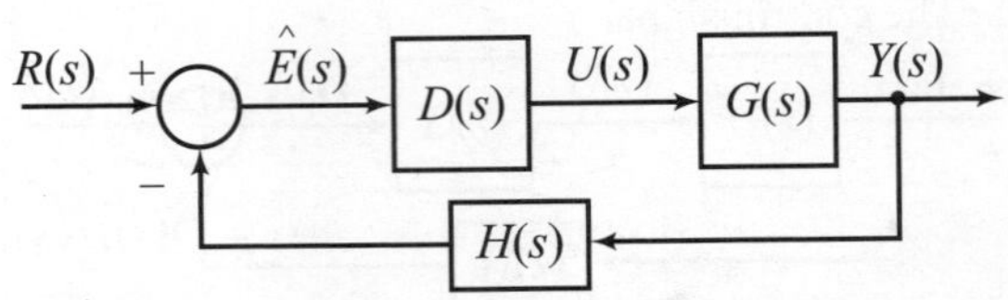

Fig. 2.38 *Block diagram without disturbance input*

and then eliminate $\hat{E}(s)$ to obtain

$$[1 + D(s)G(s)H(s)]Y(s) = D(s)G(s)R(s)$$

which leads to the desired result:

$$M(s) = \left.\frac{Y(s)}{R(s)}\right|_{W(s)=0} = \frac{D(s)G(s)}{1 + D(s)G(s)H(s)} \quad (2.118)$$

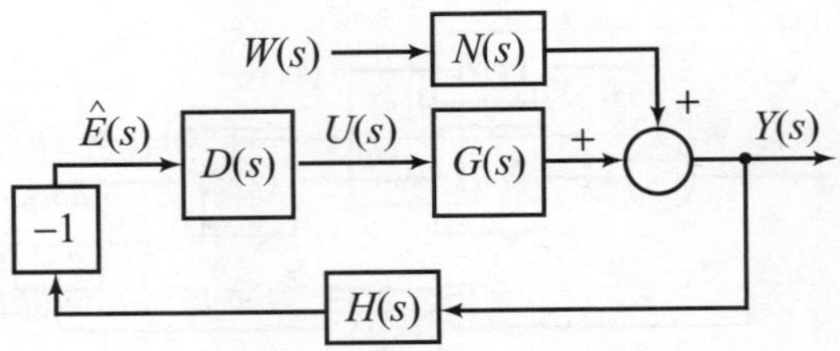

Fig. 2.39 *Block diagram without reference input*

Similarly, we obtain the *disturbance transfer function* $M_w(s)$ by setting the reference input to zero in Fig. 2.37 yielding Fig. 2.39, and then solving for $Y(s)/W(s)$. From the revised block diagram,

$$\hat{E}(s) = -H(s)Y(s)$$
$$Y(s) = G(s)D(s)\hat{E}(s) + N(s)W(s)$$

from which $\hat{E}(s)$ can be eliminated to give

$$M_w(s) = \left.\frac{Y(s)}{W(s)}\right|_{R(s)=0} = \frac{N(s)}{1 + D(s)G(s)H(s)} \quad (2.119)$$

The response to the simultaneous application of $R(s)$ and $W(s)$ is given by

$$Y(s) = M(s)R(s) + M_w(s)W(s) \quad (2.120)$$

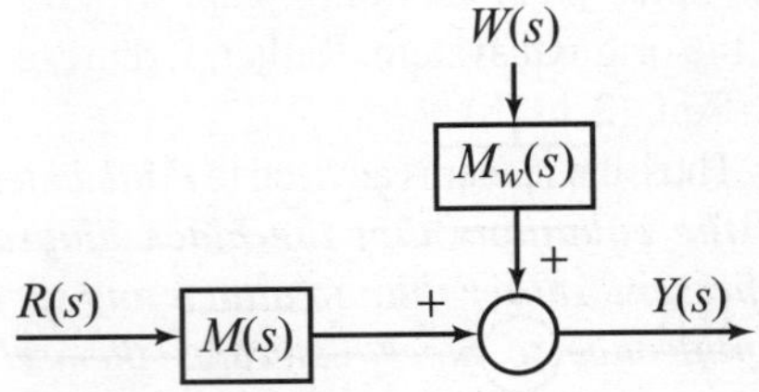

Fig. 2.40 *Reduced block diagram model for system of Fig. 2.37*

Figure 2.40 shows the reduced block diagram model of the given feedback system.

The transfer functions given by Eqns (2.118) and (2.119) are referred to as *closed-loop transfer functions*. Looking at Eqn. (2.118), we notice that the numerator is the multiplication of the transfer functions in the *forward path* between the two variables $R(s)$ and $Y(s)$. The denominator of this equation has the term $D(s)G(s)H(s)$ which is the *loop gain*. An inspection of Eqn. (2.119) shows that the numerator is again the multiplication of the transfer functions in the forward path between $W(s)$ and $Y(s)$. The denominator is the same as in the case of Eqn. (2.118).

From the examples of block diagram manipulations given above, we observe that there are two different objectives which may motivate one to manipulate a block diagram. The most obvious is to obtain the overall transfer function for the purpose of analysis (the block diagram of Fig. 2.37 was manipulated to obtain the one given in Fig. 2.40. The reader should, however, note that the CAD software available today (Simulink, for example, provides powerful tools for analysis without the need for block diagram reduction). The other objective is to manipulate a block diagram into a simpler form which is more convenient in the design process (the block diagram of Fig. 2.33 was manipulated to obtain the one given in Fig. 2.36).

The process by which we have manipulated the block diagrams is nothing more than the manipulation of a set of algebraic transform equations. We can, in fact, manipulate a complex block diagram directly by using the simple rules of block diagram reduction. The most commonly encountered block diagram operations are summarized in Table 2.3.

Example 2.5 As we shall see in the next chapter, often a feedback system will have more than one feedback path resulting in a multiple-loop configuration. The block diagram of a typical multiple-loop feedback control system is shown in Fig. 2.41 (Note that the loop $G_3(s)G_4(s)H_1(s)$ is a positive-feedback loop). Let us obtain the closed-loop transfer function of this system using block diagram manipulation rules given in Table 2.3.

Table 2.3 *Rules of block diagram algebra*

Rule	*Original diagram*	*Equivalent diagram*
1.	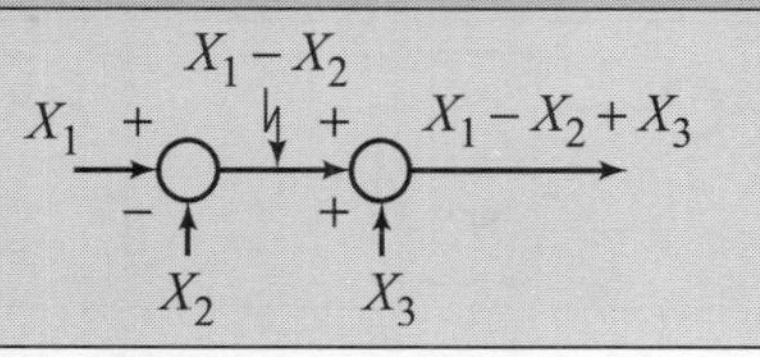	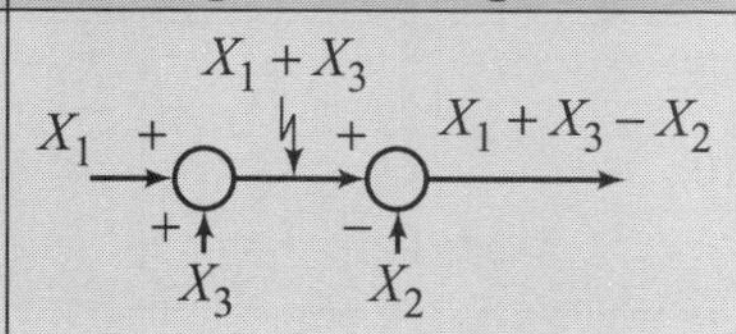
2.	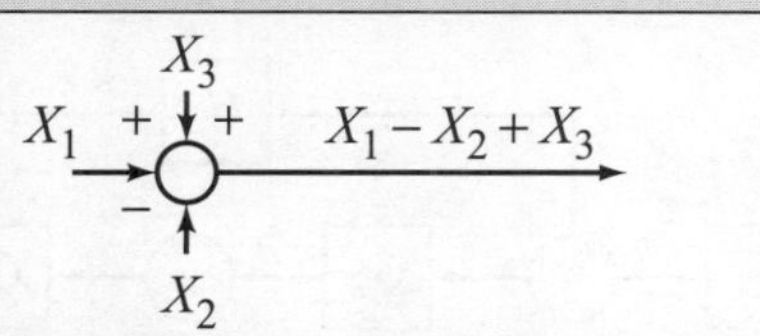	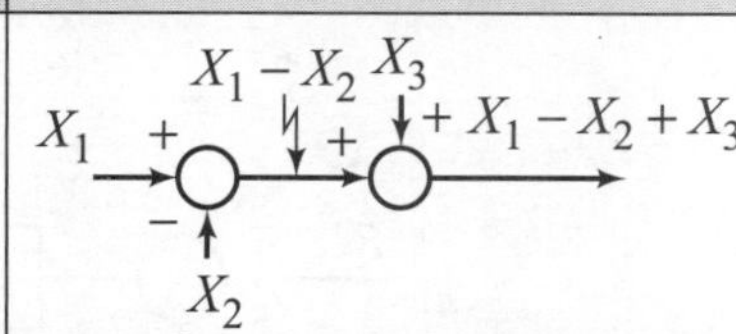
3.	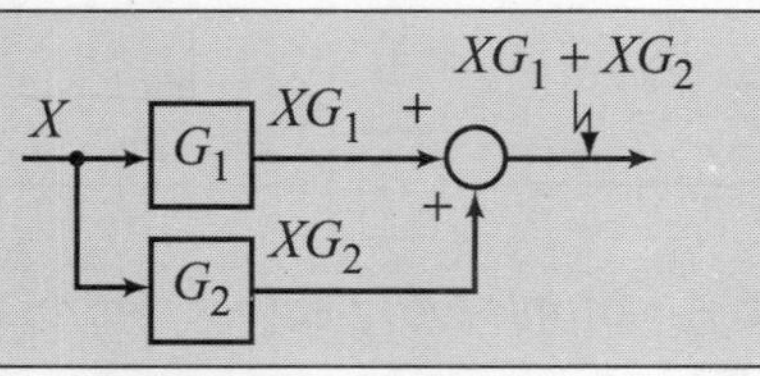	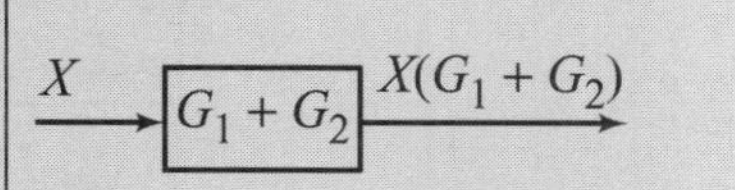
4.	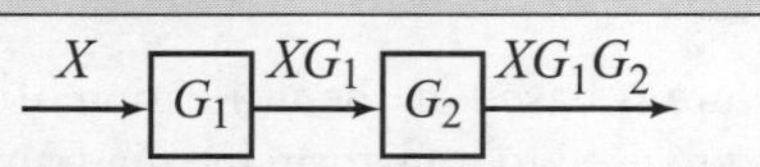	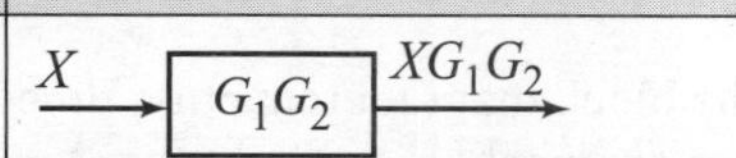
5.	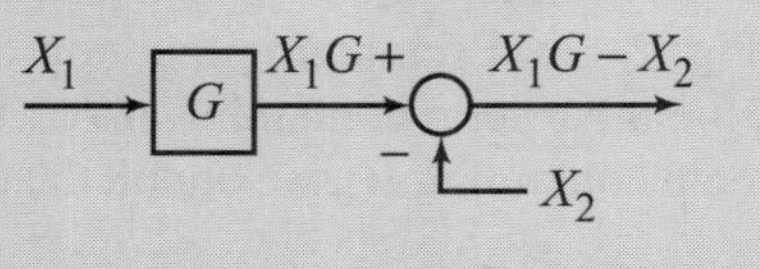	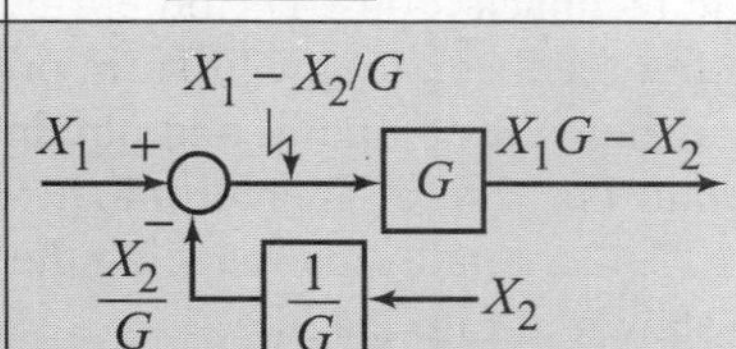

(Continued)

Table 2.3 *(Continued)*

Rule	*Original diagram*	*Equivalent diagram*
6.	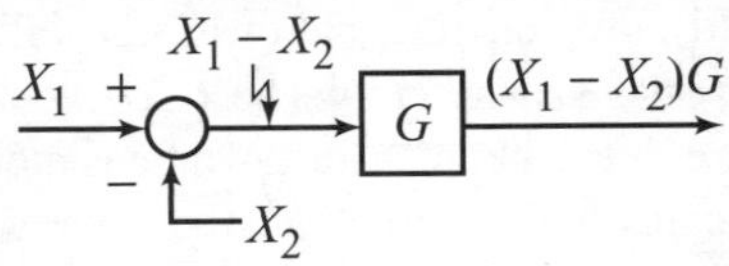	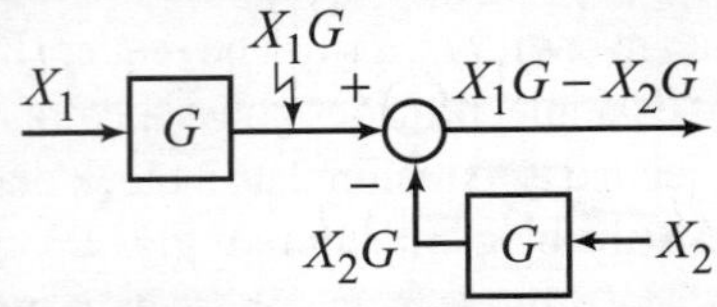
7.	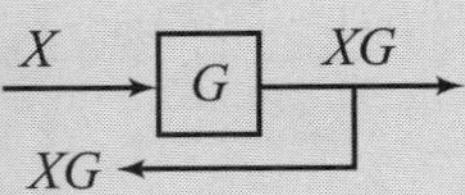	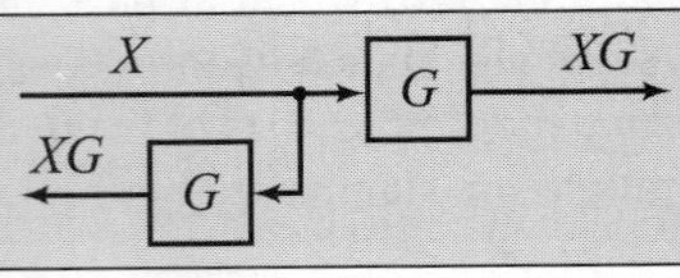
8.	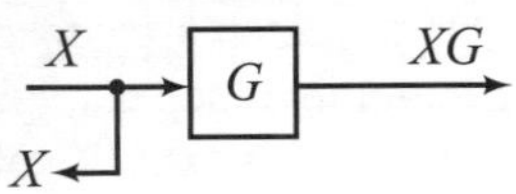	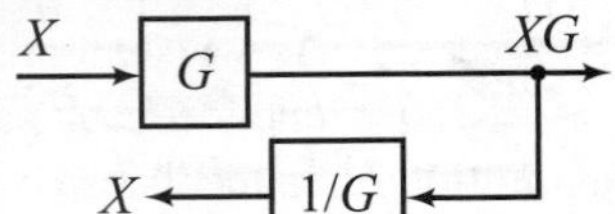
9.	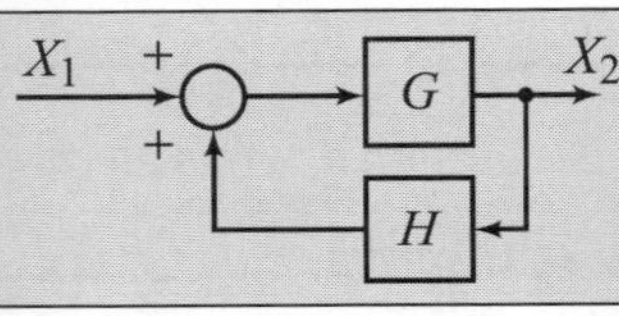	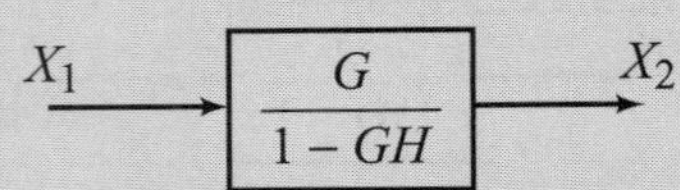
10.	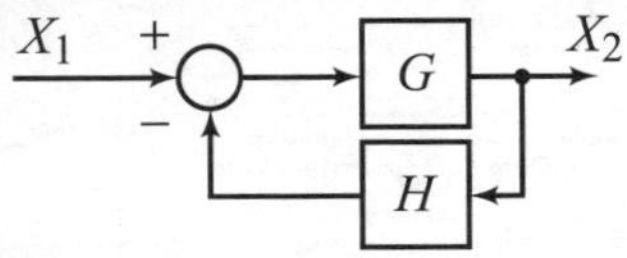	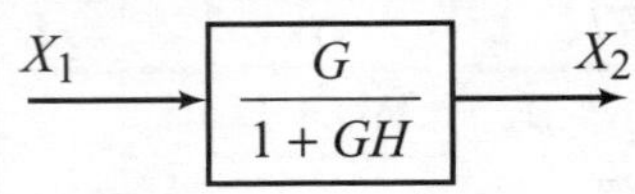

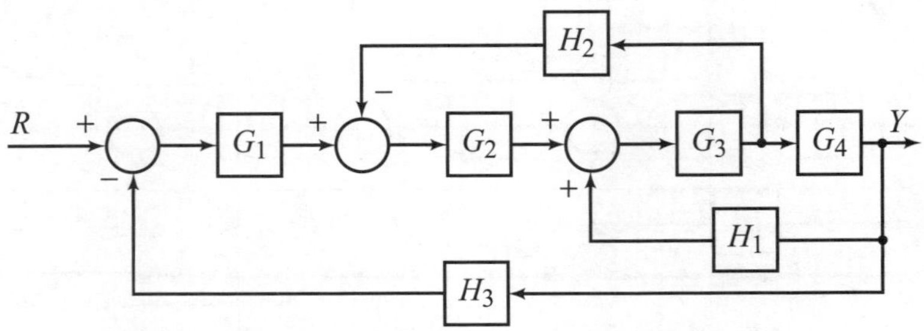

Fig. 2.41 *A multiple-loop feedback control system*

The block diagram reduction procedure is based on the utilization of rules 9 and 10, which eliminate feedback loops. The other rules are used to transform the diagram to the form ready for eliminating feedback loops. First, in order to eliminate the loop $G_3G_4H_1$, we move H_2 behind block G_4 by using rules 8 and 4, and obtain Fig. 2.42a. Eliminating the loop $G_3G_4H_1$ by rules 4 and 9, we obtain Fig. 2.42b. Then eliminating the inner loop containing H_2/G_4 by rules 4 and 10, we obtain Fig. 2.42c. Finally, by reducing the loo p containing H_3, we obtain the closed-loop transfer function as shown in Fig. 2.42d.

An alternate approach is to move the summing point of the positive feedback loop containing H_1, outside the negative feedback loop containing H_2. Application of rules 5, 4 and 1 results in Fig. 2.43a.

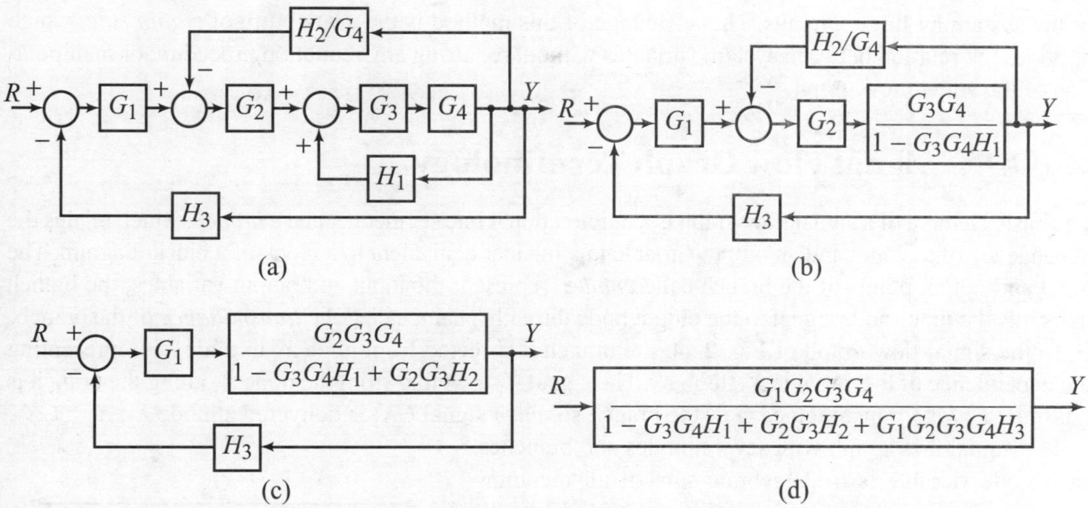

Fig. 2.42 *Block diagram reduction of the system of Fig. 2.41*

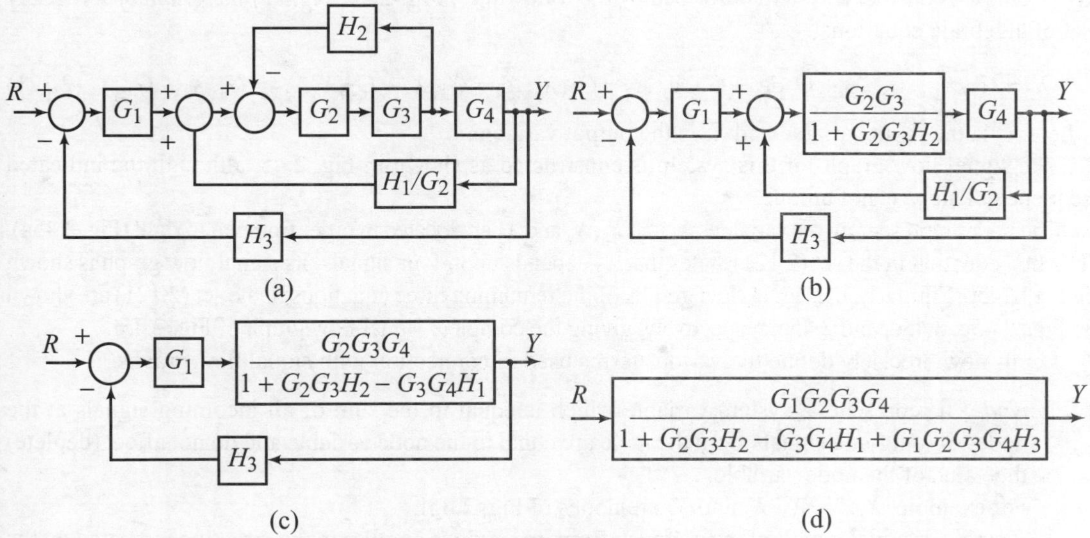

Fig. 2.43 *Block diagram reduction of the system of Fig. 2.41*

Eliminating the feedback loop containing H_2, we get Fig. 2.43b. Then, the elimination of loop containing H_1/G_2 gives Fig. 2.43c. Finally, eliminating the loop containing H_3 results in Fig. 2.43d.

2.12 SIGNAL FLOW GRAPH MODELS

Block diagrams are adequate for the representation of the interrelationships of the variables of a control system. However, for a system with reasonably complex interrelationships, the block diagram reduction procedure is cumbersome and often quite difficult to complete. An alternative method for determining the relationship between system variables, called the *signal flow graph* method, is based on a representation of the system by line segments. The advantage of this method is the availability of a *gain rule*, which provides the relation between system variables without requiring any reduction procedure or manipulation of the signal flow graph.

2.12.1 Signal Flow Graph Terminology

The basic element of a signal flow graph is a unidirectional line segment called a *branch*, which relates the dependency of an input and an output variable in a manner equivalent to a block of a block diagram. The input and output points of the branch called *nodes*, represent the input and output variables; the branch transmits the input-node signal to the output node through a gain, called the *transmittance* of the branch.

In the signal flow graph of Fig. 2.44, the branch is directed from node X_1 to node X_2, representing the dependence of the variable X_2 upon X_1. The signal X_1 traveling from the node X_1 along the branch is multiplied by the transmittance G of the branch so that a signal GX_1 is delivered at node X_2; $X_2 = GX_1$.

In a signal flow graph with several nodes and branches, each node variable is the algebraic sum of all incoming signals; and all outgoing signals from any node equal the node variable and do not affect (deplete) it. As an illustrative example, consider a system described by the following set of algebraic equations:

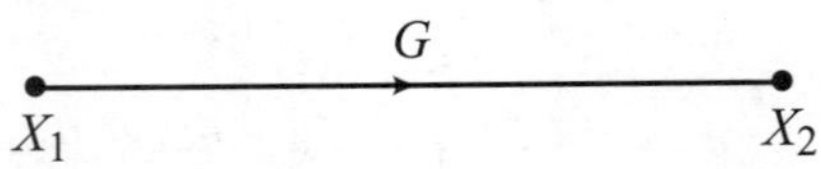

Fig. 2.44 *Signal flow graph of* $X_2 = GX_1$

$$X_2 = G_{12}X_1 + G_{32}X_3 + G_{42}X_4 + G_{52}X_5;\ X_3 = G_{23}X_2;\ X_4 = G_{34}X_3 + G_{44}X_4;\ X_5 = G_{35}X_3 + G_{45}X_4 \quad (2.121)$$

where X_1 is the input variable and X_5 is the output variable.

The signal flow graph for this system is constructed as shown in Fig. 2.45, although the indicated sequence of steps is not unique.

The nodes representing the variables X_1, X_2, X_3, X_4 and X_5 are located in order from left to right (Fig. 2.45a). The first equation in the set (2.121) states that X_2 depends upon four signals; its signal flow graph is shown in Fig. 2.45b. Similarly, the signal flow graphs of the remaining three equations in the set (2.121) are shown in Figs 2.45c, 2.45d, and 2.45e, respectively, giving the complete signal flow graph of Fig. 2.45f.

Let us now precisely define the various terms used in connection with signal flow graphs.

1. *Node:* It represents a system variable which is equal to the sum of all incoming signals at the node. All outgoing signals from the node are equal to the node variable and do not affect (deplete) the value of the node variable.
   For example, X_1, X_2, X_3, X_4 and X_5 are nodes of Fig. 2.45f.
2. *Branch:* A signal travels along a branch from one node to another in the direction indicated by the branch arrow and in the process gets multiplied by the gain or transmittance of the branch. The branch operator (gain/transmittance) is written on the signal flow graph near the branch arrow.
   For example, in Fig. 2.45f, the signal reaching node X_3 from node X_2 is given by $G_{23}X_2$ where G_{23} is the transmittance from node X_2 to node X_3.
3. *Input node or source:* It is a node with only outgoing branches (X_1 in Fig. 2.45f).
4. *Output node or sink:* It is a node with only incoming branches.
   Node X_5 in Fig. 2.45f has an outgoing branch, but after introducing an additional branch with unit transmittance as shown in Fig. 2.45g, the added node becomes an output node. This operation is equivalent to adding an equation
   $$X_5 = X_5$$
   in the set (2.121).
   However, we cannot convert a non-input node into an input node by a similar operation.

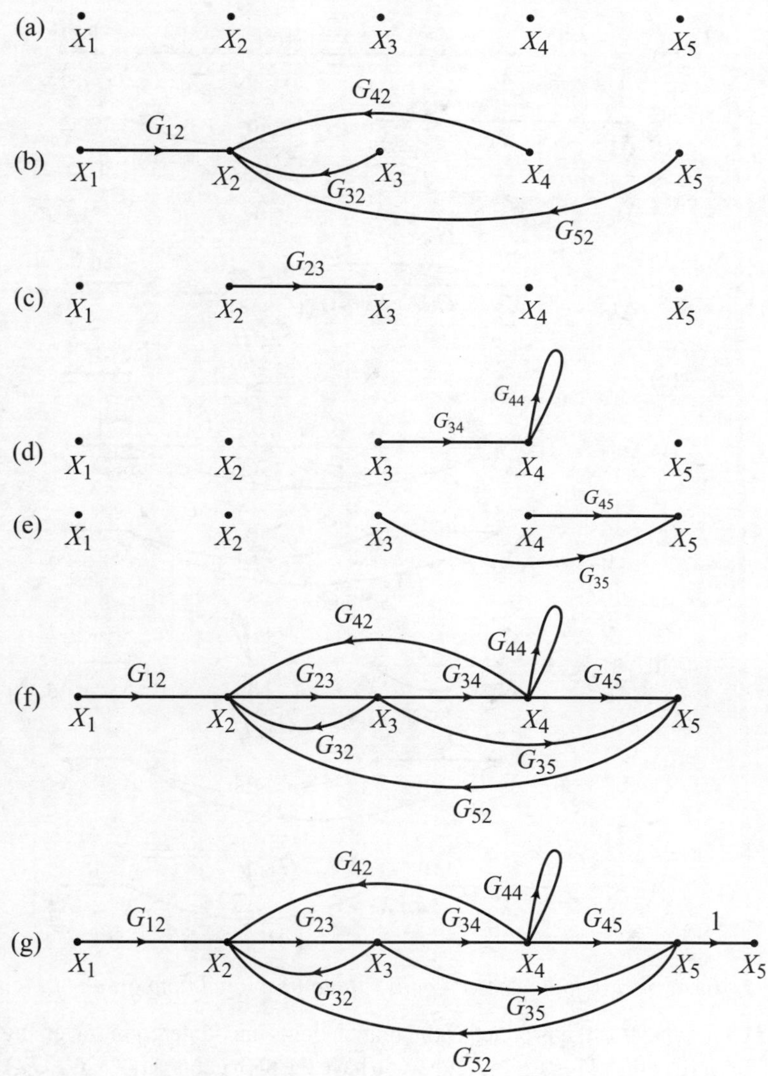

Fig. 2.45 *Construction of signal flow graph for Eqns (2.121)*

5. *Path*: It is the traversal of connected branches in the direction of the branch arrows such that no node is traversed more than once.
6. *Forward path:* It is a path from the input node to the output node.
   For the signal flow graph of Fig. 2.45g, there are two forward paths shown in Figs 2.46a and 2.46b.
7. *Loop:* It is a path which originates and terminates at the same node. The flow graph of Fig. 2.45g has five loops shown in Figs 2.46c to 2.46g.
8. *Nontouching loops:* Loops are said to be nontouching if they do not possess any common node. The flow graph of Fig. 2.45g has two possible combinations of two nontouching loops. These are shown in Figs 2.46h and 2.46i. This flow graph does not have combinations of more than two nontouching loops.

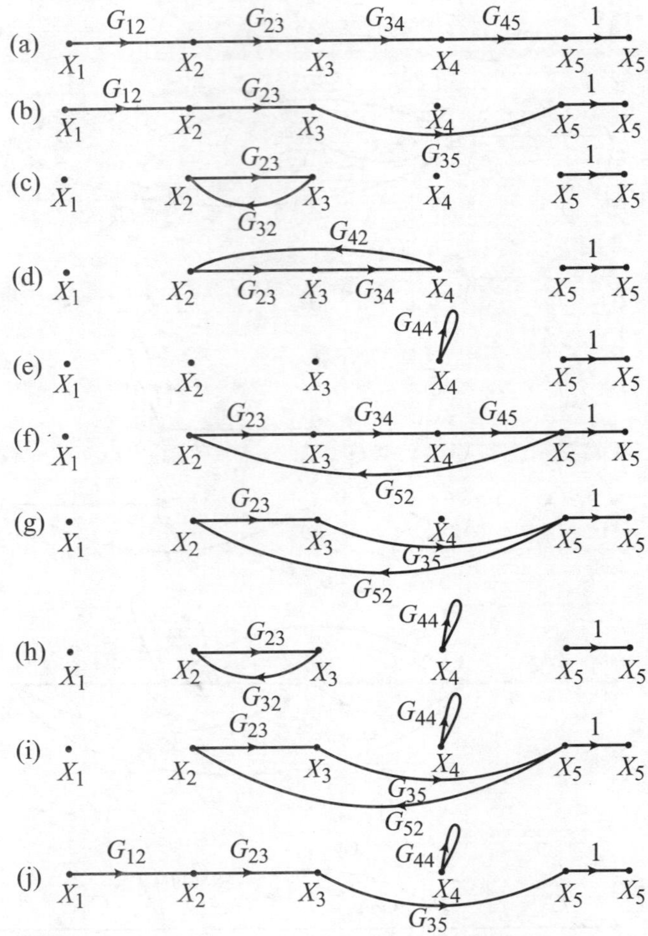

Fig. 2.46 *Application of Mason's gain rule to the signal flow graph of Fig. 2.45g*

9. *Forward path gain:* It is the product of the branch gains encountered in traversing a forward path. The two forward paths of Figs 2.46a and 2.46b have the path gains $(G_{12}G_{23}G_{34}G_{45})$ and $(G_{12}G_{23}G_{35})$, respectively.
10. *Loop gain:* It is the product of the branch gains encountered in traversing a loop.

The five individual loops of Figs 2.46c to 2.46g have loop gains $(G_{23}G_{32})$, $(G_{23}G_{34}G_{42})$, G_{44}, $(G_{23}G_{34}G_{45}G_{52})$, and $(G_{23}G_{35}G_{52})$, respectively.

2.12.2 The Mason's Gain Rule

The relationship between an input variable and an output variable of a signal flow graph is given by the net gain between the input and the output nodes and is known as the overall gain of the system. Mason's gain rule[12] for the determination of the overall system gain is given below.

[12] For the derivation, based on Cramer's rule for solving linear equations by determinants, the papers of Mason [41, 42] should be consulted.

$$M = \frac{1}{\Delta}\sum_{k=1}^{N} P_k \Delta_k = \frac{X_{out}}{X_{in}} \tag{2.122a}$$

where M = gain between X_{in} and X_{out}
X_{out} = output node variable
X_{in} = input node variable
N = total number of forward paths
P_k = path gain of the kth forward path
Δ = 1− (sum of loop gains of all individual loops) + (sum of gain products of all possible combinations of two nontouching loops) – (sum of gain products of all possible combinations of three nontouching loops) + ⋯

$$= 1 - \sum_m P_{m1} + \sum_m P_{m2} - \sum_m P_{m3} + \cdots \tag{2.122b}$$

P_{mr} = gain product of the mth possible combination of r nontouching loops.
Δ_k = the value of Δ for that part of the graph not touching the kth forward path.

Let us illustrate the use of Mason's gain rule by finding the overall gain of the signal flow graph shown in Fig. 2.45g. The following conclusions are drawn by inspection of this signal flow graph:

1. There are two forward paths with path gains

$$P_1 = G_{12}G_{23}G_{34}G_{45} \qquad \text{(Fig. 2.46a)}$$

$$P_2 = G_{12}G_{23}G_{35} \qquad \text{(Fig. 2.46b)}$$

2. There are five individual loops with loop gains

$$P_{11} = G_{23}G_{32} \qquad \text{(Fig. 2.46c)}$$

$$P_{21} = G_{23}G_{34}G_{42} \qquad \text{(Fig. 2.46d)}$$

$$P_{31} = G_{44} \qquad \text{(Fig. 2.46e)}$$

$$P_{41} = G_{23}G_{34}G_{45}G_{52} \qquad \text{(Fig. 2.46f)}$$

$$P_{51} = G_{23}G_{35}G_{52} \qquad \text{(Fig. 2.46g)}$$

3. There are two possible combinations of two nontouching loops with loop gain products

$$P_{12} = G_{23}G_{32}G_{44} \qquad \text{(Fig. 2.46h)}$$

$$P_{22} = G_{23}G_{35}G_{52}G_{44} \qquad \text{(Fig. 2.46i)}$$

4. There are no combinations of three nontouching loops, four nontouching loops, etc. Therefore,

$$P_{m3} = P_{m4} = \cdots = 0$$

Hence from Eqn. (2.122b),

$$\Delta = 1 - (G_{23}G_{32} + G_{23}G_{34}G_{42} + G_{44} + G_{23}G_{34}G_{45}G_{52} + G_{23}G_{35}G_{52}) + (G_{23}G_{32}G_{44} + G_{23}G_{35}G_{52}G_{44})$$

5. The first forward path is in touch with all the loops. Therefore, $\Delta_1 = 1$. The second forward path is not in touch with one loop (Fig. 2.46j). Therefore, $\Delta_2 = 1 - G_{44}$.

From Eqn. (2.122a), the overall gain is

$$M = \frac{X_5}{X_1} = \frac{P_1\Delta_1 + P_2\Delta_2}{\Delta} = \frac{G_{12}G_{23}G_{34}G_{45} + G_{12}G_{23}G_{32}(1-G_{44})}{\begin{bmatrix} 1-(G_{23}G_{32} + G_{23}G_{34}G_{42} + G_{44} + G_{23}G_{34}G_{45}G_{52} + G_{23}G_{35}G_{52}) \\ + G_{23}G_{32}G_{44} + G_{23}G_{35}G_{52}G_{44} \end{bmatrix}}$$

2.12.3 Block Diagram Reduction Using The Mason's Gain Rule

Due to the similarity between block diagrams and signal flow graphs, the Mason's gain rule can be used to determine the input-output relationships of either. In general, given a block diagram of a linear system, we apply the gain rule directly to it. However, in order to be able to identify all the loops and nontouching parts clearly, it may be sometimes helpful if an equivalent signal flow graph for the given block diagram is drawn before applying the gain rule.

Example 2.6 Consider the block diagram shown in Fig. 2.47a. The equivalent signal flow graph is drawn in Fig. 2.47b. Note that since a node on the signal flow graph is interpreted as a summing point of all incoming signals to the node, the negative feedbacks on the block diagram are represented by assigning negative gains to the feedback paths on the signal flow graph.

There are two forward paths with path gains

$$P_1 = G_1G_2G_3;\ P_2 = G_4$$

There are four individual loops with loop gains

$$P_{11} = -\ G_1H_1;\ P_{21} = -\ G_2H_2;\ P_{31} = -\ G_3H_3;\ P_{41} = -\ G_4H_3H_2H_1$$

There is one combination of two nontouching loops with loop gain product

$$P_{12} = G_1H_1G_3H_3$$

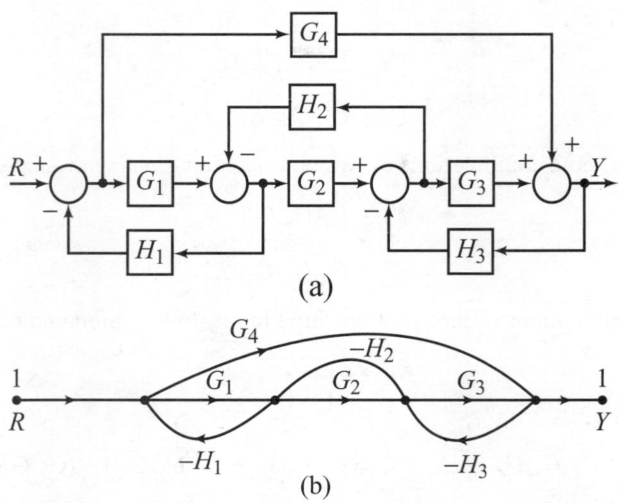

Fig. 2.47 *(a) Block diagram of a control system (b) Equivalent signal flow graph*

There are no combinations of more than two nontouching loops.
Hence, from Eqn. (2.122b),

$$\Delta = 1 - (-G_1H_1 - G_2H_2 - G_3H_3 - G_4H_3H_2H_1) + G_1H_1G_3H_3$$

The first forward path is in touch with all the loops. Therefore, $\Delta_1 = 1$. The second forward path is not in touch with one loop (Fig. 2.47b). Therefore, $\Delta_2 = 1 - (-G_2H_2)$.
From Eqn. (2.122a), the overall gain is

$$M = \frac{Y}{R} = \frac{P_1\Delta_1 + P_2\Delta_2}{\Delta} = \frac{G_1G_2G_3 + G_4 + G_4G_2H_2}{1 + G_1H_1 + G_2H_2 + G_3H_3 + G_4H_3H_2H_1 + G_1H_1G_3H_3}$$

Example 2.7 Consider another typical block diagram shown in Fig. 2.48. This feedback system has two inputs. We shall use the superposition to treat each input separately. The Mason's gain rule may be applied directly to the block diagram without transforming it into an equivalent signal flow graph.

When the disturbance input $W(s)$ is zero, the resulting block diagram has one forward path and two loops. The two loops touch each other and the forward path touches both the loops. By Mason's gain rule (Eqns (2.122))

$$M(s) = \left.\frac{Y(s)}{R(s)}\right|_{W(s)=0} = \frac{G_1(s)G_2(s)G_3(s)G_4(s)}{1 + G_3(s)H_1(s) + G_1(s)G_2(s)G_3(s)G_4(s)H_2(s)}$$

With $R(s) = 0$, the block diagram has two forward paths and two loops. The two loops touch each other and both the forward paths touch both the loops. By Mason's gain rule,

$$M_w(s) = \left.\frac{Y(s)}{W(s)}\right|_{R(s)=0} = \frac{G_5(s)G_4(s) - G_6(s)G_2(s)G_3(s)G_4(s)}{1 + G_3(s)H_1(s) + G_1(s)G_2(s)G_3(s)G_4(s)H_2(s)}$$

The response to the simultaneous application of $R(s)$ and $W(s)$ is given by

$$Y(s) = M(s)R(s) + M_w(s)W(s)$$

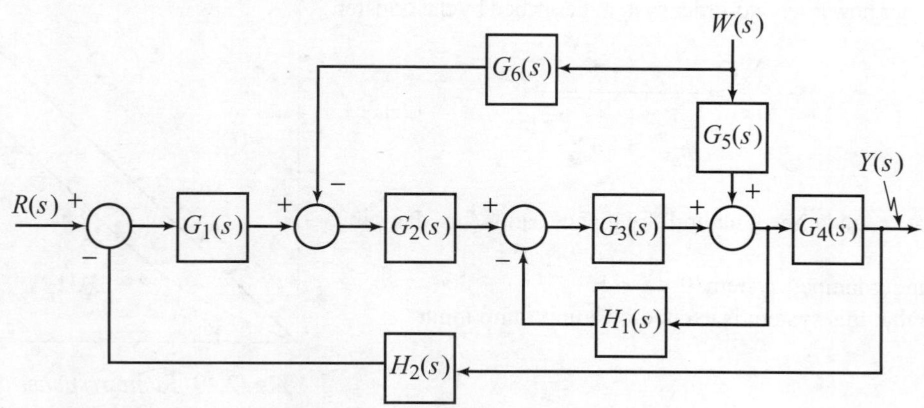

Fig. 2.48 *Block diagram of a control system*

2.13 REVIEW EXAMPLES

Review Example 2.1 Consider a first-order system described by the transfer function

$$\frac{Y(s)}{R(s)} = \frac{1}{\tau s+1} \tag{2.123}$$

when τ is the time constant of the system.

For a unit-ramp input $r(t) = t\mu(t)$, we have

$$R(s) = \frac{1}{s^2}$$

Therefore, the unit-ramp response is

$$Y(s) = \frac{1}{s^2(\tau s+1)} = \frac{1}{s^2} - \frac{\tau}{s} + \frac{\tau^2}{\tau s+1}$$

$$y(t) = t - \tau + \tau e^{-t/\tau} \tag{2.124}$$

The term $\tau e^{-t/\tau}$ is the transient response term which decays to zero with increasing time. Its rate of decay is dictated by the time constant τ. The steady-state component of the response to unit-ramp input $r(t) = t$ is

$$y_{ss} = (t - \tau) \tag{2.125}$$

which is also a ramp function.

Note that the steady-state ramp output is nothing but t delayed by τ as shown in Fig. 2.49. This phenomenon of steady-state lag in ramp response gives another interpretation of the time constant of a first-order system.

Consider now a second-order system described by the transfer function:

$$\frac{Y(s)}{R(s)} = \frac{1}{\frac{1}{\omega_n^2}s^2 + \frac{2\zeta}{\omega_n}s+1} \tag{2.126}$$

where ω_n = undamped natural frequency, and ζ = damping ratio.

For an underdamped system, $0 < \zeta < 1$.

Assume that this system is excited by a unit ramp input

$$R(s) = \frac{1}{s^2}$$

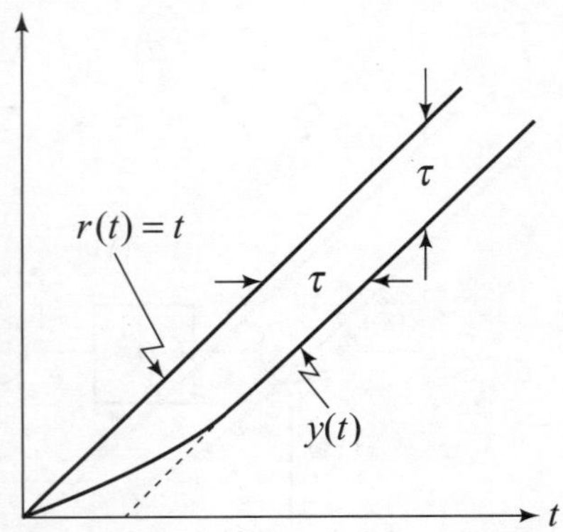

Fig. 2.49 *Ramp response of first-order systems*

This gives

$$Y(s) = \frac{\omega_n^2}{s^2(s^2 + 2\zeta\omega_n s + \omega_n^2)} = \left[\frac{1}{s^2} - \frac{2\zeta/\omega_n}{s} + \frac{2\zeta}{\omega_n}\frac{s+\zeta\omega_n}{(s+\zeta\omega_n)^2 + \omega_d^2} + \frac{2\zeta^2 - 1}{(s+\zeta\omega_n)^2 + \omega_d^2}\right]$$

where $$\omega_d = \omega_n\sqrt{1-\zeta^2}.$$

Using the table of transform pairs given in Table 2.1, we get

$$y(t) = \left[t - \frac{2\zeta}{\omega_n} + \frac{2\zeta}{\omega_n}e^{-\zeta\omega_n t}\cos\omega_d t + \frac{(2\zeta^2-1)}{\omega_d}e^{-\zeta\omega_n t}\sin\omega_d t\right]$$

$$= \left[t - \frac{2\zeta}{\omega_n} + \frac{e^{-\zeta\omega_n t}}{\omega_d}\left\{2\zeta\sqrt{1-\zeta^2}\cos\omega_d t + (2\zeta^2-1)\sin\omega_d t\right\}\right]$$

$$= t - \frac{2\zeta}{\omega_n} + \frac{e^{-\zeta\omega_n t}}{\omega_n\sqrt{1-\zeta^2}}\sin\left(\omega_d t + 2\tan^{-1}\frac{\sqrt{1-\zeta^2}}{\zeta}\right) \tag{2.127}$$

The last term in this expression is the transient response term, which decays to zero with increasing time. Its rate of decay is dictated by the time-constant $1/\zeta\omega_n$. The steady-state component of the response to unit-ramp input $r(t) = t$ is

$$y_{ss} = \left(t - \frac{2\zeta}{\omega_n}\right) \tag{2.128}$$

which is also a ramp function.

Note that the steady-state ramp output is nothing but t delayed by $2\zeta/\omega_n$ as shown in Fig. 2.50. The damped oscillations contributed by the transient component of the response are centered on the ramp.

By similar analysis, it can easily be established that the critically damped and overdamped responses ($\zeta \geq 1$) are characterized by the exponential rise toward the steady-state ramp output $\left(t - \frac{2\zeta}{\omega_n}\right)$.

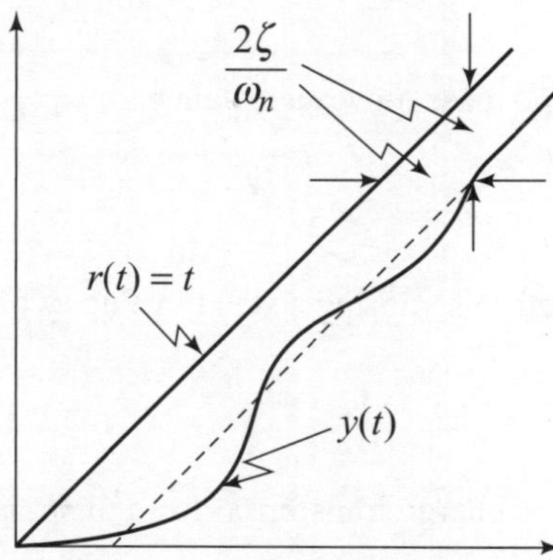

Fig. 2.50 *Ramp response of second-order underdamped systems*

Review Example 2.2 Consider the situation shown in Fig. 2.51a where a mass M is acted upon by the force $F(t)$. The system is characterized by the second-order differential equation,

$$F(t) = M\frac{d^2x(t)}{dt^2} = M\frac{dv(t)}{dt} \tag{2.129}$$

The solution of this equation requires two initial conditions $\{x(0), \dot{x}(0) = v(0)\}$, and the input $F(t)$ applied at $t = 0^+$.

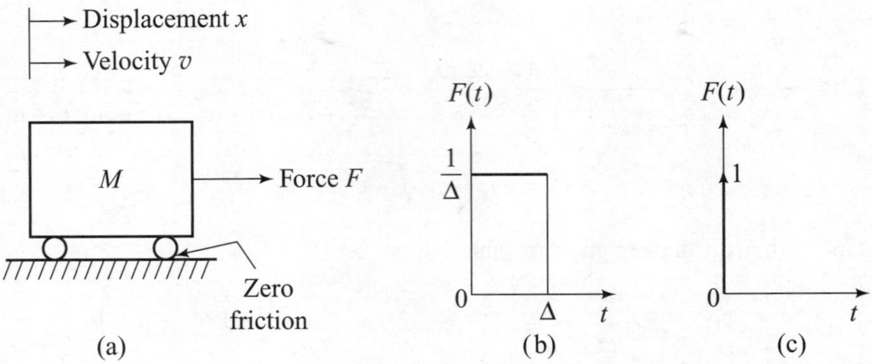

Fig. 2.51 *(a) Mass supported on frictionless wheels (b) Pulse input (c) Impulse input*

We will assume the system to be relaxed at $t = 0$, i.e.,

$$x(0) = v(0) = 0 \tag{2.129}$$

Let the force $F(t)$ be a pulse force of strength $1/\Delta$ and time duration Δ (Fig. 2.51b). For $t < \Delta$ (refer to Eqns (2.129))

$$v(t) = \frac{1}{M}\int_0^t F(\tau)d\tau = \frac{1}{M}\int_0^t \frac{1}{\Delta}d\tau = \frac{t}{M\Delta};\ x(t) = \int_0^t v(\tau)d\tau = \frac{t^2}{2M\Delta}$$

The mass moves a distance

$$x(\Delta) = \frac{\Delta}{2M}$$

before acquiring a steady velocity of

$$v(\Delta) = \frac{1}{M}$$

The energy transferred from the pulse source to the mass (in time Δ) is

$$KE = \frac{1}{2}Mv^2(\Delta) = \frac{1}{2M}$$

Consider now that the force $F(t)$ is a unit impulse $\delta(t)$ occurring at $t = 0$ (Fig. 2.51c). The velocity of the mass is given by

$$v(t) = \frac{1}{M}\int_0^t \delta(\tau)d\tau = \frac{1}{M};\ t \geq 0$$

It is the same as the steady velocity $v(\Delta)$ for the pulse case except that it is acquired now in zero distance (and of course zero time). The energy acquired by the mass is the same in the two cases.

From Figs 2.51b and 2.51c, we see that there is practically no resemblance as far as the input function is concerned and it appears at first as if the approximation of a pulse by an impulse is in doubt. However, if our aim is to approximate the response of a linear system, the impulse approximation yields useful results provided pulse width Δ is small.

Review Example 2.3 Simplified models of vehicles can be used to study the heave (vertical up and down) motion due to road profile and other disturbances. For our purposes, let us assume that the road disturbances are represented by the disturbance force $F_w(t)$. The mechanical system of Fig. 2.52a is a simplified 'one-wheeled' model of an automobile suspension system with M_1: the mass of the automobile, B: the shock absorbers, K_1: the springs, M_2: the mass of the wheels, and K_2: the elasticity of the tyres. The system objective is to reduce vibrations $y_1(t)$ due to disturbance force $F_w(t)$.

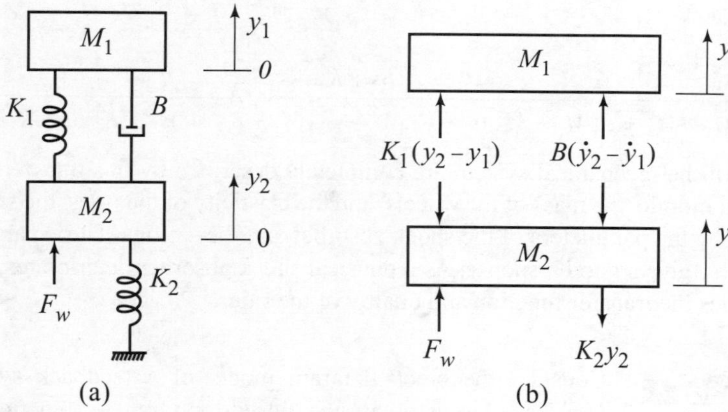

Fig. 2.52

The displacements $y_1(t)$ and $y_2(t)$ are considered with respect to positions of the vehicle and the wheels on a level road under the loaded condition. The gravitational forces, therefore, do not appear in the model being formulated.

Figure 2.52b shows the free-body diagram for the system. From this figure, we have the following differential equations describing the dynamics of the system:

$$M_1\ddot{y}_1(t) = K_1(y_2(t) - y_1(t)) + B(\dot{y}_2(t) - \dot{y}_1(t)) \tag{2.130a}$$

$$M_2\ddot{y}_2(t) = F_w(t) - K_1(y_2(t) - y_1(t)) - B(\dot{y}_2(t) - \dot{y}_1(t)) - K_2y_2(t) \tag{2.130b}$$

Taking the Laplace transform of these equations, assuming zero initial conditions, we obtain

$$M_1s^2Y_1(s) + B(sY_1(s) - sY_2(s)) + K_1(Y_1(s) - Y_2(s)) = 0 \tag{2.131a}$$

$$M_2s^2Y_2(s) + B(sY_2(s) - sY_1(s)) + K_1(Y_2(s) - Y_1(s)) + K_2Y_2(s) = F_w(s) \tag{2.131b}$$

Suppose that a transfer function is desired between $F_w(s)$ and $Y_1(s)$, i.e., between the disturbance force applied to the wheels and the resulting displacement of the automobile. This transfer function can be found by eliminating $Y_2(s)$ in Eqns (2.131). Solving Eqn. (2.131a) for $Y_1(s)$, we obtain

$$Y_1(s) = \frac{Bs + K_1}{M_1 s^2 + Bs + K_1} Y_2(s)$$

Now Eqn. (2.31b) can be solved for $Y_2(s)$:

$$Y_2(s) = \frac{1}{M_2 s^2 + Bs + K_1 + K_2} F_w(s) + \frac{Bs + K_1}{M_2 s^2 + Bs + K_1 + K_2} Y_1(s)$$

Eliminating $Y_2(s)$ from these equations, gives

$$G(s) = \frac{Y_1(s)}{F_w(s)}$$

$$= \frac{Bs + K_1}{[M_1 M_2 s^4 + B(M_1 + M_2)s^3 + (K_1 M_2 + K_1 M_1 + K_2 M_1)s^2 + K_2 Bs + K_1 K_2]}$$

The dynamics of the mechanical system are completely described by this transfer function. Given the mass of the automobile, the mass of the wheels, and the elasticity of the tyres, the smoothness of the ride is determined by the parameters of the shock absorber and the springs. These parameters are used to tune the system to give a good response. As a physical shock absorber deteriorates, the value of the parameter B changes the transfer function and quality of the ride.

Review Example 2.4 Consider the block diagram model of a feedback system, shown in Fig. 2.53. Let us obtain the closed-loop transfer function of this system using block diagram manipulation rules given in Table 2.3.

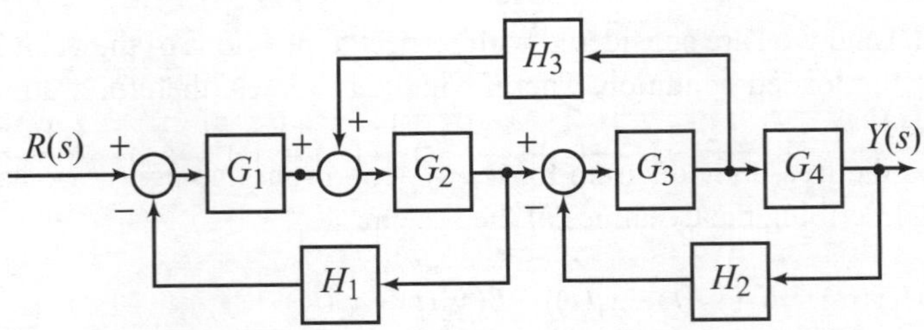

Fig. 2.53 *A block diagram model*

(i) Move the branch point between G_3 and G_4 behind block G_4 by using rules 8 and 4, and obtain Fig. 2.54a.
(ii) Move the summing point between G_1 and G_2 outside the loop containing G_1, G_2, and H_1 using rules 5 and 1, and obtain Fig. 2.54b.
(iii) By simplifying each negative feedback loop using rules 4 and 10, obtain Fig. 2.54c.
(iv) Further simplification using rules 4 and 9 results in Fig. 2.54d, from which the closed-loop transfer function can be obtained.

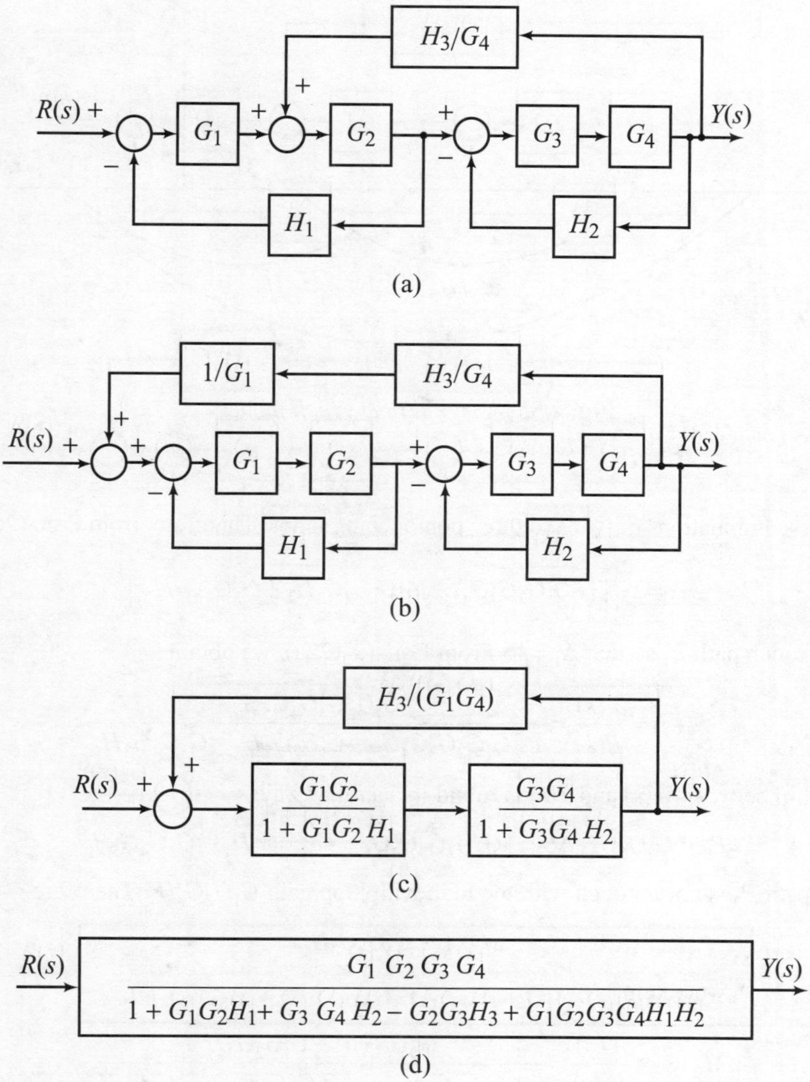

Fig. 2.54 *Successive reductions of the block diagram of Fig. 2.53*

Review Example 2.5 Consider the signal flow graph model of a feedback system shown in Fig. 2.55. Let us obtain the response $Y(s)$ to the simultaneous application of the inputs $R(s)$ and $W(s)$ using Mason's gain rule.

The forward transmittance path between $R(s)$ and $Y(s)$ is

$$P_1 = G_1G_2G_3G_4G_5G_6$$

There are three individual loops with loop gains

$$P_{11} = -\,G_1G_2G_3G_4G_5;\ P_{21} = G_1G_2G_3H_1;\ P_{31} = G_3G_4G_5H_2$$

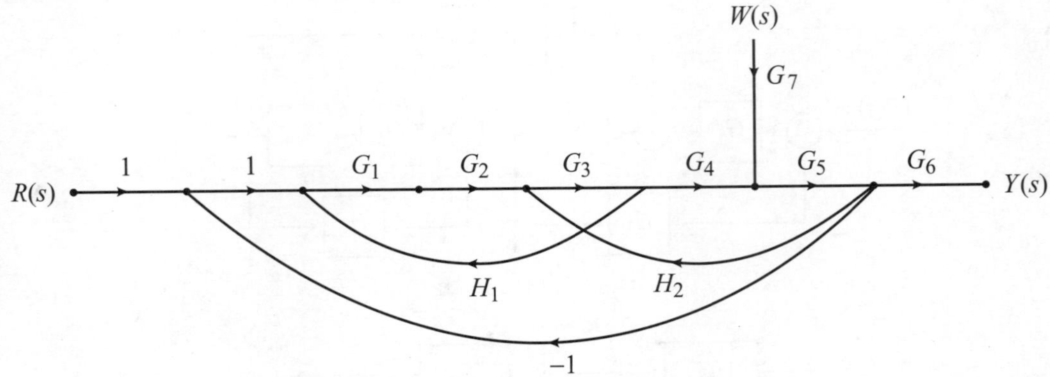

Fig. 2.55 *A signal flow graph model*

There are no combinations of two and three nontouching loops. Therefore, from Eqn. (2.122b),

$$\Delta = 1 - \left[\left(-G_1G_2G_3G_4G_5 \right) + G_1G_2G_3H_1 + G_3G_4G_5H_2 \right]$$

All the loops touch path P_1 so that $\Delta_1 = 1$. From Eqn. (2.122a), we obtain

$$M(s) = \frac{Y(s)}{R(s)} = \frac{G_1G_2G_3G_4G_5G_6}{1 + G_1G_2G_3G_4G_5 - G_1G_2G_3H_1 - G_3G_4G_5H_2}$$

The relationship between $W(s)$ and $Y(s)$ is found in a similar way:

$$P_2 = G_7G_5G_6\,; \Delta = 1 + G_1G_2G_3G_4G_5 - G_1G_2G_3H_1 - G_3G_4G_5H_2$$

The forward path P_2 is not in touch with the loop with loop gain $G_1G_2G_3H_1$. Therefore,

$$\Delta_1 = 1 - G_1G_2G_3H_1$$

From Eqn. (2.122a), we obtain

$$M_w(s) = \frac{Y(s)}{W(s)} = \frac{G_5G_6G_7(1 - G_1G_2G_3H_1)}{1 + G_1G_2G_3G_4G_5 - G_1G_2G_3H_1 - G_3G_4G_5H_2}$$

The response to simultaneous application of $R(s)$ and $W(s)$:

$$Y(s) = M(s)R(s) + M_w(s)W(s)$$

CONCLUDING REMARKS

In this chapter, we have built a mathematical platform for quantitative description of feedback control systems by developing the concept of transfer function model of a linear system; and the relationship among system variables using block diagram and signal flow graph models.

What we have learnt?

- The Laplace and Fourier transforms are very powerful tools for analysis/design of linear, time-invariant systems. Laplace transform is a mapping from time-domain (t-domain) to complex frequency domain (s-domain); and Fourier transform is a mapping from time domain to frequency domain (ω-domain). We have learnt these tools to the extent needed in the book.
- We have learnt the concept of transfer function model of a system, and found that it is nothing more than s-plane representation of differential equation model of the system obtained by application of physical laws.
- We have learnt to obtain dynamic response from a transfer function model to key test signals used in controls; and have become familiar with first-order system behavior with respect to time constant τ, and second-order system behavior with respect to the parameters: damping coefficient ζ, and undamped natural frequency ω_n.
- Frequency response model is a complete description of the system's dynamic behavior. The characteristic parameters of this form of model are: bandwidth ω_b, resonance peak M_r, and resonance frequency ω_r.
- We have learnt to represent multiple subsystems via. block diagram models and signal flow graph models; and also the techniques to reduce block diagram/signal flow graph representation to a single transfer function.

Review Questions

2.1 Define a linear time-invariant system. Explain superposition and decomposition properties which linear systems obey.

2.2 Dynamic behavior of control systems can be adequately judged and compared under application of standard test signals. Describe various standard test signals commonly used in control system design. Give time domain, s-domain, and graphical representations of the signals.

2.3 State and prove the final value theorem of the Laplace transform. What is the condition under which the theorem is valid?

2.4 Explain the reasons for dead-time element in process control loops. Show that dead time can be represented by the transfer function $e^{-s\tau_D}$. How do you approximate this transfer function by a rational function?

2.5 What is the definition of "transfer function"? What are the properties of systems whose responses can be described by transfer functions?

2.6 (a) Interconnection of various subsystems in a feedback control system is required to be nonloading. Why?

(b) Explain the use of op amp interposed as buffer amplifier between the subsystems to reduce loading effects.

2.7 (a) Given a transfer function model, how do you obtain the frequency response model? Is the frequency response model useful only for sinusoidal excitations? Justify your answer.

(b) Describe the characteristic parameters of frequency response models.

2.8 X_{in} is the input-node variable and X_{out} is the output node variable of a signal flow graph with N forward paths. Describe the Mason's gain rule for the determination of the overall gain X_{out}/X_{in} of the signal flow graph.

2.9 Unity-feedback block diagram model can be obtained only for control systems with sensor gain of unity. Do you agree with this statement? Justify your answer.

Self-Appraisal Quiz

A set of questions with multiple answer choices is given below. These questions have been designed to capture important aspects of the subject covered in the chapter. For each of these questions, identify the correct answer from the given choices. Compare your answers with the Master Key given at the end of the session. This quiz should serve the purpose of a self-appraisal test for readers.

2.1 The Laplace transform of $\mu(t-2)$ is

(A) $\frac{1}{s+2}$ (B) $\frac{1}{s-2}$ (C) $\frac{e^{-2s}}{s}$ (D) $\frac{e^{2s}}{s}$

2.2 Given that $G(s) = e^{-2s}/(s+1)$, then g(t) is

(A) $e^{-(t-2)}\mu(t-2)$ (B) $e^{-t}\mu(t-2)$ (C) $e^{-t}\mu(t+2)$ (D) $e^{-(t-2)}\mu(t)$

2.3 The inverse Laplace transform of $(s+2)/[(s+2)^2+1]$, is

(A) $e^{-t}\cos 2t$ (B) $e^{-t}\sin 2t$ (C) $e^{-2t}\cos t$ (D) $e^{-2t}\sin t$

2.4 A linear time-invariant system initially at rest, when subjected to a unit-step input, gives a response $y(t) = te^{-t}; t > 0$. The transfer function of the system is

(A) $\frac{1}{(s+1)^2}$ (B) $\frac{1}{s(s+1)^2}$ (C) $\frac{s}{(s+1)^2}$ (D) $\frac{1}{s(s+1)}$

2.5 The system shown in Fig. Q2.5 is of

(A) zero order (B) first order
(C) second order (D) third order

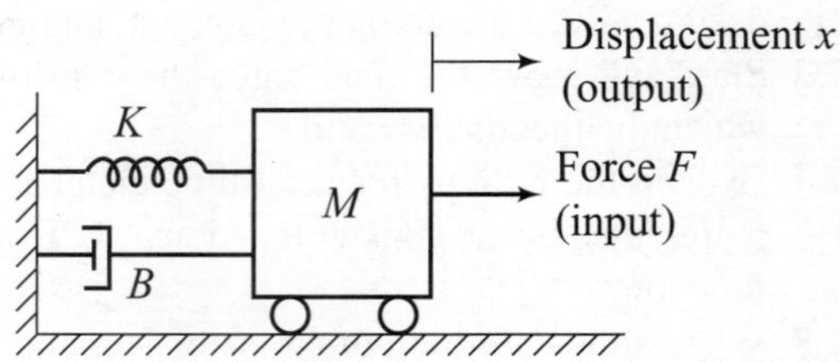

Fig. Q2.5

2.6 The impulse response of an initially relaxed linear system is $e^{-2t}\mu(t)$ where $\mu(t)$ is a unit-step function. To produce a response of $te^{-2t}\mu(t)$, the input must be equal to

(A) $2e^{-2t}\mu(t)$ (B) $\frac{1}{2}e^{-2t}\mu(t)$

(C) $e^{-2t}\mu(t)$ (D) $e^{-t}\mu(t)$

2.7 The step response of a system with transfer function $G(s) = 1/(\tau s+1)$ attains more than 98% of its final value in time t equal to

(A) τ (B) 2τ (C) 3τ (D) 4τ

2.8 The result $\mu(t) * \mu(t)$ is (* denotes convolution)

(A) $\mu^2(t)$ (B) $t\mu(t)$ (C) $t^2\mu(t)$ (D) $\delta(t)$

2.9 Time constant of the circuit of Fig. Q2.9 is

(A) 0.25 sec (B) 0.5 sec
(C) 1 sec (D) 2 sec

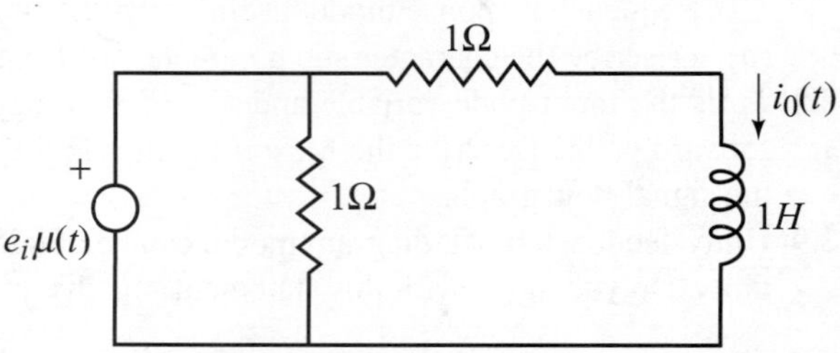

Fig. Q2.9

2.10 The response of the system of Fig. Q2.10a to an input $r(t) = 8\mu(t)$ is shown in Fig. Q2.10b. The time constant τ is equal to
(A) 0.535 msec (B) 0.32 msec (C) 0.09 msec
(D) 11.25 msec

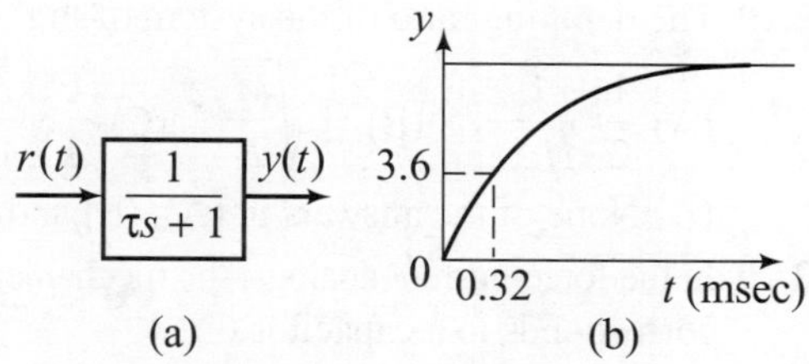

Fig. Q2.10

2.11 Closed-loop transfer function of a unity-feedback system is given by $Y(s)R(s) = 1(\tau s+1)$. Steady-state error to unit-ramp input is
(A) ∞ (B) τ (C) 1 (D) $1/\tau$

2.12 The series RLC circuit shown in Fig. Q2.12 is underdamped if

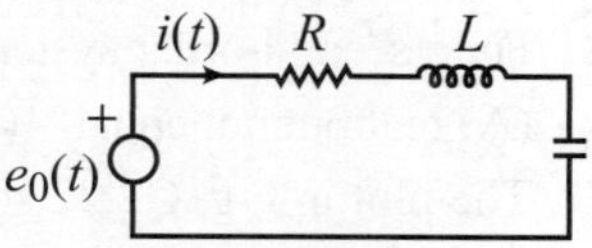

Fig. Q2.12

(A) $\left(\frac{R}{2L}\right)^2 < \frac{1}{LC}$ (B) $\left(\frac{R}{2L}\right)^2 = \frac{1}{LC}$

(C) $\left(\frac{R}{2L}\right)^2 > \frac{1}{LC}$ (D) None of the answers in (A), (B) and (C) is correct

2.13 Closed-loop transfer function of a unity-feedback system is given by $\frac{Y(s)}{R(s)} = \frac{\omega_n^2}{s^2 + 2\zeta\omega_n s + \omega_n^2}$. Steady-state error to unit-ramp input is
(A) ∞ (B) $2\zeta/\omega_n$ (C) 1 (D) $4/\zeta\omega_n$

2.14 When two networks shown in Fig. Q2.14 are cascaded in tandem, the overall transfer function $E_4(s)/E_1(s)$ is
(A) $H_1(s)H_2(s)$ (B) $H_1(s)+H_2(s)$ (C) $H_1(s)/H_2(s)$
(D) None of the answers in (A), (B), and (C) is correct

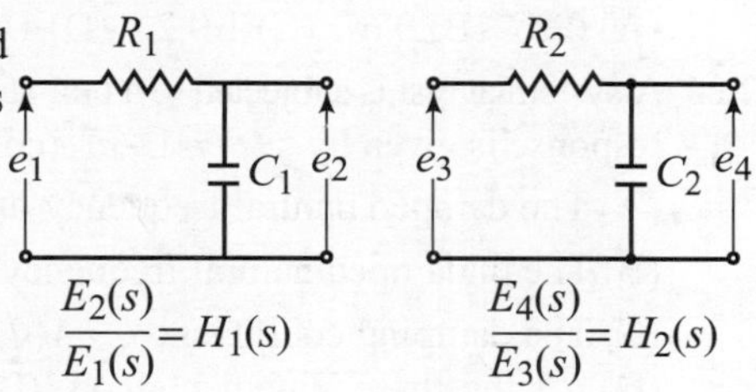

Fig. Q2.14

2.15 Dead-time model $e^{-s\tau_D}$ may be approximated by the transfer function

(A) $\dfrac{1-\frac{\tau_D}{2}s}{1+\frac{\tau_D}{2}s}$ (B) $\dfrac{1+\frac{\tau_D}{2}s}{1-\frac{\tau_D}{2}s}$ (C) $\dfrac{1-\tau_D s}{1+\tau_D s}$ (D) $\dfrac{1+\tau_D s}{1-\tau_D s}$

2.16 If a system has two real and equal characteristic roots, it is described as
(A) having no damping (B) being underdamped
(C) being critically damped (D) being overdamped

2.17 If the roots of a characteristic equaltion are given by $s_{1,2} = -3 \pm j2$, the values of damping ζ and damped natural frequency ω_d are

(A) $\left(\frac{3}{\sqrt{13}}, \sqrt{13}\right)$ (B) $\left(\frac{1}{\sqrt{13}}, 2\right)$ (C) $\left(\frac{1}{\sqrt{13}}, \sqrt{13}\right)$ (D) $\left(\frac{3}{\sqrt{13}}, 2\right)$

2.18 The system of Fig. Q2.18 is underdamped if

(A) $\frac{K}{M} > \left(\frac{B}{2M}\right)^2$ (B) $\frac{K}{M} = \left(\frac{B}{2M}\right)^2$

(C) $\frac{K}{M} < \left(\frac{B}{2M}\right)^2$ (D) None of the answers in (A), (B) and (C) is correct

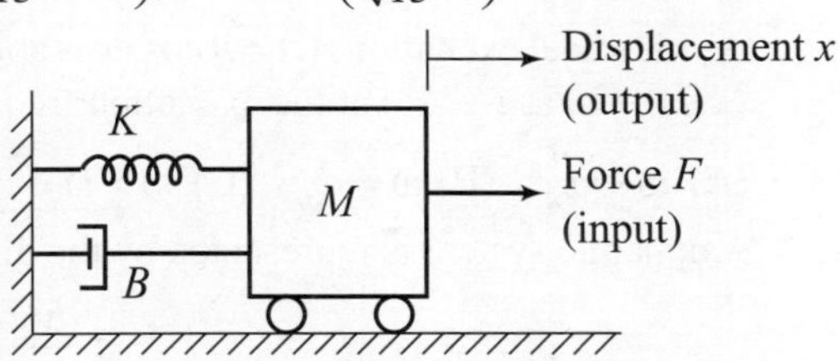

Fig. Q2.18

2.19 The damping ratio of the system of Fig. Q2.19 is

(A) $\frac{1}{2}\frac{1}{\sqrt{BM}}$ (B) $\frac{1}{2}\sqrt{\frac{M}{B}}$ (C) $\frac{1}{2}\sqrt{\frac{B}{M}}$

(D) None of the answers in (A), (B) and (C) is correct.

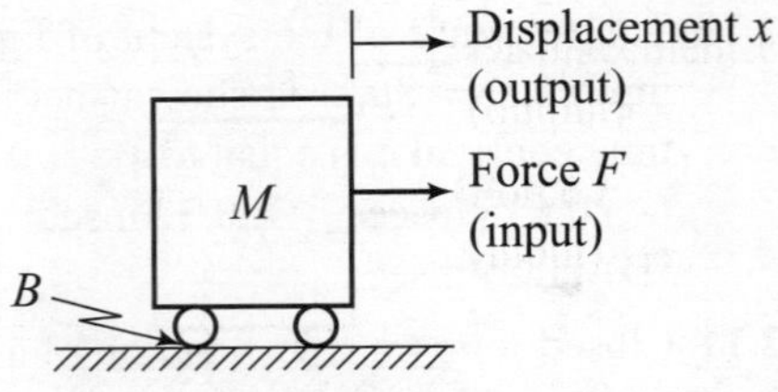

Fig. Q2.19

2.20 In the force-current analogy, the mechanical element that corresponds to a capacitor C is

(A) mass M (B) spring with stiffness K

(C) damper with coefficient B

2.21 For a second-order system with damping $\zeta > 1$, the roots of the characteristic equation are

(A) real but not equal (B) real and equal (C) complex conjugate (D) imaginary

2.22 The plot in Fig. Q2.22 shows the unit-step response of a first-order system. Transfer function of the system is

(A) $\frac{1}{5s+1}$ (B) $\frac{2}{15s+1}$ (C) $\frac{2}{3s+1}$ (D) $\frac{1}{15s+1}$

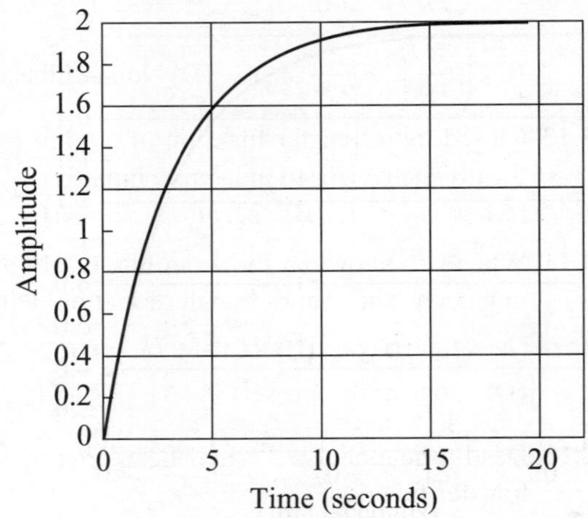

Fig. Q2.22

2.23 Three different systems have the following poles: $p_1 = -0.2$, $p_2 = -0.4$, $p_3 = -0.6$. What is the time-constant of the fastest step response?

(A) 0.2 (B) 0.6 (C) 1/0.2 (D) 1/0.6

2.24 A system at rest is subjected to a unit-step input. Its response is given by $y(t) = 1 - e^{-t}(\cos t + \sin t)$.

(A) The damped natural frequency $\omega_d = 1$

(B) The undamped natural frequency $\omega_n = \sqrt{2}$

(C) The damping coefficient $\zeta = 1/\sqrt{2}$

(D) All the three statements (A), (B), and (C) are correct.

2.25 An approximate sinusoidal transfer function of a system was determined by fitting measured excitation-reponse data into an appropriate analytical expression. This was found to be

$$G = \frac{5}{10 + j2\pi f}; f \text{ in Hz}$$

The bandwidth (in rad/sec) of the system is

(A) $\omega_b = 0.1$ (B) $\omega_b = 0.5$ (C) $\omega_b = 5$ (D) $\omega_b = 10$

2.26 The transfer function of a simple oscillator is given by

$$G(s) = \frac{Y(s)}{R(s)} = \frac{\omega_n^2}{s^2 + \omega_n^2}$$

A sinusoidal excitation $r(t) = A\cos\omega t$ is applied to this system. The system response is unbounded $(y(t) \to \infty$ as $t \to \infty)$ at the excitation frequency

(A) $\omega < \omega_n$ (B) $\omega = \omega_n$ (C) $\omega > \omega_n$

2.27 A dynamic system is represented by the differential equation

$$\tau\frac{dy(t)}{dt} + y(t) = r(t)$$

The system is given the sinusoidal input $r(t) = \sin \omega t$. The value of ω at which magnitude of the output $y(t)$ is $1/\sqrt{2}$ at steady state, is given by

(A) $\omega = \tau$ (B) $\omega = 1/\tau$ (C) $\omega = 0$

2.28 Consider the block diagram shown in Fig. Q2.28. The transfer function between $Y(s)$ and $W(s)$ is

(A) $\dfrac{D(s)G(s)N(s)}{1+D(s)G(s)H(s)}$ (B) $\dfrac{N(s)}{1+D(s)G(s)H(s)}$

(C) $\dfrac{N(s)}{1-D(s)G(s)H(s)}$ (D) None of the answers in (A), (B) and (C) is correct

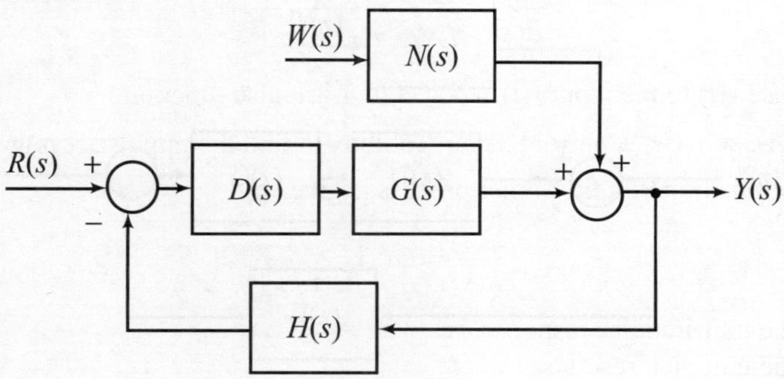

Fig. Q2.28

2.29 The signal flow diagram shown in Fig. Q2.29 represents a feedback system subjected to a disturbance $W(s)$. The transfer function relating output $Y(s)$ to $W(s)$ with reference $R(s) = 0$, is

(A) $\dfrac{G_1G_2}{1+G_1G_2}$ (B) $\dfrac{G_1G_2}{1-G_1G_2}$ (C) $\dfrac{G_2}{1+G_1G_2}$ (D) $\dfrac{G_2}{1-G_1G_2}$

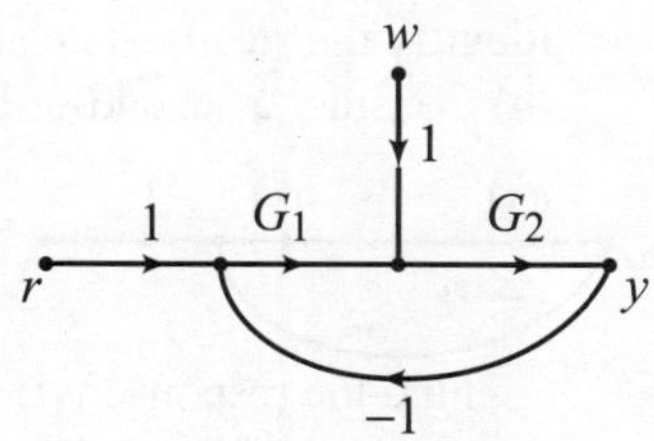

Fig. Q2.29

MASTER KEY

2.1 (C)	2.2 (A)	2.3 (C)	2.4 (C)	2.5 (C)	2.6 (C)
2.7 (D)	2.8 (B)	2.9 (C)	2.10 (A)	2.11 (B)	2.12 (A)
2.13 (B)	2.14 (D)	2.15 (A)	2.16 (C)	2.17 (D)	2.18 (A)
2.19 (D)	2.20 (A)	2.21 (A)	2.22 (C)	2.23 (D)	2.24 (D)
2.25 (D)	2.26 (B)	2.27 (B)	2.28 (B)	2.29 (C)	

Problems

2.1 Consider a system described by the linear differential equation

$$\ddot{y}(t)+7\dot{y}(t)+10y(t)=\dot{r}(t)+3r(t)$$

(a) Find the system response to an input $e^{-t}\mu(t)$ assuming zero initial conditions.

(b) Find the system response to an input $e^{-t}\mu(t)$ when the initial conditions are: $y(0) = 1, \dot{y}(0) = 1/2$.

MATLAB Exercise Use Control System Toolbox to find the system response.

2.2 Consider the linear differential equation

$$\frac{d^2y(t)}{dt^2} + 34.5\frac{dy(t)}{dt} + 1000\,y(t) = 1000\mu(t)$$

The initial values of $y(t)$ and $dy(t)/dt$ are zero. Solve the differential equation by using the Laplace transform.

MATLAB Exercise Use Control System Toolbox to solve the differential equation for $y(t)$.

2.3 Consider a system with transfer function between output and input given by

$$\frac{Y(s)}{R(s)} = G(s) = \frac{2s+1}{s^2+2s+5}$$

Find the response $y(t)$ to the input $r(t) = \delta(t)$, a unit-impulse function.

MATLAB Exercise Use Control System Toolbox to find the impulse response.

2.4 (a) The transfer function of a first-order process is given by

$$\frac{Y(s)}{R(s)} = G(s) = \frac{1}{\tau s+1}$$

(i) Find the unit-impulse response.
(ii) Find the unit-step response.
(iii) Find the unit-ramp response.
(iv) Find the response to sinusoidal input $A \sin \omega t$.

Identify the steady-state and transient components of the response in each case.

(b) Consider a second-order model

$$\frac{Y(s)}{R(s)} = G(s) = \frac{\omega_n^2}{s^2+2\zeta\omega_n s+\omega_n^2};0<\zeta<1$$

Find the response $y(t)$ to the input
(i) $r(t) = \mu(t)$, a unit-step function (ii) $r(t) = t\mu(t)$, a unit-ramp function
Find the steady-state component y_{ss} of $y(t)$ in each case.
Will the final value theorem give the correct value of y_{ss} in each case? Why?

2.5 (a) Find the response of the system

$$\frac{Y(s)}{R(s)} = \frac{K}{\tau s+1}$$

to a unit-step input $r(t)$, and discuss the effects of variations in parameters K and τ on transient behavior.

(b) Find the response of the system

$$\frac{Y(s)}{R(s)} = \frac{K\omega_n^2}{s^2+2\zeta\omega_n s+\omega_n^2}$$

to a unit-step input $r(t)$, and discuss the effects of variations in parameters K, ζ and ω_n on transient behavior.

MATLAB Exercise Using Control System Toolbox, obtain step-response plots for a range of parameter sets, and therefrom give your analysis of the effects of variations in parameters.

2.6 (a) A system with impulse response

$$g(t) = e^{-t}(1 - \sin t)$$

is subjected to unit-step input. Find the transfer function, and therefrom obtain the DC gain, and the steady-state value of the output.

(b) Determine unit-ramp response of a system having the transfer function

$$G(s) = \frac{s+1}{s+2}$$

What is the steady-state value of the response?

MATLAB Exercise Using Control System Toolbox (**lsim** command), plot the ramp response.

2.7 A second-order transfer function model is given by

$$\frac{Y(s)}{R(s)} = \frac{10000}{s^2 + 600s + 10000}$$

and the input is $r(t) = 25t\,\mu(t)$.

Show that the response is able to track the input with a steady-state delay of 60 msec. What is the value of steady-state error: $\lim_{t\to\infty} (r(t) - y(t))$?

2.8 The parameters of the mechanical system of Fig. P2.8 are

$$M = 1{,}000 \text{ kg};\ B = 10{,}000 \text{ newtons/(m/sec); and}$$

$$K = 100{,}000 \text{ newtons/m}$$

A step force of 1000 newtons is applied to the mass at $t = 0$. The initial conditions are $y(0) = \dot{y}(0) = 0$.

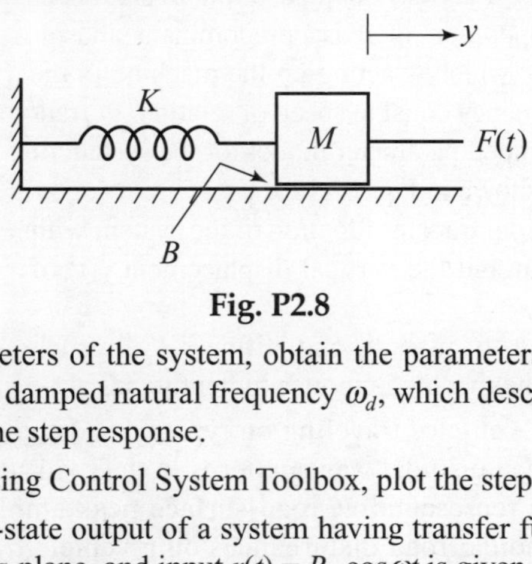

Fig. P2.8

From the physical parameters of the system, obtain the parameters: damping ratio ζ, undamped natural frequency ω_n, and damped natural frequency ω_d, which describe the dynamical behavior of the system. Also obtain the step response.

MATLAB Exercise Using Control System Toolbox, plot the step response.

2.9 (a) Prove that the steady-state output of a system having transfer function $G(s)$ with all the poles in the left half of the s-plane, and input $r(t) = R_0 \cos\omega t$ is given by

$$y_{ss}(t) = R_0|G(j\omega)|\cos(\omega t + \phi);\ \phi = \angle G(j\omega)$$

(b) A sinusoidal input $r(t) = 2 \sin 2t$ is applied to a system with transfer function

$$G(s) = \frac{2}{s(s+2)}$$

Determine the steady-state output $y_{ss}(t)$.

2.10 Consider the circuit shown in Fig. P2.10. The input voltage $e_i(t)$ is a sinusoid of frequency ω rad/sec, and $e_0(t)$ is the output. Find the sinusoidal transfer function $E_0(j\omega)/E_i(j\omega)$ and show that the circuit represents a low-pass filter. Calculate its bandwidth in Hz.

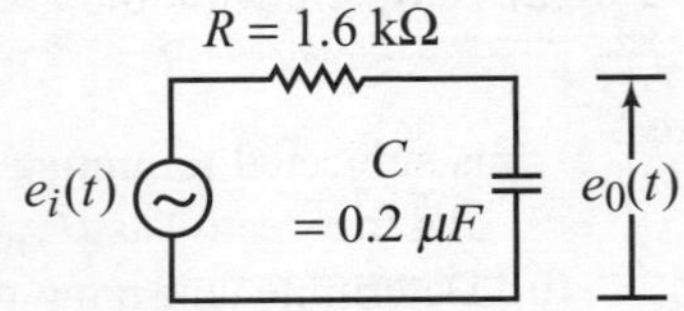

Fig. P2.10

2.11 For the circuit shown in Fig. P2.11, obtain the sinusoidal transfer function $E_0(j\omega)/E_i(j\omega)$. Show that the circuit represents a low-pass filter. Calculate its bandwidth.

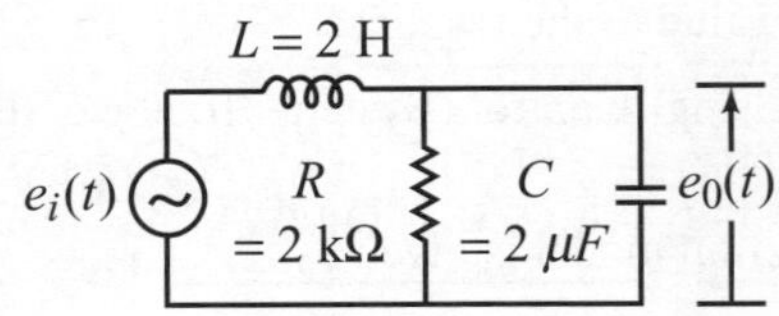

Fig. P2.11

2.12 A common cause of sinusoidal forcing in machines is the imbalance that exists to some extent in every rotating machine. The imbalances are caused by the fact that the center of mass of the rotating part does not coincide with center of rotation. Due to this imbalance, there is a predominant sinusoidal force component, $F_w(t)$, which is acting on the machine in the vertical direction at the frequency equal to speed of rotation, ω_M rad/sec of the machine [26].

A rotating machine of mass M is placed on a cantilever (Fig. P2.12a). In steady operation, there is a predominant sinusoidal force component $F(t)$, which is acting on the machine in the vertical direction at a frequency equal to speed of rotation, ω_M rad/sec, of the machine. A lumped parameter model for the system of Fig. 2.12a is of the form shown in Fig. P2.12b.

(a) Determine the sinusoidal transfer function of the system with force $F(t)$ as the input and the vertical displacement $y(t)$ of mass M as the output.

(b) Show that if the values of mass M and stiffness K are such that $\sqrt{K/M} = \omega_M$, the machine vibrates violently.

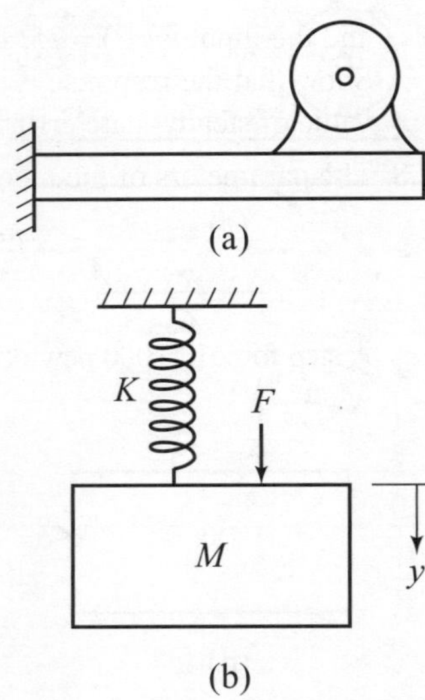

Fig. P2.12

2.13 Measurements made on vehicles traveling on typical roads have shown that some construction techniques result in sinusoidal wave patterns in the road surface; a representative wavelength λ is 15 m, and a representative road-surface peak amplitude is 7.5 cm. The forcing frequency of such sinusoidal road disturbances on a vehicle traveling with a speed of v m/sec is $\omega = 2\pi\, v/\lambda$ rad/sec.

A simplified representation of a vehicle's front wheel system consisting of a tyre of mass M and a shock absorber of damping coefficient B is shown in Fig. P2.13. Spring K_1 represents the elasticity of the tyre, while K represents that of shock absorber. We assume that the inertia of the car body is so large that the body can be considered as a fixed support.

Determine the transfer function $X(s)/Y(s)$ and therefrom obtain the peak amplitude of the tyre displacement x when the car speed is 100 km/hr.

Given: $M = 15$ kg, $B = 5.25$ knewtons/(m/sec), $K_1 = 175$ knewtons/m, and $K = 35$ knewtons/m.

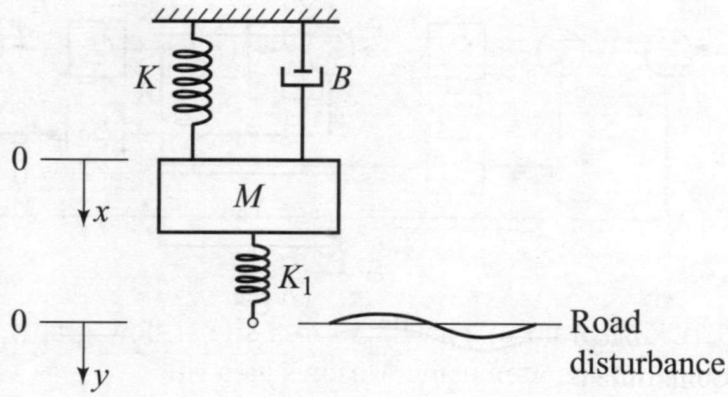

Fig. P2.13

2.14 A rotating machine of mass M is placed on a rigid concrete floor. There is an isolation pad made of elastrometric material between the machine and the floor, and is modeled as a viscous damper of coefficient B in parallel with a spring of stiffness K (Fig. P2.14a).

(a) The machine generates a disturbing force $F(t) = A \sin \omega t$. Find the force transmitted to the foundation at steady state.

(b) With reference to Fig. P2.14b, determine the machine vibration $x(t)$ due to the foundation vibration $y(t) = A \sin \gamma t$.

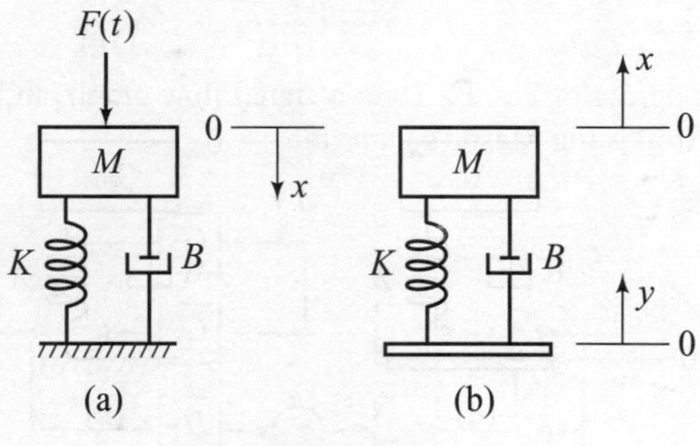

Fig. P2.14

2.15 A dynamic absorber is a device which is mounted on the vibrating area of the dynamic system. By properly selecting the parameters of the absorber, it is possible to inhibit the vibrating motions of the main system.

In Fig. P2.15, the main system and the absorber are modeled as mass-spring-damper systems with parameters (M_1,K_1,B_1,) and (M_2,K_2,B_2), respectively. This system is representative of many situations involving the vibration of machines containing unbalanced components. The parameters M_2, K_2 and B_2 may be chosen so that the main mass M_1 does not vibrate in response to foundation vibration. Obtain a set of differential equations describing the dynamics of the system.

2.16 (a) A linear feeback control system has the block diagram shown in Fig. P2.16. Using block diagram reduction rules, obtain the closed-loop transfer function $Y(s)/R(s)$.

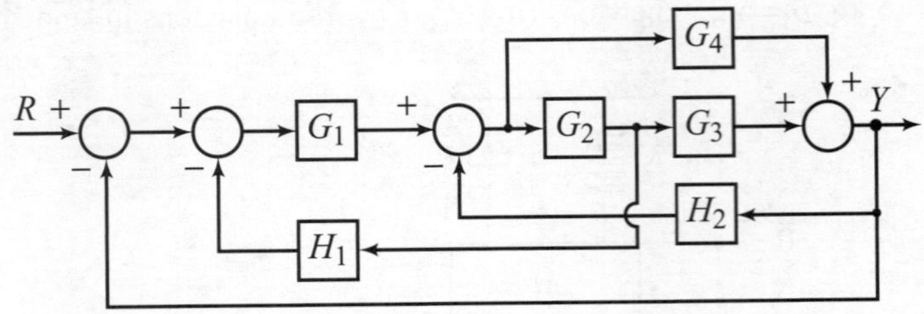

Fig. P2.16

(b) Convert the block diagram of Fig. P2.16 to a signal flow graph, and therefrom obtain the input–output transfer function using Mason's gain rule.

2.17 Using block diagram reduction rules, or Mason's gain rule applied to corresponding signal flow graph, convert the block diagram of Fig. P2.17 to a simple loop.

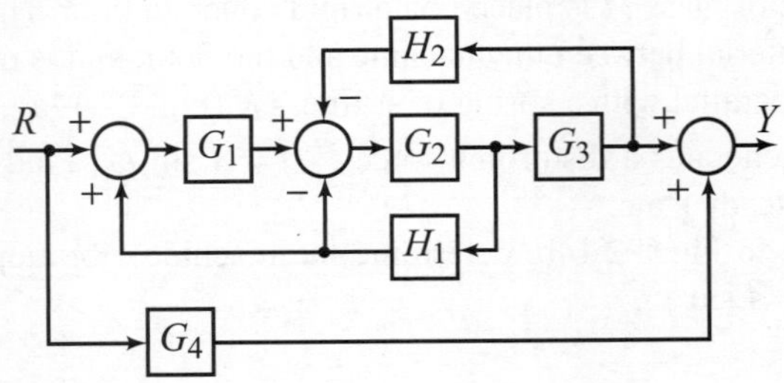

Fig. P2.17

2.18 Convert the block diagram of Fig. P2.18 to a signal flow graph, and thereform obtain the input–output transfer function using Mason's gain rule.

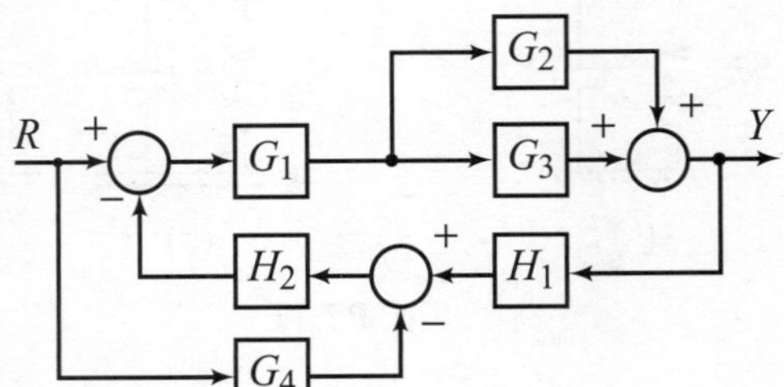

Fig. P2.18

2.19 The block diagram of a feedback control system is shown in Fig. P2.19. The output

$$Y(s) = M(s)R(s) + M_w(s)W(s)$$

Find the transfer functions $M(s)$ and $M_w(s)$

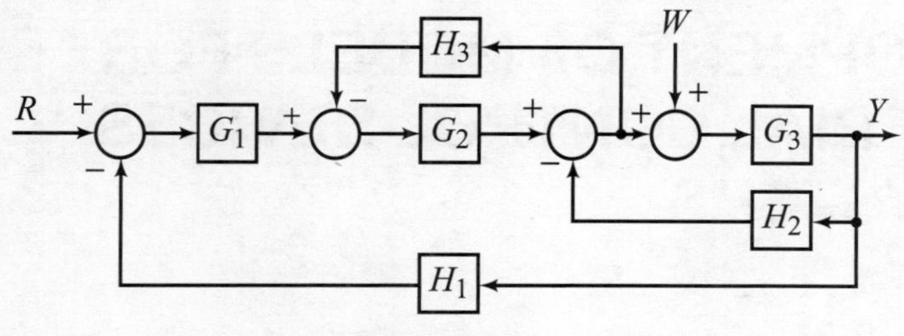

Fig. P2.19

2.20 Consider the signal flow graph model of a feedback system shown in Fig. P2.20. Obtain the closed-loop transfer function $Y(s)/R(s)$ using Mason's gain formula.

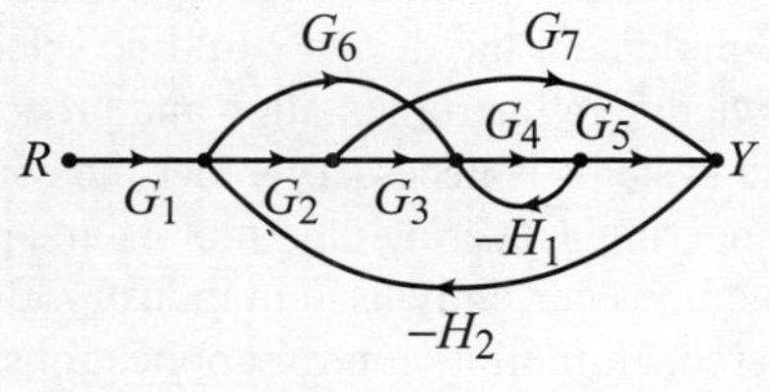

Fig. P2.20

2.21 The transfer function of a dead-time element $e^{-\tau_D s}$ may be approximated by a rational function. Two often-used approximations are given below:

(a) $e^{-\tau_D s} = \dfrac{1-\tau_D s/2}{1+\tau_D s/2}$ (b) $e^{-\tau_D s} = \dfrac{1-\tau_D s/2+\tau_D^2 s^2/12}{1+\tau_D s/2+\tau_D^2 s^2/12}$

Prove that the first approximation is correct through the first three terms, and the second approximation is correct through the first five terms of the Taylor series expansion of $e^{-\tau_D s}$.

DEVELOPMENT OF MODELS FOR INDUSTRIAL CONTROL DEVICES AND SYSTEMS

3

Preview

The learning objectives for this chapter are to acquire reasonable competence in modeling of industrial systems using judicious assumptions, and to get familiarity with available hardware so that commercially available components to implement the design could be selected. Real-life control engineering practice requires extensive knowledge in instrumentation and process dynamics. A judicious selection of industrial control devices and systems is necessary under the constraint of available time for delivery of a course on Control Engineering. Covering only motors and position/speed control systems will not capture all control design methods currently used in industry. Both the motion control and process control systems should be covered. In motion control applications, mathematical models of systems are often close enough to reality; on the other hand, industrial processes are relatively complex, and building models close enough to reality can be expensive, and also not suitable for the available design methods. Process control design based on approximate models also has to be explored. To realize this objective, the minimum learning on devices and systems includes: potentiometers, dc servomotors, dc tachogenerator, and speed/position control systems using these devices; pneumatic control valve, electropneumatic transducer, and temperature control systems using these devices. Organization of the material in this chapter helps make judicious selection of hardware and systems. The sections: AC Servo Systems, Hydraulic Devices for Motion Control, and Control of Liquid Level Processes, can be selectively studied without affecting the flow of the material that follows in later chapters.

3.1 INTRODUCTION

Control engineering is an exciting and challenging field within the engineering sciences, because by its very nature it is a multidisciplinary subject and it plays a critical role in many practical technical applications. To be more specific, control theory brings together such different fields as electrical, mechanical and chemical engineering and applied mathematics. It plays a major role in engineering applications of all kinds of complexity. This may be a robot arm, an aircraft, a satellite, a power plant or a plant for chemical industries such as a distillation column.

The most fundamental idea of control theory is to change the dynamical behavior of a plant by a controller such that the dynamics gets certain desired properties. In almost all cases, the controller design is done by building a mathematical model, which can be emulated on a computer and serves as the basis for controller synthesis. It is obvious that the model represents the real system only to a certain degree of accuracy. Despite this, the controller has to work for the real system as well as the model; in other words, it has to be "robust" against errors in the mathematical model.

Building a mathematical model for a plant, actuator, or sensor means that physical laws have to be applied to find the differential equations which govern the dynamics of the system. A decision has to be made on the level of detail to which the mathematical model has to be developed. This requires good engineering judgement and some experience. Once the differential equations which govern the system are found, a computer simulation is built for analysis. This may be done by using a simulation tool such as MATLAB's Simulink (visit MATLAB 3 in Appendix). There is no simple recipe for model building which is applicable to any thinkable situation.

In view of the complexity of the design process, modeling and design go hand in hand, in an iterative manner. In the beginning, by knowing some information about the system (e.g., intended functions, performance specifications, past experience and knowledge of related systems), we develop a model of sufficient (low to moderate) detail and complexity. By analyzing and carrying out computer simulations of the model, it will be possible to generate useful information that will guide the design process (e.g., generation of a preliminary design). In this manner, design decisions can be made, and the model can be refined using the improved design. It is unrealistic to attempt to develop a "universal model" that will incorporate all conceivable aspects of the system. The model should be as simple as possible, and may address only a few specific aspects of interest in the particular study or application.

Our discussion in this chapter is limited to the widely used control devices and systems for control of speed or position of an inertia load, and for control of temperature or pressure of a process fluid. The objective is to show the reader how the mathematical models that we shall be dealing within the remainder of the book arise from physical systems.

The devices considered in this chapter, can be broadly divided into two subgroups:

1. Information handling devices
2. Power handling devices

With reference to a generalized feedback control system model shown in Fig. 3.1 (refer to Figs 1.9 and 2.33), when we speak of a *controller*, our concern is mainly with the information handling devices; the sources of information are the reference input $r(t)$ which is derived from the command signal $y_r(t)$, and the feedback signal $b(t)$ produced by the controlled output $y(t)$. The error detector compares the feedback signal with the reference signal and generates an error signal. The physical devices[1] corresponding to $H(s)$, the error detector, and the $D(s)$ are all low-power devices; all these subsystems designed and interconnected appropriately, realize the control function/control law

$$u = f(y_r, y) \tag{3.1}$$

through the equation

$$U(s) = D(s)\left[R(s) - B(s)\right] = D(s)\left[A(s)Y_r(s) - H(s)Y(s)\right] \tag{3.2}$$

We have defined the 'control-function block $D(s)$ + error detector' as controller (Fig. 3.1). In control theory terminology, the control-function block $D(s)$ is referred to as 'compensator' (discussed later in Chapter 7). However, the term 'controller' is also in use for the control function $D(s)$. We will, therefore, not make distinction between the terms 'compensator' and 'controller'; the two will carry the same meaning at the design stage. Appropriate additional functions will be added at the implementation stage.

In the feedback-control structure represented by the block diagram of Fig. 3.1 (we will name this structure as *basic feedback-control configuration*), the controller $D(s)$ is in the forward path, and the control actions are based on the error signal. In an alternative feedback-control structure, widely used in position

[1]The block $A(s)$ may correspond to only a mathematical operation; requiring no physical devices. This was mentioned earlier in Chapter 1, and the examples in this chapter will further clarify this point.

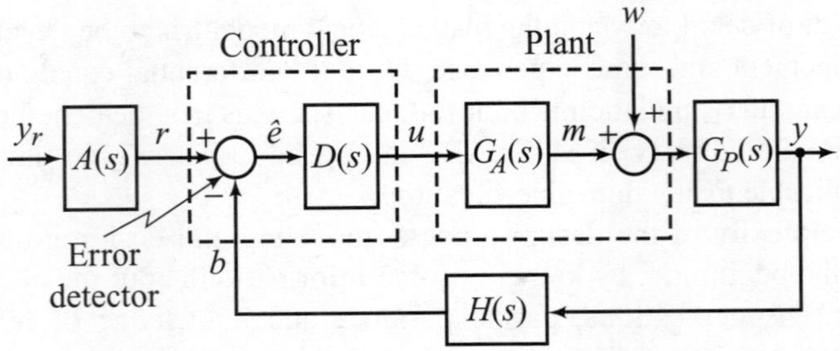

Fig. 3.1 *Basic feedback control configuration*

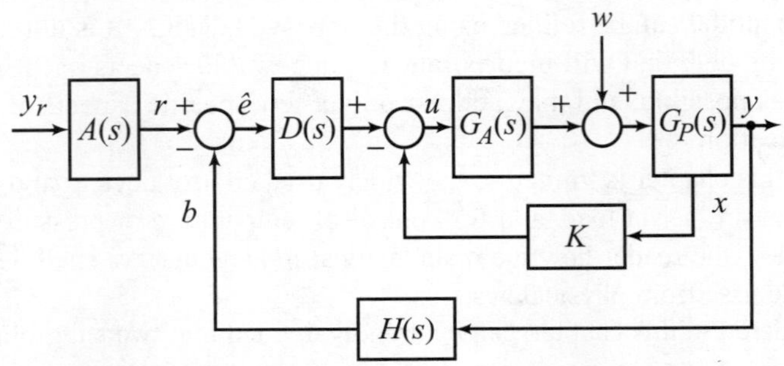

Fig. 3.2 *Minor-loop feedback control configuration*

control systems, control actions are based on some other significant variable of the process, in addition to the error signal; many a times to advantage, as we shall see in this and later chapters. The resulting *multi-loop* (loop within a loop) *control configuration* is known in the literature as *minor-loop feedback control*[2] (revisit Example 1.3). Figure 3.2 shows a block diagram representation of such a scheme.

The power handling devices—*actuators and the controlled process itself*—are represented by blocks $G_A(s)$ and $G_P(s)$ in Figs 3.1 and 3.2. We have noted earlier in Section 2.11 that a feedback loop with multiple subsystems has to obey the *no-loading* condition. The loading effects must be examined before deriving the transfer function models of subsystems. For example, in many industrial controlled systems, the process appreciably loads the actuator. In such cases, we cannot derive $G_A(s)$ and $G_P(s)$ separately; the two subsystems need to be treated as a single subsystem, and a single transfer function model $G(s)$ needs to be derived representing the dynamic behavior of the two together. Examples in this chapter will clarify this point.

3.2 DC MOTOR SPEED CONTROL

Manufacturing systems use some form of motor to actuate a process; for example, for accurate positioning in robotic welding, and for accurate velocity in conveyor systems. We will develop, in this chapter, simple models of motor *speed control* and *position control* systems.

---

[2]Extending this logic leads to *state variable feedback control scheme*, wherein feedback of all the significant variables (state variables) is required. Chapter 15 gives the details.

The applications of control systems developed in this chapter are not limited to manufacturing floors. We use feedback to control inertia load. The inertia load may consist of a small, light object such as a precision instrument in a manufacturing system, or a very large, massive object such as a radar antenna, a rigid satellite or control surfaces of an aircraft. There are numerous applications wherein one encounters the need to control mechanical motions (revisit Examples 1.3, 1.4, 1.7 and 1.8).

Basic structure of speed control/position control systems will remain the same irrespective of the application. In this section, we will develop models of subsystems of a typical speed control system; the next section will consider a position control system.

To understand the essential components of a speed control system, we consider here conveyor systems in a manufacturing process to transport goods from different areas of the production facility. A typical conveyor system (Fig. 3.3) may use a belt on rollers to carry products. The rollers are driven by a motor (*actuator*). The conveyor system may require the motor to achieve and maintain a constant speed. If the *inertia load* (conveyor belt and products)—the *Controlled Process*—presented a constant torque to the motor shaft, then no feedback control would be required. But the load does change, and as a result, feedback control is required if the motor is to achieve its purpose. A typical *feedback control system* may require the motor to achieve and maintain a constant speed (*setpoint*) while working against a load of varying torque (*disturbance*). The speed of the conveyor belt may be determined by a tachogenerator (*sensor*).

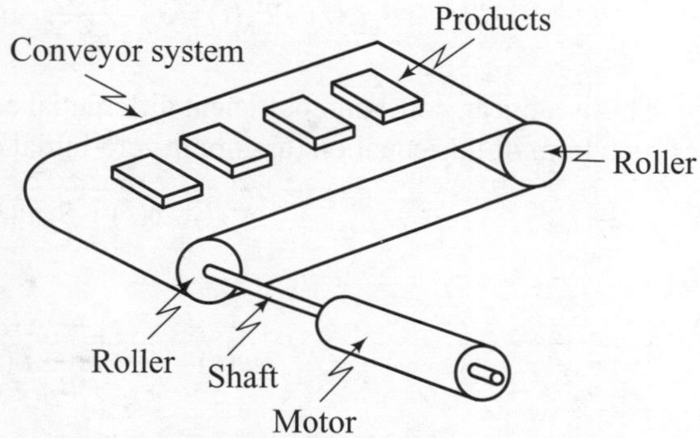

Fig. 3.3 *Conveyor belt system in a manufacturing process*

We can represent this system in a feedback-control block diagram format. We first model the controlled process, measurement, and actuator components, and then set up a feedback loop.

3.2.1 Rotational Load

For the speed control system under consideration, the controlled process is represented by the rotating shaft which moves the load (the conveyor belt and products). The input signal to this rotational system is the *torque* $T(t)$ from the motor which drives the load. The output signal of the system is the *angular velocity* $\omega(t)$ of the shaft.

The modeling procedure for mechanical rotational systems is very close to that used earlier in Section 2.6 for mechanical translational systems. The three basic parameters for rotational systems are: moment of inertia J, viscous friction with coefficient B, and the torsional spring with spring constant K. In low-power servos, the driving and driven shafts can be treated as having negligible elasticity; these rotational systems can, therefore, be modeled using two parameters: J and B. High-power servos, however, may require three parameters: J, B, and K.

A simple model of the rotational load in our example under consideration, is the moment of inertia-damper system shown in Fig. 3.4a; the parameter J_L is the moment of inertia of the load (conveyor belt and products), and the parameter B_L is the viscous friction coefficient, representing the viscous friction introduced by bearings and other sources. Torque $T(t)$ is the controlled input generated by the motor, angular velocity $\omega(t)$ of the load shaft is the output, and $T_w(t)$ represents the uncontrolled disturbance torque (due to variations in quality and quantity of products, for example).

Figure 3.4b shows the free-body diagram of the system of Fig. 3.4a. The frictional effects lead to a frictional torque $B_L\omega(t)$, that opposes the applied torque $T(t)$. The net torque $T(t) - B_L\omega(t) - T_w(t)$ drives the inertia J_L and results in angular acceleration $d\omega(t)/dt$ of the shaft. The torque equation is

$$T(t) - B_L\omega(t) - T_w(t) = J_L\frac{d\omega(t)}{dt} \quad \text{or} \quad J_L\frac{d\omega(t)}{dt} + B_L\omega(t) = T(t) - T_w(t) \tag{3.3}$$

This is a linear, constant-coefficient differential equation of first order. Taking the Laplace transform of each term of this equation (assuming zero initial conditions), we obtain

$$J_L s\omega(s) + B_L\omega(s) = T(s) - T_w(s) \tag{3.4}$$

where

$$\omega(s) = \frac{1}{J_L s + B_L}T(s) - \frac{1}{J_L s + B_L}T_w(s) \tag{3.5}$$

Figure 3.4c gives the block diagram representation of the system. The transfer function

$$G_P(s) = \frac{1}{J_L s + B_L} = \frac{1/B_L}{\frac{J_L}{B_L}s + 1} = \frac{1/B_L}{\tau s + 1} \tag{3.6}$$

where τ is the time-constant of the simple lag.

The metric system of units for mechanical rotational systems is as follows:

Torque, T : newton-m

Angular velocity, ω : rad/sec

Angular displacement, θ ($\omega = d\theta/dt$): rad

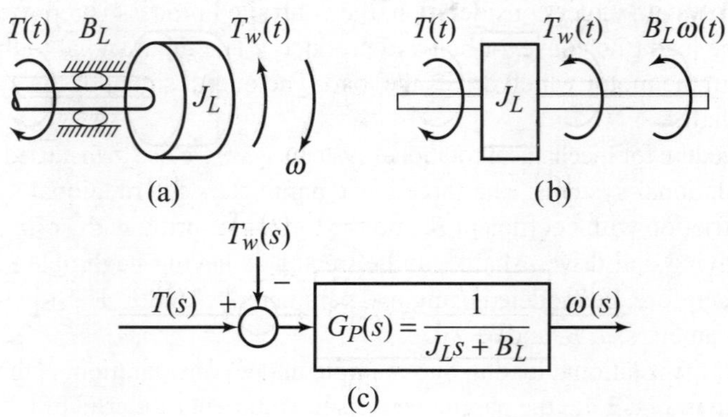

Fig. 3.4 *An inertia-damper model of the load*

Angular acceleration, $\dot{\omega}$: rad/sec^2
Moment of inertia, J : newton-m/(rad/sec^2) or kg-m^2 (3.7)
Damping coefficient, B : newton-m/(rad/sec)
Torsional spring constant, K : newton-m/rad

3.2.2 DC Servomotors

The dc motor has been a workhorse in industry for the better part of the century. It has survived because it provides good torque at all speeds and can be manufactured easily and inexpensively. It is being replaced, to some extent, by brushless dc motors and ac induction motors. However, commutator dc motors are still manufactured in great quantities because for many applications, they are still the best solution.

A sketch of the basic components of a dc motor is given in Fig. 3.5. The nonturning part (called the *stator*) has magnets which establish a field across the turning part (called the *rotor*) through the air gap between stator and rotor. The magnets may be electromagnets or, for small motors, permanent magnets. In an electromagnet motor, the stator is wound with wire and current is forced through this winding (called the *field winding*). For a constant field current i_f, the magnetic flux ϕ is constant; the magnetic flux may be varied by varying the field current.

The rotor is wound with wire and through this winding (called the *armature winding*) a current i_a is forced through the (stationary) brushes and the (rotating) commutator (The commutator comprises of a series of copper segments insulated from one another and arranged in a cylindrical shape. Riding on the commutator are appropriately placed carbon brushes which serve to conduct direct current to the armature winding. Each coil of the armature winding is made accessible to the brushes by connecting the coils to individual commutator segments). The reaction of the air-gap flux ϕ with the armature current i_a produces a torque T_M that forces the armature to rotate. The relationship among the developed torque T_M (newton-m), flux ϕ (webers) and current i_a (amps) is

$$T_M = K_{M1}\,\phi\, i_a \tag{3.8a}$$

where K_{M1} is a constant.

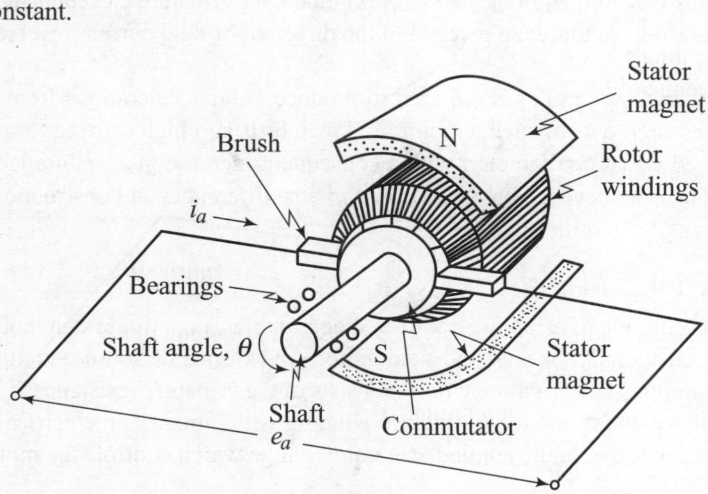

Fig. 3.5 *Sketch of a dc motor*

When using electromagnets to generate the flux, we have several options on how to wire the armature and field. In *shunt-wound* motors, we have the armature and field coils connected in parallel. *Series-wound* motors have some pairs of armature and field coils in series and some in parallel. While these motors constitute by far the largest number of dc machines, they are not very often used in closed-loop systems. *Separately excited* dc motors, wherein independent sources excite the two windings, allow smoother control of motor speed.

In servo applications, a wound-field dc motor is generally used in the linear range of magnetization; the flux ϕ is therefore proportional to field current i_f, and

$$T_M = K_{M2}\, i_f\, i_a \tag{3.8b}$$

where K_{M2} is a constant.

As the armature rotates in the magnetic field, a voltage is induced into the armature winding. This voltage is 180° out of phase with the applied armature voltage and is therefore called *back emf*. The relationship among the back emf e_b (volts), rotor velocity ω (rad/sec) and flux ϕ (webers) is

$$e_b = K_{m1}\, \phi\omega \tag{3.9a}$$

where K_{m1} is a constant.

For the wound-field motor,

$$e_b = K_{m2}\, i_f\, \omega \tag{3.9b}$$

where K_{m2} is a constant.

Equations (3.8) and (3.9) form the basis of dc motor operation.

In a permanent-magnet (PM) motor, the flux ϕ is constant; the torque T_M exerted on the motor rotor can therefore be controlled by varying the armature current (refer to Eqn. (3.8a)). If the direction of the armature current is reversed, the direction of the torque is reversed.

In a wound-field motor, the torque may be controlled by varying the armature current and/or the field current (refer to Eqn. (3.8b)). Generally, one of these is varied to control the torque while the other is held constant. In the *armature control* mode of operation, the field current is held constant and armature current is varied to control the torque. In the *field-control* mode, the armature current is maintained constant and field current controls the torque; a reversal in the direction of field current reverses the direction of the torque.

In servo applications, a dc motor is required to produce rapid accelerations from standstill. Therefore, the physical requirements of such a motor are low inertia and high starting torque. Low inertia is attained with reduced armature diameter; with a consequent increase in armature length such that the desired power output is achieved. Thus, except for minor differences in constructional features, a dc servomotor is essentially an ordinary dc motor.

Armature-Controlled dc Motor

The symbolic representation of armature-controlled dc motor as a control system component is shown in Fig. 3.6. Under consideration is a wound-field motor. An external dc source supplies a constant current i_f to the field winding. The armature circuit consists of the armature resistance R_a and the armature inductance L_a; both are due to the total armature winding which makes an electrical contact with the brushes of the commutator. e_a is the applied armature voltage which controls the motor operation, and e_b is the back emf.

On the mechanical side, the motor rotor and the attached load can be treated as inertia and viscous friction; J and B are the corresponding parameters. T_w is the disturbance load torque. The torque T_M developed by the motor drives the load with angular velocity ω.

Let us summarize the variables and parameters in the dc motor model of Fig. 3.6:

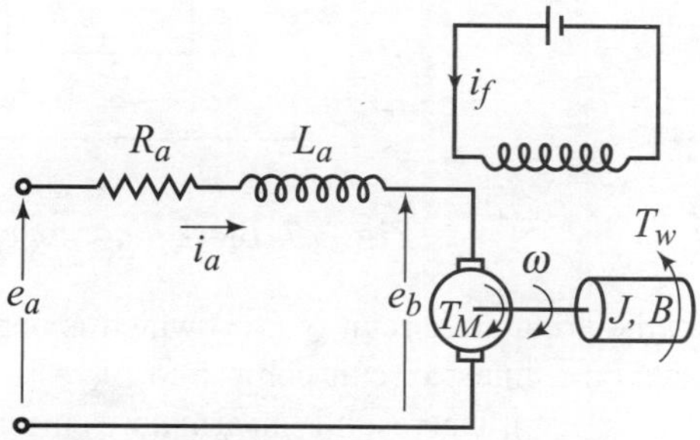

Fig. 3.6 *Armature-controlled dc motor with load*

R = armature winding resistance (ohms);
L_a = armature winding inductance (henrys);
i_a = armature current (amps);
i_f = field current (amps) = a constant;
e_a = applied armature voltage (volts);
e_b = back emf (volts);
ω = angular velocity of the motor rotor (rad/sec);
θ = angular displacement of the motor rotor (rad);
T_M = torque developed by the motor (newton-m);
J = moment of inertia of the motor rotor with attached mechanical load (kg-m^2 or (newton-m)/(rad/sec^2));
B = viscous-friction coefficient of the motor rotor with attached mechanical load ((newton-m)/(rad/sec)); and
T_w = disturbance load torque (newton-m).

Since the field current is kept constant in the armature-control mode of operation, Eqns (3.8b) and (3.9b) become

$$T_M = K_T i_a \tag{3.10a}$$

$$e_b = K_b \omega \tag{3.10b}$$

where

$$K_T \triangleq \text{torque constant; } K_b \triangleq \text{back emf constant}$$

The differential equation of the armature circuit is

$$L_a \frac{di_a}{dt} + R_a i_a + e_b = e_a \tag{3.11}$$

The torque equation is

$$J\frac{d\omega}{dt} + B\omega + T_w = T_M \tag{3.12}$$

Taking the Laplace transform of Eqns ((3.10)–(3.12)), assuming zero initial conditions, we get

$$T_M(s) = K_T I_a(s) \tag{3.13a}$$

$$E_b(s) = K_b \omega(s) \tag{3.13b}$$

$$(L_a s + R_a)\, I_a(s) = E_a(s) - E_b(s) \tag{3.13c}$$

$$(Js + B)\, \omega(s) = T_M(s) - T_w(s) \tag{3.13d}$$

Figure 3.7 shows a block diagram representation of the dc motor system. The block diagram gives a clear picture of the cause and effect relationships in the physical system of Fig. 3.6. The voltage applied

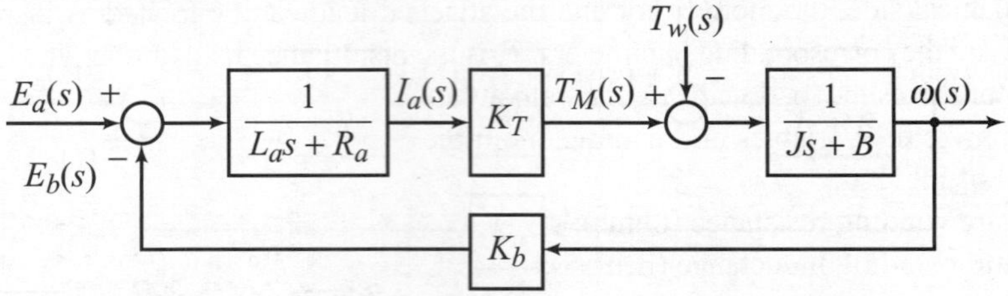

Fig. 3.7 *Block diagram of a dc motor (armature-controlled) system*

to the armature circuit is $E_a(s)$ which is opposed by the back emf $E_b(s)$. The net voltage $(E_a(s) - E_b(s))$ acts on a linear circuit comprised of resistance and inductance in series, having the transfer function $1/(sL_a + R_a)$. The result is an armature current $I_a(s)$. For the fixed field, the torque developed by the motor is $K_T I_a(s)$. The torque rotates the load at a speed $\omega(s)$ against the disturbance $T_w(s)$; the load having moment of inertia J and viscous friction with coefficient B has the transfer function $1/(Js + B)$. The back emf signal $E_b(s) = K_b\omega(s)$ is taken off from the shaft speed and fedback negatively to the summing point. Note that although a dc motor by itself is basically an open-loop system, it has a 'built in' feedback loop caused by the back emf.

The transfer function between the motor velocity $\omega(s)$ and the input voltage $E_a(s)$ obtained from the block diagram is

$$\frac{\omega(s)}{E_a(s)} = \frac{K_T}{(L_a s + R_a)(Js + B) + K_T K_b} \tag{3.14}$$

The inductance L_a in the armature circuit is usually small and may be neglected. If L_a is neglected, then the transfer function given by Eqn. (3.14) reduces to

$$\frac{\omega(s)}{E_a(s)} = \frac{K_T/R_a}{Js + B + K_T K_b/R_a} \tag{3.15}$$

The back emf constant K_b represents an added term to the viscous friction coefficient B. Therefore, the back emf effect is equivalent to an 'electric friction' which tends to improve the stability of the dc motor system.

The transfer function given by Eqn. (3.15) may be written in the form given below.

$$\frac{\omega(s)}{E_a(s)} = \frac{K_m}{\tau_m s + 1} \tag{3.16}$$

where $$K_m = \frac{K_T}{R_a B + K_T K_b} = \text{motor gain constant}$$

and $$\tau_m = \frac{R_a J}{R_a B + K_T K_b} = \text{motor time constant}$$

When the motor is used to control shaft position θ, rather than the speed ω, we get the block diagram model shown in Fig. 3.8. The transfer function between the shaft position $\theta(s)$ and the input voltage $E_a(s)$ becomes

$$\frac{\theta(s)}{E_a(s)} = \frac{K_m}{s(\tau_m s + 1)} \tag{3.17}$$

The significance of this transfer function is that the dc motor is essentially an integrating device between the input voltage and the shaft position. This is expected since if e_a is a constant input, the motor displacement will behave as the output of an integrator; that is, it will increase linearly with time. We shall see later that the integrating effect $1/s$ in Eqn. (3.17) gives a feedback position control system a better steady-state behavior than the corresponding speed control system.

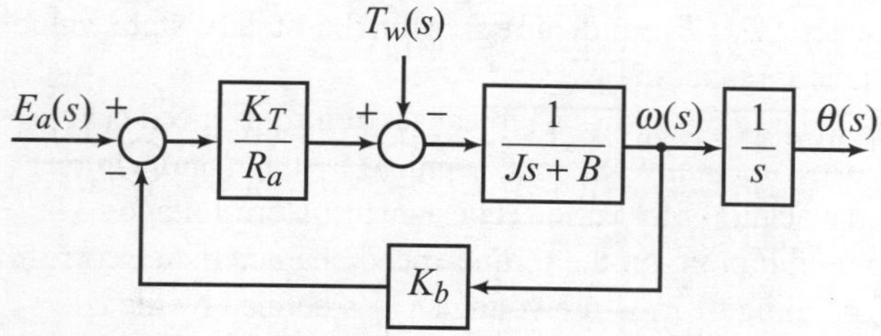

Fig. 3.8 *The dc motor (armature-controlled) model with displacement as output*

The motor torque and back emf constants are interrelated. Their relationship is deduced below. Power developed in the armature is

$$P = e_b(t)\, i_a(t)\ [(\text{volts})\ (\text{amps}) = \text{watts}]$$

$$= K_b\omega(t)i_a(t)\left[\left(\frac{\text{volts}}{\text{rad/sec}}\right)(\text{rad/sec})(\text{amp})\right]$$

In terms of torque and angular velocity,

$$P = T_M(t)\omega(t)\ [(\text{newton-m})\ (\text{rad/sec}) = \text{watts}]$$

$$= K_T i_a(t)\omega(t)\left[\left(\frac{\text{newton-m}}{\text{amp}}\right)(\text{amp})(\text{rad/sec})\right]$$

Therefore,

$$K_b\ (\text{volts/(rad/sec)}) = K_T\ (\text{newton-m/amp}) \tag{3.18}$$

This result can be used to advantage in practice as K_b can be measured more easily and with greater accuracy than K_T.

In a PM motor, the field winding is absent and a constant magnetic flux is provided by the permanent magnet. Since in the preceding development, the magnetic flux is a constant, Eqns ((3.14)–(3.18)) are also valid for a PM motor.

How does the dc motor react to the application of shaft load? What is the mechanism by which the dc motor adapts itself to supply the demanded load power? The answers to these questions can be obtained by reasoning in terms of the performance equations of the motor.

With reference to Fig. 3.7, in the steady state, the effect of inductance is zero, and the performance equations are:

$$T_M = K_T I_a;\ I_a = \frac{E_a - K_b\omega}{R_a}$$

Therefore,

$$T_M = \frac{K_T}{R_a}(E_a - K_b\omega) = -\frac{K_T K_b}{R_a}\omega + \frac{K_T}{R_a}E_a$$

where T_M, E_a, and ω represent the steady-state values of the motor torque, applied voltage, and speed, respectively.

Figure 3.9 shows a typical set of torque-speed curves (experimentally obtained/supplied by the manufacturer) of a dc motor for various constant armature voltages. The *no-load* point on the torque-speed characteristics corresponding to armature voltage E_{a3} is defined by the coordinates (ω_3, 0). Consider that a load is suddenly applied to the motor shaft. Clearly, because the developed torque at this point is only sufficient to overcome friction and not the load torque, the first reaction for the motor is to lose speed. In this way, the armature current can be increased (as seen from performance equations) so that in turn the electromagnetic torque can increase. As a matter of fact, the applied load torque causes the motor to assume the value of speed, ω_x, which yields the current sufficient to produce a developed torque, T_x to overcome the applied torque and the frictional torque. Power balance is thereby achieved, because an equilibrium condition is reached where the electromagnetic power, $E_a I_a$, is equal to the mechanical power developed, $T\omega$.

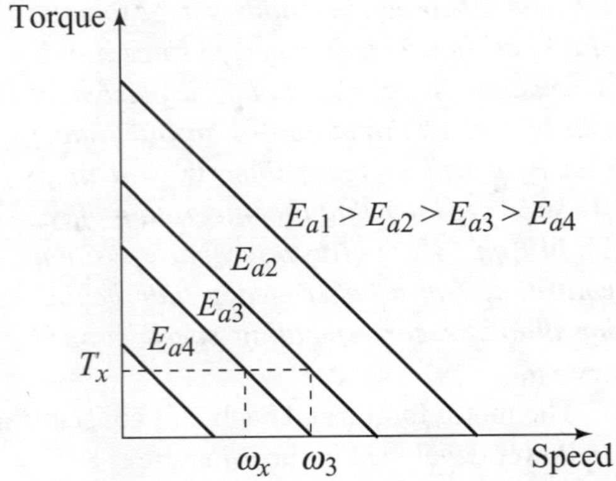

Fig. 3.9 *Typical torque-speed curves of an armature-controlled dc motor*

If the new steady-state speed ω_x is too low and the original speed ω_3 must be restored, the armature voltage must be increased. This is illustrated in Fig. 3.9; the increase in armature voltage from E_{a3} to E_{a2} restores the original no-load speed ω_3 with the applied torque.

In a dc motor system, there may be a significant nonlinear friction due to the rubbing contact between the brushes and the commutator. Also, torque saturation may occur because of the limitation of the maximum current that the motor can handle due to the heat-dissipation rating of the motor. Refer to Chapter 16 for methods for analysis of control systems with nonlinearities in the closed loop. Also see MATLAB Module 3 in Appendix, wherein Simulink tool has been employed.

Field-Controlled dc Motor

The schematic diagram of a field-controlled dc motor is shown in Fig. 3.10. In this system,

R_f = field winding resistance (ohms);
L_f = field winding inductance (henrys);
i_f = field current (amps);
i_a = armature current (amps) = a constant;
e_f = applied field voltage (volts);
ω = angular velocity of the motor rotor (rad/sec);
θ = angular displacement of the motor rotor (rad);
T_M = torque developed by the motor (newton-m);
J = moment of inertia of the motor rotor with attached mechanical load (kg-m^2 or (newton-m) /(rad/sec^2));
B = viscous-friction coefficient of the motor rotor with attached mechanical load ((newton-m)/(rad/sec)); and
T_w = disturbance load torque (newton-m).

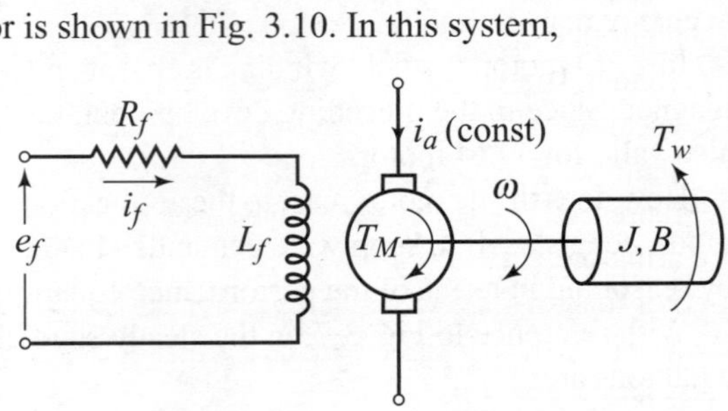

Fig 3.10 *Field-controlled dc motor with load*

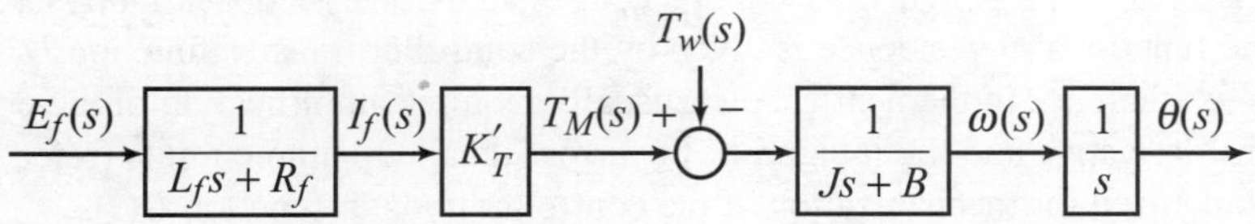

Fig. 3.11 *Block diagram of a field-controlled dc motor*

The following assumptions are made in the derivation of the transfer function of field-controlled motor:

(i) a constant current i_a is fed into the armature;
(ii) the air gap flux ϕ is proportional to field current i_f; and
(iii) the torque developed by the motor is proportional to field current;

$$T_M = K'_T i_f;\ K'_T \triangleq \text{torque constant}$$

The voltage applied to the field circuit is $E_f(s)$ which acts on a linear circuit comprising of resistance and inductance in series, having the transfer function $1/(L_f s + R_f)$. The result is a field current $I_f(s)$. For fixed armature current, the torque developed by the motor is $K'_T I_f(s)$. This torque rotates the load at a speed $\omega(s)$ against the disturbance $T_w(s)$; the load having moment of inertia J and viscous-friction coefficient B has the transfer function $1/(Js + B)$. These cause and effect relationships are described by the block diagram of Fig. 3.11. The transfer function between the motor displacement $\theta(s)$ and the input voltage $E_f(s)$, obtained from the block diagram is

$$\frac{\theta(s)}{E_f(s)} = \frac{K'_T}{s(L_f s + R_f)(Js + B)} = \frac{K'_m}{s(\tau_f s + 1)(\tau_m s + 1)} \tag{3.19}$$

where

$$K'_m = K'_T/(R_f B) = \text{motor gain constant;}$$

$$\tau_f = L_f/R_f = \text{time-constant of the field circuit; and}$$

$$\tau_m = J/B = \text{mechanical time-constant.}$$

For small size motors, field control is advantageous because only a low power servo amplifier is required while the constant armature current, which is not large, can be supplied from an inexpensive constant voltage source. This control scheme poses great practical difficulties in medium and large size motors; because of large back emf of the motor, the armature current cannot be kept constant by supplying a constant voltage to the armature. The assumption of constant armature current leads to the linear equations representing the operation of the motor. However, if instead, a constant voltage source is applied to the armature, the equation of the motor becomes nonlinear and cannot be handled by linear analysis methods. For this reason, it is a normal practice to have a fixed excitation and control the armature current.

3.2.3 Electronic Amplifiers

Broadly speaking, there are two types of electronic amplifiers used in control systems:

1. A *small signal amplifier* is one which linearly amplifies a small signal (usually voltage) and is suitable for sensor circuits to amplify weak measured signals. The operational amplifier (usually called the op amp) is in common use.

From the block diagram of Fig. 3.1, we see that a control signal in a feedback loop is a function of the error signal. The functional dependence is given by the controller transfer function $D(s)$. Mechanization of $D(s)$, as we shall see in the next chapter, usually requires amplification of the error signal, and generation of the derivative and the integral of the error signal. Op amp circuits (refer to Section 3.9) are now commonly used for mechanization of the controller transfer function $D(s)$.

2. A *power amplifier* is one which controls a large amount of power, and is therefore suitable for actuating devices.

Power amplifiers are available in three major types: the smooth transistor, the switching transistor (pulse-width modulated (PWM)), and the silicon control rectifier (SCR). Smooth transistor amplifiers use the transistors in a smooth modulating fashion and produce an output voltage that is high-power copy of the input. These amplifiers are used mainly below 1 kW applications. Switching (PWM) amplifiers use the transistors as on–off switches; the output voltage is a constant-frequency, fixed-amplitude waveform whose duty cycle is smoothly varied with input voltage. Switching amplifiers are economical in the 25 W to 5 kW range. In SCR amplifiers, the input voltage smoothly modulates the point in each power-line cycle at which the SCR (this term arises from the device as controlled rectifier constructed from silicon), also known as the *thyristor*, is made to conduct. These amplifiers are often used in high power applications.

To be precise, the transfer function of an electronic amplifier is of the form

$$G(s) = \frac{K_A}{\tau_A s + 1} \tag{3.20}$$

However, usually the time constant τ_A is negligibly small compared to the dynamics of the plant; the transfer function can therefore be approximated as

$$G(s) = K_A \tag{3.21}$$

where K_A is the proportional gain of the amplifier.

In addition to neglecting 'fast' dynamics, the model (3.21) of the amplifier further differs from reality because of its perfect linearity which assumes no limit on the magnitude of the input signal. Actually the amplifier, like other devices, exhibits saturation, limiting its output when input becomes too large (refer to Chapter 16.)

Discussion on hardware aspects of various types of electronic amplifiers is beyond the scope of this book [50].

3.2.4 DC Tachogenerator

We are interested in measuring the rotational velocity of the drive shafts for feedback control. Optical encoders are often used in manufacturing and robotics to provide a measurement of the angular movement (angular velocity/angular position). The encoder is mounted on the drive shaft and rotates at the same velocity as the shaft. The output of the encoder is a series of electrical pulses, directly proportional to the shaft velocity/position. Interfacing an encoder with a computer is straightforward (discussed later in Chapter 13). We consider here a transducer that produces continuous-time electrical signal proportional to velocity. Such a transducer facilitates interfacing with an op amp comparator/control circuit.

A tachogenerator is an electromechanical device which produces an output voltage that is proportional to its shaft speed. Figure 3.12 shows the schematic diagram of a dc tachogenerator. It comprises a stator (nonrotating part) with a permanent magnet field, a rotating armature circuit, and a commutator and brush assembly. The rotor is connected to the shaft whose speed is to be measured. The output voltage of the tachogenerator is proportional to the angular velocity of the shaft. The polarity of the output voltage is dependent on the direction of rotation of the shaft. Dynamics of a dc tachogenerator can be represented by the equation

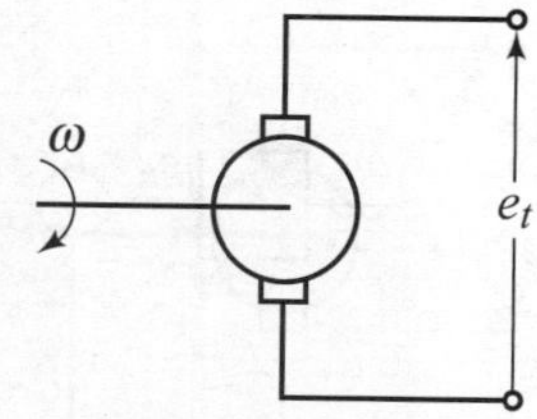

Fig. 3.12 *Schematic diagram of a dc tachogenerator*

$$e_t(t) = K_t \omega(t) \tag{3.22}$$

where e_t is the output voltage (volts), ω is the rotor velocity (rad/sec), and K_t is the sensitivity of the tachogenerator (volts per rad/sec).

The transfer function from $\omega(t)$ to $e_t(t)$ of a dc tachogenerator is of the form

$$\frac{E_t(s)}{\omega(s)} = K_t \tag{3.23}$$

3.2.5 A Speed Control System

To meet the manufacturing requirements, it is imperative that the conveyor belt (Fig. 3.3) operates at a constant speed; and has acceptable transient performance in moving from one speed setting to the next. We know that we have made several approximations while deriving models of the components that make up a speed control system. For example, nonlinear frictional effects due to brushes and commutator in dc machines have been ignored. Also we have assumed that the dc machines will be operating in the linear range of magnetization. We have neglected the inductance of the armature circuit. The tachogenerator and power amplifier models are of zero order; we have assumed that the time constants of these devices are much smaller than that of the process, and can be neglected. Also the numerical values of the parameters of the models are obtained from the design manuals of the equipment, which are usually not very precise on the parametric values. One may argue that one should attempt to take all these factors into account to develop a model. This approach will not be cost effective; the time and money spent on modeling is already much higher than that spent on developing a control design. In addition, a complex mathematical model will not necessarily lead to a better control design. The design toolkit is well developed for linear systems; but for nonlinear systems we have limited results. Robustness provided by feedback loop against external disturbances T_w and the model uncertainties, is a well-tested solution to the control problem.

Figure 3.13 shows a closed-loop block diagram for the speed control system. We have introduced a scaling block with transfer function K_t in the forward path to allow us to compare the command input with controlled output of the system (Reference voltage e_r is calculated from the command input, ω_r (setpoint), using the relation $e_r = K_t\,\omega_r$, and this value is then fed to the system through a potentiometer setting (Fig. 3.14)). Here K_b is the back emf constant of the motor, R_a is the armature resistance (armature inductance is assumed negligible), K_T is the torque constant of the moter; J and B are, respectively, moment of intertia and viscous-friction coefficient on motor shaft, and T_w is the disturbance torque on motor shaft (refer to Eqns (3.10)–(3.13), and Fig. 3.8).

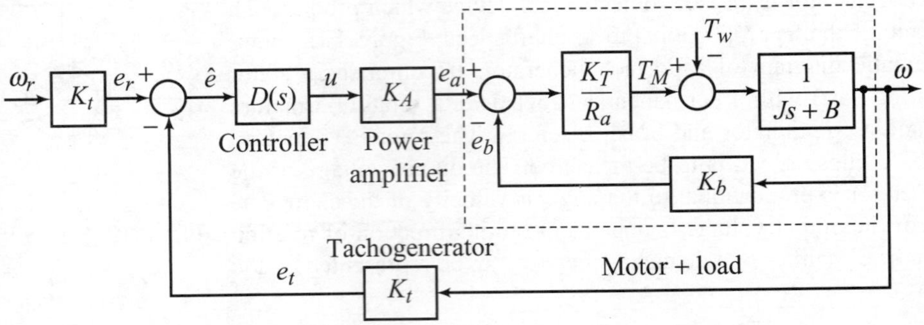

Fig. 3.13 *Closed-loop block diagram for speed control system*

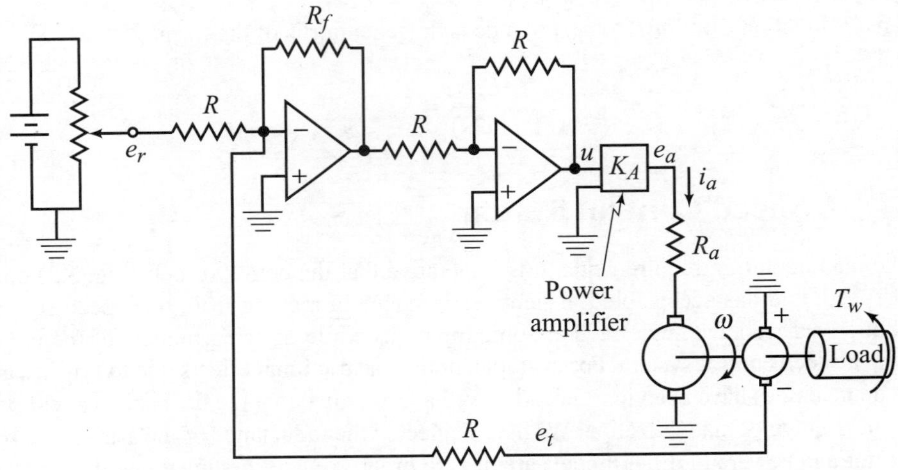

Fig. 3.14 *Schematic diagram of a speed control system*

Figure 3.14 shows schematic diagram for speed control using proportional control law. The feedback voltage e_t from the tachogenerator of sensitivity K_t volts/(rad/sec) represents the actual speed ω of the motor. The reference input voltage e_r represents the desired motor speed ω_r. An op amp circuit amplifies the error $(e_r - e_t)$ between the reference and feedback voltage signals and supplies a voltage u to a power amplifier. The output e_a of the power amplifier drives the armature-controlled dc motor which rotates in the direction to reduce the error $(e_r - e_t)$ as a consequence of the negative feedback. Reduction in $(e_r - e_t)$ results in reduction in the difference between the desired motor speed and actual motor speed. Reversal in the polarity of e_r reverses the direction of rotation of the motor.

It is seen that the control law is

$$u = \frac{R_f}{R}(e_r - e_t)$$

i.e., the control signal is proportional to the error. It will be seen in later chapters that the speed control system of Fig. 3.14 has poor disturbance-rejection property with a proportional control law.

The performance of the control system can be improved by introducing an appropriately designed controller[3] $D(s)$ in the forward path of the loop (refer to Fig. 3.13).

3.3 DC MOTOR POSITION CONTROL

In the following section, we will consider the position control systems, and the components that make up these systems.

3.3.1 Geared Drives

Electrical motors generally produce their maximum power at high speed. In other words, generally electrical motors exert rather small torques while rotating at high speeds (power = torque × angular velocity). In consequence, appropriate gearing is necessary for the electrical motors in order for these systems to drive large loads (requiring large torques) at low speeds. For example, robot arms are usually moved at low speeds, less than 1 revolution/sec, while required maximum torques range from a few newton-m to several hundred newton-m. A large gear reduction is typically required for standard servomotors.

Figure 3.15 shows a motor driving a load through a gear train which consists of two gears coupled together. The driving gear with N_1 teeth is called the primary gear and the driven gear with N_2 teeth is called the secondary gear. Angular displacements of shafts 1 and 2 are denoted by θ_1 and θ_2 respectively, with their positive directions as indicated in the figure. The moment of inertia and viscous friction of motor and gear 1 are denoted by J_1 and B_1 respectively, and those of gear 2 and load are denoted by J_2 and B_2 respectively. T_M is the torque developed by the motor and T_w is the disturbance torque on the load. Elasticity of the gear teeth and shafts is assumed negligible.

A free-body diagram is shown in Fig. 3.16. T_{12} is the torque exerted on gear 1 by gear 2, and T_{21} is the torque transmitted to gear 2 through gear 1. For the motor shaft, the differential equation is

$$J_1\ddot{\theta}_1 + B_1\dot{\theta}_1 + T_{12} = T_M \tag{3.24}$$

For the load shaft,

$$J_2\ddot{\theta}_2 + B_2\dot{\theta}_2 + T_w = T_{21} \tag{3.25}$$

Idealized characteristic of the gear train is given by the relationship $\theta_2 = (N_1/N_2)\ \theta_1$. In reality, there is always a certain amount of backlash (free play) between coupled gears. Keeping the backlash small will inevitably increase the friction between the teeth and wear out the teeth faster. On the other hand, an excessive amount of backlash may cause sustained oscillations or chattering phenomenon in control systems (refer to Chapter 16). The assumption of zero backlash is reasonable for high quality, well-adjusted gear boxes.

[3]The objective set for this chapter is the control hardware familiarity. In almost all the control system examples in the chapter, we will use an amplifier (proportional control law) as a controller, knowing very well that it may not satisfy the control requirement. The problem of design of controllers will be discussed in the next and later chapters.

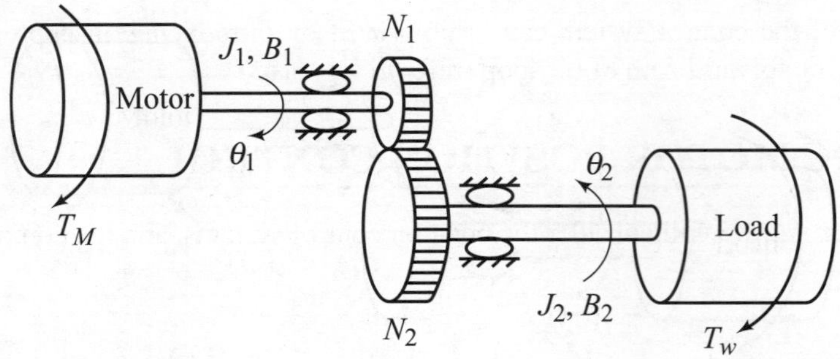

Fig. 3.15 *Motor driving a load through gearing*

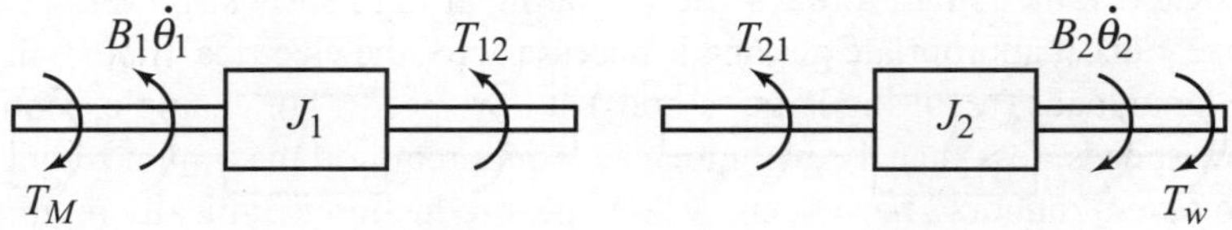

Fig. 3.16 *Free-body diagram for system of Fig. 3.15*

Let r_1 be the radius of gear 1 and r_2 be the radius of gear 2. Since the linear distance traveled along the surface of each gear is the same, $\theta_1 r_1 = \theta_2 r_2$. The number of teeth on a gear surface being proportional to gear radius, we have $(N_1/r_1) = (N_2/r_2)$. The linear forces developed at the contact point of both the gears are equal, hence $(T_{12}/r_1) = (T_{21}/r_2)$. By combining these equalities, we obtain

$$\frac{T_{12}}{T_{21}} = \frac{N_1}{N_2} = \frac{\theta_2}{\theta_1} \tag{3.26}$$

Differentiating this equation twice, we have the following relation for speed and acceleration:

$$\frac{\ddot{\theta}_2}{\ddot{\theta}_1} = \frac{\dot{\theta}_2}{\dot{\theta}_1} = \frac{N_1}{N_2} \tag{3.27}$$

From Eqns (3.26) and (3.27), it is observed that with $(N_1/N_2) < 1$, the gear train provides torque magnification and speed reduction. By proper selection of N_1/N_2, gear trains can be used to attain mechanical matching of motor to load: ***a servomotor operating at high speed and generating low torque is matched to a load to be moved at low speed but requiring high torque.***

Eliminating T_{12} and T_{21} from Eqns (3.24) and (3.25) with the help of Eqn. (3.26), we obtain

$$J_1\ddot{\theta}_1 + B_1\dot{\theta}_1 + \frac{N_1}{N_2}(J_2\ddot{\theta}_2 + B_2\dot{\theta}_2 + T_w) = T_M \tag{3.28}$$

Elimination of θ_2 from Eqn. (3.28) with the help of Eqn. (3.27), yields

$$\left[J_1 + \left(\frac{N_1}{N_2}\right)^2 J_2\right]\ddot{\theta}_1 + \left[B_1 + \left(\frac{N_1}{N_2}\right)^2 B_2\right]\dot{\theta}_1 + \frac{N_1}{N_2}T_w = T_M \tag{3.29}$$

Note that the load inertia J_2 is reflected on the motor shaft as $(N_1/N_2)^2 J_2$, load viscous friction B_2 is reflected on the motor shaft as $(N_1/N_2)^2 B_2$, and the load disturbance T_w is reflected on the motor shaft as $(N_1/N_2) T_w$.

Equation (3.29) suggests that the geared system of Fig. 3.15 is equivalent to nongeared (direct drive) system of Fig. 3.17 with equivalent moment of inertia

$$J_{1eq} = J_1 + \left(\frac{N_1}{N_2}\right)^2 J_2, \tag{3.30a}$$

equivalent friction

$$B_{1eq} = B_1 + \left(\frac{N_1}{N_2}\right)^2 B_2 \tag{3.30b}$$

and equivalent disturbance torque

$$T_{weq} = \left(\frac{N_1}{N_2}\right) T_w \tag{3.30c}$$

The problems that the mechanical gearing unavoidably possess (backlash, friction, etc.), can be solved completely using the *direct-drive technology*. An order of magnitude larger torque can be generated by direct-drive motors in a compact and lightweight body. The dc torque motors are capable of exerting much larger torque compared with regular dc motors. These motors are designed in such a way that the output torque rather than power is maximized (refer to [19] for details).

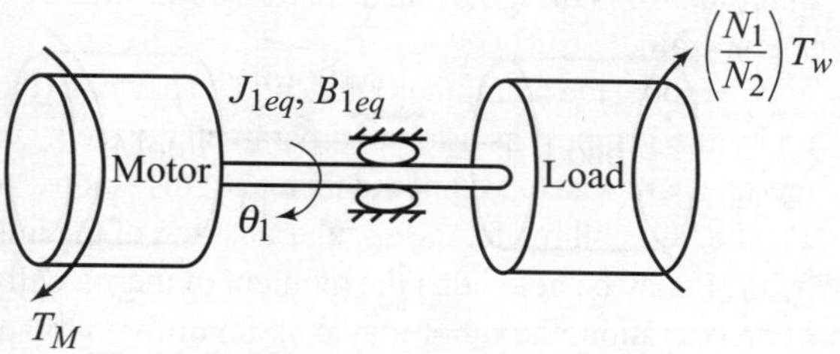

Fig. 3.17 *Equivalent direct-drive system for the geared system of Fig. 3.15*

3.3.2 Potentiometers

For conversion of a linear or an angular displacement into voltage, a potentiometer is probably the simplest device. Potentiometric transducers are relatively inexpensive and easy to apply. However, they have some limitations. The resolution (minimum change in output voltage obtained by moving the wiper, expressed as a percentage of the total applied voltage) of precision wire-wound potentiometers (constructed by winding resistance wire on a former/bobbin) ranges from 0.001 to 0.5 per cent. This discontinuous output voltage contributes to servo inaccuracy. Potentiometers are temperature sensitive, a characteristic that affects their accuracy. The wiper contact is another limiting factor, being subject to wear and dirt and potentially producing electrical noise.

In a large number of control systems, the low-amplitude high-frequency noise generated by the potentiometer does not create problems. This is because the plant usually acts as a low-pass filter. However, if the potentiometer output is differentiated to obtain a measure of velocity, the resulting signal will be completely useless.[4]

[4]Differentiation of the signal sin t contaminated by low-magnitude high-frequency noise 0.01 sin $10^3 t$ gives $(\cos t + 10 \cos 10^3 t)$; noise becomes the dominant term after differentiation.

A nonwire potentiometer (constructed by depositing conductive plastic resistance material on a former) is stepless and therefore resolution is better. However, values less than 1000 Ω are hard to obtain with nonwire type of potentiometers, while wire-wound potentiometers can be made with very low values.

3.3.3 A Position Control System

The basic structure of position control systems will remain the same irrespective of the application. We consider here a robot manipulator in a manufacturing process for load and unload operations. The motion of load gripper in a typical robot system is caused by the movements of dc motors driving the joints. Positions of the joints at any time determine the location of the gripper in the environment at that time (refer to Example 1.8).

In the following paragraphs, we will investigate the linear position control of a single-link robot manipulator, which consists of a single rigid link attached to a joint (Fig. 3.18). The joint can move the link and the gripper in two-dimensional space. The manipulator is driven by an armature controlled dc motor through a gear train.

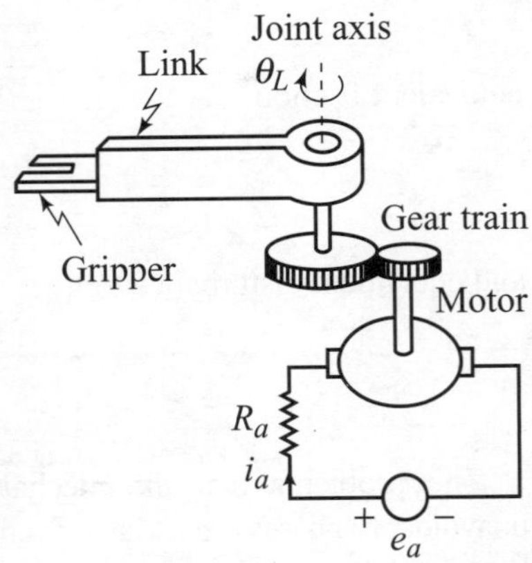

Fig. 3.18 *A single-link robot manipulator*

We model the manipulator link and its payload (the object picked up by the gripper) as a lumped parameter system with moment of inertia J_L and viscous damping factor B_L. In response to motor torque T_M, the link will rotate; the angular position of the link is designated by θ_L. It may be noted that the moment of inertia will not necessarily be constant because during the course of an operation, the robot may pick up different objects with its gripper. The variation in the payload will change the effective moment of inertia of the link. In robotic applications, J_L is usually constant during each task; however it may vary from task to task.

Consider a conceptual design of a position control system, shown in the schematic diagram of Fig. 3.19a. The operation of the system is as follows. The load with an inertia J_L and viscous friction with coefficient B_L is to be positioned at some desired angle θ_R (T_w represents the load torque disturbance). The feedback potentiometer, whose wiper arm is mechanically coupled to the load shaft, produces a voltage e_o which represents the actual load position θ_L.

The feedback voltage

$$e_o = K_P \theta_L \tag{3.31}$$

where K_P = sensitivity of the potentiometer in volts/deg.

The reference voltage e_r is calculated from the command input, θ_R (setpoint), using the relation $e_r = K_P \theta_R$, and this value is then fed to the system through a potentiometer setting.

The error voltage $(e_r - e_o)$ is amplified by an op amp circuit that supplies a voltage u to a power amplifier. The output e_a of the power amplifier drives the armature-controlled dc motor (reversal in the polarity of e_r reverses the direction of rotation of the motor). The torque developed by the motor is transmitted to the load shaft through a gear train. The transmitted torque rotates the load shaft in the direction to reduce the error $(e_r - e_o)$ as a consequence of the negative feedback. When $e_r = e_o$, the motor stops, as its drive is cut off. This must necessarily occur when $\theta_R = \theta_L$ because feedback voltage is $e_o = K_P\theta_L$, and the reference voltage (by calculation) is $e_r = K_P \theta_R$.

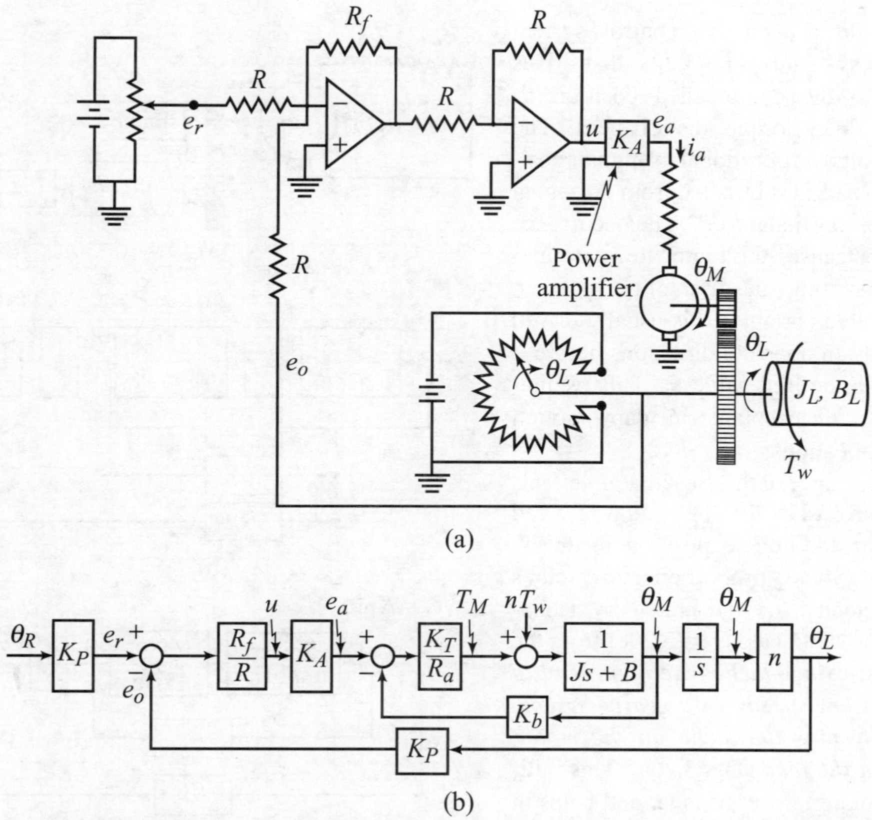

Fig. 3.19 *Closed-loop position control system (a) Schematic diagram (b) Block diagram*

A change in the load position caused by disturbances in the system (internal or external) causes the error signal $(e_r - e_o)$ to reappear, which after amplification acts on the motor; the motor torque transmitted to the load shaft forces the load to return to its original position.

A block diagram of the system is shown in Fig. 3.19b (refer to Eqns (3.10)–(3.13), Eqn. (3.30), and Fig. 3.8). Here K_b is the back emf constant of the motor, R_a is the armature resistance (armature inductance is assumed negligible), K_T is the torque constant of the motor, n is the gear ratio (n = load-shaft speed θ_L/motor-shaft speed θ_M), J and B are, respectively, equivalent moment of inertia and viscous-friction coefficient on motor shaft, and nT_w is the equivalent disturbance torque on motor shaft.

It will be seen in the next chapter that the performance of position control system of Fig. 3.19 can be improved by introducing minor-loop feedback (refer to Figs 1.12 and 1.13). A signal proportional to motor/load angular velocity, generated by a tachogenerator is fedback.

Example 3.1 Consider the system shown in Fig. 3.20a, with J_M = moment of inertia of motor shaft, B_M = coefficient of viscous friction on motor shaft, K_T = motor torque constant, and gear ratio $\dot{\theta}_L / \dot{\theta}_M = n$. The dynamics of the system is represented by the block diagram of Fig. 3.20b: $J_{eq} = J_M + n^2 J_L; B_{eq} = B_M + n^2 B_L$.

Let us build a feedback control system around the system of Fig. 3.20a, that drives the load to the commanded position θ_R inspite of load torque disturbances. The signal b_1(volts), proportional to angular position of the load, is subtracted from the signal e_r (volts) proportional to θ_R. The occurrence of the error causes the motor to rotate in a direction favoring the dissolution of error. The controlled variable θ_L is usually driven alternatively in opposite directions before it settles to the desired value θ_R. This oscillatory state is unacceptable in many motion control applications.

The occurrence of these oscillatory effects can be controlled by the application of speed feedback in addition to position feedback. A signal b_2(volts), proportional to motor's angular velocity $\dot{\theta}_M$, generated by tachogenerator, is subtracted (Fig. 3.20c) from the error signal (***Since tachogenerators tend to be noisy at low speeds, in a geared system, one usually puts the tacho on the motor, rather than the load*** (Fig. 3.20a). This initiates the braking process sooner and helps in damping out the oscillations faster. The 'loop within a loop' (velocity feedback system embedded within a position feedback system) configuration is a *minor-loop feedback control configuration* (refer to Fig. 3.2).

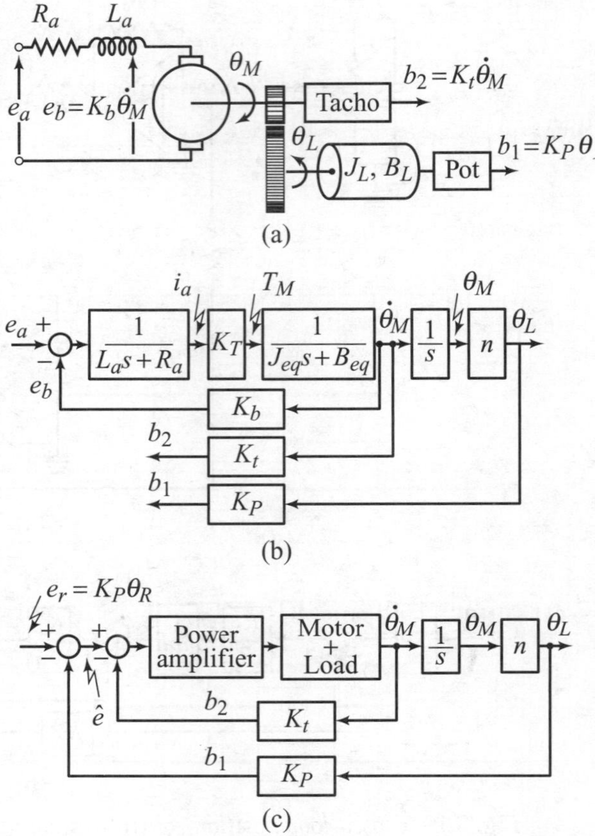

Fig. 3.20 *An armature-controlled dc motor with load*

The minor-loop feedback control configuration is an important design structure. Conceptual and analytical detailing of the structure will follow in later chapters when we take up the design issues. The objective here was simply to give information on the devices used in industrial position control systems.

3·4 AC SERVO SYSTEMS

So far in this chapter, only dc drives have been considered. AC drives for speed and position control systems provide a design alternative.

The ac servomotors possess many virtues in comparison to dc servomotors. These include significantly lower costs, weight and inertia, higher efficiency, and fewer maintenance requirements because of no commutator and brushes. However, as we shall see later, the characteristics of ac motors are quite nonlinear and these motors are more difficult to control, especially for positioning applications. The ac motors are best suited for low-power applications such as in instrument servos (e.g., control of pen in X-Y recorders) and computer-related equipment (e.g., disk drives, tape drives, printers, etc).

The dc motors are expensive because of brushes and commutators. These motors have relatively lower torque-to-volume, and torque-to-inertia ratios. However, the characteristics of dc motors are quite linear and these motors are easier to control. The dc motors have been generally used for large-power applications such as in machine tools and robotics.

Developments in technology are opening new applications for both ac and dc motors. Today, with the development of the rare-earth magnet, dc motors with very high torque-to-volume ratio at reasonable costs have become possible. Furthermore, the advances made in brush-and-commutator technology have made these wearable parts practically maintenance free. The advancements made in power electronics have made brushless dc motors quite popular in high-performance control systems [51]. Advanced manufacturing techniques have also produced dc motors with rotors of very low inertia, thus achieving very high torque-to-inertia ratios. These properties have made it possible to use dc motors in many low-power control applications that formerly used ac motors. On the other hand, three-phase induction motors with pulse-width modulated power amplifiers are currently gaining popularity in high-power control applications.

In the following section, we describe the basic characteristics of an ac servomotor, and the related components commonly used in position control systems.

3.4.1 AC Servomotor

An ac servomotor is basically a two-phase induction motor except for certain special design features. Consider a typical set of curves of speed versus torque shown in Fig. 3.21. For an induction motor used in power applications, the rotor resistance is low in order to obtain high maximum torque (Curve *A*, Fig. 3.21). However, this type of characteristic is not suitable for feedback control systems, mainly because of the positive slope on part of the curve. We will shortly see that a negative slope on the entire torque-speed curve is essential for stability throughout the operation (The dc motors inherently exhibit the desired torque-speed characteristic, Fig. 3.9). In order to obtain the desired type of characteristic, induction motors for servo applications are designed to have high rotor resistance (Curve *B*, Fig. 3.21). It should be noted that the required characteristic is purchased at a price of power which is dissipated as heat in the high rotor resistance. This decreases the efficiency of the motor and also limits its maximum power rating since the temperature rise in the motor becomes excessive for motors with high power rating.

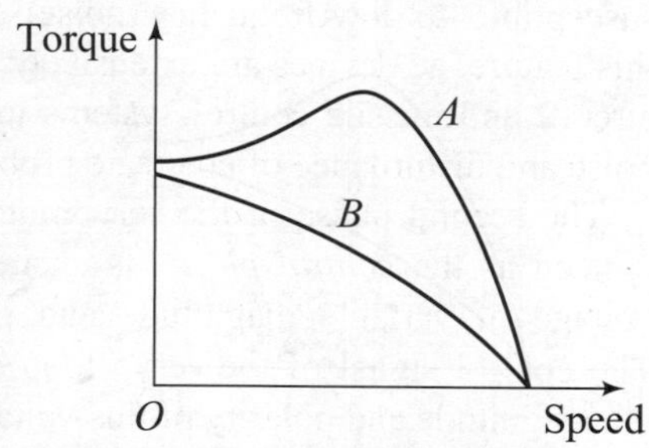

Fig. 3.21 *Torque-speed characteristics of ac motors*

In servo applications, an induction motor is required to produce rapid accelerations from standstill. Therefore, low inertia is a physical requirement of a servomotor. Thus, except for minor differences in constructional features (rotor has high resistance and low inertia), an ac servomotor is essentially a two-phase induction motor.

The stationary part (stator) of an ac motor is slotted to accommodate two field windings with their axes 90° apart. A two-phase excitation voltage is supplied to the stator windings. The stator core remains stationary, but it produces a magnetic field which rotates in space, as if the entire stator were turning. The ability of the magnetic fields produced by individual field windings, to add together or cancel makes it possible to produce a resultant field that is smoothly electrically rotating. Since the frequency of the alternating current is constant, the field rotates at a uniform rate of speed.

The rotor, when placed in a rotating magnetic field, has a current *induced*. Since the rotor current is induced by the rotating stator field, the rotor produces the same number of poles as the stator. The repulsion force between the two magnetic field vectors produces the torque.

In the dc motor, current for armature (rotor) conductors is supplied conductively by connection to a voltage source through brush and commutator. In the ac motor, the current for rotor conductors is supplied inductively by the magnetic field. Since no external connections are made to the rotor circuit, the motor can be made very robust and requires little maintenance.

Elementary analysis of the ac motor is more complex than that of the dc motor. So, rather than give the details of production of magnetic field and derivation of motor-torque equation using these details, we use the experimental data of the motor in the form of a set of torque-speed curves to obtain a model.

The symbolic representation of an ac servomotor as a control system component is shown in Fig. 3.22. One phase, designated as the *reference phase*, is excited by a constant voltage source, the frequency of which is usually 50/60, 400, or 1000 Hz. By using frequencies of 400 Hz or higher, the system can be made less susceptible to low-frequency noise. Due to this feature, ac devices are extensively used in aircraft and missile control systems in which noise and disturbance often create problems.

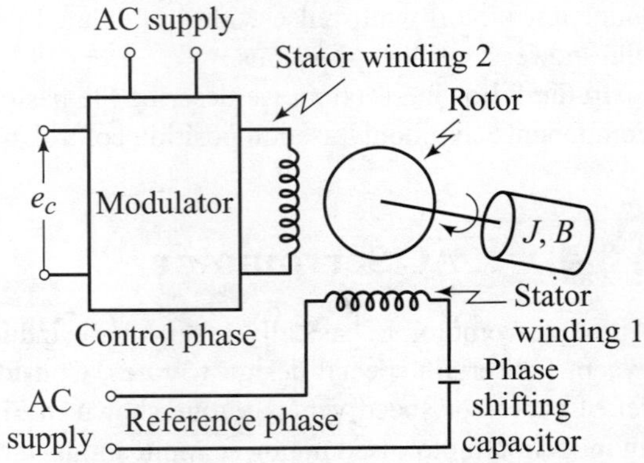

Fig. 3.22 *Symbolic representation of an ac servomotor*

The second phase of the servomotor, designated as the *control phase*, is excited by a voltage of variable magnitude and polarity. The control signal of the servo loop dictates the magnitude and polarity of this voltage.

The control signals in a servo loop are usually of low frequency, in the range of 0 to 20 Hz. For production of rotating magnetic field, the control-phase voltage must be of the same frequency as the reference-phase voltage, and in addition, the two voltages must be in time quadrature. The arrangement shown in Fig. 3.22 meets these requirements. The input signal $e_c(t)$ is applied to the control phase after modulation; the carrier frequency of the modulation process is equal to the frequency of the reference-phase supply. The 90° phase difference between the control-phase and reference-phase voltages is obtained by the insertion of a capacitor (***Note that a single-phase power supply is required to run the motor***).

Figure 3.23 shows the waveforms of a typical input signal $e_c(t)$, the carrier signal $\cos \omega_c t$ and the *suppressed-carrier modulated signal*

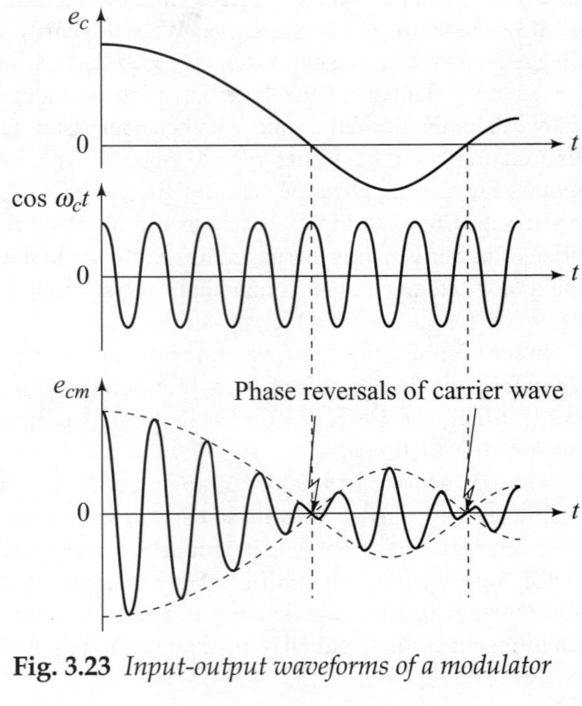

Fig. 3.23 *Input-output waveforms of a modulator*

$e_{cm}(t) = e_c(t) \cos \omega_c t$. Note that the envelope of the modulated signal e_{cm} is identical to the low-frequency signal e_c. The polarity of e_c dictates the phase of e_{cm} with respect to that of $\cos \omega_c t$. If e_c is positive, then e_{cm} and $\cos \omega_c t$ have the same phase; otherwise they have 180° phase difference:

$$\begin{aligned} e_{cm}(t) &= |e_c(t)| \cos \omega_c t && \text{for } e_c(t) > 0 \\ &= |e_c(t)| \cos(\omega_c t + \pi) && \text{for } e_c(t) < 0 \end{aligned} \tag{3.32}$$

This means that a reversal in phase of e_{cm} occurs whenever the signal e_c crosses the zero-magnitude axis. This reversal in phase causes a reversal in the direction of rotation of the magnetic field and hence a reversal in the direction of rotation of the motor shaft.

The modulator in Fig. 3.22 transforms the signal e_c into the signal e_{cm}. If we consider e_c as the input and e_{cm} in Eqn. (3.32) as the output, then the mathematical description of the modulator will be rather complicated. Recall that the reason for the introduction of e_{cm} is to match the operational frequency ranges; therefore, the information is not stored in e_{cm}, but rather in its envelope. Hence, we should consider e_c as the input and the envelope of e_{cm} as the output of the modulator. Therefore, the transfer function of the modulator is just equal to 1, for the envelope of e_{cm} is identical to e_c.

The torque-speed curves of a typical ac servomotor plotted for fixed reference-phase voltage $E_r \cos \omega_c t$ and different values of constant input voltages $e_c \le E_r$ are shown in Fig. 3.24. All these curves have negative slope. (Note that the curve for $e_c = 0$ goes through the origin; this means that when the control-phase voltage becomes zero, the motor develops a decelerating torque, causing it to stop). The curves show a large torque at zero speed. This is a requirement for a servomotor in order to provide rapid acceleration.

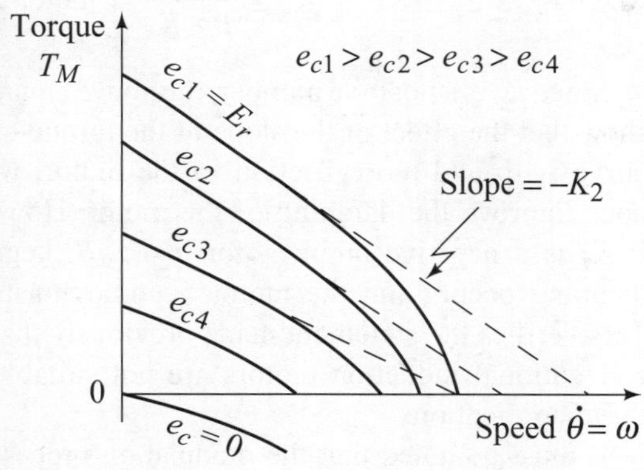

Fig. 3.24 *Torque-speed curves of an ac servomotor*

It is seen that the torque-speed curves of an ac servomotor are nonlinear except in the low-speed region. In order to derive a transfer function for the motor, some linearizing approximations are necessary. A servomotor seldom operates at high speeds; therefore, the linear portions of the torque-speed curves can be extended out to the high speed region, as shown in Fig. 3.24 by use of dashed lines. But even with this approximation, the resultant curves are still not parallel to each other. This means that for constant speeds, except near-zero speed, the torque does not vary linearly with respect to input voltage e_c. The curves in Fig. 3.25 illustrate this effect.

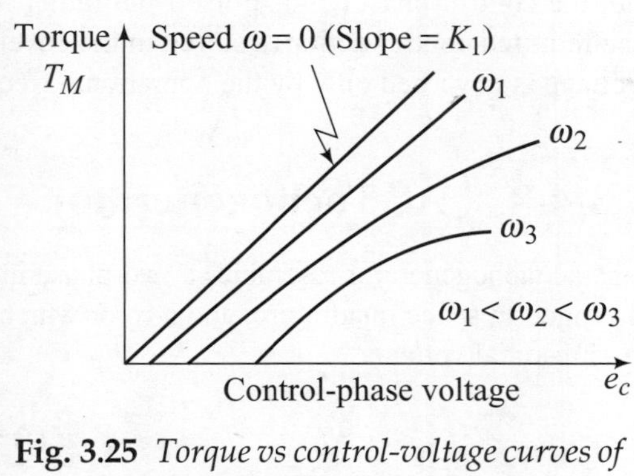

Fig. 3.25 *Torque vs control-voltage curves of an ac servomotor*

For speeds near zero, all the curves are straight lines parallel to the characteristic at rated input voltage ($e_c = E_r$), and are equally spaced for equal increments of the input voltage. Under this assumption, the torque T_M generated by the motor is represented by the equation

$$T_M = \left(\frac{\partial T_M}{\partial \dot{\theta}} \bigg|_{\substack{\dot{\theta}=0 \\ e_c=0}} \right) \dot{\theta} + \left(\frac{\partial T_M}{\partial e_c} \bigg|_{\substack{e_c=0 \\ \dot{\theta}=0}} \right) e_c = -K_2\dot{\theta} + K_1 e_c \tag{3.33}$$

where $K_2 > 0$ and $K_1 > 0$ are coefficients defining the slopes in Figs 3.24 and 3.25, respectively.

If we consider that the moment of inertia of the motor rotor with attached mechanical load is J, the viscous-friction coefficient of the motor rotor with attached mechanical load is B, and the disturbance-load torque is T_w, then we have

$$T_M = K_1 e_c - K_2\dot{\theta} = J\ddot{\theta} + B\dot{\theta} + T_w \tag{3.34}$$

Figure 3.26 shows a block diagram representation of the ac motor system. The transfer function between the shaft position $\theta(s)$ and the input voltage $E_c(s)$, obtained from the block diagram is

$$\frac{\theta(s)}{E_c(s)} = \frac{K_1}{s(Js + B + K_2)} = \frac{K_m}{s(\tau_m s + 1)} \tag{3.35}$$

where $K_m = \dfrac{K_1}{B + K_2}$ = motor gain constant and $\tau_m = \dfrac{J}{B + K_2}$ motor time constant

Since K_2 is a positive number, the above equations show that the effect of the slope of the torque-speed curve is to add more friction to the motor, which does improve the damping of the motor. However, if K_2 is a negative number, for $|K_2| > B$, negative damping occurs and the motor becomes unstable. This verifies the statement made previously that the conventional induction motors are not suitable for servo applications.

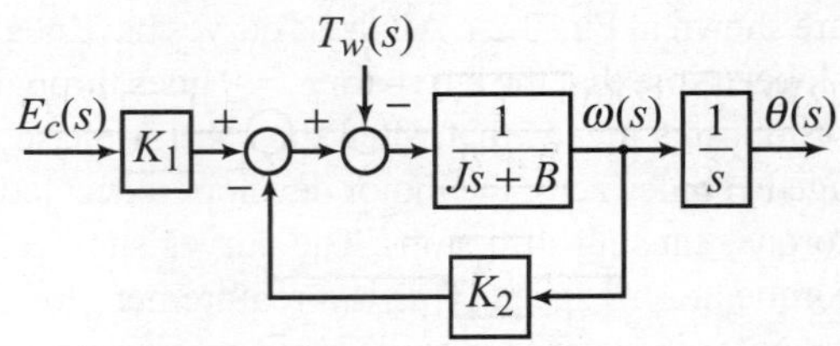

Fig. 3.26 *Block diagram of an ac motor system*

It must be noted that the modulation process for running the ac servomotor is not necessary if the control signal in the servo loop is an ac signal of frequency equal to that of reference phase supply. The transfer function model given by Eqn. (3.35) is applicable in these situations also, with the only difference that the input variable for the model is not the control-phase voltage itself but rather its envelope. This is in order as the information about the required control action is stored in the envelope of the control-phase voltage; the frequency of the voltage is governed only by the operational frequency range of the servomotor.

3.4.2 AC Tachogenerator

The ac tachogenerator resembles a two-phase induction motor in that it comprises two stator windings arranged in space quadrature, and a rotor which is not conductively connected to any external circuit. A sinusoidal voltage

$$e_r(t) = E_r \sin \omega_c t$$

is applied to the *excitation* (*reference*) *winding* (refer to Fig. 3.27). When the rotor is stationary ($\dot{\theta} = 0$), no emf is induced in the *output winding* and therefore, the output voltage will be zero. When the motor turns, a voltage at the reference frequency ω_c is induced in the output winding. The magnitude of the output voltage is proportional to the rotational speed. A change in the direction of shaft rotation causes a 180° phase shift in the output voltage. When the output voltage is in phase with the reference, the direction of rotation is said to be *positive*, and when the output voltage is 180° out of phase, the direction is said to be *negative.*

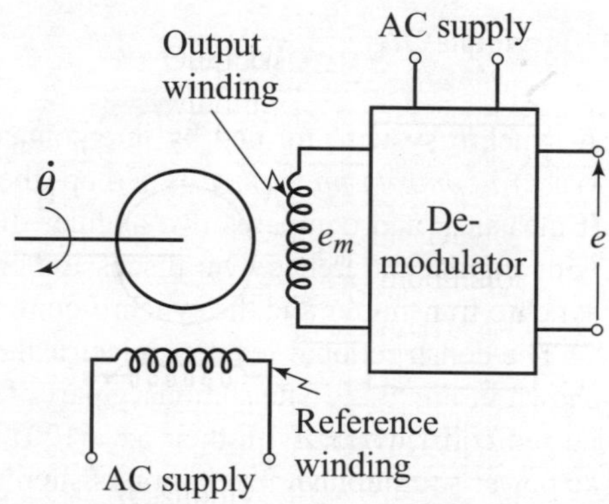

Fig. 3.27 *Schematic diagram of an ac tachogenerator*

The output voltage of an ac tachogenerator is, thus, in a modulated form. A typical waveform of the output voltage $e_m(t)$ is shown in Fig. 3.28. It can be mathematically represented as

$$e_m(t) = e(t)\sin\omega_c t$$

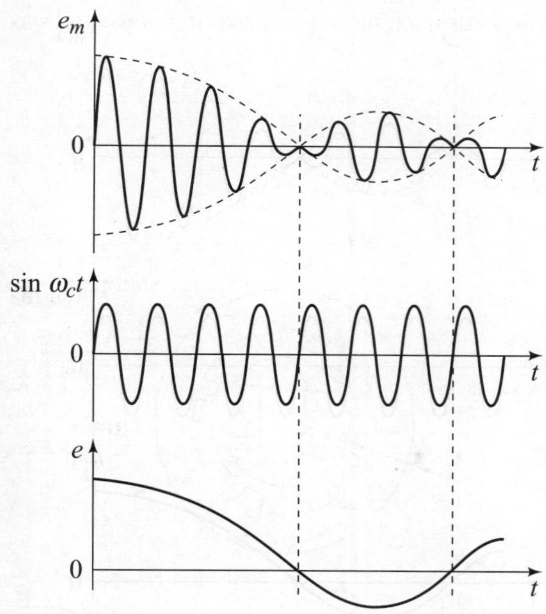

Fig. 3.28 *Waveforms illustrating the operation of a demodulator*

In the measurement system shown in Fig. 3.27, we have used a *demodulator* for signal conditioning. The input to the demodulator is $e_m(t)$ and its output is $e(t)$. Since the information about shaft speed is stored not in $e_m(t)$ itself but rather in its envelope, we should consider the envelope of $e_m(t)$ as the input of the demodulator. Therefore, the transfer function of the demodulator is just equal to 1, for the envelope of $e_m(t)$ is identical to the output $e(t)$. The output voltage of an ac tachogenerator (with signal conditioning) is therefore of the same form as that for a dc tachogenerator (refer to Eqn. (3.22)), i.e.,

$$e(t) = K_t \frac{d\theta(t)}{dt} \tag{3.36a}$$

where K_t is the sensitivity of the tachogenerator (volts per rad/sec).

The transfer function from $\theta(t)$ to $e(t)$ of an ac tachogenerator is of the form

$$\frac{E(s)}{\theta(s)} = sK_t \tag{3.36b}$$

3.4.3 Synchros

A synchro system, formed by interconnection of the devices called the *synchro transmitter* and the *synchro-control transformer*, is perhaps the most widely used error detector in feedback control systems. It measures and compares two angular displacements, and its output voltage is approximately linear with angular difference. Our discussion on synchro devices will be limited to the basic features of the synchro transmitter and the synchro-control transformer.

The constructional features, electrical circuit, and a schematic symbol of a synchro transmitter are shown in Fig. 3.29. The stationary part of the machine (stator) is slotted to accommodate three Y connected coils, wound with their axes 120° apart. The stator windings are not directly connected to the ac power source. Their excitation is supplied by the ac magnetic field produced by the rotating part of the machine (rotor).

The rotor is of a dumb-bell construction with a single winding. A single-phase excitation voltage is applied to the rotor through two slip rings. The resultant current produces a magnetic field and by the transformer action, induces voltages in the stator coils. The effective voltage induced in any stator coil depends upon the angular position of the coil's axis with respect to the rotor axis.

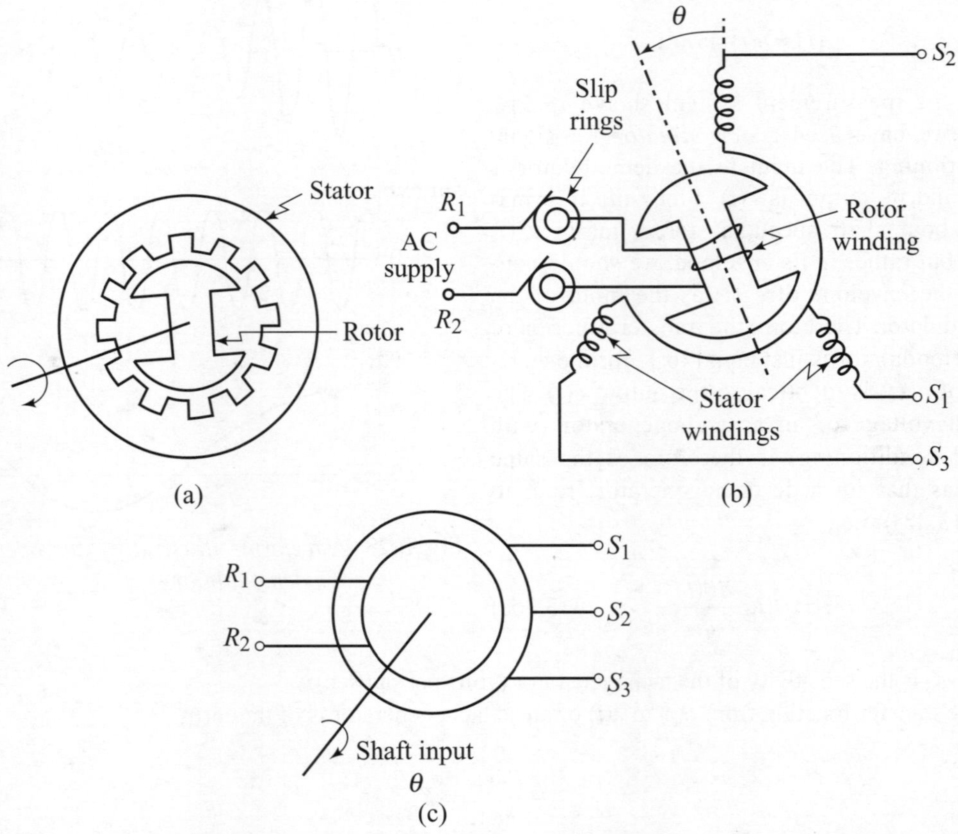

Fig. 3.29 *(a) Constructional features (b) Electrical circuit, and (c) Schematic symbol of a synchro transmitter*

Let the ac voltage applied to the rotor be

$$e_r(t) = E_r \sin \omega_c t \tag{3.37}$$

When the rotor is in the position corresponding to $\theta = 0$ in Fig. 3.29b, the voltage induced across the stator winding S_2 and the neutral is maximum and is written as

$$e_{S_2n} = KE_r \sin \omega_c t$$

where K is a proportional constant.

When the rotor is in position corresponding to $\theta = 90°$ in Fig. 3.29b, the voltage e_{S_2n} is zero. Therefore when $\theta = 0$, the coupling coefficient between the rotor winding and the stator winding S_2 is 1, and when $\theta = 90°$, the coupling coefficient equals 0. In fact, in general, the coupling coefficient between S_2 winding and rotor winding is equal to $\cos \theta$ for any position of the rotor:

$$e_{S_2n} = KE_r \cos\theta \sin \omega_c t$$

When the rotor is at 0° position with respect to the S_2 winding, it is at –120° position with respect to the S_3 winding; that is, positive (counter clockwise) rotation of 120° is required to move the rotor from the first position to the second position. Thus, if S_2 winding is taken as a reference,

$$e_{S_3n} = KE_r \cos(\theta - 120) \sin \omega_c t$$

In the same way,

$$e_{S_1n} = KE_r \cos(\theta - 240) \sin \omega_c t = KE_r \cos(\theta + 120) \sin \omega_c t$$

The three terminal voltages of the stator are (refer to trigonometric identities (2.25))

$$e_{S_1S_2} = e_{S_1n} + e_{nS_2} = e_{S_1n} - e_{S_2n} = \sqrt{3}\, KE_r \sin(\theta + 240) \sin \omega_c t \tag{3.38a}$$

$$e_{S_2S_3} = e_{S_2n} + e_{nS_3} = \sqrt{3}\, KE_r \sin(\theta + 120) \sin \omega_c t \tag{3.38b}$$

$$e_{S_3S_1} = e_{S_3n} + e_{nS_1} = \sqrt{3}\, KE_r \sin\theta \sin \omega_c t \tag{3.38c}$$

When $\theta = 0°$, maximum voltage is induced in winding S_2, and it follows from Eqn. (3.38c) that the terminal voltage $e_{S_3S_1}$ is zero. This position of the rotor is defined as the *electrical zero* of the transmitter and is used as reference for specifying the angular position of the rotor.

It should be noted that the synchro is not a three-phase machine. In a three-phase machine, there are three voltages equal in magnitude, displaced from each other by 120 electrical degrees. With the synchro, which is a single-phase device, the three stator voltages vary in magnitude, and one stator coil is in phase or 180° out of phase with another coil.

It is seen that the input to the synchro transmitter is the angular position of its rotor shaft and the output is a set of three stator coil-to-coil voltages (common connection between the stator coils is not accessible) given by Eqns (3.38). By measuring and identifying the set of voltages at the stator terminals, it is possible to identify the angular position of the rotor.[5]

[5] A synchro/digital (S/D) converter accepts the three ac voltages from the synchro as input and gives out a digital output which is a measure of the magnitude and direction of rotor movement (refer to [72] for details). There also exists a digital/synchro (D/S) transmitter to convert digital data to synchro-format voltages.

Our interest here is in synchro-error detector formed by interconnection of two synchro devices—the transmitter and the control transformer. A typical arrangement of a synchro-error detector in servo applications is to connect the stator leads of the transmitter to the stator leads of the control transformer. A servo loop using the synchro-error detector is shown in Fig. 3.30. Basically, the principle of operation of a synchro-control transformer is identical to that of the synchro transmitter, except that the rotor is cylindrically shaped so that the air-gap flux is uniformly distributed around the rotor. This feature of the control transformer minimizes the change in the rotor impedance with the rotation of the shaft; the signal conditioning circuit, therefore, sees a constant impedance.

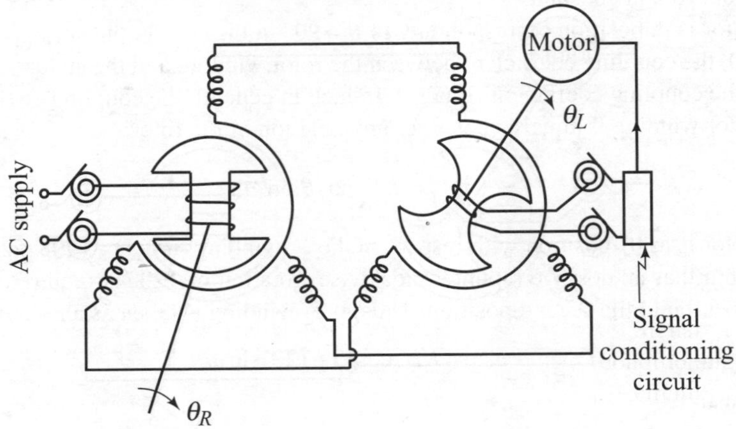

Fig. 3.30 *Servo loop using synchro-error detector*

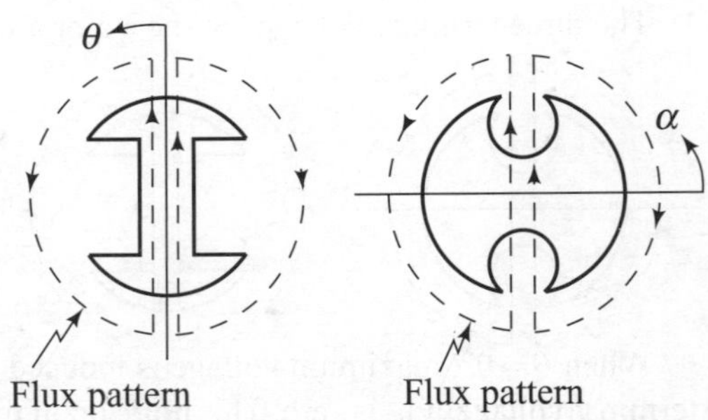

Fig. 3.31 *Rotor positions and flux patterns*

Referring to the arrangement shown in Fig. 3.30, the voltages given in Eqns (3.38a)–(3.38c) are impressed across the corresponding stator terminals of the control transformer. Due to the similarity in the magnetic construction, the flux patterns produced in the two synchros will be the same if all losses are neglected. For example, if the rotor of the transmitter is in its electrical zero position, the fluxes produced in the transmitter and in the control transformer are as shown in Fig. 3.31. When the rotor of the control transformer is in the position shown in the figure, the induced voltage at its rotor winding terminals is zero; the shafts of the two synchros are considered to be in alignment, and the control transformer is said to be in *null position*. Thus, the null position of a control transformer in a servo loop is that position of its rotor for which the output voltage on the rotor winding is zero with the transmitter in its electrical zero position.

Let the rotor of the transmitter rotate through an angle θ from its electrical zero position, and let the rotor of the control transformer rotate in the same direction through an angle α from its null position (refer to Fig. 3.31); the net angular separation of the two rotors is then $(90 - \theta + \alpha)$ and the voltage induced in the control transformer is proportional to the cosine of this angle:

$$e_m(t) = K'E_r \cos(90 - \theta + \alpha) \sin \omega_c t$$
$$= K'E_r \sin(\theta - \alpha) \sin \omega_c t$$

where K' is a proportional constant.

For small values of $\phi(t) = \theta(t) - \alpha(t)$, $e_m(t) = K'E_r\phi(t)\sin\omega_c t$. The output voltage of the synchro-error detector is thus a modulated signal with the carrier frequency ω_c equal to the frequency of the ac supply to the rotor winding of the synchro transmitter. The magnitude of the modulated carrier wave is proportional to $\phi(t) = \theta(t) - \alpha(t)$, and the phase reversals of the modulated wave depend on the sign of $\phi(t)$ (refer to Fig. 3.28).

Signal conditioning of the modulated output of the synchro-error detector by an amplifier and a demodulator gives an output voltage

$$e(t) = K_s\phi(t) \tag{3.39}$$

where K_s = sensitivity of the synchro-error detector in volts/deg.

The transfer function of the synchro-error detector is, therefore, of the form

$$\frac{E(s)}{\phi(s)} = K_s \tag{3.40}$$

It must be noted that the signal-conditioning circuit will not include the demodulation process if the synchro-output signal is connected to an ac device (e.g., ac servomotor). The transfer-function model given by Eqn. (3.40) is applicable in these situations also, with the only difference that the output variable of the model is not the synchro-output voltage itself but rather its envelope. This is quite in order as the information about the angular separation of the two shafts is stored in the envelope of the synchro-output voltage; the frequency of the voltage is governed only by the operational frequency range of the synchro device.

Compared with potentiometers, the synchros offer higher sensitivity, longer life, ruggedness, and continuous rotation capability.

3.4.4 A Position Control System

A heavy telephoto camera can be controlled by the system shown in Fig. 3.32a. The camera is driven by an ac motor through a gear train and is designed to follow the movement of the spotting scope. This system employs ac components and all the signals other than the input and output shaft positions are suppressed carrier modulated signals. Such systems are known as *carrier control systems* and are designed so that the signal cut-off frequency is much less than the carrier frequency. It is then sufficiently accurate to analyze these systems on the basis of modulating signals only.

In the system of Fig. 3.32a, a pair of synchros with sensitivity K_s is used as an error detector to generate the error signal $e = K_s(\theta_R - \theta_L)$. An ac amplifier of gain K_A is used to amplify this signal. The output e_c of the amplifier drives the ac servomotor. The torque developed by the motor is transmitted to the load shaft through a gear train with gear ratio n ($n = N_1/N_2$ = load shaft speed $\dot{\theta}_L$/motor shaft speed $\dot{\theta}_M$).

The two-phase motor develops a torque in accordance with the equation (refer to Eqn. (3.33))

$$T_M = K_1 e_c - K_2\dot{\theta}_M$$

where the constants K_1 and K_2 are given by torque-speed characteristics of the motor.

This torque drives the load (camera) of moment of inertia J_L and viscous friction with coefficient B_L, against the disturbance load torque T_w. Assume that the moment of inertia and viscous friction of the motor are represented by J_M and B_M, respectively. The equivalent moment of inertia of the motor-load combination reflected on the motor shaft is given by (refer to Eqns (3.30))

$$J = J_M + n^2 J_l$$

and the equivalent viscous-friction coefficient

$$B = B_M + n^2 B_L$$

nT_w is the equivalent disturbance torque on the motor shaft.

In terms of these parameters, the torque equation of the system becomes

$$T_M = J\ddot{\theta}_M + B\dot{\theta}_M + nT_w$$

A block diagram of the system is shown in Fig. 3.32b.

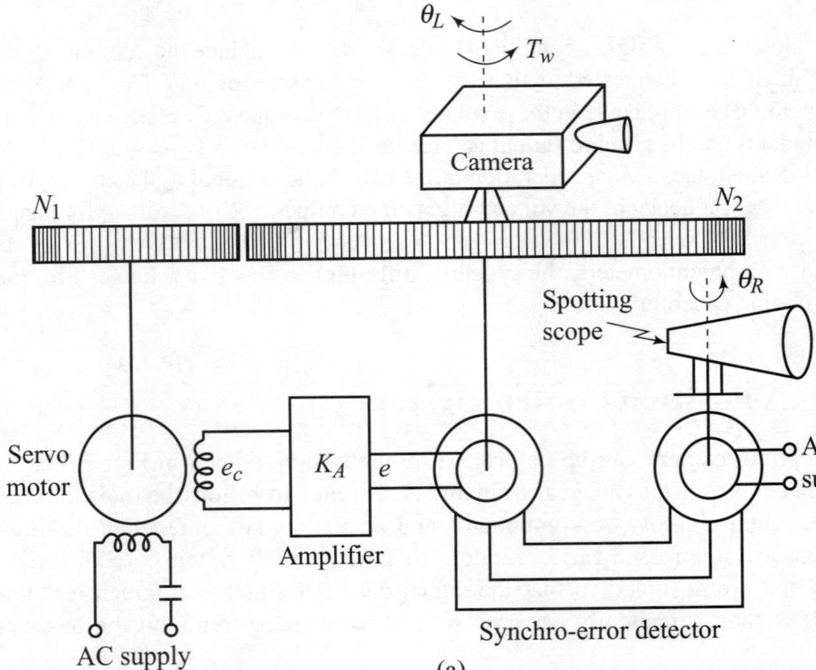

(a)

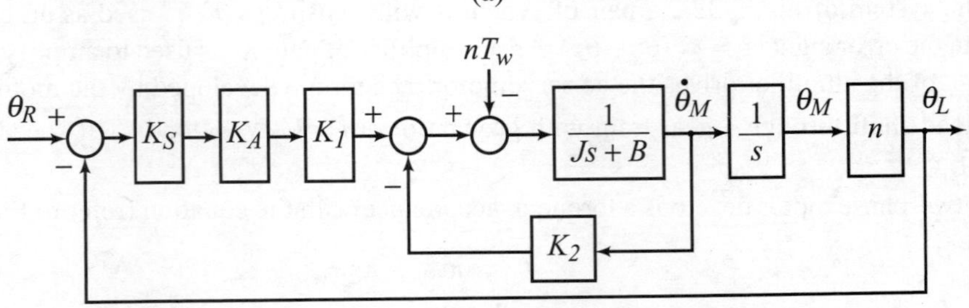

(b)

Fig. 3.32 *An ac servo system (a) Schematic diagram (b) Block diagram*

The output of the synchro-error detector in Fig. 3.32, is a carrier modulated signal. The information about the misalignment of the two shaft positions is carried by the envelope of this signal. As we shall see in Chapter 7, generation of the control signal requires processing on the information signal, i.e., the envelope of the synchro-output signal. For example, we may require a derivative and/or an integration of the information signals in addition to amplification. A demodulator provides the necessary interface between the synchro-error detector and the op amp circuit for generating the control signal. The low-frequency control signal can drive a dc motor after power amplification. Alternatively, the control signal may be modulated to drive an ac motor.

3·5 HYDRAULIC DEVICES FOR MOTION CONTROL

Hydraulically operated devices are frequently used in control systems. In these devices, power is transmitted through fluid flow under pressure. The fluid used is relatively incompressible such as petroleum-based oils or certain non-inflammable synthetic oils.

When both high speed and high power are required, hydraulic systems may be mandatory or desirable. Common hydraulic control applications are power steering and brakes in automobiles, steering mechanism of ships, aircraft control surface actuators, control of large machine tools, servo systems for cranes and hoists, robots, and rolling mills, etc.

Developments in analog and digital electronics have resulted in electronic controllers (Chapters 7 and 13) being widely accepted by the industry—currently used in almost all industrial control applications. Resulting from these developments, many hybrid systems, combining electronic and hydraulic techniques, are in use. Our selection of devices in this section is based on the current industrial practice. Classical hydraulic control schemes with mechanical feedback have not been included.

The hydraulic output devices used in control systems are generally of two types—those intended to produce rotary motion and those whose output is translational. The first type is known as the hydraulic motors and the second ones known as the hydraulic linear actuators. Our focus here is on applications where in the controlled variable is a translational position.

As an example, consider the system for machining a work piece on a lathe. The tool movement of such a system is shown in Fig. 3.33a. The carriage is moved at a constant speed and the control of the

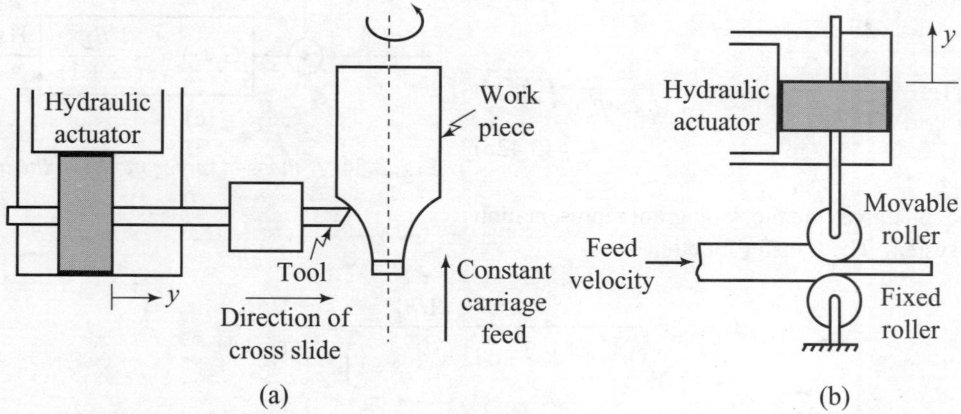

Fig. 3.33 *(a) Machining control system: (b) System for controlling the thickness of rolled steel*

tool position is in the direction of the cross slide. The load on the hydraulic actuator consists of mass M, and viscous friction with coefficient B. The disturbance acting on the actuator is the thrust force on the tool required to machine the work piece.

Another example of translational load is shown in Fig. 3.33b. The system controls the thickness of the rolled steel in a steel mill. The rolls are driven at a constant speed by drive motors. The translational motion of the movable roll is controlled by the hydraulic actuator. The load on the hydraulic actuator consists of mass M, and viscous friction with coefficient B. The disturbance acting on the actuator is the upward force on the roll which is quite significant when the slabs enter the rolls or leave the rolls.

3.5.1 Translational Load

The mass-damper system shown in Fig. 3.34a is a useful model of the load on the hydraulic actuator in industrial control systems; F_w represents the load disturbance. Let us obtain the transfer function model of this system (revisiting Section 2.6 will be helpful).

Figure 3.34b shows the free-body diagram. Applying Newton's law of motion, the force equation can be written as

$$F(t) - F_w(t) - B_L \frac{dy(t)}{dt} = M_L \frac{d^2y(t)}{dt^2} \quad \text{or} \quad M_L \frac{d^2y(t)}{dt^2} + B_L \frac{dy(t)}{dt} = F(t) - F_w(t) \tag{3.41}$$

This is a linear constant coefficient differential equation of second order.

Taking the Laplace transform of each term of this equation (assuming zero initial conditions), we obtain

$$M_L s^2 Y(s) + B_L s Y(s) = F(s) - F_w(s) \tag{3.42a}$$

Therefore,

$$Y(s) = \frac{1}{s(M_L s + B_L)} F(s) - \frac{1}{s(M_L s + B_L)} F_w(s) \tag{3.42b}$$

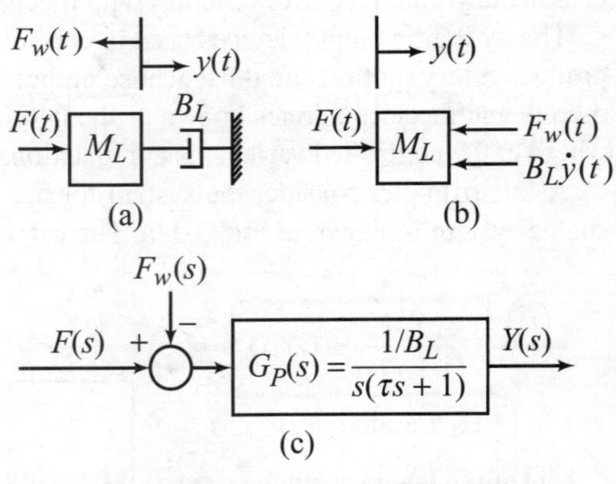

Fig. 3.34 *A mass–damper model of the load*

Figure 3.34c gives the block diagram representation of the system. The transfer function

$$G_P(s) = \frac{1}{s(M_L s + B_L)} = \frac{1/B_L}{s\left(\frac{M_L}{B_L}s + 1\right)} = \frac{1/B_L}{s(\tau s + 1)} \tag{3.43}$$

The metric system of units for mechanical translational systems is given in (2.55).

3.5.2 Hydraulic Actuator

The hydraulic linear actuator shown in Fig. 3.35 consists of a pilot valve and a power cylinder. The piston inside the power cylinder divides the cylinder into two chambers. The pilot valve—known as *spool valve* because of its shape—controls the flow rate of the hydraulic fluid to the power cylinder. It is a four-port valve: one port is connected to the hydraulic fluid supply at constant pressure, two control ports are connected to each chamber of the cylinder, and the drain port (the two drain ports are joined) is connected to the reservoir. The valve spool has two lands.

When the spool is in the neutral position ($x = 0$), the fluid flow in the power cylinder is blocked completely and the position of the piston, which moves the load, is $y = 0$. As the spool is moved to the right (x positive), the right-chamber port of the power cylinder is opened to the supply pressure P_s whereas the left-chamber port is opened to pressure P_r, which is slightly above the atmospheric pressure but we approximate it as zero gage pressure in our model. Right-chamber pressure P_1 thus rises (while left-chamber pressure P_2 falls), producing an accelerating force on the load that moves to the left (y positive). If the spool is moved to the left (x negative), the left-chamber port of the power cylinder is opened to the supply pressure P_s, whereas the right-chamber port is opened to P_r. The direction of the fluid flow in the cylinder and hence the direction of displacement y of the load is reversed. This hydromechanical system acts as a mechanical power amplifier since the power (force × velocity) applied to position the valve is much smaller than that developed at the load.

Note that when the valve spool is in its neutral position, the flow area of the control ports is not in general zero. This is because spool lands and ports cannot be manufactured with perfectly sharp edges. Thus, a practical valve will either be underlapped or overlapped (Fig. 3.36). Since an overlapped valve causes a dead zone for small spool motions (which is undesirable in control systems (refer to Chapter 16)), underlapped valves are often preferred.

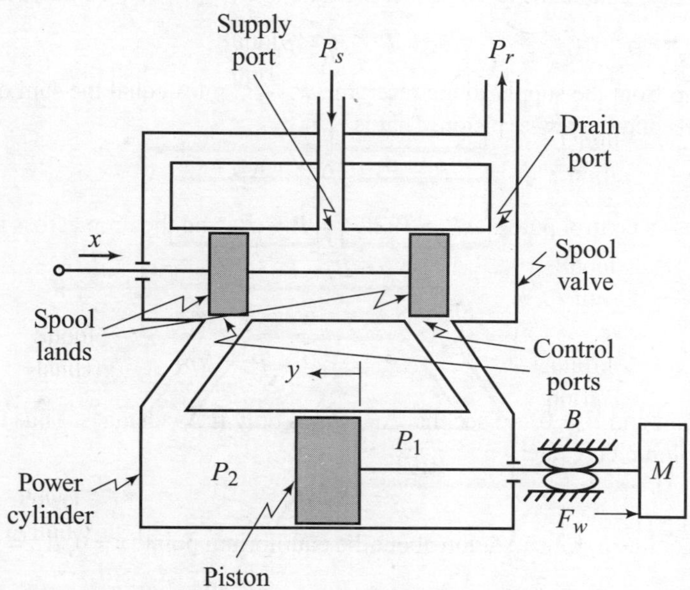

Fig. 3.35 *Hydraulic linear actuator with a load*

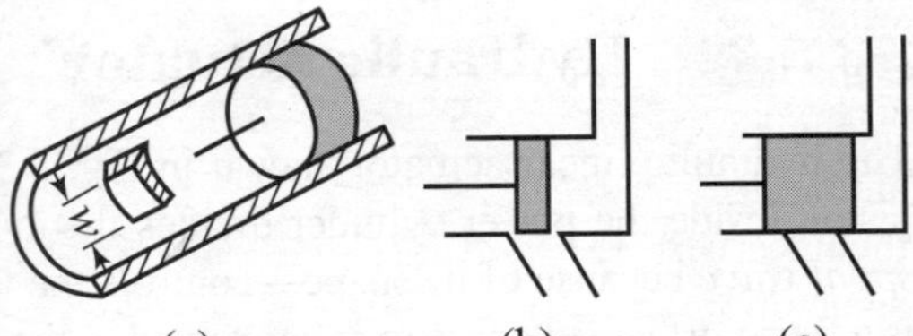

(a) (b) (c)

Fig. 3.36 *(a) Spool valve construction*
(b) Underlapped valve
(c) Overlapped valve

With an underlapped valve in the neutral position, the cylinder pressures P_1 and P_2 will approach half the supply pressure P_s. This can be explained as follows.

Fluid flow that meets with a resistance (obstruction) to flow is known as *orifice flow*. The flow through the valve ports in Fig. 3.35 meets with a controllable obstruction—flow area is adjustable. The *orifice flow relation* [9,10] can therefore be applied.

Let ΔP_v be the pressure drop across a valve port. Then

$$q = C_o A_o \sqrt{\frac{2(\Delta P_v)}{\rho}} \tag{3.44}$$

where

q = volume flow rate through the port; C_o = discharge coefficient;

A_o = uncovered area of the port; ρ = density of the fluid.

The area A_o is equal to $w \times x$, where w is the width of the control port (refer to Fig. 3.36a). Since C_o, ρ (compressibility assumed negligible), and w are constant, the flow q is a function of x and ΔP_v.

Let us now examine the neutral position of the spool valve. In this position, the fluid leakage into the right chamber of the cylinder is at the rate proportional to $(P_s - P_1)^{1/2}$, and the leakage out of the right chamber is at the rate proportional to $(P_1 - P_r)^{1/2}$. Equilibrium of leakage in and out of the right chamber of the cylinder will be achieved when $P_1 = P_s/2$. Similarly, equilibrium of leakage in and out of the left chamber will be achieved when $P_2 = P_s/2$. This situation will be taken as the equilibrium point.

The movement of the spool from its neutral position and the load-induced pressure on the power piston will disturb the system from its equilibrium position. Assume that the load changes the right-chamber pressure to $P_1 = (P_s/2) + p_1$, and the left-chamber pressure to $P_2 = (P_s/2) + p_2$; the load-induced pressure

$$\Delta p = P_1 - P_2 = p_1 - p_2$$

The pressure drop from the supply to the reservoir, $P_s - P_r$, must equal the sum of the drops across both the control ports and the power piston. That is

$$P_s - P_r = 2\Delta P_v + \Delta p$$

where the drop across a control port is $\Delta P_v = P_s - P_1 = P_2 - P_r$, and the drop across the piston is

$$\Delta p = P_1 - P_2$$

Therefore,

$$\Delta P_v = \frac{1}{2}[P_s - P_r - \Delta p]$$

Since P_s = constant and $P_r = 0$, we see that ΔP_v varies only if Δp changes. Thus the variables x and Δp determine the volume flow rate:

$$q = f(x, \Delta p) \tag{3.45}$$

This relation is nonlinear. Linearization about the equilibrium point ($x = 0$, $\Delta p = 0$, $q = 0$) gives

$$q = \left(\frac{\partial f}{\partial \Delta p} \bigg|_{\substack{\Delta p=0 \\ x=0}} \right) \Delta p + \left(\frac{\partial f}{\partial x} \bigg|_{\substack{x=0 \\ \Delta p=0}} \right) x = -K_2 \Delta p + K_1 x \tag{3.46}$$

where $K_2 > 0$ is the pressure coefficient and $K_1 > 0$ is the flow gain (refer to Eqns (2.1)).

Figure 3.37 shows this linearized relationship among q, x, and Δp. The straight lines shown are the linearized characteristic curves of the hydraulic actuator. Note that the region near the origin is most important because system operation usually occurs near this point. The flow gain K_1 and the pressure coefficient K_2 can be estimated theoretically from Eqn. (3.44), or more accurately from the experimental tests once the valve has been built.

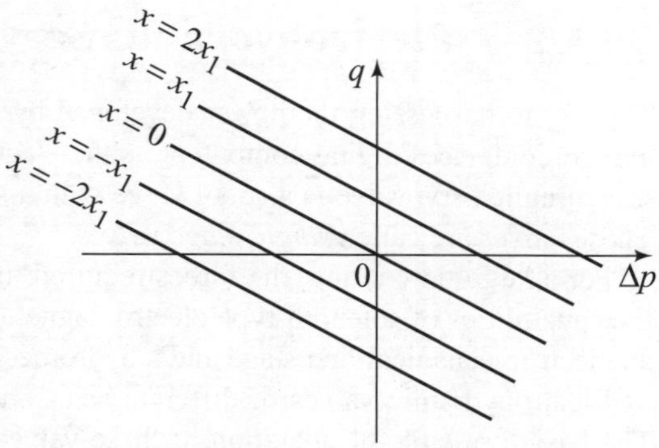

Fig. 3.37 *Linearized characteristic curves of a hydraulic actuator*

Referring to Fig. 3.35, we see that the rate of fluid flow into the power cylinder is proportional to the rate at which the piston moves, i.e.,

$$K_1 x - K_2 \Delta p = A\dot{y} \tag{3.47}$$

where A is the area of the piston.

Notice that because of high force levels and pressure differentials, there may be leakage from one chamber of the power cylinder to the other. However, translational actuators usually have piston rings that are essentially leak free. Equation (3.47) is true under the assumption of zero leakage around the piston.

The force developed by the power piston is equal to the pressure difference Δp times the piston area A. Assume that the power piston moves a load consisting of a mass M, and viscous friction with coefficient B. Then

$$A\Delta p = M\ddot{y} + B\dot{y} + F_w \tag{3.48}$$

where F_w represents the load disturbance.

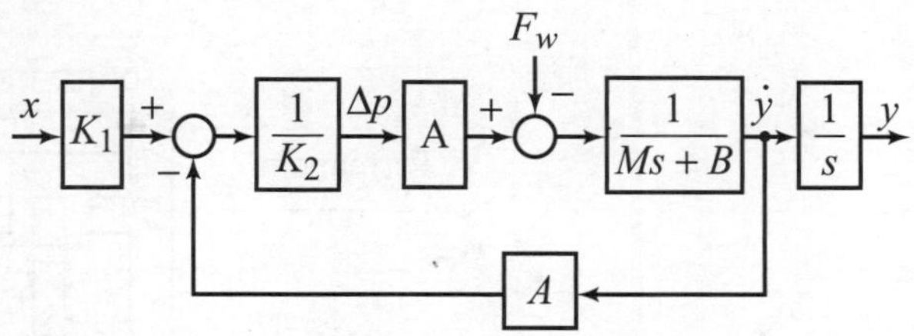

Fig. 3.38 *Block diagram of the hydraulic system of Fig. 3.35*

The hydraulic system is represented by a block diagram in Fig. 3.38, which has been drawn by rearranging Eqns ((3.47)–(3.48)):

$$\left(K_1 x - A\dot{y}\right) \times \frac{1}{K_2} = \Delta p; \; A\Delta p - F_w = M\ddot{y} + B\dot{y}$$

The transfer function between the load displacement y and the spool displacement x, obtained from the block diagram is

$$\frac{Y(s)}{X(s)} = \frac{K}{s(\tau s + 1)} \tag{3.49}$$

where $K = \dfrac{AK_1}{BK_2 + A^2}$ = gain constant of the actuator; and $\tau = \dfrac{MK_2}{BK_2 + A^2}$ = time constant of the actuator.

The controlled fluid power from a spool valve can be converted into rotating mechanical power by a hydraulic motor (refer to [49, 53]).

3.5.3 Electrohydraulic Servovalves

The key to harnessing the power developed by hydraulic actuators and controlling them with analog electronic devices/digital computers, is the electrohydraulic servovalve. In its simplest form, an electrohydraulic servovalve is a spool valve with a spool positioned by solenoid-type magnetic actuators. These valves are called *single-stage valves.*

For a large flow rating, the forces required to rapidly and accurately position the spool are beyond the capabilities of solenoid-type electromagnetic positioners. Large electrohydraulic servovalves have an electromechanical first stage and a hydromechanical second stage.

Electrohydraulic valves of different sizes and constructional features are commercially available. The basic principle of operation of these valves is illustrated by the schematic diagram in Fig. 3.39. A magnetic torque motor activates this *two-stage valve*. The armature of the torque motor, which acts as a flapper, extends into the air gap flux provided by a permanent magnet. The armature (flapper) is pivoted and its one end is positioned between two nozzles. The armature is wound with wire and current is forced through this armature winding. The reaction of the magnetic flux with armature current produces a torque that forces the armature to rotate.

In Fig. 3.39, the armature winding is energized by a balanced amplifier which has the input e. When the input is zero, the currents i_1 and i_2 are equal. Since they flow in opposite directions, the net torque is zero and the flapper is stationary at its central position between the two nozzles. If e is not zero, one of the currents increases and the other decreases in proportion to the magnitude of e. The resulting torque is proportional to the magnitude of e and its direction depends on the polarity of e.

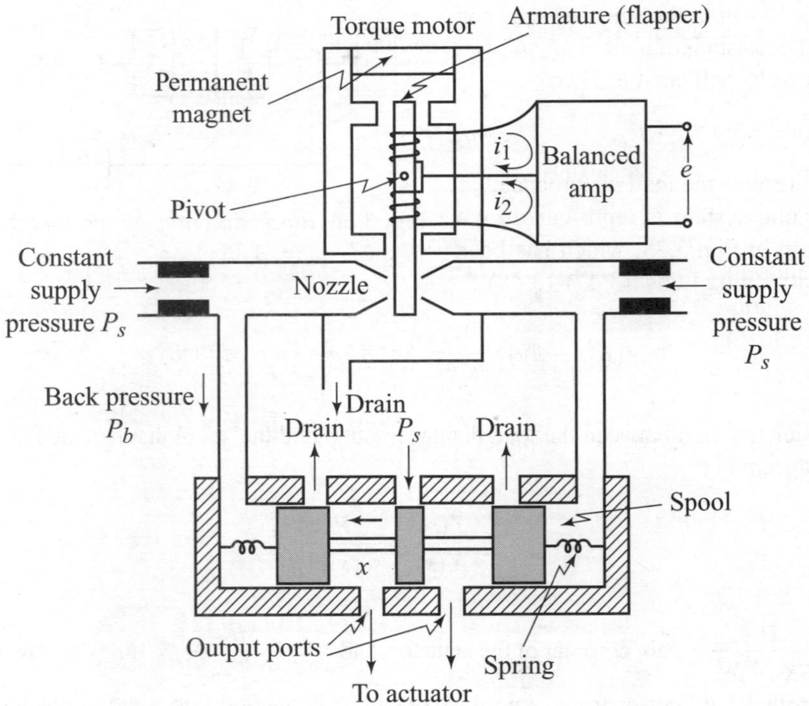

Fig. 3.39 *An electrohydraulic servovalve*

The developed torque, balanced by the torsional spring at the pivot, moves the flapper to the right or left. If the flapper moves to the right, the back pressure of the right-hand nozzle rises while that of the left-hand nozzle falls. A differential pressure between the nozzles is created which produces a force on the spool. This force, balanced by springs, results in spool displacement to the left.

A servovalve model is given by the equation

$$x = Ke \tag{3.50}$$

where x = spool displacement; e = input voltage; and K = valve gain.

This approximate model is adequate since the dynamics relating x to e is generally quite fast relative to the time constants of the hydraulic actuator interfaced to the servovalve.

3.5.4 Linear Variable Differential Transformer

A Linear Variable Differential Transformer (LVDT) measures a linear displacement. The schematic diagram of an LVDT is shown in Fig. 3.40. A primary coil is wound between two secondary coils on a hollow cylindrical former. A movable core is positioned inside the coil former. The primary is excited by an ac source. When the core is in its exact centre position ($y = 0$), the two secondaries are linked by an equal number of flux lines; the emf induced in secondary 1 will be the same as the emf induced in secondary 2. The two secondaries are so connected that their induced voltages oppose each other, and the output voltage will be zero corresponding to $y = 0$. When the core is moved to the right, fewer lines of flux link secondary 1 than secondary 2. Thus, a greater emf is induced in secondary 2 than in secondary 1 and a net voltage appears at the output. When the core is moved to the left, a net voltage will again appear at the output, but it is in anti phase to that for an equivalent displacement to the right.

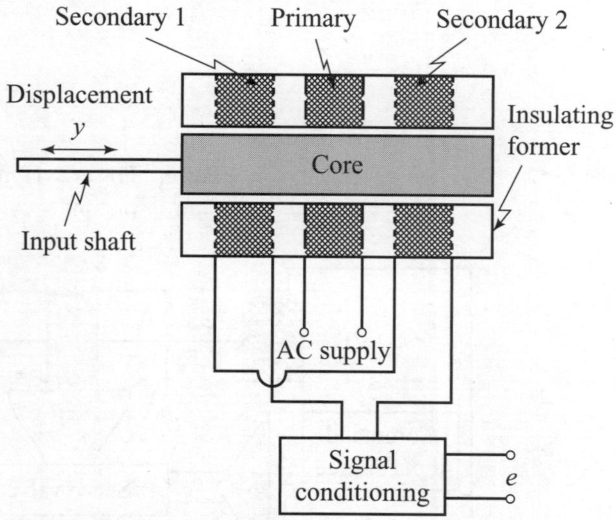

Fig 3.40 *Linear variable differential transformer (LVDT)*

The output voltage of LVDT is in a modulated form, i.e.,

$$e_m(t) = K\,y(t) \sin \omega_c t$$

where ω_c, the carrier frequency, is the same as the frequency of ac supply. The magnitude of the modulated carrier wave is proportional to the core displacement $y(t)$ with K as the constant of proportionality, and the phase reversals of the modulated carrier wave depend on the sign of $y(t)$ (refer to Fig. 3.28).

Signal conditioning by a demodulator and an amplifier gives an output voltage

$$e(t) = K_s\,y(t) \tag{3.51}$$

where K_s = sensitivity of the displacement transducer in volts/mm.

Compared with potentiometer, the LVDT offers higher sensitivity, longer life, ruggedness, and continuous motion capability.

3.5.5 A Hydraulic Servomechanism

A hydraulic servo system used to control the transverse feed of a machine tool is shown in Fig. 3.41a. The cutting tool is powerfully positioned by the hydraulic cylinder fed by the spool valve. The load on the power piston may be assumed to consist of mass M and viscous friction with coefficient B. The disturbance acting on the power piston is the thrust force F_w on the tool.

The actual tool position $y(t)$ is required to follow a time-varying command input position $y_r(t)$. The command input $y_r(t)$ may be generated by a punched tape reader (or a digital computer for computer-controlled hydraulic servo system (refer to Chapter 13)). The command input device produces a voltage $e_r = K_r y_r$. The carriage is moved at a constant speed and the control of the tool position is in the direction of cross slide. The LVDT produces a feedback voltage $e_0(t)$ proportional to the actual tool position ($e_0(t) = K_s y(t)$; K_s is a constant). The error $(e_r - e_0)$ is amplified by an amplifier of gain K_A volts/volt. The servo valve regulates the spool displacement and hence controls port opening x; $x = K_v e_a$, where K_v is the valve gain. The rate of oil flow in the power cylinder is $q = K_1 x - K_2 \Delta p$, where Δp is the differential pressure across the piston of area A, and K_1 and K_2 are constants.

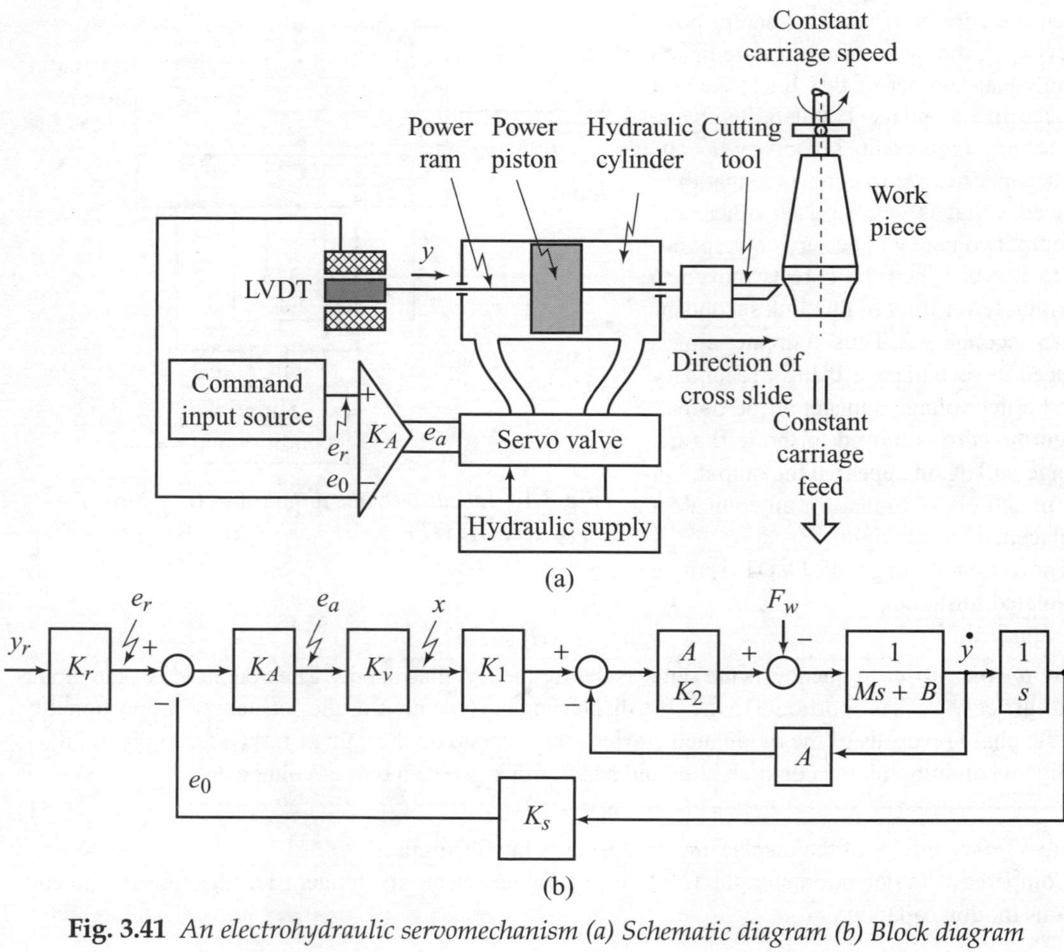

Fig. 3.41 *An electrohydraulic servomechanism (a) Schematic diagram (b) Block diagram*

All variables are deviations from an equilibrium position corresponding to a rest position of the load; the spool lands completely cover the control ports.

A block diagram of the system is shown in Fig. 3.41b (refer to Eqns ((3.47–3.49)), and Fig. 3.38).

3.6 PNEUMATIC DEVICES FOR PROCESS CONTROL

In the so-called *process industries* (chemicals, petroleum, steam power, fuel, etc.), one repeatedly encounters the need to control temperature, flow rate, liquid-level in vessels, pressure, humidity, chemical composition, and the like; such applications are generally considered *process control* applications. ***The time constants in a process control loop are generally much higher than those in a servo loop***. In Sections 1.3 and 1.4, a qualitative description of some industrial processes was given.

In servomechanism applications discussed so far, mathematical models of systems are often close enough to reality. Industrial processes are relatively complex; development of models close enough to reality can be expensive, and also not suitable for the available design methods. Simple models based on approximating assumptions appreciably deviate from reality in many cases. The design strategies for servomechanisms and process control have to take this fact into account. The model-based design of servo systems gives a fairly good performance (only a minor 'tweaking' of adjustments is needed during experimental development). In process-control field, however, adjustments needed during experimental development would be quite significant. Later chapters of the book will highlight these design issues.

Here, our concern is with the devices widely used in the process control industry. Pneumatic devices for actuation and electronic controllers (op amp/digital) for implementation of control function, are used in almost all process control applications. ***Compared to hydraulic/electromechanical devices, pneumatic devices are slow in response; the response time of these devices is, however, adequate for process control because these systems have long time constants***. The advantages of pneumatic devices include mechanical simplicity and ease of maintenance, readily available working medium—compressed air supplies, and explosion-proof characteristics.

Our selection of devices in this section is based on current industrial practice. Classical pneumatic control schemes with mechanical feedback have not been included. The focus is on hybrid systems, combining electronic and pneumatic techniques.

3.6.1 Flow Control Valve

In a large number of process control systems, the final actuator controls the flow of some fluid—liquid, gas, or steam—using flow control valves. All control valves operate by inserting a variable restriction in the flow path. The plug valve (Fig. 3.42) operates by moving a tapered plug, thereby varying the gap between the plug and the valve seat. The flow is controlled by the linear movement of the valve stem.

Pneumatic actuation of the control valves is most common in the process industries. When a valve is supplied by a manufacturer, the pneumatic actuator and the valve are usually attached to each other to form one unit. Figure 3.42 shows a unit where the pneumatic device controls the reciprocating stem of the valve as the pressure on the spring-loaded diaphragm changes[6]. The stem positions a plug in the orifice of the valve body.

[6]Electromechanical actuation of a control valve having rotating stem is another popular choice. A dc motor is configured as the position control actuator. The shaft of the motor is coupled to the rotating stem of the value through gearing. The motor and the reduction gears provide an incremental control of fluid flow.

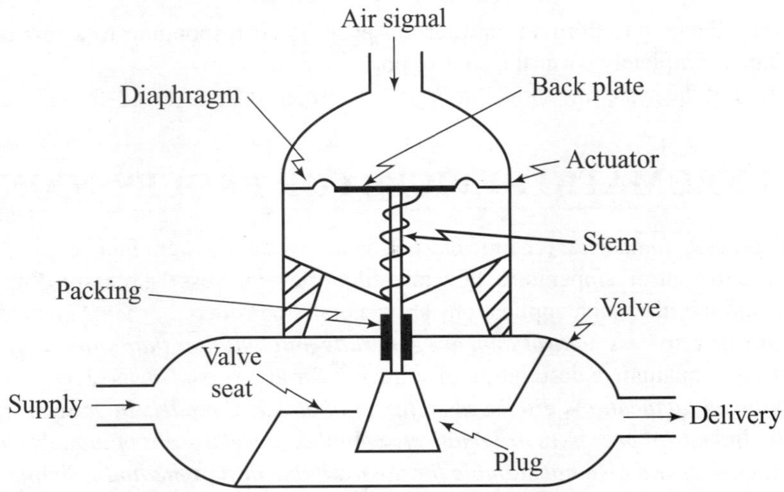

Fig. 3.42 *A pneumatic control value*

An increase in the signal pressure above the diaphragm exerts a force on the diaphragm and the back plate which causes the stem to move down until the force from the now more compressed spring balances the force on the stem. This results in an increase in the cross-sectional area for flow between the plug and the seat, thereby increasing the flow.

In general, the flow rate of fluid through the valve depends upon the upstream and downstream fluid pressures and the size of the opening through the valve. If all components (including the process) in a feedback control system are essentially linear, then the control valve flow characteristic (relationship between flow through the valve and valve stem position) should also be linear, since then the system performance (stability, speed of response, etc.) will be uniform over all system operating conditions. If other system components (usually the process) exhibit significant nonlinearities, then we can choose a valve characteristic that is the inverse of the process nonlinearity, thereby linearizing the overall system and gaining the benefits of uniform performance [49].

Control valves with linear flow characteristics are commonly used in processes where the pressure drop across the valve is fairly constant. In our process control examples, we will use control valves with the following characteristics:

1. The flow through the valve is proportional to valve stem position; and
2. The valve stem position is proportional to input signal pressure.

The standard unit of pressure is *pascal* (Pa): 1 Pa of pressure is defined as the force of 1 newton applied over an area of 1 m^2. Industrial pressures are often measured (therefore, flow control valves are often calibrated) in *pounds per square inch* (psi). Pascals can be obtained by multiplying psi by 6.8948×10^3.

The atmospheric pressure is 14.70 psi (101 kPa) at sea level. *Gage pressure* is referenced to atmospheric pressure, and *absolute pressure* is referenced to perfect vacuum. Gage pressure can be obtained by subtracting the ambient atmospheric pressure from absolute pressure. Likewise, absolute pressure can be found by adding the ambient pressure to the gage pressure. Letters *g* or *a* may be suffixed to the pressure units to clearly denote gage pressure or absolute pressure (e.g., psig, psia). The pressure variables in our discussion represent gage pressures. The atmospheric pressure has been taken as zero gage pressure.

3.6.2 Electropneumatic Transducer

The key to harnessing the power developed by pneumatic actuators and controlling them with analog electronic devices/digital computers, is the electropneumatic transducer. It receives an electrical signal and transmits a proportional output pressure.

Electropneumatic transducers of different sizes and constructional features are commercially available. The basic principle of operation of these devices is illustrated by the schematic diagram in Fig. 3.43. A magnetic torque motor activates the transducer. The armature of the torque motor, which acts as a flapper, extends into the air-gap flux provided by a permanent magnet. The signal current is passed through four coils, wired as shown, and causes a rotary torque on the armature (flapper arm). The torque is proportional to the signal current and causes a change in the flapper/nozzle gap.

Assume that an increase in signal current develops a torque that moves the flapper downwards at the nozzle end. The resulting restriction produces an increased pressure in the nozzle and in the upper chamber of the pneumatic relay. The relay responds to the increase in pressure by closing the exhaust port and opening the supply port, thereby increasing the output pressure to the control valve.

When a decreasing signal current is received, the armature rotates to uncover the nozzle, resulting in a decrease in the pressure in the nozzle and the relay. The relay diaphragm moves upward and the exhaust port opens to bleed the output pressure to atmosphere, thereby reducing the pressure applied to the control valve.

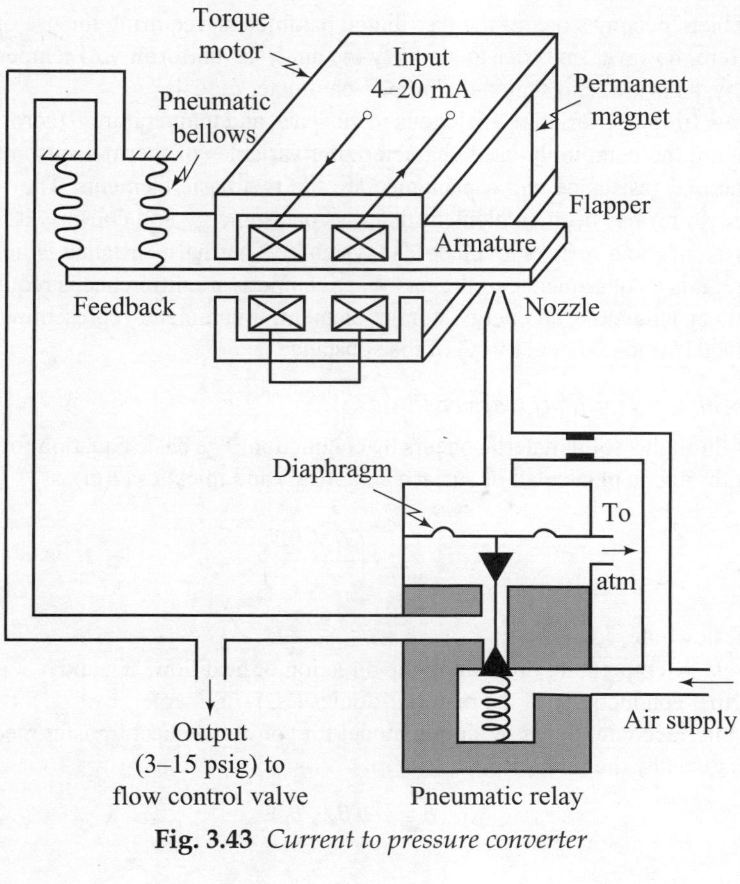

Fig. 3.43 *Current to pressure converter*

Figure 3.43 shows that the electropneumatic transducer is also itself a complete feedback system; thus it must be properly designed for stability and accuracy. If we are users (rather than designers) of such a device, we are concerned only with its overall dynamics from input current i to output pressure p. The dynamics relating p to i is generally quite fast compared to other time constants in the process control loop; so we model the electropneumatic transducer as a zero-order system with p proportional to i.

3.7 CONTROL OF THERMAL PROCESSES

Thermal processes involve the transfer of heat from one substance to another. The heat transfer can take place in different ways. We first briefly describe the heat transfer models; this is followed by development of model of a temperature control loop.

3.7.1 Heat Transfer Models

The basic requirement for the representation of thermal systems by lumped-parameter models is that the temperature of the medium be considered uniform. When the medium is small, this approximation is valid. Also when the medium consists of a body of air or liquid, the temperature can be considered uniform if there is perfect mixing of the fluid. When a single temperature does not accurately represent the thermal state of the entire medium, i.e., there is complex temperature distribution throughout the medium, the problem becomes one of the distributed parameters, requiring the use of partial differential equations. Here, however, in order to simplify the analysis, uniformity of temperature is assumed and thereby the systems are represented by lumped-parameter models.

Heat flow rate h (through variable; analogous to current), and temperature θ (across variable; analogous to voltage) are the commonly used characterizing variables of thermal systems for the purpose of modeling. Thermal resistance and capacitance are the two basic elements. The resistance that we have encountered so far has been an element that dissipates energy and converts it into heat (damper in mechanical systems, and resistor in electrical systems). Thermal resistance is not an element that dissipates energy; it is a consequence of the fact that a temperature difference is required to cause heat to flow. Thermal capacitance is an energy storage element; a parameter representing internal thermal energy (determined by molecular activity) of the substance.

Heat-Transfer Systems without Carrier Fluids

The flow of heat through a solid material occurs by conduction. The basic equation for one-dimensional heat conduction through a plane wall of surface area A(m^2) and thickness l(m) is

$$h = \sigma A\left(\frac{\theta_1 - \theta_2}{\ell}\right) \tag{3.52}$$

where h = heat flow rate, Joules/sec;

$\Delta\theta = \theta_1 - \theta_2$ = temperature drop along the direction of heat flow, °C; and

σ = thermal conductivity of the material, Joules/(°C) (m) (sec).

Many solids interface with fluids. A simple model for convective heat transfer mechanism at solid–fluid interface is given by the relation

$$h = UA(\theta_S - \theta_F) \tag{3.53}$$

where h = heat flow rate, Joules/sec (negative sign of h will indicate flow of heat from fluid to solid);

A = area of the surface (solid–fluid interface), m^2;

θ_S = temperature of the solid, °C;

θ_F = temperature of the fluid, °C; and

U = film coefficient of the solid–fluid interface, Joules/(m^2) (°C) (sec).

Using Eqns (3.52)–(3.53), we define *thermal resistance* as follows:

$$R_{CD} = \text{conductive thermal resistance} = \frac{l}{\sigma A},\ °\text{C/(Joules/sec)} \tag{3.54a}$$

$$R_{CV} = \text{convective thermal resistance} = \frac{1}{UA},\ °\text{C/(Joules/sec)} \tag{3.54b}$$

The rate of heat storage (in the form of internal energy) in a substance is governed by the relation

$$h_1 - h_2 = Mc\frac{d\theta}{dt} \tag{3.55}$$

where h_1 = heat-input rate, Joules/sec;

h_2 = heat-output rate, Joules/sec;

M = mass of the substance, kg;

c = specific heat of the substance, Joules/(kg) (°C); and

θ = temperature of the substance, °C.

Using Eqn. (3.55), we define *thermal capacitance* as follows:

$$C = Mc,\ \text{Joules/°C} \tag{3.56}$$

Heat-Transfer Systems with Carrier Fluids

Systems requiring flowing fluid as a carrier of thermal energy are very common in process industry (refer to Figs 1.15 and 1.21). In the following, we define important parameters of heat transfer system with carrier fluids.

Thermal capacitance of the carrier fluid is defined as follows:

$$C = V\rho c,\ \text{Joules/°C} \tag{3.57}$$

where

V = volume of the fluid chamber (assumed constant), m^3

ρ = fluid density (assumed constant), kg/m^3; and

c = specific heat of the fluid (assumed constant), Joules/(kg) (°C).

Carrier fluid-flow into a chamber is a source of thermal-energy input. Also, the outflowing fluid takes away thermal energy from the chamber. Carrier flow heat-input rate and heat-output rate depend upon the temperatures of the inflowing and outflowing fluids, respectively.

Let θ_i (°C) be the temperature of the inflowing fluid into a chamber and θ_0(°C) be the temperature of the outflowing fluid. Flow rate Q(m^3/sec) is assumed constant.

Heat-input rate, $h_1 = Q\rho\, c\theta_i$, Joules/sec

Heat-output rate, $h_2 = Q\rho\, c\theta_0$, Joules/sec

Net change in transportation of heat is given by

$$h_1 - h_2 = Q\rho\, c(\theta_i - \theta_0) \tag{3.58}$$

From this equation, we define the *thermal resistance* due to carrier fluid-flow as

$$R = \frac{1}{Q\rho c}, \text{°C/(Joules/sec)} \tag{3.59}$$

The metric system of units for thermal systems is summarized below.

Rate of heat flow, h : Joules/sec
Temperature, θ : °C (3.60)
Thermal capacitance, C : Joules/°C
Thermal resistance, R : °C/(Joules/sec)

3.7.2 Temperature Control of a Continuously Stirred Tank Reactor

The Continuously Stirred Tank Reactor (CSTR) sketched in Fig. 3.44, is used to heat a process stream so that its pre-mixed components achieve a uniform composition. Temperature control is important because a high temperature tends to decompose the product while a low temperature results in incomplete mixing. The tank is heated by the steam circulating through the radiator coil; the heat released for heating the tank is the latent heat of condensation. The temperature in the tank is controlled by manipulating the steam valve position.

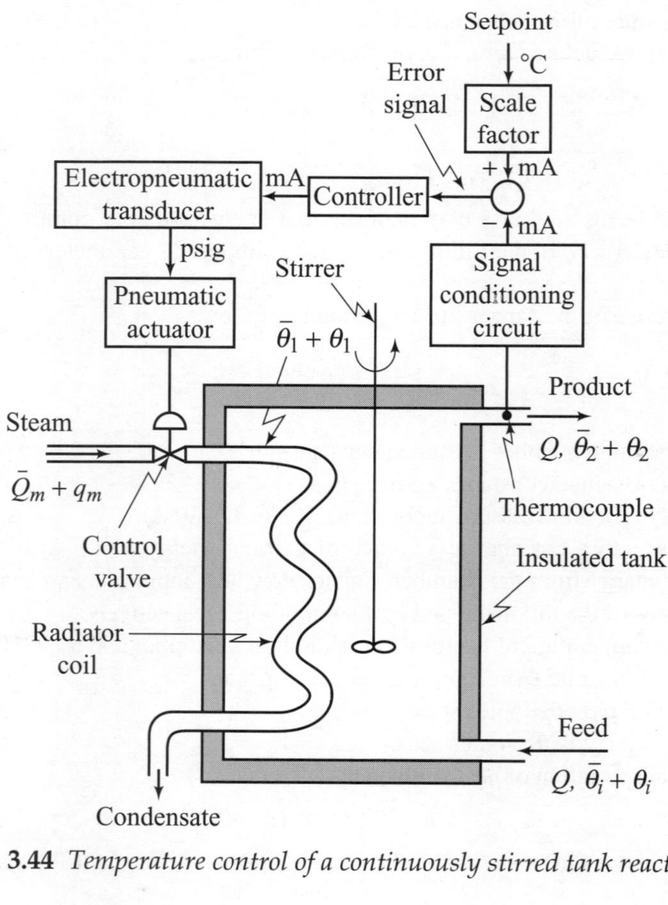

Fig. 3.44 *Temperature control of a continuously stirred tank reactor*

The control hardware required to control the temperature of a process stream leaving a CSTR is shown in Fig. 3.44. This hardware, available from manufacturers of process-control equipment, consists of the components listed below along with their respective conversions:

1. Sensor (temperature to current);
2. Transducer (current to pressure);
3. Controller (current to current);
4. Control valve (pressure to flow rate).

A thermocouple is used to measure the temperature; the signal from the thermocouple is sent to the signal-conditioning circuit, which produces an output in the range of 4–20 mA (the industrial standard current range), which is a linear function of the input. The output of the sensor is compared with the setpoint to produce an error signal. The controller converts the error to an output in the range of 4–20 mA according to the control law. The only control law we have considered so far has been proportional. In Section 3.9, other control laws will be described. The output of the controller enters the electropneumatic transducer, which produces an output in the range of 3–15 psig (the industrial standard pressure range), which is a linear function of the input. Finally, the output of the electropneumatic transducer is sent to the top of the pneumatic control valve, which adjusts the flow of the steam to the coil in the stirred tank. We will assume that the valve is linear.

Simplifying assumptions:

1. The liquid inflow and outflow rates for the tank are equal so that the liquid level in the tank is maintained constant during the operation.
2. The liquid in the tank is well stirred so that its state anywhere in the tank can be described by the temperature of the outflowing liquid.
3. The tank is well insulated so that the heat loss through its walls is negligible. Also, the heat storage capacity of the tank walls is negligible.

For a quantitative illustration, we consider here a CSTR where the process feed has a density ρ of 1090 kg/m^3 and a heat capacity c of 3350 Joules/kg-°C. The volume V of the liquid in the reactor is maintained constant at 3.4 m^3 by a constant feed rate Q of 0.425 m^3/min. Thermal capacitance C_1 of the coil mass is 503×10^3 Joules/°C. The heat transfer area A is 22.57 m^2 and the heat transfer coefficient U is 42.97×10^3 Joules/min-m^2-°C. The steam is saturated; it can be assumed that its latent heat of condensation λ is constant at 2247×10^3 Joules/kg.

At steady state, the temperature of the feed $\bar{\theta}_i$ is 55 °C. Assume that the design requirement is to maintain the tank liquid at a temperature $\bar{\theta}_2$ of 80 °C. Possible perturbations from the steady state are caused by the desired change θ_r in the temperature of the tank liquid, and the change θ_i in the feed temperature (a disturbance). These perturbation signals give rise to perturbations θ_1 and θ_2 in the temperatures of the coil mass and the tank liquid, respectively, from their steady-state values $\bar{\theta}_1$ and $\bar{\theta}_2$.

The heat balance for the coil mass:

$$C_1 \frac{d}{dt}(\bar{\theta}_1 + \theta_1) = (\bar{Q}_m + q_m)\lambda - UA[\bar{\theta}_1 + \theta_1 - (\bar{\theta}_2 + \theta_2)] \tag{3.61}$$

where

$\bar{Q}_m$ = steam flow rate (kg/min) at steady state, and
q_m = manipulation of the steam flow rate.

The heat balance for the tank liquid:

$$V\rho c\frac{d}{dt}(\bar{\theta}_2+\theta_2)=Q\rho c[\bar{\theta}_i+\theta_i]+UA[\bar{\theta}_1+\theta_1-(\bar{\theta}_2+\theta_2)]-Q\rho c[\bar{\theta}_2+\theta_2] \quad (3.62)$$

The process steady state is described by the following equations:

$$\bar{Q}_m\lambda-UA(\bar{\theta}_1-\bar{\theta}_2)=0$$

$$Q\rho c\,\bar{\theta}_i+UA(\bar{\theta}_1-\bar{\theta}_2)-Q\rho c\,\bar{\theta}_2=0$$

This gives

$$Q\rho c\,(\bar{\theta}_i-\bar{\theta}_2)+\bar{Q}_m\lambda=0$$

Since $\bar{\theta}_i=55\,°\text{C}$ and $\bar{\theta}_2=80\,°\text{C}$, the steady-state mass flow rate of steam through the coil is

$$\bar{Q}_m=\frac{0.425\times1090\times3350\times(80-55)}{2247\times10^3}=17.27\text{ kg/min}$$

At steady state, the error signal in the control configuration of Fig. 3.44 is zero. The necessary steam flow rate $\bar{Q}_m$ can be achieved with the zero error signal because the electropneumatic transducer has a zero adjustment which allows pressure to be at the middle (9 psig) of its 3–15 psig range when the input to the transducer is zero, and the valve actuator has a zero adjustment which allows the valve opening to be put anywhere desired with its input pressure at 9 psig.

Let us now derive the process model that describes perturbation dynamics about the steady state. From Eqns (3.61) and (3.62), we obtain

$$C_1\frac{d\theta_1}{dt}=q_m\lambda-UA(\theta_1-\theta_2)$$

$$V\rho c\frac{d\theta_2}{dt}=Q\rho c(\theta_i-\theta_2)+UA(\theta_1-\theta_2)$$

By Laplace transformation and manipulation of these equations, we obtain

$$(\tau_1 s+1)\,\theta_1(s)=\theta_2(s)+K_1Q_m(s)$$

$$(\tau_2 s+1)\,\theta_2(s)=K_2\theta_1(s)+K_3\theta_i(s)$$

where

$$\tau_1=\frac{C_1}{UA};\tau_2=\frac{V\rho c}{UA+Q\rho c};K_1=\frac{\lambda}{UA};$$

$$K_2=\frac{UA}{UA+Q\rho c};K_3=\frac{Q\rho c}{UA+Q\rho c}$$

Block diagram representation of these equations is shown in Fig. 3.45a.

Substituting the numerical values of various parameters, we obtain

$$\tau_1=0.52\text{ min},\ \tau_2=4.92\text{ min},\ K_1=2.32°\text{C/kg-min}$$

$$K_2=0.385°\text{C}/°\text{C},\ K_3=0.615°\text{C}/°\text{C}$$

Manipulation of the block diagram of Fig. 3.45a, with the above numerical values for various parameters gives

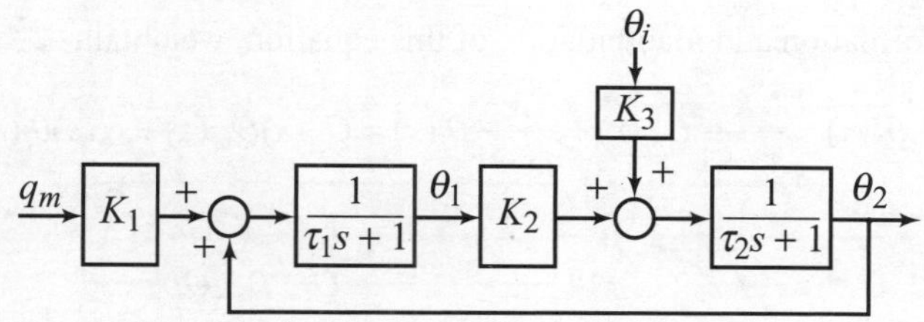

(a)

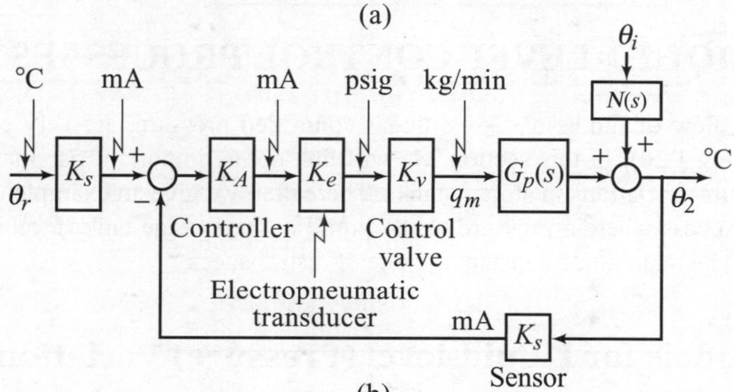

(b)

Fig. 3.45 *Block diagrams for the system of Fig. 3.44*

$$\theta_2(s) = G_p(s)\, Q_m(s) + N(s)\theta_i(s)$$

where

$$G_p(s) = \frac{0.35}{s^2 + 2.13s + 0.24};\; N(s) = \frac{0.24(0.52s+1)}{s^2 + 2.13s + 0.24}$$

The dynamics of the other components of the feedback system—the temperature sensor, the controller (amplifier), the electropneumatic transducer and the pneumatic control valve—are negligible, relative to the process time constants. All these components are fast enough to be treated as zero-order systems. Figure 3.45b shows a block-diagram representation of the temperature control loop.

The process model, $G_p(s)$, in Fig. 3.45 is a second-order model. A further approximation based on the assumption: heat storage capacity of the coil mass is negligible, leads to a first-order model for the process.

Under the assumption of $C_1 = 0$, the heat balance for the tank liquid is given by (refer to Eqn. (3.62))

$$V\rho c \frac{d}{dt}\left(\bar{\theta}_2 + \theta_2\right) = Q\rho c\left[\bar{\theta}_i + \theta_i\right] + \left(\bar{Q}_m + q_m\right)\lambda - Q\rho c\left[\bar{\theta}_2 + \theta_2\right]$$

The process state is described by the equation

$$Q\rho c(\bar{\theta}_i - \bar{\theta}_2) + \bar{Q}_m\lambda = 0$$

The process model that describes perturbation dynamics about steady state is given by

$$V\rho c \frac{d\theta_2}{dt} = Q\rho c\left(\theta_i - \theta_2\right) + q_m\lambda \qquad (3.63)$$

By Laplace transformation and manipulation of this equation, we obtain

$$\theta_2(s) = \frac{K}{\tau s+1}Q_m(s) + \frac{1}{\tau s+1}\theta_i(s) = G_p(s)Q_m(s) + N(s)\theta_i(s) \tag{3.64}$$

where $\tau = \frac{V}{Q}; K = \frac{\lambda}{Q\rho c}$

3.8 LIQUID-LEVEL CONTROL PROCESSES

Systems requiring flow of fluids in vessels under controlled pressure, are very common in process industry (refer to Fig.1.20). In this section, we will first define important parameters for models for liquid-level (pressure) variations in storage tanks. Thereafter, we give an example of feedback control of a liquid-level process wherein pressure of the liquid in the storage tank (level of the liquid in the tank) is required to be maintained constant in spite of disturbances.

3.8.1 Models for Liquid-level (Pressure) Variations in Storage Tanks

Liquids are slightly compressible at high pressures. However, approximate dynamic models based on the assumption of incompressibility of liquids are usually sufficiently accurate. Simple models developed in this section use this approximation.

Volumetric flow rate q (through variable; analogous to current) and pressure p (across variable; analogous to voltage) are the commonly used characterizing variables of hydraulic systems. The three basic elements of hydraulic systems are the resistance, the capacitance, and the inertance. The inertance represents fluid inertia and is derived from the inertia forces required to accelerate a fluid in a pipe. It is an energy storing element. Another energy storing element is the capacitance which, as we shall see, represents storage in a gravity field. In most applications of interest to us, energy storage due to the inertance element is negligible compared to that due to the capacitance element.

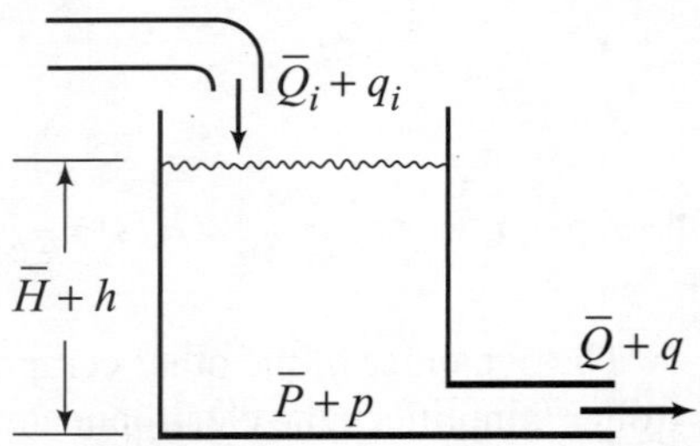

Fig. 3.46 *Capacitance of a storage tank*

The tank shown in Fig. 3.46 illustrates the concept of hydraulic capacitance. A (m^2) is the surface area of the tank's bottom, and tank's sides are vertical. If $\bar{Q}_i$ (m^3/sec) represents the steady-state inflow rate, $\bar{Q}$ (m^3/sec) represents the steady-state outflow rate, and $\bar{H}$ (m) represents the steady-state height of the liquid in the tank, then the volume balance at steady state gives

$$\bar{Q}_i = \bar{Q}; \bar{H} = \text{constant}$$

The hydrostatic pressure due to height $\bar{H}$ is

$$\bar{P} = \rho g \bar{H}, \text{ newtons/m}^2$$

where ρ = density of the liquid, kg/m^3; and
g = acceleration due to gravity, m/sec^2.

In the system of Fig. 3.46, the perturbation $q_i(t)$ in the inflow rate will change the height of the liquid in the tank by $h(t)$, the pressure across the pipe by $p(t)$ and the outflow rate by $q(t)$. The volume-balance equation becomes

$$\bar{Q}_i + q_i(t) = A\frac{d}{dt}(\bar{H} + h(t)) + \bar{Q} + q(t)$$

Perturbation from steady state is described by the equation

$$q_i(t) = A\frac{dh(t)}{dt} + q(t) \tag{3.65}$$

Since

$$p(t) = \rho g h(t) \tag{3.66}$$

it follows that

$$q_i(t) - q(t) = \frac{A}{\rho g}\frac{dp(t)}{dt} = C\frac{dp(t)}{dt} \tag{3.67}$$

The *hydraulic capacitance* due to storage in gravity field is defined as

$$C = \frac{A}{\rho g},\ \text{m}^3/(\text{newtons/m}^2) \tag{3.68}$$

In terms of hydraulic capacitance, Eqn. (3.67) may be written as

$$q_i(t) - q(t) = C\frac{dp(t)}{dt} \tag{3.69}$$

With reference to Fig. 3.46, the outflow $q(t)$ should obviously depend on $p(t)$ somehow. The greater the pressure $p(t)$, the greater the outflow rate. The concept of hydraulic resistance introduced below will help in establishing a relationship between $q(t)$ and $p(t)$.

The liquid flowing through the container in Fig. 3.46 meets with a resistance (obstruction) to flow at the outlet, due to restricted flow through the hole and friction between the liquid and the outlet pipe walls. The presence of a component such as a valve in the pipe increases the resistance.

Symbolic representation (used earlier in the book) of an obstruction to flow is shown in Fig. 3.47a. A controllable obstruction, such as a hydraulic valve with an adjustable opening, is represented as in

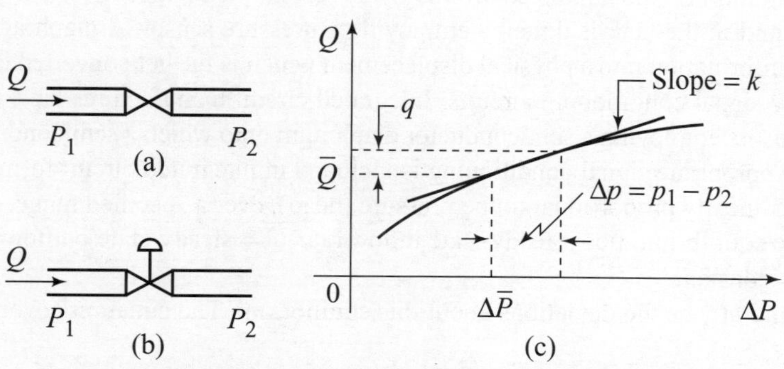

Fig. 3.47 *(a) Flow obstruction (b) Adjustable obstruction (c) Steady-state operating point*

Fig. 3.47b. The upstream pressure is P_1 and the back pressure achieved downstream is P_2. The flow rate Q depends on the pressure difference $\Delta P = (P_1 - P_2)$. This relationship is usually nonlinear; it can be obtained either experimentally or analytically.

Suppose that an experimental curve relating the flow rate Q to the pressure drop ΔP is available for an obstruction. A typical curve is shown in Fig. 3.47c. Let the slope of the curve at the equilibrium point $(\bar{Q}, \Delta\bar{P})$ be denoted by k. Then, for a small deviation from equilibrium, we obtain

$$q(t) = k(\Delta p(t)) = \frac{1}{R}(\Delta p(t)) \tag{3.70}$$

From this equation, *hydraulic resistance* is defined as

$$R = \frac{1}{k}, \left(\frac{\text{newton/m}^2}{\text{m}^3/\text{sec}}\right) \tag{3.71}$$

The value of R depends on the steady-state condition chosen for linearization.

When an equation relating the flow to the pressure drop is available, the resistance can be obtained analytically. The flow-pressure relationship (refer to Eqn. (3.44))

$$Q = f(\Delta P) \tag{3.72a}$$

may be linearized about an equilibrium point using Taylor series expansion (refer to Eqns (2.1)).

The flow rate through a valve with adjustable opening is a function of both the pressure difference ΔP across the valve and the adjustable area or equivalently the adjustable valve opening x_v:

$$Q = f(x_v, \Delta P) \tag{3.72b}$$

The metric system of units for liquid-level control systems is summarized below.

Volume flow rate, q : m^3/sec

Pressure, p : newtons/m^2

Hydraulic capacitance, C : $\text{m}^3/(\text{newtons/m}^2)$ (3.72)

Hydraulic resistance, R : $\dfrac{\text{newtons/m}^2}{\text{m}^2/\text{sec}}$

3.8.2 A Liquid-Level Control System

Figure 3.48 shows a tank containing liquid. We are interested in regulating the level of the liquid in the tank. The flow of liquid into the tank is controlled by an electropneumatic valve. The measurement of the head of the liquid in the tank is done by employing a pressure sensor. A diaphragm in the sensor converts pressure information into a physical displacement which is further converted into proportional electrical signal by signal conditioning circuits. Integrated circuit manufactures have developed composite pressure sensors employing a semiconductor diaphragm onto which a semiconductor strain gage has been grown. Appropriate signal conditioning is included in integrated circuit form, providing a dc voltage or current, linearly proportional to the pressure (head), over a specified range.

For steady-state equilibrium flow, steady-state inflow rate $\bar{Q}_i$ = steady-state outflow rate $\bar{Q}$, and the liquid height $\bar{H}$ = constant.

Let $q_i(t)$, $q(t)$ and $h(t)$ be the deviations about this equilibrium. The equation governing the change in liquid volume is

$$\frac{d}{dt}(Ah(t)) = q_i(t) - q(t) = C\frac{dh(t)}{dt} \tag{3.73}$$

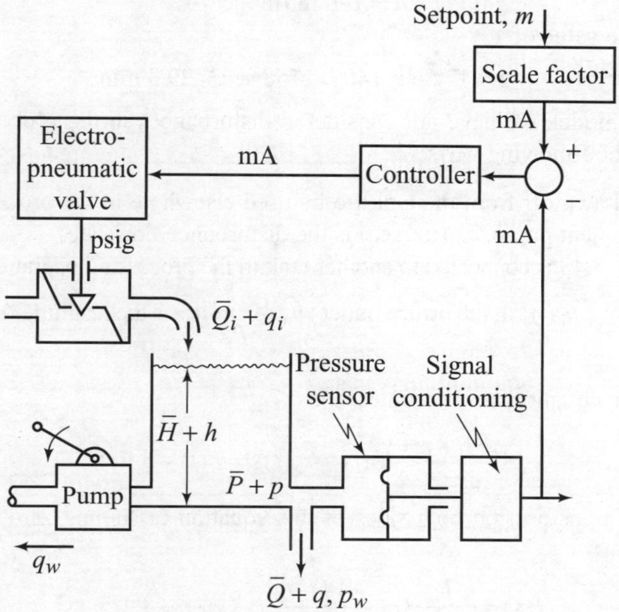

Fig. 3.48 *A liquid-level control system*

where A is the constant cross-sectional area of the tank; hydraulic capacitance[7], $C = A$.

The flow through the outlet pipe restriction is governed by the equation[7] (refer to Eqns (3.66), (3.70) – (3.71))

$$q(t) = \frac{1}{R}h(t) \tag{3.74}$$

where R (m/(m^3/sec)) is a parameter representing flow resistance through outlet pipe.

Eliminating $q(t)$ from Eqns (3.73) and (3.74), we obtain

$$A\frac{dh(t)}{dt} + \frac{1}{R}h(t) = q_i(t)$$

Defining time constant $\tau = AR$, we get

$$\tau\frac{dh(t)}{dt} + h(t) = Rq_i(t) \tag{3.75}$$

We have the following system parameters:

Radius of the tank = 2 m (Area A = 12.56 m^2)

ρ = 1000 kg/m^3; g = 9.81 m/sec^2

Experimentally obtained exit-pipe restriction parameter

[7]Relationships in the previous subsection have been derived with volumetric flow rate q as through variable and pressure p as across variable. For modeling the system of Fig. 3.48, we are using liquid height h as across variable.

$$R = 140 \text{ m/(m}^3\text{/sec)}$$

We can work out the value of τ:

$$\tau = 12.56 \times 140 = 1758 \text{ sec} = 29.3 \text{ min}$$

In developing this model, we have not considered disturbances in the process. Disturbances may enter in the model in the following ways:

1. Liquid may be drawn off from the tank (to be used elsewhere in the processing operation) by a variable displacement pump. q_w (m^3/sec) is the disturbance flow rate.
2. If the tank of Fig. 3.48 is connected to another tank in the processing operation, there will be a variable back pressure $\{p_w(t)\}$ at the orifice outlet ; $h_w = \dfrac{1}{\rho g} p_w$ is the resulting disturbance in the liquid head.

The volume balance equation now becomes

$$A\frac{dh(t)}{dt} + \frac{1}{R}(h(t) - h_w(t)) = q_i(t) - q_w(t) \tag{3.76}$$

Taking the Laplace transform on both sides of this equation (assuming zero initial conditions), we obtain

$$H(s) = \frac{R}{\tau s + 1}Q_i(s) + \frac{1/\rho g}{\tau s + 1}P_w(s) - \frac{R}{\tau s + 1}Q_w(s) \tag{3.77}$$

where

$$H(s) \triangleq \mathscr{L}[h(t)]; Q_i(s) \triangleq \mathscr{L}[q_i(t)]; P_w(s) \triangleq \mathscr{L}[p_w(t)]; Q_w(s) \triangleq \mathscr{L}[q_w(t)]$$

With the given system parameters, this equation becomes

$$H(s) = \frac{140}{29.3s + 1}Q_i(s) + \frac{1.02 \times 10^{-4}}{29.3s + 1}P_w(s) - \frac{140}{29.3s + 1}Q_w(s) \tag{3.78}$$

The system response to the controlled inflow has a gain of 140 m/(m^3/sec), and a time constant of 29.3 min. The response to pressure disturbance at the outlet has a gain of 1.02×10^{-4} m/(newton/m^2), and to disturbance flow the gain is 140 m/(m^3/sec).

We take the output signal range of pressure sensor as 4–20 mA (the industrial standard current range). The tank of our liquid-level control system is of 4 m height with a maximum fill level of 3 m.

$$\text{Sensor gain} = \frac{(20 - 4)\text{mA}}{3\text{m}} = 5.33\,\text{mA/m} \tag{3.79}$$

The time constant of the sensor is very small compared to process time constant τ; the gain 5.33 mA/m, thus, is the transfer function of the sensor.

The control valve in our system of Fig. 3.48 has a maximum capacity of 200 m^3/hr (0.056 m^3/sec). The nominal pressure of the valve is 3 to 15 psig (the industrial standard pressure range).

$$\text{Valve gain} = \frac{0.056(\text{m}^3\text{/sec})}{(15 - 3)(\text{psig})} \times \frac{(15 - 3)(\text{psig})}{16(\text{mA})} = 0.0035 \text{ (m}^3\text{/ sec)/mA} \tag{3.80}$$

Again the dynamics of the control valve are much faster than that of the process, and the valve could be treated as a zero-order system.

The block diagram of the liquid-level control loop is shown in Fig. 3.49; h_r is the change in the set-point value from the initial setting.

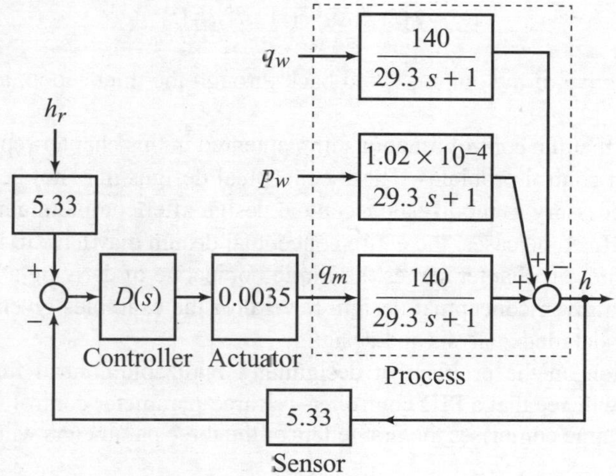

Fig. 3.49 *Block diagram for the system of Fig. 3.48*

3.9 DEVICES FOR IMPLEMENTATION OF CONTROL FUNCTIONS

The central theme of the book is to modify a system's behavior by adding feedback, and by applying a control function of the form given in Eqn. (3.1). The design of control systems is, thus, based on two tasks:

1. Determining the structure of the feedback control system; and
2. Designing a realizable/implementable control function $u = f(y_r, y)$; where y_r—the command input, and y—the controlled output, are the sources of information about the current process state, and the function $f(\cdot)$ is required to generate a control action u that drives the process towards the desired state, i.e., $y \rightarrow y_r$.

The design of vast majority of practical systems is based on the two control configurations considered in this chapter: the basic structure given by Fig. 3.1, and the minor-loop feedback structure shown in Fig. 3.2. DC speed servo of Fig. 3.13, DC position servo of Fig. 3.19, AC position servo of Fig. 3.32, hydraulic position servo of Fig. 3.41, temperature control system of Fig. 3.44, and liquid-level control system of Fig. 3.48 — all of these employ basic feedback control configuration. In position servo of Fig. 3.20, we have employed minor-loop feedback control configuration.

Revisit all of these practical control systems to see what kind of control functions have been employed. You will observe that in all the systems except the system of Fig. 3.20, the control signal u is proportional to the error signal $e = y_r - y$, i.e.,

$$u(t) = K_P\, e(t) \tag{3.81}$$

where K_P is the *proportional gain* parameter. The proportional control law has been implemented using amplifiers.

In the minor-loop feedback servo of Fig. 3.20, the control law $u = f(y_r, y)$ has been implemented as follows:

$$u(t) = K_P\, e(t) + K_D\, \dot{y}(t) \tag{3.82}$$

where $\dot{y}(t)$, the derivative of the output, is fed back through the minor loop, and K_D is the *derivative gain* parameter.

It should be noted that the control functions implemented in this chapter represent only the conceptual designs for given control problems. These conceptual designs may not be the final designs to be adopted. It may be necessary to modify a conceptual design after completing a detailed analysis of its mathematical model. In some cases, the entire conceptual design may have to be discarded in favor of another design. The present chapter serves the limited objective of deriving a mathematical model of a control system for a given conceptual design. Several of the examples given in this chapter will be reconsidered later for detailed analysis and design.

A formal discussion on the problem of designing a realizable control function will begin from Chapter 7. There we will see that a PID controller—a three-parameter control function, is widely used in industry. The PID name comprises the first letters of the three parameters which make up this control function:

P stands for **P**roportional gain parameter in the control function;
I stands for **I**ntegral gain parameter in the control function, and
D stands for **D**erivative gain parameter in the control function.

The family of PID control functions is created by various combinations of the proportional, integral, and derivative terms as required to meet specific performance requirements. The commonly used members of PID family are P, PI, PD, and PID. The PID control function is given by

$$U(s) = D(s)E(s) = \left[K_P + \frac{K_I}{s} + K_D s\right]E(s) \tag{3.83}$$

or

$$u(t) = K_P e(t) + K_I \int_0^t e(\tau)d\tau + K_D \frac{de(t)}{dt} \tag{3.84}$$

where $e(t)$ is the system error $(y_r - y)$, K_P is the *proportional gain parameter*, K_I is the *integral gain parameter*, and K_D is the *derivative gain parameter*.

Industrial representation of the PID controller often uses a *time-constant form* for the PID parameters. The time-constant form, given below, is easily derived from Eqns (3.83)/(3.84).

$$u(t) = K_c\left[e(t) + \frac{1}{T_I}\int_0^t e(\tau)d\tau + T_D \frac{de(t)}{dt}\right] \tag{3.85}$$

or

$$U(s) = K_c\left[1 + \frac{1}{T_I s} + T_D s\right]E(s) = D(s)E(s) \tag{3.86}$$

where K_c is the *controller gain* parameter, T_I is the *integral time-constant*, and T_D is the *derivative time constant*.

The control functions for other members of the PID family: P/PI/PD, easily follow from Eqns ((3.83)–(3.86)). In the conceptual designs (given in this chapter) based on basic feedback control configuration of Fig. 3.1, we have employed the control function given by Eqn. (3.81), i.e., only P-term of the PID family. In the conceptual design (given in this chapter) based on minor-loop feedback control configuration of Fig. 3.2, we have employed the control function given by Eqn. (3.82). This function is, in fact, a PD control function, with P component acting on error, and D component acting on the controlled output.

The classical control theory provides systematic methods for the design of PID control function, i.e., for designing P, I, and D parameters of the controller $D(s)$. The design aspects of control theory will be taken up from Chapter 7 onwards. Other control functions with 3 to 5 parameters will also be considered in the design process.

Today, most complex industrial processes are under computer control. A complete computer-based installation, in a multilevel viewpoint of control strategy, handles the design tasks at plant management level, production scheduling and control level, and supervisory level; in addition to controlling individual process units at the lowest level in the hierarchy [155]. It is common to delegate process control functions at the lowest level to the special individual control units, connected to the higher levels through computer system. Op Amp (Operational Amplifier) control units are in use to serve these functions.

We will take up the Op Amp realization of PID control functions in Chapter 7; Chapter 13 will highlight digital implementation.

3.10 REVIEW EXAMPLES

Review Example 3.1 Consider the temperature control system of Fig. 3.50 which is set up to produce a steady process stream at a controlled temperature. The temperature θ(°C) of the outflowing liquid is measured by a thermocouple which produces an output voltage e_t(volts) proportional to $\theta (e_t = K_t \theta)$. This voltage is subtracted from the reference voltage e_r to generate the error signal $e = (e_r - e_t)$, which in turn regulates the heating rate h by means of an SCR amplifier [5]. Basically, a solid state switch, the SCR amplifier controls the heating rate by varying the point in the ac power cycle at which the heater is connected to (or disconnected from) the power line (see Fig. 3.51). As the line frequency (50 Hz) is very high relative to the thermal system frequency response, we take h to mean the *average* power over the cycle, rather than the instantaneous electrical power, and assume that h follows the

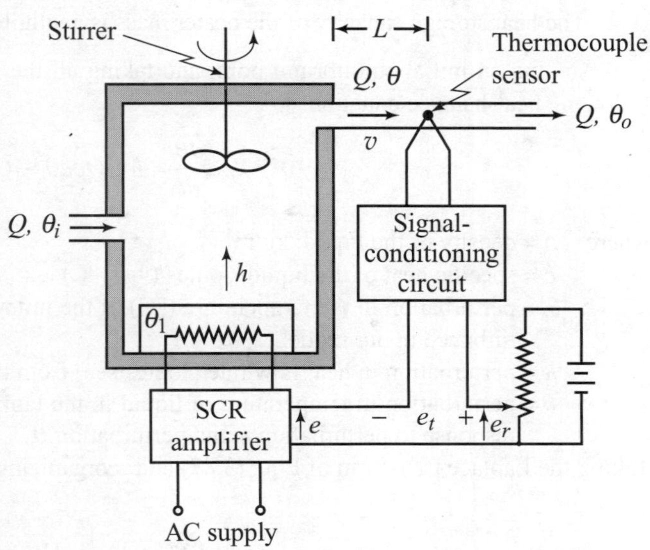

Fig. 3.50 *A temperature control system*

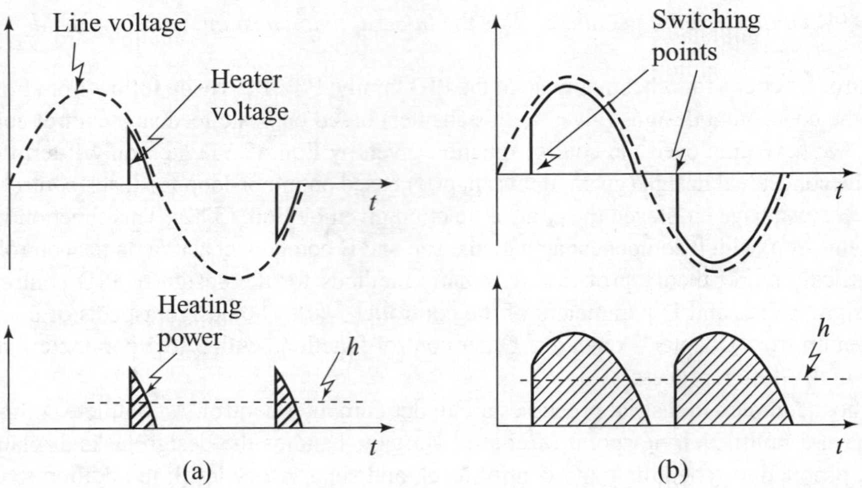

Fig. 3.51 *SCR control of heating rate (a) Low heating rate (b) High heating rate*

amplifier input voltage e instantly. The nonlinear steady-state relation (static calibration curve) between e and h can be linearized at any chosen operating point, giving heater gain K_h; $h = K_h e$ watts. Simplifying assumptions:

1. The liquid inflow and outflow rates for the tank are equal (Q m^3/sec) so that the liquid level in the tank is maintained constant (volume V m^3 of the liquid in the tank is therefore constant).
2. The liquid in the tank is well stirred so that its state can be described by the temperature θ of the outflowing liquid.
3. The tank is well insulated so that the heat loss through its walls is negligible. Also the heat storage capacity of the tank walls is negligible.
4. The heat storage capacity of the heater mass is negligible.

Assuming an initial equilibrium point and taking all the variables as perturbations, we obtain the following heat-balance equation.

$$V\rho c \frac{d\theta}{dt} = h + Q\rho c\theta_i - Q\rho c\theta \tag{3.87}$$

where ρ = density of the liquid; kg/m^3
c = specific heat of the liquid; Joules/(kg) (°C)
θ_i = perturbation in the temperature (°C) of the inflowing stream; this will be treated as a disturbance in our model.
h = perturbation in heat flow rate (Joules/sec) from the heater; and
θ = perturbation in temperature of liquid in the tank from the nominal point of operation, in response to perturbation h and perturbation θ_i.

Taking the Laplace transform of Eqn. (3.87) and reorganizing, we get

$$(s\tau + 1)\theta(s) = \frac{1}{Q\rho c}H(s) + \theta_i(s) \tag{3.88}$$

where $\tau = \frac{V}{Q} =$ process time constant.

Although we generally prefer to measure the controlled variable as directly as possible, the location of the thermocouple (which has an output of a few millivolts) in the tank encounters problems of electrical noise due to high-power SCR switching, and vibrations caused by the tank stirrer. Thus, we locate it in the pipeline downstream of the tank. Under this situation, the measured response $\theta_0(t)$ will be the same as actual response $\theta(t)$ except that it will be delayed by some amount of time. The time delay thus introduced is referred to as *dead time* or *transportation lag* to distinguish it from the delay associated with the time-constant of a simple lag. We will see in later chapters that the presence of such delays tends to cause oscillations and makes the system less stable. Dead time should be avoided to the extent possible.

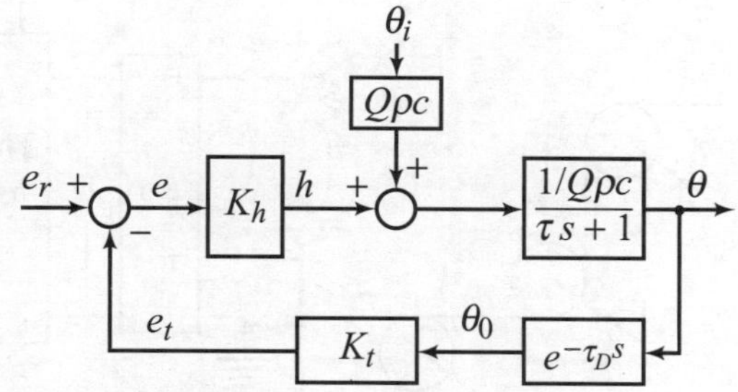

Fig. 3.52 *A block diagram for the system of Fig. 3.50*

Figure 3.50 defines the behavior of the dead-time element (pipe). The input $\theta(t)$ and the output $\theta_0(t)$ are related by

$$\theta_0 = \theta(t - \tau_D);\ t > \tau_D$$

where

$$\tau_D = \text{dead time} = \frac{\text{distance}}{\text{velocity}} = \frac{L}{v} = \frac{L}{Q/A_p}\ \text{sec};$$

L = length of pipe, m;
A_p = cross-sectional area of pipe, m^2; and
Q = flow rate, m^3/sec.

The transfer function of the dead-time element becomes

$$\frac{\theta_0(s)}{\theta(s)} = e^{-s\tau_D} \tag{3.89}$$

Using Eqns (3.88) and (3.89), we can draw the block diagram of the system as shown in Fig. 3.52.

Review Example 3.2 In a steel or paper mill, the products are moved by rollers, as shown in Fig. 3.53a. In order to maintain a prescribed uniform tension, the roller speeds are kept constant and equal to each other. This can be achieved by using the control system shown in Fig. 3.53b. Each roller is driven by an armature-controlled dc motor.

The motor speed is measured by the dc tachogenerator coupled to the motor shaft. The tachogenerator output is K_t volts/(rad/sec). This voltage is compared with the reference voltage e_r which represents the desired motor speed. The difference in voltage is amplified by a pre-amplifier of gain K_A, and then power amplified by a power amplifier.

In the thyristor-driven speed control system of Fig. 3.53b, a linear relationship between the input voltage e_c and output voltage e_a can be obtained when a proper firing control scheme is used [5].

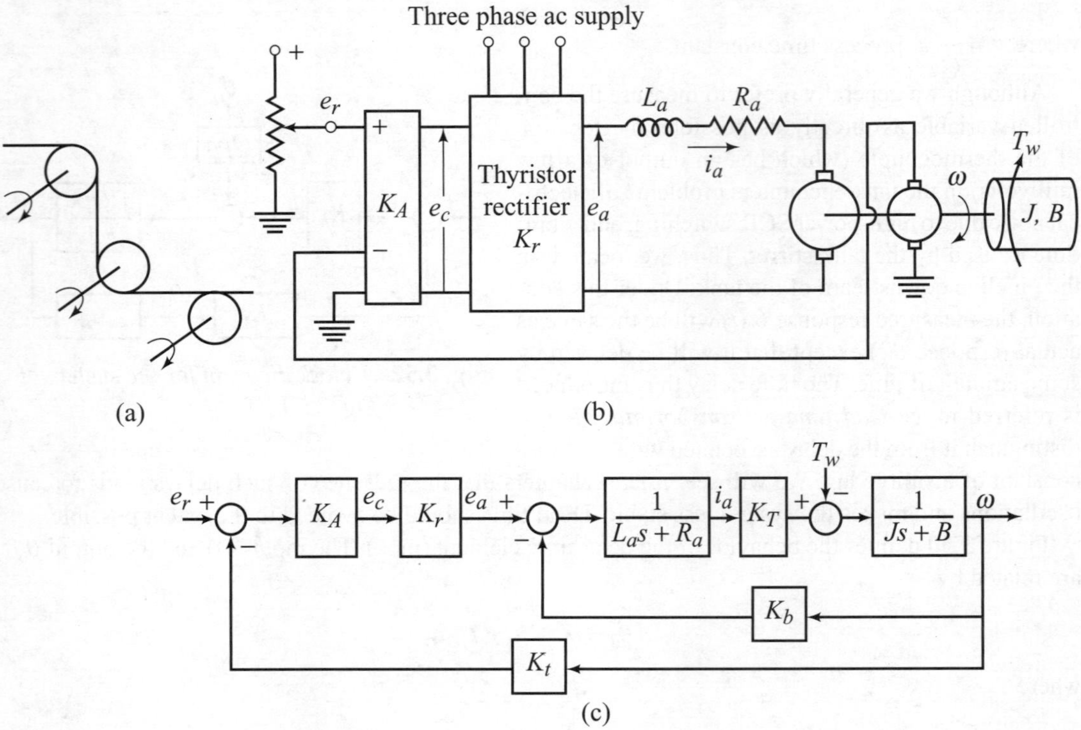

Fig. 3.53 *Speed control of rollers*

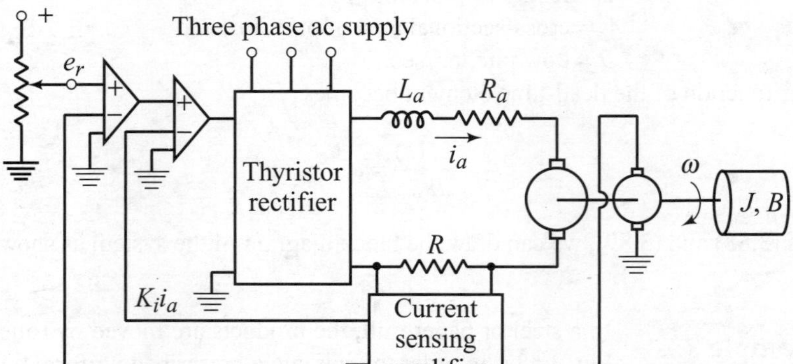

Fig. 3.54 *Speed control system with inner current-control loop*

The time constants associated with the thyristor rectifier are negligibly small. Neglecting the dynamics of the rectifier, we get

$$e_a = K_r e_c \tag{3.90}$$

where K_r is the gain of the rectifier.

The motor is separately excited with a constant field current and has a counter emf of K_b volts/(rad/sec). It produces a torque of K_T newton-m/amp of armature current. It drives a load of moment of inertia J newton-m/(rad/sec^2) and viscous friction with coefficient B newton-m/(rad/sec), against the load-torque disturbance T_w newton-m.

The functional block diagram of the speed control system is shown in Fig. 3.53c.

The thyristorized drives are usually provided with current-limiting features to prevent damage to the thyristors. A current limit can be implemented by constructing a current-control loop as shown in Fig. 3.54. A resistance R is inserted in the armature circuit to generate a voltage proportional to the armature current, which is also feedback.

CONCLUDING REMARKS

Mathematical models of physical systems are key elements in the design and analysis of control systems. The differential equations describing the dynamic behavior of these systems were utilized to construct transfer function models/block diagram models.

What we have learnt?

- The load on practical speed/position control systems is represented by moment of inertia J, and viscous friction with coefficient B.
- Due to loading effects, (motor + load) is considered a single subsystem (plant) while developing models of speed/position control systems.
- The approximate transfer function model of the plant in a speed control system is typically of first order, while that in a position control system is of second order (simple pole and an integrator).
- Gears in position control systems are used to magnify the motor torque to drive the load connected to the larger gear; the speed of the load shaft gets proportionally reduced with respect to the motor shaft.
- In a position control system, a secondary path feeding back the angular velocity of the motor shaft (in addition to the primary path feeding back the position), gives a minor-loop feedback control configuration—a practical scheme to improve the performance of the system.
- Approximate transfer function model of the process in temperature/liquid-level control system is of first order.
- In process-control field, systems are relatively slow and complex. The time constants of the processes are much higher than the time constants of servo systems. Models used for design of process control systems are, in most cases, more approximate compared to the models used for design of servo systems.

Review Questions

3.1 (a) Derive an expression for the transfer function of an armature-controlled dc servomotor. State clearly the assumptions made.

(b) How is the torque constant of the motor related to the back emf constant?

3.2 (a) Derive an expression for the transfer function of a field-controlled dc servomotor. State clearly the assumptions made.

(b) List the advantages and disadvantages of the field control compared to the armature control of a dc motor.

3.3 **(a)** Describe the construction and working of a two-phase motor suitable for use in ac servo systems. Give the torque-speed characteristics of the motor and derive the transfer function model based on the linearized characteristics. Justify the use of linear model. Appropriate symbols for variables and parameters may be used in your derivation.

(b) Compare and contrast the usefulness of ac and dc servomotors in motion control systems.

3.4 Given a two-gear system with angular displacements θ_1 and θ_2, number of teeth N_1 and N_2, and torques T_{12} (exerted on gear 1 by gear 2) and T_{21} (transmitted to gear 2 by gear 1), develop the mathematical relations between these variables and parameters. Clearly state the assumptions made.

3.5 Explain using appropriate diagrams, the construction and operation of the following transducers:
(a) AC tachogenerator **(b)** DC tachogenerator
Describe briefly using suitable block diagrams, an application of each in motion control systems.

3.6 Explain with the aid of a sketch, the construction and operation of a linear variable differential transformer (LVDT). Describe briefly using appropriate diagrams, an application of LVDT in motion control systems.

3.7 **(a)** Explain with the aid of appropriate diagrams, the construction and operation of
(i) synchro transmitter; and **(ii)** synchro-control transformer
Give a schematic diagram showing how the synchro pair (transmitter and control transformer) would be embodied in an ac position control system.

3.8 **(a)** A feedback control system is controlling the position of a single-link robot manipulator on a manufacturing floor. It comprises of a sensor, a proportional controller, and an actuator.
(i) Suggest suitable devices and describe their characteristics.
(ii) Interconnect the devices to build a feedback control system and describe the processing of signals in the loop. Give units of signals involved in the loop.

(b) Modify this feedback control system to include techogenerator damping.

3.9 Draw an electromechanical schematic diagram of a dc position control servosystem incorporating tachogenerator damping. Describe the operation of the main components used, explaining clearly how the error signal is formed.

3.10 Armature control of a dc motor is used in a closed-loop speed control system. Draw a schematic layout for the system. Describe the operation of main components used, explaining clearly how the error signal is formed.

3.11 Using appropriate diagrams, give the constructional and operational features of a hydraulic linear actuator. Derive the transfer function model of the actuator. Use appropriate symbols for variables and parameters. Clearly state the assumptions made.

3.12 Draw a schematic diagram of position control system for controlling the linear displacement of a load mass. Use potentiometers for formation of error and hydraulic linear actuator for moving the mass. Give a brief description of the equipment used in the system.

3.13 Pneumatic devices are widely used in process control systems. In a thermal system, the following hardware has been used to control the temperature of the process stream leaving a CSTR (Continuously Stirred Tank Reactor):
Sensor (temperature to current); Controller (current to current); Transducer (current to pressure); and Control valve (pressure to flow rate).
Select appropriate control hardware and draw a schematic diagram describing the complete temperature control system. Explain the constructional and operational features of the selected hardware.

3.14 Pneumatic devices are widely used in process control systems. A system uses the following hardware to regulate the level of liquid in a tank:

Sensor (pressure (liquid head) to current); Controller (current to current); Transducer (current to pressure); and Control valve (pressure to flow rate).
Select appropriate control hardware and draw a schematic diagram describing the complete liquid-level control system. Explain the constructional and operational features of the selected hardware.

Self-Appraisal Quiz

A set of questions with multiple answer choices is given below. These questions have been designed to capture the important aspects of the subject covered in the chapter. For each of these questions, identify the correct answer from the given choices. Compare your answers with the Master Key given at the end of the session. This quiz should serve the purpose of a self-appraisal test for the reader.

3.1 The gear train shown in Fig. Q3.1
(A) reduces the speed and the torque
(B) increases the speed and the torque
(C) reduces the speed and increases the torque
(D) increases the speed and reduces the torque.

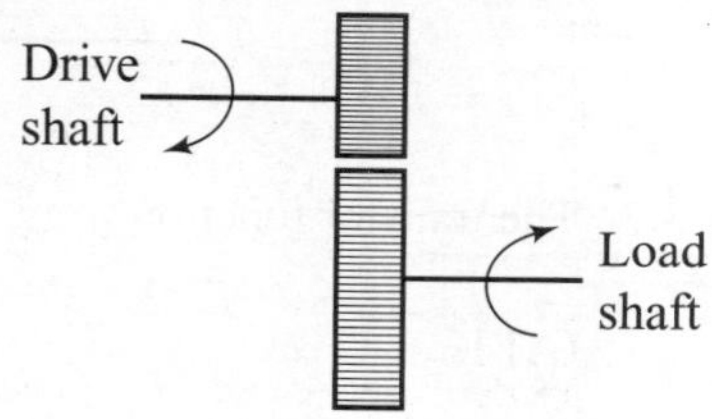

Fig. Q3.1

3.2 Effect of back emf in an armature-controlled dc servomotor is
(A) to increase effective motor friction, thereby reducing motor time constant
(B) to increase effective motor friction, thereby increasing motor time constant
(C) to increase motor inertia, thereby increasing motor time constant
(D) to increase motor inertia, thereby reducing motor time constant.

3.3 Electrical time-constant of an armature-controlled dc servomotor is
(A) equal to mechanical time constant
(B) smaller than mechanical time constant
(C) larger than mechanical time constant
(D) none of the answers in (A), (B), and (C) is correct.

3.4 Ratio of the rotor reactance X to the rotor resistance R for a two-phase servomotor
(A) is equal to that of a normal induction motor
(B) is less that that of a normal induction motor
(C) is greater than that of a normal induction motor
(D) may be less or greater than that of a normal induction motor.

3.5 A rotating load is connected to a field-controlled de motor with negligible field inductance. A test results in output steady-state speed (ω) of 2 rad/sec when an input voltage (e_f) of 100 V is applied to the motor terminals. Also the test shows that output speed reaches 0.632×2 rad/sec within 0.75 sec. The transfer function $\omega(s)/E_f(s)$ of the motor is

(A) $\dfrac{0.02}{(0.375s+1)}$ (B) $\dfrac{0.02}{0.75s+1}$ (C) $\dfrac{50}{0.75s+1}$ (D) $\dfrac{50}{(0.375s+1)}$

3.6 The transfer function from $R(s)$ to $Y(s)$ of the system of Fig. Q3.6 is ($W(s) = 0$)

(A) $\dfrac{s(\tau s+1)}{\tau s^2+(1+KK_2)s+KK_1}$ (B) $\dfrac{K_1 s(\tau s+1)}{\tau s^2+(1+KK_2)s+KK_1}$

(C) $\dfrac{KK_1}{\tau s^2+(1+KK_2)s+KK_1}$ (D) $\dfrac{s(\tau s+1+KK_2)}{\tau s^2+(1+KK_2)s+KK_1}$

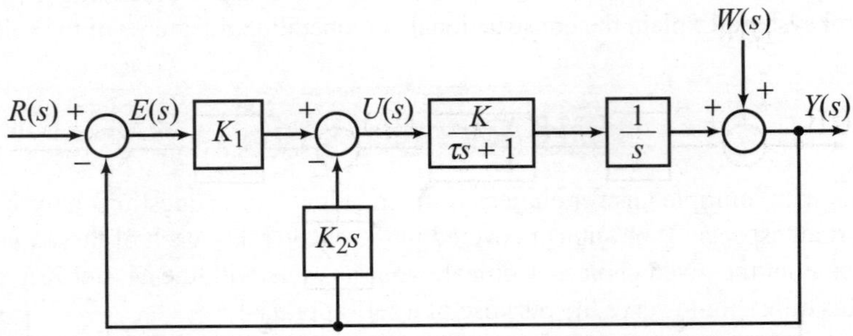

Fig. Q3.6

3.7 The transfer function from $W(s)$ to $Y(s)$ of the system of Fig. Q3.6 is ($R(s) = 0$)

(A) 1 (B) $\dfrac{s(\tau s+1)}{\tau s^2+(1+KK_2)s+KK_1}$

(C) $\dfrac{KK_1}{\tau s^2+(1+KK_2)s+KK_1}$ (D) none of the answers in (A), (B) and (C) is correct

3.8 The transfer function between $R(s)$ and $E(s)$ of the system of Fig. Q3.6 is ($W(s) = 0$)

(A) $\dfrac{s(\tau s+1+KK_2)}{\tau s^2+(1+KK_2)s+KK_1}$ (B) $\dfrac{s(\tau s+1)}{\tau s^2+(1+KK_2)s+KK_1}$

(C) $\dfrac{KK_1}{\tau s^2+(1+KK_2)s+KK_1}$ (D) $\dfrac{K_1 s(\tau s+1)}{\tau s^2+(1+KK_2)s+KK_1}$

3.9 The transfer function between $R(s)$ and $U(s)$ of the system of Fig. Q3.6 is ($W(s) = 0$)

(A) $\dfrac{s(\tau s+1)}{\tau s^2+(1+KK_2)s+KK_1}$ (B) $\dfrac{s(\tau s+1+KK_2)}{\tau s^2+(1+KK_2)s+KK_1}$

(C) $\dfrac{KK_1}{\tau s^2+(1+KK_2)s+KK_1}$ (D) $\dfrac{K_1 s(\tau s+1)}{\tau s^2+(1+KK_2)s+KK_1}$

3.10 Consider the following statements:

(i) Synchro is a three-phase machine.
(ii) Rotor of synchro transmitter is of dumb bell type while that of control transformer is cylindrical.

(A) None of the above statements is true
(B) Statement (i) is true but statement (ii) is false
(C) Statement (i) is false but statement (ii) is true
(D) Both the statements are true.

3.11 Consider the following statements:

(i) AC devices (servomotor, synchro) are normally operated at higher frequencies (400 Hz, for example) and not at 50 Hz.
(ii) The output of a synchro is demodulated before feeding it to a dc motor for actuation of inertial load.

(A) None of the above statements is true
(B) Statement (i) is true but statement (ii) is false
(C) Statement (i) is false but statement (ii) is true
(D) Both the statements are true.

3.12 Consider the following statements:
(i) Characteristics of ac sevomotors are quite linear, and these motors are easier to control.
(ii) Characteristics of dc servomotors are quite nonlinear and these motors are difficult to control.
Which of the following is the correct answer?
(A) None of the above statements is true
(B) Statement (i) is true but statement (ii) is false
(C) Statement (i) is false but statement (ii) is true
(D) Both the statements are true.

3.13 A position control system has a two-loop configuration. The minor loop is a velocity-feedback loop realized through tachogenerator. Consider the following statements:
(i) The tacho is usually mounted on motor shaft rather than the load shaft.
(ii) The minor loop is a positive-feedback loop.
(A) None of the above statements is true
(B) Statement (i) is true but statement (ii) is false
(C) Statement (i) is false but statement (ii) is true
(D) Both the statements are true.

3.14 An inertial and a frictional load are driven by a dc motor with torque T_M. The dynamic model of the system is

$$T_M(t) = J\frac{d\omega(t)}{dt} + B\omega(t)$$

The steady-state speed of the motor for step input will be doubled when
(A) inertia J is doubled
(B) friction B is doubled
(C) both the inertia J and friction B are doubled
(D) None of the answers in (A), (B), and (C) is correct.

3.15 An open-loop control system requires an operator to set a motorized valve setting c where $0 < c < 1$, so that two fluids are mixed together. One stream has a constant flow of 10 m^3/hr and the outflowing stream should be 14 m^3/hr. The configuration is shown in Fig. Q3.15, where motorized valve dynamics are given. The value of valve setting to achieve the desired steady-state outflow is
(A) 0.2 (B) 0.4 (C) 0.5 (D) 0.7

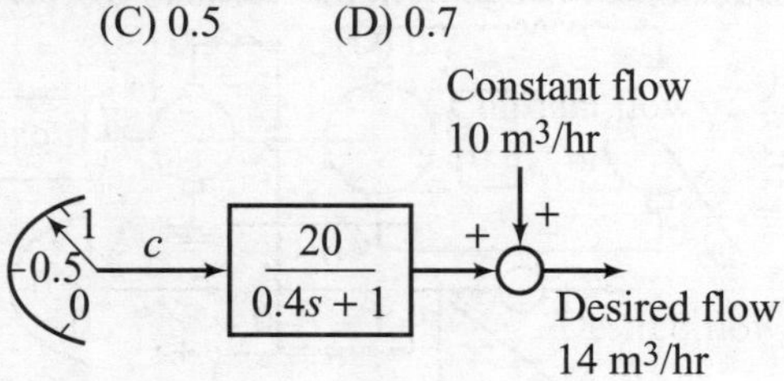

Fig. Q3.15

3.16 In a level control system, the fluid flow is related to the inflow rate by the following transfer function:

$$G(s) = \frac{\text{level } H(s)}{\text{inflow } Q(s)} = \frac{0.75}{2500s+1}$$

Initially the tank is empty. The inflow control value is suddenly opened to allow a flow rate of 0.5 m^3/sec into the tank. The level after 2500 sec is

(A) 0.474 m (B) 0.375 m (C) 0.75 m (D) 0.237 m

MASTER KEY

3.1 (C)	3.2 (A)	3.3 (B)	3.4 (B)	3.5 (B)
3.6 (C)	3.7 (B)	3.8 (A)	3.9 (D)	3.10 (C)
3.11 (D)	3.12 (A)	3.13 (B)	3.14 (D)	3.15 (A)
3.16 (D)				

Problems

3.1 (a) Consider the block diagram of a position control system shown in Fig. 3.19b. Obtain the closed-loop transfer functions $\theta_L(s)/\theta_R(s)$ of the system by block diagram manipulation. Also manipulate the block diagram to obtain $\theta_L(s)/nT_w(s)$.

(b) Consider the block diagram of a temperature control system shown in Fig. 3.45b. Obtain the transfer functions $\theta_2(s)/\theta_r(s)$, and $\theta_2(s)/\theta_i(s)$ by block diagram manipulation.

(c) Convert the block diagrams of Figs 3.19b and 3.45b to signal flow graphs and apply Mason's gain rule to obtain the transfer fucntions required in parts (a) and (b) above.

3.2 In the speed control system shown in Fig. P3.2, the generator field time constant is negligible. It is driven at constant speed giving a generated voltage of K_g volts/field amp. The separately excited motor has a back emf of K_b volts/(rad/sec). It produces a torque of K_T newton-m/amp. The motor and its load have a combined moment of inertia J kg-m^2 and negligible friction. The tachogenerator gives K_t volts/(rad/sec) and the amplifier gain is K_A amps/volt.

Draw a block diagram of the system and determine therefrom the transfer function $\omega(s)/E_r(s)$, where ω is the load speed and e_r is the reference signal which corresponds to the desired speed.

With the system initially at rest, e_r = 100 volts is suddenly applied. Determine how the load speed will change with time.

Given: K_A = 4 amps/volt; K_T = 1.5 newton-m/amp;
K_g = 50 volts/amp; K_t = 0.2 volts/(rad/sec); $R_a = 1\Omega$; J = 6 kg-m^2

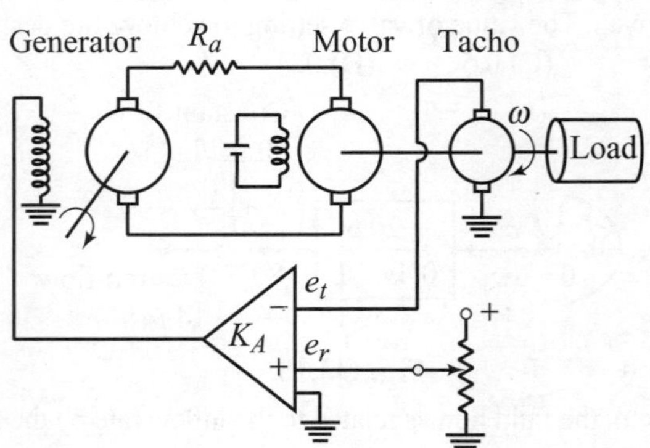

Fig. P3.2

MATLAB Exercise Use MATLAB's Simulink to obtain the step response under the following conditions.

(i) The system is linear and is driven by the generator whose gain is K_g, and field time constant is negligible.

(ii) Taking the field time constant to be half of the motor and load time-constant, determine its effect on the system's output)

3.3 Consider the positional servomechanism shown in Fig. P3.3. Assume that the input to the system is reference shaft position θ_R and the system output is the load shaft position θ_L. Draw a block diagram of the system indicating the transfer function of each block. Simplify the block diagram to obtain $\theta_L(s)/\theta_R(s)$. The parameters of the system are given below.

Sensitivity of error detector	$K_P = 10$ volts/rad
Amplifier gain	$K_A = 50$ volts/volt
Motor field resistance	$R_f = 100$ ohms
Motor field inductance	$L_f = 20$ henrys
Motor torque constant	$K_T = 10$ newton-m/amp
Moment of inertia of load	$J_L = 250$ kg-m^2
Coefficient of viscous friction of load	$B_L = 2500$ newton-m/(rad/sec)
Motor to load gear ratio	$\dot{\theta}_L/\dot{\theta}_M = 1/50$
Load to potentiometer gear ratio	$\dot{\theta}_P/\dot{\theta}_L = 1$

Motor inertia and friction are negligible.

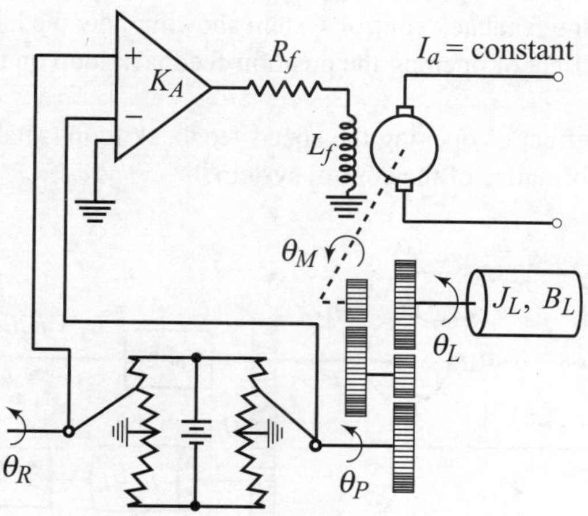

Fig. P3.3

3.4 Consider the system shown in Fig. P3.4a. The characteristic of ac motor is shown in Fig. P3.4b. The moment of inertia of the motor is $J_M = 0.003$ newton-m/(rad/sec^2). The motor drives a load through a gear train. N_1, N_2, N_3, and N_4 are the number of teeth with $N_1/N_2 = 1/2$ and $N_3/N_4 = 1/5$. The moment of inertia of the load is $J_L = 0.02$ newton-m/(rad/sec^2), and coefficient of viscous friction of the load is $B_L = 0.001$ newton-m/(rad/sec). Find the transfer function $\theta_L(s)/E_c(s)$.

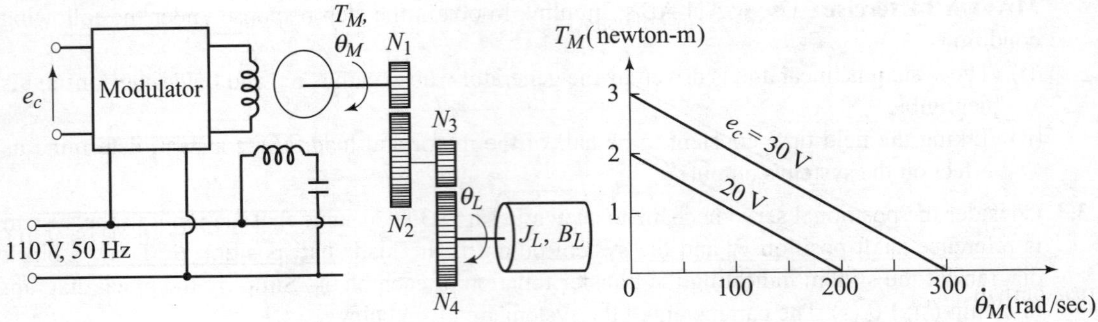

Fig. P3.4

3.5 Consider the system shown in Fig. P3.5 with $R_a = 10\,\Omega$, $L_a = 0.1$H, $K_b = 1$ volt/(rad/sec), $\dot{\theta}_L / \dot{\theta}_M = 1/2$, $K_t = 0.8$ volt/(rad/sec), $K_P = 1.5$ volt/rad.
Moment of inertia of load, $J_L = 2$ newton-m/(rad/sec^2).
Moment of inertia of motor shaft, $J_M = 0.1$ newton-m/(rad/sec^2).
Coefficient of viscous friction of load, $B_L = 0.02$ newton-m/(rad/sec).
Coefficient of viscous friction on motor shaft, $B_M = 0.01$ newton-m/(rad/sec)

(i) Find the transfer function $\theta_M(s)/E_a(s)$.

(ii) A multiloop (consisting of a speed-feedback loop and a position-feedback loop) feedback control system built around the system of Fig. P3.5 with suitable additional hardware drives the load to the commanded position θ_R inspite of load torque disturbances.
Make a sketch of the feedback control system showing how the hardware is connected.

(iii) What will be the effect of opening the position-feedback loop on the performance of the control system?

(iv) What will be the effect of opening the speed-feedback loop (retaining the position feedback intact) on the performance of the control system?

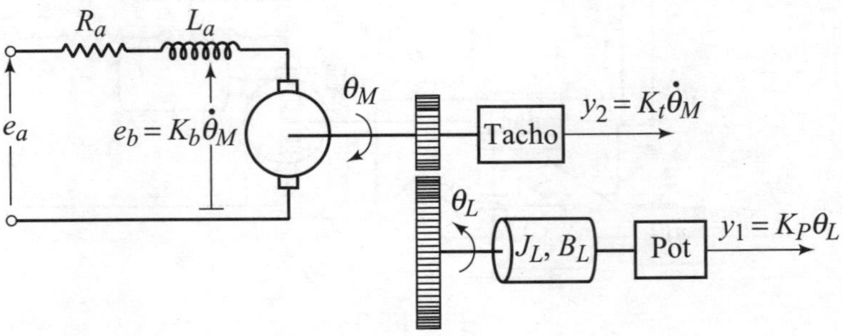

Fig. P3.5

3.6 The schematic diagram of a servo system is shown in Fig. P3.6. The two-phase servomotor develops a torque in accordance with the equation

$$T_M = K_1 e_c - K_2 \dot{\theta}_M$$

where

$$K_1 = 1 \times 10^{-5} \text{ newton-m/volt; } K_2 = 0.25 \times 10^{-5} \text{ newton-m/(rad/sec)}$$

The other parameters of the system are:

Synchro sensitivity $K_s = 1$ volt/rad
Amplifier gain $K_A = 20$ volt/volt
Tachogenerator constant $K_t = 0.2$ volt/(rad/sec)
Load inertia $J_L = 1.5 \times 10^{-5}$ kg-m^2
Viscous friction $B_L = 1 \times 10^{-5}$ newton-m/(rad/sec)

$$\dot{\theta}_M / \dot{\theta}_S = 1; \ \dot{\theta}_M / \dot{\theta}_T = 1$$

Motor inertia and friction are negligible.
Draw the block diagram of the system and therefrom obtain the transfer function $\theta_M(s)/\theta_R(s)$.

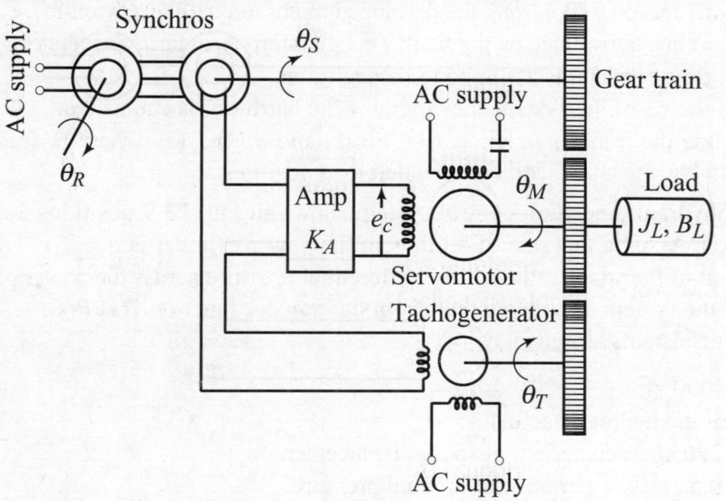

Fig. P3.6

3.7 Figure P3.7 shows a closed-loop drive system. The desired speed is ω_r and the corresponding reference voltage is e_r. The parameters describing the system are given below:

$R_a = 6\ \Omega$, $L_a = 46$ mH
$J = 0.093$ kg-m^2, $B = 0.008$ newton-m/(rad/sec)
Motor torque constant $K_T = 0.55$ newton-m/amp
Thyristor rectifier gain constant $K_r = 25$
Amplifier gain $K_A = 20$
Tachogenerator sensitivity $K_t = 0.057$ volt/(rad/sec)

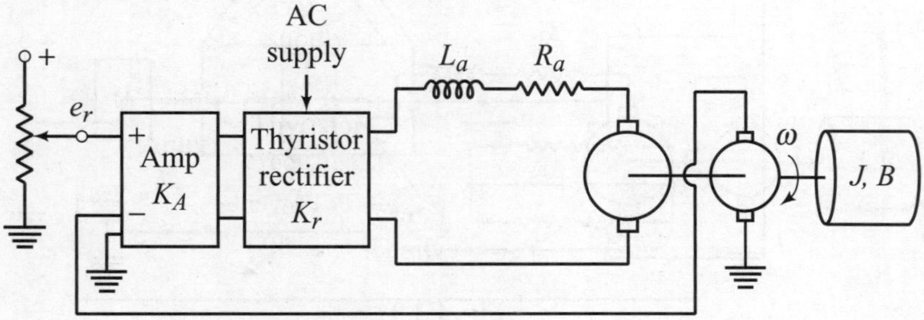

Fig. P3.7

(a) Draw a block diagram of the system that includes transfer functions of all the subsystems, and therefrom obtain $\omega(s)/E_r(s)$.

(b) The thyristorized drives are usually provided with a current-limiting feature to prevent damage to the thyristors. Modify the drive system of Fig. P3.7 to implement the current limit. Draw a block diagram of the proposed system.

3.8 A dc position control servomechanism comprises of a field-controlled dc servomotor with constant armature current, potentiometer error detector with sensitivity K_P volts/rad, and a dc amplifier of gain K_A amps/volt. The motor produces a torque of K_T newton-m/amp. Separate motor and load inertias J_M (kg-m^2) and J_L (kg-m^2), respectively, are identified, and the connecting structure is modeled by a shaft with spring constant K (newton-m/rad) and damping coefficient B (newton-m per (rad/sec)). A load disturbance T_w (newton-m) acts on J_L opposing the developed motor torque T_M (newton-m). T_M accelerates J_M and supplies the torque transmitted by the shaft. This shaft torque, in turn, accelerates J_L and supplies T_w. Motor field time constant is assumed to be negligible.

(a) Make a sketch of the system showing how the hardware is connected.

(b) Determine the transfer functions $\theta_L(s)/\theta_R(s)$ and $\theta_L(s)/T_w(s)$, where θ_L (rad) and θ_R (rad) are, respectively, the actual and commanded load positions.

3.9 The electrohydraulic position control system shown in Fig. P3.9 positions a mass M with negligible friction. Assume that rate of oil flow in the power cylinder is $q = K_1x - K_2\Delta p$ where x is the displacement of the spool and Δp is the differential pressure across the power piston. Draw a block diagram of the system and obtain therefrom the transfer function $Y(s)/R(s)$.
The system constants are given below.

Mass $M = 1000$ kg
Constants of the hydraulic actuator:
$K_1 = 200$ cm^3/sec per cm of spool displacement
$K_2 = 0.5$ cm^3/sec per unit differential pressure

Potentiometer sensitivity	$K_P = 1$ volt/cm
Power amplifier gain	$K_A = 500$ mA/volt
Linear transducer constant	$K = 0.1$ cm/mA
Piston area	$A = 100$ cm^2

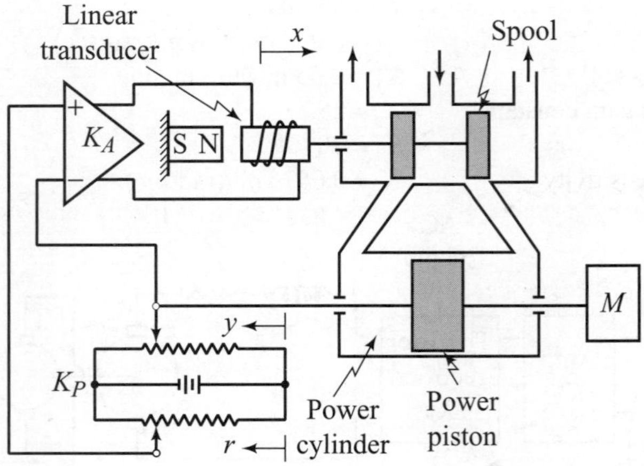

Fig. P3.9

3.10 The schematic diagram of a steel-rolling process is shown in Fig. P3.10. The steel plate is fed through rollers at a constant speed of v cm/sec. The distance between the rollers and the point where the thickness is measured is d cm. The thickness gage provides a feedback voltage $e_o(t)$ proportional to the actual thickness $y(t)$ of the plate ($e_o(t) = K_s\, y(t)$; K_s is a constant). A voltage e_r corresponding to the desired thickness is the reference input to the system. The error ($e_r - e_o$) is amplified by an amplifier of gain K_A amps/volt. The servo valve regulates the spool displacement and hence controls the port opening x; $x = K_v i$, where K_v is the valve gain. The rate of oil flow in the power cylinder is $q = K_1 x - K_2 \Delta p$ where Δp is the differential pressure across the piston of area A, and K_1 and K_2 are constants.

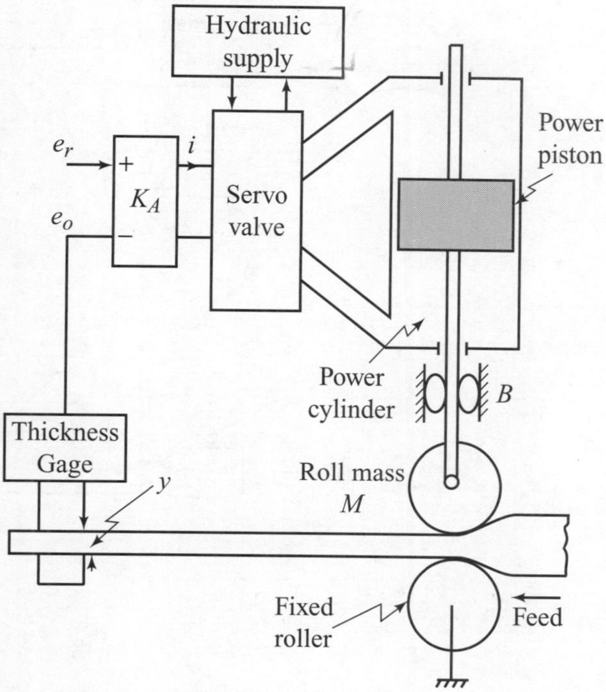

Fig. P3.10

The load on the piston is that of the roller mass M and viscous friction with coefficient B. Normal force F_w on the roller is a load disturbance.

Assuming all the variables to be deviations from an equilibrium position, draw a functional block diagram that describes perturbation dynamics of the system. Derive the transfer function $Y(s)/E_r(s)$.

3.11 Consider the liquid level control process shown in Fig. P3.11. The height h in the tank is to be controlled by adjusting the flow rate q_i into the tank. The system disturbance is the flow rate q_w drawn off from the tank (to be used elsewhere in the processing operation) by a variable displacement pump.

The command signal h_r for the desired height in the tank is set on a pot whose output is $e_r = K_s h_r$. The sensor output is $e_o = K_s h$. The differential amplifier has a gain K_a, the electropneumatic transducer has a gain K_e, and the pneumatic control valve (pneumatic actuator + flow control valve) has a gain K_v. At the initial equilibrium point, the inflow rate exactly matches the outflow rate, and the height in the tank is $\bar{H}$ with $h = h_r = 0$. A is the surface area of the tank, and R is the resistance to flow offered by the valve in the outflow path (R = incremental change in pressure across the valve/incremental change in flow through the valve).

Draw a functional block diagram of the system and therefrom derive the transfer function between h and h_r.

3.12 A system to control the temperature of an oven chamber is shown in Fig. P3.12. The temperature θ is sensed by a thermocouple whose output is $K_t\,\theta$. The desired temperature is θ_r and the corresponding reference voltage is e_r. The differential amplifier generates a signal e_A which controls

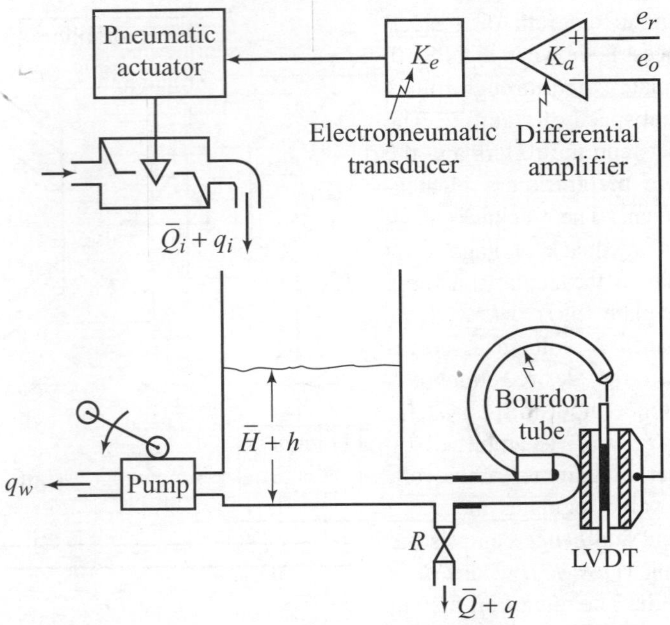

Fig. P 3.11

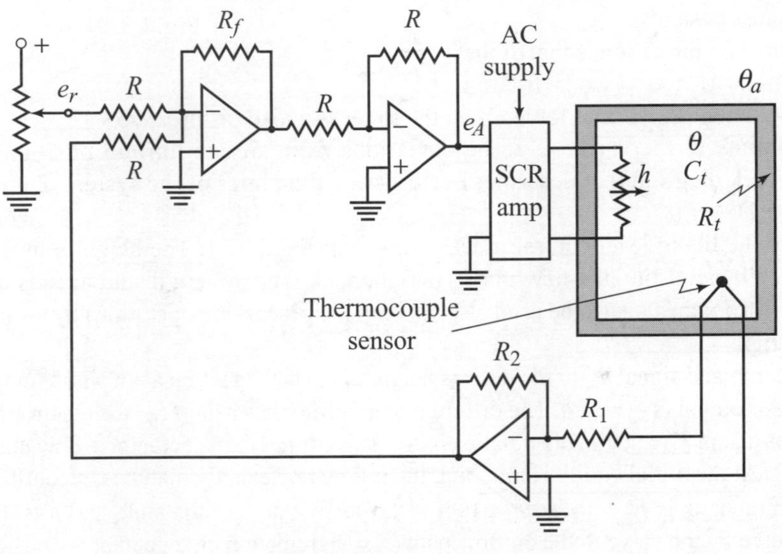

Fig. P3.12

the electrical resistance heater through an SCR amplifier and gives rise to heat flow at the rate $h = K_h e_A$ to the oven chamber. Heat also flows out of the oven due to imperfect insulation of its walls. The temperature of the ambient air is θ_a. Thermal capacitance of the medium inside the chamber is C_t and thermal resistance of the wall material is R_t. Thermal capacitances of the heater and wall are negligible.

Assuming all the variables are deviations from the equilibrium, obtain a block diagram model of the system that includes the transfer functions of all the subsystems. Derive the transfer function $\theta(s)/E_r(s)$.

3.13 In a heating system, air flows at constant volumetric flow rate Q into and out of a chamber of volume V. The density of the inflowing air is ρ, specific heat c, and temperature θ_i.

A system to control the outflow temperature is shown in Fig. P3.13. The temperature θ is sensed by a thermistor with a Wheatstone bridge whose output $e_o = K_s\theta$. The desired temperature is θ_r and the corresponding $e_r = K_s\,\theta_r$. The differential amplifier generates a signal e_A which controls electrical resistance heater through an SCR amplifier and gives rise to heat flow at the rate $h = K_h\,e_A$. The heater has temperature θ_1, thermal capacitance C_1, and thermal resistance R_1. The walls of the chamber are perfectly insulated and have negligible thermal capacitance.

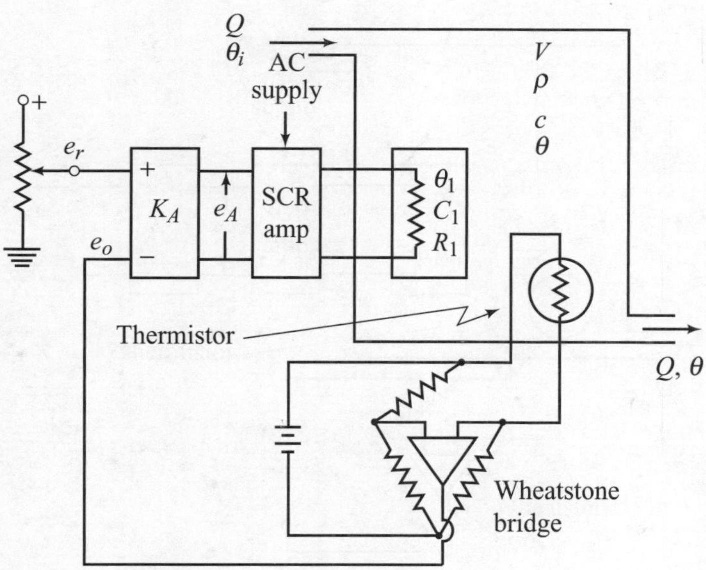

Fig. P3.13

Assume that the variables $\{\theta_r, \theta, \theta_1, \theta_i\}$ are deviations from the equilibrium position. Considering θ_i as the disturbance, draw a functional block diagram of the system that includes the transfer functions of all the subsystems. Derive the transfer functions $\theta(s)/\theta_r(s)$, and $\theta(s)/\theta_i(s)$.

3.14 **MATLAB/Simulink Exercise** (i) Consider the liquid-level control system shown in Figs 3.48 and 3.49. Create this system in the SIMULINK window, and study its response to step values of command and disturbance signals. (ii) Actuators (valves) and sensors (measurement devices) often have additional dynamic lags that can be simulated by transfer functions. Study the effect of neglecting these dynamic lags in the model of Fig. 3.49. (iii) Manipulated variables are often constrained between minimum (0 flow, for example) and maximum (fully open valve) values. A **Saturation** icon from the Simulink library can be used to simulate this behavior. For the system of Fig. 3.48, study the effect of valve saturation.

3.15 **MATLAB/Simulink Exercise** (i) Consider the temperature control system described in Figs 3.44 and 3.45. Create this system in the SIMULINK window, and study its response to step values of command and disturbance signals. (ii) Often you will want to simulate the behavior of systems that have dead time. The **Transport Delay** icon can be selected from the Simulink library. Simulate the system of Fig. 3.44 when it includes transport delay, and study the effect of dead time on the system output.

USE OF FEEDBACK FOR CONTROL OF UNCERTAIN SYSTEMS

4

Preview

Chapter 1 introduced the concept of feedback in a general and qualitative way by examining some very diverse systems that all use feedback. In Chapters 2 and 3, we built a quantitative platform; feedback control systems were represented by transfer function models, block diagram models and signal-flow graph models. Many uncertainties are involved in the development of models for designing linear control systems. In addition, uncertainties (disturbances/noise) are also introduced from outside the feedback loop. If the design performs well for substantial variations in the dynamics of the plant from that used in design, we say that the design is *robust*. The design of most industrial control systems is based on the methods that cash on the feedback property: the feedback-control designs are robust to certain extent because feedback reduces sensitivity of the closed-loop system with respect to uncertainties. Learning objectives for this chapter are to investigate quantitatively the effects of feedback on the performance of uncertain systems. Frequency-domain design, as we shall see in later chapters, provides good robustness against uncertainties because of frequency-selective characteristics of signals in control systems. Investigation of robustness properties in frequency domain is included in learning objectives set for this chapter.

4.1 INTRODUCTION

Recall the basic feedback system block diagram defined in Chapter 2, and shown in an abridged form as Fig. 4.1. When we speak of a controller $D(s)$, our concern is mainly with the *information handling devices*; the sources of information are the reference input $r(t)$ which is derived from the command $y_r(t)$, and the feedback signal $b(t)$ which is produced by the feedback path elements $H(s)$ from the controlled output $y(t)$. The *power handling devices*—actuators and the controlled process itself—are represented by $G(s)$ in the block diagram of Fig. 4.1. $N(s)$ is the plant transfer function from the disturbance input $w(t)$ to the output $y(t)$.

The controller $D(s)$ is designed to act on the actuating error signal $\hat{e}(t) = r(t) - b(t)$ to produce the control signal $u(t)$. The control logic that is physically implemented for this purpose is the *control law* or *control action*.

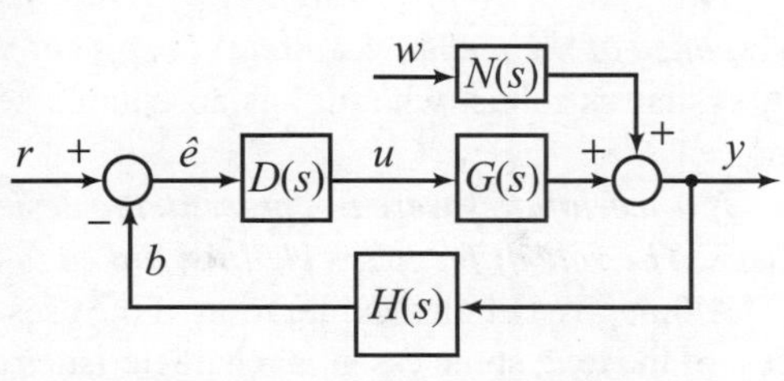

Fig. 4.1 *A non-unity feedback system*

A nonzero system error $e(t) = y_r(t) - y(t)$ results from either a change in command or a disturbance. The general function of the controller is to keep the controlled variable near its desired value. Since we never know the numerical values of the parameters of the controlled process with certainty, and some controller designs can be more sensitive to such parameter variations than others, a parameter sensitivity specification might also be included. In this chapter, we will demonstrate that by using feedback control configuration of the form shown in Fig. 4.1, we can meet these control requirements.

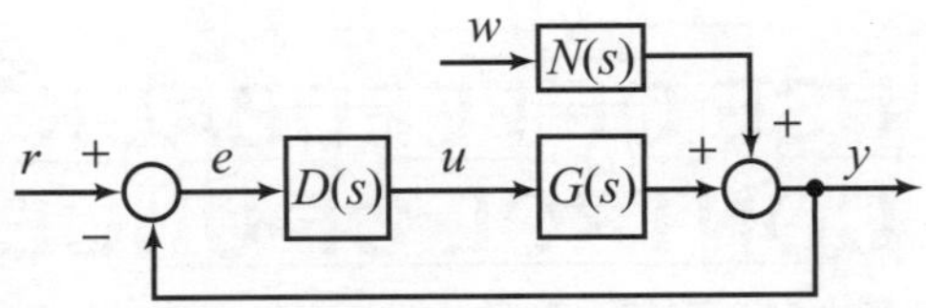

Fig. 4.2 *A unity-feedback system*

A special case of feedback systems which is of particular interest arises when $H(s) = 1$ (Fig. 4.2) and is often referred to as a *unity-feedback system*. The reference input $r(t)$, in this case, is equal to the commanded value of the controlled output. It may be noted that a system's block diagram may not appear to be of unity-feedback form when initially derived. It is often possible and convenient to manipulate the block diagram into a unity-feedback model where the system error appears explicitly as the actuating signal.

Feedback is that property of the system which permits the controlled output to be compared with the input command so that upon existence of a difference, an (actuating) error signal arises which acts to bring the two into correspondence. One possible objective in the design of a feedback control system is that the controlled output should exactly and instantaneously reproduce its input. In practice, this is not possible, since every feedback system will contain inductive- and capacitive-type components that store energy in some form. These elements and their interconnections with energy-dissipative components are responsible for the system's dynamic response characteristics. Such systems introduce dynamic lags in the response.

Once we recognize that the system dynamics cannot be ignored, we need a new design objective. A possible new design objective is to shape the dynamic response of the feedback system so that the output follows the input as closely as possible; if not instantaneously, then in as short a time as possible. To carry out control system design to meet this objective, we require mathematical models of all signals and systems within the feedback loop and the environment in which it operates. Signals and systems in the feedback loop and its environment (i) cannot be modeled precisely, (ii) may change in an unpredictable manner with time, and (iii) may be subject to significant unmeasurable disturbances.

For example, we have derived process models in the previous chapter for many industrial control systems. In no case, complete knowledge of the physical system was available. Whatever available, could not be captured mathematically because of complexity considerations. ***We derived the approximate linear models with constant parameters. These process models are an inaccurate representation of the actual physical system because of unmodeled dynamics of the system, unmodeled time delays, and changes in the parameters of the actual physical system with changes in the environmental conditions. The environment also contributes to uncertainty through unpredictable (low-frequency) disturbances (refer to Fig. 4.1), and through (high-frequency) noise entering the loop through the sensor.***

The goal of robust design is to retain assurance of system performance (the desired dynamic response of the feedback system) in spite of significant uncertainties listed above, and others, if any. A system is robust when it has acceptable performance in spite of model changes, disturbances and noise.

If uncertainties were not present, the desired system performance could be achieved without feedback. The role of feedback is, thus, linked to uncertainties in the system.

Although a detailed exploration of the uses and problems of feedback control will mostly occupy the rest of the text, some essential characteristics can be illustrated quite easily to give a helpful preview of the forthcoming developments. This is the objective of our discussion in this chapter.

4.2 THE CONTROL OBJECTIVES

In this section, we describe a control problem in general terms. Given a plant, which cannot be altered by the designer, with the following variables associated with it (Fig. 4.3):

1. A controlled variable $y(t)$, which is the variable we wish to control.
2. A command variable $y_r(t)$, which represents the prescribed value of the controlled variable.
3. An observed variable $b(t)$, which is measured by means of a sensor and which is used to obtain information about the plant during operation; the observed variable is usually contaminated with observation noise $w_n(t)$.
4. A disturbance variable $w(t)$, which influences the plant but which cannot be manipulated.
5. A control variable $u(t)$, which influences the plant and which can be manipulated.

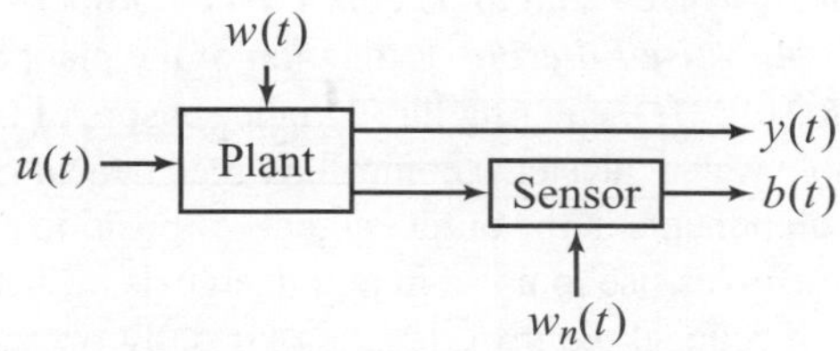

Fig. 4.3 *The plant of a control system*

The *control problem* is roughly described below.

For a given command variable, find an appropriate input $u(t)$ so that the controlled variable follows the command variable:

$$y(t) \cong y_r(t);\ t \geq t_0$$

where t_0 is the time at which the control starts.

An important class of control problems consists of those problems where the command variable is constant over long periods of time; the control objective is to force the controlled variable to reach the commanded value quickly and to maintain the controlled variable at that value thereafter. The duration of the transient period of control, which starts at the beginning of the control process and terminates when the controlled variable approximately reaches its prescribed value, will be referred to as the *settling time*. Lower the settling time, higher the *speed of response* of the process. The *steady-state accuracy* (negligible error between the controlled output and command input during steady-state period of control) is of critical importance in control systems.

In designing control systems, the following aspects must be taken into account.

1. *Stability* Roughly speaking, stability in a system implies that small changes in reference inputs, disturbance inputs, and/or initial conditions do not result in large changes in the system output. Stability is a very important property of control systems. Almost every working system is designed to be stable; we seek to improve the speed of response, steady-state accuracy, and other aspects of system performance within the constraints imposed by stability considerations. Analytical study of stability will be taken up in Chapter 5.

Feedback control systems are prone to instability. For a stable plant, the closed-loop system may not be always stable. ***Feedback control system design may, therefore, demand a trade-off between the accuracy in realizing control objectives and the stability.***

2. *Input Amplitude Constraints* For a control system designed according to linear theory and methods, it is essential that the devices used in the system operate in linear fashion under various operating conditions. If under certain circumstances, the devices are driven into the region of nonlinear characteristics, the linear design theory may describe only approximately, or quite often may even give a completely erroneous prediction of the system performance. ***Excessively large magnitudes of signals at various levels in a control system can drive the devices into the nonlinear region of operation.*** Some examples are: torque-speed curves of motors, synchro error detectors, process control valves, hydraulic servovalves, amplifiers, gears, dampers, etc. The requirement of linear operation of devices under various operating conditions imposes a constraint on the range of values over which the plant input $u(t)$ is allowed to vary. As will be seen, this constraint is of major importance and prevents us from obtaining perfect command-following systems.

3. *Disturbance Rejection* External disturbances $w(t)$ influence the plant in an unpredictable way. They are typically the uncontrolled variations in the load on a system. In systems controlling mechanical motions, load disturbances may represent forces, e.g., wind gusts on a stabilized antenna, waves on a ship, variations in payload of a robot manipulator, etc. In voltage regulating systems, variations in electrical load is a major source of disturbances. In thermal systems, the load disturbances may be caused by variations in surrounding temperature. In fluid systems, the load disturbances may result from variations in feed flow or variations in demanded flow.

Load disturbances typically vary slowly. As we shall see later in this chapter, the errors due to load disturbances tend to be corrected by proper design of feedback control.

4. *Noise Filtering* Excitation of the plant by high-frequency measurement-error signals (observation noise) $w_n(t)$, entering through the sensors of the controlled variable, influences the plant in an unpredictable way. In systems controlling mechanical motions, mechanical vibrations may lead to high-frequency components in the output signals of position and speed sensors. Continuous stirring of fluid in tank reactors gives rise to noise in output signals of thermocouple or other temperature sensors.

Feedback control depends virtually on accurate measurement of controlled variables, because the feedback action has little correcting effect for the wrong information given by the sensors. The character of the measurement errors often depends on filtering in the instruments. Effects of measurement errors on control system performance can therefore be reduced by examining the instrument and modifying the filtering so that it satisfies the requirements of the particular control problem.

5. *Sensitivity and Robustness* Robustness is an important concern in control system design. We will briefly discuss two aspects of the robustness problem.

The first aspect arises because every control system is designed on the basis of a necessarily approximate model of the plant. Since the success of the analytical approach of design is heavily dependent on the simplicity of the mathematical model, many idealizing assumptions are usually made which lead to uncertainties not only in the parameters of the model but also in the model structure. Using a linear model for an intrinsically nonlinear process is an example of deviation in model structure; omitting certain time lags in order to obtain a reduced-order model is another example. The question arises: will the physical system have a performance that is sufficiently close to the performance predicted by the idealized model? For example, if the linear model is stable, will the physical system, which, in fact, is a small perturbation (possibly nonlinear) from the linear model, still be stable?

The second aspect of the robustness appears in the following manner: the design is based on the (idealized) model which characterizes a number of nominal properties of the plant. Once the controller is implemented and the controlled plant is operating in the field, the properties of the physical system differ from the nominal properties because of environment effects, wear, aging, and other natural factors. So the question is: what is the effect of all such time dependent deviations from nominal values on the performance of the physical system in the field? Such questions can be studied by calculating the effect on the system performance due to changes in design parameters and exogenous disturbances. It is a fact that some designs, which perform nominally perfectly, are totally inadequate in the field because they are too sensitive to small perturbations in some of their parameters.

In the course of this text, we will discuss from time to time these questions of *robustness*.

The effect of differentially small deviations in the parameters of a model of known structure can be studied using sensitivity analysis. The motivation for sensitivity analysis is given by the practical experience that parameters often drift during the operation of the system in the field, but they typically remain in the vicinity of their nominal values. It is interesting to know whether such parameter variations cause severe changes of the system behavior or have merely a minor influence on the important system properties.

As we shall see later in this chapter, the errors due to parameter variations tend to be corrected by the proper design of feedback control.

The next section will demonstrate that feedback control can

(i) ***force the controlled variable to accurately follow its prescribed value in steady-state;***
(ii) ***shape the dynamics of the controlled variable;***
(iii) ***greatly reduce the effect on the controlled variable of all external disturbances except those associated with the sensor; and***
(iv) ***make the system tolerant of variations in hardware parameters other than those of the sensor.***

The use of feedback is motivated by the presence of external disturbances and model uncertainties. More strictly, the use of feedback can be regarded as unreasonable if there is no uncertainty in the system because, for the unperturbed system, open-loop control may produce the same or probably better result.

Example 4.1 In the following, we review the control objectives of heat-exchanger control system discussed in Chapter 1 (Fig. 1.15). A sketch of the exchanger is given in Fig. 4.4.

The objective is to maintain the outlet temperature of the process fluid, $\theta(t)$ °C, at its desired value or set point, θ_r°C, in the presence of disturbances in the process fluid flow, $q_m(t)$ kg/sec (= $Q\rho$; Q m^3/sec is the volumetric fluid flow and ρ kg/m^3 is the fluid density), flowing inside the tubes of the heat exchanger, and inlet temperature $\theta_i(t)$ °C. The steam flow $q_{ms}(t)$ kg/sec, condensing on the outside of the tubes, is the variable that can be adjusted to control the outlet temperature, as it determines the amount of energy supplied to the process fluid. In addition to counteracting the effects of disturbances, the control system will be called upon to follow the commands for change in the outlet temperature of the process fluid.

The first step in the design is the selection of the temperature sensor and steam-valve actuator. For temperature measurement, we select an intelligent sensor which has, in addition to the basic device such as a thermocouple, a signal processing circuit which takes the output from the thermocouple and converts it to a current signal (in the range of 4 to 20 mA) that is proportional to the temperature. The signal processing circuit includes an amplifier and a filter. For steam valve adjustments, we select a pneumatic actuator which can provide large output power. Since the valve actuator must be operated by air pressure and the sensor generates electric current signals, a current to pressure transducer is also required in the feedback loop.

The feedback control scheme works as follows: the measured controlled variable is compared with the setpoint; K_r in Fig. 4.4 represents the conversion of setpoint scale. The controller generates a control signal (in the range of 4 to 20 mA), $u(t)$, on the basis of the difference between the measurement and setpoint. The control signal is then connected

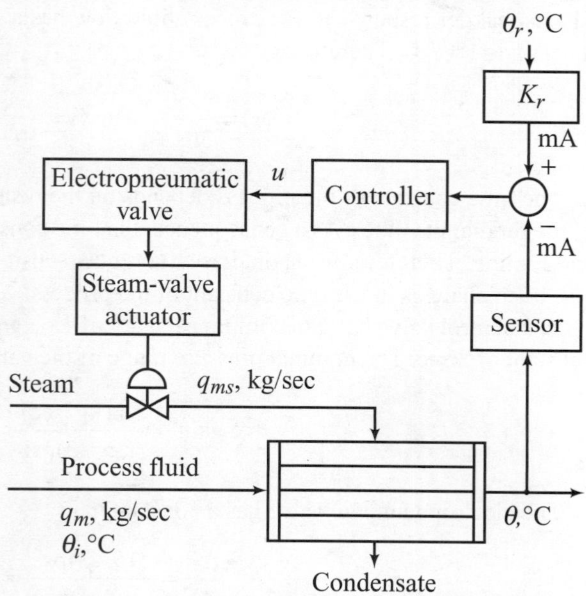

Fig. 4.4 *Feedback control loop for heat exchange*

to the pneumatic actuator of the steam valve through an electropneumatic device—a current to pressure transducer. The function of the valve actuator is to position the valve in proportion to the control signal. The steam flow is then a function of the valve position.

Once the hardware is chosen, the remainder of the design is concerned with trying to make the best use of the process. If the design is to be carried out analytically, a transfer function description of the feedback system must be found. Figure 4.5 shows the block diagram of the entire feedback loop. We now find the transfer function model of each block. Let us start with the heat exchanger.

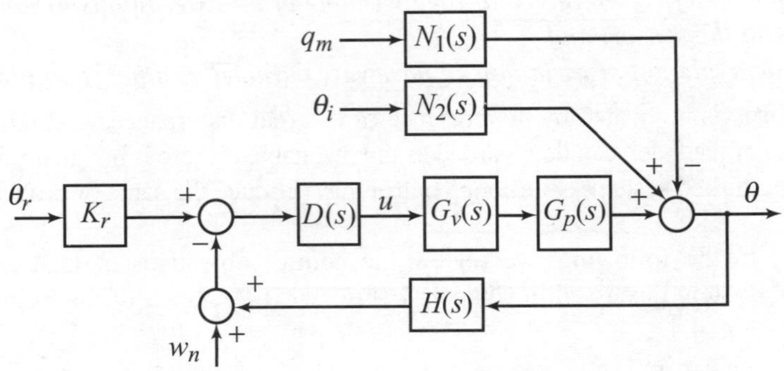

Fig. 4.5 *Block diagram of the feedback loop in Fig. 4.4*

As shown in Fig. 4.5, the heat exchanger consists of three blocks, one for each of its three inputs. $G_p(s)$ is the process transfer function relating the outlet temperature to the steam flow, $N_1(s)$ is the process transfer function relating the outlet temperature to disturbance in process fluid flow, and $N_2(s)$ is the process transfer function to disturbance in inlet temperature.

The following experimentally obtained parameters will be used in developing a model of the feedback system.

The exchanger response to the steam flow has a gain of 50 °C/(kg/sec), and a time constant of 30 sec. The exchanger response to the process fluid flow has a gain of 3 °C/(kg/sec), and to inlet temperature the gain is 1°C/°C. Therefore,

$$G_p(s) = \frac{50}{30s+1}; N_1(s) = \frac{3}{30s+1}; N_2(s) = \frac{1}{30s+1}$$

The time constant of the process depends on the residence time (tube volume/volumetric flow rate) of the fluid in the tubes. As a consequence, the time constant may vary because of flow rate fluctuations and fouling. Therefore, we should ascertain how sensitive the performance of the final system design will be to changes in the time constant of the process.

The control valve has a maximum capacity of 1.6 kg/sec of steam, linear characteristics, and a time-constant of 3 sec. The nominal pressure range of the valve is 3 to 15 psi.

$$\text{Valve gain} = \frac{1.6(\text{kg/sec})}{(15-3)\text{psi}} = \frac{1.6}{12}(\text{kg/sec})/\text{psi}$$

The electropneumatic valve has a constant gain

$$\frac{\Delta P}{\Delta I} = \frac{(15-3)\text{psi}}{(20-4)\text{mA}} = \frac{12}{16}\text{psi/mA}$$

Including the constant gain of the current-to-pressure transducer in the control-valve transfer function, $G_v(s)$, we obtain

$$G_v(s) = \frac{0.1}{3s+1}$$

The sensor has a calibrated range of 50 to 150°C and a time constant of 10 sec.

$$\text{Sensor gain} = \frac{16 \text{ mA}}{(150-50)\text{ °C}} = 0.16 \text{ mA/°C}$$

The transfer function of the sensor is given by

$$H(s) = \frac{0.16}{10s+1}$$

The setpoint scale factor

$$K_r = 0.16 \text{ mA/°C}$$

The block diagram of the temperature control loop, as initially proposed, is shown in Fig. 4.6. θ_r is the change in the setpoint value from the initial setting $\bar{\theta}_r$, θ_i is the deviation in the process fluid temperature from the nominal value $\bar{\theta}_i$, q_m is the deviation in the process fluid flow from the nominal value $\bar{Q}_m$, θ is the deviation in the outlet temperature from the nominal value $\bar{\theta}$, and q_{ms} is the deviation in the steam flow rate from the nominal value $\bar{Q}_{ms}$ (more on control objectives for a heat exchanger control loop, and their realization using PID control, will appear later in this chapter).

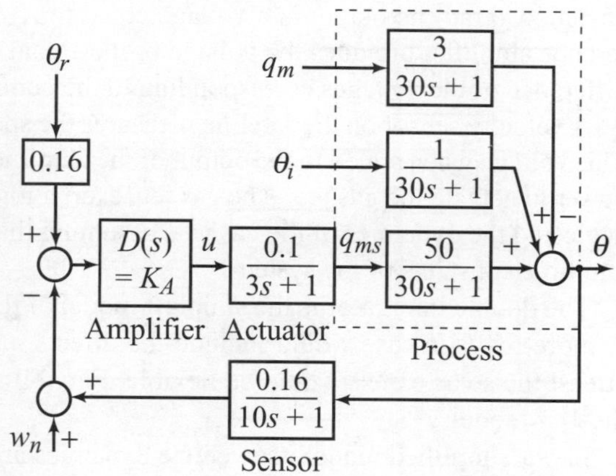

Fig. 4.6 *A model of a heat exchanger control loop*

Example 4.2 We now look at the control objectives of a satellite attitude control system. Satellites usually require attitude control so that antennas, sensors, and solar panels are properly oriented. For example, antennas are usually pointed toward a particular location on earth, while solar panels need to be oriented toward the sun for maximum power generation.

The satellite will be subjected to a variety of disturbance torques from such sources as aerodynamic drag, solar pressure, gravity gradient torques, and the rotation of various satellite components, e.g., pumps, motors and antennas. It will be assumed that the vehicle has initially been aligned to the desired attitude. Due to the above mentioned disturbance torques, it will be necessary to provide a feedback control system in order to maintain the satellite very close to the desired attitude. In addition to counteracting the effects of the disturbance torques acting on the vehicle, the control system will be called upon to change the attitude, subject to commands received from ground stations.

In this example, we shall highlight the important physical aspects of the system, taking a number of liberties with the practical aspects so as to avoid undue complication which would tend to obscure our main objectives.

So long as the angular velocity and the error angles of the vehicle are small, angular motions about the three axes are essentially uncoupled and may be controlled independently. Thus, we may design each of the three attitude-control loops independently, which is an important simplification. Figure 4.7 depicts single-axis attitude control schematic, where motion is only allowed about the axis perpendicular to the page. The angle θ that describes the satellite orientation must be measured with respect to an 'inertial' reference, that is, a reference that has no angular acceleration. The control signal comes from the gas jets that produce a torque $T(t)$ $(= F \times d)$ about the mass center.

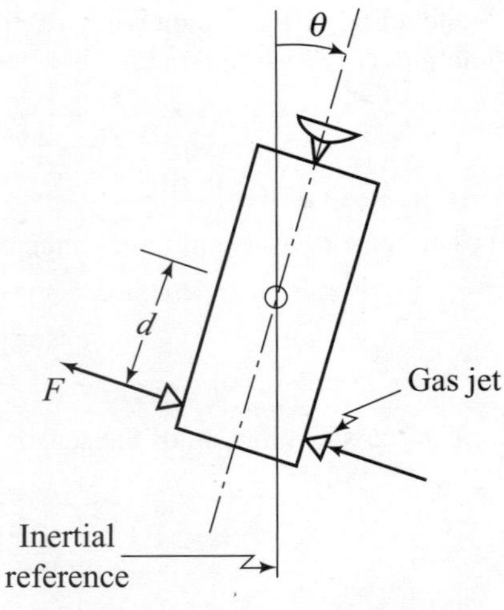

Fig. 4.7 *Satellite control schematic*

A block diagram of the control system, as initially proposed, is given in Fig. 4.8. The command signal θ_r rad, transmitted from earth, is converted to electrical form ($e_r = K_P\theta_r(t)$ volts) in the satellite. A position sensor on the satellite measures the deviation θ rad; its output is a voltage equal to $K_P\theta(t)$. A difference amplifier provides the voltage proportional to $\hat{e}(t)$: the difference of the voltages corresponding to the command signal and measured signal.

A velocity sensor on the satellite measures the speed $\dot{\theta}$ rad/sec; its output is a voltage equal to $K_V\dot{\theta}$. This voltage when added to the output of the amplifier, gives voltage $u(t)$, which must provide sufficient power to actuate the gas jets. The gas jets exert a torque $T(t) = K_T u(t)$ about the mass center. We have neglected the dynamics of the gas jets, assuming that the opening time is short compared to the important time constants of the system.

The double integrator is the simplest model of the satellite, assuming a rigid body with no friction. A more realistic case would include the effects of flexibility in the satellite attitude control, where atleast the solar panels would be flexible. Flexibility in satellite would add a set of complex poles to the $1/s^2$ model.

In our simplified model, the vehicle dynamics are governed by the differential equation,

$$J\ddot{\theta} = T(t) + T_w(t)$$

where J is the moment of inertia about the axis of rotation, and the total torque is the sum of control torque and the disturbance torque. Although the moment of inertia will vary over the life of the satellite

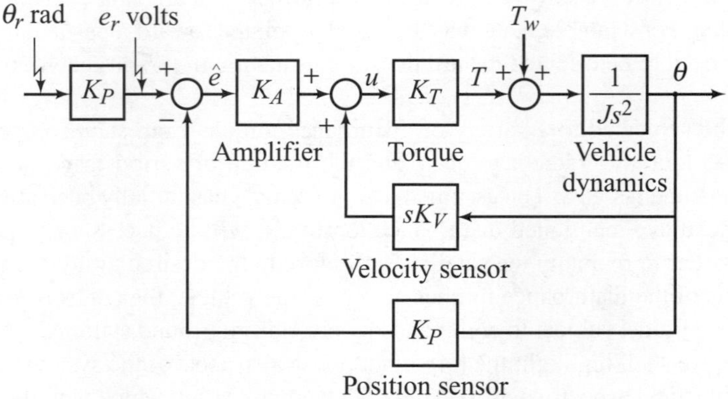

Fig. 4.8 *A model of a satellite attitude control system*

because of the consumption of fuel, it presumably changes very slowly compared with the response of the vehicle. Thus, we are justified in treating it as a constant but should ascertain how sensitive the performance of the final system design is to the anticipated changes in J. Also the effects of ignoring complex dynamics and the effects of disturbance torques should be ascertained.

4.3 FEEDBACK CONTROL SYSTEM CHARACTERISTICS

Feedback systems play an important role in control engineering practice because the use of feedback can accomplish certain desirable results that would not be readily possible with an open-loop system. Some of the potential benefits (and liabilities) of feedback will be discussed in this section. Generally, the material presented here is applicable to any system whose mathematical model takes the form of a fixed linear feedback configuration. These occur not only in automatic control systems and electronic circuits, where the feedback is introduced by design, but also in much diverse fields as physiology, economics, ecology, etc., where feedback is inherent in the natural laws governing the process. However, consistent with the major thrust of this book, our terminology, convention, and illustrations will be taken in the context of control engineering.

4.3.1 System Sensitivity

A process represented by the transfer function $G(s)$, whatever its nature, is subject to modeling errors, changing environment, aging, and other natural factors that affect the control system performance. Consider the open-loop control system of Fig. 4.9. Assume that the controller $D_o(s)$ has been designed using a given transfer function model of the process, and it meets the specifications on the system output $y(t)$. Obviously, the output will approach the desired value only if the dynamic behavior of the real process approaches that of its model. The modeling approximations and the process-behavior changes with time during operation result in an inaccurate output. There is no provision within the system for supervision of the output and no mechanism is provided to correct the output. Consider now the closed-loop control system of Fig. 4.10. This system senses the change in the output through the feedback signal and attempts to correct the output. The feedback signal gives this system the capability to act as a self-correcting mechanism.

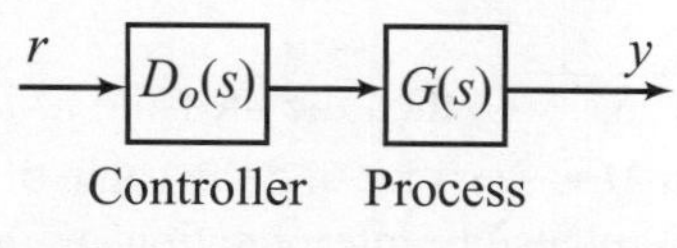

Fig. 4.9 *An open-loop control system*

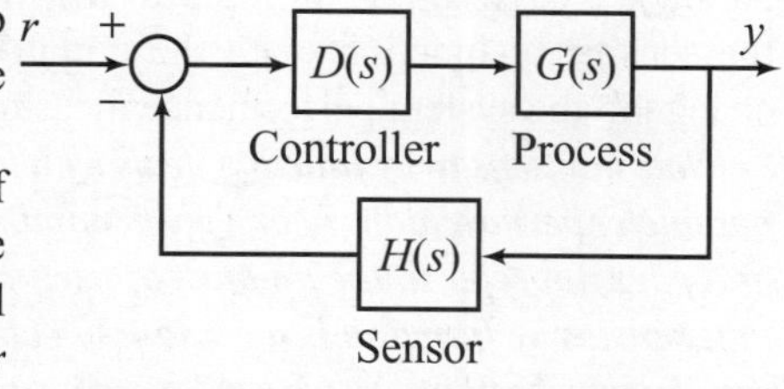

Fig. 4.10 *A closed-loop control system*

For simplicity, we assume that the structure and order of the process model are chosen exactly and they do not change with time. Then the differences between the process model and the real process behavior are manifested as *parameter errors*. Let $\theta_1, \theta_2, \ldots,$ be the parameters of the process model $G(s)$ which are subject to change.

In the following, we consider deviation in only one parameter, θ (extension of analysis to the case of deviations in more than one parameter is straightforward). If θ_n is the nominal value of the parameter, then the process model corresponding to the nominal working point may be expressed as $G(\theta_n, s)$. $G(\theta, s)$ represents a set of perturbed process models when the parameter θ changes over a specified range. We assume for the moment that $H(s)$ is not subject to parameter variations.

Sensitivity is the dependence of the system performance upon the value of the parameter θ. This property is usually investigated in terms of some specific performance measure, e.g., speed of response; or in terms of some characteristic performance function such as the frequency response of the closed-loop system. We take closed-loop transfer function $M(s)$ as the performance function characterizing the system behavior, which depends on parameter θ in addition to other parameters. Consider perturbation only in the parameter θ. For the nominal value θ_n of this parameter, we get the nominal value of the performance function as $M(\theta_n, s)$.

Parameter changes $\Delta\theta$ induce changes ΔM in the performance function. Hence, instead of $M(\theta_n, s)$, we get the new value $M(\theta_n, s) + \Delta M = M(\theta_n + \Delta\theta, s) = M(\theta, s)$.

The Taylor's series expansion (refer to Eqn. (2.1)) of $M(\theta, s)$ about the nominal point θ_n yields (assuming that the function is differentiable)

$$M(\theta_n + \Delta\theta, s) = M(\theta_n, s) + \left(\left. \frac{\partial M(\theta, s)}{\partial \theta} \right|_{\theta=\theta_n} \right) \Delta\theta + \cdots$$

If we restrict our attention to terms of at most first order, we get

$$\Delta M(\theta_n, s) = \left(\left. \frac{\partial M(\theta, s)}{\partial \theta} \right|_{\theta=\theta_n} \right) \Delta\theta$$

Expressing deviations in terms of relative changes, we obtain

$$\frac{\Delta M(\theta_n, s)}{M(\theta_n, s)} = \frac{\Delta\theta}{\theta_n} \left[\left(\left. \frac{\partial M(\theta, s)}{\partial \theta} \right|_{\theta=\theta_n} \right) \frac{\theta_n}{M(\theta_n, s)} \right] \tag{4.1}$$

System sensitivity to parameter θ is defined as the ratio of the percentage change in the system performance measure M from its nominal value to the percentage change in the parameter. S_θ^M denotes the sensitivity of M with respect to θ, and is given by

$$S_\theta^M \triangleq \lim_{\Delta\theta \to 0} \frac{\Delta M(\theta_n, s)/M(\theta_n, s)}{\Delta\theta/\theta_n} = \left[\left. \frac{\partial M(\theta, s)}{\partial \theta} \right|_{\theta=\theta_n} \right] \frac{\theta_n}{M(\theta_n, s)} \tag{4.2}$$

If $S_\theta^M = 0$, then the measure M is insensitive to the parameter θ at the nominal point θ_n. If $S_\theta^M \neq 0$, then M is sensitive and the value of S_θ^M is a measure of the degree of dependence of M on θ.

A plausible interpretation of the sensitivity function is shown in Fig. 4.11. Here we treat the relative change in the parameter θ, $\Delta\theta/\theta_n$, as an input to the process modeled by the sensitivity function, with the output being the relative change in the system performance measure.

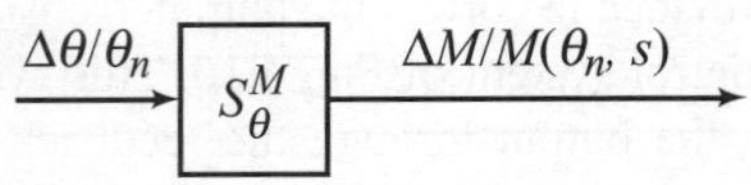

Fig. 4.11 *Interpretation of sensitivity function as a gain function*

Note that sensitivity function deals with infinitesimally small parameter perturbations near a given nominal parameter value and does not refer to the amount of perturbation. Experience on use of such functions, however, shows that sensitivity function is a reasonable characteristic of the effects of small rather than infinitesimally small perturbations. As we shall see shortly, the errors due to parameter variations tend to be corrected by the proper design of feedback control.

The sensitivity analysis of feedback systems represented by block diagram of Fig. 4.10 is frequently carried out using the sensitivity function S_G^M: the sensitivity of closed-loop transfer function $M(\theta_n, s)$ to variations in $G(\theta_n, s)$. We may define this function as

$$S_G^M \triangleq \frac{\Delta M(\theta_n, s)/M(\theta_n, s)}{\Delta G(\theta_n, s)/G(\theta_n, s)} \tag{4.3}$$

i.e., the ratio of the percentage change in the performance measure from its nominal value to percentage change in uncertain plant model from its nominal value.

Referring to Eqn. (4.1), we may express relative changes in $G(\theta_n, s)$ as

$$\frac{\Delta G(\theta_n, s)}{G(\theta_n, s)} = \frac{\Delta\theta}{\theta_n}\left[\left(\left.\frac{\partial G(\theta, s)}{\partial\theta}\right|_{\theta=\theta_n}\right)\frac{\theta_n}{G(\theta_n, s)}\right] \tag{4.4a}$$

Relative changes in (refer to Fig. 4.10)

$$M(\theta_n, s) = \frac{D(s)G(\theta_n, s)}{1 + D(s)G(\theta_n, s)H(s)} \tag{4.4b}$$

are given by

$$\frac{\Delta M(\theta_n, s)}{M(\theta_n, s)} = \frac{\Delta\theta}{\theta_n}\left[\left(\frac{\partial}{\partial\theta}\left.\left\{\frac{D(s)G(\theta, s)}{1 + D(s)G(\theta, s)H(s)}\right\}\right|_{\theta=\theta_n}\right)\frac{\theta_n}{M(\theta_n, s)}\right]$$

$$= \frac{\Delta\theta}{\theta_n}\left[\frac{D(s)}{[1 + D(s)G(\theta_n, s)H(s)]^2}\left(\left.\frac{\partial G(\theta, s)}{\partial\theta}\right|_{\theta=\theta_n}\right)\frac{\theta_n}{M(\theta_n, s)}\right]$$

$$= \frac{\Delta\theta}{\theta_n}\left[\left\{\frac{D(s)G(\theta_n, s)}{1 + D(s)G(\theta_n, s)H(s)}\right\}\left\{\frac{1}{1 + D(s)G(\theta_n, s)H(s)}\right\}\left\{\frac{1}{G(\theta_n, s)}\right\}\left(\left.\frac{\partial G(\theta, s)}{\partial\theta}\right|_{\theta=\theta_n}\right)\frac{\theta_n}{M(\theta_n, s)}\right]$$

From Eqns (4.4), it follows that

$$\frac{\Delta M(\theta_n, s)}{M(\theta_n, s)} = S(\theta_n, s)\frac{\Delta G(\theta_n, s)}{G(\theta_n, s)} \tag{4.5a}$$

with the *sensitivity function* of the feedback control given as

$$S_G^M = S(\theta_n, s) = \frac{1}{1 + D(s)G(\theta_n, s)H(s)} \tag{4.5b}$$

The sensitivity function given by Eqns (4.5) shows how relative changes in the input/output behavior of the closed-loop depend on the changes in the process transfer function. In general, the sensitivity function is a function of the Laplace transform variable s, which makes sensitivity very difficult to interpret. However, if we replace s in Eqn. (4.5b) by $j\omega$, we get the sensitivity as a frequency response. Then, we can assign meaning to the sensitivity for frequencies within the bandwidth[1] of the system. Since the system will not transmit frequencies outside its bandwidth (an approximation), the sensitivity for frequencies much greater than the system bandwidth is of little interest.

Evaluating the sensitivity in Eqn. (4.5b) as a function of frequency, yields

$$S_G^M = S(\theta_n, j\omega) = \frac{1}{1 + D(j\omega)G(\theta_n, j\omega)H(j\omega)} \tag{4.6}$$

The term *DGH* evaluated at a specific frequency is called the *loop gain* at that frequency (refer to Eqn. (2.116d)). ***At frequencies within the system bandwidth, we would like the loop gain to be***

[1] It gives the significant frequency range of the closed-loop system (refer to Section 2.10). Chapter 11 on frequency response analysis will describe this important parameter of control system performance.

as large as possible to reduce the sensitivity of the closed-loop behavior to the process parameters. Generally, one of the purposes of the controller D(s) is to allow the loop gain to be increased without destabilizing the system.

Assume now that we are dealing with an open-loop system such as the one depicted in Fig. 4.9. It is clear that for this system, we can apply the result given by Eqns (4.5) with $H(s) = 0$. Thus,

$$S_G^M = 1, \text{where } M(\theta_n, s) = D_o(s)G(\theta_n, s) \tag{4.7}$$

It can be readily seen that feedback reduces the system sensitivity to changes in process parameters.

Next we derive the sensitivity of the closed-loop transfer function to parameter changes in the sensor $H(s)$. Assume now that the parameter θ_n is a part of $H(s)$, and the process transfer function $G(s)$ is not subject to parameter variations. For this case, the closed-loop transfer function becomes

$$M(\theta_n, s) = \frac{Y(s)}{R(s)} = \frac{D(s)G(s)}{1 + D(s)G(s)H(\theta_n, s)} \tag{4.8}$$

For an infinitesimal change $\Delta\theta$ in the nominal parameter, θ_n, we obtain

$$\left.\frac{\partial M(\theta, s)}{\partial \theta}\right|_{\theta=\theta_n} = -\left[\frac{D(s)G(s)}{1 + D(s)G(s)H(\theta_n, s)}\right]^2 \left.\frac{\partial H(\theta, s)}{\partial \theta}\right|_{\theta=\theta_n} \tag{4.9}$$

From Eqns (4.8) and (4.9), it follows that

$$\frac{\Delta M(\theta_n, s)}{M(\theta_n, s)} = S_H^M \frac{\Delta H(\theta_n, s)}{H(\theta_n, s)} \tag{4.10a}$$

where

$$S_H^M = \frac{-D(s)G(s)H(\theta_n, s)}{1 + D(s)G(s)H(\theta_n, s)} \tag{4.10b}$$

This equation shows that for large values of loop gain DGH, sensitivity of the feedback system with respect to H approaches unity. ***Thus, we see that the changes in H directly affect the system output. To solve this problem, we generally can use high-quality components for the sensor.***

The use of feedback in reducing sensitivity to parameter variations is an important advantage of the feedback control systems. To have a highly accurate open-loop system, the components of the process plant must be carefully selected so that the parameters of the physical system are close to that of $G(s)$—the model used in design, under all operating conditions. On the other hand, a closed-loop system allows $G(s)$ to be less accurately specified, since the effects of parameter variations are mitigated by the use of feedback. However, a closed-loop system requires careful selection of the components of the feedback sensor. Since the plant is made up of power elements and the sensor is made up of measuring elements which operate at low power levels, the selection of accurate $H(s)$ is far less costly than that of $G(s)$ to meet the exact specifications.

Many plants, by their very nature, have parameters that vary extensively during operation. An example is an aircraft whose parameters vary over very wide ranges with speed, altitude, and so on. Hence, aircraft autopilot design must satisfy a rigorous specification on sensitivity to plant parameter variations. Feedback control is the only solution.

The price for improvement in sensitivity by use of feedback is paid in terms of* loss of *system gain. The gain of the closed-loop system shown in Fig. 4.10 is $D(s)G(s)/[1 + D(s)G(s)H(s)]$ while with $H(s) = 0$, the gain becomes $D(s)G(s)$. Hence, by use of feedback, the system gain is reduced by the same factor as by which the sensitivity of the system to parameter variations is reduced. Sufficient open-loop gain can, however, be built into a system so that we can afford to lose some gain to achieve improvement in sensitivity. We should not place a gain outside the feedback loop; the overall system will be quite sensitive to elements outside the loop.

Example 4.3 Several important speed control systems are used in steel mills for rolling steel sheets and moving the steel through the mill (Fig. 4.12). We consider here a speed control system employing an armature-controlled dc motor with negligible armature inductance, as the actuator. Figure 4.13a gives the schematic of the speed control system and Fig. 4.13b gives the operational block diagram:

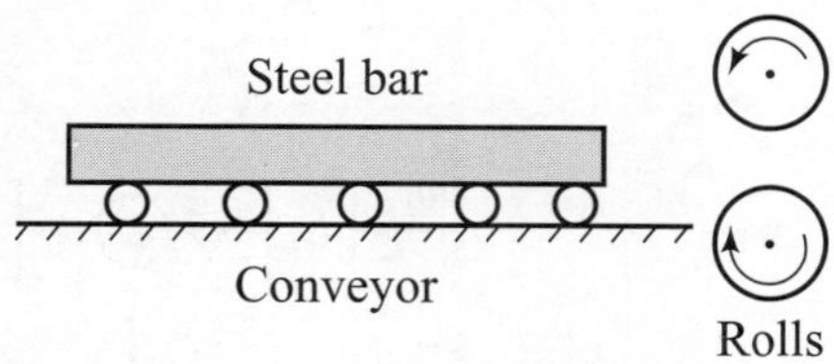

Fig. 4.12 *Rolling steel sheets*

$e_a(t)$ = armature applied voltage
R_a = armature resistance
K_b = motor back-emf constant
K_T = motor torque constant
J = moment of inertia of motor and load
B = viscous-friction coefficient of motor and load
$T_w(t)$ = disturbance torque representing loading effect on the rolls when the steel bar engages in the rolls
$\omega(t)$ = motor velocity

The transfer function of the open-loop speed control system obtained from Fig. 4.13b, is

$$M_o(s) = \frac{\omega(s)}{E_a(s)} = \frac{K}{\tau s + 1} \tag{4.11}$$

where
$$K = \frac{K_T}{BR_a + K_T K_b}; \tau = \frac{JR_a}{BR_a + K_T K_b}$$

The sensitivity of the open-loop mode of operation to variation in the constant K is (refer to Eqn. (4.2))

$$S_K^{M_o} = \frac{\partial M_o}{\partial K}\left(\frac{K}{M_o}\right) = 1 \tag{4.12}$$

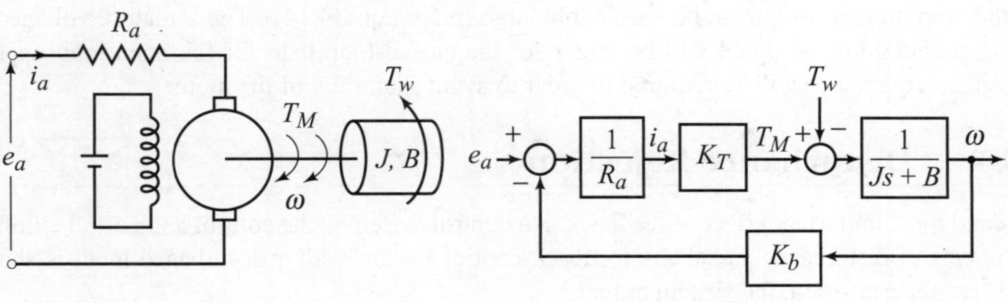

Fig. 4.13 *Open-loop speed control system*

A closed-loop speed control system is easily obtained by utilizing a tachogenerator to generate a voltage proportional to speed, as shown in Fig. 4.14. This voltage is subtracted from the reference voltage representing the desired speed, and amplified. The closed-loop transfer function is

$$\frac{\omega(s)}{\omega_r(s)} = M(s) = \frac{KK_AK_t}{\tau s + 1 + KK_AK_t} \tag{4.13}$$

where K_t = tachogenerator constant; and K_A = amplifier gain.

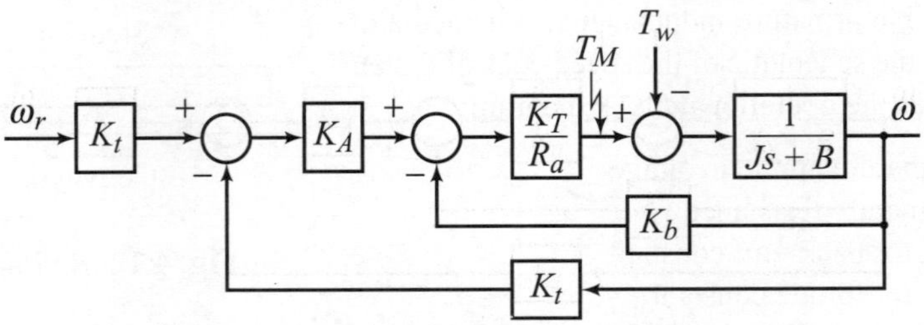

Fig. 4.14 *Closed-loop speed control system*

The sensitivity of the closed-loop system to variation in K is

$$S_K^M = \frac{\partial M}{\partial K}\left(\frac{K}{M}\right) = \frac{\tau s + 1}{\tau s + 1 + KK_AK_t} \tag{4.14}$$

For a typical application, we might have $1/\tau = 0.1$, and $(1 + KK_AK_t)/\tau = 10$. Therefore, from Eqn. (4.14) we obtain

$$S_K^M = \frac{s+0.1}{s+10}$$

We find that the sensitivity is a function of s and must be evaluated for various values of frequency. At a specific frequency, for example, $s = j\omega = j1$, the magnitude of the sensitivity of the closed-loop system is

$$\left|S_K^M\right| \cong 0.1$$

Thus, compared to that of the open-loop case, the sensitivity of the closed-loop speed control system at $\omega = 1$ is reduced by a factor of ten.

It should be noted that in order to reduce sensitivity of the closed-loop system to parameter variations, the amplifier gain K_A must be reasonably large (refer Eqn. (4.14)). The armature voltage signal and its associated torque signal will be larger for the closed-loop than for the open-loop operation. Therefore, a large motor will be required in order to avoid saturation of the motor.

4.3.2 Disturbance Rejection

The second most important effect of feedback in a control system is the control and partial elimination of the effects of disturbance signals. A feedback control system with a disturbance input is shown in Fig. 4.1. By superposition, the system output is

$$Y(s) = \frac{D(s)G(s)}{1 + D(s)G(s)H(s)}R(s) + \frac{N(s)}{1 + D(s)G(s)H(s)}W(s) \tag{4.15}$$

With $R(s) = 0$, the system output becomes

$$Y(s) = \frac{N(s)}{1 + D(s)G(s)H(s)} W(s) \tag{4.16}$$

For the open-loop system,

$$Y(s) = N(s)W(s) \tag{4.17}$$

It can be readily seen that feedback reduces the effect of disturbances on the controlled output. Replacing s in Eqn. (4.16) by $j\omega$, we get

$$\frac{Y(j\omega)}{W(j\omega)} = \frac{N(j\omega)}{1 + D(j\omega)G(j\omega)H(j\omega)}$$
$$= S(j\omega)N(j\omega)$$

where

$$S(j\omega) = \frac{1}{1 + D(j\omega)G(j\omega)H(j\omega)}$$

is the sensitivity function of the control system (defined in Eqn. (4.5b)). Recall from the preceding subsection that ***the loop gain $D(j\omega)G(j\omega)H(j\omega)$ must be made large to reduce sensitivity to variations in plant parameters. This condition also meets the requirement of reducing the effects of disturbance input on the controlled output of the system.***

For the case that loop gain is large ($DGH >> 1$), the disturbance transfer function

$$\frac{Y(j\omega)}{W(j\omega)} \cong \frac{N(j\omega)}{D(j\omega)G(j\omega)H(j\omega)}$$

An obvious way to accomplish disturbance rejection is to make $N(j\omega)$ small. Since $N(j\omega)$ is a function of the plant parameters, it cannot be modified. Alternatively, we may increase the loop gain by increasing the gain of $D(j\omega)$. Note that increasing the plant gain (if possible) will also increase the loop gain, but increasing the plant gain may also increase the gain of $N(j\omega)$. Hence, in this case, disturbance rejection may not be improved. Thus, to reject disturbances, the loop gain must be increased in such a manner that the gain from the disturbance input to the system output is not increased.

Figure 4.15 gives a block diagram of the situation wherein measurement noise $W_n(s)$ enters the system through the feedback link. The closed-loop transfer function for this disturbance is

$$\frac{Y(s)}{W_n(s)} = -\frac{D(s)G(s)}{1 + D(s)G(s)H(s)}$$

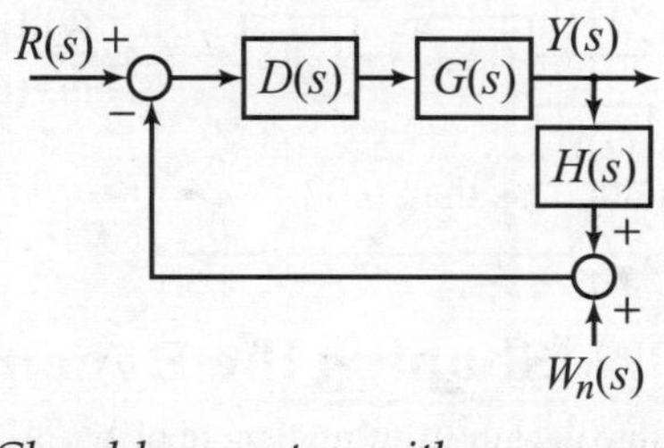

Fig. 4.15 *Closed-loop system with measurement noise*

Thus, large loop gain will lead to large output error due to measurement noise. This is in conflict with the disturbance-rejection property with respect to the configuration of Fig. 4.1. To solve this problem, we generally can examine the measuring instrument and modify the filtering so that it satisfies the requirements of a particular control system.

Example 4.4 As a specific example of a system with an unwanted disturbance, let us reconsider the speed control system for a steel rolling mill (Figs 4.12–4.14). Rolls passing steel through are subject to large load changes or disturbances. As a steel bar approaches the rolls, the rolls turn unloaded. However, when the bar engages in the rolls, the load on the rolls increases immediately to a large value. This loading effect can be approximated by a step change of disturbance torque $T_w(s)$.

Let us study the effect of load torque $T_w(s) = A/s$ in the open-loop speed control system of Fig. 4.13b and the closed-loop speed control system of Fig. 4.14.

In the open-loop case, the change in speed due to load disturbance (obtained by manipulation of the block diagram of Fig. 4.13b) is

$$\omega(s) = \left(\frac{-1}{Js + B + K_T K_b / R_a} \right) T_w(s)$$

The steady-state error in speed due to $T_w(s) = A/s$ is found by using the final value theorem:

$$\lim_{t \to \infty} \omega(t) = \lim_{s \to 0} s\omega(s) = \lim_{s \to 0} s \left[\frac{-1}{Js + B + K_T K_b / R_a} \right] \frac{A}{s}$$

$$= \frac{-A}{B + K_T K_b / R_a}$$

Manipulation of the block diagram of Fig. 4.14 gives the following relation between output speed and load disturbance (alternatively, draw a signal flow graph and use Mason's gain rule) for the closed-loop speed control system:

$$\omega(s) = \frac{-1}{Js + B + (K_T/R_a)[K_b + K_A K_t]} T_w(s)$$

Again the steady-state output is obtained by utilizing the final value theorem, and we have

$$\lim_{t \to \infty} \omega(t) = \lim_{s \to 0} s\omega(s) = \frac{-A}{B + (K_T/R_a)[K_b + K_A K_t]}$$

The ratio of the closed-loop to open-loop steady-state speed output due to an undesirable disturbance is

$$\frac{\omega_{ss}(CL)}{\omega_{ss}(OL)} = \frac{BR_a + K_T K_b}{BR_a + K_T (K_b + K_A K_t)}$$

and is usually less than 0.02.

4.3.3 Shaping the Dynamic Response

Consider the feedback control system of Fig. 4.1. The output with $W(s) = 0$ is given by

$$Y(s) = \frac{D(s)G(s)}{1 + D(s)G(s)H(s)} R(s) \tag{4.18a}$$

$$= M(s)R(s) \tag{4.18b}$$

By finding the poles and zeros of $M(s)$, we can characterize the behavior of the closed-loop system's response to a reference input.

Assume that $D(s)$, $G(s)$ and $H(s)$ are rational functions of s; they can each be expressed as ratios of their numerator and denominator polynomials of the form

$$\frac{D_1(s)}{D_2(s)} \triangleq D(s) \tag{4.19a}$$

$$\frac{G_1(s)}{G_2(s)} \triangleq G(s) \tag{4.19b}$$

and

$$\frac{H_1(s)}{H_2(s)} \triangleq H(s) \tag{4.19c}$$

Substituting $D(s)$, $G(s)$ and $H(s)$ from Eqns (4.19) into Eqns (4.18), we obtain

$$M(s) = \frac{D_1(s)G_1(s)/D_2(s)G_2(s)}{1 + D_1(s)G_1(s)H_1(s)/D_2(s)G_2(s)H_2(s)}$$

which reduces to

$$M(s) = \frac{D_1(s)G_1(s)H_2(s)}{D_2(s)G_2(s)H_2(s) + D_1(s)G_1(s)H_1(s)} \tag{4.20}$$

Having obtained the closed-loop transfer function in rational form, we are ready to make two rather important observations.

First, the finite *zeros* of $M(s)$ are given by the zeros of the plant transfer function $G(s)$, the poles of the feedback block $H(s)$, and the zeros of the controller transfer function $D(s)$. Second, the *poles* of $M(s)$ are the roots of the equation

$$D_2(s)G_2(s)H_2(s) + D_1(s)G_1(s)H_1(s) = 0 \tag{4.21}$$

which, because its left-hand side is the denominator polynomial of the transfer function, is the *characteristic equation* of the closed-loop system.

We will see in the next chapter that the poles of $M(s)$ determine the modes of closed-loop system's response; poles in the left half of the complex plane give rise to decaying modes (a requirement for system stability) and the poles in the right half of the complex plane give rise to growing modes (making the system unstable). The zeros of $M(s)$ control the magnitude of each mode and, hence, the contribution of each mode in the total response. ***Thus, we can shape the dynamic response of the closed-loop system by controlling the locations of poles and zeros of*** $M(s)$ ***through the controller*** $D(s)$.

We see from Eqn. (4.20) that the poles of a feedback system do not have a direct relationship to the open-loop poles and zeros; this tends to complicate the analysis. However, it does mean that the designer has the capability of making the closed-loop system behave quite differently from the open-loop system. In order to make effective use of this ability, it is necessary to have methods for analyzing the effects of changes in parameters of the controller $D(s)$ without having to solve Eqn. (4.20) for $M(s)$ or Eqn. (4.21) for the closed-loop poles each time a parameter is changed. Much of our subsequent work will be directed toward developing analytical tools that will allow us to take advantage of this unusual property of feedback systems.

Consider now the total response of the closed-loop system of Fig. 4.1 to both the reference and the disturbance inputs (refer to Eqn. (4.15)):

$$Y(s) = \frac{D(s)G(s)}{1 + D(s)G(s)H(s)}R(s) + \frac{N(s)}{1 + D(s)G(s)H(s)}W(s)$$
$$= M(s)R(s) + M_w(s)W(s)$$

The denominators of $M(s)$ and $M_w(s)$ are identical, implying that both the transfer functions have precisely the same poles. This situation is a consequence of the fact that the characteristic equation is a property of the closed-loop configuration and not of the particular variables that happen to be designated as input and output.

The zeros of $M(s)$ and $M_w(s)$ are different. In fact, it would be rather unfortunate if this difference did not exist because we certainly do not want the output to respond to disturbance inputs in the same way that it responds to reference inputs—generally we would prefer $M(s)$ to be unity, so that the output follows the reference input exactly, and $M_w(s)$ to be zero, so that the output would be unaffected by disturbances.

Upon examination of Fig. 4.9, one notes that the open-loop control system can easily shape the dynamics of the controlled output: the output y follows the input r exactly if

$$D_o(s) = \frac{1}{G(s)}$$

If $D_o(s)$ is realizable (i.e., a physical system with transfer function $D_o(s)$ can be constructed), it can cancel the process dynamics completely since it has the reciprocal transfer function behavior. Therefore, one may logically ask, what is the advantage of the closed-loop system in this case? Again, we return to the concepts of the sensitivity of the system to parameter changes and effects of disturbances as our answer to this question. In the open-loop control configuration of Fig. 4.9, one may implement $D_o(s)$ so that it is equal to the inverse of the nominal transfer function $G(s)$, but $G(s)$ is usually an approximate model of the process. Moreover, during the operation of the system it is inevitable that the parameters of $G(s)$ will change due to environmental changes and $D_o(s)G(s)$ will no longer be equal to unity. Also, because of the disturbances acting on the process $G(s)$ but not on the controller $D_o(s)$, the output $y(t)$ will deviate from the input $r(t)$. In contrast, a closed-loop system continually monitors the output $y(t)$, and provides an actuating signal in order to reduce the effects of parameter variations and disturbances. Thus, we find that the properties of insensitivity to model uncertainties, parameter variations, and disturbance rejection encourage the introduction of feedback.

Example 4.5 Figure 4.6 gives the block diagram description of a heat-exchanger control system (refer to Example 4.1). Assume that the actuator and the sensor devices have zero dynamic lags, i.e., the transfer function of the actuator is 0.1, and that of the sensor is 0.16. Under this assumption,

$$\frac{\theta(s)}{\theta_r(s)} = \frac{D(s)G(s)}{1 + D(s)G(s)}$$

where $$G(s) = \frac{0.8}{30s+1}.$$

If the controller $D(s)$ is an amplifier of gain K_A with negligible dynamic lag, we get

$$\frac{\theta(s)}{\theta_r(s)} = \frac{0.8K_A}{30s + 1 + 0.8K_A}$$

For positive values of K_A, the effect of the feedback is to shift the pole negatively from the open-loop location $s = -1/30$ to the location $s = -(1 + 0.8K_A)/30$; the time constant reduces from 30 to $30/(1 + 0.8K_A)$. This implies that as K_A (and hence the loop gain) increases, the system dynamics continuously becomes faster, i.e., the transient response decays more quickly. Thus, the *relative stability* (which is a quantitative measure of how fast the transients die out in the system) improves with an increase in loop gain.

Consider now the situation which is more closer to the real world: the actuator and the sensor devices in heat-exchanger control system have dynamic lags as shown in Fig. 4.6. Excessive loop gain may result in the instability of the feedback system. The instability phenomenon may be qualitatively described as follows (quantitative discussion will appear in the next chapter).

Consider the feedback system of Fig. 4.6 with a fixed desired value θ_r, and in equilibrium with the controlled variable θ at the desired value. If a process disturbance occurs, the ensuing deviation of the controlled variable will cause a correction to be applied but it will be delayed by the cumulative lags of the sensor, actuator, and process. Eventually, however, the trend of the controlled variable caused by the disturbance will be reversed by the opposition of the process manipulated input q_{ms}, returning the controlled variable toward the desired value. Now if the loop gain is high, a *strong* correction is applied and the controlled variable overshoots the desired value, causing a reversal in the algebraic sign of the system error (difference between desired value and controlled variable). Unfortunately, because of system lags, a reversal of the correction does not occur immediately, and the process manipulated input (acting on 'old' information) is now actually driving the controlled variable in the same direction as it was already going, rather than opposing its excursions, leading to a larger deviation. If the loop gain is not too large and/or system lag is not too large, it may be possible to bring the controlled variable back to its desired value. ***The combination of excessive loop gain with excessive lags always results in feedback system instability.***

4.3.4 Steady-State Accuracy

A feedback control system is valuable because the sensitivity of the system to parameter variations, and effect of disturbances can be reduced significantly. In addition, as we have seen, it provides us with the ability to adjust the transient response. However, as a further requirement, one must examine and compare the final steady-state error for an open-loop and a closed-loop system.

The error of the closed-loop system shown in Fig. 4.2 with $W(s) = 0$ is given by

$$E(s) = \frac{1}{1 + D(s)G(s)} R(s)$$

For a unit step input, $R(s) = \frac{1}{s}$, the steady-state error is given by

$$\begin{aligned} e_{ss} &= \lim_{t\to\infty} [r(t) - y(t)] = \lim_{s\to 0} sE(s) \\ &= \lim_{s\to 0} \left[\frac{1}{1 + D(s)G(s)}\right]\frac{1}{s} \\ &= \frac{1}{1 + D(0)G(0)} \end{aligned} \tag{4.22}$$

The value of $D(s)G(s)$ when $s = 0$ is often called the *dc loop gain*. Therefore, a closed-loop system with a reasonably large dc loop gain will have a small steady-state error.

Consider now the open-loop control system shown in Fig. 4.9. The error

$$E(s) = R(s) - Y(s) = [1 - D_o(s)G(s)]\, R(s)$$

For a unit-step input, the steady-state error is given by

$$e_{ss} = \lim_{s\to 0} sE(s) = 1 - D_o(0)G(0) \tag{4.23}$$

Upon examination of Eqn. (4.23), one notes that the open-loop control system can possess a zero steady-state error by simply adjusting the dc loop gain so that $D_o(0)G(0) = 1$. However, during the operation of the system, it is inevitable that the parameters of $G(s)$ will change and dc loop gain of the system will no longer be equal to unity. The steady-state error will continue to be other than zero until the system is recalibrated. By contrast, ***the feedback system of Fig. 4.2 continually monitors the steady-state error and provides an actuating signal in order to reduce the steady-state error.***

Example 4.6 The effects of feedback on steady-state accuracy can be readily demonstrated by quantitative analysis of a simplified model of a familiar system– the cruise control of an automobile (Fig. 4.16). We assume a simplified model for the car (engine plus body): $10/(10s + 1)$. Furthermore, for the range of speeds used by the system, we assume that 1° change in the throttle angle (our control variable) causes a 10 km/hr change in speed on a level road (i.e., when the disturbance w (% grade of the road) is zero). Also the grade change of 1% (up or down) is assumed to result in speed change of 5 km/hr (throttle angle change of 0.5°).

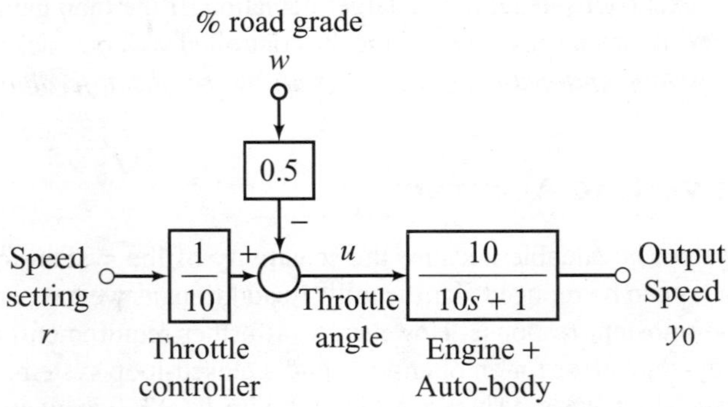

Fig. 4.16 *Block diagram of an open-loop cruise control system*

The open-loop output speed is given by

$$Y_0(s) = \frac{10}{10s+1}(0.1\,R(s) - 0.5\,W(s))$$

For $R(s) = A/s$, and $W(s) = B/s$ (A and B are constants), we have the steady-state output speed

$$y_{0ss} = \lim_{s \to 0} sY_0(s) = A - 5B$$

If $A = 50$ and road is level ($B = 0$), then the output speed will be 50 km/hr with no error. However, if $B = 1$ corresponding to 1% grade, then the speed will be 45 and we have a 5 km/hr error, which is 10% error in the speed. For a grade of 2%, the speed error would be 10 km/hr, which is an error of 20%, and so on. The example shows that there would be no error when $B = 0$, but this result depends on the controller being the exact inverse of the plant gain of 10. In practice, the plant gain is subject to change, and if it does, errors are introduced by this means also.

Assume that plant gain changes to 9, while the controller gain remains at 0.1. With $B = 0$ (i.e., on level road),

$$y_{0ss} = \lim_{s \to 0} s\left[\frac{9}{10s+1} \times \frac{0.1A}{s}\right] = 0.9A$$

$$\text{Error in speed} = \frac{0.1A}{A} = 10\%$$

The block diagram of a feedback scheme is shown in Fig. 4.17, where the controller gain has been set to 10. It has been assumed that we have an ideal speed sensor, whose block is not shown. The closed-loop output speed is given by

$$Y_c(s) = \left[\, 10\big(R(s) - Y_c(s)\big) - 0.5W(s) \,\right] \frac{10}{10s+1}$$

This gives

$$Y_c(s) = \frac{100R(s)}{10s+101} - \frac{5W(s)}{10s+101}$$

For $R(s) = A/s$, and $W(s) = B/s$, we have the steady-state output speed

$$y_{css} = \frac{100}{101}A - \frac{5}{101}B$$

Comparing y_{css} with y_{0ss}, we observe that the feedback has reduced the sensitivity of the speed error to the grade by a factor of 101, Note, however, that there is now a small speed error on level road, because even where $B = 0$,

$$y_{css} = \frac{100}{101}A = 0.99A \text{ km/hr}$$

Assume now that the plant gain changes to 9 while the controller gain remains at 10. On the level road,

$$y_{css} = \lim_{s \to 0} s\left[\frac{90}{10s+91} \times \frac{A}{s}\right] = \frac{90}{91}A$$

Error in speed $= \dfrac{(1/91)A}{A} = 1.1\%$

There is, thus, reduction in the sensitivity to plant gain variations when compared with the open-loop control scheme.

The reduction of the speed sensitivity to grade disturbances and plant gain in our example is due to the high loop gain in the feedback case. Unfortunately, there are limits to how high this gain can be made. The model of feedback system in Fig. 4.17 is over-simplified. A more realistic model will include the dynamics of throttle controller and speed sensor. In a realistic situation, increasing the loop gain can make the response worse than before, or even cause the system to become unstable (refer to Review

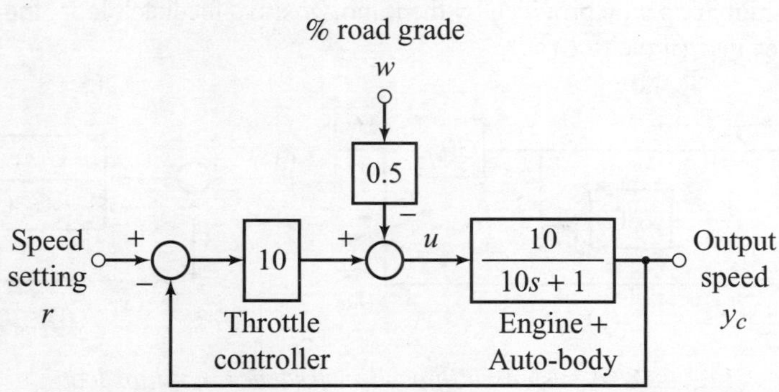

Fig. 4.17 *Block diagram of a closed-loop cruise control system*

Example 4.2). The issue of how to get the gain as large as possible to reduce the errors without making the system become unstable, is what much of feedback control design is all about.

4.3.5 The Case for High-Gain Feedback

Summarizing the benefits of negative feedback in systems with high loop gain, we can state that such systems result in:

(i) good steady-state tracking accuracy;
(ii) good disturbance signal rejection;
(iii) low sensitivity to process parameter variations; and
(iv) good relative stability, i.e., rate of decay of transients.

There are, however, factors limiting the gain:

(i) High gain may result in instability problems.
(ii) Input amplitudes limit the gain; excessively large amplitudes of control signals will drive the process to the saturation region of its operation, and the control system design based on linear model of the plant will no longer give satisfactory performance.
(iii) Measurement noise limits the gain; with high gain feedback, measurement noise appears unattenuated in the controlled output.

Therefore, in design we are faced with trade offs.

4.3.6 Positive Feedback

Positive feedback is an operation which augments an imbalance, thereby precluding stability. If a temperature controller with positive feedback were used to heat a room, it would increase the heat when the temperature was above the setpoint, and turn it off when it was below. Obviously, this property is not conducive to regulation. Sometimes, positive feedback is used in minor loops of feedback control configurations; the stability and regulation of the controlled variable is provided by the primary feedback loop. Figure 4.18a shows a feedback system with the minor loop having positive feedback. The equivalent single-loop block diagram is shown in Fig. 4.18b. If $G_m(s)$ is selected to be nearly unity, the loop gain becomes very high and the closed-loop transfer function approximates to

$$\frac{Y(s)}{R(s)} = \frac{G(s)}{1 - G_m(s) + G(s)} \cong 1$$

Thus, due to the high loop gain provided by the minor positive feedback loop, the closed-loop transfer function becomes insensitive to $G(s)$.

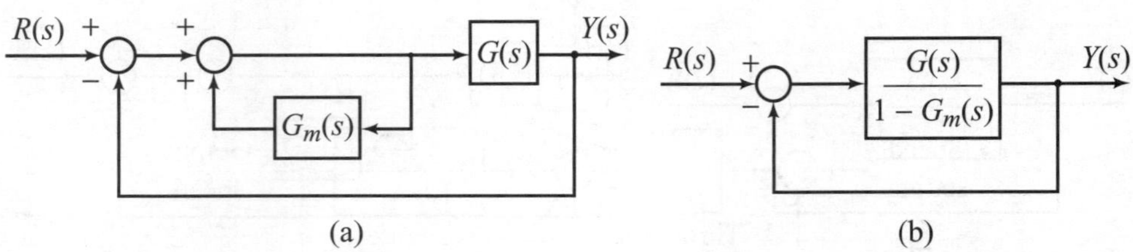

Fig. 4.18 *A system with positive feedback in minor loop*

4.4 DESIGN CONSIDERATIONS

The ultimate goal of a control-system designer is to build a system that will work in real environment. Since the real environment may change with time (components may age or their parameters may vary with temperature or other environmental conditions) or operating conditions may vary (load changes, disturbances), the control system must be able to withstand these variations. The other fact of life that the control system has to face is the issue of model uncertainty. A mathematical representation of a system often involves simplifying assumptions. Nonlinearities are either unknown, and hence unmodeled, or modeled and later ignored to simplify design. Different components of systems (sensors, amplifiers, gears, etc.) are sometimes modeled by constant gains, even though they have dynamics or nonlinearities. Dynamic structures (e.g., aircrafts, satellites, missiles) have complex dynamics in high frequencies and these may initially be ignored. Since control systems are typically designed using much simplified models of systems and other uncertainties, they must possess *robustness* property in order to operate properly in realistic situations; the proper performance demands desirable dynamic and steady-state response characteristics.

Ideally, we would like the control system to satisfy the following criteria:

1. ***The control system must be stable.***
2. ***The effects of disturbances are minimized, providing good disturbance rejection.***
3. ***Rapid, smooth responses to setpoint changes are obtained, i.e., good setpoint tracking.***
4. ***Steady-state error (offset) is eliminated.***
5. ***Excessive control action is avoided.***
6. ***The control system is robust, i.e., insensitive to changes in process conditions and to inaccuracies in the process model.***

In typical control applications, it is not possible to achieve all of these goals simultaneously because they involve inherent conflicts and trade-offs. The trade-offs must balance two important objectives, *performance* and *robustness*. A control system exhibits a high degree of performance if it provides rapid and smooth responses to disturbances and setpoint changes with little, if any, oscillation. A control system is robust if it provides satisfactory performance for a wide range of process conditions and for a reasonable degree of model inaccuracies. ***Robustness can be achieved by choosing a conservative controller (i.e., small values of gains) but this choice tends to result in poor performance.***

In the following, we study the issues of trade-off between performance and robustness from a frequency-domain perspective.

Consider the unity feedback system shown in Fig. 4.19, with the compensator $D(s)$. The system output is y, and the inputs are: r = reference (or command) input; w = disturbance input, and w_n = measurement noise.

Assume that the process model uncertainties are given by the parameters $\boldsymbol{\theta} = [\theta_1, \theta_2, \ldots, \theta_N]^T$, that vary over ranges. Corresponding to nominal parameter vector $\boldsymbol{\theta}_n = [\theta_{n1}, \theta_{n2}, \ldots, \theta_{nN}]^T$, the process transfer function is $G(\boldsymbol{\theta}_n, s)$. The closed-loop input-output behavior at the nominal working point is given by

$$M(\boldsymbol{\theta}_n, s) = \frac{Y(s)}{R(s)} = \frac{D(s)G(\boldsymbol{\theta}_n, s)}{1 + D(s)G(\boldsymbol{\theta}_n, s)} \tag{4.24}$$

The sensitivity of closed-loop system $M(\boldsymbol{\theta}_n, s)$ to variations in $G(\boldsymbol{\theta}_n, s)$ is given by (refer to Eqn.(4.5b))

$$S_G^M = S(\boldsymbol{\theta}_n, s) = \frac{1}{1 + D(s)G(\boldsymbol{\theta}_n, s)} \tag{4.25}$$

Note that

$$S(\boldsymbol{\theta}_n, s) + M(\boldsymbol{\theta}_n, s) = 1 \tag{4.26}$$

As a consequence of this relation, the choice of a particular sensitivity function $S(\boldsymbol{\theta}_n, s)$ directly implies a corresponding unique choice for the function $M(\boldsymbol{\theta}_n, s)$, and *vice versa*. This observation underlines an important factor in design; namely the trade-offs that must be made when selecting these two mutually dependent transfer functions. We call the function $M(\boldsymbol{\theta}_n, s)$, a *complementary sensitivity function*.

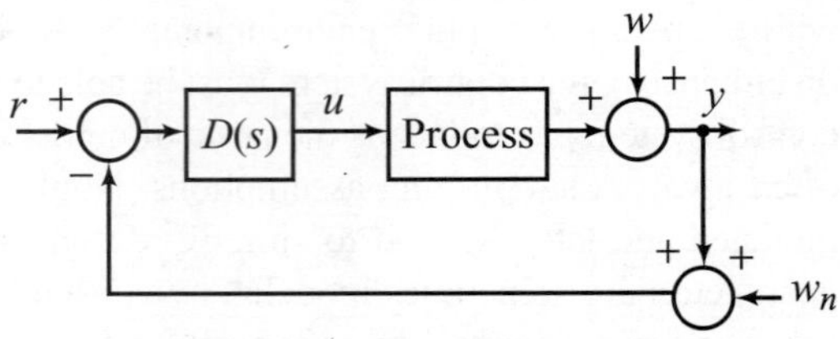

Fig. 4.19 *Block diagram of a feedback system including disturbance and measurement noise inputs.*

For the notational convenience, let us represent sensitivity function $S(\boldsymbol{\theta}_n, s)$ as $S(s)$, plant transfer function $G(\boldsymbol{\theta}_n, s)$ as $G(s)$ and closed-loop transfer function $M(\boldsymbol{\theta}_n, s)$ as $M(s)$. From the block diagram of Fig. 4.19, we may write

$$Y(s) = W(s) + D(s)G(s)[R(s) - Y(s) - W_n(s)]$$

or

$$[1 + D(s)G(s)]Y(s) = W(s) + D(s)G(s)R(s) - D(s)G(s)W_n(s)$$

Therefore, the total output of the closed-loop system is

$$Y(s) = \frac{D(s)G(s)}{1+D(s)G(s)}R(s) + \frac{1}{1+D(s)G(s)}W(s) - \frac{D(s)G(s)}{1+D(s)G(s)}W_n(s) \tag{4.27}$$

If we define the tracking error as $e = r - y$, we get

$$E(s) = \frac{1}{1+D(s)G(s)}R(s) - \frac{1}{1+D(s)G(s)}W(s) + \frac{D(s)G(s)}{1+D(s)G(s)}W_n(s) \tag{4.28}$$

Finally, the controller output (i.e., the plant input) is given by

$$U(s) = \frac{D(s)}{1+D(s)G(s)}R(s) - \frac{D(s)}{1+D(s)G(s)}W(s) - \frac{D(s)}{1+D(s)G(s)}W_n(s) \tag{4.29}$$

In the analysis to follow, it will always be assumed that the feedback system is stable; therefore S and M are stable, proper transfer functions. Equations (4.27)–(4.29) may be expressed in terms of sensitivity functions as follows.

$$Y(s) = M(s)R(s) + S(s)W(s) - M(s)W_n(s) \tag{4.30}$$

$$E(s) = S(s)R(s) - S(s)W(s) + M(s)W_n(s) \tag{4.31}$$

$$U(s) = D(s)S(s)R(s) - D(s)S(s)W(s) - D(s)S(s)W_n(s) \tag{4.32}$$

We are now ready to draw the following conclusions from these relations:

1. ***Disturbance rejection*** *S(s)* must be kept small to minimize the effect of disturbances. From the definition of S, this can be met if the loop gain is large.
2. ***Tracking*** $S(s)$ must be kept small to keep the tracking errors small.
3. ***Noise suppression*** $M(s)$ must be kept small to reduce the effects of measurement noise on the output and error. From the definition of M, this is met if loop gain is small.

4. ***Actuator limits*** $D(s)S(s)$ must be bounded to ensure that the actuating signal driving the plant does not exceed plant tolerances. Another reason for taking this into consideration is to reduce the control energy so that we can use smaller actuators (such as motors). Note that

$$D(s)S(s) = \frac{D(s)}{1+D(s)G(s)} = \frac{M(s)}{G(s)} \tag{4.33}$$

Hence, by keeping M small, we can reduce control energy.

Tracking and disturbance rejection require small sensitivity. Noise suppression and reduction in control energy require small complementary sensitivity. As these two transfer functions add up to unity, we cannot reduce both the transfer functions to zero simultaneously. We can, however, avoid the conflict by noting that in practice, command inputs and disturbances are low-frequency signals (i.e., they vary slowly with time) whereas measurement noise is a high-frequency signal. Therefore, we can meet both the objectives by keeping S small in the low-frequency range and M small for high frequencies. Figure 4.20 depicts typical plots of both the sensitivity function and the complementary sensitivity function that are consistent with our goals.

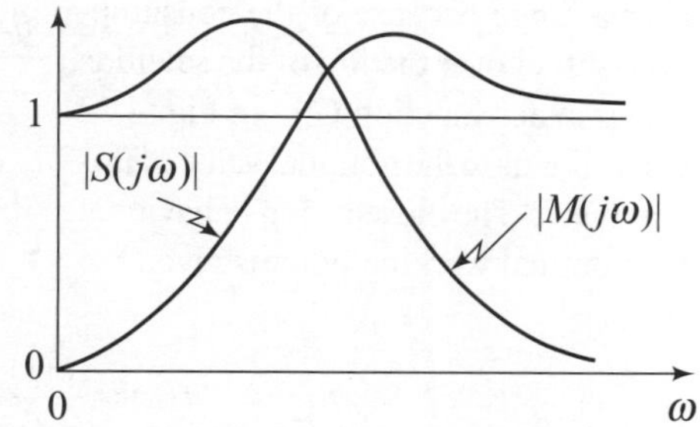

Fig. 4.20 *Typical magnitude plots of S(jw) and M(jw)*

The intermediate frequencies typically control the dynamic characteristics of the feedback system.

Putting together the effects of feedback, we arrive at a general desired shape for the loop gain $|D(j\omega)G(j\omega)|$. This is shown in Fig. 4.21.

1. The considerations of steady-state errors due to command inputs and low-frequency disturbances is an important design component. Design of acceptable steady-state errors can be thought of as placing a lower bound on the very low frequency gain ($D(0)G(0)$) of the system.
2. We have met different frequency ranges in other engineering areas (e.g., radio frequencies: 10 kHz to 1 GHz), but in most control systems, we will often work in the 0 to 100 Hz frequency range. The control signals in the feedback loop are usually of low frequency; in the range of 0 to 20 Hz. To alleviate the effects of high-frequency disturbances (e.g., sensor noise), the gain $|D(j\omega)G(j\omega)|$ at high frequencies is kept low.

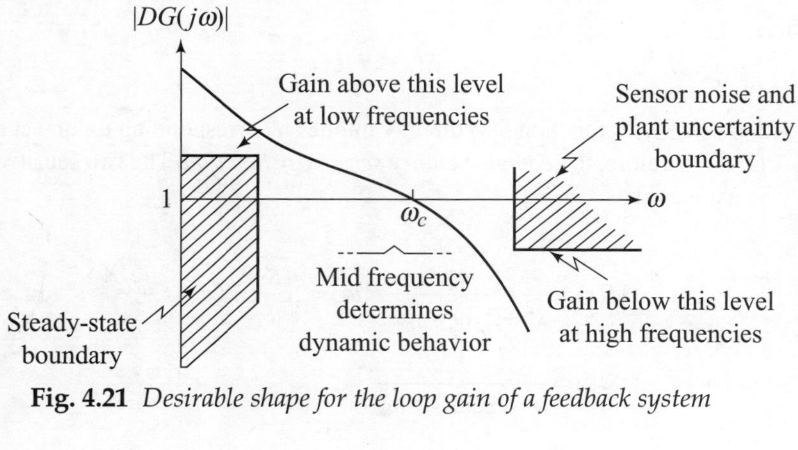

Fig. 4.21 *Desirable shape for the loop gain of a feedback system*

3. In Chapter 12, we will see that for relative stability, $|D(j\omega)G(j\omega)|$ has, for an adequate range of ω (intermediate frequency range), a smooth transition from low-frequency range to high-frequency range; and the *gain crossover* ω_c (refer to Fig. 4.21) is near the required bandwidth.

Example 4.7 Figure 4.8 shows a block diagram model of a satellite attitude control system. The output, θ (rad), is given by an equivalent block diagram of Fig. 4.22 when disturbance torque $T_w = 0$; θ_r (rad) is the reference command.

The moment of inertia J will vary over a range because of the consumption of fuel over the life of the satellite. The transfer function $G(s)$ in Fig. 4.22 corresponds to the nominal value of this parameter. The closed-loop behavior at the nominal working point is given by

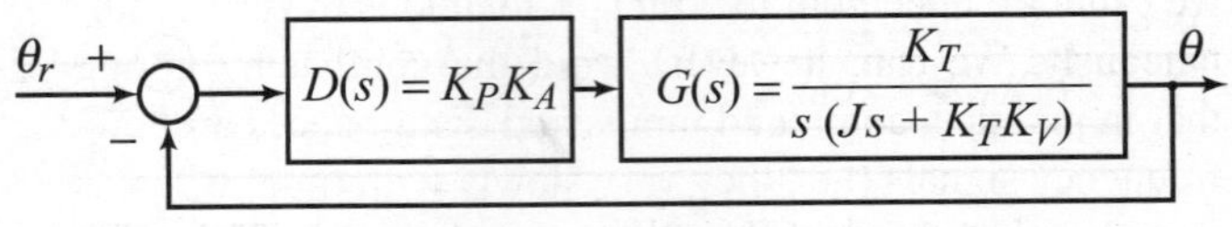

Fig. 4.22 *A model of a satellite attitude control system*

$$M(s) = \frac{\theta(s)}{\theta_r(s)} = \frac{K_P K_A K_T}{Js^2 + sK_T K_V + K_P K_A K_T}$$

$$= \frac{K/J}{s^2 + s(K_T K_V/J) + K/J}; \; K \triangleq K_P K_A K_T$$

The sensitivity of the closed-loop system to variations in $G(s)$ is given by

$$S_G^M = S(s) = \frac{1}{1 + D(s)G(s)}$$

$$= \frac{Js^2 + sK_T K_V}{Js^2 + sK_T K_V + K_P K_A K_T}$$

$$= \frac{s^2 + s(K_T K_V/J)}{s^2 + s(K_T K_V/J) + K/J}$$

Note that

$$S(s) + M(s) = 1$$

The choice of sensitivity function $S(s)$ directly implies a corresponding unique choice for $M(s)$ and *vice versa*; $M(s)$ is, therefore, the *complementary sensitivity function*. The two sensitivity functions can equivalently be expressed as

$$M(s) = \frac{\omega_n^2}{s^2 + 2\zeta\omega_n s + \omega_n^2}; \omega_n = \sqrt{\frac{K}{J}}, \; 2\zeta\omega_n = \frac{K_T K_V}{J}$$

$$S(s) = \frac{s^2 + 2\zeta\omega_n s}{s^2 + 2\zeta\omega_n s + \omega_n^2}$$

$K_V = 2\left(\sqrt{KJ}/K_T\right)$ results in $\zeta = 1$. Corresponding to this condition, plots of $S(j\omega)$ and $M(j\omega)$ are shown in Fig. 4.23. Note that

$$M(j\omega) = \frac{1}{1-\left(\frac{\omega}{\omega_n}\right)^2 + j2\left(\frac{\omega}{\omega_n}\right)}$$

$$S(j\omega) = \frac{-\left(\frac{\omega}{\omega_n}\right)^2 + j2\left(\frac{\omega}{\omega_n}\right)}{1-\left(\frac{\omega}{\omega_n}\right)^2 + j2\left(\frac{\omega}{\omega_n}\right)}$$

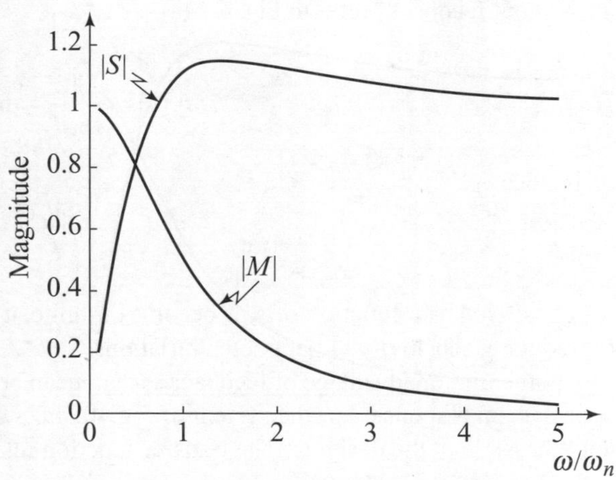

Fig. 4.23 *S(jω) and M(jω) for the second-order system in Fig. 4.22*

To control the effects of noise of high-frequency complex dynamics of the satellite, we require $M(j\omega)$ to be low, and to control the effects of variations in J, we require $S(j\omega)$ to be low. From Fig. 4.23, we observe that the frequency ω_n is an indicator on the boundary between the frequency region in which the sensitivity to J is an important design criterion, and the region in which sensitivity to noise is important. Thus, if we specify ω_n properly, we can expect the system to have an acceptable amount of robustness.

4.5 REVIEW EXAMPLES

Review Example 4.1 For the feedback system of Fig. 4.24b, suppose that the plant parameter a is nominally = 2, but is subject to small changes about its nominal value.

The closed-loop transfer function of the system is

$$M(s) = \frac{K}{s+a+K}$$

The sensitivity of $M(s)$ to changes in a is given by (refer to Eqn. (4.2))

$$S_a^M = \left(\frac{\partial M}{\partial a}\right)\frac{a}{M} = \frac{-a}{s+a+K}$$

For $a = 2$, the sensitivity is

$$S_a^M = \frac{-2}{s+2+K}$$

which can, for any s, be reduced by making K sufficiently large.

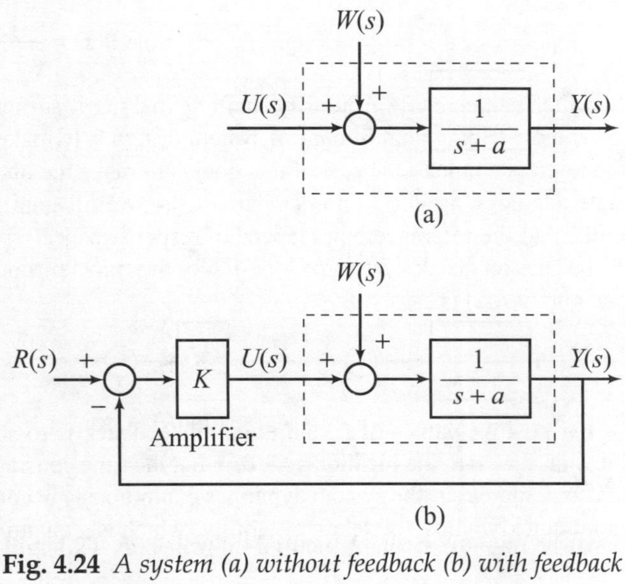

Fig. 4.24 *A system (a) without feedback (b) with feedback*

Without feedback (refer to Fig. 4.24a),

$$M_o(s) = \frac{1}{s+a} \quad \text{and} \quad S_a^{Mo} = \frac{-a}{s+a}$$

For $a = 2$,

$$S_a^{M_o} = \frac{-2}{s+2}$$

which is fixed as a function of s; it becomes infinite at $s = -2$. Feedback, thus, provides a mechanism for reducing sensitivity to parameter variations.

Another major advantage of feedback is that it can be used to reduce the effects of disturbance inputs upon system response. For the system of Fig. 4.24a, a disturbance signal, $W(s)$, affects the plant but is not accessible to the designer. The transfer function relating $Y(s)$ to $W(s)$ is

$$M_{wo}(s) = \frac{1}{s+a} = \frac{Y(s)}{W(s)}$$

For a unit-step disturbance input, the final value of the output due to the disturbance is given by

$$\lim_{t-\infty} y(t) = \lim_{s\to 0} sY(s) = \frac{1}{a}$$

If the plant is driven in the feedback arrangement of Fig. 4.24b, the transfer function relating $Y(s)$ to $W(s)$ becomes

$$M_w(s) = \frac{1/(s+a)}{1+[K/(s+a)]} = \frac{1}{s+a+K} = \frac{Y(s)}{W(s)}$$

For a unit-step disturbance input to the feedback system, the resulting steady-state output is given by

$$\lim_{s\to 0} sY(s) = \frac{1}{a+K}$$

which can be made arbitrarily small by making K sufficiently large.

The overriding goal of control system design is to make the output of the closed-loop system track the reference input. One part of this goal is to make the output converge to the input over time (steady-state accuracy aspect). Another part of the overall goal is to make sure that the output converges *quickly* to the reference input (speed of response aspect).

For the feedback system of Fig. 4.24b, the transfer function between the output $Y(s)$ and the reference input $R(s)$ is

$$M(s) = \frac{K}{s+a+K} = \frac{Y(s)}{R(s)}$$

For positive values of K, the effect of feedback is to shift the pole negatively from the open-loop location $s = -a$ to the location $s = -(a + K)$; the time constant reduces from $1/a$ to $1/(a+K)$. This implies that as K increases, the system dynamics continuously becomes faster, i.e., the transient response decays more quickly. Thus, the relative stability (which is a quantitative measure of how fast the transients die out in the system) improves with an increase in gain (Note that even for negative values of a (say $a = -1$) for which the plant is unstable, the feedback results in a stable closed-loop transfer function $M(s)$).

The error of closed-loop system of Fig. 4.24b with $W(s) = 0$ is

$$E(s) = \frac{1}{1+[K/(s+a)]}R(s) = \frac{s+a}{s+a+K}R(s)$$

The steady-state value of the error to a unit-step input is

$$\lim_{t\to\infty} e(t) = \lim_{s\to 0} sE(s) = \frac{a}{a+K}$$

Thus, by making K large relative to the magnitude of a, we can make the steady-state value of the output close to the desired reference value.

In this example, we have made one observation that an open-loop unstable system can be stabilized with feedback. Suppose the feedback system of Fig. 4.24b had been designed without neglecting smaller time constants in the loop; for example, time constant of amplifier and time constant of sensor. To check the "robustness" of our design, we must examine the stability after including these time constants in our analysis. The next example illustrates the effect of feedback on stability of higher-order systems.

Review Example 4.2 Consider the feedback system shown in Fig. 4.25. The closed-loop transfer function

$$M(s) = \frac{K(s+10)}{s(s+1)(s+10)+K} = \frac{K(s+10)}{s^3+11s^2+10s+K} = \frac{Y(s)}{R(s)}$$

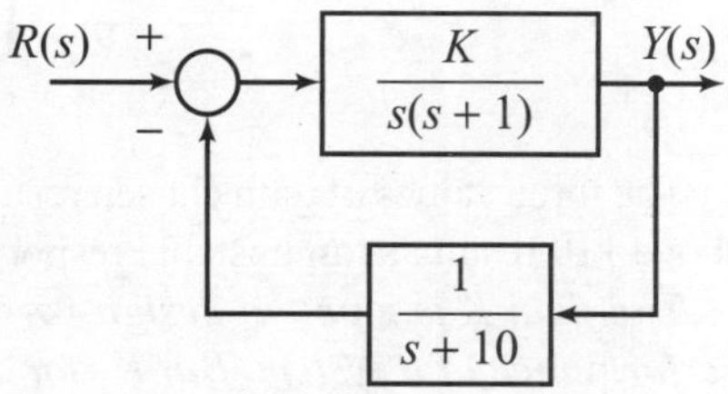

Fig. 4.25 *A feedback system*

The unit-step response of this system with $K = 2$ is shown in Fig. 4.26. For this value of K, the three closed-loop poles are real:

$$M(s) = \frac{2(s+10)}{s^3+11s^2+10s+2} = \frac{2(s+10)}{(s+0.290)(s+0.688)(s+10.02)}$$

The transient component in response $y(t)$ consists of three exponential time functions that decay to zero as $t \to \infty$.

If K is now increased to 110,

$$M(s) = \frac{110(s+10)}{s^3+11s^2+10s+110} = \frac{110(s+10)}{(s-j\sqrt{10})(s+j\sqrt{10})(s+11)}$$

The unit-step response of the system with $K = 110$ has a substantial oscillation (Fig. 4.26).

If we further increase K, say $K = 264$,

$$M(s) = \frac{264(s+10)}{s^3+11s^2+10s+264} = \frac{264(s+10)}{(s-0.5-j4.66)(s-0.5+j4.66)(s+12)}$$

The unit-step response expands sinusoidally without bound (Fig. 4.26). The system becomes unstable; certainly not a desirable response in control applications.

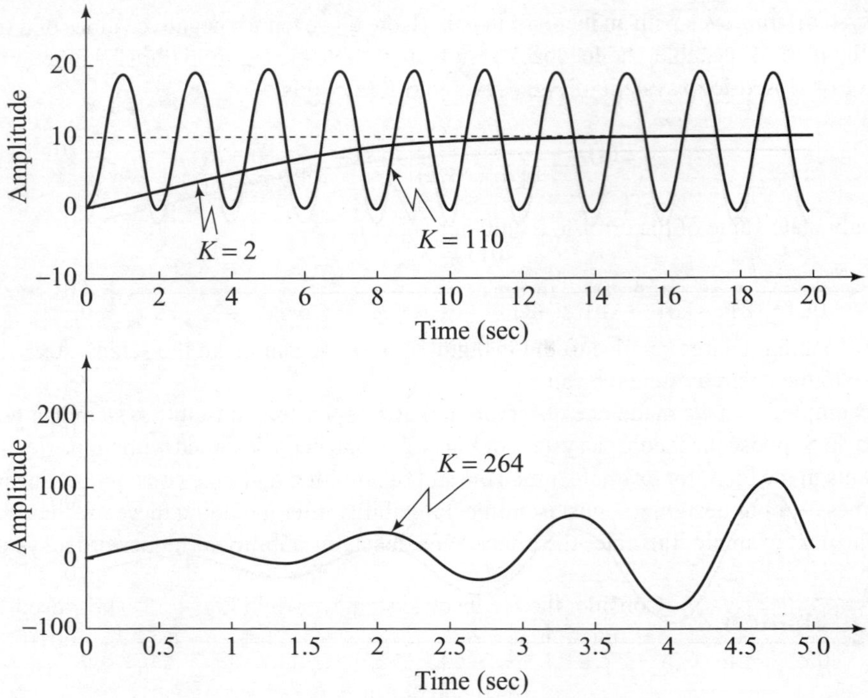

Fig. 4.26 *Effect of feedback on stability*

The three values of gain chosen result in three different behaviors. Also we see that increasing the gain above 110 results in an unstable response. These observations are generally true for all control systems.

The gain K is a potent design variable. When combined with feedback, it can markedly alter the performance of a system. But K can also cause a system with feedback to become unstable. Clearly, the feedback sword has two edges. It is up to the designer to use it wisely.

Review Example 4.3 In this example, we consider a unity-feedback temperature control system with the process transfer function

$$G(s) = \frac{K}{s+0.1}$$

where K has a nominal value of 5.0. The sensor is modeled as a pure gain H, which has a nominal value of 0.05. The loop gain DGH is varied using a controller $D(s) = K_P$. The process parameter K is subject to small changes about the nominal value. The transfer function (closed loop) of the system is

$$M(s) = \frac{K_P G(s)}{1 + K_P G(s) H} = \frac{K_P K}{s + 0.1 + 0.05 K_P K}$$

We first find the sensitivity of $M(s)$ with respect to K. From Eqn. (4.2), we get

$$S_K^M = \left(\frac{\partial M}{\partial K}\right)\frac{K}{M} = \frac{s+0.1}{s+0.1+0.05K_P K}$$

The sensitivity S_K^M is a function of the complex variable s. For $K = 5.0$, the sensitivity is

$$S_K^M(s) = \frac{s + 0.1}{s + 0.1 + 0.25K_P}$$

The sensitivity about the nominal value of K, as a function of frequency, is

$$S_K^M(j\omega) = \frac{0.1 + j\omega}{0.1 + 0.25K_P + j\omega}$$

In a like manner,

$$S_H^M(j\omega) = \frac{-0.25K_P}{0.1 + 0.25K_P + j\omega}$$

Figure 4.27 shows these sensitivity functions plotted as a function of frequency for values of the controller gain K_P of 0.4 and 1.0. These values were chosen somewhat arbitrarily.

The bandwidth of the system can be obtained from the transfer function,

$$M(j\omega) = \frac{5K_P}{0.1 + 0.25K_P + j\omega}$$

The system gain is reduced to $0.707M(j0)$ at the frequency $\omega = (0.1 + 0.25K_P)$, and thus the bandwidth is $\omega_b = (0.1 + 0.25K_P)$. For $K_P = 0.4$, the bandwidth is $\omega_b = 0.2$ rad/sec; and for $K_P = 1$, $\omega_b = 0.35$ rad/sec. Figure 4.27 shows the sensitivity plots with the bandwidths indicated therein.

We can draw the following conclusions from our analysis:

1. The sensitivity of the system to K decreases with increasing loop gain, whereas the sensitivity to H increases with increasing gain.
2. The system is very sensitive to K outside the system bandwidth, which in general, is not significant; and is very sensitive to H inside the system bandwidth, which is significant. Hence, the sensor must be constructed of high quality components.

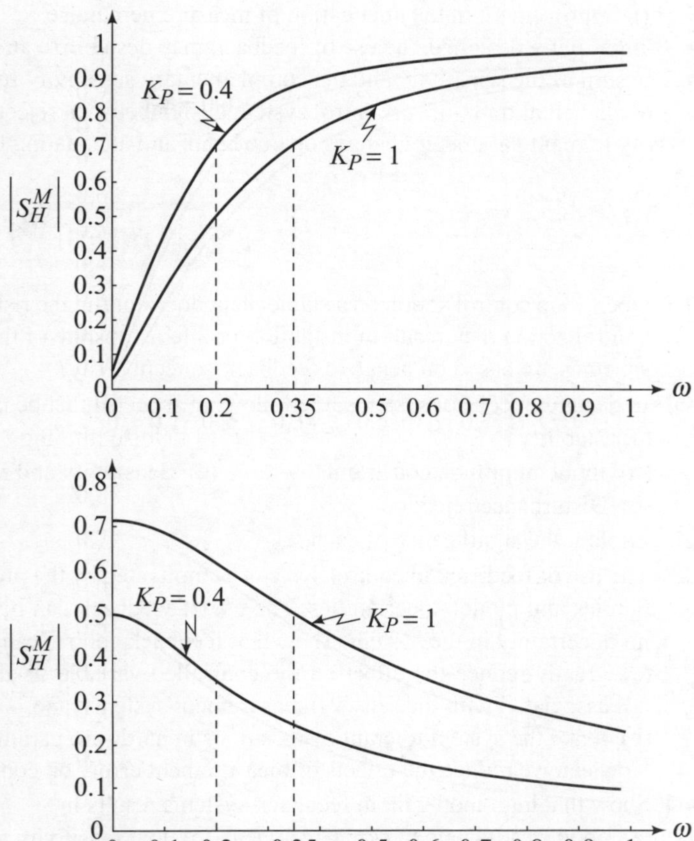

Fig. 4.27 *Sensitivity functions for the system*

CONCLUDING REMARKS

In this chapter, we introduced the concept of feedback and took a first look at its effects. Since the entire book is essentially about feedback, our learning on the topic is ongoing. The present chapter has simply laid the ground work for the following chapters.

What we have learnt?

- A closed-loop system involves the use of a comparator to calculate the error between the reference and the measured output values. By closing the loop, the overall system transfer function will change. This change can easily be manipulated (i) for control and adjustment of the transient response of the system, and (ii) for improvement in the reduction of the steady-state error of the system.
- Open-loop systems can also be used to achieve the desired transient and steady-state behavior; however, it is possible only if there are no major uncertainties which will upset the process.
- The fundamental reasons for using feedback are as follows:
  - (i) Decrease in the sensitivity of the system to variations in the parameters of the process.
  - (ii) Improvement in the rejection of the disturbances.
  - (iii) Improvement in the attenuation of measurement noise.
- If improperly designed, the use of feedback may destabilize an otherwise stable open-loop system.
- The sum of the sensitivity and the complementary sensitivity functions is always unity; this imposes a fundamental trade-off in control system design between rejecting disturbances and reducing sensitivity to plant parameter changes on one hand, and attenuating measurement noise on the other.

Review Questions

4.1 Open-loop control action on a stable plant does not run the risk of instability, but negative feedback control action may result in instability problems. Inspite of this fact, most of the industrial control schemes are based on negative feedback concept. Why?

4.2 In designing control systems, the following aspects must be taken into account.

(a) Stability
(b) Input amplitude constraints
(c) Disturbance rejection
(d) Noise filtering
(e) Sensitivity and robustness

Explain the significance of each.

4.3 The use of feedback in control systems is motivated by the presence of external disturbance uncertainties and model uncertainties. The use of feedback can be regarded as unreasonable if there is no uncertainty in the system. Show that feedback control can

(a) greatly reduce the effect on the controlled variable of all external disturbances except those associated with the sensor (measurement system); and
(b) make the system tolerant of variations in hardware parameters other than those of the sensor.

How can we reduce the effects of measurement errors on control system performance?

4.4 Show that high loop gain in feedback systems results in

(a) good steady-state tracking accuracy;
(b) good disturbance signal rejection;
(c) low sensitivity to process parameter variations; and
(d) good relative stability, i.e., rate of decay of transients.

What are the factors limiting the gain?

4.5 Design of control systems using frequency response models views the problem as a problem of *filter* design. Viewing a feedback control system as a filter, highlight the desirable characteristics of the filter in low-frequency, high-frequency, and mid-frequency ranges.

4.6 The sum of the sensitivity and the complementary sensitivity functions is always equal to unity; this imposes a fundamental trade-off in control system design. Explain the desirable properties of the two sensitivity functions using plots of these functions in the entire range of frequencies.

Self-Appraisal Quiz

A set of questions with multiple answer choices is given below. These questions have been designed to capture important aspects of the subject covered in the chapter. For each of these questions, identify the best of the given answer choices. Compare your answers with the Master Key given at the end of the session. This quiz should serve the purpose of a self-appraisal test for readers.

4.1 Feedback control systems are
(A) insensitive to both forward- and feedback-path parameter changes
(B) less sensitive to feedback-path parameter changes than to forward-path parameter charges
(C) less sensitive to forward-path parameter changes than to feedback-path parameter changes
(D) equally sensitive to forward- and feedback-path parameter charges

4.2 In the system of Fig. Q 4.2, sensitivity $M(s) = Y(s)/R(s)$ with respect to parameter K_1 is

(A) $\dfrac{1}{1+K_1K_2}$ (B) $\dfrac{1}{1+K_1G(s)}$

(C) 1 (D) None of the answers in (A), (B), and (C) is correct

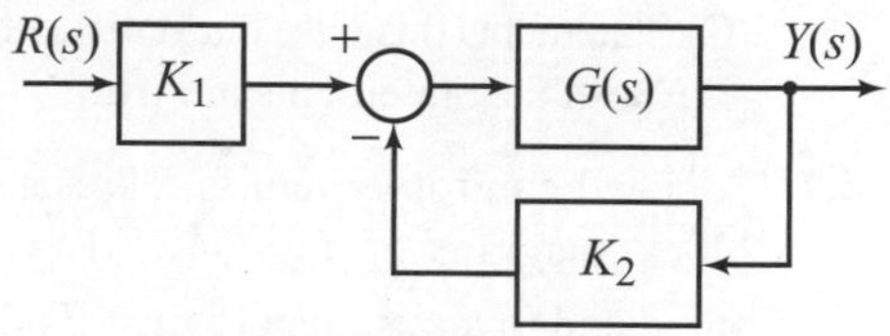

Fig. Q4.2

4.3 In the system of Fig. Q4.2, sensitivity of $M(s) = Y(s)/R(s)$ with respect to parameter K_2 is

(A) $\dfrac{1}{1+K_2K_1}$ (B) $\dfrac{1}{1+K_2G(s)}$

(C) $\dfrac{-K_2G(s)}{1+K_2G(s)}$ (D) None of the answers in (A), (B), and (C) is correct.

4.4 A speed control system is represented by the signal flow graph shown in Fig. Q4.4. The nominal value of the parameter K is 10. Sensitivity of $M(s) = \omega(s)/\omega_r(s)$ to changes in K is

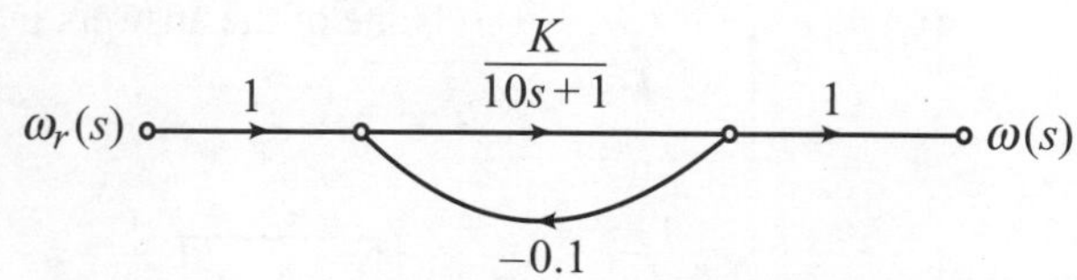

Fig. Q4.4

(A) $\dfrac{s+0.1}{s+11}$ (B) $\dfrac{s+0.1}{s+0.2}$

(C) $\dfrac{-0.1}{s+0.2}$ (D) None of the answers in (A), (B), and (C) is correct.

4.5 Sensitivity S_G^M of the closed-loop transfer function $M(s)$ with respect to open-loop transfer function $G(s)$ (Fig. Q4.5) is

(A) $\dfrac{1}{2s^2+(3K_P+1)s+3K_I}$

(B) $\dfrac{s(2s+1)}{2s^2+(K_P+1)s+3K_I}$

(C) $\dfrac{s(2s+1)}{2s^2+(K_P+1)s+K_I}$

(D) $\dfrac{s(2s+1)}{2s^2+(3K_P+1)s+3K_I}$

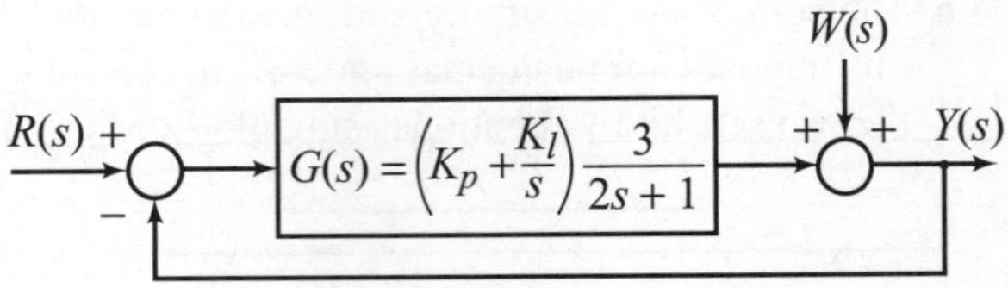

Fig. Q4.5

4.6 Consider the following statements:

(i) If an open-loop system is unstable, applying feedback will always improve its stability.

(ii) If an open-loop system is subject to parameter variations, applying feedback will always improve robustness.

Which of the following is the correct answer?

(A) None of the above statements is true

(B) Statement (i) is true but statement (ii) is false

(C) Statement (i) is false but statement (ii) is true

(D) Both the statements are true

4.7 A speed control system is represented by the block diagram of Fig. Q4.7. The system is subjected to a step disturbance $W(s)$. $\omega_{ss}^{CL}/\omega_{ss}^{OL}$ (the steady-state speed under closed-loop operation/steady-state speed under open-loop operation) is equal to

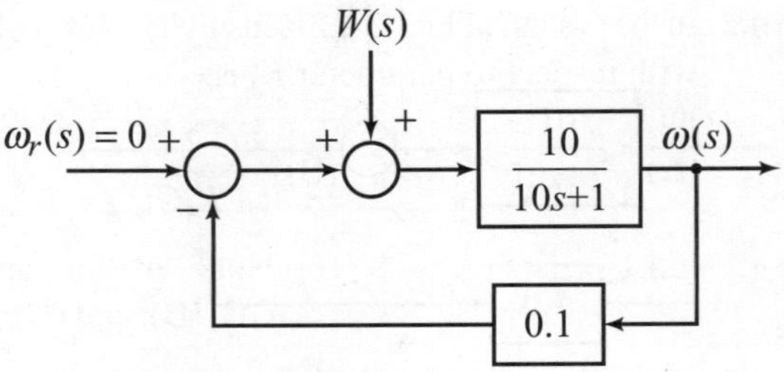

Fig. Q4.7

(A) 1/2 (B) 1

(C) 2 (D) None of the answers in (A), (B), and (C) is correct

4.8 The time response of the system of Fig. Q4.8a to an input $r(t) = 10\mu(t)$ is shown in Fig. Q4.8b. The gain K is equal to

(A) 10 (B) 8

(C) 4 (D) None of the answers in (A), (B), and (C) is correct

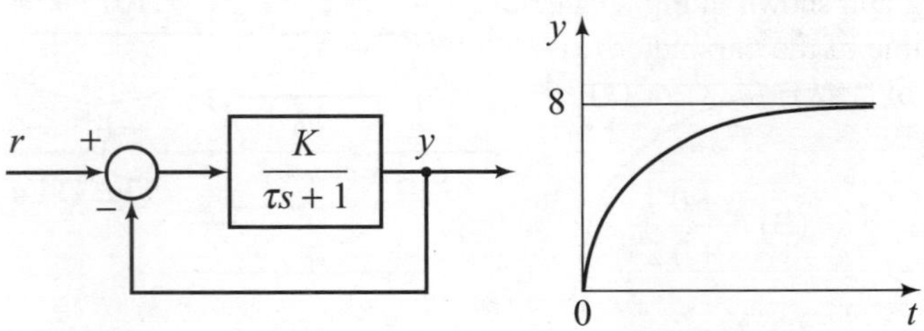

(a) (b)

Fig. Q4.8

4.9 (i) Steady-state gain (dc gain) of the system of Fig. Q4.9 is higher than for the system with $K_1 = 0$.
(ii) Bandwidth of the system of Fig. Q4.9 is higher than for the system with $K_1 = 0$.
(A) Both the statements are true
(B) Statement (i) is true but statement (ii) is false
(C) Statement (ii) is true but statement (i) is false
(D) Both the statements are false

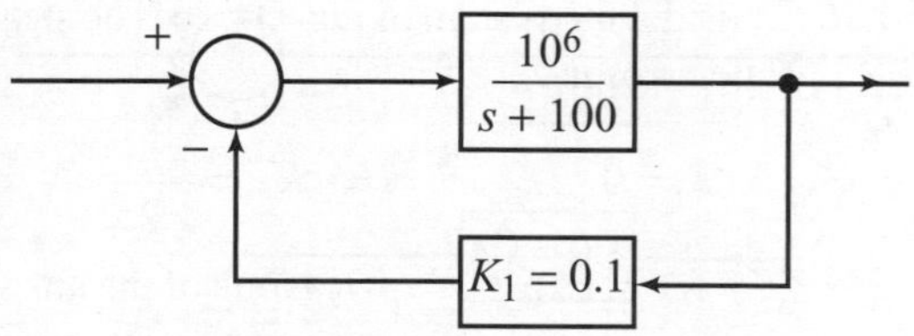

Fig. Q4.9

4.10 For the negative feedback system shown in Fig. Q4.10, the effective time constant is
(A) L/R
(B) less than L/R
(C) more than L/R
(D) zero

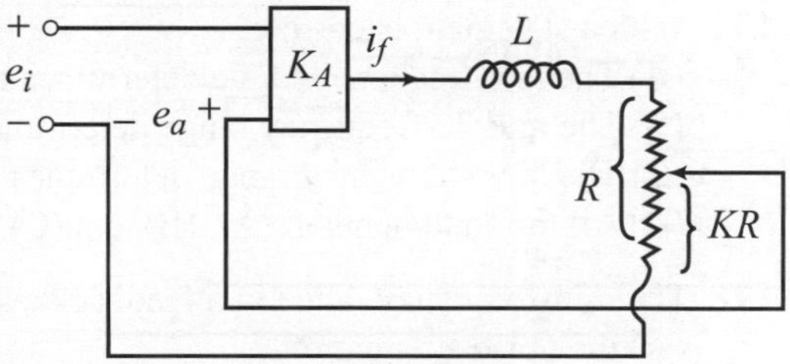

Fig. Q4.10

4.11 For what value of K is the time constant of the system of Fig. Q4.11 equal to 0.2 seconds?
(A) $K = 3$ (B) $K = 5$
(C) $K = 7$ (D) $K = 9$

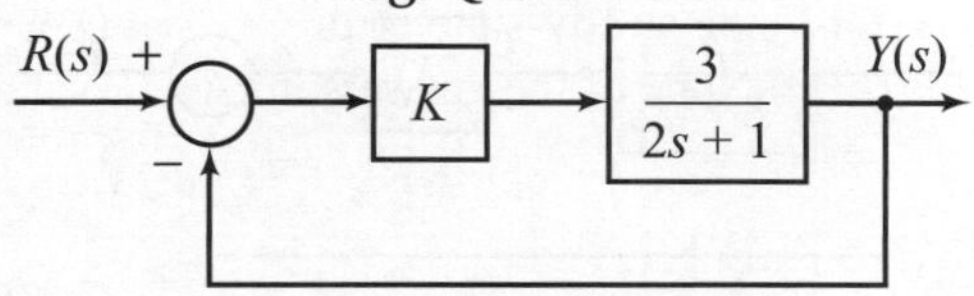

Fig. Q4.11

4.12 For what value of K is the time constant of the system of Fig. Q4.12 less than 0.2 seconds?
(A) $K > 3$ (B) $K > 5$
(C) $K > 7$ (D) $K > 9$

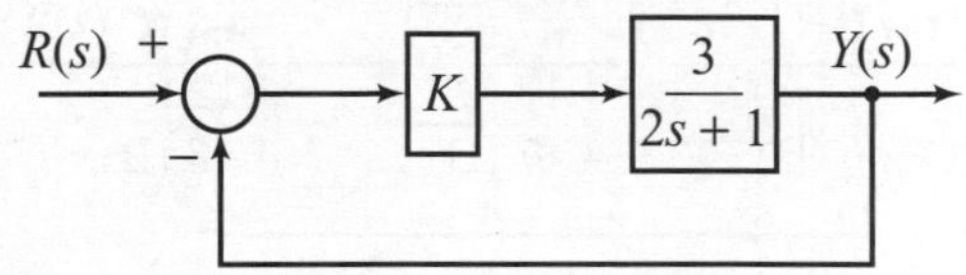

Fig. Q4.12

4.13 Consider the system of Fig. Q4.13. The K_P to move the time constant to one sixth of its open-loop value is

(A) $\frac{0.333}{2}$ (B) $\frac{0.333}{6}$

(C) $\frac{6}{0.333}$ (D) $\frac{2}{0.333}$

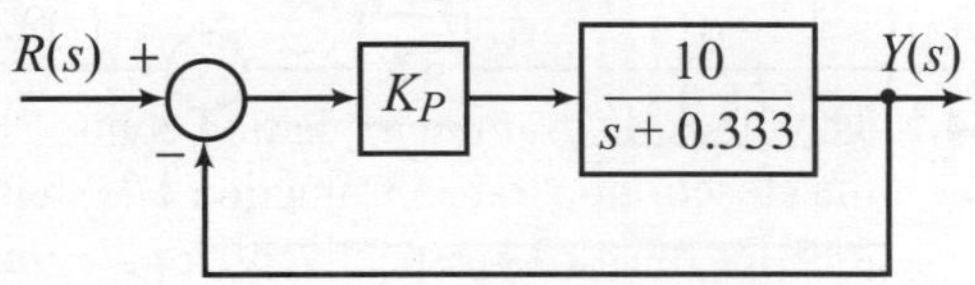

Fig. Q4.13

4.14 A first-order process plant has a time constant of 5 and a dc gain of 2. The controller gain in a unity-feedback system such that the closed-loop time constant $\tau_{CL} = 0.5$, is
(A) 4 (B) 4.5
(C) 5 (D) 5.5

4.15 The system of Fig. Q4.15 has the steady-state error

(A) $\frac{1}{1.125}$ (B) $\frac{0.1}{1.125}$

(C) $\frac{1.1}{1.125}$ (D) None of the answers in (A), (B), and (C) is correct

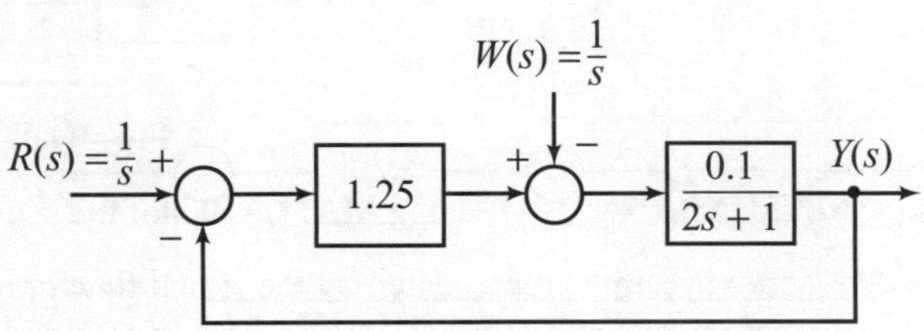

Fig. Q4.15

4.16 Consider the system of Fig. Q4.16. The steady-state error is zero for

(A) $K_f = 0$ (B) $K_f = \dfrac{1}{1.25}$

(C) $K_f = 1.25$ (D) None of the answers in (A), (B), and (C) is correct

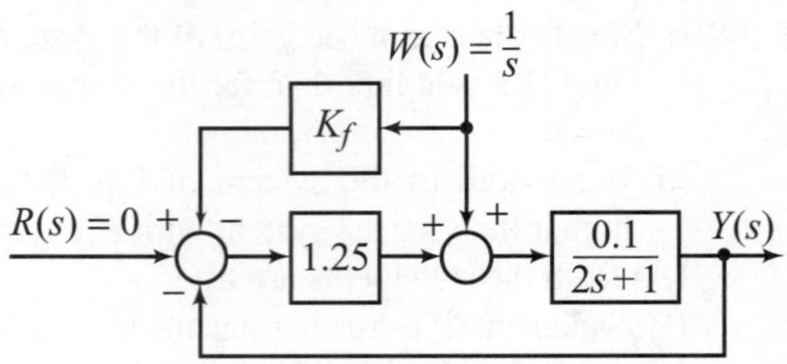

Fig. Q4.16

4.17 Which statement is correct?
(A) The low-frequency range determines the steady-state behavior of a closed-loop system.
(B) The middle-frequency range determines the closed-loop system stability.
(C) The high-frequency range determines the performance of a closed-loop system to noise.
(D) All the statements in (A), (B) and (C) are correct.

4.18 The mid-frequency range in a feedback system indicates
(A) the noise attenuation
(B) the stability
(C) the steady-state error
(D) None of the answers in (A), (B) and (C) is correct

MASTER KEY

4.1 (C)	4.2 (C)	4.3 (C)	4.4 (B)	4.5 (D)
4.6 (C)	4.7 (A)	4.8 (C)	4.9 (C)	4.10 (B)
4.11 (A)	4.12 (A)	4.13 (A)	4.14 (B)	4.15 (C)
4.16 (B)	4.17 (D)	4.18 (B)		

Problems

4.1 The block diagram of a position control system is shown in Fig. P4.1. Determine the sensitivity of the closed-loop transfer function $M(s)$ with respect to G and H, the forward path, and feedback path transfer functions respectively, for $s = j\omega = j1$.

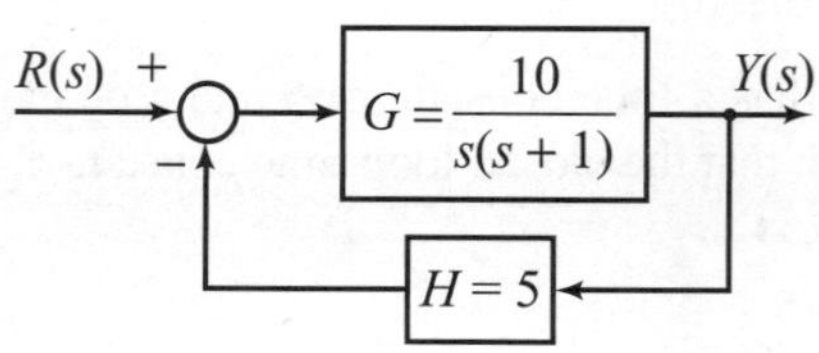

Fig. P4.1

MATLAB Exercise Using MATLAB, plot the sensitivity vs frequency graphs.

4.2 A servo system is represented by the signal flow graph shown in Fig. P4.2. The nominal values of the parameters are $K_1 = 1$, $K_2 = 5$ and $K_3 = 5$. Determine the overall transmission $Y(s)/R(s)$, and its sensitivity to changes in K_1 under steady dc conditions, i.e., $s = 0$.

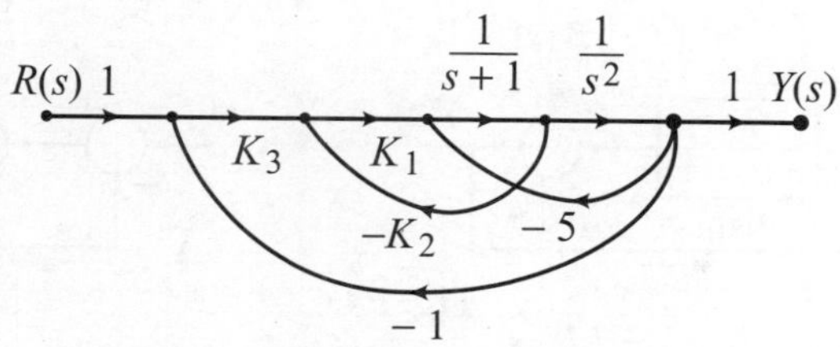

Fig. P4.2

4.3 A unity feedback system has forward path transfer function $G(s) = 20/(s + 1)$. Determine the response of open- and closed-loop systems for unit-step input.

Suppose now that parameter variations occurring during operating conditions cause $G(s)$ to modify to $G'(s) = 20/(s + 0.4)$. What will be the effect on unit-step response of open- and closed-loop systems? Comment upon the sensitivity of the two systems to parameter variations.

MATLAB Exercise After completing the hand calculations, verify your results using MATLAB.

4.4 Consider the two control systems shown in Fig. P4.4. In the open-loop control system, gain K_c is calibrated so that $K_c = 1/10$. In the closed-loop control system, gain K_p of the controller is set so that $K_p = 10$. Find the steady-state error to unit-step input in both the cases.

Suppose now that parameter variations occurring during operating conditions cause $G(s)$ to modify to $G'(s) = 11/(\tau s + 1)$. What will be the effect on the steady-state error of open- and closed-loop systems? Comment upon the sensitivity of the two systems to parameter variations.

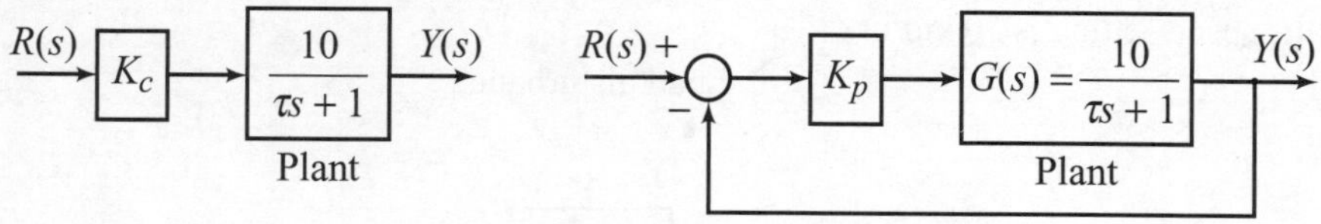

Fig. P4.4

4.5 Consider the system shown in Fig. P4.5, where $w(t)$ is a unit-step disturbance. Calculate K such that the steady-state error in $y(t)$ due to $w(t)$ is less than one percent of $w(t)$.

MATLAB Exercise Using MATLAB, plot the time response to step disturbance. Also obtain the plot in Simulink window.

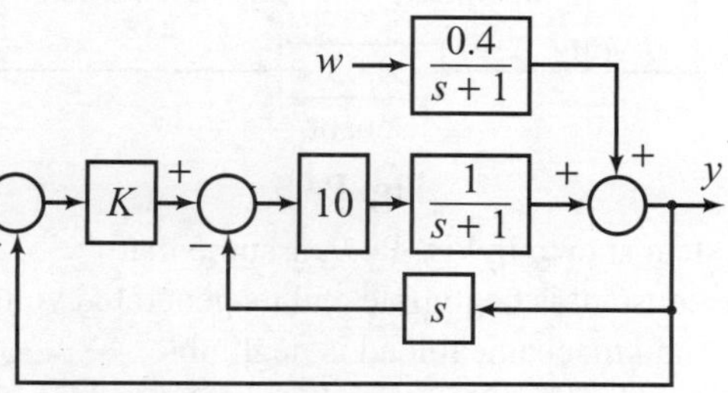

Fig. P4.5

4.6 The two control configurations shown in Fig. P4.6 have identical plants; however, external disturbances enter the two systems at different points along the forward path. Find the steady-state error in the two cases in response to step disturbance of unit magnitude.

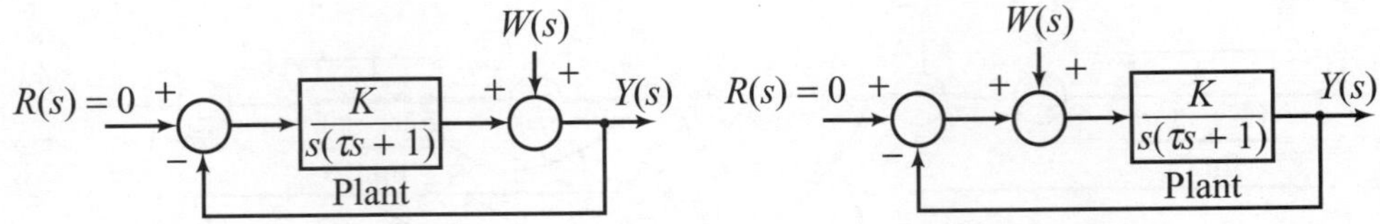

Fig. P4.6

4.7 A simplified model of cruise control of an automobile is shown in Fig. 4.17. The vehicle travels up/down a constant incline (w represents % road grade) with a constant speed setting. Sensing the error by the throttle controller sets the throttle angle. In the open-loop scheme of Fig. 4.16, the controller does not use the speed sensor; the throttle angle setting depends on the commanded speed r.

(a) For a speed setting of 50 km/hr ($R(s) = 50/s$), and level road conditions ($W(s) = 0$), determine the transient response $y(t)$ of the open-loop and closed-loop systems. Comment upon the effect of feedback on the transient response.
What is the speed in steady state in the two cases?

(b) Assume now that the road incline is 1% ($W(s) = 1/s$). Find steady-state speeds in open-loop and closed-loop systems. Comment upon the effect of feedback on steady-state performance.

4.8 An automobile speed control system will be necessary for passenger cars traveling on the automatic highways of the future. A model of a feedback control system for a standard vehicle is shown in Fig. P4.8. The load disturbance due to a percent gradient w of the highway is also shown. The engine constant $K_e = 500$.

(a) If the desired speed is $v_r = 50$ km/hr, find the constant input voltage e_r that should be applied.

(b) Determine the sensitivity of $V(s)/E_r(s)$ to changes in engine constant K_e.

(c) Determine the constant percent gradient for which the engine stalls (steady-state velocity $v_{ss} = 0$); the speed setting is 50 km/hr.

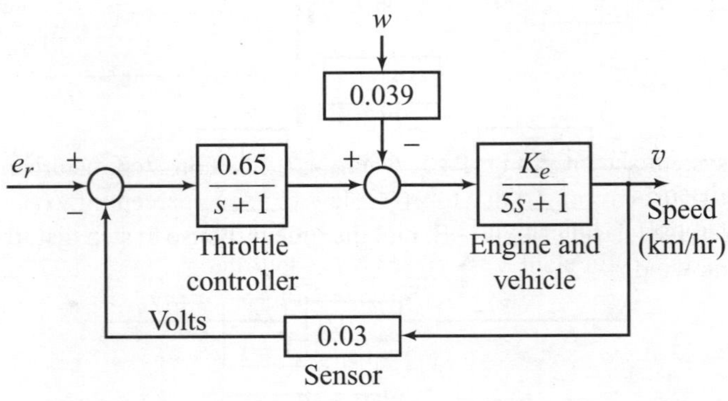

Fig. P4.8

4.9 For the speed control system shown in Fig. P4.9, assume that

(a) generator field time constant is negligible and its generated voltage is K_g volts/amp;

(b) friction of the motor and mechanical load is negligible.

(i) Find the time variation of output speed (ω_o) for a sudden command input (ω_r) of 10 rad/sec. Find also the steady-state output speed.

(ii) If the feedback loop is opened and gain K_A adjusted to give the same steady-state speed as in the case of closed loop, determine how the output speed varies with time, and compare the speed of response in the two cases.

(iii) Compare the sensitivity of $\omega_o(s)/\omega_r(s)$ to changes in amplifier gain K_A and generator speed ω_g with and without feedback.

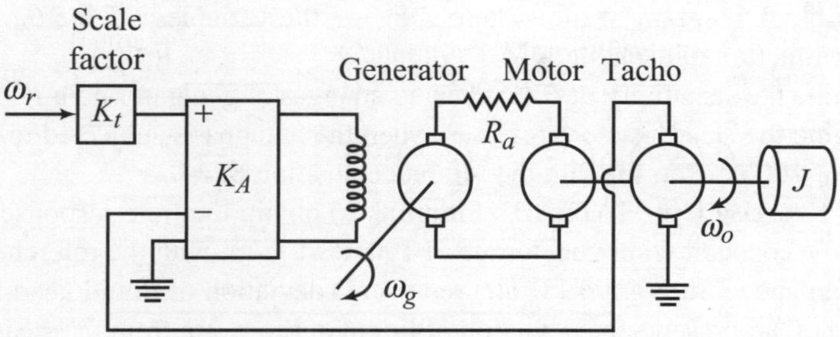

Fig. P4.9

The system parameters are given below:

Moment of inertia of motor and load	$J = 5$ kg-m^2
Motor back emf constant	$K_b = 5$ volts per rad/sec
Total armature resistance of motor and generator	$R_a = 1$ ohm
Generator gain constant	$K_g = 50$ volts/amp
Amplifier gain	$K_A = 5$ amps/volt
Tachogenerator constant	$K_t = 0.5$ volts per rad/sec

4.10 Figure P4.10 shows a control scheme for controlling the azimuth angle of a rotating antenna. The plant consists of an armature-controlled dc motor with dc generator used as an amplifier. The parameters of the plant are given below:

Motor torque constant	$K_T = 1$ newton-m/amp
Motor back emf constant	$K_b = 1$ volt/(rad/sec)
Generator gain constant	$K_g = 100$ volt/amp
Potentiometer sensitivity	$K_P = 1$ volt/rad
Amplifier gain	$K_A = 50$ volt/volt
Moment of inertia of load	$J_L = 4$ newton-m/(rad/sec^2)

$$\dot{\theta}_L/\dot{\theta}_M = \dot{\theta}_L/\dot{\theta}_C = 1/2;\ R_f = 20\ \Omega,\ L_f = 5\text{H},\ R_a = 1\Omega$$

Load friction, motor inertia and friction are negligible.

The antenna is subject to wind gust torque T_w. Experiments on the antenna show that wind exerts a maximum disturbance of 100 newton-m at the antenna.

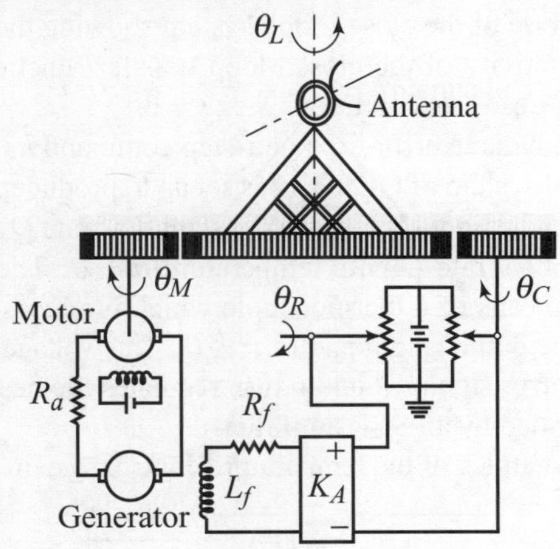

Fig. P4.10

(a) Draw a block diagram of the system showing the variables θ_R, θ_C, θ_M, θ_L, and T_w. Give the transfer function of each block.

(b) Determine the sensitivity of $\theta_L(s)/\theta_R(s)$ to changes in generator gain K_g for $s = j\omega = j0.1$.

(c) Determine the steady-state error in θ_L when the antenna is subjected to a constant wind gust torque of 100 newton-m, with the reference position $\theta_R = 0$.

MATLAB Exercise Use MATLAB's Simulink to obtain the time response when the antenna is subjected to a constant wind gust torque of 100 newton-m, with the reference position $\theta_R = 0$.

4.11 In the system shown in Fig. P4.11, $h(t)$ represents deviation of liquid head from the steady-state value $\bar{H}$ and $q(t)$ represents the deviation of liquid outflow rate from the steady-state value $\bar{Q}$. The pump controls the liquid head/outflow rate by supplying liquid at a rate $\bar{Q}_i + q_i(t)$; $q_i(t) = -K[h(t) - h_r(t)]$, where K is a constant and $h_r(t)$ is the desired change in the initial setpoint value.

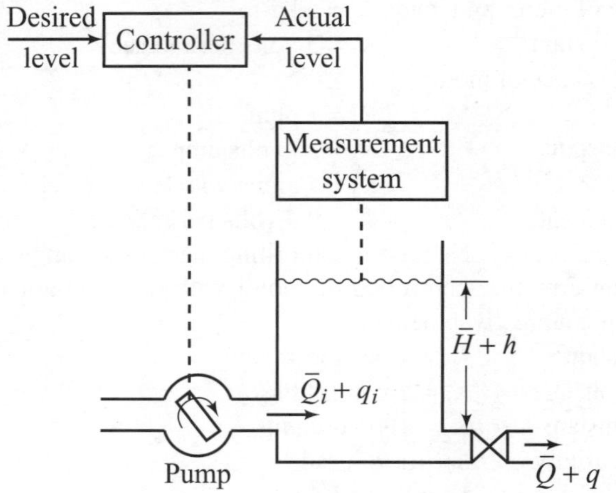

Fig. P4.11

(a) Prove that the following equation represents perturbation dynamics:

$$RC\frac{dh(t)}{dt} + h(t) = Rq_i(t)$$

where R is a constant so that $q(t) = h(t)/R$, and C = cross-sectional area of the tank.

(b) Draw a block diagram of the closed-loop system showing the variables h_r, q_i, h and q.

(c) Determine the sensitivity of the closed-loop transfer function $H(s)/H_r(s)$ to changes in the parameter R under steady dc conditions, i.e., $s = 0$.

(d) Determine the steady-state error for a unit-step command $h_r(t)$.

4.12 The temperature control system of Fig. P4.12 is set up to produce a steady stream of hot liquid at a controlled temperature. The liquid enters at a constant flow rate Q and variable temperature $\bar{\theta}_i + \theta_i$, and leaves at constant flow rate Q with temperature $\bar{\theta} + \theta$. The temperature of the outflowing liquid is measured by means of a thermocouple which produces an output voltage proportional to the temperature of the outflowing liquid ($e_t = K_t\theta$). This voltage is subtracted from the reference voltage to generate an error signal which in turn regulates the heating rate produced by electrical resistance heater controlled by an SCR amplifier.

The perturbation dynamics of the temperature control system is described by the equation

$$K(e_r - e_t) = C\frac{d\theta}{dt} + \frac{\theta - \theta_i}{R_1}$$

where

C = thermal capacitance of the liquid in the tank; (Joules/ °C)
R_1 = thermal resistance of the carrier fluid; (°C/(Joules/sec))
K = gain of the SCR amplifier-heater circuit; ((Joules/sec)/volt).

(a) Draw a block diagram of the closed-loop system showing reference voltage, e_r, disturbance in temperature, θ_i, and the controlled temperature, θ.

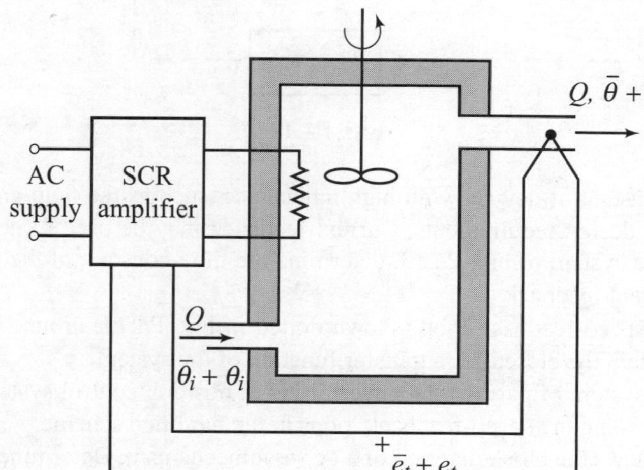

Fig. P4.12

(b) Compare the open-loop ($K_t = 0$) and the closed-loop systems for (i) sensitivity to changes in parameter K, (ii) the ability to reduce the effects of step disturbance θ_i, and (iii) the system time constants in the two cases.

(c) Modify the block diagram of part (a) to include the effect of the disturbance signal θ_a caused by changes in the ambient temperature. Find the steady change in the temperature of the out-flowing liquid for open- and closed-loop cases due to a unit-step disturbance θ_a. The surface area of the tank is A m^2, and the heat transfer coefficient of the tank-air interface is U Joules/(m^2) (°C) (sec).

4.13 Figure P4.13 shows a field-controlled dc motor with a dc amplifier. The parameters of the system are given below:

Amplifier gain	$K_A = 50$ volt/volt
Motor field resistance	$R_f = 99\ \Omega$
Motor field inductance	$L_f = 20$ H
Motor torque constant	$K_T = 10$ newton-m/amp
Moment of inertia of load,	$J = 0.5$ newton-m/(rad/sec^2)
Coefficient of viscous friction of load,	$B = 0.5$ newton-m/(rad/sec)

Motor inertia and friction are negligible.

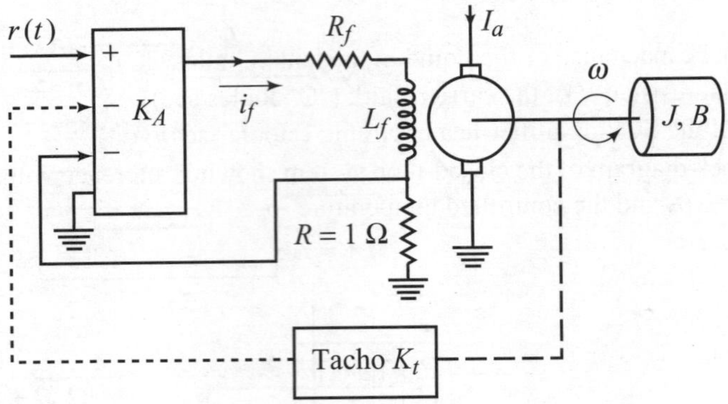

Fig. P4.13

(a) The time constant of devices with high inductance, such as the field-controlled dc motor, is too long for design requirements. Current feedback may be used to decrease the time-constant. For the system of Fig. P4.13, determine the time-constant of the field circuit with and without current feedback.

(b) Now add a speed-feedback loop (shown dotted in Fig. P4.13) around the current-feedback loop and obtain the closed-loop transfer function of the system.

(c) Modify the system of part (b) to convert it into a position control system. Use both the current-feedback and the speed-feedback loops in the modified scheme.

4.14 Figure P4.14a shows the block diagram of a dc servomechanism. Determine the sensitivity of the closed-loop transfer function $M(s)$ to the variations in K at $\omega = 5$ rad/sec. The nominal value of $K = 1$. Change the control configuration of the system to two-loop configuration with feedback of θ and $\dot{\theta}$ (see Fig. P4.14b); the closed-loop transfer function of the two-loop configuration must be equal to that of the system shown in Fig. P4.14a. Obtain sensitivity of $M(s)$ to variations in K at $\omega = 5$ for the new configuration.

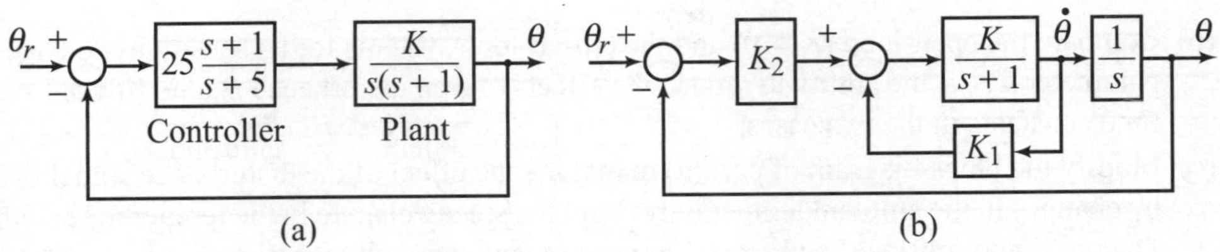

Fig. P4.14

4.15 MATLAB Exercise A unity-feedback system has open-loop transfer function

$$G(s) = \frac{10}{s(1+0.1s)(1+0.05s)}$$

Find the sensitivity of the closed-loop transfer function with respect to G. Plot the sensitivity vs frequency graph and therefrom determine the peak sensitivity and the frequency at which it occurs.

CONCEPTS OF STABILITY AND THE ROUTH STABILITY CRITERION

5

Preview

We have learnt from the previous chapter that feedback is a double-edged sword. It leads to better performance, but it can also lead to instability if improperly applied. Therefore, it becomes necessary to first find the ranges of controller parameters for which the system remains stable, and then carry out the search for optimal parameters within the limits imposed by stability considerations, to improve the performance. In this chapter, we will learn an algebraic procedure (the Routh criterion) that provides a means of determining these ranges.

5.1 INTRODUCTION

Roughly speaking, stability in a system implies that small changes in reference inputs, disturbance inputs, and/or initial conditions do not result in large changes in the system output. Stability is a very important property of control systems. Almost every working system is designed to be stable; we seek to improve the speed of response, steady-state accuracy, and other aspects of system performance within the constraints imposed by stability considerations. In this chapter, we take up analytical study of stability; the subject of system performance will be taken up in the next chapter.

We have learnt from the last chapter that feedback is a double-edged sword. It leads to better performance, but it also can lead to instability if improperly applied. For a stable plant (open-loop system), the closed-loop control system may not be always stable. Feedback control system design may, therefore, demand a trade-off between the accuracy in realizing control objectives and the stability.

For an analytical study of stability, we consider a feedback control system expressed in the form of Fig. 5.1; $R(s)$ is the input, $Y(s)$ is the output, $G(s)$ in the *forward-path transfer function* and $H(s)$ is the *feedback-path transfer function*. The *closed-loop transfer function* of the system is

$$M(s) = \frac{Y(s)}{R(s)} = \frac{G(s)}{1 + G(s)H(s)} \tag{5.1}$$

The transfer function relating $B(s)$ (the feedback signal) to $R(s)$ is the *open-loop transfer function* $G(s)H(s)$.

This is the general form in which analysis problem is normally formulated. For design purposes, the forward-path transfer function $G(s)$ is represented as a cascade of two sub-blocks—one for controller transfer function and the other for plant transfer function.

The open-loop transfer function $G(s)H(s)$ is generally known in the factored form as it is obtained by modeling the transfer functions of individual components comprising the system. Therefore, $G(s)H(s)$ can generally be expressed as

$$G(s)H(s) = \frac{K\prod_{i=1}^{m}(s+z_i)}{\prod_{j=1}^{n}(s+p_j)};\; m \le n, K > 0 \tag{5.2}$$

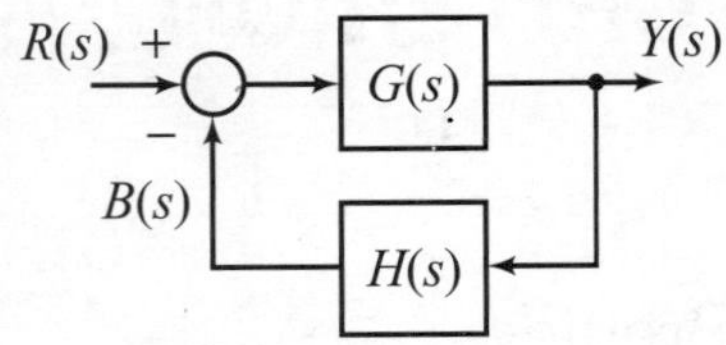

Fig. 5.1 *A feedback system*

The function $G(s)H(s)$ approaches infinity as s approaches $-p_j$: $s = -p_j$; $j = 1,\ldots, n$, are the *poles* of $G(s)H(s)$, i.e., *open-loop poles* of the feedback system of Fig. 5.1. The function $G(s)H(s)$ approaches zero as s approaches $-z_i$; $s = -z_i$; $i = 1, \ldots, m$, are the *zeros* of $G(s)H(s)$, i.e., the *open-loop zeros* of the feedback system of Fig. 5.1. K is the *open-loop gain* of the feedback system.

When the signal $B(s)$ in Fig. 5.1 is fed back (the loop is closed), the dynamic behavior of the system is governed by Eqn. (5.1). The closed-loop response

$$y(t) = \mathscr{L}^{-1}\left[\left(\frac{G(s)}{1+G(s)H(s)}\right)R(s)\right] = \mathscr{L}^{-1}\,[M(s)R(s)]$$

From Section 2.7, it follows that the response $y(t)$ consists of two components: the *transient response* and the *steady-state response*. The characteristic nature of the transient response terms (constant, exponential, oscillatory, magnitude increasing to infinity with time, etc.) is dictated by poles of $M(s)$ (zeros of $M(s)$ affect the residues in partial fraction expansion, and hence, the magnitudes of response terms) excited by the input $R(s)$. The steady-state response arises due to the *excitation poles* (poles of $R(s)$) and has the same nature as the input itself except for modification in magnitude caused by the system's behavior to the special input. Stability is a very important characteristic of the transient response. It is concerned with qualitative nature of the transient response, i.e., whether the response is bounded or unbounded. This qualitative picture emerges from the poles of $M(s)$.

The stability study of the closed-loop system will thus require the poles of $M(s)$, i.e., the closed-loop poles of the system, which are the roots of the *characteristic equation*

$$1 + G(s)H(s) = 0 \tag{5.3}$$

Substituting Eqn. (5.2) into Eqn. (5.3), we obtain the characteristic equation as

$$1 + \frac{K\prod_{i=1}^{m}(s+z_i)}{\prod_{j=1}^{n}(s+p_j)} = 0 \tag{5.4}$$

or

$$\prod_{j=1}^{n}(s+p_j) + K\prod_{i=1}^{m}(s+z_i) = 0 \tag{5.5}$$

The characteristic equation given in Eqn. (5.5) is in the conventional polynomial form:

$$\Delta(s) = a_0 s^n + a_1 s^{n-1} + \cdots + a_{n-1}s + a_n = 0 \tag{5.6}$$

where $\Delta(s)$ represents the *characteristic polynomial*.

Factoring the characteristic polynomial $\Delta(s)$ to determine closed-loop poles is one of the major problems in stability study. An obvious solution to the problem seems to be the numerical procedures.

Computer programs are available [151–154] that can be used to find the roots of a polynomial. A limitation of the numerical techniques is that all the coefficients $\{a_0, a_1, \ldots, a_n\}$ in the characteristic polynomial must be assigned numerical values. When some of these coefficients depend on the design parameters (for example, gain K in Eqn. (5.5)), we need to write a computer program that would substitute a series of values of design parameters, factor the characteristic equation for these specific values, and give the roots (closed-loop poles) required for stability study. However, this is not the best way to proceed early in design process. What we need early in the design process is a clear, *qualitative* picture of how the closed-loop poles move around as the design parameters are varied. This qualitative nature of the solution will be examined in this chapter using the Routh stability criterion, and in Chapter 8 using the root-locus analysis.

We begin with the concepts of stability and develop simple tools to examine the stability of systems. The use of these tools for the design problems then logically follows.

5.2 CONCEPTS OF STABILITY

Stability in a system implies that small changes to the system stimulus do not result in large effects on system response. In testing the stability of a dynamic system, we must carefully define the terms 'response', 'stimulus', 'small change', and 'large effect'. For example, it is possible for a system to have a well-behaved output while an internal variable (state variable) is growing without bound. If we consider only the output as the 'response', we might call such a system stable. However, if we consider the state variables to be the response, we might just as reasonably call this system unstable. The stimulus might be a signal persisting for all time, or it might be only a set of initial conditions. Finally, we might require that the response not grow if the stimulus is removed, or we might require that the response go to zero if the stimulus is removed. For systems that are nonlinear or time varying, the number of definitions of stability is large, and some of the definitions can be quite complicated in order to account for the varieties of the responses the systems can display. We will describe a few elementary results, particularly those useful in the study of linear time invariant systems.

A linear time-invariant system shown in Fig. 5.2, has two sources of excitation:

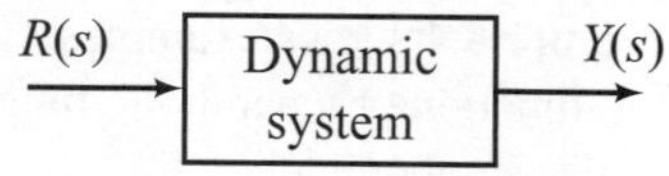

Fig. 5.2 *A linear time-invariant system*

(i) the external input $r(t)$; $t \geq 0$; and
(ii) the initial energy storage in the dynamic system, represented by the *initial conditions* of the differential equation model of the system, or equivalently the state $\mathbf{x}(t)$ at $t = 0$ of the state-variable model(described in Chapter 14) of the system.

In the stability study, we are generally concerned with the following questions:

1. If the system is *relaxed* at $t = 0$ (initial energy storage is zero, i.e., initial conditions of differential equation model are zero, or $\mathbf{x}(0) = \mathbf{0}$), will a bounded input $r(t)$; $t \geq 0$, produce a bounded output $y(t)$ for all t (regardless of what goes on inside the system) ?
2. Taking changes in the initial energy storage (i.e., $\mathbf{x}(0)$) as the stimulus, and system state $\mathbf{x}(t)$ as the response, the second question is concerned with the nature of the response to different stimulii, with the external input $r(t) = 0$, $t \geq 0$.

The first notion of stability, called *bounded-input, bounded-output (BIBO) stability*, is concerned with the boundedness of the output of a relaxed system in response to bounded input.

***A relaxed system (zero initial conditions) is said to be BIBO stable if for every bounded input** $r(t)$; $t \geq 0$, **the output** $y(t)$ **is bounded for all** t.*

The second notion of stability, called *asymptotic stability*, is based on the dynamic evolution of the state of an unforced ($r(t) = 0$; $t \geq 0$) system in response to arbitrary initial state representing initial internal energy storage.

If the state of a system with zero input** ($r(t) = 0$; $t \geq 0$) **for all possible initial states eventually decays to zero as** $t \to \infty$ **, the system is said to be asymptotically stable.

We have already prepared a platform (transfer function models of dynamic systems) helpful for BIBO stability study. However, the study of asymptotic stability requires state variable formulation of dynamic systems, which is a subject of Chapter 14. In that chapter, we will see that ***under mildly restrictive conditions (namely, the system must be both controllable and observable), the asymptotic stability ensures BIBO stability and vice versa.***

In nonlinear systems, there is no definite correspondence between the two notions. Chapter 16 highlights the problems associated with nonlinear control systems, and presents commonly used analysis tools.

In this chapter, we concern ourselves with the BIBO stability of linear time-invariant systems.

5·3 BOUNDED-INPUT BOUNDED-OUTPUT STABILITY

We first look at what is meant by *bounded* and *unbounded* signals before we address the question of system stability.

A time-domain function, $f(t)$, is said to be bounded if its magnitude does not go to infinity in the time interval $[0, \infty)$, or equivalently, there exists a real constant M such that $|f(t)| \leq M < \infty$ for all t in $[0, \infty)$.

Figure 5.3 shows examples of bounded exponential signals and bounded sinusoidal signals. The magnitude of step functions is finite even when time tends to infinity. We call these types of signals bounded.

Figure 5.4 shows examples of unbounded exponential and sinusoidal signals. The magnitude of a ramp function or a parabolic function increases without bound as time progresses. We categorize these signals as unbounded.

A linear relaxed system (initial conditions zero) is said to have BIBO stability if every bounded input results in a bounded output. Clearly, if a system is subjected to an unbounded input and produces an unbounded response, nothing can be said about stability. However, if it is subjected to a bounded input and produces an unbounded response, it is, by definition, unstable. Actually, the output of an unstable system may increase to a certain extent and then the system may break down or become nonlinear after the output exceeds a certain magnitude, so that the linear mathematical model no longer applies.

Consider the linear time-invariant system of Fig. 5.2 with input $R(s)$, initial conditions zero, and the transfer function $M(s)$; the output

$$Y(s) = M(s)R(s) \tag{5.7}$$

For a bounded input $r(t)$, the nature of the response terms contributed by the system poles (i.e., the poles of the transfer function $M(s)$) gives the nature (whether bounded or not) of the total response $y(t)$ of the system. We may answer the question of BIBO stability by examining the system poles and the response terms contributed by them.

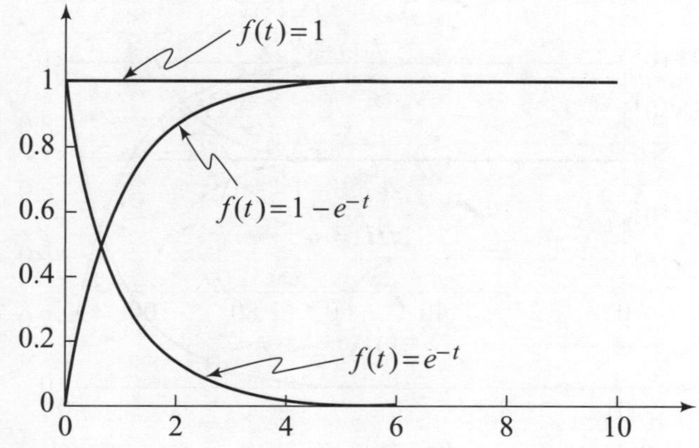

(a)

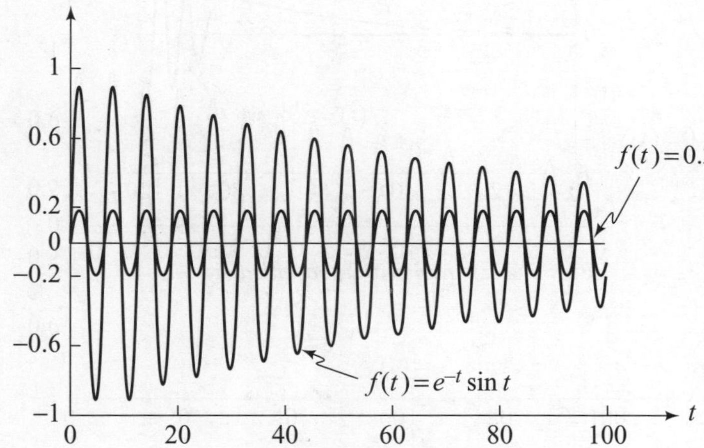

(b)

Fig. 5.3 *Examples of bounded signals*

The poles of $M(s)$ may be both real and complex-conjugate, and may have multiplicity of various orders. The transient response of the system is the summation of response from all the individual poles. If *any one* of the poles contributes an unbounded response, the total response will be dominated by this pole and the system becomes unstable. A system is therefore stable if *all* the poles individually contribute bounded components in the response.

The time function $e^{\alpha t}$ is the response function contributed by the system pole at $s = \alpha$. It is either a constant or a growing or decaying exponential depending on whether $\alpha = 0$, $\alpha > 0$, or $\alpha < 0$, respectively, as sketched in Fig. 5.5.

The complex-conjugate pole pair at $s = p = \sigma + j\omega$ and $s = p^* = \sigma - j\omega$ of $M(s)$ will give rise to the following terms in partial fraction expansion of $Y(s) = M(s)R(s)$:

$$\frac{A}{s-p} + \frac{A^*}{s-p^*}$$

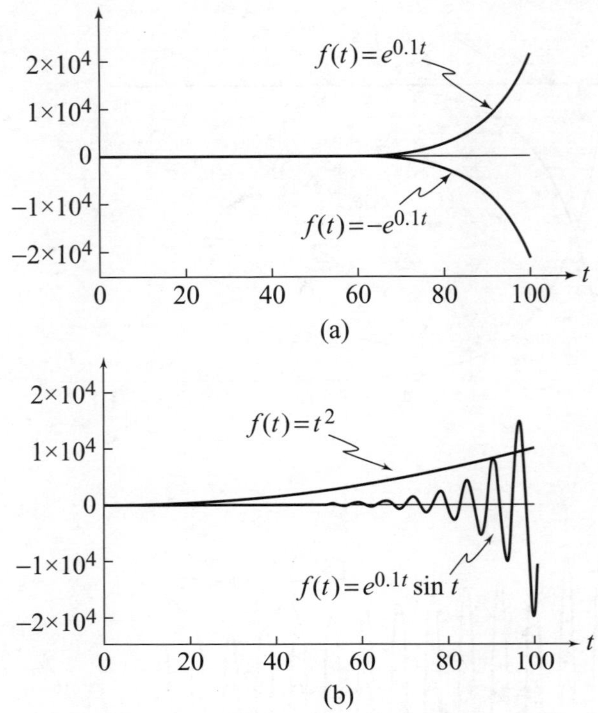

Fig. 5.4 *Examples of unbounded signals*

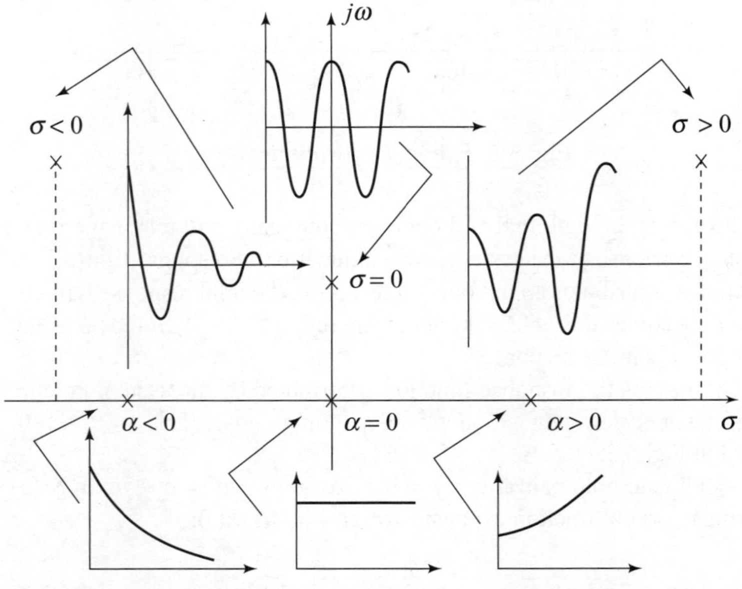

Fig. 5.5 *Time functions associated with poles in the s-plane (only one pole of a pair of complex poles is shown)*

where $A\left(=|A|\angle\phi\right)$ and A^* (complex-conjugate of A) are the residues at the poles $s = p$ and $s = p^*$, respectively. The inverse Laplace transform operation yields a time function of the form:

$$\begin{aligned} Ae^{pt} + A^* e^{p^*t} &= Ae^{pt} + (Ae^{pt})^* = 2\,\mathrm{Re}\left[Ae^{pt}\right] \\ &= 2\,\mathrm{Re}\left[|A|e^{j\phi}e^{\sigma t}e^{j\omega t}\right] = 2|A|e^{\sigma t}\,\mathrm{Re}\left[e^{j(\omega t+\phi)}\right] \\ &= 2|A|e^{\sigma t}\cos(\omega t+\phi) \end{aligned} \tag{5.8}$$

Therefore, the complex-conjugate pair of poles of $M(s)$ gives rise to a sinusoidal or oscillatory response whose envelope $2|A|e^{\sigma t}$ can be constant, growing or decaying depending on whether $\sigma = 0, \sigma > 0$, or $\sigma < 0$, respectively as sketched in Fig. 5.5.

Assume now that $M(s)$ has a real pole at $s = \alpha$ of multiplicity m. This pole will give rise to the following terms in partial fraction expansion of $Y(s) = M(s)R(s)$:

$$\frac{A_{1(m)}}{(s-\alpha)^m} + \frac{A_{1(m-1)}}{(s-\alpha)^{m-1}} + \cdots + \frac{A_{12}}{(s-\alpha)^2} + \frac{A_{11}}{(s-\alpha)}$$

where $A_{1(m)}, \ldots, A_{12}, A_{11}$ are real constants.

Response function contributed by the real pole of multiplicity m can be evaluated as follows:

Consider the transform pair

$$\frac{1}{s-\alpha} \leftrightarrow e^{\alpha t}$$

Using Eqn. (2.16),

$$-\frac{d}{ds}\left(\frac{1}{s-\alpha}\right) \leftrightarrow te^{\alpha t} \quad \text{or} \quad \frac{1}{(s-\alpha)^2} \leftrightarrow te^{\alpha t}$$

Performing differentiation once again,

$$\frac{1}{(s-\alpha)^3} \leftrightarrow \frac{1}{2!}t^2 e^{\alpha t}$$

In general,

$$\frac{1}{(s-\alpha)^m} \leftrightarrow \frac{1}{(m-1)\,!}t^{m-1}e^{\alpha t}$$

It can easily be established using final value theorem (Eqn. (2.41)) that the response function $\left[1/(m-1)!\right]\, t^{m-1}e^{\alpha t}$ equals zero as $t \to \infty$ if $\alpha < 0$. However, the response function grows without bound for $\alpha \ge 0$.

Similar conclusions can be derived in case of response functions corresponding to complex-conjugate pair of poles $(\sigma \pm j\omega)$ of multiplicity m. The limit of each response function as $t \to \infty$ equals zero if $\sigma < 0$. The case of $\sigma \ge 0$ contributes growing response functions.

This completes the investigation of the nature of response terms contributed by various types of poles of $M(s)$ in Eqn. (5.7), i.e., the roots of the characteristic equation $\Delta(s) = 0$ (refer to Eqns (5.1) and (5.6)). Observing these terms carefully leads us to the following general conclusions on BIBO stability:

1. If all the roots of the characteristic equation lie in the left half of the s-plane, then the output to bounded inputs is bounded.
2. If any root of the characteristic equation lies in the right half of the s-plane, then the output to bounded inputs grows without bound.
3. If the characteristic equation has repeated roots on the $j\omega$-axis in the s-plane, then the output to bounded inputs grows without bound.
4. If one or more nonrepeated roots of the characteristic equation are on the $j\omega$-axis of the s-plane, then the output to bounded inputs is bounded.

An exception to the above conclusions is brought out by the following observations: Consider a system with transfer function

$$M(s) = \frac{N(s)}{s(s - j\omega)(s + j\omega)}$$

The system has nonrepeated poles on the $j\omega$-axis in the s-plane. The response functions contributed by the poles at $s = 0$, and $s = \pm j\omega$ are, A_1, and $A_2 \cos(\omega t + \phi)$ respectively, where A_1 and A_2 are real constants. The terms A_1 and $A_2 \cos(\omega t + \phi)$ are bounded.

Careful examination of the input-output relation

$$Y(s) = M(s)R(s) = \frac{N(s)}{s(s - j\omega)(s + j\omega)} R(s)$$

shows that $y(t)$ is bounded for all bounded $r(t)$ unless the input has a pole matching one of the system poles on the $j\omega$-axis. For example, for a unit step inpt $r(t) = \mu(t)$,

$$R(s) = \frac{1}{s} \text{ and } Y(s) = \frac{N(s)}{s^2(s - j\omega)(s + j\omega)}$$

The response $y(t)$ is a linear combination of the terms A_3, $A_4 t$ and $A_5 \cos(\omega t + \phi)$, and therefore $y(t) \to \infty$ as $t \to \infty$. ***Such a system which has bounded output for all bounded inputs except for the input having poles matching the system poles, may be treated as acceptable or non-acceptable. We will bring the situations where the system has nonrepeated poles on the $j\omega$-axis under the class of marginally stable systems.***

Example 5.1 Consider an oscillator system with transfer function (refer to Problem 2.12)

$$G(s) = \frac{Y(s)}{R(s)} = \frac{\omega_n^2}{s^2 + \omega_n^2}$$

The system has distinct (nonrepeated) poles on $j\omega$-axis; $s = \pm j\omega_n$. For a (bounded) sinusoidal input $r(t) = A\cos\omega_n t$,

$$R(s) = \frac{As}{s^2 + \omega_n^2}$$

Excitation (distinct) poles are at $s = \pm j\omega_n$, which *match* the system poles. The response is given by

$$Y(s) = G(s)R(s) = \frac{A\omega_n^2 s}{(s^2 + \omega_n^2)^2}$$

The response function has repeated poles on $j\omega$-axis.

Using the Laplace transform property

$$t\,f(t) \Rightarrow -\frac{d}{ds}F(s)$$

we obtain

$$\frac{1}{2\omega}t\sin\omega t \Rightarrow \frac{s}{(s^2+\omega^2)^2}$$

Therefore,

$$y(t) = \frac{A\omega_n}{2}t\sin\omega_n t\,\mu(t)$$

which is unbounded ($y(t)\to\infty$ as $t\to\infty$).

Consider now the (bounded) input $r(t) = A\sin\omega_m t;\omega_n \neq \omega_n$;

$$R(s) = \frac{As}{s^2+\omega_m^2}$$

Excitation poles (distinct) are at $s = \pm j\omega_m$, which *do not match* the system poles. The response is given by

$$Y(s) = G(s)R(s) = \frac{A\omega_n^2 s}{(s^2+\omega_n^2)(s^2+\omega_m^2)}$$

Using the Laplace transform pair

$$\frac{1}{\omega_2^2-\omega_1^2}\left[\cos\omega_1 t - \cos\omega_2 t\right] \Rightarrow \frac{s}{(s^2+\omega_1^2)(s^2+\omega_2^2)}$$

we obtain

$$y(t) = \frac{A\omega_n^2}{\omega_n^2-\omega_m^2}\left[\cos\omega_m t - \cos\omega_n t\right]\mu(t)$$

which is bounded.

When the excitation frequency is equal to the frequency of the oscillator, the excitation reinforces the natural response of the oscillator; the oscillator system is in *resonance* with the excitation.

Marginal stability is useful in the design of oscillators. Linear control systems are designed to be asymptotically stable.

We will use the following terminology and tests for stability analysis of linear time-invariant systems described by the transfer function $M(s)$, with the characteristic equation $\Delta(s) = 0$.

1. ***If all the roots of the characteristic equation lie in the left half of the s-plane, the system is stable.***
2. ***If any root of the characteristic equation lies in the right half of the s-plane, the system is unstable.***
3. ***If condition 1 is satisfied except for the presence of one or more nonrepeated roots on the $j\omega$-axis in the s-plane, the system is marginally stable.***

It follows from the above discussion that stability of a linear time-invariant system can be determined by finding the roots of its characteristic polynomial. For first- and second-order polynomials, finding the roots is trivial. Algebraic methods are available for third-order polynomials, but are somewhat difficult to apply. For polynomials of order higher than three, numerical procedures are usually required. Computer Programs are available [151–154] that can be used to find the roots of a polynomial. A disadvantage of using numerical techniques is that all parameters in the system must be assigned numerical values. Finding the ranges of parameters that result in a stable system becomes difficult when a numerical technique is used to determine the stability. From the point of view of control system design, information on ranges of system parameters for stability is important.

The Routh stability criterion, which is an analytical procedure of stability study, is well suited for finding the ranges of parameters for stability, without recourse to finding the roots of characteristic polynomial.

5.4 THE ROUTH STABILITY CRITERION

The Routh stability criterion is an analytical procedure for determining if all the roots of a polynomial have negative real parts, and is used in the stability analysis of linear time-invariant systems. The characteristic equation for a linear time-invariant system is a polynomial equation, except for the case when the system contains dead-time elements. For this special class, the Routh stability criterion cannot be employed (the Nyquist stability criterion, which is covered later in Chapter 11, can be employed). However, with dead-time approximations (refer to Eqns (2.44)–(2.46)), the characteristic equation reduces to a polynomial equation and the Routh stability criterion becomes applicable.

Characteristic equation of linear time-invariant systems is of the form

$$\Delta(s) = a_0 s^n + a_1 s^{n-1} + \cdots + a_{n-1} s^1 + a_n s^0 = 0 \tag{5.9}$$

where the coefficients a_i are real numbers.

One can assume with no loss of generality that a_n, the coefficient of s^0 is not zero. Otherwise, the characteristic polynomial $\Delta(s)$ can be expressed as a power of s times a polynomial in which the coefficient of s^0 is nonzero. The power of s indicates roots at the origin of the s-plane, the number of which is evident; hence only the latter polynomial should be investigated using the Routh stability criterion. We assume in the following development that the coefficient of s^0 is not zero.

We will also assume with no loss of generality that $a_0 > 0$. In case $a_0 < 0$, it can be made positive by multiplying the characteristic equation by –1 throughout.

In order to ascertain the stability of a linear time-invariant system, it is necessary to determine if any of the roots of its characteristic equation lie in the right half of the s-plane. Assume that the roots of nth-order characteristic equation (5.9) are given by $s = r_1, r_2, \ldots, r_n$. These roots are functions of the coefficients $a_0, a_1, \ldots, a_{n-1}, a_n$.

A second-order polynomial

$$\begin{aligned} a_0 s^2 + a_1 s + a_2 &= a_0(s - r_1)(s - r_2) \\ &= a_0 s^2 - a_0(r_1 + r_2)\, s + a_0 r_1 r_2 \end{aligned} \tag{5.10}$$

A third-order polynomial

$$\begin{aligned} a_0 s^3 + a_1 s^2 + a_2 s + a_3 &= a_0(s - r_1)(s - r_2)(s - r_3) \\ &= a_0 s^3 - a_0(r_1 + r_2 + r_3)\, s^2 + a_0(r_1 r_2 + r_1 r_3 + r_2 r_3)\, s - a_0 r_1 r_2 r_3 \end{aligned} \tag{5.11}$$

Extending this expansion to the nth-order polynomial, we get

$$\begin{aligned} \Delta(s) &= a_0 s^n + a_1 s^{n-1} + \cdots + a_{n-1} s + a_n \\ &= a_0 s^n - a_0 \text{ (sum of all the roots) } s^{n-1} \\ &\quad + a_0 \text{(sum of the products of the roots taken 2 at a time) } s^{n-2} \\ &\quad - a_0 \text{(sum of the products of the roots taken 3 at a time) } s^{n-3} \\ &\quad + \cdots + a_0(-1)^n \text{(product of all the } n \text{ roots)} \end{aligned} \tag{5.12}$$

Suppose that all the roots of a polynomial are real and in the left half of the plane. Then all r_i in Eqns (5.10) and (5.11) are real and negative. Therefore, all polynomial coefficients are positive.

This characteristic also applies to the general case of Eqn. (5.12). The only case for which a coefficient can be negative is that there be atleast one root in the right half of the plane. Note also that if all the roots are in the left half of the plane, no coefficient can be zero.

If any roots of the polynomials above are complex, the roots must appear in complex-conjugate pairs, since the polynomial coefficients are assumed real. Then, as per the rules given for forming the polynomial coefficients, all imaginary parts of roots/products of roots will cancel. Therefore, if all roots occur in the left half of the plane, all coefficients of the general polynomial of Eqn. (5.12) will be positive. Presence of a negative coefficient implies that there is at least one root in the right half of the plane. A zero coefficient indicates presence of complex-conjugate roots on the $j\omega$-axis and/or one or more roots in the right half of the plane.

In summary, given a polynomial as in Eqn. (5.12):

(i) if any coefficient a_i is equal to zero, then *not all* the roots are in the left half of the plane; and
(ii) if any coefficient a_i is negative, then *at least one root* is in the right half of the plane.

We therefore conclude that the absence or negativeness of any of the coefficients of a characteristic polynomial indicates that the system is either unstable or at most, marginally stable. Thus, it is necessary for a stable system that all the coefficients of its characteristic polynomial be positive. If any coefficient is zero/negative, we immediately know that the system is not stable. The converse is, however, not true: when all the coefficients are positive, the system is not necessarily stable; there may still be roots in the right half of the plane and/or on the imaginary axis. In order for all the roots to have negative real parts, it is *necessary but not sufficient* that all of the coefficients of the characteristic equation be positive. For example, when the characteristic equation is

$$\Delta(s) = (s+2)\left(s - \frac{1}{2} - j\frac{\sqrt{15}}{2}\right)\left(s - \frac{1}{2} + j\frac{\sqrt{15}}{2}\right) = s^3 + s^2 + 2s + 8 = 0,$$

the system is unstable, and yet the polynomial possesses all positive coefficients.

The first step in analyzing the stability of a system is to examine its characteristic equation. If some of the coefficients are zero or negative, it can be concluded that the system is not stable. On the other hand, if all the coefficients of the characteristic equation are positive, the possibility of the stability of the system exists and one should proceed further to examine the sufficient conditions of stability.

A Hurwitz, and E J Routh independently published the method of investigating the sufficient conditions of stability of a system. The Hurwitz criterion is in terms of determinants and the Routh criterion is in terms of array formulation, which is more convenient to handle. We present here the Routh stability criterion[1].

The Routh stability criterion is based on ordering the coefficients of the characteristic equation

$$\Delta(s) = a_0 s^n + a_1 s^{n-1} + a_2 s^{n-2} + \cdots + a_{n-1} s + a_n = 0;\ a_0 > 0 \qquad (5.13)$$

into a schedule, called the *Routh array.*

A necessary and sufficient condition for stability is that all of the elements in the first column of the Routh array be positive. If this condition is not met, the system is unstable, and the number of sign changes in the elements of the first column of the Routh array corresponds to the number of roots of the characteristic equation in the right half of the s-plane.

The Routh stability criterion gives only the number of roots in the right half of the s-plane. It gives no information as regards the values of the roots, and also does not distinguish between real and complex roots.

[1] Proof of the Routh stability criterion involves delicate algebra and is omitted here. This algebraic criterion is derived in textbooks, such as Schwarz and Friedland [40], on linear systems.

To determine the Routh array, coefficients of the characteristic equation are arranged in two rows, beginning with the coefficients of s^n and s^{n-1}, respectively, and followed by even-numbered and odd-numbered coefficients as follows:

$$\begin{array}{llll} s^n : a_0 & a_2 & a_4 & \cdots \\ s^{n-1} : a_1 & a_3 & a_5 & \cdots \end{array}$$

The following rows are subsequently added to complete the Routh array:

$$\begin{array}{r|llll} \text{row } n : s^n & a_0 & a_2 & a_4 \cdots \\ \text{row}(n-1) : s^{n-1} & a_1 & a_3 & a_5 \cdots \\ \text{row}(n-2) : s^{n-2} & b_1 & b_2 & b_3 \cdots \\ \text{row}(n-3) : s^{n-3} & c_1 & c_2 & c_3 \cdots \\ \vdots & & & \\ \text{row } 2 : s^2 & * & a_n & \\ \text{row } 1 : s^1 & * & & \\ \text{row } 0 : s^0 & a_n & & \end{array} \tag{5.14}$$

where the elements from $(n-2)$th row onwards are computed as follows (the column with the powers of s is included as a convenient index for rows).

$$b_1 = \frac{-\det\begin{bmatrix} a_0 & a_2 \\ a_1 & a_3 \end{bmatrix}}{a_1} = (a_1a_2 - a_0a_3)/a_1$$

$$b_2 = \frac{-\det\begin{bmatrix} a_0 & a_4 \\ a_1 & a_5 \end{bmatrix}}{a_1} = (a_1a_4 - a_0a_5)/a_1 \tag{5.15}$$

$$\vdots$$

Note that the elements b_i of the s^{n-2}-row are formed from the elements of the two previous rows, s^n-row and s^{n-1}-row; determinants are formed using two elements of the first column and other elements of the successive columns. This pattern is continued until the rest of the b_i's are all equal to zero.

Elements of s^{n-3}-row are formed from the elements of the two previous rows, s^{n-1}-row and s^{n-2}-row, as follows:

$$c_1 = \frac{-\det\begin{bmatrix} a_1 & a_3 \\ b_1 & b_2 \end{bmatrix}}{b_1} = (b_1a_3 - a_1b_2)/b_1$$

$$c_2 = \frac{-\det\begin{bmatrix} a_1 & a_5 \\ b_1 & b_3 \end{bmatrix}}{b_1} = (b_1a_5 - a_1b_3)/b_1 \tag{5.16}$$

$$\vdots$$

This is continued until no more c_i elements are present. The rest of the rows are formed in this way down to the s^0-row. The complete array is triangular. Notice that the s^1-row and s^0-row contain one term each.

The elements of the first column of the Routh array must all be positive if all the roots of the characteristic equation are in the left half of the s-plane. However, if the elements of the first column are not all positive, the number of roots in the right half of the s-plane equals the number of sign changes in the first column. A pattern of +, –, + is counted as *two* sign changes, one going from + to – and the other from – to +.

It is to be noted here that in the process of generating the Routh array, the missing terms are regarded as zeros. Also, all the elements of any row can be multiplied or divided by a positive constant during the process to simplify the computational work; this modification does not change the signs of the elements of first column.

Example 5.2 Consider a fourth-order system with the characteristic equation

$$s^4 + 8s^3 + 18s^2 + 16s + 5 = 0$$

Since the equation has no missing terms and the coefficients are all positive, it satisfies the necessary condition for stability. However, the sufficient condition must still be checked. Routh array is made as follows:

s^4	1	18	5
s^3	8	16	0 (for the missing term)
s^2	$16 = \dfrac{(8)(18)-(1)(16)}{8}$	$5 = \dfrac{(8)(5)-(1)(0)}{8}$	
s^1	$\dfrac{27}{2} = \dfrac{(16)(16)-(8)(5)}{16}$	0	
s^0	5		

The elements of the first column are all positive, and hence the system is stable.

Example 5.3 Consider the characteristic equation

$$3s^4 + 10s^3 + 5s^2 + 5s + 2 = 0$$

which satisfies the necessary condition for stability since all the coefficients are positive and nonzero. The Routh array for this equation is

s^4	3	5	2
s^3	$\not{10}$ 2	$\not{5}$ 1	$\not{0}$ 0 ← divide by 5
s^2	$\dfrac{7}{2} = \dfrac{(2)(5)-(3)(1)}{2}$	$2 = \dfrac{(2)(2)-(3)(0)}{2}$	
s^1	$-\dfrac{1}{7} = \dfrac{(7/2)(1)-(2)(2)}{(7/2)}$		
s^0	2		

It may be noted that in order to simplify computational work, the s^3-row in the formation of the Routh array has been modified by dividing it by 5 throughout. The modified s^3-row is then used to complete the process of array formation.

Examining the first column of the Routh array, it is found that there are two changes in sign $\left(\text{from } \frac{7}{2} \text{ to } -\frac{1}{7} \text{ and from } -\frac{1}{7} \text{ to } 2\right)$. Therefore, the system under consideration is unstable with two poles in the right half of the s-plane.

5.4.1 Special Cases

Formation of Routh array for the polynomials of Examples 5.1–5.3 could easily be carried out using the standard procedure given in Eqns ((5.15)–(5.16)). However, the standard procedure fails if we encounter any of the following situations on our way to formation of the array:

1. A row of all zeros appears.
2. *Pivot element* (first element of a row, appearing in first column of the array) of a row is zero, but the entire row is not all zeros.

For these special situations, we have to use special techniques to complete the process of array formation.

Case I: A Row of All Zeros

When we encounter a row with all zero elements (if there is a single element in the row which is zero, it also means that the entire row is of zero elements) on our way to formation of Routh array, the process of array formation terminates prematurely because all the terms in the row next to the all-zero row become infinite.

An all-zero row indicates the existence of an even polynomial as a factor of the given characteristic polynomial. An even polynomial is one in which the exponents of s are even integers or zero only. This even-polynomial factor is called the *auxiliary polynomial.* The coefficients of the auxiliary polynomial will always be the elements in the row directly above the row of zeros in the array. The highest power in the auxiliary polynomial is the exponent that denotes the row containing its coefficients.

The roots of an even polynomial occur in pairs that are equal in magnitude and opposite in sign. Hence, these roots can be purely imaginary (Fig. 5.6a), purely real (Fig. 5.6b) or complex (Fig. 5.6c). Since complex roots must occur in conjugate pairs, any complex roots of an even polynomial must occur in groups of four, which is apparent in Fig. 5.6c. Such roots have quadrantal symmetry, that is, the roots are symmetrical with respect to both the real and the imaginary axes.

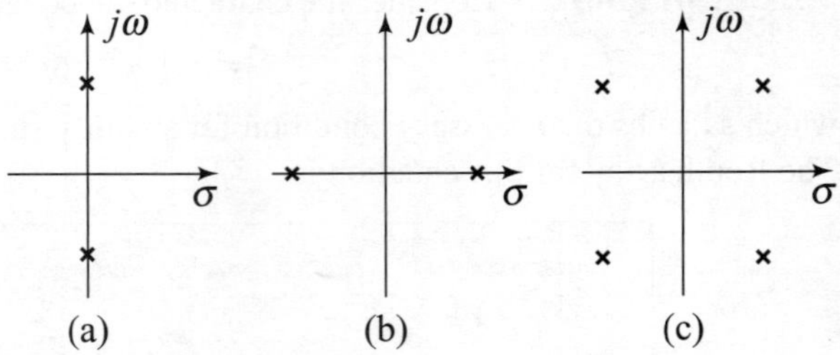

Fig. 5.6 *The roots of an even polynomial*

The *Case I* polynomial may be analyzed in either of the two ways:

1. Once the auxiliary polynomial $A(s)$ is found, it is factored from the characteristic equation, leaving a second factor which may be further analyzed by applying the Routh stability criterion. The auxiliary polynomial may be factored algebraically to directly give the roots (note that an even polynomial is easy to factorize).
2. Once the auxiliary polynomial $A(s)$ is found, it is differentiated with respect to s, and the row of zeros is replaced with the coefficients of $dA(s)/ds$. The construction of the array then continues in the usual manner, and the array is interpreted in the usual way. However, since the roots of the auxiliary polynomial are also the roots of the given characteristic equation, these must be tested separately to complete the test of stability.

Example 5.4 Consider a fifth-order system with the characteristic equation

$$\Delta(s) = s^5 + s^4 + 4\,s^3 + 24\,s^2 + 3\,s + 63 = 0$$

for which the Routh array is constructed as

s^5	1	4	3	
s^4	1	24	63	
s^3	$-\not{20}$ -1	$-\not{60}$ -3	$\not{0}$ 0	← divide by 20
s^2	$\not{21}$ 1	$\not{63}$ 3		← divide by 21
s^1	0	0		

Since a row of zeros appears prematurely, we form the auxiliary polynomial using the coefficients of the s^2-row.

$$A(s) = s^2 + 3$$

Solving the auxiliary equation

$$A(s) = s^2 + 3 = 0,$$

we get the roots at $s = \pm j\sqrt{3}$, i.e., imaginary-axis conjugate pair of roots.
In order to examine the remaining roots, we divide $\Delta(s)$ by $A(s)$ to obtain

$$\frac{\Delta(s)}{A(s)} = s^3 + s^2 + s + 21$$

Establishing Routh array for this factor, we have

s^3	1	1
s^2	1	21
s^1	-20	0
s^0	21	

The two changes in sign in the first column indicate the presence of two roots in the right half of the plane; thus the system is unstable.

Example 5.5 Consider a sixth-order system with the characteristic equation

$$\Delta(s) = s^6 + 2s^5 + 8s^4 + 12s^3 + 20s^2 + 16s + 16 = 0$$

for which the Routh array is constructed as

s^6	1	8	20	16	
s^5	~~2~~ 1	~~12~~ 6	~~16~~ 8	~~0~~ 0	← divide by 2
s^4	~~2~~ 1	~~12~~ 6	~~16~~ 8		← divide by 2
s^3	0	0			

The auxiliary polynomial

$$A(s) = s^4 + 6\,s^2 + 8$$

The derivative of the polynomial with respect to s is

$$\frac{dA(s)}{ds} = 4s^3 + 12$$

The zeros in the s^3-row are now replaced by the coefficients 4 and 12. The Routh array then becomes

s^6	1	8	12	16	
s^5	~~2~~ 1	~~12~~ 6	~~16~~ 8	0 0	← divide by 2
s^4	~~2~~ 1	~~12~~ 6	~~16~~ 8		← divide by 2
s^3	0	0			$\frac{dA(s)}{ds} = 4s^3 + 12$
	~~4~~ 1	~~12~~ 3			← divide by 4
s^2	3	8			
s^1	$\frac{1}{3}$				
s^0	8				

Since there are no sign changes in the first column of the Routh array, the polynomial $\Delta(s)/A(s)$ does not have any root in the right half of the s-plane. Solving the auxiliary equation

$$A(s) = s^4 + 6\,s^2 + 8 = 0$$

we get the roots at $s = \pm j\sqrt{2}$, and $s = \pm j2$, which are also the roots of $\Delta(s)$. Therefore, the system under consideration is marginally stable. The oscillation frequencies are $\sqrt{2}$ rad/sec, and 2 rad/sec.

Case II: A Zero Pivot Element, but the Entire Row is not All Zero

Consider now a situation wherein we encounter a row with zero pivot element while atleast one element of the row is nonzero. Because of the zero pivot element, the terms in the next row all become infinite and the array formation cannot continue.

To remedy this situation, we replace the zero pivot element by an arbitrary small number ε and then proceed with the construction of the array. The limit $\varepsilon \to 0$ is then taken to determine the changes in algebraic signs of the first column terms, yielding the information regarding the number of right-half plane roots.

Let us examine the rationale of the method. If the polynomial under test had slightly different coefficients, the troublesome zero pivot element would instead be some small nonzero number. As long as the polynomial has no roots on the imaginary axis, sufficiently small perturbations of its coefficients will not alter the number of right-half plane roots. Rather than to actually perturb the polynomial coefficients, the effect of such perturbations is indicated by replacing the zero term with ε. Therefore, for a polynomial with no imaginary-axis roots, $\varepsilon \to 0$ from positive or negative side would give identical results. ε is generally taken to be positive for convenience.

Example 5.6 Consider the polynomial

$$\Delta(s) = s^5 + 3\,s^4 + 2\,s^3 + 6\,s^2 + 6\,s + 9$$

for which the Routh array is

s^5	1	2	6	
s^4	~~3~~	~~6~~	~~9~~	← divide by 3
	1	2	3	
s^3	~~0~~	~~3~~		← replace 0 by ε
	ε	3		
s^2	~~$\frac{2\varepsilon-3}{\varepsilon}$~~	~~3~~		← replace by limiting value ($\varepsilon \to 0$)
	$\frac{-3}{\varepsilon}$	3		
s^1	~~$3+\varepsilon^2$~~			← replace by limiting value
	3			
s^0	3			

Note that in array formation, the limits were taken as $\varepsilon \to 0$ at convenient points in the calculations rather than waiting until the array was complete. This procedure simplifies the calculations and the final form of the array.

The first element in the s^2-row is $(-3/\varepsilon)$ which has a negative sign as $\varepsilon \to 0$ (in the Routh stability criterion, we are interested only in the signs of terms in the first column and not in their magnitudes). Examining the terms in the first column of the Routh array, it is found that there are two changes in sign, which implies that there are two poles in the right half of the plane.

For the case where the characteristic equation has imaginary-axis roots, replacement of the zero pivot element by ε would cause the imaginary-axis roots to move into either left or right half of the s-plane, with the result that the roots of the characteristic polynomial in the right half of the s-plane cannot be correctly determined. Therefore, to apply the ε-limiting method for a characteristic polynomial with the imaginary-axis roots, one must first extract these roots and then apply the test on the remainder polynomial with no imaginary-axis roots.

Example 5.7 Consider the polynomial

$$\Delta(s) = s^6 + s^5 + 3\,s^4 + 3\,s^3 + 3\,s^2 + 2\,s + 1$$

The Routh array is

s^6	1	3	3	1	
s^5	1	3	2		
s^4	$\cancel{0}$	$\cancel{1}$	$\cancel{1}$		← replace 0 by ε
	ε	1	1		
s^3	$\cancel{\dfrac{3\varepsilon-1}{\varepsilon}}$	$\cancel{\dfrac{2\varepsilon-1}{\varepsilon}}$			← replace by limiting value
	$-\dfrac{1}{\varepsilon}$	$-\dfrac{1}{\varepsilon}$			
s^2	$\cancel{1-\varepsilon}$	$\cancel{1}$			← replace by limiting value
	1	1			
s^1	0				

Since a row of zeros appears prematurely (elements of s^1-row are all zeros), it indicates the possibility of existence of imaginary-axis roots in the s-plane. We, therefore, need to examine the auxiliary polynomial. If the imaginary-axis roots do not exist, the usual procedure of replacing the all-zero row by coefficients of the derivative of the auxiliary polynomial is adopted. If the imaginary-axis roots are found to exist, the original polynomial is divided out by the auxiliary polynomial and the test is performed on the remainder polynomial.

For the example under consideration, the auxiliary polynomial is (let $\varepsilon \to 0$ in s^2-row):

$$A(s) = s^2 + 1 = 0$$

yielding two roots on the imaginary axis. Dividing the original polynomial $\Delta(s)$ by $A(s)$, we get

$$\frac{\Delta(s)}{A(s)} = s^4 + s^3 + 2s^2 + 2s + 1$$

The Routh array for this polynomial is

s^4	1	2	1	
s^3	1	2		
s^2	$\cancel{0}$	$\cancel{1}$		← replace 0 by ε
	ε	1		
s^1	$\dfrac{2\varepsilon-1}{\varepsilon}$			
s^0	1			

As $\varepsilon \to 0$, there are two sign changes in the first column of the array. This indicates that there are two roots in the right half of the s-plane. The original polynomial $\Delta(s)$, therefore, has two roots in the right half of the plane and two roots on the imaginary axis.

5.4.2 Relative Stability (Shifting the Origin)

The Routh stability criterion ascertains *absolute stability* of a system by determining if all the roots of the characteristic equation lie in the left half of the s-plane. Once a system is found to be absolutely stable (all characteristic roots in left half of the s-plane), it is desirable to determine its *relative stability,* which is concerned with the attributes of transient behavior of the system. We shall see in the next chapter that parameters such as overshoot, settling time, etc., are used to capture important attributes of transient response of a stable system; these parameters are, therefore, quantitative measures of relative stability. Stability margins (gain margin, phase margin) are commonly used as frequency-domain measures of relative stability. Frequency-domain analysis will appear in Chapter 10.

Although the Routh stability criterion is not as such meant for relative-stability analysis, yet in the following paragraphs, we show that the criterion can be extended for a preliminary relative-stability analysis. We know that for a real root or a complex-conjugate pair of roots on the vertical line in the s-plane defined by $s = -\sigma$, $\sigma > 0$ (refer to Fig. 5.7), the time constant $\tau = 1/\sigma$. The transient response modes corresponding to smaller time constants decay faster, and therefore the dynamic behavior of a system is dominated by the roots with larger time constants. The largest time constant associated with a system can be obtained by determining the minimum distance of the characteristic roots from the imaginary axis.

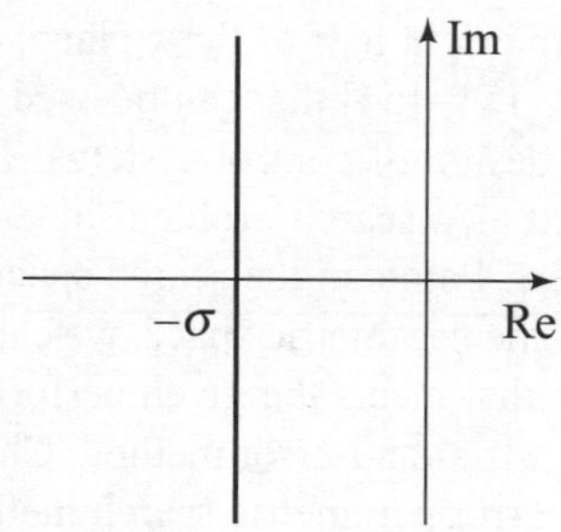

Fig. 5.7 *Shifting of the origin in the s-plane*

We substitute

$$s = \hat{s} - \sigma \tag{5.17}$$

into the characteristic equation, write a polynomial in terms of $\hat{s}$, and apply Routh stability criterion to the new polynomial in $\hat{s}$. If there are no changes of sign in the first column of the array developed for the polynomial in $\hat{s}$, it implies that all the roots of the original characteristic equation are more negative than $-\sigma$. Shifting the vertical axis on a trial-and-error basis, we can find the minimum distance of the roots from the imaginary axis, and hence the largest time constant associated with the system.

Example 5.8 Consider a third-order system with the characteristic equation

$$s^3 + 7s^2 + 25s + 39 = 0$$

which, by the Routh stability criterion, can be shown to have all its roots in the left half of the s-plane. Let us check if all the roots of this equation have real parts more negative than -1.

The substitution

$$s = \hat{s} - 1$$

results in the following polynomial in $\hat{s}$:

$$f(\hat{s}) = \hat{s}^3 + 4\hat{s}^2 + 14\hat{s} + 20$$

Forming the Routh array, we get

$\hat{s}^3$	1	14	
$\hat{s}^2$	~~4~~	~~20~~	← divide by 4
	1	5	
$\hat{s}^1$	9		
$\hat{s}^0$	5		

As the signs of all the elements of the first column of the Routh array are positive, all the roots of the original characteristic polynomial are more negative than –1.

5.5 STABILITY REGION IN THE PARAMETRIC SPACE

Our discussion so far has centered on determining whether or not a given characteristic equation has roots in the right half of the s-plane, using the Routh stability criterion. Since computer programs are available [151–154] that can be used to find the roots of a polynomial, the real value of the Routh test is in the design of control systems. The control design problem, as we will see in later chapters, can be viewed as a search problem in n-dimensional parametric space, where n is the number of design parameters. Points in the search space correspond to different design solutions. By choosing different points of the parametric space, we can generate feedback control systems with different performances. A design that meets the given performance specifications can be determined by moving in the search space by a trial-and-error method. Chapters 6–12 are devoted to system performance measures and the trial-and-error parametric search methods.

Controller design is, in fact, a constrained parametric search problem; we restrict our search to regions in the parametric space where stability is guaranteed. The first step in formulating the constrained search problem is, thus, finding the stability region in the parametric space. The Routh's test is useful for this step.

For one- or two-dimensional parametric space, visualization of stability regions is straightforward. We illustrate the concepts involved using examples.

Example 5.9 Consider the closed-loop system shown in Fig. 5.8. Let us determine the range of K for which the system is stable.

The closed-loop transfer function of the system is

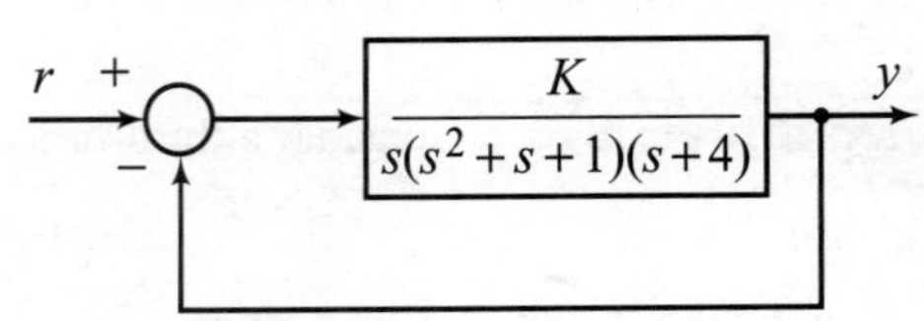

Fig. 5.8 *A feedback system*

$$\frac{Y(s)}{R(s)} = \frac{K}{s(s^2+s+1)(s+4)+K}$$

Therefore, the characteristic equation is

$$s(s^2 + s + 1)(s + 4) + K = 0$$

or

$$s^4 + 5\,s^3 + 5\,s^2 + 4\,s + K = 0$$

The Routh array for this equation is

s^4	1	5	K
s^3	5	4	
s^2	21/5	K	
s^1	$\left(\frac{84}{5}-5K\right)\Big/\left(\frac{21}{5}\right)$		
s^0	K		

Since for a stable system, the signs of elements of the first column of the Routh array should be all positive, the condition of system stability requires that

$$K > 0 \text{ and } \left(\frac{84}{5}-5K\right) > 0$$

Therefore for stability, K should lie in the range

$$\frac{84}{25} > K > 0$$

If we let $K = \frac{84}{25}$, the s^1-row of the Routh array becomes an all-zero row, resulting in the auxiliary equation,

$$\frac{21}{5}s^2 + \frac{84}{25} = 0$$

which has roots at $s = \pm j\sqrt{(4/5)} = \pm j\omega_0$. Therefore, $K = \frac{84}{25}$ will cause sustained oscillations of frequency $\omega_0 = \sqrt{(4/5)}$ rad/sec.

Example 5.10 Consider the closed-loop system shown in Fig. 5.9. A PI controller controls a second-order plant. Let us determine the range of K_c for which the closed-loop poles satisfy Re(s) < – 2.

The characteristic equation of the system is

$$s^2 + 2s + 2 + K_c\left(1+\frac{5}{s}\right)(s+3) = 0$$

or $s^3 + (2+K_c)s^2 + (2+8K_c)s + 15K_c = 0$

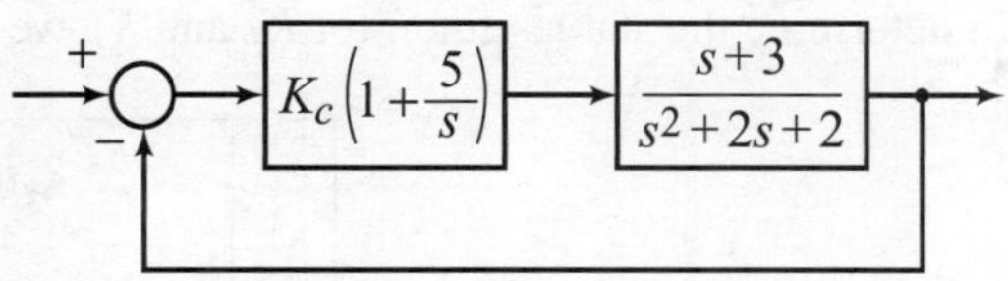

Fig. 5.9 *A feedback system*

Letting $s = \hat{s} - 2$, we have

$$(\hat{s}-2)^3 + (2+K_c)(\hat{s}-2)^2 + (2+8K_c)(\hat{s}-2) + 15K_c = 0$$

or

$$\hat{s}^3 + (K_c-4)\hat{s}^2 + (4K_c+6)\hat{s} + (3K_c-4) = 0$$

Applying the Routh's test yields the array

$\hat{s}^3$	1	$4K_c + 6$
$\hat{s}^2$	$K_c - 4$	$3K_c - 4$
$\hat{s}^1$	$\dfrac{4K_c^2 - 13K_c - 20}{K_c - 4}$	
$\hat{s}^0$	$3K_c - 4$	

If there are no sign changes in the first column of the array, then all roots satisfy Re $(\hat{s}) < 0$, that is, the roots of the original characteristic equation satisfy Re$(s) < -2$. Thus, we require that K_c satisfy all the following conditions:

$$K_c > 4;\ K_c > 4.3892 \quad \text{or} \quad K_c < -1.1392;\ K_c > \frac{4}{3}$$

The requirement $K_c < -1.1392$ is disregarded since K_c cannot be negative. Therefore, we have Re$(s) < -2$ for all closed-loop poles provided that $K_c > 4.3892$.

Example 5.11 The closed-loop control system of Fig. 5.10 involves the selection of two parameters K_P and K_I of the PI controller. Let us determine the allowable region for stability in two-dimensional parametric space.

The characteristic equation of the feedback system is

$$1 + \left(K_P + \frac{K_I}{s}\right)\frac{1}{(s+1)(s+2)} = 0$$

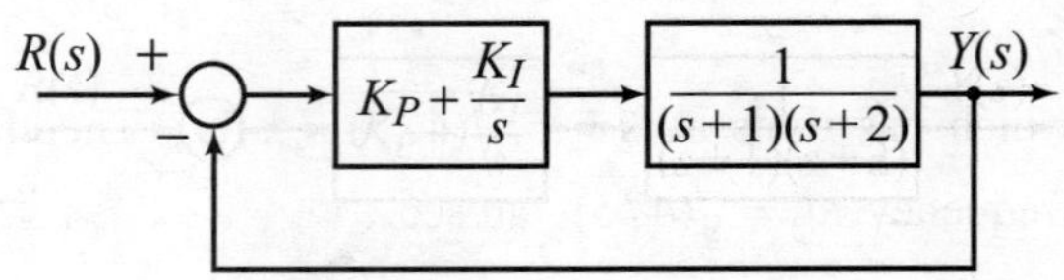

Fig. 5.10 *A feedback system*

which we may rewrite as

$$s^3 + 3s^2 + (2 + K_P)s + K_I = 0$$

To determine the stable region for K_P and K_I, we establish the Routh array as

s^3	1	$2 + K_P$
s^2	3	K_I
s^1	$(6 + 3K_P - K_I)/3$	
s^0	K_I	

The condition of system stability requires that

$$K_I > 0 \ \text{and}\ K_P > \frac{1}{3}K_I - 2$$

The allowable region is the shaded area in the (K_I, K_P) plane shown in Fig. 5.11.

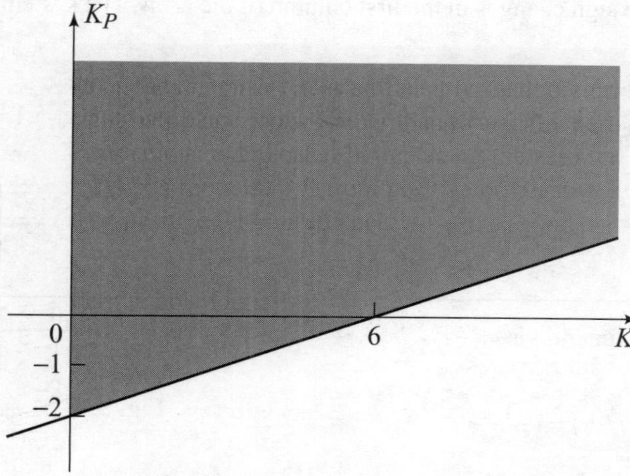

Fig. 5.11 *Region of stability in parametric space*

This example illustrates the real value of Routh's approach which provides analytical solutions. It would have been more difficult to arrive at these bounds on the parameters using numerical search.

It is probably more insightful to think about the characteristic equation of a feedback system geometrically rather than algebraically. The root locus technique developed later in Chapter 8, is intrinsically geometric as opposed to the algebraic approach of the Routh criterion. It enables us to rapidly sketch the locus of all solutions to the characteristic equation of the closed-loop transfer function. This sketch offers great insight by showing how the locations of the poles of the closed-loop transfer function change as a specific parameter is varied.

5.6 REVIEW EXAMPLES

Review Example 5.1 Consider the feedback system shown in Fig. 5.12. The characteristic equation of the system is

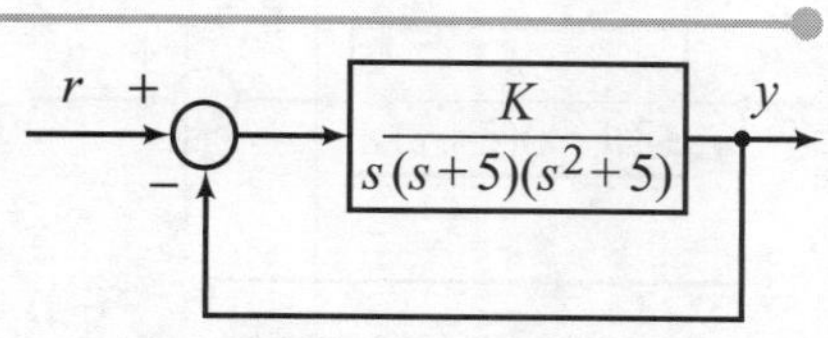

Fig. 5.12 *A feedback system*

$$\Delta(s) = s^4 + 5s^3 + 5s^2 + 25s + K = 0$$

for which the Routh array is

s^4	1	5	K	
s^3	~~5~~	~~25~~		← divide by 5
	1	5		
s^2	~~0~~	~~K~~		← replace 0 by ε
	ε	K		
s^1	$\dfrac{5\varepsilon - K}{\varepsilon}$			
s^0	1			

As $\varepsilon \to 0$, there are two sign changes in the first column of the array. The system is, therefore, unstable for all values of $K > 0$.

Review Example 5.2 A feedback control system is shown in Fig. 5.13. The process transfer function is

$$G(s) = \frac{K(s+40)}{s(s+10)}$$

and the sensor transfer function is

$$H(s) = \frac{1}{s+20}$$

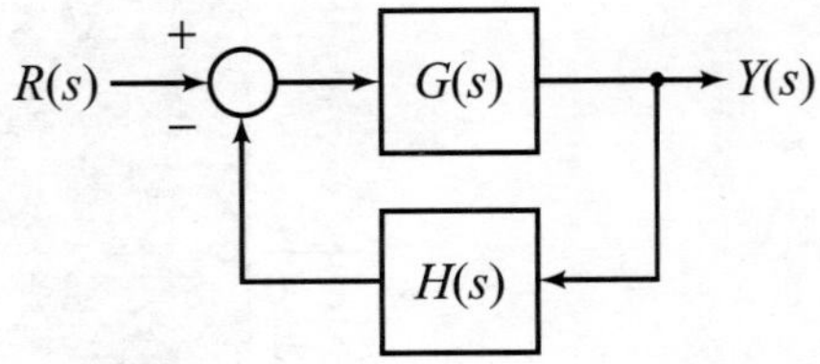

Fig. 5.13 *A feedback control system*

(a) Find the gain K that results in marginal stability. Determine the oscillation frequency.
(b) Reduce the gain to half the value found in (a) and study the relative stability of the system.

Solution: The characteristic equation of the system is

$$1 + G(s)H(s) = 0$$

This gives
$$s(s + 10)(s + 20) + K (s + 40) = 0$$
or
$$s^3 + 30\, s^2 + (K + 200)\, s + 40\, K = 0$$

for which the Routh array is

s^3	1	$K + 200$	
s^2	~~30~~ 3	~~40~~K 4K	← divide by 10
s^1	$\dfrac{600 - K}{3}$		
s^0	$4K$		

(a) $K = 600$ results in an all-zero row in the array. The auxiliary equation, for this value of K is

$$3s^2 + 2400 = 0$$

which has the roots
$$s = \pm j\sqrt{800}.$$

Therefore, $K = 600$ results in marginal stability; the oscillation frequency is $\sqrt{800}$ rad/sec.
(b) The characteristic polynomial of the system for $K = 300$ is

$$\Delta(s) = s^3 + 30\, s^2 + 500\, s + 12000$$

The substitution $s = \hat{s} - 1$ results in the following polynomial in $\hat{s}$:

$$f(\hat{s}) = \hat{s}^3 + 27\hat{s}^2 + 443\hat{s} + 11529$$

Forming the Routh array, we get

$\hat{s}^3$	1	443	
$\hat{s}^2$	~~27~~	~~11529~~	← divide by 27
	1	427	
$\hat{s}^1$	16		
$\hat{s}^0$	427		

As the signs of all the elements of the first column of the Routh array are positive, all the roots of the original characteristic polynomial $\Delta(s)$ are more negative than -1.

The substitution $s = \bar{s} - 2$ in $\Delta(s)$, results in the following polynomial:

$$g(\bar{s}) = \bar{s}^3 + 24\bar{s}^2 + 392\bar{s} + 11112$$

Forming the Routh array, we get

$\bar{s}^3$	1	392	
$\bar{s}^2$	~~24~~	~~11112~~	← divide by 24
	1	463	
$\bar{s}^1$	−71		
$\bar{s}^0$	463		

There are two sign changes in the first column of the Routh array; there are, therefore, two roots of the original characteristic polynomial that satisfy $\text{Re}(s) > -2$.

The largest time constant of the closed-loop system is greater than 0.5 and less than 1.

Review Example 5.3 Consider the feedback system shown in Fig. 5.13. The process transfer function

$$G(s) = \frac{K}{s(s+1)}$$

and the feedback transfer function is a first-order Pade approximation (refer to Eqn. (2.46)) for the transducer delay:

$$H(s) = \frac{1 - \dfrac{\tau_D}{2}s}{1 + \dfrac{\tau_D}{2}s} = \frac{1 - Ts}{1 + Ts}; \tau_D = 2T$$

The question to be studied is: what are the combinations of K and T for which the system is stable?

The characteristic polynomial of the system is

$$\Delta(s) = Ts^3 + (1 + T)s^2 + (1 - KT)\, s + K$$

The Routh array is

s^3	T	$1 - KT$
s^2	$1 + T$	K
s^1	$\dfrac{1+T-2KT-KT^2}{1+T}$	
s^0	K	

The stability boundary is given by the condition

$$1 + T - KT(2 + T) = 0$$

or

$$KT = \frac{1+T}{2+T}$$

The relationship between the maximum allowable gain K_{max} and the delay T is sketched in Fig. 5.14.

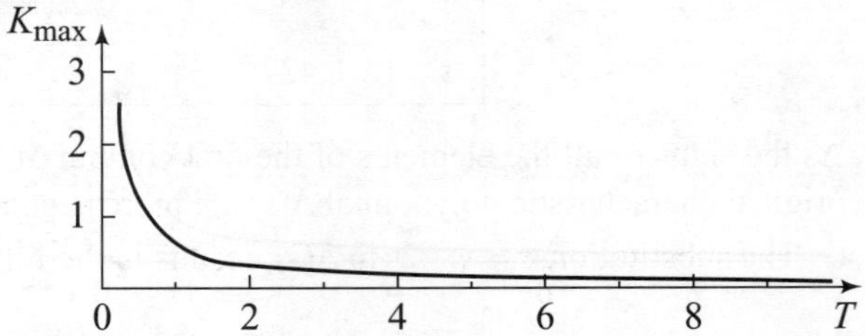

Fig. 5.14 *Maximum allowable gain versus delay*

CONCLUDING REMARKS

In this chapter, we explored the concepts of system stability from a classical viewpoint which says that a system is stable if every bounded input yields a bounded output (BIBO stability), and unstable if any bounded input yields an unbounded output.

What we have learnt?

- Stability for linear, time-invariant systems can be determined from the location of the closed-loop poles:
  1. If the poles are only in the left half of the plane, the system is stable.
  2. If any poles are in the right half of the plane, the system is unstable.
  3. If the poles are on the $j\omega$-axis and in the left half of the plane, then the system is marginally stable as long as the poles on the $j\omega$-axis are of unit multiplicity; it is unstable if there are any multiple $j\omega$ poles.
- The Routh criterion lets us find how many poles are in each of the sections of s-plane, without giving us the coordinates of the poles.
- The Routh criterion provides an analytical solution to the problem of determining the ranges of design parameters for which the system remains stable.

Review Questions

5.1 Define the terms:

(a) bounded-input, bounded-output (BIBO) stability; and

(b) asymptotic stability

State the conditions under which asymptotic stability of a linear time-invariant system ensures BIBO stability and vice versa.

5.2 (a) Define the terms: absolute stability, and relative stability.
(b) Derive conditions for absoulte stability of a linear time-invariant system described by the transfer function,

$$M(s) = \frac{b_0 s^m + b_1 s^{m-1} + \cdots + b_{m-1}s + b_m}{a_0 s^n + a_1 s^{n-1} + \cdots + a_{n-1}s + a_n}; m \le n$$

Comment upon the stability property of the system when $M(s)$ has nonrepeated poles on the imaginary axis in the s-plane.

5.3 The real value of Routh's approach lies in providing an analytical solution to the problem of determining allowable stability region in parametric space. Explain this point by addressing the stability issues in the design problem of feedback-control system of Fig. 5.15, which involves the selection of two parameters.

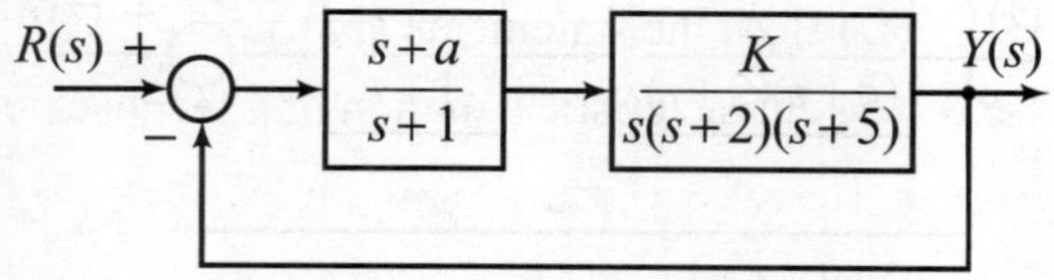

Fig. 5.15 *A feedback control system*

Self-Appraisal Quiz

A set of questions with multiple answer choices is given below. These questions have been designed to capture the important aspects of the subject covered in the chapter. For each of these questions, identify the correct answer from the given choices. Compare your answers with the Master Key given at the end of the session. This quiz should serve the purpose of a self-appraisal test for readers.

5.1 A closed-loop system is subjected to ramp input. In response, the control system output continues to rise. The system is

(A) unstable (B) stable (C) marginally stable (D) Nothing can be said about stability

5.2 The characteristic equation of the closed-loop system of Fig. Q5.2

(A) $s^2 + 11s + 10 = 0$
(B) $s^2 + 11s + 130 = 0$
(C) $s^2 + 10s + 120 = 0$
(D) $s^2 + 10s + 12 = 0$

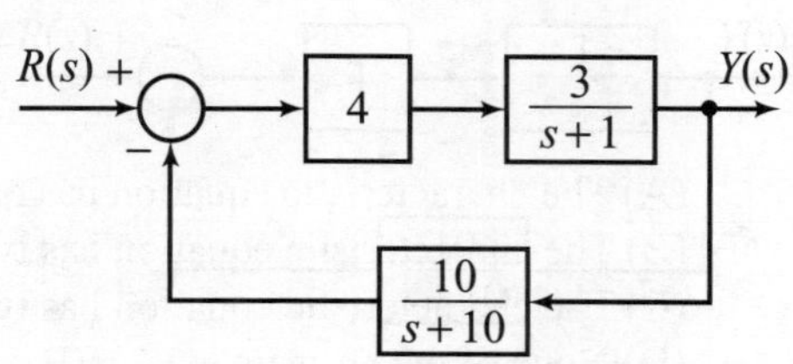

Fig. Q5.2

5.3 The condition that all the roots of the polynomial

$$\Delta(s) = a_0 s^3 + a_1 s^2 + a_2 s + a_3; a_i > 0$$

have negative real parts, is given by

(A) $a_1 a_3 > a_0 a_2$ (B) $a_1 a_0 > a_2 a_3$ (C) $a_1 a_2 > a_0 a_3$ (D) $a_2 a_0 > a_1 a_3$

5.4 Consider the following statements:
(i) All the roots of the polynomial

$$\Delta_1(s) = 3s^4 + 10s^3 + 5s^2 + 2$$

have negative real parts.

(ii) All the roots of the polynomial

$$\Delta_2(s) = s^4 + 4s^3 + 6s^2 + 4s - 5$$

have negative real parts.

Which of the following is the correct answer?

(A) None of the above statements is true

(B) Statement (i) is true but statement (ii) is false

(C) Statement (i) is false but statement (ii) is true

(D) Both the statements are true.

5.5 The characteristic equation of a feedback control system is given by

$$2s^4 + s^3 + 2s^2 + 5s + 10 = 0$$

The number of roots in the right half of the s-plane are

(A) Zero (B) 1 (C) 2 (D) 3

5.6 Consider the following statements:

(i) When a row of Routh tabulation contains all zeros before the tabulation ends, this means that the characteristic equation necessarily has roots on the imaginary axis of the s-plane.

(ii) When a row of Routh tabulation contains a zero pivot element with atleast one nonzero element, this means that the characteristic equation necessarily has roots on the imaginary axis of the s-plane.

Which of the following is the correct answer?

(A) None of the above statements is true.

(B) Statement (i) is true but statement (ii) is false

(C) Statement (i) is false but statement (ii) is true

(D) Both the statements are true.

5.7 The first two rows of Routh tabulation of a third-order system are

$$\begin{array}{c|cc} s^3 & 2 & 2 \\ s^2 & 4 & 4 \end{array}$$

(A) The characteristic equation has one root in the right half of the s-plane

(B) The characteristic equation has two roots on the $j\omega$-axis at $s = \pm j$

(C) The characteristic equation has two roots on the $j\omega$-axis at $s = \pm 2j$

(D) None of the answers in (A), (B), and (C) is correct

5.8 The polynomial

$$\Delta(s) = s^3 + 3s^2 + 2s + 6$$

has

(A) Two roots in the left half of the s-plane and one root in the right half

(B) Two roots in the right half of the s-plane and one root in the left half

(C) Two roots on $j\omega$-axis of the s-plane and one root in the right half

(D) Two roots on $j\omega$- axis of the s-plane and one root in the left half.

5.9 The characteristic equation of a feedback control system is

$$s^3 + \alpha_1 s^2 + \alpha_2 s + K\alpha_2 = 0; \alpha_1, \alpha_2, > 0,$$

where K is a variable positive scalar parameter. The system is stable for all values of K given by
(A) $K > \alpha_2$ (B) $K < \alpha_2$ (C) $K > \alpha_1$ (D) $K < \alpha_1$

5.10 The open-loop transfer function of a unity feedback system is

$$G(s) = K/[s^2(s+5)]; K > 0$$

The system is unstable for
(A) $K > 5$ (B) $K < 5$ (C) $K > 0$ (D) All the answers in (A), (B) and (C) are correct.

5.11 Consider the characteristic polynomial of a feedback system:

$$\Delta(s) = s^4 + s^3 + s^2 + s + K$$

(A) The system is stable for all $K > 0$
(B) The system is unstable for all $K > 0$
(C) $K = 8$ results in marginal stability
(D) None of the answers in (A), (B) and (C) is correct.

5.12 A unity-feedback system has open-loop transfer function

$$G(s) = K/[s(s+1)(s+2)]; K > 0$$

The value of K that results in oscillatory response to step input, is
(A) 2 (B) 6 (C) 20 (D) 60

5.13 Presence of transportion lag in the forward path of a closed-loop control system
(A) Decreases margin of stability (B) Increases margin of stability
(C) Does not affect its margin of stability

5.14 Consider a unity-feedback system with plant $G(s) = \dfrac{1}{s^2+s+1}$ and cascade controller $D(s) = K_P + \dfrac{2}{5}$.
The values of K_P for which the system is stable are
(A) $K_P > 0$ (B) unstable for all $K_P > 0$ (C) $K_P > 1$ (D) $K_P > 2$

5.15 The error function of a feedback system is

$$E(s) = \frac{(s+0.1)(s+0.5)}{s(s+0.1)(s+0.5)+0.5(s+1)}$$

The steady-state value of $e(t)$ is
(A) 0.001 (B) 0.1 (C) 0.01 (D) None of the answers in (A), (B), and (C) is correct.

5.16 Consider the system of Fig. Q5.16. The steady-state error is less than 2% for step input for
(A) $K_c > 49$ (B) any value of K_c (C) no value of K_c (D) $K_c > 50$

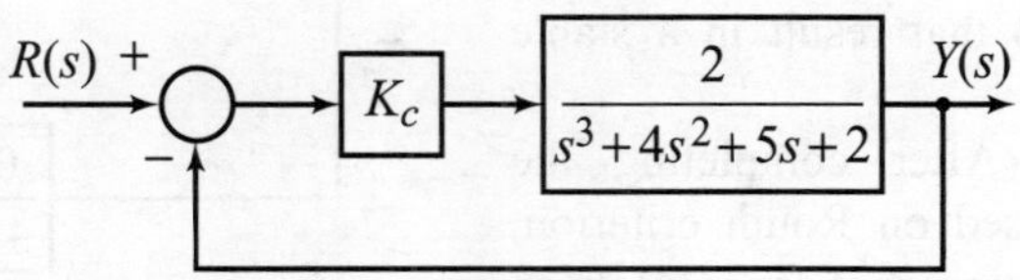

Fig. Q5.16

MASTER KEY

5.1 (D)	5.2 (B)	5.3 (C)	5.4 (A)	5.5 (C)
5.6 (A)	5.7 (B)	5.8 (D)	5.9 (D)	5.10 (D)
5.11 (B)	5.12 (B)	5.13 (A)	5.14 (C)	5.15 (D)
5.16 (C)				

Problems

5.1 Use the Routh stability criterion to determine the number of roots in the left half of the plane, the right half of the plane, and on the imaginary axis for the given characteristic equation:

(a) $s^4 + 2s^3 + 8s^2 + 4s + 3 = 0$
(b) $s^6 + 3s^5 + 2s^4 + 9s^3 + 5s^2 + 12s + 20 = 0$
(c) $s^5 + s^4 + 2s^3 + 2s^2 + 3s + 5 = 0$
(d) $s^5 + 4s^4 + 8s^3 + 8s^2 + 7s + 4 = 0$
(e) $s^4 + 9s^3 + 4s^2 - 36s - 32 = 0$
(f) $s^7 + 5s^6 + 9s^5 + 9s^4 + 4s^3 + 20s^2 + 36s + 36 = 0$

5.2 Obtain the roots of the given characteristic polynomial on the basis of the Routh array:

(a) $s^4 + 2s^3 + 11s^2 + 18s + 18 = 0$
(b) $s^5 + 2s^4 + 24s^3 + 48s^2 - 25s - 50 = 0$
(c) $s^6 + 3s^5 + 5s^4 + 9s^3 + 8s^2 + 6s + 4 = 0$

5.3 Determine whether the largest time constant of the roots of the characteristic equation given below is greater than, less than or equal to 1.0 sec.

$$s^3 + 4s^2 + 6s + 4 = 0$$

5.4 The characteristic equations for certain feedback control systems are given below. In each case, determine the range of values of K ($K > 0$) for which the system is stable.

(a) $s^4 + s^3 + s^2 + s + K = 0$ (b) $s^4 + Ks^3 + s^2 + s + 1 = 0$
(c) $s^4 + 3s^3 + 3s^2 + 2s + K = 0$ (d) $s^3 + 3Ks^2 + (K + 2)s + 4 = 0$

5.5 Determine the range of values of K ($K > 0$) such that the characteristic equation,

$$s^3 + 3(K + 1)s^2 + (7K + 5)s + (4K + 7) = 0$$

has roots more negative than $s = -1$.

MATLAB Exercise After completing the hand calculations based on Routh criterion, verify your results using root locus plots in MATLAB window.

5.6 The block diagram of a feedback control system is shown in Fig. P5.6. Determine the values of K ($K > 0$) that result in a stable closed-loop system.

MATLAB Exercise After completing the hand calculations based on Routh criterion, verify your results using root locus plots in MATLAB window.

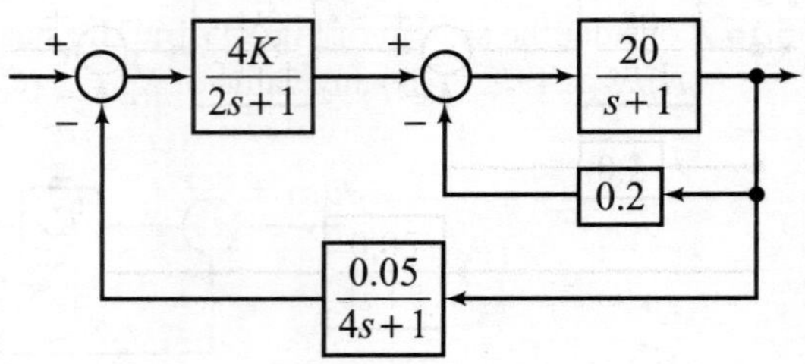

Fig. P5.6

5.7 A feedback system has open-loop transfer function

$$G(s)H(s) = \frac{Ke^{-\tau_D s}}{s(s^2+5s+9)}$$

Determine the range of K ($K > 0$) for which the system is stable when

(a) dead time $\tau_D = 0$,

(b) dead time $\tau_D = 1$, and $e^{-\tau_D s}$ is approximated by $(1 - \tau_D s)$, and

(c) dead time $\tau_D = 1$, and $e^{-\tau_D s}$ is approximated by a first-order Pade approximation:

$$e^{-\tau_D s} = \frac{1 - \frac{\tau_D}{2}s}{1 + \frac{\tau_D}{2}s}$$

5.8 Open-loop transfer functions of certain unity-feedback systems are given below. In each case, determine

(a) the range of values of K ($K > 0$) for which the system is stable,

(b) the value of K that will result in the system being marginally stable, and

(c) the location of the roots of the characteristic equation for the value of K found in (ii).

(i) $G(s) = \dfrac{K(s+13)}{s(s+3)(s+7)}$ (ii) $G(s) = \dfrac{K(s+1)}{s(s-1)(s+6)}$ (iii) $G(s) = \dfrac{K(s+2)}{s(s+5)(s^2+2s+5)}$

5.9 The open-loop transfer functions of certain unity-feedback control systems are given below. In each case, discuss the stability of the closed-loop system as a function of $K > 0$. Determine the values of K which will cause sustained oscillations in the closed-loop system. What are the corresponding oscillation frequencies?

(a) $G(s) = \dfrac{K}{(s+2)(s+4)(s^2+6s+25)}$ (b) $G(s) = \dfrac{K(s+1)}{s(s-1)(s^2+4s+16)}$

MATLAB Exercise After completing the hand calculations based on Routh criterion, verify your results using root locus plots in MATLAB window.

5.10 Determine the values of $K > 0$ and $a > 0$, so that the system shown in Fig. P5.10 oscillates at a frequency 2 rad/sec.

MATLAB Exercise After completing the hand calculations based on Routh criterion, verify your results using root locus plots in MATLAB window.

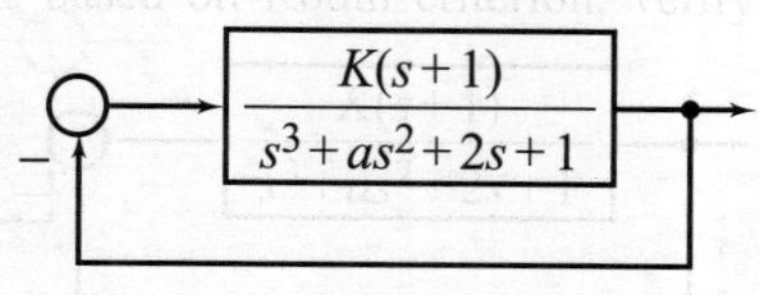

Fig. P5.10

5.11 A feedback control system is shown in Fig. P5.11. The process transfer function is

$$G(s) = \frac{K}{s(0.2s+1)}$$

and the sensor transfer function is $H(s) = 1/(0.1s + 1)$.

(a) Determine the limiting value of gain $K > 0$ for a stable system.

(b) For the gain that results in marginal stability, determine the oscillation frequency.

(c) Reduce the gain to half the value found in (b) and study the relative stability of the system by shifting the axis and using the Routh stability criterion.

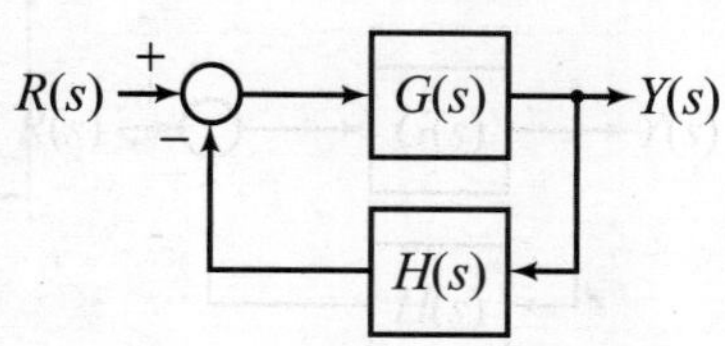

Fig. P5.11

5.12 A unity-feedback control system has the open-loop transfer function

$$G(s) = \frac{K_c(1+T_D s)}{s^2(\tau s+1)}$$

Determine the stability limits on the controller parameters $K_c > 0$ and T_D for the following cases:

(a) proportional-only control ($T_D = 0$); and

(b) proportional-derivative control ($T_D > 0$).

5.13 Shown in Fig. P5.13 is a closed-loop control system.

(a) Find the range of gain K ($K > 0$), for which the system is stable.

(b) Suppose that the gain K is set to a value of 20. In addition, suppose that the sensor has a general time constant of τ, rather than 0.1 sec given value. The transfer function of the sensor is then $1/(\tau s + 1)$. Find the allowable range for time constant $\tau > 0$ such that the system is stable.

(c) Suppose that in (b), the gain K is also variable. Find the range of gain K as a function of the sensor time constant τ such that the system is stable. The parameters K and τ may be represented in a plane with τ as the horizontal axis and K as the vertical axis. Determine the regions in the K-*vs*-τ plane in which the closed-loop system is stable, and unstable. Indicate the boundary on which the system is marginally stable.

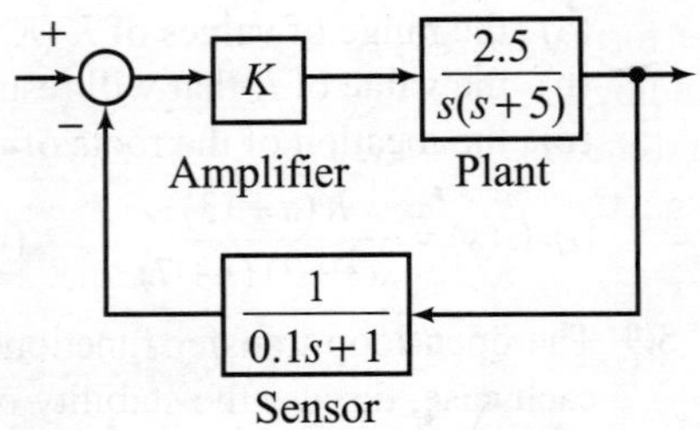

Fig. P5.13

5.14 Figure P5.14 shows a closed-loop system with a proportional-integral controller. Find the range of gain $K_c > 0$ as a function of integral time $T_I > 0$ such that the system is stable. Comment on the effect of decreasing T_I on the limit on gain K_c for stability.

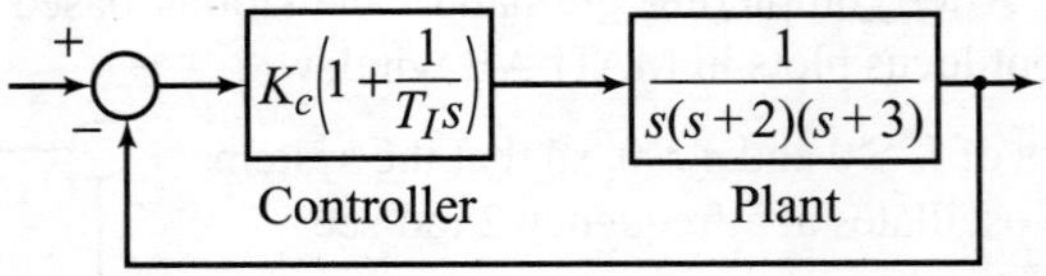

Fig. P5.14

5.15 The block diagram of a motor-control system with tachogenerator feedback is shown in Fig. P5.15. Find the range of $K > 0$ as a function of $K_t > 0$ such that the system is stable. Comment on the effect of increasing K_t on the limit on gain K for stability.

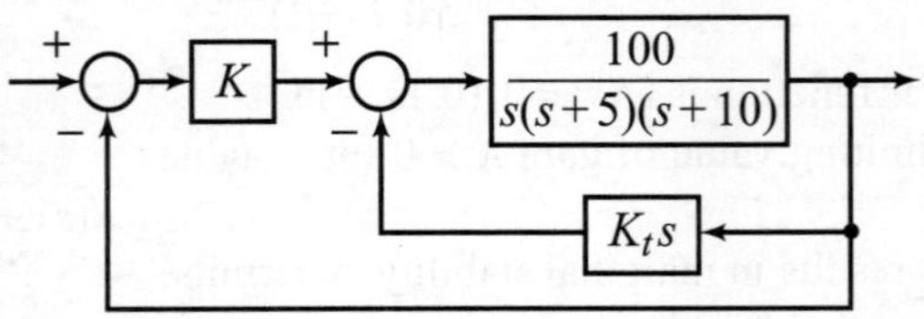

Fig. P5.15

PERFORMANCE SPECIFICATIONS ON SYSTEM TIME RESPONSE

6

Preview

We have learnt from Chapter 5 that controller design is a constrained parametric search problem; we first find ranges of the parameters of the controller for which the system remains stable, and then carry out the search for optimal parameters within the limits imposed by stability considerations, to improve performance. The first step in formulating the parametric search problem was taken in Chapter 5: finding the ranges of the controller parameters for which the feedback system stays stable. The second and the last step in formulating the design problem is to specify the desired system performance. The objective set for this chapter is to learn this important step of the design problem.

6.1 INTRODUCTION

Recall the basic feedback system block diagram defined in Chapter 2; shown as Fig. 6.1 in a form convenient for analysis. The control system objectives are that the output y follows input r, and disturbance w is ignored. Performance criteria must have to do with how well these two objectives are attained.

Clearly, the performance depends both on the system characteristics and on the nature of r and w. Precise mathematical functions for r and w will not generally be known in practice. For example, in a chamber heating system (Example 1.2), we may be quite clear that r will be constant (say, 40°C) but we cannot predict the variation of the main disturbance—outdoor temperature. In a contouring machine control (Fig. 3.41), we do not know at the design stage, all the different shapes of parts that the machine will need to produce during its lifetime. In a radar tracking system (Example 1.3), the position and speed of the target to be tracked may vary in a random fashion. ***The uncertain nature of many practical commands and disturbances makes the development of performance criteria, based on the actual r and w experienced by the real system, difficult. It is common to base performance evaluation on system response to simple 'standard' test signals such as impulse, step, ramp, parabola, and sine wave*** (refer to Section 2.3). This approach has been successful for a number of reasons.

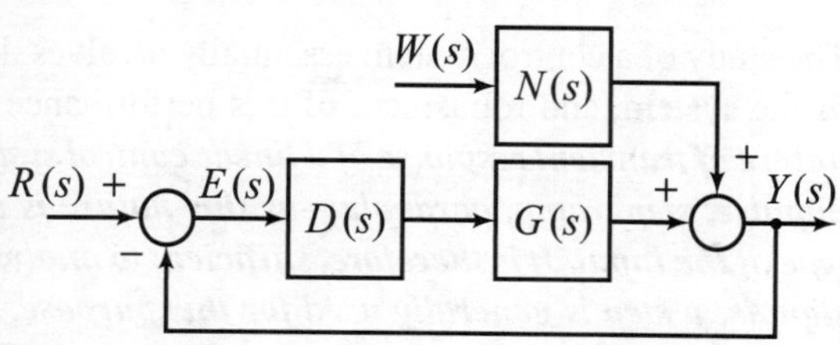

Fig. 6.1 *Block diagram representation of a feedback system*

1. The characteristics of actual signals which severely strain a control system are a sudden shock (impulse), a sudden change (step), constant velocity (ramp), and constant acceleration (parabola). Experience with the actual performance of various classes of control systems has established a good correlation between the response of systems to standard inputs and the capability of the systems to accomplish their required tasks.

2. Design is much concerned with the comparison of competitive systems. This comparison can often be made nearly as well in terms of standard inputs as for real inputs.
3. The simplicity of the form of standard inputs facilitates mathematical analysis and experimental verification.
4. For linear time-invariant systems, frequency-response curves (Section 2.9) are a complete description of the system's dynamic behavior and allow one to compute the response to any input, not just sine waves.

The standard performance criteria in common use may be classified as falling into the *time domain* or the *frequency domain*. Time-domain specifications have to do with the response to steps, ramps, parabolas, and the like; while frequency-domain specifications are concerned with certain characteristics of the system's frequency response. Both types of specifications are often applied for the same system to ensure that certain behavior characteristics will be obtained. It may be noted that all performance specifications are meaningless unless the system is absolutely stable.

The majority of analysis and design tools for feedback systems presuppose that a sufficiently accurate model of the process to be controlled has been set up. Using this model, the design is carried out to meet the time-domain and/or frequency-domain performance specifications. The analysis tools that verify the design also use the same model. This design is bound to be a disappointing one when the control system is put to real-life use. This is because in a reasonably complex system, there are always discrepancies between the model and the real process. Design methods that enable us to explicitly consider these discrepancies, i.e., designing a controller without having a precise mathematical model at hand, have yet to mature [74–79]. We need a design that is not only valid for the approximate model but does hold with certainty for a class of plant models and hence for the real process. Good robustness against modeling errors and against variations in parameters of the model due to environmental and other effects, is an important design requirement.

We will not include the robustness/insensitivity requirements in our performance specifications for design. Our design is based on the assumption that the model is fixed once the model structure and parameters are given. The design is then analysed to study its robustness properties. Qualitative guidelines will be given to improve robustness.

6.2 THE PERFORMANCE SPECIFICATIONS

The study of a control system essentially involves the evaluation of transient and steady-state responses of the system, and robustness of this performance against model uncertainties and disturbances. ***The nature of transient response of a linear control system is revealed by any of the standard test signals—impulse, step, ramp, parabola—as this nature is dependent upon system poles only and not on the type of the input. It is therefore, sufficient to analyse the transient response to one of the standard test signals; a step is generally used for this purpose. Steady-state response depends on both the system and the type of input. From the steady-state viewpoint, the 'easiest' input is generally a step since it requires only maintaining the output at a constant value once the transient is over. A more difficult problem is the tracking of a ramp input. Tracking of a parabola is even more difficult since a parabolic function is one degree faster than the ramp function. In practice, we seldom find it necessary to use a signal faster than a parabolic function; characteristics of actual signals, which the control systems generally encounter, are adequately represented by step, ramp, and parabolic functions.***

The superposition principle allows us to consider response to commands apart from response to disturbances. If both occur simultaneously, the total response is just the superposition of the two individual responses.

Robustness describes the ability of the system to satisfactorily perform in the face of inconsistencies of the model used for design and the real plant.

6.2.1 Transient Response

The transient response of a practical control system often exhibits damped oscillations before reaching steady state. In specifying the transient-response characteristics of a control system to a unit-step input, it is common to specify the following:

(i) rise time, t_r;
(ii) peak time, t_p;
(iii) peak overshoot, M_p; and
(iv) settling time, t_s.

These specifications are defined below and are shown graphically in Fig. 6.2.

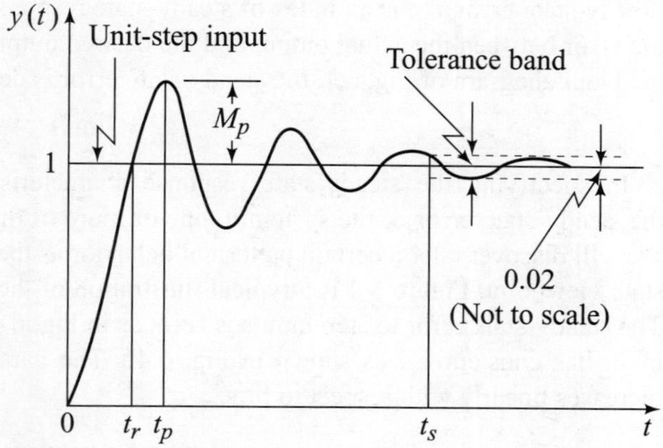

Fig. 6.2 *Typical unit-step response of a control system*

Rise Time, t_r

For underdamped systems, the rise time is normally defined as the time required for the step response to rise from 0 to 100% of its final value. For overdamped systems, the 10 to 90% rise time is commonly used.

Peak Time, t_p

It is the time required for the response to reach the first peak of the overshoot.

Peak Overshoot, M_p

It is the peak value of the response curve measured from unity. If the final steady-state value of the response differs from unity, then it is common to use *per cent peak overshoot*. It is defined by

$$\text{Per cent peak overshoot} = \frac{y(t_p) - y(\infty)}{y(\infty)} \times 100\% \tag{6.1}$$

Settling Time, t_s

The time required for the response to damp out all transients is commonly called the settling time. Theoretically, the time taken to damp out all transients may be infinity. In practice, however, the transient is assumed to be over when the error is reduced below some acceptable value. Typically, the acceptable level is set at two or five per cent of the final value. Figure 6.2 shows a *tolerance band* of ± 2%.

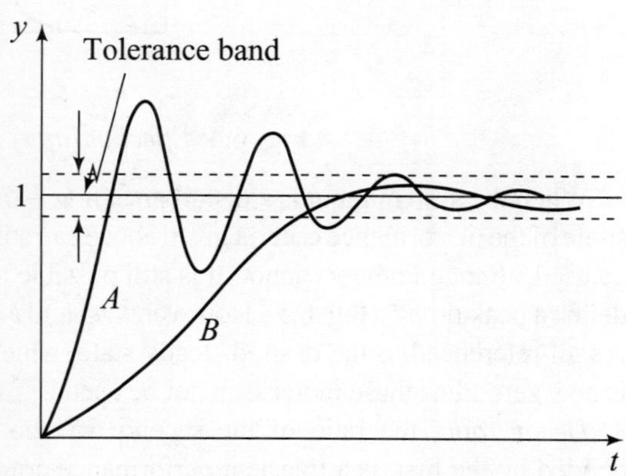

Fig. 6.3 *Significance of settling time*

Rise time and peak time are intended as speed of response criteria. Clearly, the smaller these values, the faster the system response. Peak overshoot is used mainly for relative stability. Values in excess of about 40% may indicate that the system is dangerously close to absolute instability. Settling time combines stability and speed of response aspects and is widely used. Note in Fig. 6.3 that system A (which has a much shorter rise time than system B) is, in a sense, not really faster than B since it does not settle in the neighborhood of the desired value until a later time because of the undesirable feature of overshooting (or undershooting) the final value and then having to come back in the opposite direction.

6.2.2 Steady-State Response

Steady-state error, e_{ss}, is an index of steady-state response of a system to a specified input. It indicates the error between the actual output and the desired output as time t tends to infinity. With reference to the block diagram of Fig. 6.1, the steady-state error is defined as

$$e_{ss} = \lim_{t\to\infty} [r(t) - y(t)] \tag{6.2}$$

In specifying the steady-state response characteristics of a system, it is common to specify the steady-state error of the system to one or more of the standard test signals—step, ramp, parabola. We will discover later a certain pattern of behavior as the input is made more difficult from the steady-state viewpoint. Figure 6.4 is a typical illustration of the effect of input severity on steady-state error. The steady-state error to step inputs is zero, as in Fig. 6.4a. The very same system subjected to a ramp input has constant e_{ss}, as shown in Fig. 6.4b. The parabolic input of Fig. 6.4c causes an error that increases linearly with respect to time.

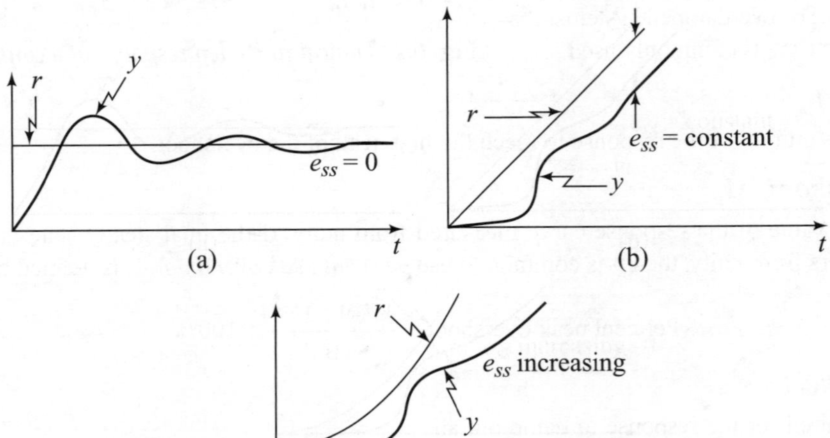

Fig. 6.4 *Effect of input severity on steady-state error*

When the system input is a disturbance w ($r \equiv 0$), some of the performance criteria given above can still be used, although others cannot. It is still possible to define a peak time t_p (Fig. 6.5). However, t_r, t_s, and M_p are all referenced to the desired steady state, which is now zero; thus these indices cannot be used.

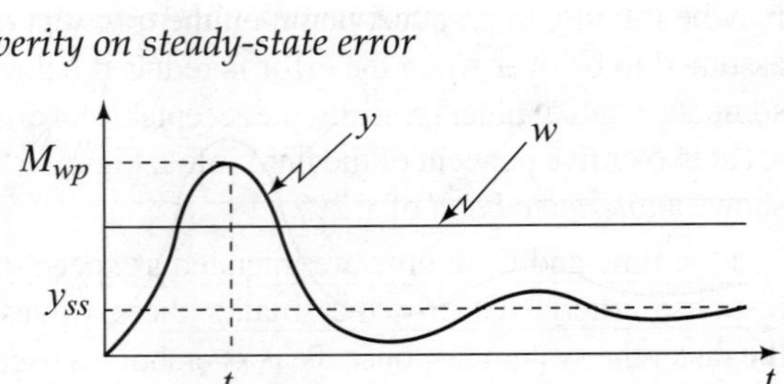

Fig. 6.5 *Time-domain specifications for disturbance input*

Decay ratio, the ratio of the second overshoot divided by the first, is a transient performance criterion often used in process control systems, the most common design value being 1/4. We could also use the number of cycles to damp the amplitude to, say, 10% of M_{wp}—the peak value of the response to disturbance input. The smaller the number of cycles, the better is the speed of response and relative stability.

When the system input is a disturbance w, the definition of steady-state error still applies and we would expect the trend of worsening error as w is changed from step to ramp to parabola, similar to the behavior of Fig. 6.4.

6.2.3 Robustness

So far the design objectives have been formulated as requirements on the closed-loop response $y(t)$ to command and disturbance inputs. These specifications on stability, speed of response, and steady-state accuracy are considered in design under the assumption that a nearly precise model of the plant is available.

The basic assumption of availability of precise model of the plant, is almost never true. There are almost always inconsistencies of the model and the real plant—called *model uncertainties*. Following are the three main reasons for model uncertainties:

1. Model uncertainties occur because of an incomplete knowledge of the system. As a typical example, the structure of the model equations can be determined by means of basic laws of physics and engineering, but the numerical values of the parameters are only known within tolerances. If the model parameters are determined experimentally, the uncertainties depend on the amount of disturbances that excited the plant during the experiment.

   Models in our examples contain parameters: mass, inertia, damping coefficient, spring constant; electrical resistance, inductance, capacitance; thermal resistance, capacitance; hydraulic resistance, capacitance; etc. The parameter values are known only to certain tolerances; hence, the discrepancy between the model and the physical system. Parameters may also vary with temperature or other environmental conditions; aging of components, etc.
2. Achievable performance of a control system depends, in part, on the quality of a model: the more stringent the requirements, the better the model must be. On the other hand, increased complexity of the model often goes hand in hand with increased complexity of design of a control system. Therefore, we use a model of no greater complexity than is needed by the performance to be achieved. This implies that the design of the control system needs to take into account the inaccuracy of the model. The use of a linearized model of a nonlinear system falls in this category, as does the use of a single linear model for all expected operating points, as opposed to a different model for each case. Different components of systems (actuators, sensors, amplifiers, gears, belts, etc.) are sometimes modeled by constant gains, even though they have dynamics or nonlinearities. Dynamic structures (e.g., aircrafts, satellites, missiles) have complicated dynamics at high frequencies, and these may initially be ignored. There is often a *truth model*, the most accurate model we have, and a simplified *design model*, used as a basis for the design. The design model may therefore have inadequate structure; though the structure of truth model (often known) may be adequate.
3. A model is based upon consideration of some physical phenomena thought to represent the salient features of the plant. In practice, a model used for control system design does not include all physical effects; secondary physical effects are excluded because of complexity and engineering study costs. Examples are parasitic capacitances, structural vibration modes in a robot whose links have some compliance, and inertance dynamics due to oil compressibility in a hydraulic

system. Since we have no truth model, we cannot precisely assess the effect of these uncertainties. Engineering judgement and experience are invoked to produce rough estimates of the bounds of the magnitude of deviations of the input-output behavior.

In any case, the model at hand yields merely an approximation of the original system performance. We call this model the *nominal model* of the plant. *Model uncertainties* are the general notion for deviations of the model dynamics from the original system behavior. Different specific notions are commonly used in literature to describe the special forms of uncertainties. Model uncertainties can be caused by *inaccuracies* or *perturbations* of the parameter values and/or the *model structure*. With respect to the knowledge available about the causes of the uncertainties, *structured* and *unstructured* uncertainties are distinguished. Uncertainties of particular parameters, or the modeling errors that occur because of imperfect structure of the model, e.g., by the use of a linear model for an intrinsically nonlinear process or by omitting high-frequency dynamics or time lags; are examples of structured uncertainties. In the case of structured uncertainties, particular points within the plant are made responsible for the uncertainties. On the contrary, in the case of unstructured uncertainties, it is only known that there are some discrepancies between the model and the real plant and, probably, how large the resulting deviations of the input-output behavior may be. Particular points within the plant responsible for the uncertainties are not known.

It is usually possible to quantify and bound the discrepancy between the plant and the model. For example, a particular parameter may be thought to lie between two limits, with a specific midrange value used in the design model (Very often, these bounds are really educated guesses). The design model then becomes a member of a *class* of models, one member of which is believed to represent the plant.

Since we do not know which member of the class represents the plant, the safe and reasonable course is to require that the control system stabilize *all* members of the class. Since that is more difficult to do than simply stabilizing a single model, we may have to give up something from the single-model performance to satisfy the additional requirement.

The deterministic way of modeling uncertain systems leads to a family $\mathcal{G}$ of plants, where each member of the family may represent the original system, but it remains unknown which member does. This can be reflected by the equation

$$\hat{G}(s) = G(s) + \Delta(s)G(s)$$

where $G(s)$ represents the nominal model of the plant, and the family $\mathcal{G}$ consists of a set of perturbed plants $\hat{G}(s)$; $\Delta(s)G(s)$ represents the error model (refer to Fig. 6.6).

Robustness describes the ability of the system to satisfactorily perform for all the models $\hat{G}(s)$ in the family $\mathcal{G}$. A *robust controller* is a feedback controller that satisfies the design requirements on stability, speed of response, and steady-state accuracy in connection with all the models of a given family $\mathcal{G}$.

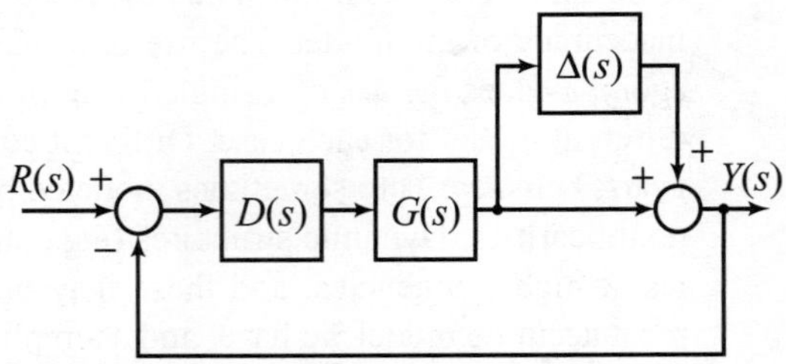

Fig. 6.6 *Block diagram model of a plant with unmodeled dynamics*

Robustness against unmodeled dynamical elements and disturbances is a 'structural' property of the feedback systems. Feedback makes it possible to improve robustness against variations of the

performance of some system component or to attenuate the effects of unmeasurable external disturbances. These are the main reasons for deliberately introducing feedback into already existing processes. The introduction of feedback is mainly motivated by the incompleteness of the knowledge of the system to be controlled and the effects of external disturbances. More strictly, the use of feedback can be regarded as unreasonable if there is no uncertainty in the system, because for the unperturbed system, open-loop control may produce the same or a probably better output.

In *linear control theory*, the robustness requirement plays a minor role in the design procedure; the design engineer focuses his attention primarily on the dynamic specifications while observing some general guidelines concerning the robustness improvement through feedback control. These qualitative robustness considerations work very well when small model uncertainties are encountered.

On the other hand, in *robust control theory*, the robustness requirement is the primary aim of design. Quantitatively specified relations between the amount of uncertainties and the properties to be attained dominate the design process. The quantitative robustness considerations become significant when model uncertainties are large [74–79].

Our focus in this book is on linear control design (based on root locus plots, Bode plots and Nichols chart) using nominal plant models for realization of specifications on stability, speed of response and steady-state accuracy. Qualitative robustness considerations will be used to evaluate the performance of the resulting controllers for real plants. In the Appendix, we will demonstrate the power of MATLAB/Simulink for evaluation of robustness properties of feedback controllers.

Example 6.1 The block diagram of a feedback control system is shown in Fig. 6.7. The design objective is to find the parameters a and K such that the steady-state error for a ramp input is less than or equal to 24% of the command. Additional design freedom is to be exploited to shape the transient response (peak overshoot for a unit step input, and the response time as measured by time to peak).

The selection of parameters a and K is obviously constrained by the requirement that the resulting system is stable. We can use Routh criterion to aid in the search for appropriate values of a and K. The characteristic equation of the feedback system is

$$1+D(s)G(s)=0$$

or

$$1+\frac{K(s+a)}{s(s+1)(s+2)(s+5)}=0$$

This gives

$$s(s+1)(s+2)(s+5)+K(s+a)=0$$

or

$$s^4+8s^3+17s^2+(K+10)s+aK=0$$

To determine the stability region for a and K, we establish the Routh array:

s^4	1	17	aK
s^3	8	$K+10$	
s^2	$\dfrac{126-K}{8}$	aK	
s^1	$\dfrac{(126-K)(K+10)-64aK}{126-K}$		
s^0	aK		

For stability, all the elements of the first column must be positive. Therefore, we require that

$$K < 126, aK > 0, \text{ and } (K+10)(126-K) - 64aK > 0$$

The region of stability for $K > 0$ is shown in Fig. 6.8. We can now find a set of (a, K) belonging to the stability region such that the steady-state specification is met.

From the block diagram of Fig. 6.7, we obtain

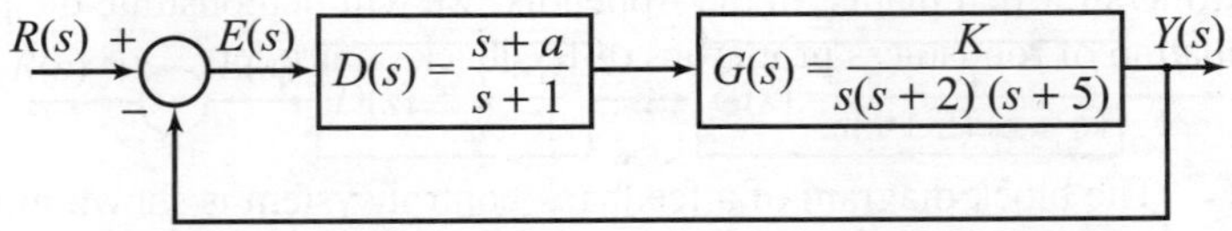

Fig. 6.7 *A feedback control system*

$$E(s) = \frac{1}{1+D(s)G(s)}R(s) = \frac{s(s+1)(s+2)(s+5)}{s(s+1)(s+2)((s+5)+K(s+a)}R(s)$$

The steady-state error of the system due to a ramp input $R(s) = 1/s^2$, is given by

$$e_{ss} = \lim_{s\to 0} sE(s) = \frac{10}{aK}$$

Given the steady-state specification $e_{ss} < 0.24$, we find that the specification is satisfied when

$$\frac{10}{aK} < 0.24 \text{ or } aK > 41.67$$

Any values of a and K that lie in the stability region in Fig. 6.8 and satisfy this inequality, will lead to an acceptable steady-state accuracy. The freedom available in the selection of a and K is to be used to satisfy the peak-overshoot and peak-time requirements on transient response, as closely as possible. This requirement on shaping the transient response can be met by selecting the parameters a and K such that the roots of the characteristic equation, and hence closed-loop poles of the system, are properly placed. So far, we have not been equipped with the procedures to carry out this portion of the design problem. The next section initiates the discussion on shaping the transient response. Simple preliminary design examples, where parametric selection is feasible analytically, will be given in this chapter.

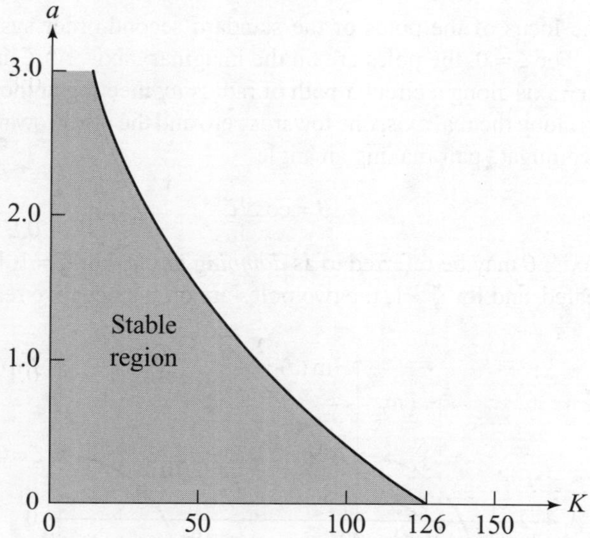

Fig. 6.8 *Stability region for the system of Fig. 6.7*

For the fourth-order system under consideration, analytical design will not be feasible; systematic trial-and-error iterative design procedures will be taken up in later chapters.

6.3 TRANSIENT RESPONSE SPECIFICATIONS IN TERMS OF POLE LOCATIONS

The design algorithms presented later require the transient specifications to be translated into a set of desired poles of the closed-loop system. This equivalently means that we have to identify a model whose response to a step input resembles the specified response. We first attempt a second-order continuous-time model.

The desired closed-loop transfer function of the system is assumed to be of the form (refer to Section 2.8)

$$\frac{Y(s)}{R(s)} = \frac{\omega_n^2}{s^2 + 2\zeta\omega_n s + \omega_n^2} \tag{6.3}$$

where ζ and ω_n are, respectively, the *damping ratio* and the *undamped natural frequency* of the system; $\zeta = 0$ for an *undamped* system, $0 < \zeta < 1$ for an *underdamped* system, $\zeta = 1$ for a *critically damped* system, and $\zeta > 1$ for an *overdamped* system.

The characteristic equation of the closed-loop system is

$$\Delta(s) = s^2 + 2\zeta\omega_n s + \omega_n^2 = 0 \tag{6.4a}$$

The characteristic roots

$$s_{1,2} = -\zeta\omega_n \pm j\omega_n\sqrt{1-\zeta^2} \tag{6.4b}$$

are the closed-loop poles of the system.

Figure 6.9 shows the locus of the poles of the standard second-order system (6.3), with ω_n held constant and ζ varying. For $\zeta = 0$, the poles are on the imaginary axis. As ζ increases, the poles move away from the imaginary axis along a circular path of radius ω_n meeting at the point $s = -\omega_n$, and then separating and traveling along the real axis, one towards zero and the other towards infinity. For $0 < \zeta < 1$, the poles are complex conjugate pair making an angle

$$\theta = \cos^{-1}\zeta \tag{6.5}$$

with the negative real axis; θ may be referred to as *damping angle.* For $\zeta = 1$, both the poles are on the negative real axis, repeated, and for $\zeta > 1$, the two poles are on the negative real axis, distinct.

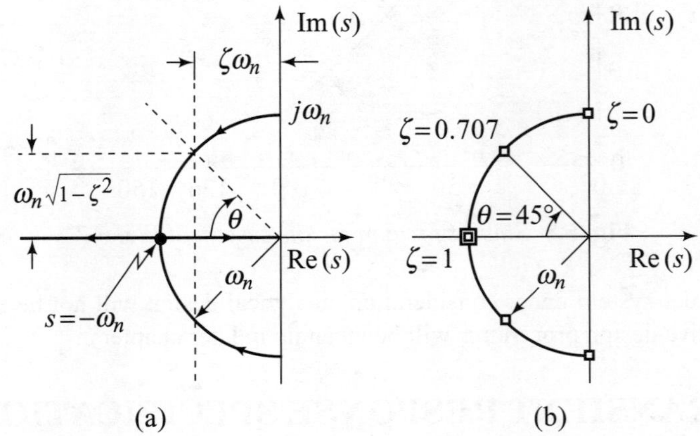

Fig. 6.9 *Pole locations for the standard second-order system*

We now study the transient response of the second-order system to unit-step input. The unit-step response has the Laplace transform (refer to Eqn. (6.3))

$$Y(s) = \frac{\omega_n^2}{s(s^2 + 2\zeta\omega_n s + \omega_n^2)} \tag{6.6}$$

For $0 < \zeta < 1$, the inverse Laplace transformation gives (refer to Eqn. (2.84))

$$y(t) = 1 - \frac{e^{-\zeta\omega_n t}}{\sqrt{1-\zeta^2}} \sin\left(\omega_n\sqrt{1-\zeta^2}\,t + \theta\right) \tag{6.7}$$

where $\theta = \cos^{-1}\zeta$

The characteristic features of this response are shown in Fig. 6.10. The response exhibits decaying oscillations of frequency $\omega_d = \omega_n\sqrt{1-\zeta^2}$; ω_d is called the *damped natural frequency.* The curves

$$\left[1 \pm \frac{e^{-\zeta\omega_n t}}{\sqrt{1-\zeta^2}}\right]$$

are the envelope curves for the step response. The response curve $y(t)$ always remains within this pair of envelope curves. The time-constant of these envelope curves is $1/\zeta\omega_n$. The speed of decay of the transient response depends on the value of the *time constant*

$$\tau = \frac{1}{\zeta\omega_n} \tag{6.8}$$

Therefore, the larger the product $\zeta\omega_n$, the faster is the decay of the transient.

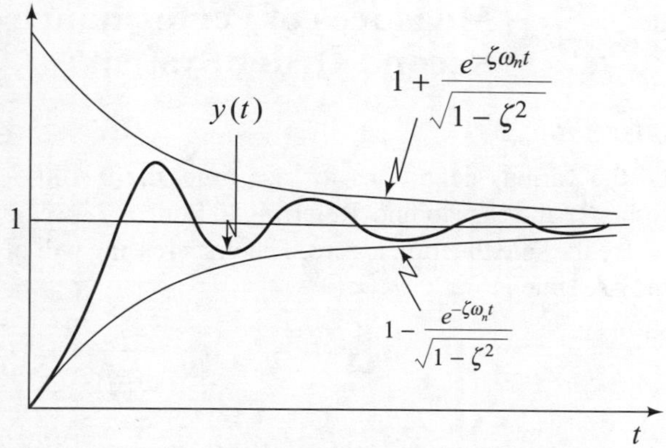

Fig. 6.10 *Typical unit-step response of underdamped standard second-order system*

The time response curves for various values of ζ plotted against normalized time $\omega_n t$ are shown in Fig. 6.11. For $\zeta = 0$, the system response exhibits continuous oscillations (refer to Eqn. (2.79)). As ζ is increased, the response becomes progressively less oscillatory till it becomes critically damped (just non-oscillatory) for $\zeta = 1$ (refer to Eqn. (2.86)), and overdamped for $\zeta > 1$ (refer to Eqn. (2.87)).

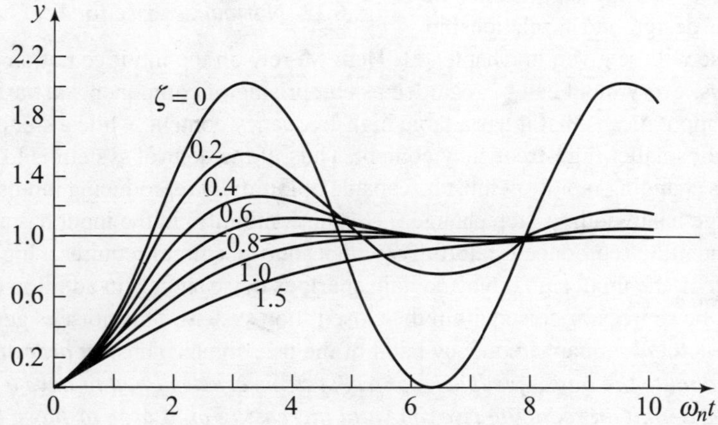

Fig. 6.11 *Unit-step response curves of standard second-order system*

Larger values of ζ give smaller values of peak overshoot but a slower rise of response. ***No overshoot at all is sometimes desirable; but usually this needlessly penalizes speed of response. Many systems are designed for 5 to 25% peak overshoot. For designs with peak overshoot less than five per cent, the price is paid in terms of speed of response, and the designs with peak overshoot more than 25% run the risk of being close to absolute instability.***

Let us obtain the expressions for rise time, peak time, peak overshoot, settling time, and steady-state error, for a second-order system whose dynamics for the underdamped case are described by Eqn. (6.7).

6.3.1 Measures of Performance of the Standard Second-Order System

Rise Time, t_r

Of the various definitions for rise time, the 0–100% rise time is meaningful for the underdamped response and easy to find. Referring to Eqn. (6.7), we observe that $y(t)$ crosses the desired value of unity where the sine function is zero. The first crossing will occur when the argument is equal to π. Therefore, the rise time is

$$t_r = \frac{\pi - \theta}{\omega_n\sqrt{1-\zeta^2}}; \theta = \cos^{-1}\zeta \tag{6.9}$$

It can be seen that the value of θ lies between $\pi/2$ and 0 for $0 < \zeta < 1$. ***For moderate overshoots of 5–25%, the damping ratio ζ is in the range 0.4–0.7*** (this we will prove shortly) and the normalized rise time $\omega_n t_r$ is approximately the same over this range, as is seen from Fig. 6.12. Thus, ***for a small value of t_r, ω_n must be large.***

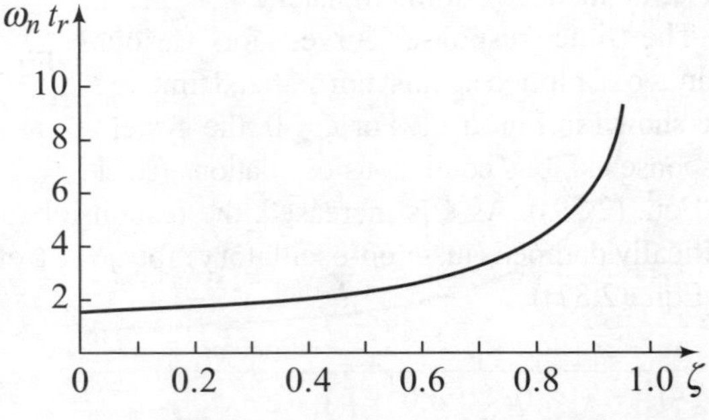

Fig. 6.12 *Normalized rise time versus damping ratio*

In frequency domain, the speed of response of a system is characterized by its *bandwidth* (refer to Section 2.10). Precise significance of bandwidth in control design, and its relationship to speed of response will be given in Chapter 11. Here we rely on the intuitive notion of bandwidth by reasoning as follows: every input can be regarded as comprising of components at various frequencies; a rapidly changing input means that it has a large high-frequency content, while a smooth, slow varying input has a relatively smaller high-frequency content. Thus, if the control system is to faithfully reproduce an input that is changing rapidly, it must be capable of faithfully reproducing inputs at high frequencies, i.e., have a large bandwidth. A step change (i.e., a discontinuity) in the input has a large amount of high-frequency content: to reproduce it faithfully (with a short rise time) requires a high bandwidth.

In a control system, the input signal may contain spurious noise signals in addition to the true signal input, or there may be sources of noise within the closed-loop system. This noise is generally in a band of frequencies above the dominant frequency band of the true signal. Thus, ***in order to reproduce the true signal and attenuate the noise, feedback control systems are designed to have a moderate bandwidth. This imposes a limit on reducing rise time and increasing ω_n. Large ω_n has a large bandwidth and will, therefore, allow the high-frequency noise signals to affect the performance.***

Peak Time, t_p

Equating the derivative of $y(t)$ in Eqn. (6.7) to zero to determine the extrema of the response, easily yields the equation,

$$\frac{\zeta\omega_n}{\sqrt{1-\zeta^2}} e^{-\zeta\omega_n t}\sin\left(\omega_n\sqrt{1-\zeta^2}\,t + \theta\right)$$

$$-\frac{e^{-\zeta\omega_n t}}{\sqrt{1-\zeta^2}}\omega_n\sqrt{1-\zeta^2}\cos\left(\omega_n\sqrt{1-\zeta^2}\,t + \theta\right) = 0$$

or
$$\frac{\omega_n e^{-\zeta\omega_n t}}{\sqrt{1-\zeta^2}}\left[\zeta \sin\left(\omega_n\sqrt{1-\zeta^2}t+\theta\right)-\sqrt{1-\zeta^2}\cos\left(\omega_n\sqrt{1-\zeta^2}t+\theta\right)\right]=0 \tag{6.10}$$

From Fig. 6.9a, we observe that $\zeta = \cos\theta$, and $\sqrt{1-\zeta^2} = \sin\theta$. Therefore, the quantity inside the brackets in Eqn. (6.10) can be expressed as

$$\sin\left(\omega_n\sqrt{1-\zeta^2}t+\theta\right)\cos\theta-\cos\left(\omega_n\sqrt{1-\zeta^2}t+\theta\right)\sin\theta=0$$

which gives (refer to Eqns (2.25)) $\sin\left(\omega_n\sqrt{1-\zeta^2 t}\right)=0$

This implies that at the extrema of the response,

$$\omega_n\sqrt{1-\zeta^2}t = m\pi; m = 0,1,2, \ldots \tag{6.11}$$

Since the peak time corresponds to first overshoot,

$$t_p = \frac{\pi}{\omega_n\sqrt{1-\zeta^2}} \tag{6.12}$$

The first undershoot will occur at $t = 2\pi\Big/\left[\omega_n\sqrt{1-\zeta^2}\right]$, the second overshoot at $t = 3\pi\Big/\left[\omega_n\sqrt{1-\zeta^2}\right]$, and so on. A plot of normalized peak time $\omega_n t_p$ vs ζ is given in Fig. 6.13.

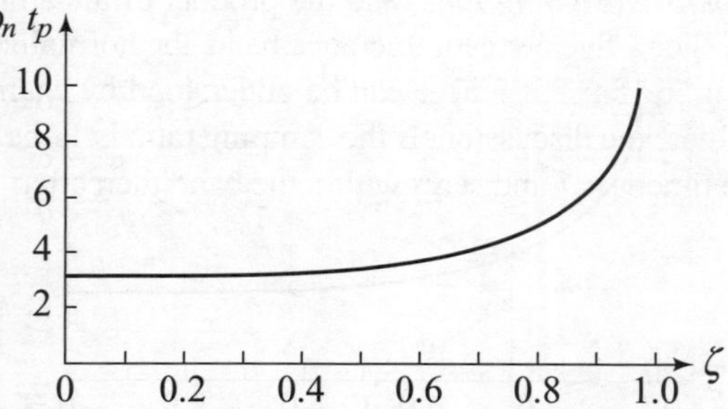

Fig. 6.13 *Normalized peak time versus damping ratio*

Peak Overshoot, M_p

From Eqns (6.7) and (6.12), we obtain

$$M_p = y(t_p) - 1 = -\frac{e^{-\zeta\omega_n t_p}}{\sqrt{1-\zeta^2}}\sin\left(\omega_n\sqrt{1-\zeta^2}t_p+\theta\right)$$

which gives
$$M_p = e^{-\pi\zeta/\sqrt{1-\zeta^2}} \tag{6.13}$$

Therefore, the per cent peak overshoot

$$= 100e^{-\pi\zeta/\sqrt{1-\zeta^2}}\,\% \tag{6.14}$$

Solving for ζ from Eqn. (6.13), we get

$$\zeta^2 = \frac{\left[\ln\left(M_p\right)\right]^2}{\left[\ln\left(M_p\right)\right]^2+\pi^2} \tag{6.15}$$

As seen from Fig. 6.14, ***the peak overshoot is a monotonically decreasing function of damping ζ and is independent of ω_n. It is, therefore, an excellent measure of system damping.***

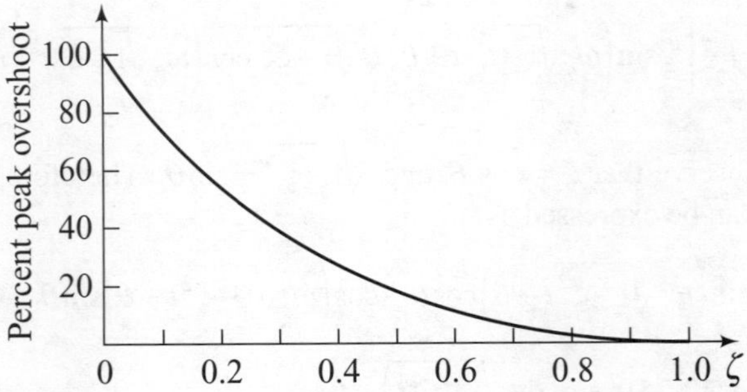

Fig. 6.14 *Per cent peak overshoot versus damping ratio*

Settling Time t_s

In the following discussion, the product of undamped natural frequency and settling time will be called the *normalized settling time*, and the product of undamped natural frequency and time is called *normalized time*. For a five per cent tolerance band, the normalized settling time *versus* damping ratio curve is shown in Fig. 6.15a. This curve can be understood by examining Fig. 6.15b which has an exaggerated scale to facilitate the discussion. If the damping ratio is 1, the response enters the tolerance band when the normalized time is 4.7 and stays within the band thereafter. This result has been obtained from Eqn. (2.86):

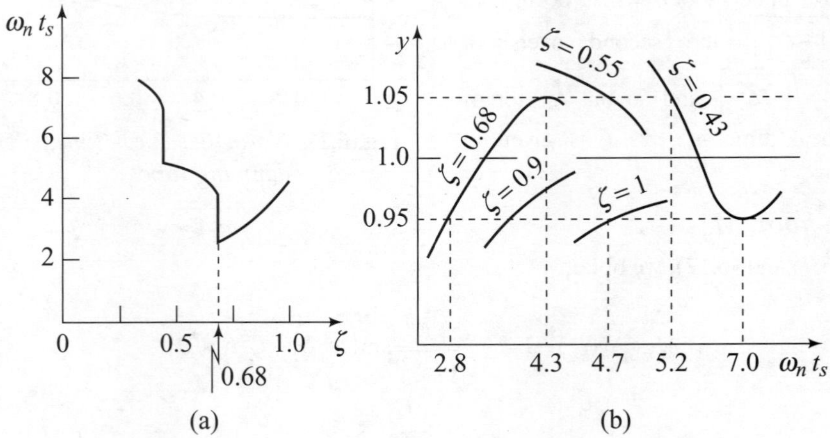

Fig. 6.15 *Normalized settling time versus damping ratio*

For a critically damped system, the response

$$y(t) = 1 - e^{-\omega_n t} - \omega_n t e^{-\omega_n t}$$

The equation

$$1 - e^{-\eta} - \eta\, e^{-\eta} = 0.95$$

gives

$$\eta = \omega_n t = 4.7$$

As the damping ratio is reduced, $\omega_n t_s$ reduces. A damping ratio of 0.68 has the smallest $\omega_n t_s$ since the response curve crosses the lower limit of the band at $\omega_n t_1 = 2.8$ and overshoots five per cent reaching the upper limit at $\omega_n t_2 = 4.3$; a small decrease in ζ from the value 0.68 will result in a jump in $\omega_n t_s$ from 2.8 to 4.3. These results have been obtained from the following equations:

$$t_2 = \frac{\pi}{\omega_n\sqrt{1-\zeta^2}}$$

$$y(t) = 1 - \frac{e^{-\zeta\omega_n t}}{\sqrt{1-\zeta^2}}\sin\left(\omega_n\sqrt{1-\zeta^2}\,t + \cos^{-1}\zeta\right)$$

When $\zeta = 0.68$, $\omega_n t_2 = 4.3$ and $y(t_2) = 1.05$. Also, when $\zeta = 0.68$, the equation $y(t_1) = 0.95$ gives $\omega_n t_1 = 2.8$. The normalized settling time increases from the value 4.3 as ζ is reduced. At a damping ratio of 0.43, the response curve crosses the upper limit of the band at $\omega_n t = 5.2$ and undershoots five per cent reaching the lower limit at $\omega_n t = 7$. Therefore, there is a jump in normalized settling time at $\zeta = 0.43$. Extending this logic, we can get the $\omega_n t_s$ vs ζ curve.

We observe that normalized settling time is minimum for $\zeta = 0.68$. This fact justifies why control systems are normally designed to be underdamped.

Analytical expression for the $\omega_n t_s$ vs ζ curve of Fig. 6.15a is difficult to obtain. We can use an approximation for the range $0 < \zeta < 0.68$ by using the envelope of the damped sinusoid shown in Fig. 6.10. Setting the time at which the lower envelope of $y(t)$ equals the value of 0.95 as the settling time t_s, we get

$$1 - \frac{e^{-\zeta\omega_n t_s}}{\sqrt{1-\zeta^2}} = 0.95$$

Solving for t_s, we obtain

$$t_s = \frac{1}{\zeta\omega_n}\ln\left(0.05\sqrt{1-\zeta^2}\right) \tag{6.16}$$

The value of $\ln\left(0.05\sqrt{1-\zeta^2}\right)$ varies between –3.0 and –3.3 as ζ varies from 0 to 0.68. A common design practice is to approximate settling time to three time constants of the envelope of the damped sinusoidal oscillations (refer to Eqn. (6.8)):

$$t_s \text{ (5\% tolerance)} \cong \frac{3}{\zeta\omega_n};\ 0 < \zeta < 0.68 \tag{6.17}$$

On similar lines, it can be established that for a two per cent tolerance band, normalized settling time is minimum at $\zeta = 0.8$. For $0 < \zeta < 0.8$, the settling time

$$t_s = -\frac{1}{\zeta\omega_n}\ln\left(0.02\sqrt{1-\zeta^2}\right) \tag{6.18}$$

The value of $\ln\left(0.02\sqrt{1-\zeta^2}\right)$ varies between – 3.91 and – 4.42 as ζ varies from 0 to 0.8. A common design practice is to approximate settling time to four time constants of the envelope of the damped sinusoidal oscillation (refer to Eqn. (6.8)):

$$t_s \text{ (2\% tolerance)} \cong \frac{4}{\zeta\omega_n};\ 0 < \zeta < 0.8 \tag{6.19}$$

Unless otherwise specified, we will use two per cent tolerance band for settling-time calculations.

Steady-state Error, e_{ss}

From Eqn. (6.7), we obtain

$$e_{ss}\Big|_{\text{unit step}} = \lim_{t\to\infty}\left[1 - y(t)\right] = 0$$

Thus, the standard second-order system has zero steady-state error to step inputs.

We now study the response of the standard second-order system to unit-ramp input $R(s) = 1/s^2$. The unit-ramp response has the Laplace transform (refer to Eqn. (6.3))

$$Y(s) = \frac{\omega_n^2}{s^2(s^2 + 2\zeta\omega_n s + \omega_n^2)} \tag{6.20}$$

For $0 < \zeta < 1$, the inverse Laplace transformation gives (refer to Eqn. (2.127))

$$y(t) = t - \frac{2\zeta}{\omega_n} + \frac{e^{-\zeta\omega_n t}}{\omega_n\sqrt{1-\zeta^2}}\sin(\omega_d t + 2\theta) \tag{6.21}$$

where $$\theta = \cos^{-1}\zeta$$

The response $y(t)$ for a typical value of ζ is plotted in Fig. 6.16. The steady-state error for unit-ramp input is given by

$$e_{ss}\Big|_{\text{unit ramp}} = \lim_{t\to\infty}[t - y(t)] = \frac{2\zeta}{\omega_n} \tag{6.22}$$

From Fig. 6.16 it is observed that the nature of transient response to ramp input is similar to that of step input (damped oscillatory) and yields no new information about the transient behavior of the system. As pointed out earlier, it is sufficient to investigate the transient response of a system to a step input only. Ramp-input response, of course, gives new information about the steady-state behavior, which may be evaluated directly by the final-value theorem (Eqn. (2.41)).

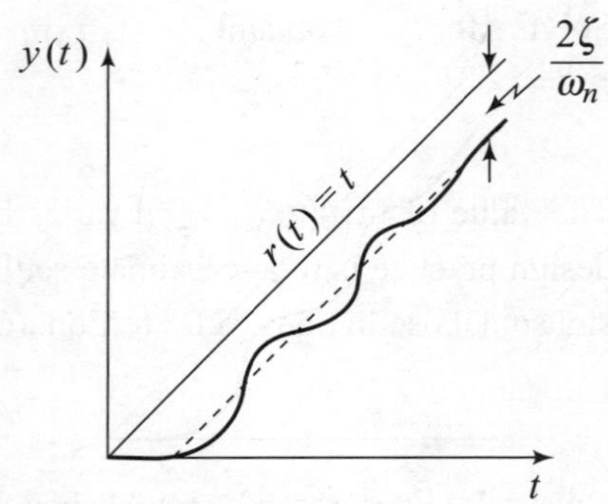

Fig. 6.16 *Typical unit-ramp response of an underdamped standard second-order system*

The error response has the Laplace transform

$$E(s) = R(s) - Y(s)$$

Using Eqn. (6.3), we have

$$E(s) = \frac{s(s + 2\zeta\omega_n)}{s^2 + 2\zeta\omega_n s + \omega_n^2}R(s) \tag{6.23}$$

For a unit-ramp input, $R(s) = 1/s^2$ and

$$E(s) = \frac{s + 2\zeta\omega_n}{s(s^2 + 2\zeta\omega_n s + \omega_n^2)}$$

$$e_{ss}\Big|_{\text{unit ramp}} = \lim_{s\to 0} s\,E(s) = \frac{2\zeta}{\omega_n} \tag{6.24}$$

To investigate the steady-state behavior of the standard second-order system to unit-parabolic input, we substitute $R(s) = 1/s^3$ in Eqn. (6.23):

$$E(s) = \frac{s + 2\zeta\omega_n}{s^2(s^2 + 2\zeta\omega_n s + \omega_n^2)}$$

Since $E(s)$ has repeated pole at $s = 0$, the error response $e(t)$ will be unbounded, and therefore

$$e_{ss}\Big|_{\text{unit parabola}} = \lim_{t\to\infty} e(t) = \infty \tag{6.25}$$

Therefore, in control systems represented by the standard second-order model, the steady-state error will be finite only if the actual signals which the control systems encounter are not faster than a ramp function.

From Eqns (6.9), (6.12), (6.13), (6.19) and (6.24), we observe that ***rise time, peak time, peak overshoot, settling time and steady-state error measures of the performance of a standard second-order system are mutually dependent, and, therefore, must be specified in a consistent manner.***

6.3.2 What is the Best Damping Ratio to Use?

A very important question to answer before leaving this section is the best value of damping ratio ζ to choose for a control system.

Selection of damping ratio for industrial control applications requires a trade-off between relative stability and speed of response. A smaller damping ratio decreases normalized rise time (refer to Fig. 6.12), but it increases the peak overshoot (refer to Fig. 6.14). The final choice of the damping ratio is subjective. Many systems are designed for damping ratio in the range of 0.4–0.7 (this corresponds to peak overshoot in the range of 5–25%). If allowed by rise-time considerations, ζ close to 0.7 is the most obvious choice because it results in minimum normalized settling time (refer to Fig. 6.15a).

If the requirement is to design an extremely accurate control system whose steady-state errors to position and velocity inputs are extremely small (e.g., to be used for navigation purposes), the transient response is not the primary performance criterion to optimize; minimum steady-state error is the major objective. With reference to Eqn. (6.24) we observe that the damping ratio ζ should be as small as possible because the steady-state error is proportional to ζ. Values close to 0.1 are not unreasonable for such applications; the disadvantage of relatively long settling time must be tolerated.

There are situations (e.g., in some robot-manipulator applications) where the overshoot cannot be tolerated. A good compromise between rise time and overshoot for such applications is $\zeta = 1$.

Example 6.2 Figure 6.17 shows block diagram of a control system whose objective is to control the position of mechanical load. The two potentiometers having sensitivity K_P convert the input and output positions into proportional electrical signals, which are in turn compared, and the error signal amplified by a factor K_A is applied to the armature circuit of a dc motor whose field winding is excited with a constant voltage. The motor is coupled to the load through a gear train of ratio n. The open-loop transfer function of the unity-feedback system is

$$D(s)G(s) = \frac{4500K_A}{s(s+361.2)} \tag{6.26}$$

The characteristic equation, $1 + D(s)G(s) = 0$, becomes

$$s^2 + 361.2s + 4500K_A = s^2 + 2\zeta\omega_n s + \omega_n^2 = 0 \tag{6.27a}$$

(Note that the system is stable for all $K_A > 0$)

Natural frequency $$\omega_n = \sqrt{4500K_A} \tag{6.27b}$$

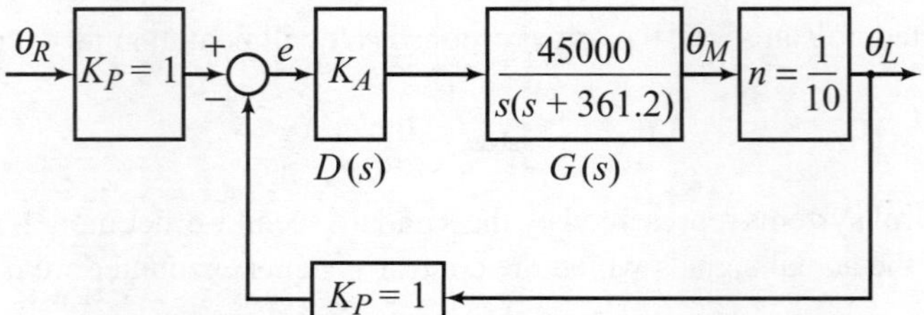

Fig. 6.17 *Block diagram of a position control system*

Damping ratio
$$\zeta = \frac{361.2}{2\sqrt{4500K_A}} = \frac{2.692}{\sqrt{K_A}} \tag{6.27c}$$

The closed-loop poles are
$$s_{1,2} = -180.6 \pm \sqrt{32{,}616.36 - 4500K_A} \tag{6.27d}$$

From the block diagram of Fig. 6.17, we obtain
$$\frac{E(s)}{\theta_R(s)} = \frac{1}{1+D(s)G(s)} = \frac{1}{1+\dfrac{4500K_A}{s(s+361.2)}} = \frac{s(s+361.2)}{s(s+361.2)+4500K_A}$$

The steady-state error of the system due to a ramp input $\theta_R(s) = 1/s^2$ is given by
$$e_{ss}\Big|_{\text{unit ramp}} = \lim_{s\to 0} sE(s) = \frac{361.2}{4500K_A} = \frac{0.0803}{K_A} \tag{6.28}$$

(i) *System with critical damping*
For $K_A = 7.247$, the damping ratio $\zeta = 1$, the natural frequency $\omega_n = 180.6$, and both the closed-loop poles are located at –180.6.
The steady-state error (Eqn. (6.28))
$$e_{ss}\Big|_{\text{unit ramp}} = 0.011\,\text{rad}$$

(ii) *System with moderate damping*
For $K_A = 14.5$, the damping ratio $\zeta = 0.707$, the natural frequency $\omega_n = 255.44$ rad/sec, and the closed-loop poles are located at $-180.6 \pm j\,180.6$. The damped natural frequency is $\omega_d = \omega_n\sqrt{1-\zeta^2} = 180.6$ rad/sec.

Rise time,
$$t_r = \frac{\pi - \cos^{-1}\zeta}{\omega_d} = \frac{\pi - 45\left(\dfrac{45}{180}\right)}{180.6} = 0.013\,\text{sec}$$

Peak time,
$$t_p = \frac{\pi}{\omega_d} = 0.0174\,\text{sec}$$

Peak overshoot,
$$M_p = e^{-\pi\zeta/\sqrt{1-\zeta^2}} \times 100 = 4.3\%$$

$$\text{Settling time } t_s \text{ (two per cent tolerance)} = \frac{4}{\zeta\omega_n} = 0.0221 \text{sec}$$

(Note that approximate expressions (6.17) and (6.19) cannot be used for calculation of settling time for systems with critical damping. However, using Fig. 6.13a we can easily establish the fact that settling time for the system with $K_A = 14.5$ is lower than for the system with $K_A = 7.247$)
The steady-state error

$$e_{ss}\Big|_{\text{unit ramp}} = 0.0055 \text{ rad}$$

Therefore, the increase in amplifier gain K_A from the value 7.247 to 14.5 has resulted in a decrease in damping from $\zeta = 1$ to $\zeta = 0.707$, decrease in settling time, and decrease in steady-state error from 0.011 to 0.0055 rad. The price has been paid in terms of overshoot. Since 4.3% peak overshoot is acceptable in many industrial control applications, moderate value of ζ results in better system performance compared to the critical value $\zeta = 1$.

(iii) *System with low damping*
For larger values of K_A, the damping ratio ζ reduces, natural frequency ω_n increases, and the real part of the poles is constant at –180.6, i.e., $\zeta\omega_n = 180.6$. As given by the approximate expression (6.19), the settling time does not change for $K_A > 14.5$. However, increase in K_A will result in reduced steady-state error to ramp input. Also, it will result in reduced rise time and hence better speed of response. The price is paid in terms of increased peak overshoot and hence poor relative stability. This means K_A can be increased at the most to a value for which the step-response overshoot is acceptable.

If 25% peak overshoot is considered acceptable, we have from Eqn. (6.15), $\zeta = 0.4$. $K_A = 45.3$ gives this value of damping. The other performance measures are:

$$\omega_n = 451.5 \text{ rad/sec}, \ \omega_d = 413.8 \text{ rad/sec}$$

$$t_r = 0.0048 \text{ sec}, \ t_p = 0.0076 \text{ sec},$$

$$e_{ss}\Big|_{\text{unit ramp}} = 0.0018 \text{ rad.}$$

Example 6.3 For the feedback control system of Fig. 6.18, select the parameters K and τ so that the following time-domain specifications will be satisfied:

(i) peak overshoot of the response to a step is less than 5%; and
(ii) the settling time to within 2% of the final value is less than 4 seconds, approaching the final value with zero error as $t \to \infty$.

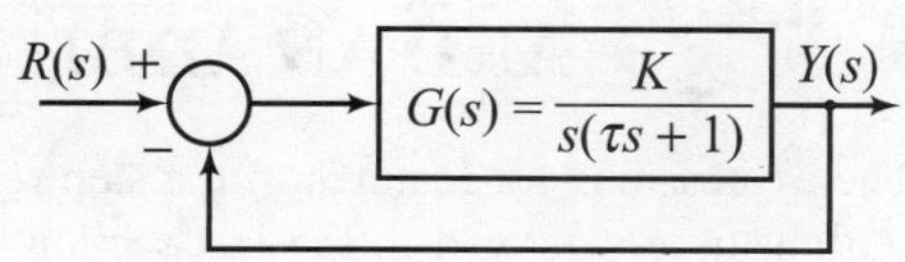

Fig. 6.18 *A feedback control system*

Solution: From Fig. 6.18, we obtain

$$E(s) = \frac{1}{1+G(s)}R(s) = \frac{s(\tau s+1)}{s(\tau s+1)+K}R(s)$$

For $R(s) = 1/s$,

$$e_{ss}\big|_{\text{unit step}} = \lim_{s\to 0} sE(s) = 0$$

Therefore, the steady-state error is zero for all values of K and τ that result in a stable system (note that the system is stable for all $K > 0$, $\tau > 0$).

The damping ratio ζ for an overshoot of 4.3% is $1/\sqrt{2}$ (This comes from Eqn. (6.13)). Because the settling time $t_s = (4/\zeta\omega_n) \le 4\,\text{sec}$, we require that $\zeta\omega_n \ge 1$. The characteristic equation of the system is

$$1 + G(s) = 1 + \frac{K}{s(\tau s+1)} = 0$$

or

$$s^2 + (1/\tau)s + K/\tau = s^2 + 2\zeta\omega_n s + \omega_n^2 = 0 \tag{6.29}$$

The characteristic roots are

$$s_{1,2} = -\zeta\omega_n \pm j\omega_n\sqrt{1-\zeta^2}$$

The requirement of $\zeta\omega_n \ge 1$ equivalently means that the real part of the characteristic roots be less than –1.

The damping ratio $\zeta = 1/\sqrt{2}$ is shown graphically as a line is Fig. 6.19. The region $\zeta\omega_n \ge 1$ is also shown in this figure. The region that will satisfy both the transient response specifications (peak overshoot and settling time) is shown shaded.

When $\zeta = 1/\sqrt{2}$, and $\zeta\omega_n = 1$ (i.e., $\omega_n = \sqrt{2}$), we have $t_s = 4$ sec and an overshoot of 4.3%. Hence, from the characteristic equation (6.29) we obtain

$$K/\tau = \omega_n^2 = 2, \quad \text{and} \quad 2\zeta\omega_n = 1/\tau = 2.$$

This gives

$$\tau = 0.5 \quad \text{and} \quad K = 1.$$

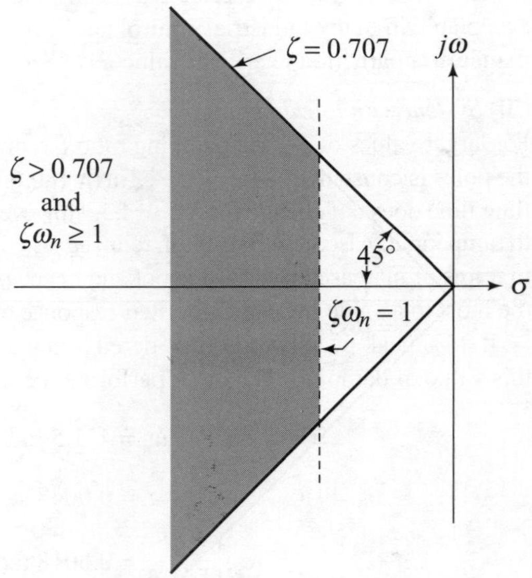

Fig. 6.19 *Specifications in terms of pole locations in the s-plane*

6.4 EFFECTS OF AN ADDITIONAL ZERO AND AN ADDITIONAL POLE

The expressions for features of the step response of a system—rise time, peak time, peak overshoot, settling time—derived in the last section, are valid only for the standard second-order system with closed-loop transfer function

$$M(s) = \frac{Y(s)}{R(s)} = \frac{\omega_n^2}{s^2 + 2\zeta\omega_n s + \omega_n^2} \tag{6.30}$$

The transfer function $M(s)$ has two poles and no zeros. For many practical systems, the closed-loop transfer function has more than two poles and/or finite zeros. In this section, we study the effects of adding poles and zeros to the standard second-order model.

Let a zero be added to $M(s)$ of Eqn. (6.30). Then we have

$$M_1(s) = \frac{Y_1(s)}{R(s)} = \frac{(s+\alpha)(\omega_n^2/\alpha)}{s^2 + 2\zeta\omega_n s + \omega_n^2} \tag{6.31}$$

The transfer function $M_1(s)$ has been normalized to have the gain at $s = 0$ equal to unity. This makes the final value of the unit-step response also 1, and thus the transfer function has been adjusted so that the system will track a step with zero steady-state error. The unit-step response of the one-zero, two-pole system may be determined from

$$Y_1(s) = \frac{(s+\alpha)(\omega_n^2/\alpha)}{s(s^2 + 2\zeta\omega_n s + \omega_n^2)} \tag{6.32}$$

$$= \frac{\omega_n^2}{s(s^2 + 2\zeta\omega_n s + \omega_n^2)} + \frac{s}{\alpha}\frac{\omega_n^2}{s(s^2 + 2\zeta\omega_n s + \omega_n^2)}$$

$$= Y(s) + \frac{s}{\alpha}Y(s) \tag{6.33a}$$

where $Y(s)$ is the Laplace transform of the unit-step response of standard second-order system (6.30).

From Eqn. (6.33a), we obtain

$$y_1(t) = y(t) + \frac{1}{\alpha}\frac{dy(t)}{dt} \tag{6.33b}$$

The effect of the added derivative term may be seen by examining Fig. 6.20a, where a case for a typical value of ζ is considered. We see from the figure that the effect of the zero is to contribute a pronounced early peak to the system's response whereby the peak overshoot may increase appreciably. From Fig. 6.20b, it is seen that the smaller the value of α, i.e., the closer the zero to the origin, the more pronounced is the peaking phenomenon. On account of this fact, *the zeros on the real axis near the origin are generally avoided in design.* However, in a sluggish system, the artful introduction of zero at the proper position can improve the transient response.

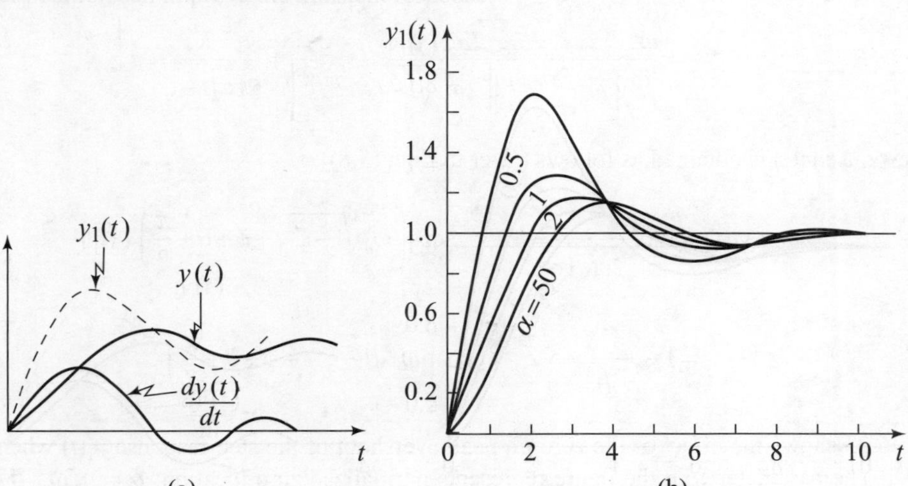

Fig. 6.20 *(a) Unit-step response of a standard second-order system with an additional zero (b) Response for several zero locations with $\zeta = 0.5, \omega_n = 1$*

We further observe from Eqn. (6.33b) that as α increases, i.e., the zero moves farther into the left half of the s-plane; its effect becomes less pronounced. For sufficiently large values of α, the effect of zero on transient response may become negligible.

The transient analysis given by Eqn. (6.33b) is very informative in case the zero is in the right-half plane. In this case, α is negative and thus the derivative term is subtracted rather than added. A typical case is sketched in Fig. 6.21; the response has become sluggish.

For a quantitative transient analysis of the standard second-order system with a finite zero, we obtain $y_1(t)$ from $Y_1(s)$ of Eqn. (6.32). Partial-fraction expansion of $Y_1(s)$ gives

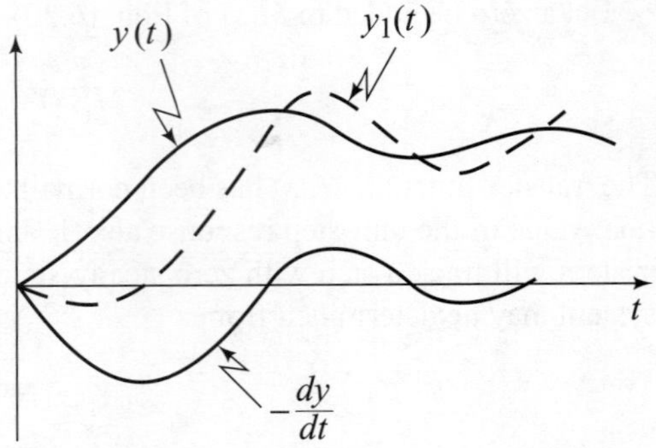

Fig. 6.21 *Unit-step response of a standard second-order system with a zero in the right-half plane*

$$Y_1(s) = \frac{1}{s} + \frac{A}{s + \zeta\omega_n - j\omega_n\sqrt{1-\zeta^2}} + \frac{A^*}{s + \zeta\omega_n + j\omega_n\sqrt{1-\zeta^2}}$$

where A^* is complex-conjugate of A.

The pole-zero pattern of $Y_1(s)$ is shown in Fig. 6.22. From the geometry of this figure, we obtain

$$A = \frac{\omega_n^2}{\alpha}\frac{s+\alpha}{s\left(s+\zeta\omega_n + j\omega_n\sqrt{1-\zeta^2}\right)}\Bigg|_{s=-\zeta\omega_n + j\omega_n\sqrt{1-\zeta^2}}$$

$$= \frac{\omega_n^2}{\alpha}\frac{Ke^{j\phi}}{\left[\omega_n e^{j(\pi-\theta)}\right]\left[2\omega_n\sqrt{1-\zeta^2}\,e^{j\pi/2}\right]} = \frac{K}{2\alpha\sqrt{1-\zeta^2}}e^{j\left(\phi+\theta+\frac{\pi}{2}\right)}$$

The response $y_1(t)$ is obtained as follows (refer to Eqn. (5.8)):

$$y_1(t) = 1 + 2\frac{K}{2\alpha\sqrt{1-\zeta^2}}e^{-\zeta\omega_n t}\cos\left(\omega_n\sqrt{1-\zeta^2}\,t + \phi + \theta + \frac{\pi}{2}\right)$$

$$= 1 + \frac{K}{\alpha\sqrt{1-\zeta^2}}e^{-\zeta\omega_n t}\cos\left(\omega_n\sqrt{1-\zeta^2}\,t + \phi + \theta + \frac{\pi}{2}\right) \tag{6.34}$$

Figure 6.23 shows the effect of the zero on peak overshoot of the step response $y_1(t)$ when $\zeta = 0.5$ and $\omega_n = 1$. The parameter η in the figure represents normalized zero location: $\alpha = \eta\zeta\omega_n$. ***When η is large, the zero is far removed from the poles and will have little effect on the response. If $\eta = 1$, the zero is at the value of the real part of the poles and has a substantial influence on the response. $\eta < 1$ results in the highly pronounced peaking phenomenon***.

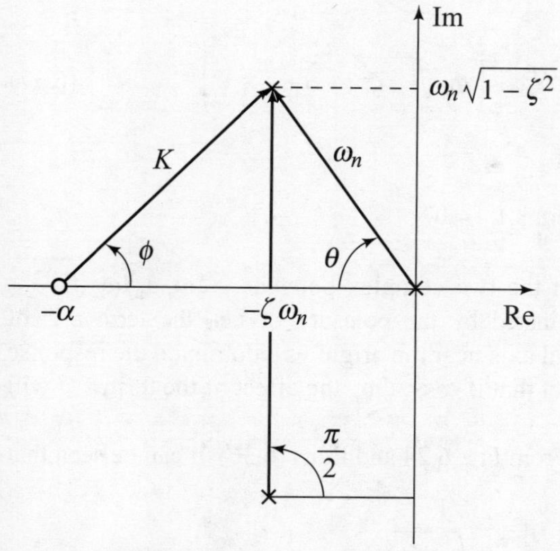

Fig. 6.22 *Pole-zero pattern of $Y_1(s)$ in Eqn. (6.32)*

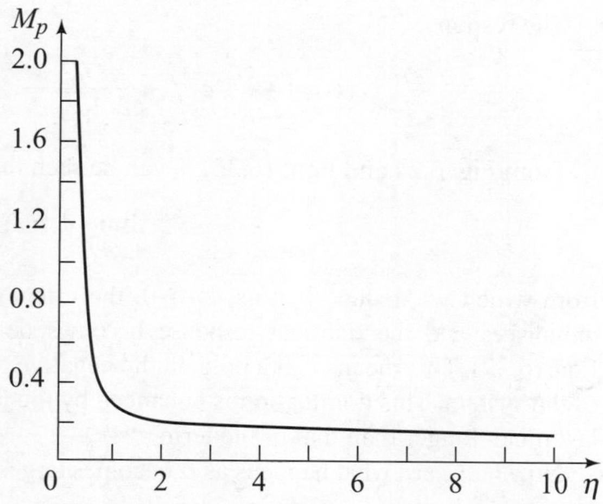

Fig. 6.23 *Plot of peak overshoot for several locations of an additional zero ($\zeta = 0.5, \omega_n = 1$)*

Now let us consider the effect of an additional pole on the response of a standard second-order system (6.30). To this end we consider a transfer function of the form

$$M_2(s) = \frac{Y_2(s)}{R(s)} = \frac{\omega_n^2\alpha}{(s+\alpha)(s^2+2\zeta\omega_n s+\omega_n^2)} \tag{6.35}$$

The transfer function $M_2(s)$ has been normalized to have the gain at $s = 0$ equal to unity. The unit-step response of this no-zero, three-pole system may be determined from

$$Y_2(s) = \frac{\omega_n^2\alpha}{s(s+\alpha)(s^2+2\zeta\omega_n s+\omega_n^2)}$$

$$= \frac{1}{s} + \frac{A_1}{s+\alpha} + \frac{A_2}{s+\zeta\omega_n - j\omega_n\sqrt{1-\zeta^2}} + \frac{A_2^*}{s+\zeta\omega_n + j\omega_n\sqrt{1-\zeta^2}}$$

For the geometry of Fig. 6.24a, we obtain

$$A_1 = \frac{\omega_n^2\alpha}{\alpha e^{j\pi} Ke^{j(\pi+\phi)} Ke^{j(\pi-\phi)}} = \frac{\omega_n^2}{K^2 e^{j\pi}} = \frac{-\omega_n^2}{K^2} \tag{6.36a}$$

From the geometry of Fig. 6.24b, we obtain

$$A_2 = \frac{\omega_n^2\alpha}{\omega_n e^{j(\pi-0)}\left(2\omega_n\sqrt{1-\zeta^2}\,e^{j\pi/2}\right)Ke^{j\phi}}$$

$$= \frac{\alpha}{2K\sqrt{1-\zeta^2}\,e^{j\left(\pi-\theta+\frac{\pi}{2}+\phi\right)}} = \frac{\alpha e^{j\left(\frac{\pi}{2}+\theta-\phi\right)}}{2K\sqrt{1-\zeta^2}} \tag{6.36b}$$

The response

$$y_2(t) = 1 - \frac{\omega_n^2}{K^2} e^{-\alpha t} + \frac{\alpha}{K\sqrt{1-\zeta^2}} e^{-\zeta\omega_n t} \cos\left(\omega_n\sqrt{1-\zeta^2}\,t + \theta + \frac{\pi}{2} - \phi\right) \tag{6.37}$$

From Fig. 6.24 and Eqns (6.36), it can be seen that

$$\lim_{\alpha\to 0} |A_1| = 1;\ \lim_{\alpha\to 0} |A_2| = 0$$

from which we deduce that as $\alpha \to 0$, the effect of the two complex poles at $-\zeta\omega_n \pm j\omega_n\sqrt{1-\zeta^2}$ diminishes and the transient response becomes dominated by the pole at $-\alpha$, i.e., the term $e^{-\alpha t}$ of Eqn. (6.37). This means that a pole on the negative real axis near the origin will dominate the response of the system. This domination is enhanced by the fact that if $\alpha \ll \zeta\omega_n$, the effect of the term $e^{-\alpha t}$ will last much longer than that of the term $e^{-\zeta\omega_n t}$.

Now let us see what happens as α becomes large. From Fig. 6.24 and Eqns (6.36), it can be seen that

$$\lim_{\alpha\to\infty} |A_1| = 0;\ \lim_{\alpha\to\infty} |A_2| = \frac{1}{2\sqrt{1-\zeta^2}}$$

from which we see that as α increases, an opposite effect is observed. The complex poles at $-\zeta\omega_n \pm j\omega_n\sqrt{1-\zeta^2}$ now dominate the response, while the contribution of the pole at $-\alpha$ diminishes in significance. Further, if $\zeta\omega_n \ll \alpha$, the effect of the term $e^{-\alpha t}$ will be over much more quickly than the term whose magnitude decreases as $e^{-\zeta\omega_n t}$. All the preceding effects may be seen quite clearly in Fig. 6.25. Figure 6.26 shows the effect of the third pole at $s = -\alpha = -\eta\zeta\omega_n$ on rise time of the step response. ***When η is large, the real pole is far removed from the complex poles and will have little effect on the response. If $\eta = 1$, the real pole is at the value of the real part of the complex pole pair and has a substantial effect on the response. $\eta < 1$ makes the system very sluggish.***

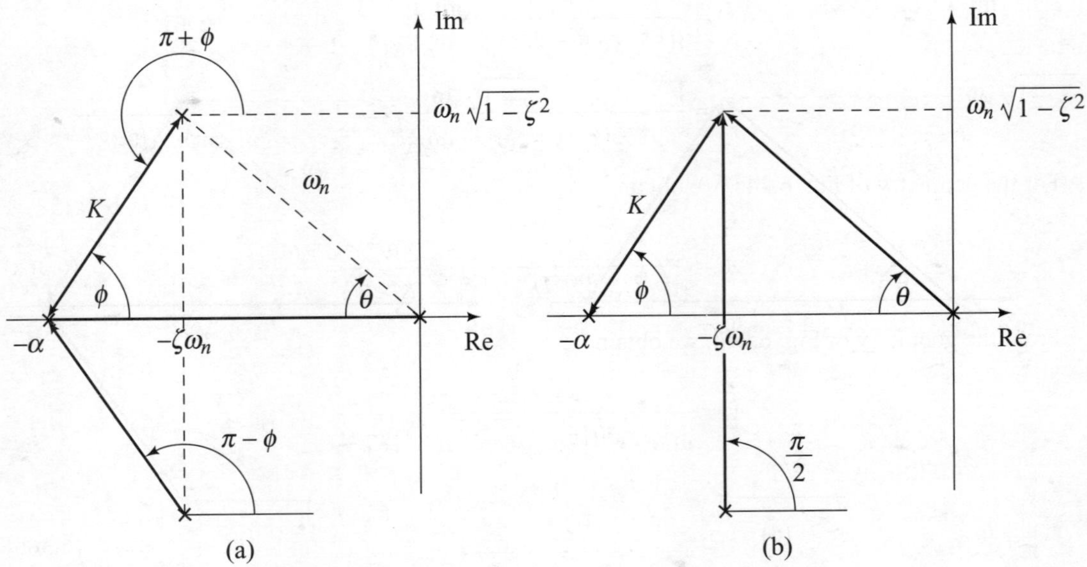

Fig. 6.24 *Pole-zero pattern of $Y_2(s)$ in Eqn. (6.35)*

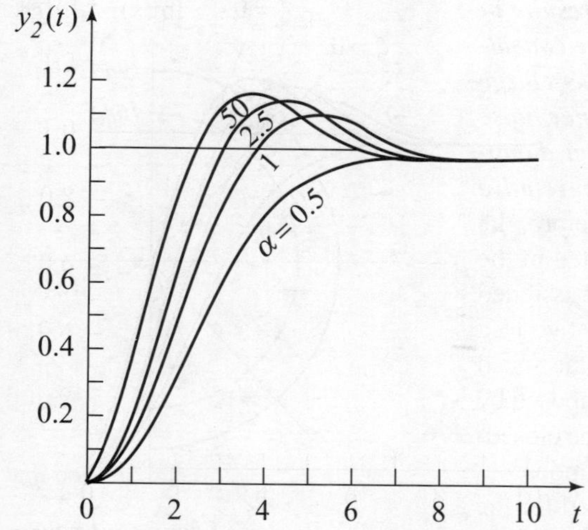

Fig. 6.25 *Unit-step response of a standard second-order system ($\zeta = 0.5$, $\omega_n = 1$) for several locations of an additional pole*

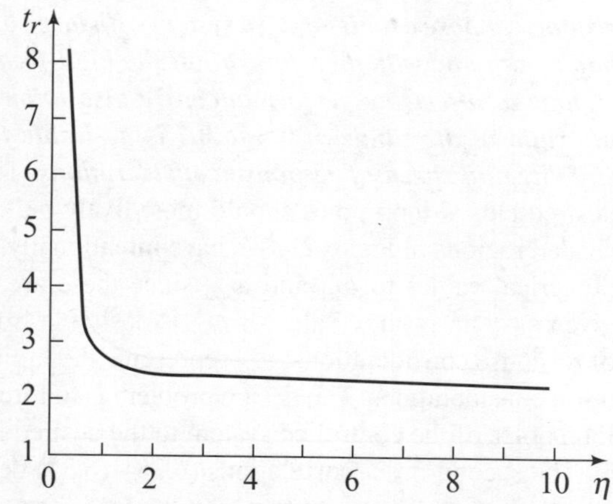

Fig. 6.26 *Plot of rise time for several locations of an additional pole ($\zeta = 0.5$, $\omega_n = 1$)*

6.5 DESIRED CLOSED-LOOP POLE LOCATIONS AND THE DOMINANCE CONDITION

From the discussion given in the preceding sections, it becomes apparent that for second-order feedback control systems with no finite zeros, the transient performance specifications can be translated into the desired locations for the two closed-loop poles. Acceptable values of damping ratio ζ and undamped natural frequency ω_n are first obtained from the consistent set of specifications on rise time, peak time, peak overshoot, and settling time using the second-order correlations given below (refer to Eqns (6.9), (6.12), (6.13) and (6.19)):

Rise time,
$$t_r = \frac{\pi - \cos^{-1}\zeta}{\omega_n\sqrt{1-\zeta^2}} \tag{6.38a}$$

Peak time,
$$t_p = \frac{\pi}{\omega_n\sqrt{1-\zeta^2}} \tag{6.38b}$$

Peak overshoot,
$$M_p = e^{-\pi\zeta/\sqrt{1-\zeta^2}} \tag{6.38c}$$

Settling time,
$$t_s = \frac{4}{\zeta\omega_n} \tag{6.38d}$$

The desired locations of the closed-loop poles are then obtained as

$$s_{1,2} = -\zeta\omega_n \pm j\omega_n\sqrt{1-\zeta^2} \tag{6.39}$$

We know that a large ω_n is desirable because it results in high speed of response. However, measurement noise considerations impose a limit on ω_n; a system with large ω_n has a large bandwidth and will, therefore, allow the high-frequency noise signals to affect the performance. We also know that damping ratio in the range of 0.4 to 0.7 is desirable from relative stability and speed of response considerations. Therefore, the desired closed-loop poles should most likely be located in the shaded regions in Fig. 6.27. We have intentionally not assigned numerical values to ω_{n1} and ω_{n2}, since these are relative to a given system. A large value of ω_{n1} is desirable from the speed of response considerations; ω_{n2} represents the limit imposed by noise considerations. The design problem is to force the closed-loop poles of the controlled system to the desired locations.

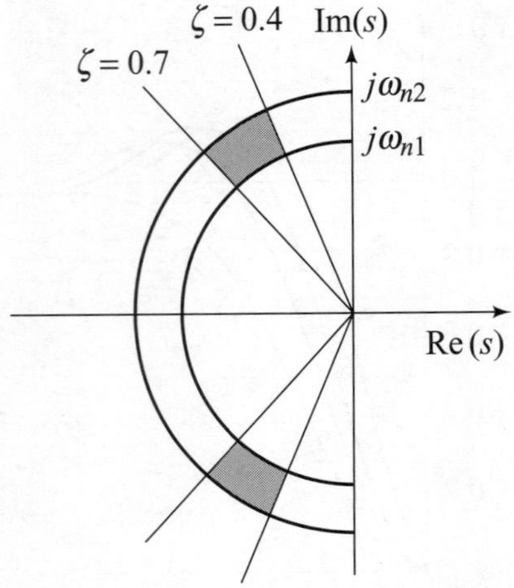

Fig. 6.27 *Region of dominant poles in the s-plane*

The second-order correlations (6.38)–(6.39) derived for the standard second-order system, can be used for a second-order system with a finite zero by appropriately accommodating the effect of the zero on the transient response. The effect of a zero in the left half s-plane is to contribute a pronounced early peak to the system's response whereby the peak overshoot may increase appreciably. The closer the zero is to the origin, the more pronounced is the peaking phenomenon. If the zero moves farther into the left half s-plane, its effect becomes less pronounced. ***If the magnitude of the zero is more than five times the magnitude of the real part of the complex-conjugate poles of the second-order system, then the zero may be regarded as insignificant insofar as the transient response is concerned*** (Fig. 6.28). However, if it is not possible to avoid a zero in the region of significant effect on the transient response, we may design the system for a higher value of ζ than the one that is acceptable, to nullify the effect of increase in overshoot due to the presence of zero. A simulation study is carried out to see whether the design is acceptable. If not, one has to re-enter the design cycle with a different and more appropriate choice of ζ.

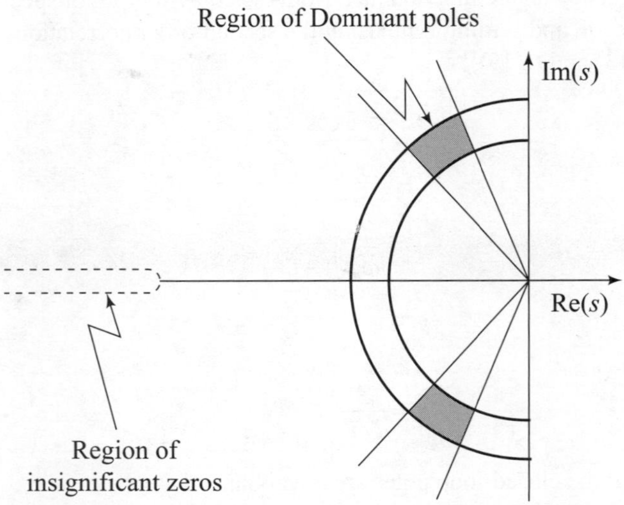

Fig. 6.28 *Regions of dominant poles and insignificant zeros in the s-plane*

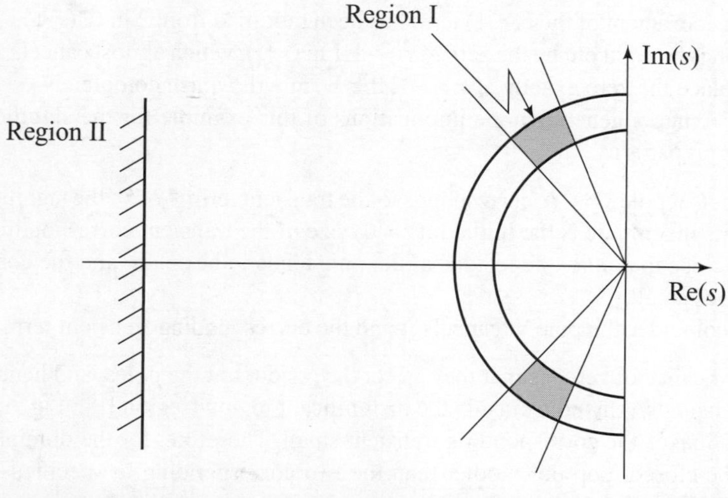

Fig. 6.29 *Regions of dominant and insignificant poles in the s-plane*

When a system is higher than second order, strictly we can no longer characterize its dynamics by damping ratio ζ and undamped natural frequency ω_n, which are defined for standard second-order systems. However, by a proper design, it may be possible to force the closed-loop poles of a higher-order system to the two regions shown in Fig. 6.29; a complex-conjugate pole pair in region I and all other poles in region II. It has been recognized in practice that ***if the magnitudes of the real parts of all the poles in region* II *are more than five times that of poles in region* I, *then the poles of region* II *may be regarded as insignificant insofar as the transient response is concerned.*** The pair of complex-conjugate poles in region I has a dominant effect on transient response; we will refer to these poles as *dominant poles* of the higher-order system. The parameters (ζ, ω_n), associated with the dominant poles, characterize the dynamics of the higher-order system.

The design procedure for the higher-order systems is as follows. The specifications of relative stability and speed of response are translated into a pair of dominant closed-loop poles using the correlations (6.38)–(6.39). The design requirement is to force two of the closed-loop poles at the specified dominant positions and all other poles in the insignificant region. If the dominance condition is not satisfied, nothing can be said about the usefulness of design. A simulation study is carried out to see whether the design is acceptable. If not, one has to re-enter the design cycle to get a different distribution of closed-loop poles.

When a system is higher than second-order and has finite zeros, the dominance condition is satisfied only when the contributions (to the transient response) of the zeros and all the poles other than the pair corresponding to specified ζ and ω_n, are insignificant. ***An artful placement of a zero can cancel the dynamics contributed by a pole; this technique of pole-zero cancellation may be utilized to meet the dominance condition.*** To illustrate this concept, we consider the following two transfer functions, which have the same poles but differ in the locations of their zeros. They are normalized to have the same gain at $s = 0$.

$$G_1(s) = \frac{2}{(s+1)(s+2)} = \frac{2}{(s+1)} - \frac{2}{(s+2)}$$

$$G_2(s) = \frac{2(s+1.1)}{1.1(s+1)(s+2)} = \frac{2}{1.1}\left\{\frac{0.1}{s+1} + \frac{0.9}{s+2}\right\} = \frac{0.18}{s+1} + \frac{1.64}{s+2}$$

Notice that the coefficient of the $(s + 1)$ term has been modified from 2 in $G_1(s)$ to 0.18 in $G_2(s)$. This dramatic reduction is brought out by the zero at $s = -1.1$ in $G_2(s)$, which almost cancels the pole at $s = -1$. Of course, if we place the zero exactly at $s = -1$, this term will vanish completely.

The following remarks generalize the implications of this example for the significance of zeros in the system pole-zero pattern.

1. The residue A_i at pole $s = -\alpha_i$ corresponds to the transient term $A_i e^{-\alpha_i t}$; the significance of the residue is that its magnitude is the initial (at $t = 0$) size of the transient corresponding to the pole.
2. If a zero is close to a pole, the residue at the pole tends to be small, and the corresponding transient becomes small.
3. If the zero coincides the pole, it cancels it and the corresponding transient term is zero.

Thus, the significance of zeros is that they affect the residues at the poles, and hence the sizes of the corresponding transients. Any pole outside the insignificant region (region II in Fig. 6.29) is acceptable if the nearby zero makes the corresponding transient small. Therefore, for the dominance condition to be satisfied, all the closed-loop poles other than the two corresponding to specified ζ and ω_n, should either lie in the insignificant region or have zeros nearby.

6.6 STEADY-STATE ERROR CONSTANTS AND SYSTEM-TYPE NUMBER

Steady-state error is a measure of system accuracy when a specific type of input is applied to a control system. The designer always strives to design the system to minimize the error for a certain anticipated class of inputs. This section considers the techniques that are available for determining the system accuracy.

Theoretically, it is desirable for a control system to have the capability of responding with zero error to all the polynomial inputs

$$r(t) = (t^k/k!)\,\mu(t) \tag{6.40}$$

of degree k; $\mu(t)$ as a unit-step input. If $k = 0$, the input is a step of unit magnitude; if $k = 1$, the input is a ramp with unit slope; if $k = 2$, the input is a parabola with unit second derivative; $k = 3$ corresponds to an input one degree faster than parabola, and so on. From the context of the common problems of mechanical motion control, the *step, ramp,* and *parabolic* inputs are called, respectively, *position, velocity* and *acceleration* inputs.

The specification of the zero-error response to all the polynomial inputs is very impractical and unrealistic. Fortunately, the requirements of practical systems are much less stringent. For example, consider the position control system for a robot manipulator. For a typical application, it would be desirable for this system to respond well to inputs of position and velocity, but not necessarily to those of acceleration. In addition, it would be desirable for this system to respond with zero error for positional-type inputs. However, a finite error could probably be tolerated for inputs of velocity.

For a quantitative analysis, we consider the unity-feedback system shown in Fig. 6.30. The input is $R(s)$, the output is $Y(s)$, and the difference between the two is the error $E(s)$. For a stable system, the *transient component* of the error response $e(t)$ will eventually die away, leaving only the *steady-state component* $e_s(t)$ whose form is dictated directly by the forcing input.

The steady-state error

$$e_{ss} \lim_{t\to\infty} e_s(t) = \lim_{t\to\infty} e(t) = \lim_{t\to\infty} [r(t) - y(t)] \tag{6.41a}$$

By the final-value theorem (Eqn. (2.41)),

$$e_{ss} = \lim_{s \to 0} sE(s) \tag{6.41b}$$

provided that $sE(s)$ has no poles on the $j\omega$-axis and in the right-half plane.

For the system shown in Fig. 6.30,

$$E(s) = R(s) - Y(s) = \left[1 - \frac{Y(s)}{R(s)}\right] R(s) = \frac{1}{1 + D(s)G(s)} R(s) \tag{6.42}$$

Substituting Eqn. (6.42) into Eqn. (6.41b), we obtain

$$e_{ss} = \lim_{s \to 0} \frac{sR(S)}{1 + D(s)G(s)} \tag{6.43}$$

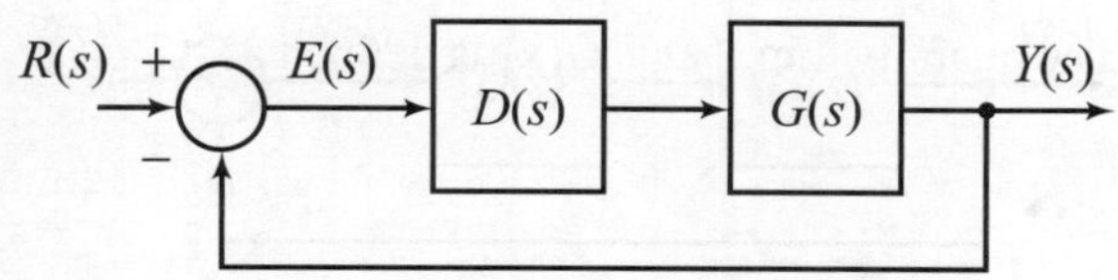

Fig. 6.30 *A unity-feedback system*

Thus, the steady state of a system with unity feedback depends on the input signal $R(s)$ and the open-loop transfer function $D(s)G(s)$. By the nature of the limit in Eqn. (6.43), we see that the result of the limit can be zero, or can be a constant different from zero, Also the limit may not exist in which case the final-value theorem does not apply; but it is easy to see from the basic definition (6.41a) that $e_{ss} = \infty$ in this case anyway because $E(s)$ will have a pole at $s = 0$ that is of order higher than one. Systems having a finite nonzero steady-state error when the reference input is a zero-order polynomial input (a step) are labeled "type-0". Similarly, a system that has finite nonzero steady-state error to a first-order polynomial input (a ramp) is called a "type-1" system; a system with finite nonzero steady-state error to a second-order polynomial input (a parabola) is called a "type-2" system, and so on.

The control engineer is usually interested in step (position), ramp (velocity), and parabola (acceleration) inputs. For a step input $r(t) = \mu(t)$, the steady-state error may be obtained by substituting $R(s) = 1/s$ into Eqn. (6.43):

$$e_{ss} = \lim_{s \to 0} \frac{1}{1 + D(s)\,G(s)} = \frac{1}{1 + \lim\limits_{s \to 0} D(s)\,G(s)}$$

The quantity $\lim\limits_{s \to 0} D(s)G(s)$ is defined as the *position error constant* and is denoted by K_p:

$$K_p = \lim_{s \to 0} D(s)G(s) \tag{6.44a}$$

Therefore, steady-state error in terms of K_p is given by

$$e_{ss} = \frac{1}{1 + K_p} \tag{6.44b}$$

The steady-state error to a velocity input $r(t) = t\mu(t)$ can be obtained by substituting $R(s) = 1/s^2$ into Eqn. (6.43):

$$e_{ss} = \lim_{s \to 0} \frac{1}{s(1 + D(s)G(s))} = \frac{1}{\lim\limits_{s \to 0} sD(s)G(s)}$$

The quantity $\lim_{s\to 0} s\,D(s)G(s)$ is defined as the *velocity error constant* and is denoted by K_v:

$$K_v = \lim_{s\to 0} sD(s)G(s) \tag{6.45a}$$

Therefore, the steady-state error in terms of K_v is given by

$$e_{ss} = 1/K_v \tag{6.45b}$$

An expression for steady-state error due to acceleration input $r(t) = \frac{1}{2}t^2\mu(t)$ can be obtained by substituting $R(s) = 1/s^3$ into Eqn. (6.43):

$$e_{ss} = \lim_{s\to 0}\frac{1}{s^2(1+D(s)G(s))} = \lim_{s\to 0}\frac{1}{s^2D(s)G(s)}$$

The quantity $\lim_{s\to 0} s^2D(s)G(s)$ is defined as the *acceleration error constant* and is denoted by K_a:

$$K_a = \lim_{s\to 0} s^2D(s)G(s) \tag{6.46a}$$

Therefore, the steady-state error in terms of K_a is

$$e_{ss} = 1/K_a \tag{6.46b}$$

As said earlier, feedback systems can be classified on the basis of their steady-state response to polynomial inputs. We can always express the open-loop transfer function $D(s)G(s)$ as

$$D(s)G(s) = \frac{K\prod_i (s+z_i)}{s^N\prod_j (s+p_j)}; z_i \neq 0, p_j \neq 0 \tag{6.47}$$

The term $1/s^N$ corresponds to N number of pure integrations in $D(s)G(s)$. As $s \to 0$, this term dominates in determining the steady-state error. Feedback systems are therefore classified in accordance with the number of pure integrators in the open-loop transfer function $D(s)G(s)$, as described below:

Type-0 System

If $N = 0$, the steady-state error to various standard inputs obtained from Eqns (6.44)–(6.46) are

$$e_{ss} = \begin{cases} \dfrac{1}{1+K_p} & \text{in response to unit-step input; } K_p = \left.\dfrac{K\prod_i(s+z_i)}{\prod_j(s+p_j)}\right|_{s=0} \\ \infty & \text{in response to unit-ramp input} \\ \infty & \text{in response to unit-parabolic input} \end{cases} \tag{6.48}$$

Thus, a system with $N = 0$ (i.e., with no integrator in $D(s)G(s)$) has a finite nonzero position error, infinite velocity and acceleration errors at steady state.

Type-1 System

If $N = 1$, the steady-state errors to various standard inputs are

$$e_{ss} = \begin{cases} 0 & \text{in response to unit-step input} \\ \dfrac{1}{K_v} & \text{in response to unit-ramp input; } K_v = \left.\dfrac{K\prod_i (s+z_i)}{\prod_j (s+p_j)}\right|_{s=0} \\ \infty & \text{in repsonse to unit-ramp input} \end{cases} \tag{6.49}$$

Thus, a system with $N = 1$ (i.e., with one integrator in $D(s)G(s)$) has zero position error, a finite nonzero velocity error, and infinite acceleration error at steady state.

Type-2 System

If $N = 2$, the steady-state errors to various standard inputs are

$$e_{ss} = \begin{cases} 0 & \text{in response to unit-step input} \\ 0 & \text{in response to unit-step input} \\ \dfrac{1}{K_a} & \text{in response to unit parabolic input; } K_a = \left.\dfrac{K\prod_i (s+z_i)}{\prod_j (s+p_j)}\right|_{s=0} \end{cases} \tag{6.50}$$

Thus, a system with $N = 2$ (i.e., with two integrators in $D(s)G(s)$) has zero position and velocity errors and a finite nonzero acceleration error at steady state.

Steady-state errors for various inputs and systems are summarized in Table 6.1.

The development above indicates that, in general, ***increased system gain** K **and/or addition of an integration in** D(s)G(s) **tend to decrease steady-state errors. However, as we have seen earlier in***

Table 6.1 *Steady-state errors for various inputs and systems*

Type of input	***Steady-state error***		
	Type-0 system	***Type-1 system***	***Type-2 system***
Unit-step	$\dfrac{1}{1+K_p}$	0	0
Unit-ramp	∞	$\dfrac{1}{K_v}$	0
Unit-parabolic	∞	∞	$\dfrac{1}{K_a}$
	$K_p = \lim_{s\to 0} D(s)G(s)$	$K_v = \lim_{s\to 0} sD(s)G(s)$	$K_a = \lim_{s\to 0} s^2 D(s)G(s)$

Chapters 4 and 5, both large system gain and the integrator in $D(s)G(s)$ have destabilizing effects on the system. Thus, a control system design is usually a trade-off between steady-state accuracy and acceptable relative stability.

For non-unity feedback system (Fig. 6.31), the difference between the input signal $R(s)$ and the feedback signal $B(s)$ is the actuating error signal $\hat{E}(s)$ which is given by

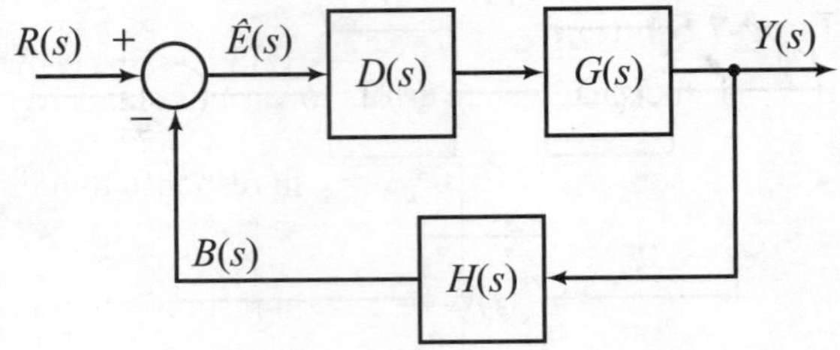

Fig. 6.31 *A non-unity feedback system*

$$\hat{E}(s) = \frac{1}{1 + D(s)G(s)H(s)} R(s) \tag{6.51}$$

Therefore, the steady-state actuating error is

$$\hat{e}_{ss} = \lim_{s \to 0} \frac{sR(s)}{1 + D(s)G(s)H(s)} \tag{6.52}$$

Example 6.4 For the feedback system of Fig. 6.32, determine the following performance measures when $K = 100$.

(i) Peak overshoot of response to $r(t)$ = unit step.
(ii) Steady-state error of response to $r(t)$ = unit step, with $w(t) = 0$.
(iii) Maximum value of response to unit step disturbance with $r(t) = 0$.
(iv) Steady-state response $y(t)$ to unit step disturbance with $r(t) = 0$.

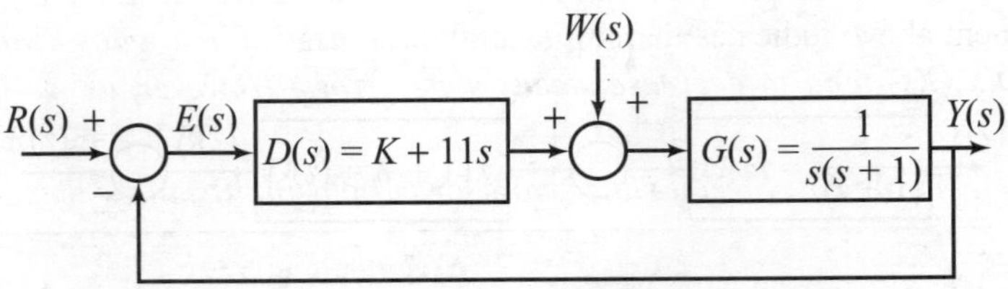

Fig. 6.32 *A feedback system*

Solution: From Fig. 6.32, we obtain

$$\begin{aligned} Y(s) &= \frac{D(s)G(s)}{1 + D(s)G(s)} R(s) + \frac{G(s)}{1 + D(s)G(s)} W(s) \\ &= \frac{K + 11s}{s^2 + 12s + K} R(s) + \frac{1}{s^2 + 12s + K} W(s) \end{aligned} \tag{6.53}$$

The characteristic equation of the system is

$$s^2 + 12s + K = s^2 + 2\zeta\omega_n s + \omega_n^2 = 0 \tag{6.54}$$

For $K = 100$, we have

$$\omega_n = 10, \ \ \zeta = 0.6$$

For $R(s) = 1/s$ and $W(s) = 0$, the response (refer to Eqn. (6.53))

$$Y(s) = M(s)R(s) = \frac{100 + 11s}{s^2 + 12s + 100}\left(\frac{1}{s}\right) \tag{6.55}$$

(i) The correlations derived in Section 6.3.1 correspond to a standard second-order system, without finite zeros. If the transfer function $M(s)$ possesses a finite zero, it can affect the transient response of the system. If the zero is far away from the poles of $M(s)$ in the left half of s-plane, it will have little effect on the transient response; if the zero is near the poles, it has a substantial influence on the response; and if the zero is close to the origin in the s-plane, it results in the highly pronounced peaking phenomenon.

In the example under consideration. The transfer function (refer to Eqn. (6.55))

$$M(s) = \frac{100 + 11s}{s^2 + 12s + 100} = \frac{\frac{\omega_n^2}{\alpha}(s + \alpha)}{s^2 + 2\zeta\omega_n s + \omega_n^2} \tag{6.56}$$

with $\omega_n = 10$, $\zeta = 0.6$, and $\alpha = 100/11$. The poles and zero of $M(s)$ are located as shown in Fig. 6.33; the zero is near the poles. The zero, therefore, will substantially affect the peak overshoot. We may determine the peak overshoot in this case using Fig. 6.23, or for better accuracy, obtain the peak overshoot M_p from the response $y(t)$ corresponding to Eqn. (6.56); MATLAB simulation of this equation has given us (Fig. 6.34) $M_p = 22\%$.

(ii) From Fig. 6.32, we see that the given system is type-1. Therefore, the steady-state error to step input with $W(s) = 0$, is zero.

(iii) Consider now the effect of disturbance with $R(s) = 0$. This is given by the equation (refer to Eqn. (6.53)).

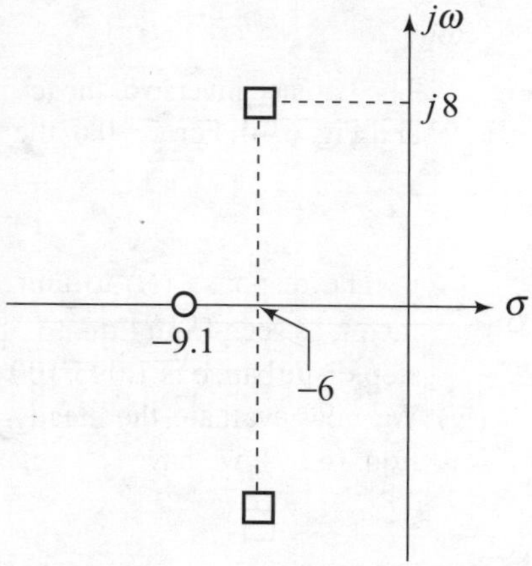

Fig. 6.33 *Closed-loop poles and zero of the system of Fig. 6.32*

$$Y(s) = \frac{1}{s^2 + 12s + 100}W(s) = \frac{1}{\omega_n^2}\left[\frac{\omega_n^2}{s^2 + 2\zeta\omega_n s + \omega_n^2}\right]W(s) \tag{6.57}$$

with $\omega_n = 10$ and $\zeta = 0.6$.

For $W(s) = 1/s$,

$$y(t) = \frac{1}{\omega_n^2}\left[\mathcal{L}^{-1}\left\{\frac{\omega_n^2}{s(s^2 + 2\zeta\omega_n s + \omega_n^2)}\right\}\right] \tag{6.58}$$

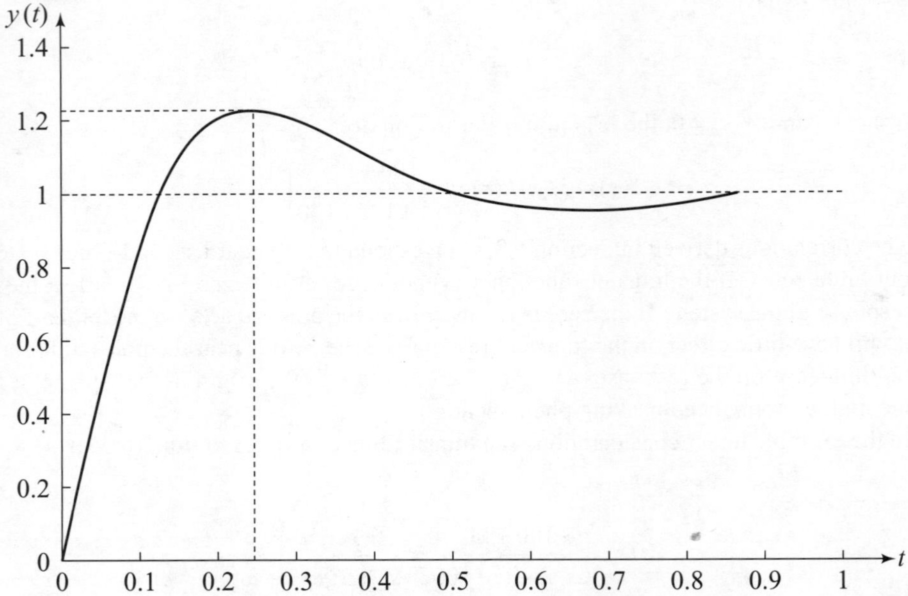

Fig. 6.34 *Response to unit-step input*

The Laplace inverse of the term inside the kinked brackets is known to us; it is given by Eqn. (6.7), and Fig. 6.10. For $\zeta = 0.6$, the maximum value of the inverse is given by

$$1 + e^{-\pi \zeta/\sqrt{1-\zeta^2}} = 1 + 0.095 = 1.095$$

The response $y(t)$ to unit step disturbance is a scaled version of this inverse; the scaling factor, as seen from Eqn. (6.58), is $1/\omega_n^2$. Therefore, the maximum value of the response to unit step disturbance is $1.095/100 = 0.01095$.

(iv) We now evaluate the steady-state response to unit step disturbance (when $R(s) = 0$). From Eqn. (6.57), we have

$$Y(s) = \frac{1}{s^2 + 12s + 100}\left(\frac{1}{s}\right)$$

$$y_{ss} = \lim_{s \to 0} sY(s) = \frac{1}{100} = 0.01$$

Example 6.5 A simplified model of a control system for a paper winding mechanism used in a paper plant is shown in Fig. 6.35. The model includes the time constants of the drive motor with load, and the gain of control amplifier and other circuitry. For a preliminary design, it is required to determine the system parameters K, τ_m and τ_e so as to meet the following design specifications:

(i) Settling time (2% tolerance) = 4/3 sec.
(ii) Peak time = $\pi/4$

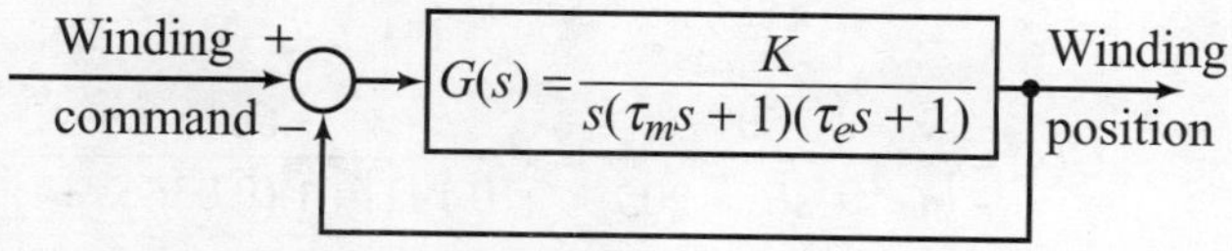

Fig. 6.35 *A simplified control system for paper winding mechanism.*

Select a value of K so that the closed-loop system behaves like a simple (second-order) damped oscillator. Determine the corresponding values of τ_m and τ_e. Estimate the percentage overshoot and steady-state error of the closed-loop system to unit ramp input.

Solution: Since the system is of third-order, the closed-loop system should have a transfer function given by

$$M(s) = \frac{K'}{(s+p)(s^2 + 2\zeta\omega_n s + \omega_n^2)}$$

If the closed-loop system is to behave like a simple damped oscillator, the real pole has to move far to the left of the complex pole pair so that the real pole could be neglected.

With the settling time $t_s = 4/3$, the closed-loop system must satisfy:

$$\frac{4}{\zeta\omega_n} = \frac{4}{3} \text{ or } \zeta\omega_n = 3$$

Also, with a peak time $t_p = \pi/4$, we have

$$\frac{\pi}{\omega_d} = \frac{\pi}{4} \text{ or } \omega_d = \omega_n\sqrt{1-\zeta^2} = 4$$

Accordingly, we have

$$\omega_n = 5,\ \zeta = 0.6$$

A rule of thumb is to move the real pole by a factor of 10 to the left of the complex pole pair. Since $\zeta\omega_n = 3$, we must then pick

$$p = 30$$

This gives

$$M(s) = \frac{K'}{(s+30)(s^2 + 6s + 25)}$$

The corresponding open-loop transfer function

$$G(s) = \frac{M(s)}{1 - M(s)} = \frac{K'}{(s+30)(s^2 + 6s + 25) - K'}$$

Since this transfer function has a free integrator (open-loop system is type-1), as given in Fig. 6.35, we must have the constant terms in the denominator cancel out; thus,

$$K' = 25 \times 30$$

Hence,

$$G(s) = \frac{750}{s(s^2 + 36s + 205)} = \frac{3.66}{s(0.141s + 1)(0.0346s + 1)}$$

So we have

$$K = 3.66, \ \tau_m = 0.141, \ \tau_e = 0.0346$$

Since $\zeta = 0.6$ for the closed-loop system, the percentage overshoot is

$$M_p = e^{-\pi\zeta/\sqrt{1-\zeta^2}} \times 100 = 9.5\%$$

System $K_v = \lim_{s \to 0} sG(s) = 3.66$

$$e_{ss}\big|_{\text{unit ramp}} = \frac{1}{3.66} = 0.25$$

6.7 REVIEW EXAMPLES

Review Example 6.1 The open-loop transfer function of a unity-feedback system is given by

$$G(s) = \frac{K}{s(\tau s + 1)}; K > 0, \tau > 0$$

By what factor should the gain K be reduced so that the peak overshoot of unit-step response of the system is reduced from 75 to 25%?

Solution: The characteristic equation of the system is

$$s^2 + \frac{1}{\tau}s + \frac{K}{\tau} = 0$$

Assume that corresponding to 75% overshoot, the gain is K_1 and damping ratio is ζ_1; and for 25% overshoot, the gain is K_2 and damping ratio is ζ_2. From the characteristic equation, we obtain

$$2\zeta_1 = \frac{1/\tau}{\sqrt{K_1/\tau}} = \frac{1}{\sqrt{K_1\tau}}; 2\zeta_2 = \frac{1}{\sqrt{K_2\tau}}$$

Therefore,

$$\left(\frac{\zeta_2}{\zeta_1}\right)^2 = \frac{K_1}{K_2} \tag{6.59}$$

Since

$$e^{-\pi\zeta/\sqrt{1-\zeta^2}} = M_p$$

we have

$$\frac{\pi\zeta_1}{\sqrt{1-\zeta_1^2}} = -\ln(0.75) = 0.29 = \frac{\pi}{\sqrt{\dfrac{1}{\zeta_1^2} - 1}} \tag{6.60}$$

and
$$\frac{\pi\zeta_2}{\sqrt{1-\zeta_2^2}} = -\ln(0.25) = 1.4 = \frac{\pi}{\sqrt{\dfrac{1}{\zeta_2^2}-1}} \tag{6.61}$$

Dividing, squaring and rearranging we get $\dfrac{\zeta_2^2}{\zeta_1^2} = 20 = \dfrac{K_1}{K_2}$.

Therefore, the gain K should be reduced by a factor of 20 so that the peak overshoot of unit-step response of the system is reduced from 75 to 25%.

Review Example 6.2 A unity-feedback system is characterized by the open-loop transfer function

$$G(s) = \frac{1}{s(0.5\,s+1)\,(0.2\,s+1)}$$

(a) Determine the steady-state errors to unit-step, unit-ramp, and unit-parabolic inputs.
(b) Determine rise time, peak time, peak overshoot, and settling time of the unit-step response of the system.

Solution: (a) The given system is a type-1 system. The error constants are obtained as follows:

$$K_p = \lim_{s\to 0} G(s) = \infty;\ K_v = \lim_{s\to 0} sG(s) = 1;\ K_a = \lim_{s\to 0} s^2G(s) = 0$$

Therefore,

$$e_{ss}\big|_{\text{unit step}} = \frac{1}{1+K_p} = 0;\ e_{ss}\big|_{\text{unit ramp}} = \frac{1}{K_v} = 1;\ e_{ss}\big|_{\text{unit parabola}} = \frac{1}{K_a} = \infty$$

(b) Since the system is of third order, we can no longer use the correlations (6.38) for rise time, peak time, peak overshoot, and settling time. These correlations were derived for a standard second-order system. For the transient analysis of the given system, we may determine the unit-step response (Eqn. (6.37)) and therefrom compute rise time, peak time, peak overshoot, and settling time.

However, if the system dynamics can be accurately represented by a pair of dominant poles, then the correlations (6.38) may be used. The characteristic equation of the given system is

$$s(0.5\,s + 1)\,(0.2\,s + 1) + 1 = 0$$

or
$$s^3 + 7s^2 + 10\,s + 10 = 0$$

Factorizing the characteristic equation, we obtain

$$(s + 5.5157)\,(s^2 + 1.4844\,s + 1.8131) = 0$$

The magnitude of the real root is more than 7 times the magnitude of the real part of the complex roots. Therefore, the complex-conjugate root pair gives the dominant closed-loop poles with

$$\omega_n = \sqrt{1.8131} = 1.3465\text{ rad/sec};\ \zeta = \frac{1.4844}{2\times 1.3465} = 0.55;\ \omega_d = \omega_n\sqrt{1-\zeta^2} = 1.1245$$

$$t_r = \frac{\pi - \cos^{-1}\zeta}{\omega_d} = \frac{\pi - 56.63 \times \dfrac{\pi}{180}}{1.1245} = 1.91 \text{ sec}; \; t_p = \frac{\pi}{\omega_d} = 2.79 \text{ sec}; \; M_p = e^{-\pi\zeta/\sqrt{1-\zeta^2}} = 0.1265;$$

$$t_s = \frac{4}{\zeta\omega_n} = 5.4 \text{ sec}$$

Review Example 6.3 A servomechanism controls the angular position θ_L of a mass through a command signal θ_R. The moment of inertia of the moving parts referred to the load shaft is 200 kg-m^2 and the motor torque at the load is 6.88×10^4 newton-m per rad of error. The damping torque coefficient referred to the load shaft is 5×10^3 newton-m per rad/sec.

(a) Determine the frequency of transient oscillation, the peak overshoot, the time to peak, and the steady-state error resulting from the application of a step input of 1 rad.
(b) Determine the steady-state error when the command signal is a constant angular velocity of 1 rev/min.
(c) Determine the steady-state error which exists when a steady torque of 1200 newton-m is applied at the load shaft.

Solution: Since the torque applied to the load must equal the reaction torque, we have

$$(200\,\ddot{\theta}_L + 5\times10^3\,\dot{\theta}_L) = 6.88\times10^4(\theta_R - \theta_L) - T_w \tag{6.62}$$

Figure 6.36 gives a block-diagram representation of this differential equation.

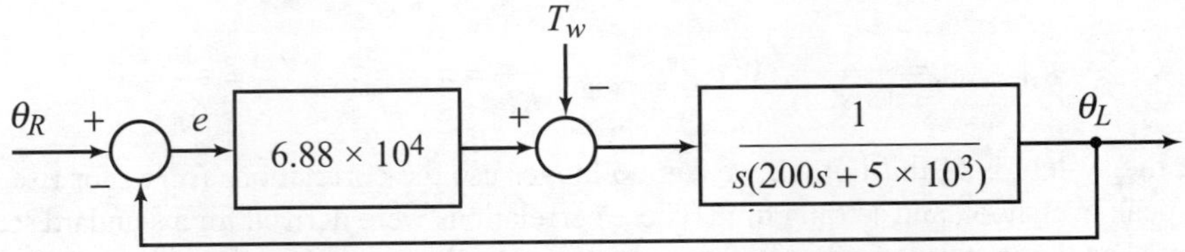

Fig. 6.36 *Block diagram representation of Eqn. (6.62)*

From the block diagram, we obtain

$$\frac{\theta_L(s)}{\theta_R(s)} = \frac{6.88\times10^4}{200\,s^2 + 5\times10^3 s + 6.88\times10^4} = \frac{344}{s^2 + 25\,s + 344}$$

$$\frac{\theta_L(s)}{T_w(s)} = \frac{-1}{200\,s^2 + 5\times10^3 s + 6.88\times10^4} \tag{6.63}$$

(a) The characteristic equation of the system is

$$s^2 + 25\,s + 344 = 0$$

Comparing with the standard characteristic equation

$$s^2 + 2\zeta\omega_n s + \omega^2_n = 0$$

we obtain $\omega_n = \sqrt{344} = 18.547;\ \zeta = \dfrac{25}{2 \times 18.547} = 0.674$

Frequency of transient oscillation, $\omega_d = \omega_n\sqrt{1-\zeta^2} = 13.7$ rad/sec

Peak overshoot, $M_p = e^{-\pi\zeta/\sqrt{1-\zeta^2}} \times 100 = 5.69\%$

Time to peak, $t_p = \pi/\omega_d = 0.229$ sec

Steady-state error, $e_{ss}|_{\text{step input}} = 0$

(b) Input = 1 rpm = $2\pi/60$ rad/sec; $\theta_R(t) = \dfrac{2\pi}{60}t$

$$e_{ss}|_{\text{unit ramp}} = \frac{2\zeta}{\omega_n} = 0.0726$$

For a ramp input of slope $2\pi/60$, steady-state error $= 0.07268 \times \dfrac{2\pi}{60} = 0.0076 \text{ rad} = 0.436 \text{ deg}$.

(c) Steady-state output in response to load torque of 1200 newton-m is given by (refer to Eqn. (6.63))

$$\theta_{Lss} = \lim_{s\to 0} s\theta_L(s)$$

$$= \lim_{s\to 0} s\left[\frac{-1}{200s^2 + 5\times 10^3 s + 6.88\times 10^4}\right]\frac{1200}{s}$$

$$= \frac{-1200}{6.88\times 10^4} \text{ rad} = -1 \text{ deg}$$

Therefore, the steady-state error for a disturbance torque of 1200 newton-m is 1 deg.

CONCLUDING REMARKS

In this chapter, we have been concerned with the definition and usefulness of quantitative measures of performance of feedback control systems. Through these quantitative measures, we specify the desired transient and steady-state response of the feedback system.

What we have learnt?

- Figures of merit, such as peak overshoot M_p, settling time t_s, and peak time t_p of the response to unit-step input specify the desired transient behaviour.
- Steady-state constants $K_p / K_v / K_a$ specify the desired steady-state behavior.
- Specifications are turned into a target closed-loop characteristic polynomial—comprising of a second-order pole pair in the quadratic factor $(s^2 + 2\zeta\omega_n s + \omega_n^2)$, with all other real/complex poles faster than this quadratic pair.
- The design problem is, then, to select the controller $D(s)$ to match this characteristic polynomial.

Review Questions

6.1 Describe a set of performance specifications characterizing desired time response, that are frequently used in the design of linear control systems using root-locus plots.

6.2 Given the following measures of step response of a feedback system with transfer function

$$\frac{Y(s)}{R(s)} = M(s) = \frac{\omega_n^2}{s^2 + 2\zeta\omega_n s + \omega_n^2}$$

(a) rise time, t_r; (b) peak time, t_p;
(c) peak overshoot; M_p; (d) settling time, t_s.

Through graphical plots, explain the correlations of these measures with ζ and ω_n. Show that rise time, peak time, and settling time measures of performance are mutually dependent, and therefore must be specified in a consistent manner.

6.3 A feedback system is higher than second-order and has finite zeros.

(a) Determine the region in the s-plane which is feasible for the system's dominant poles corresponding to a percentage overshoot $5\% \le M_p \le 10\%$, and a settling time $3 \le t_s \le 5$ seconds.

(b) Define the regions of dominant and insignificant poles, and insignificant zeros in the s-plane.

(c) Show that pole-zero cancelation may be utilized to meet the dominance condition when nondominant poles, and zeros do not lie in insignificant regions.

6.4 (a) What is system-type number? Explain the practical significance of this number.

(b) Define the error constants K_p, K_v, and K_a.

Self-Appraisal Quiz

A set of questions with multiple answer choices is given below. These questions have been designed to capture important aspects of the subject covered in the chapter. For each of these questions, identify the best of the given answer choices. Compare your answers with the Master Key given at the end of the session. This quiz should serve the purpose of a self-appraisal test for readers.

6.1 Which of the following describes the step response of the closed-loop system of Fig. Q6.1?

(A) Underdamped
(B) Critically damped
(C) Overdamped
(D) The output does not reach a steady-state value.

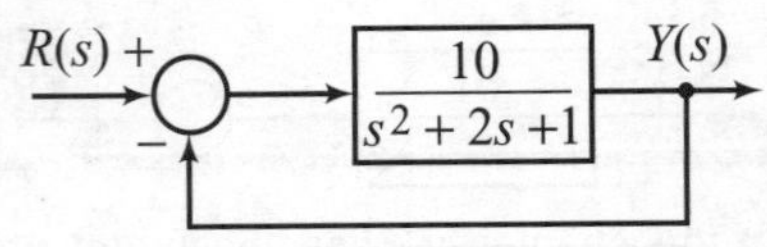

Fig. Q6.1

6.2 A unit feedback system has open-loop transfer function $G(s) = 9/[s(s+3)]$. The system has

(A) damping ratio = 1/2, and natural frequency = 9
(B) damping ratio = 1/6, and natural frequency = 3
(C) damping ratio = 1/6, and natural frequency = 9
(D) damping ratio = 1/2, and natural frequency = 3

6.3 A unity feedback system with open-loop transfer function $G(s) = 4/[s(s+p)]$ is critically damped. The value of the parameter p is

(A) 4 (B) 3 (C) 2 (D) 1

6.4 Consider the system of Fig. Q6.4. K_I that gives critical damping is
(A) 0.417 (B) 0.257 (C) 1 (D) None of the answers in (A), (B), and (C) is correct

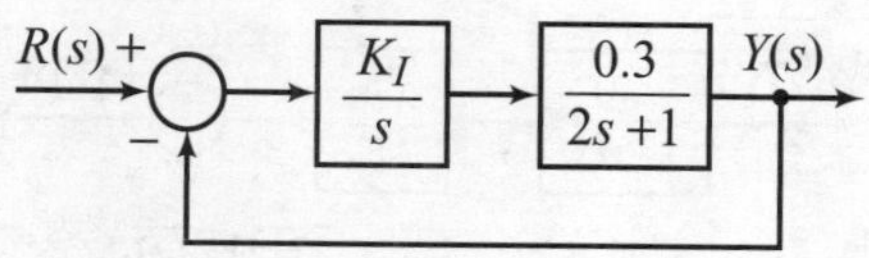

Fig. Q6.4

6.5 A unity feedback system has open-loop transfer function $G(s) = 25/[s(s+6)]$. The time t_p at which the peak of the step-input response occurs is
(A) 11/7 sec (B) 11/4 sec (C) 11/14 sec (D) 11/28 sec

6.6 Peak overshoot of step-input response of an underdamped second-order system is explicitly indicative of
(A) settling time (B) rise time (C) natural frequency (D) damping ratio

6.7 A unity feedback system has open-loop transfer function $G(s) = 25/[s(s+6)]$. The peak overshoot in the step-input response of the system is approximately equal to
(A) 5% (B) 10% (C) 15% (D) 20%

6.8 The step-input response of the system of Fig. Q6.8a is shown in Fig. Q6.8b. The value of damping ratio of the system is
(A) 0.39 (B) 0.49 (C) 0.59 (D) 0.69

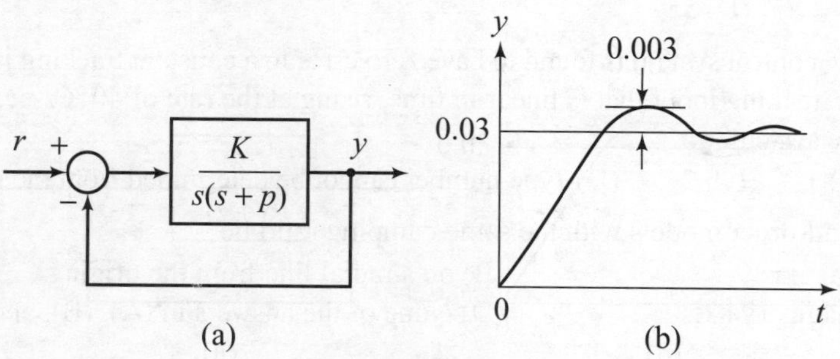

Fig. Q6.8

6.9 A unity feedback system has open-loop transfer function $G(s)$. The steady-state error is zero for
(A) step input and type-1 $G(s)$ (B) ramp input and type-1 $G(s)$
(C) step input and type-0 $G(s)$ (D) ramp input and type-0 $G(s)$

6.10 A unity feedback system with forward path transfer path transfer function $G(s) = 1/[s^2(s+1)]$ is subjected to an input $r(t) = K_1 + K_2 t + \frac{1}{2}t^2$. The steady-state error of the system is
(A) infinity (B) 1 (C) zero (D) None of the answers in (A), (B), and (C) is correct

6.11 Closed-loop transfer function of a unity feedback system is given by

$$\frac{Y(s)}{R(s)} = \frac{1}{\tau s + 1}$$

System K_v is
(A) τ (B) $1/\tau$ (C) 1 (D) ∞

6.12 Closed-loop transfer function of a unity feedback system is given by

$$\frac{Y(s)}{R(s)} = \frac{\omega_n^2}{s^2 + 2\zeta\omega_n s + \omega_n^2}$$

System K_v is

(A) $\omega_n/2\zeta$ (B) 1 (C) ∞ (D) $2\zeta/\omega_n$

6.13 Consider the position control system of Fig. Q6.13. The value of K such that the steady error is 10° for input $\theta_r = 400t\,\mu(t)$ rad/sec, is

(A) 104.5 (B) 114.5 (C) 124.5

(D) None of the answers in (A), (B), and (C) is correct

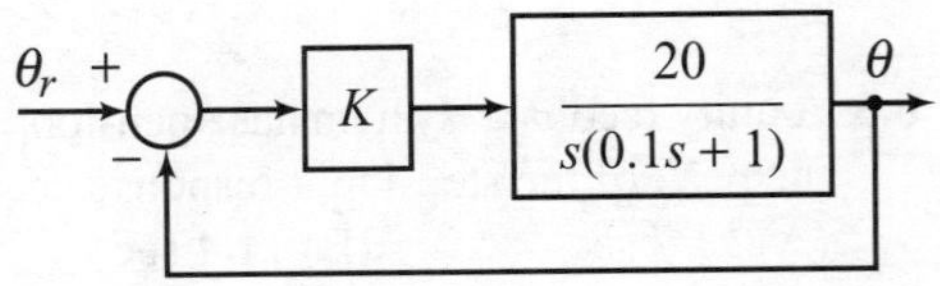

Fig. Q6.13

6.14 Consider the speed control system shown in Fig. Q6.14. Parameter variations occurring during operating conditions cause K_A to modify to $K_A' = 0.9K_A$. The value of K_A that limits the change in steady-state motor speed due to parameter variations to 0.1%, is

(A) 25 (B) 35

(C) 45 (D) 55

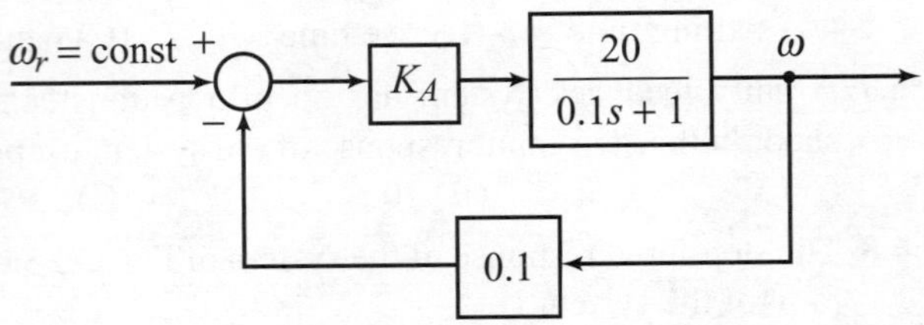

Fig. Q6.14

6.15 A temperature control system is found to have zero error to a constant tracking input, and an error of 0.5 °C to a tracking input that is linear in time, rising at the rate of 40 °C/sec. What is the type number of the system?

(A) 0 (B) 1 (C) 2 (D) Type number cannot be determined from the given information

6.16 Poles of second-order models with the same damping ratio lie

(A) on the real axis (B) on a radial line from the origin

(C) on the imaginary axis (D) None of the answers in (A), (B), and (C) is correct

6.17 The percentage overshoot of a second-order system $G(s) = \frac{K\omega_n^2}{s^2 + 2\zeta\omega_n s + \omega_n^2}$ to a step input depends only on

(A) the value of the step input (B) the value of the damping ratio

(C) the value of the gain K (D) the parameter ω_n

6.18 A unity feedback system has open-loop transfer function $G(s) = 1/s^2$. This system possesses a satisfactory steady-state error for

(A) step input signals (B) ramp input signals

(C) both the step and ramp input signals (D) none of the step and ramp input signals

6.19 Consider the following statements:

(i) Many systems are designed for peak overshoot in the range 5–25%.

(ii) Desired dominant closed-loop poles are usually a complex-conjugate pair.

(A) None of the above statements is true (B) Statement (i) is true but statement (ii) is false

(C) Statement (i) is false but statement (ii) is true (D) Both the statements are true

6.20 Figure Q6.20 shows feedback control scheme for an integrating process. The steady-state error in output y due to a unit-step disturbance, is

(A) 0 (B) 1 (C) $\frac{1}{K}$ (D) None of the answers in (A), (B), and (C) is correct

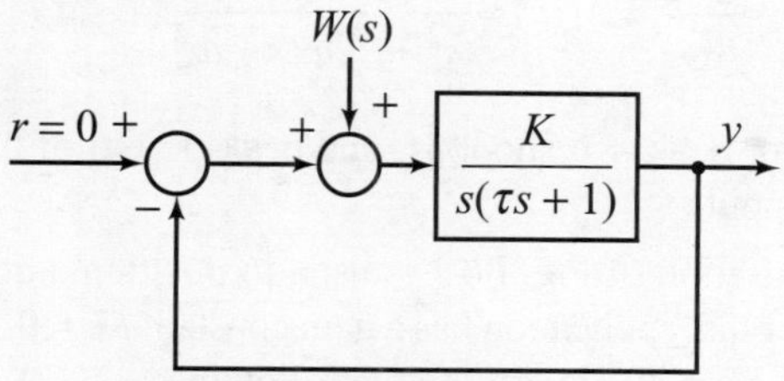

Fig. Q6.20

6.21 Consider the system of Fig. Q6.21. The steady-state offset is zero when K_P is

(A) zero (B) infinitely large (C) any value of K_P (D) no value of K_P

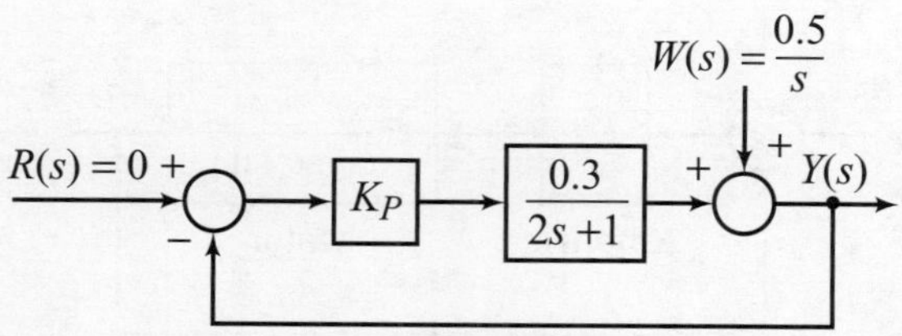

Fig. Q6.21

6.22 Consider the system of Fig. Q6.22. The steady-state offset is zero for

(A) $K_I = 0$ (B) $K_I = \infty$ (C) any value of K_I (D) no value of K_I

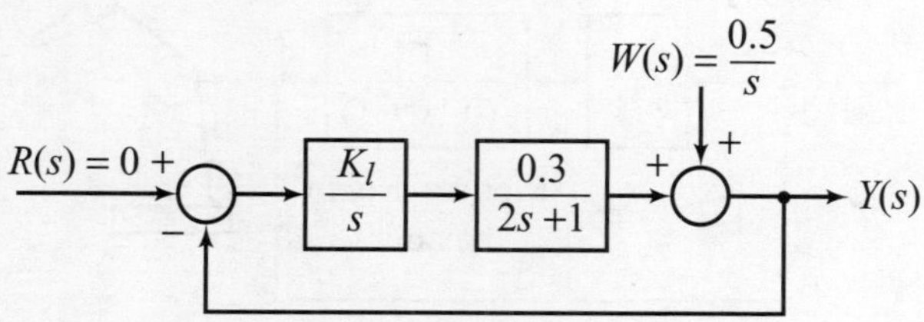

Fig. Q6.22

MASTER KEY

6.1 (A)	6.5 (C)	6.9 (A)	6.13 (B)	6.17 (B)	6.20 (B)
6.2 (D)	6.6 (D)	6.10 (B)	6.14 (D)	6.18 (D)	6.21 (B)
6.3 (A)	6.7 (B)	6.11 (B)	6.15 (B)	6.19 (D)	6.22 (C)
6.4 (A)	6.8 (C)	6.12 (A)	6.16 (B)		

Problems

6.1 (a) A feedback system has transfer function

$$\frac{Y(s)}{R(s)} = M(s) = \frac{\omega_n^2}{s^2 + 2\zeta\omega_n s + \omega_n^2}; \zeta < 1$$

Derive expressions for peak overshoot M_p and peak time t_p of the time response of the given system to unit-step input.

(b) The feedback configuration of Fig. P6.1 is used to position a device in response to an input. The motor selected for this application has a gain constant $K_m = 0.5$ and a time constant $\tau_m = 0.1$. For an amplifier gain $K_A = 20$, determine the following:

(i) Undamped natural frequency, damping ratio, and damped natural frequency.
(ii) Rise time, peak time, peak overshoot, and settling time.
(iii) Position, velocity and acceleration error constants K_p, K_v, and K_a.
(iv) Steady-state error resulting from the application of (a) unit-step input, (b) unit-ramp input, and (c) unit-parabolic input.

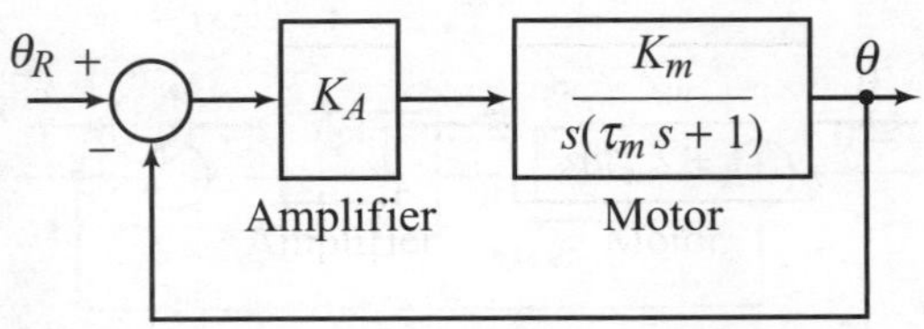

Fig. P6.1

6.2 Assume that control of one of the axes of a robot can be represented by the block diagram of Fig. P6.2.

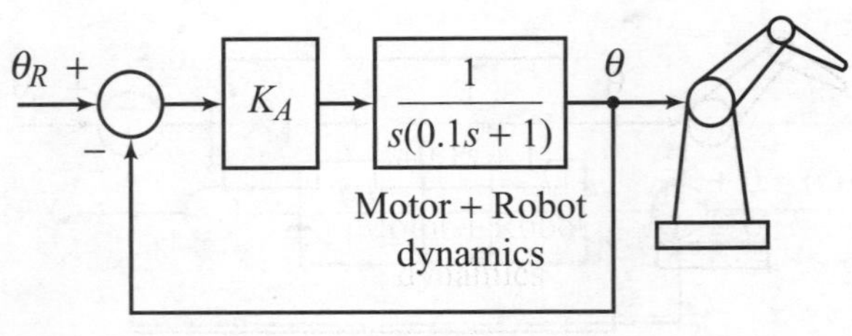

Fig. P6.2

(a) Determine the amplifier gain K_A so that the robot reaches steady-state in this axis in minimum time with no overshoot.
(b) Determine the steady-state error for an input command signal of $\theta_R = (5 + t)\mu(t)$, with the value of K_A obtained in part (a).

6.3 A ground-based tracking radar is used to track aircraft targets. Assume that azimuth axis position control loop of the tracking radar can be represented by the block diagram of Fig. P6.3a.

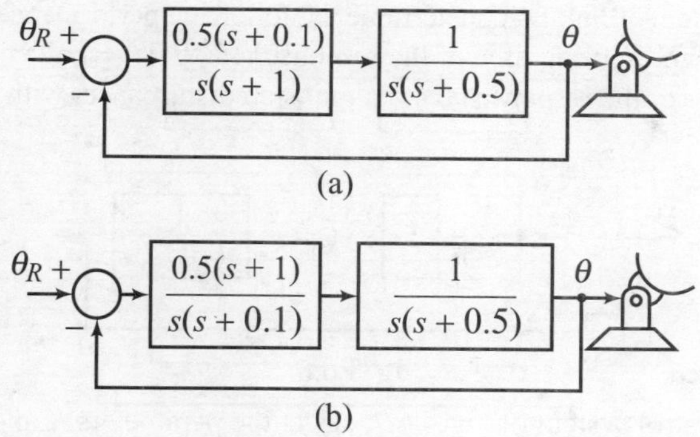

Fig. P6.3

(a) Determine the steady-state errors of the tracker for the following inputs caused by the aircraft motion.

(i) $\theta_R = 10t$

(ii) $\theta_R = 10t + 0.1t^2$

(b) Compare the steady-state performance of system of Fig. P6.3b with that of Fig. P6.3a.

6.4 The system of Fig. P6.4 has a constant reference input $r = 10$ and a constant disturbance input $w = 4$. Find the steady-state error.

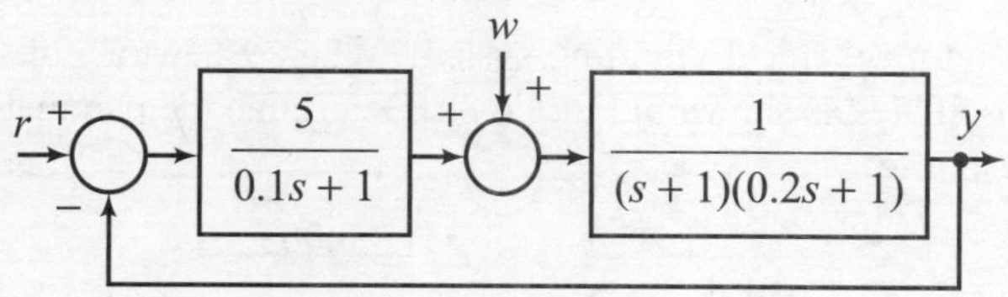

Fig. P6.4

6.5 Consider the heat-exchanger temperature control loop shown in Fig. P6.5.

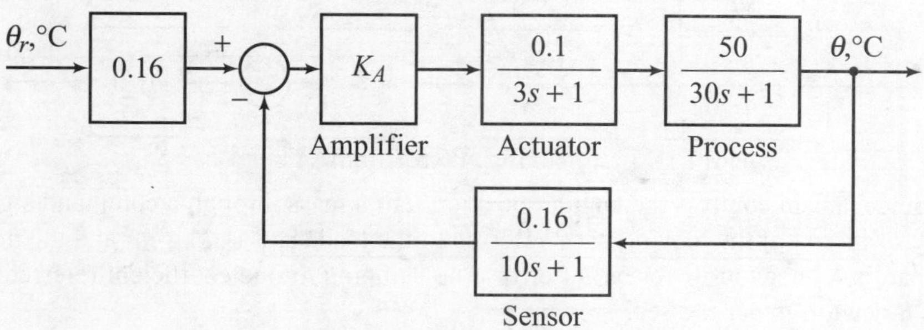

Fig. P6.5

(a) Find the range of values of $K_A > 0$ for which the system is stable.

(b) The acceptable value of steady-state error is 1°C for the step input of 10 °C. Find the value of K_A that meets this specification on static accuracy.

6.6 For the feedback system of Fig. P6.6, determine the following performance measures when $K = 60$.

(i) Peak overshoot and settling time of the response $y(t)$ to $r(t)$ = unit step, with $w(t) = 0$.

(ii) Maximum value of the response $y(t)$ to a unit step disturbance, with $r(t) = 0$.

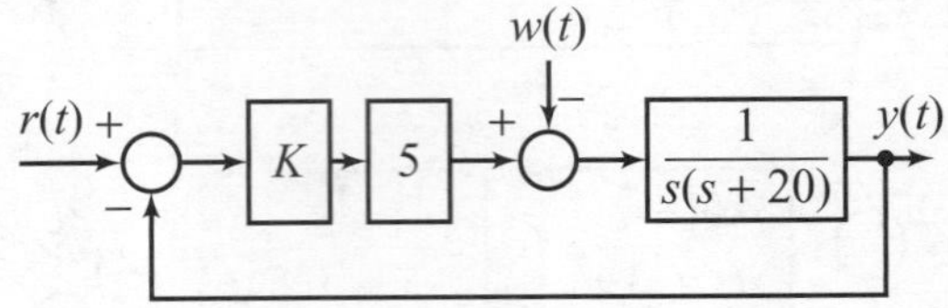

Fig. P6.6

6.7 For the feedback control system of Fig. P6.7, select the parameters K and p so that the following specifications will be satisfied:

(i) Peak overshoot of the response to a step input ≤ 9.5%.

(ii) Steady-state error to unit step disturbance ≤ 0.01.

Determine the steady-state error of the resulting system to unit ramp command.

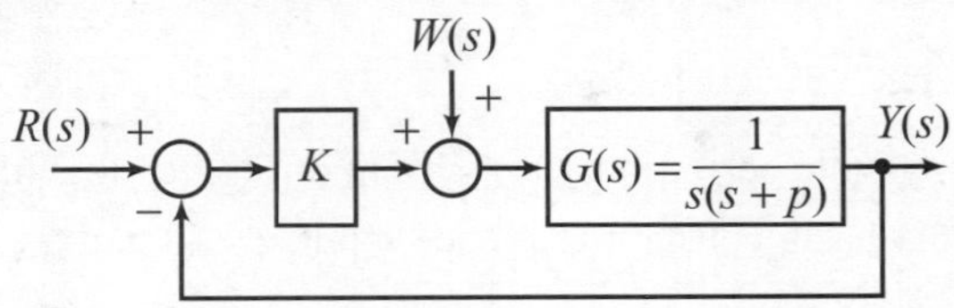

Fig. P6.7

6.8 Figure P6.8a shows a mechanical vibrating system. When 9 newtons of force is applied to the system, the mass oscillates as shown in Fig. P6.8b. Determine the parameters M, B, and K of the system from the response.

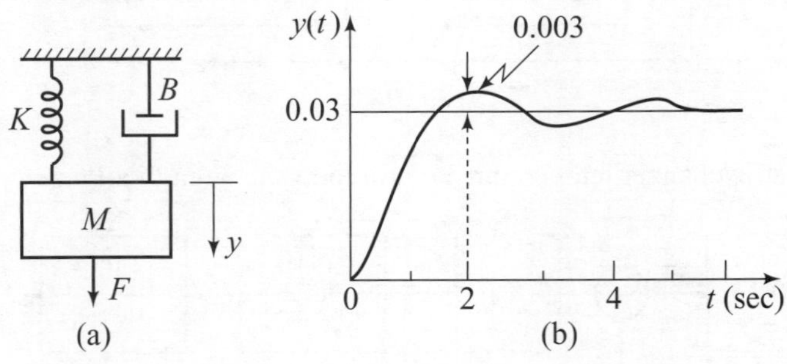

Fig. P6.8

6.9 A servomechanism controls the angular position θ of a mass through a command signal θ_R. The moment of inertia of the moving parts referred to the load shaft is J kg-m^2 and the motor torque at the load is K_T newton-m per rad of error. The damping torque coefficient referred to the load shaft is B newton-m per rad/sec.

The following tests were conducted on the position control system.

(i) Unit-step response was recorded and peak overshoot measured therefrom has a value of 25%.

(ii) A constant velocity input of 1 rad/sec produced a steady-state error of 0.04 rad.

(ii) With the command θ_R held fixed, a torque of 10 newton-m applied at the load shaft produced a steady-state error of 0.01 rad.

From the above data, determine the parameters J, B and K_T of the system.

6.10 The speed of a flywheel, driven by an electric motor, is to be controlled from the setting of an input potentiometer using a closed-loop control system. The moment of inertia of the flywheel and the motor is 100 kg-m^2, and the motor torque at the flywheel is 45 newton-m for a speed error of 1 rad/sec. Frictional torque is 5 newton-m when the flywheel velocity is 1 rad/sec. With the system at rest, the input pot setting is suddenly increased from zero to 50 rev/min. Determine an expression for the subsequent flywheel angular velocity as a function of time, and calculate steady-state velocity error of the flywheel.

MATLAB Exercise Plot the angular velocity as a function of time using MATLAB.

6.11 A flywheel driven by an electric motor is automatically controlled to follow the movement of a handwheel. The moment of inertia of the flywheel and the motor is 150 kg-m^2 and the motor torque applied to the flywheel is 2400 newton-m per radian of misalignment between the flywheel and the handwheel. The viscous friction is equivalent to a torque of 600 newton-m per rad/sec. If the handwheel is suddenly turned through 60° when the system is at rest, determine an expression for the subsequent angular position of the flywheel as a function of time, and calculate the peak overshoot of the response.

MATLAB Exercise Plot the angular position as a function of time using MATLAB.

6.12 The schematic of a servo system is shown in Fig. P6.12. The two-phase servomotor develops a torque in accordance with the equation

$$T_M = K_1 e_c - K_2 \dot{\theta}_M$$

where $K_1 = 2 \times 10^{-3}$ newton-m /volt, and $K_2 = 0.5 \times 10^{-3}$ newton-m per rad/sec.
The other parameters of the system are: load inertia, $J_L = 2.5$ kg-m^2; coefficient of load friction, $B_L = 250$ newton-m per rad/sec; motor to load gear ratio, $(\dot{\theta}_L / \dot{\theta}_M) = 1/50$ motor to synchro gear ratio, $(\dot{\theta}_S / \dot{\theta}_M) = 1/1$, and sensitivity of synchro error detector, K_s = 100 volts/rad. The motor inertia and friction are negligible.

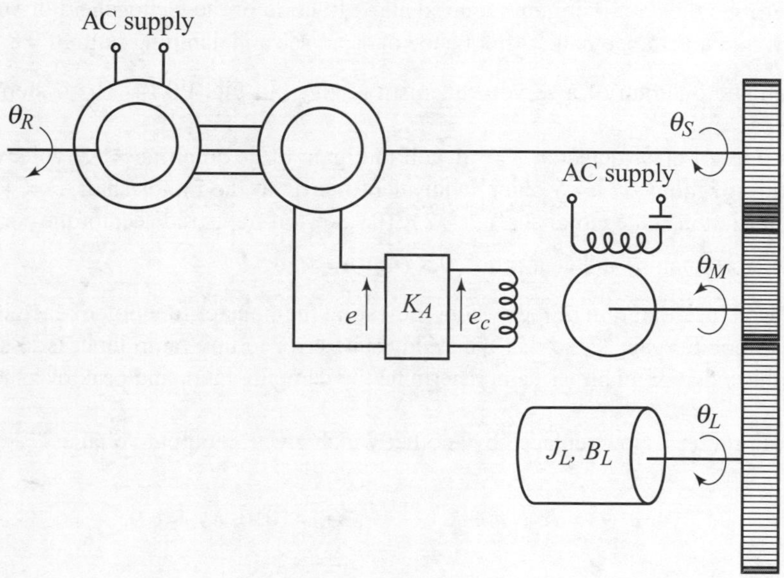

Fig. P6.12

(a) Draw a block diagram of the system indicating the transfer function of each block.
(b) If the input shaft is driven at a constant speed of π rad/sec, determine the value of the amplifier gain K_A such that the steady-state deviation between input and output positions is less than five degrees. For this value of amplifier gain, determine the peak overshoot and settling time of the system.
(c) The amplifier is now replaced by another which gives an output voltage of

$$e_c(t) = K_A e(t) + K_A \int e(t)dt; \; K_A = 904.5$$

Determine the steady-state error to ramp input $\theta_R(t) = \pi t$
(d) A simulation study on the system of part (c) shows that the transient behavior of the third-order system resembles that of the standard second-order system with characteristic equation

$$s^2 + 99.4508\, s + 3514.6322 = 0$$

Give reasons. Also determine the peak overshoot and settling time of the system.

6.13 In a dc position control servomechanism, the load is driven by a motor supplied with constant armature current. The motor field current is supplied from a dc amplifier, the input to which is the difference between the voltages obtained from input and output potentiometers. A tachogenerator is coupled to the motor shaft and a fraction K of tachogenerator output is fed back to produce a stabilizing effect. The following particulars refer to the system: moment of inertia of motor, $J_M = 2 \times 10^{-3}$ kg-m^2; moment of inertia of load, $J_L = 5$ kg-m^2; motor to load gear ratio, $(\dot{\theta}_L / \dot{\theta}_M) = 1/50$; load to potentiometer gear ratio, $\dot{\theta}_P / \dot{\theta}_L = 1/1$; motor torque constant, $K_T = 2$ newton-m/amp; tachogenerator constant, $K_t = 0.2$ volt per rad/sec; sensitivity of error detector, $K_P = 0.6$ volt/rad, and amplifier gain $= K_A$ amps/volt.
(a) Make a sketch of the system showing how the hardware is connected.
(b) Draw a block diagram of the system indicating the transfer function of each block.
(c) Determine the amplifier gain required and a fraction of the tachogenerator voltage fed back to give an undamped natural frequency of 4 rad/sec and damping ratio of 0.8.

6.14 The schematic diagram of a servomechanism is given in Fig. P6.14. The system constants are given below.

Sensitivity of error detector, $K_P = 1$ volt/rad; gain of dc amplifier $= K_A$ volts/volt; resistance of armature of motor, $R_a = 5$ ohms; equivalent inertia at the motor shaft, $J = 4 \times 10^{-3}$ kg-m^2; equivalent friction at the motor shaft, $B = 2 \times 10^{-3}$ newton-m per rad/sec; torque constant of motor, $K_T = 1$ newton-m/amp; and gear ratio $(\dot{\theta}_L / \dot{\theta}_M), n = 1/10$.

(a) Draw a block diagram of the system indicating the transfer function of each block.
(b) Find amplifier gain K_A so that the steady-state error to unit-ramp input is less than 0.01 rad. For this value of amplifier gain, determine the damping ratio, and peak overshoot to unit-step input.
(c) The amplifier is now replaced by another which gives an output voltage

$$e_a(t) = K_A e(t) + K_D \frac{de(t)}{dt}; \; K_A = 1010, \; K_D = 6.95$$

Determine the steady-state error to unit-ramp input, and damping ratio of the system. Also calculate peak overshoot corresponding to this damping ratio using correlation for standard second-order systems.

(d) A simulation study of your design in part (c) shows a different value of peak overshoot. Give reasons.

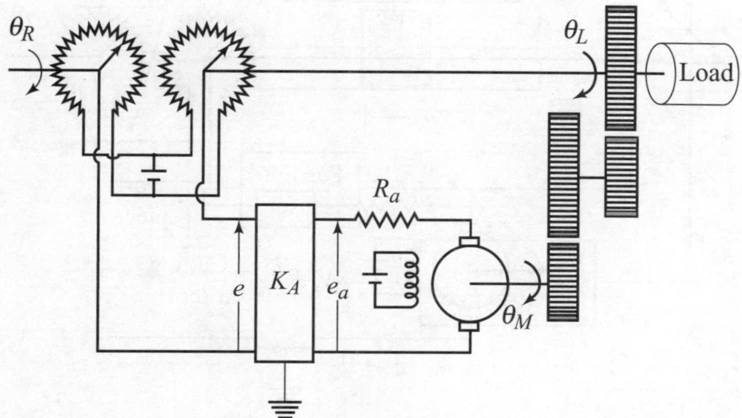

Fig. P6.14

MATLAB Exercise Use MATLAB's Simulink to obtain the step response under the following conditions:

(i) The system is linear and is driven by amplifier whose gain K_A is given by the solution to part (b).

(ii) The amplifier saturates at ± 5 volts. Describe the effect of the saturation on the system's output (Drag **Saturation** icon from Simulink library).

(iii) The amplifier with gain K_A drives the gear train that has backlash (As the motor reverses direction, the output shaft remains stationary while the motor begins to reverse. When the gears finally connect, the output shaft itself begins to turn in the reverse direction; refer to Chapter 16). The deadband width of the backlash is 0.15. Describe the effect of backlash on the system's output (Drag **Backlash** icon from Simulink library).

(iv) Deadzone (refer to Chapter 16) is present when the motor cannot respond to small voltages. The response begins when the input voltage to the motor exceeds a threshold. The deadzone width is ± 2. Describe the effect of deadzone on the system output (Drag **Deadzone** icon from Simulink library).

6.15 Hydraulic actuators are used to drive cutting tools in machining applications (Fig. P6.15a). The reference input e_r is a voltage representing the desired position y_r of the cutting tool as function of time. It can be generated by a cam drive, a punched paper tape, or a computer program. A model of a feedback control system is shown in Fig. P6.15b. The load disturbance due to reactive force F_w of the cutting tool is also shown.

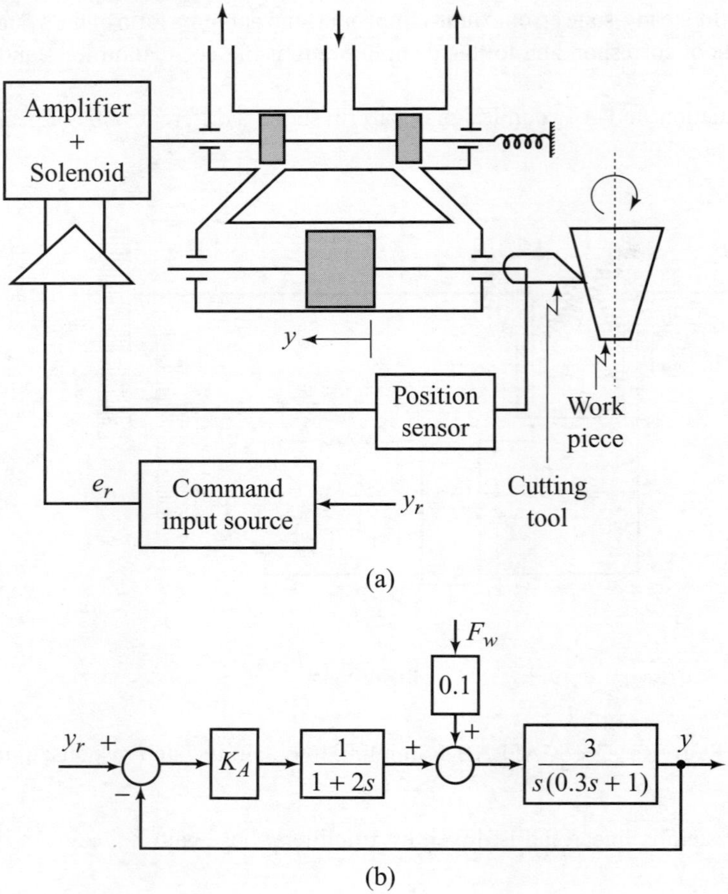

Fig. P6.15

(a) When $F_w = 0$, determine the steady-state tracking error for a constant velocity input of 0.03 mm/sec, as a function of K_A.

(b) Determine the steady-state error in $y(t)$ introduced by a unit-step disturbance, as a function of K_A.

(c) Find the minimum values of error in part (a) and that in part (b) which can be obtained by the control configuration of Fig. P6.15b.

6.16 A tank-level control system is shown in Fig. P6.16a. If the level is not correct, an error voltage e is developed and a signal $K_P e + K_D \dot{e}$ is applied to a servomotor. This in turn adjusts a valve through appropriate gearing and thereby restores balance by adjusting the inlet flow rate. A model of the feedback control system is shown in Fig. P6.16b.

(a) What is the system type number?

(b) Find the ranges of values of K_P and K_D for which the steady-state error to unit-ramp disturbance is less than 0.1.

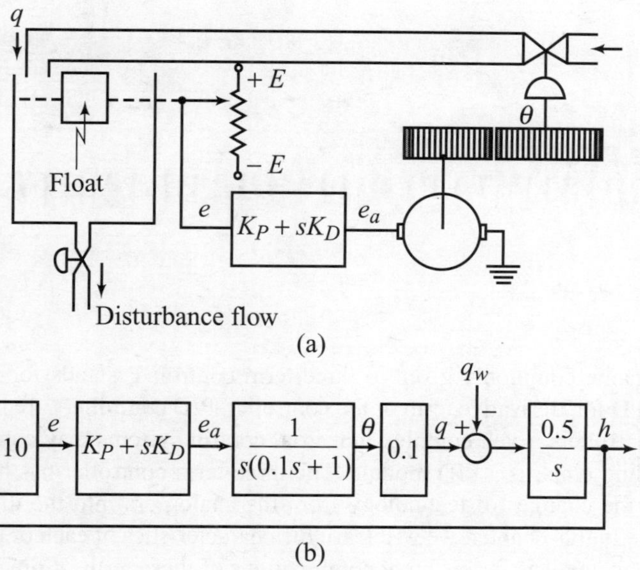

Fig. P6.16

6.17 Figure P6.17 shows a thermal system in which hot air is circulated to keep the temperature of a chamber constant. The transfer function of the plant can be adequately represented by

$$\frac{\theta(s)}{U(s)} = \frac{1}{s+1}$$

For $K_1 = 3$ and $K_2 = 19.237$, determine the peak overshoot and settling time of the response of the feedback system to unit step input.

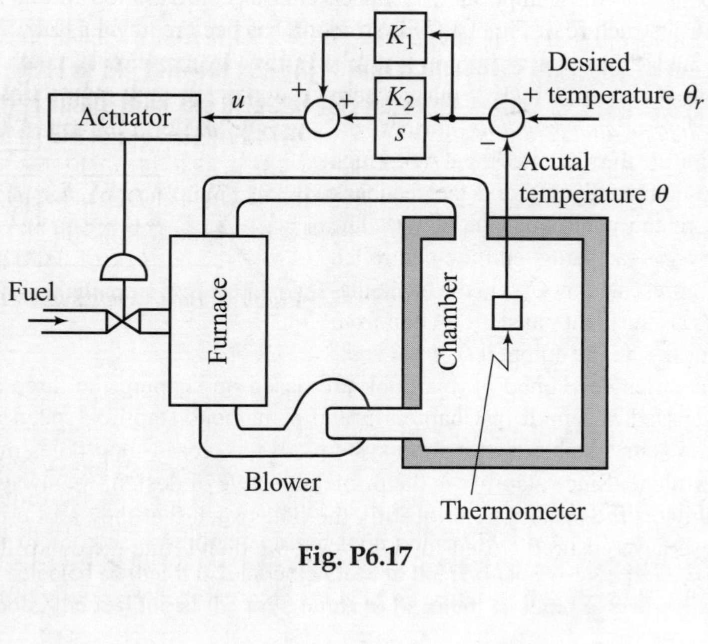

Fig. P6.17

PID CONTROL

7

Preview

PID control is a name commonly given to three-term control; **P** stands for **P**roportional term, **I** for **I**ntegral term, and **D** for **D**erivative term of the controller. PID controllers are probably the most widely used industrial controllers. Even complex industrial control systems may comprise a control function whose main building block is a PID module. The three-term controller has had a long history of use and has survived the changes of technology from the analog era into the digital control system age quite satisfactorily. In this chapter, we will learn the characteristics of each of the terms P, I, and D; and perform closed-loop analysis for different combinations of these terms. Some implementation aspects of the PID controller will also be reported in this chapter.

7.1 INTRODUCTION TO DESIGN AND COMPENSATION

The design of a vast majority of industrial control systems is not based on some mathematical synthesis procedure, but is based on the knowledge of the control configurations that have been devised and practiced—since this storehouse is what one draws on when conceiving a design for a new application. The first, and probably the most important, feedback control configuration in this storehouse is the basic feedback control structure of Fig. 3.1. This structure has been redrawn as Fig. 7.1 in an abridged form (refer to Figs 2.33–2.36).

When we speak of a controller $D(s)$, our concern is mainly with the *information handling devices*; the sources of information are the reference input $r(t)$, which is derived from the command $y_r(t)$, and the feedback signal produced from the controlled output $y(t)$. The *power handling devices*—actuators and the controlled process itself—are represented by $G(s)$ in the block diagram of Fig. 7.1. $N(s)$ is the plant transfer function from the disturbance input $w(t)$ to the output $y(t)$.

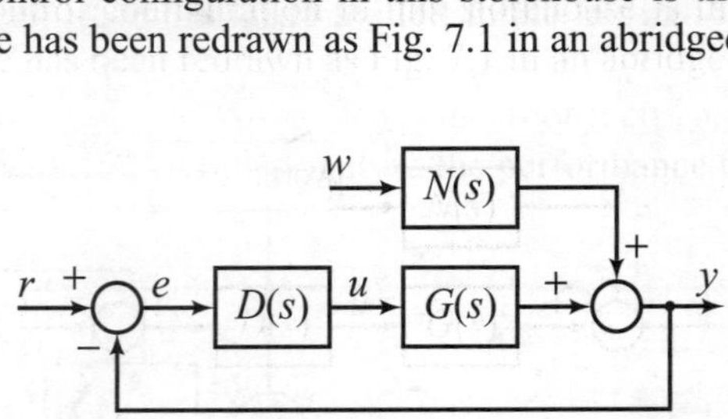

Fig. 7.1 *Basic feedback control configuration*

The design procedures developed in this book are based on shaping the three aspects of a system's response discussed at length in Chapters 5 and 6, namely, stability, speed of response, and steady-state accuracy. For a stable closed-loop system, all the closed-loop poles must strictly lie in the left half of the s-plane. Since stability is the prime objective of design, we always ensure that our controller $D(s)$ stabilizes the system by relocating the poles such that they all lie in the left half of the s-plane. Once we have ensured closed-loop stability, we then refine the controller to achieve the desired performance.

The performance is often specified in terms of figures of merit, such as peak overshoot, settling time, steady-state error constant, . . . , which we have already met; and bandwidth, phase margin, gain margin, . . . , which we will meet later in this book. One of the ideas which has been emerging from the previous chapters is that of turning the specifications into a target closed-loop characteristic polynomial and then selecting the controller $D(s)$ to match this polynomial. An appropriate choice of a target closed-loop polynomial is to use a second-order pole pair in the quadratic factor $(s^2 + 2\zeta\omega_n s + \omega_n^2)$ and all other real/complex poles faster than this quadratic pair, so that the response of the higher-order system is dominated by the second-order response characteristic. Overshoot, settling time or other performance indices can be used in this step.

At several places in the preceding chapters, we have used gain setting to improve the performance of a feedback system. Merely by gain setting, it may be possible to meet the given specifications of simple control systems. However, in most practical cases, the gain adjustment does not provide the desired result. As is usually the case, increasing the gain reduces the steady-state error but results in oscillatory transient response or even instability. Under such circumstances, it is possible to introduce appropriate corrective action, integral or derivative, to force the plant to meet the given specifications. The job of integral or derivative actions is to *compensate* for the deficiency in the performance. The subsystem $D(s)$ in the feedback loop (Fig. 7.1) is thus a *compensator*. The transfer function $D(s)$ will, in the sequel, be referred to as a controller transfer function or a compensator transfer function; *both the terms (controller/compensator) will point to D(s) in the feedback loop.*

7.1.1 Compensator Design: A Parametric Search Problem

The compensator $D(s)$ widely used in industry, is basically a three-parameter control function given by Eqns (3.83)/(3.86). As we will see later in the book, control functions/compensators with more than three parameters are also important in industrial practice to handle the situations wherein the three-parameter compensator fails to yield acceptable performance. ***The compensator design problem can thus be viewed as a search problem in n-dimensional parametric space, where n is the number of design parameters of D(s). Points in the search space correspond to different selections of compensators. By choosing different points of the parametric space, we can generate feedback control systems with different performances. A compensator that meets the given performance specifications can be determined by moving in this search space on trial-and-error basis.*** *Classical control theory* provides systematic methods for solving this trial-and-error search problem. In this context, the two most successful practical methods—the *root-locus method* and the *frequency-response method*—will be presented in Chapters 9 and 12, respectively.

In the *minor-loop feedback control configuration* of Fig. 7.2, the trial-and-error parametric search process involves the parameters of $D(s)$ and the minor-loop feedback parameter K. The design aspects of this control configuration will also appear in Chapters 9 and 12.

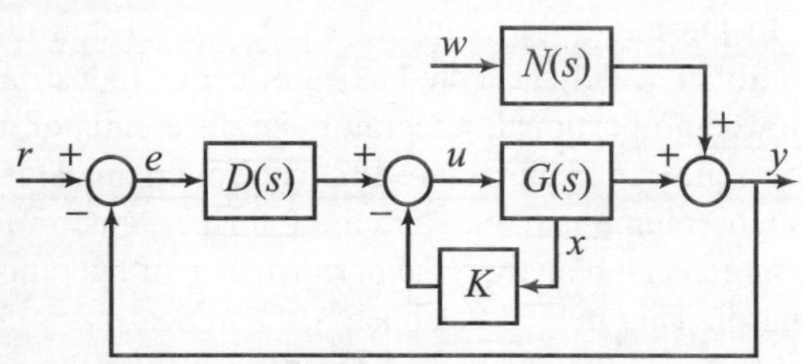

Fig. 7.2 *Minor-loop feedback control configuration*

7.1.2 Alternative Approaches to Control Design

There is no guarantee that a compensator in the feedback control structure of Fig. 7.1/Fig. 7.2 will meet all the system specifications. When the failure occurs after repeated trial-and-error designs, alternatives beyond the search in a preselected search space must be attempted. These include the use of components with improved dynamics and/or totally new system concepts (change from electric drive to hydraulic drive, for example). The changes in control structure of Fig. 7.1/Fig. 7.2 may also yield acceptable performance. For example, one may include feedforward control (in addition to feedback) for improving the performance (refer to Section 1.3).

Another way could be supplying the controller with more complete information about the dynamical status of the process. This equivalently means changing the structure from 'output feedback' to 'state feedback'. In *state feedback controller*, we use internal dynamics of the system as opposed to only output information in the classical structures. These developments are relatively new— whereas classical control dates from 1940 onwards, state variable methods emerged from around 1960 and underwent many refinements before reaching the highly developed level available today. We will take up two design approaches based on state-feedback control—*pole-placement design*, and *linear quadratic optimal control*—in Chapter 15. The strengths and weaknesses of these design approaches will be dealt with in detail.

Though state variable methods provide new fundamental insights about control theory, these methods could not replace the classical approach to design of controllers. In fact, the classical approach is going strong; most of the designs of industrial control systems today are based on this approach. Our major focus in this book is on classical methods of control design. Up to Chapter 13, we will talk about these design methods only.

This chapter concentrates on basic features of the well-known classical compensation[1] schemes. Implementation of the compensators using operational amplifiers is also within the scope of this chapter. From next chapter onwards, we will be concerned with compensator design problem—a trial-and-error search problem in *n*-dimensional parametric space.

7.2 PID CONTROL FUNCTIONS

PID or *three-term controllers* are widely used in industry. The PID name comprises the first letters of the three terms which make up this controller:

P stands for the **P**roportional term in the controller,
I stands for the **I**ntegral term in the controller, and
D stands for the **D**erivative term in the controller.

In the mid-1900s, several technological and analysis concepts came together to make the PID controller widely accepted as an effective industrial controller. The electronic amplifiers of the time allowed proportional, integrator and differentiator units to be made fairly easily with good reliability. The ability to manufacture PID controllers made it the first really successful controller product with a high volume market. Also the Laplace transform formalism had begun to be understood and used by engineers to analyse the performance of operational amplifiers, and hence the performance of PID control. This helped establish a theoretical basis for PID control design. The three-term controller has survived the changes of technology from the analog era into the digital control system age quite satisfactorily.

---

[1] The term 'compensation' is normally used to indicate the process of increasing accuracy and speeding up the response. The term 'design' incompasses the entire process of basic system selection and modifications; compensation is an important aspect of overall system design. We may not strictly follow this distinction while using the terms 'design' and 'compensation'; the context, however, will make the purpose clear.

Figure 7.3a shows the three links for the PID control:

$$U(s) = \left[K_P + \frac{K_I}{s} + K_D s\right]E(s) \tag{7.1a}$$

or

$$u(t) = K_P e(t) + K_I \int_0^t e(\tau)d\tau + K_D \frac{de(t)}{dt} \tag{7.1b}$$

where K_P is the *proportional gain*, K_I is the *integral gain*, K_D is the *derivative gain*; the controller operates on the error signal $e(t)$ to provide control signal $u(t)$.

The time-constant form of PID control (Fig. 7.3b) is as follows:

$$u(t) = K_c\left[e(t) + \frac{1}{T_I}\int_0^t e(\tau)d\tau + T_D \frac{de(t)}{dt}\right] \tag{7.2a}$$

or

$$U(s) = K_c\left[1 + \frac{1}{T_I s} + T_D s\right]E(s) \tag{7.2b}$$

where K_c is the *controller gain*, T_I is the *integral* or *reset time*, and T_D is the *derivative* or *rate time*.

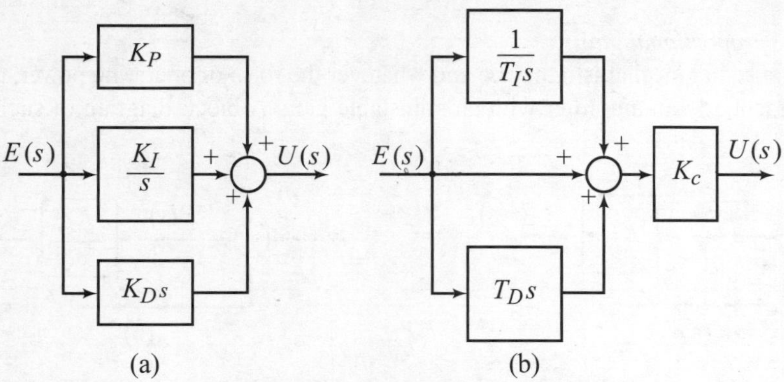

Fig. 7.3 *PID controller blocks*

If we look at real control applications, we find that the feedback signal is the sum of measured output (controlled) signal and a measurement noise component. The error signal $e(t)$ thus contains a corrupting high-frequency noise signal. This noise component has important implications for the use of the derivative term in PID controllers, as is seen below.

$$D(s) = \frac{U(s)}{E(s)} = K_P + \frac{K_I}{s} + K_D s$$

$$D(j\omega) = K_P + \frac{K_I}{j\omega} + K_D \times j\omega \tag{7.3}$$

The derivative term produces amplification of high-frequency signals. The consequence of the possibility of measurement noise being present within the feedback loop is that we do not, in practical

applications, apply the derivative directly to the measured output of the process. Instead, we introduce a low-pass filter in the* D*-term. A low-pass filter has the effect of attenuating high-frequency signals.

We incorporate a low-pass filter of the form $1/(\tau_D s + 1)$ into the derivative term as follows:

$$D(s) = K_c\left[1 + \frac{1}{T_I s} + \frac{T_D}{\tau_D s + 1}\right] \tag{7.4}$$

This introduces one additional design parameter τ_D. In some of the process control applications, the design procedure is carried out for three parameters: K_c, T_I and T_D; with the filter term $\tau_D = T_D/N$, and the value of N is usually selected to be in the range $5 \le N \le 20$. This amounts to handling the control design and filter design problems independently.

7.2.1 P/I/PI/PD/PID Functions

The family of PID control functions is created by various combinations of proportional, integral and derivative terms as required to meet performance specifications. The commonly used members of PID family are P, I, PI, PD, and PID.

For a controller with only *proportional* (P) term, the relationship between the control variable $u(t)$ and the error signal $e(t)$ is

$$u(t) = K_P e(t) \tag{7.5a}$$

or

$$U(s) = K_P E(s) \tag{7.5b}$$

where K_P is the *proportional gain.*

Whatever the actual mechanism may be and whatever the form of operating power, the proportional controller is essentially an amplifier with an adjustable gain. A block diagram of such a controller is shown in Fig. 7.4a.

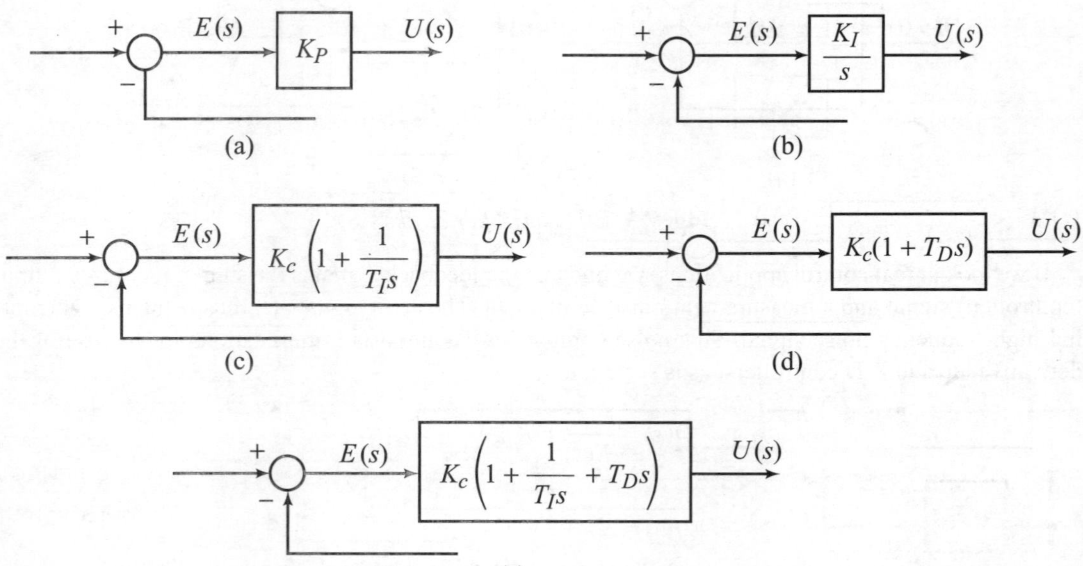

Fig. 7.4 *(a) Proportional control (b) Integral control (c) PI control (d) PD control (e) PID control*

In the *integral* (*I*) control mode, also called the *reset control*, the change in the control signal is proportional to the integral of the error:

$$u(t) = K_I \int_0^t e(t)dt \tag{7.6a}$$

or

$$U(s) = \frac{K_I}{s} E(s) \tag{7.6b}$$

where K_I is the *integral gain*.

Figure 7.4b shows the block diagram of such a controller.

Integral control can be used by itself or in combination with other control modes; indeed *proportional plus integral* (PI) control is the most common mode. The control action of a PI controller is defined by the following equations:

$$U(s) = K_P\, E(s) + \frac{K_I}{s} E(s) = K_c\left(1 + \frac{1}{T_I s}\right) E(s) \tag{7.7a}$$

$$u(t) = K_c\left[e(t) + \frac{1}{T_I}\int_0^t e(t)dt\right] \tag{7.7b}$$

where T_I is the *integral* or *reset time*, and K_c is the *controller gain*.

Figure 7.4c shows the block diagram of a PI controller.

The *derivative* (*D*) control mode, also called the *rate control*, is described by the following equations:

$$u(t) = K_D \frac{d}{dt} e(t) \tag{7.8a}$$

or

$$U(s) = K_D s E(s) \tag{7.8b}$$

where K_D is the *derivative gain*.

Derivative control mode is normally not used alone because the control produces no corrective effort for any constant error, and would, therefore, allow uncontrolled steady-state errors. A combination of proportional and derivative controls is a practical mode of control for industrial processes. The control action of a *proportional plus derivative* (PD) controller is defined by the following equations:

$$U(s) = K_P\, E(s) + K_D\, sE(s) = K_c(1 + T_D s)E(s) \tag{7.9a}$$

$$u(t) = K_c\left[e(t) + T_D \frac{de(t)}{dt}\right] \tag{7.9b}$$

where T_D is the *derivative* or *rate time*.

Figure 7.4d shows the block diagram of a PD controller.

A combination of the three basic modes—proportional, integral and derivative—results in a *proportional plus integral plus derivative* (PID) controller described by the following equations:

$$U(s) = K_c\left(1 + \frac{1}{T_I s} + T_D s\right) E(s) \tag{7.10a}$$

$$u(t) = K_c\left[e(t) + \frac{1}{T_I}\int_0^t e(t)dt + T_D \frac{de(t)}{dt}\right] \tag{7.10b}$$

Figure 7.4e shows the block diagram of such a controller.

In the block diagrams of Fig. 7.4, the controller is in the forward path and therefore the control actions are based on the error signal. This, however, is not a restriction. The control actions may be taken on controlled output variable; many a times to advantage, as we shall see in this chapter. The resulting *multiloop* (loop within a loop) *control configuration* has been defined earlier as *minor-loop feedback control* (Fig. 7.2).

7.3 CHARACTERISTICS OF PROPORTIONAL MODE OF CONTROL

We use a temperature control system to develop an understanding of proportional control characteristics.

The system shown in Fig. 7.5 is designed to maintain the chamber temperature θ at desired value θ_r in the face of disturbances in outdoor temperature. We will assume an initial equilibrium operating point (all variables steady) at which steam inflow $\bar{Q}_m$ kg/sec exactly matches the heat loss to the surroundings with chamber temperature $\bar{\theta}$ = set-point value $\bar{\theta}_r$. The necessary steam flow rate $\bar{Q}_m$ can be achieved because the electropneumatic transducer has a zero adjustment which allows pressure $\bar{P}$ to be set at the middle (9 psig) of its 3 to 15 psig range when $e_A = 0$, and the valve positioner has a zero adjustment which allows the valve opening to be put anywhere desired, with the pressure $\bar{P}$ at 9 psig. With these initial 'trimming' adjustments made, all the variables named in Fig. 7.6 are considered as small perturbations away from the initial steady state.

The thermocouple output is $e_t = K_t\theta$. This simple model is justifiable if the dynamics of the sensor are negligible relative to the process time constant τ_p. The electronic amplifier is obviously fast enough to be treated as zero-order system. The dynamics of the electropneumatic transducer and valve positioner

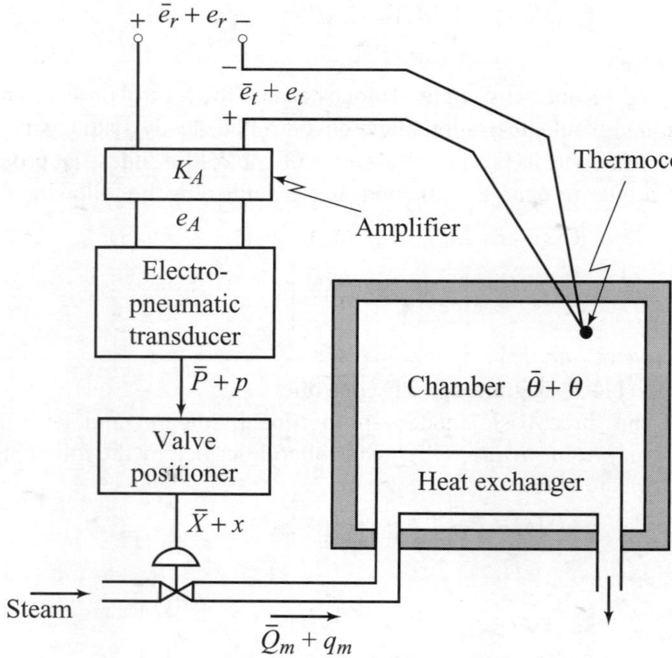

Fig. 7.5 *Heating system with proportional control*

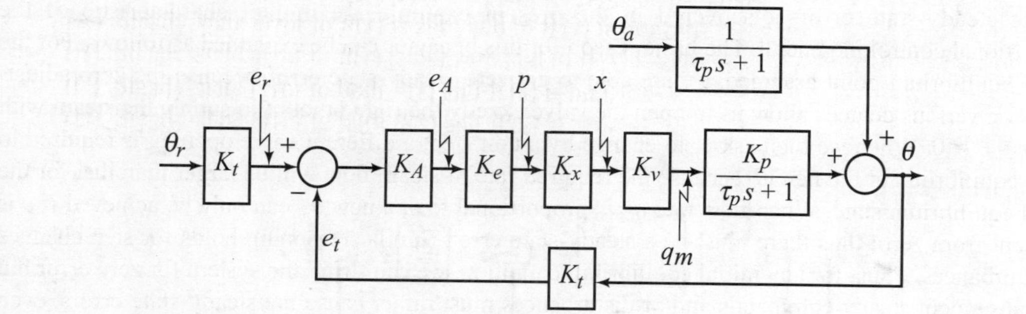

Fig. 7.6 *A block diagram for the system of Fig. 7.5*

have been neglected in the block diagram of Fig. 7.6, assuming their response to be very fast relative to τ_p. Also the relation between q_m and x is assumed statically linear and dynamically instantaneous. The process model itself has been approximated as a first-order model (refer to Section 3.7.2).

Having justified the block diagram of Fig. 7.6, we now use it directly to obtain the transfer function of the closed-loop system, relating θ to θ_r and θ_a:

$$[(K_t\theta_r(s) - K_t\theta(s))K_AK_eK_xK_v]\frac{K_p}{\tau_p s+1} + \left(\frac{1}{\tau_p s+1}\right)\theta_a(s) = \theta(s)$$

or

$$(\tau_p s + 1)\theta(s) + K_tK_AK_eK_xK_vK_p\theta(s) = K_tK_AK_eK_xK_vK_p\theta_r(s) + \theta_a(s)$$

or

$$(\tau s+1)\theta(s) = \frac{K}{1+K}\theta_r(s) + \frac{1}{1+K}\theta_a(s) \tag{7.11}$$

where $\tau \triangleq$ closed-loop system time constant $= \dfrac{\tau_p}{1+K}$;

$K \triangleq$ system loop gain $= K_tK_AK_eK_xK_vK_p$

For the linear model of Eqn. (7.11), superposition holds and we can investigate the effects of command θ_r and disturbance θ_a separately. For θ_r, a step input of size unity (perturbation θ_a is held at zero), we get

$$\theta(t) = \frac{K}{1+K}(1-e^{-t/\tau}) \tag{7.12}$$

Figure 7.7 shows the first-order system response with speed determined by τ and the steady-state error given by

$$e_{ss} = 1 - \theta_{ss} = \frac{100}{K+1}\%$$

Since $\tau = \tau_p/(K + 1)$, we see that ***both the speed of response and the steady-state error are improved if we increase loop gain*** K. For step disturbance input $\theta_a(s) = 1/s$ ($\theta_r = 0$), we get

$$\theta(t) = \frac{1}{(K+1)}(1-e^{-t/\tau}) \tag{7.13}$$

and again increasing K improves both the speed of response and the steady-state error.

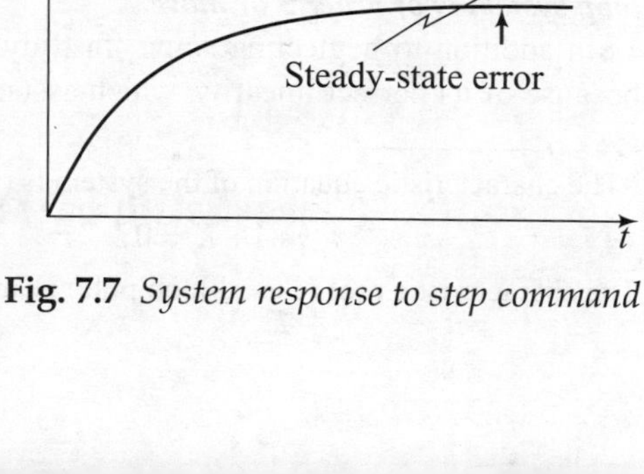

Fig. 7.7 *System response to step command*

The steady-state errors discovered in the present example are in fact characteristic of the proportional control in general. The basic reason for this behavior can be explained as follows. For the initial equilibrium point assumed, we are able to get zero steady-state error because the zero adjustments on various devices allow us to open the valve exactly the right amount to supply the steam with $e = \theta_r - \theta = 0$. When we then ask θ_r to change by a unit step, a different valve opening is required to reach equilibrium at the new θ_r because the required rate of steam flow will be larger than that for the initial equilibrium state. When flow rate q_m is proportional to e, a new q_m can only be achieved if e is different from zero; thus there must be a steady-state error. Similar reasoning holds for step changes in disturbances. Thus, for any initial equilibrium condition, we can 'trim' the system for zero error but any subsequent steady commands and/or disturbances must rather cause the steady-state errors, even when proportional control is at work.

The steady-state error due to step commands can theoretically be eliminated from proportional control systems by intentionally 'misadjusting' θ_r. For example, for a step input θ_r of size $(1 + K)/K$ with $\theta_a = 0$, we get from Eqn. (7.11):

$$\theta(t) = 1 - e^{-t/\tau};\ \theta_{ss} = \text{steady-state value of } \theta(t) = 1$$

Thus, if we want θ to change by a unit step, we ask for a step input of size $(1 + K)/K$.

This trick does in fact work, but depends upon our knowledge about K. If K is not accurately known or varies, then the trick is not 100% effective. Also, the trick does not help in reducing steady-state errors caused by disturbances. In the next section, we will see that the *integral mode* of feedback control removes steady-state errors due to both commands and disturbances and does not depend on the knowledge of specific numerical values for system parameters.

Equations (7.11)–(7.13) indicate that the following aspects of systems behavior are improved by increasing the loop gain:

(i) steady-state tracking accuracy;
(ii) disturbance signal rejection; and
(iii) relative stability, i.e., rate of decay of the transients.

In addition, it can easily be established that increase in loop gain decreases the sensitivity of the system to parameter variations. Thus, all aspects of system behavior are improved by high-gain proportional control. This is, however, true up to a point but cannot be carried to extremes because the system will go unstable. Note that the model of Fig. 7.6 gives no warning of instability[2]. This defect in our model is caused by our neglect of dynamics in some of the system components. We originally neglected these dynamics relative to τ_p, but in the closed-loop system, response speed is determined by τ and not τ_p. Equation (7.11) shows that τ decreases as K is increased and, therefore, at some point the initially neglected dynamics are no longer negligible. Thus, the model of Fig. 7.6 correctly predicts system behavior, as long as loop gain is not made 'too large'; it is not valid for stability predictions. If we want to make valid stability predictions, we must include enough dynamics in our system model. We have learnt in Chapter 5 that ***instability will occur for large values of loop gain when the closed-loop system is of order 3 or more.***

In addition to neglecting some 'fast' dynamics, the model of Fig. 7.6 further differs from reality because of its perfect linearity, which assumes no limit on the magnitude of corrective control action

[2] The characteristic equation of the system is (refer to Eqn. 7.11)

$$\tau_p s + 1 + K = 0$$

The characteristic root (closed-loop pole) is always in the left half of complex plane.

that can be applied by the system. Actually the amplifier, the electropneumatic transducer, and valve positioner all exhibit saturation, limiting their output when input becomes too large. In fact, all aspects of closed-loop system behavior suffer from saturation.

Example 7.1 We consider here a heating control system with the process model [63]

$$Y(s) = G(s)U(s) = \frac{0.3}{2s+1}U(s)$$

The input signal $U(s)$ is the power in kW from the heater and the output signal $Y(s)$ is the resulting temperature. The time constant of the process is 2 hrs. The sensor has a very short time constant in comparison to the process time constant; so we can model the system as a unity feedback system (Fig. 7.8). We let the step reference signal be $R(s) = 1/s$, and let the constant load disturbance effect be modeled as $W(s) = 0.5/s$. System dynamics is governed by the equation

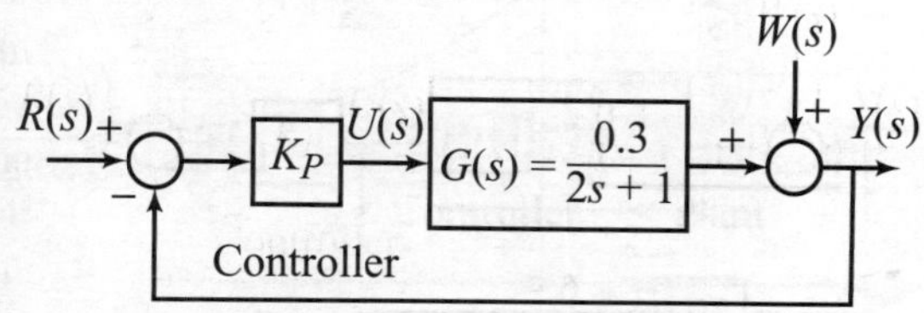

Fig. 7.8 *Proportional control of a system*

$$Y(s) = \frac{0.3K_P}{2s+1+0.3K_P}R(s) + \frac{2s+1}{2s+1+0.3K_P}W(s) \quad (7.14\text{a})$$

$$= M(s)R(s) + M_w(s)W(s) \quad (7.14\text{b})$$

Figure 7.9 shows the effects of increasing the values of K_P. We see that increasing proportional gain causes a faster closed-loop system response. The other effect of increasing the size of proportional gain is that the offset from the desired output level is reduced.

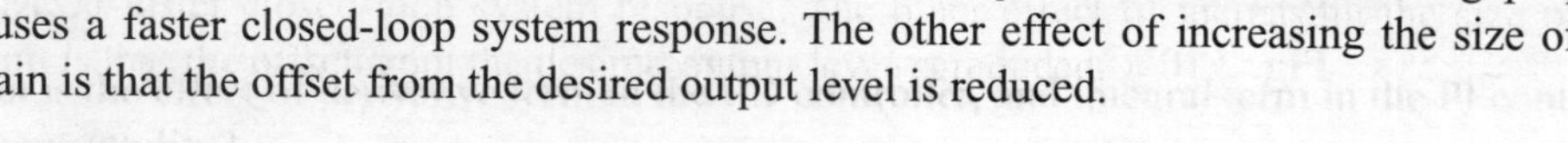

Fig. 7.9 *Unit-step responses for increasing proportional gain*

Figure 7.10 is a plot of control signal $u(t)$ for increasing values of K_P. The control signal becomes increasingly large as the proportional gain increases.

Figure 7.11 shows the disturbance rejection response curves for increasing proportional gain K_P. We deduce from this figure that increasing the proportional gain reduces the steady-state offset from zero.

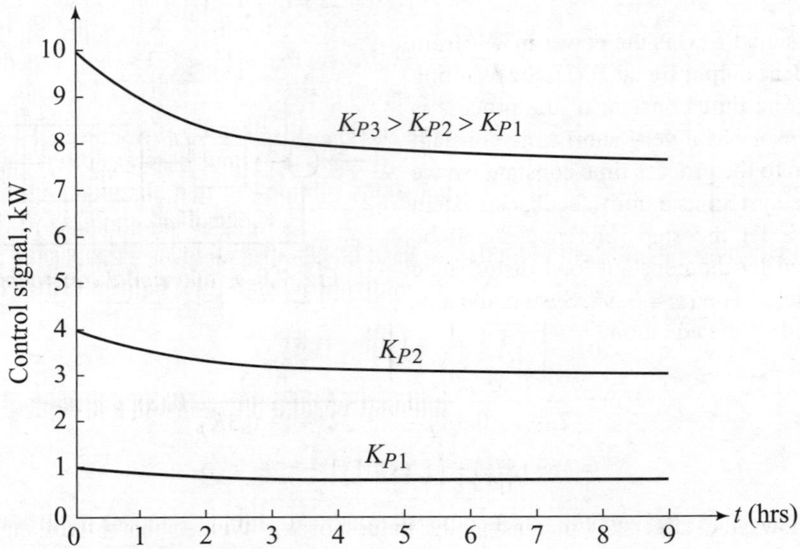

Fig. 7.10 *Control signals for increasing proportional gain*

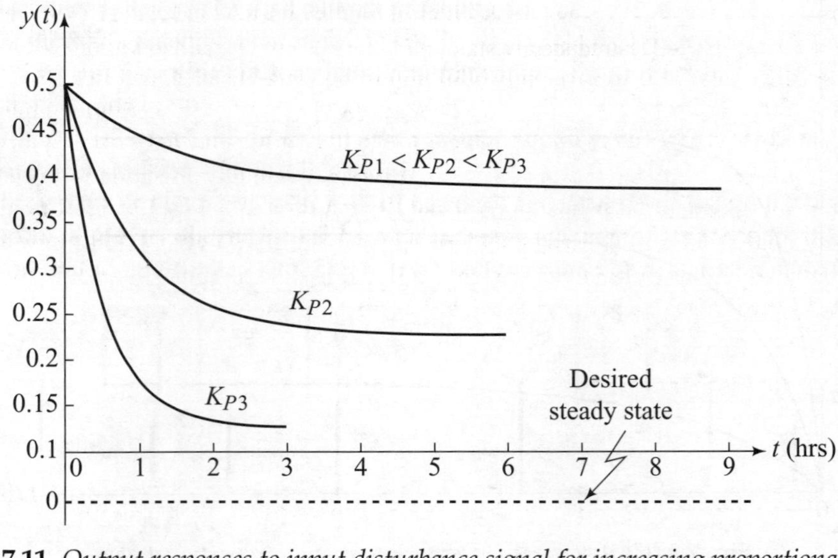

Fig. 7.11 *Output responses to input disturbance signal for increasing proportional gain*

Three observations can be made:

1. ***Increasing the size of the proportional gain K_P has two effects on the system output. The speed of response to both the reference input and the disturbance is increased, and the offset from the desired output due to both the reference input and the disturbance is reduced. The offset can only be completely removed by allowing the proportional gain to become infinitely large.***
2. ***Increasing the proportional gain to achieve the desired effects of faster output response and significantly reduced steady-state offset causes the controller signal $u(t)$ to become increasingly large. This means that the demanded output from the actuator could be too large, possibly reaching physical system limits. Thus, too large a proportional gain may not be able to achieve the theoretical improvement because the hardware of the actuator cannot actually deliver the required control signals.***
3. ***For all values of proportional gain $K_P > 0$, the closed-loop pole of $M(s)$ (or of $M_w(s)$), is in the left half of s-plane (refer to Eqns (7.14)) and the closed-loop system is stable. This analysis is, of course, true if the first-order model represents the dynamics of the feedback loop. If the time constants of sensor, amplifier, actuator, etc., which have been neglected relative to the process time constant τ, become significant at high loop gain, the order of the system would go high and the system would run the risk of instability for high values of K_P.***

7.4 CHARACTERISTICS OF INTEGRAL MODE OF CONTROL

Consider the closed-loop system of Fig. 7.12a. Assume that the initial 'trimming' (zero adjustments) of the control system has been done to obtain $y = r$, with $e = u = 0$. In response to a change in the set-point, the controller must change the flow of energy and/or material to the process in order to achieve equilibrium at the new operating conditions. If the controller is based on proportional law, then in the new steady state, a nonzero error e_{ss} must exist to get a nonzero value of control signal u_{ss} (Fig. 7.12b). The new set-point is, say, 10. It will result in a response y_{ss} of less value, say 9.9. The steady-state error is 0.1 and the steady value of the control signal is $0.1K_P$ where K_P is proportional gain. The operator must then 'reset' the setpoint to bring the output to the desired value of 10.

We need a controller that automatically brings the output to 10 with a setpoint of 10, i.e., it gives a steady control signal of value $0.1K_P$ with system error $e_{ss} = 0$. An integrator is capable of exactly this type of behavior (Fig. 7.12c). Integral control is thus a means of removing steady-state errors without the need for manual reset and, in fact, it is sometimes called *automatic reset*.

Although integral control is very useful for removing or reducing steady-state errors, it has the undesirable side-effects of degrading stability. Conversion of the proportional control system of Fig. 7.6 to integral control will bring out some general results about integral mode. Suppose we replace the amplifier block K_A by K_I/s (refer to Eqns (7.6)). The closed-loop system is then described by the equation

$$\left[(K_t\theta_r(s) - K_t\theta(s))\frac{K_I}{s}K_eK_xK_v\right]\frac{K_p}{\tau_p s+1} + \frac{1}{\tau_p s+1}\theta_a(s) = \theta(s)$$

or
$$(\tau_p s+1)\theta(s) + \frac{K_tK_IK_eK_xK_vK_p}{s}\theta(s) = \frac{K_tK_IK_eK_xK_vK_p}{s}\theta_r(s) + \theta_a(s) \tag{7.15}$$

or
$$(\tau_p s^2 + s + K)\theta(s) = K\theta_r(s) + s\theta_a(s)$$

where
$$K \triangleq \text{loop gain} = K_tK_IK_eK_xK_vK_p$$

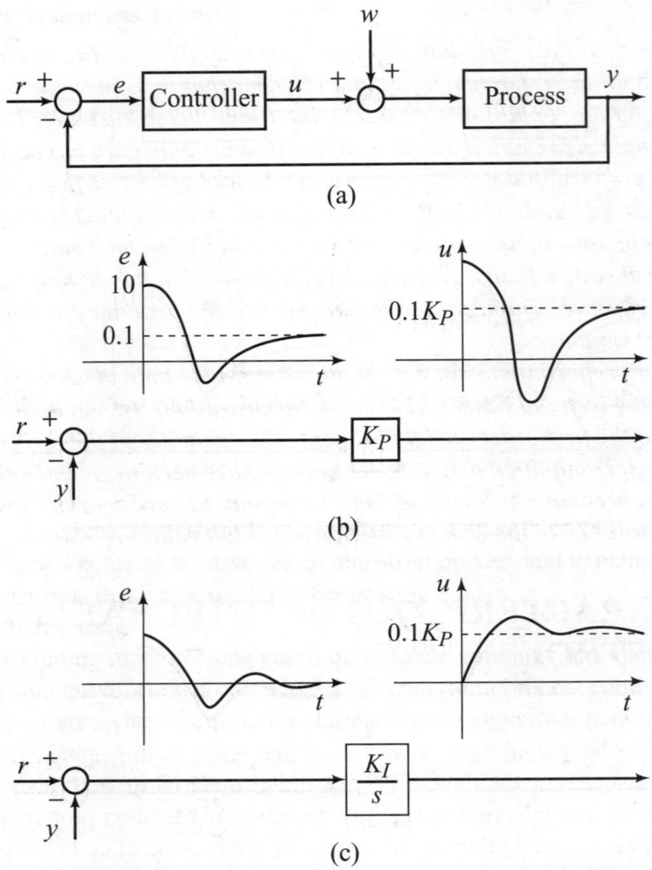

Fig. 7.12 *Comparison of proportional and integral control (graphs not to scale)*

For θ_r, step input of size unity (perturbation θ_a held at zero), we get

$$\theta(s) = \left(\frac{K}{\tau_p s^2 + s + K}\right)\frac{1}{s}$$

By final-value theorem (refer to Eqn. (2.42)), the steady value of the output is unity, and therefore steady-state error is zero. For step disturbance input $\theta_a(s) = 1/s$ ($\theta_r = 0$), we get

$$\theta(s) = \left(\frac{s}{\tau_p s^2 + s + K}\right)\frac{1}{s};\ \theta_{ss} = \text{steady-state value of output} = 0$$

Thus, step changes of θ_r and θ_a give zero steady-state errors regardless of the value of loop gain K. The steady-state performance of the system with constant inputs is thus insensitive to parameter variations.

We need to consider what integral control does to the dynamic response. For this, we need to look at the characteristic equation corresponding to Eqn. (7.15), which is

$$\tau_p s^2 + s + K = 0$$

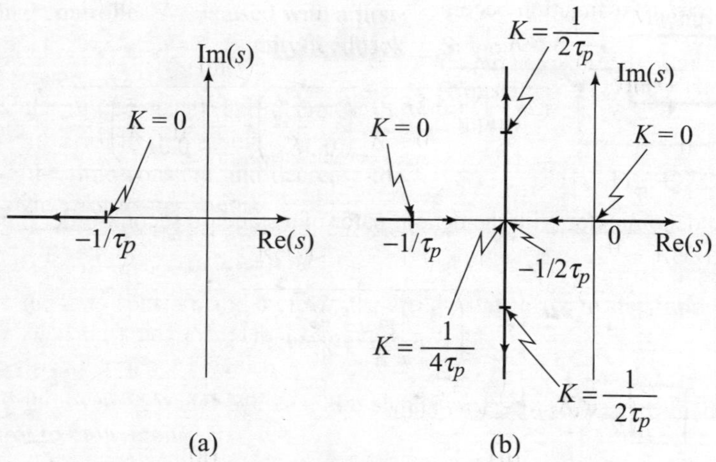

Fig. 7.13 *Destabilizing effect of integral control*

The roots of this equation, which are readily computed, vary as gain K is varied. In fact, it is informative to compute these roots and plot them as a function of K. By quadratic formula of algebra, the roots are

$$s = \frac{-1 \pm \sqrt{1 - 4K\tau_p}}{2\tau_p}$$

and are plotted in Fig. 7.13b for changing values of the gain K. When the gain is zero, the roots are at $-1/\tau_p$ and at 0. As the gain is raised, the roots move closer until there is a double root at $-1/2\tau_p$, and then for higher gains, the roots become complex conjugate.

Recall the case of proportional control; from Eqn. (7.11), we obtain the following characteristic equation:

$$\frac{\tau_p}{1+K}s + 1 = 0$$

The root of this equation, $s = -\dfrac{1+K}{\tau_p}$, moves increasingly to the left in the complex plane for increasing values of K (Fig. 7.13a) which implies a decreasing time constant for the closed-loop system; the larger the loop gain, faster is the rate of decay of transients. With integral control, the rate of decay of transients is dictated by the real root closer to the imaginary axis or real part of the pair of complex conjugate roots. There is obviously a limit on the speed of response; also the response will be oscillatory for large values of K rather than exponential as with proportional control. Due to our too simple dynamic modeling, instability for large values of loop gain is not predicted.

Example 7.2 Reconsider the heating control system that was studied in Example 7.1 for proportional control [63]. We replace the proportional controller K_P by integral controller K_I/s.

The closed-loop transfer function becomes

$$M(s) = \frac{0.3K_I}{s(2s+1)+0.3K_I} = \frac{1}{\left(\frac{2}{0.3K_I}\right)s^2 + \left(\frac{1}{0.3K_I}\right)s + 1}$$

If we associate the denominator with the denominator of a standard second-order transfer function:

$$\frac{1}{\left(\frac{1}{\omega_n^2}\right)s^2 + \left(\frac{2\zeta}{\omega_n}\right)s + 1}$$

and equate coefficients of powers of s, we obtain

$$\frac{1}{\omega_n^2} = \frac{2}{0.3K_I}; \frac{2\zeta}{\omega_n} = \frac{1}{0.3K_I}$$

giving

$$\omega_n = \sqrt{\frac{0.3K_I}{2}}; \zeta = \frac{1}{2\sqrt{0.6K_I}}$$

Letting $K_I = 1$, gives $\omega_n = 0.39$ rad/sec and $\zeta = 0.65$. Figure 7.14 shows closed-loop system response for different values of integral gain K_I. Disturbance rejection response curves are shown in Fig. 7.15. The largest peak occurs for the case with the largest value of K_I, i.e., the most underdamped pole locations. Also note that there is no offset in desired steady state.

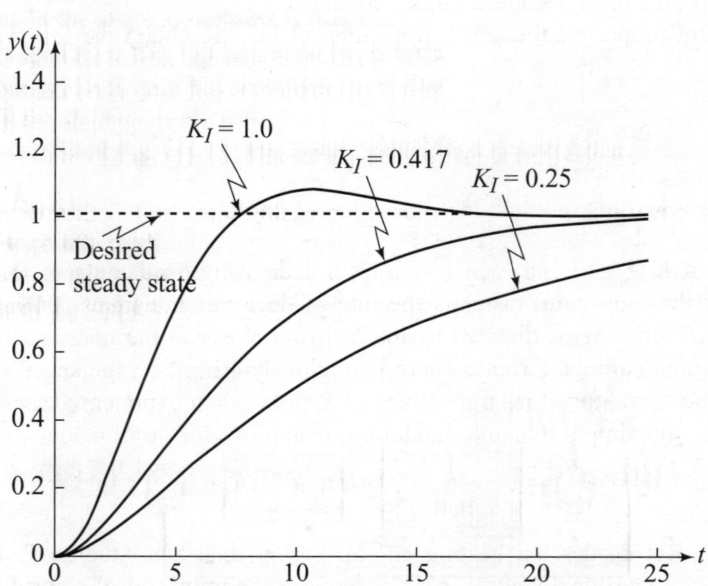

Fig. 7.14 *Outputs of closed-loop system for different values of integral gain*

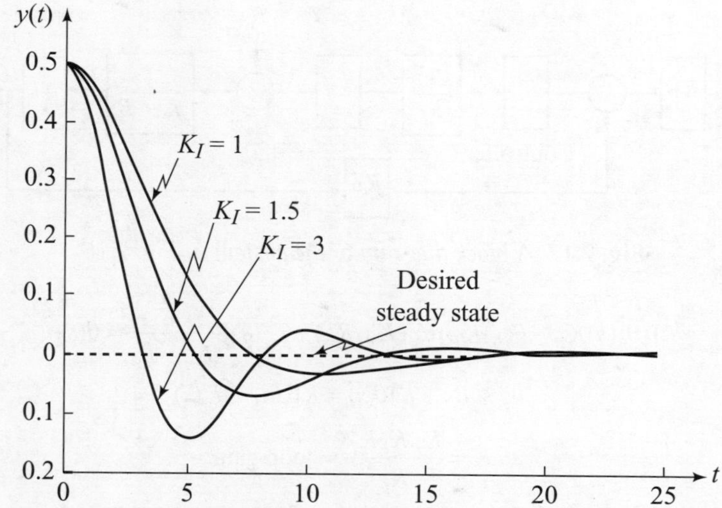

Fig. 7.15 *System output responses with different values of integral gain for disturbance input signal*

Example 7.3

Figure 7.16 shows an electromechanical servo wherein motor torque T_M controls position θ of inertia J which is subject to torque disturbance T_w. The field time-constant $\tau_f = L_f/R_f$ is negligibly small; K_T is the motor torque constant and K_P is the potentiometer sensitivity.

Notice that integrating effect naturally appears in the system plant. Let us examine whether we need integral control for such a plant. The block diagram in Fig. 7.17 shows the integrating plant with proportional control. The following equation easily follows from this block diagram:

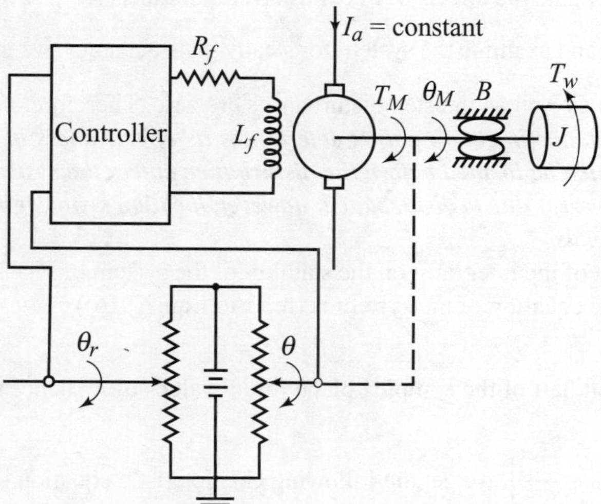

Fig. 7.16 *An electromechanical servo*

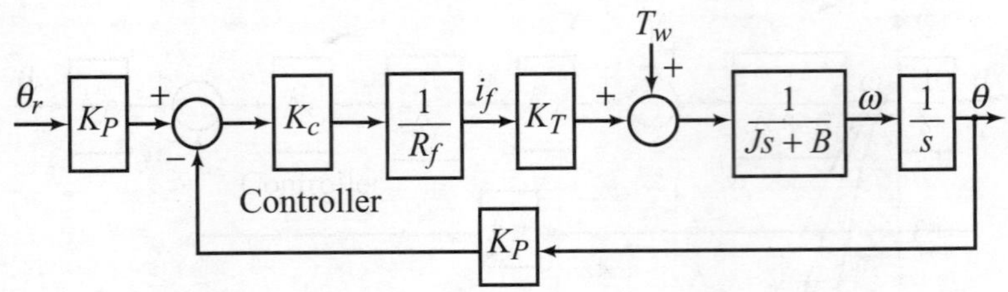

Fig. 7.17 *A block diagram of the system in Fig. 7.16*

$$[(\theta_r(s)K_P - \theta(s)K_P)K_c(K_T/R_f) + T_w(s)]\frac{1}{Js^2+Bs} = \theta(s)$$

or
$$(Js^2 + Bs + K)\theta(s) = K\theta_r(s) + T_w(s) \tag{7.16}$$

where
$$K \triangleq \frac{K_P K_c K_T}{R_f} = \text{loop gain}$$

For a unit-step command input with $T_w = 0$, we have $\theta(s) = \left(\frac{K}{Js^2+Bs+K}\right)\frac{1}{s}$

By final-value theorem, the steady-state value of the output is given by

$$\theta_{ss} = \lim_{s\to 0} s\left(\frac{K}{Js^2+Bs+K}\right)\frac{1}{s} = 1$$

and therefore, the steady-state error, $e_{ss} = \theta_r - \theta_{ss}$, is zero.

For a unit step disturbance with $\theta_r = 0$, we have $\theta(s) = \left(\frac{1}{Js^2+Bs+K}\right)\frac{1}{s}$ and the steady-state value of the output is $\theta_{ss} = \frac{1}{K}$.

The plant integrator, thus, gives zero steady-state error for a step command but not for a step disturbance. Let us investigate the effects of PI control (refer to Eqns (7.7)). If we replace the controller gain K_c by $K_c\left(1+\frac{1}{T_I s}\right)$, and examine the system for steady-state accuracy, we find that the steady-state errors to step commands as well as the step disturbances are zero. Therefore, ***the location of the integrator relative to disturbance injection points determines its effectiveness in removing steady-state errors. The integrator must be located before the disturbance enters the system if it is to be effective in removing steady-state error due to disturbance; however, location is not significant for steady-state errors caused by commands.***

Let us study the effect of the integrator on the stability of the system of Fig. 7.17. With proportional control, the characteristic equation of the system is (refer to Eqn. (7.16))

$$Js^2 + Bs + K = 0$$

which has roots in the left half of the complex plane for all values of K; hence the system is stable for all values of loop gain.

Replacing K_c by $K_c\left(1+\frac{1}{T_I s}\right)$, we get the following characteristic equation:

$$Js^3 + Bs^2 + Ks + K/T_I = 0$$

It is a third-order characteristic equation and, as we have seen in Chapter 5, large values of loop gain K lead to instability.

Although controllers with a single integrator are most common, double (and occasionally triple) integrators are useful for the more difficult steady-state error problems; however, they require careful stability augmentation. ***Conventionally, the number of integrators in the forward path has been called the system type number; type-0 systems have no integrators, type-1 have one, and so forth*** (refer to Chapter 6).

7.5 CHARACTERISTICS OF DERIVATIVE MODE OF CONTROL

Consider the closed-loop system shown in Fig. 7.18. The controller is given the information on system error $e(t)$ and has the task of changing control variable $u(t)$ so as to keep $e(t)$ close to zero. If the control law is a proportional control action, then the value of u at time t_1 will be same as that at t_2. However, a stronger corrective effort seems appropriate at t_1 and a lesser one at t_2, since at t_1 the error is e_{12} and *increasing*, whereas at t_2 it is again e_{12} but *decreasing*. That is, the controller must sense not only the ordinate of the error curve but also its trend or slope. The slope is clearly de/dt; so to mechanize this desirable control function, ***we need a controller sensitive to error derivative. Such a controller can, however, not be used alone since it acts as open circuit for steady errors of any size, as at t_3; thus a combination of proportional plus derivative* (PD)*, for example, makes sense*** (refer to Eqns (7.9)).

The noise-accentuating characteristics of derivative operations may often require use of approximate (low-pass filtered) derivative signals (refer to Eqn. (7.4)). This prevents realization of ideal performance but significant system improvement is still often possible. Here we will explore idealized versions of derivative controls in order to make clear their basic characteristics. Later chapters will develop their practical implementation in a more comprehensive way.

To investigate the use of PD control, we perform an assessment using a process whose transfer function is of second order (As we shall see in later chapters, for the first-order processes of the type shown in Fig. 7.8, usually a proportional plus integral (PI) controller yields an acceptable solution). Figure 7.19 shows our typical block diagram for tracking and disturbance rejection analysis.

The usual block diagram analysis yields

$$Y(s) = \left(\frac{K(K_P + K_D s)}{\tau s^2 + (1 + KK_D)s + KK_P} \right) R(s) + \frac{s(\tau s + 1)}{\tau s^2 + (1 + KK_D)s + KK_P} W(s) \tag{7.17a}$$

$$= M(s)R(s) + M_w(s)W(s) \tag{7.17b}$$

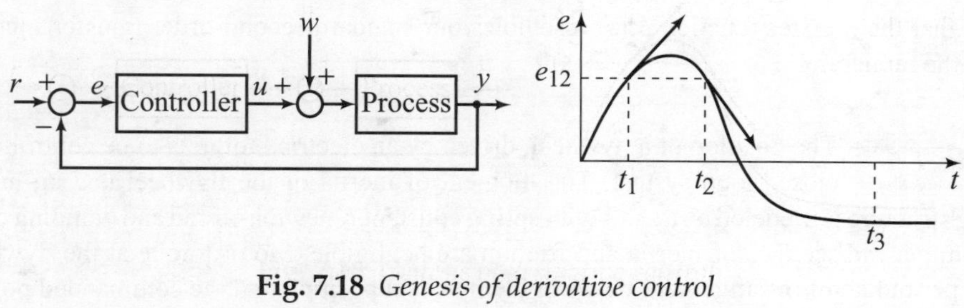

Fig. 7.18 *Genesis of derivative control*

The closed-loop stability of the system will be given by the denominator of $M(s)$ and $M_w(s)$, which is seen to be

$$\tau s^2 + (1 + KK_D)s + KK_P = 0$$

or

$$s^2 + \left(\frac{1 + KK_D}{\tau}\right)s + \frac{KK_P}{\tau} = s^2 + 2\zeta\omega_n s + \omega_n^2 = 0$$

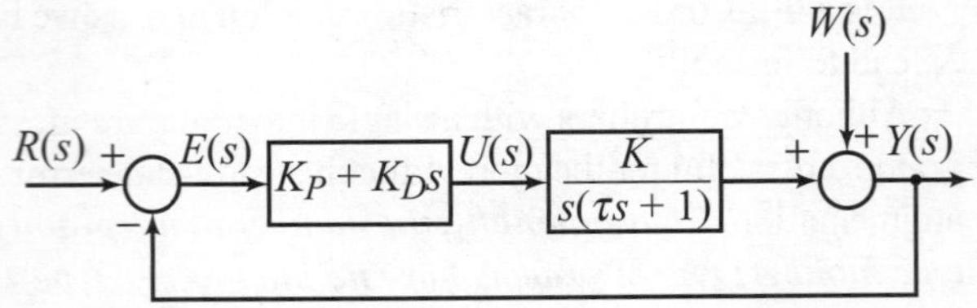

Fig. 7.19 *PD control of a second-order system*

From this we get the design equations

$$\frac{KK_P}{\tau} = \omega_n^2 \text{ and } \frac{1 + KK_D}{\tau} = 2\zeta\omega_n$$

Thus, ***the design utility of* PD *control is that it can be used to meet the closed-loop stability requirements and shape the transient response of the closed loop*.**

If we look at the form of $M(s)$ carefully:

$$M(s) = \frac{K(K_P + K_D s)}{\tau s^2 + (1 + KK_D)s + KK_P}, \tag{7.18}$$

we note that the numerator has a zero and so $M(s)$ is not quite the standard second-order system form, which does not have a zero. In some cases, this zero will restrict the speed of response obtainable by the controller or cause excessive overshoot (refer to Chapter 6).

Consider an alternative PD Control structure shown in Fig. 7.20; the derivative term is now acting on the output signal and not on the error signal.

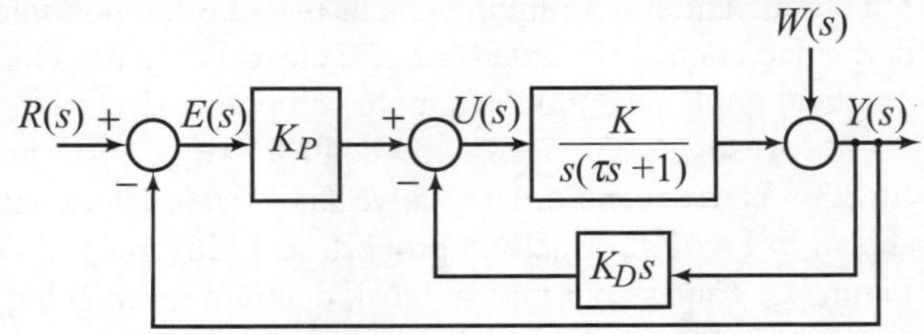

Fig. 7.20 *An alternative PD control structure*

The block diagram analysis gives

$$Y(s) = \left(\frac{KK_P}{\tau s^2 + (1 + KK_D)s + KK_P}\right)R(s) + \left(\frac{s(\tau s + 1)}{\tau s^2 + (1 + KK_D)s + KK_P}\right)W(s) \tag{7.19a}$$

$$= M(s)R(s) + M_w(s)W(s) \tag{7.19b}$$

We see that the transfer function $M(s)$ resembles our standard second-order transfer function with no zero in the numerator.

Example 7.4 The position of a flywheel, driven by an electric motor, is to be controlled using a closed-loop system. The moment of inertia of the flywheel is J kg-m^2, and the connecting structure is modeled by a shaft with spring constant K newton-m/rad and damping coefficient B newton-m per rad/sec. Motor inertia and friction are negligible. Motor torque at the flywheel is T_M newton-m per radian of misalignment between the flywheel position and the commanded position.

Figure 7.21 shows a model of this position control system [63]. The reference signal is a unit step ($R(s) = 1/s$) and we model the constant load disturbance by $W(s) = 0.5/s$. A PD controller with a fixed proportional gain $K_P = 0.2$ is used. The plant has low damping.

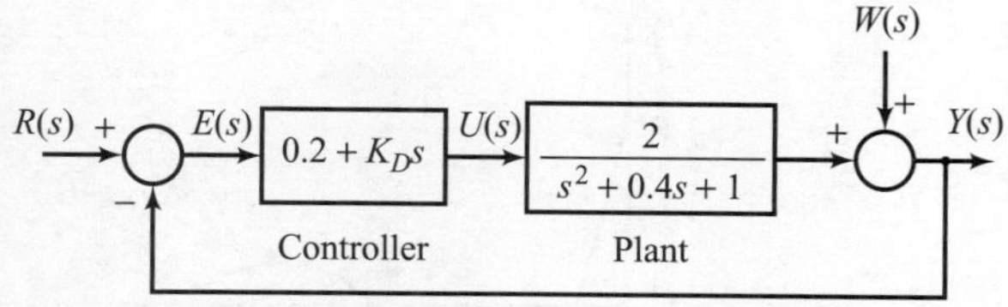

Fig. 7.21 *Closed-loop system with PD control*

$$M(s) = \frac{Y(s)}{R(s)} = \frac{0.4 + 2K_D s}{s^2 + (0.4 + 2K_D)s + 1.4}$$

The characteristic equation is

$$\Delta(s) = s^2 + (0.4 + 2K_D)s + 1.4 = 0$$

We establish a link with the standard form for a second-order system:

$$\Delta(s) = s^2 + 2\zeta\omega_n s + \omega_n^2 = 0$$

Comparing the two characteristic equations, we obtain

$$\omega_n = 1.1832\,;\ \zeta = 0.169 + 0.845K_D;\ K_D > 0$$

We can tune K_D to achieve a desired value for ζ. Output responses for different values of K_D are shown in Fig. 7.22. The steady-state offset is (1– 0.2846) = 0.7154.

Note that steady-state tracking is independent of different values of derivative gain K_D.

We now examine the disturbance rejection performance.

$$\frac{Y(s)}{W(s)} = M_w(s) = \frac{s^2 + 0.4s + 1}{s^2 + (0.4 + 2K_D)s + 0.4}$$

Output responses to a disturbance input $W(s) = 0.5/s$ for different values of K_D are shown in Fig. 7.23. Again we note that steady-state offset is independent of the value of K_D.

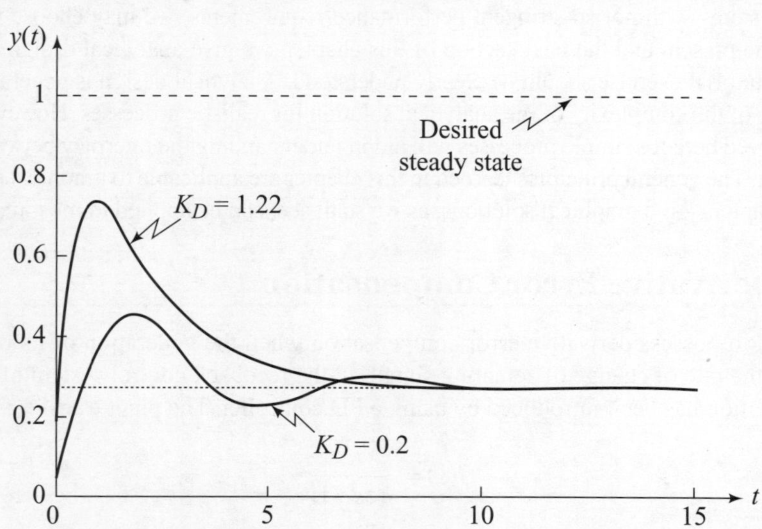

Fig. 7.22 *Output responses to unit-step input*

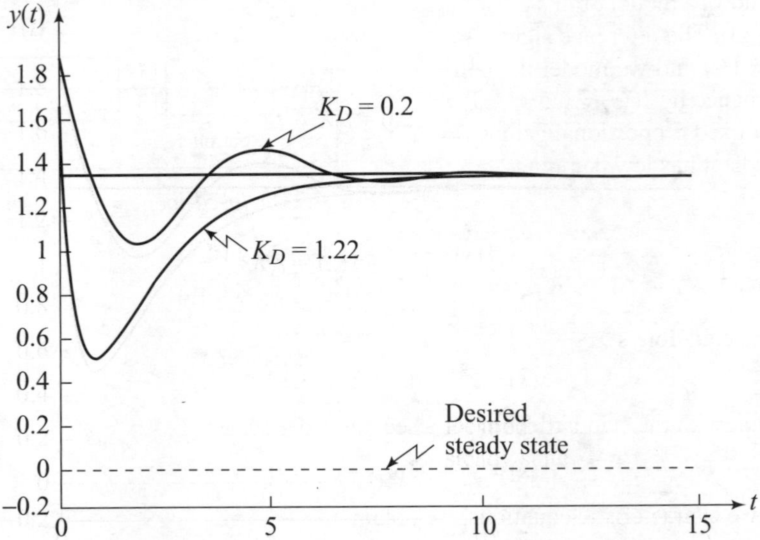

Fig. 7.23 *Output responses to constant disturbance*

7.6 ANALYTICAL DESIGN FOR PD CONTROL

In the previous three sections of this chapter, we learnt about the characteristics of the individual P, I and D terms of the three-term controller, and their effects on the closed-loop system performance. This knowledge will help us select the appropriate combination PI/PD/PID to match the effects we wish the controller to achieve. Trial-and-error parametric search for these controllers to meet the performance specifications will begin from the next chapter. We will employ graphical methods—the root locus plots and the frequency response plots—to carry out the trial-and-error parametric search.

For simple systems with not so stringent performance requirements, we may choose the parameters analytically. In the present and the next section of this chapter, we give analytical design examples. The objective of this tutorial exercise should be clearly understood. Analytical design is generally *not done in practice*, because of the complexity of the analytical solution for realistic processes. However, the analytical solutions derived here for simple processes will aid in understanding the interplay between the process and the controller. The general principles learned in this chapter are applicable to graphical solutions introduced in later chapters—the graphical solutions, as we shall see, can be applied to most realistic systems.

7.6.1 Derivative Error Compensation

A system is said to possess derivative error compensation when the generation of its output depends in some way on the rate of change of actuating signal. In the feedback control system of Fig. 7.24, this type of compensation has been introduced by using a PD controller. The plant transfer function is

$$G(s) = \frac{K}{s(\tau s + 1)} \tag{7.20}$$

The controller transfer function is

$$D(s) = K_P + K_D s \qquad (7.21)$$

The control signal is

$$u(t) = K_P e(t) + K_D \frac{de(t)}{dt} \qquad (7.22)$$

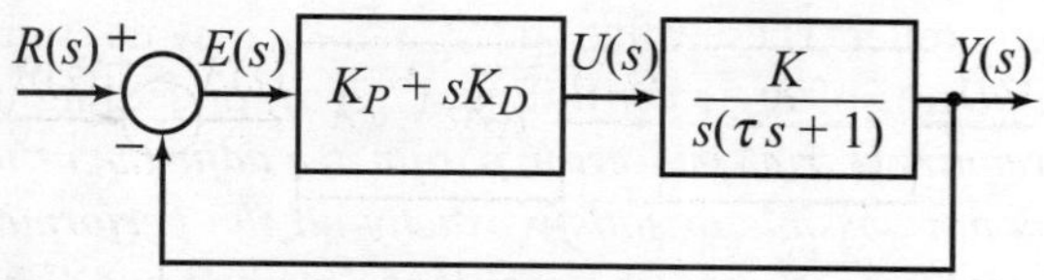

Fig. 7.24 *Derivative error compensation scheme*

The open-loop transfer function of the overall system is

$$D(s)G(s) = \frac{K(K_P + K_D s)}{s(\tau s + 1)} \qquad (7.23)$$

which shows that the PD control is equivalent to adding a simple zero at $s = -K_P/K_D$ to the open-loop transfer function.

The proper selection of performance specifications is the most important step in control system design. The desired behavior is specified in terms of transient response measures and steady-state error. The steady-state error is usually specified in terms of error constants for specific inputs while the transient response is specified by measures of relative stability and speed of response; the relative stability is measured in terms of damping ζ or peak overshoot M_p, while the speed of response is measured in terms of rise time, settling time or natural frequency.

For the system under consideration, velocity error constant

$$K_v = \lim_{s \to 0} sD(s)G(s) = KK_P \qquad (7.24a)$$

Therefore, the steady-state error to velocity input

$$e_{ss} = \frac{1}{K_v} = \frac{1}{KK_P} \qquad (7.24b)$$

By comparison with the standard characteristic equation

$$s^2 + 2\zeta\omega_n s + \omega_n^2 = 0$$

the characteristic equation of the system is

$$\tau s^2 + (1 + KK_D)s + KK_P = 0$$

or

$$s^2 + \left(\frac{1 + KK_D}{\tau}\right)s + \frac{KK_P}{\tau} = 0 \qquad (7.25)$$

The natural frequency and damping of the compensated system are given by

$$\omega_n = \sqrt{\frac{KK_P}{\tau}} \qquad (7.26a)$$

$$\zeta = \frac{1}{2}\left[\frac{1 + KK_D}{\sqrt{KK_P \tau}}\right] \qquad (7.26b)$$

It is obvious from the above equations that only two of the three specifications, i.e., K_v, ζ and ω_n, can be met exactly by a suitable choice of K_P and K_D. For a specified K_v, the proportional gain K_P of the PD controller gets fixed (refer to Eqn. (7.24a)). The closed-loop damping can now be raised to the desired level by a suitable choice of K_D (refer to Eqn. (7.26b)) which is the gain of derivative term of the PD controller. These values of K_P and K_D may not satisfy the requirement on speed of response, specified in terms of ω_n (refer to Eqn. (7.26a)) or settling time $t_s = 4/(\zeta\omega_n)$. ***Practically, most control systems require a trial-and-error parameter adjustment to achieve at least an acceptable performance if it is not possible to satisfy exactly all the performance specifications.*** Two commonly used trial-and-error procedures—the root-locus method and the frequency-response method—will be presented in Chapters 9 and 12.

We should point out here that since Eqn. (7.23) no longer represents a standard second-order system, the transient response is also affected by the zero of the transfer function at $s = -K_P/K_D$. ***In general, if the value of*** K_D ***is very large, the zero will be very close to the origin in the s-plane, the overshoot could be increased substantially, and the damping ratio*** ζ ***is then no longer an accurate estimate on the maximum overshoot of the output.*** Simulation of the compensated system is necessary to check the design. If the transient response is not acceptable, we have to re-enter the design cycle to make necessary changes in design.

Example 7.5 Figure 7.25 shows block diagram of a control system whose objective is to control the position of mechanical load. The two potentiometers having unity sensitivity convert the input and output positions into proportional electrical signals, which are in turn compared, and the error signal amplified by a factor K_A, is applied to the armature circuit of a dc motor whose field winding is excited with a constant voltage. The motor is coupled to the load through a gear train of ratio n. The block diagram also shows a minor feedback loop which corresponds to a tachogenerator connected in the system to improve damping. In this example, we consider proportional control only (tachogenerator disconnected); K_A is the only design parameter. The other control scheme (minor-loop feedback) for this system will be considered in the next example.

When switch 'S' is open (tachogenerator disconnected), the block diagram of Fig. 7.25 reduces to that given in Fig. 7.26.

The position control system of Fig. 7.26 has open-loop transfer function

$$D(s)G(s) = \frac{4500\, K_A}{s(s+361.2)}$$

Let us consider an application wherein static accuracy requirement is very high; steady-state error to unit-ramp input is required to be less than 0.025 deg (0.000436 rad).

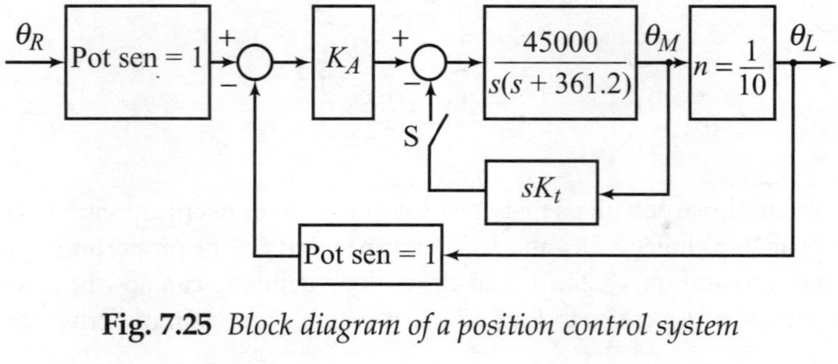

Fig. 7.25 *Block diagram of a position control system*

Required $$K_v = \frac{1}{0.000436} = 2293.6$$

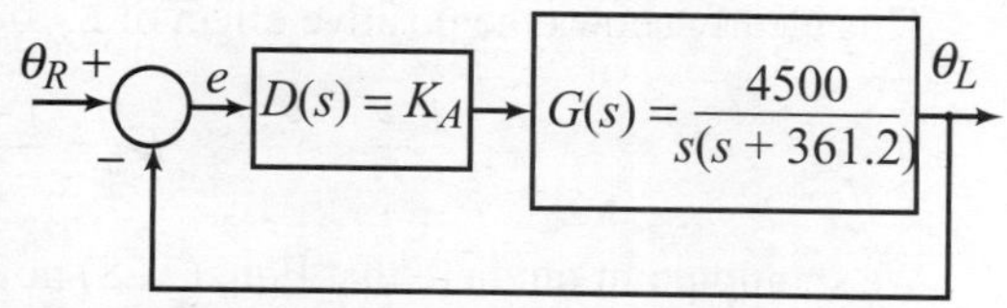

Fig. 7.26 *Simplified block diagram of the system in Fig. 7.25 with $K_t = 0$*

This requirement is satisfied if

$$K_A = \frac{2293.6 \times 361.2}{4500} = 184.1$$

For this value of K_A, the characteristic equation of the system becomes

$$s^2 + 361.2s + 828450 = 0$$

Therefore,

$$\omega_n = \sqrt{828450} = 910.2; \zeta = \frac{361.2}{2 \times 910.2} = 0.198; M_p = e^{-\pi\zeta/\sqrt{1-\zeta^2}} \times 100 = 53\%$$

Relative stability is obviously very poor. To improve damping and peak overshoot while maintaining K_v at 2293.6, we propose the replacement of amplifier with gain K_A by a PD controller (refer to Fig. 7.26)

$$D(s) = K_P + K_D s$$

With the PD controller, the open-loop transfer function becomes

$$D(s)G(s) = \frac{4500(K_P + K_D s)}{s(s+361.2)} \tag{7.27}$$

The closed-loop transfer function

$$\frac{\theta_L(s)}{\theta_R(s)} = \frac{4500(K_P + K_D s)}{s^2 + (361.2 + 4500K_D)s + 4500K_P} \tag{7.28}$$

Velocity error constant

$$K_v = \frac{4500K_P}{361.2} = 2293.6$$

when $$K_P = 184.1$$

For this value of K_P, the characteristic equation of the system becomes

$$s^2 + (361.2 + 4500K_D)s + 828450 = 0$$

Therefore,

$$\omega_n = \sqrt{828450} = 910.2$$

$$\zeta = \frac{361.2 + 4500\,K_D}{2 \times 910.2} = 0.198 + 2.472\,K_D$$

This clearly shows the positive effect of K_D on damping. For critical damping,

$$K_D = \frac{1-0.198}{2.472} = 0.324$$

We should point out here that Eqn. (7.28) no longer represents a standard second-order system; the transient response is also affected by the zero of the transfer function at $s = -K_P/K_D$. In general, if K_D is large, zero will be close to the origin in the s-plane, the overshoot will be increased substantially and damping ratio ζ no longer gives an accurate estimate on the peak overshoot of the output. In the present case, the increase in the overshoot is acceptable as seen from the unit-step response curves in Fig. 7.27. Without the PD controller, the response has a peak overshoot of 53%. With the PD controller, the peak overshoot is about four per cent. Although K_D was chosen for critical damping, the overshoot is due to the zero at $s = -K_P/K_D = -568.2$ (the two closed-loop poles are located at $s = -\omega_n = -910.2$ when $\zeta = 1$).

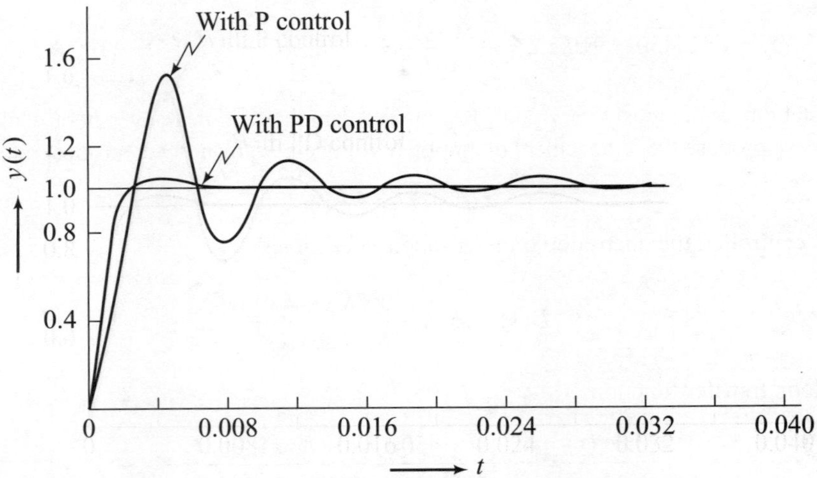

Fig. 7.27 *Unit-step responses for system of Example 7.5*

7.6.2 Derivative Output Compensation

The philosophy of using the derivative of the actuating error signal to improve the damping of a system can also be extended to the output signal. Figure 7.28 shows the block diagram of a second-order system with a secondary path feeding back the derivative of the output.

Eliminating the minor feedback loop, we obtain the following forward-path transfer function:

$$G(s) = \frac{K\,K_P}{s(\tau s+1+K\,K_D)} \qquad (7.29)$$

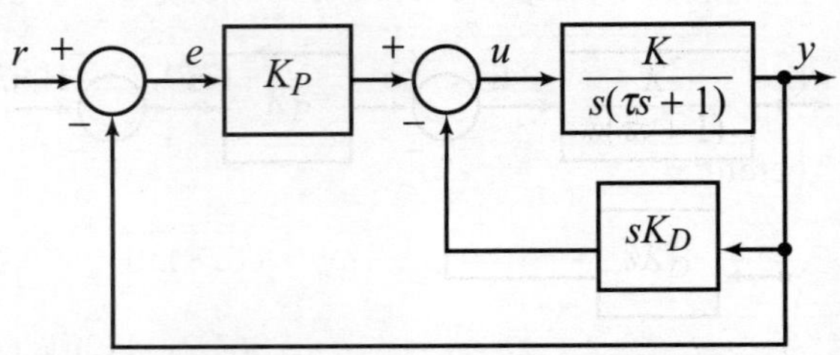

Fig. 7.28 *Derivative output compensation scheme*

The velocity error constant

$$K_v = \lim_{s \to 0} sG(s) = \frac{K K_P}{1 + K K_D} \tag{7.30}$$

The closed-loop transfer function is given by

$$\frac{Y(s)}{R(s)} = \frac{K K_P}{\tau s^2 + (1 + K K_D)s + K K_P} \tag{7.31}$$

and the characteristic equation is

$$s^2 + \frac{(1 + K K_D)}{\tau} + \frac{K K_P}{\tau} = 0 \tag{7.32}$$

The natural frequency and damping of the compensated system are given by

$$\omega_n = \sqrt{\frac{K K_P}{\tau}} \tag{7.33a}$$

$$\zeta = \frac{1}{2}\left[\frac{1 + K K_D}{\sqrt{K K_P \tau}}\right] \tag{7.33b}$$

For a specified K_v and ζ, we can write from Eqns (7.30) and (7.33b),

$$\zeta K_v = \frac{1}{2}\left[\sqrt{\frac{K K_P}{\tau}}\right] \tag{7.34}$$

from which K_P can be determined. Using this value of K_P, K_D is obtained from Eqn. (7.30). Thus, we notice that by a suitable choice of K_P and K_D, we can simultaneously meet the specifications on K_v and ζ. We will have to resort to trial-and-error methods of compensator design if the resulting speed of response is not acceptable.

Example 7.6 Consider the position control system of Fig. 7.25. When switch 'S' is open (tachogenerator disconnected), the forward-path transfer function of the system becomes

$$G(s) = \frac{4500\, K_A}{s(s + 361.2)}$$

Assume that the steady-state error to unit-ramp input is required to be less than 0.025 deg (0.000436 rad).

Required $$K_v = \frac{1}{0.000436} = 2293.6$$

In Example 7.5, we have seen that the requirement on K_v is satisfied if $K_A = 184.1$. For this value of K_A

$$\omega_n = 910.2,\ \zeta = 0.198$$

Relative stability is obviously very poor.

Closing the switch 'S' will result in a minor feedback loop which represents the feeding back of the derivative of the output signal using a tachogenerator. Eliminating the minor feedback loop, we obtain the following forward-path transfer function:

$$G(s) = \frac{4500\,K_A}{s(s + 361.2 + 45000\,K_t)} \tag{7.35}$$

The closed-loop transfer function

$$\frac{\theta_L(s)}{\theta_R(s)} = \frac{4500\;K_A}{s^2 + (361.2 + 45000\,K_t)s + 4500\,K_A} \tag{7.36}$$

Velocity error constant

$$K_v = \frac{4500\;K_A}{361.2 + 45000\;K_t} \tag{7.37}$$

The characteristic equation of the system is

$$s^2 + (361.2 + 45000K_t)s + 4500K_A = 0 \tag{7.38a}$$

from which we obtain

$$\omega_n = \sqrt{4500\;K_A} \tag{7.38b}$$

$$\zeta = \frac{361.2 + 45000\;K_t}{2\sqrt{4500\;K_A}} \tag{7.38c}$$

From these equations, it is apparent that the tachogenerator feedback increases the damping of the system; it, however, reduces the system K_v.

With $K_A = 184.1$, we get critical damping when $K_t = 0.0324$. For these values of K_A and K_t, we obtain from Eqn. (7.37):

$$K_v = 455.4$$

Therefore,

$$e_{ss}|_{\text{unit ramp}} = 0.0022 \text{ rad} = 0.126 \text{ deg.}$$

With the amplifier gain K_A set at the value 184.1, the tachogenerator feedback (K_t = 0.0324) increases the damping ratio from 0.198 to 1; it also increases the steady-state error to unit-ramp input from 0.025 deg to 0.126 deg.

Equations (7.37) and (7.38c) show that specified values of K_v and ζ can be realized by appropriate choice of K_A and K_t. For example, $K_v = 2293.6$ (steady-state error to unit-ramp input = 0.025 deg) and $\zeta = 1$ are realized by

$$K_A = 4676.4 \text{ and } K_t = 0.196$$

This choice of K_A gives $\omega_n = 4587.35$ rad/sec. Hardware considerations (e.g., amplifier saturation) and bandwidth considerations (large ω_n corresponds to large bandwidth) impose a limit on K_A which in turn imposes a limit on achievable accuracy/relative stability.

7·7 ANALYTICAL DESIGN FOR PI CONTROL

In an integral error compensation scheme, the output response depends in some manner upon the integral of the actuating signal. In the feedback control system of Fig. 7.29, this type of compensation has been introduced by using a PI controller. The plant transfer function is

$$G(s) = \frac{K}{s(\tau s+1)} \quad (7.39\text{ a})$$

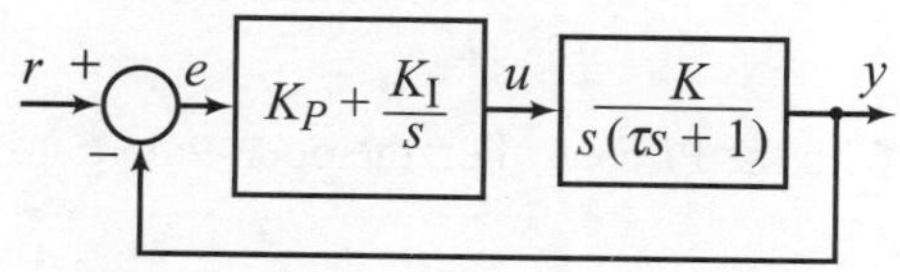

Fig. 7.29 *Integral error compensation scheme*

The controller transfer function is

$$D(s) = K_P + \frac{K_I}{s} \quad (7.39\text{ b})$$

The control signal is

$$u(t) = K_P e(t) + K_I \int_0^t e(\tau)d\tau$$

The open-loop transfer function of the overall system is

$$D(s)G(s) = \frac{K(K_P s + K_I)}{s^2(\tau s+1)} \tag{7.40}$$

Clearly, the PI control adds a zero at $s = -K_I/K_P$, and a pole at $s = 0$ to the open-loop transfer function. ***One obvious effect of the integral control is that it increases the order of the system by one; it may therefore be less stable than the original second-order system or even become unstable if the parameters K_P and K_I are not properly chosen.***

Integral control increases the type of the system by one. The PI controller in Fig. 7.29 converts a type-1 system to type-2. Therefore, the steady-state error of the compensated system is always zero for velocity input. The problem is to choose a proper combination of K_P and K_I so that the transient response is satisfactory.

The characteristic equation of the closed-loop system is

$$\tau s^3 + s^2 + KK_P s + KK_I = 0 \tag{7.41}$$

Applying Routh's test to this equation yields the result that the system is stable for $0 < K_I < K_P/\tau$. This means that the system zero at $s = -K_I/K_P$ cannot be placed too far to the left in the left half of s-plane, or the system would be unstable.

A viable method of designing a PI controller is to determine the ranges of parameters K_P and K_I that result in a pair of complex-conjugate closed-loop poles and a real closed-loop pole. The zero at $s = -K_I/K_P$ is then selected so that it is relatively close to the real closed-loop pole. We know that if the zero is close to a pole, the residue at the pole tends to be small so that the corresponding transient is small. The dynamics of the system is then dominated by only the complex-conjugate pair of closed-loop poles. By trial-and-error parameter adjustments, it is possible to almost cancel a closed-loop pole with a zero, and simultaneously achieve the desired transient performance. Trial-and-error procedures of parameter adjustment are given in Chapters 9 and 12.

Example 7.7 Reconsider the design problem of Example 7.5. Assume that we require zero steady-state error to ramp inputs. To realize this objective, we replace the amplifier with gain K_A (refer to Fig. 7.26) by a PI controller

$$D(s) = K_P + \frac{K_I}{s}$$

With the PI controller, the open-loop transfer function becomes

$$D(s)G(s) = \frac{4500(K_P s + K_I)}{s^2(s+361.2)} \tag{7.42}$$

The closed-loop transfer function

$$\frac{\theta_L(s)}{\theta_R(s)} = \frac{4500(K_P s + K_I)}{s^3 + 361.2 s^2 + 4500\, K_P s + 4500\, K_I} \tag{7.43}$$

From Eqn. (7.42), we see that with the PI controller, the system has become a type-2 system; therefore, the steady-state error to ramp inputs is zero. To study the effect of PI controller on transient response, we consider the characteristic equation,

$$s^3 + 361.2\, s^2 + 4500 K_P s + 4500 K_I = 0 \tag{7.44}$$

Applying Routh's test to this equation yields the result that the system is stable for $0 < K_I < 361.2 K_P$. This means that the system zero at $s = -K_I/K_P$ cannot be placed too far to the left in the left half of s-plane.

Design to determine suitable values of K_P and K_I can be carried out using the root-locus and frequency-response methods discussed in the following chapters. Here we analyze the performance of the system for a typical design:

$$K_P = 14.728;\ K_I = 147.28$$

For these values of K_P and K_I, the characteristic equation becomes

$$s^3 + 361.2\, s^2 + 66276\, s + 662760 = 0$$

or

$$(s + 10.59)\,(s + 175.30 + j178.41)\,(s + 175.30 - j178.41) = 0$$

The pole at $s = -10.59$ is insignificant as far as the transient response is concerned, because the zero at $s = -K_I/K_P = -10$ is close to this pole. The transient response is therefore dominated by the quadratic factor

$$(s + 175.30 + j\,178.41)\,(s + 175.30 - j\,178.41) = 0$$

or

$$s^2 + 350.6\, s + 62560.22 = 0$$

The unit-step response of the third-order system, therefore, resembles a second-order system with

$$\omega_n = \sqrt{62560.22} = 250.12 \text{ rad/sec; and } \zeta = \frac{350.6}{2\times 250.12} = 0.7$$

The unit-step response of the PI-compensated system with $K_P = 14.728$ and $K_I = 147.28$ is shown in Fig. 7.30. Notice that although the damping is adequate, the integral control causes the step response to have long rise and settling times.

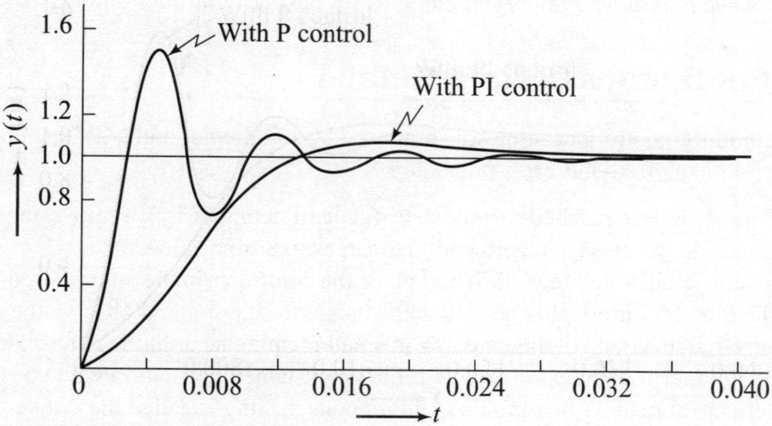

Fig. 7.30 *Unit-step responses for the system of Example 7.7*

7.8 ON-LINE CONTROLLER TUNING

As we shall see in later chapters, sophisticated methods are available to develop a controller that will meet steady state and transient response specifications, for both tracking input references and rejecting disturbances. These methods require that the designer has a dynamic model of the system in the form of transfer function/frequency response/state-space equations.

In servomechanism applications, mathematical models close enough to reality are often easily built; therefore, controller parameters for optimum response can be fairly well predicted at the design stage. In the process-control field, however, the industrial processes are relatively slow and complex for which, in most cases, only approximate models are built. A model close to reality may become too complex to be of practical value for the available control-design procedures. In the process-control field, it is common for the controller to be installed with little modeling and design being done before hand. Only preliminary controller settings based on process knowledge, control objectives and prior experience are specified. The preliminary settings may also be obtained from the control-design procedures based on approximate models. After a controller is installed, the preliminary settings often prove to be satisfactory. But for critical control loops, these settings may have to be adjusted in order to achieve satisfactory control. This on-line adjustment is referred to as *on-line controller tuning*.

On-line controller tuning, as we will shortly see, involves plant testing, often on a trial-and-error basis. The tuning, therefore, can be quite tedious and time-consuming. Consequently, good initial controller settings are very desirable to reduce the required time and effort. Ideally, the preliminary settings from the control system design can be used as initial field settings. If the preliminary settings are not satisfactory, alternative settings can be obtained from simple experimental tests. If necessary, the settings can be fine-tuned by a modest amount of trial-and-error.

A great number of manufacturers are making available in the market off-the-shelf process controllers with features that permit plant testing and controller tuning: adjusting the setpoint, transferring between manual (open-loop) and auto (closed-loop) control modes, adjusting the controlled output from the control-action unit (adjusting the parameters K_c, T_I, and T_D of the PID controller (refer to Eqns (7.2)), and displaying the controlled variable, setpoint and control signal. Various plant testing and PID-control tuning methods of interest to us, can easily be implemented using these controllers [155].

In this section, we present two important methods for on-line controller tuning [6].

7.8.1 Continuous Cycling Method

This classic method for controller tuning was proposed by J G Ziegler and N B Nichols in 1942 [8]. It is based on the following trial-and-error procedure:

1. After the process has reached steady-state (at least temporarily), remove the derivative and integral modes by setting T_D to zero and T_I to largest possible value.
2. Set K_c equal to small value (e.g., 0.5) and place the controller in the auto (closed-loop) mode.
3. Disturb the loop by introducing a small step change in setpoint and observe the response of the controlled variable. Gradually increase K_c in small increments until the controlled variable starts oscillating. The term *continuous cycling* refers to sustained oscillation with constant amplitude. The numerical value of K_c that produces continuous cycling is called the *ultimate gain*, K_{cu}. The period of the corresponding sustained oscillation is referred to as the *ultimate period*, T_u (refer to Fig. 7.31).
4. Calculate the PID controller settings using the Ziegler–Nichols tuning relations given in Table 7.1.
5. Evaluate the controller settings by introducing a small setpoint change and observing the closed-loop response. Fine-tune the settings, if necessary.

Typically, determination of K_{cu} involves trial-and-error. For $K_c < K_{cu}$, the closed-loop response is usually overdamped or slightly oscillatory. Continuous cycling occurs when $K_c = K_{cu}$. For $K_c > K_{cu}$, the closed-loop system is unstable and will theoretically have unbounded response. But in practice, saturation in the loop prevents the response from becoming unbounded and produces continuous cycling instead. To eliminate the risk of using gain corresponding to this cycling as K_{cu} for the purpose of determining controller settings, it is very important that saturation be avoided during the experimental tests.

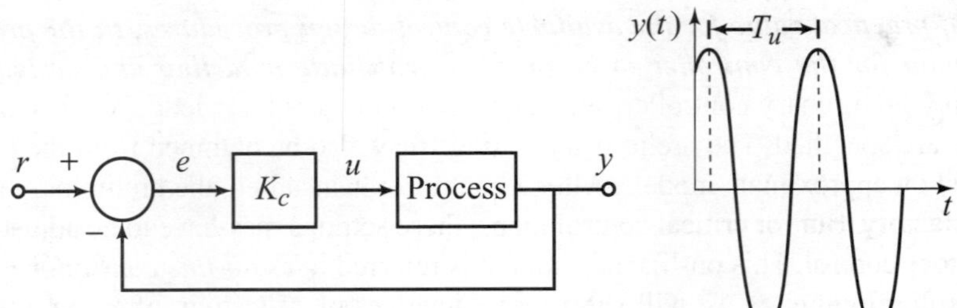

Fig. 7.31 *Determination of the ultimate gain and period*

Table 7.1 *QDR tuning formulas based on ultimate gain and period*

Controller	*Gain*	*Integral time*	*Derivative time*
P	$K_c = 0.5K_u$	–	–
PI	$K_c = 0.45K_u$	$T_I = T_u/1.2$	–
PID	$K_c = 0.75K_u$	$T_I = T_u/1.6$	$T_D = T_u/10$

Quarter-Decay Ratio Response

The tuning relations reported by Ziegler and Nichols were determined empirically to provide closed-loop responses that have a ¼ (quarter) decay ratio. This means that the transient in closed-loop step response decays to a quarter of its value after one period of oscillation, as shown in Fig. 7.32. Table 7.1 provides a safety margin of two for K_c because it is equal to one-half of the stability limit, K_u. When integral action is added to form a PI controller, K_c is reduced from $0.5K_u$ to $0.45K_u$. The stabilizing effect of derivative action allows K_c to be increased to $0.75K_u$ for PID control.

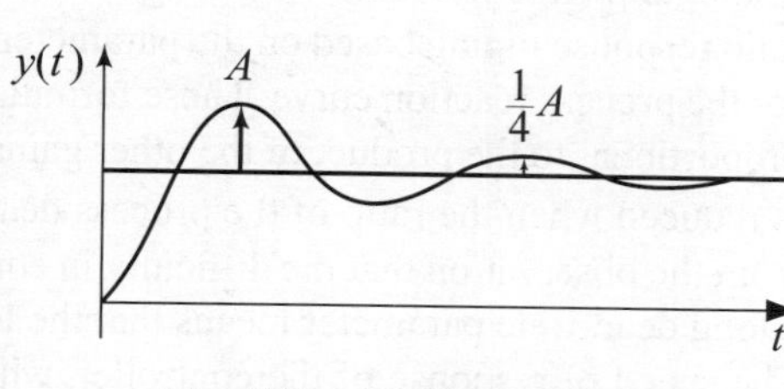

Fig. 7.32 *QDR response*

PI and PID tuning parameters that produce quarter-decay response are not unique. For each setting of the integral and derivative terms, there will usually be a setting for the gain K_c that produces quarter-decay response. The settings given in Table 7.1 are the figures based on experience; these settings have produced fast response for most industrial loops.

7.8.2 Process Reaction Curve Method

Ziegler and Nichols proposed a second on-line tuning technique based on a single step test. The experimental procedure is quite simple. After the process has reached steady state (at least approximately), the controller is placed in the manual (open-loop) mode. Then a small step change in the control signal (e.g., 3 to 5%) is introduced. The step change in control should be large enough for the consequent change in the process output variable to be distinguishable from usual fluctuations due to noise and disturbances but not so large that the response will be distorted by process nonlinearities. The process output is recorded for a period from the introduction of the step change in input until the process reaches a new steady state.

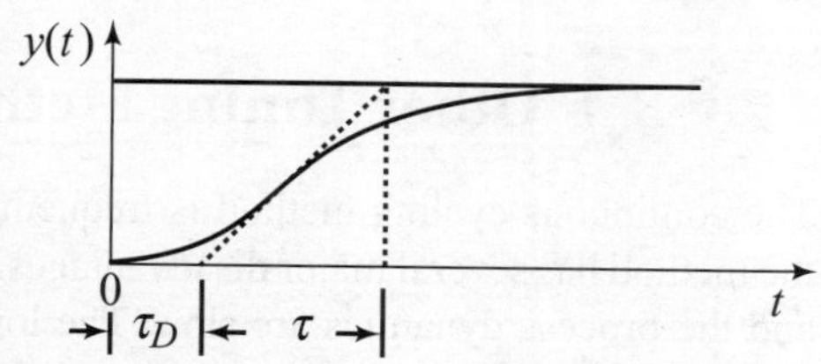

Fig. 7.33 *Process reaction curve*

Zeigler and Nichols recognized that the open-loop step responses of a large number of processes exhibit a *process reaction curve* like that shown in Fig. 7.33. The S-shape of the curve can be approximated by the step response of

$$\frac{Y(s)}{U(s)} = \frac{Ke^{-s\tau_D}}{\tau s+1} \tag{7.45}$$

K = the process steady-state gain,
τ_D = the effective process dead time; and
τ = the effective process time constant.

The constants in Eqn. (7.45) can be determined from the experimentally obtained step response $y(t)$ to step input $u(t) = A\mu(t)$. The steady-state value of the response $y_{ss} = KA$. This gives the parameter K. If a tangent is drawn at the inflection point of the reaction curve (refer to Fig. 7.33), then the slope of the line is KA/τ, and the intersection of the tangent line with the time axis identifies the dead-time τ_D.

Besides the formulas for quarter-decay ratio response tuning based on the ultimate gain and period of the loop (refer to Table 7.1), Ziegler and Nichols also developed tuning formulas for quarter-dacay ratio response tuning based on the parameters K, τ and τ_D of the first-order plus dead-time model (7.45) for the process reaction curve. These formulas are given in Table 7.2. The controller gain K_c is inversely proportional to the product of the other gains in the feedback loop, i.e., $K_c \propto 1/K$. The controller gain is reduced when the ratio of the process dead time to its time constant (τ_D/τ) increases. This follows from the observation that the difficulty in controlling the loop increases when τ_D/τ increases. Having a long dead-time parameter means that the loop is difficult to control only if the time constant is short. The speed of response of the controller, which is determined by integral and derivative times (T_I and T_D), must match the speed of response of the process. The formulas in Table 7.2 match the response speeds by relating the integral and derivative times to the process dead time.

The formulas in Table 7.2 were developed empirically for the most common range of τ_D/τ, which is between 0.1 and 0.3.

Table 7.2 *QDR tuning formulas based on process reaction curve*

Controller	*Gain*	*Integral time*	*Derivative time*
P	$K_c = \tau/K\tau_D$	–	–
PI	$K_c = 0.9\tau/K\tau_D$	$T_I = 3.33\ \tau_D$	–
PID	$K_c = 1.5\tau/K\tau_D$	$T_I = 2.5\ \tau_D$	$T_D = 0.4\ \tau_D$

7.8.3 Other Tuning Methods

The continuous cycling method is frequently recommended by control system vendors, even though, the method has several major disadvantages. It can be quite time-consuming if several trials are required and the process dynamics are slow. The long experimental tests may result in reduced production or poor product quality. Also, in many applications, continuous cycling is objectionable because the process is pushed to the stability limits. Consequently, if external disturbances or process changes occur during the test, unstable operation could result.

Astrom and Hagglund [11] have developed an attractive alternative to the continuous cycling method. In their relay auto-tuning method, a simple experimental test is used to determine K_{cu} and T_u. For this test, the feedback controller is temporarily replaced by an on-off controller (or *relay*). After the control loop is closed, the controlled variable exhibits a sustained oscillation that is characteristic of an on-off control (refer to Chapter 16).

The process reaction curve method (the step test method) also avoids the disadvantages of the continuous cycling method; only a single experimental test is necessary. However, the experimental test is performed under open-loop conditions. Thus, if a significant disturbance occurs during the test, no corrective action is taken. Consequently, the process can be upset, and the test results may be misleading.

Closed-loop versions of the step test method have been proposed [7]. Typically, a step change in the setpoint is introduced while the process is controlled using proportional-only control. Then the parameters in the simple transfer function models can be estimated from the closed-loop step response.

The early controller tuning relations published by Ziegler and Nichols were developed to provide closed-loop responses that have a ¼ decay ratio. Because a response with ¼ decay ratio is considered to be excessively oscillatory for many process control applications, other tuning relations (for example, tuning relations based on *integral error criteria*) have been reported in the literature [12].

7.9 IMPLEMENTATION OF PID CONTROLLERS

Today, most complex industrial processes are under computer control. A complete computer-based installation, in a multilevel viewpoint of control strategy, handles the design tasks at plant management level, production scheduling and control level, and supervisory level, in addition to controlling individual process units at the lowest level in the hierarchy [155]. It is common to delegate process control functions at the lowest level to the special individual control units, connected to the higher levels through computer system. Op Amp (Operational Amplifier) control units are in use to serve these functions.

We take up the op amp realization of PID control functions in the present section. Chapter 13 will highlight digital implementation.

Basics of operational amplifiers have been given in Section 2.6.2. Revisiting this section will be helpful.

7.9.1 Simple Op Amp Circuit Models

Figure 7.34 shows seven simple op amp circuits. In the following, we derive transfer function for each circuit. From the transfer function models, it will become obvious that these circuits are used for amplification, integration, and differentiation of a given signal.

Figure 7.34*a*

$$\frac{E_0(s)}{E_i(s)} = -\frac{R_2}{R_1} \tag{7.46a}$$

The circuit is thus a constant gain amplifier. The gain can be made adjustable by using a potentiometer for one of the resistances.

Note the phase relationship between input and output; the circuit of Fig. 7.34a is called an *inverting amplifier*.

Figure 7.34*b*

A circuit with $R_1 = R_2$ is an *inverter*; it simply inverts the sign of the input voltage without changing its magnitude. It can be used in cascade with other elements to maintain proper sign relations, as shown in Figure 7.34b. For this circuit,

$$\frac{E_0(s)}{E_i(s)} = \frac{R_2}{R_1} \tag{7.46b}$$

Figure 7.34*c*

$$\frac{E_0(s)}{E_i(s)} = -\frac{1/sC}{R} = -\frac{1}{RCs} \tag{7.47a}$$

The circuit is thus an *integrator*.

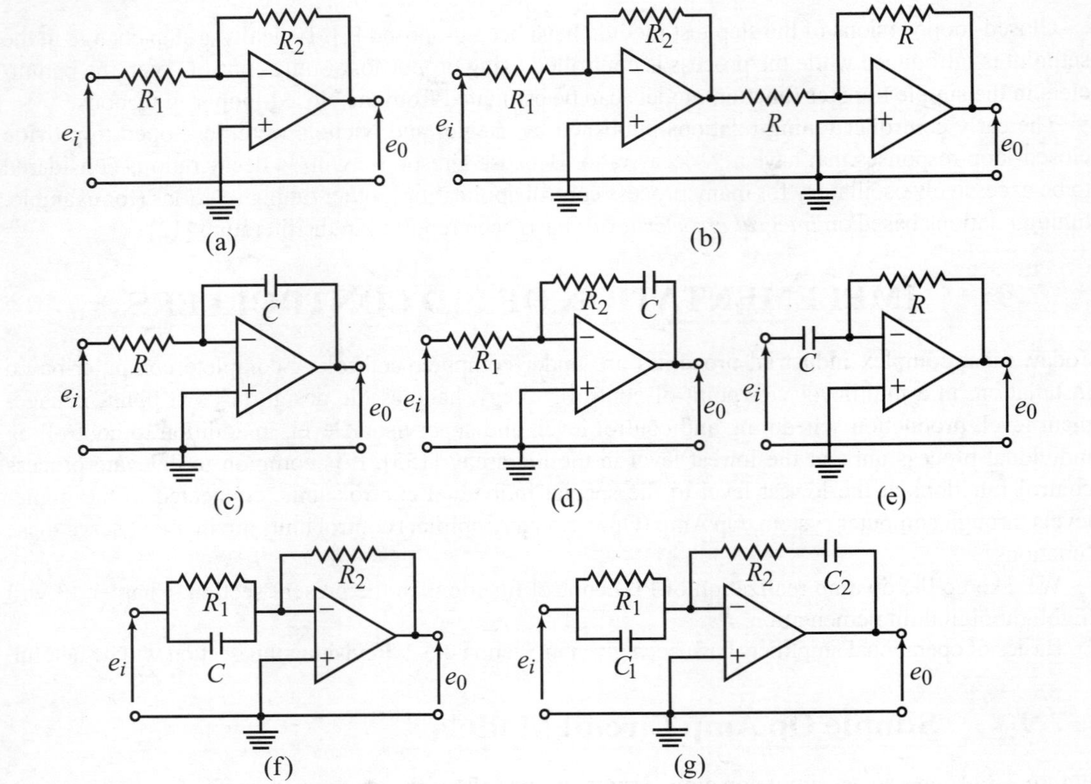

Fig. 7.34 *Commonly used op amp circuits*

Equation (7.47a) is true if the initial voltage on the capacitor is zero. Inverse Laplace transformation of this equation gives

$$e_0(t) = -\frac{1}{RC}\int_0^t e_i(\tau)d\tau \tag{7.47b}$$

Figure 7.34*d*

$$\frac{E_0(s)}{E_i(s)} = -\frac{R_2 + 1/sC}{R_1} = -\frac{R_2}{R_1} - \frac{1}{R_1Cs} \tag{7.48a}$$

$$e_0(t) = -\frac{R_2}{R_1}e_i(t) - \frac{1}{R_1C}\int_0^t e_i(\tau)d\tau \tag{7.48b}$$

The circuit thus processes the input signal by 'proportional + integral' action.

Figure 7.34*e*

$$\frac{E_0(s)}{E_i(s)} = -\frac{R}{1/sC} = -RCs \tag{7.49a}$$

$$e_0(t) = -RC\frac{de_i(t)}{dt} \tag{7.49b}$$

The circuit is thus a *differentiator*.

Figure 7.34f

$$\frac{E_0(s)}{E_i(s)} = -\frac{R_2}{\dfrac{R_1(1/sC)}{R_1 + 1/sC}} = -\frac{R_2}{R_1}(R_1Cs+1) \tag{7.50a}$$

$$e_0(t) = -\frac{R_2}{R_1}e_i(t) - R_2C\frac{de_i(t)}{dt} \tag{7.50b}$$

The circuit thus processes the input signal by 'proportional + derivative' action.

Figure 7.34g

$$\frac{E_0(s)}{E_i(s)} = -\frac{R_2 + 1/sC_2}{\dfrac{R_1(1/sC_1)}{R_1 + 1/sC_1}} = -\frac{(R_2C_2s+1)(R_1C_1s+1)}{R_1C_2s} = -\left[\frac{R_2C_2R_1C_1s^2 + R_2C_2s + R_1C_1s + 1}{R_1C_2s}\right]$$

$$= -\left[\frac{R_1C_1 + R_2C_2}{R_1C_2} + \frac{1}{R_1C_2s} + R_2C_1s\right] \tag{7.51a}$$

$$e_0(t) = -\frac{R_1C_1 + R_2C_2}{R_1C_2}e_i(t) - \frac{1}{R_1C_2}\int_0^t e_i(\tau)d\tau - R_2C_1\frac{de_i(t)}{dt} \tag{7.51b}$$

The circuit thus processes the input signal by 'proportional + integral + derivative' action.

7.9.2 Practical Op Amp Circuits

In the following, we outline the limitations of some of the circuits of Fig. 7.34, and give 'practical' circuits used in implementing compensator transfer functions of feedback control systems [44].

Gain-limited Integrator

Figure 7.34c is the ideal op amp implementation of an integrator. The transfer function of this circuit is given by Eqn. (7.47a), and is reproduced below.

$$\frac{E_0(s)}{E_i(s)} = -\frac{1}{RCs} \tag{7.52}$$

The ideal integrator has infinite dc (zero frequency) gain. For applications requiring finite dc gain (In Chapter 9, we will discuss feedback-control applications), resistor R_1 connected in parallel with the capacitor C is used. Figure 7.35a shows this implementation. The resistor R_1 in conjunction with resistor R limits the maximum gain of the integrator.

The integrator gain now becomes

$$\frac{E_0(s)}{E_i(s)} = \frac{R_1/R}{R_1Cs+1} \tag{7.53}$$

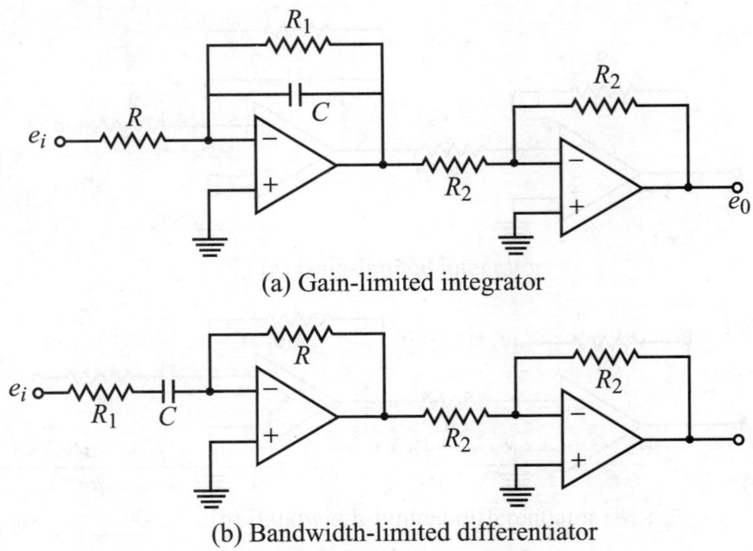

Fig. 7.35 *Practical Op Amp Circuits*

This (low frequency) gain-limited integrator can be thought of as a low-pass filter with a dc gain equal to R_1/R and a time-constant of R_1C.

In Fig. 7.35a, one amplifier is set up as an integrator with an output which is of opposite polarity to that of the input. To reverse the inverted signal, another amplifier has been added. The two amplifiers form a non-inverting integrator, resulting in a positive output for a positive input.

Bandwidth-limited Differentiator

Figure 7.34e is the ideal op amp implementation of a differentiator. The transfer function of the circuit is given by Eqn. (7.49a), and is reproduced below.

$$\frac{E_0(s)}{E_i(s)} = -RCs \tag{7.54}$$

The ideal differentiator has gain that increases with an increase in frequency and becomes infinite as frequency approaches infinity.

In feedback control applications, a significant amount of noise that occupies the high end of the frequency spectrum, gets superimposed on the sensor feedback signal which is processed by a PID controller. Since the gain of the derivative term increases linearly with frequency, the noise signal tends to be greatly amplified. To reduce the controller sensitivity to noise, amplifier gain needs to be limited. Without using a complex filtering scheme, it can be achieved by adding a resistor in series with the input capacitor. This network changes the amplifier characteristics, resulting in a limited high-frequency gain.

A schematic diagram of the modified differentiator is shown in Fig. 7.35b. Resistor R_1 in conjunction with resistor R, limits the maximum amplifier gain. As the frequency increases, the impedance of the input capacitor becomes smaller. Because the amplifier gain is controlled by the ratio of the feedback and input impedances, with the input impedance becoming smaller with an increase in frequency, the amplifier gain continues to increase. The resistor R_1 in series with the capacitor sets a limit on the minimum value of input impedance.

The derivative gain now becomes

$$\frac{E_0(s)}{E_i(s)} = \frac{RCs}{R_1Cs+1} \tag{7.55}$$

This practical differentiator can be thought of as a derivative block followed by a low-pass filter with a time constant equal to R_1C. The result is a bandwidth-limited derivative action.

Example 7.8 Consider the network shown in Fig. 7.36. The transfer function for this network can be obtained as follows:

The equivalent impedance of R_1 and C_1 is

$$Z_1(s) = \frac{R_1}{R_1C_1s+1}$$

The equivalent impedance of R_2 and C_2 is

$$Z_2(s) = \frac{R_2}{R_2C_2s+1}$$

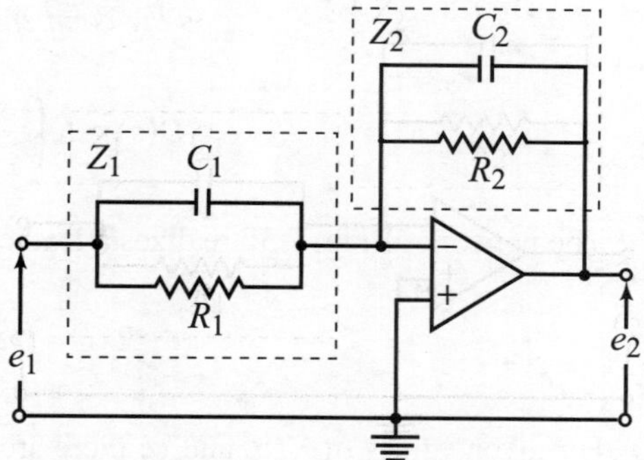

Fig. 7.36 *Operational amplifier circuit*

The transfer function between the output $E_2(s)$ and the input $E_1(s)$ is

$$\frac{E_2(s)}{E_1(s)} = -\frac{Z_2(s)}{Z_1(s)} = -\frac{R_2}{R_1}\left(\frac{R_1C_1s+1}{R_2C_2s+1}\right) \tag{7.56}$$

Notice that the transfer function in Eqn. (7.56) contains a minus sign. If this sign is not acceptable in the actual application, a sign inverter may be connected to either the input or the output of the network in Fig. 7.36. An example is shown in Fig. 7.37. The sign inverter has the transfer function

$$\frac{E_3(s)}{E_2(s)} = -\frac{R_4}{R_3}$$

Hence, the network shown in Fig. 7.37 has the following transfer function:

$$\frac{E_3(s)}{E_1(s)} = \frac{R_2R_4}{R_1R_3}\left(\frac{R_1C_1s+1}{R_2C_2s+1}\right) = \frac{R_4C_1}{R_3C_2}\left(\frac{s+\frac{1}{R_1C_1}}{s+\frac{1}{R_2C_2}}\right) \tag{7.57}$$

Controllers (also called compensators) are sometimes formulated with a transfer function of the form (this form, as we shall see in later chapters, effectively realizes PD control law and is called *lead compensator*)

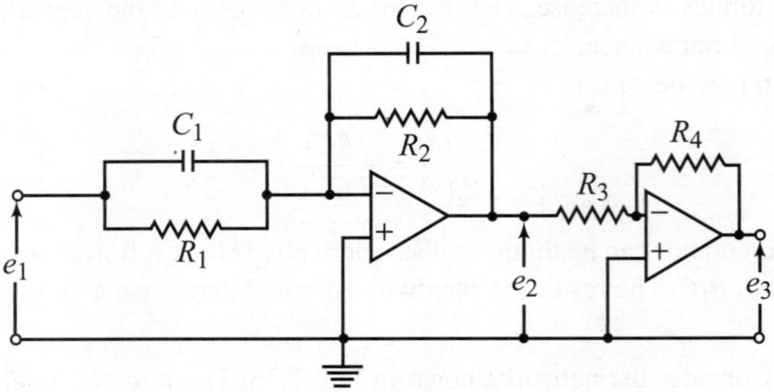

Fig. 7.37 *Operational amplifier circuit used as a lead or a lag compensator*

$$D(s) = K\left(\frac{s+1/\tau}{s+1/\alpha\tau}\right); \alpha < 1, K > 0, \tau > 0 \tag{7.58}$$

The network of Fig. 7.37 realizes a lead compensator if $R_1C_1 > R_2C_2$, as seen from Eqn. (7.57).

$$R_1C_1 = 1/\tau; R_2C_2 = 1/\alpha\tau; \frac{R_4C_1}{R_3C_2} = K$$

For given values of K, τ, and α; there are six op amp circuit parameters: R_1, R_2, R_3, R_4, C_1, and C_2; which are required for realization of the compensator. The values of the six circuit parameters are determined from the three compensator parameters using the three equations given above. There is an additional degree of freedom in the choice of the values of the circuit parameters which is used to set the impedance level of the network.

Another common formulation of compensators is of the form (this form, as we shall see in later chapters, effectively realizes PI control law and is called *lag compensator*)

$$D(s) = K\left(\frac{s+1/\tau}{s+1/\beta\tau}\right); \beta > 1; K > 0, \tau > 0 \tag{7.59}$$

The network of Fig. 7.37 realizes a lag compensator if $R_1C_1 < R_2C_2$, as seen from Eqn. (7.57).

A *lag-lead compensator* (we shall encounter such designs in later chapters)

$$D(s) = K\left(\frac{s+1/\tau_2}{s+1/\beta\tau_2}\right)\left(\frac{s+1/\tau_1}{s+1/\alpha\tau_1}\right); \beta > 1, \alpha < 1, K > 0, \tau_1 > 0, \tau_2 > 0 \tag{7.60}$$

may be realized by cascading two active networks as shown in Fig. 7.38. This form, as we shall see in later chapters,effectively realizes PID control law.

In addition to realizing a given compensator transfer function, the impedance level of the op amp circuit is required to be set to acceptable values to reduce the loading effects in feedback loop.

It should be noted that many alternative circuit designs for a given compensation requirement are possible. Some practical circuits [44] are given in Problems 7.18–7.23. Choosing an optimal design is an important task for the control engineer (The topic of Operational-Amplifier circuit design is not covered in this book).

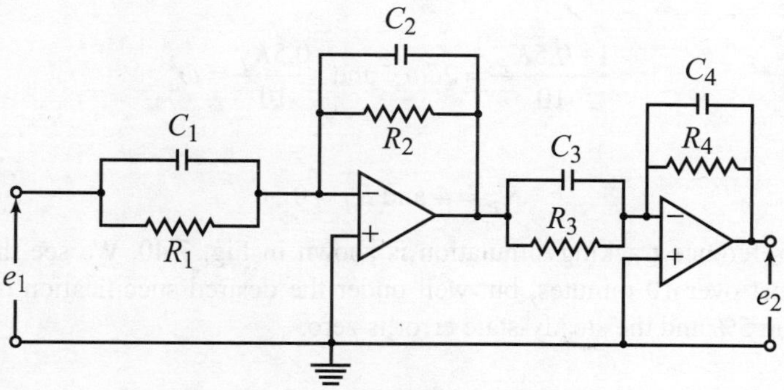

Fig. 7.38 *Operational amplifier circuit used as a lag-lead compensator*

7.10 REVIEW EXAMPLES

Review Example 7.1 We consider here the temperature control loop in a processing plant (Fig. 7.39) [63]. The controller is PI type:

$$\frac{U(s)}{E(s)} = D(s) = K_P + \frac{K_I}{s}$$

and the process model is

$$\frac{Y(s)}{U(s)} = G(s) = \frac{0.5}{10s+1}$$

with steady-state gain of 0.5 °C for 1% input change; and time constant of the process is 10 minutes.

It is desired to select K_P and K_I to ensure no steady-state offset to step reference signals, to achieve a settling time of less than 20 minutes (5% tolerance band), and to have only a little overshoot.

PI control ensures zero steady-state offset to step reference signals. For t_s (5% tolerance) = 20 minutes, $\zeta\omega_n = 3/t_s = 0.15$. Little overshoot implies that the closed-loop damping ratio should be close to unity; so we choose $\zeta = 0.9$. Therefore, $\omega_n = 0.15/\zeta = 0.167$.

From Fig. 7.39,

$$\frac{Y(s)}{R(s)} = \frac{D(s)G(s)}{1+D(s)G(s)} = \frac{0.5(K_P s + K_I)}{10s^2 + (1+0.5K_P)s + 0.5K_I}$$

The closed-loop characteristic equation is

$$10s^2 + (1+0.5K_P)s + 0.5K_I = 0$$

The specified standard second-order characteristic equation is

$$s^2 + 2\zeta\omega_n s + \omega_n^2 = 0;\ \zeta = 0.9, \omega_n = 0.167$$

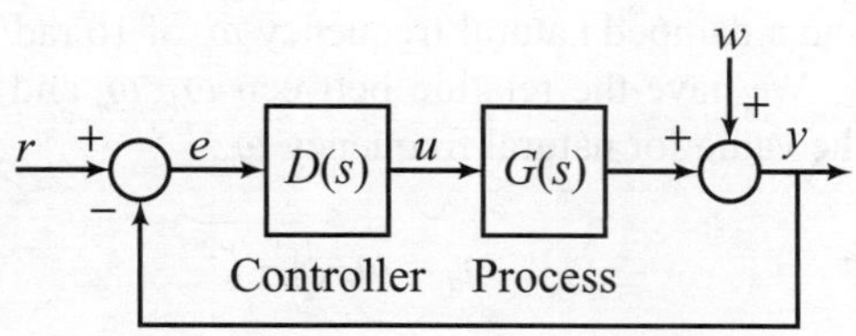

Fig. 7.39 *A temperature control loop*

We need to solve

$$\frac{1+0.5K_P}{10} = 2\zeta\omega_n \text{ and } \frac{0.5K_I}{10} = \omega_n^2$$

The solution is

$$K_P = 4 \text{ and } K_I = 0.56$$

The result of a reference tracking simulation is shown in Fig. 7.40. We see that the value for t_s (5% tolerance) is just over 10 minutes, but well under the desired specification of 20 minutes. The overshoot is less than 5% and the steady-state error is zero.

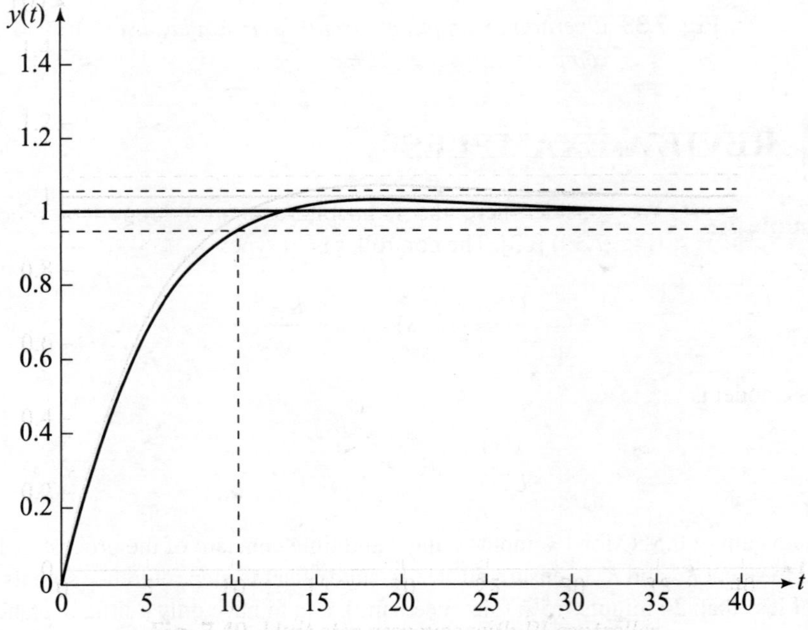

Fig. 7.40 *Unit step response with PI controller*

Review Example 7.2 The position control loop based on a dc motor actuator is a very common student laboratory experiment. The loop uses a type-1 model and demonstrates how derivative control can be used to tune system damping.

A design for proportional plus derivative control on error is required as shown in Fig. 7.41 [63]. The system time constant is given as 1 sec and the control specification is for closed-loop damping $\zeta = 0.6$ and a damped natural frequency ω_d of 10 rad/sec.

We have the relation between ω_d, ω_n and ζ to find the value for natural frequency ω_n:

$$\omega_d = \omega_n\sqrt{1-\zeta^2}$$

For $\omega_d = 10$ and $\zeta = 0.6$, we obtain $\omega_n = 12.5$.

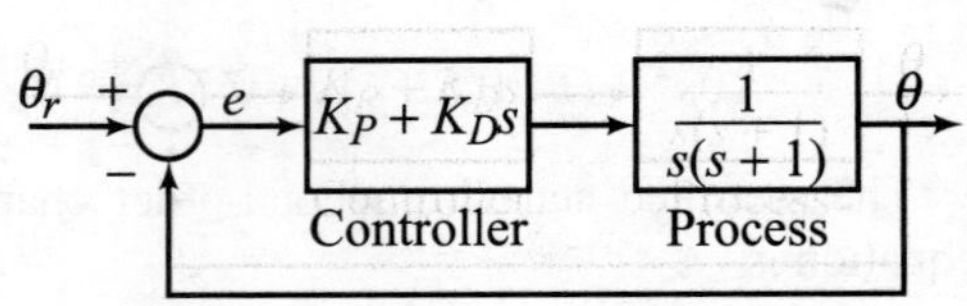

Fig. 7.41 *A position control loop*

This gives us the target closed-loop characteristic equation:

$$s^2 + 2\zeta\omega_n s + \omega_n^2 = s^2 + 15s + 156.25 = 0$$

The characteristic equation of the system of Fig. 7.41 is obtained as follows.

$$1 + (K_P + K_D s)\left(\frac{1}{s(s+1)}\right) = 0$$

This gives a polynomial equation

$$s^2 + (1 + K_D)s + K_P = 0$$

This is identical to target characteristic equation when $K_D = 14$ and $K_P = 156.25$.

An alternative design with proportional control on error and derivative control on measured variable θ is as shown in Fig. 7.42. This structure also realizes the target characteristic equation when $K_D = 14$ and $K_P = 156.25$.

Figure 7.43a shows the response of the system of Fig. 7.41 to a reference input of $\theta_r = 0.1$ occurring at $t = 0.5$ second; while Fig. 7.43b shows the response for system of Fig. 7.42. The overshoot in the second case has reduced. The corresponding control signal trace is shown in Fig. 7.44. Clearly, even with the more practical structure having proportional control on error and derivative control on measured variable, we have a control signal input which is not very desirable. It has a spike, and has a large magnitude despite the fact that the reference change is only a small one. The reason of spike is high proportional gain. Going for a PID control structure is one solution to this problem.

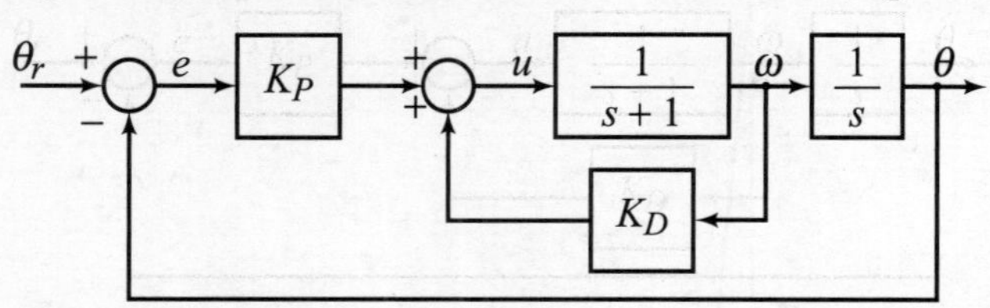

Fig. 7.42 *Position control system with minor-loop feedback*

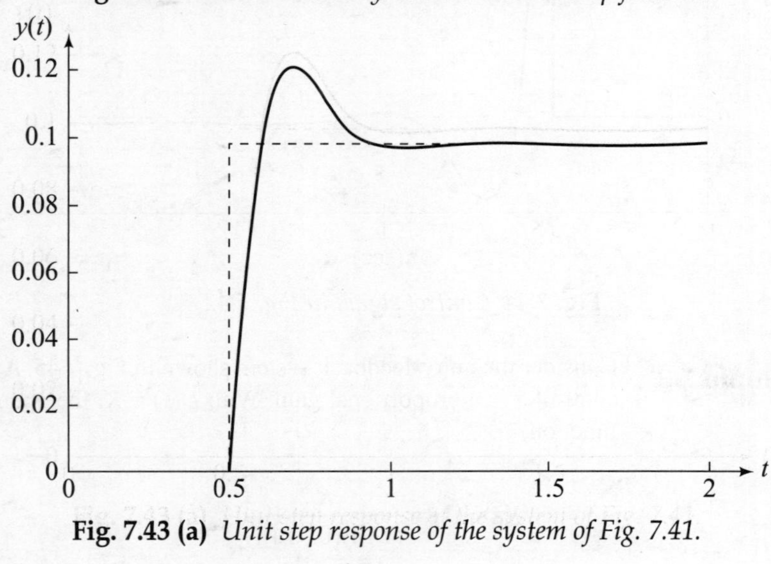

Fig. 7.43 (a) *Unit step response of the system of Fig. 7.41.*

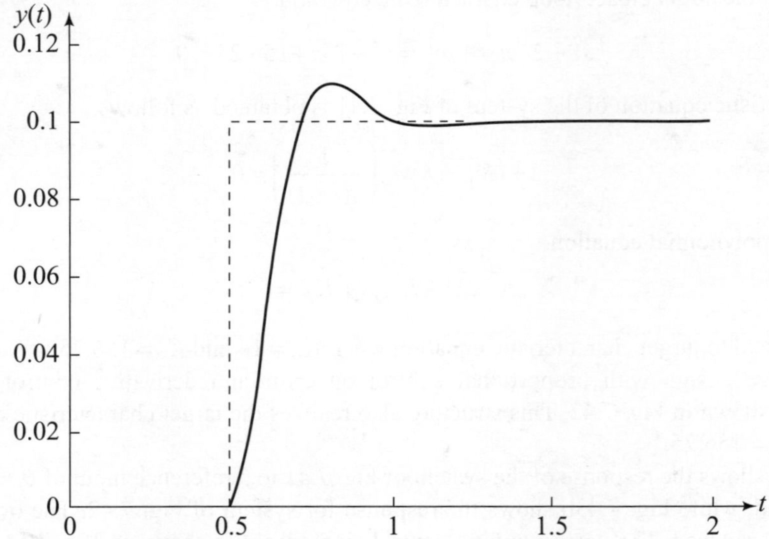

Fig. 7.43(b) *Unit step response of the system of Fig. 7.42*

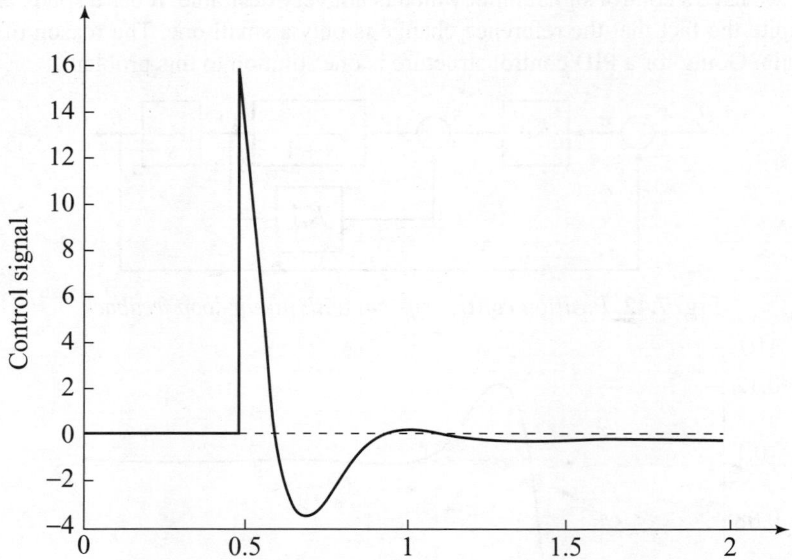

Fig. 7.44 *Control signal in Fig. 7.42*

Review Example 7.3

Consider the unity-feedback system shown in Fig. 7.45. Assume that the controller is a proportional gain. With $D(s) = K$, the open-loop transfer function,

$$D(s)G(s) = \frac{K}{(s+1)(0.5s+1)}$$

It is a type-0 system. The position error constant $K_p = K$, and

$$e_{ss}|_{\text{unit step}} = \frac{1}{1+K}$$

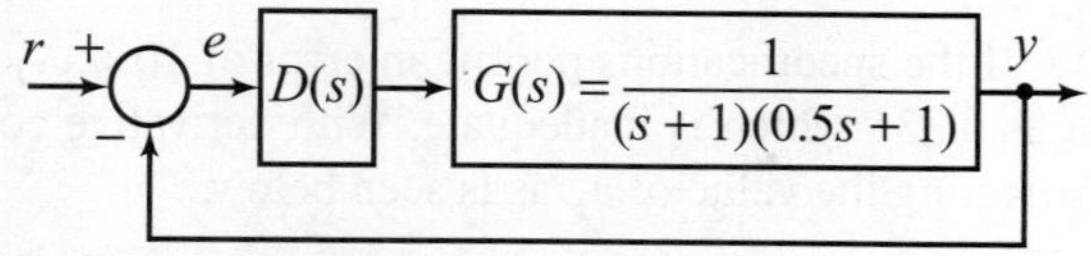

Fig. 7.45 *A unit-feedback system*

Improvement in steady-state accuracy with higher gain is evident. Relative stability considerations, however, impose a limit on the gain K. Large values of gain give rise to large overshoots.

The closed-loop transfer function of the system is

$$\frac{Y(s)}{R(s)} = \frac{2K}{s^2+3s+2+2K}$$

The characteristic equation is

$$s^2+3s+2+2K = s^2+2\zeta\omega_n s+\omega_n^2 = 0$$

$\zeta = 0.5$ corresponds to peak overshoot $M_p = 16.3\%$. The limiting value of K for $M_p \le 16.3\%$ is obtained from the following equations:

$$\omega_n = \sqrt{2+2K}\,;\, 2\times 0.5\times\omega_n = 3$$

Solving for K, we get $K = 3.5$

The steady-state error for this value of K is

$$e_{ss}|_{\text{unit step}} = \frac{1}{1+3.5} = 0.2222 = 22.22\%$$

An error of 22.22% is too large. We can reduce the error by increasing K, but this will result in an overshoot more than 16.3%. The objective of reducing the steady-state error without increasing overshoot can be realized using a PD controller:

$$D(s) = K_P + K_D s$$

With the PD controller, the open-loop transfer function becomes

$$D(s)G(s) = \frac{K_P + K_D s}{(s+1)(0.5s+1)}$$

The closed-loop transfer function

$$\frac{Y(s)}{R(s)} = \frac{2(K_P + K_D s)}{s^2+(3+2K_D)s+2+2K_P}$$

For $K_D = 0$, the characteristic equation is

$$s^2+3s+2+2K_P = 0 = s^2+2\zeta\omega_n s+\omega_n^2$$

To achieve $\zeta = 0.5$, we require $K_P = 3.5$. The steady-state error for unit-step input is $1/(1 + K_P) =$ 0.2222 or 22.22%.

If the specifications permit an error of 10%, $K_P = 9$ is required. Then for $K_D = 0$, the damping ratio ζ is 0.335 which is inadequate. With derivative control, the damping can be increased to 0.5 without affecting the value of e_{ss} as is seen below.

$$\omega_n = \sqrt{2+2K_P}\ ;\ 2\zeta\omega_n = 3 + 2K_D$$

With $K_P = 9$, $\zeta = 0.5$ is obtained when K_D statisfies the following relations:

$$\omega_n = \sqrt{2+18} = 4.47; 3 + 2K_D = 2 \times 0.5 \times \omega_n$$

This gives $\qquad K_D = 0.736$

The zero of the closed-loop transfer function is at $s = -K_P/K_D = -9/0.736 = -12.23$. The magnitude of the zero is about 5.5 times the magnitude of the real part of the poles ($\zeta\omega_n = 2.235$). So the unit-step response of the system with the PD controller should closely resemble the response of a standard second-order system with $\zeta = 0.5$ and $\omega_n = 4.47$.

If static accuracy demand is very high, PI control would probably be necessary. With $D(s) = K_P + K_I/s$, the open-loop transfer function becomes

$$D(s)G(s) = \frac{sK_P + K_I}{s(s+1)(0.5s+1)}$$

The system has become a type-1 system and the steady-state error to step inputs is zero.
The velocity error constant $K_v = K_I$, and

$$e_{ss}\big|_{\text{unit ramp}} = \frac{1}{K_I}$$

Thus, the steady-state error to ramp inputs reduces as the integral gain factor K_I is increased (note that the steady-state error to ramp inputs is infinity for the system with only proportional control or PD control). As one would expect, this improvement in static accuracy is limited by considerations of relative stability. The closed-loop transfer function

$$\frac{Y(s)}{R(s)} = \frac{sK_P + K_I}{0.5s^3 + 15s^2 + (1+K_P)s + K_I}$$

shows a third-order characteristic equation. Design to determine suitable values of K_P and K_I can be carried out using the root-locus and frequency-response methods discussed in the following chapters. This will show that careful design is necessary to avoid oscillatory or slow system dynamic behavior.

Review Example 7.4 The system shown in Fig. 7.46 is a unity-feedback control system with a minor rate-feedback loop.

(a) In the absence of rate feedback ($\alpha = 0$), determine the peak overshoot of the system to unit-step input, and the steady-state error resulting from a unit-ramp input.
(b) Determine the rate-feedback constant α which will decrease the peak overshoot of the system to unit-step input to 1.5%. What is steady-state error to unit-ramp input with this setting of the rate-feedback constant?

(c) Illustrate how in the system with rate feedback, the steady-state error to unit-ramp input can be reduced to the same level as in part (a), while the peak overshoot to unit-step input is maintained at 1.5%.

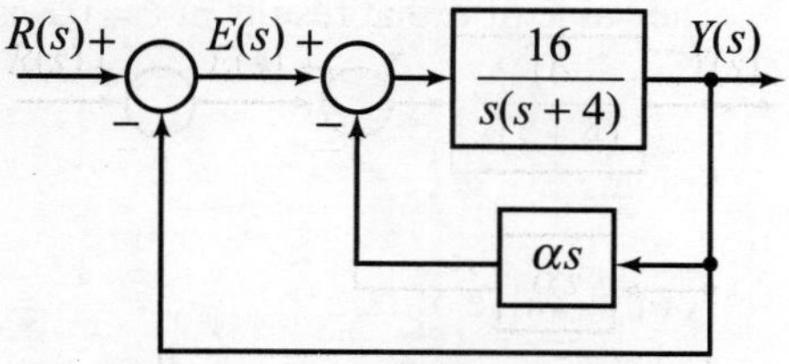

Fig. 7.46 *A unity-feedback system with a minor rate-feedback loop*

Solution: (a) With $\alpha = 0$, the forward-path transfer function,

$$G(s) = \frac{16}{s(s+4)}$$

Velocity error constant $K_v = \lim_{s \to 0} sG(s) = 4$

$$e_{ss}\big|_{\text{unit ramp}} = \frac{1}{K_v} = 0.25$$

The characteristic equation of the system is

$$s^2 + 4s + 16 = 0$$

Comparison with the standard characteristic equation,

$$s^2 + 2\zeta\omega_n s + \omega_n^2 = 0$$

gives $\omega_n = 4$, and $\zeta = 0.5$

Therefore, $M_p = e^{-\pi\zeta/\sqrt{1-\zeta^2}} \times 100 = 16.3\%$

(b) With rate feedback, the forward-path transfer function is obtained by eliminating the inner loop:

$$G(s) = \frac{\dfrac{16}{s(s+4)}}{1+\dfrac{16\alpha s}{s(s+4)}} = \frac{16}{s(s+4+16\alpha)} \tag{7.61}$$

$$K_v = \frac{16}{4+16\alpha} = \frac{4}{1+4\alpha}$$

The characteristic equation now takes the form

$$s^2 + (4 + 16\alpha)s + 16 = 0 \tag{7.62}$$

Corresponding to 1.5% overshoot, the damping ratio ζ is given by

$$\zeta^2 = \frac{[\ln(0.015)]^2}{[\ln(0.015)]^2 + \pi^2}$$

Solving, we get $\zeta = 0.8$.

The value of α that results in $\zeta = 0.8$ is given by the following equation (refer to Eqn. (7.62)):

$$2 \times 0.8 \times 4 = 4 + 16\alpha$$

Solving, we get $\alpha = 0.15$.

For this value of α, $e_{ss}|_{\text{unit ramp}} = \dfrac{1}{K_v} = \dfrac{1+4\alpha}{4} = 0.4$

(c) The system of Fig. 7.46 may be extended by including an amplifier K_A as shown in Fig. 7.47. The forward-path transfer function now becomes

$$G(s) = \frac{16K_A}{s(s+4+16\alpha)} \quad (7.63)$$

$$K_v = \frac{16K_A}{4+16\alpha} = \frac{4K_A}{1+4\alpha}$$

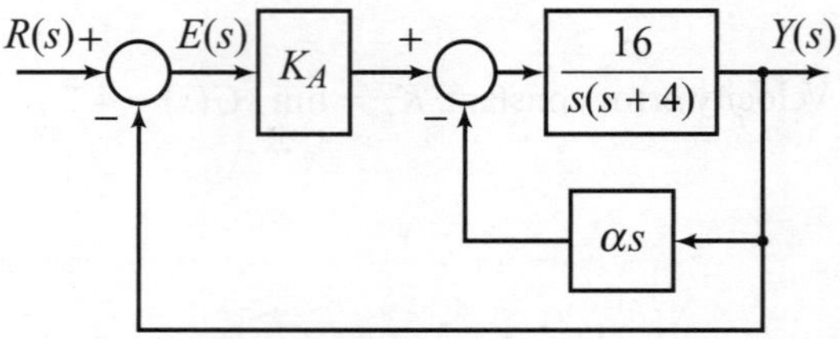

Fig. 7.47 *Amplifier added to the system shown in Fig. 7.46*

The required $K_v = 4$. Therefore,

$$\frac{4K_A}{1+4\alpha} = 4 \quad (7.64)$$

The characteristic equation of the extended system is

$$s^2 + (4 + 16\alpha)\, s + 16K_A = 0 \quad (7.65)$$

The requirement of $\zeta = 0.8$ gives rise to the following equation:

$$2 \times 0.8 \times \sqrt{16\,K_A} = 4 + 16\alpha \quad (7.66)$$

From Eqns (7.64) and (7.66), we obtain

$$K_A = 2.56;\ \alpha = 0.39$$

CONCLUDING REMARKS

PID controllers are probably the most widely used industrial controllers. This chapter concentrated on the basic characteristics of the individual modes of control—P, I, and D; and simple closed-loop analysis for the different combinations of these modes of control.

What we have learnt?

- Increasing the proportional gain speeds up the transient response (reference tracking response and disturbance rejection response), and decreases (but does not entirely eliminate) the output offset from the desired constant reference value, or the output offset from zero value due to the constant disturbance. Increasing the gain may lead to instability. Increasing the gain generally increases the size of control signal; this may lead to saturation or limiting problems with the system actuators.

- The presence of integral action in a controller usually leads to a wider range of closed-loop transient responses (both speeds and magnitudes), sometimes unstable ones. It generally slows down the response. The presence of integral action eliminates constant offset signals in steady-state for both reference tracking and disturbance rejection responses.
- The presence of derivative action in a controller leads to wide range of responses (both reference tracking and disturbance rejection) and can be used to tune response damping; generally speeds up the response. Derivative control has *no effect* on constant offsets in either reference tracking or disturbance rejection responses.
- The classical approach treats the design problem as a parameter-search problem. The design of PI/PD/PID controller is a simpler search problem with 2 to 3 trial parameters; the filtering and other signal conditioning are independently designed. The lag/lead/lag-lead compensator design involves 3 to 5 parameters; these control functions are, in fact, nothing but PI/PD/PID controllers with filtering and other signal conditioning embedded in the design procedure.
- In process-control field (control of pressure (liquid-level in a storage tank), temperature, etc), systems are relatively (with respect to position/speed control systems) slow and complex, and in most cases, the time delay in signal flow between subsystems results in dead-time effects. Even where a pure dead-time element is not present, the complexity of the process (which will typically contain several first-order lags) will often result in a response which has the appearance of a pure dead-time element. Modeling of such complex systems is a very difficult task. However, many years of experience has proved that the controller based on approximate process models (open-loop process reaction curve characterized by three parameters: the gain, the dead-time, and the time-constant; and closed-loop model characterized by ultimate gain and ultimate period) are quite versatile. These models can be experimentally determined by performing a step test. The crudity of such a model precludes the use of sophisticated design techniques (developed in Chapters 9 and 12). These models are useful in the simple 'experimental design' approach called *controller tuning* where a controller based on approximate process characterization is first installed, and then controller parameters are tuned by experiment.
- Increasingly controllers are implemented using op amps and embedded computers. We have learnt about op amp realization in this chapter; the digital implementation will appear later in Chapter 13.

Review Questions

7.1 Consider a unity feedback system with forward-path transfer function $D(s)G(s)$, where $G(s)$ is the transfer function of the plant and $D(s)$ corresponds to controller. Show that, in general, increased system gain and/or addition of an integrator in $D(s)G(s)$ tend to decrease steady-state errors; however, both large system gain and the integrator in $D(s)G(s)$ have destabilizing effects on the system.

7.2 A unity feedback system has a plant

$$G(s) = \frac{K}{s(\tau s + 1)}$$

with a cascade controller

$$D(s) = K_c(1 + T_D s)$$

Describe the effects of K_c and T_D on steady-state error, settling time, and peak overshoot of the system response.

7.3 A unity feedback system has a plant

$$G(s) = \frac{K}{s(\tau s + 1)}$$

with a cascade controller

$$D(s) = K_c\left(1 + \frac{1}{T_I s}\right)$$

Describe the effects of K_c and T_I on steady-state error, settling time, and peak overshoot of the system response.

7.4 (a) Show with the help of examples that introduction of derivative mode of control in a feedback system with proportional control, makes the system response less oscillatory. What is its effect on steady-state accuracy?

(b) Derivative control may be applied to the system error or the controlled variable. Which scheme do you prefer? Why?

(c) When do we introduce integral mode in a feedback system with proportional-derivative control?

7.5 Sketch a typical process reaction curve for an industrial process and explain how it can be used to obtain a first-order plus dead-time model of the process.

7.6 (a) What is QDR response?

(b) Describe two tuning methods, one based on ultimate gain and period and the other based on process reaction curve. How do we make a selection for a process control application?

7.7 Sketch op amp circuits capable of providing outputs which approximate to

(a) proportional plus integral of the input;

(b) proportional plus derivative of the input; and

(c) proportional plus integral plus derivative of the input.

Self-Appraisal Quiz

A set of questions with multiple answer choices is given below. These questions have been designed to capture important aspects of the subject covered in the chapter. For each of these questions, identify the best of the given answer choices. Compare your answers with the Master Key given at the end of the session. This quiz should serve the purpose of a self-appraisal test for readers.

7.1 It is desired that the output θ changes by unit step with zero error at steady state (Fig. Q7.1). We can achieve it by injecting

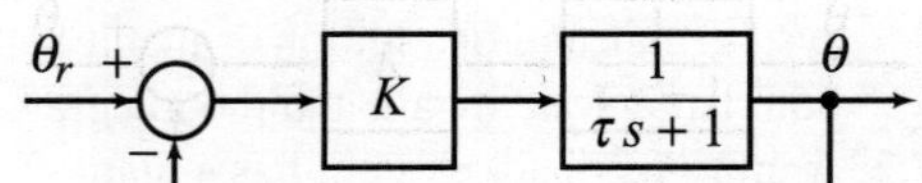

Fig. Q7.1

(A) θ_r = unit step

(B) θ_r of step size $(1 + K)/K$

(C) θ_r of step size $K/(K + 1)$

(D) None of the answers in (A), (B), and (C) is correct

7.2 A feedback control system of a machine tool is shown in Fig. Q7.2. The setpoint value in order to make the steady-state response value equal to the desired value of y_r, is equal to

(A) y_r (B) $0.9\, y_r$

(C) $1.11\, y_r$ (D) None of the statements in (A), (B) or (C) is correct.

7.3 A proportional controller, K_P, is used with a first-order system $G(s) = K/(\tau s + 1)$ in unity-feedback structure. Increasing K_P will

(A) increase the time constant and decrease the steady-state error to step inputs

(B) decrease the time constant and decrease the steady-state error to step inputs

(C) decrease the time constant and increase the steady-state error to step inputs

(D) increase the time constant and increase the steady-state error to step inputs

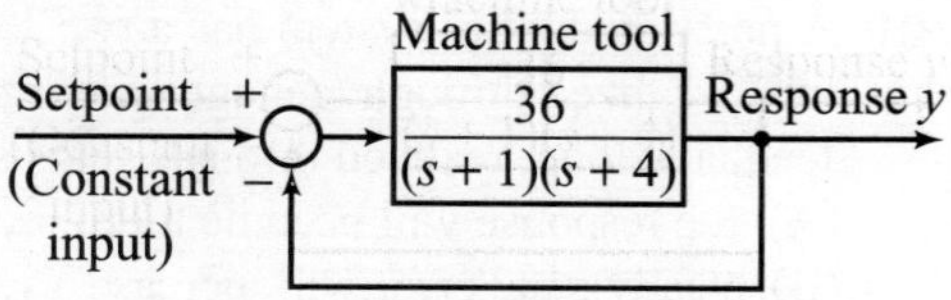

Fig. Q7.2

7.4 Consider the system of Fig. Q7.4. Increasing proportional gain will

(A) increase the overshoot and decrease the steady-state error to ramp inputs

(B) decrease the overshoot and increase the steady-state error to ramp inputs

(C) increase the overshoot and increase the steady-state error to ramp inputs

(D) decrease the overshoot and decrease the steady-state error to ramp inputs

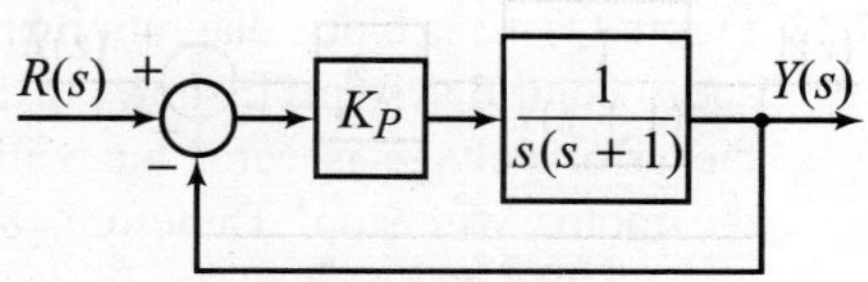

Fig. Q7.4

7.5 A dc motor model is shown in Fig. Q7.5. A unity-feedback speed control and a unity-feedback position control system are built using this model (with proportional control only). Steady-state error to step input is

(A) zero in both speed and position controls

(B) zero in speed control but not in position control

(C) zero in position control but not in speed control

(D) non-zero in both speed and position controls

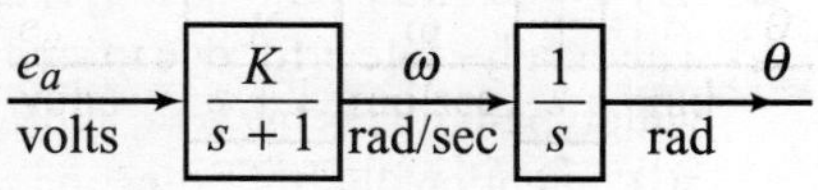

Fig. Q7.5

7.6 If there is a constant error in the output signals, derivative control will

(A) reduce the error to zero

(B) reduce the error but not necessarily to zero

(C) have no effect on the error

(D) increase the error

7.7 An integral controller K_I/s is used with a first-order system $G(s) = K/(\tau s + 1)$ in unity-feedback structure. Increasing K_I will

(A) decrease damping with no change in steady-state offset to step commands.

(B) decrease damping with decrease in steady-state offset to step commands.

(C) increase damping with no change in steady-state offset to step commands.

(D) increase damping with decrease in steady-state offset to step commands.

7.8 Derivative error compensation is employed in feedback control systems to

(A) increase the effective damping in the system

(B) decrease the effective damping in the system

(C) improve the steady-state response of the system

7.9 Tachogenerator feedback is sometimes used in position control systems to

(A) increase the effective damping in the system

(B) decrease the effective damping in the system

(C) improve the steady-state response of the system

7.10 A position control system has a two-loop configuration. The minor loop is a velocity feedback loop, realized through tachogenerator mounted on the motor shaft. The major loop is a position feedback loop, realized through gears and potentiometer. When the minor loop is broken

(A) the response will become more oscillatory
(B) steady-state error will increase
(C) Both the answers in (A) and (B) are correct
(D) None of the answers in (A) and (B) is correct

7.11 Integral error compensation is employed in feedback control systems to

(A) improve damping
(B) improve speed of response
(C) reduce steady-state error

7.12 Figure Q7.12 is block diagram representation of a speed control system. It shows that when $\omega = \omega_r$, the error voltage is zero; but with zero voltage the motor will stop. Therefore, $\omega = \omega_r$ cannot be achieved at steady state. To achieve $\omega = \omega_r$ at steady state,

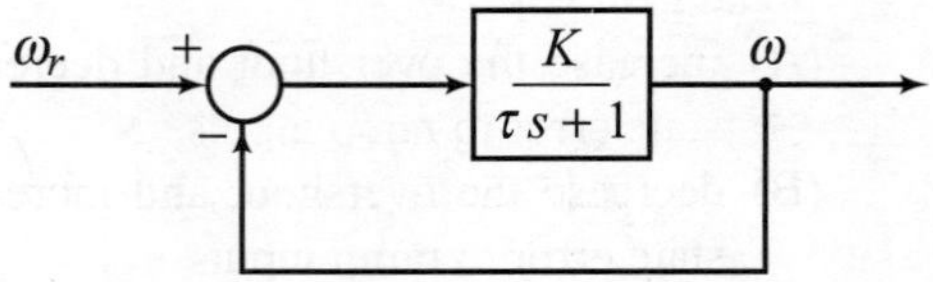

Fig. Q7.12

(A) give input voltage larger than that corresponding to the desired value of ω at steady state
(B) use integral control to provide voltage input to the motor when $\omega = \omega_r$
(C) Both the options in (A) and (B) can yield the desired performance
(D) None of the options in (A) and (B) can yield the desired performance

7.13 The steady-state error due to step commands can be eliminated from proportional control systems with type-0 plants by

(i) intentionally 'misadjusting' reference input
(ii) introducing integral mode in the controller

(A) None of the above statements is true
(B) Statement (i) is true but statement (ii) is false
(C) Statement (i) is false but statement (ii) is true
(D) Both the statements are true

7.14 Consider the system of Fig. Q7.14. The steady-state offset is zero when

(A) $K_P = \infty, K_D = \infty$
(B) $K_P = 0, K_D =$ any value
(C) $K_P = \infty, K_D =$ any value
(D) no values of K_P and K_D

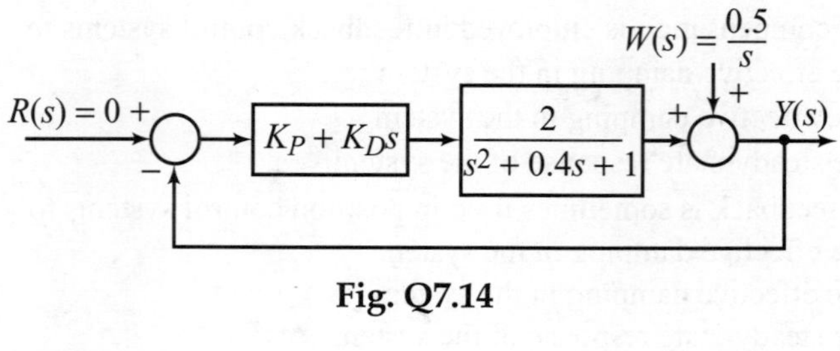

Fig. Q7.14

7.15 Consider the system of Fig. Q7.15. $R(s)$ and $W(s)$ are step signals.
(A) Steady-state error to $R(s)$ is zero and to $W(s)$ is nonzero
(B) Steady-state error to $R(s)$ is nonzero and to $W(s)$ is zero
(C) Steady-state error to both $R(s)$ and $W(s)$ is nonzero
(D) Steady-state error to both $R(s)$ and $W(s)$ is zero

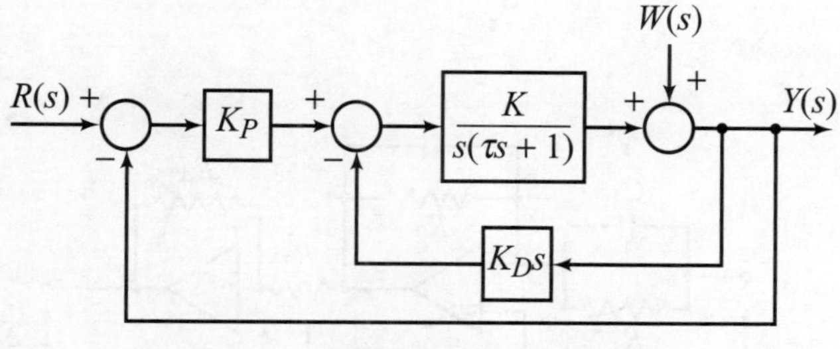

Fig. Q7.15

7.16 Consider the system of Fig. Q7.16. The steady-state error is less than 2% to step inputs for
(A) any value of K_I
(B) no value of K_I
(C) $K_I > 50$
(D) None of the answers is (A), (B), and (C) is correct

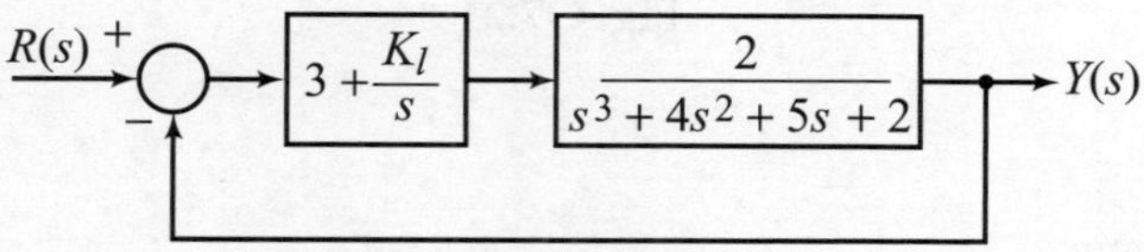

Fig. Q7.16

7.17 Consider the system of Fig. Q7.16. The steady-state error is less than 2% to ramp inputs for
(A) any value of K_I
(B) no value of K_I
(C) $K_I > 50$
(D) None of the answers is (A), (B), and (C) is correct

7.18 A type-1plant is changed to type-2 feedback system by the following cascade control action:
(A) PD (B) PI (C) Either PD or PI (D) Neither PD nor PI

7.19 A process was subjected to a step input of magnitude 0.8, and the reaction curve shown in Fig. Q7.19 was obtained. The parameters K, τ_D and τ of the process model $G(s) = Ke^{-s\tau_D}/(\tau s+1)$ are, respectively,
(A) 0.64, 7, 62
(B) 0.64, 7, 55
(C) 0.8, 7, 62
(D) 0.8, 7, 55

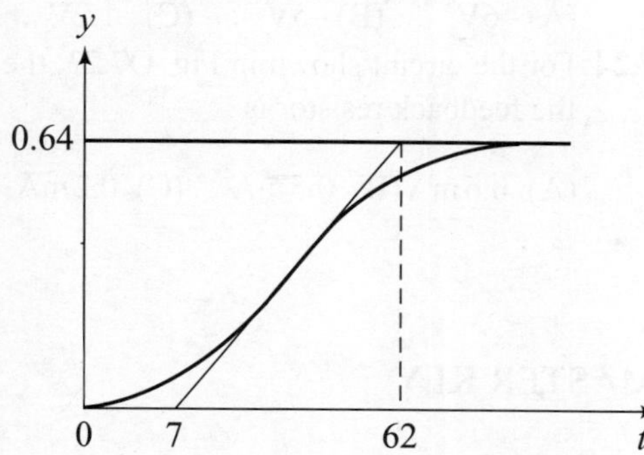

Fig. Q7.19

7.20 The characteristic equation of a unity feedback system is

$$s^3 + 3s^2 + 2s + K = 0$$

The parameters ultimate gain K_{cu}, and ultimate period T_u, of the system are
(A) 6, 4.44 sec (B) 4.44, 6 sec (C) Neither (A) nor (B) is correct.

7.21 Consider the op amp circuit shown in Fig. Q7.21. It can be used as
(A) a lead compensator only
(B) a lag compensator only
(C) either a lead or a lag compensator
(D) neither a lead nor a lag compensator

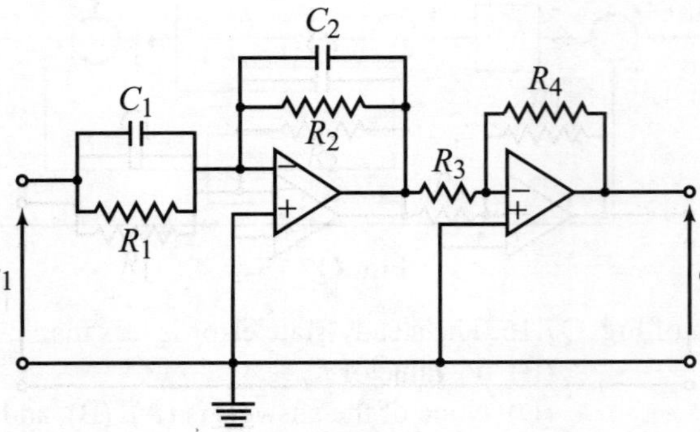

Fig. Q7.21

7.22 For an ideal op amp:
(i) the differential voltage across the input terminals is zero;
(ii) the current into the input terminals is zero;
(iii) the input resistance is zero;
(iv) the output resistance is zero.
Which statement is not true?
(A) (i) (B) (ii) (C) (iii) (D) (iv)

7.23 For the circuit shown in Fig. Q7.23, voltage e_0 is
(A) –6V (B) –5V (C) –1.2V (D) –0.2V

7.24 For the circuit shown in Fig. Q7.23, the current through the feedback resistor is
(A) 0.6 mA (B) 0.5 mA (C) 0.2 mA (D) $\frac{1}{12}$ mA

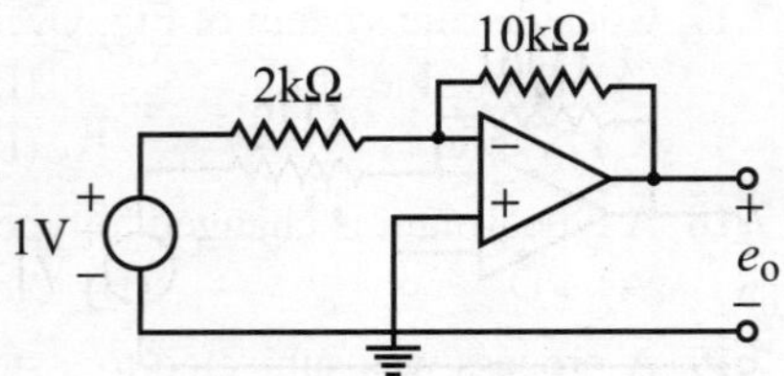

Fig. Q7.23

MASTER KEY

7.1 (B)	7.6 (C)	7.11 (C)	7.16 (D)	7.21 (C)
7.2 (C)	7.7 (A)	7.12 (C)	7.17 (B)	7.22 (C)
7.3 (B)	7.8 (A)	7.13 (D)	7.18 (B)	7.23 (B)
7.4 (A)	7.9 (A)	7.14 (C)	7.19 (D)	7.24 (B)
7.5 (C)	7.10 (A)	7.15 (D)	7.20 (A)	

Problems

7.1 (a) A unity feedback temperature control loop in a processing plant has open-loop data such that there is a 5°C output temperature to a 10% change in input signal level, and a measured time constant of 10 minutes. Derive a first-order process model. Find the closed-loop time constant, and steady-state error to unit-step reference input.

(b) It is desired to cascade a proportional controller to ensure faster response and better steady-state behavior. Find a proportional gain to achieve a closed-loop time constant of 3 minutes. What is the effect of this proportional controller on steady-state behavior?

7.2 A temperature control system has the block diagram given in Fig. P7.2. The input signal is a voltage and represents the desired temperature θ_r. Find the steady-state error of the system when θ_r is a unit step function and (i) $D(s) = 1$, (ii) $D(s) = 1 + \frac{0.1}{s}$, (iii) $D(s) = 1 + 0.3s$. What is the effect of the integral term in the PI controller, and derivative term in the PD controller on the steady-state error ?

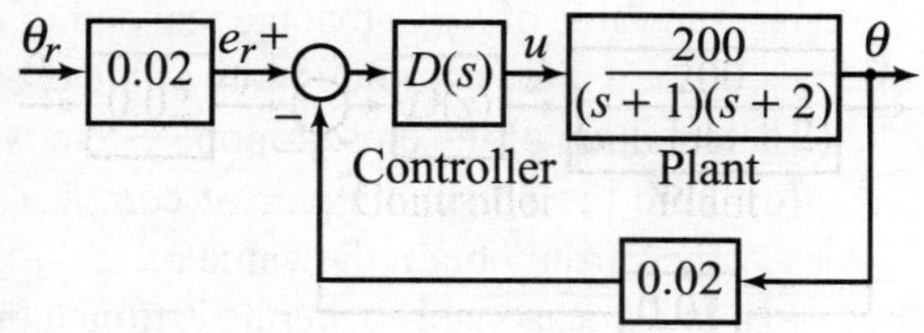

Fig. P7.2

7.3 Consider the closed-loop system of Fig. P7.3. Compute the characteristic equation and therefrom determine the stability of the system when

(a) $D(s) = 1$, (b) $D(s) = 1 + 2s$, and

(c) $D(s) = 1 + \frac{2}{s}$.

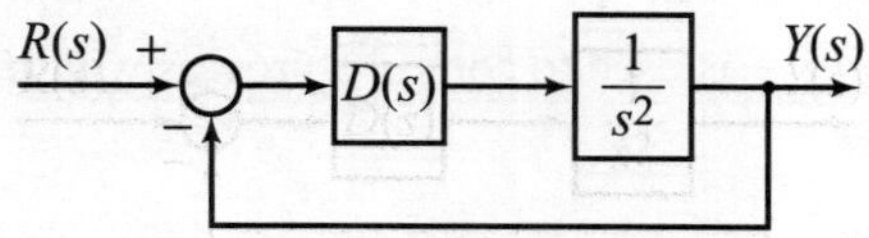

Fig. P7.3

What is the effect of derivative term in the PD controller, and integral term in the PI controller on system stability?

7.4 Figure P7.4 shows a position control system.

(a) Compute the values of K_c and T_D so that the characteristic equation of the closed-loop system will have roots at $-1 \pm j1.732$.

(b) We know that it is advantageous to avoid derivative action on the error signal if it is possible to obtain equivalent effect with an alternative control configuration. Suggest an alternative control scheme that results in the same characteristic roots of closed-loop system as in part (a) and avoids derivative action on the error signal.

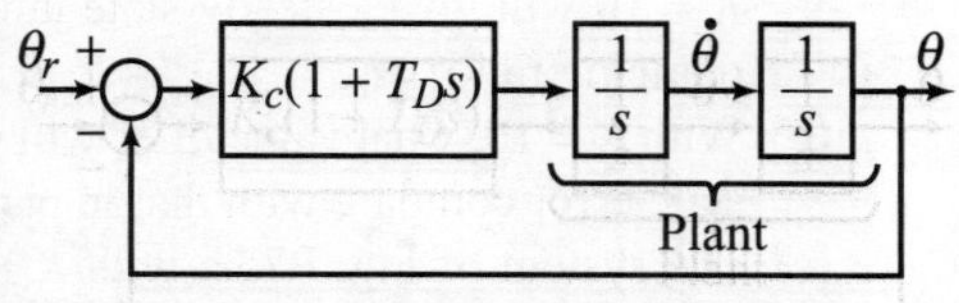

Fig. P7.4

7.5 Consider the control system in Fig. P7.5 in which a proportional compensator is employed. A specification on the control system is that the steady-state error must be less than two per cent for constant inputs.

(a) Find K_P that satisfies this specification.

(b) If the steady-state criterion cannot be met with a proportional compensator, use a dynamic compensator $D(s) = 3 + \frac{K_I}{s}$. Find the range of K_I that satisfies the requirement on steady-state error.

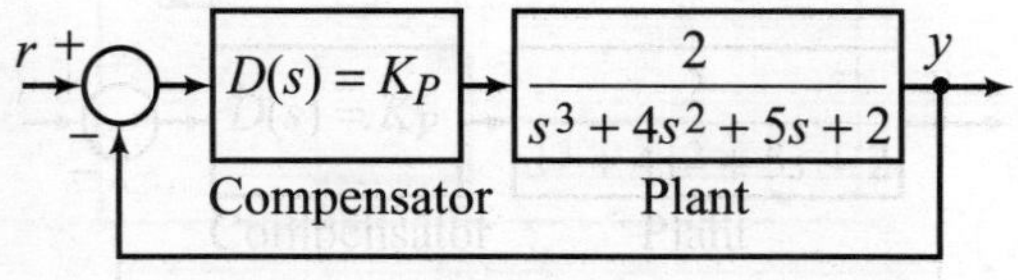

Fig. P7.5

7.6 A control system with a PD controller is shown in Fig. P7.6.

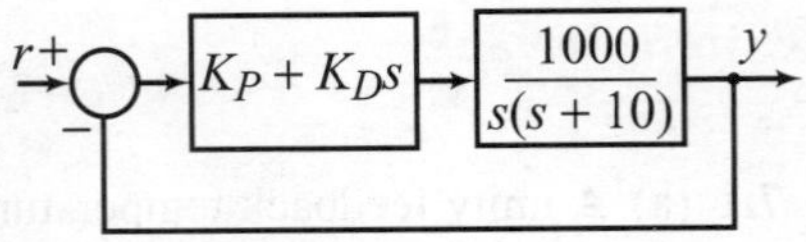

Fig. P7.6

(a) Find the values of K_P and K_D so that the velocity error constant K_v is 1000, and the damping ratio is 0.5.

(b) Is damping ratio ζ a reasonably accurate estimate of peak overshoot of the unit-step response? If not, why?

MATLAB Exercise Find the peak overshoot using MATLAB, and compare it with one estimated from ζ value.

7.7 A control system with a type-0 process and a PI controller is shown in Fig. P7.7.

(a) Find the value of K_I so that the velocity error constant K_v is 10.

(b) With the value of K_I found in part (a), determine the value of K_P so that the real part of the conjugate pair of complex roots of the characteristic equation of the closed-loop system is –1.

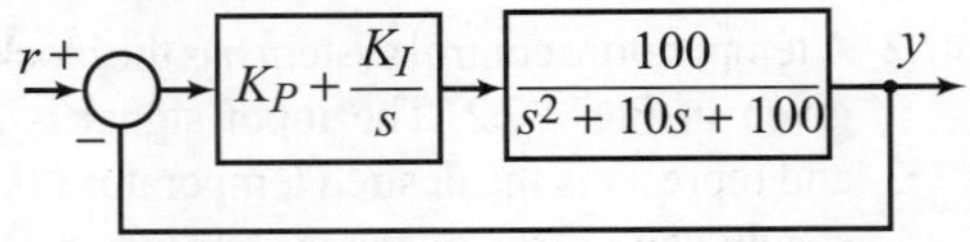

Fig. P7.7

(c) Find the conjugate pair of complex roots and therefrom obtain the value of ζ.

(d) Is ζ a reasonably accurate estimate on peak overshoot of the unit-step response? Give reasons.

7.8 A three-term controller with transfer function

$$D(s) = 20\left[1 + \frac{1}{T_I s} + T_D s\right]$$

is used to control a process with transfer function

$$G(s) = \frac{4}{s^2 + 8s + 80}$$

The closed-loop system is a unity-feedback structure with $D(s)$ and $G(s)$ in cascade.

(a) If integral action is not employed, find the derivative time T_D required to make the closed-loop damping ratio unity.

(b) A simulation study on your design in part (a) shows overshoot in unit-step response. Give reasons.

(c) If the value of derivative time is maintained as in part (a), determine the minimum value of reset time T_I that can be used without instability arising.

7.9 (a) For the motor position servo shown in Fig. P7.9a without the rate feedback ($K_t = 0$), show that $K = 10$ will give a steady-state unit-ramp following error of 0.1. What is the corresponding damping ratio?

(b) With $K = 10$, what value of K_t will give a system damping ratio of 0.5? How does the steady-state error compare with that in part (a)?

(c) The system of Fig. P7.9a with $K = 10$ has been extended by including the amplifier K_A as shown in Fig. P7.9b. Determine whether it is now possible to achieve both the steady-state error of 0.1 and damping ratio of 0.5. If so, find the values of K_t and K_A required.

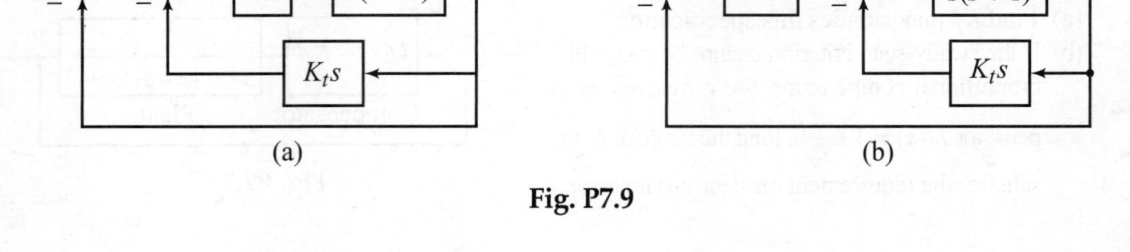

Fig. P7.9

7.10 The block diagram of a control system is shown in Fig. P7.10.

(a) Find the values of K_1 and K_2 so that peak overshoot is 10% and settling time is 0.05 sec.

(b) For the values of K_1 and K_2 obtained in part (a), find the step, ramp and parabolic error constants.

MATLAB Exercise Using MATLAB, plot the time response to step command. Also obtain the plot in simulink window.

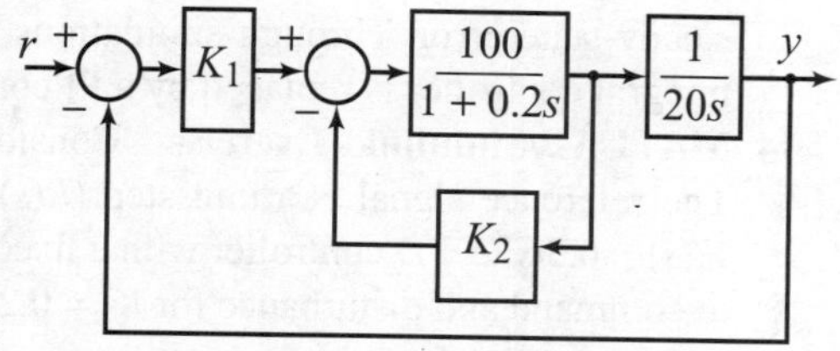

Fig. P7.10

7.11 The block diagram of a motor position servo with a load disturbance torque T_w is shown in Fig. P7.11. Find K and K_t to obtain a system damping ratio 0.5 and 5% steady-state error for step input T_w.

MATLAB Exercise Using MATLAB, plot the time responses to step command and step disturbance, respectively. Also obtain these plots in Simulink window.

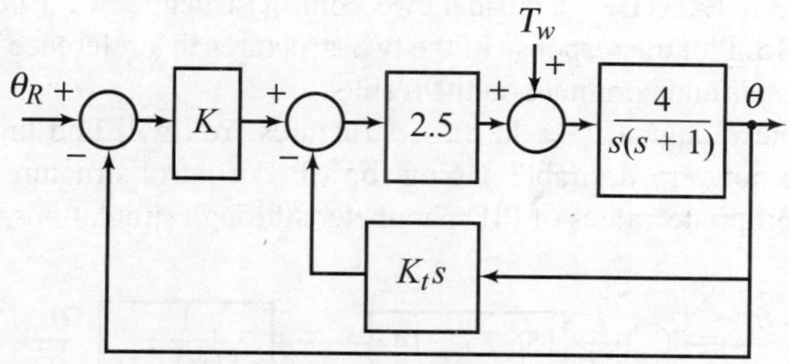

Fig. P7.11

7.12 MATLAB/Simulink Exercise Consider a heating control system with the plant model

$$Y(s) = \frac{0.3}{2s+1}U(s)$$

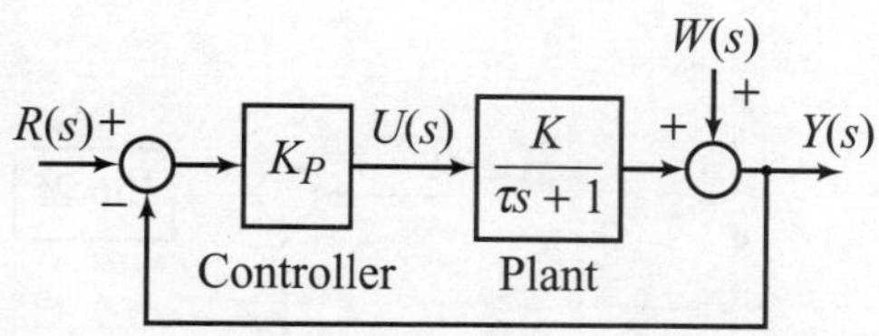

Fig. P7.12

The input signal $U(s)$ is the power in kW from the heater and the output signal $Y(s)$ is the resulting temperature. The time constant of the process is 2 hrs. The sensor has a very short time constant in comparison to the process time constant; so we can model the system as a unity feedback system (Fig. P7.12). The step reference signal $R(s) = 1/s$, and the constant load disturbance effect is modeled as $W(s) = 0.5/s$.

Using MATLAB/Simulink, plot the time response to reference and disturbance inputs for different values of K_P. Also plot the signal $u(t)$ for these values of K_P. Comment upon the effects of increasing the values of K_P on transient response and steady-state error, and also on the required control signal.

7.13 MATLAB/Simulink Exercise Reconsider the heating control system that was simulated in Problem 7.12 for proportional control. We replace the proportional controller K_P by integral controller K_I/s. Plot the closed-loop system response to command and disturbance signals for $K_I = 0.25$, 0.417, and 1.8. Comment upon the effects of increasing K_I on transient response and steady-state error. Through simulations, examine the effect on transient performance when the integral controller is replaced by a PI controller $K_P + K_I/s$, with fixed proportional gain $K_P = 4$.

7.14 MATLAB/Simulink Exercise Consider a position control system shown in Fig. P7.14. The reference signal is a unit step ($R(s) = 1/s$) and we model the constant load disturbance by $W(s) = 0.5/s$. A PD controller with a fixed proportional gain $K_P = 0.2$ is used. Plot output response to command and disturbance for $K_D = 0.2$, and 1.22. Comment on the effect of K_D on transient and steady-state performance.

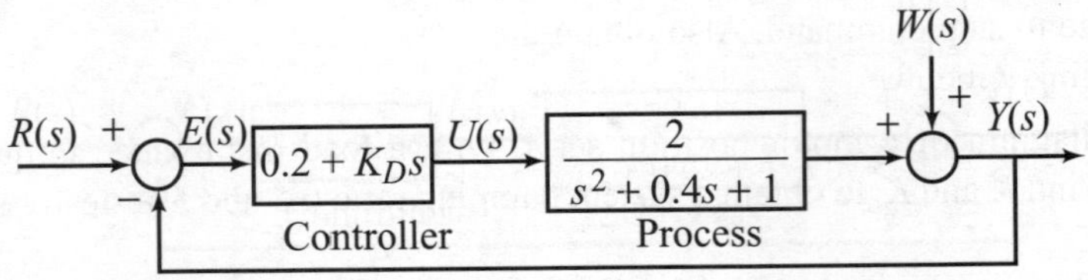

Fig. P7.14

7.15 MATLAB/Simulink Exercise Consider two control structures of a position control system, shown in Figs P7.15. Plot the response of the two structures to a reference input of $\theta_r = 0.1$ occurring at $t = 0.5$ seconds and comment on the results.

Observe the control signal trace in these structures. You will find high control-signal magnitudes which are not very desirable. Going for a PID control structure is one solution to this problem. Find appropriate values of PID parameters through simulations.

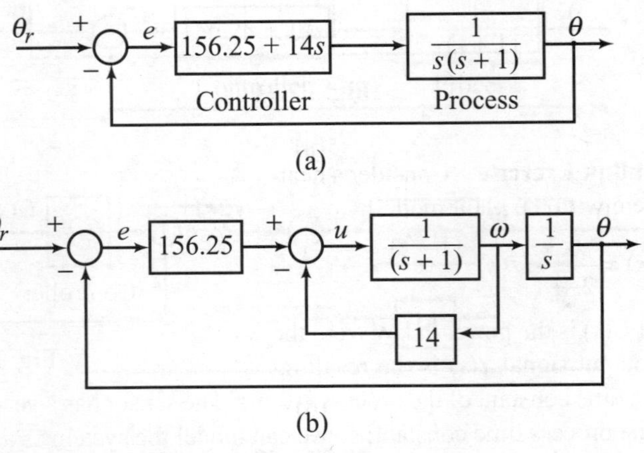

Fig. P7.15

7.16 From a test conducted on a temperature control loop, it is determined that the controller gain required to cause sustained oscillations is 1.2 and the period of oscillations is 4.5 min. Determine the QDR tuning parameters for

(a) a PI controller

(b) a PID controller

7.17 The following table gives the measured open-loop response of a process using unit-step input. Approximate this response curve by a first-order plus dead-time model.

Time (min) t	0	0.5	1.0	1.5	2.0	2.5	3.0	3.5
Response y	0	0.009	0.056	0.15	0.28	0.428	0.572	0.7
t	4.0	4.5	5.0	5.5	6.0	7.0		
y	0.802	0.8876	0.924	0.953	0.967	0.973		

(a) Tune a proportional-only controller for QDR response.
(b) Tune a PI controller for QDR response.

7.18 Show that the parallel arrangement of 'constant gain amplifier' and 'differentiator' op amp circuits shown in Fig. P7.18 is a realization of the transfer function

$$D(s) = K_P + K_D s$$

of a 'proportional + derivative' compensator. Determine the parameters K_P and K_D in terms of the parameters of the circuit.

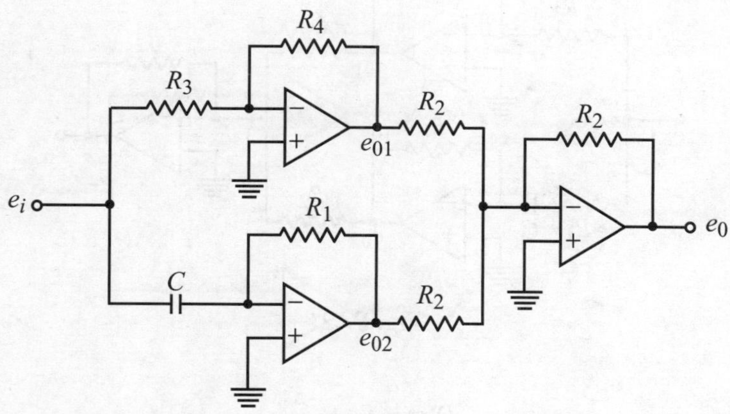

Fig. P7.18

7.19 Show that the parallel arrangement of 'constant gain amplifier' and 'integrator' op amp circuits shown in Fig. P7.19 is a realization of the transfer function

$$D(s) = K_P + K_I / s$$

of a 'proportional + integral' compensator.
Determine the parameters K_P and K_I in terms of the parameters of the circuit.

7.20 Consider the PD circuit of Fig. P7.18, cascaded with the PI circuit of Fig. P7.19 (In principle, it makes no difference whether the PD element or the PI element comes first). Show that the cascaded circuit realizes the transfer function

$$D(s) = K_P + K_I/s + K_D s$$

of a 'proportional + integral + derivative' compensator.

7.21 (a) Show that the parallel arrangement of 'constant gain amplifier' and 'bandwidth-limited differentiator' op-amp circuits shown in Fig. P7.21 is a realization of the transfer function

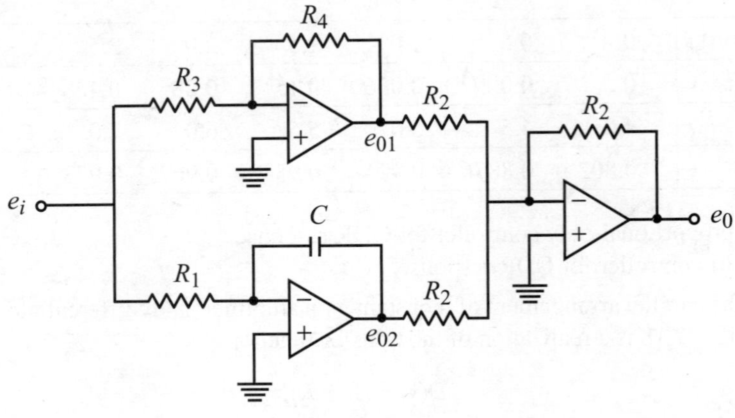

Fig. P7.19

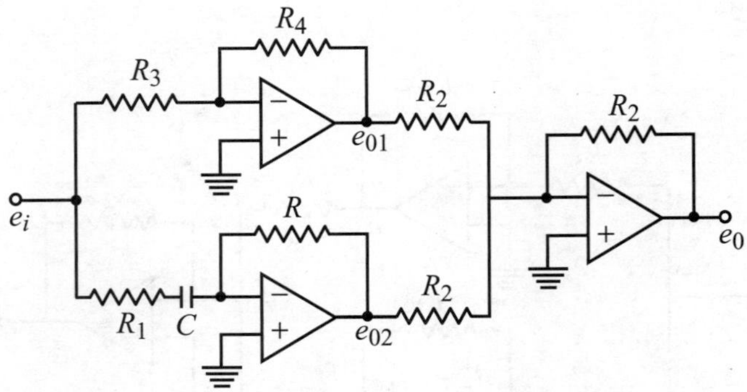

Fig. P7.21

$$D(s) = K_2 + \frac{K_1 s}{\tau_1 s + 1}$$

of a 'proportional + bandwidth-limited derivative' compensator.
Determine the parameters K_2, K_1, and τ_1 in terms of the parameters of the circuit.

(b) Compensators are sometimes formulated with a transfer function of the form (called *lead compensator*)

$$D(s) = K_c\left(\frac{\tau s + 1}{\alpha\tau s + 1}\right); \alpha < 1, K_c > 0, \tau > 0$$

Prove that this transfer function can be obtained with the parallel arrangement of op amp circuits shown in Fig. P7.21. Determine K_c, τ and α in terms of K_2, K_1, and τ_1.

7.22 (a) Show that the parallel arrangement of 'constant gain amplifier and gain-limited integrator' op amp circuits shown in Fig. P7.22 is a realization of the transfer function

$$D(s) = K_2 + \frac{K_1}{\tau_1 s + 1}$$

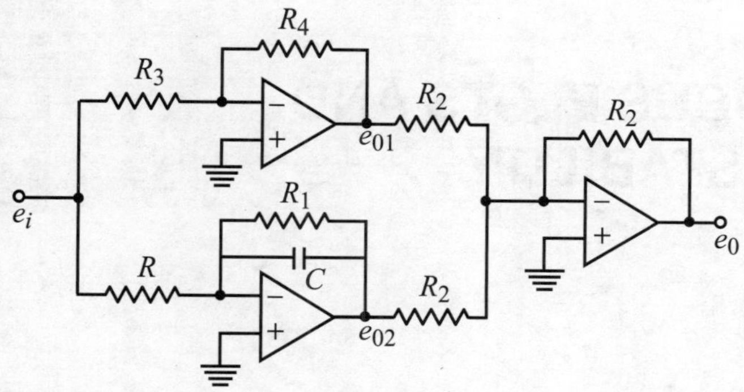

Fig. P7.22

of a 'proportional + gain-limited integral' compensator.
Determine the parameters K_2, K_1, and τ_1 in terms of the parameters of the circuit.

(b) Compensators are sometimes formulated with a transfer function of the form (called *lag compensator*)

$$D(s) = K_c\left(\frac{\tau s+1}{\beta\tau s+1}\right); \beta > 1, K_c > 0, \tau > 0$$

Prove that this transfer function can be obtained with the parallel arrangement of op amp circuits shown in Fig. P7.22. Determine K_c, τ and β in terms of K_2, K_1 and τ_1.

7.23 Consider the 'proportional + bandwidth-limited derivative' circuit of Fig. P7.21, cascaded with 'proportional + gain-limited integral circuit of Fig. P7.22. Show that the cascaded circuit realizes the transfer function

$$D(s) = K_c\left(\frac{\tau_1 s+1}{\alpha\tau_1 s+1}\right)\left(\frac{\tau_2 s+1}{\beta\tau_2 s+1}\right); \alpha < 1, \beta > 1, K_c > 0, \tau_1 > 0, \tau_2 > 0$$

of a lag-lead compensator.

ROOT LOCUS PLOTS AND SYSTEM STABILITY

8

Preview

In Chapter 5, we examined the characteristic equation of a feedback system algebraically using Routh stability criterion to determine ranges of certain parameters ensuring a stable operation. For stable systems, the relative stability in terms of minimum distance of the characteristic roots from the imaginary axis, and hence the largest time constant associated with the system was also considered. It is probably more insightful to think about the characteristic equation geometrically. The root locus technique, developed in this chapter, is intrinsically geometric as opposed to the algebraic approach of Routh criterion. It enables us to rapidly sketch the locus of all solutions to the characteristic equation. This sketch (root locus plot) offers great insight on the relative stability and the transient response performance of a closed-loop control system by showing how the locations of the closed-loop poles change as a specific parameter is varied. The learning objective for this chapter is to examine this graphical tool in detail with respect to its role in control system analysis and design. The next chapter extends the concept of root locus to the design of compensators.

8.1 INTRODUCTION

From the point of view of control system design, information on range of system parameters for stability is important. The Routh criterion was introduced in Chapter 5 for stability analysis, and we observed that the criterion is well suited for finding ranges of parameters for stability. Extension of the Routh criterion for *relative stability* analysis—finding minimum distance of the characteristic roots from the imaginary axis, and hence the largest time constant associated with the system, was also demonstrated.

The approach of stability analysis by the Routh criterion is purely algebraic. We can gain some insight into the system behavior by determining how the largest time constant would vary as a design parameter is varied. However, the criterion lacks the ability to give an overall picture of how the closed-loop response would vary with changes in the parameter.

We have seen in Chapter 6 that the transient response (a measure of relative stability) of a closed-loop system is directly related to the location of the closed-loop roots of the characteristic equation,

$$\Delta(s) = a_0 s^n + a_1 s^{n-1} + \cdots + a_{n-1}s + a_n = 0$$

An obvious solution to the relative stability analysis is factoring the characteristic polynomial $\Delta(s)$ using numerical techniques, determining its roots and therefrom the time functions that will show up in the transient response of the closed-loop system to a reference input. The trouble with the characteristic polynomial $\Delta(s)$ is that all of its coefficients are not available in numeric form; the design parameters are buried in the coefficients, and the only way to factor this polynomial is to use some iterative search technique—writing a computer program that would substitute a series of values for the parameters, factor the polynomial for these specific values of the parameters, and give the roots.

However, this is not the best way to proceed early in the design process. What we need early in the design process is a clear *qualitative* picture of how the closed-loop poles move around as the design parameters are varied. This qualitative nature of the solution will be examined in this chapter using the root-locus analysis. The *root locus technique* is intrinsically geometric as opposed to the algebraic approach of Routh criterion.

The root locus technique was introduced by Walter R Evans in 1948, and has been developed and utilized extensively in control engineering practice. Useful rules for graphically constructing the *root locus plot* enable us to rapidly sketch the locus of all solutions to the characteristic equation of the closed-loop system when a design parameter is varied. The sketch is usually only *approximate,* but even so, it offers great insight by showing how the location of the poles of the closed-loop transfer function change as design parameter is varied.

This chapter first introduces the basic concept of the root locus method and then presents useful rules for graphically constructing the root loci. What we learn in this chapter will be subjected to a test in the next chapter on design using root locus plots.

8.2 THE ROOT LOCUS CONCEPT

We introduce the root locus concept through an example. Consider the second-order system shown in Fig. 8.1, which represents a typical position control system. The plant consists of servomotor and load, driven by a power amplifier with gain K. The open-loop transfer function of the system is

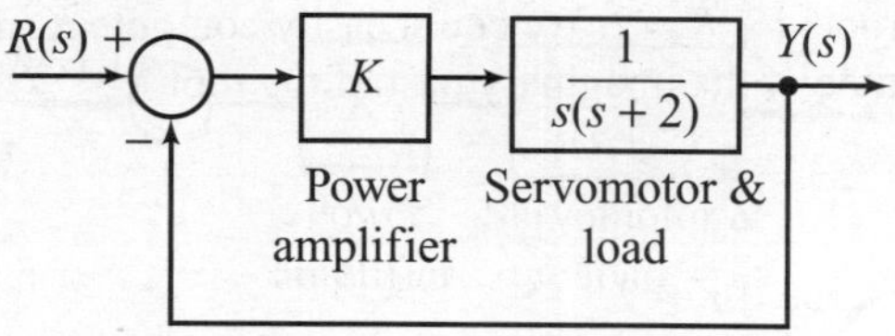

Fig. 8.1 *A position control system*

$$G(s) = \frac{K}{s(s+2)} \tag{8.1}$$

The open-loop poles marked × in Fig. 8.2 are at $s = 0$ and $s = -2$. The closed-loop transfer function of the system is

$$M(s) = \frac{G(s)}{1+G(s)} = \frac{K}{s^2+2s+K} \tag{8.2}$$

The characteristic equation is

$$\Delta(s) = s^2 + 2s + K = 0 \tag{8.3}$$

The second-order system under consideration is always stable for positive values of K. The relative stability of the system depends on the location of the characteristic roots (closed-loop poles)

$$s_{1,2} = -1 \pm \sqrt{1-K} \tag{8.4}$$

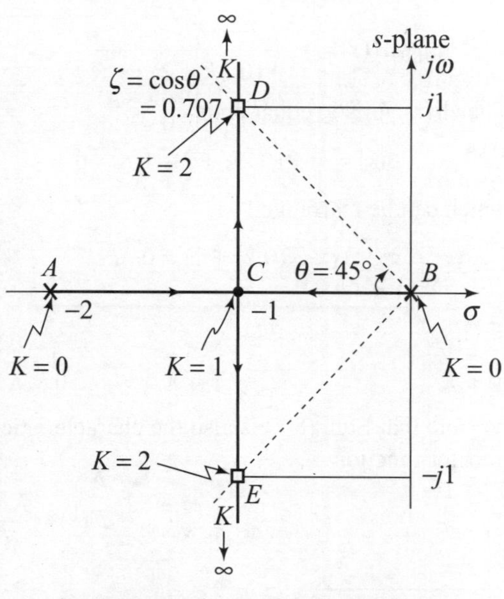

Fig. 8.2 *Root locus plot for Eqn. (8.3)*

and hence on the choice of the parameter K. As K is varied from zero to infinity, the characteristic roots move in the s-plane as shown in Fig. 8.2. At $K = 0$, the root s_1 is equal to the open-loop pole at $s = 0$, and the root s_2 is equal to the open-loop pole at $s = -2$. As K increases, the roots move toward each other. The two roots meet at $s = -1$ for $K = 1$. As K is increased further, the roots break away from the real axis, become complex conjugate, and since the real part of both the roots remains fixed at $s = -1$, the roots move along the line $\sigma = -1$.

A root locus of a system is a plot of the roots of the system characteristic equation (poles of the closed-loop transfer function) as some parameter of the system is varied.

The two branches[1] A-C-E and B-C-D of the plot of Fig. 8.2 are thus two *root loci* of the system of Fig. 8.1. Each root locus starts at an open-loop pole with $K = 0$ and terminates at infinity as $K \to \infty$. Each root locus gives one characteristic root (closed-loop pole) for a specific value of K.

The root locus plot gives us a great deal of information concerning the transient behavior of the system as the gain K is varied. We can see from the plot that for $0 < K < 1$, the roots are real and distinct and the system is overdamped. For $K = 1$, the roots are real and repeated and the system is critically damped. For $K > 1$, the roots are complex conjugate and the system is underdamped with the value of ζ decreasing as K increases. From the viewpoint of design, we see that by the choice of gain K, we can cause a characteristic root at any point on the root locus. For example, suppose that a system specification requires damping ζ equal to 0.707. The dashed lines in Fig. 8.2 correspond to $\zeta = 0.707$. The points when the root loci cross the dashed lines have been marked □. These points correspond to characteristic roots for $K = 2$. We could easily compute K in this case because we know that if $\zeta = 0.707$, the magnitude of the imaginary part of the root

$$\left|\sqrt{1-K}\right| = 1$$

is true if $K = 2$.

Consider now the second-order system shown in Fig. 8.3 which represents a typical position control system with velocity feedback. It is a multiloop system with closed-loop transfer function

$$M(s) = \frac{K}{s^2 + (0.2K + 2)\,s + K} \tag{8.5}$$

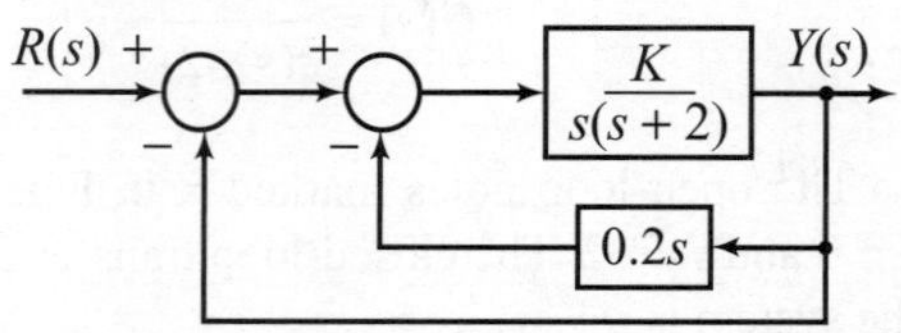

Fig. 8.3 *A position control system with velocity feedback*

The characteristic equation is

$$\Delta(s) = s^2 + (0.2K + 2)s + K = 0 \tag{8.6}$$

which can be rearranged as

$$s^2 + 2s + K(0.2s + 1) = 0$$

or

$$1 + \frac{0.2K(s+5)}{s(s+2)} = 0 \tag{8.7}$$

Note that Eqn. (8.7) is also the characteristic equation of a unity feedback system with (open-loop) transfer function

$$F(s) = \frac{K'(s+5)}{s(s+2)};\ K' = 0.2K \tag{8.8}$$

[1]Treating the two branches A-C-D and B-C-E as two root loci will give identical results.

which has two poles at $s = 0$ and $s = -2$, and a zero at $s = -5$. In Fig. 8.4, the open-loop poles (poles of $F(s)$) are indicated by × and the open-loop zero (zero of $F(s)$) is indicated by ○.

Let us examine the nature of characteristic roots

$$s_{1,2} = -(0.1K + 1) \pm \sqrt{(0.1K + 1)^2 - K} \tag{8.9}$$

As K is varied from zero to infinity, the characteristic roots move in the s-plane as shown in Fig. 8.4. At $K = 0$, the root s_1 is equal to the open-loop pole at $s = 0$ and the root s_2 is equal to the open-loop pole at $s = -2$. As K increases, the roots move towards each other. The two roots meet at $s = -1.127$ for $K = 1.27$. As K is increased further, the roots break away from the real axis and become complex conjugate. It turns out that the complex roots move along a circle[2] of radius $\sqrt{15}$ with center at $s = -5$.

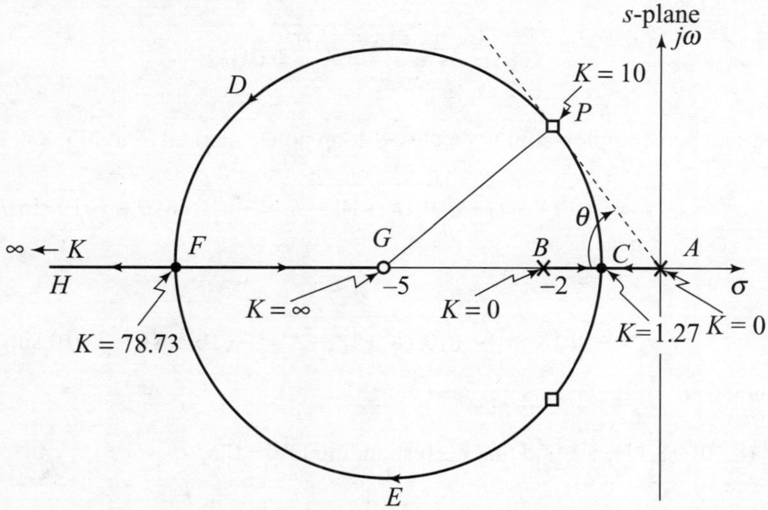

Fig. 8.4 *Root locus plot for Eqn. (8.6)*

The two complex-conjugate roots meet at $s = -8.873$ for $K = 78.73$. As K is increased further, the roots break away and depart in opposite directions along the real axis.

From Eqn. (8.9), we obtain

$$s_{1,2} = -(0.1K + 1) \pm (0.01K^2 - 0.8K + 1)^{1/2}$$

$$= -(0.1K + 1) \pm 0.1K\left(1 - \frac{80}{K} + \frac{100}{K^2}\right)^{1/2}$$

For large K,

$$s_{1,2} \cong -(0.1K + 1) \pm 0.1K\left(1 - \frac{80}{K}\right)^{1/2}$$

$$\cong -(0.1K + 1) \pm 0.1K\left(\left(1 - \frac{40}{K}\right)\left(1 - \frac{40}{K}\right)\right)^{1/2}$$

$$\cong -0.1K - 1 \pm 0.1K \mp 4$$

[2] This can easily be proved as we shall see later in Example 8.1.

Therefore, as $K \to \infty$, $s_1 \to -5$ and $s_2 \to -\infty$.
Figure 8.4 shows the two root loci: *A-C-D-F-G* and *B-C-E-F-H*.

As with the earlier example of Fig. 8.2, valuable information can be obtained from the root locus plot of Fig. 8.4. As K increases, the root loci move away from imaginary axis and the system becomes more stable. The least damped complex-conjugate poles are obtained by drawing straight line AP tangential to the circular locus as shown in Fig. 8.4.

By geometry,

$$\zeta_{min} = \cos\theta = \frac{AP}{AG} = \frac{\sqrt{(AG)^2 - (GP)^2}}{AG}$$

$$= \frac{\sqrt{25-15}}{5} = \frac{\sqrt{10}}{5} = 0.632$$

The least damped complex-conjugate closed-loop poles, marked □ in Fig. 8.4, are

$$s_1 = -(0.1K+1) + \sqrt{(0.1K+1)^2 - K} = -\sqrt{10}\cos\theta + j\sqrt{10}\sin\theta$$

$$= -2 + j\,2.45 \tag{8.10}$$

$$s_2 = -(0.1K+1) - \sqrt{(0.1K+1)^2 - K} = -\sqrt{10}\cos\theta - j\sqrt{10}\sin\theta$$

$$= -2 - j\,2.45 \tag{8.11}$$

From Eqns (8.10)–(8.11), we find that corresponding to $\zeta = 0.632$,

$$0.1K + 1 = 2$$

is true if $K = 10$. Therefore, $K = 10$ results in the least damping.

The two root locus plots we have presented, illustrate a number of properties which will be derived and discussed in the next section. First, a root locus plot is always symmetrical with respect to the real axis. This is so because the roots must always be complex conjugate. Second, a root locus plot is made of as many branches as there are poles in the open-loop transfer function, and each branch starts from one of the open-loop poles. The branches either go off to infinity or end on an open-loop zero.

The two cases we have discussed involved only a modest amount of analysis. The effort rapidly grows with the order of the system. The sketching rules given in the next section serve the purpose of avoiding direct solution for the roots. As one can imagine, repeated formation of the characteristic polynomial and factoring by hand calculator is a tedious job. Programs are now available [151–154] to generate the root locus plots; some of them based on repeated factoring and some that mechanize the sketching rules. Whatever the package available to us does, it will be useful to know its basis and its limitations. The hand-sketching process trains our intuition and helps us to avoid buying computer-generated trash results.

A graphical procedure for sketching root locus plots was developed by W R Evans in 1948. In the following, we present the locus equations which form the basis of the Evans method; guidelines for sketching root loci will be given in the next section.

8.2.1 The Locus Equations

Most control systems can be expressed in the form of Fig. 8.5, where the closed-loop transfer function is

$$M(s) = \frac{Y(s)}{R(s)} = \frac{G(s)}{1+G(s)H(s)} \tag{8.12}$$

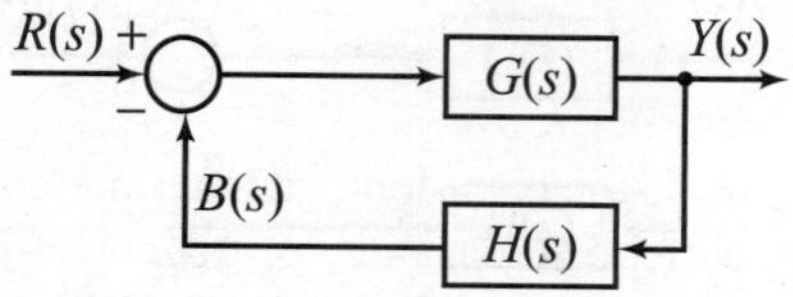

Fig. 8.5 *A feedback system*

The characteristic equation of the system is

$$1 + G(s)H(s) = 0 \tag{8.13}$$

The open-loop transfer function (refer to Eqn. (2.116c)) $G(s)H(s)$ is generally known in the factored form as it is obtained by modeling the transfer function of individual components comprising the system. Therefore, $G(s)H(s)$ can generally be expressed in either of the factored forms given below:

Pole-zero Form

$$G(s)H(s) = \frac{K\prod_{i=1}^{m}(s+z_i)}{\prod_{j=1}^{n}(s+p_j)}; \; m \le n, \, K \ge 0 \tag{8.14}$$

The function $G(s)H(s)$ approaches infinity as s approaches $-p_j$; $s = -p_j; j=1,2,\ldots, n$ are the poles of $G(s)H(s)$. The function $G(s)H(s)$ approaches zero as s approaches $-z_i; s=-z_i; i=1,2,\ldots,m$ are the *zeros* of $G(s)H(s)$.

For a rational function of s, i.e., a ratio of two polynomials of s, the total number of poles equals the total number of zeros when the s-plane includes the points at infinity. The function $G(s)H(s)$ given by Eqn. (8.14) has $(n - m)$ zeros at infinity, since

$$\lim_{s\to\infty} G(s)H(s) = \lim_{s\to\infty} \frac{K}{s^{n-m}} = 0$$

Thus, the function $G(s)H(s)$ in Eqn. (8.14) has n poles at $s = -p_j$, *m finite zeros at* $s = -z_i$, and $(n - m)$ zeros at infinity.

Time-constant Form

$$G(s)H(s) = \frac{K'\prod_{i=1}^{m}(\tau_{zi}s+1)}{\prod_{j=1}^{n}(\tau_{pj}s+1)}; \; m \le n, \, K \ge 0 \tag{8.15}$$

The pole-zero and time-constant forms are interrelated by the following expressions:

$$-z_i = -\frac{1}{\tau_{zi}}; i=1,2,\ldots,m; \; -p_j = -\frac{1}{\tau_{pj}}; j=1,2,\ldots,n \tag{8.16}$$

$$K = \frac{K' \prod_{j=1}^{n} p_j}{\prod_{i=1}^{m} z_i}$$

K' is the open-loop gain in time-constant form, K is the open-loop gain in pole-zero form, $-z_i$ are m open-loop zeros, and $-p_j$ are n open-loop poles. In all physically realizable systems, $n \geq m$.

The pole-zero form of Eqn. (8.14) is more convenient for drawing root locus and will be used throughout this chapter. ***The open-loop gain K is considered variable for drawing the root locus of the system.***

Substituting Eqn. (8.14) into Eqn. (8.13), we obtain the characteristic equation,

$$1 + = \frac{K \prod_{i=1}^{m}(s + z_i)}{\prod_{j=1}^{n}(s + p_j)} = 0 \tag{8.17}$$

or

$$\prod_{j=1}^{n}(s + p_j) + K \prod_{i=1}^{m}(s + z_i) = 0 \tag{8.18}$$

The characteristic equation given in (8.18) is in conventional polynomial form. The characteristic equation given in (8.17) is not in conventional form; however, the values of s for which it is satisfied will be same as those for which Eqn. (8.18) is satisfied provided that the numerator and denominator polynomials in Eqn. (8.17) have no common factors.

If the variable parameter is other than open-loop gain, the characteristic equation can be rearranged to obtain the form (8.17) with the parameter of interest appearing as the multiplying factor. The characteristic equation of a multiloop control configuration can also be brought to the form (8.17) by block diagram manipulation rules or by Mason's gain formula. In our study of the root locus technique, we therefore, consider a general characteristic equation,

$$1 + F(s) = 0 \tag{8.19a}$$

with

$$F(s) = \frac{K \prod_{i=1}^{m}(s + z_i)}{\prod_{j=1}^{n}(s + p_j)}; m \leq n \tag{8.19b}$$

where $K \geq 0$ is the *root locus gain*.

From Eqns (8.19) it is seen that the roots of the characteristic equation occur only for those values of complex variable s where

$$F(s) = \frac{K \prod_{i=1}^{m}(s + z_i)}{\prod_{j=1}^{n}(s + p_j)} = -1 \tag{8.20}$$

It may be noted that in the characteristic equation in the form (8.20), the design parameter K is no longer buried in the coefficients of the characteristic polynomial as in (8.18); it becomes a *scaling factor* for a complex number.

8.2.2 Magnitude and Angle Conditions

Since s is a complex variable, Eqn. (8.20) is converted into two Evans conditions given below:

Magnitude Criterion

$$|F(s)| = \frac{K\prod_{i=1}^{m}|s+z_i|}{\prod_{j=1}^{n}|s+p_j|} = 1 \tag{8.21a}$$

Angle Criterion

$$\angle F(s) = \sum_{i=1}^{m}\angle s+z_i - \sum_{j=1}^{n}\angle s+p_j = \pm 180°(2q+1); q = 0,1,2,... \tag{8.21b}$$

Equations (8.21) imply that the roots of 1 + $F(s)$ = 0 are those values of s at which the magnitude of $F(s)$ equals 1 and the angle of $F(s)$ equals an odd multiple of 180°. Figure 8.6 illustrates graphical evaluation of $|s_0 + p_j|$, $|s_0 + z_i|$, $\angle s_0 + p_j$ and $\angle s_0 + z_i$ at a point s_0 in the s-plane.

It should be appreciated that not all values of s that satisfy Eqn. (8.21a) also satisfy Eqn. (8.21b), while for every s that satisfies Eqn. (8.21b), there exists a K satisfying Eqn. (8.21a). ***The root locus, thus, is a plot of the points in the complex plane where the angle criterion (8.21b) is satisfied. The value of gain corresponding to a root, i.e., a point on the root locus, can be obtained from the magnitude criterion (8.21a).***

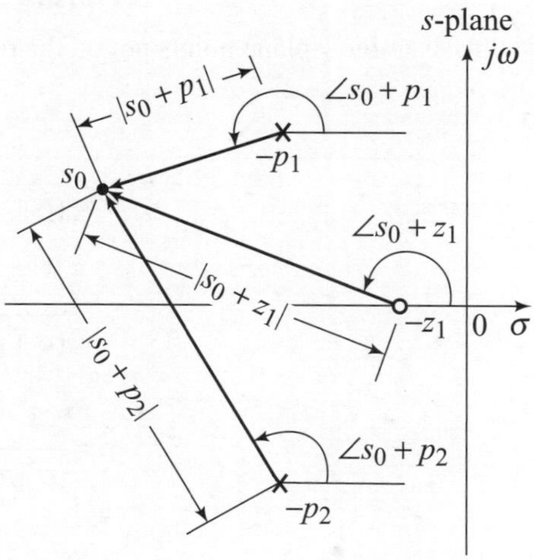

Fig. 8.6 *Evaluation of* $|F(s)|$ *and* $\angle F(s)$

Example 8.1 Consider the second-order characteristic equation

$$1 + F(s) = 0$$

with

$$F(s) = \frac{K(s+b)}{(s+a_1)(s+a_2)}; K \ge 0$$

having pole-zero configuration shown in Fig. 8.7a.

We will search for all the points in the s-plane for which the angle criterion (8.21b) is satisfied. A plot of these points is the root locus of the given characteristic equation.

We begin our search process with the real axis of the s-plane. Let us examine any point s_0 in between the two poles (refer to Fig. 8.7a). As this point is joined by phasors to the poles at $s = -a_1$ and $s = -a_2$, and the zero at $s = -b$, it is easily seen that

(i) the pole at $s = -a_1$ contributes an angle of 180°,
(ii) the pole at $s = -a_2$ contributes an angle of 0°, and
(iii) the zero at $s = -b$ contributes an angle of 0°.

Therefore, the angle criterion (8.21b) is satisfied and the point s_0 is on the root locus. Scanning all the points on the real axis, yields the real-axis root locus segments shown in Fig. 8.7b by dark lines.

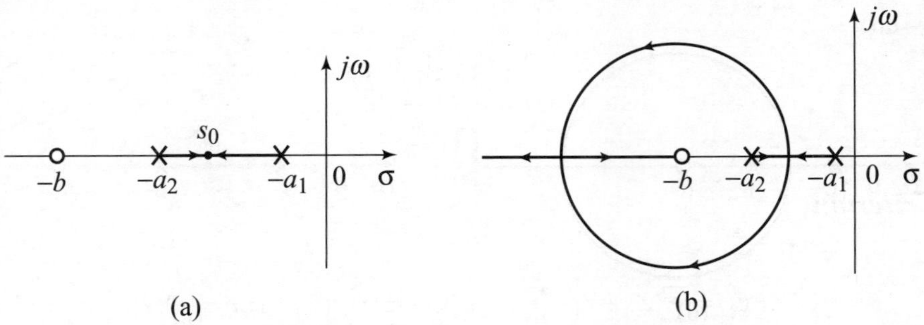

Fig. 8.7 *Search for points on root loci*

To examine s-plane points not on the real axis, we consider a representative point $s = \sigma + j\omega$. At this point

$$F(s) = \frac{K(\sigma + j\omega + b)}{(\sigma + j\omega + a_1)(\sigma + j\omega + a_2)}$$

$$= \frac{K(\sigma + b + j\omega)}{(\sigma + a_1)(\sigma + a_2) + j\omega(\sigma + a_1 + \sigma + a_2) - \omega^2}$$

$$= \frac{K(\sigma + b + j\omega)}{(\sigma + a_1)(\sigma + a_2) - \omega^2 + j\omega(2\sigma + a_1 + a_2)}$$

$$\angle F(s) = \tan^{-1}\frac{\omega}{\sigma + b} - \tan^{-1}\frac{\omega(2\sigma + a_1 + a_2)}{(\sigma + a_1)(\sigma + a_2) - \omega^2}$$

$$= \tan^{-1}\left\{\frac{\dfrac{\omega}{\sigma + b} - \dfrac{\omega(2\sigma + a_1 + a_2)}{(\sigma + a_1)(\sigma + a_2) - \omega^2}}{1 + \dfrac{\omega}{\sigma + b}\left[\dfrac{\omega(2\sigma + a_1 + a_2)}{(\sigma + a_1)(\sigma + a_2) - \omega^2}\right]}\right\}$$

$\angle F(s)$ is a multiple of 180° if

$$\frac{\dfrac{\omega}{\sigma + b} - \dfrac{\omega(2\sigma + a_1 + a_2)}{(\sigma + a_1)(\sigma + a_2) - \omega^2}}{1 + \dfrac{\omega}{\sigma + b}\left[\dfrac{\omega(2\sigma + a_1 + a_2)}{(\sigma + a_1)(\sigma + a_2) - \omega^2}\right]} = 0$$

Manipulation of this equation gives

$$(\sigma + b)^2 + \omega^2 = (b - a_1)(b - a_2)$$

This is the equation of a circle with center at $(-b, 0)$ and radius $= \sqrt{(b-a_1)(b-a_2)}$. It can easily be verified that at every point on this circle in the *s*-plane, $\angle F(s)$ is $\pm 180°$. Every point on the circle, therefore, satisfies the angle criterion (8.21b).

Let us assume the following numerical values for open-loop poles and zeros:

$$a_1 = 1, a_2 = 2, b = 5$$

The root locus plot for this case is shown in Fig. 8.8. The two root loci are *A-C-D-F-G* and *B-C-E-F-H*; the locus *A-C-D-F-G* starts at open-loop pole $s = -2$ with $K = 0$ and terminates at open-loop zero $s = -5$ with $K = \infty$; and the locus *B-C-E-F-H* starts at open-loop pole $s = -1$ with $K = 0$ and goes off to infinity along the negative real axis as $K \to \infty$.

The value of K corresponding to least damping may be obtained by applying the magnitude criterion (8.21a) at the point P where OP is a straight line tangential to the circular locus.

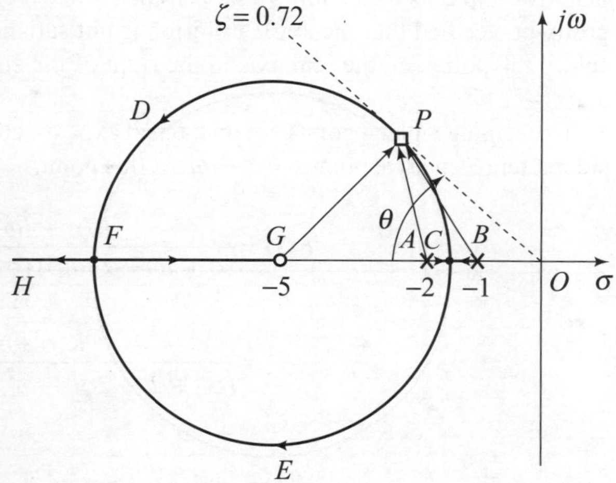

Fig. 8.8 *A root locus plot (Example 8.1)*

$$K = \frac{\text{Length of the vector } AP \times \text{Length of the vector } BP}{\text{Length of the vector } GP}$$

$$= \frac{2.6 \times 2.9}{3.45} \cong 2.2; \zeta_{\min} = \cos\theta = 0.72$$

The closed-loop poles corresponding to this value of damping are (read from Fig. 8.8): $s_{1,2} = -2.6 \pm j2.5$.

Example 8.2 Consider the second-order characteristic equation:

$$1 + F(s) = 0$$

with

$$F(s) = \frac{K(s+b)}{(s+\alpha+j\beta)(s+\alpha-j\beta)}; K \ge 0$$

having pole-zero configuration shown in Fig. 8.9.

Let us examine any point s_0 on the negative real axis to the left of the zero at $s = -b$. As this point is joined by phasors to the two poles at $s = -\alpha \pm j\beta$ and the zero at $s = -b$, it is easily seen that (i) the zero contributes an angle of 180°, and (ii) the net angle contribution of complex-conjugate pole pair is zero. Therefore, angle criterion (8.21b) is satisfied for all the points on the negative real axis to the left of the zero at $s = -b$. It can easily be verified that the angle criterion is not satisfied for all the points on the real axis to the right of the zero at $s = -b$.

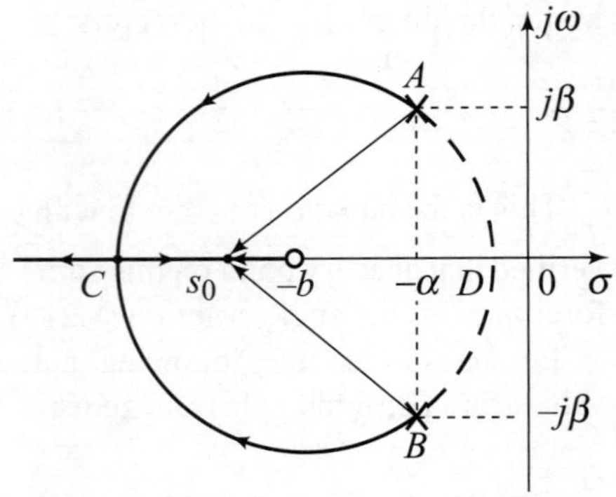

Fig. 8.9 *Search of points on root loci*

To examine s-plane points not on the real axis, we consider a representative point $s = \sigma + j\omega$. At this point,

$$F(s) = \frac{K(\sigma + j\omega + b)}{(\sigma + j\omega + \alpha + j\beta)(\sigma + j\omega + \alpha - j\beta)}$$

$$= \frac{K(\sigma + b + j\omega)}{(\sigma^2 + \alpha^2 + 2\sigma\alpha - \omega^2 + \beta^2) + j2\omega(\sigma + \alpha)}$$

$$\angle F(s) = \tan^{-1}\frac{\omega}{\sigma + b} - \tan^{-1}\frac{2\omega(\sigma+\alpha)}{(\sigma^2 + \alpha^2 + 2\sigma\alpha - \omega^2 + \beta^2)}$$

$$= \tan^{-1}\left\{\frac{\dfrac{\omega}{\sigma+b} - \dfrac{2\omega(\sigma+\alpha)}{\sigma^2+\alpha^2+2\sigma\alpha-\omega^2+\beta^2}}{1 + \dfrac{\omega}{\sigma+b}\left[\dfrac{2\omega(\sigma+\alpha)}{(\sigma^2+\alpha^2+2\sigma\alpha-\omega^2+\beta^2}\right]}\right\}$$

$\angle F(s)$ is a multiple of 180° if

$$\frac{\dfrac{\omega}{\sigma+b} - \dfrac{2\omega(\sigma+\alpha)}{(\sigma^2+\alpha^2+2\sigma\alpha-\omega^2+\beta^2}}{1 + \dfrac{\omega}{\sigma+b}\left[\dfrac{2\omega(\sigma+\alpha)}{\sigma^2+\alpha^2+2\sigma\alpha-\omega^2+\beta^2}\right]} = 0$$

Manipulation of this equation gives

$$(\sigma + b)^2 + \omega^2 = (\alpha - b)^2 + \beta^2$$

This is the equation of a circle with center at $(-b, 0)$ and radius $= \sqrt{(\alpha - b)^2 + \beta^2}$. It can easily be verified that at every point of the *A-C-B* section of the circle, $\angle F(s)$ is ± 180°; and at every point of the *A-D-B* section of the circle, $\angle F(s) = 0°$. Every point of the *A-C-B* section, therefore, satisfies the angle criterion (8.21b).

Let us assume the following numerical values for the open-loop poles and zeros:

$$\alpha = 1, \beta = \sqrt{2}, b = 2$$

The root locus plot for this case is shown in Fig. 8.10. The two root loci are *A-C-E* and *B-C-F*; the locus *A-C-E* starts at open-loop pole at $s = -1 + j\sqrt{2}$ with $K = 0$ and terminates at open-loop zero at $s = -2$ with $K = \infty$, and the locus *B-C-F* starts at open-loop pole at $s = -1 - j\sqrt{2}$ with $K = 0$ and goes off to infinity along the negative real axis as $K \to \infty$.

The value of K corresponding to a damping ratio of 0.7 may be obtained by applying the magnitude criterion (8.21a) at the point P in Fig. 8.10.

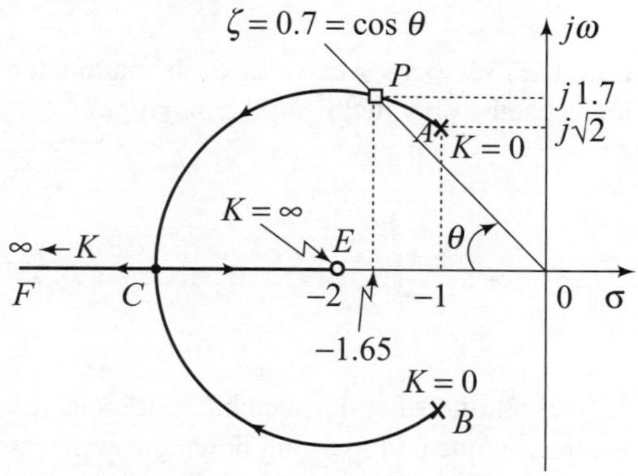

Fig. 8.10 *A root locus plot (Example 8.2)*

$$K = \frac{\text{Length of the vector } AP \times \text{Length of the vector } BP}{\text{Length of the vector } EP}$$

$$= \frac{0.7 \times 3.2}{1.73} \cong 1.3$$

The closed-loop poles corresponding to this value of K are (read from Fig. 8.10): $s_{1,2} = -1.65 \pm j1.7$.

In systems of Examples 8.1 and 8.2, the root loci in the complex plane are a part of a circle. Circular root loci may occur in systems that involve two poles and one zero, two poles and two zeros, or one pole and two zeros. Even in such systems, whether or not circular root loci occur depends on the locations of poles and zeros involved.

It is important to note that easily interpretable equations for the root locus can be derived for simple systems only. For complicated systems having many poles and zeros, any attempt to derive equations for the root loci is discouraged. Such derived equations are very complicated and their configuration in the complex plane is difficult to visualize. The next section provides a solution to this problem.

8.3 GUIDELINES FOR SKETCHING ROOT LOCI

The purpose of the root locus is to show in graphical form the general trend of the roots of the characteristic equation,

$$1 + F(s) = 0 \tag{8.22a}$$

where

$$F(s) = \frac{K\prod_{i=1}^{m}(s + z_i)}{\prod_{j=1}^{n}(s + p_j)}; m \le n \tag{8.22b}$$

as the parameter K is varied from zero to infinity. Every point $s = \sigma + j\omega$ in the complex plane that satisfies the *angle criterion* (refer to Eqn. (8.21b))

$$\angle F(s) = \sum_{i=1}^{m} \angle s + z_i - \sum_{j=1}^{n} \angle s + p_j = \pm 180° \, (2q+1); \, q = 0, 1, 2, ... \qquad (8.23a)$$

is on the root locus. The value of the parameter K (*root locus gain*) corresponding to a point on the root locus can be obtained from the *magnitude criterion* (refer to Eqn. (8.21a))

$$|F(s)| = \frac{K \prod_{i=1}^{m} |s + z_i|}{\prod_{j=1}^{n} |s + p_j|} = 1 \qquad (8.23b)$$

We can, in principle, sketch a root locus for a given $F(s)$ by measuring $\angle F(s)$ at all the points of the complex plane and marking down those places where we find $\angle F(s)$ equal to an odd multiple of 180°. This trial-and-error procedure of sketching the root locus is illustrated in Fig. 8.6. The poles and zeros of $F(s)$ are located on the s-plane on a graph sheet and a trial point s_0 is selected. The lines joining each of the poles $(-p_j)$ and zeros $(-z_i)$ to s_0 represent the phasors $(s_0 + p_j)$ and $(s_0 + z_i)$ respectively, as shown in Fig. 8.6. The angles of these phasors can be measured by placing a protector at the pole and zero locations, and the angle values are substituted in the angle criterion given by Eqn. (8.23a). The trial point s_0 is suitably shifted till the angle criterion is satisfied to a desired degree of accuracy. A number of points on the s-plane are determined in this manner and a smooth curve drawn through these points gives the root locus with variable K.

The value of K for a particular root location s_0 can be obtained from the magnitude criterion given by Eqn. (8.23b) by substituting the phasor magnitudes read to scale. It is to be noted that the same scale must be employed for both the real and imaginary axes of the complex plane.

From the magnitude criterion of Eqn. (8.23b),

$$K = \frac{\prod_{j=1}^{n} |s_0 + p_j|}{\prod_{i=1}^{m} |s_0 + z_i|} \qquad (8.24a)$$

With reference to Fig. 8.6, we can write

$$K = \frac{\begin{bmatrix} \text{Product of phasor lengths (read to scale)} \\ \text{from } s_0 \text{ to poles of } F(s) \end{bmatrix}}{\begin{bmatrix} \text{Product of phasor lengths (read to scale)} \\ \text{from } s_0 \text{ to zeros of } F(s) \end{bmatrix}} \qquad (8.24b)$$

If we were to use the trial-and-error method just described, the search for all the root locus points in the s-plane that satisfy the angle criterion (8.23a) would in general be a very tedious task. In order to help reduce the tedium of the trial-and-error procedure, certain rules have been developed for making a quick approximate sketch of the root locus. This approximate sketch provides a guide for the selection of trial points such that a reasonably accurate root locus can be obtained by a few trials. Further, the approximate root locus sketch is very useful in vizualizing the effects of variation of system parameter K, the effects of shifting pole-zero locations and of bringing in a new set of poles and zeros.

With the availability of digital computers and efficient root-finding techniques, the trial-and-error method has become obsolete. Nevertheless, the analyst should still have an understanding of the properties of the root loci in order to interpret the computer results correctly and apply the root locus techniques for the design of linear control systems. Presentation of guidelines for hand-sketching an approximate root locus plot is, therefore, in order.

8.3.1 Rules for Constructing Root Loci

To help in the rough drafting of the general form of a root locus, construction rules have been devised. We shall discuss several of them, first stating the rule, then giving a justification for its validity, and finally citing an example if appropriate. We assume that $F(s)$ for which the root locus is desired, is of the form given by Eqn. (8.22b), i.e., with m zeros at $s = -z_i$, and n poles at $s = -p_j$; $m \le n$. The m zeros and n poles of $F(s)$ will be referred to as open-loop zeros and open-loop poles, respectively[3].

Rule 1 Number of Root Loci

The root locus plot consists of n root loci as K varies from 0 to ∞. The loci are symmetric with respect to the real axis.

The characteristic equation (8.22a) can be written as

$$\prod_{j=1}^{n}(s+p_j)+K\prod_{i=1}^{m}(s+z_i)=0$$

This equation has degree n. Thus, for each real K, there are n roots. As the roots of this equation are continuous functions of its coefficients, the n roots form n continuous loci as K varies from 0 to ∞.

Since the complex roots of the characteristic equation always occur in conjugate pairs, the n root loci, must be symmetrical about the real axis.

Rule 2 Starting and Terminal Points of Root Loci

As K increases from zero to infinity, each root locus originates from an open-loop pole with K = 0 and terminates either on an open-loop zero or on infinity with K = ∞. The number of loci terminating on infinity equals the number of open-loop poles minus zeros.

The characteristic equation (8.22a) can be written as

$$\prod_{j=1}^{n}(s+p_j)+K\prod_{i=1}^{m}(s+z_j)=0 \tag{8.25}$$

When $K = 0$, this equation has roots at $-p_j$ ($j = 1, 2, \ldots, n$), which are the open-loop poles. The root loci, therefore, start at the open-loop poles.

The characteristic equation (8.25) can be rearranged as

$$\frac{1}{K}\prod_{j=1}^{n}(s+p_j)+\prod_{i=1}^{m}(s+z_i)=0$$

[3] The equation $1 + F(s) = 0$ can be interpreted as the characteristic equation of a unity-feedback system with open-loop transfer function equal to $F(s)$.

When $K = \infty$, this equation has roots at $-z_i(i = 1, 2, \ldots, m)$, which are the open-loop zeros. Therefore, m root loci terminate on the open-loop zeros.

In case $m < n$, the open-loop transfer function has $(n - m)$ zeros at infinity. Examining the magnitude criterion (8.23b) in the form

$$\frac{\prod_{i=1}^{m}|s+z_i|}{\prod_{j=1}^{n}|s+p_j|} = \frac{1}{K}$$

we find that this is satisfied by $s \rightarrow \infty\, e^{j\phi}$ as $K \rightarrow \infty$. Therefore, $(n - m)$ root loci terminate on infinity.

Rule 3 Asymptotes to Root Loci

The $(n - m)$ root loci which tend to infinity do so along straight line asymptotes radiating out from a single point $s = -\sigma_A$ on the real axis (called the centroid), where

$$-\sigma_A = \frac{\sum(\text{real parts of open-loop poles}) - \sum(\text{real parts of open-loop zeros})}{n-m} \tag{8.26a}$$

These $(n - m)$ asymptotes have angles

$$\phi_A = \frac{(2q+1)180°}{n-m};\ q = 0, 1, \ldots, (n-m-1) \tag{8.26b}$$

We justify the rule by using the pole-zero pattern shown in Fig. 8.11a. For a point s_0 very far away from the origin, the poles and zeros can be considered to cluster at the same point, say $-\sigma_A$ as shown in Fig. 8.11b.

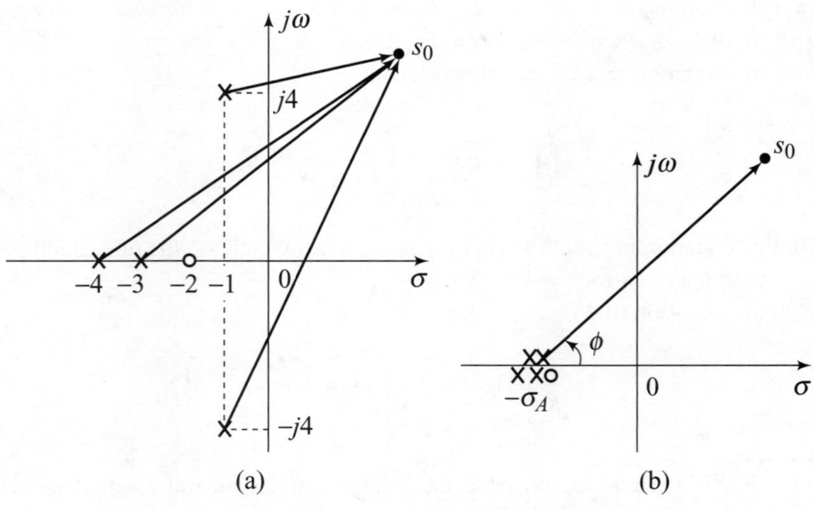

Fig. 8.11 *Asymptotes to root loci*

Consequently, the characteristic equation in (8.22a) can be approximated by

$$1+\frac{K\prod_{i=1}^{m}(s+z_i)}{\prod_{j=1}^{n}(s+p_j)} \cong 1+\frac{K}{(s+\sigma_A)^{n-m}}=0 \tag{8.27}$$

In other words, all m finite zeros are canceled by poles, and only $(n-m)$ poles are left at $-\sigma_A$. Now we compute the relationship among z_i, p_j, and $-\sigma_A$. From Eqn. (8.27), we obtain

$$\frac{\prod_{j=1}^{m}(s+p_j)}{\prod_{i=1}^{m}(s+z_j)} \cong (s+\sigma_A)^{n-m}$$

or

$$\frac{s^n+\left(\sum_{j=1}^{n}p_j\right)s^{n-1}+\cdots+\prod_{j=1}^{n}p_j}{s^m+\left(\sum_{i=1}^{m}z_i\right)s^{m-1}+\cdots+\prod_{i=1}^{m}z_i} \cong (s+\sigma_A)^{n-m}$$

which implies, by direct division and expansion,

$$s^{n-m}+\left(\sum_{j=1}^{n}p_j-\sum_{i=1}^{m}z_j\right)s^{n-m-1}+\cdots$$

$$\cong s^{n-m}+(n-m)\sigma_A s^{n-m-1}+\cdots$$

Equating the coefficients of s^{n-m-1} yields

$$(n-m)\sigma_A=\sum_{j=1}^{n}p_j-\sum_{i=1}^{m}z_i$$

or

$$-\sigma_A=\frac{\sum_{j=1}^{n}(-p_j)-\sum_{i=1}^{m}(-z_i)}{n-m}$$

This establishes Eqn. (8.26a).

For the point s_0 in Fig. 8.11b to be on the root locus,

$$-(n-m)\phi=-(2q+1)\,180°;\ q=0,1,\ldots$$

or

$$\phi=\frac{(2q+1)180°}{n-m};q=0,1,\ldots$$

The $(n - m)$ angles given by this equation divide 360° equally and are symmetric with respect to real axis. The $(n - m)$ root loci, therefore, tend to infinity along $(n - m)$ asymptotes radiating out from $s = -\sigma_A$ given by Eqn. (8.26a) at angles ϕ_A given by Eqn. (8.26b).

The pole-zero map of Fig. 8.11a corresponds to

$$F(s) = \frac{K(s+2)}{(s+1+j4)(s+1-j4)(s+3)(s+4)} \tag{8.28a}$$

The root locus plot of

$$1 + F(s) = 0 \tag{8.28b}$$

will consist of four root loci, each starting from an open-loop pole with $K = 0$. One root locus will terminate on open-loop zero with $K = \infty$. The other three loci will terminate on infinity as $K \to \infty$ along the asymptotes radiating out from $s = -\sigma_A$

where $$-\sigma_A = \frac{-1-1-3-4-(-2)}{4-1} = -7/3$$

at angles 60°, 180°, and 300°, respectively. Figure 8.12 shows the asymptotes. Note that asymptotes are developed for large s; thus the root loci will *approach* them for large s or large K.

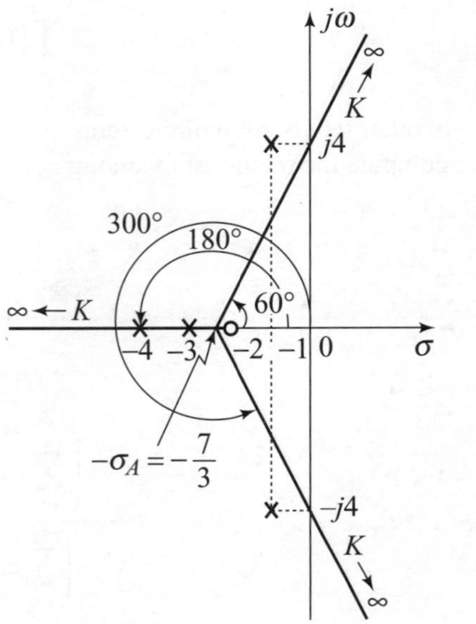

Fig. 8.12 *Asymptotes for the characteristic equation (8.28)*

Rule 4 On-Locus Segments of the Real Axis

A point on the real axis lies on the locus if the number of open-loop poles plus zeros on the real axis to the right of this point is odd.

For the fourth-order example under consideration (refer to Eqns (8.28)), open-loop pole-zero configuration is shown in Fig. 8.13a. Let us examine any point s_0 on the real axis. As this point is joined by phasors to all the open-loop poles and zeros, it is easily seen that (i) the poles and zeros on the real axis to the right of this point contribute an angle of 180° each, (ii) the poles and zeros on the real axis to the left of this point contribute an angle of 0° each, and (iii) the net angle contribution of a complex conjugate pole or zero pair is always zero.

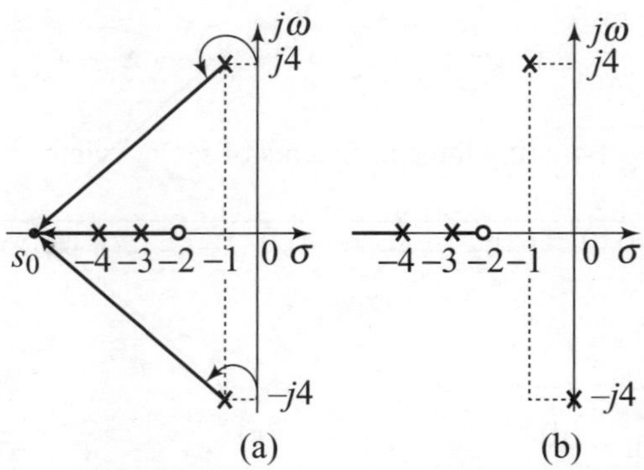

Fig. 8.13 *On-locus segments of the real axis*

Therefore,

$$\angle F(s) = (m_r - n_r)180° = \pm(2q + 1)180°, q = 0, 1, 2, \ldots$$

where

m_r = number of open-loop zeros on the real axis to the right of s_0; and

n_r = number of open-loop poles on the real axis to the right of s_0.

We therefore see that for a point on the real axis, the angle criterion is met only if $(n_r - m_r)$ or $(n_r + m_r)$ is odd and hence the rule. If this rule is satisfied at any point on the real axis, it continues to be satisfied as the point is moved on either side unless the point crosses a real axis pole or zero. By use of this additional fact, the real axis can be divided into segments *on-locus* and *not-on-locus*; the dividing points being the real open-loop poles and zeros. The on-locus segments of the real axis alternate as shown in Fig. 8.13b.

Rule 5 On-Locus Points of the Imaginary Axis

The intersections (if any) of root loci with the imaginary axis can be determined by use of the Routh criterion.

Segments of root loci can exist in the right half of s-plane. This signifies instability. The points at which the root loci cross the imaginary axis define the stability limits. Basic application of the Routh array (Chapter 5) determines the gains at the stability limit. By substituting this value of gain in the auxiliary equation, the value of $s = j\omega_0$ at the stability limit is evaluated.

For the fourth-order example under consideration (refer to Eqns (8.28)), the characteristic equation is

$$1+\frac{K(s+2)}{(s+1+j4)(s+1-j4)(s+3)(s+4)}=0$$

which is equivalent to

$$s^4 + 9s^3 + 43s^2 + (143 + K)s + 204 + 2K = 0 \tag{8.29}$$

The Routh array for the characteristic polynomial is

s^4	1	43	$204 + 2K$
s^3	9	$143 + K$	
s^2	$\dfrac{244-K}{9}$	$204 + 2K$	
s^1	$\dfrac{18368-61K-K^2}{244-K}$		
s^0	$204 + K$		

For Eqn. (8.29) to have no roots on $j\omega$-axis or in the right half s-plane, the elements in the first column of Routh array must all be of the same sign. Therefore, the following inequalities must be satisfied:

$$244 - K > 0$$

$$18368 - 61K - K^2 > 0$$

$$204 + K > 0$$

These inequalities are satisfied if K is less than 108.4, which means that the critical value of K which corresponds to the roots on the $j\omega$-axis is 108.4.

The value of $K = 108.4$ makes all the coefficients of s^1-row of the Routh array zero. For this value of K, the auxiliary equation formed from the coefficients of the s^2-row, is given by

$$\left(\frac{244-K}{9}\right)s^2+(204+2K)=0$$

For $K = 108.4$, the roots of the above equation lie on the $j\omega$-axis and are given by

$$s = \pm j5.28$$

Thus, the root loci intersect the $j\omega$-axis at $s = \pm j5.28$ and the value of K corresponding to these roots is 108.4.

Rule 6 Angle of Departure from Complex Poles

The angle of departure, ϕ_p, of a locus from a complex open-loop pole is given by

$$\phi_p = 180° + \phi \tag{8.30}$$

where ϕ is the net angle contribution at this pole of all other open-loop poles and zeros.

For the fourth-order example under consideration (refer to Eqns (8.28)), the characteristic equation is

$$1 + F(s) = 1 + \frac{K(s+2)}{(s+1+j4)(s+1-j4)(s+3)(s+4)} = 0$$

The pole-zero map of $F(s)$ is shown in Fig. 8.14. Let s_0 be an arbitrary point on the root locus starting from the pole at $s = -1 + j4$. Let the phase from this pole to s_0 be denoted by ϕ_p. The net angle contribution of all other open-loop poles and zeros at the point s_0 is

$$\phi = \theta_2 - (\theta_1 + \theta_3 + \theta_4)$$

Therefore, the total phase of $F(s)$ at s_0 is $\phi - \phi_p$. In order for s_0 to be on the root locus, the total phase must be $\pm 180°$. Thus, we have $\phi_p = 180° + \phi$. This is the angle of departure from the complex open-loop pole.

If the point s_0 is very close to the pole at $-1 + j4$, then the vectors drawn from the zero and all other open-loop poles to s_0 can be approximated by the vectors drawn to the pole at $-1 + j4$, i.e., we consider the point s_0 to be at $-1 + j4$ for measurement of angles θ_1, θ_2, θ_3, and θ_4. Under this approximation, $\theta_1 = 90°$, $\theta_2 = 76°$, $\theta_3 = 63°$, and $\theta_4 = 53°$. Therefore,

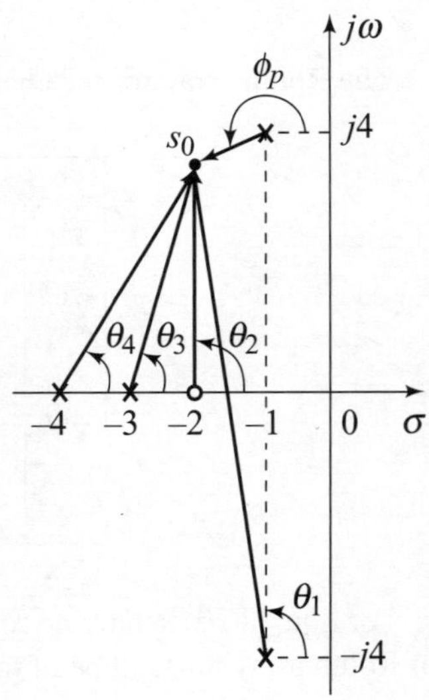

Fig. 8.14 *Angle of departure from complex poles*

$$\phi = \theta_2 - (\theta_1 + \theta_3 + \theta_4) = -130°;\ \phi_p = 180° + \phi = 50°$$

We can now sketch a rough root locus plot for the fourth-order example under consideration. The plot is shown in Fig. 8.15. The information given in Figs 8.12–8.14 has been used to construct this plot.

There are four open-loop poles; so there are four loci. One locus departs from the real pole at –3 and terminates on the zero at –2 along the real axis. The second locus departs from another real pole at –4 and moves along the asymptote on the negative real axis. The third locus departs from the complex pole at $-1 + j4$ with a departure angle of $\phi_p = 50°$, and moves toward the asymptote radiating out from $-\sigma_A = -7/3$ at an angle $+ 60°$; it crosses the imaginary axis at $j\omega_0 = j5.28$. Using the symmetry property, the fourth locus is obtained immediately by reflection about the real axis.

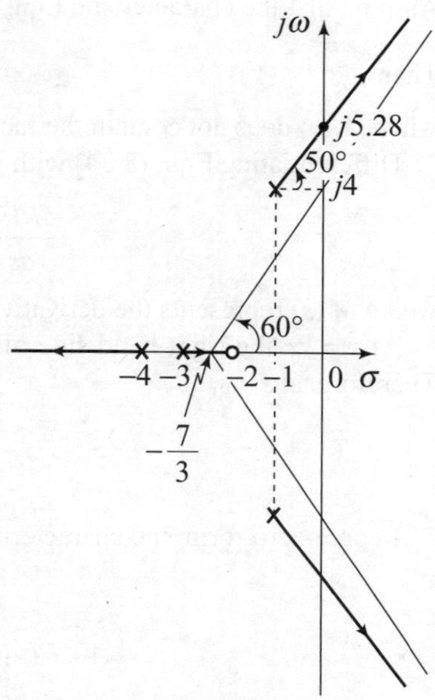

Fig. 8.15 *Root locus plot for the characteristic equation (8.28)*

Rule 7 Angle of Arrival at Complex Zeros

The angle of arrival, ϕ_z, of a locus at a complex zero is given by

$$\phi_z = 180° - \phi \tag{8.31}$$

where ϕ is the net angle contribution at this zero of all other open-loop poles and zeros.

Consider the characteristic equation

$$1 + F(s) = 1 + \frac{K(s^2+1)}{s(s+2)} = 0 \tag{8.32}$$

The pole-zero map of $F(s)$ is shown in Fig. 8.16. The open-loop poles are located at $s = 0$ and $s = -2$ and the open-loop zeros are located at $s = \pm j1$. Let s_0 be an arbitrary point on the root locus terminating on the zero at $s = j1$. Let the phase from this zero to s_0 be denoted by ϕ_z (refer to Fig. 8.16a).

If the point s_0 is very close to the zero at $j1$, then the vectors drawn from the other zero at $-j1$ and open-loop poles at 0 and -2 to s_0 can be approximated by vectors drawn to the zero at $j1$. Under this approximation, the net angle contribution at s_0 of the zero at $-j1$ and open-loop poles at 0 and -2 is given by (refer to Fig. 8.16b)

$$\phi = 90° - 90° - 26.5° = -26.5°$$

In order for s_0 to be on the root locus, the total phase must be $\pm 180°$. Thus, we have $\phi_z = 180° - \phi = 206.5°$.

The complete root locus plot for roots of Eqn. (8.32) is shown in Fig. 8.16b; it can easily be proved (refer to Example 8.2) that the complex root branches lie on a circle with centre at (1/2, 0) and radius equal to $\sqrt{5}/2$.

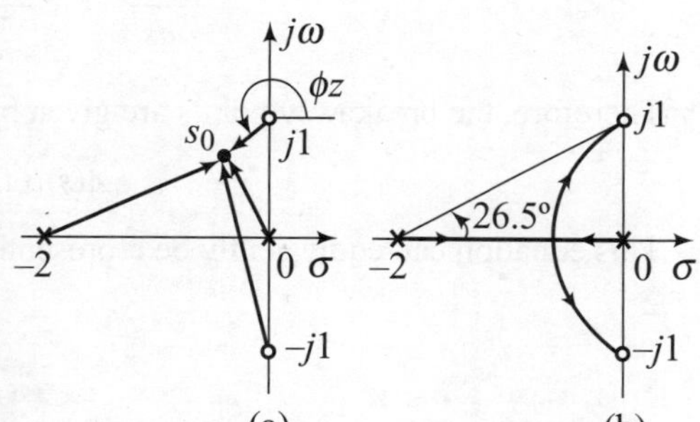

Fig. 8.16 *Angle of arrival at complex zeros*

Rule 8 Locations of Multiple Roots

Points at which multiple roots of the characteristic equation occur (breakaway points of root loci) are the solutions of

$$\frac{dK}{ds} = 0 \tag{8.33a}$$

where

$$K = -\frac{\prod_{j=1}^{n}(s+p_j)}{\prod_{i=1}^{m}(s+z_i)} \tag{8.33b}$$

Assume that the characteristic Eqn. (8.22a) has a multiple root at $s = s_0$ of multiplicity r.

Then $$1 + F(s) = (s - s_0)^r M(s) \tag{8.34}$$

where $M(s)$ does not contain the factor $(s - s_0)$.

Differentiating Eqn. (8.34) with respect to s, we have

$$\frac{dF(s)}{ds} = (s - s_0)^{r-1}[rM(s) + (s - s_0)M'(s)] \tag{8.35}$$

where $M'(s)$ represents the derivative of $M(s)$.

At $s = s_0$, the right-hand side of this equation is zero because it has a factor $(s - s_0)^{r-1}$ and $r \geq 2$. Therefore, at $s = s_0$,

$$\frac{dF(s)}{ds} = 0$$

In pole-zero form, the characteristic equation may be written as

$$1 + F(s) = 1 + \frac{K\prod_{i=1}^{m}(s + z_i)}{\prod_{j=1}^{n}(s + p_j)} = 1 + \frac{KB(s)}{A(s)} = 0 \tag{8.36a}$$

Taking the derivative of this equation with respect to s, we get

$$\frac{dF(s)}{ds} = K\frac{A(s)B'(s) - A'(s)B(s)}{[A(s)]^2} = 0$$

Therefore, the breakaway points are given by the roots of

$$A(s)B'(s) - A'(s)B(s) = 0 \tag{8.36b}$$

This equation can equivalently be represented as

$$\frac{dK}{ds} = 0 \tag{8.36c}$$

where $$K = -\frac{A(s)}{B(s)} = -\frac{\prod_{j=1}^{n}(s + p_j)}{\prod_{i=1}^{m}(s + z_i)}$$

Consider the characteristic equation

$$1 + \frac{K(s+2)(s+3)}{s(s+1)} = 1 + \frac{KB(s)}{A(s)} = 0 \tag{8.37}$$

Figure 8.17 shows open-loop poles and zeros on the complex plane. Root loci segments exist on the negative real axis between 0 and –1 and between –2 and –3. At $K = 0$, the roots are at $s = 0$ and $s = -1$. As K increases, the two real roots move away from poles $s = 0$ and $s = -1$ toward each other inside the real-axis segment [–1, 0]. As K continues to increase, the two real roots will eventually coalesce into a repeated real root and then break away from the real axis into two complex conjugate roots. Such a point is called a *breakaway point.*

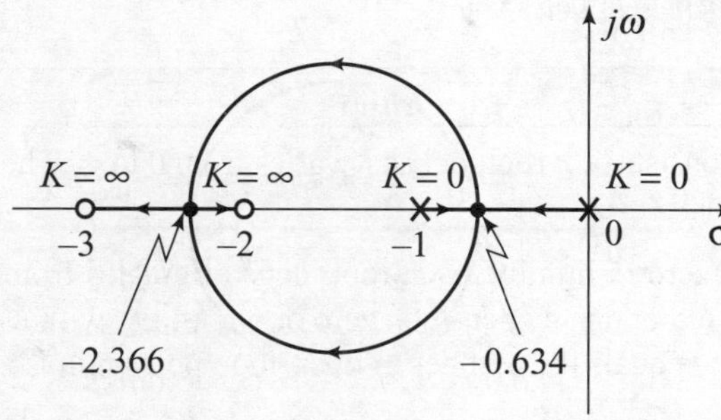

Fig. 8.17 *Root locus plot for characteristic Eqn. (8.37)*

Similarly, as K approaches infinity, one root will approach zero at $s = -2$ along the negative real axis and another will approach zero at $s = -3$ along the negative real axis. As the root loci are continuous, the two complex-conjugate roots will approach the real axis somewhere inside the segment [–3, –2] and then depart in opposite directions along the real axis. Such a point is also called a *breakaway point*.

From Eqns (8.36)–(8.37), we have

$$B(s) = (s + 2)\,(s + 3) = s^2 + 5s + 6;\; B'(s) = 2s + 5$$

$$A(s) = s(s + 1) = s^2 + s;\; A'(s) = 2s + 1$$

and
$$A(s)B'(s) - A'(s)B(s) = (s^2 + s)\,(2s + 5) - (2s + 1)\,(s^2 + 5s + 6)$$

$$= -\,3s^2 - 12s - 6 = 0$$

Its roots are –0.634 and –2.366. Thus, the root locus plot has two breakaway points: $s = -0.634$ and $s = -2.366$.

The complete root locus plot for roots of Eqn. (8.37) is shown in Fig. 8.17; it can easily be proved (refer to Example 8.1) that the complex root branches lie on a circle with center at (–1.5, 0) that passes through the breakaway points.

It is important to note that the condition for the breakaway point given by Eqns (8.33) is *necessary* but not *sufficient*. In other words, all breakaway points on the root locus must satisfy Eqns (8.33) but not all solutions of Eqns (8.33) are breakaway points. To be a breakaway point, the solution of Eqns (8.33) must satisfy the characteristic equation (8.22) for some real K. This and many other important aspects of breakaway points will be illustrated through examples in the next section.

8.3.2 Summary of the Rules

Except for extremely complex cases, the rules described above should be adequate for making a reasonably accurate sketch of the root locus plot. For quick reference, these rules are summarized in Table 8.1.

Table 8.1 *Rules for construction of Root Locus Plot of* $1+F(s)=0; F(s)=\dfrac{K\prod_{i=1}^{m}(s+z_i)}{\prod_{j=1}^{n}(s+p_j)}; K \leq 0, n \geq m$
z_i*: m open-loop zeros;* p_j*: n open-loop poles*

Rules
(i) The root locus plot consists of n root loci as K varies from 0 to ∞. The loci are symmetric with respect to the real axis.
(ii) As K increases from zero to infinity, each root locus originates from an open-loop pole with $K = 0$ and terminates either on an open-loop zero or on infinity with $K = \infty$. The number of loci terminating on infinity equals the number of open-loop poles minus zeros.
(iii) The $(n - m)$ root loci which tend to infinity, do so along straight-line asymptotes radiating out from a single point $s = -\sigma_A$ on the real axis (called the centroid), where $-\sigma_A = \dfrac{\sum(\text{real parts of open-loop poles}) - \sum(\text{real parts of open-loop zeros})}{n-m}$ These $(n - m)$ asymptotes have angles $\phi_A = \dfrac{(2q+1)180°}{n-m}; q = 0, 1, ..., (n-m-1)$
(iv) A point on the real axis lies on the locus if the number of open-loop poles plus zeros on the real axis to the right of this point is odd. By use of this fact, the real axis can be divided into segments *on-locus* and *not-on-locus*; the dividing points being the real open-loop poles and zeros.
(v) The intersections (if any) of root loci with the imaginary axis can be determined by use of the Routh criterion.
(vi) The angle of departure, ϕ_p, of a root locus from a complex open-loop pole is given by $\phi_p = 180° + \phi$ where ϕ is the net angle contribution at this pole of all other open-loop poles and zeros.
(vii) The angle of arrival, ϕ_z, of a locus at a complex zero is given by $\phi_z = 180° - \phi$ where ϕ is the net angle contribution at this zero of all other open-loop poles and zeros.
(viii) Points at which multiple roots of the characteristic equation occur (breakaway points of root loci) are the solutions of $\dfrac{dK}{ds} = 0$

(Continued)

Table 8.1 (*Continued*)

where	$K = -\dfrac{\prod_{j=1}^{n}(s+p_j)}{\prod_{i=1}^{m}(s+z_i)}$
(ix) The gain K at any point s_0 on a root locus is given by $$K = \frac{\prod_{j=1}^{n}\lvert s_0+p_j\rvert}{\prod_{i=1}^{m}\lvert s_0+z_i\rvert} = \frac{\text{[Product of phasor lengths (read to scale) from } s_0 \text{ to poles of } F(s)]}{\text{[Product of phasor lengths (read to scale) from } s_0 \text{ to zeros of } F(s)]}$$	

8.4 SELECTED ILLUSTRATIVE ROOT LOCI

The root locus concept was introduced in Section 8.2 and locus equations were derived in Examples 8.1 and 8.2 for two simple characteristic equations. Since for complex cases, it is not possible to derive easily interpretable locus equations, point-by-point plotting of the root loci becomes necessary. In Section 8.3, we presented some construction rules which help in making a quick rough sketch of the root locus plot.

In order to gain experience with the root-locus method, it is helpful to sketch a number of locus plots and visualize the various alternative shapes a locus of a given complexity can take. In this section, we will go through the construction rules for several cases selected to illustrate important features of root loci.

Example 8.3 Consider a feedback system with the characteristic equation,

$$1+\frac{K}{s(s+1)(s+2)}=0; K \geq 0 \tag{8.38}$$

The open-loop poles are located at $s = 0$, –1 and –2, while there are no finite open-loop zeros. The pole-zero configuration is shown in Fig. 8.18.

Rule (i) (refer to Table 8.1) tells us that the root locus plot consists of three root loci as K is varied from 0 to ∞.

Rule (ii) tells us that the three root loci originate from the three open-loop poles with $K = 0$ and terminate on infinity with $K = \infty$.

Rule (iii) tells us that the three root loci tend to infinity along asymptotes radiating out from

$$s = -\sigma_A = \frac{\sum \text{real parts of poles} - \sum \text{real parts of zeros}}{\text{number of poles} - \text{number of zeros}} = \frac{-2-1}{3} = -1$$

with angles $$\phi_A = \frac{(2q+1)180°}{\text{number of poles} - \text{number of zeros}}; q = 0, 1, \ldots$$

$$= \frac{(2q+1)180°}{3}; q = 0, 1, 2$$
$$= 60°, 180°, 300°$$

The asymptotes are shown by dotted lines in Fig. 8.18.

Rule (iv) tells us that the segments of the real axis between 0 and –1, and between –2 and –∞ lie on the root locus. On-locus segments of the real axis are shown by thick lines in Fig. 8.18.

From Fig. 8.18, it is seen that out of the three loci, one is a real-root locus which originates from $s = -2$ and terminates on $-\infty$. The other two loci originate from $s = 0$ and $s = -1$, and move on the real axis approaching each other as K is increased. These two loci must, therefore, meet on the real axis. The characteristic equation has a double root at such a point. As the gain K is further increased, the root loci breakaway from the real axis to give a complex conjugate pair of roots.

Rule (v) tells us to check by use of the Routh criterion if the two root loci breaking away from the real axis will intersect the imaginary axis. The system characteristics equation is

$$s(s + 1)(s + 2) + K = 0$$

or
$$s^3 + 3s^2 + 2s + K = 0$$

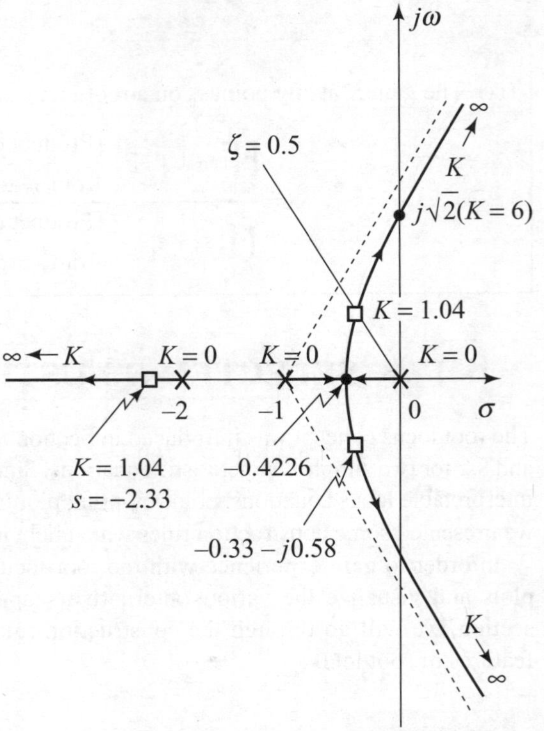

Fig. 8.18 *Root locus plot for the characteristic equation (8.38)*

The Routh array becomes

s^3	1	2
s^2	3	K
s^1	$(6 - K)/3$	
s^0	K	

For all the roots of the characteristic equation to lie to the left of the imaginary axis, the following conditions must be satisfied:

$$K > 0; (6 - K)/3 > 0$$

Therefore, the critical value of K which corresponds to the location of roots on the $j\omega$-axis is 6. This value of K makes all the coefficients of s^1-row of Routh array zero. The auxiliary equation formed from the coefficients of the s^2-row is given by

$$3s^2 + K = 3s^2 + 6 = 0$$

The roots of this equation lie on the $j\omega$-axis and are given by

$$s = \pm j\sqrt{2}$$

Thus, the complex-root branches intersect the $j\omega$-axis at $s = \pm j\sqrt{2}$, and the value of K corresponding to these roots is 6.

The characteristic equation under consideration does not require application of rules (vi) and (vii). Rule (viii) is used below to determine the breakaway points.

From the characteristic equation of the system,

$$K = -s(s+1)(s+2) = -(s^3 + 3s^2 + 2s)$$

Differentiating this equation, we get

$$\frac{dK}{ds} = -(3s^2 + 6s + 2)$$

The roots of the equation $dK/ds = 0$, are $s_{1,2} = -0.4226, -1.5774$

From Fig. 8.18, we see that the breakaway point must lie between 0 and –1. Therefore, $s = -0.4226$ corresponds to the breakaway point. The other solution of the equation $dK/ds = 0$ is $s = -1.5774$. This solution is not on the root locus (it does not satisfy Eqn. (8.38) for any $K \geq 0$) and, therefore, does not represent a breakaway point. The derivative condition $dK/ds = 0$ is, therefore, necessary but not sufficient to indicate a breakaway, or multiple-root situation.

If two loci breakaway from a breakaway point, then their tangents will be 180° apart. In general, as we shall see in other illustrative examples, if r loci breakaway from a breakaway point, then their tangents will be $360°/r$ apart, i.e., the tangents will equally divide 360°.

With the information obtained through the use of these rules, the root locus is sketched in Fig. 8.18 from where it is seen that for $K > 6$, the system has two closed-loop poles in the right half of the s-plane and is thus unstable.

It is important to note that the root locus plot given in Fig. 8.18 is only a rough sketch; it gives qualitative information about the behavior of the closed-loop system when the parameter K is varied. For quantitative analysis, the loci or specific segments of the loci need be constructed with sufficient accuracy.

Suppose it is required to determine dominant closed-loop poles with damping ratio $\zeta = 0.5$. A closed-loop pole with $\zeta = 0.5$ lies on a line passing through the origin and making an angle $\cos^{-1}\zeta = \cos^{-1}0.5 = 60°$ with the negative real axis. The point of intersection of the ζ-line with the rough sketch of the root locus plot gives the first guess. In the region where the rough root locus sketch intersects the ζ-line, a trial-and-error procedure is adapted along the ζ-line by applying the angle criterion to determine the point of intersection accurately. From Fig. 8.18, we see that the point $-0.33 + j0.58$, which lies on the ζ-line, satisfies the angle criterion. Therefore, the dominant closed-loop poles of the system are $s_{1,2} = -0.33 \pm j0.58$.

The value of K that yields these poles is found from the magnitude criterion as follows:

$$K = \{|s|\,|s+1|\,|s+2|\}_{s=-0.33+j0.58}$$

= Product of the distances from the open-loop poles to the point $-0.33 + j0.58$

$= 0.667 \times 0.886 \times 1.768 = 1.04$

The third closed-loop pole will lie on the third locus which is along the negative real axis. We need to guess a test point, compute a trial gain, and correct the guess until we found the point where $K = 1.04$. Using this trial-and-error procedure, we find from Fig. 8.18 that the point $s = -2.33$ corresponds to $K = 1.04$. Therefore, the third closed-loop pole is at $s = -2.33$.

The closed-loop transfer function of the system under consideration is

$$M(s) = \frac{1.04}{(s+0.33-j0.58)\,(s+0.33+j0.58)\,(s+2.33)}$$

Example 8.4 Consider a feedback system with the characteristic equation,

$$1 + \frac{K}{s(s+3)\,(s^2+2s+2)} = 0;\ K \geq 0 \tag{8.39}$$

The open-loop poles are located at $s = 0$, -3, $-1 + j1$ and $-1 - j1$, while there are no finite open-loop zeros. The pole-zero configuration is shown in Fig. 8.19.

Rule (i) tells us that the root locus plot consists of four root loci as K is varied from 0 to ∞.

Rule (ii) tells us that the four root loci originate from the four open-loop poles with $K = 0$ and terminate on infinity with $K = \infty$.

Rule (iii) tells us that the four root loci tend to infinity along asymptotes radiating out from

$$s = -\sigma_A = \frac{-3-1-1}{4} = -1.25$$

with angles $\phi_A = \dfrac{(2q+1)180^\circ}{4}; q = 0,1,2,3$

$$= 45^\circ, 135^\circ, 225^\circ, 315^\circ$$

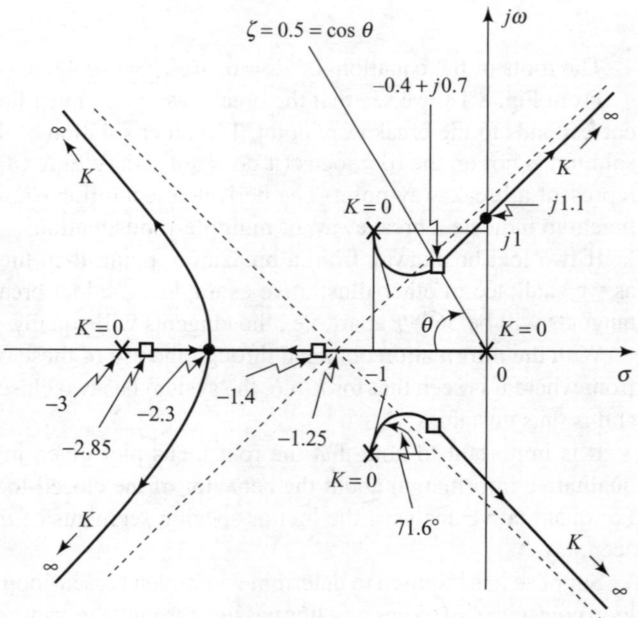

Fig. 8.19 *Root locus plot for the characteristic equation (8.39)*

The asymptotes are shown by dotted lines in Fig. 8.19.

Rule (iv) tells us that root loci exist on the real axis for $-3 \leq s \leq 0$; shown by thick lines on the real axis in Fig. 8.19.

Rule (v) tells us to check by use of the routh criterion if the root loci will intersect the imaginary axis. The system characteristic equation is

$$s(s+3)\,(s^2+2s+2) + K = 0$$

or

$$s^4 + 5s^3 + 8s^2 + 6s + K = 0$$

The Routh array becomes

s^4	1	8	K
s^3	5	6	
s^2	34/5	K	
s^1	$\dfrac{\{(204/5)-5K\}}{34/5}$		
s^0	K		

Examination of the elements in the first column of the Routh array reveals that the root loci will intersect the imaginary axis at a value of K given by

$$(204/5) - 5K = 0$$

from which $K = 8.16$.

The auxiliary equation, formed from the coefficients of the s^2-row when $K = 8.16$, is

$$(34/5)s^2 + 8.16 = 0$$

from which $s = \pm j1.1$.

Therefore, purely imaginary closed-loop poles of the system are located at $s = \pm j1.1$ as shown in Fig. 8.19; the value of K corresponding to these poles is 8.16.

Rule (vi) tells us that a root locus leaves the pole at $s = -1 + j1$ at angle ϕ_p, given by (refer to Fig. 8.19)

$$\phi_p = 180° + [-135° - 90° - 26.6°] = -71.6°$$

The characteristic equation under consideration does not require application of rule (vii).

Rule (viii) is used below to determine the breakaway points.

From the characteristic equation of the system,

$$K = -s(s+3)(s^2+2s+2) = -(s^4 + 5s^3 + 8s^2 + 6s)$$

Differentiating this equation, we get

$$\frac{dK}{ds} = -4(s^3 + 3.75s^2 + 4s + 1.5)$$

The solutions to the cubic

$$s^3 + 3.75s^2 + 4s + 1.5 = 0 \tag{8.40}$$

are the possible breakaway points.

From Fig. 8.19, we see that a breakaway point must lie between 0 and –3 on the real axis. By trial-and-error procedure, we find that $s = -2.3$ satisfies Eqn. (8.40) to a reasonable accuracy. MATLAB Module 8 in the Appendix may be used to obtain the plot whenever the problem becomes a little complex. Experience of hand-sketching will be helpful in interpreting the MATLAB results. The other two solutions of Eqn. (8.40) are $s_{1,2} = -0.725 \pm j0.365$.

It can easily be checked that $-0.725 \pm j0.365$ are not the root locus points as the angle criterion is not met at these points.

The root locus plot, therefore, has only one breakaway point at $s = -2.3$. Tangents to the two loci breaking away from this point will be 180° part.

With the information obtained through the use of these rules, the root locus plot is sketched in Fig. 8.19 from where it is seen that for $K > 8.16$, the system has two closed-loop poles in the right half of s-plane and is thus unstable.

Dominant roots of the characteristic equation with damping $\zeta = 0.5$ are determined as follows.

A ζ-line is drawn in the second quadrant at an angle of $\theta = \cos^{-1}0.5 = 60°$ with the negative real axis. By trial-and-error procedure, it is found that the point $s = -0.4 + j0.7$, which lies on the ζ-line, satisfies the angle criterion. Therefore, the dominant roots of the characteristic equation are $s_{1,2} = -0.4 \pm j0.7$. The value of gain K at the dominant root is

$$K = \text{Product of distances from open-loop poles to the point } -0.4 + j0.7$$

$$= 0.84 \times 1.86 \times 2.74 \times 0.68 = 2.91$$

The other two roots of the characteristic equation are obtained from the loci originating from the poles at $s = 0$ and $s = -3$. From Fig. 8.19, it is seen that at the double-root location $s = -2.3$, the value of K is 4.33:

$$K = \{|s||s+3|\,|s+1+j1|\,|s+1-j1|\}_{s=-2.3}$$

$$= 2.3 \times 0.7 \times 1.64 \times 1.64 = 4.33$$

Therefore, the points corresponding to $K = 2.91$ must lie on the real-root segments of these loci. Using the trial-and-error procedure, it is found that the points $s = -1.4$ and $s = -2.85$ have gain $K = 2.91$. Thus the closed-loop transfer function of the system under consideration is

$$M(s) = \frac{2.91}{(s+0.4+j0.7)(s+0.4-j0.7)(s+1.4)(s+2.85)}$$

Example 8.5 In this example, we illustrate multiple roots of multiplicity more than two. Consider the root locus of $1 + F(s)$,

where
$$F(s) = \frac{K(s+1)}{s^2(s+9)};\ K \ge 0$$

The pole-zero configuration is shown in Fig. 8.20.

Rule (i) tells us that the root locus plot consists of three root loci as K is varied from 0 to ∞.

Rule (ii) tells us that the three root loci originate from open-loop poles with $K = 0$; one of these terminates on the zero at $s = -1$ with $K = \infty$, and the other two terminate on infinity with $K = \infty$.

Rule (iii) tells us that the two root loci tend to infinity along asymptotes radiating out from

$$s = -\sigma_A = \frac{-9-0-(-1)}{3-1} = -4$$

with angles

$$\phi_A = \frac{(2q+1)180°}{3-1}; q = 0, 1$$
$$= 90°, 270°$$

The asymptotes are shown by dotted lines in Fig. 8.20.

Rule (iv) tells us that root loci exist on the real axis for $-9 \le s \le -1$; shown by thick lines on the real axis in Fig. 8.20.

Rule (v) tells us to check by use of the Routh criterion if the root loci will intersect the imaginary axis. It is found that the root loci do not cross the imaginary axis.

The characteristic equation under consideration does not require application of rules (vi) and (vii).

Rule (viii) is used below to determine the breakaway points.

From the characteristic equation of the system,

$$K = -\frac{s^2(s+9)}{s+1} = -\frac{s^3+9s^2}{s+1}$$

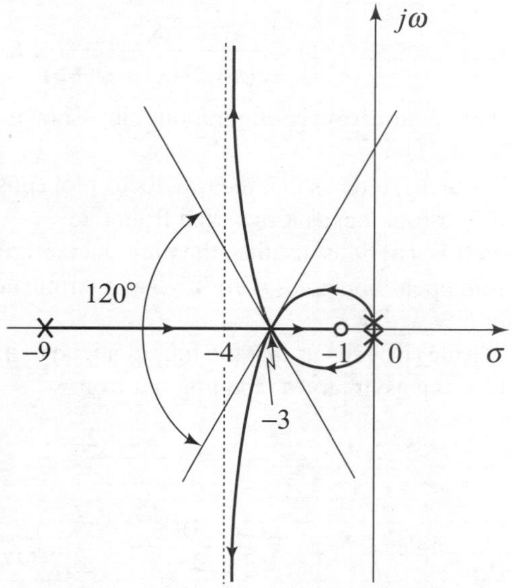

Fig. 8.20 *Root locus plot (Example 8.5)*

Differentiating this equation, we get

$$\frac{dK}{ds} = \frac{(s^3+9s^2)-(s+1)(3s^2+18s)}{(s+1)^2}$$

$$= \frac{-2s(s+3)^2}{(s+1)^2}$$

Possible points of multiple roots are at $s = 0, -3, -3$, which are on the locus as seen in Fig. 8.20.

One feature of the breakaway point at $s = -3$ needs special attention. At $s = -3$, we have double root in the derivative dK/ds; which means that not only $dK/ds = 0$ at this point but d^2K/ds^2 is also zero. This equivalently means that (refer to Eqns (8.36)) $d^2F(s)/ds^2 = 0$ at $s = -3$ which is possible only if $1 + F(s)$ has a root of multiplicity three at $s = -3$ (refer to Eqn. (8.34)). Therefore, the characteristic equation has three roots at $s = -3$; the three loci originating from the open-loop poles will approach this point and then break away. Tangents to the three loci breaking away from $s = -3$ will be $360°/3 = 120°$ apart.

Tangents to the two loci originating from $s = 0$ will be 180° apart.

With the information obtained through the use of these rules, the root locus is sketched in Fig. 8.20. (The specific examples discussed in this section, serve the objective of explaining various features of root locus plots. If a problem becomes too complex for hand-sketching, the help of MATLAB Module 8 in the Appendix may be taken).

Example 8.6 In this example, we illustrate multiple roots off the real axis. Consider the root locus of $1 + F(s)$, where $$F(s) = \frac{K}{s(s+2)(s^2+2s+5)}; K \geq 0$$

The pole-zero configuration is shown in Fig. 8.21.

Rule (i) tells us that the root locus plot consists of four root loci as K is varied from 0 to ∞.

Rule (ii) tells us that the four loci originate from open-loop poles with $K = 0$ and terminate on infinity with $K = \infty$.

Rule (iii) tells us that the four loci tend to infinity along asymptotes radiating out from

$$s = -\sigma_A = \frac{-1-1-2}{4} = -1$$

with angles $$\phi_A = \frac{(2q+1)180°}{4}; q = 0, 1, 2, 3$$

$$= 45°, 135°, 225°, 315°$$

The asymptotes are shown by dotted lines in Fig. 8.21.

Rule (iv) tells us that root loci exist on the real axis for $-2 \leq s \leq 0$; shown by thick line on the real axis in Fig. 8.21.

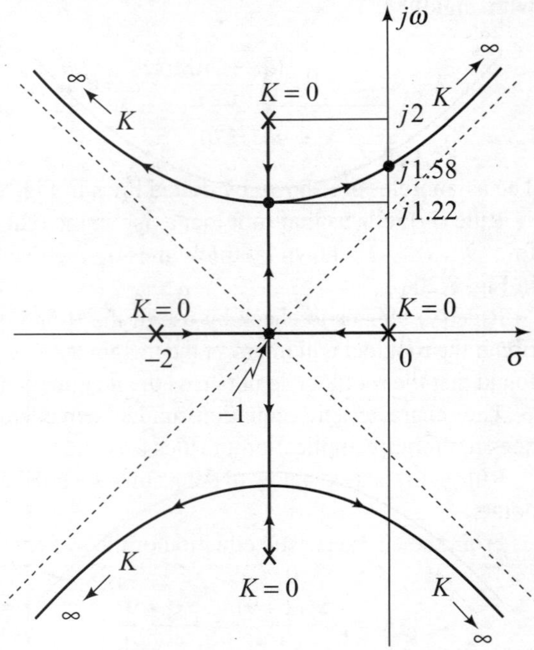

Fig. 8.21 *Root locus plot (Example 8.6)*

Rule (v) tells us to check, by using the Routh criterion, if the root loci will intersect the imaginary axis.

The characteristic equation is

$$s(s + 2)(s^2 + 2s + 5) + K = 0$$

or $$s^4 + 4s^3 + 9s^2 + 10s + K = 0$$

The Routh array becomes

s^4	1	9	K
s^3	4	10	
s^2	26/4	K	
s^1	$\frac{\{(260/4) - 4K\}}{26/4}$		
s^0	K		

Examination of the elements in the first column of the Routh array reveals that the root loci will intersect the imaginary axis at a value of K given by

$$(260/4) - 4K = 0$$

from which $K = 16.25$.

The auxiliary equation formed from the elements of s^2-row when $K = 16.25$ is

$$(26/4)s^2 + 16.25 = 0$$

from which $$s = \pm j1.58.$$

Therefore, the root loci cross the imaginary axis at $s = \pm j1.58$; the value of K corresponding to these roots is 16.25.

Rule (vi) tells us that a root locus leaves the pole at $s = -1 + j2$ at angle ϕ_p, given by (refer to Fig. 8.21)

$$\phi_p = 180° + [-116.6° - 90° - 63.4°] = -90°$$

In fact, we can observe at once that along the line $s = -1 + j\omega$, the angles contributed by the poles at $s = 0$ and $s = -2$ always add to 180°, and the angles contributed by the poles at $s = -1 + j2$ and $s = -1 - j2$ are respectively, –90° and 90°; so the entire line from one complex pole to the other is on the locus *in this special case*.

The characteristic equation under consideration does not require application of rule (vii).

Rule (viii) is used below to determine the breakaway points.

From the characteristic equation of the system,

$$K = -s(s+2)(s^2 + 2s + 5) = -(s^4 + 4s^3 + 9s^2 + 10s)$$

Differentiating this equation, we get

$$\frac{dK}{ds} = -4(s^3 + 3s^2 + 4.5s + 2.5)$$

We could find solutions to the cubic

$$s^3 + 3s^2 + 4.5s + 2.5 = 0,$$

but from the application of rule (vi) we know that the entire line $s = -1 + j\omega$ is on the locus; so there must be a breakaway at $s = -1$. Therefore, $(s + 1)$ is one of the factors of the cubic. We can now easily show that

$$(s^3 + 3s^2 + 4.5s + 2.5) = (s+1)(s+1+j1.22)(s+1-j1.22)$$

Since the solutions $-1 \pm j1.22$ are on the line between the complex poles, they are true points of multiple roots on the locus.

The complete root locus plot is sketched in Fig. 8.21. Notice that here we have complex multiple roots of multiplicity 2. Segments of the root loci come together at $-1 \pm j1.22$ and break away; tangents to the loci breaking away will be 180° apart.

8.5 THE ROOT LOCUS OF SYSTEMS WITH DEAD-TIME

Let us consider the problem of constructing the root locus for systems where a dead time τ_D is involved in the loop function. Figure 8.22 shows in block diagram form two systems involving delay in the forward path and the feedback path. In either case, the characteristic equation is given by

$$1 + KG(s)H(s)e^{-\tau_D s} = 0; K \geq 0 \qquad (8.41)$$

Equation (8.41) is not a polynomial and thus does not fall in the same class as our other examples so far. How would one plot the root locus corresponding to this equation? There are two basic approaches: direct application, and approximation of the phase criterion.

The direct application approach demands the use of angle and magnitude criteria directly on the given characteristic equation without any approximation. These criteria are not changed if the process is nonrational.

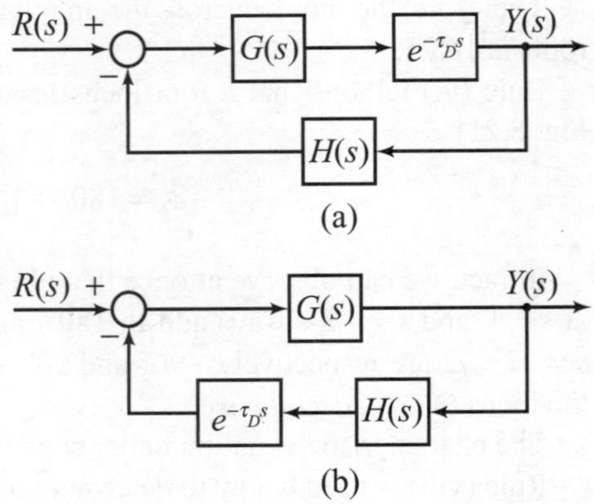

Fig. 8.22 *Closed-loop systems with time delays involved*

Angle criterion

$$\angle G(s)H(s)e^{-\tau_D s} = \pm 180°(2q+1);\ q = 0, 1, \ldots \qquad (8.42a)$$

Magnitude criterion

$$|KG(s)H(s)e^{-\tau_D s}| = 1 \qquad (8.42b)$$

One can formulate a root-locus problem as searching for locations where the angle criterion (8.42a) is satisfied. The value of the root locus gain K at any point on the root locus can then be determined by using the magnitude criterion (8.42b). It may however be noted that the rules given in Table 8.1 are not applicable for such a problem.

In the second approach, we reduce the given problem close to the one we have previously solved by approximating the nonrational function $e^{-\tau_D s}$ with a rational function. Since we are concerned with control systems (systems with low frequencies), we want an approximation that will be good for $s = 0$ and nearby. The most common means used to find such an approximation is attributed to Pade, and is based on matching the series expansion of $e^{-\tau_D s}$ with that of a rational function; a ratio of numerator polynomial of degree p and denominator polynomial of degree q. For our purposes, we will consider only the case $p = q = 1$ (refer to Eqn. (2.46)):

$$e^{-\tau_D s} \cong \frac{1-\dfrac{\tau_D}{2}s}{1+\dfrac{\tau_D}{2}s} = \frac{-[s-(2/\tau_D)]}{s+(2/\tau_D)} \qquad (8.43)$$

With this approximation, the characteristic Eqn. (8.41) becomes

$$1 - KG(s)H(s)\,[s - (2/\tau_D)]\,/\,[s + (2/\tau_D)] = 0 \qquad (8.44)$$

Examining Eqn. (8.44), we note that the equation is of the form

$$1 - F(s) = 0 \qquad (8.45)$$

where

$$F(s) = \frac{K\prod_{i=1}^{m}(s+z_i)}{\prod_{j=1}^{n}(s+p_j)}$$

This equation again is not of the standard form (8.22) used earlier to derive construction rules of Table 8.1. Therefore, the construction rules of Table 8.1 cannot be used directly even in the approach based on approximation of nonrational function $e^{-\tau_D s}$. However, only a minor modification in the rules of Table 8.1 is required to deal with characteristic equations of the form (8.45). Our discussion will be limited to second approach; the approach based on approximation given by Eqn. (8.43).

In a manner similar to the development of the root-locus method in Section 8.2, we require that the root locus of the characteristic equation (8.45) satisfy the equations:

Angle criterion

$$\angle F(s) = \sum_{i=1}^{m}(s+z_i) - \sum_{j=1}^{n}(s+p_j) = \pm q(360°);\ q = 0, 1, 2, \ldots \tag{8.46a}$$

Magnitude criterion

$$|F(s)| = \frac{K\prod_{i=1}^{m}|s+z_i|}{\prod_{j=1}^{n}|s+p_j|} = 1 \tag{8.46b}$$

The locus of roots follows a zero-degree locus (Eqn. (8.46a)) in contrast with the 180° locus considered previously. However, the root locus rules of Table 8.1 may be altered to account for the zero-degree phase angle requirement, and then the root locus may be obtained as in the preceding sections.

Rule (i) : No change

Rule (ii) : No change

Rule (iii) : No change in centroid calculation. Angles of asymptotes are now given by the relation

$$\phi_A = \frac{q(360°)}{n-m};\ q = 0, 1, \ldots, (n-m-1) \tag{8.47}$$

Rule (iv) : A point on the real axis lies on the locus if the number of open-loop poles plus zeros on the real axis to the right of this point is even.

Rule (v) : No change

Rule (vi) : The angle of departure, ϕ_p, of a locus from a complex open-loop pole is given by

$$\phi_p = \phi \tag{8.48}$$

where ϕ is the net angle contribution at this pole of all other open-loop poles and zeros.

Rule (vii) : The angle of arrival, ϕ_z, of a locus at a complex zero is given by

$$\phi_z = -\phi \tag{8.49}$$

where ϕ is the net angle contribution at this zero of all other open-loop poles and zeros.

Rule (viii) : No change

Rule (ix) : No change.

Example 8.7 Suppose that the plant in Fig. 8.22 is $G(s) = K/s$, the sensor $H(s) = 1$ and the dead time $\tau_D = 1$ sec.

The characteristic equation of the system becomes

$$1+\frac{Ke^{-s}}{s}=0; K\geq 0$$

With the approximation $e^{-s} \cong \dfrac{1-s/2}{1+s/2}$, the characteristic equation becomes

$$1-\frac{K(s-2)}{s(s+2)}=0 \qquad (8.50)$$

Examining the real-axis segments of the root loci shown in Fig. 8.23, it is obvios that there will be two breakaway points. These points are the solutions of the equation $\dfrac{dK}{ds}=0$; where $K=\dfrac{s(s+2)}{(s-2)}$.

The solutions are $s = -0.83, 4.83$.

The characteristic equation (8.50) can equivalently be expressed as

$$s^2 + (2 - K)s + 2K = 0$$

Obviously, $K = 2$ results in oscillations of frequency $\omega_0 = 2$. Therefore, the root loci intersect the imaginary axis at $s = \pm j2$; the value of K corresponding to these points is 2.

The complete root locus plot is sketched in Fig. 8.23. It can, in fact, be shown that the complex root branches of the root loci form a circle (refer to Example 8.1).

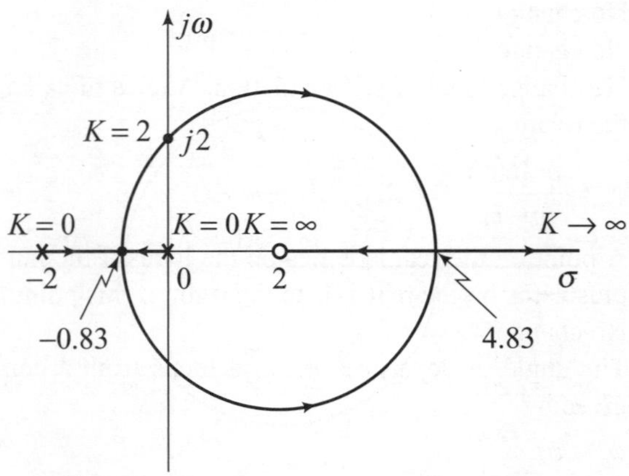

Fig. 8.23 *Root locus plot for the characteristic equation (8.50)*

8.6 REVIEW EXAMPLES

Review Example 8.1 Several open-loop pole-zero configurations and their corresponding approximate root loci are given in Table 8.2. Computing (where applicable) (i) the large-gain asymptotes, (ii) the angles of departure from complex poles, (iii) the angles of arrival at complex zeros, (iv) the breakaway points, and (v) the frequency at which the root loci cross the imaginary axis, is left as an exercise for the reader. This exercise will help attain good practice on hand-sketching rules for constructing root-locus plots.

Table 8.2 *Open-loop pole-zero configurations and the corresponding root loci (equal axes scales)*

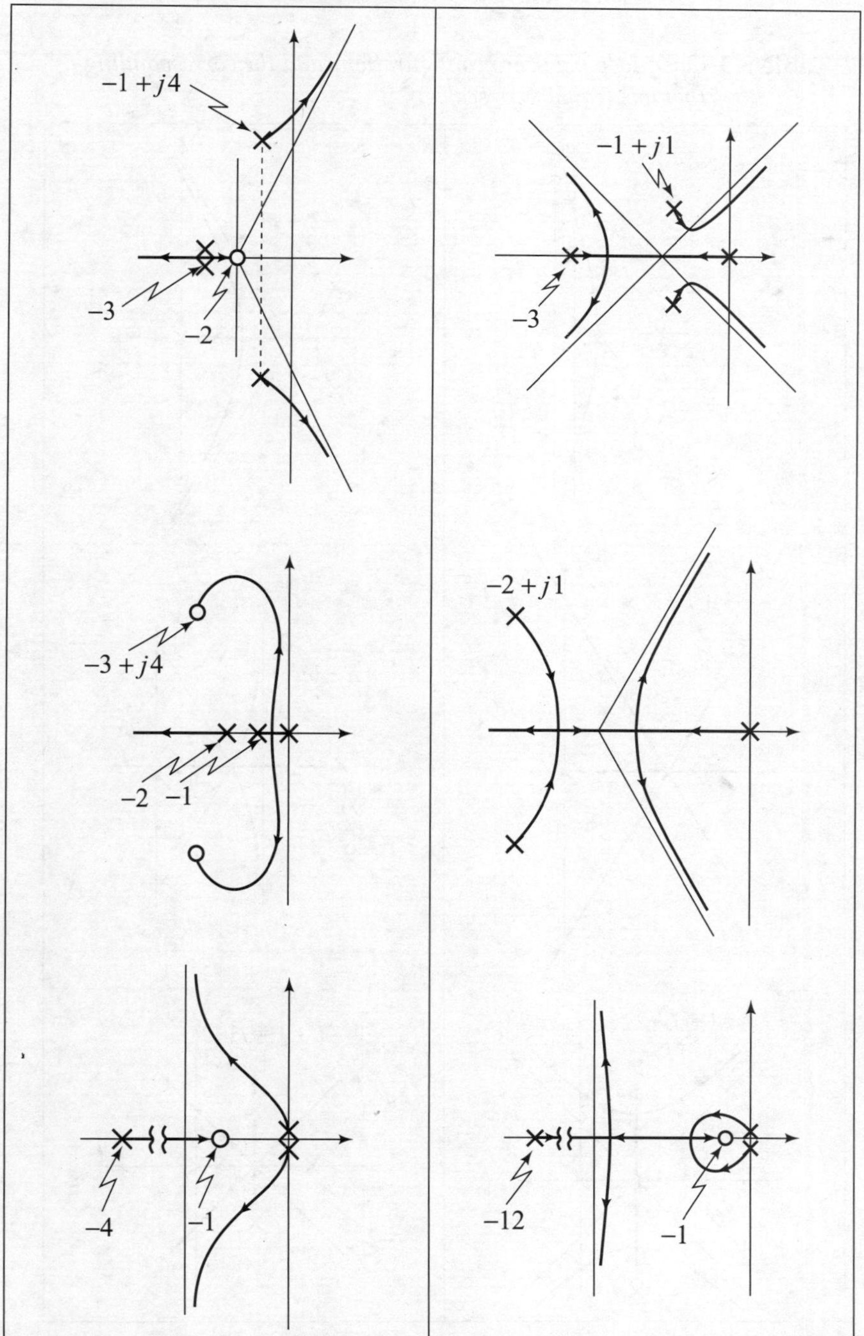

Review Example 8.2 Several open-loop pole-zero configurations and their corresponding approximate root loci were given in Table 8.2. Some additional examples are tabulated in Table 8.3. These examples have been selected to show multiple roots of various orders in root-locus plots. Estimating the locations of multiple roots (if any) and the order of multiplicity of the roots, is left as an exercise for the reader.

Table 8.3 *Open-loop pole-zero configurations and the corresponding root loci (equal axes scales)*

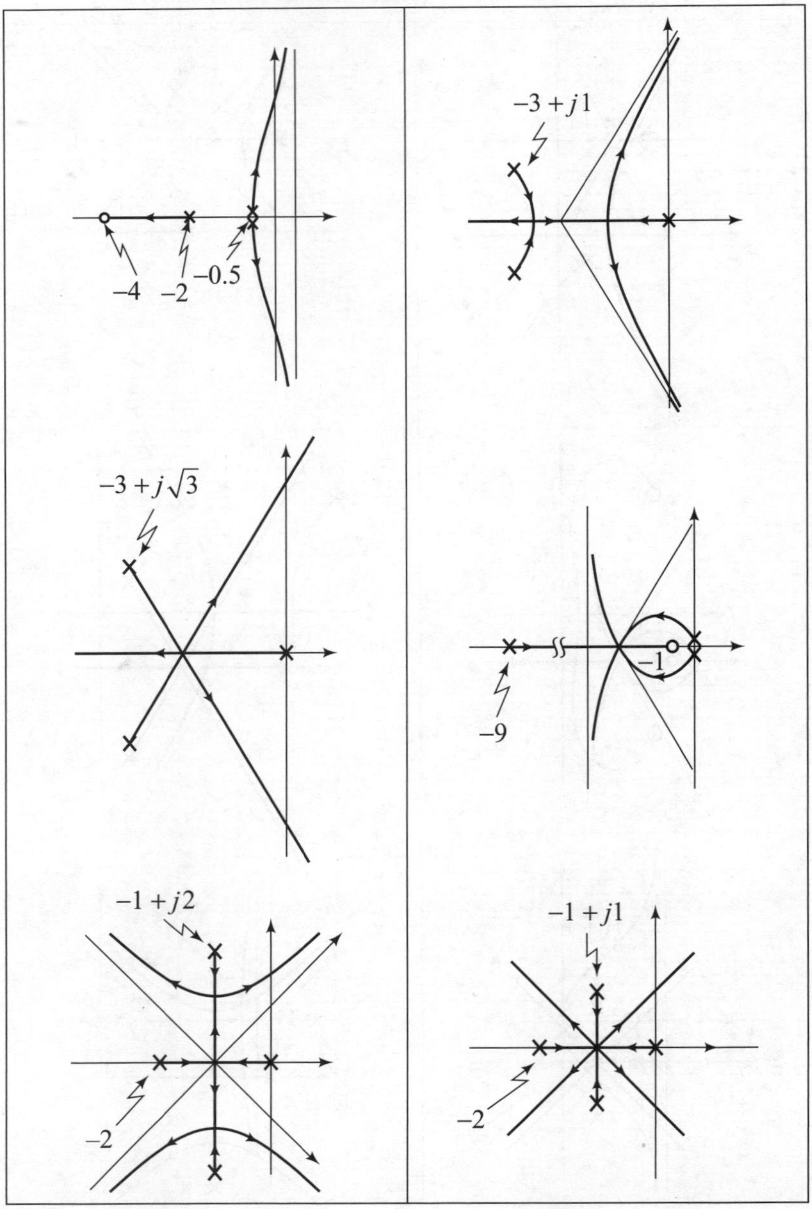

Review Example 8.3 The root locus technique discussed in this chapter is limited to only one variable parameter in K. In many design problems, the effects of varying several parameters need to be investigated. For example, it would be useful to investigate the effects on the characteristic equation roots when the parameters K_1 and K_2 of the feedback system of Fig. 8.24 take on various values.

The characteristic equation of the system of Fig. 8.24 is

$$s^3 + 10s^2 + K_1K_2s + K_1 = 0 \qquad (8.51)$$

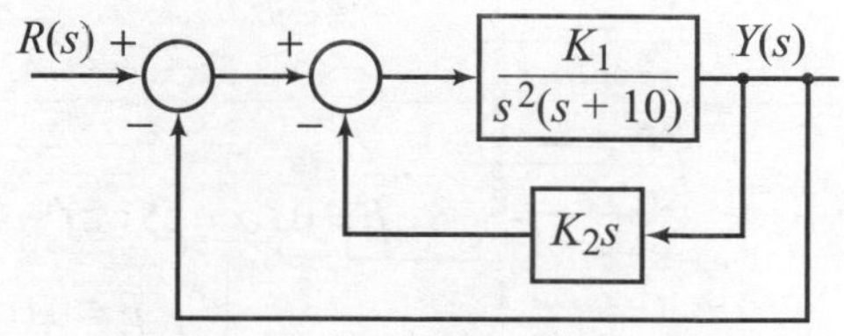

Fig. 8.24 *A feedback system*

where $\alpha = K_1$ and $\beta = K_1K_2$ are the variable parameters, which vary from 0 to ∞.

In terms of the parameters α and β, the characteristic equation becomes

$$s^3 + 10s^2 + \beta s + \alpha = 0 \qquad (8.52)$$

The effect of varying β from zero to infinity is determined from the root locus equation,

$$1 + \frac{\beta s}{s^3 + 10s^2 + \alpha} = 0 \qquad (8.53)$$

The open-loop poles of this root locus equation are the roots of the characteristic equation,

$$s^3 + 10s^2 + \alpha = 0 \qquad (8.54)$$

We must first evaluate the effect of varying α from zero to infinity by using Eqn. (8.54), rewritten as

$$1 + \frac{\alpha}{s^2(s+10)} = 0 \qquad (8.55)$$

Then, after evaluating the effect of α, a value of α is selected and used with Eqn. (8.53) to evaluate the effect of β. This two-step method of evaluating the effect of α and then β may be carried out by two root locus procedures—two root locus plots are generated; one as α varies, and then the second as β varies for a selected value of α.

We may instead, plot the root loci for variation of one parameter (α) and then embed on it the root loci for the variation of the other parameter (β). This family of loci, often called *root contours*, illustrates the total effect of varying both α and β from zero to infinity, on the roots of the characteristic equation of the system.

The root locus plot of Eqn. (8.55) as a function of α is shown in Fig. 8.25. The roots of these loci, indicated by slashes, become the open-loop poles for the root locus plot of Eqn. (8.53). This plot, embedded on the root locus plot of Eqn. (8.53) for several selected values of α is also shown in Fig. 8.25 (broken thick lines).

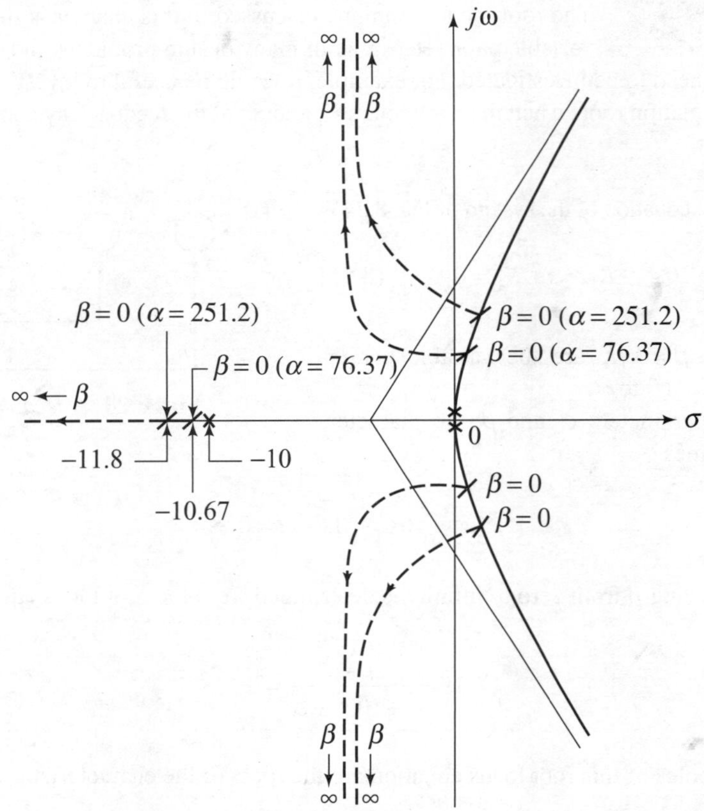

Fig. 8.25 *Root contours for the characteristic equation (8.52)*

CONCLUDING REMARKS

The root locus method is a powerful and useful approach for analysis and design of control systems, and will continue to be one of the most important procedures of control engineering.

What we have learnt?

- The root locus is the locus of roots of the characteristic equation of a closed-loop system as a specific parameter is varied from zero to infinity. We looked at the ways to sketch the root locus rapidly, even for higher-order systems.
- We have seen in Chapter 6 that the transient performance of a closed-loop system is directly related to the location of the roots of the closed-loop characteristic equation. Therefore, a root locus sketch gives a qualitative information about changes in the transient response as system parameters are varied. Approximate root locus plots can be used to analyze the initial design of a system and determine suitable alterations in parameter values.
- If the system has a variable loop gain, then the location of the closed-loop poles depends on the loop gain chosen. Setting the gain is the first step in adjusting the system for satisfactory performance. For many practical cases, however, the adjustment of the gain alone may not provide sufficient alteration

of the system behavior to meet the given specifications. It is then necessary to redesign the system by modifying the structure and/or by incorporating additional control devices. The next chapter deals with the redesign problem using root locus plots.

- The reader may wonder why hand calculations for plotting root loci have been discussed in the present chapter when powerful MATLAB commands are available (refer to MATLAB Module 8 in the Appendix). In fact, through the systematic formulation of set of rules, we are looking for the clearest and simplest explanation of the dynamic phenomena of the system. The rules of the root locus give us a clear and precise understanding of the endless patterns that can be created by an infinite set of characteristic equations. We could eventually learn to design without these rules, but our level of skill would never be as high or our understanding as great. This is true of other analysis techniques also such as Bode plots, Nyquist plots, Nichols chart, and so on, covered later in the book.

Review Questions

8.1 (a) Give an account of the meaning of the terms 'centroid' and 'breakaway points' with reference to root locus plots. Give the equations to determine their values.
(b) What is the way to find out whether or not a root locus plot crosses the (i) imaginary axis (ii) real axis? If it does cross, how to determine the point at which it crosses and what is the significance of that point?

8.2 How can we analyze the transient response of a feedback system from its root locus plot? In particular, indicate on a root locus plot (i) how to determine settling time and peak overshoot for a given value of root locus gain, and (ii) how to find the gain for given values of settling time and peak overshoot. Under what conditions, your analysis will approximately match the simulation results?

8.3 Give examples of feedback systems which, for all values of loop gain K, are (i) stable (ii) unstable (iii) marginally stable. Sketch the corresponding root locus plots.

Self-Appraisal Quiz

A set of questions with multiple answer choices is given below. These questions have been designed to capture important aspects of the subject covered in the chapter. For each of these questions, identify the best of the given answer choices. Compare your answers with the Master Key given at the end of the session. This quiz should serve the purpose of a self-appraisal test for readers.

8.1 A unit feedback system has open-loop poles at $s = -2 \pm j2, s = -1$, and $s = 0$; and a zero at $s = -3$. The angles made by the root-locus asymptotes with the real axis, and the point of intersection of the asymptotes are, respectively,

(A) (60°, –60°, 180°) and –3/2 (B) (60°, –60°, 180°) and –2/3
(C) (45°, –45°, 180°) and –2/3 (C) (45°, –45°, 180°) and –4/3

8.2 Which of the root locus plots shown in Fig. Q8.2 is the correct plot for a unity feedback system with open-loop poles at $s = -1 \pm j1$, and a zero at $s = -2$?

(A) Fig. Q8.2a (B) Fig. Q8.2b (C) Fig. Q8.2c (D) Fig. Q8.2d

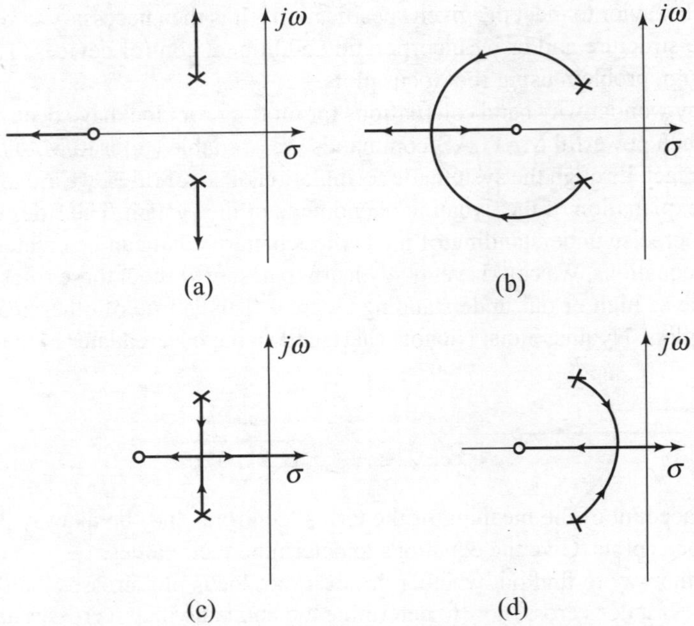

Fig. Q8.2

8.3 Which of the root locus plots shown in Fig. Q8.3 is the correct plot for a unity feedback system with open-loop transfer function $G(s) = K/[s^2(s+5)]$?

(A) Fig Q8.3a (B) Fig. Q8.3b (C) Fig. Q8.3c

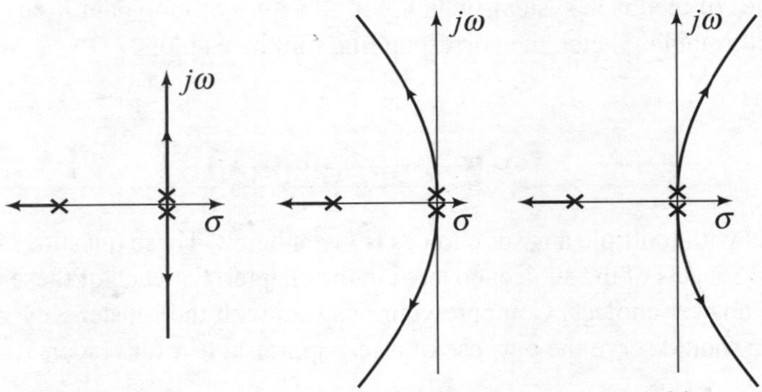

Fig. Q8.3

8.4 A unity feedback system has open-loop transfer function $G(s) = K(s+1)(s+2)/[s(s+3)(s+4)]$. For $K = 10$, the closed-loop poles are

(A) all real and distinct
(B) one real and two complex conjugate
(C) all real and repeated
(D) None of the answers in (A), (B), and (C) is correct

8.5 Root locus plot of a feedback system as gain K is varied, is shown in Fig. Q8.5. The system response to step input is non-oscillatory for

(A) $0 < K < 0.4$
(B) $0.4 < K < 6$
(C) $6 < K < \infty$
(D) None of the answers in (A), (B), and (C) is correct

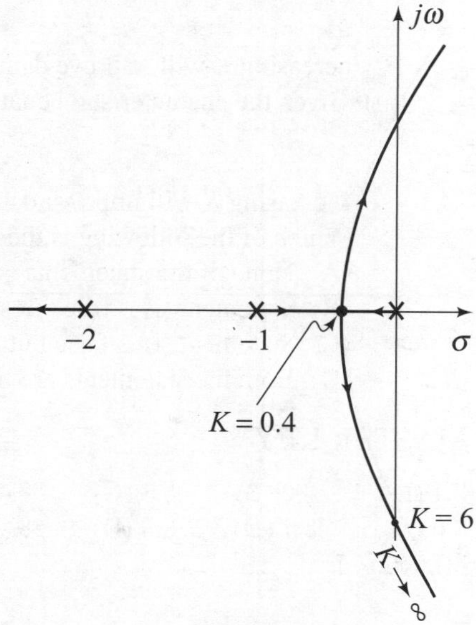

Fig. Q8.5

8.6 Consider the root locus plot shown in Fig. Q 8.5.

(i) Adding a zero between $s = -1$ and $s = -2$ would move the root locus to the left.
(ii) Adding a pole at $s = 0$ would move the root locus to the right.

Which of the following is the correct answer?

(A) None of the above statements is true
(B) Statement (i) is true but statement (ii) is false
(C) Statement (i) is false but statement (ii) is true
(D) Both the statements are true

8.7 Consider the root locus plot of unity-feedback system with open-loop transfer function,

$$G(s) = \frac{K(s+5)}{s(s+2)(s+4)(s^2+2s+2)}$$

The meeting point of the asymptotes on the real axis occurs at

(A) –1.2 (B) –0.85 (C) –1.05 (D) –0.75

8.8 Consider the root locus plot of unity-feedback system with open-loop transfer function,

$$G(s) = \frac{K(s+5)}{s(s+2)(s+4)(s^2+2s+2)}$$

The breakaway point of the root loci on the real axis occurs at

(A) –3 (B) –4.5
(C) –5.5 (D) None of the answers in (A), (B), and (C) is correct

8.9 The root locus plot of the characteristic equation

$$1 + KF(s) = 0$$

is given in Fig. Q8.9. The value of K at $s = \pm j1$ is

(A) 4 (B) 1
(C) 10 (D) None of the answers in (A), (B), and (C) is correct

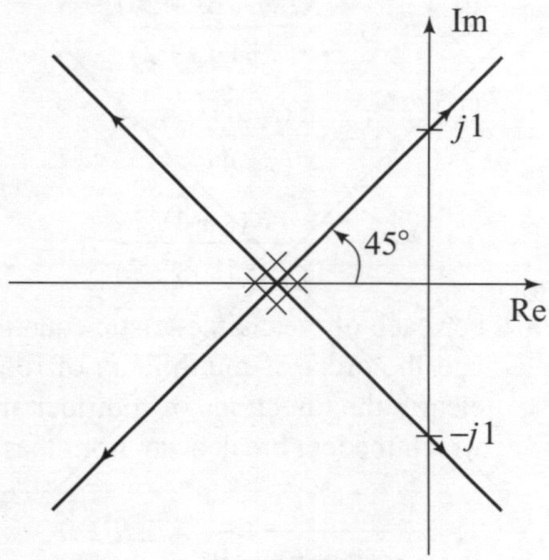

Fig. Q8.9

8.10 In a root locus plot,

(i) there is only one intersect of the asymptotes and it is always on the real axis;
(ii) the breakaway points always lie on the real axis.

Which of the following is the correct answer?

(A) None of the statements is true
(B) Statement (i) is true but statement (ii) is false
(C) Statement (i) is false but statement (ii) is true
(D) Both the statements are true

8.11 Consider the following statements:

(i) Given the characteristic equation

$$s^3 + 3s^2 + 2s + K = 0$$

Increasing K will improve damping.

(ii) Given the characteristic equation

$$s^2 + 2s + K(0.2s + 1) = 0$$

Increasing K will improve damping.

Which of the following is the correct answer?

(A) None of the statements is true
(B) Statement (i) is true but statement (ii) is false
(C) Statement (i) is false but statement (ii) is true
(D) Both the statements are true

MASTER KEY

8.1 (B)	8.2 (B)	8.3 (C)	8.4 (A)	8.5 (A)
8.6 (D)	8.7 (D)	8.8 (D)	8.9 (A)	8.10 (B)
8.11 (C)				

Problems

8.1 For each of the open-loop transfer functions $F(s)$ given below, sketch the general shape of the root locus plot of the characteristic equation $1+F(s) = 0$, as the gain K is varied from 0 to ∞. Where applicable, the plot should include (i) the large-gain asymptotes, (ii) the angles of departure from complex poles of $F(s)$, (iii) the angles of arrival at complex zeros of $F(s)$, (iv) breakaway points, and (v) the frequency at which the root loci cross the $j\omega$-axis.

(a) $F(s) = \dfrac{K(s+2)}{(s+3)^2(s+1+j4)(s+1-j4)}$

(b) $F(s) = \dfrac{K}{s(s+3)(s^2+2s+2)}$

(c) $F(s) = \dfrac{K(s^2+6s+25)}{s(s+1)(s+2)}$

(d) $F(s) = \dfrac{K}{s(s^2+4s+5)}$

(e) $F(s) = \dfrac{K(s+1)}{s^2(s+4)}$

(f) $F(s) = \dfrac{K(s+1)}{s^2(s+12)}$

(g) $F(s) = \dfrac{K(s+4)}{(s+0.5)^2(s+2)}$

8.2 For each of the characteristic equations given below, estimate locations of multiple roots (if any), and the order of multiplicity of roots. Roughly sketch the root locus plot in each case showing clearly the directions of root loci approaching the points of multiple roots, and the directions at which the loci break away from these points.

(i) $1 + \dfrac{K}{s(s^2+6s+10)} = 0; K \ge 0$

(ii) $1 + \dfrac{K}{s(s^2+6s+12)} = 0; K \ge 0$

(iii) $1+\dfrac{K(s+1)}{s^2(s+9)}=0; K\ge 0$ (iv) $1+\dfrac{K}{s(s+2)(s^2+2s+5)}=0; K\ge 0$

(v) $1+\dfrac{K}{s(s+2)(s^2+2s+2)}=0; K\ge 0$

8.3 A unity-feedback control system has an open-loop transfer function

$$G(s)=\frac{K}{s(s+1)(s+2)}; K\ge 0$$

Make a rough sketch of the root locus plot of the system, explicitly identifying the centroid, the asymptotes, the breakaway points and $\pm j\omega$-axis crossover points. By a trial-and-error application of the angle criterion, locate a point on the locus that gives dominant closed-loop poles with $\zeta = 0.5$. Determine the value of K at this point. Corresponding to this value of K, determine the closed-loop transfer function of the system.

MATLAB Exercise Verify your rough sketch with a MATLAB plot. Turn in your hand sketch and the MATLAB plot on the same scale. Also use MATLAB to obtain the closed-loop step response and validate or refute your second-order approximation.

8.4 A unity-feedback control system has an open-loop transfer function,

$$G(s)=\frac{K}{s(s+4)(s^2+8s+32)}; K\ge 0$$

Make a rough sketch of the root locus plot of the system, explicitly identifying the centroid, the asymptotes, the breakaway points, the departure angles from poles of $G(s)$, and the $\pm j\omega$-axis crossover points.

By a trial-and-error application of the angle criterion, locate a point on the locus that gives dominant closed-loop poles with $\zeta = 0.707$. Determine the value of K at this point. Corresponding to this value of K, roughly locate the other two closed-loop poles and comment upon the dominance condition.

MATLAB Exercise Verify your rough sketch with a MATLAB plot. Turn in your hand sketch and the MATLAB plot on the same scale. Also check the accuracy of the second-order approximation using MATLAB to simulate the system.

8.5 A unity-feedback control system has an open-loop transfer function,

$$G(s)=\frac{K}{s(s^2+8s+32)}; K\ge 0$$

Make a rough sketch of the root locus plot of the system, explicitly identifying the centroid, the asymptotes, the departure angles from complex poles of $G(s)$, and the $\pm j\omega$-axis crossover points.

By a trial-and-error application of the angle criterion, locate a point on the locus that gives dominant closed-loop poles with $\zeta = 0.5$. Determine the value of K at this point. Corresponding to this value of K, roughly locate the third closed-loop pole and comment upon the dominance condition.

MATLAB Exercise Verify your rough sketch with a MATLAB plot. Turn in your hand sketch and the MATLAB plot on the same scale. Also check the accuracy of the second-order approximation using MATLAB to simulate the system.

8.6 Draw the root locus plot for a control system with unity feedback having the forward-path transfer function.

$$G(s)=\frac{K}{s(s+1)(s+5)};K\geq 0$$

Give all relevant characteristics of the curves that are useful in establishing the locus.

Find the least value of K to give an oscillatory response and the greatest value of K that can be used before continuous oscillations occur. Find the frequency of the continuous oscillations when K is just large enough to give this condition.

MATLAB Exercise Attempt this problem using MATLAB dialogs.

8.7 Prove that a combination of two poles $s=-a_1$ and $s=-a_2$ and one zero $s=-b$ to the left of both of them on the real axis, results in a root locus whose complex root branches form a circle centered at the zero with radius given by $\sqrt{(b-a_1)(b-a_2)}$; the root locus gain varying from 0 to ∞.

8.8 Consider a unity-feedback system with a forward path transfer function,

$$G(s)=\frac{K(s+3)}{s(s+2)};K\geq 0$$

Show that part of the root locus is a circle.

Construct the root locus and determine the damping ratio for maximum oscillatory response. What is the value of K at this point of the locus?

8.9 Prove that the combination of two complex-conjugate poles $s_{1,2}=-\alpha\pm j\beta$ and a real zero $s=-b$ results in a root locus whose complex root branches lie on a circle centered at the zero with radius $=\sqrt{(\alpha-b)^2+\beta^2}$; the root locus gain varies from 0 to ∞.

8.10 Consider a unity-feedback system with open-loop transfer function,

$$G(s)=\frac{K(s^2+1)}{s(s+2)};K\geq 0$$

Sketch the root locus plot. Show that the complex root branches lie on a circle.

8.11 Consider a unity-feedback system with a forward path transfer function,

$$G(s)=\frac{K(s+4)}{(s+2)(s-1)};K\geq 0$$

Draw a root locus plot and find the value of K that results in $\zeta=0.707$ and $t_s<4$ sec. Determine the peak overshoot, the settling time, and the position error for this value of K.

MATLAB Exercise Attempt this problem using MATLAB dialogs.

8.12 Sketch the root locus plot for the system shown in Fig. P8.12 as K varies from 0 to ∞. Show that the complex root branches of the plot form a circle.

Find the value of K that results in (i) damping ratio $\zeta=1$, and (ii) minimum steady-state error to ramp inputs.

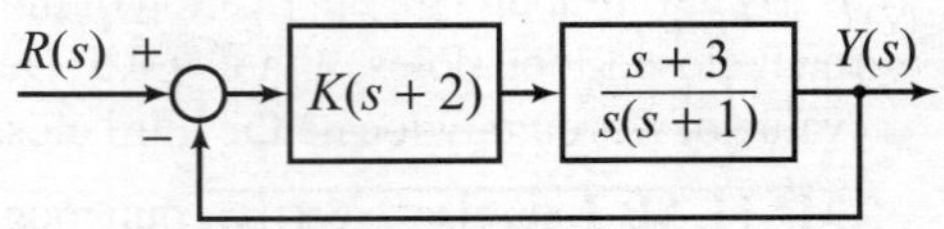

Fig. P8.12

8.13 Consider the control loop shown in Fig. P8.13. Sketch the root locus plot as K varies from 0 to ∞. Find the value of K that yields an oscillatory response of the control loop with a damping ratio of 0.707.

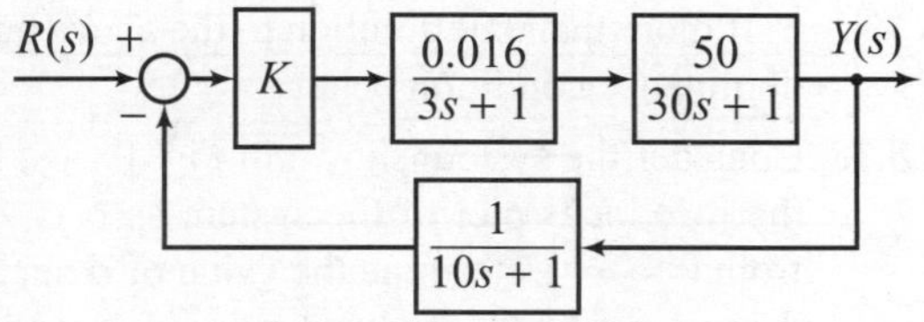

Fig. P8.13

MATLAB Exercise Attempt this problem using MATLAB dialogs.

8.14 Figure P8.14 represents a position servo with an appreciable sensor time constant.

(a) Plot root loci and find $K > 0$ for a damping ratio of 0.5 for unity feedback (i.e., assuming an ideal feedback sensor).

(b) If the actual sensor has in fact an appreciable time constant, what will be the actual damping ratio if the value of K found in part (a) is used? An approximate answer will suffice.

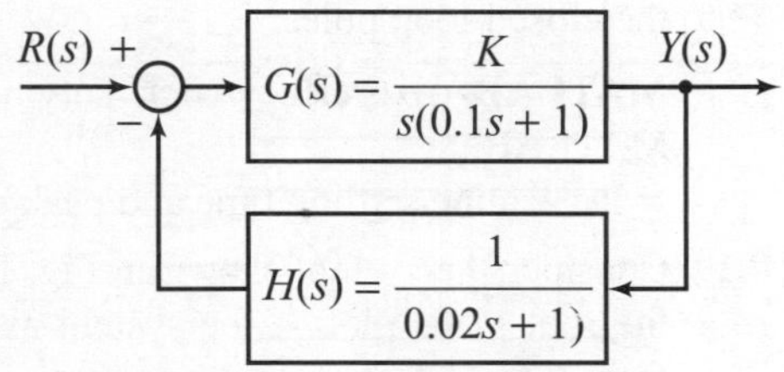

Fig. P8.14

8.15 Consider the system shown in Fig. P8.15. Sketch the root locus plot of the system as gain K varies from 0 to ∞. Determine the value of K such that the damping ratio ζ of the dominant closed-loop poles is 0.5. For the selected value of K, find all the closed-loop poles.

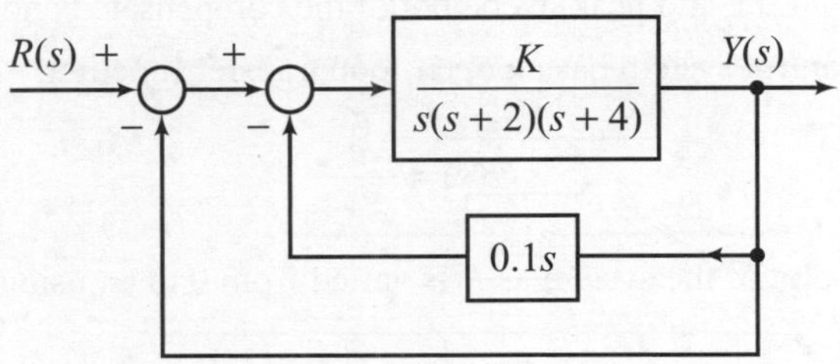

Fig. P8.15

MATLAB Exercise Attempt this problem using MATLAB dialogs.

8.16 Consider the servo system shown in Fig. P8.16. Sketch the root locus plot of the system as the velocity feedback constant K_t varies from 0 to ∞. Determine the value of K_t such that the closed-loop poles have a damping ratio of 0.7.

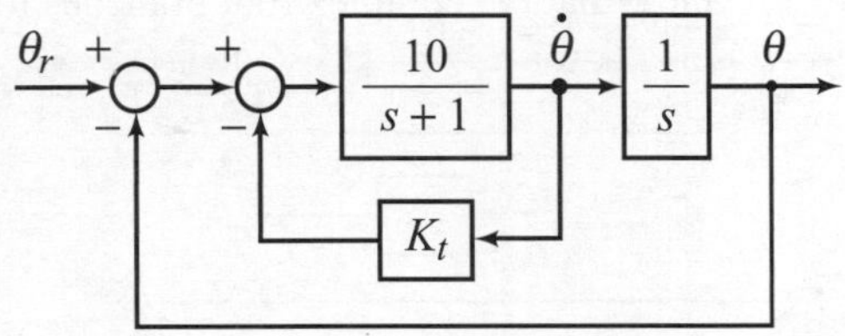

Fig. P8.16

MATLAB Exercise Attempt this problem using MATLAB dialogs.

8.17 Consider the system shown in Fig. P8.17. Draw the root locus plot of the system as K_t is varied from 0 to ∞. Determine the value of K_t such that damping ratio of the dominant closed-loop poles is 0.4.

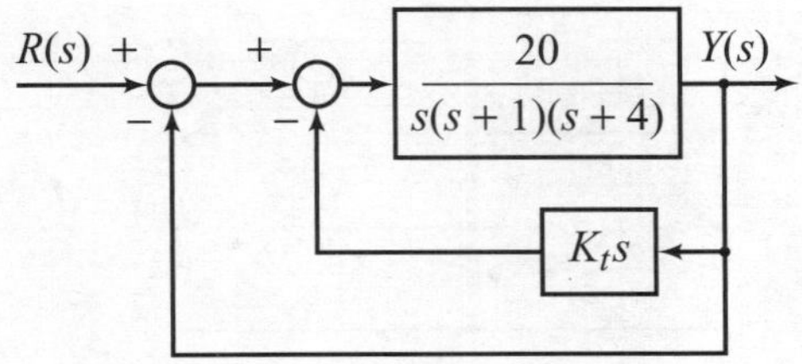

Fig. P8.17

If more than one solution to the above problem is possible, then select a solution that meets the dominance condition better.

8.18 Consider the system shown in Fig. P8.18. Draw the root locus plot of the system as α is varied from 0 to ∞. Determine the value of α such that damping ratio ζ of the dominant closed-loop poles is 0.5. For the selected value of α, find all the closed-loop poles.

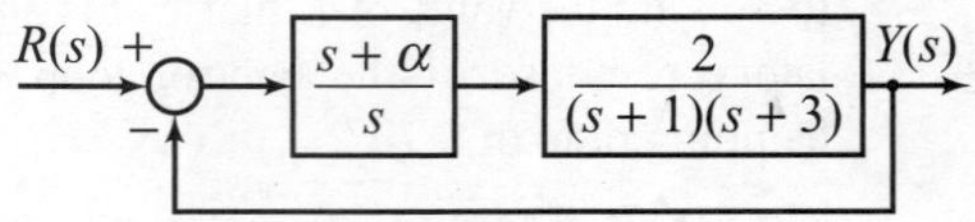

Fig. P8.18

MATLAB Exercise After solving the problem using hand sketches, attempt it using MATLAB.

Determine settling time and peak overshoot of the compensated and uncompensated systems.

8.19 Consider the system shown in Fig. P8.19. Draw the root locus plot of the system as α is varied from 0 to ∞. Determine the value of α such that damping ratio ζ of the dominant closed-loop poles is 0.5. For the selected value of α, find all the closed-loop poles.

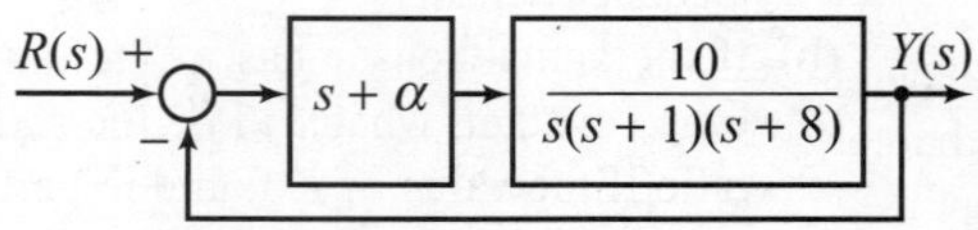

Fig. P8.19

MATLAB Exercise After solving the problem using hand sketches, attempt it using MATLAB.

Determine settling time and peak overshoot of the compensated and uncompensated systems.

8.20 A unity-feedback control system has an open-loop transfer function.

$$G(s) = \frac{Ke^{-s}}{s+1}$$

Sketch a root locus plot of the system as K is varied from 0 to ∞, using the approximation

$$e^{-s} = \left(1 - \frac{1}{2}s\right) \Big/ \left(1 + \frac{1}{2}s\right).$$

Show that the complex root branches form a circle.

COMPENSATOR DESIGN USING ROOT LOCUS PLOTS

9

Preview

The basic characteristics of the transient performance of a closed-loop system are closely related to the closed-loop poles. If the system has a variable loop gain, then the location of the closed-loop poles depends on the value of the loop gain chosen. The root locus technique, introduced in the previous chapter, is a graphic tool that helps us know how the closed-loop poles move in the s-plane as the loop gain is varied. From the design viewpoint, simple gain adjustment may move the closed-loop poles of some systems to desired locations while simultaneously satisfying the steady-state performance requirements. If the gain adjustment alone does not yield the desired result, modifying the feedback control structure and/or incorporating additional control devices (compensators) may become necessary. The objective in this chapter is to learn procedures for compensator design using root locus plots. The procedures for lead and lag compensator design, and PID control design will be developed. In Chapter 12, we will look at another method of design based on frequency response plots.

9.1 INTRODUCTION

Recall the basic feedback system block diagram defined in Chapter 2 and shown in abridged form as Fig. 9.1a. When we speak of a controller $D(s)$, our concern is mainly with the *information handling devices;* the sources of information are the reference input $r(t)$, which is derived from the command $y_r(t)$, and the feedback signal $b(t)$, which is produced by the feedback path elements $H(s)$ from the controlled output $y(t)$. The *power handling devices*—actuators and the controlled process itself—are represented by $G(s)$ in the block diagram of Fig. 9.1a.

It is often possible and convenient to manipulate the block diagram into a unity-feedback model shown in Fig. 9.1b, where the reference input $r(t)$ is equal to the commanded value of the controlled output; the system error appears explicitly as the actuating signal.

This is the first chapter devoted exclusively to control system design. The design procedures developed in this chapter are based on shaping the three aspects of a system's response discussed at length in Chapters 5 and 6, namely, stability, speed, and steady-state accuracy of the closed-loop response. For a

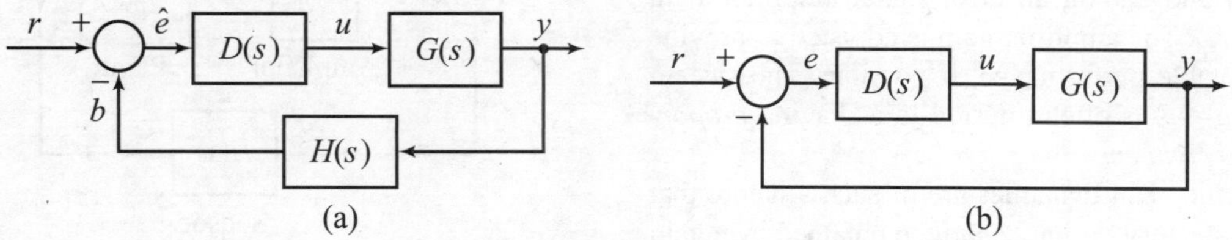

Fig. 9.1 *(a) A non-unity feedback system (b) A unity feedback system*

stable closed-loop system, all the closed-loop poles must strictly lie in the left half of the s-plane. Since stability is the prime objective of design, we always ensure that our controller $D(s)$ stabilizes the system by relocating the poles such that they all lie in the left half of the s-plane. Once we have ensured closed-loop stability, we then refine the controller (without violating the constraint on stability) to achieve the desired performance.

The performance is often specified in terms of figures of merit, such as peak overshoot, settling time, steady-state error constant, ….., which we have already met; and bandwidth, phase margin, gain margin, ….., which we meet later in this book. One of the ideas which has been emerging from the previous chapters is that of turning the specifications into a target closed-loop characteristic polynomial and then selecting the controller $D(s)$ to match this polynomial. An appropriate choice of a target closed-loop polynomial is to use a second-order pole pair in the quadratic factor $\left(s^2 + 2\zeta\omega_n s + \omega_n^2\right)$ and all other real/complex poles faster than this quadratic pair so that the response of the higher-order system is dominated by the second-order response characteristic. Overshoot, settling time or other performance indices can be used in this step.

At several places in the preceding chapters, we have used gain setting to improve the performance of a feedback system. Merely by gain setting, it may be possible to meet the given specifications of simple control systems. However, in most practical cases, the gain adjustment does not provide the desired result. As is usually the case, increasing the gain reduces the steady-state error but results in oscillatory transient response or even instability. Under such circumstances, it is possible to introduce appropriate corrective action, integral or derivative, to force the plant to meet the given specifications. The job of integral or derivative actions is to *compensate* for the deficiency in the performance. The subsystem $D(s)$ in the feedback loop (Fig. 9.1) is thus a *compensator.* The transfer function $D(s)$ will, in the sequel, be referred to as a controller transfer function or a compensator transfer function; both the terms (controller/compensator) will point to $D(s)$ in the feedback loop.

The selection of a compensator $D(s)$ is basically a search problem in n-dimensional parametric space where n is the number of design parameters of $D(s)$. Points in the search space correspond to different selections of compensators. By choosing different points of the parametric space, we can generate feedback systems with different performances. A compensator that meets the given performance specifications can be determined by moving in this search space on trial-and-error basis. Classical control theory provides systematic methods for solving this trial-and-error search problem. The two most successful practical methods, the *root-locus method* and the *frequency-response method*, will be presented in the current chapter and in Chapter 12, respectively.

9.2 RESHAPING THE ROOT LOCUS

A simple control-system configuration of Fig. 9.2 is created by providing feedback around the plant $G_p(s)$ and adding an error-signal amplifier with gain K_A. The amplifier gain is adjusted to provide acceptable performance, if possible. The system of Fig. 9.2 is often referred to as having *proportional compensation.*

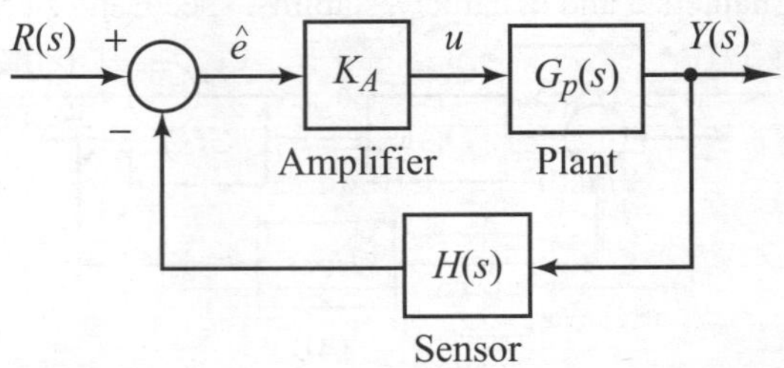

Fig. 9.2 *A simple control system configuration*

If the plant dynamics are of such a nature that a satisfactory design *cannot* be obtained by a gain adjustment alone, some dynamic compensation for the plant dynamics is required. A simple control-system configuration that includes dynamic compensation is shown in Fig. 9.3; the compensator $D_c(s)$ is placed in cascade with the forward-path transfer function $G_p(s)$. The system of Fig. 9.3 is often referred to as having *cascade compensation.*[1]

The dynamic compensator will generally also have a gain K that has to be adjusted to improve steady-state accuracy, i.e., the transfer function $D_c(s)$ is typically of the form

$$D_c(s) = \frac{K\prod_{i=1}^{m}(s+z_i)}{\prod_{j=1}^{n}(s+p_j)} \tag{9.1a}$$

$$= KD\,(s) \tag{9.1b}$$

For the purpose of simplifying the design procedure, we normally associate the adjustable gain parameter K with the plant transfer function and then design the dynamic compensator $D(s)$ given by Eqns (9.1) based on the new plant transfer function $KG_p(s)$. At the implementation stage, the gain K is associated with the compensator and the transfer function $D_c(s)$ is realized.

The configuration of Fig. 9.4a will accordingly be referred to as the *uncompensated system* and that of Fig. 9.4b as *compensated system.*

We introduce root-locus design through selected design problems.

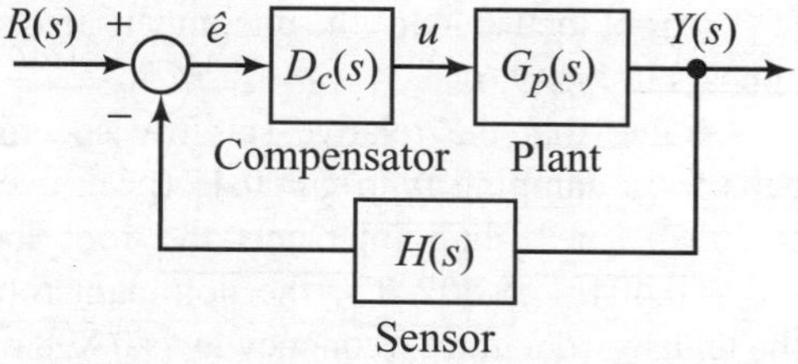

Fig. 9.3 *A control system with dynamic compensation*

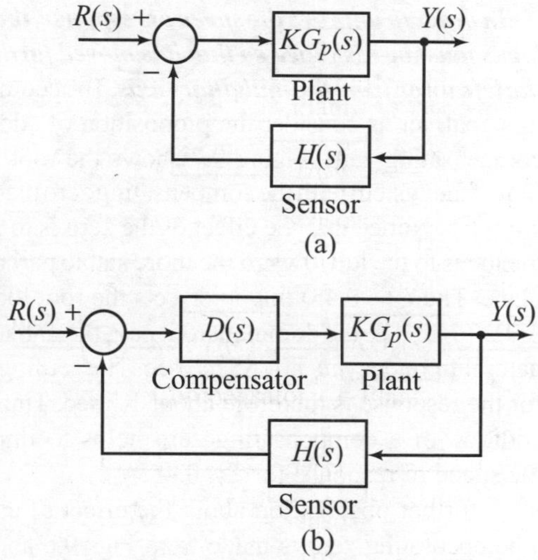

Fig. 9.4 *(a) Uncompensated system (b) Compensated system*

9.2.1 Design Problem 1

Consider the system of Fig. 9.5 which has an open-loop transfer function

$$G(s) = \frac{K_A}{s(s+1)\left(\frac{1}{5}s+1\right)} \tag{9.2}$$

From Eqn. (9.2), we have

$$G(s) = \frac{5K_A}{s(s+1)(s+5)} = \frac{K}{s(s+1)(s+5)} \tag{9.3}$$

[1] The term 'cascade compensation' used in Chapters 9–12 refers to the use of compensator (controller) in cascade (series) with the forward-path transfer function. It may be pointed out here that in process control field, the term 'cascade control' is frequently used for feedback configurations having 'loop within a loop'.

The root locus plot of the uncompensated system appears in Fig. 9.6.

Assume that the relative stability specification calls for a damping ratio $\zeta = 0.45$ (peak overshoot $\cong$ 20%). The ζ-line intersects the root locus at $s = -0.404 + j0.802$. For the dominant root pair, the undamped natural frequency ω_n is 0.898 rad/sec. The settling time ($= 4/\zeta\omega_n$) for the response is about 9.9 sec. The uncompensated system, therefore, has sluggish transient response.

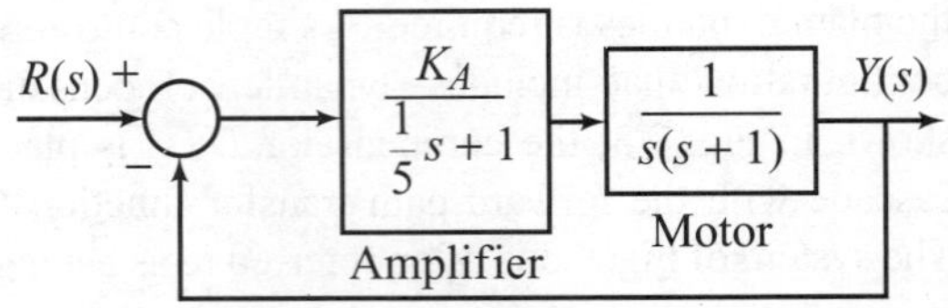

Fig. 9.5 *A position control system*

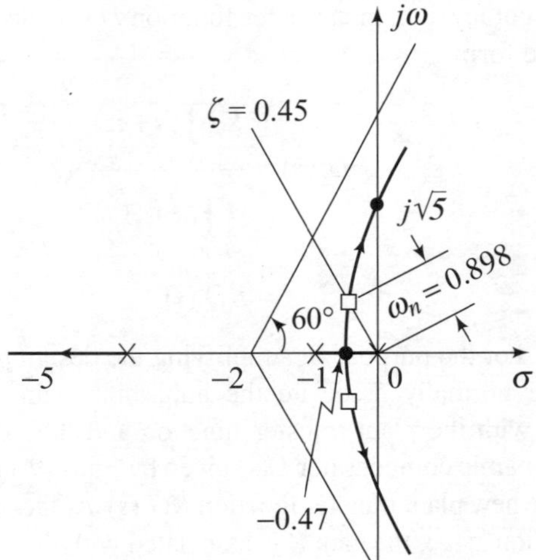

Fig. 9.6 *Root locus plot of* $1+\frac{K}{s(s+1)(s+5)}$

In order to improve the speed of response, the root locus must be reshaped so that it is moved farther to the left, away from the imaginary axis. To accomplish this goal, let us consider the proposition of adding a compensating zero. Figure 9.7 shows the root locus plot of the system with the compensating zero placed at $s = -1.5$. Notice that the effect of the zero is to move the locus to the left, toward the more stable part of the plane. The ζ (= 0.45) line intersects the root locus at $-2.2 + j4.3$. For the dominant root pair, the undamped natural frequency ω_n is 4.85 rad/sec. The settling time for the response is therefore about 1.7 sec. Thus, the addition of a compensating zero helps to improve the speed of response for $\zeta = 0.45$.

A further observation about the effect of adding a compensating zero is made from Figs 9.6 and 9.7. Without the compensating zero, the real root in addition to the dominant complex-root pair is always located to the left of the open-loop pole at $s = -5$. On the other hand, in the presence of compensating zero, this root moves over to the right of $s = -5$ and the dominance condition is therefore weakened. Since the real root (which is the closed-loop pole) will be located close to the compensating zero, the amplitude of the response term due to this root will be small. Also, the time constant of the real root will be comparable to that of complex dominant root pair; therefore, its effect on settling time may be tolerable.

An acceptable transient response can generally be obtained by trial-and-error adjustments of the compensating zero on the real axis. It can easily be verified that if we place the compensating zero too far to the left (say, near the open-loop pole at $s = -5$), the dominance condition is improved but complex branches of the root locus are pressed toward the right and the dynamic response becomes poor. If, on the other hand, the zero is placed close to the origin (say, to the right of the open-loop pole at $s = -1$), the complex branches are advantageously far into the left, but a real root close to the origin appears. This root will also be close to the compensating zero; therefore amplitude of the response term due to this root may be small. However, since the real root will have a large time constant compared to that of dominant complex root pair, its effect on settling time may be appreciable.

No general rule can be laid down for selecting the location of compensating zero that satisfactorily meets all the design requirements. Trial-and-error procedure is the best tool in hand. However, the following *guidelines* can be of considerable help:

Place the compensating zero on the real axis in the region below the desired closed-loop pole location such that it lies close to the left of any open-loop pole in this region. This technique ensures that the closed-loop pole on the real axis caused by the introduction of the compensating zero will be located close to it with a time constant comparable to that of desired dominant closed-loop pole-pair, and would therefore make a negligible contribution to system dynamics. In case, the uncompensated system does not have any open-loop pole in the region below the desired closed-loop poles, the dominance condition must be checked by locating the real-axis closed-loop pole created by the compensating zero.

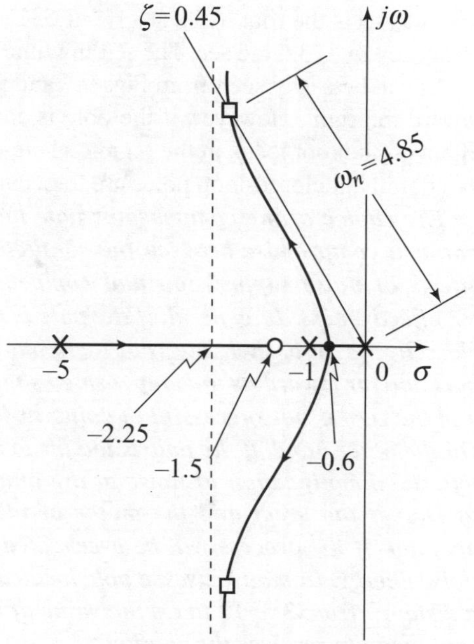

Fig. 9.7 *Root locus plot of* $1+\dfrac{K(s+1.5)}{s(s+1)(s+5)}$

The addition of a zero in the open-loop transfer function can be achieved by placing a compensator

$$D(s) = s + z = s + 1/\tau \qquad (9.4a)$$

in cascade with the forward-path transfer function. This results in (refer to Fig. 9.3)

$$D_c(s) = K(s + 1/\tau) \qquad (9.4b)$$

It is seen that $D_c(s)$ given by Eqn. (9.4b) is a *proportional plus derivative (PD) controller:*

$$D_c(s) = \frac{K}{\tau}(1+\tau s) = K_c(1+T_D S)$$

$$= \frac{K}{\tau} + Ks = K_P + K_D s$$

An op amp circuit realization of this controller is of the form shown in Fig. 9.8 (refer to Fig. P7.18 for circuit details).

The problem with PD compensation is that the inevitable high-frequency noise present at the input to the $D_c(s)$ will be greatly amplified. The usual method for overcoming this problem is to add a pole to the compensator transfer function which limits the high-frequency gain. The pole must of course be added such that it has relatively negligible effect on the root locus in the region where the two dominant closed-loop poles are to occur.

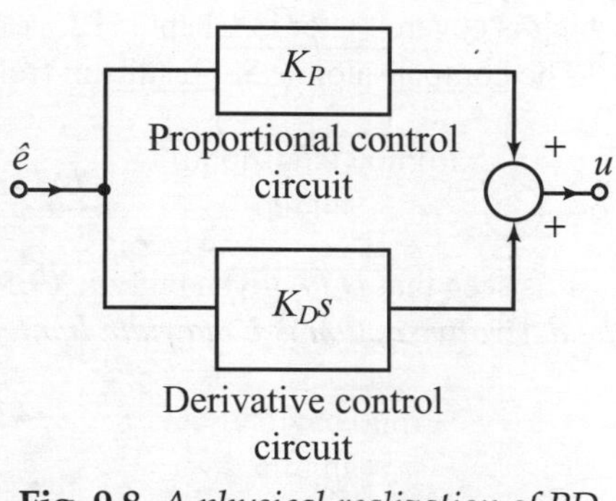

Fig. 9.8 *A physical realization of PD controller*

Figure 9.9 shows the root locus plot in the presence of both a compensating zero and a pole. The root locus near the dominant closed-loop poles gets somewhat modified by the presence of the compensating pole. The effect on the transient response of the system is not very pronounced. The $\zeta(= 0.45)$

line intersects the root locus at $s = -1.531 + j3.039$. For the dominant root-pair, the undamped natural frequency ω_n is 3.4 rad sec. The settling time for the response is about 2.6 sec.

As can be easily seen from Figs 9.7 and 9.9, the effect of the compensator pole is to press the locus toward the right. However, if the pole is added far away from the $j\omega$-axis, it has relatively negligible effect on the root locus in the region where the two dominant closed-loop poles are to occur.

The choice of the compensator pole location is a compromise between the conflicting effects of noise suppression and compensator effectiveness. In general, if the pole is too close to the zero, then the root locus moves back too far toward its uncompensated shape, and the zero is not successful in doing its job. On the other hand, if the pole is too far to the left, the magnification of noise at the output of D(s) is too great and the motor or other actuator of the process will be overheated by noise energy. In many cases, a pole located at a distance from 3 to 10 times the value of the zero location resolves the conflict.

We see from the above discussion that the compensating pole must be located to the left of the compensating zero; both the pole and the zero confined to the left half of the s-plane. The transfer function of such a first-order compensator then becomes

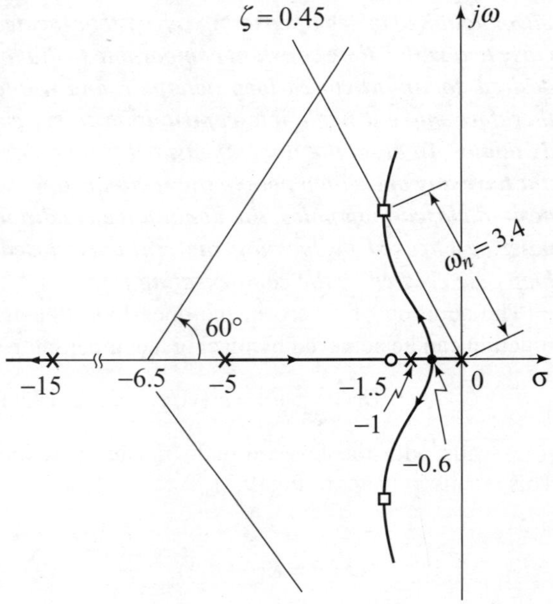

Fig. 9.9 *Root locus plot of* $1+\dfrac{K(s+1.5)}{s(s+1)(s+5)(s+15)}$

$$D(s) = \frac{s+z}{s+p} = \frac{s+1/\tau}{s+1/\alpha\tau}; \alpha = \frac{z}{p} < 1, \tau > 0 \tag{9.5a}$$

Note that $\alpha < 1$ ensures that the pole is located to the left of the zero. The contribution of $D(s)$ to the angle criterion of the root locus is always positive. This type of compensator is called *lead compensator*. The name 'lead compensator' has a clearer meaning with respect to frequency-domain design, which is covered later in Chapter 12.

The compensator (9.5a) results in (refer to Fig. 9.3)

$$D_c(s) = K\left(\frac{s+1/\tau}{s+1/\alpha\tau}\right); \alpha < 1, \tau > 0 \tag{9.5b}$$

It is seen that $D_c(s)$ given by Eqn. (9.5b) is a form of *proportional plus derivative controller wherein the derivative action is bandwidth limited.*

$$D_c(s) = K\alpha + \frac{K\alpha\tau(1-\alpha)s}{\alpha\tau s+1}$$

$$= K_P + \frac{K_D s}{\tau_D s+1}$$

An op amp realization of this controller is of the form shown in Fig. 9.10 (refer to Fig. P7.21 for circuit details).

The following is the summary of the results obtained for uncompensated and compensated systems.

The system of Fig. 9.5 has an open-loop transfer function,

$$G(s) = \frac{K}{s(s+1)(s+5)}$$

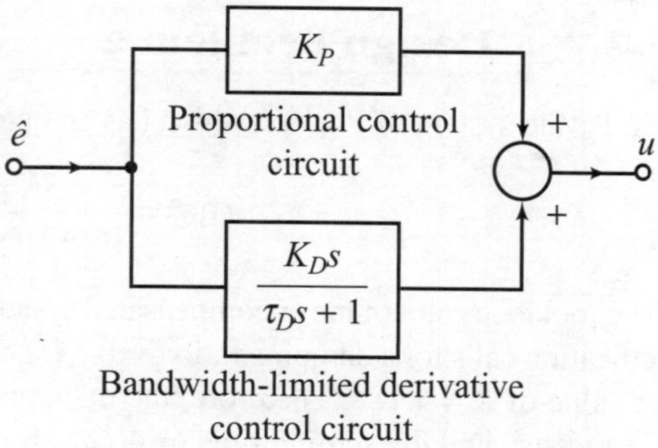

Fig. 9.10 *A physical realization of lead compensator*

The root locus plot of this uncompensated system is shown in Fig. 9.6. A damping ratio $\zeta = 0.45$ yields the following pertinent data:

Dominant roots,	$s_{1,2} = -0.404 \pm j0.802$
Root locus gain at s_1,	$K = \|s_1\|\ \|s_1 + 1\|\ \|s_1 + 5\| = 4.188$
Velocity-error constant,	$K_v = \lim_{s\to 0} sG(s) = \frac{K}{5} = 0.838$
Undamped natural frequency,	$\omega_n = 0.898$ rad/sec
Settling time ($= 4/\zeta\omega_n$),	$t_s = 9.9$ sec

Figure 9.9 is the root locus of lead-compensated system with open-loop transfer function

$$D(s)G(s) = \frac{K(s+1.5)}{s(s+1)(s+5)(s+15)}$$

A damping ratio $\zeta = 0.45$ yields the following pertinent data:

Dominant roots,	$s_{1,2} = -1.531 \pm j3.039$
Root locus gain at s_1,	$K = \frac{\|s_1\|\|s_1+1\|\|s_1+5\|\|s_1+15\|}{\|s_1+1.5\|} = 220$
Velocity-error constant,	$K_v = \lim_{s\to 0} sD(s)G(s) = \frac{K(1.5)}{5\times 15} = 4.4$
Undamped natural frequency,	$\omega_n = 3.4$ rad/sec
Settling time	$t_s = 2.6$ sec

The positive angle contributed by the lead compensator shifts the root locus towards the left in the s-plane; this results in improvement in the dynamic response of the system. ***The lead compensator also helps to increase system error constant, though to a limited extent.***

9.2.2 Design Problem 2

Consider the system of Fig. 9.5 which has an open-loop transfer function

$$G(s) = \frac{5K_A}{s(s+1)(s+5)} = \frac{K}{s(s+1)(s+5)} \tag{9.6}$$

The root locus plot of the uncompensated system appears in Fig. 9.6. Suppose that the relative stability specification calls for a damping ratio $\zeta = 0.45$. The ζ-line intersects the root locus at $s = -0.404 + j0.802$ for a value of $K = 4.188$. Therefore, the undamped natural frequency ω_n of the closed-loop system is 0.898 rad/sec, and the settling time ($= 4/\zeta\omega_n$) for the response is 9.9 sec. We assume that the dynamic response of the uncompensated system with $K = 4.188$ is satisfactory.

For the uncompensated system with $K = 4.188$, the velocity-error constant is

$$K_v = \lim_{s\to 0} sG(s) = \lim_{s\to 0} s\left\{\frac{K}{s(s+1)(s+5)}\right\} = \frac{K}{5} = 0.898$$

and thus with a unit-ramp input (1 rad/sec), the steady-state error

$$e_{ss} = \frac{1}{K_v} = 1.114 \text{ rad}$$

which is too large. It is therefore required to raise the system K_v. To accomplish this, consider the proposition of adding a *proportional plus integral (PI) controller* in cascade with the plant transfer function in the forward path. A PI controller has transfer function of the form

$$D_c(s) = K_c\left(1 + \frac{1}{T_I s}\right) = K_c\left(\frac{s + 1/T_I}{s}\right) = K_c\frac{(s+z)}{s} \tag{9.7}$$

The PI controller, thus, adds a pole at $s = 0$ ***and a zero at*** $s = -z$ ***to the open-loop transfer function. System type increases by one because of the added open-loop pole at the origin. Hence, for stable design, the corresponding error constant is infinity and steady-state error is zero.***

An op amp circuit realization of the PI controller given by Eqn. (9.7) is of the form shown in Fig. 9.11 (refer to Fig. P7.19 for circuit details).

Let us now investigate the effects of the pole-zero pair of the PI controller on transient response of the system. ***It can easily be verified that the pole alone would move the root locus to the right, thereby making the system unstable. The zero must be near the pole in order to minimize the effect on the transient response.*** Figure 9.12 shows the root locus plot of the system with the compensating zero placed at $s = -0.01$; the compensating pole being at the origin. Notice that the angle contribution of the compensating pole-zero pair at the desired closed-loop pole $-0.404 \pm j0.802$ is negligibly small, and therefore, the root locus in the vicinity of this point is not appreciably disturbed.

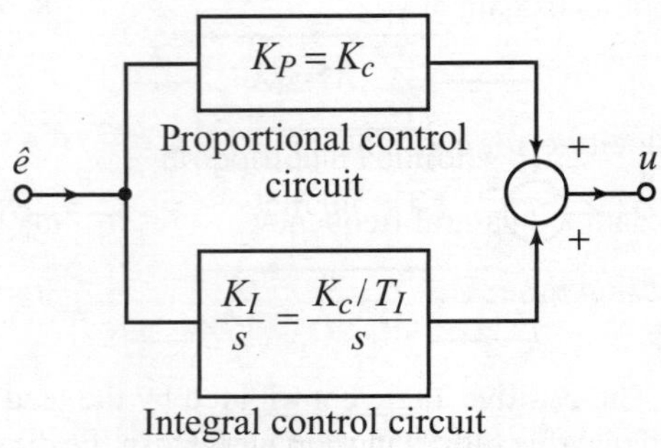

Fig. 9.11 *A physical realization of PI controller*

A further observation about the effect of adding pole-zero pair is made from Figs 9.6 and 9.12. Without the compensating pole-zero pair, the root locus has a real root in addition to the desired dominant complex root-pair $-0.404 \pm j0.802$. The real root is always located to the left of the open-loop pole at $s = -5$, and therefore, its effect on transient response of the system will be negligible. On the other hand, in the presence of compensating pole-zero pair, there are two real roots in addition to the desired complex root pair; one to the left of open-loop pole at $s = -5$ and the other on the real-axis segment between -0.02 and -0.01. The additional real root contributed by the compensating pole-zero pair is too close to the origin and therefore, will have a very large time-constant, affecting the settling time appreciably. There is, however, a zero very close to this real root, and therefore, the magnitude of the response term contributed by this root will be negligibly small. ***The closed-loop pole-zero dipole introduced by* PI *controller will not contribute significantly to the transient response of the closed-loop system.***

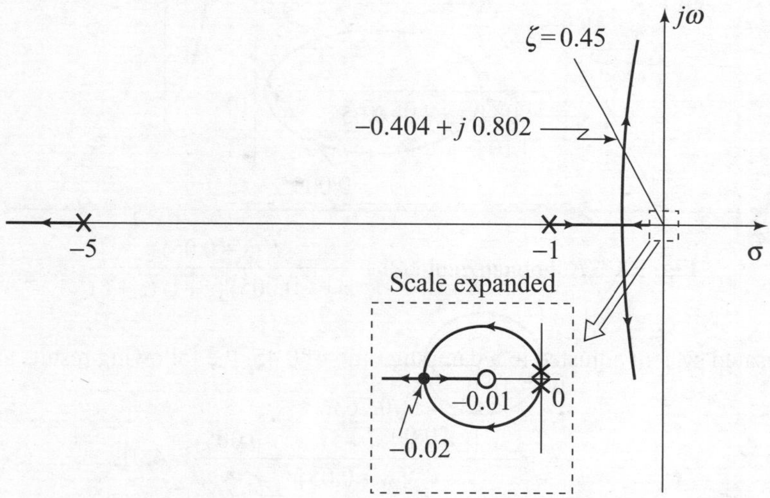

Fig. 9.12 *Root locus plot of* $1+\dfrac{K(s+0.01)}{s^2(s+1)(s+5)}$

It is desirable to place the compensating pole-zero combination as far to the left as possible without causing major shifts in the dominant root locations and giving required improvement in the steady-state response. The compensating pole-zero pair near the origin (the pole not necessarily at the origin) will give rise to a closed-loop pole-zero dipole shifted relatively to the left compared to the case of* PI *controller; therefore, the time constant of the additional root will be relatively less, resulting in a smaller effect on the settling time of the system. In system response, a long tail of small amplitude introduced by closed-loop dipole will decay relatively faster (refer to Example 9.4 for more details).

It can easily be verified that if the compensating pole-zero pair is placed too far to the left (say, on the real-axis segment to the left of desired dominant complex closed-loop poles), no significant improvement in the steady-state response is possible.

Figure 9.13 shows the root locus plot in the presence of compensating pole at $s = -0.005$ and the compensating zero at $s = -0.05$. The angle contributed by the compensator pole-zero pair at $s = -0.404 + j0.802$, the desired dominant root, is about 2.6° and is therefore acceptable. For the damping ratio

$\zeta = 0.45$, the new locus is close to the locus of uncompensated system. The new dominant roots are $s_{1,2} = -0.384 \pm j0.763$; thus, the roots are essentially unchanged.

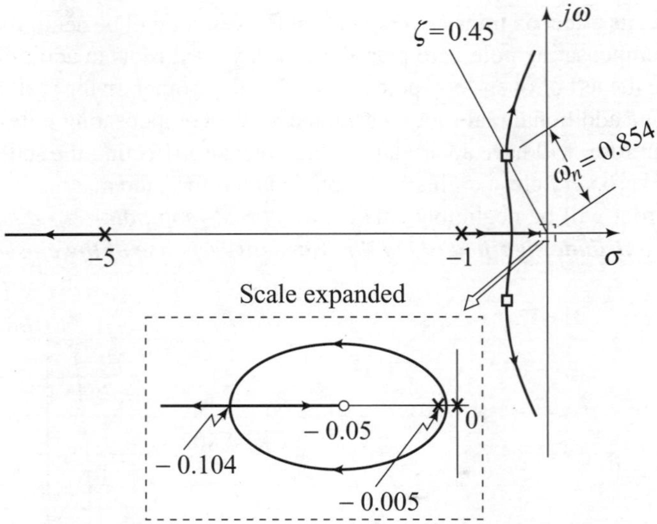

Fig. 9.13 *Root locus plot of* $1+\dfrac{K(s+0.05)}{s(s+0.005)(s+1)(s+5)}$

For the compensated system adjusted to a damping ratio of 0.45, the following results are obtained:

Dominant roots, $\quad s_{1,2} = -0.384 \pm j0.763$

Root locus gain at s_1: $\quad K = \dfrac{|s_1||s_1+1||s_1+5||s_1+0.005|}{|s_1+0.05|} = 4.01$

Velocity-error constant, $\quad K_v = \lim_{s\to 0} s\left\{\dfrac{K(s+0.05)}{s(s+1)(s+5)(s+0.005)}\right\}$

$$= \frac{K(0.05)}{5\times 0.005} = 8.02$$

Undamped natural frequency $\quad \omega_n = 0.854$ rad/sec.

A comparison of the uncompensated system and the compensated system shows that the undamped natural frequency has decreased slightly from 0.898 to 0.854 rad/sec, but there is considerable increase in velocity-error constant: from 0.838 to 8.02. ***This increase in K_v is in fact, approximately equal to $|z|/|p|$ where $s = -z$ is the location of the compensating zero and $s = -p$ is the location of the compensating pole.***

We see from the above discussion that the compensating zero must be located to the left of the compensating pole; both the zero and the pole confined to the left half of the s-plane. The transfer function of such a first-order compensator then becomes

$$D(s) = \frac{s+z}{s+p} = \frac{s+1/\tau}{s+1/\beta\tau};\ \beta = \frac{z}{p} > 1,\ \tau > 0 \qquad (9.8a)$$

Note that $\beta > 1$ ensures that the pole is located to the right of the zero, i.e., nearer the origin than the zero. The contribution of $D(s)$ to the angle criterion of the root locus is always negative. This type of compensator is called *lag compensator*. The name 'lag compensator' has a clearer meaning with respect to frequency-domain design, which is covered later in Chapter 12.

The compensator (9.8a) results in (refer to Fig. 9.3)

$$D_c(s) = K\left(\frac{s+1/\tau}{s+1/\beta\tau}\right);\ \beta>1,\ \tau>0 \tag{9.8b}$$

It is seen that $D_c(s)$ given by Eqn. (9.8b) is a form of *proportional plus integral controller wherein the integral action is low-frequency gain limited* :

$$D_c(s) = K + \frac{K(\beta-1)}{\beta\tau s+1}$$

$$= K_P + \frac{K_I}{\tau_I s+1}$$

An op amp realization of the controller is of the form shown in Fig. 9.14 (refer to Fig. P7.22 for circuit details).

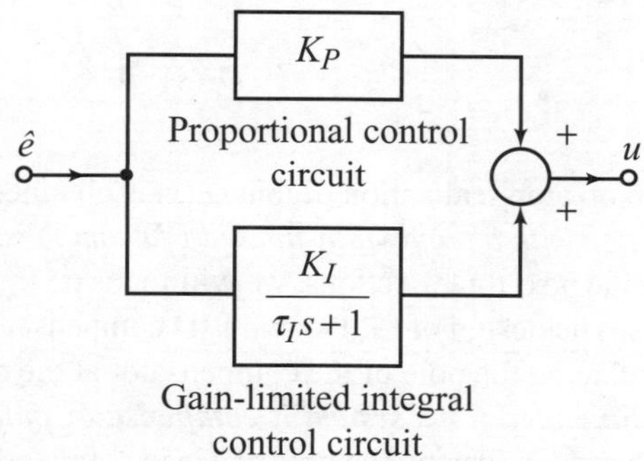

Fig. 9.14 *A physical realization of lag compensator*

9.2.3 Reprise

The purpose of reshaping the root locus generally falls into one of the following categories:

1. A given system is stable and its transient response is satisfactory, but its steady-state error is too large. Thus, the corresponding error constant must be increased while nearly preserving the transient response. This can be achieved by a cascade PI/lag compensator.
2. A given system is stable, but its transient response is unsatisfactory. Thus, the root locus must be reshaped so that it is moved farther to the left, away from the imaginary axis. This can be achieved by a cascade PD/lead compensator.
3. A PD/lead-compensated system has satisfactory transient response but its steady-state error is too large. The steady-state error can be decreased by cascading a PI/lag compensator to the PD/lead-compensated system. This is equivalent to using a second-order compensator.

Consider first the cascade of PI and PD compensators:

$$D_c(s) = \left(K_1 + \frac{K_i}{s}\right)(K_2 + K_d s)$$

$$= K_1K_2 + K_iK_d + \frac{K_1K_i}{s} + K_2K_d s$$

$$= K_P + \frac{K_I}{s} + K_D s$$

Thus, *cascading a PI compensator with a PD compensator is completely equivalent to a PID compensator*; an op amp realization of this compensator is of the form shown in Fig. 9.15 (refer to Fig. P7.20 for circuit details).

Consider now the cascade of lag and lead compensators:

$$D_c(s) = K\left(\frac{s+1/\tau_2}{s+1/\beta\tau_2}\right)\left(\frac{s+1/\tau_1}{s+1/\alpha\tau_1}\right);\ \beta>1,\ \alpha<1,\ \tau_2>0,\ \tau_1>0$$

This is a second-order compensator, called the *lag-lead* compensator.

Expressed explicitly in the form of proportional, integral, and derivative control actions, $D_c(s)$ looks like

$$D_c(s) = \left(K_{P1} + \frac{K_I}{\tau_I s+1}\right)\left(K_{P2} + \frac{K_D s}{\tau_D s+1}\right)$$

An op amp realization of this $D_c(s)$ is obtained by *cascading* a (low frequency) *gain-limited PI compensator with a bandwidth-limited PD compensator* (refer to Fig. P7.23 for circuit details).

In the next three sections, we will focus on the design procedures for lag, lead, and lag-lead compensators. The design of PI, PD and PID compensators follows from these procedures in a straight forward way. Placing the pole of the compensator at the origin gives us a PI compensator (***A* PI *compensator is recommended for a system if compensator pole at the origin does not degrade closed-loop stability appreciably***). Placing the pole of a lead compensator at $-\infty$ (equivalently, designing a lead compensator by appropriately placing a zero only) gives us a PD compensator (***A* PD *compensator is recommended if the feedback loop is free from high-frequency noise problems. This is unlikely in real-life applications***). An example of PID compensation will be given later in Section 9.6.

The systems, which are type-2 or higher, are usually absolutely unstable. For these types of systems, clearly PD/lead compensators are required, as only a PD/lead compensator increases the margin of stability. In type-1 and type-0 systems, stable operation is always possible if the gain is sufficiently reduced. In such cases, any of the three compensators, namely, PI/lag, PD/lead, or PID/lag-lead, may be used to obtain the desired performance. The particular choice is based upon transient and steady-state response characteristics of uncompensated system.

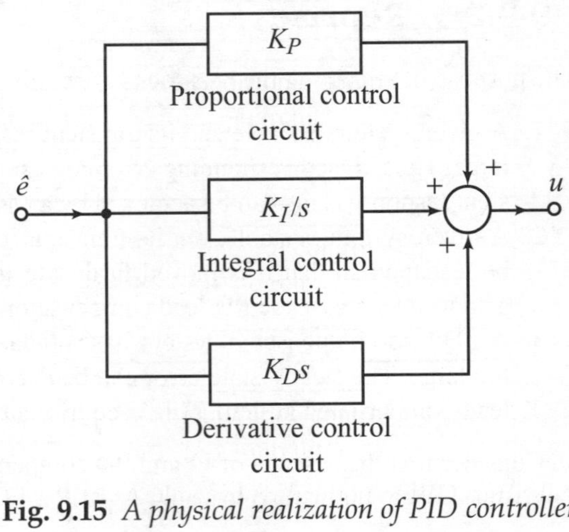

Fig. 9.15 *A physical realization of PID controller*

9.3 CASCADE LEAD COMPENSATION

As has been pointed out in the previous section, lead compensation is generally used to improve the transient response of a system. In this section, we will consider how the root locus may be used to design a compensator so as to realize a desired improvement in transient response.

Consider a unity-feedback system with a forward-path transfer function $G(s)$. The transfer function $G(s)$ is assumed to be unalterable except for the adjustment of gain. Let the system dynamic response specifications be translated into the desired location s_d for the dominant complex closed-loop poles as shown in Fig. 9.16.

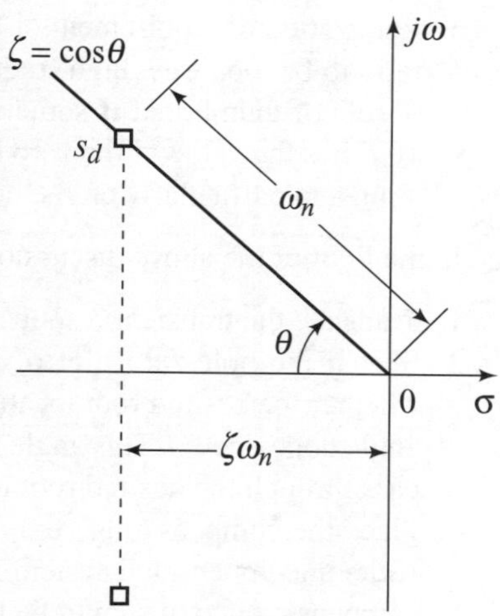

Fig. 9.16 *Desired dominant closed-loop poles*

If the angle criterion at s_d is not met, i.e., $\angle G(s_d) \neq \pm 180°$, the uncompensated root locus with variable open-loop gain will not pass through the desired root location, indicating the need for compensation. Let us introduce a first-order lead compensator with zero and pole in the left half of the complex plane:

$$D(s) = \frac{s+z}{s+p} = \frac{s+1/\tau}{s+1/\alpha\tau}; \alpha = \frac{z}{p} < 1, \tau > 0 \tag{9.9}$$

in cascade with the forward-path transfer function $G(s)$ as shown in Fig. 9.17.

The compensator $D(s)$ has to be so designed that the compensated root locus passes through s_d. In terms of the angle criterion, this requires that

$$\angle D(s_d)G(s_d) = \angle D(s_d) + \angle G(s_d) = \pm 180°$$

or
$$\angle D(s_d) = \phi = \pm 180° - \angle G(s_d) \tag{9.10}$$

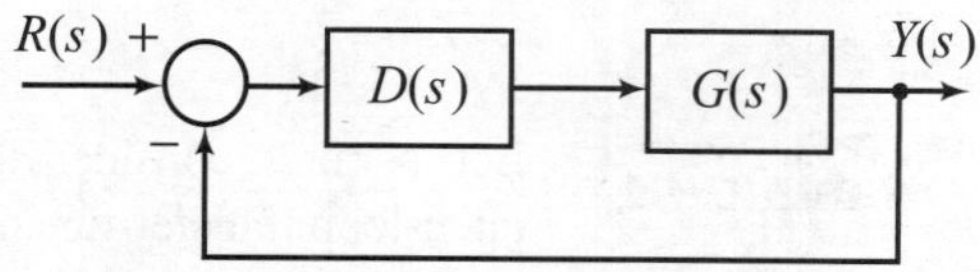

Fig. 9.17 *A feedback system with cascade compensator*

Thus, for the root locus of the compensated system to pass through the desired root location, the lead compensator pole-zero must contribute an angle ϕ given by Eqn. (9.10) and shown in Fig. 9.18. For a given angle ϕ for lead compensation, there is no unique location for the pole-zero pair. Out of many possible solutions, we select the one that satisfies the following additional requirements:

1. The least-damped complex poles of the resulting closed-loop transfer function correspond to the desired dominant poles, and all other closed-loop poles are either located very close to the open-loop zeros or are relatively far away from the $j\omega$-axis. This ensures that the poles other than the dominant poles make a negligible contribution to the system dynamics.
2. Changing the locations of the compensator pole and zero can produce a range of values for the gain while maintaining the desired root s_d. The pole-zero pair location that produces maximum gain (and hence results in minimum steady-state error) is preferred.

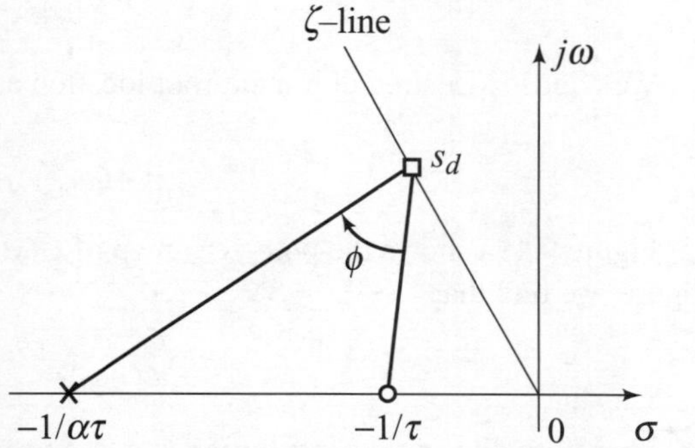

Fig. 9.18 *Angle contribution of lead compensator*

3. Realization of requirements 1 and 2 may demand a large value of the ratio p/z. If we choose this ratio to be too large, high frequency noise problems can, and probably will, become significant. A rule of thumb that is sometimes used is to limit this ratio to 10 (pole is located at a distance 10 times the value of the zero location), but this depends on the physical system under consideration and particularly on the noise present.

In the light of the above discussion, the steps for lead compensator design are summarized below.

1. Translate the transient response specifications into a pair of complex dominant roots.
2. Sketch the pole-zero plot of the uncompensated system and determine using angle criterion whether the desired root locations can be realized by gain adjustment alone. If not, calculate the angle deficiency ϕ. This angle must be contributed by the lead compensator if the root locus is to pass through the desired root locations.
3. Place the compensating zero on the real axis in the region below the desired dominant root s_d. If the uncompensated system has an open-loop pole on the real axis in the region below s_d, the compensating zero should lie to the left of this pole.
4. Determine the compensator pole location so that the total angle contributed by compensator pole-zero pair at s_d is ϕ.
5. Evaluate the gain at s_d and then calculate the error constant.
6. Repeat the steps if transient performance is unsatisfactory because of a weak dominance condition and/or if the error constant is not satisfactory.

Example 9.1 Let us first consider the example of a unity-feedback type-2 system with open-loop transfer function

$$G(s) = \frac{K}{s^2}$$

It can be seen from its root locus plot that the closed-loop poles always lie on the $j\omega$-axis. As explained earlier, lead compensator is the only choice for this system.

It is desired to compensate the system so as to meet the following transient response specifications:
Settling time, $t_s \leq 4$ sec
Peak overshoot for step input $\leq 20\%$
These specifications imply that

$$\zeta \geq 0.45, \quad \text{and} \quad \zeta\omega_n \geq 1$$

We choose a desired dominant root location as

$$s_d = -\zeta\omega_n + j\omega_n\sqrt{1-\zeta^2} = -1 + j2$$

Figure 9.19a shows the pole-zero map of $G(s)$, and the desired dominant root location s_d. From this figure, we find that

$$\angle G(s_d) = -2 \times 116° = -232°$$

Therefore, the angle contribution at s_d required of the lead compensator pole-zero pair is

$$\phi = -180° - (-232°) = 52°$$

Place the zero of the compensator on the real-axis segment directly below s_d, at $s = -z = -1$. Join the compensator zero to s_d and locate the compensator pole by making an angle of $\phi = 52°$, as shown in Fig. 9.19a. The location of the pole is found to be at $s = -p = -3.6$.

With the addition of the cascade lead compensator

$$D(s) = \frac{s+1}{s+3.6},$$

the open-loop transfer function of the system becomes

$$D(s)G(s) = \frac{K(s+1)}{s^2(s+3.6)}$$

The gain K is evaluated by measuring the vector lengths from the poles and zeros to the root location s_d:

$$K = \frac{(2.23)^2(3.25)}{2} = 8.1$$

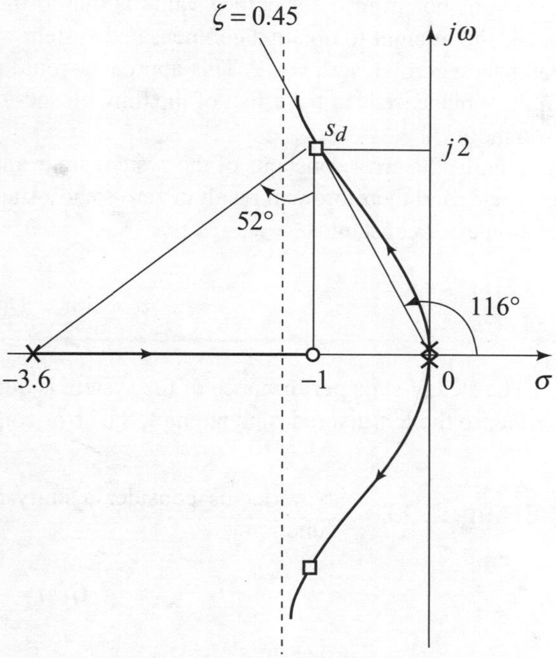

Fig. 9.19(a) *Cascade lead compensation (Example 9.1)*

Therefore, the open-loop transfer function of the compensated system is

$$D(s)G(s) = \frac{8.1(s+1)}{s^2(s+3.6)} \tag{9.11}$$

The root locus plot of the compensated system is shown in Fig. 9.19a. The third closed-loop pole introduced because of the lead compensator lies on the real-axis root locus between $s = -3.6$ and $s = -1$. Due to this closed-loop pole, there will be an exponential term in the transient response in addition to the oscillatory term due to the complex pair of closed-loop poles. The specifications for the transient response were originally expressed in terms of the overshoot and the settling time of the system. These specifications were translated on the basis of an approximation of the system by a second-order system, to an equivalent ζ and ω_n, and therefrom to the desired closed-loop pole locations. However, the original specifications will be satisfied only if the poles selected are dominant. Often the designer will simulate the final design and obtain the actual transient response of the system. If the peak overshoot and settling time indices of the transient response do not compare well with the specified values, the system is redesigned.

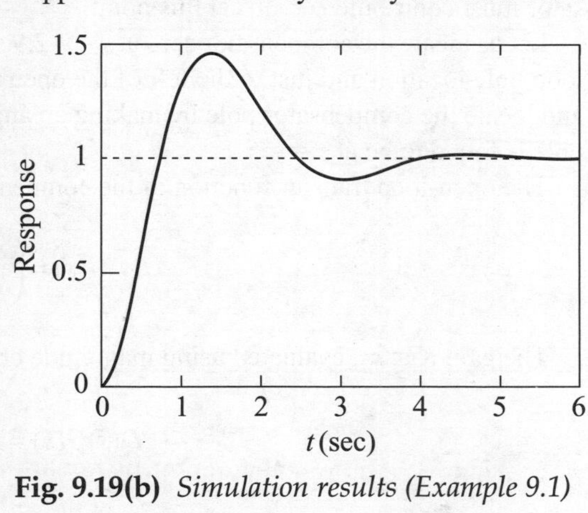

Fig. 9.19(b) *Simulation results (Example 9.1)*

Computer simulation (MATLAB Module 9 in the Appendix) of the unity-feedback system with open-loop transfer function (9.11) resulted in a peak overshoot of 46.7% and a settling time of 3.78 sec for a step input (Fig. 9.19b). The difference in the overshoot from the specified value is due to the third closed-loop pole, which is not negligible. A second attempt to obtain a compensated system with an overshoot of 20% would utilize a compensator with zero at say $s = -2$. This approach would move the third closed-loop pole farther to the left in the s-plane, reduce the effect of the third closed-loop pole on the transient response, and reduce the overshoot.

Finally, the error constants of the system are evaluated. We find that the system under consideration is type-2, and therefore will result in zero steady-state error for step and ramp input signals. The acceleration error constant is

$$K_a = \lim_{s\to 0} s^2 D(s)G(s) = \frac{8.1}{3.6} = 2.25$$

The steady-state performance of the system is quite satisfactory. With the compensator zero shifted farther to the left in a redesign attempt, the error constant K_a will increase marginally.

Example 9.2 Now let us consider a unity-feedback type-1 system with open-loop transfer function

$$G(s) = \frac{K}{s(s+2)}$$

It is desired to design a cascade lead compensator such that the dominant closed-loop poles provide a damping ratio $\zeta = 0.5$ and have an undamped natural frequency $\omega_n = 4$ rad/sec.

Our specifications provide the location of the desired dominant closed-loop poles. An s-plane diagram showing the desired pole location s_d and the open-loop poles of the uncompensated transfer function is given in Fig. 9.20. The desired closed-loop pole is at

$$s_d = -\zeta\omega_n + j\omega_n\sqrt{1-\zeta^2} = -2 + j3.46$$

The angle of $G(s)$ at the desired closed-loop pole, obtained from Fig. 9.20, is

$$\angle G(s_d) = -210°$$

Thus, if we need to force the root locus to go through the desired closed-loop pole, the lead compensator must contribute $\phi = 30°$ at this point.

Let us locate the compensator zero at $s = -2.9$, i.e., in the region below the desired dominant closed-loop pole location and just to the left of the open-loop pole at $s = -2$. Join the compensator zero to s_d and locate the compensator pole by making an angle $\phi = 30°$ as shown in Fig. 9.20. The location of the pole is found to be at $s = -5.4$.

The open-loop transfer function of the compensated system becomes

$$D(s)G(s) = \left(\frac{s+2.9}{s+5.4}\right)\left[\frac{K}{s(s+2)}\right]$$

The gain K at s_d, evaluated using magnitude criterion, is 18.7. It follows that

$$D(s)G(s) = \frac{18.7(s+2.9)}{s(s+2)(s+5.4)}$$

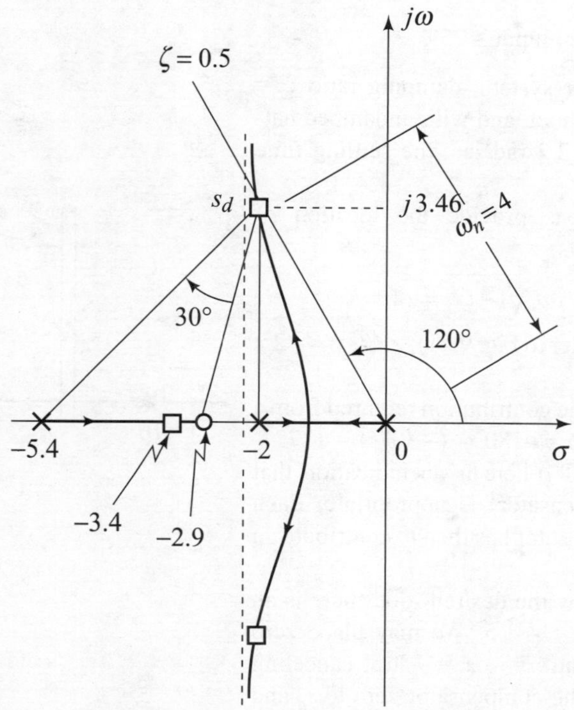

Fig. 9.20 *Root locus plot (Example 9.2)*

The root locus plot of the compensated system is shown in Fig. 9.20. The third closed-loop pole is at $s = -3.4$, which is close to the added zero at $s = -2.9$. Therefore, the effect of this pole on the transient response is expected to be small. This should be verified by simulation study on the closed-loop transfer function

$$\frac{Y(s)}{R(s)} = M(s) = \frac{18.7(s+2.9)}{(s+2+j3.46)(s+2-j3.46)\,(s+3.4)}$$

Simulation study (MATLAB Module 9 in the Appendix) shows that the response has a peak overshoot of 20.8% and a settling time of 2.05 sec for a step input.

The velocity-error constant of the compensated system is

$$K_v = \lim_{s \to 0} s\,D(s)G(s) = \frac{18.7 \times 2.9}{2 \times 5.4} = 5.02 \text{ sec}^{-1}$$

Although a certain change in the value of K_v can be made by altering the pole-zero location of the lead compensator, if a large increase in the value of K_v is desired, then we must alter the lead compensator to a lag-lead compensator.

Example 9.3 A unity-feedback system with open-loop transfer function

$$G(s) = \frac{K}{s^2(s+1.5)}$$

is to be compensated to meet the following specifications:

Settling time, $t_s < 5$ sec
Peak overshoot for step input ≅ 25%.

For a second-order system, damping ratio $\zeta = 0.45$ gives 20% overshoot, and with undamped natural frequency $\omega_n = 2.2$ rad/sec, the settling time $t_s = 4/\zeta\omega_n = 4$ sec.

These specifications provide the location of desired dominant roots:

$$s_{1,2} = -\zeta\omega_n \pm j\omega_n\sqrt{1-\zeta^2} = -1 \pm j2$$

At $s_d = -1 + j2$ (refer to Fig. 9.21), $\angle G(s_d) = -2 \times 116° - 75° = -307°$

Therefore, the angle contribution required from a lead compensator is $\phi = -180° - (-307°) = 127°$.

The large value of ϕ here is an indication that a double lead compensator[2] is appropriate. Each section of the compensator has then to contribute an angle of 63.5° at s_d.

In the region below the desired root, there is an open-loop pole at $s = -1.5$. We may place zero of the compensator at $s = -z = -1.5$, canceling the plant pole. Join the compensator zero to s_d and locate the compensator pole by making an angle of $\phi = 63.5°$, as shown in Fig. 9.21. The location of the pole is found to be at $s = -10$.

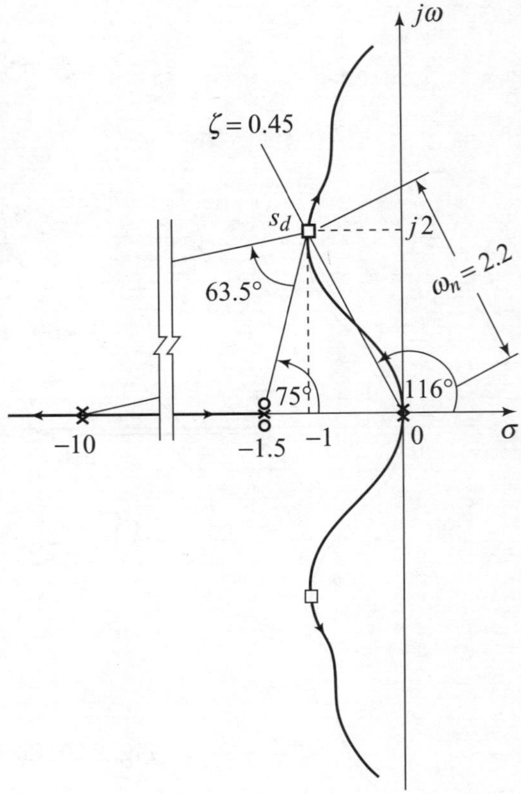

Fig. 9.21 *Root locus plot (Example 9.3)*

With the cascade double lead compensator $D(s) = \dfrac{(s+1.5)^2}{(s+10)^2}$, the open-loop transfer function of the compensated system becomes

$$D(s)G(s) = \frac{K(s+1.5)^2}{s^2(s+1.5)(s+10)^2} = \frac{K(s+1.5)}{s^2(s+10)^2}$$

The gain K at s_d, evaluated using magnitude criterion, is 247. It follows that

$$D(s)G(s) = \frac{247(s+1.5)}{s^2(s+10)^2}$$

A rough root locus plot of the compensated system is shown in Fig. 9.21. By locating all the closed-loop poles and carrying out a simulation study on the closed-loop system, we can determine whether the dominance of the complex pole-pair is ensured, i.e., whether the actual transient response compared well with the specified one. Simulation study shows that the response has a peak overshoot of 53.1% and a settling time of 3.96 sec for a step input. More attempts on compensator design are, therefore, required.

[2] An important consideration governing this choice is the inherent noise in control system. Refer to Example 12.2.

9.4 CASCADE LAG COMPENSATION

As pointed out earlier in Section 9.2, a lag compensator improves the steady-state behavior of a system, while nearly preserving its transient response. In this section, we consider how the root locus may be used to design a compensator so as to realize a desired improvement in steady-state response.

Consider a unity-feedback system with a forward-path transfer function,

$$G(s) = \frac{K\prod_{i=1}^{m}(s+z_i)}{s^r \prod_{j=r+1}^{n}(s+p_j)} \tag{9.12}$$

We assume that at a certain value of K, this system has satisfactory transient response, i.e., its root locus plot passes through (or is close to) the desired closed-loop pole location s_d indicated in Fig. 9.22. It is required to improve the system error constant K_p, K_v or K_a (depending on whether the system is type-0, -1 or -2, respectively) to a specified value without impairing its transient response. This requires that after compensation, the root locus should continue to pass through s_d, while the error constant at s_d is raised to the specified value. To accomplish this, consider adding a lag compensator pole-zero pair close to origin, with zero to the left of the pole. If pole and zero of this pair are located close to each other, it will contribute a negligible angle λ at s_d such that s_d continues to lie on the root locus of the compensated system. Figure 9.22 shows the location of such a lag compensator pole-zero pair.

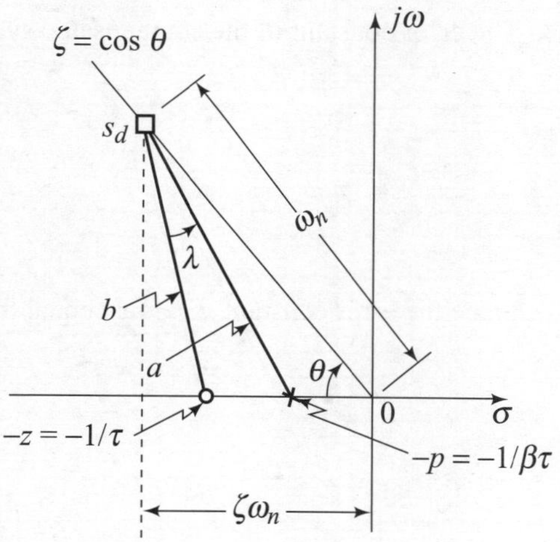

Fig. 9.22 *Locating the lag compensator pole-zero pair*

The gain of the uncompensated system at s_d is given by

$$K^{uc}(s_d) = \frac{|s_d|^r \prod_{j=r+1}^{n}|s_d+p_j|}{\prod_{i=1}^{m}|s_d+z_i|} \tag{9.13}$$

For the system with a cascade lag compensator

$$D(s) = \frac{s+z}{s+p} = \frac{s+1/\tau}{s+1/\beta\tau}; \beta = \frac{z}{p} > 1, \tau > 0 \tag{9.14}$$

the system gain at s_d is (refer to Fig. 9.22)

$$K^{c}(s_d) = \frac{|s_d|^r \prod_{j=r+1}^{n}|s_d+p_j|}{\prod_{i=1}^{m}|s_d+z_i|}\left(\frac{a}{b}\right) \tag{9.15}$$

where $$a = |s_d + 1/\beta\tau|, \text{ and } b = |s_d + 1/\tau|.$$

Since the compensator pole and zero are located close to each other, they are nearly equidistant from s_d, i.e., $a \cong b$. It, therefore, follows from Eqns (9.13) and (9.15) that

$$K^c(s_d) \cong K^{uc}(s_d)$$

The error constant K_e (which is equal to K_p, K_v, or K_a depending on the type of the system) of the uncompensated system is given by

$$K_e^{uc} = K^{uc}(s_d)\frac{\prod_{i=1}^{m} z_i}{\prod_{j=r+1}^{n} p_j}$$

The error constant of the compensated system is given by (refer to Fig. 9.22)

$$K_e^c = K^c(s_d)\frac{\prod_{i=1}^{m} z_i}{\prod_{j=r+1}^{n} p_j}\left(\frac{z}{p}\right) \cong K_e^{uc}\left(\frac{z}{p}\right)$$

Since the error constant K_e^c must equal the desired value K_e^d, we have

$$K_e^d \cong K_e^{uc}\left(\frac{z}{p}\right)$$

or

$$\beta = \left(\frac{z}{p}\right) \cong \frac{K_e^d}{K_e^{uc}} \tag{9.16}$$

Thus, the β-parameter of the lag compensator (refer to Eqn. (9.14)) is nearly equal to the ratio of the desired error constant to the error constant of the uncompensated system.

Since the lag compensator does contribute a small negative angle λ at s_d, the compensated root locus will not pass through s_d. The effect of the small lag angle λ is to press the root locus slightly to the right. The intersection of the compensated root locus with the specified ζ-line will be at s_d' having slightly lower value of ω_n compared to the one corresponding to s_d. Also the actual error constant will somewhat fall short of the specified value if β obtained from Eqn. (9.16) is used.

The effects of lag angle λ can be anticipated and counteracted by taking design parameter ω_n to be somewhat larger than the specified value. Also in the design process, we may choose a β somewhat larger than that given by Eqn. (9.16).

In the light of the above discussion, the steps for lag compensator design are summarized below.

1. Draw the root locus plot for the uncompensated system.
2. Translate the transient response specifications into a pair of complex dominant roots and locate these roots on the uncompensated root locus plot. It is assumed that acceptable transient performance can be obtained by gain adjustment; the desired dominant root will, therefore, lie on (or close to) the uncompensated root locus.

3. Calculate the gain of the uncompensated system at the dominant root s_d and evaluate the corresponding error constant.
4. Determine the factor by which the error constant of the uncompensated system should be increased to meet the specified value. Choose the β-parameter of lag compensator to be somewhat greater than this factor.
5. Select zero ($s = -z$) of the compensator sufficiently close to the origin. As a guide rule, we may construct a line making an angle of 10° (or less) with the desired ζ-line from s_d; the intersection of this line with the real axis gives location of the compensator zero.
6. The compensator pole can then be located at $s = -p = -z/\beta$. It is important to note that the compensator pole-zero pair should contribute an angle λ less than 5° at s_d so that the root locus plot in the region of s_d is not appreciably changed and hence satisfactory transient behavior of the system is preserved.

Example 9.4 Reconsider the uncompensated system of Example 9.2, where the uncompensated open-loop transfer function is

$$G(s) = \frac{K}{s(s+2)}$$

The system is to be compensated to meet the following specifications:

Damping ratio, $\zeta = 0.707$
Settling time, $t_s \le 5$ sec
Velocity-error constant, $K_v \ge 4$

The uncompensated root locus is a vertical line at s = – 1 and results in a root on the $\zeta = 0.707$ line at $s_d = -1 + j1$, as shown in Fig. 9.23a. Settling time corresponding to this root is $4/\zeta\omega_n = 4$ sec. The gain at s_d is $K = 2$. Therefore, the uncompensated system with the gain adjusted to a value of 2 meets the transient response specifications. Slight reduction in ω_n due to the small lag angle of the compensator required to improve steady-state performance, is also permissible.

The velocity-error constant of the uncompensated system is

$$K_v^{uc} = \lim_{s \to 0} sG(s) = K/2 = 1$$

The desired velocity-error constant is

$$K_v^d \ge 4$$

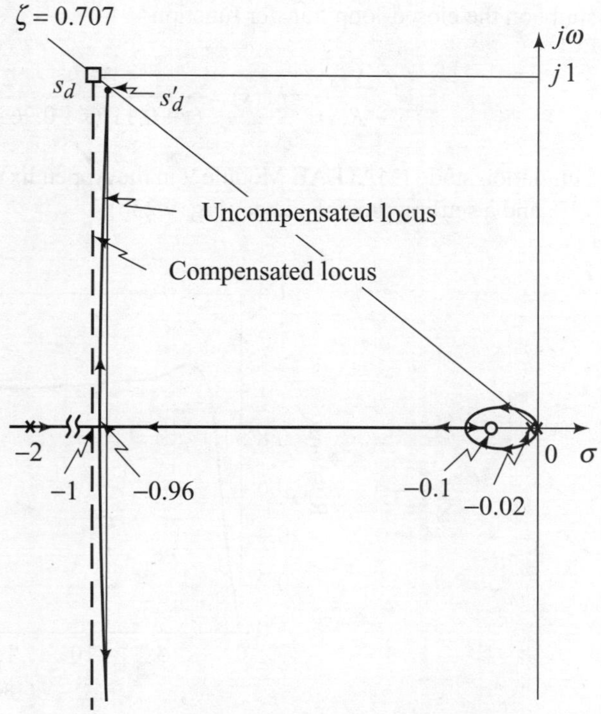

Fig. 9.23(a) *Root locus plot (Example 9.4)*

Thus, the velocity-error constant must be increased by at least a factor of 4. Lag compensator with $\beta = 5$ is expected to meet this requirement. We might set compensator zero at $s = -z = -0.1$. Note that $|z|$ is much smaller compared to $|s_d| = \sqrt{2}$. The compensator pole is located at $s = -p = -z/\beta = -0.02$. The difference of the angles from compensator zero and pole at the desired root s_d is about 3°; therefore, $s_d = -1 + j1$ will lie close to compensated root locus.

The open-loop transfer function of the compensated system becomes

$$D(s)G(s) = \left(\frac{s+0.1}{s+0.02}\right)\left[\frac{K}{s(s+2)}\right]$$

The root locus plot of the compensated system is shown in Fig. 9.23a. It intersects the $\zeta = 0.707$ line at $s_d' = -0.96 + j0.96$. The gain K at s_d', evaluated by magnitude criterion, remains approximately equal to 2. It follows that

$$D(s)G(s) = \frac{2(s+0.1)}{s(s+0.02)(s+2)}$$

From Fig. 9.23a, we find that corresponding to $K = 2$, the third closed-loop pole is at $s = -0.11$. Note that the real pole has a time constant of $\tau = 1/0.11 = 9.1$ sec, which would appear to make the system so slow as to be unusable. However, there is a zero close to this additional pole. The net effect is that the settling time will increase because of the third pole, but the amplitude of the response term contributed by this pole will be very small. In system response, a long tail of small amplitude will appear which may not degrade the performance of the system appreciably. This should be verified by simulation study on the closed-loop transfer function,

$$\frac{Y(s)}{R(s)} = M(s) = \frac{2(s+0.1)}{(s+0.11)(s+0.96+j0.96)(s+0.96-j0.96)}$$

Simulation study (MATLAB Module 9 in the Appendix) shows that the response has peak overshoot of 12% and a settling time of 15 sec (Fig. 9.23b).

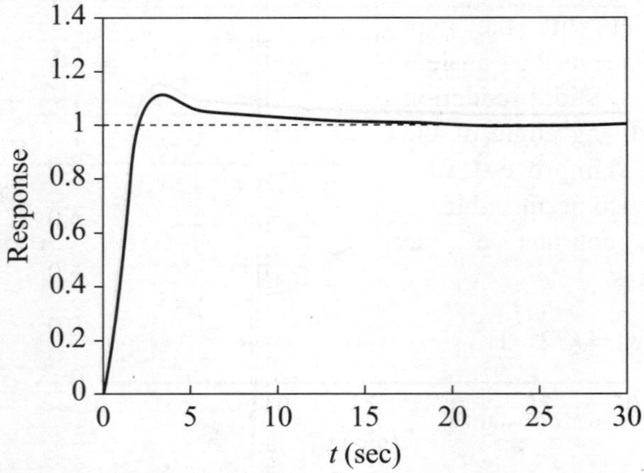

Fig. 9.23(b) *Simulation results (Example 9.4)*

9.5 CASCADE LAG-LEAD COMPENSATION

In the preceding sections, we have seen that the lead compensator is suitable for systems having unsatisfactory transient response. However, it provides only a limited improvement in steady-state response. If the steady-state behavior is highly unsatisfactory, the lead compensator may not be the answer. On the other hand, for systems with satisfactory transient response but unsatisfactory steady-state response, the lag compensator is found to be a good choice.

When both transient and steady-state responses are unsatisfactory, we must draw upon the combined powers of lag and lead compensators in order to meet the specifications. The transfer function for the lag-lead compensator is of the form

$$D(s)=\left(\frac{s+z_2}{s+p_2}\right)\left(\frac{s+z_1}{s+p_1}\right) \tag{9.17a}$$

$$=\underbrace{\left(\frac{s+1/\tau_2}{s+1/\beta\tau_2}\right)}_{\text{Lag Section}}\underbrace{\left(\frac{s+1/\tau_1}{s+1/\alpha\tau_1}\right)}_{\text{Lead Section}};\beta=\frac{z_2}{p_2}>1,\alpha=\frac{z_1}{p_1}<1,\tau_2>0,\tau_1>0 \tag{9.17b}$$

In designing a lag-lead compensator, we frequently choose $\beta = 1/\alpha$ (this is not necessary; we can, of course, choose $\beta \neq 1/\alpha$). The reason for this choice is that with the reduction in number of design parameters, the design problem becomes simpler. With $\beta = 1/\alpha$, the transfer function (9.17) of the lag-lead compensator reduces to

$$D(s)=\underbrace{\left(\frac{s+1/\tau_2}{s+\alpha/\tau_2}\right)}_{\text{Lag Section}}\underbrace{\left(\frac{s+1/\tau_1}{s+1/\alpha\tau_1}\right)}_{\text{Lead Section}};\alpha<1\ \tau_2,\tau_1>0 \tag{9.18}$$

We first design the lead section to realize the required ζ and ω_n for the dominant closed-loop poles. The error constant is then determined for the lead-compensated system, say it is K_e^{lc}. If K_e^{lc} is sufficiently high to give the desired steady-state response, we need not go further. Lead compensation is all that is required. If, however, K_e^{lc} is not large enough, we consider adding the lag section with $\beta = 1/\alpha$. This results in an error constant

$$K_e^{llc} \cong \frac{1}{\alpha}K_e^{lc} \tag{9.19}$$

If K_e^{llc}, as given in Eqn. (9.19), is not large enough to meet steady-state specifications, we choose $\beta > 1/\alpha$ and redesign the lag section.

Example 9.5 Consider a plant with transfer function

$$G(s)=\frac{4}{s(s+0.5)}$$

We design a feedback system to meet the following specifications:

1. Damping ratio of dominant closed-loop poles, $\zeta = 0.5$.
2. Undamped natural frequency of dominant closed-loop poles, $\omega_n = 5$ rad/sec.
3. Velocity-error constant = 80 sec.$^{-1}$

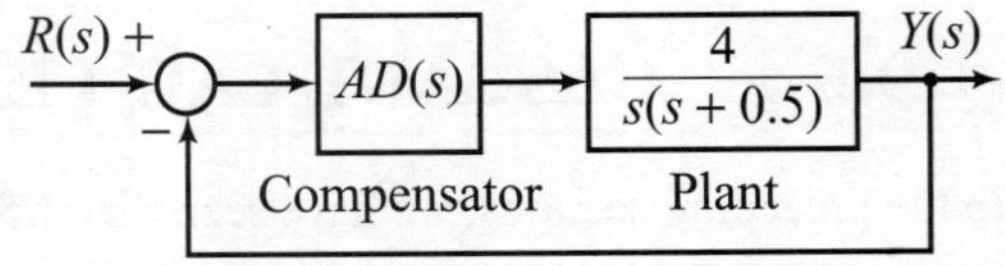

Fig. 9.24 *Control configuration (Example 9.5)*

We choose the unity-feedback control configuration shown in Fig. 9.24. Root locus plot of the system with $A > 0$ and $D(s) = 1$ is shown by dashed line in Fig. 9.25. The $\zeta = 0.5$ line intersects the root locus at the point $s = -0.25 + j0.433$. The undamped natural frequency corresponding to this root is 0.5 rad/sec. Therefore, the configuration in Fig. 9.24 with $D(s) = 1$ cannot meet the specifications.

We introduce a cascade lead compensator. From the performance specifications we find that the dominant closed-loop poles must be at $s = -2.5 \pm j4.33$. Consider the desired location,

$$s_d = -2.5 + j4.33$$

Since (refer to Fig. 9.25) $\angle G(s_d) = -120° - 115° = -235°$, the lead compensator must contribute 55° so that the root locus passes through the desired root s_d. To design the lead compensator, we first determine the location of the compensator zero. There are many possible choices, but we shall here choose the zero at $s = -z = -0.5$ so that it cancels the open-loop pole at $s = -0.5$. Once the compensator zero is chosen, the compensator pole can be located such that the angle contribution of the pole-zero pair is 55°. By simple graphical analysis (refer to Fig. 9.25), we find that the pole must be located at $s = -p = -5$. Thus, with the lead compensator

$$D_1(s) = \frac{s + 1/\tau_1}{s + 1/\alpha\tau_1} = \frac{s + 0.5}{s + 5}, \ \alpha = 0.1,$$

the open-loop transfer function of the system becomes

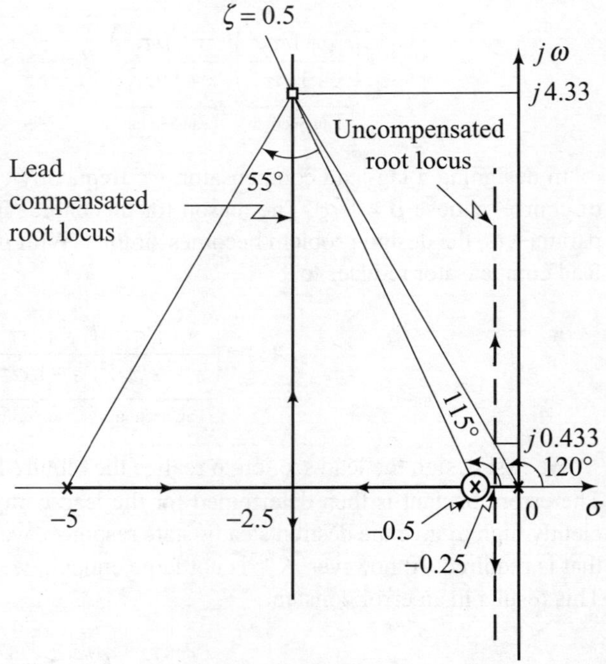

Fig. 9.25 *Root locus plot (Example 9.5)*

$$D_1(s)G(s) = \frac{K(s + 0.5)}{s(s + 0.5)(s + 5)} = \frac{K}{s(s + 5)}; \ K = 4A$$

The value of K obtained by magnitude criterion, is

$$K = |s_d||s_d + 5| \cong 25$$

Root locus plot of the lead-compensated system is shown in Fig. 9.25. Let us evaluate the steady-state performance of this system,

$$K_v = \lim_{s \to 0} s D_1(s)G(s) = \frac{K}{5} = 5$$

The desired K_v is 80. A large increase in K_v cannot be obtained by adjustment of pole-zero pair of the lead compensator. We, therefore, must alter the lead compensation to a lag-lead compensation.

Let us design a lag compensator

$$D_2(s) = \frac{s+1/\tau_2}{s+1/\beta\tau_2}; \beta > 1$$

for the lead-compensated system. If we choose $\beta = 1/\alpha = 10$, the velocity-error constant will increase by a factor of about 10, i.e., the velocity-error constant of the lag-lead compensated system will become about 50. The choice $\beta = 1/\alpha$, therefore, does not meet our requirements on steady-state accuracy.

The velocity-error constant of the lead-compensated system is $K_v^{lc} = 5$.

The desired velocity-error constant is $K_v^d = 80$.

Thus, the velocity-error constant must be increased by a factor of 16. The lag compensator with $\beta = 16$ will meet this requirement. We now choose the pole-zero pair of the lag compensator. We want that

$$|D_2(s_d)| = \left|\frac{s_d + z_2}{s_d + p_2}\right| \cong 1$$

$$-5° < \angle D_2(s_d) = \angle s_d + z_2 - \angle s_d + p_2 < 0°$$

$z_2 = 0.2$, and $p_2 = z_2/\beta = 0.2/16 = 0.0125$ meet these requirements.

The open-loop transfer function of the lag-lead compensated system becomes

$$D_2(s)D_1(s)G(s) = \frac{25(s+0.2)}{s(s+0.0125)(s+5)}$$

$$= \frac{25}{4}\frac{(s+0.2)(s+0.5)}{(s+0.0125)(s+5)}\left[\frac{4}{s(s+0.5)}\right]$$

Therefore, (refer to Fig. 9.24)

$$AD(s) = \frac{25}{4}\frac{(s+0.2)(s+0.5)}{(s+0.0125)(s+5)}$$

Root locus analysis of the lag-lead compensated feedback system is left as an exercise for the reader.

9.6 CASCADE PID COMPENSATION

The design of a controller is basically a search problem is n-dimensional parametric space where n is the number of design parameters of the controller. Points in the search space correspond to different selections of controllers. By choosing different points of the parametric space, we can generate feedback systems with different performances. A controller that meets the given performance specifications can be determined by moving in this search space on trial-and-error basis.

The design of PI/PD/PID compensator is a simpler problem with 2 to 3 trial parameters. The filtering and other signal conditioning may be independently designed. The lag/lead/lag-lead compensator design involves 3 to 5 parameters.

In the case of design wherein the primary requirement is to improve transient performance, a* PD *compensator may be attempted. This corresponds to a search problem in two-dimensional parametric space. This compensation will, however, be useful if the feedback loop is free from high-frequency noise problem. This is unlikely in real-life applications. Therefore, the total design problem includes the design of a filter in addition to the design of a* PD *compensator. An alternative to this strategy is the design of a lead compensator. This corresponds to a search problem in three-dimensional parametric space. Lead compensator may meet the filtering requirement in addition to satisfying the specifications on transient performance.

Consider now the problem of design wherein the primary requirement is to improve steady-state performance. A* PI *compensator may be attempted. This corresponds to a search problem in two-dimensional parametric space. The compensator will, however, be useful if its pole at the origin does not degrade closed-loop stability appreciably. The lag compensator, which is a search problem in three-dimensional parametric space, is an alternative strategy; it may yield an acceptable improvement in the steady-state performance without appreciably degrading the closed-loop stability.

The design of PI, PD and PID compensators follows in a straight-forward way from the procedures of the design of lag, lead and lag-lead compensators detailed in earlier sections. An illustrative example follows.

Example 9.6 A unity feedback system has forward path transfer function,

$$G(s) = \frac{K(s+8)}{(s+3)(s+6)(s+10)}$$

Our goal is to design a cascade PID compensator that improves the transient as well as steady-state performance of the closed-loop system.

Let us fix specifications for design. Say, we want to achieve a transient response that has no more that 20% overshoot with peak time $t_p = 0.2$ sec. The steady-state error to step input is required to be zero. The root locus plot of the uncompensated system is shown in Fig. 9.26. The $M_p = 20\%$ ($\zeta = 0.456$) line intersects the root locus at the point $s = -5.4 + j\,10.5$. (It is observed that the effect of the third closed-loop pole will be nearly canceled by a zero.) The damped natural frequency corresponding to this root is 10.5 rad/sec:

$$t_p = \frac{\pi}{10.5} = 0.6 \text{ sec}$$

The value of K obtained by magnitude criterion is 119.66.

The position-error constant

$$K_p = \lim_{s \to 0} G(s) = \frac{119.66 \times 8}{3 \times 6 \times 10} = 5.32$$

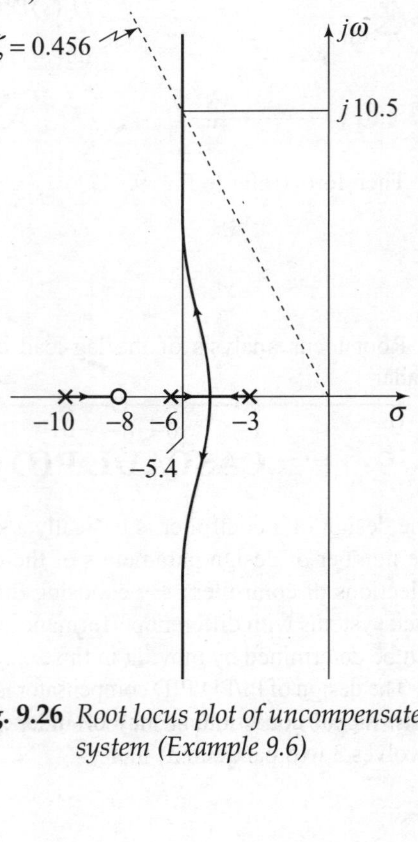

Fig. 9.26 *Root locus plot of uncompensated system (Example 9.6)*

Steady-state error to unit-step input,

$$e_{ss} = \frac{1}{1+K_p} = 0.158$$

Therefore, the uncompensated system (gain adjustment only) cannot meet the specifications.

We introduce a cascade PD compensator. From the performance specifications we find that the dominant closed-loop poles must be at $s_d = -8.16 \pm j\,15.92$ (The imaginary part of the compensated dominant poles is $\omega_d = \pi/t_p = \pi/0.2 = 15.92$; and the real part is $-\zeta\omega_n = -\zeta\omega_d/\sqrt{1-\zeta^2} = -8.16$).

We search for compensator's zero location so that the root locus of the compensated system passes through the desired dominant pole location.

The sum of the angles from uncompensated system's poles and zeros to the desired compensted dominant pole is – 198.5°. Thus the contribution required from the compensator zero is 198.5 – 180 = 18.5° Assuming that compensator zero is located at $s = -z_c$ as shown in Fig. 9.27, we obtain

$$\frac{15.92}{z_c - 8.16} = \tan(18.5°)$$

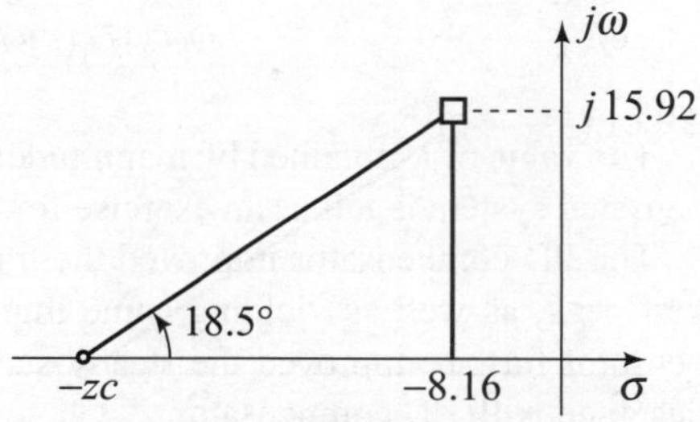

Fig. 9.27

This gives

$$z_c = 55.75$$

Thus, the PD compensator is

$$D_1(s) = s + 55.75$$

With this compensator, the open-loop transfer function of the system becomes

$$D_1(s)G(s) = \frac{K(s+8)(s+55.75)}{(s+3)(s+6)(s+10)}$$

Root locus plot of the PD-compensated system is shown in Fig. 9.28. The root locus passes through the design point s_d. It is observed that the effect of the third closed-loop pole will be nearly canceled by a zero.

The value of K obtained by the magnitude criterion is 5.24. Let us evaluate the steady-state performance of this system.

$$K_p = \lim_{s\to 0} D_1(s)G(s) = \frac{5.24 \times 8 \times 55.75}{3 \times 6 \times 10}$$

$$e_{ss} = \frac{1}{1+K_p} = 0.072$$

The PD-compensated system, thus, gives reduction in peak time with improvement in steady-state error over uncompensated system. The desired steady-state error is zero. We, therefore, alter the PD compensation to a PID compensation.

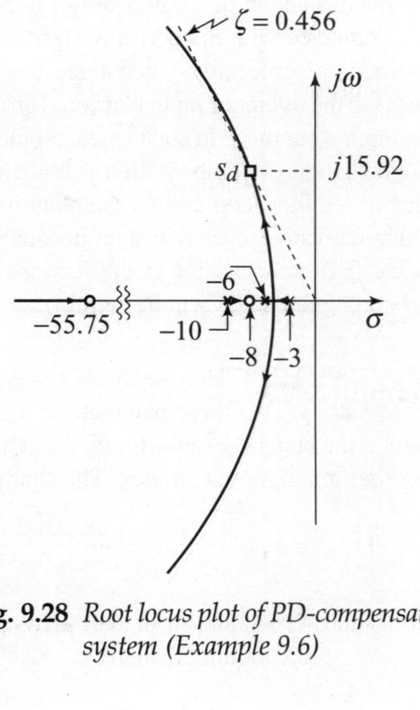

Fig. 9.28 *Root locus plot of PD-compensated system (Example 9.6)*

Let us design a PI compensator for the PD-compensated system. Any PI compensator will work as long as its zero is placed close to the origin to ensure that the compensator pole-zero pair will not appreciably change the transient response of the PD-compensated system. (The placement of the zero of the PI compensator is not entirely arbitrary. If K_v were specified, the location of the zero would have influenced the steady-state error to ramp inputs).

Choosing

$$D_2(s) = \frac{s+0.5}{s}$$

the open-loop transfer function of the PID-compensated system becomes

$$D_2(s)D_1(s)G(s) = \frac{K(s+0.5)(s+8)(s+55.75)}{s(s+3)(s+6)(s+10)}$$

The value of K obtained by magnitude criterion is 4.4. Complete root locus analysis of the PID compensated system is left as an exercise for the reader.

The PD compensator improved the transient response by decreasing the time required to reach the first peak, as well as yielding some improvement in the steady-state error. The complete PID compensator further improved the steady-state error without appreciably changing the transient response designed with PD compensator.

9.7 MINOR-LOOP FEEDBACK COMPENSATION

The feedback control schemes discussed in the preceding sections have all utilized a compensator $D(s)$ placed in the forward path in cascade with the plant transfer function $G(s)$. The compensator $D(s)$ typically involves PI terms (lag compensator), PD terms (lead compensator), or PID terms (lag-lead compensator).

Derivative action of a compensator need not be implemented with the derivative term in the forward path, especially when there are step changes in the reference input that could produce plant input saturation. In such cases, a practical implementation of compensation is realized by means of a minor loop around the plant implementing derivative term, with a proportional (P) or a lag (PI) compensator in the forward path (Revisiting Section 7.5 will be helpful).

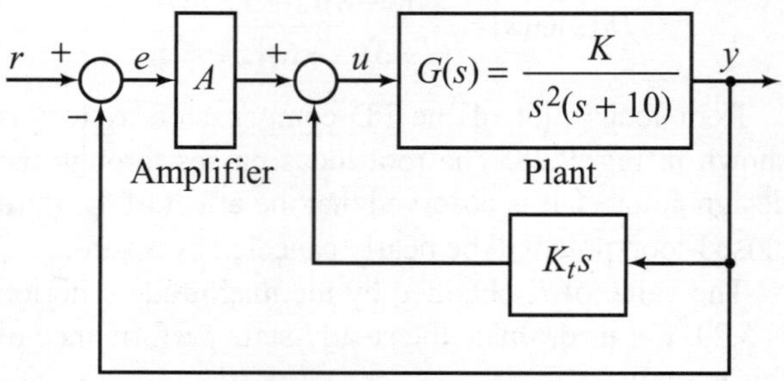

Fig. 9.29 *Rate feedback compensation*

Example 9.7 The feedback system shown in Fig. 9.29 provides the opportunity for varying three parameters: A, K, and K_t. The larger the number of variables inserted into a system, the better the opportunity for achieving a specified performance. However, the design problem becomes much more complex. The characteristic equation of the given system is (refer to Fig. 9.29)

$$1+\frac{AG(s)}{1+sK_tG(s)} = 1+\frac{AK}{s^2(s+10)+KK_ts} = 0 \tag{9.20}$$

Rearranging, we get

$$s^3 + 10\,s^2 + KK_t\,s + AK = 0$$

The parameters to be selected are $\alpha = AK$, and $\beta = KK_t$. The characteristic equation in terms of α and β parameters is

$$s^3 + 10s^2 + \beta s + \alpha = 0$$

We may plot a family of root loci (the root contours) to study the total effect of both α and β on the roots of the characteristic equation of the given system (refer to Review Example 8.3).

By appropriate partitioning of the characteristic equation, we can use single-parameter root loci for the design of three parameters (A, K, K_t) of the given system.

Consider the following partitioning:

$$1 + \frac{KK_t(s + A/K_t)}{s^2(s+10)} = 0 \tag{9.21}$$

This equation shows that the net effect of rate feedback is to add a zero at $s = -A/K_t$. Now the design procedure calls for a search for suitable location for this additional zero to satisfy the specifications.

Assume that the system is to be compensated to meet the following specifications:

Percent peak overshoot $\le 10\%$

Settling time ≤ 4 sec

The damping ζ and natural frequency ω_n of the dominant roots, obtained from the relations

$$e^{-\pi\zeta/\sqrt{1-\zeta^2}} \cong 0.1; \quad \text{and} \quad \frac{4}{\zeta\omega_n} = 4$$

are

$$\zeta = 0.6,\ \omega_n = 1.67 \text{ rad/sec.}$$

The desired dominant roots are then given as $s_d = -1 \pm j1.34$.

From Fig. 9.30, we see that the angle contribution of the open-loop poles at s_d is $-2(127°) - 8° = -262°$. Therefore, for the point s_d to be on the root locus, the compensating zero should contribute an angle of $\phi = -180° - (-262°) = 82°$. The location of the compensating zero is thus found to be at $s = -1.2$. The open-loop transfer function of the compensated system becomes

$$F(s) = \frac{KK_t(s+1.2)}{s^2(s+10)} \tag{9.22}$$

The root locus plot of the compensated system is shown in Fig. 9.30. It is now observed that for all values of KK_t, the system is stable. The value of KK_t at s_d is found to be 17.4. The third closed-loop pole is at $s = -8$ which is far away from the imaginary axis compared to dominant closed-loop poles. Hence, its effect on the transient behavior is negligible.

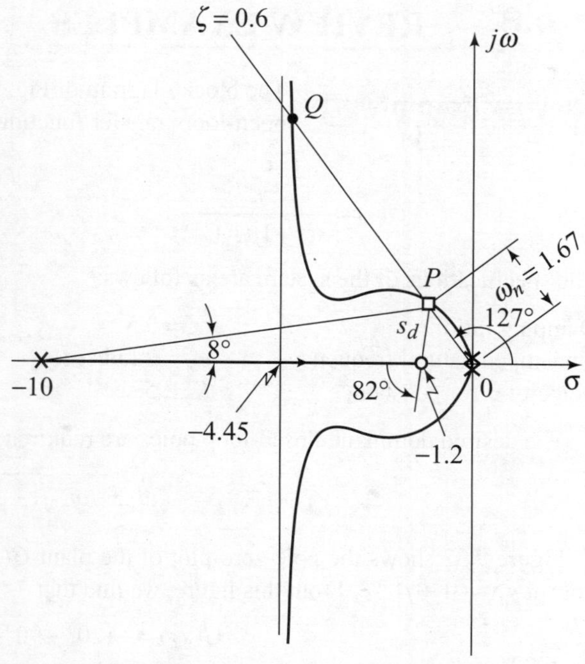

Fig. 9.30 *Root locus plot for the characteristic equation (9.21)*

From Fig. 9.30, we observe that the root locus intersects the $\zeta = 0.6$ line at two points, P and Q. Thus, two values of KK_t will give the damping ratio ζ of the closed-loop poles equal to 0.6. $KK_t = 17.4$ at point P results in a pair of dominant complex-conjugate closed-loop poles which satisfy the transient response specifications. The third closed-loop pole is at $s = -8$ which has insignificant contribution to system dynamics. At point Q, the dominance condition will be weakened; the third closed-loop pole on the real axis will force a slow overdamped response. It is important to point out that the zero at $s = -1.2$ is not the zero of the system of Fig. 9.29. It is the open-loop zero of $F(s)$ (refer to Eqn. 9.22) which was introduced in the process of partitioning the characteristic equation such that the adjustable variable KK_t appears as a multiplying factor.

Let us now investigate the steady-state behavior of the compensated system. From Eqn. (9.20), the velocity-error constant K_v is given by

$$K_v = \lim_{s\to 0} s\left[\frac{AK}{s^2(s+10)+KK_t s}\right] = \frac{AK}{KK_t} = A/K_t = 1.2$$

If this value of K_v is acceptable, then the design is complete. Otherwise, the design procedure given above cannot satisfy the steady-state and transient specifications simultaneously. A trade-off between the steady state and transient behavior becomes necessary.

We have three parameters and two equations:

$$A/K_t = 1.2;\ KK_t = 17.4$$

If we select $K_t = 0.4$, then the other two parameters become: $A = 0.48$ and $K = 43.5$. K can be decreased with no change in performance by increasing A and K_t; e.g., $K_t = 0.8$ gives $A = 0.96$ and $K = 21.75$.

9.8 REVIEW EXAMPLES

Review Example 9.1 The block diagram of Fig. 9.31 represents a position control system. The open-loop transfer function of the uncompensated system is

$$G(s) = \frac{K}{s(s+1)(s+4)}$$

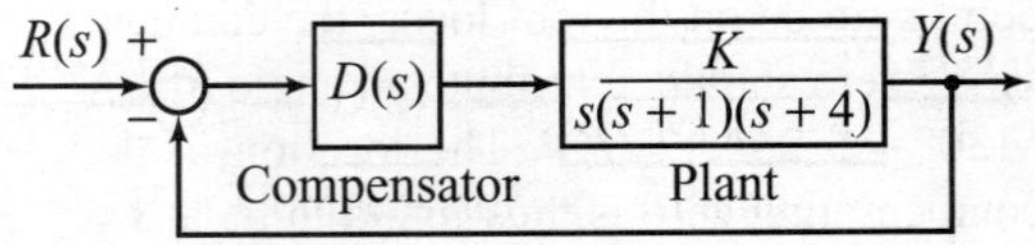

Fig. 9.31 *A position control system*

The specifications of the system are as follows:

Damping ratio, $\zeta = 0.5$
Undamped natural frequency, $\omega_n = 2$ rad/sec
Velocity-error constant, $K_v \ge 5\ \text{sec}^{-1}$

The desired dominant closed-loop poles are required to be located at

$$s_{1,2} = -\zeta\omega_n \pm j\omega_n\sqrt{1-\zeta^2} = -1 \pm j1.73$$

Figure 9.32 shows the pole-zero plot of the plant $G(s) = \dfrac{K}{s(s+1)(s+4)}$, and a desired closed-loop pole at $s_d = -1 + j1.73$. From this figure, we find that

$$\angle G(s_d) = -120° - 90° - 30° = -240°$$

The point s_d, therefore, does not lie on the uncompensated root locus and transient response specifications cannot be met by gain adjustment alone. We propose to add a cascade lead compensator to meet the transient response specifications. In order for the root locus to go through s_d, the angle contribution from the lead compensator pole-zero pair must be

$$\phi = -180° - \angle G(s_d) = -180° - (-240°) = 60°$$

Further, it is observed that the open-loop pole at $s = -1$ lies directly below the desired closed-loop pole location. It appears that the best transient response due to the complex closed-loop pole will occur when the zero of the compensator cancels the open-loop pole at $s = -1$; there will be no real closed-loop pole near the imaginary axis in this case. This ideal situation, however, cannot be realized practically because of imperfect cancelation. A real closed-loop pole near the imaginary axis will occur but a zero will be very close to it; the dominance condition will, therefore, be ensured.

Join the compensator zero to s_d and locate the compensator pole making an angle of $\phi = 60°$ as shown in Fig. 9.32. The location of the pole is found to be at $s = -4$.

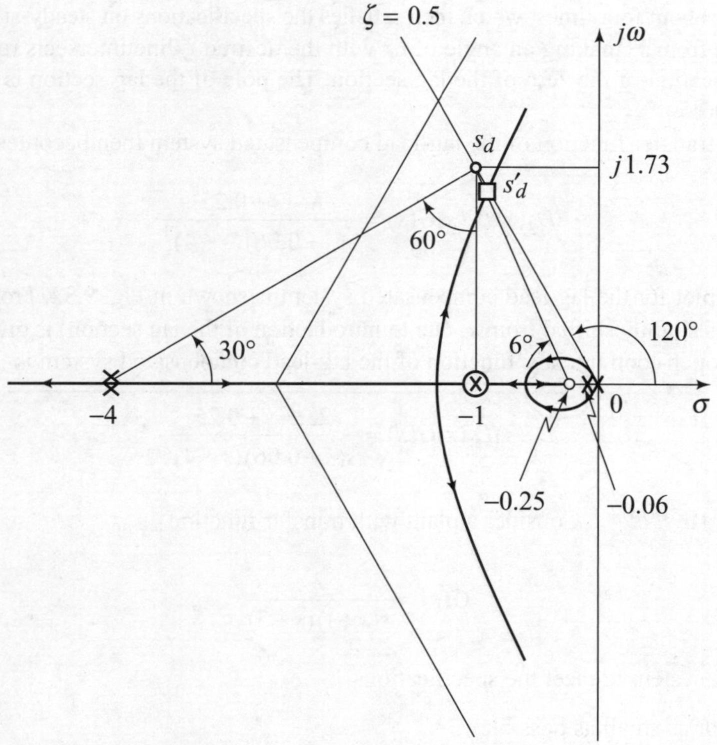

Fig. 9.32 *Root locus plot (Review Example 9.1)*

With the addition of the cascade lead compensator

$$D_1(s) = \frac{s+1}{s+4} = \frac{s+1/\tau_1}{s+1/\alpha\tau_1}; \tau_1 = 1, \alpha = \frac{1}{4}$$

the open-loop transfer function of the system becomes

$$D_1(s)G(s) = \frac{K(s+1)}{s(s+1)(s+4)^2} = \frac{K}{s(s+4)^2}$$

The value of K at s_d is found to be 24. Therefore,

$$D_1(s)G(s) = \frac{24}{s(s+4)^2}$$

The velocity-error constant of the lead compensated system is

$$K_v^{lc} = \lim_{s\to 0} s\,D_1(s)G(s) = 1.5$$

This does not meet the specified $K_v \geq 5$. A lag section

$$D_2(s) = \frac{s+1/\tau_2}{s+\alpha/\tau_2} = \frac{s+1/\tau_2}{s+1/4\tau_2}$$

will increase K_v by about four times, which then satisfies the specifications on steady-state performance.

The line drawn from s_d making an angle of 6° with the desired ζ-line intersects real axis at – 0.25, which gives the location of the zero of the lag section. The pole of the lag section is then found to be at – 0.25/4 ≅ – 0.06.

The open-loop transfer function of the lag-lead compensated system then becomes

$$D_2(s)D_1(s)G(s) = \frac{K^c(s+0.25)}{s(s+0.06)(s+4)^2}$$

The root locus plot for the lag-lead compensated system is shown in Fig. 9.32. From this figure, the gain at s_d' (which is slightly shifted from s_d due to introduction of the lag section) is given by $K^c = 23.5$.

Therefore, the open-loop transfer function of the lag-lead compensated system is

$$D_2(s)D_1(s)G(s) = \frac{23.5(s+0.25)}{s(s+0.06)(s+4)^2}$$

Review Example 9.2 Consider a plant with transfer function

$$G(s) = \frac{2}{s(s+1)(s+5)}$$

Design a feedback system to meet the specifications:

1. Velocity error as small as possible
2. $\zeta = 0.707$
3. Settling time < 4.5 seconds

Solution

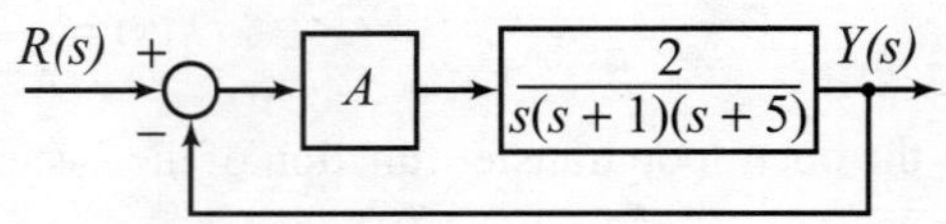

Fig. 9.33 *A control configuration (Review Example 9.2)*

As a first try, we choose a unity-feedback system shown in Fig. 9.33. The root locus plot of this system for $A > 0$ is shown in Fig. 9.34. In order for the feedback system to have settling time less than 4.5 seconds, all the closed-loop poles must lie on the left hand side of the vertical line passing through the point $-4/t_s = -4/4.5 = -0.9$.

From the root loci in Fig. 9.34 we see that this is not possible for any $A > 0$. Therefore, the configuration in Fig. 9.33 cannot meet the specifications.

As a next try, we introduce an additional tachogenerator feedback as shown in Fig. 9.35. The characteristic equation of the system becomes

$$1 + \frac{AG(s)}{1 + sK_t G(s)} = 0 \tag{9.23}$$

where $G(s) = \dfrac{2}{s(s+1)(s+5)}$.

Rearranging, we get

$$s^3 + 6s^2 + (5 + 2K_t)\, s + 2A = 0 \tag{9.24}$$

The parameters to be selected are $a = 2A$, and $K = 2K_t$. The characteristic equation in terms of a and K parameters, is

$$s^3 + 6s^2 + 5s + Ks + a = 0$$

which may be rearranged as a root locus equation:

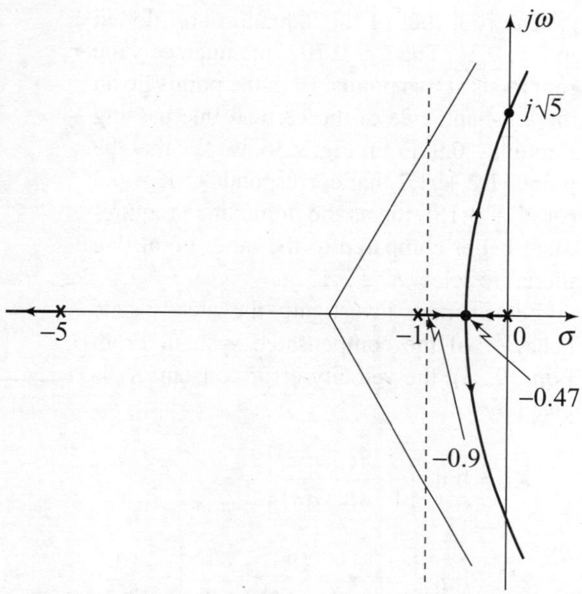

Fig. 9.34 *Root locus plot for the system of Fig. 9.33*

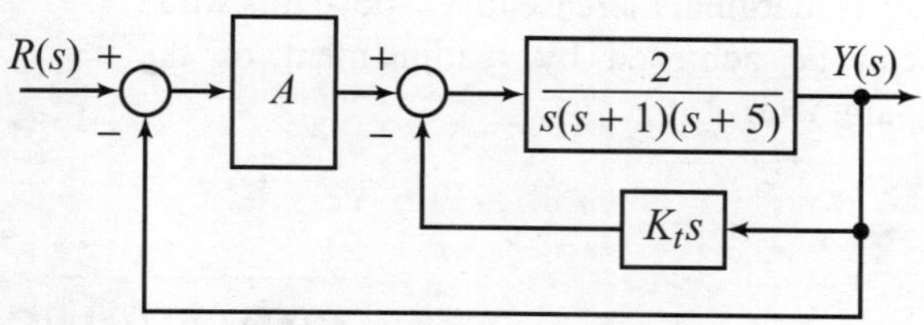

Fig. 9.35 *Tachogenerator feedback control configuration (Review Example 9.2)*

$$1 + \frac{Ks}{s^3 + 6s^2 + 5s + a} = 0 \tag{9.25}$$

We will carry out the iterative design by choosing a value for a, and then plotting the root loci of Eqn. (9.25) as K is varied from zero to infinity. Our first trial value is $a = 10$ ($A = 5$). The open-loop poles of the root locus Equation (9.25) may be obtained by computer simulation or by sketching a root locus plot of the equation (refer to Review Example 8.3):

$$1 + \frac{a}{s(s^2 + 6s + 5)} = 0$$

For $a = 10$, Eqn. (9.25) becomes

$$1 + \frac{Ks}{(s + 5.42)(s + 0.29 + j1.33)(s + 0.29 - j1.33)} = 0$$

The root loci of this equation are plotted in Fig. 9.36. The $\zeta = 0.707$ line intersects the root locus at two points. Both the points lie on the left-hand side of the vertical line passing through – 0.9. From Fig. 9.36, we see that the point $-1.2 + j1.2$ that corresponds to $K = 6.3$ (or $K_t = 3.15$), meets the dominance requirement better compared to the other point. We therefore select $K_t = 3.15$.

Let us now investigate the steady-state behavior of the compensated system. From Eqn. (9.23), the velocity-error constant K_v is given by

$$K_v = \lim_{s\to 0} s\left[\frac{AG(s)}{1+sK_tG(s)}\right]$$

$$= \lim_{s\to 0} s\left[\frac{10}{s(s+1)(s+5)+6.3s}\right] = \frac{10}{11.3}$$

A marginal increase in K_v above this value can be achieved by readjustment of the parameter A.

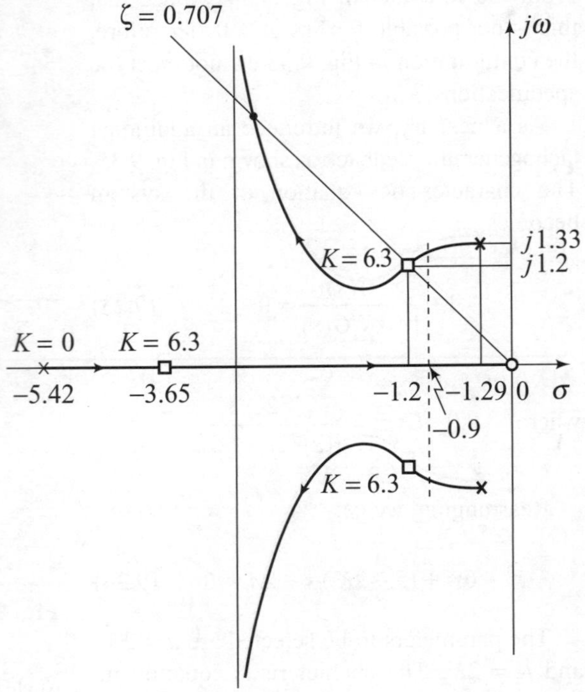

Fig. 9.36 *Root locus plot for the system of Fig. 9.35*

CONCLUDING REMARKS

In this chapter, we have considered alternative approaches to the design of feedback control systems using root locus plots.

What we have learnt?

- The root locus is reshaped by adding additional open-loop poles and zeros via a lead/lag/lead-lag compensator or a PID control function in a selected feedback control structure—a classical control configuration with/without minor-loop feedback. The additional poles and zeros must be located by trial and error so that the second-order approximations used in the design are valid. All closed-loop poles besides the dominant second-order pair, must yield a response that is much faster than the desired response. Thus, these poles must be at least 5 times as far from the imaginary axis as the dominant pair. Further, any zeros of the system must be close to a pole for pole-zero cancelation or far from the dominant pole pair. The resulting closed-loop system can then be approximated by two dominant poles. If second-order approximations cannot be justified, then a simulation is required to make sure the design is within tolerance.
- The dominant pole pair controls the transient response of the closed-loop system. Best results for transient response design are achieved through lead compensator/PD control function. PD control function adds an open-loop zero to compensate the transient response. Lead compensator, on the other hand, adds an open-loop pole along with an open-loop zero, giving better filtering.

- Best results for steady-state response design are achieved through lag compensator/PI control function. PI control function adds an open-loop pole at the origin, thereby increasing the system type. A lag compensator, on the other hand, places the pole off the origin but near it. Both the methods add a zero very close to the pole in order to minimize the effect on the transient response.
- We can correct both the transient response and the steady-state error with lag-lead compensator/PID control function.
- A minor-loop feedback control configuration with controller gain in forward path and derivative gain in feedback path, has also been investigated using root locus plots.

Review Questions

9.1 Discuss the compensation characteristics of cascade lag and lead compensators using root locus plots. Show that
(a) lead compensation is suitable for systems having unsatisfactory transient response, and it provides a limited improvement in steady-state performance; and
(b) lag compensation is suitable for systems with satisfactory transient response but unsatisfactory steady-state response.

9.2 In many control systems, it is desirable to use a feedback compensator forming a minor loop around the plant, in addition to a cascade gain compensator which is used to adjust the overall performance. Discuss compensation characteristics of the minor-loop feedback scheme.

Self-Appraisal Quiz

A set of questions with multiple answer choices is given below. These questions have been designed to capture important aspects of the subject covered in the chapter. For each of these questions, identify the best of the given answer choices. Compare your answers with the Master Key given at the end of the session. This quiz should serve the purpose of a self-appraisal test for readers.

9.1 The transfer function of a lag compensator is

$$D(s) = \frac{1+\alpha\tau s}{1+\tau s}; \tau > 0$$

The value of α is given by
(A) $\alpha = 1$ (B) $\alpha > 1$ (C) $\alpha < 1$ (D) α is any constant

9.2 A compensator is of the form

$$D(s) = K_c\left(\frac{\alpha\tau s+1}{\tau s+1}\right); \alpha > 1, K_c > 0, \tau > 0$$

This compensator may be realized using op amp circuits for
(A) proportional plus integrator action
(B) proportional plus gain-limited integrator action
(C) proportional plus derivative action
(D) proportional plus bandwidth-limited derivative action

9.3 A compensator is of the form

$$D(s) = K_c\left(\frac{\alpha\tau s+1}{\tau s+1}\right); \alpha < 1, K_c > 0, \tau > 0$$

This compensator may be realized using op amp circuits for
(A) proportional plus integrator action
(B) proportional plus gain-limited integrator action
(C) proportional plus derivative action
(D) proportional plus bandwidth-limited derivative action

9.4 A compensator is of the form

$$D(s) = K_2 + \frac{K_1}{\tau_1 s+1}; K_2 > 0,\ K_1 > 0, \tau_1 > 0$$

This is a
(A) lag compensator
(B) lead compensator
(C) lag-lead compensator
(D) None of the answers in (A), (B), and (C) is correct

9.5 A compensator has a transfer function

$$D(s) = K_2 + \frac{K_1 s}{\tau_1 s+1}; K_2 > 0, K_1 > 0, \tau_1 > 0$$

This is a
(A) lag compensator
(B) lead compensator
(C) lag-lead compensator
(D) None of the answers in (A), (B), and (C) is correct

9.6 Consider the following statements:
(i) The effect of compensating pole is to pull the root locus towards left.
(ii) The effect of compensating zero is to press the locus towards right.
(A) None of the above statements is true.
(B) Statement (i) is true but statement (ii) is false.
(C) Statement (i) is false but statement (ii) is true.
(D) Both the statements are true.

9.7 Consider the following statements:
(i) For a noise-less system, PD compensation is a better choice than lead compensation for transient performance.
(ii) PI compensation is a better choice than lag compensation for steady-state performance, if both yield acceptable relative stability.
(A) None of the above statements is true.
(B) Statement (i) is true but statement (ii) is false.
(C) Statement (i) is false but statement (ii) is true.
(D) Both the statements are true.

9.8 Consider the following statements:
(i) Replacing PI compensator with lead compensator will increase system sensitivity to noise.
(ii) Replacing PI compensator with lag compensator will result in smaller negative effect on transient response.

(A) None of the above statements is true.
(B) Statement (i) is true but statement (ii) is false.
(C) Statement (i) is false but statement (ii) is true.
(D) Both the statements are true.

9.9 A long tail of small amplitude may appear in the output response with
(A) PI compensation (B) Lag compensation
(C) Either of the two (D) Neither of the two

9.10 A closed-loop pole-zero dipole near the origin will appear with
(A) PD compensation (B) PI compensation
(C) Either of the two (D) Neither of the two

9.11 A process control system has type-0 plant. For an accurate step-command following, a recommended cascade compensation scheme for the system employs.
(A) a lag compensator
(B) a lead compensator
(C) either a lag or a lead compensator
(D) neither a lag nor a lead compensator

MASTER KEY

9.1 (C)	9.2 (D)	9.3 (B)	9.4 (A)	9.5 (B)
9.6 (A)	9.7 (D)	9.8 (C)	9.9 (C)	9.10 (B)
9.11 (A)				

Problems

9.1 A unity-feedback system has an open-loop transfer function,

$$G(s) = \frac{K}{s(s+1)(s+5)}$$

Draw the root locus plot and determine the value of K that yields a damping ratio of 0.3 for the dominant closed-loop poles. A phase-lag compensator having a transfer function,

$$D(s) = \frac{10s+1}{100s+1}$$

is now introduced in tandem. Find the new value of K that gives the same damping ratio for the dominant closed-loop poles. Compare the velocity-error constant and settling time of the original and the compensated systems.

MATLAB Exercise You have approximated the closed-loop system by a second-order system, assuming the dominance condition is satisfied. Generate root locus plots of the original and compensated systems in MATLAB window and determine all the closed-loop poles for the two systems. Also generate step responses in the MATLAB window and find settling time and steady-state error in the two cases. Comment whether dominance condition was satisfied.

9.2 A unity-feedback system has forward path transfer function

$$G(s) = \frac{K}{s(s+1)}$$

It is required that the closed-loop poles be located at $s_{1,2} = -1.6 \pm j4$ using a lead compensator with transfer function $D(s) = \dfrac{s+2.5}{s+\alpha}$.

Determine the values of α and K to locate the closed-loop poles as required. What is the location of the third closed-loop pole?

MATLAB Exercise Attempt this problem using MATLAB dialogs.

9.3 Consider the feedback system of Fig. P9.3. Note that the plant is open-loop unstable.

(a) Plot the root loci to determine whether the system can be stabilized by proportional control, i.e., $D(s) = K$.

(b) If not, could the unstable pole of $G(s)$ be canceled by a zero of $D(s)$ to stabilize the system, and if not, why not?

(c) Choose a compensator $D(s) = \dfrac{K(s+1)}{s+8}$ and sketch root loci. How does this phase-lead compensator affect the transient as well as steady-state behavior?

MATLAB Exercise The pole of the physical plant may be different from the one given by its model. Cancelation of the pole of the model by compensator zero may not yield stable closed-loop when the compensator is cascaded to the physical plant. Show this by generating root locus plots in MATLAB window with compensator zero at $s = 2$, and the plant pole varying between $s = 1.5$ and $s = 2.5$.

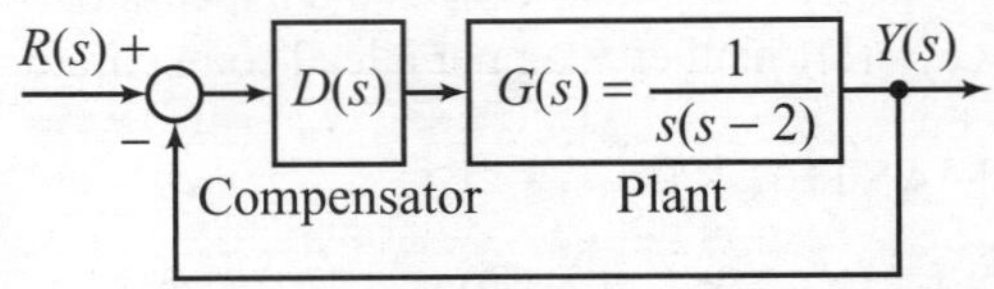

Fig. P9.3

9.4 A unity-feedback servo system has open-loop transfer function,

$$G(s) = \frac{100(1+10s)}{s(1+0.2s+0.25s^2)}$$

A second-order cascade compensator

$$D(s) = \frac{s^2+0.8s+4}{(s+0.384)(s+10.42)}$$

cancels the complex poles of $G(s)$. Using root-locus analysis, compare the transient performance of compensated and uncompensated systems. What is the effect of $D(s)$ on steady-state performance?

9.5 Consider the type-1 system of Fig. P9.5. We would like to design the compensator $D(s)$ to meet the following specifications:

(i) Damping ratio $\zeta = 0.707$

(ii) Settling time ≤ 2 sec

(a) Show that the proportional control is not adequate.

(b) Show that proportional plus derivative control will work.

(c) For $D(s) = K_P + K_D s$, find gains K_P and K_D to meet the design specifications.

MATLAB Exercise After solving the problem using hand sketches, attempt it using MATLAB. Determine settling time and peak overshoot of the compensated and uncompensated systems.

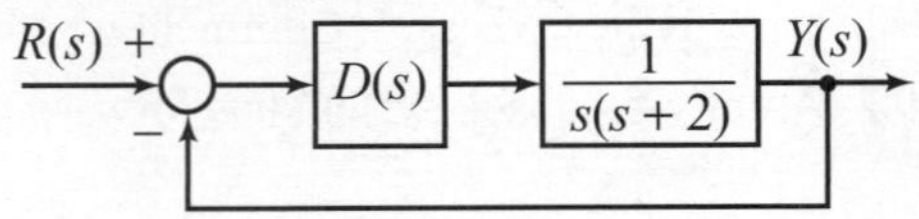

Fig. P9.5

9.6 Determine the values of K, z, and p of the system shown in Fig. P9.6, so that (i) the dominant closed-loop poles have damping ratio $\zeta = 0.45$ and (ii) velocity-error constant $K_v = 20$.

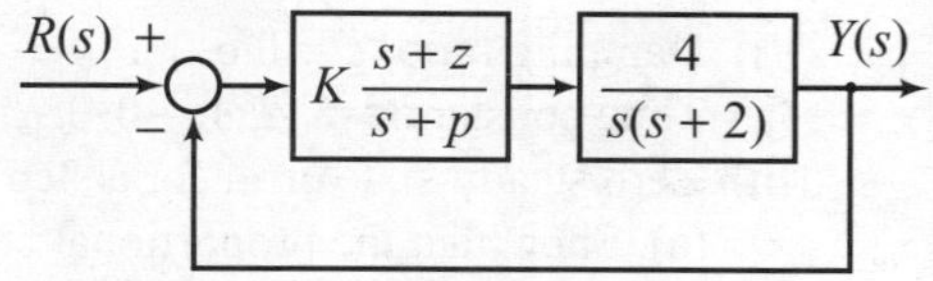

Fig. P9.6

MATLAB Exercise Verify your solution using MATLAB, and if required refine your design. Due to uncertainties in the load, the time-constant of the plant may vary by 10%. Check the robustness of your design by determining transient and steady-state performance for various values of this time-constant.

9.7 The controlled plant of a unity-feedback system is

$$G(s) = \frac{K}{s(s+1)(s+5)}$$

It is desired to compensate the system so as to meet the following transient response specifications:
Settling time, $t_s \le 3$ sec
Peak overshoot for step input ≤ 20 %

Design a suitable cascade compensator (choose the compensator zero so as to cancel the plant pole at $s = -1$). What is the velocity-error constant of the compensated system?

MATLAB Exercise Attempt this problem using MATLAB dialogs.

9.8 The controlled plant of a unity-feedback system is

$$G(s) = \frac{K}{s(s+1)}$$

It is desired to compensate the system so as to meet the following transient response specifications:

Damping ratio, $\zeta = 0.707$
Settling time, $t_s = 1.4$ sec

Design a suitable first-order or second-order (two first-order sections in cascade) lead compensator for the system.

9.9 The controlled plant of a unity-feedback system is

$$G(s) = \frac{1.06}{s(s+1)(s+2)}$$

Sketch a root locus plot of the system, and determine the dominant closed-loop poles and the velocity-error constant.

It is desired to increase the velocity-error constant to about 5 sec^{-1} without appreciably changing the location of the dominant closed-loop poles. Design a first-order cascade compensator $D(s) = \dfrac{K(s+z)}{s+p}$ to meet this specification.

MATLAB Exercise Attempt this problem using MATLAB dialogs.

9.10 Consider the type-0 system of Fig. P9.10. We would like to design the compensator $D(s)$ to meet the following specifications:

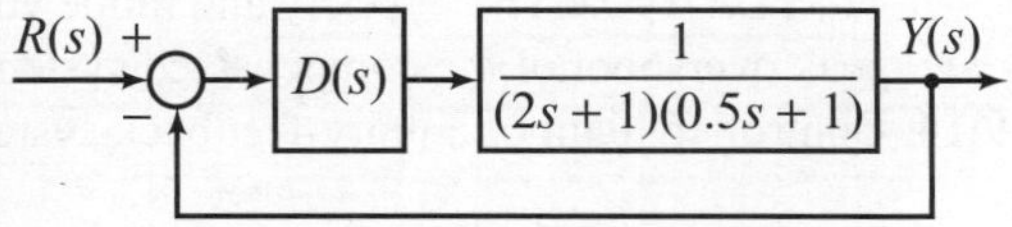

Fig. P9.10

(i) Damping ratio, $\zeta = 0.6$
(ii) Time constant $\tau = 1/\zeta\omega_n = 1/0.75$
(iii) Zero steady-state error for a step input.
(a) Show that the proportional control is not adequate.
(b) Show that proportional plus integral control will work.
(c) For $D(s) = K_P + \frac{K_I}{s}$, find gains K_P and K_I to meet the design specifications.

MATLAB Exercise After solving the problem using hand sketches, attempt it using MATLAB. Determine settling time and peak overshoot of the compensated and uncompensated systems.

9.11 The controlled plant of a unity-feedback system is

$$G(s) = \frac{K}{s(s+10)(s+20)}$$

Determine the value of K so that the damping ratio of the dominant closed-loop poles is 0.6. For this value of K, determine the velocity-error constant K_v.

It is desired to increase K_v by a factor of 10. It is also desired to keep the damping ratio of the dominant closed-loop poles at 0.6. A small change in the undamped natural frequency of the dominant closed-loop poles is permissible.
(a) Design a suitable cascade lead, lag, or lag-lead compensator to realize the given objectives.
(b) Design a suitable cascade PD, PI, or PID compensator to realize the given objectives.

MATLAB Exercise Attempt this problem using MATLAB dialogs.

9.12 The controlled plant of a unity-feedback system is

$$G(s) = \frac{K}{s(s+10)^2}$$

It is specified that velocity-error constant of the system be equal to 20, while the damping ratio of the dominant roots be 0.707.
(a) Design a suitable cascade lead, lag, or lag-lead compensator to realize the given objectives.
(b) Design a suitable cascade PD, PI, or PID compensator to realize the given objectives.

MATLAB Exercise Attempt this problem using MATLAB dialogs.

9.13 Controlled plant of a unity-feedback system is

$$G(s) = \frac{K}{s(s+1)(s+5)}$$

It is desired to compensate the system so as to meet the following specifications:

Damping ratio of dominant roots = 0.45
Undamped natural frequency of dominant roots = 3.5 rad/sec
Velocity-error constant = 30 sec^{-1}

Design a cascade lag-lead compensator to meet these objectives (choose the zero of the lead section of the compensator to cancel the plant pole at $s = -1$).

MATLAB Exercise Verify and refine your design using MATLAB. Determine settling time and peak overshoot of the compensated system.

9.14 Controlled plant of a unity-feedback system is

$$G(s) = \frac{1}{s^2+1}$$

(a) Design a PD controller such that the dominant closed-loop poles are located at $-1 \pm j\sqrt{3}$. What is the position-error constant of the compensated system?

(b) It is desired to reduce the steady-state error to step inputs to zero. Design a PID controller that meets the requirements on both the transient and the steady-state performance.

9.15 A unity-feedback system is characterized by the open-loop transfer function

$$G(s) = \frac{K}{s(s+3)(s+9)}$$

(a) Determine the value of K if 20% overshoot to a step input is desired.

(b) For the above value of K, determine the settling time and velocity-error constant K_v.

(c) Design a suitable cascade lead, lag or lag-lead compensator that will give approximately 15% overshoot to a unit-step input, while the settling time is decreased by a factor of 2.5 with $K_v \geq 20$.

(d) Design a suitable PD, PI, or PID compensator to realize the objectives given in (c).

MATLAB Exercise You have employed second-order correlations to solve this problem using hand sketches. Verify your solution using MATLAB, and if required refine your design.

9.16 For a plant with transfer function $G(s) = \dfrac{K}{s(s+2)}$, a feedback system is to be designed to satisfy the following specifications:

(a) Steady-state error for a ramp input $\leq 35\%$ of input slope.

(b) Damping ratio of dominant roots ≥ 0.707.

(c) Settling time of the system ≤ 3 sec.

The structure of the control system is shown in Fig. P9.16. Select the parameters K and K_t to meet the given specifications.

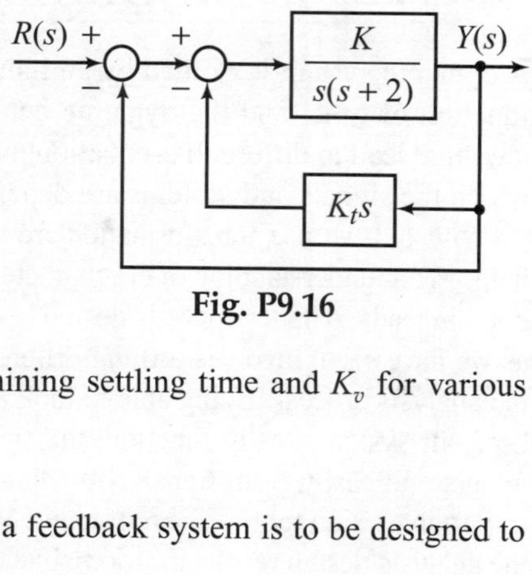

Fig. P9.16

MATLAB Exercise The parameter K_t is subject to variations by 10%. Check by simulation the robustness of the closed-loop system to these variations by determining settling time and K_v for various values of K_t.

9.17 For a plant with transfer function $G(s) = \dfrac{K}{s(s+1)(s+5)}$, a feedback system is to be designed to satisfy the following specifications:

(i) Dominant closed-loop poles at $-1 \pm j2$.

(ii) Ramp error constant, $K_a \geq 1.5$.

The structure of the control system is shown in Fig. P9.17.

(a) Select the parameters A, K, and K_t to meet the given specifications.

(b) Find the closed-loop transfer function of the system.

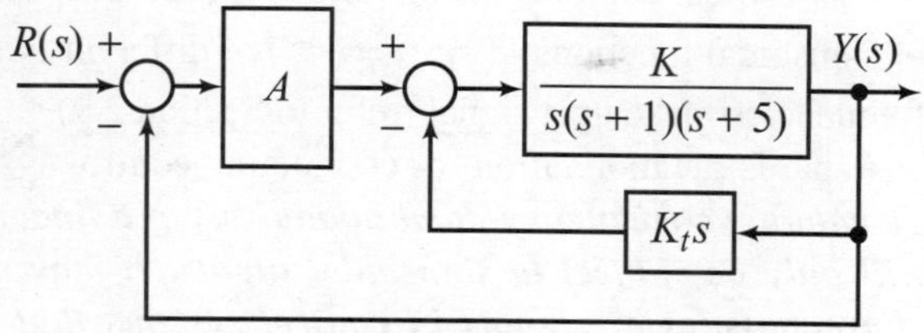

Fig. P9.17

NYQUIST/BODE FREQUENCY RESPONSE PLOTS AND SYSTEM STABILITY

10

Preview

In Chapters 5 and 8, we discussed stability and developed various tools (Routh criterion, root locus analysis) to determine stability and to assess relative stability. We continue that discussion in this chapter by showing how frequency response methods can be used to investigate stability. The root locus method was used earlier to study relative stability of feedback control systems as a function of certain design parameters. We realized this objective in *s*-domain by using transfer function models of subsystems in the feedback loop, and root locus plots of closed-loop characteristic equation when a system parameter is varied from zero to infinity. Here in this chapter, we use sinusoidal transfer function models and Nyquist/Bode plots to study relative stability of feedback systems.

10.1 INTRODUCTION

A dynamic system is excited by an input, which is a function of time; the response of the system is a function of time; and the dynamic behavior (system response to excitation) is governed by physical laws that lead to differential-equation model of the system. Time '*t*' is, thus, the independent variable on which the signals and systems are dependent.

Why did we use transformation from *t*-domain to *s*-domain? Transformation is beneficial only if it helps in our understanding of the characteristics of signals and systems and/or understanding in transformed domain leads to better analysis/design results involving these signals and systems. Laplace transformation, as we have seen, involves turning differential equations into algebraic equations, thus greatly facilitating the analysis process. Being able to look at signals and systems in the *s*-domain helps us understand better how our systems really function; the stability, the transient and steady-state response characteristics are understood easily from transfer function models. This understanding has given birth to one of the most important design tools—the root locus. We use Laplace transformation from *t*-domain to *s*-domain, obtain the analysis/design results in *s*-domain cashing on better understanding, and then apply the inverse Laplace transform to the result to transform it back to the *t*-domain.

We now turn our attention to *Fourier transformation,* which is a process of decomposition of a signal into sinusoidal components of various frequencies. A time-varying signal may be resolved into a sum of sinusoidal functions $A_i \sin(\omega_i t + \phi_i)$ of different frequencies ω_i. Fourier transformation tells us what frequencies make up a signal (refer to Section 2.9).

What is the motivation for conducting control systems analysis and design in frequency domain? ***To conduct a frequency-domain analysis of a linear control system does not imply that the system will only be subject to sinusoidal inputs. It may never be. In fact, it is the frequency-selective characteristics of signals in control systems that provide the motivation for transformation to frequency domain.***

A control system is required to process two kinds of input signals: command (reference) signals, represented by $r(t)$; and disturbances represented by w and w_n (refer to Fig. 10.1). The signals represented by w_n (noise), are generally of high frequencies, while r and w generally represent low-frequency signals. Therefore, a feedback control design is required to stop (attenuate) the high-frequency components of the input and 'selectively' pass the low-frequency components. Thus, ***the alternate viewpoint of design reduces the control-design problem to an appropriate design of a low-pass filter.***

In Chapter 9, the shaping of the response $y(t)$ of a control system was done by the root locus method, which is a graphical tool that manipulates the controller transfer function $D(s)$ to adjust the closed-loop poles of the system (refer to Fig. 10.1) giving desired shape to the output $y(t)$. ***In frequency domain, we can visualize the response y(t) in terms of its frequencies rather than an output which changes with time. We can shape y(t) by manipulating its frequencies.*** Frequency-selective characteristics of signals in control systems make this design approach simple and mostly better. By using feedback loop and a suitable controller within the loop, we reshape the frequency response of the closed-loop system; and this frequency response is translated back into time domain through inverse Fourier transformation, to see if the desired $y(t)$ has been achieved. Complex computations of inverse Fourier transformation (refer to Section 2.9), however, make this iterative design procedure difficult. There is a way out. Using the relationship (2.99) between Fourier transform and Laplace transform of stable transfer functions, we will convert the frequency response model of the closed-loop system to s-domain and apply inverse Laplace transformation to study the output $y(t)$. Another important factor in favor of design based on frequency-reponse models is the robustness against model uncertainties. As we have seen in Chapter 3, transfer function models obtained from application of physical laws are based on many approximating assumptions. The modeling errors introduce uncertainties, and affect the robustness of the performance of the resulting control systems. The frequency-response models for stable plants can be obtained directly from experimental measurements (discussed later in Section 10.8); therefore uncertainties related to modeling errors will be minimized. The design based on experimentally obtained frequency-response models is expected to be relatively more robust.

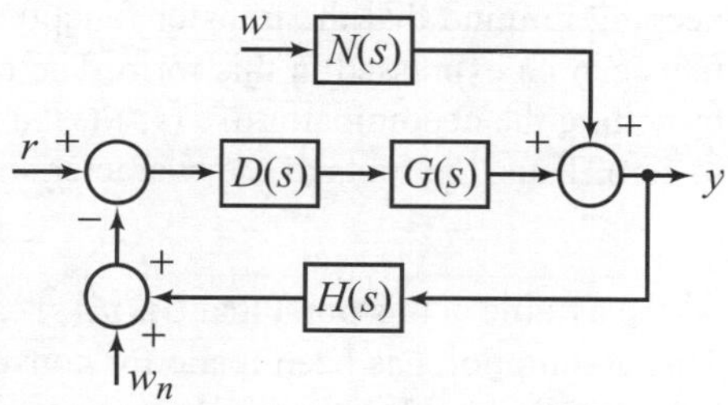

Fig. 10.1 *A feedback control system*

A plant for which the feedback system is to be designed may, however, be an unstable plant. A practical viewpoint in favor of ignoring this kind of limitation is that: ***how many unstable plants have we come across in process industry/monufacturing floors?*** Since most of the plants in industrial control systems are stable, the robustness aspect of frequency response based design continues to benefit the industry. Theoretically, however, we may argue that if a plant is unstable, create a stable feedback system for the plant, make the frequency response measurements on this feedback system, and there from extract the frequency response data for the plant using the relationship between the closed-loop transfer function $M(j\omega)$ and the plant transfer $G(j\omega)$.

The question: why frequency-domain design?, will get naturally answered in a better way as our maturity of understanding the control system characteristics in frequency domain increases through the present and the subsequent two chapters.

The present chapter is concerned with the stability analysis in frequency domain. A frequency-domain stability criterion was developed by H Nyquist in 1932 and remains a fundamental approach to the investigation of stability of linear control systems. In addition to answering the question of absolute stability, this criterion also gives some useful results on relative stability. The relative stability measures given by the Nyquist stability criterion are, in fact, central to the great importance of frequency-domain design methods.

We first develop the Nyquist stability criterion. Two common types of frequency-domain plots (Nyquist plots, Bode plots) are then introduced, and using these plots, absolute and relative stability are investigated on the basis of the Nyquist stability criterion. The use of the Nyquist criterion in frequency-domain design procedures will appear in the next two chapters.

10.2 DEVELOPMENT OF THE NYQUIST STABILITY CRITERION

In order to investigate the stability of a control system, we consider the closed-loop transfer function,

$$\frac{Y(s)}{R(s)} = \frac{G(s)}{1+G(s)H(s)} \tag{10.1}$$

keeping in mind that the transfer functions of both the single-loop and the multiple-loop control systems can be expressed in this form. The characteristic equation of the closed-loop system is obtained by setting the denominator of $Y(s)/R(s)$ to zero, which is same as setting the numerator of $1 + G(s)H(s)$ to zero. Thus, the roots of the characteristic equation must satisfy

$$1 + G(s)H(s) = 0 \tag{10.2}$$

We assume at this point that $G(s)H(s)$ can be expressed by a ratio of finite algebraic polynomials in s. This assumption has been made for convenience; the Nyquist criterion applies to more general situations, as shown subsequently. Let

$$G(s)H(s) = \frac{K(s+z_1')(s+z_2')\cdots(s+z_m')}{(s+p_1)(s+p_2)\cdots(s+p_n)}; m \le n \tag{10.3}$$

$G(s)H(s)$ is the product of plant, compensator, and sensor transfer functions; its pole and zero locations are assumed to be known since these transfer functions are generally available in factored form. Substituting for $G(s)H(s)$ from Eqn. (10.3) into Eqn. (10.2), we obtain

$$\begin{aligned} 1+G(s)H(s) &= 1+\frac{K(s+z_1')(s+z_2')\cdots(s+z_m')}{(s+p_1)(s+p_2)\cdots(s+p_n)} \\ &= \frac{(s+p_1)(s+p_2)\cdots(s+p_n)+K(s+z_1')(s+z_2')\cdots(s+z_m')}{(s+p_1)(s+p_2)\cdots(s+p_n)} \\ &= \frac{(s+z_1)(s+z_2)\cdots(s+z_n)}{(s+p_1)(s+p_2)\cdots(s+p_n)} \end{aligned} \tag{10.4}$$

It is apparent from Eqn. (10.4) that the poles of $1 + G(s)H(s)$ are identical to those of $G(s)H(s)$, i.e., the open-loop poles of the system; and the zeros of $1 + G(s)H(s)$ are identical to the roots of the characteristic equation, i.e., the closed-loop poles of the system. ***For the closed-loop system to be stable, the zeros of*** $1 + G(s)H(s)$ ***must lie in the left half of the s-plane.*** It is important to note that even if some of the open-loop poles lie in the right half s-plane, all the zeros of $1 + G(s)H(s)$, i.e., the closed-loop poles may lie in the left half s-plane, meaning thereby that an open-loop unstable system may lead to closed-loop stable operation.

Recall that we have introduced two methods of checking whether or not all zeros of $1 + G(s)H(s)$ have negative real parts. The zeros of $1 + G(s)H(s)$ are roots of the polynomial

$$(s+p_1)(s+p_2)\cdots(s+p_n)+K(s+z_1')(s+z_2')\cdots(z+z_m')$$

and we may apply the Routh test. Another method is to plot the root loci of

$$\frac{K(s+z_1')(s+z_2')\cdots(s+z_m')}{(s+p_1)(s+p_2)\cdots(s+p_n)} = -1$$

as a function of K. In this section, we shall introduce yet another method of checking whether or not all zeros of $1 + G(s)H(s)$ lie inside the open left half s-plane. The method, called the *Nyquist stability criterion*, is based on the *principle of argument* in the theory of complex variables [32]. The basic concept used in the Nyquist criterion is explained below.

Consider a rational function,

$$1+G(s)H(s) = Q(s) = \frac{s+z_1}{s+p_1}$$

The pole-zero map of $Q(s)$ is shown in Fig. 10.2a. This figure also shows a closed contour Γ_1. A point or an area is said to be *enclosed* by a closed path if it is found to lie to the right of the path when the path is traversed in the clockwise direction. The pole $-p_1$ and the zero $-z_1$ are therefore enclosed by the contour Γ_1 in Fig. 10.2a.

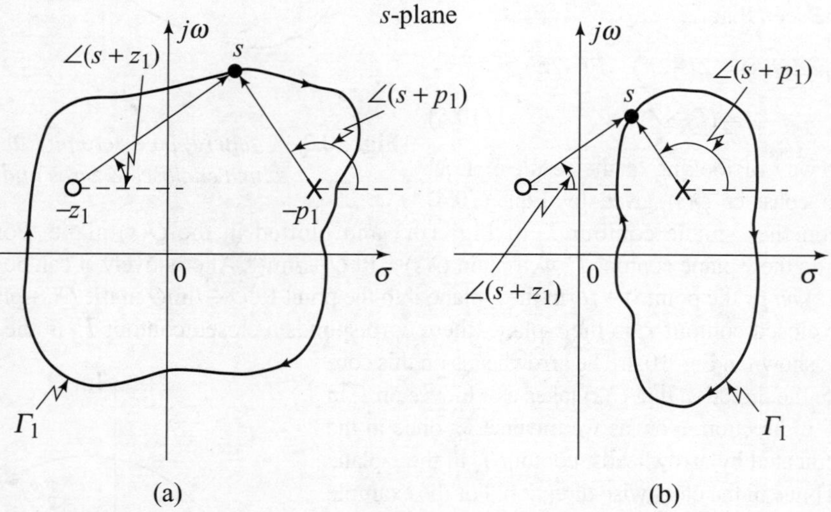

Fig. 10.2 *(a) A contour which encloses both the pole and the zero*
(b) A contour which encloses only the pole

Now we see that as the point s follows the prescribed path (i.e., clockwise direction) on the s-plane contour Γ_1, both $\angle(s+z_1)$ and $\angle(s+p_1)$ decrease continuously. For one clockwise traversal of Γ_1, we have

$$\delta_{\Gamma_1}\angle(s+z_1) = -2\pi$$
$$\delta_{\Gamma_1}\angle(s+p_1) = -2\pi$$

where $\delta_{\Gamma_1}\angle$ indicates the change in angle as Γ_1 is traversed.

$$\delta_{\Gamma_1}\angle Q(s) = \delta_{\Gamma_1}\angle(s+z_1) - \delta_{\Gamma_1}\angle(s+p_1)$$
$$= -2\pi - (-2\pi) = 0$$

Thus, the change in $\angle Q(s)$ as s traverses Γ_1 (which is a completely general contour in the s-plane except that it encloses both $-z_1$ and $-p_1$) once in clockwise direction, is zero. Using these arguments on the contour Γ_2 in the s-plane which encloses only $-p_1$ and not $-z_1$ (Fig. 10.2b), we find that (for the zero $s = -z_1$ not enclosed by the contour Γ_2, $\delta_{\Gamma_2}\angle(s+z_1) = 0$ for one clockwise traversal of Γ_2)

$$\delta_{\Gamma_2}\angle Q(s) = 0 - (-2\pi) = 2\pi$$

We thus see that a change in $\angle Q(s)$ that one gets in traversing a contour in the s-plane, is strictly a function of how many poles and zeros of $Q(s)$ are enclosed by the s-plane contour. Let us consider a general $Q(s) = 1 + G(s)H(s)$, given by Eqn. (10.4), for which a typical pole-zero plot might be made as shown in Fig. 10.3. The figure also shows an s-plane contour Γ_s which encloses Z zeros and P poles of $Q(s)$. Note that Γ_s does not go through any of the poles or zeros of $Q(s)$. If we make one clockwise traversal of the contour Γ_s, it may be seen that

$$\delta_{\Gamma_s}\angle Q(s) = Z(-2\pi) - P(-2\pi)$$

$$= (P - Z)\,2\pi \qquad (10.5)$$

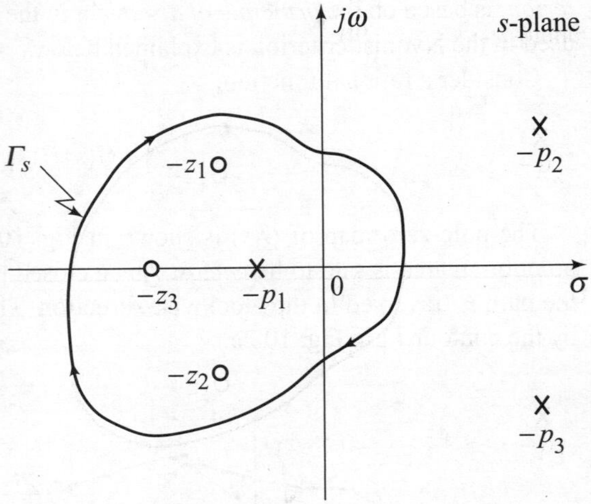

Fig. 10.3 *A general pole-zero plot with contour Γ_s which encloses Z zeros and P poles*

Another way of looking at the result of Eqn. (10.5) is to consider $Q(s)$ given by Eqn. (10.4) evaluated on the s-plane contour Γ_s of Fig. 10.3 and plotted in the $Q(s)$-plane. For every point $s = \sigma + j\omega$ on the s-plane contour Γ_s, we obtain $Q(s) = \text{Re}Q + j\text{Im}Q$. Alternatively, it can be stated that the function $Q(s)$ *maps* the point $\sigma + j\omega$ in the s-plane into the point $\text{Re}Q + j\text{Im}Q$ in the $Q(s)$-plane. It follows that for the closed contour Γ_s in the s-plane, there corresponds a closed contour Γ_Q in the $Q(s)$-plane. A typical Γ_Q is shown in Fig. 10.4. The arrowheads on this contour indicate the direction that $Q(s)$ takes as s moves on Γ_s in the clockwise direction. Now as we traverse Γ_Q once in the direction indicated by arrowheads, contour Γ_s in the s-plane is traversed once in the clockwise direction. For the example of Fig. 10.4, traversing Γ_Q once in the clockwise direction gives a change in $\angle Q(s)$ of -4π, since the origin is encircled twice in the negative direction. Thus, Eqn. (10.5) gives

$$\delta_{\Gamma_s}\angle Q(s) = 2\pi(P - Z) = -4\pi$$

or

$$Z - P = 2$$

In general,

$$Z - P = N \qquad (10.6)$$

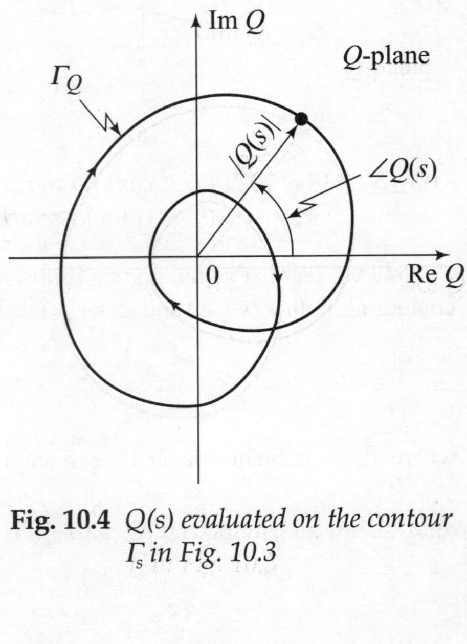

Fig. 10.4 *Q(s) evaluated on the contour Γ_s in Fig. 10.3*

where N is the number of clockwise encirclements of the origin that Γ_Q makes in the $Q(s)$-plane. It may be noted that we are not interested in the exact shape of the $Q(s)$-plane contour. The important fact that concerns us is the encirclements of the origin by the $Q(s)$-plane contour.

The relation (10.6) between the enclosure of poles and zeros of $Q(s)$ by the s-plane contour and the encirclements of the origin by the $Q(s)$-plane contour is commonly known as the *principle of argument* which may be stated as follows.

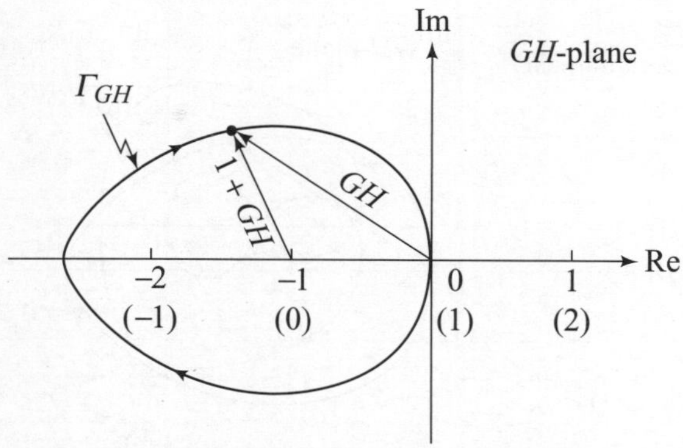

Fig. 10.5 *Evaluating* $1 + G(s)H(s)$ *from the* $G(s)H(s)$ *map*

Let $Q(s)$ be a ratio of polynomials in s. Let P be the number of poles and Z be the number of zeros of $Q(s)$ which are enclosed by a simple closed contour in the s-plane, multiplicity accounted for. Let the closed contour be such that it does not pass through any poles or zeros of $Q(s)$. The s-plane contour then maps into the $Q(s)$-plane contour as a closed curve.

The number N of clockwise encirclements of the origin of the $Q(s)$-plane, as a representative point s traces out the entire contour in the s-plane in the clockwise direction, is equal to $Z - P$.

Since
$$Q(s) = 1 + G(s)H(s),$$

we can obtain $Q(s)$-plane contour from the $G(s)H(s)$-plane contour simply by adding +1 to each point of $G(s)H(s)$-plane contour. Figure 10.5 shows a typical $G(s)H(s)$-plane contour Γ_{GH}. The effect of adding +1 to each point of Γ_{GH} to obtain $Q(s)$-plane contour Γ_Q is accomplished simply by adding +1 to the scale of the real axis, as shown by the numbers in parentheses in Fig. 10.5. It is seen that $-1 + j0$ point of the $G(s)H(s)$ map Γ_{GH} corresponds to the origin of the $Q(s)$ map Γ_Q. For convenience, we designate the $-1 + j0$ point of the $G(s)H(s)$-plane as the *critical point*. Thus, the encirclements of the origin by the contour Γ_Q is equivalent to the encirclements of the critical point $-1 + j0$ by the contour Γ_{GH}. In the light of these observations, we can express Eqn. (10.6) as follows:

$$Z - P = N \tag{10.7}$$

where

Z = number of zeros of $1 + G(s)H(s)$ enclosed by the s-plane contour Γ_s;

P = number of poles of $G(s)H(s)$ enclosed by the s-plane contour Γ_s; and

N = number of clockwise encirclements of the critical point $-1 + j0$ made by the $G(s)H(s)$-plane contour Γ_{GH}.

In general N can be positive ($Z > P$), zero ($Z = P$), or negative ($Z < P$). $N > 0$ corresponds to N *net* encirclements of the critical point $-1 + j0$ in clockwise direction by the Γ_{GH} contour. $N = 0$ indicates zero *net* encirclements of the critical point by the Γ_{GH} contour. $N < 0$ corresponds to N *net* encirclements of the critical point in counterclockwise direction by the Γ_{GH} contour.

A convenient way of determining N ***with respect to the critical point*** $-1 + j0$ ***of the*** $G(s)H(s)$***-plane is to draw a radial line from this point. The number of net intersections of the radial line with the*** Γ_{GH} ***contour gives the magnitude of*** N. Figure 10.6 gives several examples of the method of determining N.

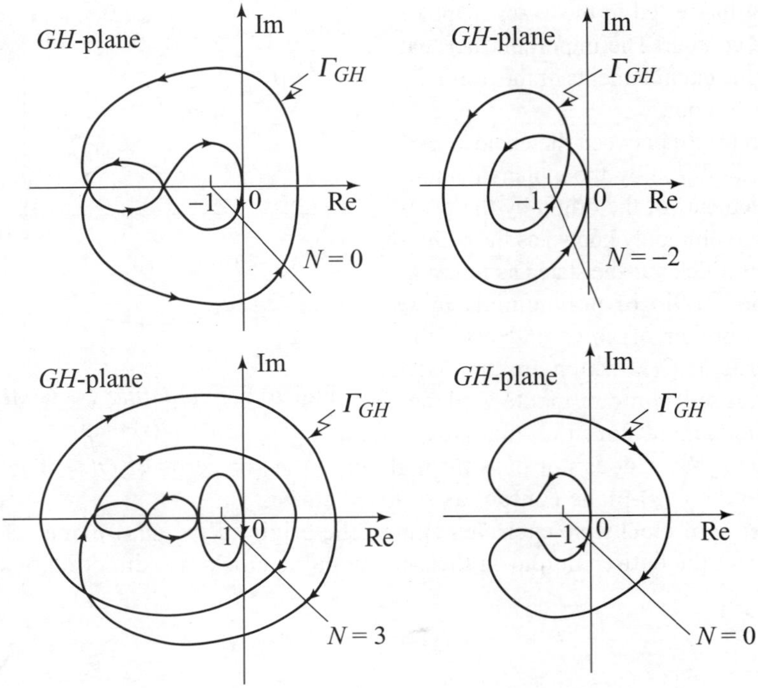

Fig. 10.6 *Examples of determination of N*

10.2.1 The Nyquist Contour

At this point the reader may place himself/herself in the position of Nyquist many years ago, confronted with the problem of stability of the closed-loop system that has the transfer function of Eqn. (10.1), which is equivalent to determining whether or not the function $1 + G(s)H(s)$ has zeros in the right half s-plane. Apparently, Nyquist discovered that the principle of argument of the complex-variable theory could be applied to solve the stability problem if the s-plane contour Γ_s is taken to be one that encloses the entire right half of the s-plane. Of course, as an alternative, Γ_s can be chosen to enclose the entire left half of the s-plane, as the solution is a relative one.

Figure 10.7 illustrates a Γ_s contour that encloses the entire right half of the s-plane. Such a contour is called the *Nyquist contour*. It is directed clockwise and comprises of an infinite line segment C_1 along the $j\omega$-axis and arc C_2 of infinite radius.

Along C_1, $s = j\omega$ with s varying from $-j\infty$ to $+j\infty$.

Along C_2, $s = Re^{j\theta}$ with $R \to \infty$ and θ varying from $+90°$ through $0°$ to $-90°$.

As the s-plane contour Γ_s must avoid all poles of $1 + G(s)H(s)$, modifications in the Nyquist contour defined in Fig. 10.7 are required when $G(s)H(s)$, and therefore $1 + G(s)H(s)$, has one or more poles on the imaginary axis (the zeros of $1 + G(s)H(s)$ are unknown to us; we will shortly see the implications of s-plane contour passing through the zeros of $1 + G(s)\,H(s)$). The basic trick, of course, is to take a small detour around the imaginary axis poles. Figure 10.8 illustrates a modified Nyquist contour when

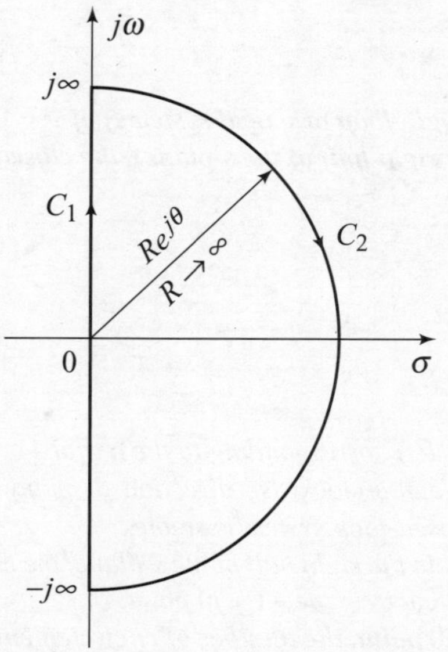

Fig. 10.7 *The Nyquist contour*

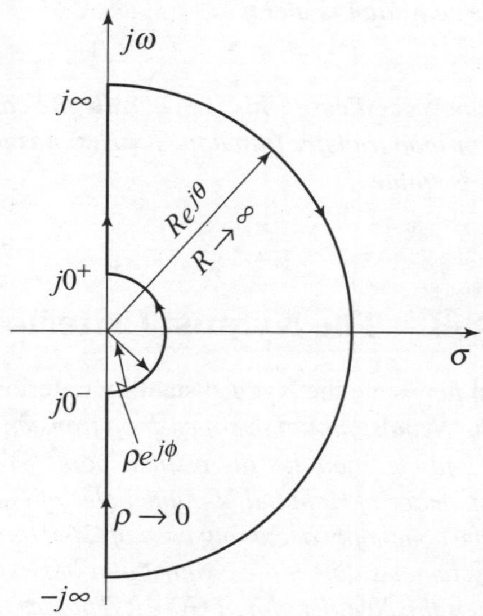

Fig. 10.8 *Indented Nyquist contour*

$G(s)H(s)$ has a pole at $s = 0$. Along the semicircular indent around the pole at the origin, $s = \rho e^{j\phi}$ with $\rho \to 0$ and ϕ varying from –90° through 0° to +90°. The indented Nyquist contour in Fig. 10.8 does not enclose the pole at the origin. Of course, as an alternative, the Nyquist contour may be indented to enclose the pole at the origin.

10.2.2 The Nyquist Plot

The Nyquist stability criterion is a direct application of the principle of argument when the s-plane contour Γ_s is the Nyquist contour. In principle, once the Nyquist contour is specified, the stability of a closed-loop system can be determined by plotting the $G(s)H(s)$ locus when s takes on values along the Nyquist contour, and investigating the behavior of the $G(s)H(s)$ plot with respect to the critical point $-1 + j0$. The $G(s)H(s)$ plot that corresponds to the Nyquist contour is called the *Nyquist plot* of $G(s)H(s)$.

With the additional dimension added to the stability problem, we define N, P, and Z as follows:

N = number of clockwise encirclements of the critical point $-1 + j0$ made by the $G(s)H(s)$ locus of the Nyquist plot.

P = number of poles of $G(s)H(s)$, and therefore of $1 + G(s)H(s)$, enclosed by the Nyquist contour. Then (refer to Eqn. (10.7))

Z = number of zeros of $1 + G(s)H(s)$ enclosed by the Nyquist contour

$$= N + P \tag{10.8a}$$

Closed-loop stability requires that

$$Z = 0 \tag{10.8b}$$

This condition is met if

$$N = -P \tag{10.8c}$$

In the special case (this is generally the case in most single-loop practical systems) of $P = 0$ ***(i.e., the open-loop transfer function*** $G(s)H(s)$ ***has no poles in the right half of the*** s***-plane), the closed-loop system is stable if***

$$N = 0 \tag{10.8d}$$

10.2.3 The Nyquist Criterion

We can now state the Nyquist stability criterion as follows:

If the Nyquist plot of the open-loop transfer function G(s)H(s) corresponding to the Nyquist contour in the s-plane encircles the critical point $-1 + j0$ *in the counterclockwise direction as many times as the number of right half s-plane poles of G(s)H(s), the closed-loop system is stable.*

In the commonly occurring case of G(s)H(s) with no poles in the right half of the s-plane, the closed-loop system is stable if the Nyquist plot of G(s)H(s) does not encircle the $-1 + j0$ *point.*

When the Nyquist plot of $G(s)H(s)$ ***passes through*** $-1 + j0$ ***point, the number of encirclements*** N ***is indeterminate. This corresponds to the condition where*** $1 + G(s)H(s)$ ***has zeros on the imaginary axis. A necessary condition for applying the Nyquist criterion is that the Nyquist contour must not pass through any poles or zeros of*** $1 + G(s)H(s)$***. When this condition is violated, the value for*** N ***becomes indeterminate and the Nyquist stability criterion cannot be applied.***

10.3 SELECTED ILLUSTRATIVE NYQUIST PLOTS

The following examples serve to illustrate the application of the Nyquist criterion to the stability study of control systems.

Example 10.1 Consider a single-loop feedback control system with the open-loop transfer function given by

$$G(s)H(s) = \frac{K}{(\tau_1 s+1)+(\tau_2 s+1)}; \tau_1, \tau_2 > 0, K > 0 \tag{10.9}$$

$G(s)H(s)$ has no poles in the right-half s-plane; therefore, stability is assured if the Nyquist plot of $G(s)H(s)$ does not encircle the $-1 + j0$ point.

The Nyquist plot of $G(s)H(s)$, as we know, is the mapping of the Nyquist contour in the s-plane onto $G(s)H(s)$-plane. In Fig. 10.9a, the Nyquist contour has been divided into three sections: C_1, C_2, and C_3. Section C_1 is defined by $s = j\omega$, $0 \le \omega < \infty$; section C_2 is defined by $s = j\omega$, $-\infty < \omega \le 0$; and section C_3 is defined by $s = Re^{j\theta}$, $R \to \infty$, and θ varies from $+90°$ through $0°$ to $-90°$.

Mapping of Section C_1 *onto G(s)H(s)-plane*

Substituting $s = j\omega$ into $G(s)H(s)$, we obtain

$$G(j\omega)H(j\omega) = \frac{K}{(j\omega\tau_1+1)(j\omega\tau_2+1)} \tag{10.10}$$

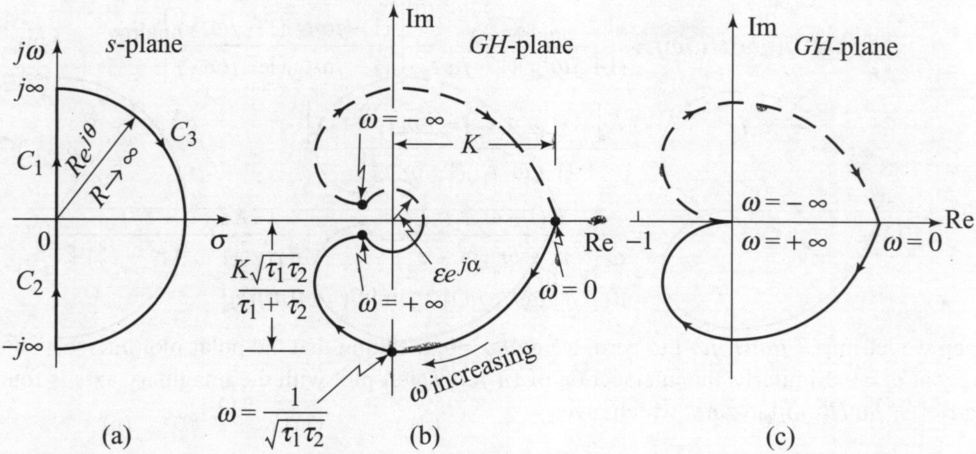

Fig. 10.9 *(a) Nyquist contour; (b) Nyquist plot of G(s)H(s) of Eqn. (10.9); (c) Simplified Nyquist plot*

A plot of $G(j\omega)H(j\omega)$ on polar coordinates as ω is varied from 0 to ∞ is the map of section C_1 on the $G(s)H(s)$-plane. This plot is called the *polar plot* of sinusoidal transfer function[1] $G(j\omega)H(j\omega)$.

For application of the Nyquist stability criterion, an exact polar plot of $G(j\omega)H(j\omega)$ is not essential. Often a rough sketch is adequate for stability analysis. The general shape of the polar plot of $G(j\omega)H(j\omega)$ may be determined from the following information:

1. The behavior of the magnitude and phase of $G(j\omega)H(j\omega)$ at $\omega = 0$ and $\omega = \infty$.
2. The points of intersection of the polar plot with the real and imaginary axes, and the values of ω at these intersections.
3. A few on-locus points in second and third quadrants in the vicinity of the critical point $-1 + j0$.

For $G(j\omega)H(j\omega)$ given by Eqn. (10.10), the magnitude and phase at $\omega = 0$ and $\omega = \infty$ are calculated as follows:

$$|G(j\omega)H(j\omega)|_{\omega=0} = K; \angle G(j\omega)H(j\omega)\Big|_{\omega=0} = 0°$$

$$|G(j\omega)H(j\omega)|_{\omega\to\infty} = \left|\frac{K}{j^2\omega^2\tau_1\tau_2}\right|_{\omega\to\infty} = 0$$

$$\angle G(j\omega)H(j\omega)\Big|_{\omega\to\infty} = \angle\left(\frac{K}{j^2\omega^2\tau_1\tau_2}\right) = -180°$$

The intersections of the polar plot with the axes of $G(s)H(s)$-plane can easily be ascertained by identifying the real and imaginary parts of $G(j\omega)H(j\omega)$.

[1] This plot also gives the *open-loop frequency response* of the system if all the poles of $G(s)H(s)$ lie in the left-half s-plane (refer to Eqns. (2.94–2.99)). The frequency response of stable open-loop systems obtained experimentally (refer to Section 10.8) can also be used in the stability analysis using Nyquist criterion. For $G(s)H(s)$ that is not stable, the sinusoidal transfer function $G(j\omega)H(j\omega)$ represents a map of a section of the Nyquist contour on $G(s)H(s)$-plane; and using this map, the stability of the resulting closed-loop system can be determined using Nyquist criterion.

$$G(j\omega)H(j\omega)=\frac{K}{(1+j\omega\tau_1)(1+j\omega\tau_2)}\left[\frac{(1-j\omega\tau_1)(1-j\omega\tau_2)}{(1-j\omega\tau_1)(1-j\omega\tau_2)}\right]$$

$$=\frac{K\left[(1-\omega^2\tau_1\tau_2)-j\omega(\tau_1+\tau_2)\right]}{(1+\omega^2\tau_1^2)(1+\omega^2\tau_2^2)}$$

$$=\frac{K(1-\omega^2\tau_1\tau_2)}{\omega^4\tau_1^2\tau_2^2+\omega^2(\tau_1^2+\tau_2^2)+1}-j\frac{K\omega(\tau_1+\tau_2)}{\omega^4\tau_1^2\tau_2^2+\omega^2(\tau_1^2+\tau_2^2)+1}$$

$$=\text{Re}[G(j\omega)H(j\omega)]+j\text{Im}[G(j\omega)H(j\omega)]$$

When we let Im[$G(j\omega)H(j\omega)$] to zero, we get $\omega = 0$, meaning that the polar plot intersects the real axis only at $\omega = 0$. Similarly, the intersection of $G(j\omega)H(j\omega)$-plot with the imaginary axis is found by setting Re[$G(j\omega)H(j\omega)$] to zero, which gives

$$\omega=\frac{1}{\sqrt{\tau_1\tau_2}}$$

$$\left|G(j\omega)H(j\omega)\right|_{\omega=\frac{1}{\sqrt{\tau_1\tau_2}}}=\left.\frac{K\omega(\tau_1+\tau_2)}{\omega^4\tau_1^2\tau_2^2+\omega^2(\tau_1^2+\tau_2^2)+1}\right|_{\omega=\frac{1}{\sqrt{\tau_1\tau_2}}}$$

$$=\frac{K\sqrt{\tau_1\tau_2}}{\tau_1+\tau_2}$$

Based on this information, a rough sketch of the polar plot can easily be made as shown in Fig. 10.9b (the portion of the locus from $\omega = 0$ to $\omega = +\infty$).

Mapping of Section C_2 onto G(s)H(s)-plane

Given $G(j\omega)H(j\omega)$, $0 \le \omega < \infty$, $G(j\omega)H(j\omega)$ for $-\infty < \omega \le 0$ is constructed by realizing

$$G(-j\omega)H(-j\omega)=[G(j\omega)H(j\omega)]^*$$

where * denotes conjugate. Thus, given the $G(j\omega)H(j\omega)$-locus for $0 \le \omega < \infty$, we get the $G(j\omega)H(j\omega)$-locus for the negative $j\omega$-axis by simply taking the mirror image about the real axis of the locus for positive ω. The dashed portion of the locus in Fig. 10.9b from $\omega = -\infty$ to $\omega = 0$ is thus the map of section C_2 of Nyquist contour.

Mapping of Section C_3 onto G(s)H(s)-plane

Along section C_3, $s = Re^{j\theta}$; $R \to \infty$ and θ varies from +90° through 0° to −90°. Substituting into $G(s)H(s)$, we obtain

$$G(s)H(s)\Big|_{s=Re^{j\theta}}=\frac{K}{(\tau_1Re^{j\theta}+1)(\tau_2Re^{j\theta}+1)}$$

$$=\frac{K}{\tau_1\tau_2R^2e^{j2\theta}}=\frac{K}{\tau_1\tau_2R^2}e^{-j2\theta}$$

The infinitesimal semicircular locus around the origin (refer to Fig. 10.9b), represented by

$$G(s)H(s)=\varepsilon\, e^{j\alpha}$$

with $\varepsilon \to 0$ and α varying from $-180°$ through $0°$ to $+180°$, is thus the map of section C_3 of Nyquist contour.

For application of the Nyquist stability criterion, exact shape of the Nyquist plot around the origin of $G(s)H(s)$-plane is of no interest to us. We can therefore take $\varepsilon = 0$; the resulting Nyquist plot is shown in Fig. 10.9c. This plot does not encircle the critical point $-1 + j0$ for any positive values of K, τ_1, and τ_2. Therefore, the system is stable for all positive values of K, τ_1, and τ_2.

Example 10.2 Consider now a feedback system whose open-loop transfer function is given by

$$G(s)H(s) = \frac{K}{s(\tau s+1)};\ K > 0, \tau > 0 \tag{10.11}$$

We note here that for this type-1 system, there is one pole on the imaginary axis, precisely at the origin. The Nyquist contour is shown in Fig. 10.10a, where a semicircular detour about the origin is indicated. The Nyquist contour has been divided into four sections: C_1, C_2, C_3 and C_4. Section C_1 is defined by $s = j\omega$, $0 < \omega < \infty$; section C_2 is defined by $s = j\omega$, $-\infty < \omega < 0$; section C_3 is defined by $s = Re^{j\theta}$, $R \to \infty$, and θ varies from $+90°$ through $0°$ to $-90°$; and section C_4 is defined by $s = \rho e^{j\phi}$, $\rho \to 0$, and ϕ varies from $-90°$ through $0°$ to $+90°$.

Mapping of section C_1 onto the $G(s)H(s)$-plane is given by the polar plot of sinusoidal transfer function,

$$G(j\omega)H(j\omega) = \frac{K}{j\omega(j\omega\tau+1)} \tag{10.12}$$

This sinusoidal transfer function can be expressed as

$$G(j\omega)H(j\omega) = -\frac{K\tau}{1+\omega^2\tau^2} - j\frac{K}{\omega(1+\omega^2\tau_2)}$$

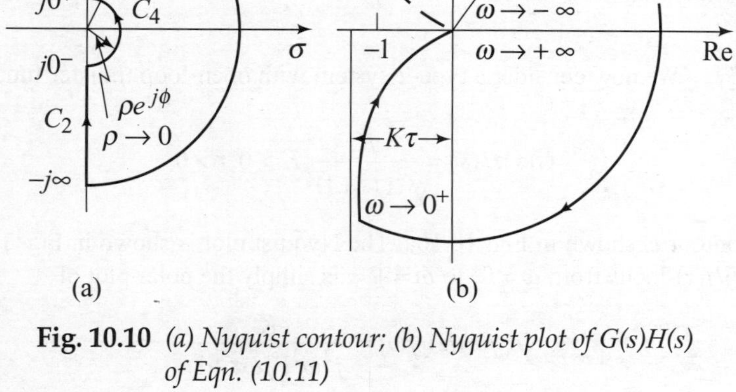

Fig. 10.10 *(a) Nyquist contour; (b) Nyquist plot of G(s)H(s) of Eqn. (10.11)*

The low-frequency portion of the polar plot becomes

$$\lim_{\omega \to 0} G(j\omega)H(j\omega) = K\tau - j\infty$$

and the high-frequency portion is (refer to Eqn. (10.12))

$$\lim_{\omega \to \infty} G(j\omega)H(j\omega) = \frac{K}{\omega^2 \tau} \angle \left(\frac{K}{j^2 \omega^2 \tau} \right) \Bigg|_{\omega \to \infty} = 0\angle -180°$$

The general shape of the polar plot is shown in Fig. 10.10b. The $G(j\omega)H(j\omega)$-plot is asymptotic to the vertical line passing through the point $-K\tau + j0$.

Mapping of section C_2 onto $G(s)H(s)$-plane is obtained simply by taking the mirror image about the real axis of the polar plot of $G(j\omega)H(j\omega)$.

Mapping of section C_3 onto the $G(s)H(s)$-plane is obtained as follows:

$$G(s)H(s)\Big|_{s=Re^{j\theta}} = \frac{K}{Re^{j\theta}(\tau Re^{j\theta}+1)} = \frac{K}{\tau R^2} e^{-j2\theta}$$

The $G(s)H(s)$-locus thus turns at the origin with zero radius from –180° through 0° to +180°.

Mapping of section C_4 onto the $G(s)H(s)$-plane is obtained as follows. Along C_4, $s = \rho e^{j\phi}$; $\rho \to 0$, and ϕ varies from –90° through 0° to +90°.

Substituting into $G(s)H(s)$, we obtain

$$G(s)H(s)\Big|_{s=\rho e^{j\phi}} = \frac{K}{\rho e^{j\phi}(\tau\rho e^{j\phi}+1)} = \frac{K}{\rho} e^{-j\phi}$$

The value K/ρ approaches infinity as $\rho \to 0$, and $-\phi$ varies from +90° through 0° to –90° as s moves along section C_4 of the Nyquist contour. Thus, the infinitesimal semicircular indent around the origin in the s-plane maps into a semicircular arc of infinite radius on the $G(s)H(s)$-plane as shown in Fig. 10.10b.

The complete Nyquist plot for $G(s)H(s)$ given by Eqn. (10.11) is shown in Fig. 10.10b. In order to investigate the stability of this second-order system, we first note that the number of poles of $G(s)H(s)$ in the right-half s-plane is zero. Therefore, for this system to be stable, we require that the Nyquist plot of $G(s)H(s)$ does not encircle the critical point $-1+j0$. Examining Fig. 10.10b, we find that irrespective of the value of the gain K and the time-constant τ, the Nyquist plot does not encircle the critical point, and the system is always stable.

Example 10.3 We now consider a type-2 system with open-loop transfer function,

$$G(s)H(s) = \frac{K}{s^2(\tau s+1)}; K > 0, \tau > 0 \tag{10.13}$$

The Nyquist contour is shown in Fig. 10.10a. The Nyquist plot is shown in Fig. 10.11. Clearly, the portion of the $G(s)H(s)$-locus from $\omega = 0^+$ to $\omega = +\infty$ is simply the polar plot of

$$G(j\omega)H(j\omega) = \frac{K}{j^2\omega^2(j\omega\tau+1)} \tag{10.14}$$

Note that

$$G(j\omega)H(j\omega)\Big|_{\omega\to 0^+} = \left(\frac{K}{\omega^2}\Big|_{\omega\to 0^+}\right)\angle -180^\circ = \infty\angle \pm 180^\circ$$

$$G(j\omega)H(j\omega)\Big|_{\omega\to \infty} = \left(\frac{K}{\omega^3\tau}\Big|_{\omega\to +\infty}\right)\angle -270^\circ = 0\angle -270^\circ \text{ or } +90^\circ$$

It can easily be examined that the locus of $G(j\omega)H(j\omega)$ does not intersect the real or imaginary axis; it lies in the second quadrant for all values of ω.

The $G(s)H(s)$-locus turns at the origin with zero radius from –270° through 0° to +270°. This corresponds to the locus from $\omega = +\infty$ to $\omega = -\infty$. The portion of the $G(s)H(s)$-locus from $\omega = -\infty$ to $\omega = 0^-$ is the mirror image about the real axis of the locus for positive ω. From $\omega = 0^-$ to $\omega = 0^+$, $s = \rho e^{j\phi}$, $\rho \to 0$ and ϕ varies from –90° through 0°, to +90°. This gives

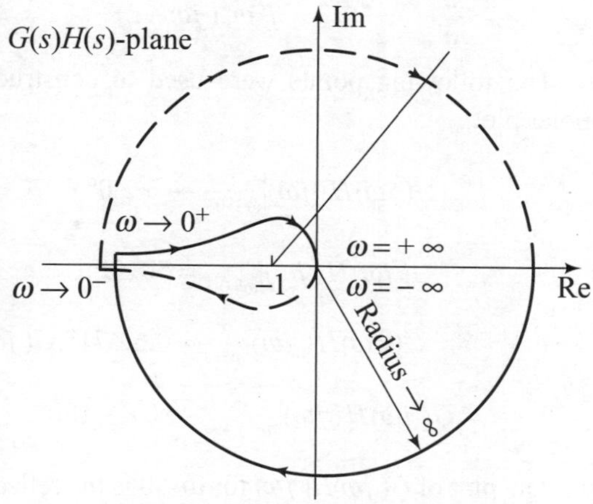

Fig. 10.11 *Nyquist plot[2] for G(s)H(s) of Eqn. (10.13)*

$$G(s)H(s)\Big|_{s=\rho e^{j\phi}} = \frac{K}{\rho^2 e^{j2\phi}(\tau\rho e^{j\phi}+1)} = \frac{K}{\rho^2}e^{-j2\phi}$$

The portion of the $G(s)H(s)$ locus from $\omega = 0^-$ to $\omega = 0^+$ is a circular arc of infinite radius ranging from +180° at $\omega = 0^-$ through 0° to –180° at $\omega = 0^+$ (+180° → + 90° → 0° → – 90° → –180°).

From Fig. 10.11, we observe that the Nyquist plot encircles the critical point $-1 + j0$ twice in the clockwise direction, i.e., $N = 2$. Since P = number of poles of $G(s)H(s)$ in right-half s-plane = 0, we have from Eqn. (10.8a), $Z = 2$; i.e, there are two roots of the closed-loop system in the right-half s-plane and the system, irrespective of the gain K and time constant τ, is unstable.

Example 10.4 Consider now an open-loop system with the transfer function,

$$G(s)H(s) = \frac{s+1}{s^2(s-2)} \tag{10.15}$$

Let us determine whether the system is stable when the feedback path is closed.

[2] For $K = 1$, $\tau = 1$;

ω	0.01	0.05	0.1	0.5	1
$\lvert GH\rvert$	4×10^6	600	98.51	3.57	0.7
$\angle GH$	179.97	177.68	174.2	153.4	135

From the transfer function of the open-loop system it is observed that there is one open-loop pole in the right half of the s-plane. Therefore, $P = 1$.

The Nyquist contour is shown in Fig. 10.10a. The Nyquist plot is shown in Fig. 10.12. Clearly, the portion of the $G(s)H(s)$-locus from $\omega = 0^+$ to $\omega = +\infty$ is simply the polar plot of

$$G(j\omega)H(j\omega) = \frac{j\omega + 1}{j^2\omega^2(j\omega - 2)} \qquad (10.16)$$

The following points were used to construct the polar plot:

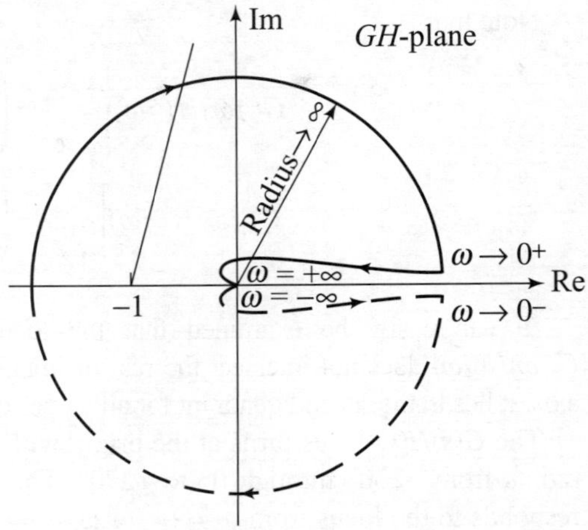

Fig. 10.12 *Nyquist plot for G(s)H(s) of Eqn. (10.15)*

$$G(j\omega)H(j\omega)\big|_{\omega \to 0^+} \to \infty\angle 0^\circ$$

$$G(j\omega)H(j\omega)\big|_{\omega = 0.1} = 50\angle 9^\circ$$

$$G(j\omega)H(j\omega)\big|_{\omega = 1} = 0.6\angle 71^\circ; G(j\omega)H(j\omega)\big|_{\omega = 10} = 0.01\angle 162^\circ$$

$$G(j\omega)H(j\omega)\big|_{\omega \to +\infty} \to 0\angle -180^\circ$$

The plot of $G(j\omega)H(j\omega)$ for $\omega < 0$ is the reflection, with respect to the real axis, of the plot for $\omega > 0$. Every point on the large semicircle in Fig. 10.10a, is mapped by $G(s)H(s)$ into the origin of the $G(s)H(s)$-plane. A point $s = \rho e^{j\phi}$; $\rho \to 0$, ϕ varying from -90° through 0° to $+90^\circ$ is mapped by $G(s)H(s)$ into

$$G(s)H(s)\Big|_{s=\rho e^{j\phi}} = \frac{s+1}{s^2(s-2)}\Big|_{s=\rho e^{j\phi}} = -\frac{1}{2\rho^2 e^{j2\phi}} = \frac{1}{2\rho^2}\angle(180^\circ - 2\phi)$$

The small semicircle in Fig. 10.10a is mapped by $G(s)\,H(s)$ into a circular arc of infinite radius ranging from 360° at $\omega = 0^-$ through 180° to 0° at $\omega = 0^+$ ($360^\circ \to 270^\circ \to 180^\circ \to 90^\circ \to 0^\circ$).

Figure 10.12 indicates that the critical point $-1 + j0$ is encircled by the Nyquist plot once in the clockwise direction. Therefore, $N = 1$, and (refer to Eqn. (10.8a))

Z = number of zeros of $1 + G(s)H(s)$ enclosed by the Nyquist contour
$= N + P = 2$

Hence, the feedback system is unstable with two poles in the right half of the s-plane.

Example 10.5 Consider a feedback system with the following open-loop transfer function,

$$G(s)H(s) = \frac{K}{s(s+3)(s+5)} \qquad (10.17)$$

Let us investigate the stability of this system for various values of K.

First set $K = 1$ and sketch the Nyquist plot for the system using the contour shown in Fig. 10.10a. For all points on the imaginary axis,

$$G(j\omega)H(j\omega) = \left.\frac{K}{s(s+3)(s+5)}\right|_{\substack{K=1 \\ s=j\omega}} = \frac{-8\omega^2 - j(15\omega - \omega^3)}{64\omega^4 + \omega^2(15-\omega^2)^2} \tag{10.18}$$

At $\omega = 0$, $G(j\omega)H(j\omega) = -0.0356 - j\infty$.

Next, find the point where the Nyquist plot intersects the negative real axis. Setting the imaginary part of Eqn. (10.18) equal to zero, we find $\omega = \sqrt{15}$. Substituting this value of ω back into Eqn. (10.18), yields the real part of -0.0083.

Finally, at $\omega = \infty, G(j\omega)H(j\omega) = G(s)H(s)|_{s\to j\infty} = \frac{1}{(j\omega)^3} = 0\angle{-270°}$.

The Nyquist plot for $G(s)H(s)$ of Eqn. (10.17) with $K = 1$ is shown in Fig. 10.13. Application of the Nyquist criterion shows that with $K = 1$, the closed-loop system is stable.

If we were to increase the gain by a factor $1/0.0083 = 120.48$, all points on the $G(j\omega)H(j\omega)$-locus would increase in magnitude by this factor along radial lines centered at the origin. The length of the vector $G(j\sqrt{15})H(j\sqrt{15})$ would be $0.0083 (1/0.0083) = 1$, and, therefore, the curve $G(j\omega)H(j\omega)$ would go through the critical point; the closed-loop system would be at the limit of instability. Hence, for stability, K must be less than 120.48, i.e., the stability range is $0 < K < 120.48$.

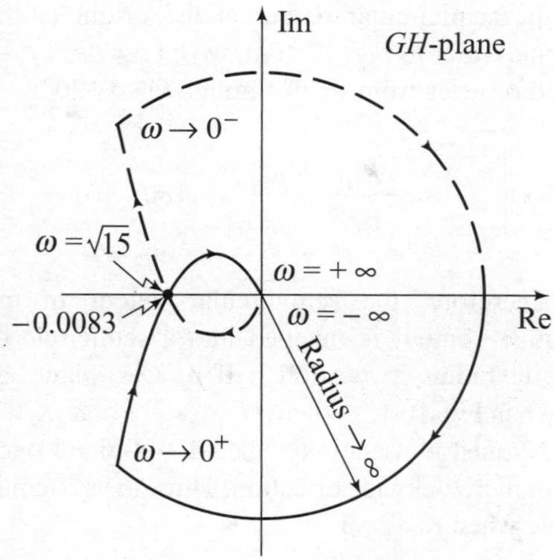

Fig. 10.13 *Nyquist plot for G(s)H(s) of Eqn. (10.17)*

Example 10.6 Let us consider the control system shown in Fig. 10.14 and determine its stability. The characteristic equation of the system is

$$s(s-1) + K_A(1 + K_t s) = 0$$

We rearrange it in the form convenient for Nyquist plotting:

$$1 + \frac{K_A(1+K_t s)}{s(s-1)} = 0$$

or

$$1 + G(s) = 0$$

$$G(s) = \frac{K_A(1+K_t s)}{s(s-1)} \tag{10.19}$$

The open-loop transfer function has one pole in the right half of the s-plane, and, therefore, $P = 1$. In order for the closed-loop system to be stable, we require $N = -1$ (i.e., the Nyquist plot must encircle the critical point once in the counterclockwise direction).

Several important values of $G(j\omega)$-locus are given below:

$$G(j\omega)\big|_{\omega\to 0^+} = \infty\angle+90°;\ G(j\omega)\big|_{\omega\to+\infty} = 0\angle-90°$$

$$G(j\omega) = \frac{K_A(1+K_t j\omega)}{-\omega^2 - j\omega}$$

$$= \frac{-K_A(\omega^2+\omega^2 K_t) + j(\omega - K_t\omega^3)K_A}{\omega^2+\omega^4} \qquad (10.20)$$

$$G(j\omega)\big|_{\omega^2=1/K_t} = -K_A K_t$$

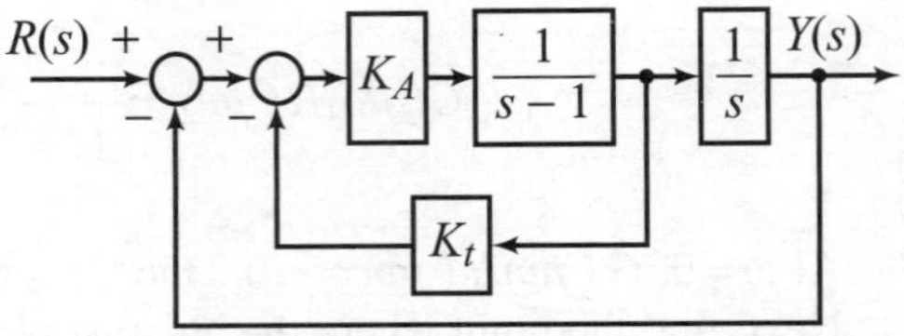

Fig. 10.14 *A feedback control system*

At the semicircular detour at the origin in the s-plane (refer to Fig. 10.10a), we let $s = \rho e^{j\phi}$; $\rho \to 0$ and ϕ varies from $-90°$ through $0°$ to $+90°$.

$$G(s)\Big|_{s=\rho e^{j\phi}} = \frac{K_A}{-\rho e^{j\phi}} = \frac{K_A}{\rho}\angle(-180° - \phi)$$

Therefore, the semicircular detour in the Nyquist contour is mapped into a semicircle of infinite radius in the left half of $G(s)$-plane, as shown in Fig. 10.15. When $-K_AK_t < -1$ or $K_AK_t > 1$, the Nyquist plot encircles the $-1 + j0$ point once in counterclockwice direction. Thus, the system is stable when $K_AK_t > 1$.

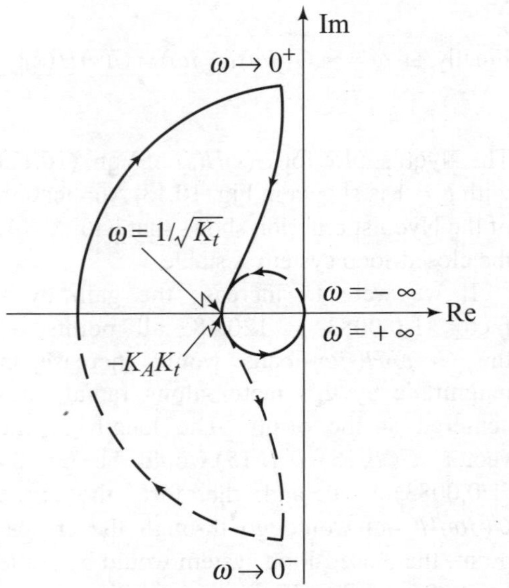

Fig. 10.15 *Nyquist plot[3] for G(s) of Eqn. (10.19)*

10.4 STABILITY MARGINS

In the preceding sections, the Nyquist plot was used to detemine if a system was stable or unstable. Of course, in general, a usable system must be stable. However, there are concerns beyond simple stability, for two reasons. First, a stable system must also have, among other characteristics, an acceptable transient response. Also, the model that is used in the analysis and design of a control system is never exact. Hence, the model may indicate that a system is stable, whereas in fact the physical system is unstable. Generally, we require not only that a system be stable but also that it be stable by some margin of safety.

Measures of degree of stability of a closed-loop system that has open-loop transfer function with no poles in right-half s-plane, can be conveniently created through Nyquist plot. The stability information for such a system becomes obvious by the inspection of the Nyquist plot of its open-loop transfer

[3] For $K_A = 1$, $K_t = 3$;

ω	0.02	0.05	0.09	1
$\|G\|$	146.3	74	37.16	2.23
$\angle G$	91.57	93.1	96.14	153.56

function $G(s)H(s)$, since the stability criterion is merely the non-encirclement of the critical point $-1+j0$. It can be intuitively imagined that as the Nyquist plot gets closer to the critical point, the system tends towards instability.

Consider two different systems whose dominant closed-loop poles are shown on the s-plane in Figs 10.16a and 10.16b. Obviously, system A is 'more stable' than system B, since its dominant closed-loop poles are located comparatively away to the left from the $j\omega$-axis. The Nyquist plots of open-loop transfer functions of systems A and B are shown in Figs 10.16c and 10.16d, respectively. Comparison of the closed-loop pole locations of these systems with their corresponding Nyquist plots reveals that as a Nyquist plot moves closer to the critical point $-1+j0$, the system closed-loop poles move closer to the $j\omega$-axis and hence the system becomes relatively less stable and *vice versa*.

Thus, the distance between the Nyquist plot of open-loop transfer function $G(s)H(s)$ and the critical point $-1+j0$ can be used as a measure of degree of stability of the closed-loop system. Generally, the larger the distance, the more stable the system (the larger the distance, the more robust the system–refer to subsection 11.3.3 (Fig. 11.6)). The distance can be found by drawing a circle touching the Nyquist plot as shown in Figs 10.16c and 10.16d. Such a distance, however, is not convenient for analysis and design. The measures gain margin and phase margin, are commonly used to quantify the closeness of a Nyquist plot to the critical point.

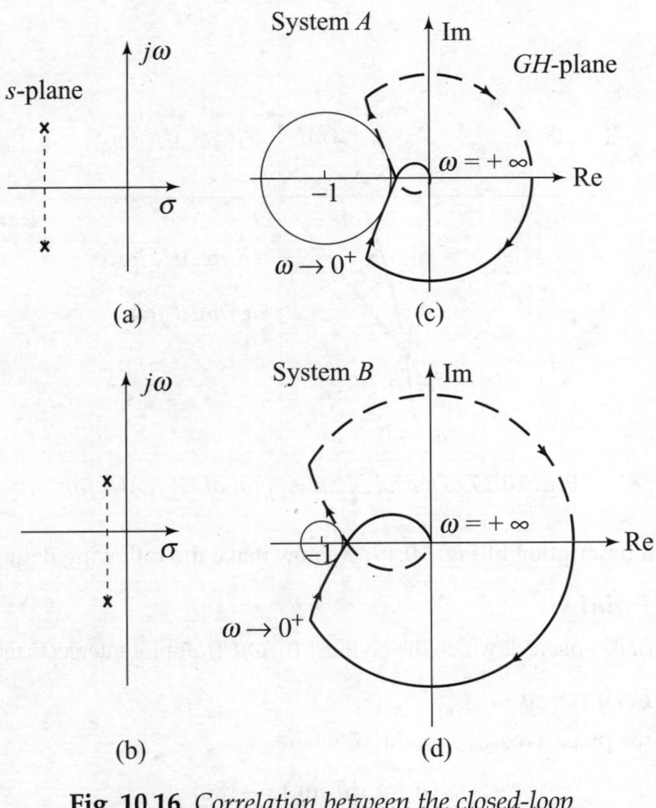

Fig. 10.16 *Correlation between the closed-loop s-plane root locations and open-loop frequency-response curves*

10.4.1 Gain Margin

Figure 10.17 shows a 'typical' Nyquist plot of $G(j\omega)H(j\omega)$; $\omega \geq 0$. $G(s)H(s)$ is known to have no poles in the right half of the s-plane. A closed-loop system with the $G(s)H(s)$ as open-loop transfer function, is stable.

The point at which the Nyquist plot crosses the negative real axis is marked A, and the frequency at that point is designated ω_ϕ. The point A is thus the tip of the vector $G(j\omega_\phi)H(j\omega_\phi)$ with magnitude $|G(j\omega_\phi)H(j\omega_\phi)|$ and angle 180°. It is seen from Fig. 10.17 that $|G(j\omega_\phi)H(j\omega_\phi)| < 1$, and therefore the tip of the vector $G(j\omega_\phi)H(j\omega_\phi)$ does not reach the critical point $-1+j0$.

If we were to increase the gain of $G(j\omega)H(j\omega)$ by a factor $1/|OA|$ without altering its phase, all points on the Nyquist plot of $G(j\omega)H(j\omega)$ would increase in magnitude by this factor along radial lines centered at the origin. The length of the vector $G(j\omega_\phi)H(j\omega_\phi)$ would be $|OA|(1/|OA|) = 1$, and therefore the curve $G(j\omega)H(j\omega)$ would go through the critical point; the closed-loop system would be at the limit of stability. We call the multiplying factor $1/|OA|$ the *gain margin* because *it is the factor by which the gain can be increased to drive the system to the verge of instability*.

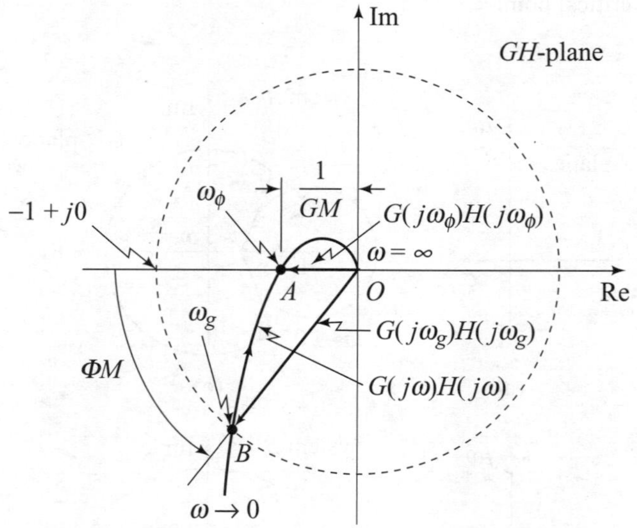

Fig. 10.17 *Typical Nyquist plot of $G(j\omega)H(j\omega)$*

From the graphical description of Fig. 10.17, we now make the following definitions:

Phase Crossover Point

It is a point on the $G(s)H(s)$-plane at which the Nyquist $G(j\omega)H(j\omega)$-plot intersects the negative real axis.

Phase Crossover Frequency ω_ϕ

It is the frequency at the phase crossover point, or where

$$\angle G(j\omega_\phi)H(j\omega_\phi) = -180° \tag{10.21}$$

Gain margin The gain margin GM of the closed-loop system that has $G(s)H(s)$ as its open-loop transfer function is defined as the number

$$GM = \frac{1}{|G(j\omega_\phi)H(j\omega_\phi)|} \tag{10.22}$$

For example, if $|G(j\omega_\phi)H(j\omega_\phi)| = 0.5$, then the $GM = 2$; the loop gain magnitude could be increased by a factor of 2 before the closed-loop system would go unstable. ***For stable systems, GM is always a number greater than one.***

10.4.2 Phase Margin

A unit circle centered at the origin has been drawn in Fig. 10.17 in order to identify point B at which Nyquist $G(j\omega)H(j\omega)$-plot has unity magnitude; the frequency at point B has been designated ω_g. The point B is thus the tip of the vector $G(j\omega_g)H(j\omega_g)$ with magnitude $|G(j\omega_g)H(j\omega_g)| = 1$ and angle $\angle G(j\omega_g)H(j\omega_g)$. It is seen from Fig. 10.17 that the tip of the vector $G(j\omega_g)H(j\omega_g)$ does not reach the critical point $-1+j0$.

If we were to decrease the phase of $G(j\omega)H(j\omega)$ without altering its gain, all points on the Nyquist $G(j\omega)H(j\omega)$-plot would rotate clockwise about the origin an angular amount equal to the increase in phase lag. In the light of Fig. 10.17, we see that the tip of the vector $G(j\omega_g)H(j\omega_g)$ would be placed at the critical point if the increase in phase lag of $G(j\omega)H(j\omega)$ were made equal to the angle ΦM. We call the angle ΦM, *measured positively in the counterclockwise direction from the negative real axis to the vector $G(j\omega_g)H(j\omega_g)$, the phase margin* because it is the angle by which the phase of $G(j\omega)H(j\omega)$ can be decreased to drive the system to the verge of instability.

From the graphical description of Fig. 10.17, we now make the following definitions:

Gain Crossover Point

It is a point on the $G(s)H(s)$-plane at which the Nyquist $G(j\omega)H(j\omega)$-plot has unity magnitude.

Gain Crossover Frequency ω_g

It is the frequency at the gain crossover point, or where

$$|G(j\omega_g)H(j\omega_g)| = 1 \tag{10.23}$$

Phase margin The phase margin ΦM of the closed-loop system that has $G(s)H(s)$ as its open-loop transfer function is defined as

$$\Phi M = 180° + \angle G(j\omega_g)H(j\omega_g) \tag{10.24}$$

For example, if $\angle G(j\omega_g)H(j\omega_g) = -135°$, then $\Phi M = 45°$; an additional phase lag of 45° could be associated with $G(j\omega)H(j\omega)$ before the closed-loop system would go unstable. ***For stable systems, ΦM is always positive.***

Consider now an unstable closed-loop system with Nyquist $G(j\omega)H(j\omega)$-plot of Fig. 10.18 ($G(s)H(s)$ is known to have no poles in the right half of the s-plane). Applying the definition of gain margin to this plot we find that $G(j\omega_\phi)H(j\omega_\phi)$ is a number greater than one and therefore GM is a number less than one. Applying the definition of phase margin, we observe that the vector $G(j\omega_g)H(j\omega_g)$ lies in the second quadrant. The angle measured from negative real axis to this vector is negative (counterclockwise measurement is taken as positive by definition) and therefore, the phase margin ΦM is negative.

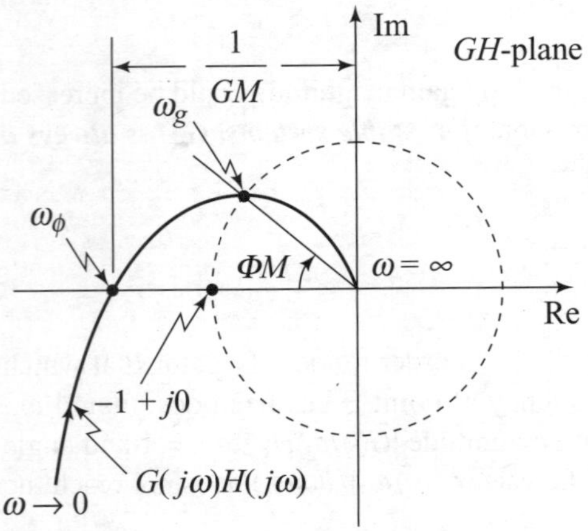

Fig. 10.18 *Nyquist plot of an unstable closed-loop system*

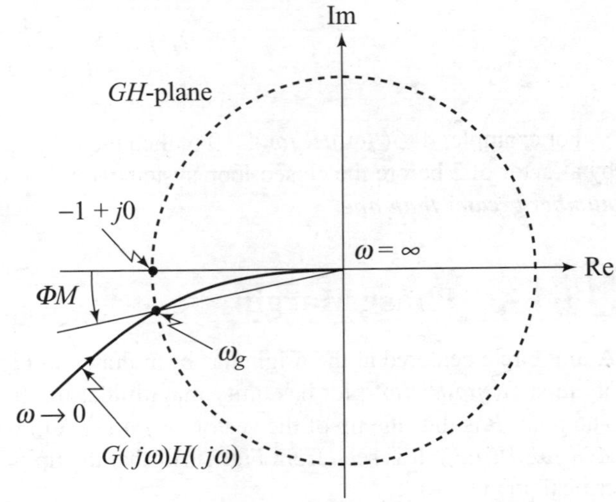

Fig. 10.19 *Infinite GM, poor ΦM*

In general, the gain margin and the phase margin are mutually independent. For example, it is possible for $G(j\omega)H(j\omega)$ to reflect an excellent gain margin but a poor phase margin. Such is the case depicted in Fig. 10.19, where the $GM = \infty$ but the $\Phi M < 15°$. In such a case, the phase crossover frequency ω_ϕ is undefined. Conversely, $G(j\omega)H(j\omega)$ can reflect a poor gain margin but an excellent phase margin, as is the case depicted in Fig. 10.20, where the $\Phi M = \infty$ but the GM is only slightly greater than 1. In such a case, the gain crossover frequency ω_g is undefined.

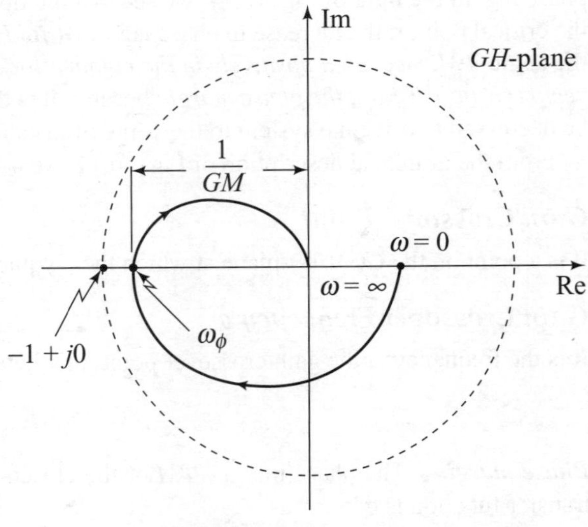

Fig. 10.20 *Poor GM, infinite ΦM*

In the light of these observations, gain and phase margins should be used in conjunction with one another when determining the degree of stability of a closed-loop system. Together, they represent a measure of the distance of $G(j\omega)H(j\omega)$ from the critical point $-1 + j0$. ***When we cannot define gain margin*** ($|G(j\omega)| \to 0$ ***as*** $\angle G(j\omega) \to -180°$)***, then only phase margin is used to determine the degree of stability. Only gain margin becomes the measure of degree of stability when phase margin cannot be defined.***

10.4.3 Some Constraints and Cautions

Gain margin and phase margin are valid measures of relative stability of a closed-loop system only if its open-loop transfer function $G(s)H(s)$ ***has no poles in the right half of the s-plane.*** If the open-loop transfer function has poles in the right half *s*-plane, it is safer to examine the complete Nyquist plot and/or root locus for interpreting relative stability of the closed-loop system.

It should be observed that the sketches of Figs 10.17–10.20 each cross the negative real axis or the unit circle atmost once. For systems with numerous zeros and poles, the Nyquist plot may cross the negative real axis several times, or it may cross the unit circle more than once. For such curves, several values of gain margin and phase margin are defined, as shown in Figs 10.21 (a) and (b). Caution is suggested when interpreting the stability of such systems. It is safer not to depend on the gain margin and phase margin for relative stability analysis.

It should also be observed that the sketches of Figs 10.12–10.20 correspond to systems where increasing gain leads to instability. There are certain practical situations wherein increase in gain can make the system stable. Consider, for example, unity-feedback system with open-loop transfer function,

$$G(s) = \frac{K(s+10)^2}{s^3}$$

This is a system for which increasing gain causes a transition from instability to stability (The Routh criterion shows that the closed-loop system is unstable for $K < 5$ and stable for $K > 5$). The Nyquist plot in Fig. 10.21c has been drawn for the stable value $K = 7$. From this figure, we observe that the phase margin is positive and the gain margin is less than one. According to the rules for stability discussed earlier, these two margins yield conflicting signals on system stability. Caution is suggested when interpreting the stability of such systems. It is safer not to depend on the gain margin and phase margin for relative stability analysis.

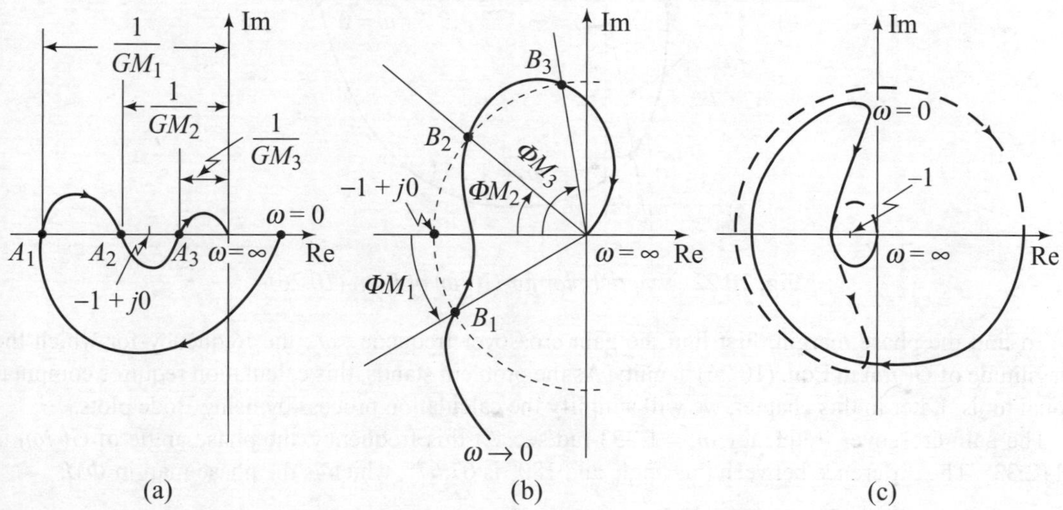

Fig. 10.21 *Nyquist plots requiring careful interpretation of GM and ΦM*

Example 10.7 A unity-feedback system has open-loop transfer function,

$$G(s) = \frac{6}{(s^2 + 2s + 2)(s + 2)} \tag{10.25}$$

Since the open-loop poles are only in the left half of the s-plane, the Nyquist criterion tells us that we want no encirclement of $-1+j0$ point for stability. In such cases, closed-loop stability can be evaluated from Nyquist $G(j\omega)$-plot.

$$G(j\omega)\big|_{\omega=0} = 1.5\angle 0°; G(j\omega)\big|_{\omega\to\infty} = 0\angle 90°$$

$$G(j\omega) = \frac{6}{(s^2+2s+2)(s+2)}\bigg|_{s=j\omega}$$

$$= \frac{6[4(1-\omega^2) - j\omega(6-\omega^2)]}{16(1-\omega^2)^2 + \omega^2(6-\omega^2)^2} \qquad (10.26)$$

Setting the imaginary part to zero, we find the phase crossover frequency $\omega_\phi = \sqrt{6}$ rad/sec. At this frequency, the real part of $G(j\omega)$ in Eqn. (10.26) is calculated to be – 0.3 (refer to Fig. 10.22). Thus, the gain can be increased by (1/0.3) before the real part becomes –1. The gain margin is $GM = 3.33$.

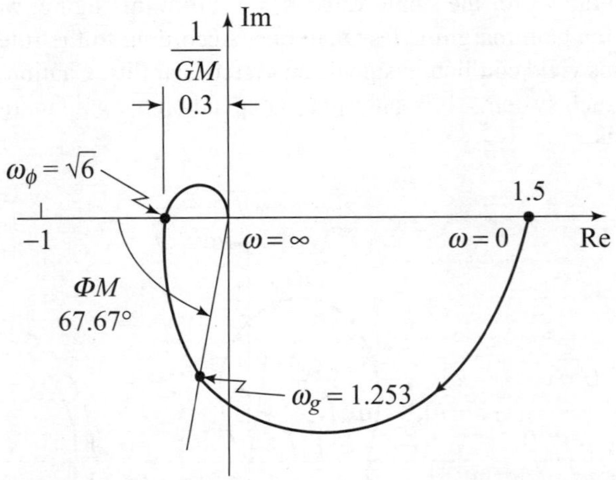

Fig. 10.22 *Nyquist plot for G(jω) of Eqn. (10.26)*

To find the phase margin, first find the gain crossover frequency ω_g: the frequency for which the magnitude of $G(j\omega)$ in Eqn. (10.26) is unity. As the problem stands, this calculation requires computational tools. Later in this chapter, we will simplify the calculation process by using Bode plots.

The gain crossover frequency $\omega_g = 1.253$ rad/sec. At this frequency, the phase angle of $G(j\omega)$ is –112.33°. The difference between this angle and 180° is 67.67°, which is the phase margin ΦM.

10.5 THE BODE PLOTS

A sinusoidal transfer function may be represented by two separate plots; one giving the magnitude *versus* frequency, and the other the phase angle *versus* frequency. A *Bode plot* (named after Hendrick W Bode) consists of two graphs: one is a plot of the logarithm of the magnitude of the sinusoidal transfer function, and the other is a plot of the phase angle in degrees; both are plotted against the frequency in logarithmic scale.

In a Bode plot, the logarithmic magnitude of sinusoidal transfer function $G(j\omega)$ is represented as 20 log $|G(j\omega)|$, where the base of the logarithm is 10. The unit used in this representation of the magnitude is *decibel* (one-tenth of a 'Bel': a unit named after Alexander Graham Bell), usually abbreviated as dB.

The main advantage of using the logarithmic plot is that the multiplication of magnitudes can be converted into addition. Consider a typical rational transfer function $G(s)$ factored in the time-constant form:

$$G(s) = \frac{K(1+s\tau_a)(1+s\tau_b)\cdots}{s^N(1+s\tau_1)(1+s\tau_2)\cdots[1+(2\zeta/\omega_n)s+s^2/\omega_n^2]\cdots} \tag{10.27}$$

This transfer function has zeros at $s = -1/\tau_a, -1/\tau_b, \ldots$, N poles at the origin, real poles at $s = -1/\tau_1, -1/\tau_2, \ldots$, and complex poles at $-\zeta\omega_n \pm j\omega_n\sqrt{1-\zeta^2}$. We note that the constant multiplier $K = \lim_{s\to 0} s^N G(s)$, where N is the type number for the system with open-loop transfer function $G(s)$ given by Eqn. (10.27). Thus for a type-0 system, $K_p = K$. For a type-1 system, $K_v = K$, and for a type-2 system, $K_a = K$. If $G(s)$ has complex zeros, then even though it is not shown in Eqn. (10.27), a quadratic term of the form $[1 + (2\zeta/\omega_n)s + s^2/\omega_n^2]$ will also appear in the numerator.

The magnitude of $G(j\omega)$ in dB is obtained by multiplying the logarithm to the base 10 of $|G(j\omega)|$ by 20.

$$\begin{aligned}|G(j\omega)|_{\text{dB}} &= 20\log|G(j\omega)| \\ &= 20\log K + 20\log|1+j\omega\tau_a| + 20\log|1+j\omega\tau_b| + \cdots \\ &\quad -20N\log|j\omega| - 20\log|1+j\omega\tau_1| - 20\log|1+j\omega\tau_2| - \cdots \\ &\quad -20\log|1+j2\zeta\omega/\omega_n - \omega^2/\omega_n^2| - \cdots\end{aligned} \tag{10.28a}$$

The phase of $G(j\omega)$ is

$$\begin{aligned}\angle G(j\omega) &= \tan^{-1}\omega\tau_a + \tan^{-1}\omega\tau_b + \cdots - N(90°) - \tan^{-1}\omega\tau_1 - \tan^{-1}\omega\tau_2 - \cdots \\ &\quad -\tan^{-1} 2\zeta\omega\omega_n/(\omega_n^2 - \omega^2) - \cdots\end{aligned} \tag{10.28b}$$

Equations (10.28) express the magnitude and the phase of $G(j\omega)$ as linear combinations of relatively simple terms. The curves for each term can be added together graphically to get the curves for the complete transfer function.

The logarithmic scale used for frequency in Bode plots, has some interesting properties. The frequency in a typical control system application varies over many (two/three) powers of ten so that most information would be compressed near the origin if a linear scale were used. The logarithmic scale is nonlinear, that is, the distance between 1 and 2 is greater than the distance between 2 and 3 and so on. As a result, ***use of this scale enables us to cover a greater range of frequencies; both low and high frequency behavior of a system can be adequately displayed in one plot.*** The other interesting property of the use of log scale for frequency, as we shall see shortly, is that the straight-line approximation of the Bode plots becomes possible. This property allows the plotting by hand that is quick and yet sufficiently accurate for control-system design. Most control-system designers will have access to computer programs that will diminish the need for hand plotting; however, it is still important to develop good intuition so that erroneous computer results are quickly identified, and for this one needs the ability to check results by hand plotting.

The phase angle graph of the Bode plot, $\angle G(j\omega)$ *versus* log ω, has a linear scale for phase in degrees and a logarithmic scale for frequency ω. The magnitude graph of the Bode plot, $|G(j\omega)|_{dB}$ versus log ω, has a linear scale for magnitude in dB and a logarithmic scale for frequency ω.

In general, a rational transfer function $G(s)$ can contain just four simple types of factors:

1. Real constant: $K > 0$
2. Poles or zeros at the origin of order N: $(s)^{\mp N}$
3. Poles or zeros at $s = -1/\tau$ of order q: $(1 + s\tau)^{\mp q}$
4. Complex poles or zeros of order r: $(1 + 2\zeta s/\omega_n + s^2/\omega_n^2)^{\mp r}$

Equations (10.28) indicate one of the unique characteristics of Bode plot—each of the four types of factors listed can be considered as a separate plot; the individual plots are then algebraically added to yield the total plot of the given transfer function. Bode plots of the four types of factors are, therefore, the basic building blocks for the construction of the Bode plot of the given transfer function.

10.5.1 Magnitude Plot: Straight-Line Approximation

We shall now discuss how the $|G(j\omega)|_{dB}$ *versus* log ω graph can be sketched with very little effort using straight-line approximation. We first examine the basic building blocks, and then use these blocks to build the magnitude plot of the given transfer function.

The dB *versus* log ω graph can be sketched on a linear rectangular coordinate graph paper. However, use of semilog graph paper is more convenient as it eliminates the need to take logarithm of many numbers. The logarithmic scale of the semilog graph paper is used for frequency and the linear scale is used for dB.

The units used to express frequency bands or frequency ratios are the *octave* and the *decade*. An octave is a frequency band from ω_1 to ω_2 where $\omega_2/\omega_1 = 2$. There is an increase of one decade from ω_1 to ω_2 when $\omega_2/\omega_1 = 10$. Semilog graph papers come in two, three, four or five cycles, indicating the range of coverage in decades (refer to Fig. 10.23). ***It may be noted that we cannot locate the point $\omega = 0$ on the log scale since*** log 0 = – ∞.

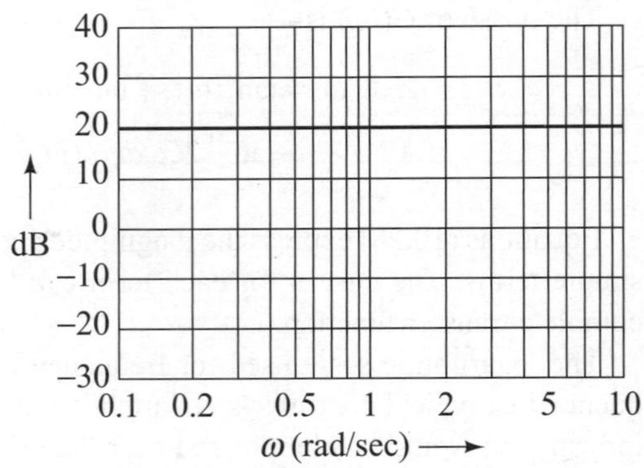

Fig. 10.23 *Bode magnitude plot of K*

Real Constant K

Since the constant K is frequency invariant, the plot of

$$\text{dB} = 20\log K \tag{10.29}$$

is a horizontal straight line. The magnitude plot for $K = 10$ is shown in Fig. 10.23 in semilog coordinates. The horizontal axis is log ω; labeled ω on logarithmic scale. The vertical axis is magnitude in decibels; labeled dB on linear scale.

Poles or Zeros at the Origin

The factor $1/(j\omega)$ appearing in a transfer function $G(j\omega)$ has the magnitude

$$\text{dB} = 20\log\left|\frac{1}{j\omega}\right| = -20\log\omega \tag{10.30}$$

The dB versus log ω plot of Eqn. (10.30) is a straight line with a slope of −20 dB per unit change in log ω. A unit change in log ω means

$$\log(\omega_2/\omega_1) = 1 \text{ or } \omega_2 = 10\omega_1$$

This range of frequencies is called a *decade*. Thus, the slope of Eqn. (10.30) is −20 dB/decade. The range of frequencies $\omega_2 = 2\omega_1$ is called an *octave*. Since −20 log 2 = −6 dB, the slope of Eqn. (10.30) could also be expressed as −6 dB/octave. The plot of Eqn. (10.30) is shown in Fig. 10.24a; it intersects the 0-dB axis at $\omega = 1$.

For an Nth-order pole at the origin, the magnitude is

$$\text{dB} = 20\log\left|\frac{1}{(j\omega)^N}\right| = -20N\log\omega \tag{10.31}$$

The magnitude plot is still a straight line that intersects the 0-dB axis at $\omega = 1$, but the slope is now $-20N$ dB/decade.

For the case that a transfer function has a zero at the origin, the magnitude of the term is given by

$$\text{dB} = 20\log|j\omega| = 20\log\omega \tag{10.32}$$

and the plot is the negative of that for a pole at the origin; it is a straight line with a slope of + 20 dB/decade that intersects the 0-dB axis at $\omega = 1$ (Fig. 10.24b). For the case of an Nth-order zero at the origin, it is seen that the plot is the negative of that for the Nth-order pole.

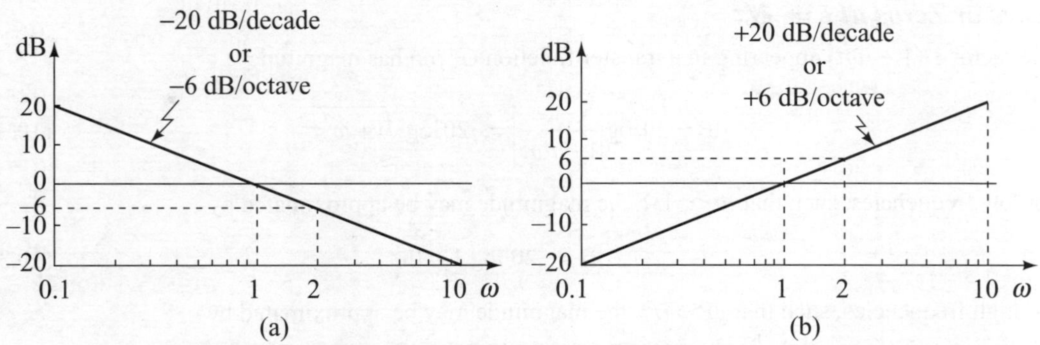

Fig. 10.24 *Bode magnitude plot of (a) pole; (b) zero at the origin*

Example 10.8 Consider the transfer function,

$$G(s) = \frac{K}{s^N}$$

There are two terms in $G(s)$: the constant K, and the repeated pole at the origin. We could add the Bode magnitude plots of the two terms to get the Bode magnitude plot of $G(s)$. However, treating K/s^N as a single term turns out to be more convenient, as is seen below.

$$\text{dB} = 20\log\left|\frac{K}{(j\omega)^N}\right| = -20N\log\omega + 20\log K \tag{10.33}$$

With log ω as abscissa, the plot of Eqn. (10.33) is a straight line having a slope $-20N$ dB/decade and passing through 20 log K dB when log $\omega = 0$, i.e., when $\omega = 1$, as shown in Fig. 10.25. Further, the plot has a value of 0 dB at the frequency of

$$20N \log \omega = 20 \log K \text{ or } \omega = (K)^{1/N}$$

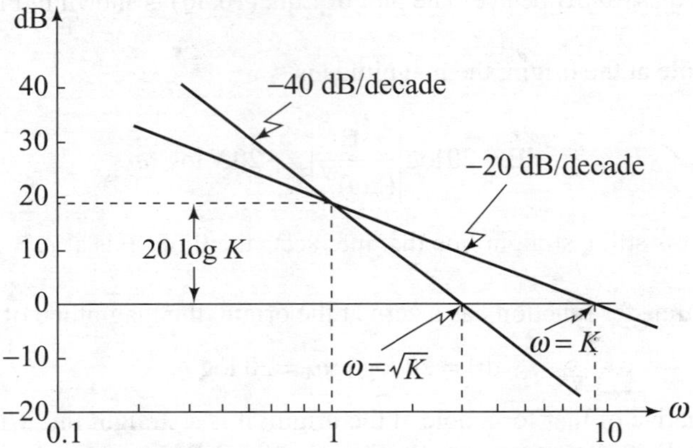

Fig. 10.25 *Bode magnitude plot of K/s^N*

Poles or Zeros at $s = -1/\tau$

The factor $1/(1 + j\omega\tau)$ appearing in a transfer function $G(j\omega)$ has magnitude

$$\text{dB} = 20\log\left|\frac{1}{1+j\omega\tau}\right| = -20\log\sqrt{1+\omega^2\tau^2} \tag{10.34}$$

For low frequencies, such that $\omega << 1/\tau$, the magnitude may be approximated by

$$\text{dB} = -20 \log 1 = 0 \tag{10.35}$$

For high frequencies, such that $\omega >> 1/\tau$, the magnitude may be approximated by

$$\text{dB} = -20 \log(\omega\tau) = -20 \log \omega - 20 \log \tau \tag{10.36}$$

The plot of Eqn. (10.35) is a straight line coincident with the 0 dB axis. The plot of Eqn. (10.36) is also a straight line having a slope of −20 dB/decade intersecting the 0 dB axis at $\omega = 1/\tau$, as shown in Fig. 10.26. We call the straight line approximations the *asymptotes*. The low-frequency approximation is called the *low-frequency asymptote* and the high-frequency approximation is called the *high-frequency asymptote*. The frequency $\omega = 1/\tau$ at which the two asymptotes meet is called the *corner frequency* or *break frequency*.

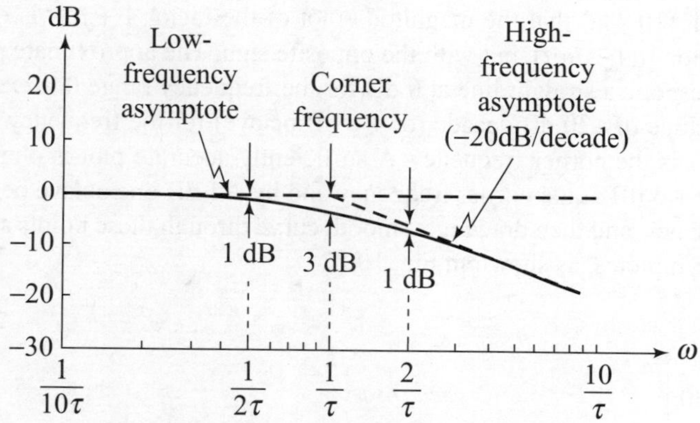

Fig. 10.26 *Bode magnitude plot of $1/(1 + j\omega\tau)$*

Though the asymptotic approximations hold good for $\omega << 1/\tau$ and $\omega >> 1/\tau$, with some loss of accuracy these could be extended for $\omega \le 1/\tau$ and $\omega \ge 1/\tau$. Therefore, the dB *versus* log ω curve of $1/(1 + j\omega\tau)$ can be approximated by two asymptotes, one a straight line at 0 dB for the frequency range $0 < \omega \le 1/\tau$ and the other, a straight line with a slope −20 dB/decade (or − 6dB/octave) for the frequency range $1/\tau \le \omega < \infty$. The corner frequency divides the plot into two regions, a low-frequency region and a high-frequency region.

The error in the magnitude plot caused by the use of asymptotes can easily be calculated. The maximum error occurs at the corner frequency $\omega = 1/\tau$ and is given by (refer to Eqns (10.34) and (10.35))

$$-20 \ \log\sqrt{1+1} + 20 \ \log 1 = -3.01 \text{ dB}$$

The error at the frequency one octave below the corner frequency ($\omega = 1/(2\tau)$) is (refer to Eqns (10.34) and (10.35))

$$-20 \ \log \sqrt{1+\tfrac{1}{4}} + 20 \ \log 1 = -0.97 \text{ dB}$$

The error at the frequency one octave above the corner frequency ($\omega = 2/\tau$) is (refer to Eqns (10.34) and (10.36))

$$-20 \ \log \sqrt{1+4} + 20 \ \log 2 = -0.97 \text{ dB}$$

The error at the frequency one decade below the corner frequency ($\omega = 1/(10\tau)$) is

$$-20 \ \log\sqrt{1+\tfrac{1}{100}} + 20 \ \log 1 = -0.04 \text{ dB}$$

The error at the frequency one decade above the corner frequency ($\omega = 10/\tau$) is

$$-20 \ \log \sqrt{1+100} + 20 \ \log 10 = -0.04 \text{ dB}$$

Similarly, the errors at other frequencies may be calculated.

In practice, a sufficiently accurate magnitude plot is obtained by correcting the asymptotic plot by −3 dB at the corner frequency and by −1 dB one octave below and one octave above the corner frequency, and then drawing a smooth curve through these points approaching the low and high frequency asymptotes, as shown in Fig. 10.26.

It is seen from Eqn. (10.28a) that the magnitude plot of the factor $1 + j\omega\tau$ is exactly of the same form as that of the factor $1/(1 + j\omega\tau)$, but with the opposite sign. The approximate plot consists of two straight line asymptotes, one a straight line at 0 dB for the frequency range $0 < \omega \leq 1/\tau$, and the other a straight line with a slope of +20 dB/decade (or + 6 dB/octave) for the frequency range $1/\tau \leq \omega < \infty$. The frequency $\omega = 1/\tau$ is the corner frequency. A sufficiently accurate plot is obtained by correcting the asymptotic plot by + 3 dB at the corner frequency and by + 1 dB one octave below and one octave above the corner frequency, and then drawing a smooth curve through these points approaching the low and high frequency asymptotes, as shown in Fig. 10.27.

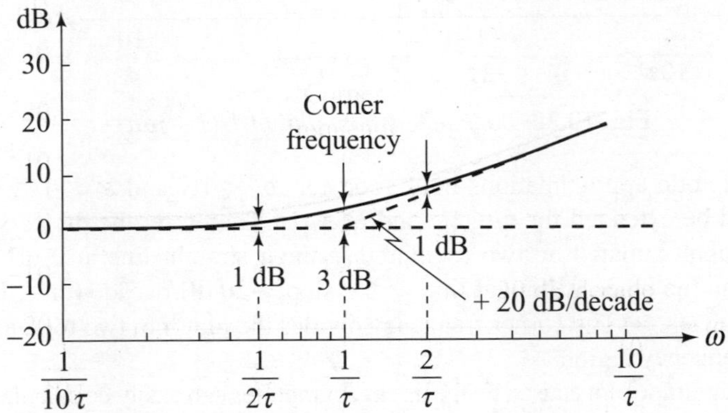

Fig. 10.27 *Bode magnitude plot of $(1 + j\omega\tau)$*

For the case when a given transfer function involves terms like $(1+ j\omega\tau)^{\mp q}$, a similar asymptotic construction may be made. The corner frequency is still at $\omega = 1/\tau$ and the asymptotes are straight lines. The low-frequency asymptote is a straight line at 0 dB and the high-frequency asymptote has a slope of $\mp 20q$ dB/decade (or $\mp 6q$ dB/octave). The error involved in asymptotic approximation is q times that for $(1+ j\omega\tau)^{\mp 1}$; $\mp 3q$ dB at the corner frequency and $\mp q$ dB one octave below and one octave above the corner frequency.

Example 10.9 Let us draw the Bode magnitude plot for the transfer function,

$$G(s) = \frac{200(s+1)}{(s+10)^2}$$

The rearrangement of the transfer function in the time-constant form gives

$$G(s) = \frac{2(1+s)}{(1+s/10^2)}$$

Therefore, the sinusoidal transfer function in the time-constant form is given by

$$G(j\omega) = \frac{2(1+ j\omega)}{(1+ j\omega/10)^2}$$

Our approach is to first construct an asymptotic plot and then apply corrections to it to get an accurate plot. The corner frequencies of the asymptotic plot in order of their occurrence as frequency increases are

(i) $\omega_{c1} = 1$, due to zero at $s = -1$;
(ii) $\omega_{c2} = 10$, due to double pole at $s = -10$

At frequencies less than ω_{c1}, the first corner frequency, only the factor $K = 2$ is effective.

Pertinent characteristics of each factor of the transfer function are given in Table 10.1. By sketching the asymptotes for each factor and then algebraically adding them, we obtain the asymptotic plot for the transfer function $G(j\omega)$. We can, however, draw the composite asymptotic plot directly, as is outlined below.

Step 1: We start with the factor $K = 2$. Its magnitude plot is the asymptote 1; a horizontal straight line at the magnitude of 6 dB (Fig. 10.28).

Step 2: Let us now add to the asymptote 1, the plot of the factor $(1 + j\omega)$ corresponding to the lowest corner frequency $\omega_{c1} = 1$. Since this factor contributes zero dB for $\omega \le \omega_{c1} = 1$, the resultant plot up to $\omega = 1$ is the same as that of the asymptote 1. For $\omega > \omega_{c1} = 1$, this factor contributes +20 dB/decade such that the resultant plot of the two factors is the asymptote 2 of slope +20 dB/decade passing through (6 dB, 1 rad/sec) point. At $\omega = \omega_{c2} = 10$, the resultant plot has a magnitude of 26 dB as shown in Fig. 10.28.

Step 3: We now add to the resultant plot of Step 2, the plot of the factor $1/(1 + j\omega/10)^2$ corresponding to the corner frequency $\omega_{c2} = 10$. Since this factor contributes 0 dB for $\omega \le \omega_{c2} = 10$, the resultant plot up to $\omega = 10$ is the same as that of step 2. For $\omega > \omega_{c2} = 10$, this factor contributes –40 dB/decade such that the resultant plot of the three factors is the asymptote 3 of slope (+20) + (–40) = –20 dB/decade passing through (26 dB, 10 rad/sec) point. Figure 10.28 shows the asymptotic magnitude plot of given $G(j\omega)$.

Table 10.1 *Asymptotic approximation table for Bode magnitude plot of* $2(1 + j\omega)/(1 + j\omega/10)^2$

Factor	*Corner frequency*	*Asymptotic magnitude characteristic*
2	None	Constant magnitude of + 6 dB
$1 + j\omega$	$\omega_{c1} = 1$	Straight line of 0 dB for $\omega \le \omega_{c1}$; straight line of +20 dB/decade for $\omega \ge \omega_{c1}$
$1/(1 + j\omega/10)^2$	$\omega_{c2} = 10$	Straight line of 0 dB for $\omega \le \omega_{c2}$; straight line of – 40 dB/decade for $\omega \ge \omega_{c2}$

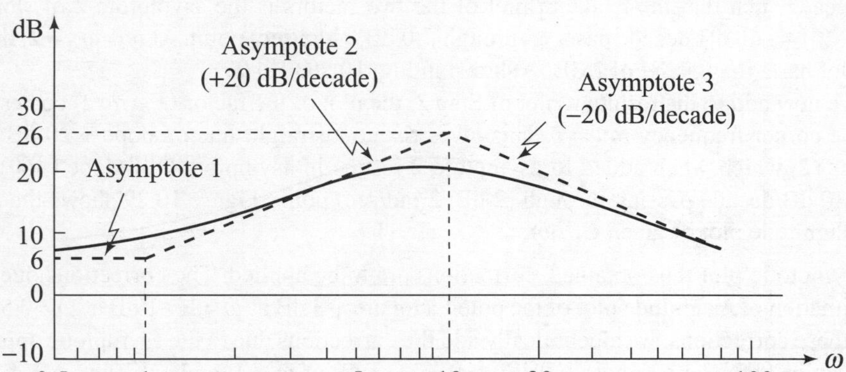

Fig. 10.28 *Bode magnitude plot for Example 10.9*

To the asymptotic plot thus obtained, corrections are to be applied. The corrections at each corner frequency and at an octave above and below the corner frequency are usually sufficient. The corner frequency $\omega_{c1} = 1$ corresponds to the first-order factor $(1 + j\omega)$: the corrections are +3 dB at $\omega = 1$, +1 dB at $\omega = 0.5$ and +1 dB at $\omega = 2$. The corner frequency $\omega_{c2} = 10$ corresponds to second-order factor $1/(1 + j\omega/10)^2$: the corrections are –6 dB at $\omega = 10$, –2 dB at $\omega = 5$ and –2 dB at $\omega = 20$. Table 10.2 lists the net corrections.

Table 10.2 *Corrections to the asymptotic magnitude plot of* $2(1 + j\omega)/(1 + j\omega/10)^2$

Frequency, ω	0.5	1	2	5	10	20
Net correction	+1 dB	+3 dB	+1 dB	–2 dB	–6 dB	–2 dB

The corrected asymptotic magnitude plot of the given transfer function is shown in Fig. 10.28.

Example 10.10 Consider the transfer function, $G(j\omega) = \dfrac{10(1 + j\omega/2)}{(j\omega)^2(1 + j\omega)}$

The corner frequencies of the asymptotic Bode magnitude plot of $G(j\omega)$ in order of their occurrence as frequency increases are

(i) $\omega_{c1} = 1$, due to simple pole,
(ii) $\omega_{c2} = 2$, due to simple zero.

At frequencies less than ω_{c1}, only the factor $10/(j\omega)^2$ is effective.

Asymptotic magnitude plot of $G(j\omega)$ is shown in Fig. 10.29. The plot is obtained following the steps given below:

Step 1: We start with the factor $10/(j\omega)^2$ corresponding to double pole at the origin. Its magnitude plot is the asymptote 1, having a slope of –40 dB/decade and passing through the point 20 log 10 = 20 dB at $\omega = 1$. Asymptote 1 intersects the 0 dB line at $\omega = \sqrt{10}$.

Step 2: Let us now add to the asymptote 1, the plot of the factor $1/(1 + j\omega)$ corresponding to the lowest corner frequency $\omega_{c1} = 1$. Since this factor contributes zero dB for $\omega \le 1$, the resultant plot up to $\omega = 1$ is the same as that of asymptote 1. For $\omega > 1$ this factor contributes –20 dB/decade such that the resultant plot of the two factors is the asymptote 2 of slope (–40) + (–20) = –60 dB/decade passing through (20 dB, 1 rad/sec) point. At $\omega = \omega_{c2} = 2$, the resultant plot has a magnitude of 2 dB as shown in Fig. 10.29.

Step 3: We now add to the resultant plot of Step 2, the plot of the factor $(1 + j\omega/2)$ corresponding to the corner frequency $\omega_{c2} = 2$. This gives rise to a straight line of slope + 20 dB/decade for $\omega > 2$, which when added to asymptote 2 results in asymptote 3 of slope (–60) + (+20) = –40 dB/decade passing through (2 dB, 2 rad/sec) point. Figure 10.29 shows the asymptotic magnitude plot of given $G(j\omega)$.

To the asymptotic plot thus obtained, corrections are to be applied. The corrections due to asymptotic approximation of magnitude plot of the pole factor are: –3 dB at $\omega = 1$, –1 dB at $\omega = 0.5$ and –1 dB at $\omega = 2$. To these corrections, we algebraically add the corrections due to the asymptotic approximation of magnitude plot of the zero factor: + 3 dB at $\omega = 2$, +1 dB at $\omega = 1$, and +1 dB at $\omega = 4$. Table 10.3 lists the net corrections.

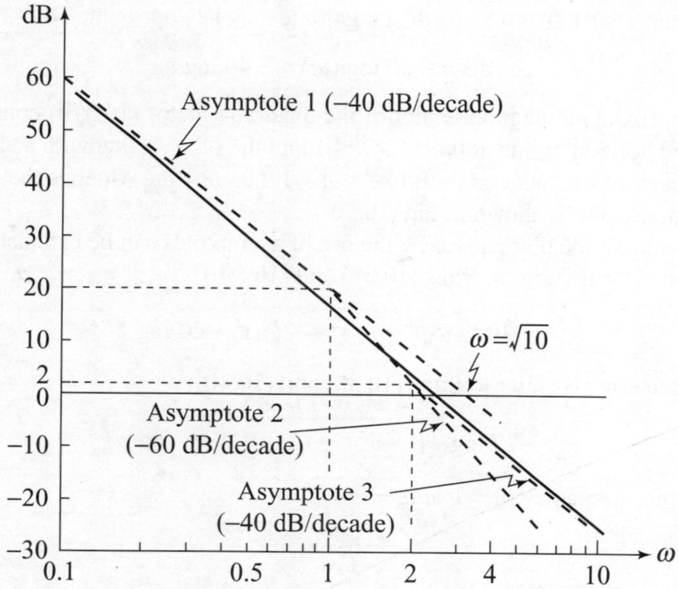

Fig. 10.29 *Bode magnitude plot for Example 10.10*

Table 10.3 *Corrections to the asymptotic magnitude plot of* $10(1 + j\omega/2)/(j\omega)^2(1 + j\omega)$

Frequency, ω	0.5	1	2	4
Net correction	−1 dB	−2 dB	+2 dB	+1 dB

The corrected asymptotic magnitude plot of the given transfer function is shown in Fig. 10.29.

Complex Poles or Zeros

Transfer functions of control systems often possess quadratic factors of the form,

$$\frac{1}{1 + j(2\zeta/\omega_n)\omega - \omega^2/\omega_n^2}; 0 < \zeta \le 1 \quad (10.37a)$$

This term is slightly more complicated since besides being a function of ω as before, it is also a function of the variable ζ. As may be expected, the shape of the Bode plot depends strongly upon what value of damping ratio ζ is being considered.

In normalized form, the quadratic factor (10.37a) may be written as

$$\frac{1}{(1 + j2\zeta u - u^2)} \quad (10.37b)$$

where $u = \omega/\omega_n$ is the normalized frequency. The magnitude of this factor is

$$\text{dB} = 20 \log\left|\frac{1}{1 - u^2 + j2\zeta u}\right| = -20 \log\sqrt{(1-u^2)^2 + (2\zeta u)^2} \quad (10.38)$$

For low frequencies, such that $u << 1$, the magnitude may be approximated by

$$\text{dB} = -20 \log 1 = 0 \quad (10.39)$$

For high frequencies, such that $u >> 1$, the magnitude may be approximated by

$$\text{dB} = -20 \log (u^2) = -40 \log u \tag{10.40}$$

Therefore, an approximate magnitude plot of the quadratic factor (10.37b) consists of two straight line asymptotes, one horizontal line at 0 dB for $u \le 1$ and the other, a line with a slope −40 dB/decade for $u \ge 1$. The two asymptotes meet on 0 dB line at $u = 1$, i.e., $\omega = \omega_n$, which is the corner frequency of the plot. The asymptotic plot is shown in Fig. 10.30.

The error in the magnitude plot caused by the use of asymptotes can be calculated as follows:

For $0 < u \le 1$, the error is (refer to Eqns (10.38) and (10.39))

$$-20 \log \sqrt{(1-u^2)^2 + (2\zeta u)^2} + 20 \log 1 \tag{10.41a}$$

and for $1 \le u < \infty$, the error is (refer to Eqns (10.38) and (10.40))

$$-20 \log \sqrt{(1-u^2)^2 + (2\zeta u)^2} + 40 \log u \tag{10.41b}$$

The error at the corner frequency ($u = 1$ or $\omega = \omega_n$) is

$$-20 \log 2\zeta \tag{10.42}$$

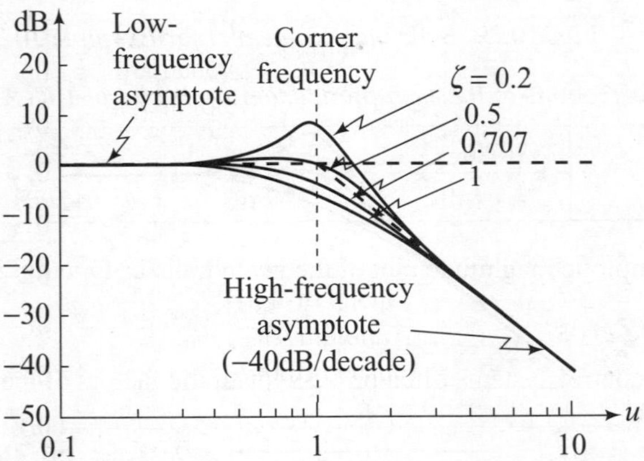

Fig. 10.30 *Bode magnitude plot of $1/(1 + j2\zeta u - u^2)$*

For $\zeta = 1$, the error at the corner frequency is $-20 \log 2 \cong -6$ dB. Note that for $\zeta = 1$, Eqn. (10.37) has two equal real poles:

$$\left.\frac{1}{(1 + j2\zeta u - u^2)}\right|_{\zeta=1} = \frac{1}{(1 + ju)^2} = \frac{1}{(1 + j\omega/\omega_n)^2}$$

From expression (10.42) we see that at the corner frequency, the error is negative for $\zeta > 0.5$ and positive for $\zeta < 0.5$. The corrected magnitude curves for various values of ζ are shown in Fig. 10.30. From this figure, it is seen that for small values of ζ, the magnitude curves have pronounced peak at a frequency slightly less than the corner frequency. In fact, smaller the value of ζ, higher the peak. We shall discuss this peak in some detail.

Consider Fig. 10.31, which shows what might be a typical plot of (refer to Eqn. (10.38))

$$\text{dB} = -20 \log \sqrt{(1-u^2)^2 + (2\zeta u)^2}$$

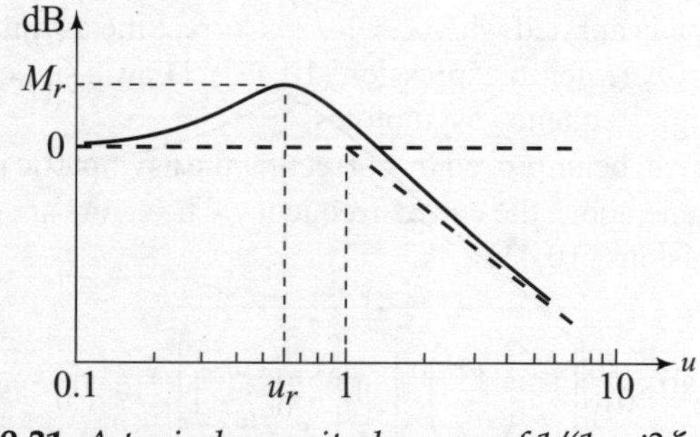

Fig. 10.31 *A typical magnitude curve of $1/(1 + j2\zeta u - u^2)$*

The peak value occurs at normalized frequency which is labeled u_r ($u_r = \omega_r/\omega_n$) and the height of the peak is M_r. This peak will occur at a u where $\sqrt{(1-u^2)^2 + (2\zeta u)^2}$ is at a minimum. The minimum can be found by simple differentiation.

$$\frac{d}{du}\left[\sqrt{(1-u^2)^2+(2\zeta u)^2}\right]_{u=u_r} = \frac{\frac{1}{2}[-4(1-u_r^2)u_r + 8\zeta^2 u_r]}{[(1-u_r^2)^2+(2\zeta u_r)^2]^{1/2}} = 0$$

which gives

$$4u_r^3 - 4u_r + 8\zeta^2 u_r = 0$$

or

$$u_r = \sqrt{1-2\zeta^2} \tag{10.43}$$

The corresponding peak value M_r is then

$$M_r = -20\log\sqrt{(1-u_r^2)^2+(2\zeta u_r)^2}$$

$$= -20\log 2\zeta\sqrt{1-\zeta^2}\ \text{dB} \tag{10.44}$$

Note that as ζ goes to zero, the peak value M_r goes off to infinity (refer to Eqn. (10.44)) and frequency u_r approaches the corner frequency (refer to Eqn. (10.43)). For $0 < \zeta < 1/\sqrt{2}$, u_r is less than the corner frequency. For $\zeta \ge 1/\sqrt{2}$, the errors given by expressions (10.41) are negative for all frequencies and therefore magnitude curves do not have a peak.

For given values of ζ and ω_n, an exact dB versus logω curve is obtained from the following expression (refer to Eqn. (10.38)):

$$\text{dB} = -20\ \log\sqrt{\left[1-\left(\frac{\omega}{\omega_n}\right)^2\right]^2 + \left(\frac{2\zeta\omega}{\omega_n}\right)^2} \tag{10.45}$$

The following procedure is useful for hand plotting:

Draw two straight line asymptotes: one a 0 dB line for $0 < \omega \le \omega_n$ and the other a -40 dB/decade line for $\omega_n \le \omega < \infty$. The two asymptotes meet at the corner frequency $\omega = \omega_n$. Correct the asymptotic plot by M_r dB at the frequency ω_r where (refer to Eqns (10.44) and (10.43))

$$M_r = -20\log 2\zeta\sqrt{1-\zeta^2}\ \text{dB} \tag{10.46a}$$

$$\omega_r = \omega_n\sqrt{1-2\zeta^2}\ \text{rad/sec} \tag{10.46b}$$

Of course, this correction is applied when $\zeta < 1/\sqrt{2}$. Correct the asymptotic plot by –20 log (2ζ) dB at the corner frequency $\omega = \omega_n$ (refer to expression (10.42)). Draw a smooth curve through these points approaching the low and high frequency asymptotes.

The accuracy of the plot can be improved by correcting the asymptotic plot at some other frequencies also; say one octave below and above the corner frequency. The errors are calculated from the following expressions (refer to expressions (10.41)):

$$-20\log\sqrt{\left[1-\left(\frac{\omega}{\omega_n}\right)^2\right]^2+\left[2\zeta\left(\frac{\omega}{\omega_n}\right)\right]^2} \quad \text{for } 0 < \omega \le \omega_n \tag{10.47a}$$

$$-20\log\sqrt{\left[1-\left(\frac{\omega}{\omega_n}\right)^2\right]^2+\left[2\zeta\left(\frac{\omega}{\omega_n}\right)\right]^2}+40\log\left(\frac{\omega}{\omega_n}\right) \quad \text{for } \omega_n \le \omega < \infty \tag{10.47b}$$

The magnitude plot of the factor $[(1+j(2\zeta/\omega_n)\omega-\omega^2/\omega_n^2]$ is exactly of the same form as that of the factor $1/[1+j(2\zeta/\omega_n)\omega-\omega^2/\omega_n^2]$, but with the opposite sign. Similar asymptotic construction may be made for the case when a given transfer function involves terms like $[1+j(2\zeta/\omega_n)\omega-\omega^2/\omega_n^2]^{\mp r}$.

Example 10.11 Let us draw the Bode magnitude plot of the transfer function,

$$G(s)=\frac{200(s+2)}{s(s^2+10s+100)}$$

The rearrangement of the transfer function in time-constant form gives

$$G(s)=\frac{4(1+s/2)}{s[1+s/10+(s/10)^2]}$$

Therefore, the sinusoidal transfer function in time-constant form is given by

$$G(j\omega)=\frac{4(1+j\omega/2)}{j\omega(1+j\omega/10-\omega^2/100)} \tag{10.48}$$

The corner frequencies of the asymptotic plot of $G(j\omega)$, in order of their occurrence as frequency increases are

(i) $\omega_{c1} = 2$, due to zero at $s = -2$;
(ii) $\omega_{c2} = 10$, due to pair of complex conjugate poles with $\zeta = 0.5$ and $\omega_n = 10$.

At frequencies less than ω_{c1}, only the factor $4/(j\omega)$ is effective.

Asymptotic magnitude plot of $G(j\omega)$ is drawn in Fig. 10.32. The plot is obtained following the steps given below:

Step 1: We start with the factor $4/(j\omega)$ corresponding to the pole at the origin. Its magnitude plot is the asymptote 1 having a slope of –20 dB/decade passing through the point 20 log 4 = 12 dB at $\omega = 1$. Asymptote 1 intersects the 0 dB line at $\omega = 4$. At $\omega = \omega_{c1} = 2$, the magnitude is 6 dB.

Step 2: We now add to the asymptote 1, the plot of the factor $(1 + j\omega/2)$ corresponding to the lowest corner frequency $\omega_{c1} = 2$. Since this factor contributes 0 dB for $\omega \le 2$, the resultant plot up to $\omega = 2$ is the same as that of asymptote 1. For $\omega > 2$, this factor contributes +20 dB/decade such that the resultant plot of the two factors is the asymptote 2 of slope (−20) + (+20) = 0 dB/decade passing through (6 dB, 2 rad/sec) point. At $\omega = \omega_{c2} = 10$, the resultant plot has a magnitude of 6 dB as shown in Fig. 10.32.

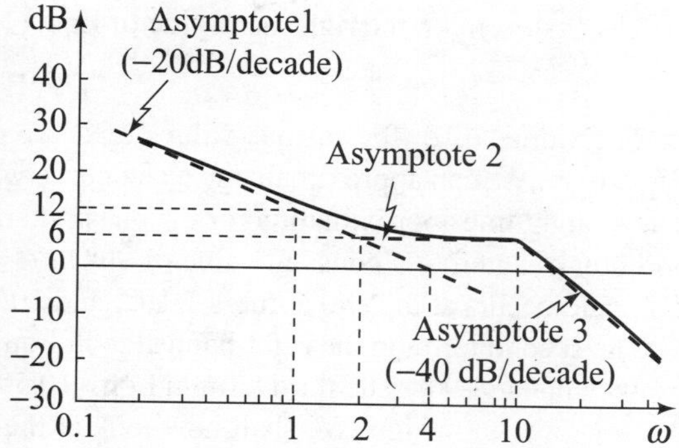

Fig. 10.32 *Bode magnitude plot for Example 10.11*

Step 3: We now add to the resultant plot of Step 2, the plot of the factor $1/(1 + j\omega/10 - \omega^2/100)$ corresponding to the corner frequency $\omega_{c2} = \omega_n = 10$. This gives rise to a straight line of slope – 40 dB/decade for $\omega > 10$ which when added to asymptote 2 results in asymptote 3 with a slope 0 + (– 40) = –40 dB/decade passing through the point (6 dB, 10 rad/sec).

To the asymptotic plot thus obtained, corrections are to be applied. Table 10.4 lists the corrections.

Table 10.4 *Corrections to the asymptotic plot of* $4\,(1 + j\omega/2)/(j\omega)(1 + j\omega/10 - \omega^2/100)$

Frequency	*Correction*	*Remarks*
$\omega = 2$	+ 3 dB	
$\omega = 1$	+ 1 dB	
$\omega = 4$	+ 1 dB	
$\omega = 10$	0 dB	Error = – 20 log $2\zeta = 0$
$\omega_r = 10/\sqrt{2}$	$-20\ \log(\sqrt{3}/2)$ dB	$M_r = -20\ \log 2\zeta\sqrt{1-\zeta^2}$

The corrected asymptotic plot of the given transfer function is shown in Fig. 10.32.

10.5.2 Phase Plot

The phase angle at any frequency can be obtained as the algebraic sum of the phase angles due to various factors in the transfer function, in a manner similar to that used for obtaining magnitude plots (refer to Eqns (10.28)). To illustrate the procedure, we consider the transfer function,

$$G(j\omega) = \frac{K(1 + j\omega\tau_1)}{j\omega(1 + j\omega\tau_2)(1 + (j2\zeta/\omega_n)\omega - \omega^2/\omega_n^2} \tag{10.49}$$

The phase of $G(j\omega)$ is

$$\begin{aligned}\angle G(j\omega) &= \tan^{-1}\omega\tau_1 - 90° - \tan^{-1}\omega\tau_2 - \tan^{-1} 2\zeta\omega\omega_n/(\omega_n^2 - \omega^2 \\ &= \phi_1 + \phi_2 + \phi_3 + \phi_4\end{aligned} \tag{10.50}$$

The first term on the right-hand side of Eqn. (10.50) is due to the real zero, and is given by

$$\phi_1 = \tan^{-1}\omega\tau_1$$

The values of ϕ_1 for various values of ω are given in Table 10.5, and these values are plotted in Fig. 10.33. We can approximate the exact curve with the straight line construction shown in Fig. 10.33. The straight line approximation for the phase characteristic breaks from 0° at the frequency $\omega = 1/10\tau_1$, and breaks again to a constant value of 90° at $\omega = 10/\tau_1$. The maximum error in the approximation is 5.7°, and occurs at the two corners $1/10\tau_1$ and $10/\tau_1$.

The second term on the right-hand side in Eqn. (10.50) is due to the pole at the origin and is identicaclly equal to –90°. The third term in Eqn. (10.50) is due to the real pole and is similar to the first term but for reversal in sign. We shall now look at the fourth term, which is due to the complex-conjugate pair of poles, and is given by

$$\phi_4 = -\tan^{-1}\frac{2\zeta\omega_n\omega}{\omega_n^2 - \omega^2}$$

Table 10.5 *Phase characteristics of $(1 + j\omega\tau_1)$*

ω (rad/sec)	*Exact value*	*Straight line approximation*
$0.05/\tau_1$	2.9°	0°
$0.1/\tau_1$	5.7°	0°
$0.2/\tau_1$	11.3°	13.5°
$0.5/\tau_1$	26.6°	31.5°
$0.8/\tau_1$	38.7°	40.6°
$1/\tau_1$	45°	45°
$2/\tau_1$	63.4°	58.5°
$5/\tau_1$	78.7°	76.5°
$8/\tau_1$	82.9°	85.6°
$10/\tau_1$	84.3°	90°
$20/\tau_1$	87.1°	90°

For $\omega << 0.1\omega_n$, we have $\phi_4 = 0$, and for $\omega >> 10\omega_n$, we get $\phi_4 = 180°$. At the intermediate frequency $\omega = \omega_n$, we have $\phi_4 = 90°$. The plots of the asymptotic approximation and the actual curve for various values of ζ are shown in Fig. 10.34. It is seen that for small values of ζ, the actual phase angle changes very rapidly for ω near the undamped natural frequency ω_n.

The overall phase angle is now obtained as the algebraic sum of the phase angle due to each factor. The addition, however, is not as simple as in the case of the magnitude plot due to the fact

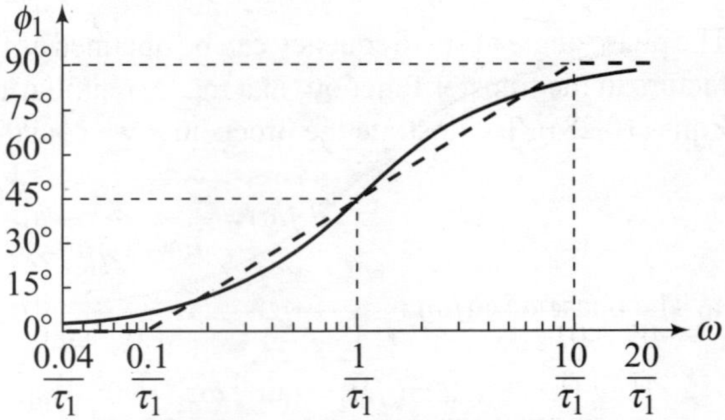

Fig. 10.33 *Phase characteristics of $(1 + j\omega\tau_1)$*

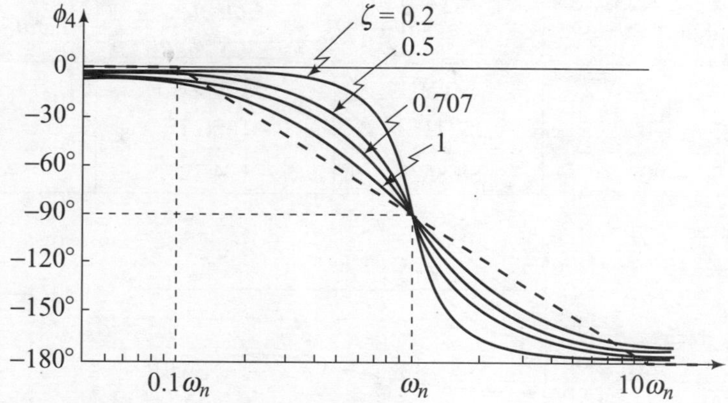

Fig. 10.34 *Phase characteristics of* $1/[1+(j2\zeta/\omega_n)\omega-\omega^2/\omega_n^2]$

that the straight line approximation consists of three parts over four decades. The only exception arises when the various poles and zeros have 'large' separation, so that frequency range of phase angle plot of one factor does not overlap with the frequency range of phase angle plot of any other factor. In this case, the phase angle plots for the various factors can be superimposed on each other to obtain the overall phase.

For a general case, it is usually advisable to make a table of phase angle against frequency for each factor and then obtain the total phase as the algebraic sum of these. Better accuracy is obtained by using the exact values in the table instead of those obtained through the asymptotic approximation. The computational effort to achieve this accuracy is not heavy.

Example 10.12 Consider the transfer function given in Eqn. (10.48) for which the magnitude plot was obtained in Fig. 10.32. We shall now plot the phase curve.

Solution: The phase angle due to each factor for different values of ω is given in Table 10.6. The plot of the total phase is shown in Fig. 10.35.

Table 10.6 *Phase angle due to different factors of the transfer function given by Eqn. (10.48)*

ω (rad/sec)	*Pole at the origin*	*Zero at s = − 2*	*Complex conjugate poles*	*Total phase angle, φ*
0.1	− 90°	2.86°	− 0.57°	− 87.71°
0.2	− 90°	5.71°	− 1.15°	− 85.44°
0.5	− 90°	14.04°	− 2.87°	− 78.33°
1	− 90°	26.57°	− 5.77°	− 69.20°
2	− 90°	45.00°	− 11.77°	− 56.77°
5	− 90°	68.20°	− 33.69°	− 55.49°
10	− 90°	78.69°	− 90.00°	− 101.32°

(continued)

ω (rad/sec)	Pole at the origin	Zero at s = – 2	Complex conjugate poles	Total phase angle, ϕ
20	– 90°	84.29°	– 146.31°	– 152.02°
50	– 90°	87.71°	– 168.23°	– 170.52°
100	– 90°	88.85°	– 174.22°	– 175.68°

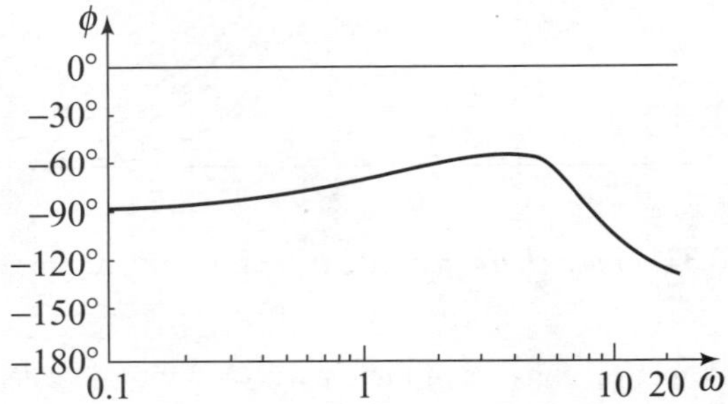

Fig. 10.35 *Bode phase plot for Example 10.12*

10.6 STABILITY MARGINS ON THE BODE PLOTS

The general objective of the frequency-domain design procedures is to shape the Nyquist plot such that the critical point $-1+j0$ is avoided with some reasonable stability margins, thereby providing acceptable closed-loop response characteristics. Direct use of the Nyquist plot for design is not particularly convenient, since changes in parameters other than gain require extensive plot revisions. Fortunately, general characteristics of Nyquist plots can be visualized with reasonable accuracy in most cases of interest from Bode plots, and Bode plots are easily constructed and modified.

Since the Bode plot corresponds to only the positive portion of the $j\omega$-axis from $\omega = 0$ to $\omega = \infty$ in the s-plane, its use in stability studies is limited to the determination of gain and phase crossover points and the corresponding phase and gain margins. Recall that gain margin and phase margin are valid measures of relative stability of a closed-loop system only if the system has open-loop transfer function with no poles in right half of the s-plane.

Since the straight line approximation of the Bode plot is relatively easier to construct, the data necessary for the other frequency-domain plots, such as the Nyquist plots, can be easily generated from the Bode plot.

Example 10.13 A unity-feedback system has open-loop transfer function,

$$G(s) = \frac{K}{(s+2)(s+4)(s+5)}; K = 200 \tag{10.51a}$$

Let us determine the stability of this system implementing the Nyquist stability criterion using Bode plots.

Solution: We begin by sketching the Bode magnitude and phase plots. In the time-constant form, the transfer function $G(s)$ is expressed as

$$G(s) = \frac{5}{(1+s/2)(1+s/4)(1+s/5)} \tag{10.51b}$$

The sinusoidal transfer function

$$G(j\omega) = \frac{5}{(1+j\omega/2)(1+j\omega/4)(1+j\omega/5)}$$

Bode magnitude and phase plots for $G(j\omega)$ are shown in Fig. 10.36.

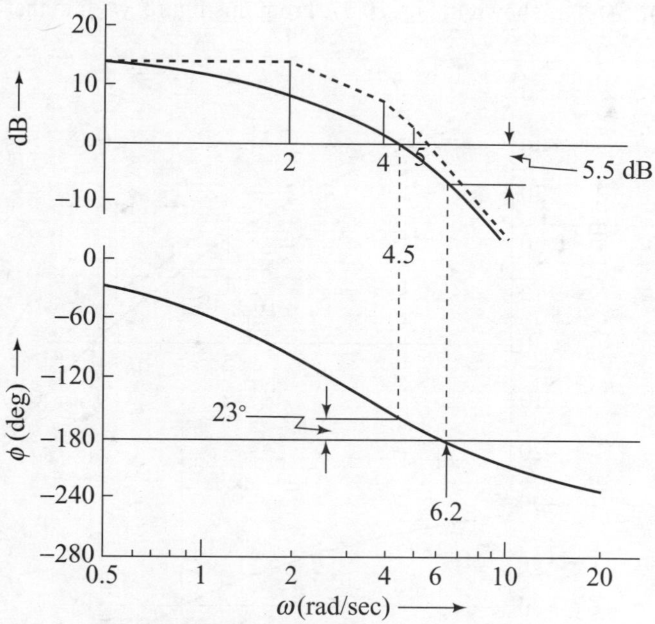

Fig. 10.36 *Bode plot for Example 10.13*

The Nyquist criterion for this example tells us that we want zero encirclements of the $-1+j0$ point for stability. Thus, we recognise that on Bode magnitude plot, the magnitude must be less than 0 dB (which corresponds to unity gain of $G(j\omega)$ on polar plane) at that frequency where the phase is $-180°$. Accordingly, we see that at a frequency of 6.2 rad/sec where the phase is $-180°$, the magnitude is -5.5 dB. Therefore, an increase in gain of $+5.5$ dB is possible before the system becomes unstable. Since the magnitude plot was drawn for a gain of 200 (refer to Eqn. (10.51a)), + 5.5 dB ($20 \log 1.885 \cong 5.5$) represents the allowed increase in gain above 200. Hence, the gain for instability is $200 \times 1.885 = 377$. The closed-loop system is stable for $0 < K < 377$.

Next we show how to evaluate the gain and phase margins using Bode plots. ***The gain margin is found by using the phase plot to find the phase crossover frequency ω_ϕ where the phase angle is $-180°$. At this frequency, look at the magnitude plot to determine the gain margin, GM, which is the gain required to raise the magnitude curve to*** 0 dB. From Fig. 10.36, we find that

$$\omega_\phi = 6.2 \text{ rad/sec}; \; GM = 5.5 \text{ dB}$$

The phase margin is found by using the magnitude curve to find the gain crossover frequency ω_g***, where the gain is*** 0 dB. ***On the phase curve at that frequency, the phase margin,*** ΦM***, is the difference between the phase value and*** $-180°$. From Fig. 10.36, we find that

$$\omega_g = 4.5 \text{ rad/sec}; \; \Phi M = -157° - (-180°) = 23°$$

Example 10.14 Consider now a unity-feedback system with open-loop transfer function,

$$G(s) = \frac{K}{s(1+0.1s)(1+0.2s)}; K = 100$$

The Bode plot of $G(j\omega)$ is shown in Fig. 10.37. From this figure, we find that

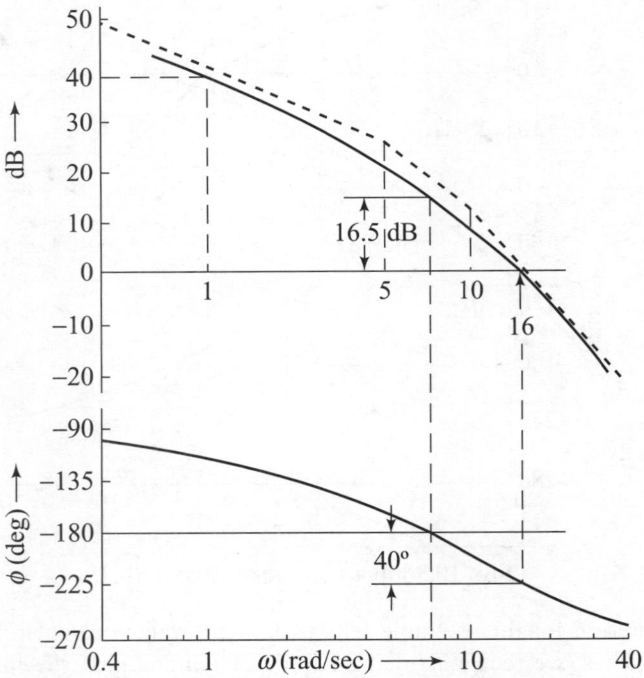

Fig. 10.37 *Bode plot for Example 10.14*

ω_ϕ = phase crossover frequency = 7 rad/sec
GM = gain margin = −16.5 dB
ω_g = gain crossover frequency = 16 rad/sec
ΦM = phase margin = −40°

The closed-loop system is thus unstable.

If we were to decrease the gain by 16.5 dB, all points on dB vs ω plot would decrease in magnitude by this amount without affecting the phase plot. The closed-loop system would be at the limit of instability. The gain K must be less than $100 \times x$ $(16.5 = 20 \log x)$, i.e., the stability range is $0 < K < 668$.

10.7 STABILITY ANALYSIS OF SYSTEMS WITH DEAD-TIME

Figure 10.38 shows the block diagram of a system with dead-time elements in the loop. Open-loop transfer function of the system is

$$G(s)H(s) = G_1(s)e^{-s\tau_D} \tag{10.52}$$

where $G_1(s)$ is a rational function.

Root locus analysis of systems with dead-time elements is possible, but the method is quite complex. In Section 8.5, root locus analysis was carried out using the following approximation:

$$e^{-s\tau_D} \cong \frac{1 - s\tau_D/2}{1 + s\tau_D/2}$$

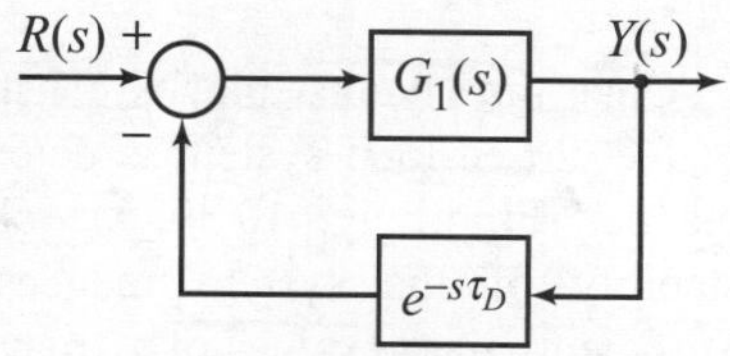

Fig. 10.38 *A closed-loop system with dead-time element in the loop*

This approximation works fairly well as long as the dead-time τ_D is small in comparison to the system time constants.

The frequency-domain graphical methods provide a simple yet exact approach to handle the dead-time problem since the factor $e^{-s\tau_D}$ ***is readily interpreted in terms of either the Nyquist or the Bode plot.***

10.7.1 Nyquist Plots of Systems with Dead-Time

Assume that $G_1(s)$ in Fig. 10.38 is an integrator. Then the open-loop transfer function of the system, becomes

$$G(s)H(s) = \frac{e^{-s\tau_D}}{s}$$

The sinusoidal transfer function is

$$G(j\omega)H(j\omega) = \frac{e^{-j\omega\tau_D}}{j\omega}$$

Clearly,

$$|G(j\omega)H(j\omega)| = 1/\omega;\ \angle G(j\omega)H(j\omega) = -\ \omega\tau_D - \pi/2$$

$$G(j\omega)H(j\omega) = \frac{1}{j\omega}(\cos\omega\tau_D - j\sin\omega\tau_D)$$

$$= -\frac{\sin\omega\tau_D}{\omega} - j\frac{\cos\omega\tau_D}{\omega}$$

$$\lim_{\omega\to 0}\,[G(j\omega)H(j\omega)] = -\tau_D - j\infty$$

Since the magnitude decreases monotonically, and the phase angle also decreases monotonically indefinitely, the polar plot of the given transfer function will spiral into the $\omega \to \infty$ point at the origin, as shown in Fig. 10.39. A curious reader may like to find some of the intersections with real and imaginary axes. For the first intersection with the real axis, set

$$\text{Im}\,[G(j\omega)H(j\omega)] = 0 \text{ to obtain } \omega\tau_D = \pi/2.$$

$$\text{Re}[G(j\omega)H(j\omega)]\Big|_{\omega=\frac{\pi}{2\tau_D}} = -\frac{1}{\omega}\Big|_{\omega=\frac{\pi}{2\tau_D}} = -2\tau_D/\pi$$

Once the Nyquist plot is constructed, the stability of the closed-loop system is determined in the usual manner. From Fig. 10.39, we observe that relative stability of the system reduces as τ_D increases. Sufficiently large values of τ_D may drive the system to instability.

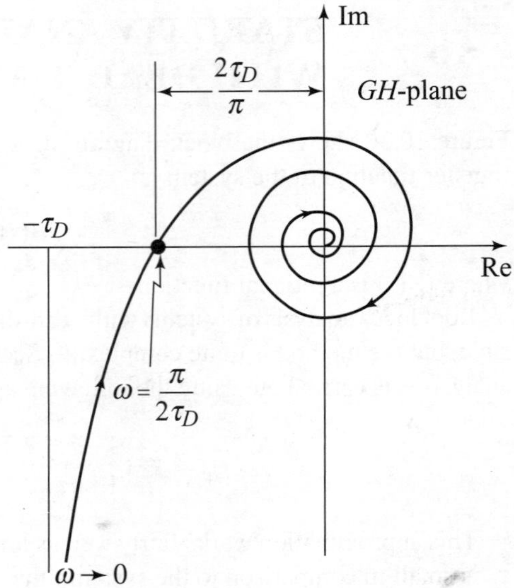

Fig. 10.39 *Polar plot of* $e^{-s\tau_D}/s$

The Critical Trajectory

Thus far, we have used the point $-1+j0$ of the $G(s)H(s)$-plane as the critical point for stability analysis using Nyquist criterion. We can extend the critical-point idea into a trajectory if necessary.

The roots of the characteristic equation of the system shown in Fig. 10.38 satisfy (refer to Eqn. (10.52))

$$1 + G_1(s)e^{-s\tau_D} = 0$$

or

$$G_1(s)e^{-s\tau_D} = -1 \tag{10.53}$$

The right-hand side of the last equation points to the fact that $-1+j0$ is the critical point for stability analysis of the closed-loop system. Equation (10.53) can be written as

$$G_1(s) = -e^{+s\tau_D} \tag{10.54}$$

The corresponding condition for the system without dead-time is

$$G_1(s) = -1 \tag{10.55}$$

Comparison of Eqns (10.54) and (10.55) reveals that the effect of dead-time is simply to shift the critical stability point $-1+j0$ to $-e^{s\tau_D}$, which describes a *critical trajectory*. When $\omega\tau_D = 0$, the trajectory starts at the point $-1+j0$, and as $\omega\tau_D$ increases, the critical point traces out a circle with unit radius centered at the origin of $G_1(s)$-plane, in the counterclockwise direction.

Example 10.15 Let us determine, with the help of Nyquist stability criterion, the maximum value of τ_D for stability of the closed-loop system of Fig. 10.38 with

$$G_1(s) = \frac{1}{s(s+1)(s+2)}$$

From Eqn. (10.54), we have with $s = j\omega$,

$$G_1(j\omega) = \frac{1}{j\omega(j\omega+1)(j\omega+2)} = -e^{+j\omega\tau_D}$$

Figure 10.40 shows the $G_1(j\omega)$ locus together with the critical trajectory of $-e^{+j\omega\tau_D}$. The frequency at which the $G_1(j\omega)$ locus intersects the critical trajectory is found by setting the magnitude of $G_1(j\omega)$ to unity, i.e.,

$$|G_1(j\omega)| = \left|\frac{1}{-3\omega^2 + j\omega(2-\omega^2)}\right| = 1$$

which gives $\omega = 0.446$ rad/sec.
Since

$$\angle G_1(j\omega)\Big|_{\omega=0.446} = \angle -e^{j\omega\tau_D}\Big|_{\omega=0.446}$$

we obtain (refer to Fig. 10.40)

$$53.4\ (\pi/180) = 0.932 \text{ rad} = 0.446\ \tau_D \text{ rad}$$

which gives

$$\tau_D = 2.09 \text{ sec}$$

From Fig. 10.40, we observe that the critical point on the critical trajectory is encircled by the $G_1(j\omega)$ locus for $\omega\tau_D > 0.932$ ($\tau_D > 2.09$) and is not encircled for $\omega\tau_D < 0.932$ ($\tau_D < 2.09$); and hence, we conclude that the system under discussion is stable if $\tau_D < 2.09$ sec.

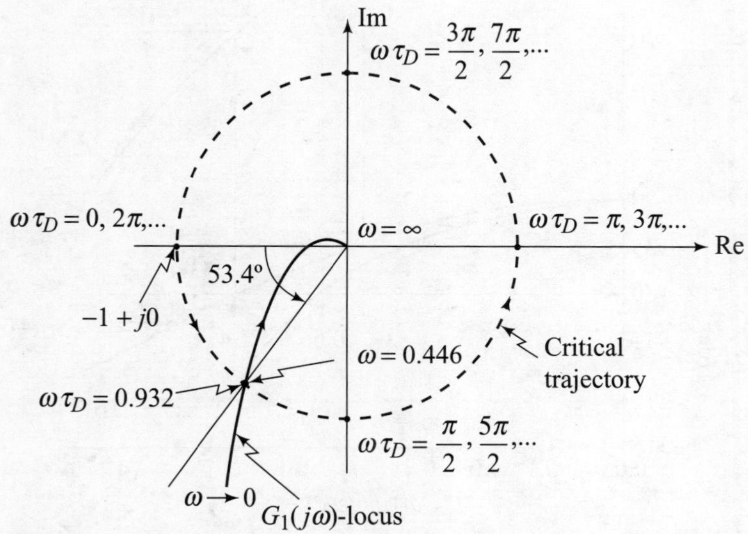

Fig. 10.40 *Critical trajectory and $G_1(j\omega)$-locus for Example 10.15*

10.7.2 Bode Plots of Systems with Dead-Time

The stability analysis of systems with dead time can be conducted easily using the Bode plots. Since $20 \log |e^{-j\omega\tau_D}| = 0$, the magnitude plot of a system is unaffected by the presence of dead time. The dead time, of course, contributes a phase angle of $-(\omega\tau_D \times 180°)/\pi$, thereby causing the modification of the phase plot. An illustrative example follows.

Example 10.16 Consider the closed-loop system of Fig. 10.38 with

$$G_1(s) = \frac{K}{s(s+1)(s+2)} = \frac{K/2}{s(1+s)(1+s/2)}$$

Figure 10.41 shows the Bode plot of $G_1(j\omega)e^{-j\omega\tau_D}$ with $K = 1$ and $\tau_D = 0$. We find from this figure that

ω_g = gain crossover frequency = 0.446 rad/sec
ΦM = phase margin = 53.4°
ω_ϕ = phase crossover frequency = 1.4 rad/sec
GM = gain margin = 16 dB

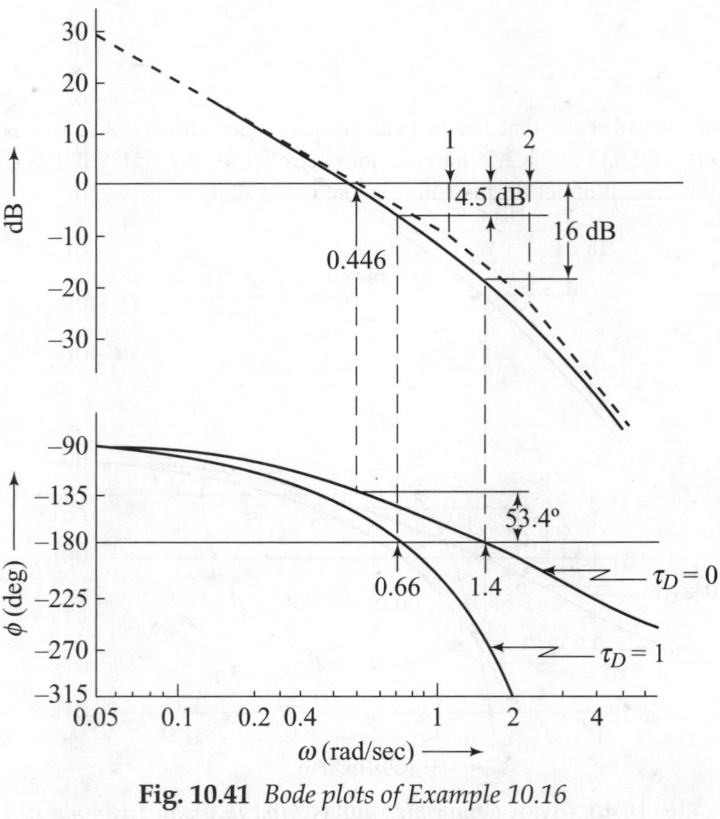

Fig. 10.41 *Bode plots of Example 10.16*

The effect of dead time is to add the phase shift of $-(\omega\tau_D \times 180)/\pi$ degrees to the phase curve while not affecting the magnitude curve. The adverse effect of dead time on stability is apparent since the negative phase shift caused by the dead time increases rapidly with the increase in ω. Let us set $\tau_D = 1$ sec and find the critical value of K for stability. Figure 10.41 also shows Bode plot of $G_1(j\omega)e^{-j\omega\tau_D}$ with $K = 1$ and $\tau_D = 1$. The magnitude curve is unchanged; the phase curve drops with the increase in ω, and the phase crossover frequency is now at 0.66 rad/sec. The gain margin is 4.5 dB. Note that 20 log 1.67 = 4.5 dB; therefore, the critical value of K for stability when $\tau_D = 1$ is 1.67.

Let us now determine the critical value of dead time, τ_D, for stability with $K = 1$. Since the phase margin with $\tau_D = 0$ and $K = 1$ is 53.4°, a phase lag of 53.4° can be introduced by dead time before instability sets in. The critical value of dead time is, therefore, given by the relation,

$$\omega_g \tau_D \times 180/\pi = 53.4$$

which gives

$$\tau_D = 2.09 \text{ sec}$$

10.8 FREQUENCY RESPONSE MEASUREMENTS

In this section we will describe how a system responds to sinusoidal input—called the system's *frequency response*—and how the frequency response measurements can be used for design purposes.

To introduce the ideas, we consider a *stable* linear time-invariant system shown in Fig. 10.42. The input and output of the system are described by $x(t)$ and $y(t)$, respectively. The response of the system to sinusoidal input

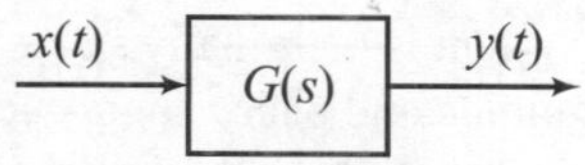

Fig. 10.42 *A stable linear time-invariant system*

$$x(t) = X \sin \omega t$$

will consist of two components: the transient component and the steady-state component (refer to Section 2.9). The transient component will disappear as time goes by. The steady-state output that results after the transient dies out, is of the form

$$y(t) = A(\omega) \times \sin(\omega t + \phi(\omega))$$

which is a sine wave of the same frequency as the input; it differs from the sinusoidal input only in the amplitude and the phase angle. Both the amplitude and the phase angle change when we use a different frequency ω.

The output-input amplitude ratio $A(\omega)$, and the phase angle $\phi(\omega)$ between the output and input sinusoids, are given by the relations

$$A(\omega) = \frac{\text{Amplitude of the sinusoidal output}}{\text{Amplitude of sinusoidal input}} = |G(s = j\omega)|$$

$$\phi(\omega) = \text{Phase angle of the sinusoidal output} - \text{phase angle of the sinusoidal input}$$
$$= \angle G(s = j\omega)$$

The variation of $A(\omega)$ and $\phi(\omega)$ with frequency ω is, by definition, the *frequency response* of the system.

If the system equations are known, complete with numerical values, we can derive its transfer function $G(s)$ and therefrom the *sinusoidal transfer function* $G(j\omega)$. Frequency response of the system can be computed from $G(j\omega)$.

If the equations and/or parameter values are not known, but the physical system or its components exist and are available for test, we can generate the frequency-response data by experimental measurements:

1. A sinusoidal input of known amplitude and frequency is applied to the system.
2. The system output is allowed to 'settle' into a steady-state pattern.
3. The amplitude and relative phase of the sinusoidal output are measured and recorded.
4. This procedure is repeated for values of ω spanning the frequency range of interest.

Obviously, no steady-state response exists for an unstable system; so a frequency-response test will directly indicate instability.

Raw measurements of the output amplitude and phase of a stable plant undergoing a sinusoidal input are sufficient to analyse feedback system stability. No intermediate processing of the data is necessary because Nyquist plot/Bode plot can be sketched directly from experimental measurements.

Sometimes it is desirable to obtain an approximate model, in terms of a transfer function, directly from the frequency-response measurements. Determination of a transfer function $G(s)$ from measured data is an *identification problem*.

For a stable system, $|G(j\omega)|$ and $\angle G(j\omega)$ can be obtained by measurement. This is also possible if the system is marginally stable with a pole at $s = 0$. Experimental data are used to obtain the exact log-magnitude and phase angle curves of the Bode plot. Asymptotes are drawn on the exact log-magnitude curves by utilizing the fact that their slopes must be multiples of ± 20 dB/decade. From these asymptotes, the system type and approximate time constants are determined.

Careful interpretation of the phase angle curve is necessary to identify whether the transfer function is a *minimum phase* or a *nonminimum phase* transfer function. Let us first define these terms.

Consider the following transfer functions:

$$G_1(j\omega) = \frac{1 - j\omega\tau}{(1 + j\omega\tau_1)(1 + j\omega\tau_2)}; G_2(j\omega) = \frac{1 + j\omega\tau}{(1 + j\omega\tau_1)(1 + j\omega\tau_2)}$$

It can easily be deduced that the two transfer functions have the same magnitude characteristics:

$$|G_1(j\omega)| = |G_2(j\omega)|;\ \omega \geq 0$$

However, the phase characteristics are different for the two cases as illustrated in Fig. 10.43. The transfer function with a zero in the right half of the s-plane undergoes a net change in phase, when evaluated for frequency inputs between zero and infinity, which is greater than that for the transfer function with the zero in the left half of the s-plane; magnitude plots of two transfer functions being the same. Based on the phase shift characteristics, transfer functions are classified as minimum phase and nonminimum phase.

The range of phase shift of a *minimum phase transfer function* is the least possible corresponding to a given magnitude curve, whereas the range of phase shift of *nonminimum phase transfer function* is greater than the minimum possible for the given magnitude curve.

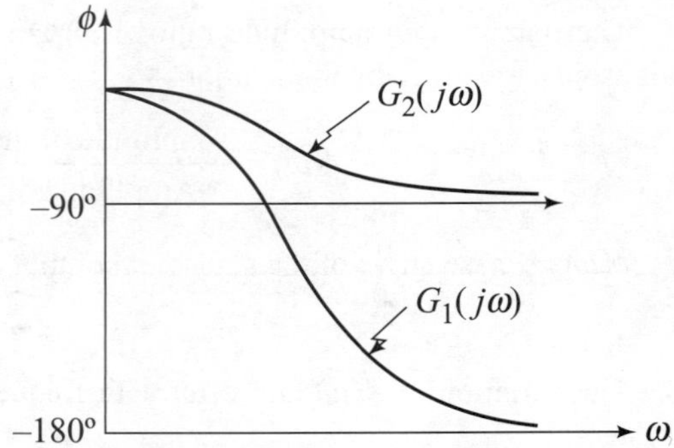

Fig. 10.43 *Phase angle characteristics of minuimum phase and nonminimum phase transfer functions*

***It can easily be established that a proper rational transfer function is a minimum phase transfer function if all of its zeros lie in the left half of the s-plane. It is a nonminimum phase transfer function if it has one or more zeros in the right half of the s-plane*[4].**

Using the basic definition, we can easily establish the minimum phase or nonminimum phase character of irrational transfer functions as well. A nonminimum phase irrational transfer function of practical importance is one with $e^{-\tau_D s}$ as its factor, e.g.,

$$G(j\omega) = \frac{(1 + j\omega\tau)e^{-j\omega\tau_D}}{(1 + j\omega\tau_1)(1 + j\omega\tau_2)}$$

With these observations, we can go back to the identification problem. If the Bode plot is obtained by frequency response measurements, then, except for a possible pole at $s = 0$, all the poles must lie in the left half of the s-plane. However, zeros in the right half of the s-plane and/or presence of dead-time elements cannot be ruled out. Therefore, care must be taken in interpreting the phase angle plot to detect the presence of right-half plane zeros and/or dead-time elements. We illustrate the identification procedure through examples.

Example 10.17

Table 10.7 gives experimentally-obtained frequency-response data of a system. From this data, let us determine the approximate transfer function model of the system.

Table 10.7 *Experimental frequency response data*

f(Hz)	*ω(rad/sec)*	*Gain(dB)*	*Phase(deg)*
60	377	– 7.75	– 155
50	314	– 4.3	– 150
40	251	– 0.2	– 145
35	219	0.75	– 140
25	157	5.16	– 135
20	126	7.97	– 120
16	100	10.5	– 110
10	63	15.0	– 100
7	44	16.9	– 85
2.5	16	20.4	– 45
1.3	8	21.6	– 30
0.22	1.38	24.0	– 5
0.16	1.0	24.1	0

[4] Nonminimum phase systems are particularly difficult to control and require special attention.

Bode plot shown in Fig. 10.44 is drawn using the data of Table 10.7. First we approximate the magnitude plot by the three straight dashed lines shown. They intersect at $\omega = 15$ and $\omega = 150$. We begin with the leftmost part of the magnitude plot. The low frequency asymptote is a horizontal line at 25 dB. It indicates that the transfer function represents a type-0 system with a gain K given by

$$20 \log K = 25 \text{ or } K = 17.8$$

At $\omega = 15$, the slope becomes –20 dB/decade. Thus, there is a pole with corner frequency 15. At $\omega = 150$, the slope becomes –40 dB/decade, a decrease of 20 dB/decade. Therefore, there is another pole with corner frequency 150.

The transfer function must be of the form: $G(s) = \dfrac{17.8}{(1+s/15)(1+s/150)}$.

The phase characteristics calculated from this transfer function are in fair agreement with experimentally obtained characteristics shown in Fig. 10.44. Dead-time element is, therefore, not present.

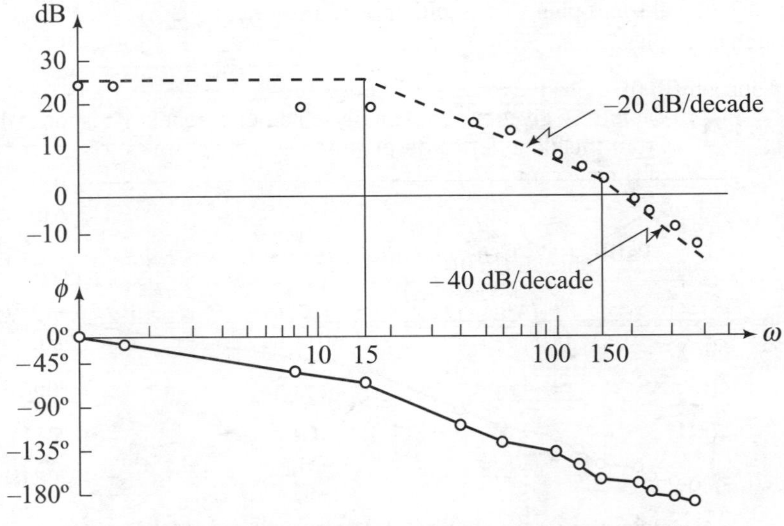

Fig. 10.44 *Experimental data in Bode coordinate system*

Example 10.18 Let us find the transfer function of the experimentally obtained Bode plot shown in Fig. 10.45. First we approximate the magnitude plot by the three dashed lines shown. They intersect at $\omega = 1$ and $\omega = 10$. In the low-frequency range, there is an asymptote with slope – 20 dB/decade; it indicates the presence of a factor of the form $K/(j\omega)$ in the transfer function. At $\omega = 1$, the magnitude of the asymptote is 40 dB. Therefore,

$$20 \log K = 40$$

This gives $$K = 100$$

At $\omega = 1$, the slope becomes – 40 dB/decade, a decrease of 20 dB/decade; thus there is a pole with corner frequency $\omega = 1$. At $\omega = 10$, the slope becomes – 20 dB/decade, an increase of 20 dB/decade; there is a zero with corner frequency $\omega = 10$. The transfer function must be of the form,

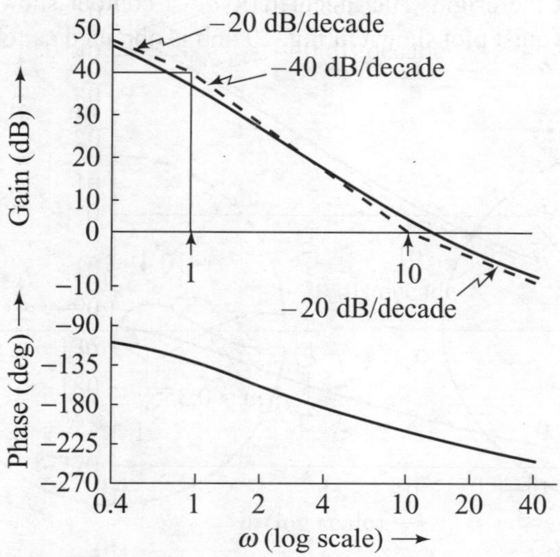

Fig. 10.45 *Experimentally obtained Bode plot*

$$G(s) = \frac{100\left(1 \pm \frac{s}{10}\right)}{s(1+s)}$$

We have assumed that the Bode plot of Fig. 10.45 is obtained by frequency response measurements; the pole in the right half of the s-plane is, therefore, not possible. However, a zero in the right half of the s-plane may exist, i.e., the system under experimental analysis may be a nonminimum-phase system.

Now we use the phase plot to determine the sign of the zero factor. If the sign of zero $(1 \pm 0.1s)$ is positive, the zero will introduce positive phase into $G(s)$ or, equivalently, the phase of $G(s)$ will increase as ω passes through the corner frequency at 10. This is not the case as seen from Fig. 10.45; thus we have $(1 - 0.1s)$, and the transfer function of the Bode plot is

$$G(s) = \frac{100(1-0.1s)}{s(1+s)}$$

The phase characteristics calculated from this transfer function are in fair agreement with experimentally obtained characteristics shown in Fig. 10.45.

10.9 REVIEW EXAMPLES

Review Example 10.1 Consider a feedback system with an open-loop transfer function,

$$G(s)H(s) = \frac{1+4s}{s^2(1+s)(1+2s)}$$

which has a double pole at the origin. The indented Nyquist contour shown in Fig. 10.46a does not enclose these poles. The Nyquist plot shown in Fig. 10.46b is obtained as follows:

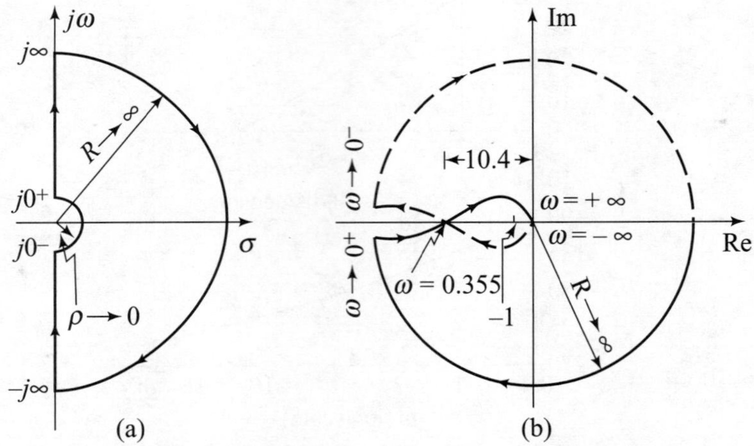

Fig. 10.46 *Nyquist contour and the corresponding mapping*[5] *for* $G(s)H(s) = (4s + 1)/[s^2(s + 1)(2s + 1)]$

The portion of the $G(s)H(s)$-locus from $\omega = 0^+$ to $\omega = +\infty$ is simply the polar plot of

$$G(j\omega)H(j\omega) = \frac{1 + j4\omega}{(j\omega)^2(1 + j\omega)(1 + j2\omega)}$$

The following points were used to construct the polar plot:

1. $G(j\omega)H(j\omega)\big|_{\omega\to 0^+} \to \infty\angle -180^\circ$
2. The $G(j\omega)H(j\omega)$-locus intersects the real axis at a point where

$$\angle G(j\omega)H(j\omega) = -180^\circ$$

or
$$-180^\circ - \tan^{-1}\omega - \tan^{-1}2\omega + \tan^{-1}4\omega = -180^\circ$$

or
$$\tan^{-1}4\omega - \tan^{-1}\omega = \tan^{-1}2\omega$$

Therefore,
$$\tan(\tan^{-1}4\omega - \tan^{-1}\omega) = \tan(\tan^{-1}2\omega)$$

or
$$\frac{4\omega - \omega}{1 + 4\omega^2} = 2\omega$$

This gives
$$\omega = \frac{1}{2\sqrt{2}} = 0.355 \text{ rad/sec}$$

$$\left|G(j\omega)H(j\omega)\right|\Big|_{\omega=0.355} = 10.4$$

5

ω	0.01	0.1	0.35	0.36
$\lvert GH \rvert$	4×10^6	104.09	11.4	10.3
$\angle GH$	–179.97	–175.2	–179.35	179.64

3. $G(j\omega)H(j\omega)\big|_{\omega\to+\infty} \to 0\angle-270°$

The plot of $G(j\omega)H(j\omega)$ for $\omega < 0$ is the reflection with respect to the real axis, of the plot for $\omega > 0$. Every point on the large semicircle in Fig. 10.47a, is mapped by $G(s)H(s)$ into the origin of the $G(s)H(s)$-plane. A point $s = \rho e^{j\phi}$; $\rho \to 0$, ϕ varying from – 90°, through 0° to + 90°, is mapped by $G(s)H(s)$ into

$$G(s)H(s)\Big|_{s=\rho e^{j\phi}} = \frac{1+4s}{s^2(1+s)(1+2s)}\Big|_{s=\rho e^{j\phi}} = \frac{1}{\rho^2 e^{j2\phi}} = \frac{1}{\rho^2}\angle-2\phi$$

The small semicircle in Fig. 10.46a is mapped by $G(s)H(s)$ into a circular arc of infinite radius ranging from 180° at $\omega = 0^-$ through 0° to –180° at $\omega = 0^+$ (180° → 90° → 0° → – 90° → – 180°).

Figure 10.46b indicates that the critical point $-1 + j0$ is encircled by the Nyquist plot twice in the clockwise direction. Therefore, $N = 2$. From the given transfer function it is seen that no pole of $G(s)H(s)$ lies in the right half of the s-plane, i.e., $P = 0$. Therefore (refer to Eqn. (10.8a))

Z = number of zeros of $1 + G(s)H(s)$ enclosed by the Nyquist contour

$$= N + P = 2$$

Hence, the feedback system is unstable with two poles in the right half of the s-plane.

Review Example 10.2 The frequency response test data for the forward-path elements of a unity-feedback control system are given in Table 10.8. These data are sufficient to analyse feedback system stability, as is seen below.

Figure 10.47 shows the frequency response data of Table 10.8 on Bode magnitude and phase plots. From these plots, we find that

phase crossover frequency $\omega_\phi = 48$ rad/sec,
gain crossover frequency $\omega_g = 3$ rad/sec,
gain margin $GM = 28$ dB, and
phase margin $\Phi M = 49°$.
The feedback system is therefore stable.

It may be desirable to obtain a transfer function that approximates the experimental amplitude and phase characteristics. A series of straight-line asymptotes fitted to the experimentally obtained

Table 10.8 *Experimental frequency response data*

ω (rad/sec)	0.5	1	2	4	10	20	30
Gain (dB)	19.75	13	5	– 4	– 17	– 24	– 26
Phase (deg)	– 101	– 111	– 125	– 134	– 130	– 127	– 134
ω(rad/sec)	40	50	60	70	80	90	100
Gain (dB)	– 27	– 29	– 34	– 39	– 43	– 47	– 50
Phase (deg)	– 154	– 189	– 218	– 235	– 243	– 249	– 252

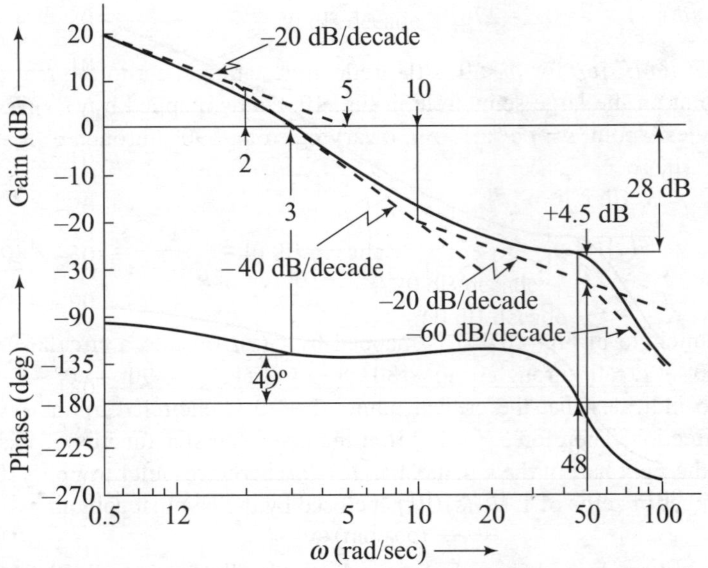

Fig. 10.47 *Experimentally obtained magnitude and phase characteristics*

magnitude curve are shown in Fig. 10.47. The low frequency asymptote has a slope of −20 dB/decade and when extended, intersects the 0-dB axis at $\omega = 5$. Therefore, the asymptote is a plot of the factor $5/(j\omega)$. The corner frequencies are found to be located at $\omega = 2$, $\omega = 10$ and $\omega = 50$. At the first corner frequency, the slope of the curve changes by −20 dB/decade and at the second corner frequency, it changes by + 20 dB/decade. Therefore, the transfer function has factors $1/(1 + j\omega/2)$ and $(1 + j\omega/10)$ corresponding to these corner frequencies. At $\omega = 50$, the curve changes by a slope of – 40 dB/decade. At this frequency, the error between actual and approximate plots is + 4.5 dB. The change of slope by – 40 dB/decade indicates the presence of either double pole on real axis or a pair of complex conjugate poles. The error of + 4.5 dB at the corner frequency and the peak occurring at a frequency less than the corner frequency indicate the presence of a quadratic factor with $\zeta < 0.5$. In fact, we can calculate the approximate value of ζ using expression (10.42):

$$-20 \log 2\zeta = 4.5$$

This gives $\zeta = 0.3$

Therefore, the transfer function has a quadratic factor

$$\frac{1}{1 + j2\zeta(\omega/50) + (j\omega/50)^2} \quad \text{where } \zeta = 0.3$$

Thus, an approximate transfer function of the forward-path elements of the feedback system becomes

$$G(j\omega) = \frac{5(1 \pm j\omega/10)}{j\omega(1 + j\omega/2)[1 + j0.6(\omega/50) + (j\omega/50)^2]}$$

Now we use the phase plot to determine the sign of the zero factor. The phase characteristics calculated from the transfer function are in fair agreement with experimentally obtained characteristics shown in Fig. 10.47, when the sign of the zero factor is positive. Therefore, the transfer function of the Bode plot of Fig. 10.47 is

$$G(s) = \frac{5(1+s/10)}{s(1+s/2)\left[1+0.6s/50+(s/50)^2\right]}$$

CONCLUDING REMARKS

The root locus analysis and design discussed in Chapters 8 and 9 seems to yield satisfactory solution to the control-design problems. Then why look for an alternative like frequency response analysis and design? We started the dialogue in this chapter to answer this question, and our answer will mature through the discussion in the next two chapters.

What we have learnt?

- The frequency response of a system can be represented either as a polar plot or as separate magnitude and phase diagrams. As a polar plot, the magnitude response is the length of the vector drawn from the origin to a point on the curve, whereas the phase response is the angle of that vector; frequency is implicit and is shown at each point on the polar curve. Separate magnitude and phase diagrams present the data with frequency explicitly enumerated along the abscissa.
- The Nyquist criterion, which uses polar plots (called the Nyquist diagram), sets forth the theoretical foundation from which the open-loop sinusoidal transfer function can be used to determine stability of the closed-loop system. The Nyquist criterion also provides us with the relative stability measures—gain margin and phase margin.
- Separate magnitude and phase diagrams are called Bode plots when magnitude curve is a plot of log-magnitude versus log-frequency, and the phase curve is a plot of phase angle versus log-frequency. The advantage of Bode plots over the Nyquist diagram is that they can easily be drawn using asymptotic approximations to the actual curve.
- Since Bode plots simplify the computations considerably, and also give better interpretation of filtering characteristics of feedback systems, we prefer using Bode plots. Based on Nyquist stability criterion, relative stability measures can easily be calculated from Bode plots; we can also investigate the effect of variation of loop gain parameter on the relative stability of the feedback system.
- As with the root locus plots, it will be tempting to rely exclusively on MATLAB to obtain Bode, Nyquist and other frequency-response plots. We strongly recommend that MATLAB should be treated as one tool in the tool kit that can be used to design and analyze control systems. There is no substitute for a clear understanding of the underlying theory.

Review Questions

10.1 (a) Using the principle of argument, derive the Nyquist stability criterion.
(b) What is the interpretation of stability analysis when Nyquist plot crosses the $-1+j0$ point?

10.2 (a) Give an account of meaning of the terms 'gain margin' and 'phase margin' with reference to Nyquist plots.
(b) How can a frequency response be represented by a Bode plot? Indicate what gain and phase margins are in this context.
(c) Why do we use logarithmic scale for frequency in Bode plots?

10.3 Give the properties of minimum-phase and nonminimum-phase transfer functions.
Describe a method of identification of transfer function model of a system using frequency-response measurements.

10.4 Give examples of Nyquist plots with
(i) poor ΦM, infinite GM (ii) poor GM, infinite ΦM

Self-Appraisal Quiz

A set of questions with multiple answer choices is given here. These questions have been designed to capture important aspects of the subject covered in the chapter. For each of these questions, identify the best of the given answer choices. Compare your answers with the Master Key given at the end of the session. This quiz should serve the purpose of a self-appraisal test for readers.

10.1 Figure Q10.1 shows the Nyquist plot of a unity feedback system having open-loop transfer function $G(s)$ with one pole in right half of s-plane. The feedback system is
(A) stable
(B) unstable
(C) marginally stable

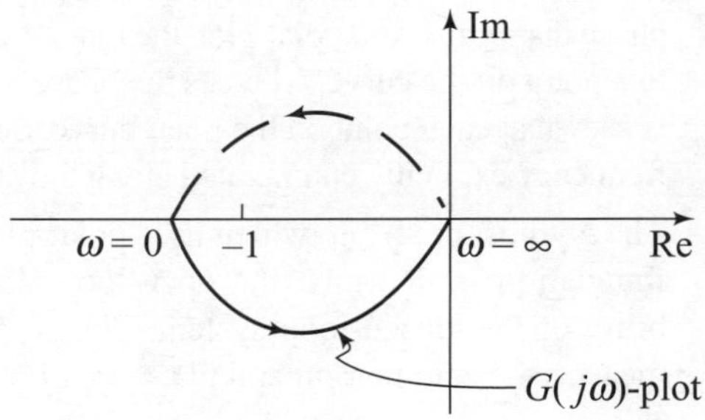

Fig. Q10.1

10.2 Figure Q10.2 shows the Nyquist plot of a unity feedback system having open-loop transfer function $G(s)$ with one pole in right half of s-plane. The feedback system is
(A) stable
(B) unstable
(C) marginally stable

10.3 Polar plot $G(j\omega) = 1/[j\omega(1 + j\omega\tau)]$
(A) crosses the negative real axis
(B) crosses the negative imaginary axis
(C) crosses the positive imaginary axis
(D) None of the answers in (A), (B), and (C) is correct

10.4 Which of the polar plots shown in Fig. Q10.4 is the correct plot for $G(j\omega) = 1/[(j\omega)^2(1 + j\omega\tau)]$?
(A) Fig. Q10.4a (B) Fig. Q10.4b
(C) Fig. Q10.4c (D) Fig. Q10.4d

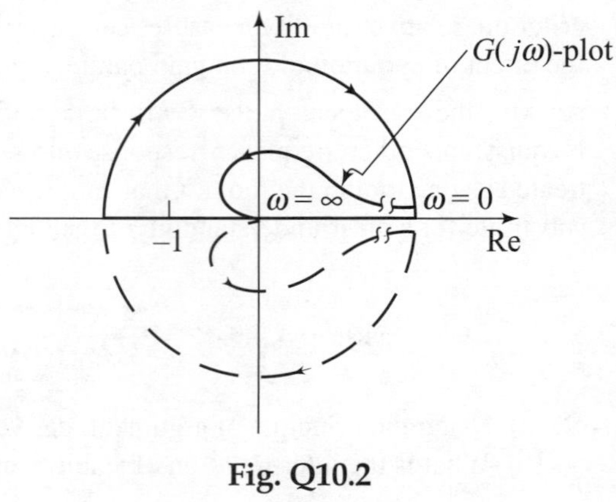

Fig. Q10.2

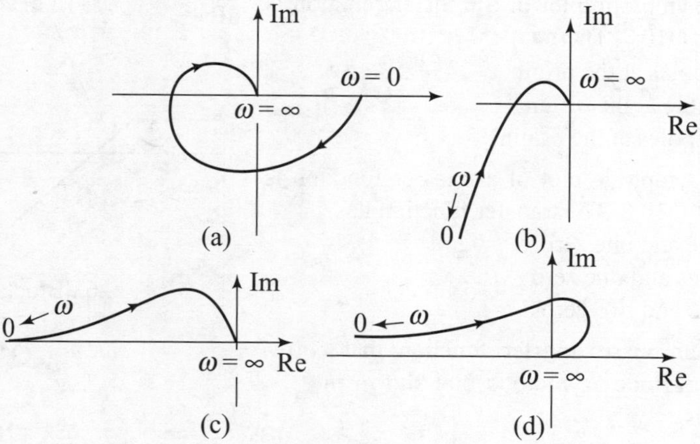

Fig. Q10.4

10.5 The transfer function of a system is $G(s) = 10(1+0.2s)/(1+0.5s)$. The phase shifts at $\omega = 0$ and $\omega = \infty$, will be respectively

(A) 90° and 0° (B) −180° and 180°
(C) −90° and 90° (D) None of the answers in (A), (B), and (C) is correct

10.6 Which of the Bode asymptotic plots shown in Fig. Q10.6 is the correct plot for $G(s) = K/[s^2(s + 5)]$?

(A) Fig. Q10.6a (B) Fig. Q10.6b (C) Fig. Q10.6c (D) Fig. Q10.6d

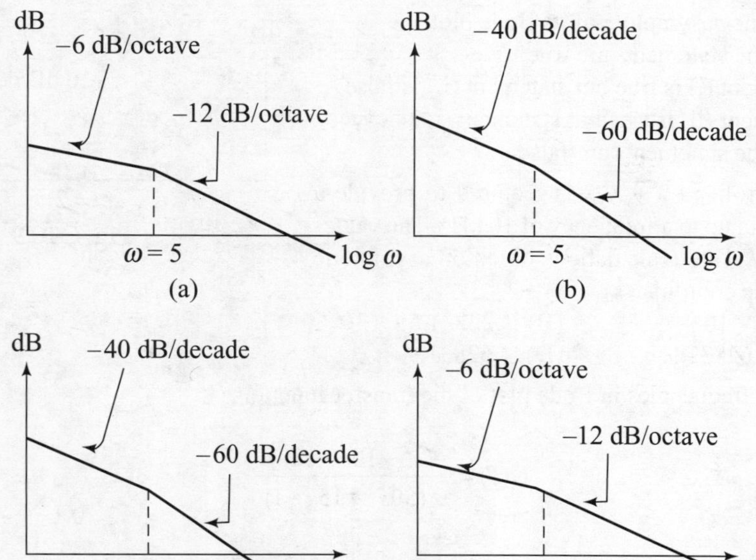

Fig. Q10.6

10.7 The Bode asymptotic plot of a transfer function is given in Fig. Q10.7. There
(A) are no poles at the origin
(B) is one pole at the origin
(C) are two poles at the origin

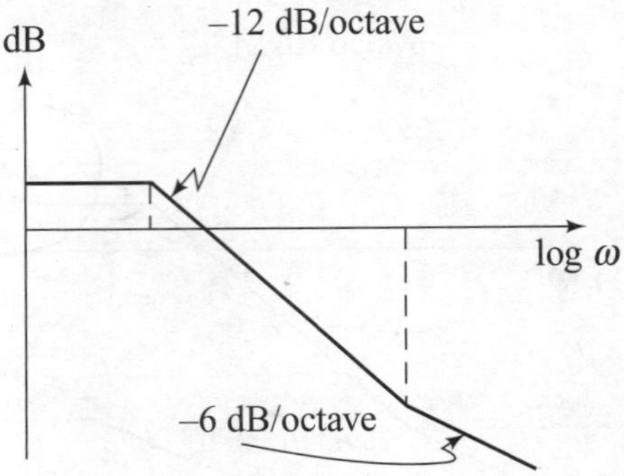

Fig. Q10.7

10.8 The Bode asymptotic plot of a transfer function is given in Fig. Q10.8. The transfer function has
(A) one pole and one zero
(B) two poles and one zero
(C) one pole and two zeros

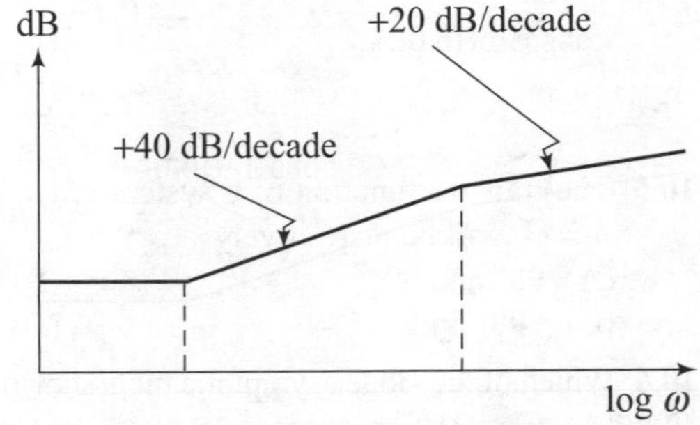

Fig. Q10.8

10.9 The minimum-phase transfer function that corresponds to the Bode asymptotic plot shown in Fig. Q10.9, is

(A) $\dfrac{1}{2s+1}$ (B) $2s+1$

(C) $\dfrac{1}{\frac{1}{2}s+1}$ (D) $\frac{1}{2}s+1$

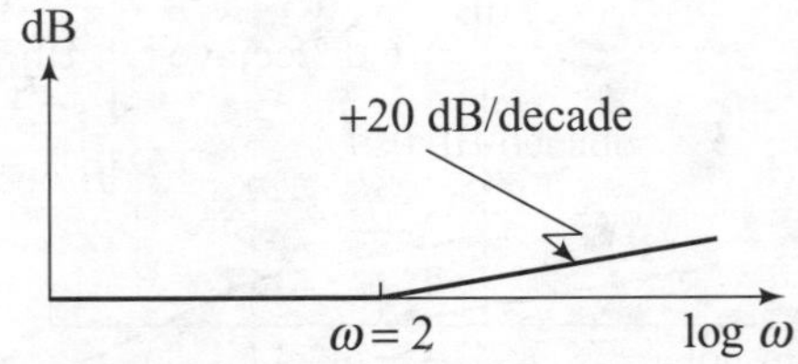

Fig. Q10.9

10.10 (i) Type number of a transfer function dictates high-frequency asymptote of its Bode plot.
(ii) Order of a transfer function dictates low-frequency asymptote of its Bode plot.
(A) Both the statements are true
(B) Statement (i) is true but statement (ii) is false
(C) Statement (ii) is true but statement (i) is false
(D) Both the statements are false

10.11 A PD controller $K_P + K_D s$ is required to provide a unity dc gain up to a frequency of 10 kHz. The values of K_P and K_D are (calculations based on asymptotic plot in Bode coordinates)
(A) $1, 1.59 \times 10^{-5}$ (B) $1.222, 1.59 \times 10^{-5}$
(C) 1.222, 62831.9 (D) 1,62831.9

10.12 The corner frequencies in Bode plot of the transfer function,

$$G(s) = \frac{6(s^2 + 10s + 100)}{s^2(50s^2 + 15s + 1)}$$

are
(A) 1,50 (B) 100,1
(C) 10,1 (D) None of the answers in (A), (B), and (C) is correct

10.13 The low-frequency asymptote in Bode plot of

$$G(s) = \frac{6(s^2 + 10s + 100)}{s^2(50s^2 + 15s + 1)}$$

has a slope of
(A) 0 dB/decade
(B) −20 dB/decade
(C) −40 dB/decade
(D) None of the answers in (A), (B) and (C) is correct

10.14 The high-frequency asymptote in Bode plot of

$$G(s) = \frac{6(s^2 + 10s + 100)}{s^2(50s^2 + 15s + 1)}$$

has a slope of
(A) 0 dB/decade
(B) −20 dB/decade
(C) −40 dB/decade
(D) None of the answers in (A), (B), and (C) is correct

10.15 If a system's gain is given as $|G(j\omega_0)| = 0$ dB, what will happen to an input signal at the frequency ω_0?
(A) It will be amplified
(B) It will be attenuated
(C) There will be no output signal
(D) None of the answers in (A), (B) and (C) is correct

10.16 An engineer applies the input $r(t) = 2 \sin t$ to a chemical process and measures the output as $y(t) = 0.4 \sin(t - 1.55)$. What are the gain and phase of the system?
(A) (0.2, −89°)
(B) (0.4, −1.55°)
(C) (0.2, −1.55°)
(D) (0.4, −89°)

10.17 A current to pressure transducer has a gain of 0.316 and a phase of −60° at $\omega = 10$ rad/sec. An input signal of $r(t) = 5\cos(10t - \pi/2)$ is injected. What is the output signal?
(A) $1.58\cos(20t - 5\pi/6)$
(B) $1.58\cos(10t - \pi/6)$
(C) $1.58\cos(10t + \pi/6)$
(D) None of the answers in (A), (B), and (C) is correct

10.18 Bode magnitude plot of a system has −20 dB gain at low frequencies. The system is
(A) type-0
(B) type-1
(C) type-2
(D) Nothing can be deduced about type number from the given information

10.19 A PI controller $K_P + K_I/s = 100 + 0.01/s$ has a high-frequency gain of 40 dB. The values of K_P and K_I that reduce high-frequency gain to 20 dB are (calculations based on asymptote plot in Bode coordinates)
(i) (10,0.01) (ii) (10,0.001)
(A) Both the statements are true
(B) Statement (i) is true but (ii) is false
(C) Statement (i) is false but (ii) is true
(D) Both the statements are false

10.20 The frequency response of the system $G(s) = e^{-\tau_D s}/s$ at frequency $\omega = \pi/(2\tau_D)$ gives magnitude and phase of
(A) $2\tau_D/\pi$, −90°
(B) $2\tau_D/\pi$, −180°
(C) $2\tau_D/\pi$, 0°
(D) None of the answers in (A), (B) and (C) is correct

10.21 A PD controller has transfer function $(0.1 + 0.01s)$. The frequency at which the magnitude is 20 dB is (calculations based on asymptotic plot is Bode coordinates)
(A) 10 (B) 1000 (C) 1 (D) 100

10.22 A unity feedback system has open-loop transfer function $G(s)$. Polar plot of $G(j\omega)$ is shown in Fig. Q10.22. The gain margin (GM) and the phase margin (ΦM) of the feedback system are
(A) $GM = -0.3$; $\Phi M = 112.33°$
(B) $GM = 0.3$; $\Phi M = 112.33°$
(C) $GM = 3.33$; $\Phi M = 67.67°$
(D) None of the answers in (A), (B), and (C) is correct

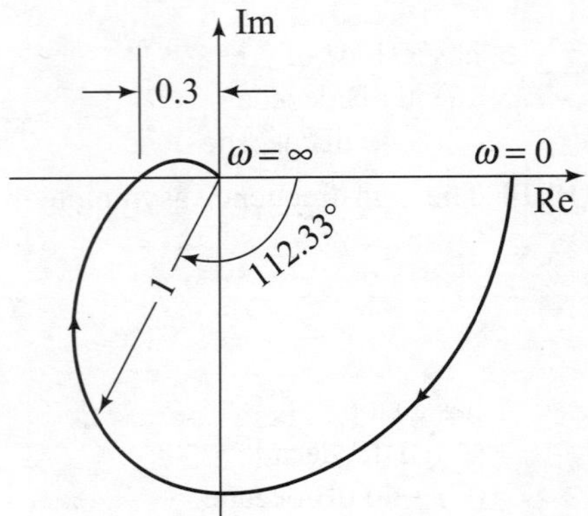

Fig. Q10.22

10.23 A unity feedback system has open-loop transfer function $G(s) = K/[s(1 + s\tau)]$. The gain margin of the feedback system is
(A) ∞
(B) 0
(C) 1
(D) None of the answers in (A), (B), and (C) is correct

10.24 A unity feedback system has open-loop transfer function $G(s)$. Bode plot of $G(j\omega)$ is shown in Fig. Q10.24. The feedback system has
(A) positive phase margin and negative gain margin
(B) positive phase margin and positive gain margin
(C) negative phase margin and negative gain margin
(D) negative phase margin and positive gain margin

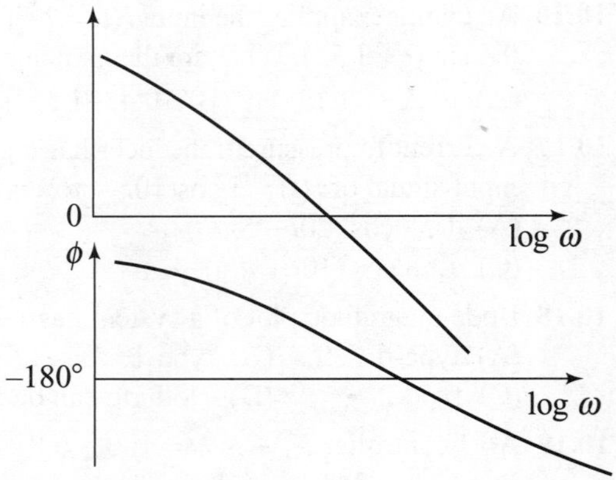

Fig. Q10.24

10.25 Frequency response of a system is given in the following table:

Frequency (rad/sec)	*Gain*	*Phase (deg)*
0.1	5	0
1	1.5	−90
10	0.25	−180
100	0.1	−235

Gain margin is
(A) 4 (B) 0.25
(C) 5 (D) 1.5

10.26 Frequency response of a system is given in the following table:

Frequency (rad/sec)	*Gain*	*Phase (deg)*
1	10	–50
10	2	–100
100	1	–155
1000	0.1	–235

Phase margin is

(A) –25° (B) +25°

(C) –55° (D) +55°

10.27 Nyquist plot and Bode magnitude plot of two systems are given in Fig. Q10.27

(A) Both the systems are type-0

(B) Both the systems are type-1

(C) System I is type-0 and system II is type-1

(D) System II is type-0 and system I is type-1

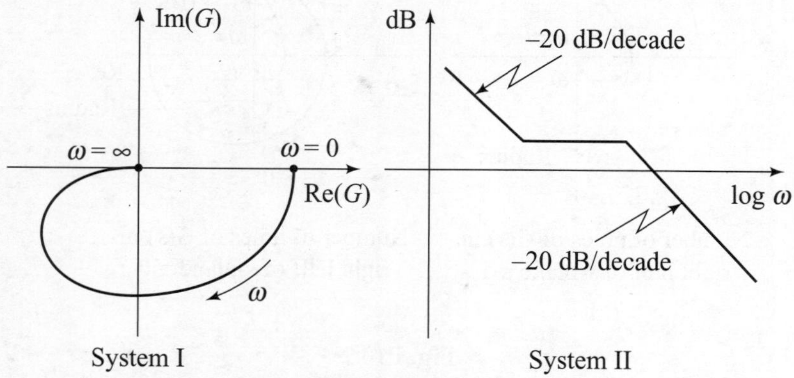

Fig. Q10.27

MASTER KEY

10.1 (A)	10.2 (B)	10.3 (D)	10.4 (C)	10.5 (D)
10.6 (B)	10.7 (A)	10.8 (C)	10.9 (D)	10.10 (D)
10.11 (A)	10.12 (D)	10.13 (C)	10.14 (C)	10.15 (D)
10.16 (A)	10.17 (D)	10.18 (A)	10.19 (A)	10.20 (B)
10.21 (B)	10.22 (C)	10.23 (A)	10.24 (B)	10.25 (A)
10.26 (B)	10.27 (C)			

Problems

10.1 Sketch the general shapes of the polar plots for

(a) $G(j\omega)=\dfrac{1}{(1+j\omega\tau_1)(1+j\omega\tau_2)}$ (b) $G(j\omega)=\dfrac{1}{j\omega(1+j\omega\tau)}$

(c) $G(j\omega)=\dfrac{1}{(j\omega)^2(1+j\omega\tau)}$ (d) $G(j\omega)=\dfrac{1}{j\omega(1+j\omega\tau_1)(1+j\omega\tau_2)}$

(e) $G(j\omega)=\dfrac{1}{(j\omega)^2(1+j\omega\tau_1)(1+j\omega\tau_2)}$

based on magnitude and phase calculations at (i) $\omega=0$, (ii) $\omega=\infty$, (iii) the point of intersection (if any) with the real axis, and (iv) the point of intersection (if any) with the imaginary axis.

10.2 Check the stability of unity-feedback systems whose Nyquist plots are shown in Fig. P10.2. $G(s)$ represents open-loop transfer function.

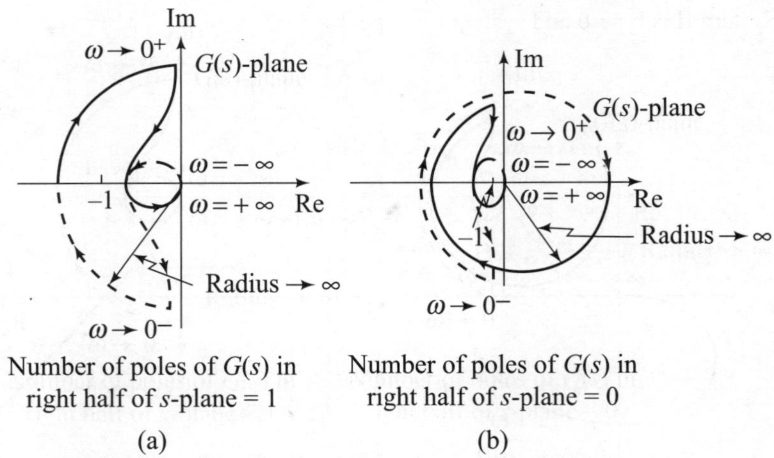

Fig. P10.2

10.3 Consider the unity-feedback system shown in Fig. P10.3a. The polar plot of $G(s)$ is of the form shown in Fig. P10.3b. Assuming that the Nyquist contour in s-plane encloses the entire right half of s-plane, draw a complete Nyquist plot in the $G(s)$-plane. Then answer the following questions:

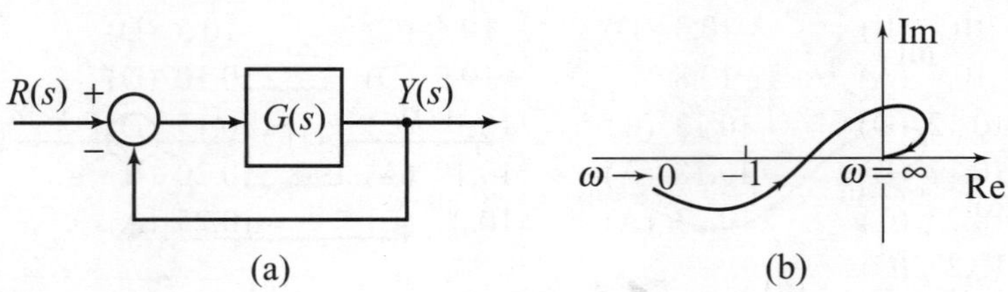

Fig. P10.3

(i) If $G(s)$ has no poles and no zeros in the right half of s-plane, is the closed-loop system stable?

(ii) If $G(s)$ has one pole and no zeros in the right half of s-plane, is the closed-loop system stable?

(iii) If $G(s)$ has one zero and no poles in the right half of s-plane, is the closed-loop system stable?

10.4 By use of the Nyquist criterion, determine whether the closed-loop systems having the following open-loop transfer functions are stable or not. If not, how many closed-loop poles lie in the right half of the s-plane?

(a) $G(s)H(s)=\dfrac{180}{(s+1)(s+2)(s+5)}$ (b) $G(s)H(s)=\dfrac{2}{s(s+1)(2s+1)}$

(c) $G(s)H(s)=\dfrac{s+2}{(s+1)(s-1)}$ (d) $G(s)H(s)=\dfrac{s+2}{s^2}$

(e) $G(s)H(s)=\dfrac{1}{s^2+100}$ (f) $G(s)H(s)=\dfrac{2(s+3)}{s(s-1)}$

MATLAB Exercise After completing the hand sketches, verify your results using MATLAB. Turn in your hand sketches and MATLAB-based plots on the same scale.

10.5 The stability of a closed-loop system with open-loop transfer function,

$$G(s)H(s)=\frac{K(\tau_2 s+1)}{s^2(\tau_1 s+1)};K,\tau_1,\tau_2>0$$

depends on the relative magnitudes of τ_1 and τ_2. Draw Nyquist plots and therefrom determine stability of the system when (i) $\tau_1<\tau_2$, and (ii) $\tau_1>\tau_2$.

10.6 Use the Nyquist criterion to determine the range of values of $K>0$ for the stability of the system in Fig. P10.6 with

(a) $G(s)=\dfrac{8}{(s+1)(s^2+2s+2)}$

(b) $G(s)=\dfrac{4(1+s)}{s^2(1+0.1s)}$

(c) $G(s)=\dfrac{4(1+0.1s)}{s^2(1+s)}$

(d) $G(s)=\dfrac{e^{-0.8s}}{s+1}$

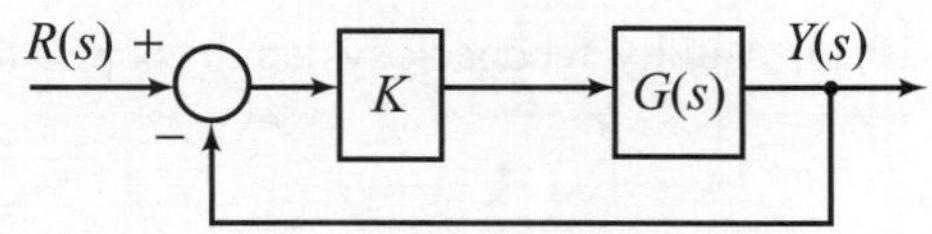

Fig. P10.6

MATLAB Exercise (i) After completing the hand sketches, verify your results using MATLAB. Turn in your hand sketches and MATLAB-based plots on the same scale. (ii) Also verify your results using MATLAB-based root locus plots. Make suitable approximation for the deadtime in Problem 10.6d.

10.7 Sketch the Nyquist plot for a feedback system with open-loop transfer function,

$$G(s)H(s)=\frac{K(s+3)(s+5)}{(s-2)(s-4)};K>0$$

Find the range of values of K for which the system is stable.

10.8 Sketch the Nyquist plot for a feedback system with open-loop transfer function,

$$G(s)H(s) = \frac{K(1+0.5s)(s+1)}{(1+10s)(s-1)}; K > 0$$

Find the range of values of K for which the system is stable.

10.9 Consider the control system shown in Fig. P10.9. Using the Nyquist criterion, determine the range of gain $K > 0$ for stability of the system.

MATLAB Exercise Make a rough sketch to decide on appropriate frequency range. Then generate a MATLAB-based plot over this frequency range. Also verify your result using MATLAB-based root locus plot.

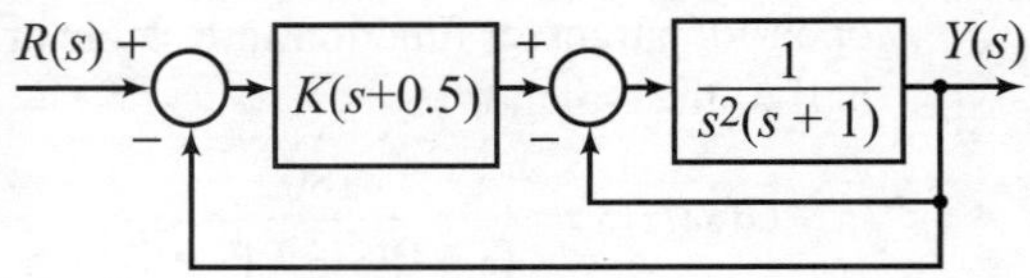

Fig. P10.9

10.10 Consider the feedback systems shown in Figs P10.10a and P10.10b. Sketch the Nyquist plot in each case and therefrom determine the maximum value of K for stability.

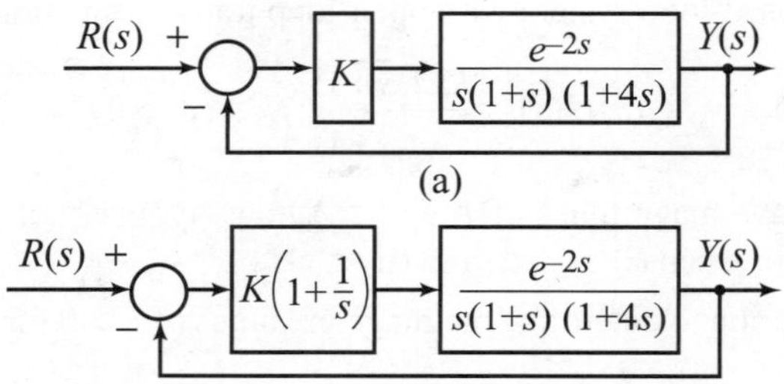

(b)

Fig. P10.10

10.11 A unity-feedback system has open-loop transfer function,

$$G(s) = \frac{e^{-s\tau_D}}{s(s+1)}$$

Draw the Nyquist plot of $G_1(j\omega) = \dfrac{1}{j\omega(j\omega+1)}$, together with the critical trajectory of $e^{-j\omega\tau_D}$

Using these curves, obtain the maximum value of deadtime τ_D in seconds for the closed-loop system to be stable.

10.12 Using the Nyquist plot, determine gain crossover frequency, phase crossover frequency, gain margin, and phase margin of feedback system with open-loop transfer function,

$$G(s)H(s) = \frac{10}{s(1+0.2s)(1+0.02s)}$$

MATLAB Exercise Using MATLAB dialogs, sketch the Nyquist plot with selection of appropriate frequency range from the Bode plot of the system. From the Nyquist plot, determine the required performance indices. Also determine the indices using Bode plot.

10.13 Sketch the Nyquist plot for a feedback system with open-loop transfer function,

$$G(s)H(s) = \frac{K(s+10)^2}{s^3}; K > 0$$

(a) Show that the feedback system is stable for $K > 5$.
(b) Determine the gain margin and phase margin when $K = 7$.

10.14 Sketch the Bode asymptotic plots showing the magnitude in dB as a function of log frequency for the transfer functions given below. Determine the gain crossover frequency in each case.

(a) $G(s) = \dfrac{25}{(s+1)(0.1s+1)(0.05s+1)}$

(b) $G(s) = \dfrac{50(0.2s+1)}{s(s+1)(0.02s+1)}$

(c) $G(s) = \dfrac{500(0.2s+1)(0.1s+1)}{s^2(s+1)(0.02s+1)}$

(d) $G(s) = \dfrac{50(0.05s+1)}{s(0.1s+1)(0.02s+1)((s/200)^2+(0.02s/200)+1)}$

10.15 Using Bode plot, determine gain crossover frequency, phase crossover frequency, gain margin, and phase margin of a feedback system with open-loop transfer function,

(a) $G(s) = \dfrac{10}{s(0.1s+1)}$ (b) $G(s) = \dfrac{10}{s(0.1s+1)^2}$ (c) $G(s) = \dfrac{20(0.2s+1)}{s(0.5s+1)}$

(d) $G(s) = \dfrac{20(0.2s+1)e^{-0.1s}}{s(0.5s+1)}$ (e) $G(s) = \dfrac{40}{(s+2)(s+4)(s+5)}$ (f) $G(s) = \dfrac{10}{s^2(0.2s+1)}$

MATLAB Exercise Generate Bode plots in MATLAB window and determine the performance indices.

10.16 Use Bode plots to determine the stability of the system shown in Fig. P10.16 for the two cases: (i) $K = 10$, and (ii) $K = 100$.

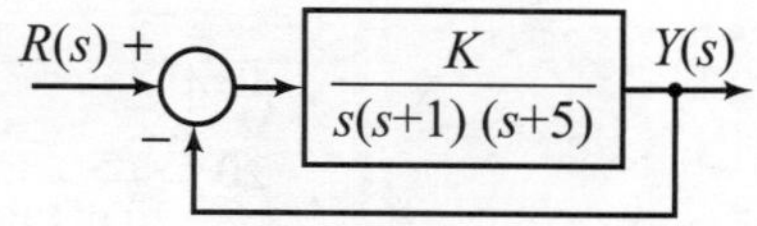

Fig. P10.16

MATLAB Exercise Determine the stability using both the Bode plots and the root locus plots, generated using MATLAB.

10.17 Use Bode plot to determine the range of K within which a unity-feedback system with open-loop transfer function $G(s)$ is stable. Given:

(a) $G(s) = \dfrac{K}{(s+2)(s+4)(s+5)}$ (b) $G(s) = \dfrac{K}{s(1+0.2s)(1+0.02s)}$

(c) $G(s) = \dfrac{Ke^{-s}}{s(s+1)(s+2)}$

MATLAB Exercise Determine the range using both the Bode plots and the root locus plots, generated using MATLAB. Note that the transfer function for Problem 10.17c has deadtime; it may be approximated by a rational function for generating root locus plot.

10.18 Open-loop transfer function of a closed-loop system is

$$G(s)H(s) = \frac{10e^{-s\tau_D}}{s(0.1s+1)(0.05s+1)}$$

(a) Find the gain margin and phase margin when $\tau_D = 0$.

(b) Find the gain margin and phase margin when $\tau_D = 0.04$ sec. Comment upon the effect of dead time.

(c) Determine the maximum value of τ_D for the closed-loop system to be stable.

MATLAB Exercise Use MATLAB environment in dialog mode to solve this problem.

10.19 The experimental frequency response data of certain systems presented on Bode plots and asymptotically approximated are shown in Fig. P10.19. Find the transfer function in each case (systems are known to have minimum-phase characteristics).

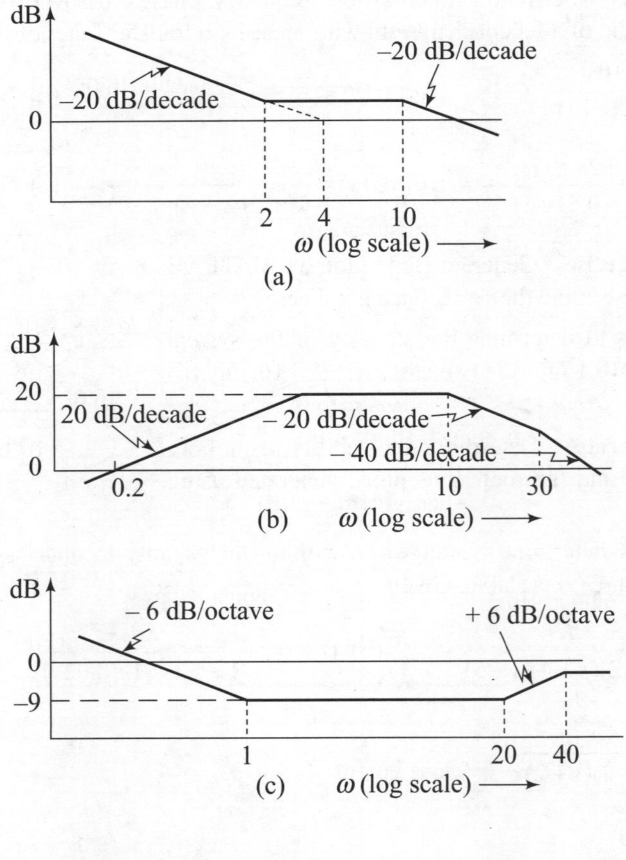

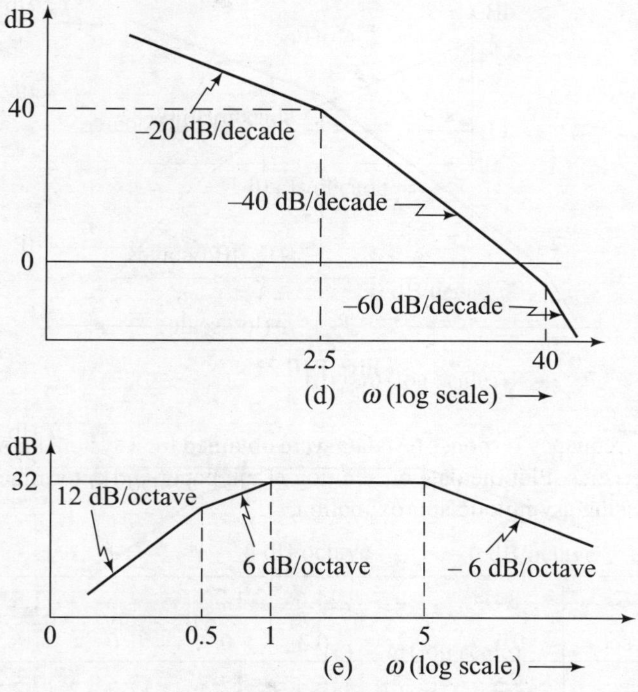

Fig. P10.19

10.20 Consider a minimum-phase system whose asymptotic amplitude frequency response is depicted in Fig. P10.20.

(a) Determine the transfer function $G(s)$ of the system.

(b) Determine the two gain crossover frequencies ω_{g1} and ω_{g2}.

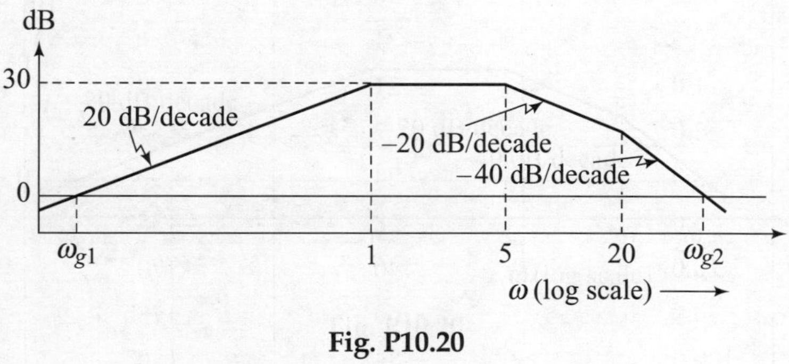

Fig. P10.20

10.21 The asymptotic amplitude frequency response of the open path of a feedback system is shown in Fig. P10.21. Determine the gain margin of the system. The system is known to have minimum-phase characteristics.

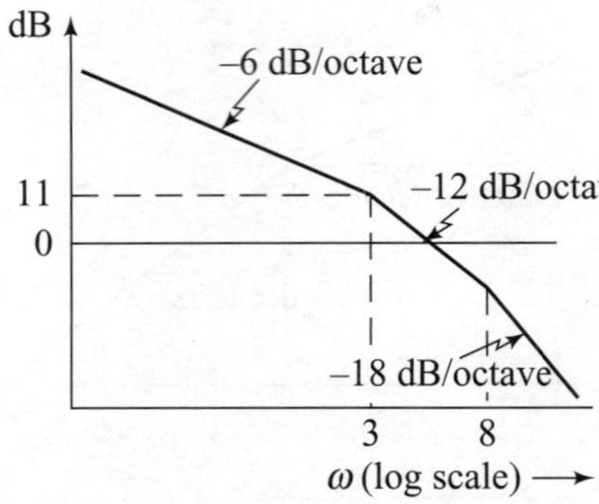

Fig. P10.21

10.22 The following frequency response test data were obtained for a system known to have minimum-phase characteristics. Plot the data on semilog graph paper and determine the transfer function of the system using asymptotic approximation.

Gain (dB)	34	28	24.6	14.2	8	1.5	–3.5	–7.2
Frequency (rad/sec)	0.1	0.2	0.3	0.7	1.0	1.5	2.0	2.5
Gain (dB)	–12.5	–14.7	–16.0	–17.5	–17.5	–17.5		
Frequency (rad/sec)	4.0	5.0	6.0	9.0	20	35		

10.23 Determine an approximate transfer function model of a system for which experimental frequency response data is given in the following table:

ω	*dB*	*Phase*
0.1	– 20	0
0.5	– 21	0
1.0	– 21	– 9°
2.0	– 22	– 54°
3.0	– 24	– 90°
5.0	– 28	– 135°
10.0	– 40	– 170°
30.0	– 60	– 178°
100.0	– 84	– 180°

PERFORMANCE SPECIFICATIONS ON SYSTEM FREQUENCY RESPONSE

11

Preview

In a way, this chapter runs parallel to Chapter 6; here we look at performance specifications on frequency response, while in Chapter 6, the focus was on performance specifications on time response. The characteristic parameters that quantify the frequency response characteristics of a feedback system are the *bandwidth* ω_b, the *resonance peak* M_r, and the *resonance frequency* ω_r (refer to Section 2.10). The parameters that characterize the relative stability of a feedback system are the *phase margin* and the *gain margin*—given by open-loop sinusoidal transfer function on Bode plot, when examined in the light of Nyquist criterion. These are the performance measures used in frequency-domain design procedures. This chapter is primarily concerned with the description of these performance measures—bandwidth, resonance peak, resonance frequency, phase margin, gain margin; and their correlation with time-domain performance measures—rise time, settling time, peak overshoot, etc.

11.1 INTRODUCTION

In practice, the performance of a control system is more realistically measured by the time-domain characteristics. The reason is that the performance of most control systems is judged based on time responses to typical test signals—step function, ramp function, and parabolic function. The performance is often specified in terms of figures of merit, such as peak overshoot, settling time, steady-state error constant, …., which we have already discussed in Chapter 6. In Chapter 9, we have introduced design techniques based on root locus plots that use these performance specifications.

The advantage of root locus design is that it is very intuitive. The root locus plot provides the designer with a picture of all potential closed-loop pole locations. This picture, when combined with the performance specifications, makes it an easy matter to find a good compensator. The overriding principle is the use of *near* pole/zero cancelation to make the closed-loop transfer function as close as possible to the standard second-order transfer function. The assumption is that the response for near cancelation will be essentially the same as for perfect cancelation. We have not provided rigorous analysis to support this assumption, but the examples in Chapter 9 suggest that the assumption is generally true. With this assumption, the specifications on peak overshoot, settling time, etc., of the closed-loop time response are translated into the parameters ζ and ω_n of the standard second-order transfer function and a compensator that matches the closed-loop transfer function with the standard second-order transfer function is selected.

In Chapter 10, we initiated the development of an alternative approach of design based on frequency response models. The crux of the approach is that the frequency response of a closed-loop system can be used to predict the time-domain transient and steady-state responses. This important result was established in Section 2.9 through Fourier transformation. It was proved that ***the frequency response***

model of a feedback system is a complete description of the system's dynamic behavior and allows us to compute the response to any input, not just sine waves.

The correlation between transient response and frequency response through the Fourier integral forms an important basis of the design procedures that use frequency response models. The desired time-domain behavior is interpreted in terms of frequency-response characteristics, the design is carried out in frequency domain, and the frequency response is translated back into the time domain. Conceptually, it sounds all right, but there are practical difficulties associated with the procedure based on Fourier transformation and inverse Fourier transformation. Complex computations of Fourier transformation and its inverse make the iterative design procedure, based on translating the desired time-domain behavior to frequency domain, difficult. Fortunately, there is a way out:

Through the design examples developed in the next chapter, we will see that the design procedures based on frequency response analysis are essentially pole cancelation techniques. In many cases, the resulting closed-loop system closely approximates the standard second-order transfer function. Direct and simple correlations exist between the transient response measures and frequency-response measures of performance for the standard second-order system of Fig. 11.1. The design based on these correlations proceeds very fast, though it is not exact. Sound engineering skill and judgment are necessary to estimate the accuracy of such a procedure. It is, in fact, essential to check the exact response after the design is completed.

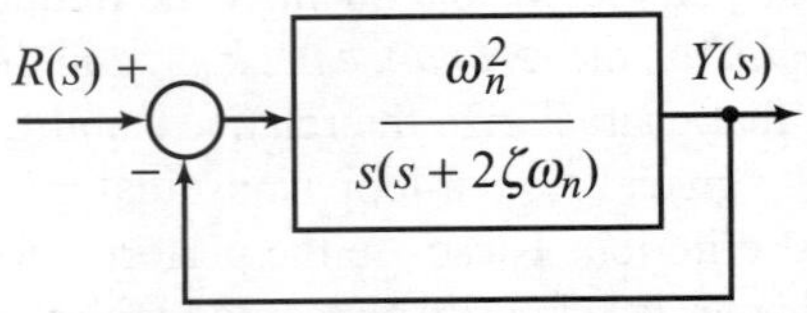

Fig. 11.1 *Standard second-order feedback system*

11.2 CLOSED-LOOP FREQUENCY RESPONSE

Recall the basic feedback system block diagram defined in Chapter 2; shown as Fig. 11.2 in a form convenient for analysis. The controlled process (and the actuator) is represented by $G(s)$, and $N(s)$ is the plant transfer function from the disturbance input to the output. The control system objectives are that output y follow input r, and ignore disturbance w.

For the single-loop control-system configuration of Fig. 11.2, the closed-loop transfer function is

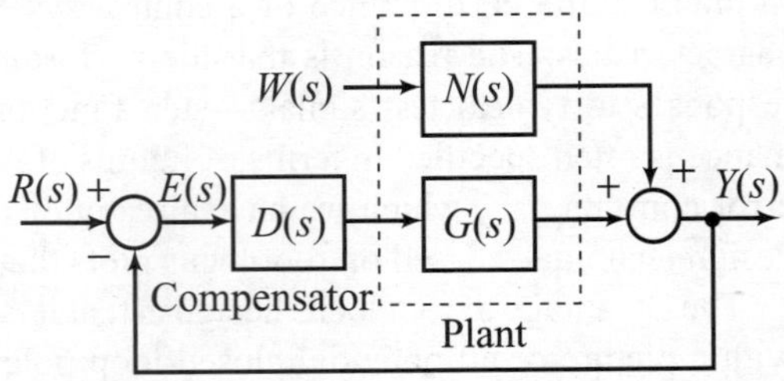

Fig. 11.2 *A feedback system*

$$M(s) = \frac{Y(s)}{R(s)} = \frac{D(s)G(s)}{1+D(s)G(s)} \tag{11.1}$$

For a *stable* closed-loop system, the sinusoidal transfer function

$$M(j\omega) = \frac{Y(j\omega)}{R(j\omega)} = \frac{D(j\omega)G(j\omega)}{1+D(j\omega)G(j\omega)} \tag{11.2}$$

completely characterizes the system's dynamic behavior and allows one to compute the response to any input through inverse Fourier transformation (refer to Section 2.9). Therefore, frequency response curves—magnitude frequency response $|M(j\omega)|$ vs ω, and phase frequency response

$\angle M(j\omega)$ vs ω—obtained analytically from Eqn. (11.2) or obtained experimentally, give a complete characterization of the dynamics of the system.

The next chapter on design in frequency domain will demonstrate that in many ways, the design of a feedback control system is quite similar to a filter design, and the control system is regarded as a signal processor. In fact, if the *ideal filter* of Fig. 11.3 (refer to Section 2.10) were physically realizable, it would be ideal as the frequency response of a feedback control system. The output of such a system would follow all inputs with frequencies up to ω_{BW}, without any error. Therefore, ***the design of a feedback control system is an iterative procedure that approximates the frequency response of the feedback system to an ideal lowpass filter as closely as possible***. Shaping the frequency-response characteristics is, thus, the objective of frequency-response design methods.

Typical plots of $|M(j\omega)|$ and $\angle M(j\omega)$ vs ω, showing frequency-domain characteristics of a feedback control system, are shown in Fig. 11.4. We find that the control-system response deviates substantially from the *ideal* one shown in Fig. 11.3. This deviation is measured in terms of *height* and *width* of the control-system frequency response.

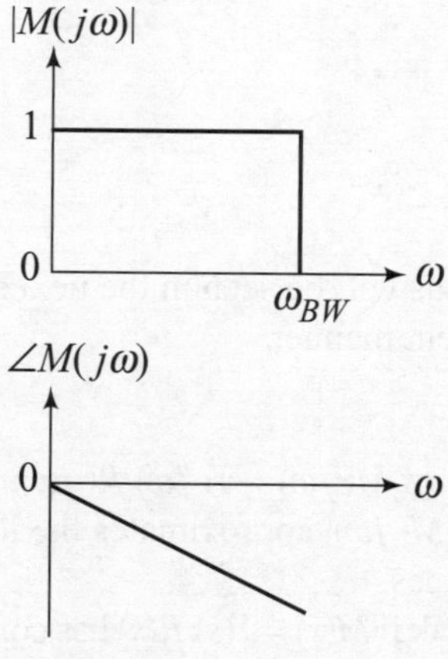

Fig. 11.3 *Ideal lowpass filter*

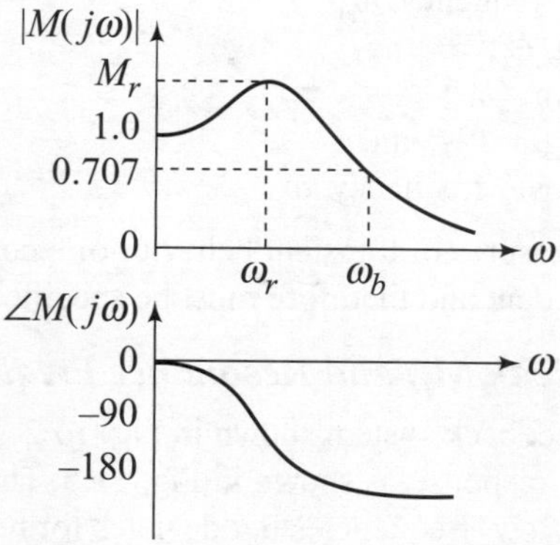

Fig. 11.4 *Frequency-domain characteristics of a feedback control system*

The frequency response of the feedback system shown in Fig. 11.4 exhibits *resonance peak*, M_r, in its amplitude characteristic at *resonance frequency*, ω_r. It determines the height of the response curve. Also it has a gradual transition between passband and rejection, unlike the ideal filter which has a sharp cutoff at ω_{BW}. A commonly used convention is to represent the width of the response curve by the *bandwidth*, ω_b, defined as the range of positive frequencies over which the amplitude drops not lower than $1/\sqrt{2} = 0.707$ times the value at zero frequency. The phase characteristic of the frequency response of a control system is also of importance, as it plays an important role in the relative stability (phase margin, gain margin) of closed-loop systems, as was observed in the previous chapter.

11.3 PERFORMANCE SPECIFICATIONS IN FREQUENCY DOMAIN

The study of a control system essentially involves the evaluation of transient and steady state responses of the system, and robustness of this performance against model uncertainties and disturbances.

11.3.1 Transient Response

The transient response of a practical control system (which often exhibits damped oscillations before reaching steady state) is revealed by rise time t_r, peak time t_p, peak overshoot M_p, and setting time t_s of the system response to a unit-step input (refer to Section 6.2). Rise time and peak time are intended as speed of response criteria; smaller these values, faster will be the system response. Peak overshoot is used mainly for relative stability. Values in excess of 40% may indicate the system is dangerously close to absolute instability. Settling time combines stability and speed of response aspects and is widely used. All these measures of transient behavior of a control system are mutually dependent and therefore must be specified in a consistent manner.

In the design of linear control systems using frequency-domain methods (revisiting Sections 2.9 and 2.10 will be helpful), it is necessary to define a set of specifications so that the quality of the transient response can be described using frequency-response characteristics. It is common to specify the following for this purpose:

(i) resonance peak, M_r;
(ii) resonance frequency, ω_r;
(iii) bandwidth, ω_b;
(iv) gain margin, *GM*;
(v) phase margin ΦM; and
(vi) gain crossover frequency, ω_g.

All these measures of transient behavior of a control system, as will be seen in the next section, are mutually dependent and therefore must be specified in a consistent manner.

Resonance Peak M_r, and Resonance Frequency ω_r

Consider the feedback system shown in Fig. 11.2. A typical plot of $M(j\omega) = Y(j\omega)/R(j\omega)$, the closed-loop frequency response, is shown in Fig. 11.5. The magnitude $|M(j\omega)|$ approximates the ideal 1.0 for some range of 'low' frequencies but deviates for higher frequencies.

We know that $|M(j\omega)|$ *versus* ω curve will have a pronounced peak if $M(s) = Y(s)/R(s)$ has complex poles with very small damping ratio. In fact, smaller the value of damping ratio, higher will be the peak. The peak occurs at a frequency slightly less than the natural frequency of the complex poles (refer to Figs 10.30 and 10.31). The maximum value M_r of $|M(j\omega)|$ is termed the *resonance peak*, and the frequency ω_r at which it occurs is the *resonance frequency* (refer to Fig. 11.5).

Resonance peak M_r is a very good measure of relative stability of a closed-loop system; a large M_r corresponds to a large overshoot in transient response. Many systems are designed to exhibit resonance peak in the range 1.2 to 1.4.

The resonance frequency ω_r is a measure of speed of response; the higher the ω_r the faster the system.

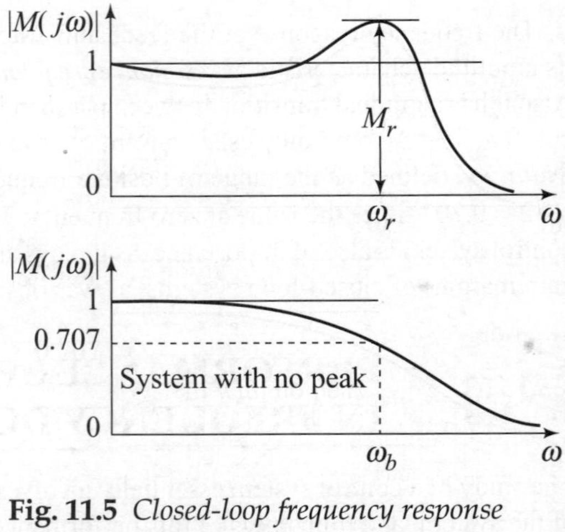

Fig. 11.5 *Closed-loop frequency response criteria*

Bandwidth ω_b

For systems that exhibit no peak (refer to Fig. 11.5), which is sometimes the case, the ***bandwidth ω_b is used for the speed of response specification***. It can, of course, be specified even if there is a peak. (Bandwidth of the closed-loop frequency response is the frequency at which the magnitude M is $1/\sqrt{2} = 0.707$ times its value at zero frequency).

Gain Margin GM, and Phase Margin ΦM

Alternative measures of relative stability are stability margins. We have seen in Chapter 10 that *both* a good *gain margin* and a good *phase margin* are needed for robust stability; neither is sufficient by itself. General numerical design goals for gain margin and phase margin cannot be given since systems that satisfy other specific performance criteria may exhibit a wide range of gain and phase margins. It is possible, however, to give useful lower bounds; ***gain margin should usually exceed* 2.5** (20 log 2.5 dB) ***and phase margin should exceed* 30°.**

Gain Crossover Frequency ω_g

Of particular importance for the speed of response analysis is the *gain crossover frequency* ω_g (refer to Section 10.4). ***Raising the gain crossover frequency increases the speed of response.***

With respect to disturbances, if we set $R \equiv 0$ and let W be a sine wave (refer to Fig. 11.2), we can measure or calculate $Y(j\omega)/W(j\omega)$, which should of course ideally be zero for all frequencies. For low frequencies, the feedback action is effective in keeping Y near the ideal value of zero.

11.3.2 Steady-State Response

The steady-state behavior of a practical control system is revealed by the steady-state error of the system to one or more of the standard test signals—step, ramp, parabola. The specification of the zero-error response to all the three test signals is very impractical and unrealistic. Fortunately, the requirements of practical system are much less stringent. A finite error is usually tolerated as the severity of the input increases (refer to Section 6.6).

11.3.3 Robustness

So far the design objectives have been formulated as requirements on the frequency response of the closed-loop system. These specifications on stability, speed of response, and steady-state accuracy are considered in design under the assumption that a nearly precise model of the plant is available.

The basic assumption of the availability of precise model of the plant is almost never true. There are almost always inconsistencies of the model and the real plant—called *model uncertainties* (refer to Section 6.2). It is usually possible to quantify and bound the discrepancy between the plant and the model. The design model then becomes a member of a class $\mathcal{G}$ of models, one member of which is believed to represent the plant.

Since we do not know which member of the class $\mathcal{G}$ represents the plant, the safe and the reasonable course is to require that the control system stabilize *all* members of the class $\mathcal{G}$. Since that is more difficult to do than simply stabilizing a single (*nominal*) model, we may have to give up something from the single-model performance to satisfy the additional requirement.

Robustness describes the ability of the system to satisfactorily perform for all the models in class $\mathcal{G}$. A *robust controller* is a feedback controller that satisfies the design requirements on stability, speed of response, and steady-state accuracy in connection with all the models of a given family $\mathcal{G}$. The explicit consideration of the model uncertainties poses novel problems in the analysis and design of feedback systems. The problem of designing controllers for uncertain systems necessitates new concepts of modeling, analysis and design.

Since to fully understand the ramifications of modern *robust control theory* would take us on a lengthy and hard mathematical journey, we will content ourselves with the understanding of the intent of robust control, and its relation to the design techniques developed in this book. This is a worthwhile exercise that will at least give us an appreciation of the implications of robust control.

From the viewpoint of single-input, single-output systems, the contributions of robust control theory are primarily an extension of Bode's work on sensitivity analysis plus a sounder theoretical basis for the loop-shaping techniques that had been in use for the past 60 years. Many of these results have been given more rigorous mathematical form in modern robust control theory.

Sensitivity optimization is currently receiving considerable attention. For the system of Fig. 11.2, the sensitivity function is (refer to Eqn. (4.6))

$$|S(j\omega)| = \frac{1}{|1 + D(j\omega)G(j\omega)|} \tag{11.3a}$$

Qualitative considerations for optimization of $|S(j\omega)|$ are as follows (refer to Section 4.4).

1. Small $|S(j\omega)|$ to minimize the effects of disturbances, and to keep the tracking errors small.
2. Small complementary sensitivity $|M(j\omega)|$ $(M(j\omega) = 1 - S(j\omega))$ to minimize the effects of measurement noise, and to reduce control energy.

Since disturbances and command signals are usually of low frequency, and the noise is of high frequency, these design considerations imply that the sensitivity function must be kept small for low frequencies, while for high frequencies, a large value is in order.

Quantitatively, the least requirement on sensitivity function is that $|S(j\omega)| < 1$, i.e., we are looking for a controller such that the closed-loop system performance is less affected by uncertainties than a nominally equivalent open-loop control system for which $|S(j\omega)| = 1$ (refer to Eqn. (4.17)). Since the tolerance on the value of $|S(j\omega)|$ varies with the values of frequency ranges, our aim is, in fact, to minimize a *weighted sensitivity* index such as

$$|W(j\omega)S(j\omega)| < 1 \ \forall\, \omega \qquad \text{or} \qquad |S(j\omega)| < \frac{1}{|W(j\omega)|} \ \forall\, \omega \tag{11.3b}$$

where $W(j\omega)$ is the transfer function that can be tuned by the control engineer, guided by a thorough understanding of the physical capabilities and limitations of the process, and the tolerance on errors. For good disturbance attenuation, $|S(j\omega)|$ should be made small over the frequency range of disturbances. This can be ensured by choosing large $W(j\omega)$ in this frequency band. The requirement of alleviation of the effects of sensor noise dictates that the complementary sensitivity at high frequencies be kept low. The weighted sensitivity optimization must meet these constraints. Over the intermediate frequencies, the required stability margins tune the shape of $W(j\omega)$.

From Eqns (11.3), we obtain

$$|W(j\omega)| < |1 + D(j\omega)G(j\omega)| \ \forall\, \omega$$

Figure 11.6a shows graphical interpretation of this inequality.

Figure 6.6 shows a block diagram of a plant with unmodeled dynamics. The closed-loop transfer function between input $R(s)$ and output $Y(s)$ is

$$\frac{Y(s)}{R(s)} = \frac{D(s)G(s)[1 + \Delta(s)]}{1 + D(s)G(s)[1 + \Delta(s)]}$$

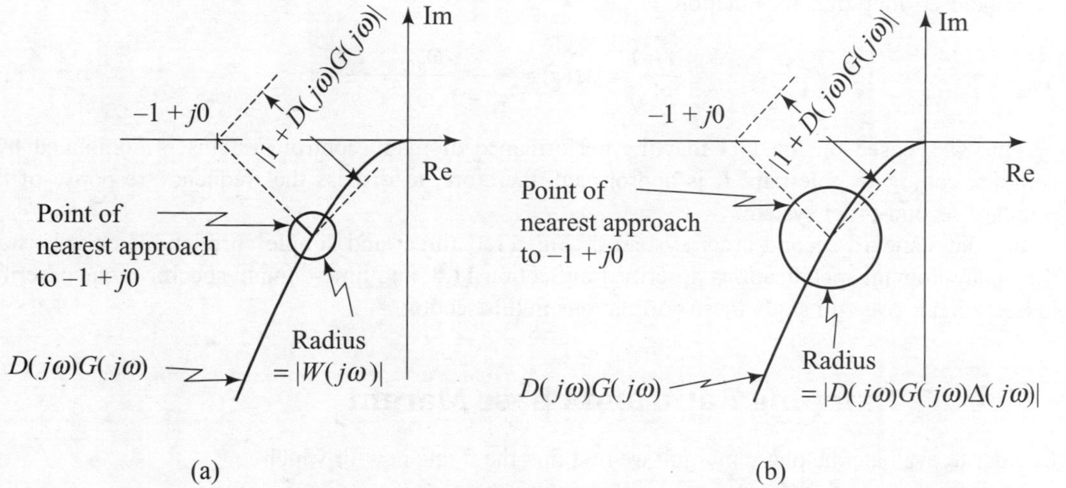

Fig. 11.6 *Qualitative robustness considerations*

where $\Delta(s)$ represents the unknown dynamics of the plant and $G(s)$ is its nominal plant model. We assume that the number of poles of $D(s)G(s)$ in the right half of s-plane is zero. The nominal closed-loop system is designed to be stable—the Nyquist plot of $D(j\omega)G(j\omega)$ in the complex plane does not encircle $-1 + j0$ point. The feedback system with perturbed plant is stable if the Nyquist plot of $D(j\omega)G(j\omega)[1 + \Delta(j\omega)]$ does not encircle $-1 + j0$ point. A sufficient mathematical condition for robust closed-loop stability is $|D(j\omega)G(j\omega)\Delta(j\omega)| < |1 + D(j\omega)G(j\omega)|\ \forall\omega$. This will ensure that $D(j\omega)G(j\omega)\ [1 + \Delta(j\omega)]$ does not reach the $-1 + j0$ point because of unknown dynamics of the plant. Graphical interpretation of this inequality is given in Fig. 11.6b.

In fact, building of uncertainty models $W(j\omega)$ and $\Delta(j\omega)$ itself is subject to uncertainties; the industry, by and large, continue to rely on frequency-domain design methods with the gain margin and phase margin specifications for robustness. These specifications indirectly relate to the requirements dictated by $W(j\omega)$ and $\Delta(j\omega)$. Uncertainty models are not used implicitly in the conventional design process; however, the robustness properties may be checked by simulation for a given uncertainty information. Re-entering the design cycle may become necessary if the robustness requirements are not satisfied. In the Appendix, we will demonstrate the power of MATLAB/Simulink for evaluation of robustness properties of feedback controllers.

Frequency-domain design provides good robustness against uncertainties; design based on Bode plots (Chapter 12) is probably the best method to use for cascade compensation of linear time-invariant systems. However, the root locus method can be helpful in seeing what the Bode design is accomplishing.

11.4 FREQUENCY RESPONSE OF A STANDARD SECOND-ORDER SYSTEM

In Chapter 6, the transient performance of a control system was discussed by the examination of the standard second-order system of Fig. 11.1, with open-loop transfer function

$$G(s) = \frac{\omega_n^2}{s(s + 2\zeta\omega_n)} \tag{11.4}$$

and closed-loop transfer function

$$\frac{Y(s)}{R(s)} = M(s) = \frac{\omega_n^2}{s^2 + 2\zeta\omega_n s + \omega_n^2} \tag{11.5}$$

This was based on the fact that the performance of many control systems is dominated by a complex-conjugate pole pair. It is appropriate, therefore, to discuss the frequency response of this standard second-order system.

For the standard second-order system of Fig. 11.1, direct and simple correlations exist between frequency-domain specifications described in Section 11.3, and time-domain specifications described in Section 6.2. We will study these correlations in this section.

11.4.1 Damping Ratio and Phase Margin

In order to evaluate the phase margin, we first find the frequency for which

$$|G(j\omega)| = \frac{\omega_n^2}{\left|-\omega^2 + j2\zeta\omega_n\omega\right|} = 1 \tag{11.6}$$

The frequency, ω_g, that satisfies Eqn. (11.6) is given by

$$\omega_n^2 = \omega_g\sqrt{\omega_g^2 + (2\zeta\omega_n)^2}$$

or

$$\omega_g^4 + 4\zeta^2\omega_n^2\omega_g^2 - \omega_n^4 = 0$$

The roots of this equation follow by applying the quadratic formula in terms of ω_g^2:

$$\omega_g^2 = \omega_n^2\left(-2\zeta^2 \pm \sqrt{4\zeta^4 + 1}\right)$$

For ω_g to be real-valued, the positive root must be used so that

$$\omega_g = \omega_n\sqrt{\sqrt{4\zeta^4 + 1} - 2\zeta^2} \tag{11.7}$$

The phase angle of $G(j\omega)$ at this frequency is

$$\angle G(j\omega_g) = -90° - \tan^{-1}\frac{\omega_g}{2\zeta\omega_n}$$

$$= -90° - \tan^{-1}\frac{\sqrt{\sqrt{4\zeta^4 + 1} - 2\zeta^2}}{2\zeta} \tag{11.8}$$

The difference between the angle of Eqn. (11.8) and $-180°$ is the phase margin ΦM. Thus,

$$\Phi M = 90° - \tan^{-1}\frac{\sqrt{\sqrt{4\zeta^4 + 1} - 2\zeta^2}}{2\zeta}$$

$$= \tan^{-1}\frac{2\zeta}{\sqrt{\sqrt{4\zeta^4+1}-2\zeta^2}} \quad (11.9)$$

Figure 11.7 shows phase margin, ΦM, in degrees, *versus* damping ratio ζ. The dashed line in the figure shows a straight line approximation to the function

where
$$\zeta = \frac{\Phi M}{100} \quad (11.10)$$

It is clear that the approximation only holds for phase margin below about 70°.

For the standard second-order system under consideration, we cannot define gain margin because $|G(j\omega)| \to 0$ as $\angle G(j\omega) \to -180°$.

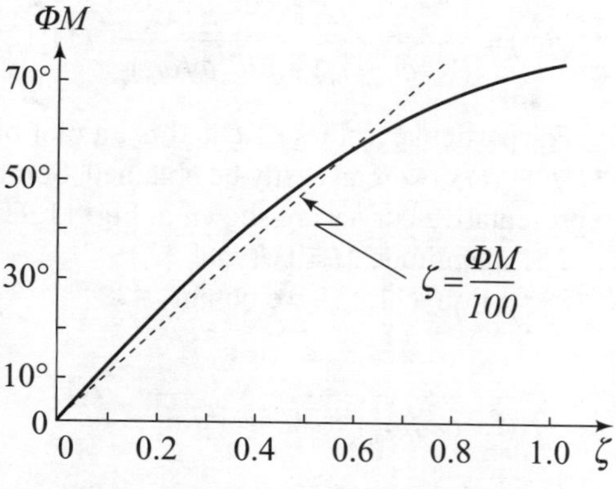

Fig. 11.7 *Phase margin versus damping ratio*

11.4.2 Response Speed and Gain Crossover Frequency

Equation (11.7), reproduced below, relates the gain crossover frequency to the undamped natural frequency ω_n and damping ratio ζ.

$$\omega_g = \omega_n\sqrt{\sqrt{4\zeta^4+1}-2\zeta^2} \quad (11.11)$$

Figure 11.8 shows a plot of ω_g/ω_n versus ζ.

We know that when a root-locus design holds the damping ratio constant, both the rise time and the settling time decrease as the undamped natural frequency increases. Therefore, ***when a frequency-domain design holds the phase margin constant while increasing the gain crossover frequency, the resulting rise time and settling time would reduce in the time domain.***

It should be remembered that the use of the second-order system of Fig. 11.3 as a standard for reference is based on the assumption of dominance. We are assuming that the contributions of all additional poles and zeros may be neglected. There will certainly be cases where this assumption is not valid. It is not practical to develop correlations for exceptions; so the transient response should be determined, usually by simulation, as a check.

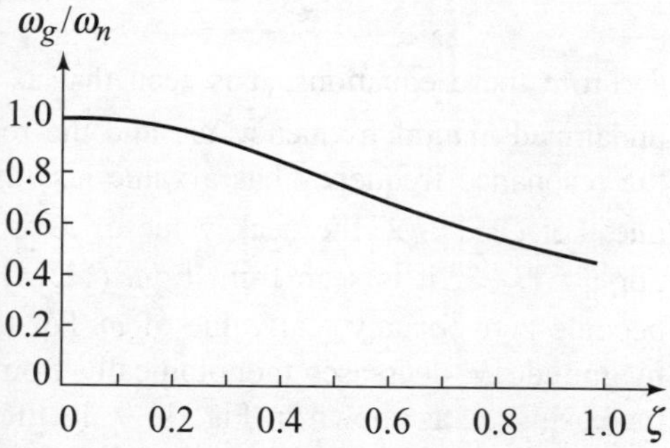

Fig. 11.8 ω_g/ω_n *versus damping ratio*

11.4.3 Damping Ratio and Resonance Peak

To evaluate resonance peak and the frequency at which it occurs, we consider the closed-loop transfer function (11.5) with $s = j\omega$:

$$M(j\omega) = \frac{1}{(1-\omega^2/\omega_n^2) + j2\zeta(\omega/\omega_n)} \quad (11.12)$$

For particular values of ζ and ω_n, a plot of $|M(j\omega)|$ *versus* ω can easily be obtained. Some representative curves are shown in Fig. 11.9.

The magnitude $M \triangleq |M(j\omega)|$

From Eqn. (11.12), we obtain

$$M^2 = \frac{1}{(1-\omega^2/\omega_n^2)^2 + 4\zeta^2(\omega^2/\omega_n^2)} \quad (11.13)$$

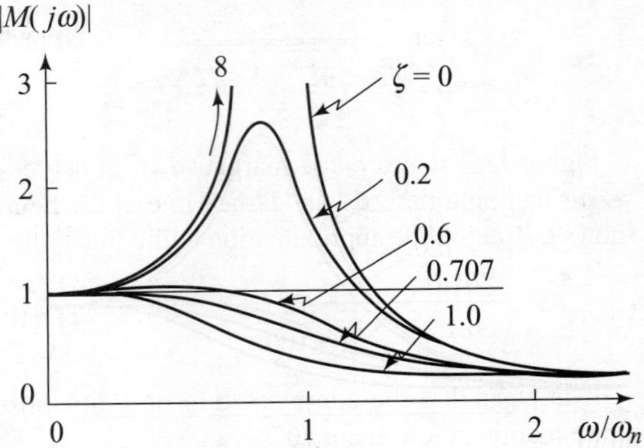

Fig. 11.9 *M versus* ω/ω_n

To find the peak value of M and the frequency at which it occurs, Eqn. (11.13) is differentiated with respect to frequency and set equal to zero:

$$\frac{dM^2}{d\omega} = -\frac{-4(1-\omega^2/\omega_n^2)(\omega/\omega_n^2) + 8\zeta^2(\omega/\omega_n^2)}{[(1-\omega^2/\omega_n^2)^2 + 4\zeta^2(\omega^2/\omega_n^2)]^2} = 0 \quad (11.14)$$

The frequency ω_r at which the value M exhibits a peak, as found from Eqn. (11.14), is

$$\omega_r = \omega_n\sqrt{1-2\zeta^2} \quad (11.15)$$

This value of frequency is substituted into Eqn. (11.13) to yield resonance peak

$$M_r = \frac{1}{2\zeta\sqrt{1-\zeta^2}} \quad (11.16)$$

From these equations, it is seen that as ζ approaches zero, resonance frequency ω_r approaches undamped natural frequency ω_n, and the resonance peak M_r goes off to infinity. For $0 < \zeta < 1/\sqrt{2}$, the resonance frequency has a value less than ω_n and the resonance peak has a value greater than one. For $\zeta = 1/\sqrt{2}$, the peak value of M is one at zero frequency. For $\zeta \geq 1/\sqrt{2}$, it is seen from Eqn. (11.14) that $dM^2/d\omega$ does not become zero for any real value of ω. For the range $\zeta \geq 1/\sqrt{2}$, the magnitude M decreases monotonically from $M = 1$ at $\omega = 0$ with increasing ω, as shown in Fig. 11.9. It, therefore, follows that for $\zeta \geq 1/\sqrt{2}$, there is no resonance peak as such and the greatest value of M equals one. This limiting value of ζ for peaking on the magnitude curve of closed-loop frequency response should not be confused with the overshoot on the step response, where there is overshoot for $0 < \zeta < 1$.

It is important to note that for the standard second-order system, resonance peak M_r is a function of damping ratio only. Figure 11.10 illustrates the relationship between M_r and ζ.

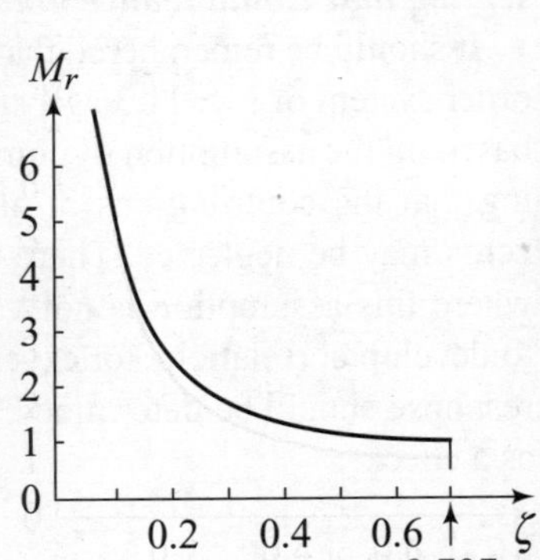

Fig. 11.10 M_r *versus* ζ

11.4.4 Response Speed and Resonance Frequency

Equation (11.15), reproduced below, relates the resonance frequency ω_r to the undamped natural frequency ω_n and damping ratio ζ.

$$\omega_r = \omega_n\sqrt{1-2\zeta^2}\,;\ \zeta < 1/\sqrt{2} \tag{11.17}$$

Figure 11.11 shows a plot of ω_r/ω_n versus ζ.

From Eqns (11.16) and (11.17), we see that ***when a frequency-domain design holds M_r constant while increasing the resonance frequency, the resulting rise time and settling time would reduce in time domain.***

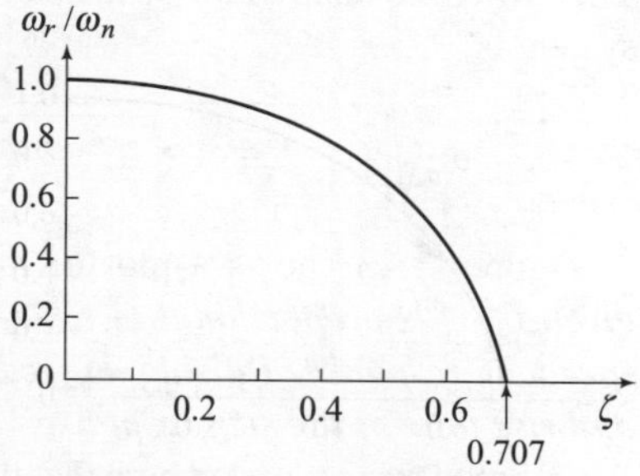

Fig. 11.11 *ω_r/ω_n versus ζ*

11.4.5 Response Speed and Bandwidth

Another relationship between the frequency response and time response is between the speed of time response (as measured by settling time, and rise time), and the *bandwidth* of the closed-loop frequency response, which is defined as the *frequency ω_b, at which the magnitude M is $1/\sqrt{2} = 0.707$ times its value at zero frequency* (see Fig. 11.12a). ***In Bode coordinate system, the bandwidth ω_b is the frequency at which the magnitude falls off by 3 dB from its low-frequency value*** (see Fig. 11.12b).

The bandwidth of the standard second-order system can be found by finding that frequency for which $M = 1/\sqrt{2}$ in Eqn. (11.13):

$$\frac{1}{(1-\omega_b^2/\omega_n^2)^2 + 4\zeta^2(\omega_b^2/\omega_n^2)} = \frac{1}{2}$$

Thus,

$$(1-\omega_b^2/\omega_n^2)^2 + 4\zeta^2(\omega_b^2/\omega_n^2) = 2$$

or

$$(\omega_b^4/\omega_n^4) - 2(1-2\zeta^2)(\omega_b^2/\omega_n^2) - 1 = 0$$

Solving for ω_b^2/ω_n^2, we get

$$(\omega_b^2/\omega_n^2) = (1-2\zeta^2) \pm \sqrt{4\zeta^4 - 4\zeta^2 + 2}$$

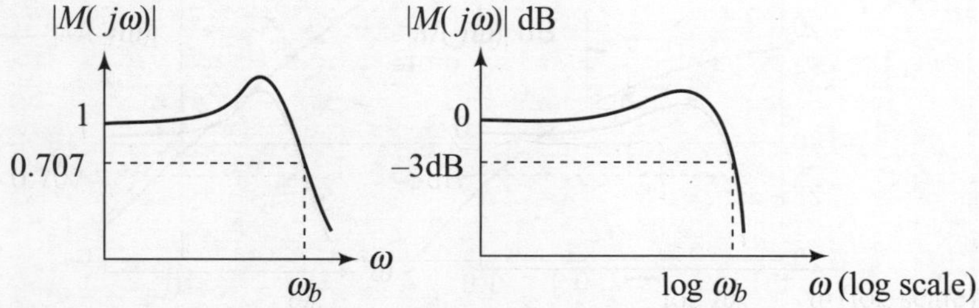

Fig. 11.12 *Definition of bandwidth*

In the last equation the plus sign should be chosen, since ω_b/ω_n must be a positive real quantity for any ζ. Therefore, the bandwidth of the standard second-order system is

$$\omega_b = \omega_n\sqrt{\left[(1-2\zeta^2)+\sqrt{4\zeta^4-4\zeta^2+2}\right]} \qquad (11.18)$$

Figure 11.13 shows a plot of ω_b/ω_n *versus* ζ. ***For given ζ, ω_b is proportional to ω_n, and is a measure of speed of response. Raising ω_b reduces settling time and rise time of the step response.***

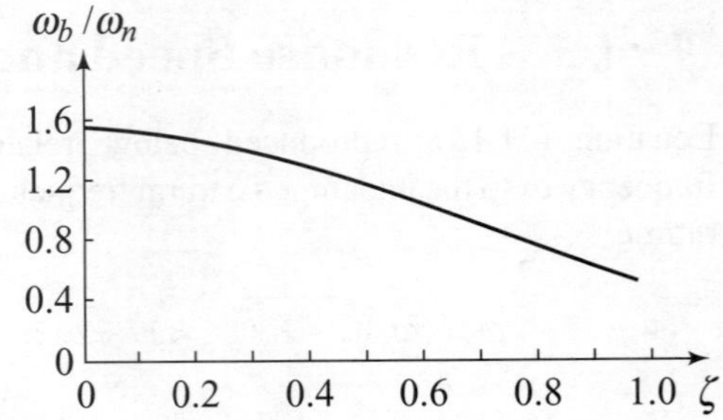

Fig. 11.13 ω_b/ω_n *versus* ζ

The reader is reminded here that the correlations between time-domain and frequency-domain performance specifications arrived at in this section apply only to the standard second-order system of Fig. 11.3. When other second-order or higher-order systems are involved, the correlations are different and more complex. It is not practical to develop correlations for all the cases.

Application of correlations developed for standard second-order system to other second-order and higher-order systems is based on the assumption of dominance. There will certainly be cases where this assumption is not valid; therefore, the transient response must be determined, usually by simulation, as a check.

Example 11.1 The transient behavior of the system of Fig. 11.14 can be estimated from Bode plots of $G(j\omega)$, as is seen below.

$$G(j\omega) = \frac{0.5e^{-j\omega}}{j\omega(1+j\omega)(1+0.1j\omega)}$$

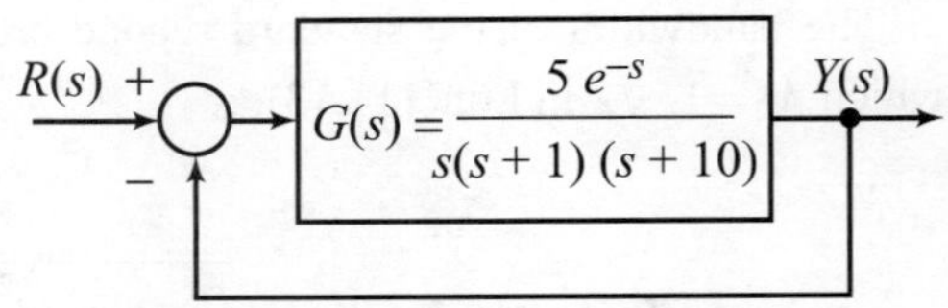

Fig. 11.14 *A feedback system*

Bode plots for this transfer function are shown in Fig. 11.15. Note that the magnitude plot is not affected by the dead time; however, the phase plot is affected.

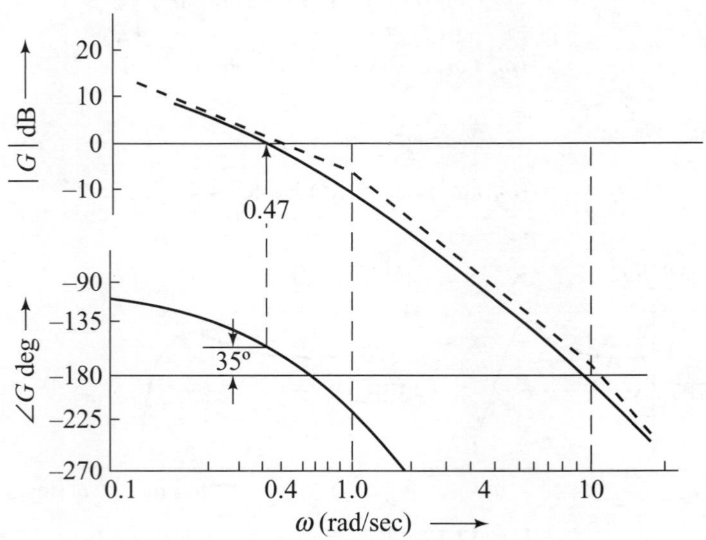

Fig. 11.15 *Bode plot for G(s) of Fig. 11.14*

The dead time yields a decreased phase margin since at any frequency, the phase angle is more negative. Using a second-order approximation, this decrease in phase margin implies a lower damping ratio and a more oscillatory response for the closed-loop system.

The magnitude curve of Fig. 11.15 crosses the 0 dB axis at a frequency of 0.47 rad/sec; the corresponding phase angle is –145°. Therefore, the gain crossover frequency $\omega_g = 0.47$ rad/sec, and the phase margin $\Phi M = -145° - (-180°) = 35°$.

The values of ζ and ω_n to be used for approximating the transient behavior of the system of Fig. 11.14 can be determined from Eqns (11.9)–(11.11) or Figs 11.7–11.8. The values obtained from these second-order approximations are

$$\zeta = 0.33;\ \omega_n = 0.52 \text{ rad/sec}$$

An estimate of the transient behavior of the system of Fig. 11.14 to step input is therefore obtained as

$$M_p = \text{peak overshoot} = 100\, e^{-\pi\zeta/\sqrt{1-\zeta^2}} = 33\%$$

$$t_s = \text{settling time} = 4/\zeta\omega_n = 23.3 \text{ sec}$$

Computer simulation (Fig. 11.16) shows $M_p \cong 38\%$ and $t_s \cong 24$ sec.

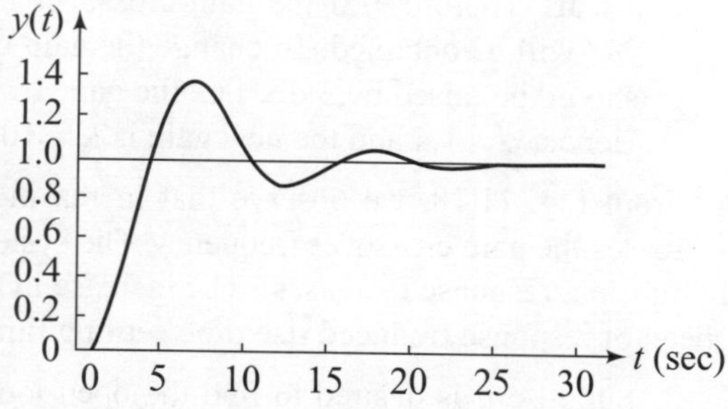

Fig. 11.16 *Step response for system of Fig. 11.14*

Example 11.2 From the correlation between damping ratio (equivalently, peak overshoot in step response) and phase margin developed for standard second-order systems, we observe that if we can vary the phase margin, we can vary the peak overshoot in the step response of the system. Looking at Fig. 11.17, we see that if we desire a phase margin represented by CD, we would have to lower the magnitude curve by AB. Thus, a simple gain adjustment can be used to design phase margin and hence percentage overshoot in step response. Such a design is acceptable if both the resulting gain crossover frequency (equivalently, speed of response) and gain margin are acceptable. Dynamic compensation (discussed in Chapter 12) will become necessary in cases wherein the design for specified phase margin by gain adjustment results in poor gain crossover frequency and/or gain margin.

Gain adjustments can also be used to design gain margin. The resulting phase margin and gain crossover frequency are then examined to see whether the design is acceptable. Similarly, we could adjust the gain for specified gain crossover frequency and then examine the resulting gain margin and phase margin.

Consider a unity-feedback system with open-loop transfer function,

$$G(s) = \frac{K}{s(1+0.1s)(1+0.05s)}$$

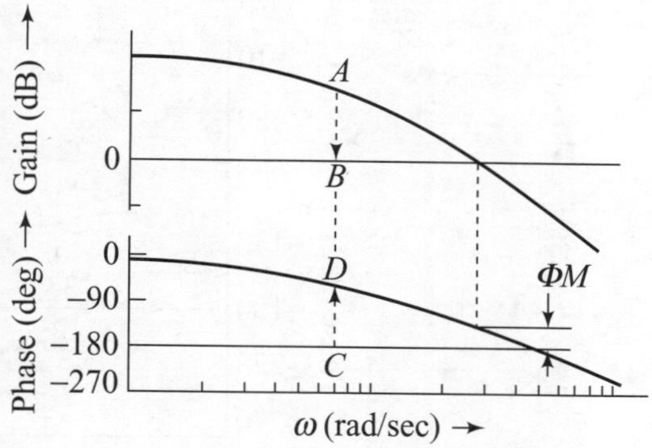

Fig. 11.17 *Bode plot showing gain adjustment for desired phase margin*

Figure 11.18 shows the Bode plot of $G(j\omega)$ with open-loop gain $K = 10$. From this figure, it is found that gain margin $GM = 10$ dB, phase margin $\Phi M = 33°$, and gain crossover frequency $\omega_g = 7.4$ rad/sec. The steady-state error for unit-ramp input for this type-1 system is $e_{ss} = 1/K = 1/10$.

Let us examine the effects of gain adjustments.

1. Suppose that to reduce the steady-state error, a phase margin of 24° will be accepted. To obtain phase margin of 24°, we must first determine the frequency at which the phase angle of $G(j\omega)$ is –156°, and then adjust the value of K so that the gain at this frequency is 0 dB. From the Bode plot of Fig. 11.18, we find that for $\omega = 9$ rad/sec, we get this phase angle and a gain of –3 dB. Therefore, if the gain crossover frequency is changed to 9 rad/sec, a phase margin of 24° will be obtained. To change the gain crossover frequency to 9 rad/sec, the magnitude curve should be raised by 3 dB, i.e., the gain 10 must be raised by a factor of α given by $20 \log \alpha = 3$. Hence, $\alpha = 1.4$ and the new gain is $K = 10 \times 1.4 = 14$.

From Fig. 11.18, we observe that increasing K reduces the phase margin and gain margin, and increases the gain crossover frequency. The system thus becomes relatively less stable (peak overshoot in transient response increases). The increase in gain crossover frequency will generally result in better speed of response (reduced rise time/settling time).

2. Suppose it is desired to find the open-loop gain for a gain margin of 20 dB. To determine the gain for a gain margin of 20 dB, first we determine the phase crossover frequency and the gain at this frequency. From the Bode plot of Fig. 11.18 we find that phase crossover frequency $\omega_\phi = 14$ rad/sec and the gain at ω_ϕ is –10 dB. Hence, for the desired gain margin, we must decrease the gain by 10 dB. It means that magnitude curve must be lowered by 10 dB, i.e., the gain K must be reduced by a factor of β, given by $20 \log(1/\beta) = -10$. Hence, $\beta = 3.16$ and the new gain is $K = 10/3.16 = 3.16$.

From Fig. 11.18 we observe that reducing K increases the phase margin and gain margin, and reduces the gain crossover frequency.

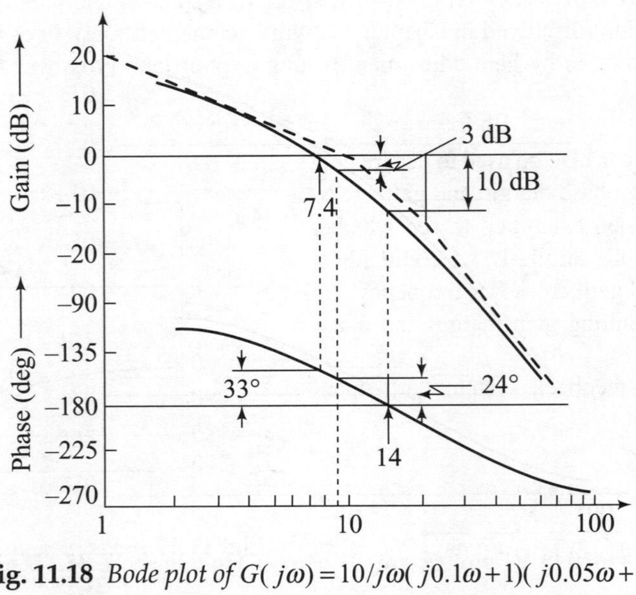

Fig. 11.18 *Bode plot of* $G(j\omega) = 10/j\omega(j0.1\omega + 1)(j0.05\omega + 1)$

11.5 CONSTANT-*M* CIRCLES

A control system is a closed-loop system. By Nyquist stability criterion, we predict its stability using plots of the open-loop transfer function. Closed-loop performance measures such as gain crossover frequency, phase margin, and gain margin are also determined using plots of open-loop transfer functions.

Other useful measures of performance of a feedback system are resonance peak, resonance frequency, and bandwidth of the closed-loop frequency response. Fortunately, these closed-loop performance indices, for a *stable* feedback system, can also be determined using plots of the open-loop transfer function, as is seen below.

Consider a single-loop unity-feedback system shown in Fig. 11.19. The closed-loop transfer function is

$$M(s) = \frac{G(s)}{1+G(s)}$$

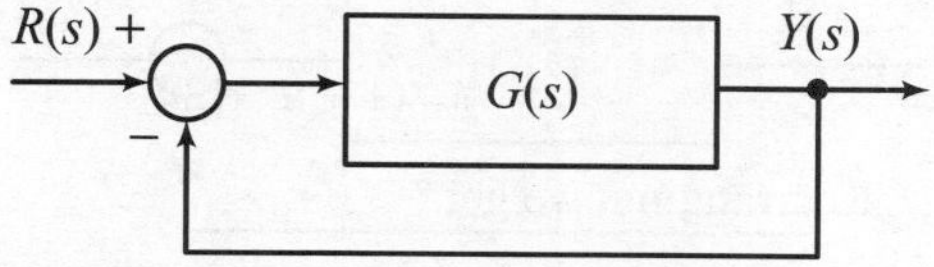

Fig. 11.19 *A unity-feedback system*

The frequency response of the closed-loop transfer function is given by

$$M(j\omega) = \frac{G(j\omega)}{1+G(j\omega)} \tag{11.19}$$

Figure 11.20 shows a typical plot of $G(j\omega)$ in polar coordinates. Consider the point A on this plot associated with frequency ω_1. $G(j\omega_1)$ is a vector from the origin to the point A; and $1 + G(j\omega_1)$ is a vector from $-1 + j0$ point to the point A. One point on the closed-loop frequency response can be calculated by measuring lengths and angles of these vectors:

$$|M(j\omega_1)| = \frac{|G(j\omega_1)|}{|1+G(j\omega_1)|}$$

$$\angle M(j\omega_1) = \angle G(j\omega_1) - \angle(1 + G(j\omega_1))$$

Obviously, this calculation can be repeated at many points on the $G(j\omega)$-curve, and the entire closed-loop frequency response can be calculated in this manner.

It is easier to prepare nomographs so that the results can be read off without additional calculations. Consider the graphical configuration of Fig. 11.20 as a problem in geometry and trigonometry, without regard to frequency response. The lines for $G(j\omega_1)$ and $1 + G(j\omega_1)$ are simply vectors which have lengths and angles associated with them when drawn from the origin and the $-1 + j0$ point, respectively, to a common point A in the complex plane.

$$G(j\omega_1) = \text{Re } G(j\omega_1) + j\text{Im } G(j\omega_1) = u + jv$$

$$1 + G(j\omega_1) = (1 + u) + jv$$

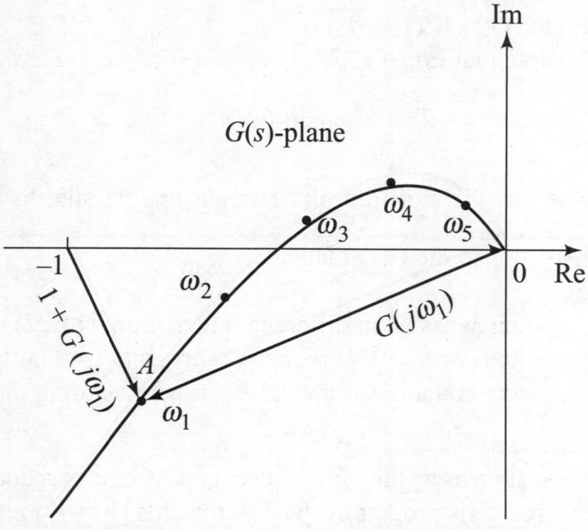

Fig. 11.20 *Polar relationship between open-loop and closed-loop frequency responses*

By definition,

$$\frac{|G(j\omega_1)|}{|1+G(j\omega_1)|}=\frac{|u+jv|}{|(1+u)+jv|}=M,\ \text{a number} \tag{11.20}$$

Are there any other points in the plane such that the same value of M would be associated with vectors drawn to them? The answer to this question is *yes*. In fact, the locus of constant-M points on the polar plane is a circle. This can be proved as follows.

From Eqn. (11.20), we have

$$M^2=\frac{u^2+v^2}{(1+u^2)+v^2}$$

or

$$M^2(1+u^2+2u+v^2)=u^2+v^2$$

Rearranging, we get

$$(1-M^2)u^2+(1-M^2)v^2-2\,M^2\,u=M^2 \tag{11.21}$$

This equation is conditioned by dividing through by $(1-M^2)$ and adding the term $[M^2/(1-M^2)]^2$ on both sides. We have

$$u^2+v^2-\frac{2M^2}{1-M^2}u+\left(\frac{M^2}{1-M^2}\right)^2=\frac{M^2}{1-M^2}+\left(\frac{M^2}{1-M^2}\right)^2$$

which is finally simplified to

$$\left(u-\frac{M^2}{1-M^2}\right)^2+v^2=\left(\frac{M}{1-M^2}\right)^2 \tag{11.22}$$

For a given M, Eqn. (11.22) represents a circle with the center at $(M^2/(1-M^2), 0)$. The radius of the circle is $r = M/(1-M^2)$.

Note that Eqn. (11.22) is invalid for $M = 1$. For $M = 1$, Eqn. (11.21) gives

$$u=-\frac{1}{2} \tag{11.23}$$

which is the equation of a straight line parallel to the imaginary axis and passing through the point $\left(-\frac{1}{2},0\right)$ in the $G(s)$-plane.

When M takes on different values, Eqn. (11.22) describes in the $G(s)$-plane a family of circles that are called *constant-M circles*. A representative family of constant-M circles is shown in Fig. 11.21. The following conclusions are easily drawn regarding these circles:

Case I: $M > 1$

As M increases, the radii of constant-M circles reduce monotonically, and the centers located on negative real axis progressively shift towards $(-1+j0)$ point. $M=\infty$ circle has a radius zero and is centered at $(-1+j0)$.

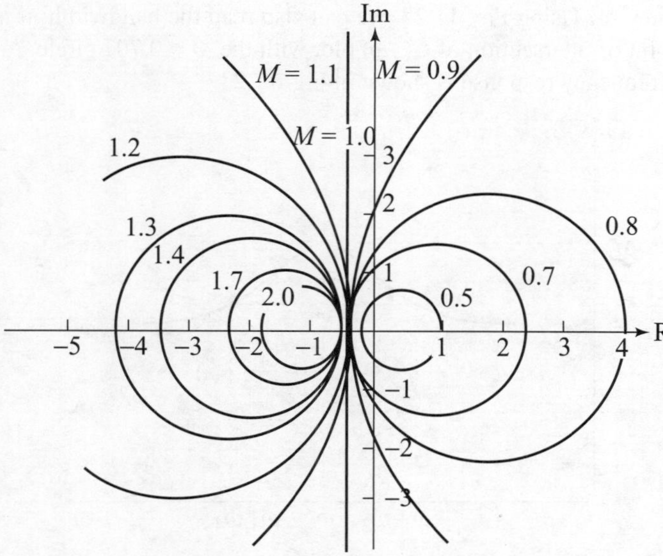

Fig. 11.21 *Constant-M circles*

Case II: M = 1
It follows that $M = 1$ circle is of infinite radius with center at infinity on the real axis, i.e., it is a straight line parallel to the imaginary axis of the $G(s)$-plane. It intersects the real axis at $-1/2$.

Case III: M < 1
As M decreases, the radii of constant-M circles reduce monotonically, and centers located on positive real axis shift towards the origin. $M = 0$ circle has a radius zero and is centered at the origin.

The constant-M circles are useful when overlaid on the polar plot of $G(j\omega)$ as shown in Fig. 11.22. The intersection of $G(j\omega)$ plot with M_1 circle at frequencies ω_1 and ω_4 indicates that

$$\frac{G(j\omega_1)}{1+G(j\omega_1)} = M(j\omega_1) = M_1$$

$$\frac{G(j\omega_4)}{1+G(j\omega_4)} = M(j\omega_4) = M_1$$

Similarly, the intersection of $G(j\omega)$ plot with M_2 circle at frequencies ω_2 and ω_3 indicates that

$$M(j\omega_2) = M(j\omega_3) = M_2$$

Note that the circle tangent to the $G(j\omega)$ plot provides us with the peak magnitude M_r of the closed-loop frequency response and the

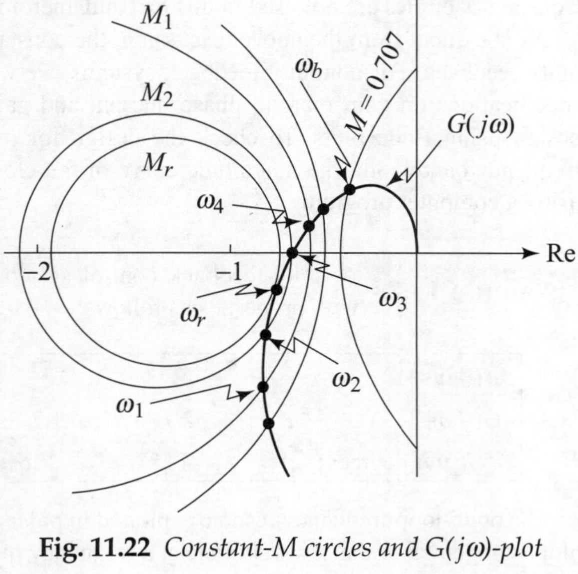

Fig. 11.22 *Constant-M circles and G(jω)-plot*

corresponding frequency ω_r. Using Fig. 11.22, we can also read the bandwidth; it is the frequency ω_b associated with the point of intersection of $G(j\omega)$ plot with the $M = 0.707$ circle. A typical magnitude curve of closed-loop frequency response is shown in Fig. 11.23.

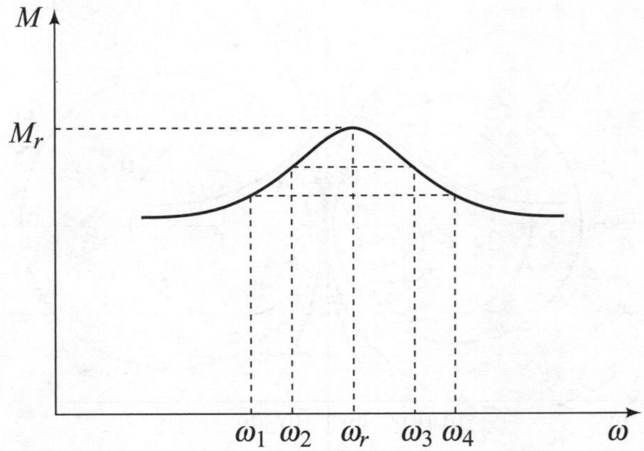

Fig. 11.23 *Closed-loop frequency response*

It is also possible to plot loci of constant phases of $M(j\omega)$ on the $G(s)$-plane. The plot consists of a family of circles called *constant-N circles*; $N = \tan \alpha$, $\alpha = \angle M(j\omega)$. When overlaid on the $G(j\omega)$ plot, the constant-N circles give the phase curve of closed-loop frequency response in a manner similar to that used for obtaining magnitude curve.

It may be noted that instead of using constant-M and constant-N circles, we can compute closed-loop frequency response directly from Eqn. (11.19) by use of a computer program. However, as we shall see shortly, the graphical method allows the design of a system with specified values of M_r, ω_r and/or ω_b. These performance measures are associated with only the magnitude curve of closed-loop frequency response and can be read off directly from a plot of $G(j\omega)$ superimposed on constant-M circles. Constant-N circles are not used in this text and, therefore, will not be discussed.

As is evident from the above discussion, the constant-M circles can be used only for systems with unity feedback. For nonunity feedback systems, we will translate the performance requirements into specifications on gain margin, phase margin, and gain crossover frequency, and then carry out the design using Bode plots. To check the design for requirements on resonance peak and resonance frequency/bandwidth, the magnitude curve of the closed-loop frequency response may be generated using a computer program.

Example 11.3 A unity-feedback control system has an open-loop frequency response $G(j\omega)$ with properties as follows:

ω (rad/sec)	2	3	4	5	6
$\lvert G(j\omega)\rvert$	2.7	1.7	0.97	0.63	0.4
$\angle G(j\omega)$(degrees)	–115	–126	–138	–150	–163

The open-loop frequency response, plotted in polar coordinates, is shown in Fig. 11.24. The $G(j\omega)$-plot is tangential to the $M = 1.4$ circle at a frequency $\omega = 4$ rad/sec. Therefore, the resonance peak $M_r = 1.4$, and the resonance frequency $\omega_r = 4$ rad/sec.

The values of ζ and ω_n to be used for approximating the transient behavior of the system can be determined from Eqns (11.16)–(11.17) or Figs 11.10–11.11. Rearranging Eqn. (11.16), we obtain

$$\zeta^4 - \zeta^2 + (1/4M_r^2) = 0 \quad (11.24)$$

Since $M_r = 1.4$, we have $\zeta = 0.39$ and 0.92. A damping ratio larger than 0.707 yields no peak above zero frequency. Thus, use select $\zeta = 0.39$. Using Eqn. (11.17) with $\omega_r = 4$ and $\zeta = 0.39$, we get $\omega_n = 4.8$ rad/sec.

The bandwidth of a system is defined as the frequency at which the magnitude of the closed-loop frequency response is 0.707 of its magnitude at $\omega = 0$. From Fig. 11.24, we see that $M = 0.707$ circle intersects the $G(j\omega)$-plot at $\omega = 6$ rad/sec. This frequency can be taken as the bandwidth if it is assumed that the closed-loop gain is unity at $\omega = 0$.

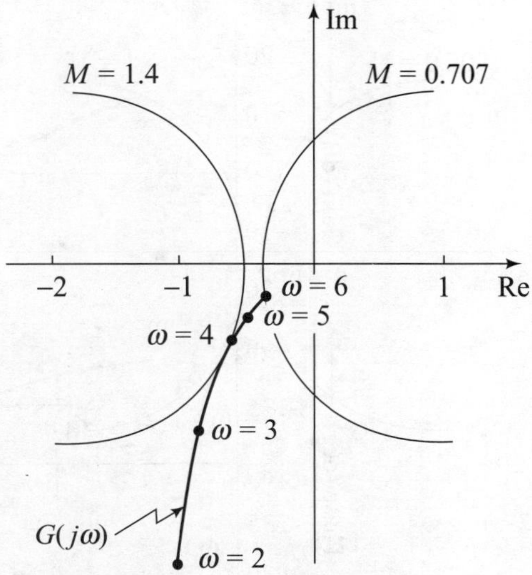

Fig. 11.24 *Constant-M circles and G(jω)-plot for Example 11.3*

11.6 THE NICHOLS CHART

In the previous section, we have seen that closed-loop frequency response of a unity-feedback system (refer to Fig. 11.19) can easily be determined from polar plot of open-loop transfer function $G(j\omega)$ superimposed on constant-M circles. Use of constant-M circles on polar plane is straightforward but seldom used because Nyquist analysis is normally carried out on the Bode plots and it is more efficient to use a nomograph compatible with Bode data. The *Nichols Chart*, introduced by N.B. Nichols in the 1940s, is such a nomograph.

The Nichols chart consists of curves that are maps of the constant-M and constant-N circles on a new coordinate system. The Nichols coordinate system plots $|G(j\omega)|$ in dB as ordinate *versus* $\angle G(j\omega)$ in degrees as abscissa, with origin at $|G(j\omega)| = 0$ dB and $\angle G(j\omega) = -180°$. This origin is the map of the critical point $-1 + j0$ of the polar coordinate (Nyquist) plot. Figure 11.26 illustrates the coordinate system (not the nomograph) with a plot of Bode data from Fig. 11.25. Note that the phase and gain margins are readily identified.

The constant-M and constant-N circles in polar coordinates do not remain circles when mapped onto dB *versus* phase plane. A typical constant-M circle is shown in Fig. 11.27. The magnitude ρ and angle λ of a vector drawn to any point on this circle are shown. The equation of the M circle is (refer to Eqn. (11.22))

$$\left(u - \frac{M^2}{1-M^2}\right)^2 + v^2 = \left(\frac{M}{1-M^2}\right)^2$$

where $u = \rho \cos \lambda$, and $v = \rho \sin \lambda$. Combining these equations produces

$$\left(\rho \cos \lambda - \frac{M^2}{1-M^2}\right)^2 + \rho^2 \sin^2 \lambda = \left(\frac{M}{1-M^2}\right)^2$$

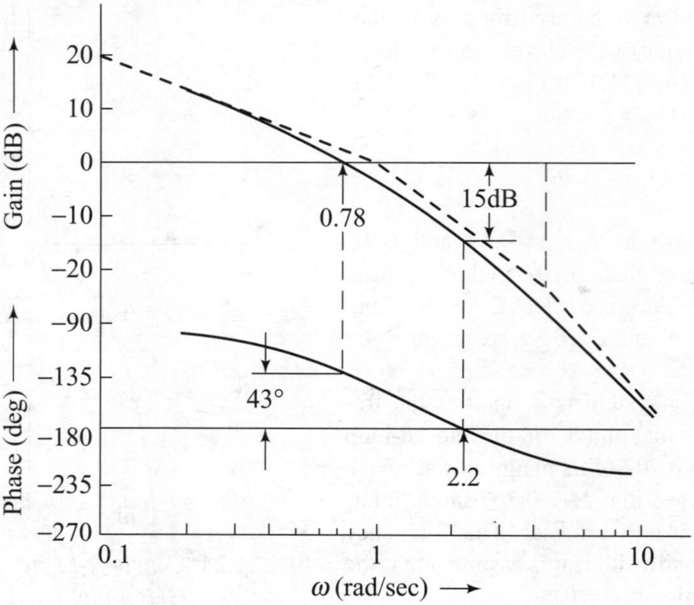

Fig. 11.25 *Frequency-response plot in a Bode coordinate system*

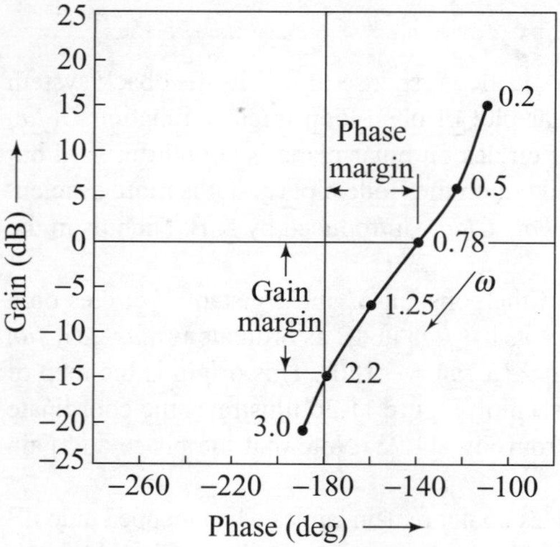

Fig. 11.26 *Frequency-response plot in a Nichols coordinate system*

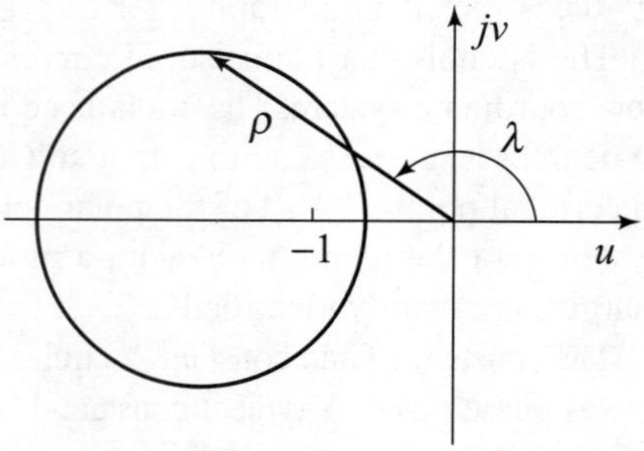

Fig. 11.27 *A constant-M circle*

or

$$\rho^2 - 2\cos\lambda\left(\frac{M^2}{1-M^2}\right)\rho - \frac{M^2}{1-M^2} = 0 \tag{11.25}$$

For any value of M, a series of values of λ can be inserted in Eqn. (11.25) to solve for ρ. This magnitude must be changed to dB. The data thus generated when plotted in Nichols coordinate system, gives a constant-M locus. When M takes on different values, Eqn. (11.25) describes in the Nichols system a family of loci. A representative family of constant-M loci is shown in Fig. 11.28. The constant-N circles can be transferred on Nichols chart in a similar manner, if needed.

Note that the $M = 1$ curve is asymptotic to $-90°$ and $-270°$ lines of Nichols coordinate system. The curves for $M > 1$ are closed curves inside the limits of $-90°$ and $-270°$ lines. The curves for $M < 1$ cross these limits. The curves are symmetrical about $-180°$ line.

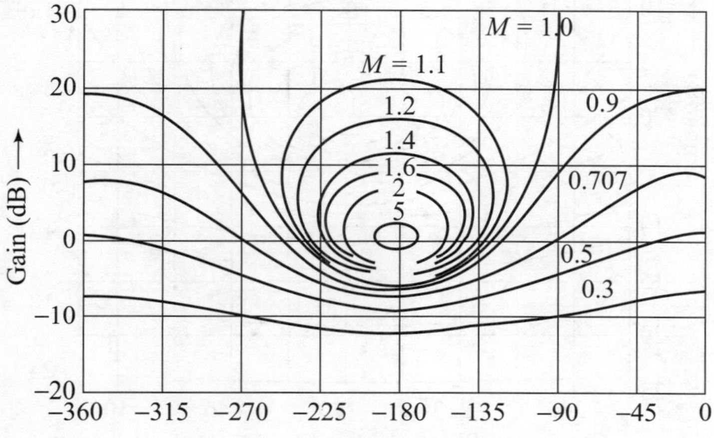

Fig. 11.28 *Constant-M loci in Nichols coordinate system*

The Nichols charts were commercially available until the middle 1970s. These charts cannot be purchased any more. They can, however, be easily generated with the help of a computer. An expanded section of the Nichols chart with selected constant-M loci is shown in Fig. 11.29; the values of M are given in dB.

Superimposing the dB versus phase angle curve of $G(j\omega)$ shown in Fig. 11.26, on the Nichols chart we obtain the plot of Fig. 11.30. The intersection of $G(j\omega)$-curve with $M = 1$ dB locus at frequencies ω_1 and ω_2 indicates that

$$\frac{|G(j\omega_1)|}{|1+G(j\omega_1)|} = |M(j\omega_1)| = 1 \text{ dB}$$

$$\frac{|G(j\omega_2)|}{|1+G(j\omega_2)|} = |M(j\omega_2)| = 1 \text{ dB}$$

Similarly, the intersections of $G(j\omega)$-curve with other constant-M loci determine the magnitude (in dB) of the closed-loop frequency response for various frequencies. The data so generated when plotted with magnitude in dB as ordinate and frequency ω (log scale) as abscissa, gives the Bode magnitude plot of the closed-loop frequency response.

Note that the constant-M locus tangent to the $G(j\omega)$-curve provides us with the peak magnitude M_r (in dB) of the closed-loop frequency response and the corresponding frequency ω_r. We can also directly read off the bandwidth from the Nichols chart; it is the frequency associated with the point of intersection of $G(j\omega)$-curve with $M = -3$ dB locus.

The Nichols chart of Fig. 11.30 has 2 dB and 3 dB closed-loop gain curves; the M_r value lies between 2 dB and 3 dB. One must do a little interpolation to determine M_r. From Fig. 11.30 we see that an approximate value of M_r is 2.5 dB and it occurs at a frequency $\omega_r = 0.8$ (the frequency ω_r corresponding to the point A has been determined from Fig. 11.25).

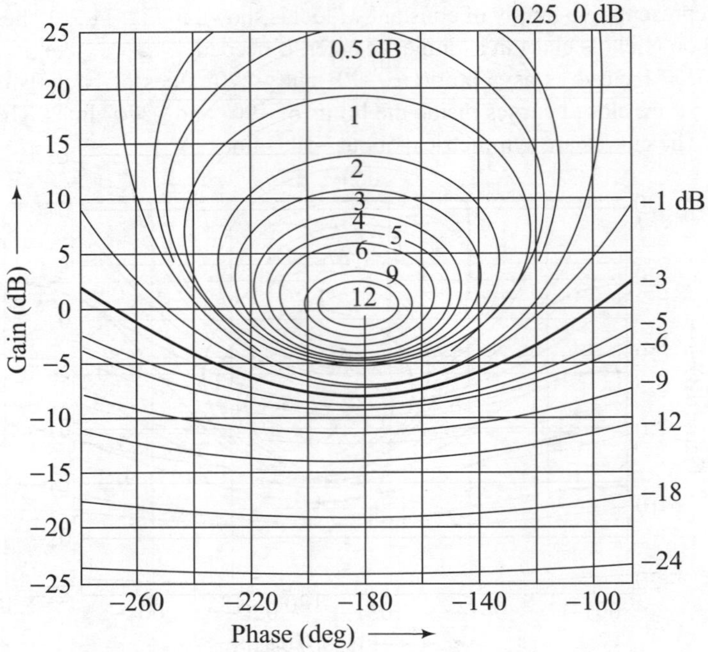

Fig. 11.29 *An expanded section of the Nichols chart*

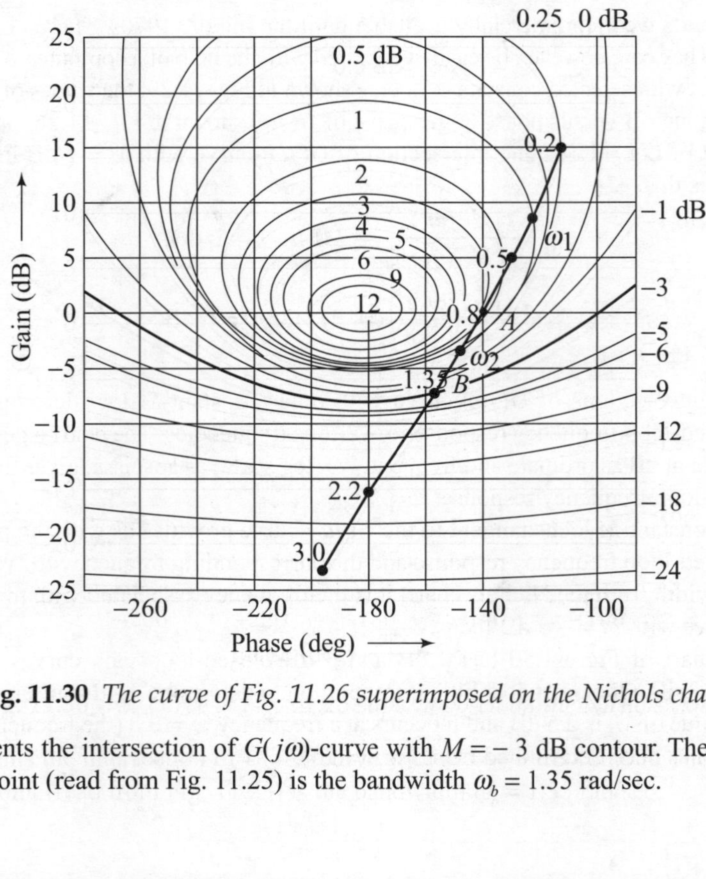

Fig. 11.30 *The curve of Fig. 11.26 superimposed on the Nichols chart*

Point *B* represents the intersection of $G(j\omega)$-curve with $M = -3$ dB contour. The frequency corresponding to this point (read from Fig. 11.25) is the bandwidth $\omega_b = 1.35$ rad/sec.

Example 11.4 The concept of constant-M circles will now be applied to the design of control systems. In obtaining acceptable performance, the adjustment of gain is usually the first consideration. The adjustment may be based on a desirable value for the resonance peak. In what follows, we shall demonstrate how the Nichols chart can be used to determine the gain K so that the system will have a desired value of M_r. Such a design is usually acceptable if the resulting bandwidth ω_b is satisfactory.

Let us reconsider the system discussed in Example 11.2. The system has an open-loop transfer function

$$G(j\omega) = \frac{K}{j\omega(j0.1\omega+1)(j0.05\omega+1)}; K = 10$$

Figure 11.18 shows the Bode plot of $G(j\omega)$. Transferring the Bode data to a Nichols chart, we get the dB versus phase angle plot of $G(j\omega)$. This is shown in Fig. 11.31. From the figure, it is seen that

$$M_r = 5.3 \text{ dB, and } \omega_r = 8.14 \text{ rad/sec}$$

The dB versus phase angle plot of $G(j\omega)$ intersects the –3 dB M-locus at the point A. The frequency parameter at this point can be easily determined by transferring the open-loop dB data at A to the Bode plot of Fig. 11.18. We find the frequency to be 12.6 rad/sec. Therefore, bandwidth ω_b = 12.6 rad/sec.

Let us examine the effects of gain adjustments. Suppose it is desired to find the open-loop gain for resonance peak of 2 dB. From Fig. 11.31, it is found that the dB versus phase angle plot of $G(j\omega)$ must be moved downwards by 3.5 dB in order that it be tangent to the 2 dB M-locus. It means that the gain $K = 10$ must be reduced by a factor of β, given by $20 \log (1/\beta) = -3.5$. Hence, $\beta = 1.5$ and the new gain is $K = 10/1.5 = 6.33$.

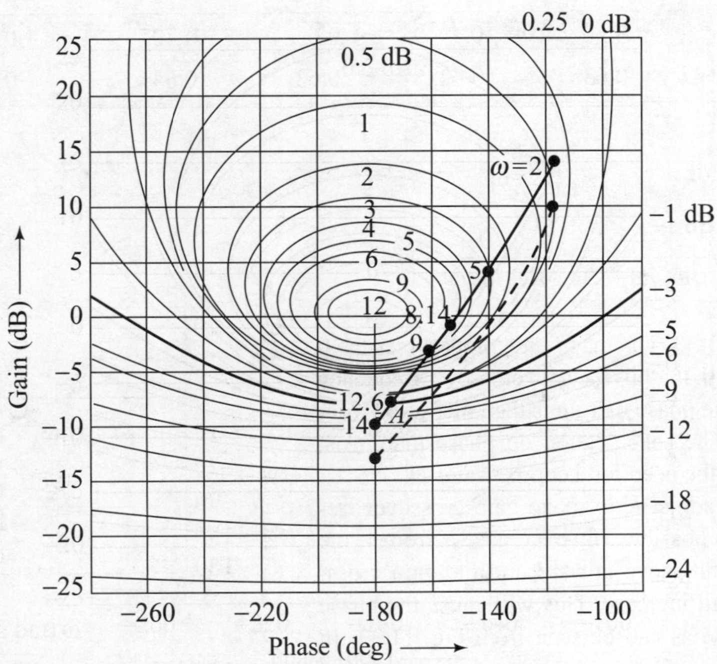

Fig. 11.31 *The Nichols chart for Example 11.4*

From Fig. 11.31, we observe that reducing K reduces the resonance peak M_r; it also reduces the bandwidth ω_b. The system thus becomes relatively more stable (peak overshoot in transient response reduces). The reduced bandwidth will generally result in increased rise time/settling time of the transient response.

11.7 REVIEW EXAMPLES

Review Example 11.1 Consider the unity-feedback system shown in Fig. 11.32. Let the compensator $D(s)$ be simply a constant K. Find a gain K such that

(i) phase margin $\geq 60°$, and
(ii) bandwidth is as large as possible.

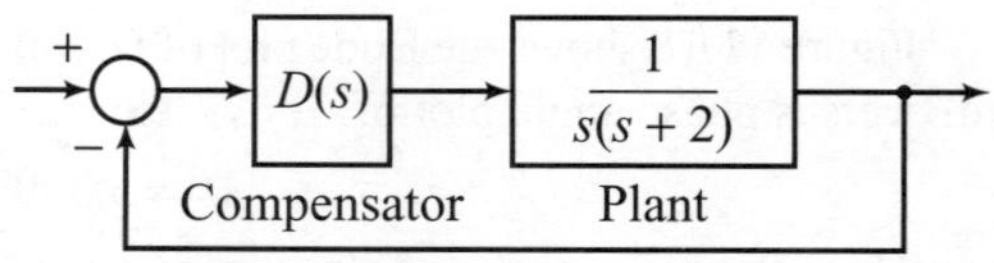

Fig. 11.32 *Unity-feedback system*

Solution

The system under consideration is a standard second-order system. Correlations derived in Section 11.4 are applicable without any approximation. Using Eqns (11.11) and (11.18), we obtain the ratio of gain crossover frequency ω_g and bandwidth ω_b as a function of ζ. Table 11.1 shows this ratio for a range of values of ζ. It is quite interesting to note that the ratio is rather constant. Raising the gain crossover frequency increases the bandwidth of the system.

Table 11.1 *Bandwidth-gain crossover frequency correlation*

ζ	0.3	0.4	0.6	0.707	0.8
ω_g/ω_b	0.63	0.62	0.63	0.64	0.68

The Bode plot of

$$G(j\omega) = \frac{1}{j\omega(j\omega+2)} = \frac{0.5}{j\omega(j0.5\omega+1)}$$

is shown in Fig. 11.33. The magnitude plot crosses the 0 dB line roughly at 0.46; thus the gain crossover frequency is 0.46 rad/sec. The phase margin is then measured from the plot as 77°. The specification on phase margin is, thus, met without the need for a compensator $D(s)$.

Now we shall adjust K to make gain crossover frequency as large as possible. If we increase K from 1, the magnitude plot will move upward and the gain crossover frequency will increase. This will cause the phase margin to decrease as can be seen from Fig. 11.33. In other words, as the gain crossover frequency increases, the phase margin will decrease. The largest permissible

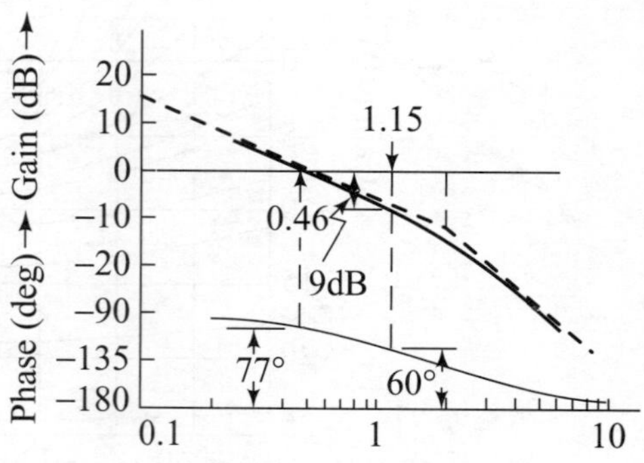

Fig. 11.33 *Bode plot of $G(j\omega) = 0.5/j\omega(j0.5\omega + 1)$*

gain crossover frequency corresponds to a phase margin of 60°. It is seen from Fig. 11.33 that if magnitude plot is shifted up by 9 dB, then 1.15 rad/sec becomes the new gain crossover frequency and the phase margin is 60°. Therefore, the required gain K is given by 20 log $K = 9$, which implies $K = 2.8$.

We may determine the bandwidth of the gain-compensated closed-loop system using the Nichols chart. Transferring the Bode data from Fig. 11.33 corresponding to $K = 2.8$ to the Nichols coordinate system, we obtain Fig. 11.34. The –3dB M-locus has been constructed using Table 11.2, obtained from the Nichols chart of Fig. 11.29. The dB versus phase angle curve of $KG(j\omega)$ intersects the –3dB M-locus at point A. The frequency parameter at this point can be easily determined by transferring the open-loop dB data at A to the Bode plot of Fig.11.33. We find this frequency to be 2; therefore the bandwidth $\omega_b = 2$ rad/sec.

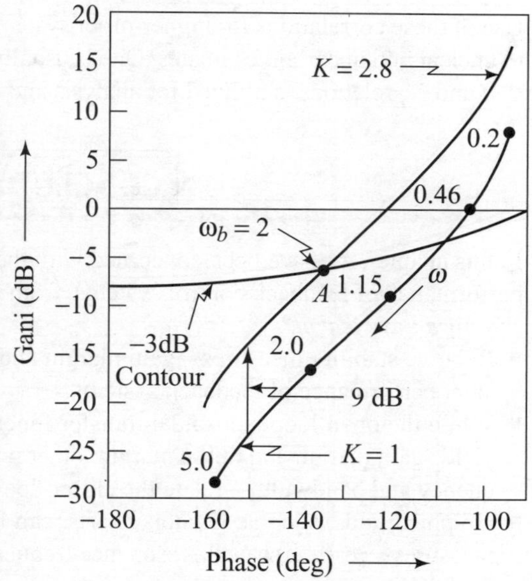

Fig. 11.34 *Determination of bandwidth*

Table 11.2 *Data for construction of –3dB M-locus of the Nichols chart*

Degrees	– 90	– 100	– 120	– 140	– 160	– 180	– 200	– 220
dB	0	– 1.5	– 4.18	– 6.13	– 7.28	– 7.66	– 7.28	– 6.13

Note that the above design problem could also be solved analytically using correlations developed in Section 11.4; we opted for the graphical tools to illustrate a design procedure applicable to second-order as well as higher-order systems.

Review Example 11.2 Let us consider a unity-feedback system with open-loop transfer function

$$G(s) = \frac{0.64}{s(s^2 + s + 1)}$$

The $G(j\omega)$ plot in polar coordinate system is shown in Fig. 11.35. From this figure, we find that

$$\Phi M = 30°, M_r = 2.8$$

On the basis of phase margin, we estimate the system damping ratio as $\zeta \cong 0.30$ (refer to Eqn. (11.10)), and on the basis of resonance peak we find that $\zeta \cong 0.175$ (refer to Eqn. (11.16)). The apparent conflict discovered in this example clearly indicates that a designer must use the frequency-domain to time-domain correlations with caution. The correlations arrived at in Section 11.4 apply to standard second-order systems.

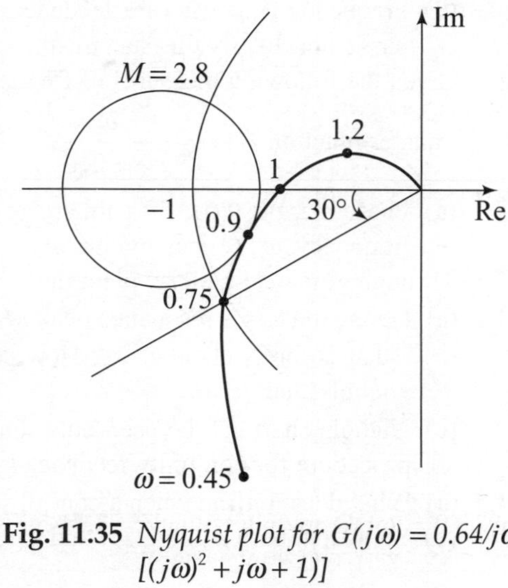

Fig. 11.35 *Nyquist plot for $G(j\omega) = 0.64/j\omega [(j\omega)^2 + j\omega + 1)]$*

Use of these correlations for higher-order systems where dominance condition is not satisfied, can lead to unclear and uncertain estimates. One is usually safe if the lesser of the two values of ζ resulting from ΦM and M_r relations is utilized for analysis and design purposes.

CONCLUDING REMARKS

In this chapter, we have been concerned with the definition and usefulness of quantitative measures of performance of feedback control systems, from the perspective of frequency response characteristics.

What we have learnt?

- Relative stability measures—gain margin and phase margin—can be utilized as indices of the transient performance of feedback systems.
- While the open-loop sinusoidal transfer function, when examined in the light of Nyquist criterion, yields gain margin and phase margin, other performance measures—resonance peak, resonance frequency and bandwidth—relate the closed-loop frequency response to the transient response.
- Graphical aids, such as Nichols charts, can be used to determine closed-loop frequency response measures—resonance peak, resonance frequency, and bandwidth—from open-loop sinusoidal transfer function.
- Setting the loop gain is the first step in adjusting the feedback system for satisfactory performance. We have learnt this preliminary design exercise in this chapter from the perspective of frequency-domain behavior. For many practical cases, however, the adjustment of the gain alone may not provide sufficient alteration of the system behavior to meet the given specifications. It is then necessary to redesign the system by modifying the structure and/or by incorporating additional control devices. The next chapter deals with the redesign problem using frequency response plots.

Review Questions

11.1 (a) Describe a set of performance specifications characterizing desired frequency response, that are frequently used in the design of linear control systems using Bode plots.

(b) Frequency response of a feedback control system should be similar to that of an ideal low pass filter. Justify this statement.

11.2 Given the following measures of frequency response of a unity feedback system with open-loop transfer function $G(s) = \dfrac{\omega_n^2}{s(s+2\zeta\omega_n)}$:

(a) Phase margin, ΦM; (b) gain crossover frequency, ω_g; (c) resonance peak, M_r; (d) resonance frequency, ω_r; (e) bandwidth, ω_b.

Through graphical plots, explain the correlation of these functions with ζ and ω_n.

11.3 (a) Define the terms resonance peak M_r, and the bandwidth ω_b of a closed-loop control system.

(b) What is the Nichols chart? How can we identify resonance peak and bandwidth from the Nichols chart?

(c) Nichols chart can be used only for systems with unity feedback. Suggest a suitable design procedure for non-unity feedback systems.

11.4 (a) What does it mean when a control system is described as being robust?

(b) Show that both a good GM and a good ΦM are needed for robust stability.

Self-Appraisal Quiz

A set of questions with multiple answer choices is given below. These questions have been designed to capture important aspects of the subject covered in the chapter. For each of these questions, identify the best of the given answer choices. Compare your answers with the Master Key given at the end of the session. This quiz should serve the purpose of a self-appraisal test for readers.

11.1 The sinusoidal transfer function of a feedback control system is given by

$$G(j\omega) = \frac{\omega_n^2}{(\omega_n^2 - \omega^2 + j2\zeta\omega_n\omega)}; \quad 0 < \zeta < \frac{1}{\sqrt{2}}$$

If a sinusoidal excitation $\cos(\omega_n\sqrt{1-2\zeta^2})t$ is applied to the system, the steady-state response magnitude is

(A) $\dfrac{1}{2\zeta\sqrt{1-\zeta^2}}$ (B) $\dfrac{1}{2\zeta}$

(C) $e^{-\pi\zeta/\sqrt{1-\zeta^2}}$ (D) None of the answers in (A), (B) or (C) is correct.

11.2 A dynamic system is represented by the transfer function,

$$\frac{Y(s)}{R(s)} = G(s) = \frac{\omega_n^2}{s^2 + 2\zeta\omega_n s + \omega_n^2}; \quad 0 < \zeta < \frac{1}{\sqrt{2}}$$

The system is given the sinusoidal input $r(t) = \sin \omega t$

The value of ω at which the output $y(t)$ is maximum at steady state is given by

(A) $\omega < \omega_n$ (B) $\omega = \omega_n$ (C) $\omega > \omega_n$

11.3 Consider the following statements for an underdamped second-order system:

(i) Peak overshoot in step-input response reduces as damping is increased from 0.2 to 0.6

(ii) Resonance peak in frequency response reduces as damping is increased from 0.2 to 0.6.

Which of the following is the correct answer?

(A) None of the above statements is true (B) Statement (i) is true but statement (ii) is false.

(C) Statement (i) is false but statement (ii) is true. (D) Both the statements are true.

11.4 Undamped natural frequency ω_n and resonance frequency ω_r of a unity feedback system with open-loop transfer function,

$$G(s) = \frac{\omega_n^2}{s(s + 2\zeta\omega_n)}; \zeta < \frac{1}{\sqrt{2}},$$

are related as

(A) $\omega_n = \omega_r$ (B) $\omega_n > \omega_r$

(C) $\omega_n < \omega_r$ (D) None of the answers in (A), (B), and (C) is correct.

11.5 undamped natural frequency ω_n and bandwidth ω_b of a unity feedback system with open-loop transfer function,

$$G(s) = \frac{\omega_n^2}{s(s + 2\zeta\omega_n)}; \zeta < \frac{1}{\sqrt{2}},$$

are related as

(A) $\omega_n = \omega_b$ (B) $\omega_n > \omega_b$
(C) $\omega_n < \omega_b$ (D) None of the answers in (A), (B), and (C) is correct.

11.6 For a unity feedback system with open-loop transfer function $G(s) = \dfrac{\omega_n^2}{s(s + 2\zeta\omega_n)}, \zeta < 0.7;$

(i) phase margin is explicitly indicative of damping ratio;
(ii) resonance peak is explicitly indicative of damping ratio.

Which of the following answers is correct?

(A) None of the above statements is true (B) Statement (i) is true but statement (ii) is false
(C) Statement (i) is false but statement (ii) is true. (D) Both the statements are true

11.7 Open-loop transfer function of a unity feedback system is $G(s) = K/[s(s+5)]$. The gain K that results in a phase margin of 50° is approximately

(A) 15 (B) 20 (C) 25 (D) 30

11.8 Consider a unity-feedback system with open-loop transfer function $G(s) = \dfrac{K}{s(s+1)}$. Increase in gain K will cause

(A) the gain crossover frequency to reduce (B) the gain crossover frequency to increase
(C) the system to respond more slowly (D) None of the statements in (A), (B), and (C) is correct.

11.9 The closed-loop dynamics of a system is of second-order. To improve the damping, we should

(A) decrease the phase margin (B) increase the phase margin
(C) decrease the gain margin (D) increase the gain margin

11.10 The closed-loop dynamics of a system is of second-order. The resonance peak of the closed-loop frequency response is related to

(i) overshoot of time response
(ii) damping ratio

(A) Both the statements are true (B) Statements (i) is true but (ii) is false
(C) Statement (i) is false but (ii) is true (D) Both the Statements are false

11.11 A unity-feedback system has open-loop transfer function,

$$G(s) = \frac{\omega_n^2}{s(s + 2\zeta\omega_n)}$$

(i) The peak overshoot of step-input response of $G/(1 + G)$ is indicative of damping
(ii) The resonance peak of frequency response of $G/(1 + G)$ is indicative of damping.

(A) Both the statements are true (B) Statement (i) is true but (ii) is false
(C) Statement (i) is false but (ii) is true (D) Both the statements are false

11.12 For the system $1/(s^2 + 0.1s + 1)$, the resonance frequency ω_r and natural frequency ω_n are related as

(A) $\omega_r = \omega_n$ (B) $\omega_r > \omega_n$
(C) $\omega_r < \omega_n$ (D) None of the answers in (A), (B), and (C) is correct.

11.13 Bode plot characteristic which is the best indicator of the closed-loop step response rise time is

(A) phase margin (B) gain margin
(C) phase crossover frequency (D) gain crossover frequency

11.14 The vertical axis of the Nichols chart represents

(A) open-loop gain (B) open-loop phase
(C) closed-loop gain (D) closed-loop phase

11.15 *M*-contours on the Nichols chart represent

(A) open-loop gain (B) open-loop phase
(C) closed-loop gain (D) closed-loop phase

11.16 Consider the following statements

(i) The resonance peak of the frequency response of $G/(1 + G)$ is given by the *M*-contour on the Nichols chart which is tangent to the Nichols plot of G.
(ii) The resonance peak of the frequency response of $G/(1 + G)$ is the peak value of the frequency response of $G/(1 + G)$ on Bode magnitude plot.

(A) Both the statements are true (B) Statements (i) is true but (ii) is false
(C) Statement (i) is false but (ii) is true (D) Both the statements are false

11.17 The bandwidth of $G/(1 + G)$

(i) is the frequency at which the –3 dB *M*-contour on Nichols chart intersects the Nichols plot of G.
(ii) is the frequency at which Bode magnitude plot of $G/(1 + G)$ has a gain of –3 dB

(A) Both the statements are true (B) Statements (i) is true but (ii) is false
(C) Statement (i) is false but (ii) is true (D) both the statements are false

11.18 The gain margin of a system can be found from the Nichols chart by examining

(A) the intersection of *M*-contours with the –180° phase axis
(B) the intersection of *M*-contours with the 0 dB magnitude axis
(C) the intersection of the open-loop plot with the 0 dB magnitude axis.

MASTER KEY

11.1 (A)	11.2 (A)	11.3 (D)	11.4 (B)	11.5 (C)	11.6 (D)
11.7 (C)	11.8 (B)	11.9 (B)	11.10 (A)	11.11 (A)	11.12 (C)
11.13 (D)	11.14 (A)	11.15 (C)	11.16 (A)	11.17 (A)	11.18 (D)

Problems

11.1 Consider a unity-feedback system with open-loop transfer function

$$G(s) = \frac{\omega_n^2}{s(s + 2\zeta\omega_n)}$$

Show that the phase margin (relative stability measure in frequency domain) is given by

$$\Phi M = \tan^{-1}\frac{2\zeta}{\sqrt{\sqrt{4\zeta^4+1}-2\zeta^2}}$$

11.2 For the standard second-order system

$$\frac{Y(s)}{R(s)} = \frac{\omega_n^2}{s^2+2\zeta\omega_n s+\omega_n^2},$$

show that the bandwidth is given by

$$\omega_b = \omega_n\sqrt{\left[1-2\zeta^2)+\sqrt{4\zeta^4-4\zeta^2+2}\right]}$$

11.3 Show that when $0 < \zeta < 0.707$, the frequency response of a system defined by a transfer function,

$$M(s) = \frac{\omega_n^2}{s^2+2\zeta\omega_n s+\omega_n^2}$$

has a peak amplitude given by

$$M_r = \frac{1}{2\zeta\sqrt{1-\zeta^2}}$$

which occurs at the resonance frequency

$$\omega_r = \omega_n\sqrt{1-2\zeta^2}$$

11.4 Consider the feedback system shown in Fig. P11.4.

(a) Find the values of K and a to satisfy the following frequency-domain specifications:

$M_r = 1.04$; $\omega_r = 11.55$ rad/sec

(b) For the values of K and a determined in part (a), calculate the settling time and bandwidth of the system.

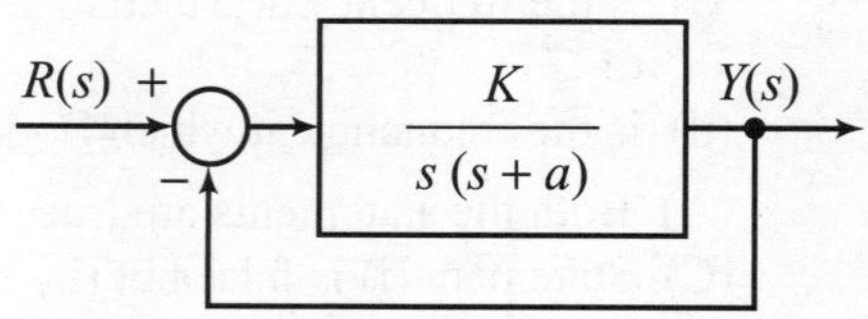

Fig. P11.4

MATLAB Exercise Use MATLAB to obtain resonance peak, resonance frequency, settling time, and bandwidth for the values of K and a obtained in part (a).

11.5 Unit-step response data of a second-order system is given below. Obtain the corresponding frequency response indices (M_r, ω_r, ω_b) for the system.

t(sec)	0	0.05	0.10	0.15	0.20	0.25	0.30	0.35	0.40	0.45	0.50
$y(t)$	0	0.25	0.8	1.08	1.12	1.02	0.98	0.98	1.0	1.0	1.0

11.6 Phase margin for a second-order system is given by

$$\Phi M = \tan^{-1}\frac{2\zeta}{\sqrt{\sqrt{4\zeta^4+1}-2\zeta^2}}$$

Write the approximate expression for ΦM for low values of ζ. Using the approximate expression, find the value of gain K such that the system shown in Fig. P11.6. has a phase margin of ϕ_m degrees.

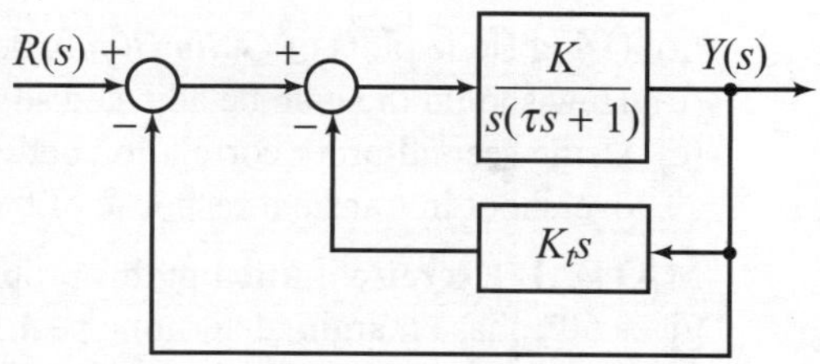

Fig. P11.6

11.7 The open-loop transfer function of a unity-feedback system is given by

$$G(s) = \frac{K}{s(\tau_1 s+1)(\tau_2 s+1)}$$

Derive an expression for gain K in terms of τ_1 and τ_2 and specified gain margin G_m.

11.8 Feedback system of Fig. P11.8 has open-loop transfer function,

$$G(s)H(s) = \frac{199}{s(s+1.71)(s+100)}$$

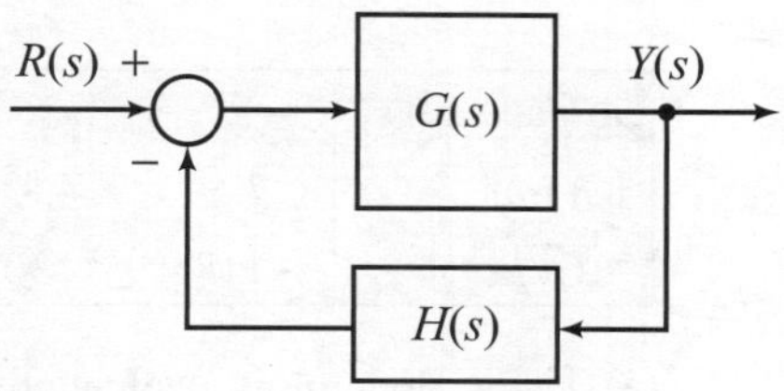

Fig. P11.8

(a) Using Bode plots, determine the gain crossover frequency and phase margin.
(b) Using second-order correlations between time-domain and frequency-domain measures, estimate the peak overshoot and settling time of the transient response to step input.

MATLAB Exercise Use MATLAB to obtain gain crossover frequency, phase margin, peak overshoot and settling time.

11.9 The Bode magnitude plot of a system (known to have minimum-phase characteristics) is shown in Fig. P11.9. Deduce approximate Bode phase plot of the system.

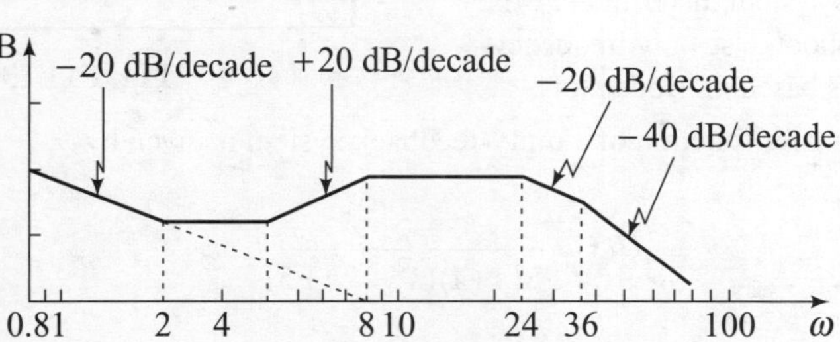

Fig. P11.9

Determine phase margin and gain margin of the system when operating in the closed-loop mode with unity feedback. From the frequency-domain performance measures, estimate the peak overshoot in transient response of the feedback system to step input.

11.10 Suppose that for the system of Fig. P11.8,

$$G(s)H(s) = \frac{4}{s(s+1)(s+2)}$$

(a) Using Bode plots of $G(j\omega)H(j\omega)$, determine the phase margin of the system.
(b) How should the gain be adjusted so that phase margin is 50°?
(c) Using second-order correlation between phase margin and damping ratio, estimate the peak overshoot in transient response of the system to step input.

MATLAB Exercise Attempt this problem using MATLAB dialogs. For the value of gain that gives 50° phase margin, determine peak overshoot in the MATLAB window.

11.11 Figure P11.11 shows block diagram of a control system. Determine the gain K such that the phase margin is 50°. What is the gain margin in this case?

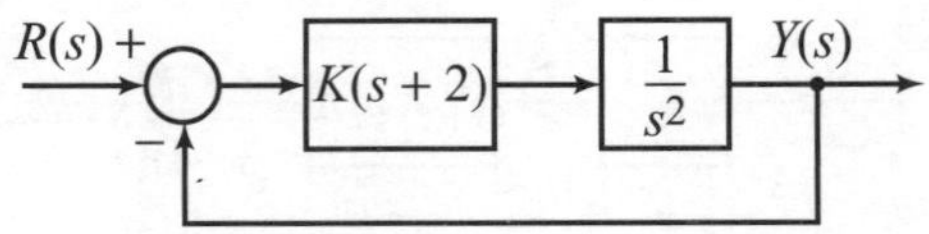

Fig. P11.11

MATLAB Exercise Attempt this problem using MATLAB dialogs.

11.12 A unity-feedback system has the following open-loop frequency response.

ω (rad/sec)	2	3	4	5	6	8	10
$\|G(j\omega)\|$	7.5	4.8	3.15	2.25	1.70	1.00	0.64
$\angle G(j\omega)$	– 118°	– 130°	– 140°	–150°	–157°	– 170°	– 180°

(a) Using Bode plots, evaluate the gain margin and phase margin of the system.
(b) Determine the change in gain required so that the gain margin of the system is 20 dB.
(c) Determine the change in gain required so that the phase margin of the system is 60°.

11.13 For the position control system shown in Fig. P11.13, find the value of the preamplifier gain K to yield a 9.48% overshoot in the transient response for a step input. For the gain-compensated system, determine t_p: the time to peak overshoot. Use only frequency-domain techniques based on Bode plots.

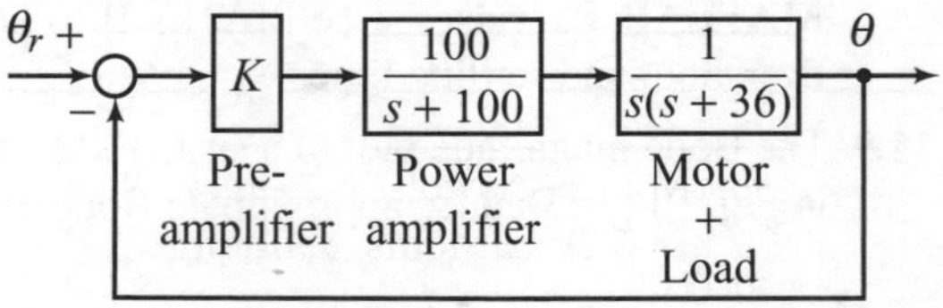

Fig. P11.13

11.14 The open-loop transfer function of a unity-feedback system is given by

$$G(s) = \frac{5}{s(s+1)(s^2+2s+5)}$$

(a) Using Bode plots, evaluate the gain margin, the phase margin, and the gain crossover frequency.
(b) How should the gain be adjusted in the system so that the gain margin becomes 10 dB?
(c) How should the gain be adjusted so that the gain crossover frequency becomes 1.27 rad/sec?
(d) How should the gain be adjusted so that the phase margin becomes 45°?

MATLAB Exercise Attempt this problem using MATLAB dialogs.

11.15 The open-loop transfer function of a unity-feedback system is

$$G(s) = \frac{K}{s(s+2)(s+10)}$$

Determine the value of K so that the system is stable with (a) a phase margin $\geq 45°$, and (b) gain crossover frequency as large as possible.

MATLAB Exercise Attempt this problem using MATLAB dialogs.

11.16 (a) The open-loop system with transfer function,

$$G(s) = \frac{5e^{-s}}{s(s+1)(s+10)}$$

is used in unity-feedback configuration. Estimate the peak overshoot in transient response to step input. Use frequency-domain techniques based on Bode plots.

(b) Repeat part (a) for the system without dead-time.

11.17 The open-loop system with transfer function,

$$G(s) = \frac{31.5e^{-s\tau_D}}{(s+1)(30s+1)[(s^2/9)+(s/3)+1]}$$

is used in unity-feedback configuration.

(a) Find phase margin when $\tau_D = 0$.

(b) Find phase margin when $\tau_D = 1$. Adjust the gain to obtain phase margin of 30°.

(c) Compare steady-state errors of the two cases with the same phase margin.

11.18 A unity-feedback system has open-loop transfer function,

$$G(s) = \frac{50}{s(s+3)(s+6)}$$

(a) Using constant-M circles and polar plot of $G(j\omega)$, determine resonance peak M_r and bandwidth ω_b of the closed-loop frequency response.

(b) Using second-order correlations between time-domain and frequency-domain measures, estimate the peak overshoot and settling time of the transient response to step input.

11.19 A unity-feedback control system has the following open-loop frequency response:

ω (rad/sec)	2	3	4	5	10
$\lvert G(j\omega)\rvert$	2.7	1.7	0.97	0.63	0.4
$\angle G(j\omega)$	– 115°	– 126°	– 138°	– 150°	– 163°

(a) Plot the open-loop frequency response in polar coordinates.

(b) Find the resonance peak M_r and the resonance frequency ω_r of the closed-loop frequency response.

(c) From the frequency-domain performance measures, estimate the peak overshoot and settling time of the transient response of the feedback system to step input.

11.20 The open-loop transfer function of a unity-feedback system is

$$G(s) = \frac{11.7}{s(1+0.05s)(1+0.1s)}$$

(a) Draw Bode plot of $G(j\omega)$, and therefrom determine gain crossover frequency, phase crossover frequency, gain margin and phase margin.

(b) Transfer the Bode data to a Nichols chart and determine the resonance peak, resonance frequency and bandwidth.

MATLAB Exercise Attempt this problem using MATLAB dialogs.

11.21 The open-loop frequency response of a unity-feedback system is given below.

ω (rad/sec)	1	2	3	4	5
$\|G(j\omega)\|$(dB)	12.6	3.92	– 2.22	– 7	– 11
$\angle G(j\omega)$(degree)	– 127°	– 152°	– 168°	– 179°	– 188°

Transfer the frequency-response data to a Nichols chart and therefrom determine the gain margin, phase margin, resonance peak, resonance frequency and bandwidth.

11.22 A unity-feedback system has open-loop transfer function,

$$G(s) = \frac{4}{s(s+1)(s+2)}$$

(a) Using Bode plots of $G(j\omega)$, determine the phase margin of the system.
(b) How should the gain be adjusted so that phase margin is 50°?
(c) Determine the bandwidth of the gain-compensated system.

The –3dB contour of the Nichols chart may be constructed using the following table:

Phase, degrees	0	– 30	– 60	– 90	– 120	– 150	–180	– 210
Magnitude, dB	7.66	6.8	4.18	0	– 4.18	– 6.8	–7.66	– 6.8

(d) Using second-order correlations between frequency-domain measures (phase margin, bandwidth) and time-domain measures (damping ratio, undamped natural frequency), estimate the peak overshoot and settling time of the transient response of gain-compensated system to step input.

11.23 A unity-feedback system has open-loop transfer function,

$$G(s) = \frac{54}{(1+0.1s)\,(s^2+8s+25)}$$

(a) Using Nichols chart, determine M_r and ω_r.
(b) What value of ζ and ω_n should be used to approximate the transient behavior of the system.

MATLAB Exercise Determine M_r and ω_r from the Nichols chart in the MATLAB window. Use Second-order correlations to determine ζ and ω_n (peak overshoot and settling time). Compare these values of peak overshoot and settling time with the ones obtained from the step response of the given system generated using MATLAB.

11.24 For the system shown in Fig. P11.24, the transfer function $G(s)$ is

$$G(s) = \frac{100}{(1+0.1s)\,(s^2+8s+25)}$$

(a) Using Nichols chart, determine the factor K by which the gain of the system should be changed so that the resulting system will have an M_r of 1.4.

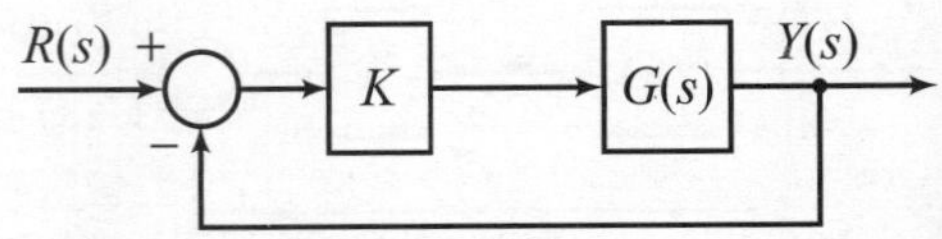

Fig. P11.24

(b) Find the bandwidth of the gain-compensated system.

MATLAB Exercise Attempt this problem using MATLAB dialogs.

11.25 The open-loop transfer function of a unity-feedback system is

$$G(s) = \frac{Ke^{-0.1s}}{s(1+0.1s)(1+s)}$$

By use of Bode plots and/or Nichols chart, determine the following:

(a) The value of K so that the gain margin of the system is 20 dB.

(b) The value of K so that the phase margin of the system is 60°.

(c) The value of K so that the resonance peak M_r of the system is 1 dB. What are the corresponding values of ω_r and ω_b?

(d) The value of K so that the bandwidth ω_b of the system is 1.5 rad/sec.

MATLAB Exercise Attempt this problem using MATLAB dialogs.

11.26 The following experimental results were obtained from an open-loop frequency response test of a unity-feedback control system:

ω	4	5	6	8	10
Gain	0.66	0.48	0.36	0.23	0.15
Phase angle	– 134°	– 143°	– 152°	– 167°	– 180°

Plot the locus of the loop transfer function in Nichols coordinates (dB versus angle) and measure the gain and phase margins.

By what factor should the gain be increased so that the resonance peak in closed-loop frequency response is 1.4, and what would then be the gain and phase margins?

11.27 The open-loop frequency response of a unity-feedback system is given below.

ω (rad/sec)	0.8	1.0	1.2	1.4	1.6
Re[$KG(j\omega)$]	−3.5	−2.9	−2.3	−2.0	−1.2
Im[$KG(j\omega)$]	−4.4	−3.2	−1.9	−1.2	−0.5

Determine the change in gain K required to make the resonance peak $M_r = 1.4$. For this value of gain, determine the phase margin of the system. Using second-order correlation between resonance peak and damping ratio, determine the value of ζ that approximates the transient behavior of the gain-compensated system. Repeat the calculation for ζ using second-order correlation between phase margin and damping ratio, and comment on the result.

COMPENSATOR DESIGN USING BODE PLOTS

12

Preview

This chapter runs parallel to Chapter 9; we look at frequency-domain design method, which is an alternative to root locus method. In the root locus technique, the performance specifications are translated into desired locations for closed-loop poles, and the root locus plot helps us know how the closed-loop poles move in the s-plane as the parameters of lead and lag compensators/PID control function are varied. In frequency-domain design, on the other hand, performance specifications are translated into characteristic parameters of desired frequency response and stability margins; the technique helps us know the effects of variation of parameters of lead and lag compensators/PID control function on filtering characteristics and the resulting closed-loop stability. Frequency-domain design provides good robustness against uncertainties because of frequency-selective characteristics of signals in control systems. A feedback control system is typically required to stop (or attenuate) the high-frequency components of the input, such as sensor noise and/or poorly known or changing high-frequency resonances; and pass the low-frequency components—the command signals which the control system is required to faithfully reproduce. This viewpoint on design is realized more easily using the Bode plots. Bode design approach is probably the best method to use for cascade compensation of linear time-invariant systems.

The two methods—the root locus design and the frequency response design—do not compete with each other; they complement each other. In fact, a practical design in industry is carried out using both of these graphical tools; the root locus plot is helpful in seeing what the Bode plot design is accomplishing.

12.1 INTRODUCTION

The design of feedback control systems in industry is probably accomplished using frequency-domain methods more often than other methods. The primary reason for the popularity of frequency-domain design is that the effects of disturbances, sensor noise and plant uncertainties are relatively easy to visualize and access in the frequency domain. Disturbance rejection can be specified in terms of low-frequency gain, and sensor-noise rejection can be specified in terms of high-frequency attenuation. The intermediate frequencies typically control the stability margins.

The robust control design methods developed over the last two decades seek to quantify uncertainties in the modeling of physical systems to ensure that the compensated system is always stable. It goes without saying that this has obviously always been the goal of control system designers. The recent advances are more mathematically rigorous than the work done during the 1940s and 1950s by Bode, Nichols, and others. From one point of view, the recent work on robust control serves to illustrate and corroborate the frequency-domain design techniques that have been developed over the last half century. There are strong connections between the classical design techniques presented in this chapter, and the modern robust control theory (refer to Section 11.3). In fact building of uncertainty model itself is subject to uncertainties; the industry, by and large, continue to rely on gain margin and phase margin specifications for robustness.

Through the design examples developed in this chapter, we will see that Bode plot design is essentially a pole cancelation technique. In many cases, the resulting overall system closely approximates a standard second-order transfer function. The fact that Bode plot design approach can in many cases yield a closed-loop system that approximates standard second-order transfer function, accounts in large part for its power.

Our design strategy in this chapter is similar to the one we have used earlier in root locus design. Our real interest is how the closed-loop system behaves in the time domain but we predict this behavior by examining the open-loop frequency response. We have seen in Chapter 10 that open-loop frequency response data on Bode plots, when examined in the light of Nyquist criterion, yield stability margins (gain margin and phase margin) of the closed-loop system. In Chapter 11, we have seen that open-loop frequency response data on Nichols chart, when examined in the light of *M*-contours, yield frequency-response (resonance peak, resonance frequency, bandwidth) of the closed-loop systems. We have also examined the correlations between the frequency response measures—phase margin, resonance peak, resonance frequency, bandwidth—with the time response measures—peak overshoot, settling time, etc., through standard second-order system.

The design procedure calls for shaping the Bode plot/Nichols chart using a compensator, to match it to the specified closed-loop frequency response. The frequency-domain design thus involves "loop shaping" techniques. These techniques make it possible to build control systems with bandwidths sufficient to meet performance requirements, but no wider (Recall the dangers of building control systems with wide bandwidths using root-locus design).

From the 1940s to 1960, the Nichols chart was in wide use for loop shaping. Some very good designers of that era preferred it to the Bode plot as a design tool. Some designers prefer to use the Bode magnitude and phase plots. The advantage here is that the frequency information is more readily available. Our preference is for Bode plots as design tools. We will use Nichols chart only for evaluation of the closed-loop frequency response from open-loop data.

The objective of the frequency-response design is to reshape, by means of a compensator, the frequency-response plot of the basic system and achieve the performance specifications in terms of resonance peak, resonance frequency, band width, gain margin, phase margin, etc. We can obtain the transient-response parameters from the frequency-response parameters using the correlations developed in the previous chapter. These correlations are, however, approximate and give only qualitative information about the transient response.

The root-locus design provides very good indicators of the transient response of a system. However, we do need a relatively accurate model of the system to obtain these benefits from the root locus plots. In fact, various design techniques supplement each other, rather than one method excluding others; the control engineer must therefore be familiar with all the design methods.

This chapter is concerned with the changes that can be made in the frequency-response characteristics of a given system by various types of compensators. Design procedures to obtain improvement in the system performance are described.

12.2 RESHAPING THE BODE PLOT

The performance of a closed-loop system may be described in terms of the following frequency-domain specifications:

1. Phase margin ΦM/resonance peak M_r—indicative of relative stability.
2. Gain crossover frequency ω_g/bandwidth ω_b/resonance frequency ω_r—indicative of speed of response.
3. Error constants $K_p/K_v/K_a$—indicative of steady-state error to step/ramp/parabolic input.

In case the transient-response specifications are given in time domain, we must first translate these into frequency domain to carry out frequency-domain compensation. This translation is carried out by using the explicit correlations between the two domains for the standard second-order system. These correlations are valid approximations for higher-order systems dominated by a pair of complex conjugate poles. The correlations, developed earlier in chapter 11, are reproduced below for convenience of use:

$$\Phi M = \tan^{-1}\frac{2\zeta}{\sqrt{\sqrt{4\zeta^4+1}-2\zeta^2}} \cong 100\zeta \tag{12.1}$$

$$\omega_g = \omega_n\sqrt{\sqrt{4\zeta^4+1}-2\zeta^2} \tag{12.2}$$

$$M_r = \frac{1}{2\zeta\sqrt{1-\zeta^2}} \tag{12.3}$$

$$\omega_r = \omega_n\sqrt{1-2\zeta^2} \tag{12.4}$$

$$\omega_b = \omega_n\sqrt{\left[(1-2\zeta^2)+\sqrt{4\zeta^4-4\zeta^2+2}\right]} \tag{12.5}$$

After completing the compensation design in the frequency domain, one must recheck the time response specifications by simulating the compensated system. Based on the results of this check, it may sometimes be necessary to repeat the complete design process.

The design procedure usually starts with the gain adjustments to meet the performance specifications. If the desired specifications are not met by gain adjustments alone, compensating devices must be used. These devices alter the shape of the frequency-domain plot to meet the performance specifications.

The compensating device may be inserted into the system either in cascade with the forward portion of the loop as shown in Fig. 12.1 or as part of minor feedback loop as shown in Fig. 12.2. We have seen in Chapter 9 that cascade compensation is usually concerned with the addition of PD/lead, PI/lag, or PID/lag-lead device, and that the minor-loop feedback compensation is primarily concerned with the addition of rate feedback.

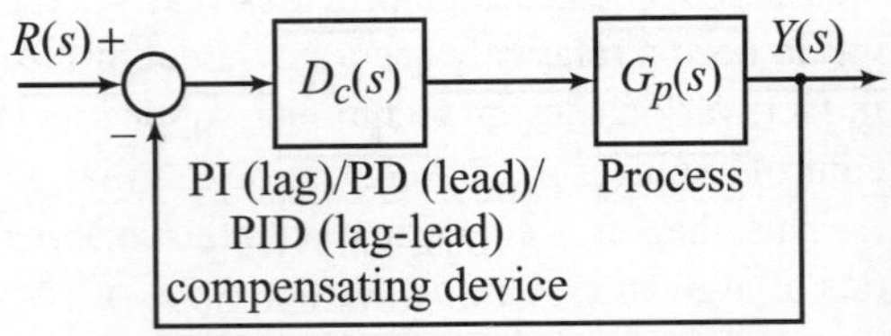

Fig. 12.1 *Illustration of cascade compensation*

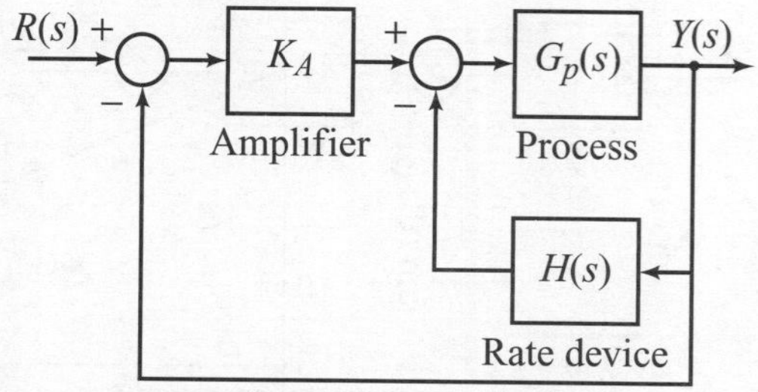

Fig. 12.2 *Illustration of minor-loop feedback compensation*

The decision to use a minor-loop feedback scheme (refer to Section 9.7) is sometimes a matter of convenience, sometimes a matter of necessity, and for some problems it can be shown that minor-loop feedback scheme will do a better job. For example, the use of velocity feedback for damping is very effective and consequently quite popular. Servomotors with in-built tachogenerators are shelf items; the use of velocity feedback can be very convenient. When designing compensation for process controls, it becomes necessary to take derivative action on controlled variable and not on error variable in order to avoid the derivative kick (violent response to sudden command changes (such as steps) because of differentiating action) on setpoint changes. Finally, in systems subjected to frequent load disturbances, feedback compensation may be preferred as it provides greater stiffness against load disturbances. Besides all these factors, the available components, and designer's experience and preferences influence the choice between a cascade and a feedback compensation scheme.

A basic point of view that is helpful in developing techniques for analysis and design is to realize that either method, cascade or feedback compensation, ultimately replaces the forward-path transfer function $G_p(s)$ with a new transfer function $G_f(s)$. The design process then can be thought of as converting a $G_p(s)$ that is not suitable, to $G_f(s)$ that is more satisfactory for the proposed system.

In the following section, we will develop through selected design problems, the guidelines for cascade compensation design to satisfy the given frequency-domain specifications. Based on these guidelines, the parameter values of minor-loop compensation scheme can easily be determined by trial-and-error using MATLAB (refer to the Appendix).

12.2.1 Design Problem 1

Consider the system of Fig. 12.1 with the plant transfer function,

$$G_p(s) = \frac{1}{s(s+2)}$$

We will assume that the phase margin ΦM and the velocity error constant K_ϑ are our major design specifications. Impose the requirements of (i) $K_v \ge 10$, and (ii) $\Phi M \ge 60°$.

First we examine whether or not adjustment of gain alone (i.e., $D_c(s) = K$) can achieve the design.

$$K_v = \lim_{s\to 0} s\,KG_p(s) = \frac{K}{2}$$

In order to meet the K_v specification, we require that $K \ge 20$. We choose $K = 20$ and sketch the Bode plot[1] of

$$G(s) = KG_p(s) = \frac{20}{s(s+2)} = \frac{10}{s(0.5s+1)}$$

[1] The slope of the low-frequency asymptote determines the system type. A type-0 system has a low-frequency asymptote with slope 0 dB/decade, a type-1 system has a low-frequency asymptote with slope −20 dB/decade, while for a type-2 system, the slope is −40 dB/decade.

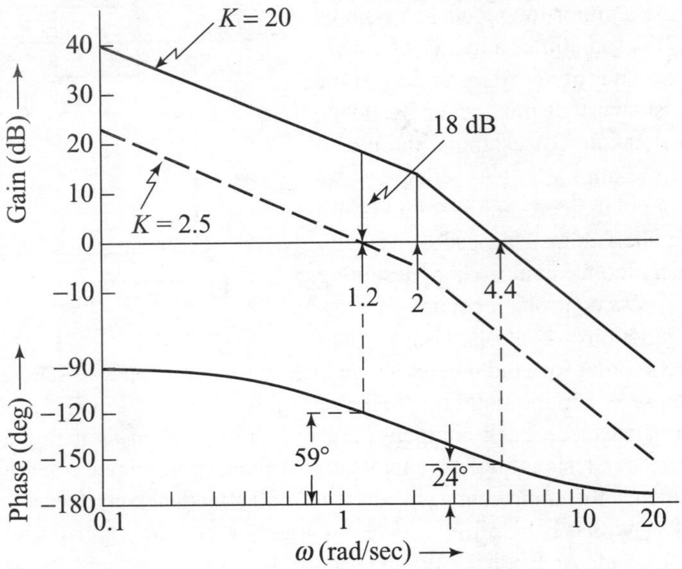

Fig. 12.3 *Reshaping of the Bode plot by gain adjustment*

in Fig. 12.3. We plot only the asymptotes. We see that the gain crossover frequency roughly equals 4.4 rad/sec, and the phase margin roughly equals 24°. This phase margin does not meet the specification. If K is the only parameter that can be adjusted to achieve the design, the only way to increase the phase margin is to decrease K. Decreasing K to 2.5 results in gain crossover frequency of 1.2 rad/sec and phase margin of 59° (refer to Fig. 12.3). This, however, will violate the K_v specification. Thus for this problem, both the steady-state and transient performance requirements cannot be met by adjusting K alone.

The steady-state performance of the system is governed by low-frequency characteristics of the transfer function, and the transient performance of the system is governed by high-frequency[2] characteristics. To simultaneously satisfy the transient and the steady-state performance requirements, the Bode plot of $G(j\omega)$ ***must be reshaped so that the high-frequency portion of the plot satisfies the phase margin requirement and the low-frequency portion satisfies the*** K_v ***requirement***. When we inspect the Bode plot of Fig. 12.3 for $K = 20$, we see that we can meet the design objectives by adding more phase to the system in the positive direction in the high-frequency range around gain crossover frequency 4.4 rad/sec, while keeping the low-frequency region relatively unaltered. This reshaping of the Bode plot can be achieved using a cascade compensator $D_c(s)$ having frequency-response characteristics shown in Fig. 12.4. Figure 12.5 serves to illustrate the principle of this compensation. This figure shows Bode plots of gain-compensated system $G(j\omega) = 10/j\omega(j0.5\omega + 1)$ and $D_c(j\omega)G_p(j\omega)$ where $D_c(j\omega)$ is given by Fig. 12.4. The gain-compensated system has a phase margin of 24° and gain crossover frequency of 4.4 rad/sec. Using compensator $D_c(j\omega)$, the phase angle plot is raised for higher frequencies. Since the positive phase in the compensator is accompanied by an additional gain at high frequencies, the gain crossover frequency increases when this compensation is employed. From Fig. 12.5 we see that the compensated system has gain crossover frequency of 6.8 rad/sec, phase margin of 62° and $K_v = 10$.

[2] It is, in fact, the mid-frequency range with respect to the frequency ranges given in Fig. 4.21; the higher frequencies need to be blocked by proper bandwidth realization.

The transfer function of the compensator having frequency-response characteristics shown in Fig. 12.4, is of the form

$$D_c(s) = K\left(\frac{\tau s+1}{\alpha\tau s+1}\right); \alpha<1, \tau>0 \qquad (12.6)$$

Note that $\alpha < 1$ ensures that the compensator adds positive phase to the system over some appropriate frequency range. Because of this property, the first-order compensator given by Eqn. (12.6) is called a *lead compensator* (also refer to Eqns (9.5)).

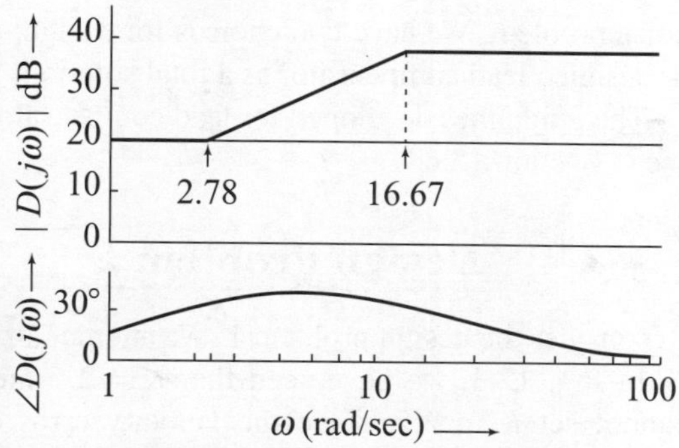

Fig. 12.4 *Bode plot of a lead compensator*

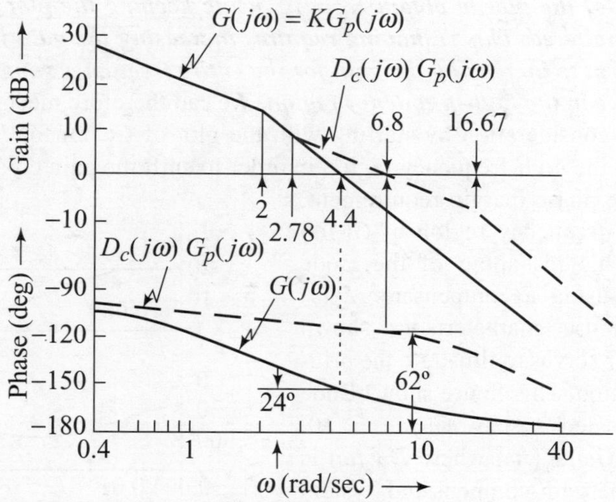

Fig. 12.5 *Reshaping of the Bode plot by lead compensation*

It is seen that $D_c(s)$ given by Eqn. (12.6) is a form of proportional plus derivative controller wherein the derivative action is bandwidth limited (refer to Fig. 9.10):

$$D_c(s) = K\alpha + \frac{K\alpha\tau(1-\alpha)s}{\alpha\tau s+1} \qquad (12.7a)$$

$$= K_P + \frac{K_D s}{\tau_D s+1}$$

The design of compensator (controller) is basically a search problem in n-dimensional space where n is the number of design parameters of the compensator. The design of a lead compensator is a problem with three trial parameters; K, α, and τ. A lead compensator may meet the filtering requirement in addition to satisfying the specifications on transient response.

For feedback loops free from high-frequency noise problem, a PD compensator

$$D_c(s) = K_P + sK_D \qquad (12.7b)$$

with only two trial parameters: K_P and K_D, is more appropriate choice. With the reduction in trial parameters, the design problem becomes simpler. However, real-life applications do have high-frequency noise problem. We have two choices for design; design a PD compensator, and a filter independently; or design a lead compensator as a total solution.

The guidelines developed for lead compensator are equally helpful for PD compensation, as we will see in Section 12.6.

12.2.2 Design Problem 2

Reconsider the design problem 1. We attempt an alternative solution.

In Fig. 12.3, we have seen that $K = 2.5$ meets the phase margin requirement. Apparently, we cannot set K to 2.5, since the velocity error constant would only be 1.25, and the steady-state performance requirement would not be satisfied. When we inspect the Bode plot of Fig. 12.3 for $K = 2.5$, we see that ***we can meet both the phase margin and the K_v requirements by reshaping the low-frequency portion of the plot to obtain $K_v = 10$, while keeping the plot near gain crossover frequency relatively unaltered. This reshaping requires increasing the gain in the low-frequency region. This is equivalent to increasing the gain for the entire frequency range and then attenuating the magnitude curve in the high-frequency region.*** We can therefore meet the design objectives for the problem under consideration by starting with the plot of $G(j\omega)$ for $K = 20$, reducing the magnitude of $G(j\omega)$ in the high-frequency region in order to shift the gain crossover frequency to a lower value to meet the phase margin requirement, while keeping the low-frequency region of $G(j\omega)$ relatively unaltered. This reshaping of the Bode plot can be achieved using a compensator $D_c(s)$ having frequency-response characteristics shown in Fig. 12.6. Figure 12.7 serves to illustrate the principle of this compensation. This figure shows Bode plots of gain-compensated system $G(j\omega) = 10/j\omega\,(j0.5\omega + 1)$ and $D_c(j\omega)G_p(j\omega)$ where $D_c(j\omega)$ is given by Fig. 12.6. The gain-compensated system has a phase margin of 24° and gain crossover frequency of 4.4 rad/sec. Using compensator $D_c(j\omega)$, the gain crossover frequency reduces to 0.95 rad/sec, and the requirements on both the ΦM and the K_v are met.

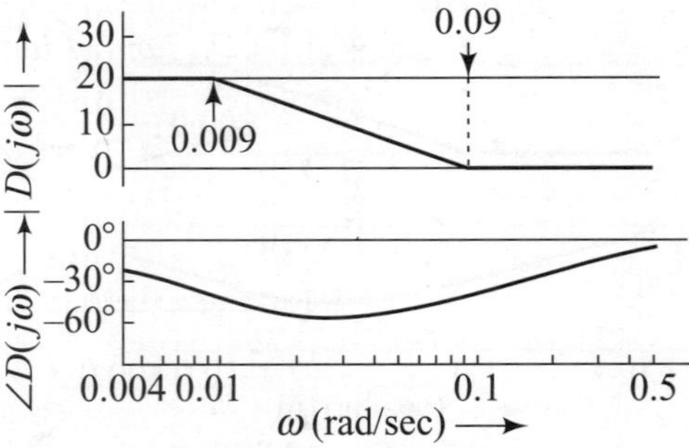

Fig. 12.6 *Bode plot of a lag compensator*

It should be noted that the attenuation characteristic of the compensator is accompanied by phase lag that tends to offset the increase in phase margin achieved due to reduction in gain crossover frequency. The loss in phase margin due to the phase lag characteristic of the compensator can be minimized by placing the frequency band over which the compensator has large phase lag much below the new gain crossover frequency. From Fig. 12.6 we see that the upper corner frequency of the lag compensator is 0.09 rad/sec, which is nearly one decade below the new gain crossover frequency 0.95 rad/sec.

In general, it may not be possible to satisfy all the design specifications by simply using a lead or a lag compensator. Situations often arise when both the low-frequency and the high-frequency portions of the Bode plot of the controlled process need to be reshaped by using a *lag-lead compensator*, which is basically a lag compensator and a lead compensator in cascade.

It should be pointed out that the design of linear control systems is not an exact science. A number of compensation schemes and parameter values may satisfy the given set of performance specifications. Some degree of trial-and-error and fine tuning is inevitable. Therefore, a CAD facility for frequency-domain analysis should be invaluable (refer to MATLAB Module 12 in the Appendix).

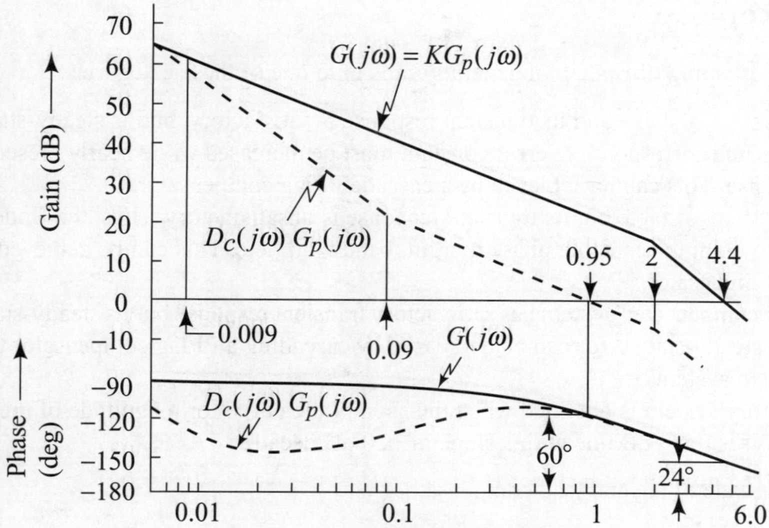

Fig. 12.7 *Reshaping of the Bode plot by lag compensation*

The transfer function of the compensator having frequency-response characteristics shown in Fig. 12.6 is of the form

$$D_c(s) = K\left(\frac{\tau s+1}{\beta\tau s+1}\right); \beta > 1, \tau > 0 \tag{12.8}$$

The compensator gets its name *lag-compensator* for a reason similar to the lead compensator; it introduces phase lag (also refer to Eqns (9.8)).

It is seen that $D_c(s)$ given by Eqn. (12.8) is a form of proportional plus integral controller wherein the integral action is low-frequency gain limited (refer to Fig. 9.14):

$$D_c(s) = K + \frac{K(\beta-1)}{\beta\tau s+1} = K_P + \frac{K_I}{\tau_I s+1} \tag{12.9a}$$

The design of a lag compensator is a problem with three trial parameters; K, β, and τ. For feedback loops wherein the primary requirement is to improve steady-state performance, a PI compensator

$$D_c(s) = K_P + \frac{K_I}{s} \tag{12.9b}$$

with only two trial parameters may be attempted. The limit on the low-frequency gain has now been relaxed. This compensator will, however, be useful if relaxing the limit does not degrade closed-loop stability appreciably.

The guidelines developed for lag compensator are equally helpful for PI compensation, as we will see in Section 12.6.

12.2.3 Reprise

The purpose of reshaping of Bode plot generally falls in to one of these categories:

1. A given system is stable and its transient response is satisfactory, but its steady-state error is too large. Thus, the corresponding error constant must be increased while nearly preserving the transient response. This can be achieved by a cascade PI/lag compensator.
2. A given system is stable but its transient response is unsatisfactory. Thus, the Bode plot must be reshaped to obtain acceptable phase margin, bandwidth, etc. This can be achieved by a cascade PD/lead compensator.
3. A PD/lead compensated system has satisfactory transient response but its steady-state error is too large. The steady-state error can be decreased by cascading a PI/lag compensator to the PD/lead compensated system.
4. A satisfactory system is obtained when the asymptotic curve for magnitude of the compensated system crosses the 0 dB line with a slope of −20 dB/decade.

Consider a cascade of lag and lead compensators:

$$D_c(s) = K\left(\frac{\tau_2 s+1}{\beta\tau_2 s+1}\right)\left(\frac{\tau_1 s+1}{\alpha\tau_1 s+1}\right);\ \beta>1,\ \alpha<1;\ \tau_2,\tau_1>0 \tag{12.10}$$

This is a second-order *lag-lead compensator*. Expressed explicitly in the form of proportional, integral, and derivative actions, $D_c(s)$ looks like

$$D_c(s) = \left(K_{P1} + \frac{K_I}{\tau_I s+1}\right)\left(K_{P2} + \frac{K_D s}{\tau_D s+1}\right) \tag{12.11a}$$

It is thus a cascade of gain-limited PI compensator with a bandwidth-limited PD compensator. Consider now the cascade of PI and PD compensators:

$$\begin{aligned} D_c(s) &= \left(K_1 + \frac{K_i}{s}\right)(K_2 + K_d s) \\ &= K_1K_2 + K_iK_d + \frac{K_1K_i}{s} + K_2K_d s \\ &= K_P + \frac{K_I}{s} + K_D s \end{aligned} \tag{12.11b}$$

Thus cascading a PI compensator with a PD compensator is completely equivalent to a PID compensator. In general, we can represent a dynamic compensator by the transfer function $D_c(s)$, given by

$$D_c(s) = \frac{K\prod_{i=1}^{m}(\tau_{zi}s+1)}{\prod_{j=1}^{n}(\tau_{pj}s+1)};\ m \le n,\ K>0 \tag{12.12a}$$

$$= KD(s) \tag{12.12b}$$

The gain K is *zero-frequency gain* of the dynamic compensator $D_c(s)$, that has to be adjusted to improve steady-state accuracy.

For the purpose of simplifying the design procedure, we normally associate the adjustable gain parameter K with the plant transfer function and then design the compensator $D(s)$ given by Eqns (12.12) based on new plant transfer function $KG_p(s)$. At the implementation stage, the gain K is associated with the compensator and the transfer function $D_c(s)$ is realized. The configuration of Fig. 12.8a will accordingly be referred to as the *uncompensated system*, and that of Fig. 12.8b as *compensated system*.

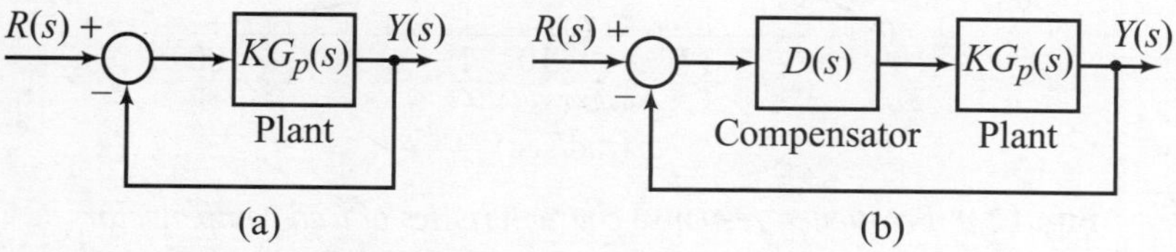

Fig. 12.8 *(a) Uncompensated system (b) Compensated system*

12.3 CASCADE LEAD COMPENSATION

The primary function of a lead compensator is to reshape the frequency-domain plot by providing sufficient phase lead angle over some appropriate frequency range to offset the excessive phase lag associated with the plant.

12.3.1 Lead Compensator Frequency Response

Let us first look at the frequency-response characteristics of a lead compensator and derive some valuable relationships that will help us in the design process. The transfer function of a lead compensator with unity zero-frequency gain is of the form

$$D(s) = \frac{\tau s + 1}{\alpha \tau s + 1}; \alpha < 1, \tau > 0 \tag{12.13}$$

The sinusoidal transfer function is of the form

$$D(j\omega) = \frac{1 + j\omega\tau}{1 + j\omega\alpha\tau} \tag{12.14}$$

The Bode plot of $D(j\omega)$, shown in Fig. 12.9, has two corner frequencies at $\omega = 1/\tau$ and $\omega = 1/\alpha\tau$. The phase lead mainly occurs within and around the two corner frequencies. ***Phase lead of the compensator is the property that is utilized to give a system sufficient phase margin.***

The phase lead at any frequency ω is given by

$$\phi = \tan^{-1}\omega\tau - \tan^{-1}\omega\alpha\tau \tag{12.15}$$

Differentiating with respect to ω, we obtain

$$\frac{d\phi}{d\omega} = \frac{\tau}{1 + (\omega\tau)^2} - \frac{\alpha\tau}{1 + (\omega\alpha\tau)^2}$$

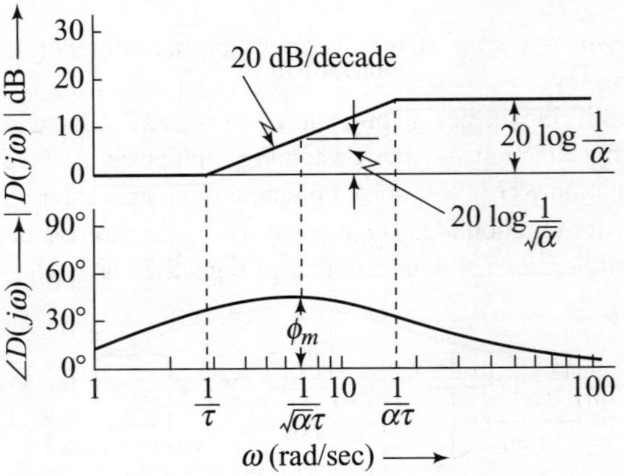

Fig. 12.9 *Frequency response characteristics of lead compensators*

Setting $d\phi/d\omega$ equal to zero, we find that the frequency ω_m at which the maximum phase lead ϕ_m occurs, is

$$\omega_m = \frac{1}{\tau\sqrt{\alpha}} = \sqrt{\frac{1}{\tau}\left(\frac{1}{\alpha\tau}\right)} \tag{12.16}$$

As seen from Eqn. (12.16), ω_m is the geometric mean of the two corner frequencies of the compensator. At $\omega = \omega_m$, the maximum phase lead ϕ_m is given by (refer to Eqn. (12.15))

$$\phi_m = \tan^{-1}\frac{1}{\sqrt{\alpha}} - \tan^{-1}\sqrt{\alpha}$$

Making use of the relation

$$\tan(\theta_1 - \theta_2) = \frac{\tan\theta_1 - \tan\theta_2}{1 + \tan\theta_1 \tan\theta_2},$$

we obtain

$$\tan\phi_m = \frac{1}{\sqrt{\alpha}} - \sqrt{\alpha} = \frac{1-\alpha}{\sqrt{\alpha}}$$

or

$$\sin\phi_m = \frac{1-\alpha}{1+\alpha} \tag{12.17}$$

This gives α in terms of ϕ_m as

$$\alpha = \frac{1-\sin\phi_m}{1+\sin\phi_m} \tag{12.18}$$

The magnitude of $D(j\omega)$ at $\omega = \omega_m$, the frequency of maximum phase lead is

$$|D(j\omega_m)| = \frac{|1 + j\omega_m\tau|}{|1 + j\omega_m\alpha\tau|} = \frac{1}{\sqrt{\alpha}} \tag{12.19}$$

An important consideration governing the choice of α is the inherent noise in control systems. The noise is generally of high frequency nature relative to the frequencies of control signals that appear in the loop. We see from Fig. 12.9 that in a lead compensator, the high-frequency noise signals are amplified by a factor $(1/\alpha) > 1$, while the low-frequency control signals undergo unit amplification (0 dB gain). This means that ***the lead compensator boosts the noise signal level relative to the control signal, and the lower the value of*** α***, the higher this boost.*** Thus, if noise is a problem in a particular system, one must be careful about choosing a value of α for a lead compensator. It is recommended that the value of α should not be less than 0.07. A common choice is $\alpha = 0.1$. Limiting the value of α amounts to limiting an increase in gain crossover frequency of the system due to addition of a lead compensator.

We are now ready to enumerate a design procedure for lead compensation.

12.3.2 Design Procedure

To illustrate the lead compensator design procedure, we reconsider the unity-feedback system of design problem 1, which was discussed earlier in Section 12.2. The design problem is to obtain the unity zero-frequency gain compensator

$$D(s) = \frac{\tau s + 1}{\alpha \tau s + 1}; \alpha < 1, \tau > 0$$

for the plant with adjustable gain,

$$G(s) = \frac{K}{s(s+2)}$$

so that the closed-loop system has (i) $\Phi M \geq 60°$, and (ii) $K_v \geq 10$.

Since the lead compensator has negligible effect at low frequencies, we set the gain of the uncompensated system to a value that satisfies the steady-state error requirement. It is seen that $K = 20$ meets the K_v specification. The Bode plot of $G(j\omega)$ for $K = 20$ is shown in Fig. 12.10. From the plot we see

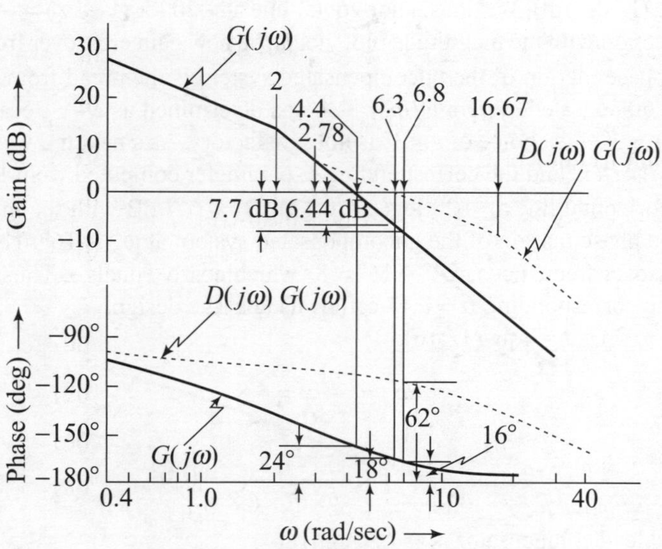

Fig. 12.10 *Cascade lead compensation*

that the uncompensated system has gain crossover frequency $\omega_g = 4.4$ rad/sec and phase margin = 24°. The required phase margin is 60°. If it were possible to simply add phase without affecting the magnitude, we would only need an additional 36° phase at the gain crossover frequency of $\omega_g = 4.4$. However, we know that insertion of lead compensator will maintain the same low-frequency gain but will affect the gain in high-frequency range; the gain crossover frequency will increase to a value $\omega'_g > \omega_g$. As the phase of most control processes decreases with increase in frequency, the phase of $G(j\omega)$ at ω'_g will be less than that at ω_g. Let this inevitable phase drop be equal to ε degrees. Therefore, to realize the specified phase margin, the lead compensator must provide a lead angle of 36° + ε at new gain crossover frequency ω'_g.

The value of ε depends upon ω'_g, which in turn depends upon the parameters of the lead compensator. In the trial-and-error design procedure, we make a guess on the value of ε and design a lead compensator. If the characteristics of the feedback system are not acceptable after the design, then a redesign with a different value of ε may be necessary. The initial guess on ε depends upon the slope of the magnitude plot of the uncompensated system in the region of gain crossover frequency ω_g. For a slope of –40 dB/decade, as in the present example, ε = 3°–5° is a good guess. The guessed value may have to be as high as 15 to 20° for a slope of –60 dB/decade.

In order to provide a phase lead angle of 36° + ε at ω'_g with the largest value of parameter α (this is desirable from signal/noise considerations), the frequency ω_m at which maximum phase lead occurs, must be made to coincide with ω'_g. Thus, we get

$$\omega_m = \omega'_g;\ \phi_m = 36° + \varepsilon = 36° + 3° = 39°$$

The α-parameter of the compensator can then be computed from Eqn. (12.18).

$$\alpha = \frac{1-\sin 39°}{1+\sin 39°} = \frac{0.37}{1.63} = 0.227$$

Since at ω_m, the compensator provides a gain of 20 $\log(1/\sqrt{\alpha})$ dB (see Fig. 12.9), the new gain crossover frequency $\omega'_g = \omega_m$ can be determined as that frequency at which the uncompensated system has a gain of $-20\log(1/\sqrt{\alpha})$ dB. We draw a horizontal line at $-20\log(1/\sqrt{\alpha}) = -6.44$ dB as shown in Fig. 12.10; its intersection with the magnitude plot yields the new gain crossover frequency as $\omega'_g = 6.3$. The corresponding phase margin of the uncompensated system is then read from the plot as 18°. The phase margin of uncompensated system at $\omega_g = 4.4$ was determined as 24°. Because 24° – 18° = 6°, which is larger than ε = 3°, the choice of ε = 3° is not satisfactory. As a next try, we choose ε = 7°. Then we have ϕ_m = 36° + 7° = 43° and the corresponding α-parameter computed from Eqn. (12.18) is 0.17. Intersection of the horizontal line at $-20\log(1/\sqrt{0.17})$ dB = –7.7 dB with the magnitude plot yields $\omega''_g = 6.8$ rad/sec. The phase margin of the uncompensated system at ω''_g is 16°. The phase drop due to increase in gain crossover frequency is 24° – 16° = 8° which nearly equals ε. Thus the choice of ε = 7° is satisfactory and the corresponding $\alpha = 0.17$ can be used in the design.

Now we let $\omega_m = \omega''_g$. Using Eqn. (12.16),

$$\omega_m = \frac{1}{\tau\sqrt{\alpha}} = 6.8$$

which implies

$$\tau = \frac{1}{\sqrt{0.17}\times 6.8} = 0.36$$

Thus, the required lead compensator is

$$D(s) = \frac{\tau s+1}{\alpha\tau s+1} = \frac{0.36s+1}{0.06s+1}$$

The open-loop transfer function of the compensated system becomes

$$D(s)G(s) = \frac{10(0.36s+1)}{s(0.5s+1)(0.06s+1)}$$

The frequency-response characteristics of $D(s)G(s)$ in Fig. 12.10 can be seen to yield a phase margin of 62°. The compensated system, therefore, meets both the steady-state and the relative stability requirements.

The lead-compensation design procedure given above has been tailored to the set of two specifications: phase margin (measure of relative stability), and error constant (measure of steady-state performance). The resulting compensated system must be evaluated with respect to other performance requirements. It may become necessary to repeat the design procedure several times to meet all the specifications within acceptable bounds.

In many design problems, relative stability is adequately specified by the phase margin alone. However, if gain margin is also specified, we must check the gain margin from the Bode plot of $D(s)G(s)$ and repeat the design if the requirement is not satisfied. For the problem under consideration, specification on gain margin is not useful because the phase never crosses the −180° line and the gain margin is always infinite.

Gain crossover frequency read from the Bode plot of $D(s)G(s)$ is an estimate of speed of response of the compensated system. We have seen that introduction of cascade lead compensator increases the gain crossover frequency and, consequently, the speed of response of the resulting system.

Bandwidth of the feedback system may be determined by transferring the data from Bode plot to Nichols chart. ***Since the gain crossover frequency is a rough measure of the bandwidth of a feedback system, we can say that the lead compensation increases the system bandwidth. It must, of course, be noted that too large an increase in system bandwidth will include some of the noise frequencies; it is undesirable and must be guarded against in lead compensation design.***

A summary of the lead compensation design procedure for designing compensator $D(s) = (\tau s + 1)/(\alpha\tau s + 1)$ for the plant $G(s)$ with adjustable gain K, is given below. This design procedure will apply to many cases; however, the designer should keep in mind that the specific procedure followed in any particular design may need to be tailored to its particular set of specifications.

1. Sketch the Bode plot of the uncompensated system with the gain K set according to the steady-state error requirement. Measure the gain crossover frequency ω_g, and phase margin of the uncompensated system.
2. Using the relation given below, determine the additional amount of phase lead needed at the new gain crossover frequency ω'_g to realize the specified phase margin.

   Additional phase lead required = specified phase margin − phase margin of the uncompensated system at $\omega_g + \varepsilon$.

   ε is a margin of safety required by the fact that the gain crossover frequency will increase due to compensation. A guess is made on the value of ε depending upon the slope of the magnitude plot in the region of gain crossover frequency. For a slope of –40 dB/decade, ε = 3°–5° is a good guess. The guessed value may have to be as high as 15 to 20° for a slope of −60 dB/decade.
3. Set the maximum phase of the lead compensator,

   $$\phi_m = \text{additional phase lead required,}$$

   and compute parameter α of the compensator using the following equation:

   $$\alpha = \frac{1-\sin\phi_m}{1+\sin\phi_m}$$

4. Locate the frequency at which the uncompensated system has a gain of $-20\log(1/\sqrt{\alpha})$ dB. Select this frequency as the new gain crossover frequency ω_g'. Measure the phase margin of the uncompensated system at ω_g'. If the difference in phase margins of $G(j\omega)$ at ω_g and ω_g' is less than ε, go to the next step. Otherwise, choose a larger ε in Step 2 and repeat Steps 3 and 4.
5. Set $\omega_m = \omega_g'$ and compute parameter τ of the compensator using the following equation:

$$\omega_m = \frac{1}{\tau\sqrt{\alpha}}$$

6. Sketch the frequency-response plots of compensated system in Bode and Nichols coordinate systems. Check any additional specifications on system performance, e.g., gain margin, bandwidth, etc. Redesign for another choice of ω_g', till all specifications are met. It may be noted that the additional specifications can only be met if they are consistent.

Example 12.1 Consider a type-1 unity-feedback system with an open-loop transfer function,

$$G(s) = \frac{K}{s(s+1)}$$

It is desired to have the velocity error constant $K_v = 10$. Furthermore, we desire that the phase margin of the system be atleast 45°.

The first step in the design procedure is to sketch the Bode plot of the uncompensated system with the gain K set according to the steady-state error requirement. $K = 10$ meets the K_v specification. The Bode plot of

$$G(j\omega) = \frac{10}{j\omega(j\omega+1)}$$

is drawn in Fig. 12.11. The gain crossover frequency of the uncompensated system is $\omega_g = 3.16$ rad/sec, and the phase margin is 18°.

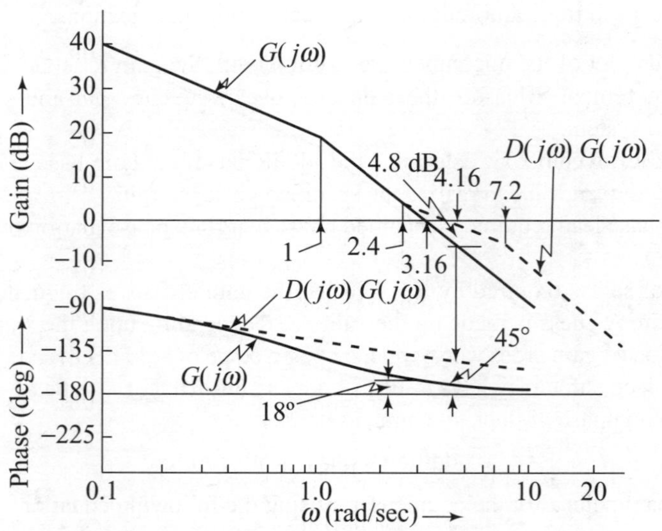

Fig. 12.11 *Bode plot for Example 12.1*

If cascade lead compensation scheme is employed to increase the phase margin to the specified value, the additional phase lead required at the new gain crossover frequency ω'_g is given by the following relation:

Additional phase lead required
= specified phase margin – phase margin of the uncompensated system at $\omega_g + \varepsilon$
= 45° – 18° + 3° = 30°

Setting the maximum phase ϕ_m of the lead compensator equal to the additional required phase lead of 30°, we obtain the α parameter of the lead compensator:

$$\alpha = \frac{1-\sin\phi_m}{1+\sin\phi_m} = \frac{1-\sin 30°}{1+\sin 30°} = 0.333$$

The uncompensated system has a gain of $-20\log(1/\sqrt{\alpha}) = -4.8$ dB at $\omega'_g = 4.16$ rad/sec. Setting ω_m, the frequency at which the lead compensator has maximum phase ϕ_m, equal to ω'_g, we obtain the τ parameter of the lead compensator:

$$\omega_m = \frac{1}{\tau\sqrt{\alpha}} = \omega'_g = 4.16$$

This gives $\tau = 0.416$

Therefore, the lead compensator is

$$D(s) = \frac{\tau s+1}{\alpha\tau s+1} = \frac{0.416s+1}{0.139s+1}$$

and the compensated open-loop transfer function becomes

$$D(s)G(s) = \frac{10(0.416s+1)}{s(s+1)(0.139s+1)}$$

The frequency-response plots of uncompensated and compensated systems in Bode and Nichols coordinate systems are shown in Figs 12.11 and 12.12, respectively. From these plots we observe the following:

1. Phase margin is increased from 18° to 45°.
2. Resonance peak is reduced from 10 dB to 3 dB.
3. Bandwidth is increased from 5 rad/sec to 7 rad/sec.

Thus, in general, the effect of the lead compensation is to increase the margin of stability and speed of response (the design has been done here with the straightline asymptotic plots rather than the exact curves for the magnitude).

To reinforce these observations, we include the root locus and step response of the design just completed. To draw the root locus, we rearrange $D(s)G(s)$ in the form convenient for root-locus sketching:

$$D(s)G(s) = \frac{10(0.416s+1)}{s(s+1)(0.139s+1)} = \frac{29.928(s+2.404)}{s(s+1)(s+7.194)}$$

The root locus for this design is shown in Fig. 12.13a, with the root locations marked for root-locus gain 29.928. Note that the complex poles are at $-2.55 \pm j3.85$, and the real-axis pole is at -3.1; it is close to the zero at $s = -2.404$. Reshaping of Bode plot has resulted in near-cancelation compensation.

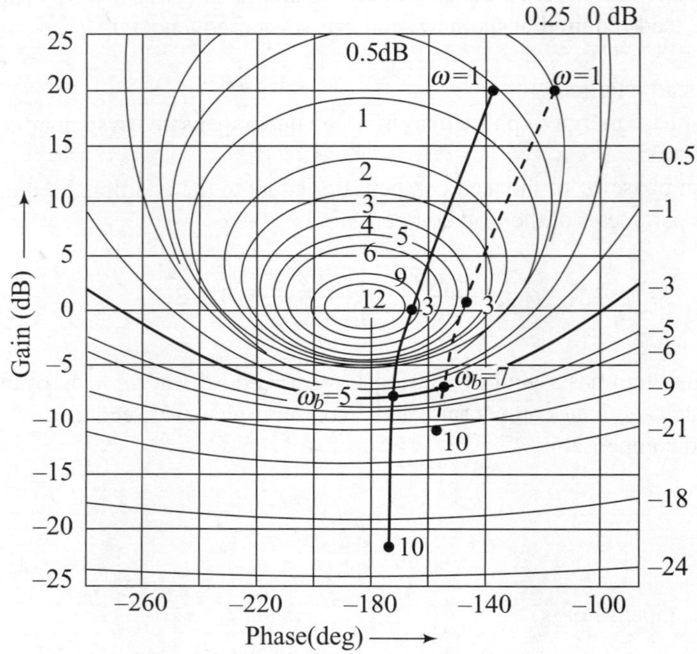

Fig. 12.12 *Nichols chart for Example 12.1*

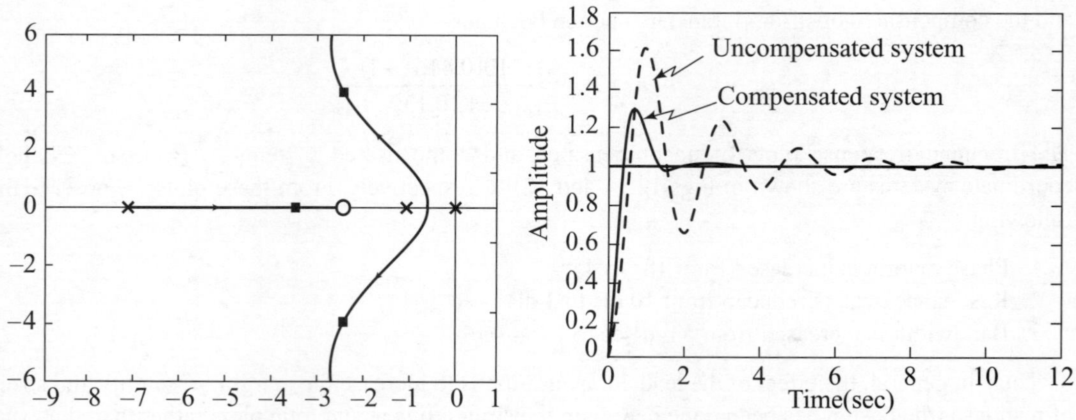

Fig. 12.13 *(a) Root locus for Example 12.1 (b) Step responses for Example 12.1*

For a phase margin of 40°, the value of ζ is 0.4 and the corresponding peak overshoot based on second-order approximation is 25.38 %.

The unit step response of the compensated closed-loop system shown in Fig. 12.13b, has an overshoot of 28.8%. This correlates reasonably well with the per cent overshoot predicted by the formula $\zeta = 0.01 \times \Phi M$.

From the response curves of uncompensated and compensated systems shown in Fig. 12.13b, we observe that the speed of response of the compensated system (t_s = 1.8 sec) has increased due to the higher gain crossover incurred when we included the lead compensator. Peak overshoot has been reduced due to increase in phase margin.

Example 12.2 Consider now a unity-feedback control system with open-loop transfer function,

$$G(s) = \frac{K}{s^2(0.2s+1)}$$

Assume that the system is required to be compensated to meet the following specifications:

1. Acceleration-error constant $K_a = 10$.
2. Phase margin $\geq 35°$.

From the steady-state requirement, we set $K = 10$. The Bode plot of $G(j\omega)$ when $K = 10$ is shown in Fig. 12.14. As observed from this plot, the gain crossover frequency $\omega_g = 3.16$ rad/sec, and the phase margin is $-32°$, which means that the uncompensated system is unstable. In fact, the uncompensated system is unstable for all values of K.

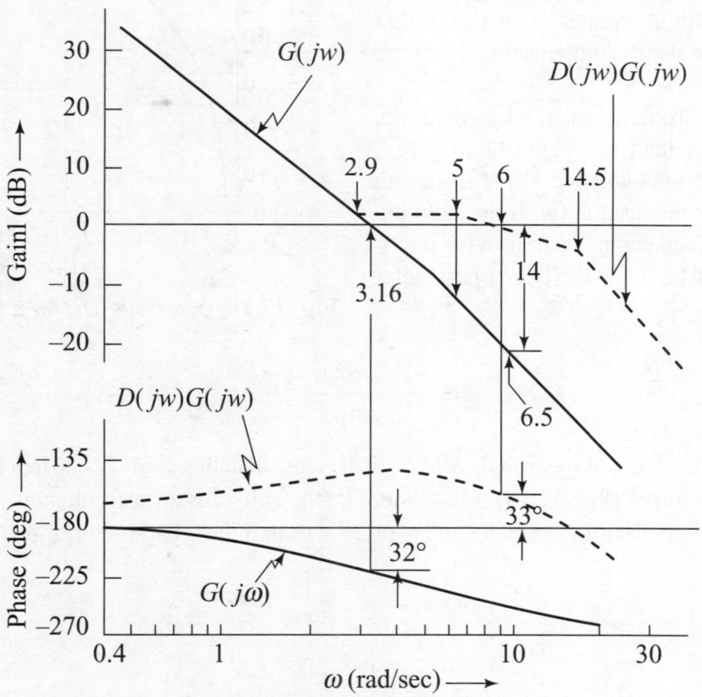

Fig. 12.14 *Bode plot for Example 12.2*

The phase lead needed at the new gain crossover frequency ω'_g is obtained as follows:

Additional phase lead required

= specified phase margin – phase margin of the uncompensated system at $\omega_g + \varepsilon$

$= 35° + 32° + 15° = 82°$

A large value of ε has been selected in this example because of rapid decrease of $\angle G(j\omega)$ in the region of gain crossover frequency ω_g.

Setting the maximun phase ϕ_m of the lead compensator equal to the additional required phase lead of 82°, we obtain the α parameter of the compensator as

$$\alpha = \frac{1-\sin 82^\circ}{1+\sin 82^\circ} = 0.005$$

Since the high-frequency noise signals are amplified by a factor of $1/\alpha$, the use of $\alpha = 0.005$ may not be acceptable.

Figure 12.15 shows the plot of ϕ_m *versus* $1/\alpha$, from which it is observed that to obtain phase leads more than 60° ($\alpha \cong 0.07$), $1/\alpha$ increases rather sharply, quite out of proportion to the increase in phase lead. Therefore, for phase leads greater than 60°, it is advisable to use two or more cascaded stages of lead compensator (described by Eqn. (12.13)) with moderate values of α rather than a single-stage lead compensator with too small a value of α.

For the design problem under consideration, using a single-stage lead compensator to give a phase lead of 82° is not advisable. We shall design a two-stage lead compensator (with an isolating amplifier), so that each stage has to provide maximum phase lead of 82/2 = 41°. The α parameter of each stage is

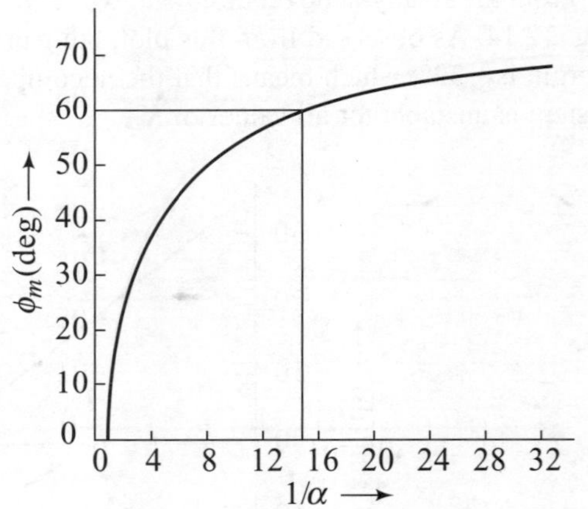

Fig. 12.15 ϕ_m *versus* $1/\alpha$ *for a lead compensator*

$$\alpha = \frac{1-\sin 41^\circ}{1+\sin 41^\circ} \cong 0.2$$

From the Bode plot of uncompensated system, it is seen that the gain $2(-20\log(1/\sqrt{\alpha})) = -14$ dB occurs at $\omega'_g = 6.5$ rad/sec (Fig. 12.14). This should be the gain crossover frequency of the system with single-stage lead compensator. Choosing $\omega_m = \omega'_g$, we obtain the τ parameter of one stage:

$$\omega_m = \frac{1}{\tau\sqrt{\alpha}} = \omega'_g = 6.5$$

This gives $$\tau = 0.344$$

By adding another identical stage of lead compensator, the open-loop transfer function of the compensated system becomes

$$D(s)G(s) = \frac{10(0.344s+1)^2}{s^2(0.2s+1)(0.069s+1)^2}$$

The Bode plot of $D(j\omega)G(j\omega)$ is shown in Fig. 12.14, from which it is found that the phase margin of the compensated system is 33°. ***This indicates that the system has become stable but the desired phase margin is not yet fully achieved. This is because of the excessive lag of the fixed part of the system at the new gain crossover frequency.*** If this phase margin is not acceptable, then the compensator should be redesigned with a higher value of ε.

Simulation of the closed-loop system corresponding to the design just completed shows a peak overshoot of 46.4% and settling time of 1.1 sec. The phase margin specification, when converted to time-domain measure on the basis of second-order correlations, gives $\zeta = 0.35$ (peak overshoot 30.92%). The actual response does not correlate well with the one predicted by second-order approximation. This time-response analysis dictates redesign if the actual performance is not in the acceptable range.

12.3.3 Reprise

1. The steady-state error of a feedback system is not affected by introduction of a unity zero-frequency gain lead compensator.
2. The phase of the open-loop transfer function in the vicinity of the gain crossover frequency is increased by lead compensation. This improves the phase margin of the closed-loop system.
3. Lead compensation increases the gain crossover frequency and bandwidth; consequently, the speed of response of the closed-loop system is increased.
4. The value of the α parameter of the lead compensator should not be smaller than 0.07, nor should the maximum phase lead ϕ_m be more than 60°, because such choices will need an additional gain of excessive value. If more than 60° phase lead is required, two (or more) lead networks may be used in series, with an isolating amplifier.
5. If the uncompensated system has low stability margin, the additional phase lead required to obtain a certain desired phase margin may be excessive. This requires a relatively small value of α parameter of the compensator, which, as a result, will give rise to large bandwidth for the compensated system. Such a compensator may not be acceptable because of considerations of noise transmission. As we shall see later in this chapter, a lead compensator cascaded with a lag compensator may realize a large phase margin while limiting the increase in bandwidth of the closed-loop system.
6. If the phase angle of the uncompensated system decreases rapidly near the gain crossover frequency, lead compensation becomes less effective because the shift in the gain crossover frequency to the right makes it difficult to provide enough phase lead at the new gain crossover frequency.
7. The steady-state performance of a feedback system can be improved to any desired level by appropriately raising the open-loop gain K. Raising K results in amplification of signals at all frequencies. Introduction of a lead compensator provides additional gain for high-frequency signals, thereby boosting the noise signal level relative to the control signal. This imposes a limit on the improvement of steady-state error by raising gain K in a lead-compensated system.

12.4 CASCADE LAG COMPENSATION

The primary function of a lag compensator is to reshape the frequency-domain plot by providing sufficient attenuation over some appropriate frequency range. The phase lag characteristic is of no consequence in lag compensation.

12.4.1 Lag Compensator Frequency Response

The transfer function of a lag compensator with unity zero-frequency gain is of the form

$$D(s) = \frac{\tau s + 1}{\beta \tau s + 1} \quad \beta > 1, \tau > 0 \tag{12.20}$$

The sinusoidal transfer function is of the form

$$D(j\omega) = \frac{1 + j\omega\tau}{1 + j\omega\beta\tau} \tag{12.21}$$

The Bode plot of $D(j\omega)$, shown in Fig. 12.16, has two corner frequencies at $\omega = 1/\beta\tau$ and $\omega = 1/\tau$. ***The phase lag mainly occurs within and around the two corner frequencies. It must be recognized here that any phase lag at the gain crossover frequency of the compensated system is undesirable. To prevent detrimental effects of phase lag due to the lag compensator, the corner frequencies of the lag compensator must be located substantially lower than the gain crossover frequency of the compensated system.***

Figure 12.16 also shows that in the high-frequency range, the lag compensator has an attenuation of 20 log β dB, ***which is the property that is utilized to give a system sufficient phase margin.***

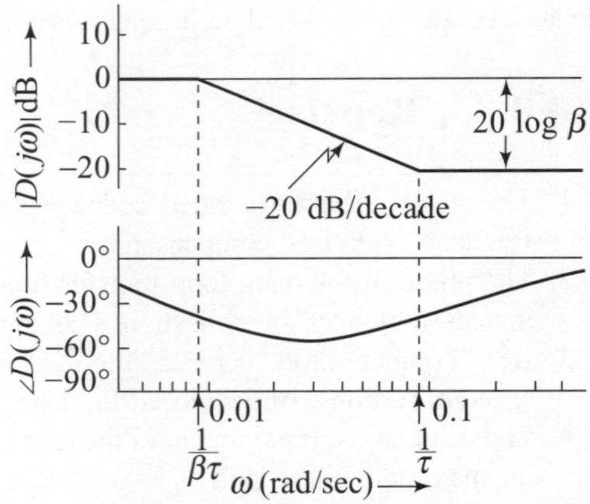

Fig. 12.16 *Frequency response characteristics of lag compensators*

The addition of a lag compensator results in an improvement in the ratio of control signal to noise signal in the loop. The high-frequency noise signals are attenuated by a factor $\beta > 1$, while the low-frequency control signals undergo unit amplification (0 dB gain). The choice of β is usually restricted because a very large β will appreciably reduce the gain crossover frequency and consequently speed of response of the system. A typical choice of β is 10.

We are now ready to enumerate a design procedure for lag compensation.

12.4.2 Design Procedure

To illustrate the lag-compensator design procedure, we reconsider the unity-feedback system of design problem 2, which was discussed earlier in Section 12.2. The design problem is to obtain the unity zero-frequency gain compensator

$$D(s) = \frac{\tau s + 1}{\beta\tau s + 1};\ \beta > 1, \tau > 0$$

for the plant with adjustable gain,

$$G(s) = \frac{K}{s(s+2)}$$

so that the closed-loop system has (i) $\Phi M \geq 60°$, and (ii) $K_v \geq 10$.

The steady-state error requirement sets $K = 20$. The Bode plot of $G(j\omega)$ for $K = 20$ is shown in Fig. 12.17. From this plot we see that the uncompensated system has gain crossover frequency $\omega_g = 4.4$ rad/sec and phase margin = 24°. The required phase margin of 60° can be achieved by lowering the gain crossover frequency to a region where $G(j\omega)$ has more favorable phase characteristics. This can be achieved by introducing a lag compensator.

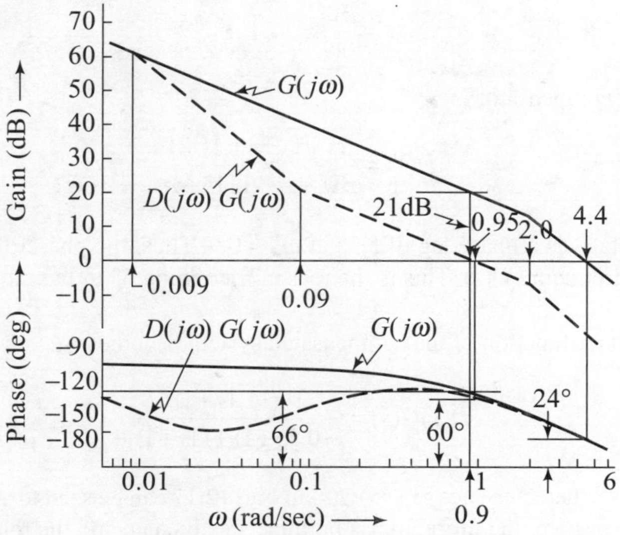

Fig. 12.17 *Cascade lag compensation*

First we search for a new gain crossover frequency that has a phase margin of 60° plus 6° (this will be explained later), or 66°. This can be found by drawing a horizontal line with 66° phase margin. Its intersection with the phase plot yields the new gain crossover frequency. It is read from Fig. 12.17 as $\omega'_g = 0.9$. We then draw a vertical line upward to the magnitude plot. We see that the magnitude at $\omega = \omega'_g$ is 21dB. To bring the magnitude curve down to 0 dB at ω'_g, the lag compensator must provide 21 dB attenuation at ω'_g. From Fig. 12.16, we see that a lag compensator provides an attenuation of 20 log β for $\omega > 1/\tau$, the upper corner frequency. Therefore, a lag compensator with upper corner frequency $1/\tau < \omega'_g$, and the β parameter given by

$$20 \log \beta = 21 \tag{12.22}$$

will provide the required attenuation. From Eqn. (12.22), we get $\beta = 11$. We take $\beta = 10$ as the initial trial value.

Once the value of β is determined, it is necessary only to select the proper value of τ to complete the design. If the upper corner frequency $1/\tau$ is close to ω'_g, the phase margin at ω'_g will be greatly reduced by the phase lag of the compensator. If the upper corner frequency is very far away from ω'_g (on the left-hand side), then the phase margin will be hardly affected by the phase lag of the compensator. Therefore, it seems that we should choose τ so that $1/\tau$ is as small as possible. However, this will cause a different problem. If $1/\tau$ is very small, then $D(s)$ has a pole very close to the origin. If we plot the root locus of $1 + D(s)G(s)$, then we can see that the closed-loop system will have a pole very close to the origin and the corresponding time constant will be very large. Due to this additional closed-loop pole, the speed of response of the resulting system may become slow. The magnitude of the response term contributed by the additional pole will not be too large because of the presence of compensator zero close to the additional closed-loop pole.

Usually, as a general guideline, the upper corner frequency, $1/\tau$, of the compensator should be approximately one octave to one decade below the new gain crossover frequency ω'_g. Taking

$$\frac{1}{\tau} = \frac{\omega'_g}{10}$$

we obtain
$$\tau = \frac{10}{0.9} = 11.1$$

Thus, the required lag compensator is

$$D(s) = \frac{\tau s + 1}{\beta \tau s + 1} = \frac{11.1s + 1}{111s + 1}$$

This lag compensator has a phase lag of 5.7° at $\omega_g' = 0.9$. Thus, the phase of $G(s)$ will be reduced by roughly 6° after introducing $D(s)$. This is the reason for adding 6° to the required phase margin in determining ω_g'.

The open-loop transfer function of the compensated system becomes

$$D(s)G(s) = \frac{10(11.1s + 1)}{s(0.5s + 1)(111s + 1)}$$

The frequency-response characteristics of $D(s)G(s)$ in Fig. 12.17 can be seen to yield a phase margin of 60°. The compensated system, therefore, meets both the steady-state and the relative stability requirements.

A summary of the lag-compensation design procedure for designing compensator $D(s) = (\tau s + 1)/(\beta \tau s + 1)$ for the plant $G(s)$ with adjustable gain K, is given below.

1. Sketch the Bode plot of the uncompensated system with the gain K set according to the steady-state error requirement. Measure the gain crossover frequency ω_g, and the phase margin of the uncompensated system.
2. Determine the frequency at which the phase angle of the uncompensated system is

   $$-180° + \text{specified phase margin} + \varepsilon$$

   Select this frequency as new gain crossover frequency ω_g'.
   ε is a margin of safety required by the fact that the compensator will contribute phase lag at ω_g'. A guess is made on the value of ε depending upon the value of ω_g'. If we can place upper corner frequency, $1/\tau$, of the compensator far below ω_g', a small value of ε is taken. If ω_g' is fairly low, the upper corner frequency $1/\tau$ cannot be taken far to its left to avoid large time constants. In such cases, phase lag contribution of the compensator at ω_g' will be considerable; a large value of ε is taken to compensate for this detrimental effect.

   Allow for ε from 5° to 15° for phase lag contribution of the compensator at ω_g'.
3. To bring the magnitude curve down to 0 dB at ω_g', the lag compensator must provide the amount of attenuation equal to the value of gain of the uncompensated system at ω_g'. Measure the gain of the uncompensated system at ω_g' and equate it to 20 log β. Calculate therefrom, the parameter β of the compensator.
4. Choose the upper corner frequency of the compensator one octave to one decade below ω_g'. This gives the parameter τ of the compensator.
5. Calculate the phase lag of the compensator at ω_g'. If it is less than ε, go to the next step. Otherwise, choose a larger ε in Step 2 and repeat Steps 3 and 4.
6. Sketch the frequency-response plots of compensated system in Bode and Nichols coordinate systems. Check any additional specification on system performance. Redesign for another choice of $1/\tau$ till all specifications are met. It may be noted that the additional specifications can only be met if they are consistent.

Example 12.3 Consider a type-1 unity-feedback system with an open-loop transfer function,

$$G(s)=\frac{K}{s(s+1)}$$

It is desired to have the velocity error constant $K_v = 10$. Furthermore, we desire that the phase margin of the system be atleast 45°. Earlier in Example 12.1, a lead compensator was used to realize these specifications. We now attempt a lag compensation scheme.

The first step in the design procedure is to sketch the Bode plot of the uncompensated system with the gain K set according to the steady-state error requirement. $K = 10$ meets the K_v specification. The Bode plot of

$$G(j\omega)=\frac{10}{j\omega(j\omega+1)}$$

is drawn in Fig. 12.18. The gain crossover frequency of the uncompensated system is $\omega_g = 3.16$ rad/sec, and the phase margin is 18°.

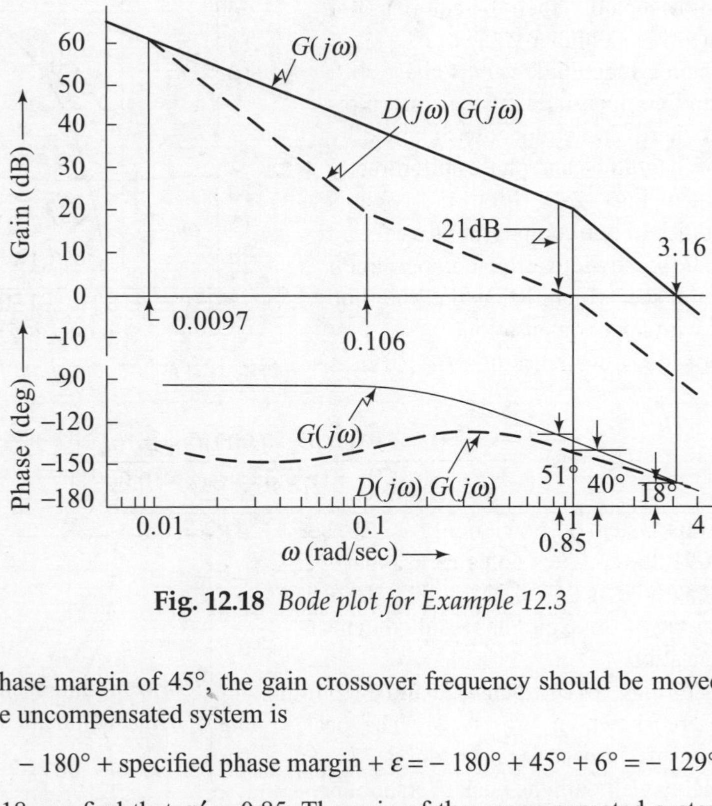

Fig. 12.18 *Bode plot for Example 12.3*

To realize a phase margin of 45°, the gain crossover frequency should be moved to ω_g' where the phase angle of the uncompensated system is

$$-180° + \text{specified phase margin} + \varepsilon = -180° + 45° + 6° = -129°$$

From Fig. 12.18, we find that $\omega_g' = 0.85$. The gain of the uncompensated system at ω_g' is 21 dB. Therefore, to bring the magnitude curve down to 0 dB at ω_g', the lag compensator must provide an attenuation of 21 dB. The β parameter of the lag compensator can now be calculated.

$$20 \log \beta = 21$$

This gives $\beta = 11$

Placing the upper corner frequency of the compensator three octaves below ω'_g, we obtain

$$\frac{1}{\tau} = \frac{\omega'_g}{(2)^3} = \frac{0.85}{8}$$

which gives $\tau = 9.4$

The lower corner frequency is $\frac{1}{\beta\tau} = \frac{1}{103.4}$. Therefore, the lag compensator $D(s) = \frac{9.4s+1}{103.4s+1}$, and the compensated open-loop transfer function,

$$D(s)G(s) = \frac{10(9.4s+1)}{s(s+1)(103.4s+1)}$$

The Bode plot of the compensated system is shown in Fig. 12.18. It is seen that the phase margin of the compensated system is approximately 40°. If this phase margin is not acceptable, then the compensator should be redesigned with a higher value of ε.

In Fig. 12.19, the log-magnitude *versus* phase shift plots of original and compensated systems are plotted on the Nichols chart. These plots are obtained by taking the values of magnitude and phase shift directly from the Bode plot of Fig. 12.18. From the Nichols chart, the bandwidth without compensation is found to be 5 rad/sec. The bandwidth after compensation is found to be 1.5 rad/sec, which means that the time response is slower with lag compensation.

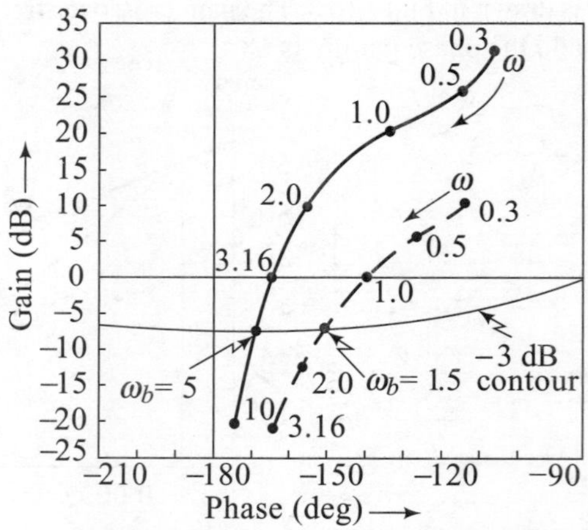

Fig. 12.19 *Nichols chart for Example 12.3*

To draw the root locus, we rearrange $D(s)G(s)$ as follows:

$$D(s)G(s) = \frac{10(9.4s+1)}{s(s+1)(103.4s+1)} = \frac{0.9091(s+0.1064)}{s(s+1)(s+10.0097)}$$

The root locus for this system is shown in Fig. 12.20. For root locus gain 0.9091, the complex poles are at $-0.445 \pm j0.784$, and the real pole is at -0.12, with a zero at $s = -0.1064$. Reshaping of Bode plot has resulted in near-cancelation compensation.

The step response curves of compensated and uncompensated systems are shown in Fig. 12.21. The peak overshoot of compensated closed-loop system is 27.3%; it correlates reasonably well with the peak overshoot predicted by the formula $\zeta = 0.01 \times \Phi M$ (25.38%). We also note from Fig. 12.21, that the compensated system becomes slower corresponding to the decrease in gain

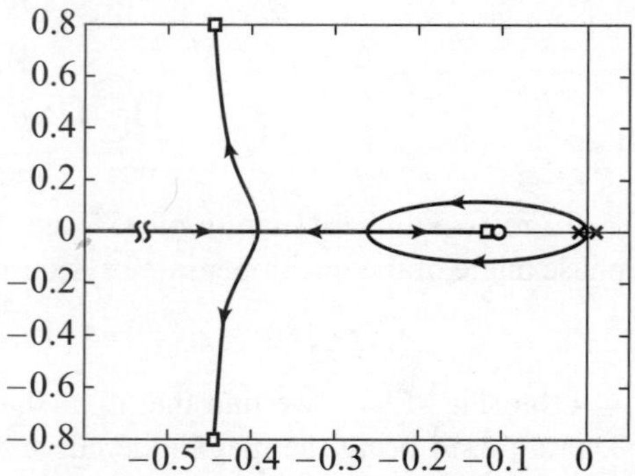

Fig. 12.20 *Root locus for Example 12.3*

crossover, giving a lower bandwidth for the system. The introduction of the slow root from the lag compensator has caused the response to require 15.6 sec settling time; the system 'creeps' towards the steady state. Peak overshoot has been reduced due to increase in phase margin.

A redesign might lower the phase margin, which may make the system response overshoot a bit; but this may reduce the creeping effect. In many applications, the long tail (creeping) in the system response is acceptable because the amplitude of its deviation from steady-state value is low.

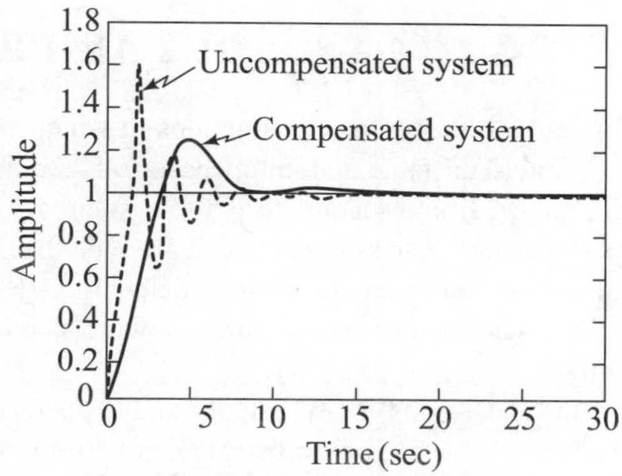

Fig. 12.21 *Step responses for Example 12.3*

Note that Examples 12.1 and 12.3 meet an identical set of specifications for the same plant in very different ways. In the first case, the specifications are met with a lead compensation and a bandwidth $\omega_b = 7$ rad/sec, results. In the second case, the same specifications are met with a lag compensator and $\omega_b = 1.5$ rad/sec, results. Clearly, had there been specifications for rise time or bandwidth, they would have influenced the choice of compensation (lead or lag). Likewise, if the slow settling (creeping) to the steady-state value was a problem, it might have suggested the use of lead compensation instead of lag.

12.4.3 Reprise

1. We utilize the attenuation characteristic of the lag compensator rather than its phase lag characteristic. The phase lag characteristic has no use for compensation purposes.
2. The steady-state error of a feedback system is not affected by introduction of a unity zero-frequency gain lag compensator.
3. Lag compensation reduces high-frequency gain for better phase margin.
4. Lag compensation decreases the gain crossover frequency and bandwidth; consequently, the closed-loop system becomes more sluggish.
5. Attenuation of high-frequency gain by lag compensation improves signal/noise ratio of the system.
6. If the uncompensated system has low stability margin, the high-frequency attenuation required to obtain a certain desired phase margin may be excessive. Such a compensator may not be acceptable because of consideration of speed of response. As we shall see later in this chapter, a lag compensator cascaded with a lead compensator may realize a large phase margin while limiting the loss of speed of response of the closed-loop system.
7. Steady-state performance of a feedback system can be improved to any desired level by appropriately raising the open-loop gain K. Raising K results in amplification of signals at all frequencies. Introduction of a lag compensator results in attenuation of the high-frequency signals, thereby reducing the noise signal level relative to the control signal. This feature of lag compensation allows a large value of K for significant improvement in steady-state performance of the feedback system.

12.5 CASCADE LAG-LEAD COMPENSATION

In general, there are two situations in which compensation is required. In the first case, the system is absolutely unstable and compensation is required to stabilize it, as well as to achieve a specified performance. In the second case, the system is stable but compensation is required to obtain the desired performance. The systems which are type-2 or higher, are usually absolutely unstable. The phase curve of such systems may lie entirely below the $-180°$ line in the Bode coordinate system. Under such situations, lag compensation cannot provide stability and required performance; lead compensation is the only answer.

In type-1 and type-0 systems, stable operation is always possible if the gain is sufficiently reduced. For such systems, phase margin can be improved to any level for a specified error constant by either lead or lag compensation. Decision on the choice of lead or lag compensation is based on performance requirements other than the relative stability and steady-state error. The lead compensation results in increased bandwidth and, consequently, faster speed of response. For higher-order systems and for systems with large specified error constants, large lead angles are required for compensation, resulting in excessively large bandwidth which is undesirable from noise transmission point of view. For such systems, lag compensation is preferred. The lag compensation, on the other hand, reduces system bandwidth and, consequently, slows down the speed of response.

For large specified error constant and moderately large desired bandwidth, it may not be possible to meet the specifications through either lead or lag compensation. Under such circumstances, we can use a lag-lead compensator.

12.5.1 Lag-Lead Compensator Frequency Response

As the name implies, the lag-lead compensator is basically just a lag compensator and a lead compensator in cascade. The transfer function of a lag-lead compensator with unity zero-frequency gain is of the form

$$D(s) = \underbrace{\left(\frac{\tau_1 s+1}{\beta\tau_1 s+1}\right)}_{\text{Lag section}}\underbrace{\left(\frac{\tau_2 s+1}{\alpha\tau_2 s+1}\right)}_{\text{Lead section}};\ \beta>1,\ \alpha<1;\ \tau_1,\tau_2>0 \tag{12.23}$$

The lead section alters the frequency-domain plot by adding phase lead angle and increasing the phase margin at the gain crossover frequency. The lag section provides attenuation near and above the gain crossover frequency and thereby allows an increase of gain in the low-frequency range to improve the steady-state performance.

For applications where independent choice of α and β is not essential (with the reduction in number of design parameters, the design problem becomes simpler), a lag-lead compensator of the form

$$D(s) = \frac{(\tau_1 s+1)(\tau_2 s+1)}{(\beta\tau_1 s+1)\left(\dfrac{1}{\beta}\tau_2 s+1\right)};\ \beta>1\ \tau_1,\ \tau_2>0 \tag{12.24}$$

may be used. The sinusoidal transfer function of this compensator is

$$D(j\omega) = \frac{(1+j\omega\tau_1)(1+j\omega\tau_2)}{(1+j\omega\beta\tau_1)(1+j\omega\tau_2/\beta)} \tag{12.25}$$

Figure 12.22 shows the Bode plot of $D(j\omega)$. Note that the phase shift is a function of input frequency; phase angle varies from lag to lead as the frequency is increased. Thus, phase lag and phase lead occur in different frequency bands.

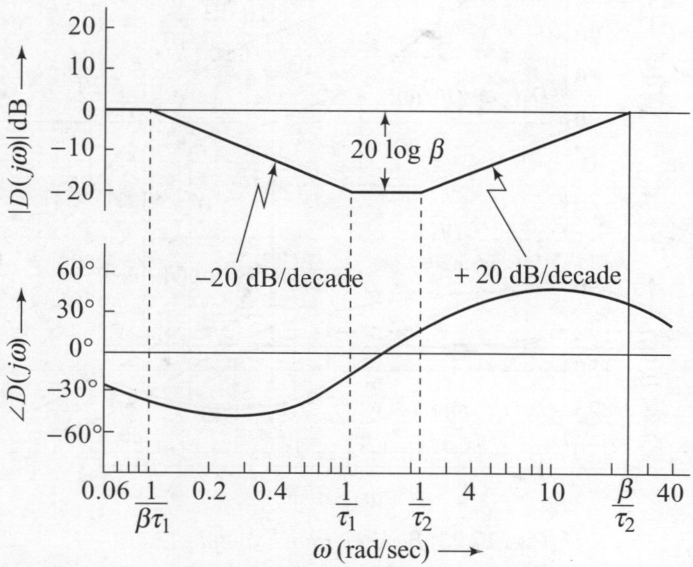

Fig. 12.22 *Frequency response characteristics of lag-lead compensators*

12.5.2 Design Procedure

To start with design, we check the phase margin and bandwidth of the uncompensated system with gain adjusted to meet the error constant specification. If the bandwidth is smaller than the specified value, lead compensation may be tried. However, if the bandwidth is larger than acceptable, lead compensation would not be desirable; so we try lag compensation, provided that the uncompensated system is not absolutely unstable. If the lag compensator design results in too low a bandwidth, the need for a lag-lead compensation is indicated.

Let us demonstrate the design procedure with an example.

Example 12.4 Consider the unity-feedback system whose open-loop transfer function is

$$G(s) = \frac{K}{s(0.1s+1)(0.2s+1)}$$

The system is to be compensated to meet the following specifications:

1. Velocity-error constant $K_v = 30$.
2. Phase margin $\Phi M \geq 50°$.
3. Bandwidth $\omega_b = 12$ rad/sec.

It easily follows that $K = 30$ satisfies the specification on K_v. The Bode plot of $G(j\omega)$ with $K = 30$ is shown in Fig. 12.23 from which it is found that the uncompensated system has a gain crossover frequency $\omega_g = 11$ rad/sec and phase margin $= -23°$.

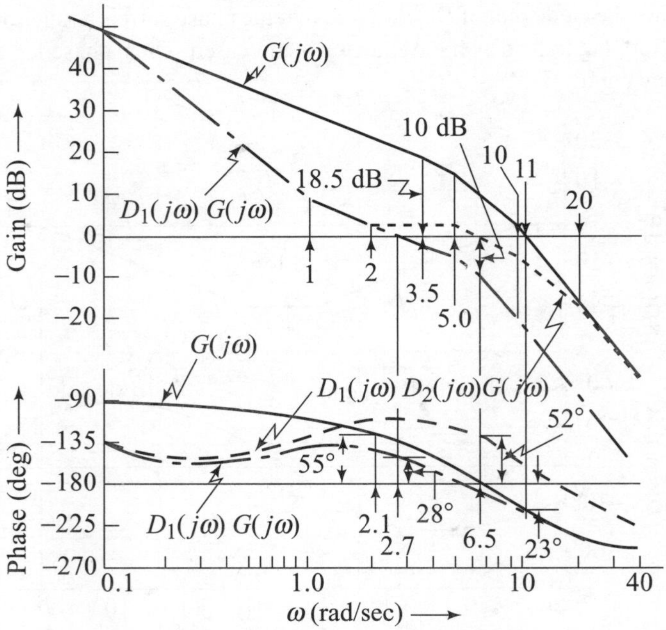

Fig. 12.23 *Bode plot for Example 12.4*

If lag compensation is employed for this system, the bandwidth will decrease sufficiently so as to fall short of the specified value of 12 rad/sec, resulting in a sluggish system. This fact can be verified by designing a lag compensator. If on the other hand, lead compensation is attempted, the bandwidth of the resulting system will be much higher than the specified value; the closed-loop system will be sensitive to noise which is undesirable. This fact can also be verified by designing a lead compensation scheme. Owing to the large phase lead needed at the gain crossover frequency of the compensated system, two stages of lead compensator described by Eqn. (12.13), may be used.

Let us design a lag-lead compensator to overcome the difficulties mentioned above. Since a full lag compensator would reduce the system bandwidth excessively, the lag section of the lag-lead compensator must be designed to provide partial compensation only. The lag section design, therefore, proceeds by making a choice of the new gain crossover frequency which must be higher than the crossover frequency if the system were fully lag compensated. Full lag compensation demands that the gain crossover frequency should be shifted to a point where the phase angle of the uncompensated system is

$$-180° + \text{specified phase margin} + \varepsilon = -180° + 50° + 5° = -125°$$

From Fig. 12.23, we find that this requirement is met at 2.1 rad/sec. For the design of lag section of lag-lead compensator, the selected gain crossover frequency should be higher than 2.1. The choice is made as $\omega'_g = 3.5$ rad/sec to start with.

From Fig. 12.23, we find that the gain of the uncompensated system at ω'_g is 18.5 dB. Therefore, to bring the magnitude curve down to 0 dB at ω'_g, the lag section must provide an attenuation of 18.5 dB. This gives the β parameter of the lag section as

$$20 \log \beta = 18.5$$

or

$$\beta = 8.41, \text{ say } 10$$

Let us now place the upper corner frequency of the lag section at 1 rad/sec. This gives the lag-section transfer function,

$$D_1(s) = \frac{\tau_1 s+1}{\beta\tau_1 s+1} = \frac{s+1}{10s+1}$$

Bode plot of lag-section compensated system is shown in Fig. 12.23; the phase margin is 28°.

We now proceed to design the lead section. We choose $\alpha = 1/\beta = 0.1$. The maximum lead provided by the lead section is therefore (refer to Eqn. (12.17)),

$$\phi_m = \sin^{-1}\left(\frac{1-\alpha}{1+\alpha}\right) = 55°$$

The lag-compensated system has a gain of $-20 \log \left(1/\sqrt{\alpha}\right) = -10$ dB at $\omega_g'' = 6.5$ rad/sec. Setting ω_m, the frequency at which the lead section has maximum phase ϕ_m, equal to ω_g'', we obtain

$$\omega_m = \frac{1}{\tau_2\sqrt{\alpha}} = \omega_g'' = 6.5$$

This gives

$$\tau_2 = 0.5$$

The transfer function of the lead section, therefore, is

$$D_2(s) = \frac{\tau_2 s+1}{\alpha\tau_2 s+1} = \frac{0.5s+1}{0.05s+1}$$

Combining the transfer functions of the lead and lag sections, we obtain

$$D(s) = \frac{(s+1)(0.5s+1)}{(10s+1)(0.05s+1)}$$

The open-loop transfer function of the lag-lead compensated system is

$$D(s)G(s) = \frac{30(s+1)(0.5s+1)}{s(0.1s+1)(0.2s+1)(10s+1)(0.05s+1)}$$

The Bode plot of $D(j\omega)G(j\omega)$ is shown in Fig. 12.23. The phase margin is found to be 52°.

The magnitude *versus* phase angle curve of the lag-lead compensated system is drawn on the Nichols chart in Fig. 12.24, from which the bandwidth of the system is found to be 9.6 rad/sec.

The design does not meet the specifications laid down. We must, therefore, redesign by adjusting the initial choice of β and τ_1. This is left as an exercise for readers.

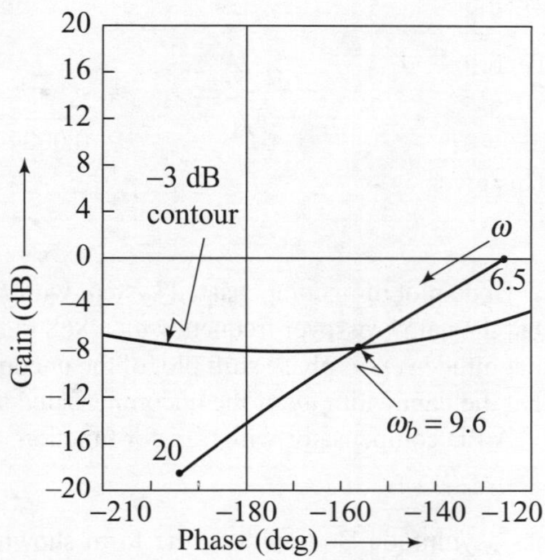

Fig. 12.24 *Nichols chart for Example 12.4*

12.6 CASCADE PID COMPENSATION

The design of a controller is basically a search problem in n-dimensional parameter space, where n is the number of design parameters. Lesser the value of n, simpler is the design problem.

In the case of the design wherein the primary requirement is to improve transient response, we have recommended a lead compensator, which has three trial parameters. An alternative strategy is to design a PD compensator, which has two trial parameters. However, an independent design of a filter, to take care of high-frequency noise problem, will also be required.

When the primary requirement is to improve steady-state performance, we have recommended a lag compensator, which has three trial parameters. An alternative strategy is to design a PI compensator, which has two trial parameters. This compensator, however, will be useful if the pole at origin does not degrade closed-loop stability appreciably.

The design of PI, PD, and PID compensators follows in a straight forward way from the procedures of design of lag, lead and lag-lead compensators. Illustrative examples follow.

Example 12.5 A unity feedback system has forward-path transfer function,

$$G(s) = \frac{4500K}{s(s+361.2)}$$

Our goal is to design a cascade PD compensator so that the closed-loop system meets the following specifications:

Steady-state error due to unit-ramp input ≤ 0.000443

Phase margin $\Phi M \geq 80°$

Bandwidth $\omega_b \leq 2000$ rad/sec

We want to achieve a steady-state response that has no more than 0.000443 error due to unit-ramp input. This, as we know, can be achieved by simple gain adjustment. We therefore first evaluate the 'uncompensated system' that meets the steady-state accuracy requirement.

Desired

$$K_v = \frac{1}{0.000443}$$

Therefore,

$$\frac{1}{0.000443} = \lim_{s\to 0} \frac{s(4500K)}{s(s+361.2)}$$

This gives

$$K = 181.17$$

Bode plot of uncompensated system with $K = 181.17$ is shown in Fig. 12.25, from which it is found that the gain crossover frequency $\omega_g = 868$ rad/sec and the phase margin $\Phi M = 22.6°$. In Fig. 12.26, the magnitude *versus* phase shift plot of the uncompensated system is drawn on the Nichols chart. It is seen that the bandwidth ω_b of the uncompensated system is 1400 rad/sec.

A PD compensator with transfer function,

$$D(s) = 1 + T_D s$$

has asymptotic Bode plot of the form shown in Fig. 12.27. The logical way to approach the design problem is to first examine how much additional phase is needed to realize a ΦM of 80°. Since the

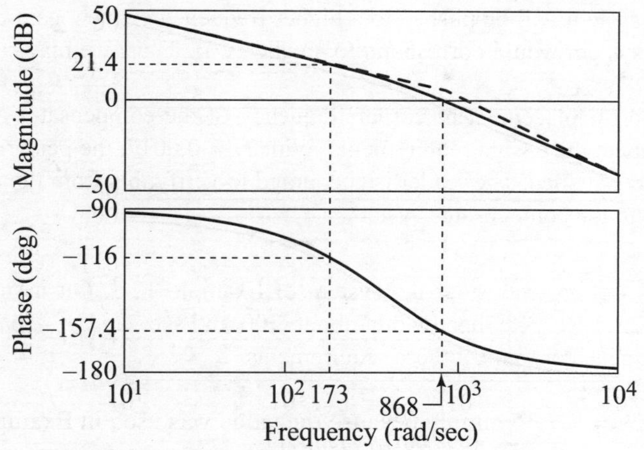

Fig. 12.25 *Bode plot for Example 12.5*

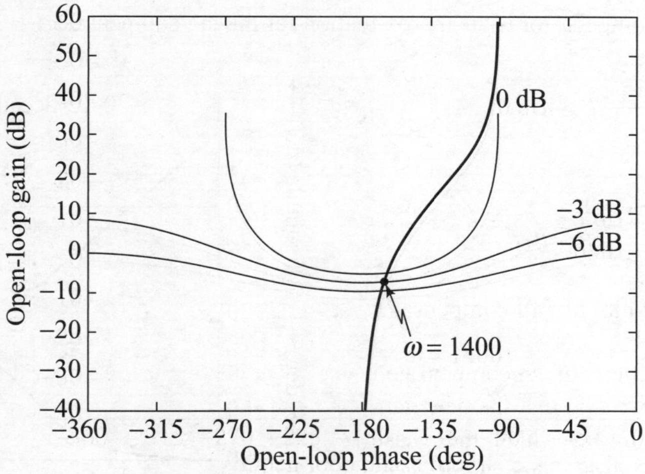

Fig. 12.26 *Nichols chart for Example 12.5*

uncompensated system with the gain set to meet the steady-state requirement has ΦM of only 22.6°, the PD compensator must provide an additional phase of 80 – 22.6 = 57.4°. The additional phase must be placed at the desired gain crossover frequency of the compensated system in order to realize a ΦM of 80°.

Referring to the asymptotic Bode plot of PD compensator shown in Fig. 12.27, we see that the additional phase at frequencies > $1/T_D$ is always accompanied by a gain in the magnitude curve. As a result, the gain crossover

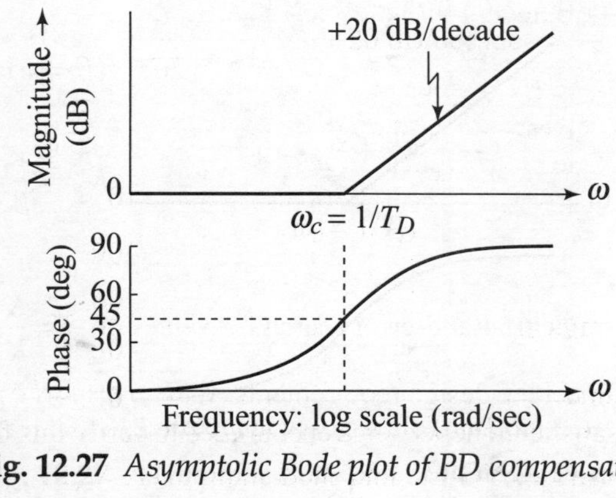

Fig. 12.27 *Asymptolic Bode plot of PD compensator*

of the compensated system will be pushed to a higher frequency. At new gain crossover, the phase of the uncompensated system would correspond to smaller ΦM. Thus, we may run into the problem of diminishing returns.

Simple trial-and-error placement of corner frequency of the compensator around the gain crossover frequency can meet the design requirements. With $T_D = 0.0017$, the performance requirements in frequency domain are satisfied. The reader is encouragd to verify this from the Bode plot and Nichols chart corresponding to the compensated system.

Example 12.6 Let us reconsider the system of Example 12.5. Let us now specify ΦM to be atleast 64°, bandwidth about 300 rad/sec, and e_{ss} (parabolic input) ≤ 0.2. We attempt a PI compensator to meet these requirements.

Initially we take $K = 181.17$ simply because the value was used in Example 12.5. We will tune the total loop gain later to meet the steady-state performance requirement. The uncompensated system with $K = 181.17$ has $\Phi M = 22.6°$, gain crossover frequency $\omega_g = 868$ rad/sec, and bandwidth $\omega_b = 1400$ rad/sec (refer to Figs 12.25 and 12.26).

We know that PI compensator is an approximation for the lag compensator:

$$D(s) = \frac{\tau s + 1}{\beta \tau s + 1}; \beta > 1$$

$$\cong \frac{\tau s + 1}{\beta \tau s} = \frac{1}{\beta} + \frac{1}{\beta \tau s} = K_P + \frac{K_I}{s}$$

The asymptotic Bode plot of a PI compensator is shown in Fig. 12.28.

From the Bode plot of uncompensated system (Fig. 12.25), we find that the new gain crossover frequency ω'_g at which the phase margin is 64°, is 173 rad/sec. The magnitude of $G(j\omega)$ at this frequency is 21.4 dB. Thus, the PI compensator should provide an attenuation of 21.4 dB at $\omega'_g = 173$ rad/sec.

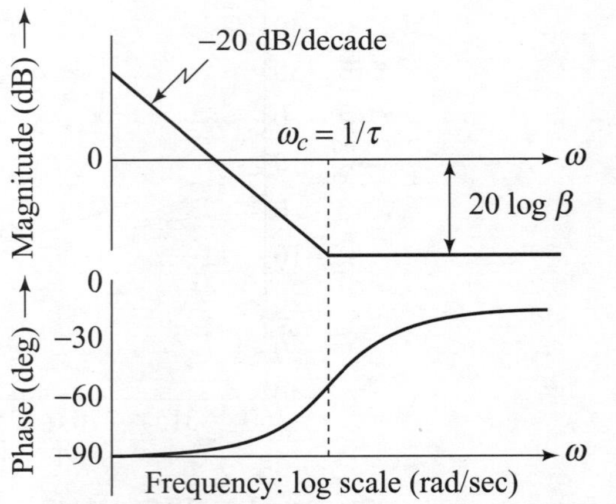

Fig. 12.28 *Asymptatic Bode plot of PI compensator*

$$20 \log(\beta) = 21.4; \beta = 1/0.0851.$$

This gives

$$K_P = \frac{1}{\beta} = 0.0851$$

A simple trial-and-error search for values of $\frac{1}{\tau} = \frac{K_I}{K_P}$ that are sufficiently small (low-frequency range) can meet the design requirements. With $1/\tau = 1$, the performance requirements in frequency domain are all satisfied. The reader is encouraged to verify this from the Bode plot and Nichols chart corresponding to the compensated system.

Let us now evaluate the steady-state performance.

$$D(s)G(s) = \frac{4500 \times 181.17 \times 0.0851(s+1)}{s^2(s+361.2)}$$

$$K_a = \frac{4500 \times 181.17 \times 0.0851}{361.2} = 192.0793$$

$$e_{ss} = 0.0052$$

Steady-state requirement is thus satisfied.

It should be noted that ΦM of the system can be improved further by increasing the value of β above 1/0.0851. However, the bandwidth of the system will be reduced. For example, for $\beta = 1/0.04$ and $1/\tau = 1$, the ΦM is increased to 75.7° but ω_b is reduced to 127 rad/sec.

12.7 REVIEW EXAMPLES

Review Example 12.1 Consider a unity-feedback type-1 system with open-loop transfer function,

$$G(s) = \frac{K}{s(s+1)(s+4)}$$

The system is to be compensated to meet the following specifications:

Velocity-error constant	$K_v = 5$
Peak overshoot	$M_p = 25\%$
Settling time	$t_s = 10$ sec

Peak overshoot of 25% and settling time of 10 sec imply $\zeta = 0.4$ and $\omega_n = 1$ for the closed-loop dominant poles (refer to Eqns (6.14) and (6.19)). Equation (12.1) yields a 40° phase margin for a damping ratio of 0.4, and from Eqn. (12.5) we find that with $\zeta = 0.4$ and $\omega_n = 1$, a closed-loop bandwidth of $\omega_b = 1.375$ rad/sec is required.

In order to meet the specification of $K_v = 5$, K must be set at 20, yielding

$$G(s) = \frac{5}{s(s+1)(0.25s+1)}$$

Bode plot of uncompensated system for $K = 20$ is shown in Fig. 12.29, from which it is found that the gain crossover frequency $\omega_g = 2.2$ rad/sec and the phase margin is –4°. If lead compensation is employed for this system, the gain crossover frequency of the compensated system will be more than 2.2 rad/sec and the bandwidth will exceed the specified value, making the system sensitive to noise. We, therefore, try lag compensation.

It is seen from the Bode plot that neglecting the phase-lag contribution of the compensator, the specified phase margin of 40° is obtained if the gain cross over frequency is 0.8 rad/sec. Since this is fairly low, the upper corner frequency of the lag compensator cannot be taken far to its left to avoid large time-constants. This indicates that phase-lag contribution of the compensator at the new gain

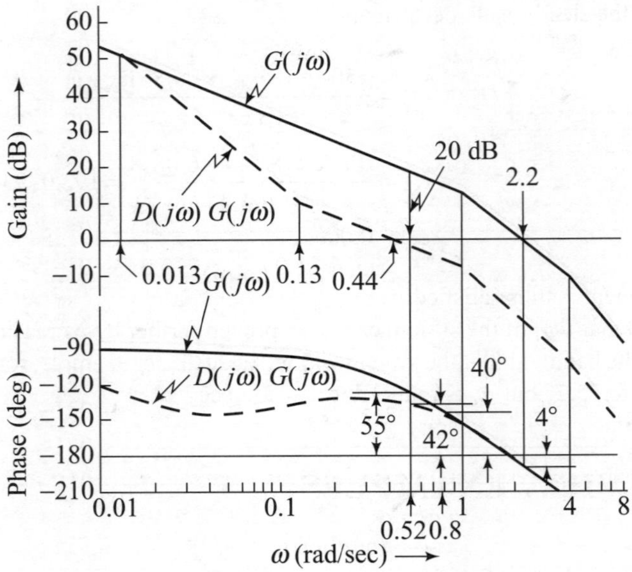

Fig. 12.29 *Bode plot for Review Example 12.1*

crossover frequency ω_g will be considerable and may be guessed at $\varepsilon = 15°$. The uncompensated system must, therefore, have a phase angle of

$$-180° + \text{specified phase margin} + \varepsilon = -180° + 40° + 15° = -125°$$

at ω_g', which is found to be 0.52 rad/sec.

The gain of the uncompensated system at ω_g' is 20 dB. Therefore, to bring the magnitude curve down to 0 dB at ω_g', the lag compensator must provide an attenuation of 20 dB. The β parameter of the lag compensator can now be calculated.

$$20 \log \beta = 20$$

This gives

$$\beta = 10$$

Placing the upper corner frequency of the compensator two octaves below ω_g', we have

$$\frac{1}{\tau} = \frac{\omega_g'}{(2)^2} = 0.13 \text{ rad/sec} = \frac{1}{7.7}$$

The lower corner frequency of the compensator is then fixed at

$$\frac{1}{\beta\tau} = 0.13 \text{ rad/sec} = \frac{1}{77}$$

The transfer function of the lag compensator is then

$$D(s) = \frac{7.7s + 1}{77s + 1}$$

Phase-lag introduced by the lag compensator at ω'_g is

$$\tan^{-1}(7.7\omega'_g) - \tan^{-1}(77\omega'_g) = 76° - 88° = -12°$$

Therefore, the safety margin of ε = 15° is justified.

The open-loop transfer function of the compensated system becomes

$$D(s)G(s) = \frac{5(7.7s+1)}{s(s+1)(0.25\,s+1)(77\,s+1)}$$

The Bode plot of $D(j\omega)G(j\omega)$ is shown in Fig. 12.29, from where the phase margin of the compensated system is found to be 42°.

In Fig. 12.30, the magnitude *versus* phase shift plot of the compensated system is drawn on the Nichols chart. It is seen that the bandwidth of the compensated system is 1 rad/sec. We must redesign by adjusting the initial choice of β and τ if an improvement in bandwidth is required.

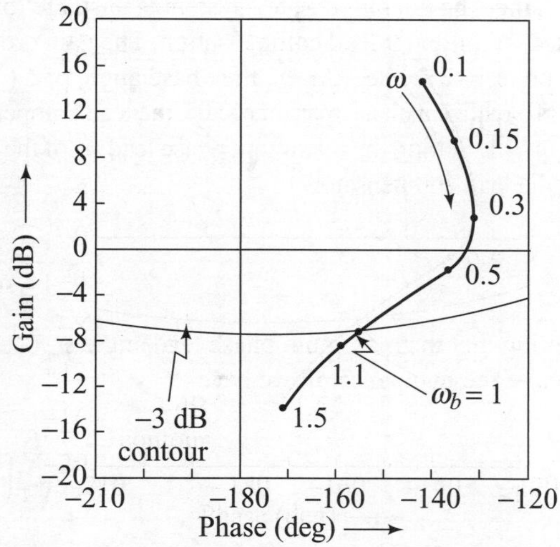

Fig. 12.30 *Nichols chart for Review Example 12.1*

Simulation of the closed-loop system corresponding to the design just completed gives peak overshoot of 31.7% and setting time of 18.7 sec.

Review Example 12.2 Consider the system shown in Fig. 12.31. Design a suitable compensator such that the closed-loop system will have a phase margin of 50°, a gain margin of not less than 10 dB and a gain crossover frequency of 1 rad/sec.

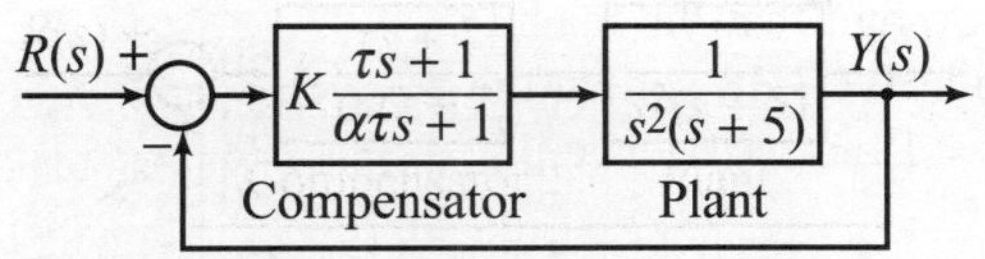

Fig. 12.31 *Feedback system for Review Example 12.2*

Solution The design problem may be reformulated as follows:

Design the unity zero-frequency gain compensator

$$D(s) = \frac{\tau s+1}{\alpha\tau s+1}$$

for the plant with adjustable gain,

$$G(s) = \frac{K}{s^2(s+5)}$$

so that the closed-loop system meets the performance specifications.

Since the uncompensated system is unstable for all values of K, lag compensation scheme cannot be used. We attempt lead compensation. The gain crossover frequency of the compensated system is given to be $\omega_g = 1$ rad/sec. At ω_g, the phase angle of $G(j\omega)$ is $-191.3°$.

To realize a phase margin of 50°, the lead compensator must provide a phase lead of $50° + 11.3° = 61.3°$ at $\omega_g = 1$. Setting the maximum phase lead ϕ_m of the compensator equal to 61.3°, we obtain the α parameter of the lead compensator:

$$\alpha = \frac{1 - \sin\phi_m}{1 + \sin\phi_m} = 0.065$$

Noting that the maximum phase lead angle ϕ_m occurs at the geometric mean of the two corner frequencies of the compensator, we have

$$\omega_m = \sqrt{\frac{1}{\tau}\left(\frac{1}{\alpha\tau}\right)} = 1 = \frac{1}{\tau\sqrt{\alpha}}$$

This gives $\tau = 3.9$

Therefore, the lead compensator $D(s) = \dfrac{\tau s + 1}{\alpha\tau s + 1} = \dfrac{3.9s + 1}{0.25s + 1}$

and the compensated open-loop transfer function

$$D(s)G(s) = \frac{K(3.9s + 1)}{s^2(s + 5)(0.25s + 1)} = \frac{0.2K(3.9s + 1)}{s^2(0.2s + 1)(0.25s + 1)}$$

A Bode plot of $D(j\omega)G(j\omega)/(0.2K)$ is shown in Fig. 12.32. From this plot we find that the magnitude curve must be lowered by 11.7 dB so that the magnitude equals 0 dB at $\omega = 1$ rad/sec. Hence,

$$20 \log (0.2K) = -11.7 \quad \text{or} \quad 0.2K = 0.26$$

which yields $K = 1.3$

The Bode plot of compensated system (Fig. 12.32) shows that the system has phase margin of 50° and gain margin of 4.3 + 11.7 = 16 dB. Hence, the design specifications are satisfied.

In this design problem, both the gain margin and phase margin have been specified. The two together qualitatively give a measure of the robustness property of the closed-loop system (refer to Section 11.3).

Simulation of the closed-loop system corresponding to the design just completed gives peak overshoot of 25.5% and settling time of 9.59 sec.

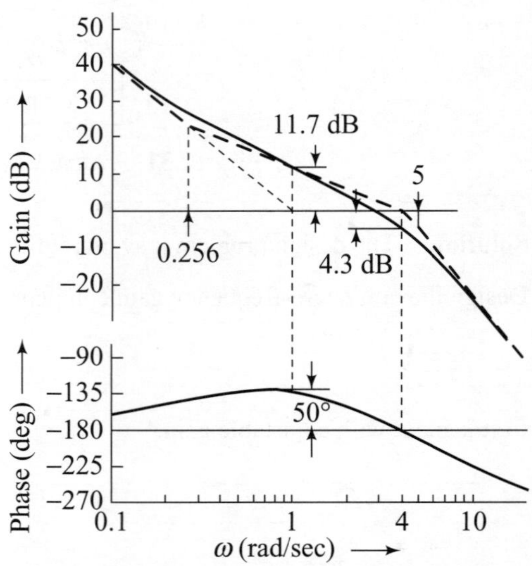

Fig. 12.32 *Bode plot for Review Example 12.2*

Review Example 12.3 The transfer function of the controlled process of a unity-feedback system is

$$G(s) = \frac{1.57e^{-0.785s}}{s(s+1)}$$

The Bode plot of $G(j\omega)$ is shown in Fig. 12.33. From the plot we see that the phase margin of the uncompensated system is approximately −5° and gain crossover frequency is $\omega_g = 1.1$ rad/sec.

Let us assume that the design specification calls for a phase margin of at least 45°. From Fig. 12.33 we see that the phase of $G(j\omega)$ decreases rapidly beyond 1 rad/sec. Therefore, a cascade lead compensator may not be effective in improving the phase margin of the system. A cascade lag compensator seems to be more appropriate for this system.

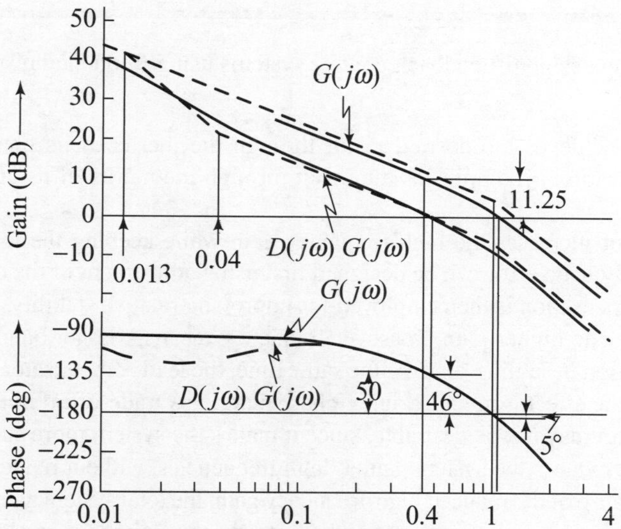

Fig. 12.33 *Bode plot for Review Example 12.3*

To realize a phase margin of 45°, the gain crossover frequency should be moved to ω'_g where the phase angle of the uncompensated system is

$$-180° + \text{specified phase margin} + \varepsilon = -180° + 45° + 5° = -130°$$

From Fig. 12.33 we find that $\omega'_g = 0.4$ rad/sec. The gain of the uncompensated system at ω'_g is 11.25 dB. Therefore, to bring the magnitude curve down to 0 dB at ω'_g, the lag compensator must provide an attenuation of 11.25 dB. The β parameter of the lag compensator can now be calculated.

$$20 \log \beta = 11.25$$

This gives

$$\beta = 3.65$$

Placing the upper corner frequency of the compensator one decade below ω'_g, yields

$$\frac{1}{\tau} = \frac{\omega'_g}{10} = 0.04 \quad \text{or} \quad \tau = 25$$

Therefore, the lag compensator $D(s) = \dfrac{25s+1}{91.25s+1}$

The Bode plot of the compensated system transfer function $D(j\omega)G(j\omega)$ is sketched in Fig. 12.33. The phase margin of the compensated system is as per specifications.

CONCLUDING REMARKS

This chapter covered the design of feedback control systems using frequency response plots.

What we have learnt?

- Lead compensation achieves the desired result through the merits of its phase-lead contribution, whereas lag compensation accomplishes the result through the merits of its attenuation property at high frequencies.
- The lead compensator increases the high-frequency gain while keeping the low-frequency gain the same. Thus, the steady-state error can be designed first using adjustment of the overall gain (at all frequencies). Lead compensation is then employed for improving relative stability. It yields a higher gain crossover frequency. The higher gain crossover frequency means a larger bandwidth. A large bandwidth means a reduction in settling time. At the same time, the lead compensator increases phase angle at high frequencies; the effect is to yield higher phase margin. A trade-off: if noise signals are present, then a large bandwidth may not be desirable, since it makes the system more susceptible to noise.
- The lag compensator reduces the system gain at high frequencies, without reducing the system gain at low frequencies. Because of the reduced high-frequency gain, the total system gain can be increased, and thereby low-frequency gain can be increased and the steady-state accuracy can be improved. Lag compensation consists of setting the total system gain to meet the steady-state requirement and then reducing the high-frequency gain to create stability and meet the phase-margin requirement for the transient response. Since the system bandwidth is reduced, the lag-compensated feedback system has a slower speed to respond; however, any high-frequency noises involved in the system can be attenuated.
- A lag-lead compensator combines the advantages of both the lag and the lead compensation. First, the lag compensation is designed to yield the proper steady-state error with improved stability. Next, the lead compensation is designed to speed up the transient response.
- Good phase margin and hence damping is obtained when crossover occurs at essentially a slope of −20 dB/decade.
- A PD compensator is, in effect, a lead compensator. It has reduced number of trial parameters, making the design problem simpler. It enhances the improvement in relative stability. The price paid is in terms of increase in the system's sensitivity to noise.
- A PI compensator is, in effect, a lag compensator with a reduced number of trial parameters. It enhances the improvement in steady-state accuracy. The compensator will, however, be useful if reduction in parameters does not degrade closed-loop stability appreciably.

Review Questions

12.1 Discuss the compensation characteristics of cascade lag and lead compensators using Bode plots. Show that

(a) lead compensation is suitable for systems having unsatisfactory transient response, and it provides a limited improvement in steady-state performance; and

(b) lag compensation is suitable for systems with satisfactory transient response but unsatisfactory steady-state response.

12.2 Sketch op amp circuits capable of providing

(a) phase lead;

(b) phase lag; and

(c) phase lead and lag over different frequency bands.

Determine transfer function model of each circuit and draw Bode plot for each model.

12.3 The transfer function of a lag-lead compensator is given by

$$D(s) = \underbrace{\left[\frac{\tau_1 s+1}{\beta\tau_1 s+1}\right]}_{\text{Lag Section}}\underbrace{\left[\frac{\tau_2 s+1}{\alpha\tau_2 s+1}\right]}_{\text{Lead Section}};\beta>1,\alpha<1,\tau_1,\tau_2>0$$

Give an op amp circuit that realizes this $D(s)$.

12.4 Sketch op amp circuits capable of realizing

(a) PD control action;

(b) PI control action; and

(c) PID control action.

Determine transfer function model of each circuit and draw Bode plot for each model.

Self-Appraisal Quiz

A set of questions with multiple answer choices is given below. These questions have been designed to capture important aspects of the subject covered in the chapter. For each of these questions, identify the correct answer. Compare your answers with the Master Key given at the end of the session. This quiz should serve the purpose of a self-appraisal test for readers.

12.1 A motion control system has type-2 plant. A recommended cascade compensation scheme for this system employs

(A) a lag compensator

(B) a lead compensator

(C) either a lag or a lead compensator

(D) neither a lag nor a lead compensator

12.2 Consider the following statements:

(i) Cascade lead compensation decreases the bandwidth.

(ii) Cascade lag compensation increases the bandwidth.

Which of the following is the correct answer?

(A) None of the above statements is true
(B) Statement (i) is true but statement (ii) is false
(C) Statement (i) is false but statement (ii) is true
(D) Both the statements are true

12.3 Which of the following control schemes will be employed for systems with high demands on speed of response, margin of stability and steady-state accuracy?
(A) Cascade lag
(B) Cascade lead
(C) Lead network in minor-loop feedback
(D) Cascade lag-lead

12.4 Maximum phase-lead of the compensator $D(s) = (0.5s + 1)/(0.05s + 1)$, is
(A) 52° at 4 rad/sec
(B) 52° at 10 rad/sec
(C) 55° at 12 rad/sec
(D) None of the answers in (A), (B), and (C) is correct

12.5 In control systems
(i) reduction in bandwidth results in sluggish response;
(ii) reduction in bandwidth results in better signal/nose ratio.
Which of the following is the correct answer?
(A) None of the above statements is true
(B) Statement (i) is true but statement (ii) is false
(C) Statement (i) is false but statement (ii) is true
(D) Both the statements are true

12.6 Consider the following statements:
(i) Lead compensation is suitable for systems having unsatisfactory transient response. It also provides a limited improvement in steady-state response.
(ii) Lag compensation is suitable for systems with satisfactory transient response but unsatisfactory steady-state response.
Which of the following is the correct answer?
(A) None of the above statements is true
(B) Statement (i) is true but statement (ii) is false
(C) Statement (i) is false but statement (ii) is true
(D) Both the statements are true

12.7 Consider the following statements:
(i) The natural frequency ω_n must be large for good performance; however, bandwidth considerations impose a limit on increasing ω_n.
(ii) When design considerations impose a limit on ω_b, a PD compensator with a filter is suitable.
(A) None of the above statements is true
(B) Statement (i) is true but statement (ii) is false
(C) Statement (i) is false but statement (ii) is true
(D) Both the statements are true

12.8 The compensation scheme highly sensitive to noise is
(A) PD
(B) PI
(C) Both PD and PI
(D) Neither PD nor PI

12.9 (i) Lead compensation achieves the desired result through phase-lead contribution.
(ii) Lag compensation achieves the desired result through phase-lag contribution.
(A) Both the statements are true.
(B) Statement (i) is true and statement (ii) is false.
(C) Statement (i) is false and statement (ii) is true.
(D) Both the statements are false

12.10 (i) PD compensator is, in effect, a lag compensator with relatively higher sensitivity to noise.
(ii) PI compensator is, in effect, a lead compensator with relatively higher negative effect on transient response.
(A) Both the statements are true.
(B) Statement (i) is true and statement (ii) is false
(C) Statement (i) is false and statement (ii) is true
(D) Both the statements are false.

MASTER KEY

12.1 (B)	12.2 (A)	12.3 (D)	12.4 (D)	12.5 (D)
12.6 (D)	12.7 (D)	12.8 (A)	12.9 (B)	12.10 (D)

Problems

12.1 The open-loop transfer function of a control system is

$$G(s)H(s) = \frac{1}{s(1+0.5s)(1+2s)}$$

(a) Determine approximate values of gain margin and phase margin.
(b) If a lag compensator with transfer function $D(s) = K_c(1 + 3s)/(1 + 5s)$ is inserted in the forward path, find the value of K_c to keep the gain margin unchanged.

MATLAB Exercise After solving the problem using hand sketches, attempt it using MATLAB. Determine settling time and peak overshoot of the compensated and uncompensated systems.

12.2 The open-loop transfer function of a control system is

$$G(s)H(s) = \frac{10}{s(1+0.5s)(1+0.1s)}$$

(a) Draw the Bode plot and determine gain crossover frequency, and phase and gain margins.
(b) A lead compensator with transfer function $D(s) = (1 + 0.23s)/(1 + 0.023s)$ is now inserted in the forward path. Determine the new gain crossover frequency, and phase and gain margins. Comment upon the effects of lead compensation on system performance.

MATLAB Exercise After completing the hand sketches, verify your results using MATLAB. Determine the bandwidth, resonance peak and resonance frequency of the uncompensated and compensated systems.

12.3 The block diagram of Fig. P12.3 represents a position control system. You have been given the task of designing the compensation $D(s)$.

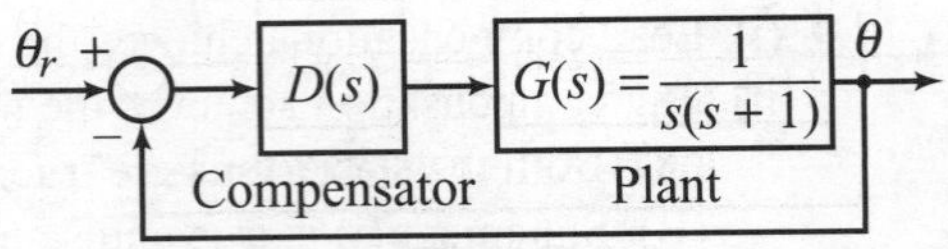

Fig P12.3

(a) Consider first $D(s) = K$. What value of K would provide velocity error constant $K_v = 12$?

(b) For the value of K in part (a), determine phase margin ΦM, bandwidth ω_b, resonance peak M_r, and resonance frequency ω_r.

(c) Next consider $D(s) = K\dfrac{\tau s+1}{\alpha\tau s+1}$. With the parameter K set for K_v of 12, choose $\tau > 0$, and $\alpha < 1$ so that ΦM is greater than 40°.

(d) For the value of $D(s)$ in part (c), determine ΦM, ω_b, M_r, and ω_r.

MATLAB Exercise Verify and refine your design using MATLAB. Determine settling time and peak overshoot of the compensated system.

12.4 Consider the system shown in Fig. P12.3. It is desired to obtain a phase margin greater than 45°, and steady-state error less than 0.1 for a unit ramp input.

(a) Design a cascade lead compensator that meets the given specifications.

(b) Design a cascade lag compensator that meets the given specifications.

(c) Since an identical set of specifications can be met by lead as well as lag compensation, how do we make a choice of the compensation scheme?

MATLAB Exercise Verify and refine your designs using MATLAB. Determine settling time and peak overshoot of the compensated systems.

12.5 A unity-feedback control system has open-loop transfer function

$$G(s) = 1/s^2$$

The specifications for the system are:
Settling time $t_s \le 4$ sec
Peak overshoot for a step input $\le 20\%$

(a) Design a suitable cascade lead, lag, or lag-lead compensator to realize the given objectives.

(b) Design a suitable PD, PI, or PID compensator to realize the given objectives.

MATLAB Exercise Convert the specifications to frequency domain using second-order correlations. After completing the design by hand sketches, verify the design using MATLAB. Obtain the time response of the compensated system and check whether time-domain requirements are satisfied. If found necessary, refine the design.

12.6 An uncompensated control system with unity feedback has a plant transfer function,

$$G(s) = \frac{K}{s(s+2)}$$

It is desired to have a steady-state error for a ramp input less than five per cent of the magnitude of ramp. It is also desired that the phase margin of the system be at least 45° and the closed-loop bandwidth greater than 10 rad/sec. Design a cascade lead compensator to yield the desired specifications.

MATLAB Exercise Verify and refine your design using MATLAB. Determine settling time and peak overshoot of the compensated system.

12.7 For the system of Problem 12.6, design a cascade lag compensator to yield the desired specifications, with the exception that a bandwidth equal to or greater than 2 rad/sec will be acceptable.

MATLAB Exercise Attempt this problem using MATLAB dialogs.

12.8 An uncompensated control system with unity feedback has a plant transfer function

$$G(s) = \frac{K}{s(1+0.1\,s)(1+0.2\,s)}$$

The system must satisfy the following performance specifications:

(a) The magnitude of the steady-state error of the system due to a unit ramp function input is 0.01.

(b) Phase margin $\geq 40°$.

Use two identical cascaded lead compensators to compensate the system. Justify the use of two-stage lead compensator.

MATLAB Exercise Attempt this problem using MATLAB dialogs. Include root-locus analysis and simulation study of your design. Redesign if found necessary.

12.9 Consider the problem of design of angular control system of a robot's joint. A block diagram of the system is shown in Fig. P12.9. The performance specifications of the system are as follows:

(a) The steady-state error due to ramp input should be less than or equal to one per cent.

(b) The phase margin of the system should be greater than 45°.

Design a cascade lead compensator to yield the desired specifications. Determine the resonance peak M_r, and the bandwidth ω_b of the compensated system.

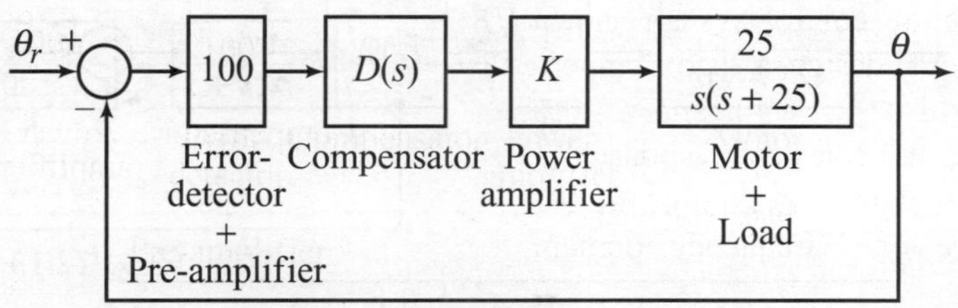

Fig. P12.9

MATLAB Exercise (i)You have neglected the time constant of the power amplifier. Check the robustness of your design with respect to this modeling approximation. Determine the range of amplifier time constant for which your design gives acceptable performance (ii) Set up block diagram of the system in Simulink window and study the saturation effects of the amplifier. Determine the linear range of the amplifier for which your design gives acceptable performance.

12.10 For the system of Problem 12.9, design a cascade lag compensator to yield the desired specifications. Determine the resonance peak M_r, and the bandwidth ω_b of the compensated system.

MATLAB Exercise Attempt this problem using MATLAB dialogs. Include root-locus analysis and simulation study of your design. Redesign if found necessary.

12.11 A unity-feedback system has an open-loop transfer function,

$$G(s) = \frac{5}{s(s+1)(0.5s+1)}$$

It is desired to obtain the phase margin of atleast 40° and the gain margin of atleast 10 dB without sacrificing the velocity-error constant of the system.

(a) Design a suitable cascade lead, lag, or lag-lead compensator to realize the given objectives. Give reasons for your choice of the compensation scheme.

(b) Design a suitable PD, PI or PID compensator to realize the given objectives. Give reasons for your choice of the compensation scheme.

MATLAB Exercise Attempt this problem using MATLAB dialogs. Include root-locus analysis and simulation study of your design. Redesign if found necessary.

12.12 The block diagram of Fig. P12.12 represents a position control system. You have been given the task of designing the compensation $D(s)$ to meet the following specifications:

Velocity-error constant, $K_v = 30$.
Phase margin, $\Phi M \geq 50°$.
Bandwidth, $\omega_b = 12$ rad/sec.

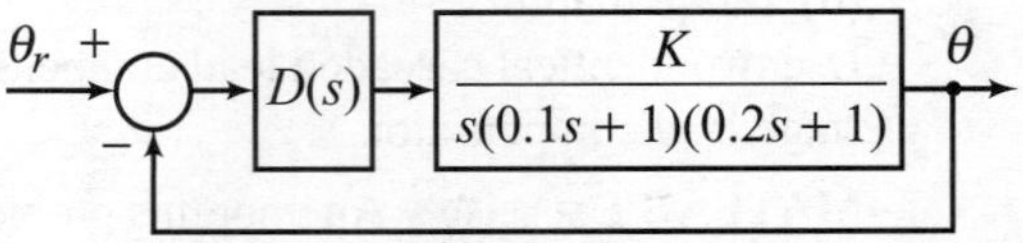

Fig. P12.12

(a) Show that if lead compensation is employed to meet the K_v and ΦM requirements, the bandwidth will be larger than the specified value, resulting in a system sensitive to noise.

(b) Show that if lag compensation is attempted to meet the K_v and ΦM requirements, the bandwidth will fall short of the specified value, resulting in a sluggish system.

(c) Design a suitable lag-lead compensator that meets the specifications on K_v, ΦM, and ω_b.

MATLAB Exercise Attempt this design problem using MATLAB dialogs. Many iterations may be required to meet the design objectives.

12.13 For the position control system shown in Fig. P12.13, design a lead compensator to yield a $K_v = 40$, and a 20% overshoot in the transient response for a step input, with a peak time of 0.1 second. Use only frequency domain methods.

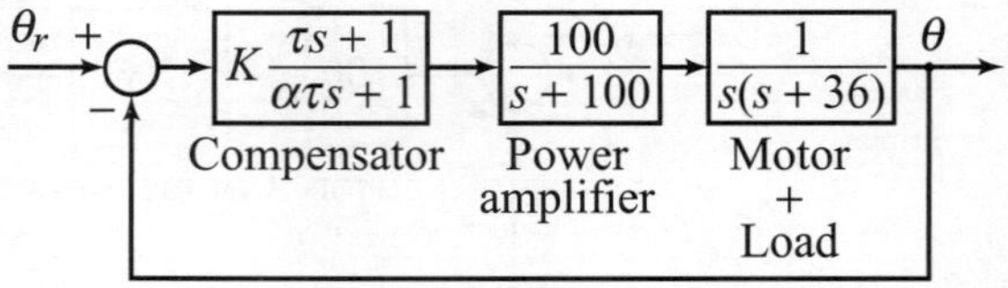

Fig. P12.13

MATLAB Exercise You have employed second-order correlations to solve this problem using hand sketches. Verify your solution using MATLAB, and if required refine your design. Due to uncertainties in the load, the time constant of Motor + Load subsystem may vary by 10%. Check the robustness of your design determining transient and steady-state performance for various values of this time-constant.

12.14 The open-loop transfer function of a unity-feedback system is

$$G(s) = \frac{K}{s(s+10)^2}$$

It is specified that the velocity error constant of this system be equal to 20, while the damping ratio of the dominant roots be equal to 0.707.

(a) It is not advisable to compensate this system by a single-stage cascade lead compensator. Why?

(b) Design a cascade lag compensator to meet the given performance specifications. Use only frequency-domain methods.

(c) Determine the bandwidth of the compensated system.

The –3 dB contour of the Nichols chart may be constructed using the following table:

Phase, degrees	0	–30	–60	–90	–120	–150	–180	–210
Magnitude, dB	7.66	6.8	4.18	0	–4.18	–6.8	–7.66	–6.8

12.15 A unity-feedback type-0 system with dead time has a forward path transfer function,

$$G(s)=\frac{10e^{-0.02s}}{(0.5s+1)(0.1s+1)(0.05s+1)}$$

Design a suitable compensation scheme so that the system acquires a damping ratio of 0.4 without loss of steady-state accuracy. Use only frequency-domain methods.
Estimate the bandwidth and the settling time of the compensated system.

MATLAB Exercise You have employed second-order correlations to solve this problem using hand sketches. Verify your solution using MATLAB, and if required refine your design.

12.16 Consider the servo system shown in Fig. P12.16.

(a) Sketch the Bode plot of $G(s)$ when $D(s) = 1$. Find the phase margin of the uncompensated system.

(b) The second-order compensator

$$D(s)=\frac{s^2+0.8s+4}{(s+0.384)(s+10.42)}$$

cancels the complex poles of the process transfer function $G(s)$. Find the phase margin of the compensated system. What is the effect of $D(s)$ on the steady-state performance of the system?

(c) The closed-loop system is required to have a phase margin of at least 70° without sacrificing the steady-state performance of the system. If the cancelation-compensation proposed in part (b) cannot meet the phase-margin requirement, add a stage of lead compensator to further improve the relative stability of the system.

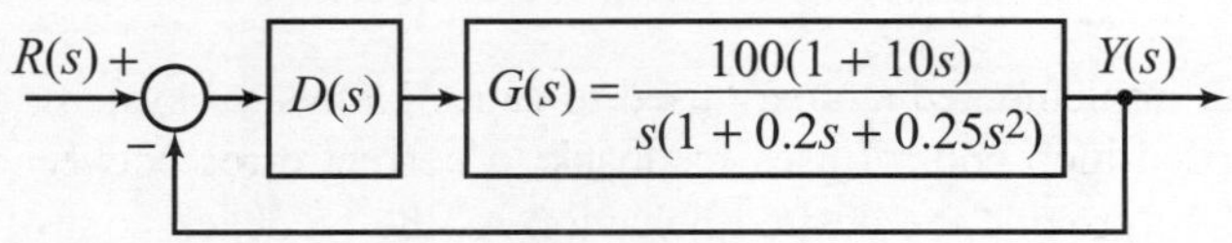

Fig. P12.16

12.17 Consider the position control system of Fig. P12.17.

Show that $K_t = \frac{1}{8}$ provides approximately 25° phase margin. Find K_v of the compensated system.

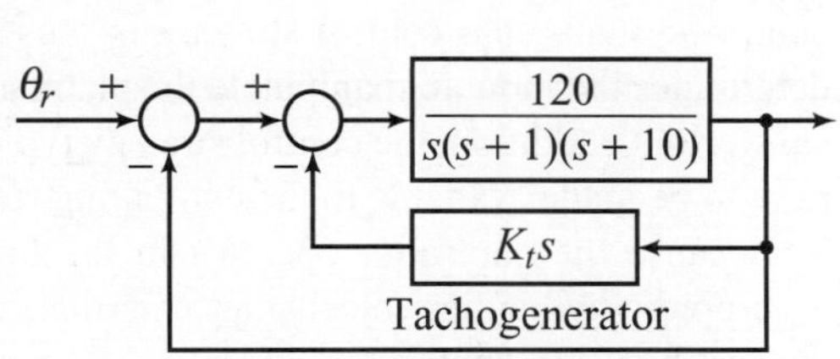

Fig. P12.17

MATLAB Exercise The parameter K_t is subject to variations by 10%. Check by simulation the robustness of the closed-loop system to these variations by determining phase margin and K_v for various values of K_t.

DIGITAL CONTROL SYSTEMS

13

Preview

The controllers we have studied so far were described by integro-differential equations, which were assumed to be built using analog electronics. However, most control systems today use digital computers (usually microprocessors) to implement the controllers. The intent of this chapter is to expand on the transform-domain analysis and design methods covered in the book so far, for analysis and design of control systems that will be implemented using a digital computer. The state variable methods for analysis and design of digital control systems are covered in the next two chapters.

The arrangement of topics for the material on digital control systems is a little different from the conventional one. Typically, a presentation on digital control systems starts with transform-domain design and then carries over to state space—a detailed account of state variable methods for discrete-time systems is given. Since the state variable methods for discrete-time systems run quite parallel to those for continuous-time systems, a full-blown repetition followed in conventional approach, is, in fact, not required. A natural way of introducing state variable methods to students with no background on the subject, is to give the treatment for continuous-time systems, followed by brief parallel presentation for discrete-time systems. This approach has been followed in the next two chapters.

Our treatment in this book is a limited version of a complex subject; covered in more detail in reference [155].

13.1 INDUSTRIAL AND EMBEDDED CONTROL

A control system is an implemented strategy used to cause a physical system, or *plant*, to behave in a desired manner. Closed-loop control uses feedback to correct error between the plant output and a reference input, *i.e.*, to provide the desired behavior. As the feedback control strategy increases in complexity, it becomes more difficult to apply analog components for its implementation. In modern control systems, the control strategy is thus typically implemented in software. A digital computer determines the input to manipulate the plant; and this requires facilities to apply this input to the physical world. In addition, the control strategy typically relies on measured values of the plant behavior that have to be made available to the computing resources.

Because the controller operates in the low-power electronics domain and the plant operates in high-power hydraulics, mechanics, thermal, and other physical domains, transducers are needed to convert between controllers and plant. These transducers are used either by *actuators*, to drive the plant with controller-computed values, or by *sensors*, to provide measurements to the low-power electronics domain. In addition to the transformation between high- and low-power domains, transformations between discrete-time and continuous-time behavior are required. The plant dynamics change continuously with time. The controller, however, has a discrete clock that governs its behavior, and so its values change only at discrete points in time. To ensure data integrity, the sensor must include a mechanism to sample continuous data from the plant at discrete points in time, while the actuators need to provide a continuous value between the time points of discrete-time data from the controller.

Computer interfacing for data acquisition consists of analog-to-digital (A/D) conversion of input (to the controller) analog signals. Prior to the conversion, the analog signal has to be conditioned to meet the input requirements of the A/D converter. Signal conditioning consists of amplification (for sensors generating very low power signals), and filtering (to limit the amount of noise in the signal). Conversion of a digital signal to on analog signal (D/A) at the output (of the controller) is to be carried out to send this signal to an actuator which requires an analog signal. The signal has to be amplified by a transistor or solid-state relay or power amplifier. Most manufacturers of electronic instrumentation devices are producing signal conditioners as modules. The immersion of the computing power into the physical world has changed the scene of control system design. A comprehensive theory of digital control has been developed.

The broad space of control applications can be roughly divided into two categories: *industrial control* and *embedded control*. Industrial or process control applications are those in which control is used as part of the process of creating or producing an end-product. The control system is not a part of the actual end-product itself. Examples include the manufacture of pharmaceuticals and the refining of oil. In the case of industrial process control, the control system must be robust and reliable, since the processes typically run continuously for days, weeks or years.

Embedded control applications are those in which the control system is a component of the end-product itself. For example, Electronic Control Units (ECUs) are found in a wide variety of products including automobiles, airplanes, and home appliances. Most of these ECUs implement different feedback control tasks, for instance, engine control, traction control, anti-lock braking, active stability control, cruise control, and climate control. While embedded control systems must also be reliable, cost is a more significant factor, since the components of the control system contribute to the overall cost of manufacture of the product. In this case, much more time and effort is usually spent in the design phase of the control system to ensure reliable performance without requiring any unnecessary excess of processing power, memory, sensors, actuators, etc., in the digital control system. Our focus in this book has been on industrial control applications.

Perhaps more than any other factor, the development of microprocessors has been responsible for the explosive growth of the computer industry. A microprocessor is a computer on a chip. While early microprocessors required many additional components in order to perform any useful task, the increasing use of large-scale integration (LSI) and very large-scale integration (VLSI) semiconductor fabrication techniques has led to the production of microcomputers, where all of the required circuitry is embedded on one or a small number of integrated circuits. A further extension of the integration is the single chip microcontroller, which adds analog and binary I/O, timers and counters, so as to be able to carry out real-time control functions with almost no additional hardware. The major types of microcontrollers are (i) embedded 8-bit microcontrollers, (ii) 16-to-32-bit microcontrollers, and (iii) digital signal processors (DSPs). Because of DSP's special architecture, it is more useful than a general-purpose microprocessor for high-speed processing applications.

Why would any one want to use a microcontroller when there exists a powerful PC that can be programmed to perform almost any computer-related task? The main reasons for selecting microcontrollers over PCs are the cost and the size. For many applications, a moderate amount of local computer-processing power is required. A PC would not be a good choice for such applications. Cost-effective microcontrollers are ideal for such applications. Attractive compactness of microcontrollers is another factor which dictates the choice.

Today, most complex industrial processes are under computer control. We call a complete instalation (hardware and software) a SCADA (*Supervisory Control and Data Acquisition*) system. SCADA systems adopt a multilevel or hierarchical view point of control strategy. The multilevel approach subdivides the system into a hierarchy of simpler control design problems. On the lowest level of control (*direct process control level*), closed-loop control functions based on directives from the next 'higher' level are handled. The higher levels are: *supervisory level, production scheduling and control level,* and *plant management level*; the plant management level being the 'highest' hierarchical level of the plant automation system. The direct process control functions are delegated to special control units connected to the SCADA systems. Hierarchical, distributed, flexible and extremely powerful *Computer Integrated Process System* (CIPS) is now a technical reality.

A quiet revolution is going on in the manufacturing world as well, that is changing the look of the factory. Computers are controlling and monitoring the manufacturing processes. The high degree of automation that until recently was reserved for mass production only, is applied now also to small batches. This requires a change from hard automation in the production line to a *flexible manufacturing system* which can be more readily rearranged to handle new market requirements. Flexible manufacturing systems, combined with automatic assembly and product inspection on one hand, and CAD/CAM systems (Computer-Aided Design/Computer-Aided Manufacturing) on the other, are the basic components of the modern *Computer Integrated Manufacturing System* (CIMS). In a CIMS, the production flow, from the conceptual design to the finished product, is entirely under computer control and management. CIMS also adapts a multilevel or hierarchical view point of control strategy which subdivides the system into a hierarchy of simpler control design problems. The lowest level of this structure contains stand-alone control systems of manufacturing processes and industrial robots. This includes all types of machine tools, welding, electrochemical machining, etc. The stand-alone control systems have a proper interfacing with the higher levels of hierarchy: *supervisory level*, *production scheduling* and *control level*, and *plant management level*.

The strength of PC-based solutions comes from being able to provide an extremely wide range of measurements and record, and analyze the resultant data with high real-time speed. The CIPS and CIMS employ such solutions. Plug-in (in expansion slots) data acquisition and control boards (called add-on boards) are available for many popular computers.

PC-based monitoring and control systems using textual and graphical programming are today in use. The user familiarity with a computer-aided package like MATLAB allows easy adaptation of textual programming. Graphical programming permits the utilization of graphical components for assembling systems, and simplifies the development of monitoring and control programs. LabVIEW Real-Time provides a single software development tool that can implement the functions of PC-based control systems.

With the option of calculating the control signal in software (microcontrollers) rather than hardware (op amps), a digital controller gives the designer much more flexibility in making modifications to the control law after the hardware design (actuators, sensors, etc.) is fixed. In many applications, hardware and software designs can proceed almost independently.

In general, the design of control systems using a microcontroller requires a sophisticated use of new concepts such as z-transform. It is however quite straight forward to translate an analog design into a digital equivalent. Digital controller differs from analog control in that the signals must be *sampled* and *quantized.* A continuous-time signal to be used in digital logic needs to be sampled first (with sampling period T sec), and then each sample needs to be converted into a quantized digital number by an A/D converter. Once the digital computer has calculated the proper next control signal value, this value needs to be converted back to an analog signal by a D/A converter in order to be applied to the actuator of the process.

In Sections 13.2–13.6 of this chapter, we will describe a simple technique for finding a sampled equivalent to a given analog compensator. Finding a discrete equivalent to a given analog compensator is equivalent to finding a recurrence equation for the samples of the controller which will approximate the transfer function of the analog device. The method requires the sampling period T sec. being short enough so that the reconstructed (output of D/A converter) control signal is close to the signal that the original analog compensator would have produced. The rest of the chapter deals with direct digital design using z-transforms. This route of design, as we shall see, relaxes the need for a very short sampling interval; thereby giving us flexibility in terms of selection of digital hardware for implementation of design.

13.2 USE OF DIGITAL COMPUTER AS A COMPENSATOR DEVICE

In recent years, there has been a rapid increase in the use of digital computers in control systems. In fact, many industrial control systems include digital computers as an integral part of their operation. Also the availability of low-cost microprocessors and microcomputers has established a new trend towards the use of digital computers even in small-scale control systems, including those for automobiles and household appliances.

13.2.1 Advantages Offered by Digital Control

1. The accuracy of analog control systems is often limited. For example, if an analog compensator is to be built using a resistor with resistance 980.5 ohms and a capacitor with capacitance 81.33 microfarads, it would be difficult and expensive to obtain components with exactly these values. In digital control systems, the accuracy of a digital compensator can be increased simply by increasing the number of bits.
2. Digital signals are coded in sequences of ‘0’ and ‘1’, which represent ranges of voltages. This representation is less susceptible to noise and drift of power supply.
3. Due to the growth of very large scale integration (VLSI) technology, the price of digital systems has been constantly decreasing during the last decade. Now the use of a digital computer or microprocessor is cost effective even for small control systems.
4. An important advantage offered by digital control is in flexibility of modifying controller characteristics, or of adapting the controller if plant dynamics change with operating conditions. The ability to ‘redesign’ the controller by changing software (rather than hardware) is an important feature of digital control over analog control.
5. If a digital computer is used, it can be used not only as a compensator but also to collect data, to carry out complex computations, and to monitor the status of the controlled system. To enable the computer to meet the variety of demands imposed on it, it is time shared among its tasks.
6. Implementation of nonlinear and/or time-varying control laws, and adaptive control schemes was earlier constrained by the limitations of analog controllers and high cost of digital computers. However, with the advent of inexpensive digital computers with virtually limitless computing power, advanced control techniques are now being put to practice.
7. The study of emerging applications shows that artificial-intelligence (AI) will affect design and application of control systems as profoundly as the impact of microprocessor has been in the last decade. It is clear that future generation control systems will have a significant AI component; the list of applications of computer-based control will continue to expand.

13.2.2 Implementation Problems in Digital Control

The main problems associated with the implementation of digital control are related to the effects of sampling and quantization.

Most processes that we are called upon to control operate in continuous time. This implies that we are dealing largely with an analog environment; to this environment we must interface the digital computer through which we seek to influence the process.

The interface is accomplished by a system of the form shown in Fig. 13.1. It is a cascade of the A/D conversion system followed by a computer followed by a D/A conversion system. The A/D conversion process involves deriving samples of the analog signal at discrete instants of time separated by sampling period T sec. The D/A conversion process involves reconstructing a continuous-time signal from the samples given out by the digital computer.

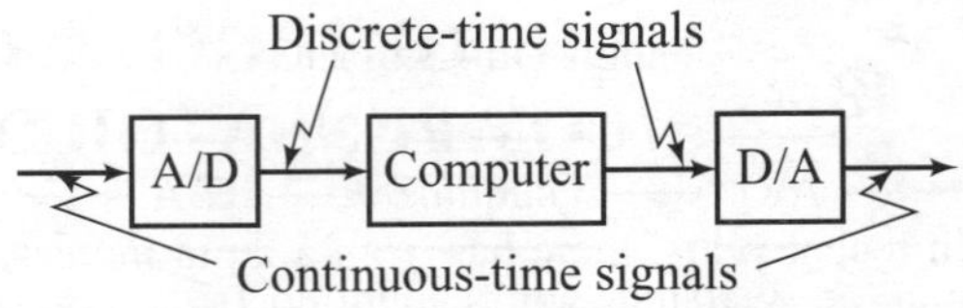

Fig. 13.1 *Discrete-time processing of continuous-time signals*

Quantization Effects

The conversion of signals from analog into digital form and *vice versa* is performed by electronic devices (A/D and D/A converters) of finite resolution. A device of n-bit resolution has 2^n quantization levels and the analog signal gets tied to these finite number of quantization levels in the process of conversion to digital form. Therefore, by the sheer act of conversion, a valuable part of information about the signal is lost.

Furthermore, any computer employed as a real-time controller must perform all of the necessary calculations with a limited precision, thus introducing a truncation error after each arithmetic operation has been performed. As the computational accuracy is normally much higher than the resolution of real converters, further truncation must take place before the computed data are converted into the analog form. The repeated process of approximate conversion—computation—conversion may be costly, if not disastrous, in terms of control system performance.

Sampling Effects

Selection of sampling period T is a fundamental problem in digital control systems. The *sampling theorem* (refer to Section 13.10) states that ***a continuous-time signal whose frequency spectrum is bounded by upper limit ω_m, can be completely reconstructed from its samples when the sampling frequency $\omega_s = 2\pi/T > 2\omega_m$. There are two problems associated with the use of sampling theorem in practical control systems.***

1. ***Frequency spectra of real signals do not possess strictly defined bandwidth ω_m. There are almost always frequency components outside the 3 dB-bandwidth ω_b. Therefore, the selection of the sampling frequency ω_s using the sampling theorem on the basis of system bandwidth ($\omega_b = \omega_m$) will result in information loss due to sampling.***
2. ***Ideal lowpass filter needed for distortionless reconstruction of the continuous-time signal from its samples, is not physically realizable. Practical devices, such as D/A converters, introduce distortions.***

Thus the process of sampling and reconstruction affects the amount of information available to the plant, and degrades control system performance. The ill-effects of sampling can be reduced, if not eliminated completely, by sampling at a very high rate. As a signal is sampled more frequently, adjacent samples have more or less similar magnitudes. In order to realize the beneficial effects of shorter sampling, longer word lengths are needed to resolve the differences between adjacent samples. Excessively fast sampling ($T \to 0$) may, however, result in numerical ill-conditioning in implementation of recursive control algorithms.

With the availability of low-cost, high-performance digital computers and interfacing hardware, the implementation problems in digital control do not pose a serious threat to its usefulness; the advantages of digital control outweigh its degrading effects for most of the applications.

13.3 CONFIGURATION OF THE BASIC COMPUTER-CONTROL SCHEME

Figure 13.2 depicts a block diagram of a computer-controlled system showing a configuration of the basic control scheme. The basic elements of the system are shown by the blocks.

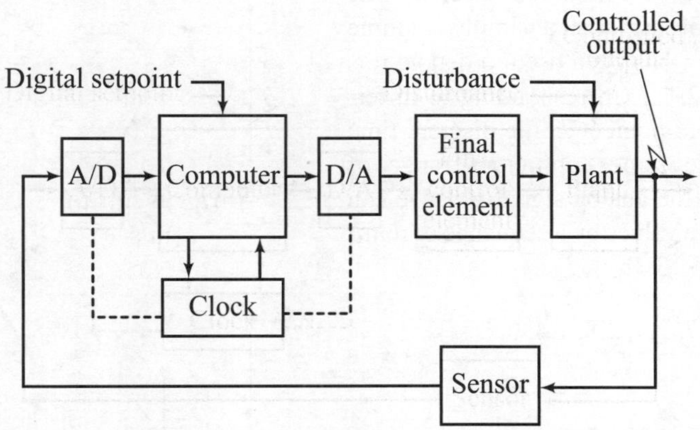

Fig. 13.2 *Configuration of the basic digital control scheme*

The analog feedback signal coming from the sensor is converted into digital form by an A/D converter. The A/D converter converts a voltage (or current) amplitude at its input into a binary code representing a quantized amplitude value closest to the amplitude of the input.

The digital computer processes the sequence of numbers by means of an algorithm and produces a new sequence of numbers. Since data conversions and computations take time, there will always be a delay when a control law is implemented using a digital computer. The delay, which is called the *computational delay*, degrades the control system performance. It should be minimized by a proper choice of hardware and by a proper design of software for the control algorithm. Floating-point operations take a considerably longer time to perform (even when carried out by arithmatic co-processor) than the fixed-point ones. We, therefore, try to execute fixed-point operations whenever possible.

The D/A conversion system in Fig. 13.2 converts the sequence of numbers in numerical code into a piecewise continuous-time signal. The output of the D/A converter is fed to the plant through the actuator (final control element) to control its dynamics.

The basic control scheme of Fig. 13.2 assumes a uniform sampling operation, i.e., only one sampling rate exists in the system and the sampling period is constant. The real-time clock in the computer synchronizes all the events of A/D conversion $\to$ computation $\to$ D/A conversion.

The overall system in Fig. 13.2 is hybrid in nature; the signals are in a *sampled form* (*discrete-time signals/digital signals*) in the computer and in continuous-time form in the plant. Such systems have traditionally been called sampled-data control systems; we will use this term as a synonym to *computer control systems/digital control systems*.

13.4 PRINCIPLES OF SIGNAL CONVERSION

Figure 13.3a shows an analog signal $y(t)$—it is defined at a continuum of times, and its amplitudes assume a continuous range of values. Such a signal cannot be stored in digital computers; the signal, therefore, must be converted to a form that will be accepted by digital computers. One very common way to do this is to record sample values of this signal at equally spaced instants. If we sample the signal every 10 msec, for example, we obtain the *discrete-time signal* sketched in Fig. 13.3b. This *sampling interval* corresponds to a *sampling rate* of 100 samples/sec. The choice of sampling rate is an important one, since it determines how accurately the discrete-time signal can represent the original signal.

Notice that the time axis of the discrete-time signal in Fig. 13.3b is labeled simply 'sample number' and index k has been used to denote this number; k = 0, 1, 2, ... Corresponding to different values of sample number k, the discrete-time signal assumes the same continuous range of values assumed by the analog signal $y(t)$. We can represent the sample values by a sequence of numbers y_s (refer to Fig. 13.3b):

$$y_s = \{1.7, 2.4, 2.8, 1.4, 0.4, ...\}$$

In general,

$$y_s = \{y(k)\}, 0 \le k < \infty$$

where $y(k)$ denotes the kth number in the sequence.

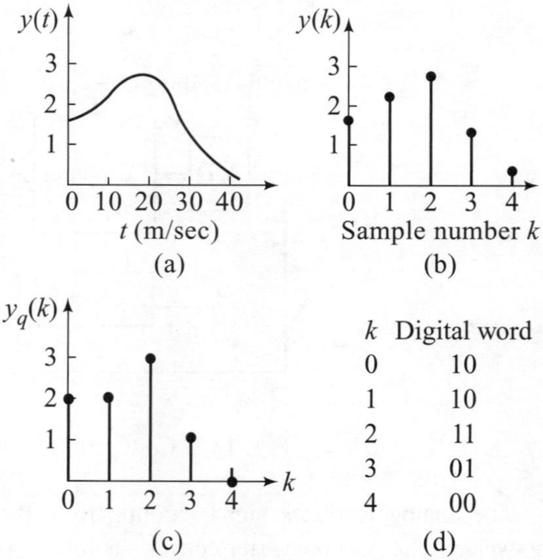

Fig. 13.3 *Sampling, quantization, and coding of an analog signal*

The sequence defined above is a *one-sided sequence*; $y_s = 0$ for $k < 0$. In digital control applications, we normally encounter one-sided sequences.

Although, strictly speaking, $y(k)$ denotes the kth number in the sequence, the notation given above is often unnecessarily cumbersome, and it is convenient and unambiguous to refer to $y(k)$ itself as a sequence.

Throughout our discussion on digital control, we will assume *uniform sampling*, i.e., sample values of the analog signal are extracted at equally spaced sampling instants. If the physical time corresponding to the sampling interval is T seconds, then the kth sample $y(k)$ gives the value of the discrete-time signal at $t = kT$ seconds. We may, therefore, use $y(kT)$ to denote a sequence wherein the independent variable is physical time.

The signal of Fig. 13.3b is defined at discrete instants of time; the sample values are, however, tied to a continuous range of numbers. Such a signal, in principle, can be stored in an infinite-bit machine because a finite-bit machine can store only a finite set of numbers.

A simplified hypothetical two-bit machine can store four numbers given in Table 13.1.

Table 13.1

Binary number	*Decimal equivalent*
00	0
01	1
10	2
11	3

The signal of Fig. 13.3b can therefore be stored in such a machine if the sample values are quantified to four *quantization levels*. Figure 13.3c shows a quantized discrete-time signal for our hypothetical machine. We have assumed that any value in the interval [0.5, 1.5) is rounded to 1, and so forth. The signals for which both time and amplitude are discrete, are called *digital signals*.

After sampling and quantization, the final step required in converting an analog signal to a form acceptable to digital computer is *coding* (or *encoding*). The encoder maps each quantized sample value into a digital word. Figure 13.3d gives a coded digital signal corresponding to the analog signal of Fig. 13.3a for our hypothetical two-bit machine.

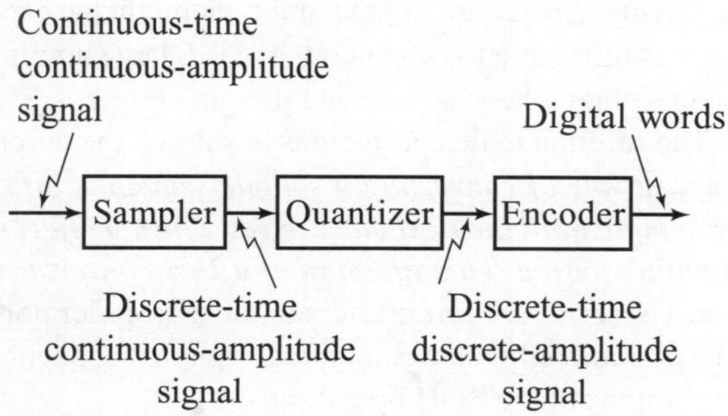

Fig. 13.4 *Operations performed by an A/D converter*

The device that performs the sampling, quantization and coding is an *analog-to-digital* (A/D) *converter*. Figure 13.4 is a block diagram representation of the operations performed by an A/D converter.

It may be noted that the quantized discrete-time signal of Fig. 13.3c and the coded signal of Fig. 13.3d carry exactly the same information. For the purpose of an analytical study of digital systems, we will use the quantized discrete-time form for digital signals.

The number of binary digits carried by a device is its *word length*, and this is obviously an important characteristic related to the *resolution* of the device—the smallest change in the input signal that will produce a change in the output signal. The A/D converter that generates signals of Fig. 13.3 has two binary digits and thus four quantization levels. Any change, therefore, in the input over the interval (0.5, 1.5) produces no change in the output. With three binary digits, 2^3 quantization levels can be obtained, and the resolution of the converter would improve.

The A/D converters in common use have word length of 8 to 16 bits. For an A/D converter with a word length of eight-bits, an input signal can be resolved to one part in 2^8, or 1 in 256. If the input signal has a range of 10 V, the resolution is 10/256, or approximately 0.04 V. Thus, the input signal must change by at least 0.04 V in order to produce a change in the output.

With the availability of converters with resolution ranging from 8 to 16 bits, the quantization errors do not pose a serious threat in computer control of industrial processes. In our treatment of the subject, we assume quantization errors to be zero. This is equivalent to assuming infinite-bit digital devices. Thus, we treat digital signals as if they are discrete-time signals with amplitudes assuming a continuous range of values. In other words, we make no distinction between the words 'discrete-time' and 'digital'.

A typical topology of a single-loop digital control system is shown in Fig. 13.2. It has been assumed that the measuring transducer and the actuator (final control element) are analog devices requiring, respectively, A/D and D/A conversion at the computer input and output. The D/A conversion is a process of producing an analog signal from a digital signal and is, in some sense, the reverse of the sampling process discussed above.

The *digital-to-analog* (D/A) *converter* performs two functions; first, generation of output samples from the binary-form digital signals produced by the machine, and second, conversion of these samples to analog form. Figure 13.5 is a block diagram representation of the operations performed by a D/A converter. The decoder maps each digital word into a sample value of the signal in discrete-time form. It is usually not possible to drive a load, such as a motor, with these samples. In order to deliver sufficient energy, the sample amplitude might have to be so large that it is infeasible to be generated. Also, large amplitude signals might saturate the system being driven.

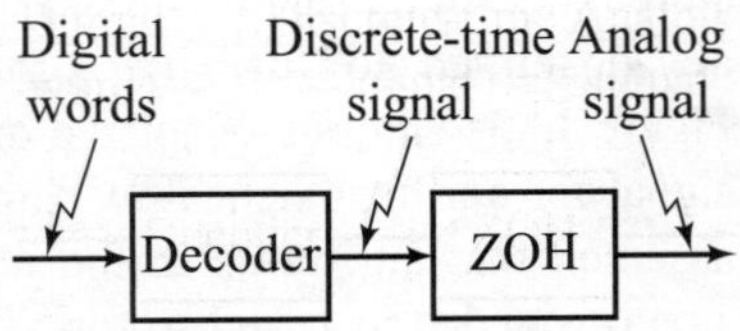

Fig. 13.5 *Operations performed by a D/A converter*

The solution to this problem is to smooth the output samples to produce a signal in analog form. ***The simplest way of converting a sample sequence into a continuous-time signal is to hold the value of the sample until the next one arrives. The net effect is to convert a sample to a pulse of duration T, the sampling period. This function of a D/A converter is referred to as a zero-order hold (ZOH) operation.*** The term *zero order* refers to the zero-order polynomial used to extrapolate between the sampling times. Figure 13.6 shows a typical sampled sequence produced by the decoder, and the analog signal resulting from the ZOH operation.

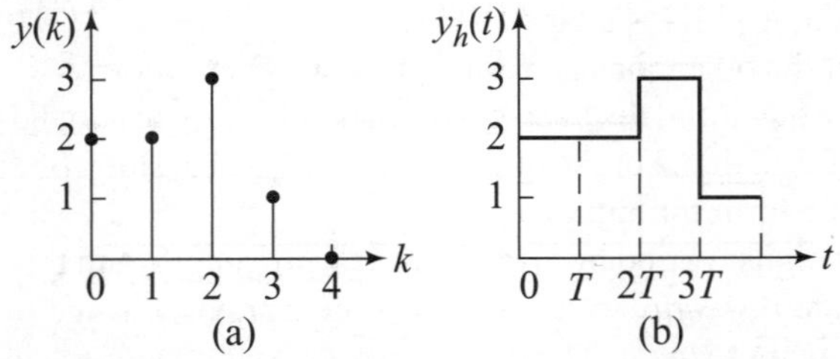

Fig. 13.6 *(a) Sampled sequence; (b) Analog output from ZOH*

Figure 13.7 represents a model of a unity-feedback digital control system. This representation is useful for the purpose of analytical study of digital control systems. ***The sampler which samples continuous-time signal*** $e(t)$ ***every*** T ***secs, is a model of A/D converter, and the ZOH which constructs a continuous-time signal from a discrete-time output of digital control algorithm, is a model of D/A converter.***

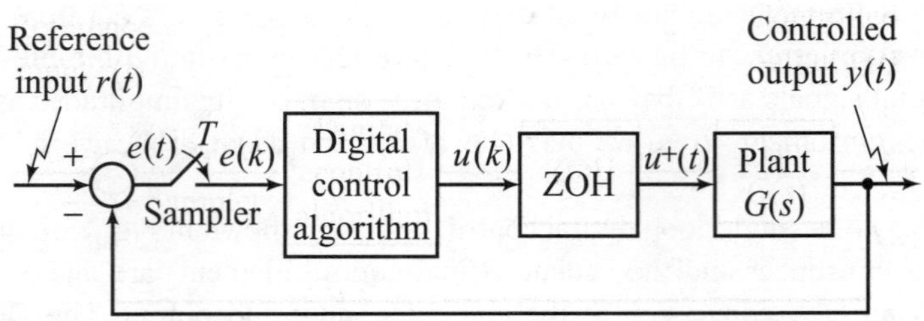

Fig. 13.7 *Analytical model of a unity-feedback digital control system*

13.5 SHAFT-ANGLE ENCODERS FOR DIGITAL MEASUREMENT OF SHAFT POSITION/SPEED

Recall the configuration of the basic digital control scheme given in Fig. 13.2. In most cases, the measuring transducer (sensor) and the actuator (final control element) are analog devices requiring, respectively, A/D and D/A conversion at the computer input and output. There are, of course, exceptions; sensors which combine the functions of the transducer and the A/D converter, and actuators which combine the functions of the D/A converter and the final control element are available. In the control scheme of Fig. 1.14, the control signal from the digital computer is processed by SCR trigger control circuit to control the firing angle of the SCR (D/A conversion and the power conversion by the final control element are embedded in the scheme). In the control scheme of Fig. 1.12, the measurement of shaft position and processing of signals (calculation of control signal) are digital in nature (A/D function and the transducer function are embedded in the scheme).

In this section, we present the functional details of *shaft-angle encoders* for digital measurement of shaft position/speed.

The digital measurement of shaft position requires conversion from the analog quantity 'shaft angle' to a binary number. One way of doing this would be to change shaft angle to a voltage using a potentiometer, and then to convert it to a binary number through an electronic A/D converter. This is perfectly feasible but is not sensible because

(i) high quality potentiometers of good accuracy are expensive and subject to wear; and
(ii) the double conversion is certain to introduce more errors than a single conversion would do.

We can go straight from angle to number using an optical angular *absolute-position encoder*. It consists of a rotary disk made of a transparent material. The disk is divided into a number of equal angular sectors depending on the resolution required. Several tracks, which are transparent in certain sectors but opaque in others, are laid out. Each track represents one digit of a binary number. Detectors on these tracks sense whether the digit is a '1' or a '0'. Figure 13.8 gives an example. Here the disk is divided into eight 45° sectors. To represent eight angles in binary code requires three digits ($2^3 = 8$), hence there are three tracks. Each track has a light source sending a beam on the disk, and on the opposite side a photoelectric sensor receiving this beam. Depending upon the angular sector momentarily facing the sensors, they transmit a bit pattern representing the angular disk position. For example, if the bit pattern is 010, then Sector IV is facing the sensors.

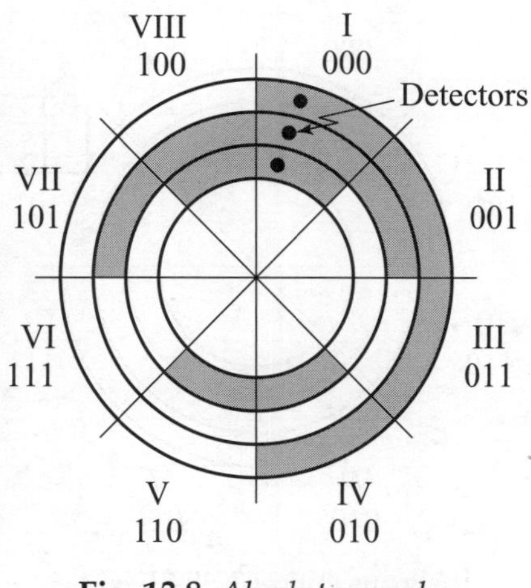

Fig. 13.8 *Absolute encoder*

Figure 13.8 is an example of an 'absolute encoder', so called because for a given angle the digital output must always be the same. Note that a cyclic (Gray) binary code is normally used on absolute encoders (in cyclic codes, only one bit changes between adjacent numbers). If a natural binary-code pattern were used, a transition from, say 001 to 010, would produce a race between the two right-hand bits. Depending on which photosensor responded faster, the output would go briefly through 011 or 000. In either case, a momentary false bit pattern would be sent. Cyclic codes avoid such races. A cyclic code can be converted into natural binary code by either hardware or computer software.

Encoders similar to Fig. 13.8 have been widely used. But they have certain disadvantages:

1. The resolution obtainable with these encoders is limited by the number of tracks on the encoder disk. The alignment of up to ten detectors and the laying out of ten tracks is still quite difficult and thus expensive.
2. The resulting digital measurement is in a cyclic code and must usually be converted to natural binary before use.
3. The large number of tracks and detectors inevitably increases the chance of mechanical and/or electrical failure.

For these reasons, another form of encoder is commonly used today and is known as the *incremental encoder*. The basis of an incremental encoder is a single track served by a single detector and laid out in equal segments of '0' and '1' as in Fig. 13.9. As the track moves relative to the detector, a pulse train is generated, and can be fed to a counter to record how much motion has occurred. With regard to this scheme of measurement, the following questions may be raised.

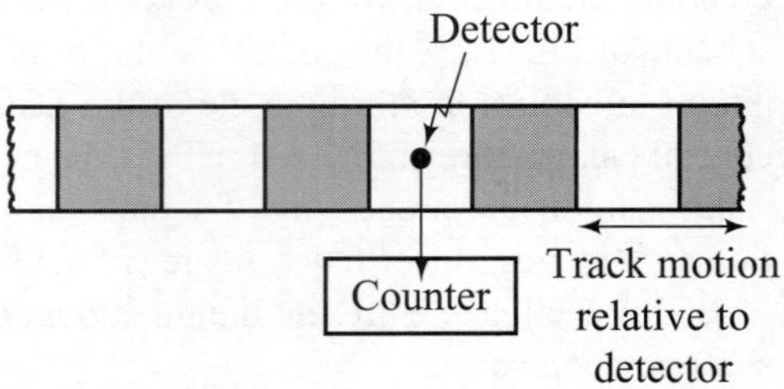

Fig. 13.9 *Incremental encoder*

1. How do we know which direction the motion was?
2. If we can record only the distance moved, how do we know where we were?

The answer to the first question involves the addition of a second detector. Figure 13.10a shows two detectors, spaced one half of a segment apart. As the track moves relative to the detectors (we assume

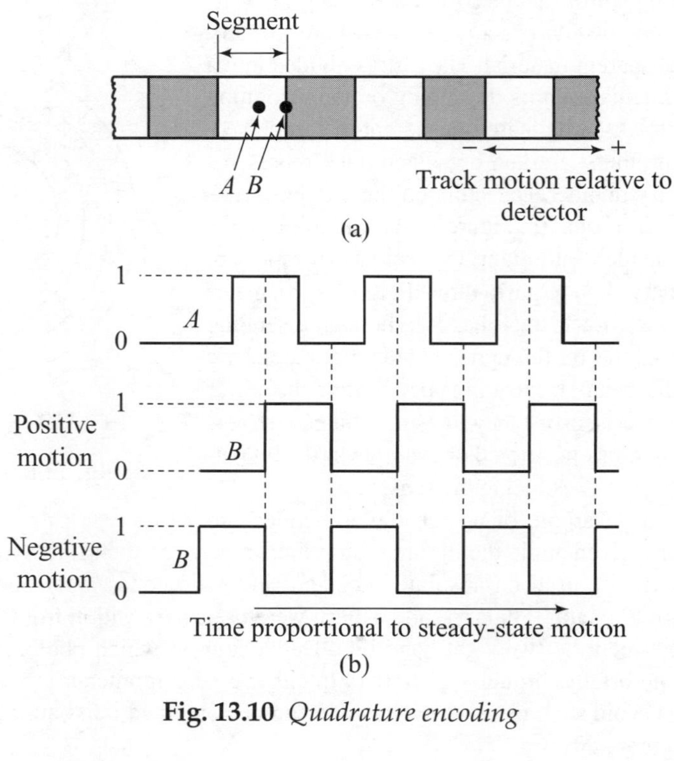

Fig. 13.10 *Quadrature encoding*

at a constant rate), the detector outputs vary with time as shown in the waveforms of Fig. 13.10b. We can see that the relative 'phasing' of the A and B signals depends upon the direction of motion and so gives us a means of detecting the direction.

For example, if signal B goes from '0' to '1' while signal A is at '1', the motion is positive. For the same direction, we see that B goes from '1' to '0' whilst A is at '0'. For negative motion, a similar but different pair of statements can be made. By application of some fairly simple logic, it is possible to control a reversible counter as is indicated in Fig. 13.11.

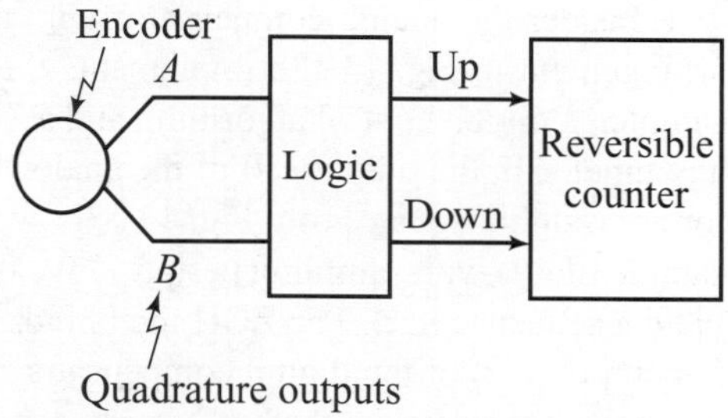

Fig. 13.11 *Direction-detecting circuit for an incremental encoder*

This method of direction sensing is referred to as *quadrature encoding*. The detectors are one half of a segment apart, but reference to the waveforms of Fig. 13.10 shows that there are two segments to one cycle; so the detectors are one-quarter of a cycle apart, and hence the name.

The solution to the second problem also requires an additional detector working on a datum track as shown in Fig. 13.12. The datum resets the counter every time it goes by.

We have thus three detectors in an incremental encoder. But this is still a lot less than on an absolute encoder.

In an analog system, speed is usually measured by a tachogenerator attached to the motor shaft. This is because the time differentiation of analog position signal presents practical problems.

In a digital system, however, it is relatively easy to carry out step-by-step calculation of the 'slope' of the position/time curve. We have the position data in digital form from the shaft encoder, so the rest is fairly straightforward.

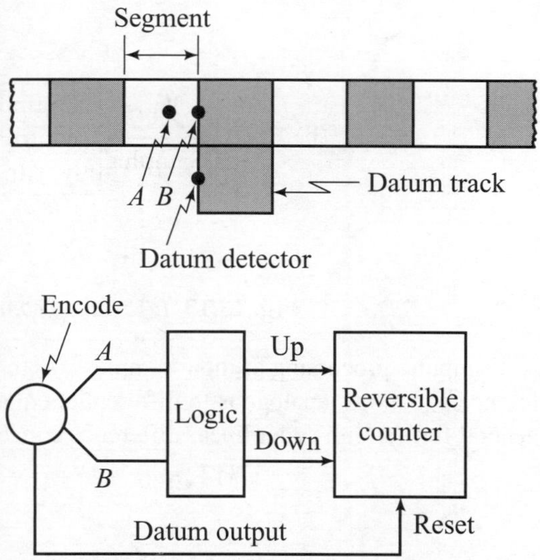

Fig. 13.12 *Datum detector in an incremental encoder*

13.6 DIGITAL IMPLEMENTATION OF ANALOG COMPENSATORS

Plants of control systems are mostly analog systems. However, because digital compensators have many advantages over analog ones, we may be asked to design digital compensators to control analog plants. There are two approaches to carrying out the design. The first approach uses the design methods discussed in the preceding chapters to design an analog compensator and then transforms it into a digital one. The second approach first transforms analog plants into digital plants and then carries out design using digital techniques. We consider in this section the first approach. The rest of the chapter deals with second approach.

Consider the analog compensator with a proper transfer function $D(s)$ shown in Fig. 13.13a. The arrangement in Fig. 13.13b implements the analog compensator digitally. It consists of three parts: a sampler, a digital control algorithm, and a ZOH. The problem is to find a digital algorithm such that for any input $e(t)$, the output $u(t)$ of the analog compensator and the output $u^+(t)$ of the digital compensator are roughly equal. From Fig. 13.13b, we see that the output of the sampler is $e(k) \triangleq e(t = kT)$, the sample of $e(t)$ with sampling period T. We then search for a digital algorithm which operates on $e(k)$ to yield a sequence $u(k)$. The ZOH then holds the value of u constant until the arrival of next data. Thus, the output $u^+(t)$ of the digital compensator is stepwise as shown. The output of analog compensator is generally not stepwise; the best we can achieve is that $u^+(t)$ approximately equals $u(t)$.

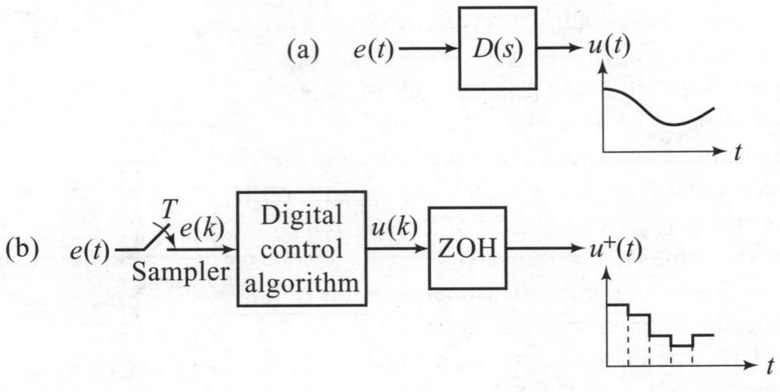

Fig. 13.13 *(a) Analog compensator; (b) Digital compensator*

Computer processing of input samples $e(k)$ to produce output samples $u(k)$ may be described by difference equations, analogous to differential equations that characterize continuous-time systems. The general form of nth-order linear difference equation relating output $u(k)$ to input $e(k)$ is given below.

$$u(k) + a_1u(k-1) + \cdots + a_nu(k-n)$$
$$= b_0e(k) + b_1e(k-1) + \cdots + b_me(k-m) \tag{13.1}$$

The coefficients a_i and b_j are real constants; m and n are integers with $m \le n$.

We will consider the general linear difference equation in the following form:

$$u(k) + a_1u(k-1) + \cdots + a_nu(k-n)$$
$$= b_0e(k) + b_1e(k-1) + \cdots + b_ne(k-n) \tag{13.2}$$

There is no loss of generality in this assumption; the results for $m = n$ can be used for the case of $m < n$ by setting appropriate b_j coefficients to zero.

If the input is assumed to be switched on at $k = 0$ ($e(k) = 0$ for $k < 0$), then the difference equation model (13.2) gives the output at instant '0' in terms of the past values of the output; $u(-1), u(-2), \ldots, u(-n)$, and the present input $e(0)$. Thus the initial conditions of the model (13.2) are $\{u(-1), u(-2), \ldots, u(-n)\}$.

Since the difference equation model (13.2) represents a time-invariant system, the choice of the initial point on the time scale is simply a matter of convenience in analysis. Shifting the origin from $k = 0$ to $k = n$, we get the equivalent difference equation model:

$$u(k+n) + a_1u(k+n-1) + \cdots + a_nu(k)$$
$$= b_0e(k+n) + b_1e(k+n-1) + \cdots + b_0e(k) \tag{13.3}$$

Substituting $k = 0$ in Eqn. (13.3), we observe that the output at instant 'n' is expressed in terms of n values of the past outputs: $u(0), u(1), \ldots, u(n-1)$ and in terms of inputs: $e(0), e(1), \ldots, e(n)$. If k is incremented to take on values $k = 0, 1, 2, \ldots$, *etc*., the $u(k)$; $k = n, n + 1, \ldots$, can easily be generated by the iterative procedure. Given $\{u(-1), u(-2), \ldots, u(-n)\}$, the initial conditions $\{u(0), u(1), \ldots, u(n-1)\}$ of the model (13.3) can be determined by successively substituting $k = -n, -n + 1, \ldots, -2, -1$ in Eqn. (13.3).

Example 13.1 Consider the computer processing of the input samples $e(k) \triangleq e(kT)$ to generate output sequence $u(k) \triangleq u(kT)$ as per the algorithm described by second-order difference equation,

$$u(k + 2) - 3u(k + 1) + 2u(k) = 2e(k + 1) - e(k) \tag{13.4}$$

The input is applied at $k = 0$ ($e(k) = 0$ for $k < 0$). We assume zero initial conditions, i.e., $u(k) = 0$ for $k < 0$.

If we write Eqn. (13.4) as

$$u(k) = 3u(k - 1) - 2u(k - 2) + 2e(k - 1) - e(k - 2)$$

then its response to the given input sequence $e(k)$; $k \geq 0$, can be computed recursively as

$$\begin{aligned} u\,(0) &= 0 \\ u\,(1) &= 3u\,(0) + 2e\,(0) \\ u\,(2) &= 3u\,(1) - 2u\,(0) + 2e\,(1) - e\,(0) \\ u\,(3) &= 3u\,(2) - 2u\,(1) + 2e\,(2) - e\,(1) \\ &\vdots \end{aligned}$$

and so forth.

The discretization problem, therefore, amounts to conversion of the transfer function

$$\frac{U(s)}{E(s)} = D(s)$$

to an equivalent difference equation. There is no exact solution to this problem because $D(s)$ responds to the complete history of $e(t)$, whereas the digital control algorithm has access only to the samples $e(k)$. In a sense, various discretization procedures simply make different assumptions about what happens to $e(k)$ between the sample points. A simple discretization approach is based on numerical integration scheme:

$$\int_{(k-1)T}^{KT} e(t)dt \cong \frac{T}{2}[e(k) + e(k-1)] \tag{13.5}$$

where $e(k) \triangleq e(t = kT)$.

In Eqn. (13.5), the area in question has been approximated by that of a trapezoid formed by the base T and vertices $e(k)$ and $e(k-1)$, as shown in Fig. 13.14. The area can also be approximated by the rectangle of amplitude $e(k-1)$ or $e(k)$ to give

$$\int_{(k-1)T}^{kT} e(t)dt \cong T[e(k-1)] \tag{13.6}$$

or

$$\int_{(k-1)T}^{kT} e(t)dt \cong T[e(k)] \tag{13.7}$$

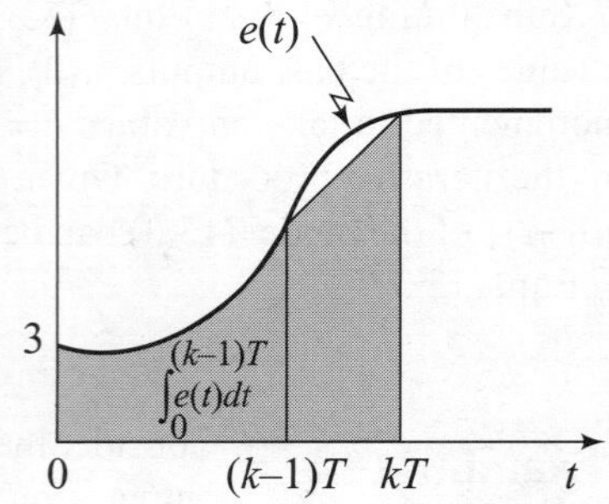

Fig. 13.14 *Trapezoidal rule for integral approximation*

In this chapter, we will employ *trapezoidal rule for integration*, given by Eqn. (13.5). Refer to [155] for the characteristics of rectangular rules for integration, given by Eqns (13.6) and (13.7).

Example 13.2 The simplest formula for the PID or three-mode controller is the addition of the proportional, integral, and derivative modes:

$$u(t) = K_c\left[e(t) + \frac{1}{T_I}\int_0^t e(t)dt + T_D\frac{de(t)}{dt}\right] \tag{13.8}$$

where

u = controller output signal
e = error (controller input) signal
K_c = controller gain
T_I = integral or reset time
T_D = derivative or rate time

For the digital realization of the PID controller, it is necessary to approximate each mode in Eqn. (13.8) using the sampled values of $e(t)$.

The proportional mode requires no approximation since it is purely static part:

$$u_P(k) = K_c e(k) \tag{13.9}$$

The integral mode may be approximated by trapezoidal rule for integration:

$$u_I(k) = \frac{K_c}{T_I}\int_0^{kT} e(t)\,dt$$

$$= \frac{K_c}{T_I}\int_0^{(k-1)T} e(t)\,dt + \frac{K_c}{T_I}\int_{(k-1)T}^{kT} e(t)\,dt$$

$$\cong u_I(k-1) + \left(\frac{K_c}{T_I}\right)\frac{T}{2}[e(k) + e(k-1)] \tag{13.10}$$

The derivative-mode term in Eqn. (13.8) is

$$u_D(t) = K_c T_D \frac{de(t)}{dt}$$

Therefore,

$$K_c T_D\, e(t) = \int_0^t u_D(t)\, dt$$

In the derivative-mode term, the values of u and e are reversed with respect to integral-mode term, and the consistent approximation can be written down at once from Eqn. (13.10):

$$K_c T_D e(k) \cong K_c T_D e(k-1) + \frac{T}{2}[u_D(k) + u_D(k-1)]$$

or

$$u_D(k) = (K_c T_D)\frac{2}{T}[e(k) - e(k-1)] - u_D(k-1) \tag{13.11}$$

Bringing all the three modes together results in the following PID algorithm:

$$u(k) = u_P(k) + u_I(k) + u_D(k) \tag{13.12}$$

13.6.1 Sampling Rate Selection

The selection of the best sampling rate for a digital control system is the result of a compromise among many factors. The basic motivation to lower the sampling rate is cost. A decrease in sampling rate means more time is available for the control calculations; hence slower computers can be used for a given control function or more control capability can be achieved from a given computer. Either way, the cost per function is lowered. For systems with A/D converters, less demand on conversion speed will also lower cost. These economic arguments indicate that the best engineering choice is the slowest possible sampling rate that still meets all performance specifications.

With excessive reduction in sampling rate, a number of potentially degrading effects start to become significant. For a particular application, one or more of these degrading effects set the lower limit for the sampling rate. The process dynamics, the type of algorithm and control requirements, and the characteristics of input and noise signals, all interact to set the maximum usable value for sampling period T.

How to find an adequate sampling interval, has been widely discussed in the literature. Some commonly used empirical rules are reported below.

Empirical Rules for the Selection of Sampling Rate

Practical experience and simulation results have produced a number of useful approximate rules for the specification of minimum sampling rate/maximum sampling interval.

1. The recommendations in Table 13.2 for the most common process variables follow from the experience of process industries:
2. Fast-acting electromechanical systems require much shorter sampling intervals, perhaps down to a few milliseconds.

Table 13.2

Type of variable	*Sampling time (sec)*
Flow	1–3
Level	5–10
Pressure	1–5
Temperature	10–20

3. A rule of thumb says that select a sampling period that is much shorter than any of the time constants in the continuous-time plant to be controlled digitally. Sampling interval equal to one-tenth of the smallest time constant or the inverse of the largest real pole (or real part of complex pole) has been recommended.
4. For complex poles with imaginary part ω_d, the frequency of transient oscillations corresponding to the poles is ω_d. A convenient rule suggests sampling at the rate of 6 to 10 times per cycle. Thus, if the largest imaginary part in the poles of the continuous-time plant is 1 rad/sec, which corresponds to transient oscillations with a frequency of 1/6.28 cycles per second, $T = 1$ sec may be satisfactory.
5. Rules of thumb based on open-loop plant model are full of danger under the conditions where high closed-loop performance is forced from a plant with a low open-loop performance. The rational choice of the sampling rate should be based on an understanding of its influence on the closed-loop performance of the control system. It seems reasonable that the highest frequency of interest should be closely related to the 3 dB-bandwidth of the closed-loop system. The selection of sampling rates can then be based on the bandwidth of the closed-loop system. ***Reasonable sampling rates are* 10 *to* 30 *times the bandwidth,* ω_b.**
6. Another rule of thumb based on closed-loop performance is to select sampling interval T equal to or less than one-tenth of the desired settling time.

Example 13.3 Consider the analog control system shown in Fig. 13.15. The specifications for the system are that the damping ratio ζ of the pair of dominant closed-loop poles be 0.7, and the natural frequency ω_n be 0.3 rad/sec. The specifications can be met with

$$D(s) = \frac{0.81(s+0.2)}{s+2}$$

as can be verified by the root locus plot.

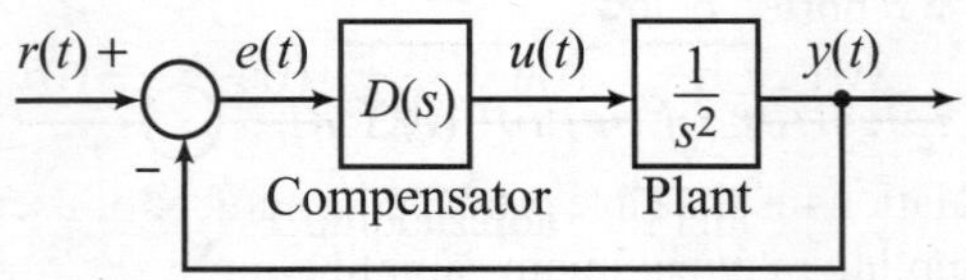

Fig. 13.15 *An analog control system*

To digitize this $D(s)$, we first need to select a sampling rate. For a system with $\omega_n = 0.3$ rad/sec, a very 'safe' sampling rate would be a factor of 20 faster than ω_n, yielding $\omega_s = 0.3 \times 20 = 6$ rad/sec, or 1 Hz.

Thus, let us pick $T = 1$ sec. Using the trapezoidal rule for integration, a digital control algorithm is obtained as follows:

$$D(s) = \frac{0.81(s+0.2)}{s+2} = \frac{U(s)}{E(s)} \tag{13.13}$$

gives $$\dot{u}(t) + 2u(t) = 0.81\,\dot{e}(t) + 0.162\,e(t)$$

or $$\dot{u}(t) = -2u(t) + 0.81\,\dot{e}(t) + 0.162e(t)$$

Therefore, $$u(t) = u(0) - 2\int_0^t u(\tau)d\tau + 0.81e(t) - 0.81e(0) + 0.162\int_0^t e(\tau)\,d\tau$$

Using the trapezoidal rule for integration, the following recursive algorithm follows from this equation.

$$u(k) = u(k-1) - 2\left[\frac{u(k)+u(k-1)}{2}\right]T + 0.81e(k) - 0.81e(k-1) + 0.162\left[\frac{e(k)+e(k-1)}{2}\right]T$$

or $$u(k) = 0.4455e(k) - 0.3645e(k-1) \tag{13.14}$$

where $e(k) = r(k) - y(k)$

This completes the digital algorithm design. The complete digital control system is shown in Fig. 13.16.

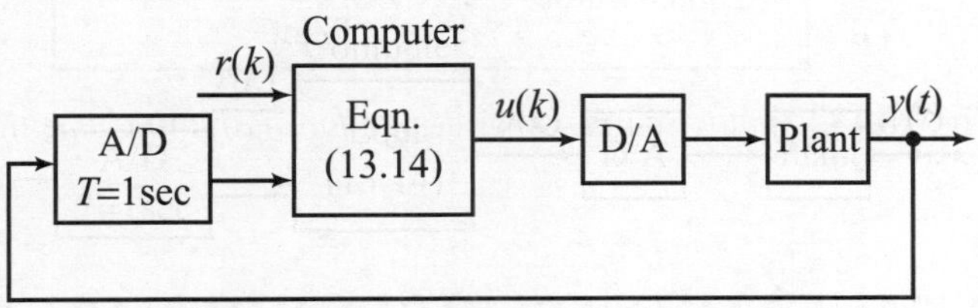

Fig. 13.16 *A digital control system*

13.6.2 Applicability Limits of Digital Implementation of Analog Compensators

When a continuous-time signal is sampled at regular intervals of time and is then reconstructed by holding the sampled values constant for each sampling interval, the reconstructed signal is effectively delayed by approximately one half of the sampling interval, as shown in Fig. 13.17.

In digital control configuration of Fig. 13.7, the ZOH holds the output of the digital controller constant between updates, thus introducing a delay of $T/2$ sec into the loop. This dead time introduces a phase lag, and thus reduces the stability margin in a closed-loop system. If an exact discrete-time analysis or a simulation of a digital control system obtained by discretization of analog design were performed, and the discretization determined for a wide range of sampling rates, the expected results are: ***the system would be unstable for sampling rates lower than approximately*** $5 \times \omega_b$***, and the damping would be substantially degraded for sampling rates lower than*** $10 \times \omega_b$***. At sampling rates of the order of*** $20 \times \omega_b$***, this design method will yield reasonable results, and can be used with confidence for sampling rates of*** $30 \times \omega_b$ ***or higher.***

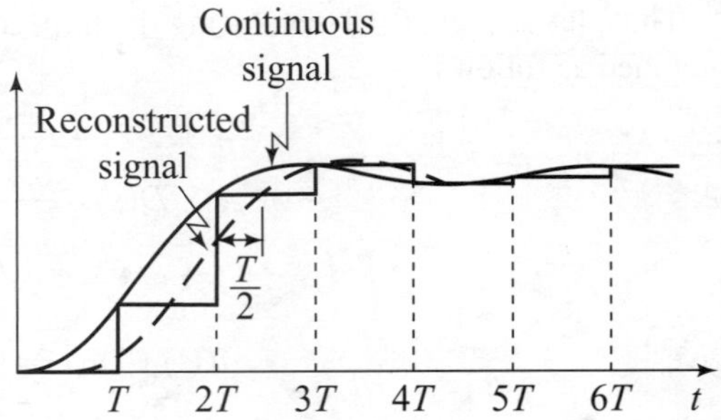

Fig. 13.17 *Time delay introduced by sampling and reconstruction*

A simple way to refine the design is to account for the lagging effect of the ZOH in the analog design itself. We introduce the transfer function,

$$G_h(s) = e^{-sT/2} \tag{13.15}$$

in analog control loop (refer to Fig. 13.18) in anticipation of the conversion from an analog controller to a digital controller. Note that $G_h(s)$ is a pure time delay with dead time of $T/2$ and a dc gain of unity. Considering $G_h(s)G(s)$ as the plant model, we design the analog compensator $D(s)$. Discretization of this $D(s)$ will allow for higher values of sampling interval.

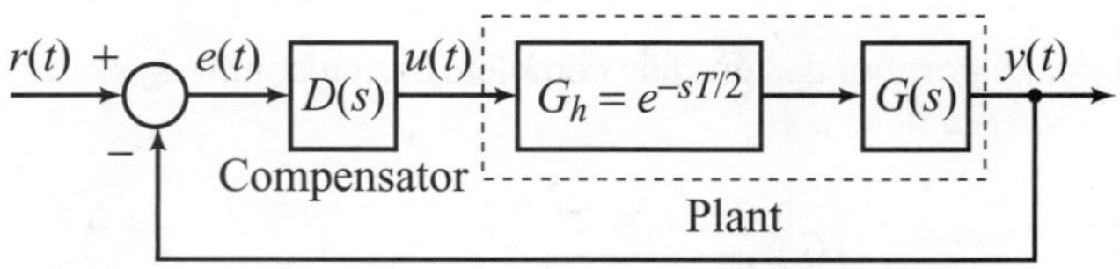

Fig. 13.18 *Analog compensator design for digital implementation*

13.7 FORMULATION OF DIRECT DIGITAL CONTROL DESIGN PROBLEM

There are two basic routes to the design of digital controllers for continuous-time processes. The route: *continuous-time modeling → continuous-time control design → discrete-time approximation of the controller*, was considered in earlier section of this chapter. This approach is based on the fact that a digital system with a high sampling rate approximates to an analog system. However, the use of high sampling rates wastes computer power, and can lead to problems of arithmetic precision, etc. One is, therefore, driven to find methods of design which take account of the sampling process. The rest of the chapter is devoted to the following route, which takes sampling process into account: *continuous-time modeling → discrete-time approximation of the model → discrete-time control design.*

An actual design process is often a combination of the two methods. First iteration to a digital design may be obtained using discretization of an analog design. Then the result is tuned up using direct digital analysis and design.

Consider an analytical model of a unity-feedback digital control system shown in Fig. 13.7. This representation is useful for direct digital control design. The sampler which samples continuous-time signals $e(t)$ every T secs to generate the sequence $e(k)$, is a model of A/D converter, and the zero-order hold(ZOH) which constructs a continuous-time signal $u^+(t)$ from a discrete-time output $u(k)$ of the digital computer (digital control algorithm), is a model of D/A converter. Direct digital control design begins with sufficiently accurate mathematical models of sampling and hold operations.

13.7.1 A Mathematical Model of Sampling Operation

We assume that the sampler in Fig. 13.7 is an ideal switch that closes and reopens instantaneously. Of course, no switch opens and closes instantaneously, but the modern A/D converter comes very close. A/D converters with a conversion rate of 100 kHz are fairly inexpensive, and conversion rates in the megahertz range are available (In control applications, the sampling rate is usually less than 100 Hz). If the conversion rate is 100 kHz, the conversion is completed in less than one one-thousand of the typical sample period. Thus, the conversion rate of A/D converter is close enough to instantaneous.

To establish a relationship of the sequence $e(k)$ to the continuous-time function $e(t)$ from which this sequence is derived, we take the following approach:

We treat each sample of the sequence $e(k)$ as an impulse function of strength equal to the value of the sample (Impulse function $A\delta(t - t_0)$ is an impulse of strength A occurring at $t = t_0$). The idea is to give a mathematical description to periodic samples of a continuous-time function in such a way that we can analyze the samples and the function simultaneously using the same tool (Laplace transform). The sequence $e(k)$ can be viewed as a train of impulses, represented by the continuous-time function $e^*(t)$:

$$e^*(t) = e(0)\delta(t) + e(1)\delta(t - T) + e(2)\delta(t - 2T) + \cdots$$

$$= \sum_{k=0}^{\infty} e(k)\,\delta(t - kT) \tag{13.16}$$

Typical signals $e(t)$, $e(k)$ and $e^*(t)$ are shown in Fig. 13.19. The sampler of Fig. 13.7 can thus be viewed as an 'impulse modulator' with the carrier signal

$$\delta_T(t) = \sum_{k=0}^{\infty} \delta(t - kT) \tag{13.17}$$

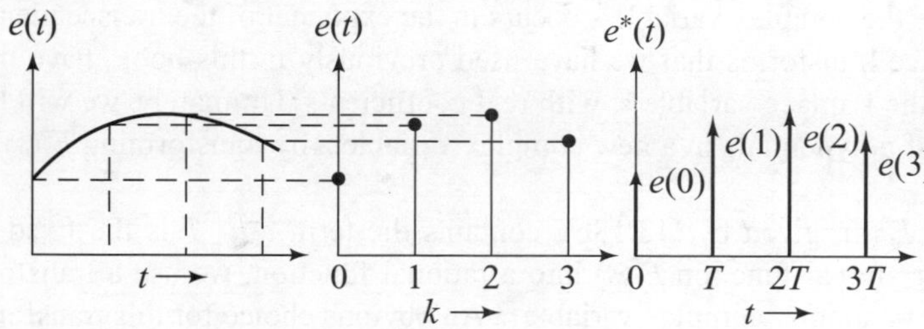

Fig. 13.19 *Conversion of a continuous-time signal to a sequence*

and modulating signal $e(t)$. The modulation process is schematically represented in Fig. 13.20a, and the *impulse train* $\delta_T(t)$ in Fig. 13.20b:

$$e^*(t) = e(t)\delta_T(t) \tag{13.18a}$$

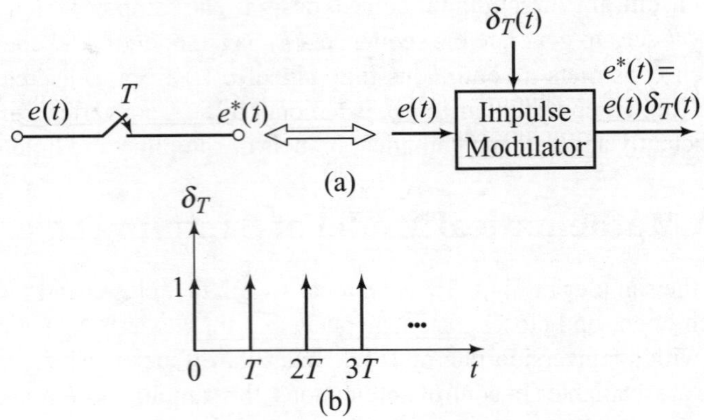

Fig. 13.20 *(a) Representation of sampling as impulse modulator; (b) Representation of impulse train*

We will eliminate the delta function by simply taking the Laplace transform of $e^*(t)$ to obtain (refer to Eqn. (2.2))

$$\mathscr{L}[e^*(t)] \triangleq E^*(s)$$

$$= \int_0^\infty e(t)\delta_T(t)e^{-st}\,dt$$

$$= \sum_{k=0}^\infty \int_0^\infty e(t)\delta(t-kT)e^{-st}\,dt$$

$$= \sum_{k=0}^\infty e(kT)e^{-skT} \tag{13.18b}$$

We see that by applying Laplace transform to $e^*(t)$, we have sifted out the values $e(0)$, $e(T)$, $e(2T)$,..., $e(kT)$, ... These are simply the values of $e(t)$ at the output of the A/D converter.

The expression (13.18b) for $E^*(s)$ represents a Laplace transform, but it is not a transform that is easy to use because the complex variable s occurs in the exponent of the transcendental function e. By contrast, the Laplace transforms that we have used previously in this book, have mostly been ratios of polynomials in the Laplace variable s, with real coefficients. Ultimately, we will be able to achieve these same ratios of polynomials in a new complex variable z by transforming $E^*(s)$ to reach what we call the z-plane.

The expression $E^*(s)$, given by (13.18b), contains the term e^{-Ts}; T is the fixed sampling period. To transform the irrational function $E^*(s)$ into a rational function, we use a transformation from the complex variable s to another complex variable z. An obvious choice for this transformation is

$$z = e^{Ts} \tag{13.19a}$$

although $z = e^{-Ts}$ would be just as acceptable.
Solving for s in Eqn. (13.19a), we obtain

$$s = \frac{1}{T} \ln z \tag{13.19b}$$

The relationship between s and z in Eqns (13.19) may be defined as the *z-transformation*.
When Eqns (13.19) are substituted in Eqn. (13.18b), we have

$$E^*(s = \frac{1}{T} \ln z) = E(z) = \sum_{k=0}^{\infty} e(kT)z^{-k} = \sum_{k=0}^{\infty} e(k)z^{-k} \tag{13.20}$$

The mapping (13.20) is often used as the definition of *z-transform*.

13.7.2 A Mathematical Model of Zero-Order Hold

We have replaced the D/A converter with a zero-order hold (ZOH) in the analytical model of Fig. 13.7. A ZOH clamps the output signal to a value equal to that of the input signal at the sampling instant. If a discrete-time signal $u(k)$ is applied at $t = kT$, the ZOH immediately puts out a value of $u(k)$ and holds it for T units of time. No physical device behaves exactly like this, but a modern D/A converter can convert a digital input to an analog signal in roughly 10^{-5} sec. This is nearly the ideal behavior (instantaneous response).

In an actual operation, the digital output $e(k)$ of the A/D converter (refer to Fig. 13.7) is processed by the digital control algorithm, and the computer outputs digital signal $u(k)$ which is converted to an analog signal $u^+(t)$ by the D/A converter (refer to Fig. 13.13). For mathematical modeling, we have assumed that the sampler (A/D converter) generates impulse signals $e^*(t)$ (refer to Fig. 13.19). To continue with this school of thought, we may assume that the computer algorithm has a transfer function $D^*(s)$, and has been designed to process impulse signals $e^*(t)$ and outputs impulse signals $u^*(t)$, which are then converted to $u^+(t)$ by the ZOH as shown in Fig.13.21.

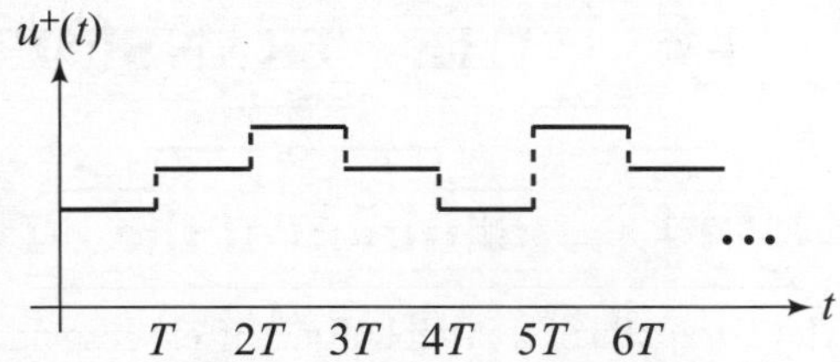

Fig. 13.21 *Output of a zero-order hold*

We can, therefore, write the ZOH output as (step function $\mu(t - t_0)$ is a step of strength unity occurring at $t = t_0$)

$$u^+(t) = \sum_{k=0}^{\infty} u(kT)\left[\mu(t - kT) - \mu(t - \overline{k+1})T\right] \tag{13.21a}$$

Taking the Laplace transform, we have (refer to Eqn. (2.43))

$$U^+(s) = \sum_{k=0}^{\infty} u(kT)\left[\frac{e^{-skT}}{s} - \frac{e^{-s(k+1)T}}{s}\right]$$

$$= \left[\frac{1-e^{-sT}}{s}\right]\left[\sum_{k=0}^{\infty} u(kT)e^{-skT}\right] \tag{13.21b}$$

The second factor in Eqn. (13.21b) is the Laplace transform of the input signal $u^*(t)$ (refer to Eqn. (13.18b)). We can, therefore, write Eqn. (13.21b) as

$$U^+(s) = G_{h0}(s)U^*(s)$$

where

$$G_{h0}(s) = \frac{1-e^{-sT}}{s} \tag{13.22}$$

is the transfer function of zero-order hold.

It should be remembered that $U^*(s)$ ***is not present in the physical system but appears in the mathematical model as a result of Eqn. (13.22). Also the sampler with model given by Eqns (13.18) is not present in the physical system. However, the combination shown in Fig. 13.22, does accurately model the sampling and reconstruction operation in a physical system.***

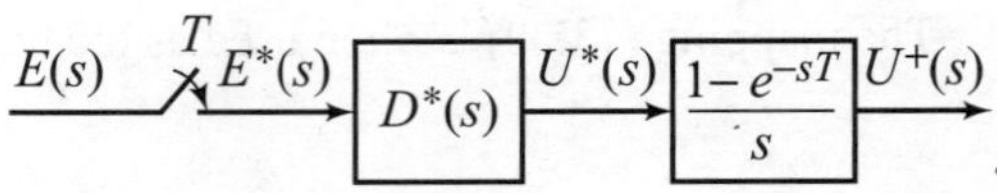

Fig. 13.22 *Mathematical representation of sampler and zero-order hold*

If we make the substitution $s = \ln z/T$, we get the mathematical representation of sampled-data system on the z-plane. This transformation enables us to preserve, in large part, the frequency-domain techniques developed in the previous chapters. The design approaches that follow in this chapter use z-transform as the background tool. In Chapter 15, we will solve the identical problems using state-space formulation.

13.8 THE z-TRANSFORM

13.8.1 Definition of the z-Transform

The z-transform is a technique of expressing a sequence $x(k)$ as a sum of functions of the form z^{-k}, where z is a complex variable. Recall the mapping (13.20) introduced in the previous section. The expression,

$$X(z) = \sum_{k=0}^{\infty} x(k)z^{-k}$$

is used as the definition of the z-transform of sequence $x(k)$, denoted symbolically as $\mathscr{Z}[x(k)]$:

$$X(z) \triangleq \mathscr{Z}[x(k)] = \sum_{k=0}^{\infty} x(k)z^{-k};\ x(k) = 0 \text{ for } k < 0 \tag{13.23}$$

where the variable z is a complex variable:

$$z = a + jb = r\,e^{j\Omega} \tag{13.24}$$

For a fixed value of r, the locus of z is a circle of radius r in the complex z-plane. Circle of radius unity in the complex z-plane will be of specific interest to us. This circle is called a *unit circle* (Fig. 13.23).

The summation in Eqn. (13.23) does not converge for all functions; and when it does, it does so for restricted values of z in the z-plane. Therefore, for each sequence for which the summation converges, there is an associated *convergence region* in the complex z-plane (Examples of z-plane convergence regions will shortly follow).

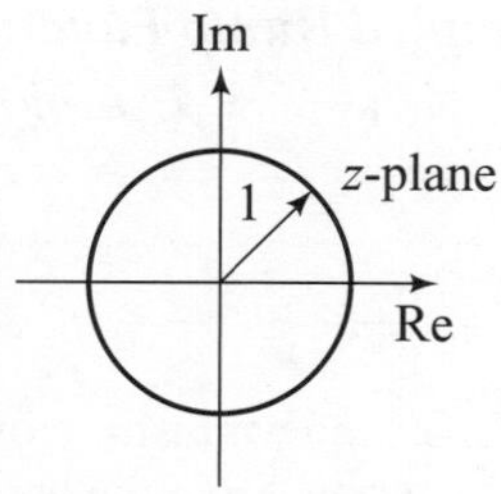

Fig. 13.23 *Unit circle is z-plane*

13.8.2 z-Transforms of Functions Useful for Control System Analysis

In this section, our goal is to find the z-transform of the functions we will subsequently need for our analysis of control systems.

Unit Sample Sequence

The unit sample sequence (*discrete-time impulse*) contains only one nonzero element and is defined by

$$\delta(k) = \begin{cases} 1 \text{ for } k = 0 \\ 0 \text{ otherwise} \end{cases} \tag{13.25a}$$

The z-transform of this elementary signal is

$$\mathscr{Z}[\delta(k)] = \sum_{k=0}^{\infty} \delta(k) z^{-k} = z^0 = 1; |z| > 0 \tag{13.25b}$$

Unit Step Sequence

The unit step sequence is defined as

$$\mu(k) = \begin{cases} 1 \text{ for } k \geq 0 \\ 0 \text{ otherwise} \end{cases} \tag{13.26a}$$

The z-transform is

$$\mathscr{Z}[\mu(k)] = \sum_{k=0}^{\infty} \mu(k) z^{-k} = \sum_{k=0}^{\infty} z^{-k}$$

Using the geometric series formula

$$\sum_{k=0}^{\infty} x^k = \frac{1}{1-x}; |x| < 1,$$

this becomes

$$\mathscr{Z}[\mu(k)] = \frac{1}{1-z^{-1}} = \frac{z}{z-1} \tag{13.26b}$$

Note that this equation holds only if the infinite sum converges, that is if $|z^{-1}| < 1$ or $|z| > 1$. Thus, the region of convergence is the area outside the unit circle in the z-plane.

Sampled Ramp Function

Discrete form of $y(t) = t\mu(t)$ is

$$y(k) = \begin{cases} kT \text{ for } k \geq 0 \\ 0 \text{ otherwise} \end{cases} \tag{13.27}$$

Given the z-transform of $x(k)$, we can obtain z-transform of $\{kT\,x(k)\}$ using the following property of z-transforms. From the basic definition (13.23),

$$\frac{dX(z)}{dz} = \sum_{k=0}^{\infty} \frac{d}{dz}[x(k)z^{-k}] = \sum_{k=0}^{\infty}(-k)\,x(k)z^{-k-1}$$

Consequently,

$$-z\frac{dX(z)}{dz} = \sum_{k=0}^{\infty} k\; x(k)z^{-k}$$

or

$$\mathscr{Z}[kT\; x(k)] = -Tz\frac{dX(z)}{dz} \tag{13.28}$$

For $x(k) = \mu(k)$, we obtain

$$\mathscr{Z}[kT\mu(k)] = \mathscr{Z}[y(k)] = Y(z) = -Tz\frac{d}{dz}\left(\frac{z}{z-1}\right) \tag{13.29}$$

$$= -Tz\left[\frac{-z}{(z-1)^2} + \frac{1}{z-1}\right] = \frac{Tz}{(z-1)^2}; |z| > 1$$

Sampled Exponential Function

Discrete form of $x(t) = e^{-at}$ is

$$x(k) = \begin{cases} e^{-a\,kT} \text{ for } k \geq 0 \\ 0 \text{ otherwise} \end{cases} \tag{13.30a}$$

Then

$$X(z) = \sum_{k=0}^{\infty} e^{-a\,kT}z^{-k} = \sum_{k=0}^{\infty}(e^{aT}z)^{-k} = \sum_{k=0}^{\infty}\alpha^{-k}; \alpha = e^{aT}$$

$$= \frac{1}{1-\alpha^{-1}} = \frac{e^{aT}z}{e^{aT}z-1} = \frac{z}{z-e^{-aT}} \tag{13.30b}$$

This equation holds only if the infinite sum converges, that is if $|e^{-aT}z^{-1}| < 1$ or $|z| > |e^{-aT}|$. This result holds for both real and complex a.

The transform of sampled sinusoid $A\cos(\omega kT + \phi)$ can be found by expanding the sequence into complex exponential components. Given the z-transform of $A\cos(\omega kT + \phi)$, we can obtain z-transform of $Ae^{-a\,kT}\cos(\omega kT + \phi)$ using the following property of z-transforms:

$$\mathscr{Z}[e^{-a\,kT}x(k)] = \sum_{k=0}^{\infty} x(k)e^{-a\,kT}z^{-k} = \sum_{k=0}^{\infty} x(k)(z\,e^{aT})^{-k} = X(ze^{aT}) \tag{13.31}$$

We summarize the z-transforms we have derived to this point plus some additional transforms in Table 13.3. The table lists commonly encountered functions $x(t)$; $t \geq 0$ and z-transforms of sampled version of these functions, given by $x(kT)$. We also include in this table the Laplace transforms $X(s)$ corresponding to the selected functions $x(t)$. Whenever a function in s is given, one approach for finding the corresponding z-transform is to convert $X(s)$ into $x(t)$ and then find z-transform of $x(kT)$. Another approach is to expand $X(s)$ into partial fractions and directly use a z-transform table to find the z-transforms of the expanded terms. Table 13.3 will be helpful for this second approach.

All the transforms listed in the table can easily be derived from first principles. It may be noted that in this transform table, regions of convergence have not been specified. In our applications of systems analysis, which involve transformation from time-domain to z-domain and inverse transformation, the variable z acts as a dummy operator. If transform pairs for sequences of interest to us are available, we are not concerned with the region of convergence.

Table 13.3 ***Abbreviated table of z-transforms***

$X(s)$	$x(t);\ t \geq 0$	$x(k);\ k = 0, 1, 2 \ldots$	$X(z)$
–	–	$\delta(k)$	1
$\frac{1}{s}$	$\mu(t)$	$\mu(k)$	$\frac{z}{z-1}$
$\frac{1}{s+a}$	e^{-at}	e^{-akT}	$\frac{z}{z-e^{-aT}}$
$\frac{1}{s^2}$	t	kT	$\frac{Tz}{(z-1)^2}$
$\frac{1}{(s+a)^2}$	$t\,e^{-at}$	kTe^{-akT}	$\frac{Te^{-aT}z}{(z-e^{-aT})^2}$
$\frac{a}{s(s+a)}$	$1-e^{-at}$	$1-e^{-akT}$	$\frac{(1-e^{-aT})z}{(z-1)(z-e^{-aT})}$
$\frac{\omega}{s^2+\omega^2}$	$\sin \omega t$	$\sin \omega kT$	$\frac{(\sin \omega T)z}{z^2-(2\cos \omega T)z+1}$
$\frac{s}{s^2+\omega^2}$	$\cos \omega t$	$\cos \omega kT$	$\frac{z^2-(\cos \omega T)z}{z^2-(2\cos \omega T)z+1}$
$\frac{\omega}{(s+a)^2+\omega^2}$	$e^{-at}\sin \omega t$	$e^{-akT}\sin \omega kT$	$\frac{(e^{-aT}\sin \omega T)z}{z^2-(2e^{-aT}\cos \omega T)z+e^{-2aT}}$
$\frac{s+a}{(s+a)^2+\omega^2}$	$e^{-at}\cos \omega t$	$e^{-akT}\cos \omega kT$	$\frac{z^2-(e^{-aT}\sin \omega T)z}{z^2-(2e^{-aT}\cos \omega T)z+e^{-2aT}}$

13.8.3 Useful Operations with z-Transforms

Shifting Theorem: Advance (forward shift)

z-transformation of difference equations written in terms of advanced versions of the input and output variables (refer to Eqn. (13.3)) requires the following results:

$$\mathscr{Z}[y(k+1)] = \sum_{k=0}^{\infty} y(k+1)z^{-k} = z\sum_{k=0}^{\infty} y(k+1)z^{-(k+1)}$$

Letting $k + 1 = m$, yields

$$\mathscr{Z}[\,y(k+1)] = z\sum_{m=1}^{\infty} y(m)z^{-m} = z\left[\sum_{m=0}^{\infty} y(m)z^{-m} - y(0)\right]$$

$$= zY(z) - zy(0) \tag{13.32a}$$

Use of this result in a recursive manner results in the general transform formula:

$$\mathscr{Z}[\,y(k+n)] = z^{n}Y(z) - z^{n}y(0) - z^{n-1}y(1) - \cdots - z^{2}y(n-2) - zy(n-1) \tag{13.32b}$$

Shifting Theorem: Delay (backward shift)

z-transformation of difference equations written in terms of delayed versions of input and output variables (refer to Eqn. (13.2)) requires the following results:

$$\mathscr{Z}[\,y(k-1)] = \sum_{k=0}^{\infty} (k-1)z^{-k} = y(-1)z^{0} + y(0)z^{-1} + y(1)z^{-2} + \cdots$$

$$= z^{-1}\left[\sum_{k=0}^{\infty} y(k)z^{-k}\right] + y(-1) = z^{-1}Y(z) + y(-1) \tag{13.33a}$$

If $y(k) = 0$ for $k < 0$, we have

$$\mathscr{Z}[y(k-n)] = z^{-n}Y(z) \tag{13.33b}$$

Remember that multiplication of the z-transform $X(z)$ by z has the effect of advancing the signal $x(k)$ by one sampling period, and that multiplication of the z-transform $X(z)$ by z^{-1} has the effect of delaying the signal $x(k)$ by one sampling period. In control engineering and signal processing, $X(z)$ is frequently expressed as a ratio of polynomials in z^{-1}; z^{-1} is interpreted as the *unit delay operator.*

13.8.4 z-Transform Inversion

We will obtain the inverse z-transform in exactly the same way as we obtained the inverse Laplace transform (Chapter 2), namely, by partial fraction expansion. The reason that the partial fraction expansion method works is that we frequently encounter transforms that are rational functions, i.e., ratio of two polynomials in z with real coefficients:

$$X(z) = \frac{b_0 z^m + b_1 z^{m-1} + \cdots + b_m}{z^n + a_1 z^{n-1} + \cdots + a_n}; m \le n \tag{13.34a}$$

$$= \frac{K \prod_{i=1}^{m} (z - \beta_i)}{\prod_{j=1}^{n} (z - \alpha_j)} \tag{13.34b}$$

where the β_i and α_j are either real or complex-conjugate pairs.

The z-transform *zeros* are the values of z for which the transform is zero; $z = \beta_i$ are the zeros of $X(z)$. The z-transform *poles* are the values of z for which the transform is infinite; $z = \alpha_j$ are the poles of $X(z)$.

The transform pairs encountered using the partial fraction expansion technique will usually be those found in Table 13.3. Those not found in the table can easily be derived by using basic properties of z-transformation. The partial fraction expansion method for z-transforms is very straightforward and similar in most respects to the partial fraction expansion in Laplace transforms. We illustrate the method with some examples.

Example 13.4 We observe that the transforms of the elementary functions (see Table 13.3) contain a factor of z in the numerator. To ensure that the partial fraction expansion will yield terms corresponding to those tabulated, it is customary to first expand the function $Y(z)/z$ if $Y(z)$ has one or more roots at the origin, and then multiply the resulting expansion by z. For instance, if $Y(z)$ is given:

$$Y(z) = \frac{2z^2 - 1.5z}{z^2 - 1.5z + 0.5} = \frac{z(2z - 1.5)}{(z - 0.5)(z - 1)},$$

we are justified in writing

$$\frac{Y(z)}{z} = \frac{2z - 1.5}{(z - 0.5)(z - 1)} = \frac{A_1}{z - 0.5} + \frac{A_2}{z - 1}$$

Constants A_1 and A_2 can be evaluated by applying the conventional partial fraction expansion rules.

$$A_1 = (z - 0.5)\left[\frac{Y(z)}{z}\right]_{z=0.5} = 1$$

$$A_2 = (z - 1)\left[\frac{Y(z)}{z}\right]_{z=1} = 1$$

$$\frac{Y(z)}{z} = \frac{1}{z - 0.5} + \frac{1}{z - 1} \text{ or } Y(z) = \frac{z}{z - 0.5} + \frac{z}{z - 1}$$

Using the transform pairs from Table 13.3,

$$y(k) = (0.5)^{kT} + 1;\ k \geq 0$$

Example 13.5 When $Y(z)$ does not have one or more zeros at the origin, we expand $Y(z)$, instead of $Y(z)/z$, into partial fractions and utilize shifting theorem given by Eqn. (13.33b) to obtain inverse z-transform.

Let us consider an example:

$$Y(z) = \frac{10}{(z-1)(z-0.2)} = \frac{12.5}{z-1} - \frac{12.5}{z-0.2}$$

Notice that the inverse z-transform of $1/(z-1)$ is not available in Table 13.3. However, using the shifting theorem, we find that

$$\mathscr{Z}^{-1}\left[\frac{1}{z-1}\right] = \mathscr{Z}^{-1}\left[z^{-1}\left(\frac{z}{z-1}\right)\right] = \begin{cases} 1; k = 1, 2, \ldots \\ 0; k \leq 0 \end{cases}$$

Also

$$\mathscr{Z}^{-1}\left[\frac{1}{z-0.2}\right] = \mathscr{Z}^{-1}\left[z^{-1}\left(\frac{z}{z-0.2}\right)\right] = \begin{cases} (0.2)^{k-1};\ k=1,2,\ldots \\ 0 \qquad ;\ k \leq 0 \end{cases}$$

Therefore,

$$y(k) = \mathscr{Z}^{-1}[Y(z)] = \begin{cases} 12.5[1-(0.2)^{k-1}];\ k = 1, 2, \ldots \\ 0 \qquad ;\ k \leq 0 \end{cases}$$

which can be rewritten as

$$y(k) = 12.5[1-(0.2)^{k-1}]\mu(k-1)$$

13.8.5 Final Value Theorem

The final value theorem is concerned with the evaluation of $y(k)$ as $k \to \infty$ assuming, of course, that $y(k)$ does approach a limit. Using partial fraction expansion for inverting z-transforms, it is a simple matter to show that $y(k)$ ***approaches a limit as*** $k \to \infty$ ***if all the poles of*** $Y(z)$ ***lie inside the unit circle*** ($|z| < 1$) ***in the complex z-plane. The unit circle boundary is, however, excluded except for a single pole at*** $z = 1$. This is because purely sinusoidal signals whose transforms will have poles on the unit circle, do not settle to a constant value as $k \to \infty$. Multiple poles at $z = 1$ are also excluded because, as we have already seen in the table of z-transforms, these correspond to unbounded signals like ramps. A more compact way of phrasing these conditions is to say that $(z-1)Y(z)$ ***must be analytic on the boundary and outside the unit circle in the complex z-plane***. The final value theorem states that when this condition on $(z-1)Y(z)$ is satisfied, then

$$\lim_{k\to\infty} y(k) = \lim_{z\to 1}(z-1)\,Y(z) \tag{13.35}$$

The proof is as follows:

$$\mathscr{Z}[y(k+1)-y(k)] = \lim_{m\to\infty}\sum_{k=0}^{m}[y(k+1)-y(k)]z^{-k}$$

or

$$zY(z) - zy(0) - Y(z) = \lim_{m\to\infty}\sum_{k=0}^{m}[y(k+1)-y(k)]z^{-k}$$

Letting $z \to 1$ on both sides,

$$\lim_{z\to 1}[(z-1)Y(z)] = y(0) + \lim_{z\to 1}\lim_{m\to\infty}\sum_{k=0}^{m}[y(k+1-y(k)]z^{-k}$$

Interchanging the order of limits on the right-hand side, we have

$$\lim_{z\to 1}[(z-1)\,Y(z)] = y(0) + \lim_{m\to\infty}\sum_{k=0}^{m}[y(k+1)-y(k)] = y(\infty)$$

13.9 CLOSED-LOOP SAMPLED-DATA SYSTEMS: TRANSFER FUNCTION MODELS AND DYNAMIC RESPONSE

The very first step in the analytical study of a system is to set up mathematical equations to describe the system. Because of different analytical methods used, or because of different questions asked, we may often set up different mathematical models to describe the same system. In this chapter, our focus is on transfer function models; state variable representation will be given in later chapters.

13.9.1 Transfer Function Models of Digital Control Algorithms

Computer processing of input samples $e(k)$ to produce output samples $u(k)$ (refer to Fig. 13.7) is generally described by difference equations of the form (13.2) or (13.3). Given the model of the form (13.3), we can shift the origin from $k = n$ to $k = 0$ to obtain a model of the form (13.2), reproduced below.

$$u(k) + a_1u(k-1) + \cdots + a_nu(k-n) = b_0e(k) + b_1e(k-1) + \cdots + b_ne(k-n)$$

We assume that the input is switched on at $k = 0$ ($e(k) = 0$ for $k < 0$) and the initial conditions $u(-1)$, $u(-2)$,..., $u(-n)$ are all zero. The z-transform operation on Eqn. (13.2) gives (using the shifting theorem given by Eqn. (13.33b))

$$U(z) + a_1z^{-1}U(z) + \cdots + a_nz^{-n}U(z) = b_0E(z) + b_1z^{-1}E(z) + \cdots + b_nz^{-n}\,E(z)$$

or

$$(1 + a_1z^{-1} + \cdots + a_nz^{-n})\,U(z) = (b_0 + b_1z^{-1} + \cdots + b_nz^{-n})E(z)$$

Therefore,

$$D(z) = \frac{U(z)}{E(z)} = \frac{b_0 + b_1z^{-1} + \cdots + b_nz^{-n}}{1 + a_1z^{-1} + \cdots + a_nz^{-n}} \tag{13.36a}$$

This transfer function can be equivalently expressed as

$$D(z) = \frac{U(z)}{E(z)} = \frac{b_0 z^n + b_1 z^{n-1} + \cdots + b_n}{z^n + a_1 z^{n-1} + \cdots + a_n} \tag{13.36b}$$

13.9.2 Transfer Function Model of Plant

Consider the sampled-data system of Fig. 13.7. The plant is a continuous-time system with transfer function $G(s)$, and the ZOH operation has been approximated by a transfer function model $G_{h0}(s) = ((1 - e^{-sT})/s)$. The two are in cascade and are not separated by a sampler; we can treat the two together as a single system with transfer function $G_{h0}(s)G(s)$.

We therefore consider $G_{h0}G(s) \triangleq G_{h0}(s)G(s)$ as the continuous-time system with input $u(k)$ and the output $y(t)$ as shown in Fig. 13.24a. Note that the input is a discrete-time signal and the output is a continuous-time signal. If, however, we are satisfied with finding the output at the sampling instants and not in between, the representation of the sampled-data system of Fig. 13.24a can be greatly simplified. Our assumption is visually described in Fig. 13.24b where the output is conceptually sampled in synchronization with the input by a *mathematical sampler T(M)*. The block represented by dotted lines has now discrete-time signals both at the input and the output and therefore a z-domain transfer function model can be derived.

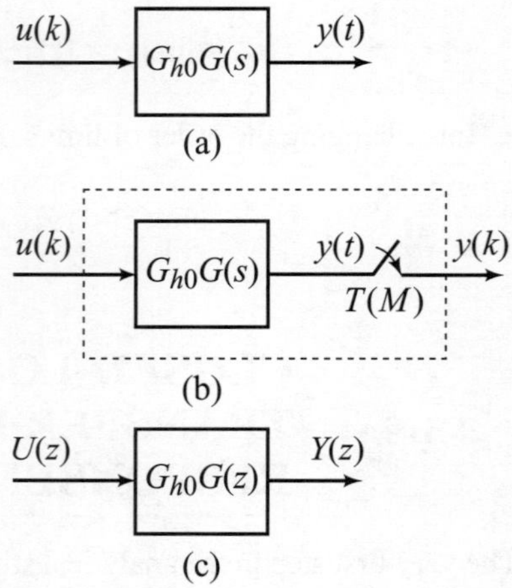

Fig. 13.24 *Sampled-data systems*

$$G_{h0}G(s) = \left(\frac{1 - e^{-sT}}{s}\right) G(s) = (1 - e^{-sT})\frac{G(s)}{s}$$

Let $$\mathscr{L}^{-1}\left[\frac{G(s)}{s}\right] = g_1(t), \text{ and } \mathscr{Z}[g_1(kT)] = G_1(z)$$

Then $$\mathscr{L}^{-1}\left[e^{-sT}\frac{G(s)}{s}\right] = g_1(t - T), \text{ and } \mathscr{Z}[g_1(kT - T)] = z^{-1}G_1(z)$$

Therefore,

$$G_{h0}G(z) \triangleq \mathscr{Z}[G_{h0}G(s)] = \mathscr{Z}\left[\frac{G(s)}{s} - e^{-sT}\frac{G(s)}{s}\right] = (1 - z^{-1})\mathscr{Z}\left[\frac{G(s)}{s}\right] \tag{13.37}$$

Expanding $\frac{G(s)}{s}$ into partial fractions, $\mathscr{Z}\left[\frac{G(s)}{s}\right]$ can be found by using the entries in Table 13.3. Figure 13.24c gives the z-domain equivalent of Fig. 13.24b.

Example 13.6 Let $$G(s) = \frac{ae^{-\tau_D s}}{s+a} \tag{13.38}$$

The dead time τ_D can be expressed as

$$\tau_D = NT + \Delta T \tag{13.39a}$$

where N is an integer and $0 < \Delta < 1$. We can find $G_{h0}G(z)$ using this formulation (refer to the companion book [155]). However, given the capabilities of modern microprocessors, we can always adjust the sampling interval slightly so that

$$\tau_D = NT \tag{13.39b}$$

This assumption would simplify the model $G_{h0}G(z)$, and in most cases we make this adjustment in the sampling frequency. Under this assumption (using Table 13.3),

$$G_{h0}G(z) = (1-z^{-1})\mathscr{Z}\left[\frac{G(s)}{s}\right] = (1-z^{-1})\mathscr{Z}\left[\frac{ae^{-NTs}}{s(s+a)}\right]$$

$$= (1-z^{-1})z^{-N}\mathscr{Z}\left[\frac{a}{s(s+a)}\right] = (1-z^{-1})z^{-N}\mathscr{Z}\left[\frac{1}{s} - \frac{1}{s+a}\right]$$

$$= \frac{z-1}{z(z^N)}\left[\frac{z}{z-1} - \frac{z}{z-e^{-aT}}\right] = \frac{1-e^{-aT}}{z^N(z-e^{-aT})} \tag{13.40}$$

13.9.3 Transfer Function Model of Closed-loop Sampled-Data System

Consider now the closed-loop sampled-data system of Fig. 13.7, reproduced in Fig. 13.25a. It has $r(t)$ as its input and $y(t)$ is the output. The system of Fig. 13.25b is functionally equivalent to that of Fig. 13.25a; the manipulation of the block diagram from Fig. 13.25a to Fig. 13.25b helps us to use the relations previously established in this section. The following equations easily follow:

$$U(z) = D(z)E(z) \tag{13.41a}$$

$$Y(z) = G_{h0}G(z)U(z) \tag{13.41b}$$

Since $$e(kT) = r(kT) - y(kT), \tag{13.41c}$$

we have $$E(z) = R(z) - Y(z)$$

Combining Eqns (13.41a), (13.41b), and (13.41c), gives

$$\frac{E(z)}{R(z)} = \frac{1}{1 + D(z)G_{h0}G(z)} \tag{13.42}$$

$$M(z) = \frac{Y(z)}{R(z)} = \frac{D(z)G_{h0}G(z)}{1 + D(z)G_{h0}G(z)} \tag{13.43}$$

Figure 13.25c gives the z-domain equivalent of Fig. 13.25b.

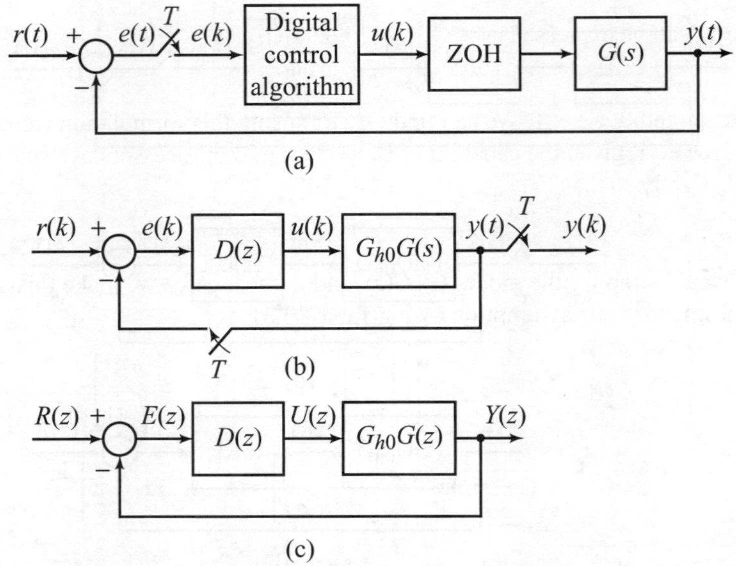

Fig. 13.25 *Closed-loop sampled-data system*

Example 13.7 Consider the sampled-data feedback system of Fig. 13.26a where the sensor dynamics is represented by transfer function $H(s)$. Figure 13.26b shows a functionally equivalent system. The following equations easily follow:

$$E(z) = R(z) - B(z) \tag{13.44a}$$

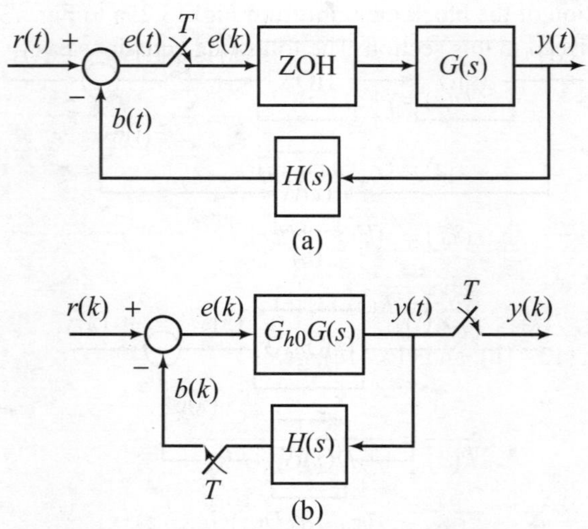

Fig. 13.26 *A non-unity feedback sampled-data system*

$$Y(z) = G_{h0}G(z)E(z) = \mathscr{Z}[G_{h0}(s)G(s)]E(z) \tag{13.44b}$$

$$B(z) = G_{h0}GH(z)E(z) = \mathscr{Z}[G_{h0}(s)G(s)H(s)]E(z) \tag{13.44c}$$

Equations (13.44a), (13.44b), and (13.44c) give

$$\frac{E(z)}{R(z)} = \frac{1}{1+G_{h0}GH(z)} \tag{13.45a}$$

$$M(z) = \frac{Y(z)}{R(z)} = \frac{G_{h0}G(z)}{1+G_{h0}GH(z)} \tag{13.45b}$$

Having become familiar with the technique, from now onwards we may directly write z-domain relationships without manipulating the block diagrams of sampled-data systems.

13.9.4 Dynamic Response

A standard problem in control engineering is to find the response, $y(k)$, of a system given the input, $r(k)$, and a model of the system. With z-transforms, we have a means for easily computing the response of linear time-invariant systems to quite general inputs.

Given a general relaxed, closed-loop sampled-data system and an input signal $r(k)$, the procedure for determining the output $y(k)$ is given by the following steps:

STEP1: Determine the transfer function $M(z)$ of the sampled-data system.
STEP2: Determine the z-transform of the input signal; $R(z) = \mathscr{Z}[r(k)]$.
STEP3: Determine the z-transform of the output; $Y(z) = M(z)R(z)$.
STEP4: Break-up $Y(z)$ by partial-fraction expansion.
STEP5: Invert $Y(z)$ to get $y(k)$: look-up the components of $y(k)$ in a table of transform pairs and combine the components to get the total solution in the desired form.

Example 13.8 A sampled-data system is described by the transfer function,

$$M(z) = \frac{Y(z)}{R(z)} = \frac{1}{z^2 + a_1 z + a_2}; \; a_1 = -\frac{3}{4}, \; a_2 = \frac{1}{8}$$

Find the response $y(k)$ to the input (i) $r(k) = \delta(k)$, (ii) $r(k) = \mu(\text{k})$

Solution

The transfer function $M(z)$ expressed as a ratio of polynomials in z^{-1}:

$$M(z) = \frac{z^{-2}}{1 - \frac{3}{4}z^{-1} + \frac{1}{8}z^{-2}}$$

Since z^{-1} is a unit delay operator, we can immediately write the corresponding difference equation:

$$y(k) - \frac{3}{4}y(k-1) + \frac{1}{8}y(k-2) = r(k-2)$$

The difference equation can be solved easily by means of recursion procedure. The recursion procedure is quite simple and convenient for digital computations. In the following, we obtain the closed-form solutions.

(i) The z-transform of $\delta(k)$ is (refer to Table 13.3): $\mathscr{Z}[\delta(k)] = 1$

Letting $R(z) = 1$, we obtain

$$Y(z) = \frac{1}{z^2 - \frac{3}{4}z + \frac{1}{8}} = \frac{1}{\left[z - \frac{1}{4}\right]\left[z - \frac{1}{2}\right]} = \frac{4}{z - \frac{1}{2}} - \frac{4}{z - \frac{1}{4}}$$

The response $y(k)$ is therefore (refer to Table 13.3)

$$y(k) = \mathscr{Z}^{-1}\left[z^{-1}\left(\frac{4z}{z - \frac{1}{2}}\right)\right] - \mathscr{Z}^{-1}\left[z^{-1}\left(\frac{4z}{z - \frac{1}{4}}\right)\right] = 4\left(\frac{1}{2}\right)^{k-1} - 4\left(\frac{1}{4}\right)^{k-1} \quad ; k \ge 1$$

The discrete-time impulse input, thus, excites the system poles without creating any additional response term.

(ii) The z-transform of $\mu(k)$ is (refer to Table 13.3)

$$\mathscr{Z}[\mu(k)] = \frac{z}{z-1}$$

Letting $R(z) = \frac{z}{z-1}$, we obtain

$$Y(z) = \frac{1}{\left(z - \frac{1}{4}\right)\left(z - \frac{1}{2}\right)}\left[\frac{z}{z-1}\right] = \frac{z}{\left(z - \frac{1}{4}\right)\left(z - \frac{1}{2}\right)(z-1)}$$

$$= \underbrace{\frac{\frac{16}{3}z}{z - \frac{1}{4}} + \frac{-8z}{z - \frac{1}{2}}}_{\text{System poles}} + \underbrace{\frac{\frac{8}{3}z}{z-1}}_{\text{Excitation pole}}$$

The inverse transform operation gives (refer to Table 13.3)

$$y(k) = \left[\underbrace{\frac{16}{3}\left(\frac{1}{4}\right)^k - 8\left(\frac{1}{2}\right)^k}_{\text{Transient response}} + \underbrace{\frac{8}{3}(1)^k}_{\text{Steady-state response}}\right]$$

The *transient response* terms correspond to system poles excited by the input $\mu(k)$. These terms vanish as $k \to \infty$. The second response term arises due to the excitation pole, and has the same nature as the input itself except for a modification in magnitude caused by the system's behavior to the specified input. Since the input exists as $k \to \infty$, the second response term does not vanish and is called the *steady-state response* of the system.

The steady-state response can be quickly obtained without doing the complete inverse transform operation by use of the final value theorem (Eqn. (13.35)):

$$\lim_{k\to\infty} y(k) = \lim_{z\to 1}(z-1)Y(z)$$

if $(z-1)Y(z)$ has no poles on the boundary and outside of the unit circle in the complex z-plane.

13.10 s-PLANE TO z-PLANE MAPPING

13.10.1 The $z = e^{sT}$ Mapping

For continuous-time systems, we saw in earlier chapters that certain behaviors result from different pole locations in the s-plane: oscillatory behavior for poles near the imaginary axis, exponential decay for poles on the negative real axis, and unstable behavior for poles with positive real part. A similar kind of association would also be useful to know when designing sampled-data systems.

The mapping (13.19a):

$$z = e^{sT};\ T = \text{sampling period} \tag{13.46}$$

defines z in terms of s, and is therefore a mapping from the s-plane to the z-plane. The equivalent characteristics in the z-plane are related to those in the s-plane by the mapping (13.46). Consider for example, a continuous-time signal

$$f(t) = e^{-at};\ t > 0$$

which has the Laplace transform

$$F(s) = \frac{1}{s+a}$$

that corresponds to the pole at $s = -a$. The z-transform of $f(kT)$ is (refer to Table 13.3)

$$F(z) = \mathscr{Z}[e^{-akT}] = \frac{z}{z - e^{-aT}}$$

which corresponds to a pole at $z = e^{-aT}$. This means that a pole at $s = -a$ in the s-plane corresponds to a pole at $z = e^{-aT}$ in the z-plane. The relationship $z = e^{sT}$ is, in fact, true in general for the poles of Laplace transform and z-transform functions, as is demonstrated by the entries in Table 13.3.

In Fig. 13.27a, the s-plane has been divided into strips of width ω_s where $\omega_s = 2\pi/T$ is the sampling frequency. The primary strip extends from $\omega = -\frac{\omega_s}{2}\left(= \frac{-\pi}{T}\right)$ to $+\frac{\omega_s}{2}\left(= \frac{\pi}{T}\right)$, and the complementary strips extend from $-\frac{\omega_s}{2}\left(= \frac{-\pi}{T}\right)$ to $-\frac{3\omega_s}{2}\left(= \frac{-3\pi}{T}\right)$, ..., for negative frequencies, and from $\frac{\omega_s}{2}\left(= \frac{\pi}{T}\right)$ to $\frac{3\omega_s}{2}\left(= \frac{3\pi}{T}\right)$, ..., for positive frequencies.

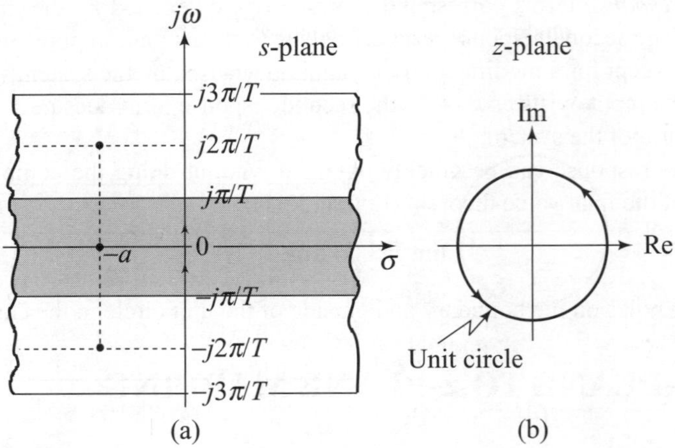

Fig. 13.27 *Mapping of the s-plane to the z-plane using the $z = e^{sT}$ transformation*

It can easily be demonstrated that the primary strip, shown shaded in Fig. 13.27a, maps into the entire z-plane. ***The points in the shaded strip belonging to the left half of the s-plane map into the interior of the unit circle; the points in the strip belonging to jω-axis of the s-plane map onto the boundary of the unit circle; and the points in the strip belonging to the right half of the s-plane map into the exterior of the unit circle in the complex z-plane* (*Fig. 13.27b*).**

An important observation on the mapping $z = e^{sT}$ is that each of the complementary strips maps into the entire z-plane the way the primary strip maps. Consider poles at $s = -a \pm j2n\pi/T$; $n = 1, 2,...$ (For $n = 1$, the poles lie in the complementary strip that extends from π/T to $3\pi/T$, as shown in Fig. 13.27a). The corresponding pole locations in the z-plane are $z = e^{-aT}e^{\pm j2n\pi} = e^{-aT}$, which is the mapping of the pole $s = -a$ in the primary strip. This means that ***with the mapping $z = e^{sT}$, we cannot distinguish the poles in the primary strip from the poles in the secondary strips. Thus, the largest frequency we can distinguish is $\omega = \pi/T$, which is half the sampling frequency $\omega_s = 2\pi/T$.***

While sampling a continuous-time signal to produce the discrete-time sequence, we want to ensure that all the information in the original signal is retained in the samples. There will be no information loss if we can exactly recover the continuous-time signal from the samples. To determine the condition under which there is no information loss, let us consider the continuous-time signal to be band-limited signal with maximum frequency ω_m. From Fig. 13.27 we observe that there is no information loss if

$$\frac{\pi}{T} > \omega_m \text{ or, } \omega_s > 2\omega_m$$

This, in fact, is the *sampling theorem*: Let $x(t)$ be a band-limited signal with $X(j\omega) = 0$ for $|\omega| > \omega_m$. Then $x(t)$ is uniquely determined from its samples $x(kT)$ if the sampling frequency $\omega_s\left(= \frac{2\pi}{T}\right) > 2\omega_m$, i.e., the ***sampling frequency must be at least twice the highest frequency present in the signal.***

13.10.2 Stability on the z-Plane

From Fig. 13.27, the following conclusions can easily be drawn (Revisiting Chapter 5 will be helpful).

A sampled-data control system with closed-loop transfer function $M(z)$ is

(i) ***stable if all the poles of M(z) are inside the unit circle;***

(ii) ***unstable if any pole of M(z) is outside the unit circle and/or there are poles of multiplicity greater than one on the unit circle; and***

(iii) ***marginally stable if the poles on the unit circle are of multiplicity one and all other poles are inside the unit circle.***

The determination of stability hinges upon our ability to determine whether the roots of the denominator of the closed-loop transfer function $M(z)$ are in the stable region of the z-plane. The problem for high-order systems is complicated because of the fact that the denominator of $M(z)$ is in polynomial form, not in factored form. Tabular methods for determining stability for high-order sampled-data systems exist (Jury's stability test is a well-known tabular method). For want of space, these methods are not covered in this introductory chapter on digital control (refer to the companion book [155]).

The approach of stability analysis by tabular methods (such as Jury's criterion) is purely algebraic. The methods lack the ability to give an overall picture of how the closed-loop response would vary with changes in gain. The key to achieving this global picture is to rethink the solution in terms of geometry. The root locus technique is intrinsically geometric as opposed to algebraic approach of the tabular methods. In Chapter 8, we have discussed relative stability analysis of continuous-time systems by the root-locus technique. The analysis was centered around some particular patterns in the s-plane: (i) loci of points in the s-plane with constant damping ratio ζ; and (ii) loci of points in the s-plane with constant undamped natural frequency ω_n. Correlating these patterns of the s-plane with particular features of system behavior on the z-plane will be of great value in relative stability analysis of sampled-data systems using the root locus technique. The correlations based on the mapping $z = e^{sT}$ are described below, and the root locus analysis will appear in the next section.

Complex-conjugate pair of s-plane poles:

$$s_{1,2} = -\zeta\omega_n \pm j\omega_n\sqrt{1-\zeta^2} = -\zeta\omega_n \pm j\omega_d \tag{13.47a}$$

is transformed to

$$z_{1,2} = e^{-\zeta\omega_n T} e^{\pm j\omega_n T\sqrt{1-\zeta^2}} = re^{\pm j\theta} \tag{13.47b}$$

We know from earlier chapters that as ω_n is varied for constant ζ, the poles given by (13.47a) lie along a radial line in the s-plane (Fig.13.28a). In the z-plane, the magnitude of z (i.e., the distance to the origin) is $r = e^{-\zeta\omega_n T}$ and the angles with the positive real axis of the z-plane, measured positive in the counterclockwise direction, are $\theta = \omega_n T\sqrt{1-\zeta^2}$. As ω_n increases for a constant ζ, the magnitude of z decreases and the phase angle increases; constant ζ locus is a logarithmic spiral in the z-plane (Fig.13.28c). This, in fact, is one-to-one map of constant-ζ line in the primary strip.

Consider now the locus of the poles in Eqn. (13.47a) when ζ is varied but ω_d is held constant (Fig. 13.28b). We assume that ω_d is in the primary strip. Note that we are considering constant-ω_d locus instead of constant-ω_n locus. As we shall see, this helps us to extract the information on system behavior using a simple straight line locus. The constant ω_n locus is nonlinear [155].

From Eqns (13.47),

$$z_{1,2} = e^{-\zeta\omega_n T} e^{\pm j\omega_d T} = re^{\pm j\theta}$$

For a constant ω_d, the angle θ is fixed. Therefore, a radial line at an angle $\theta°$ with the real axis is a constant ω_d locus (Fig. 13.28d):

$$\omega_d T = \frac{\theta° \times \pi}{180} \text{ rad} \tag{13.47c}$$

Figure 13.28e illustrates the translation of specifications on ζ and ω_n to the characteristic root locations in the z-plane. The z-plane poles for constant ζ lie along a spiral, and the z-plane poles for constant ω_d (value of ω_d derived from specified values of ζ and ω_n) lie along a straight line. The intersection of the two curves provides the z-plane pole locations for specified values of ζ and ω_n.

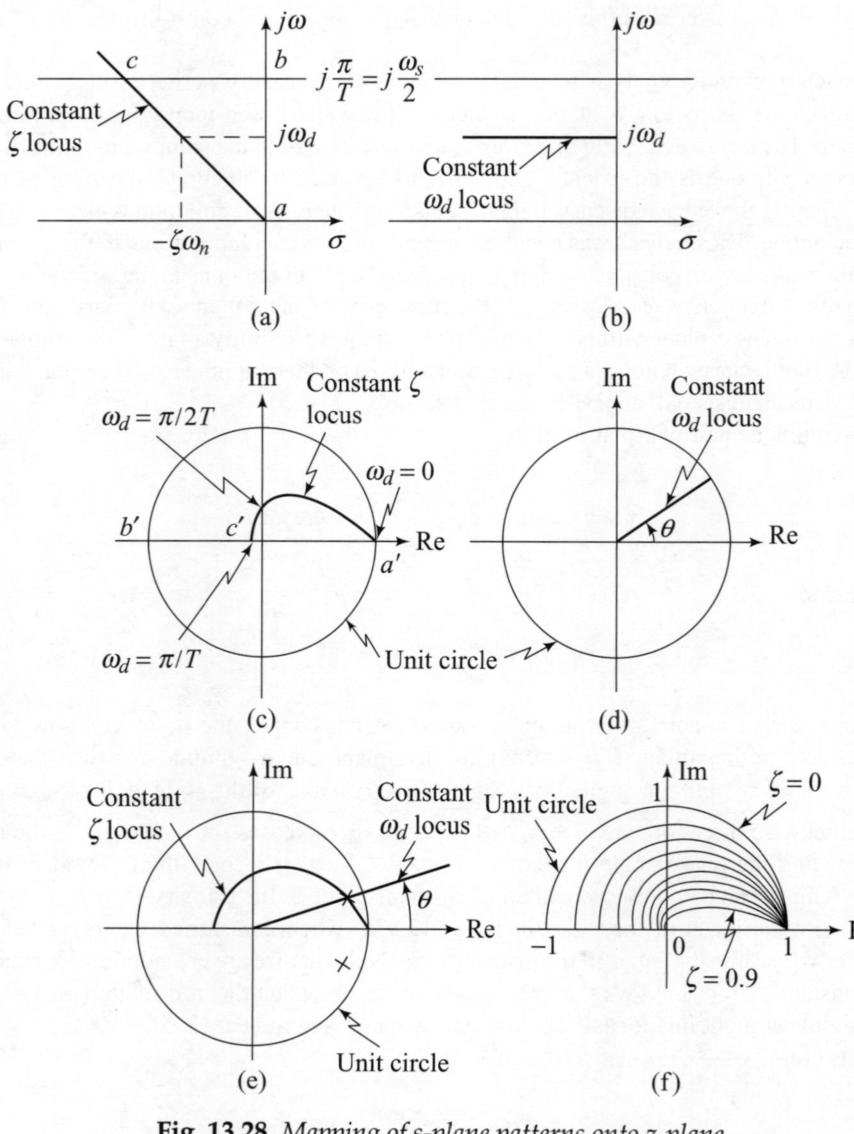

Fig. 13.28 *Mapping of s-plane patterns onto z-plane*

For a given ζ, the constant ζ curve on the z-plane may be constructed using Eqn. (13.47b). Table 13.4 gives the real and imaginary coordinates of points on some constant ζ curves. Figure 13.28f shows a plot of constant ζ curves. Constant ω_d curves are radial lines which are relatively easy to construct for given values of ω_d.

Table 13.4 *Coordinates of points on some constant ζ curves*

$\zeta = 0.3$	Re	0.932	0.735	0.360	0	–0.259	–0.380	–0.373
	Im	0.164	0.424	0.623	0.610	0.448	0.220	0
$\zeta = 0.4$	Re	0.913	0.689	0.317	0	–0.201	–0.276	–0.254
	Im	0.161	0.398	0.549	0.504	0.347	0.160	0
$\zeta = 0.5$	Re	0.891	0.640	0.273	0	–0.149	–0.191	–0.163
	Im	0.157	0.370	0.473	0.404	0.259	0.110	0
$\zeta = 0.6$	Re	0.864	0.585	0.228	0	–0.104	–0.122	–0.095
	Im	0.152	0.338	0.395	0.308	0.180	0.070	0
$\zeta = 0.7$	Re	0.830	0.519	0.179	0	–0.064	–0.067	–0.046
	Im	0.146	0.299	0.310	0.215	0.111	0.039	0
$\zeta = 0.8$	Re	0.780	0.431	0.124	0	–0.031	–0.026	–0.015
	Im	0.138	0.249	0.215	0.123	0.053	0.015	0

13.10.3 The Bilinear Mapping

In Section 13.6, we discussed a numerical integration scheme: trapezoidal rule for integration, for performing the discretization of analog compensators. Armed with an understanding of the z-transform, we now revisit this discretization method.

Assume we are given an analog compensator $D(s)$, shown in Fig. 13.13a. We wish to find a set of difference equations (or $D(z)$) for the digital implementation of that compensation in Fig. 13.13b. We rephrase the problem as one of finding the best $D(z)$ in the digital implementation shown in Fig. 13.29a to match the analog system represented by $D(s)$ in Fig. 13.29b. The technique followed in Section 13.6 is to approach the problem as one of numerical integration. Suppose

$$\frac{U(s)}{E(s)} = D(s) = \frac{1}{s}$$

Therefore,

$$u(kT) = \int_{0}^{kT-T} e(t)\,dt + \int_{kT-T}^{kT} e(t)\,dt$$

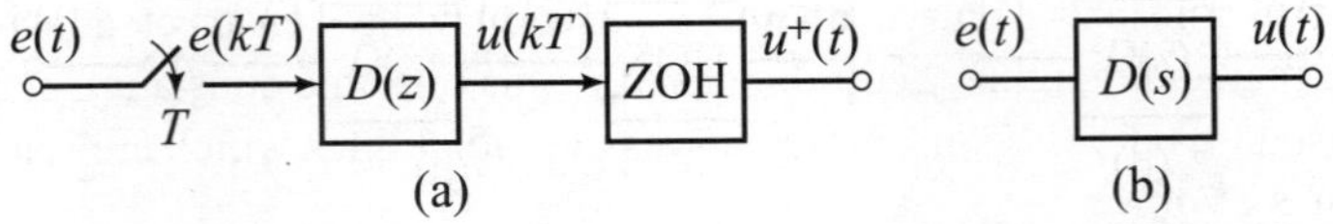

Fig. 13.29 *(a) Digital compensator; (b) Analog compensator*

which can be rewritten as

$$u(kT) = u(kT - T) + (\text{area under the } e(t) \text{ curve over } t = kT - T \text{ to } t = kT) \tag{13.48}$$

where T is the sampling period.

The trapezoidal rule for integration approximates $e(t)$ by a straight line between the two samples (Fig. 13.14). Writing $u(kT)$ as $u(k)$ and $u(kT - T)$ as $u(k - 1)$ for short, we approximate Eqn. (13.48) to

$$u(k) = u(k-1) + \frac{T}{2}[e(k-1) + e(k)] \tag{13.49}$$

z-transformation of this equation gives

$$\frac{U(z)}{E(z)} = \frac{T}{2}\left(\frac{1+z^{-1}}{1-z^{-1}}\right) = \frac{1}{\dfrac{2}{T}\left(\dfrac{1-z^{-1}}{1+z^{-1}}\right)} = D(z)$$

For $D(s) = \dfrac{a}{s+a}$, applying the same integration procedure yields

$$D(z) = \frac{a}{\dfrac{2}{T}\left(\dfrac{1-z^{-1}}{1+z^{-1}}\right) + a}$$

In fact, substituting

$$s = \frac{2}{T}\left(\frac{1-z^{-1}}{1+z^{-1}}\right) \tag{13.50}$$

for every occurrence of s in any $D(s)$ yields a $D(z)$ based on the trapezoidal integration rule. This is called the *bilinear mapping*.

13.11 TRANSFORM DESIGN OF DIGITAL CONTROLS

The transform design of digital compensators is very similar to the design of analog compensators discussed is earlier chapters; in fact, all the same rules apply.

13.11.1 Design Using Root-Locus Plots

A closed-loop sampled-data system with a digital control algorithm used to improve the performance is shown in Fig. 13.25. The closed-loop transfer function is (refer to Eqn. (13.43))

$$\frac{Y(z)}{R(z)} = M(z) = \frac{D(z)G_{h0}G(z)}{1 + D(z)G_{h0}G(z)} \tag{13.51}$$

The transient behavior of the closed-loop system is given by the factors in the denominator of Eqn. (13.51), that is, the roots of the characteristic equation,

$$1 + D(z)G_{h0}G(z) = 0 \tag{13.52}$$

The root-locus technique used for continuous-time systems (Chapter 8) to find roots of a polynomial in s apply equally well and without modification to the polynomial in z; however, the interpretation of the results is quite different as we saw in Fig. 13.28. A major difference is that the stability boundary is now the unit circle instead of the imaginary axis.

We use Table 8.1 with s replaced by z for sketching the root locus plot for sampled-data systems. All the rules in the table apply for sampled-data systems without modification. For stability analysis and design of digital compensator, we superimpose a unit circle on the root locus plot and determine the characteristics of the closed-loop sampled-data systems. Examples that follow will illustrate the procedure.

Transient Accuracy

The specifications on peak overshoot, settling time, etc., of the closed-loop transient response are translated into the parameters ζ and ω_n of the standard second-order transfer function (refer to Chapter 6), the poles of the standard second-order transfer function are mapped onto the z-plane (revisit Section 13.10), and a digital compensator that matches the dominant roots of the characteristic equation (13.52) with the z-plane poles of the standard second-order transfer function is selected.

On the z-plane root locus, a finite zero/additional pole has only a small effect when located in the range 0 to –1. This can be seen from Fig. 13.28. For $0 < \zeta < 1$, the farthest s-plane point in the primary strip is given by the intersection of the ζ line and the boundary of the primary strip (point c in Fig. 13.28a). For $\zeta = 0$, this point moves to $j\omega_s/2$ (point b in Fig. 13.28a), and as $\zeta \rightarrow 1$, the point tends to infinity along the negative real axis. The point b in Fig. 13.28a maps to the point b' in Fig. 13.28c (–1 point on the z-plane); the point c in Fig. 13.28a maps to the point c' in Fig. 13.28c (a point on the negative real axis of the z-plane between –1 and 0); and as $\zeta \rightarrow 1$, the point on the z-plane tends to the origin (Fig.13.28f).

Steady-State Accuracy

The steady-state behavior of the closed-loop sampled-data system of Fig. 13.25 may be determined from the relation (refer to Eqn. (13.42))

$$E(z) = \frac{R(z)}{1 + D(z)G_{h0}G(z)} \tag{13.53}$$

The final value theorem given by Eqn. (13.35) states that when $(z - 1)E(z)$ is analytic,

$$e_{ss}^{*} = \lim_{z \to 1} (1 - z^{-1})E(z) \tag{13.54a}$$

where e_{ss}^{*} is the final sampled value of $e(t)$, i.e., the final value of $e(kT)$.

Using the final value theorem on Eqn. (13.53), we find that the sampled steady-state error for the system of Fig. 13.25 is

$$e_{ss}^{*} = \lim_{z \to 1} (z-1)\frac{R(z)}{1+D(z)G_{h0}G(z)} \tag{13.54b}$$

For a unit step input (from Table 13.3),

$$R(z) = \frac{z}{z-1}$$

Substituting into Eqn. (13.54b), we have

$$e_{ss}^{*} = \frac{1}{1+\lim_{z \to 1} D(z)G_{h0}G(z)} = \frac{1}{1+K_p} \tag{13.55a}$$

where the *position-error constant*

$$K_p = \lim_{z \to 1} [D(z)G_{h0}G(z)] \tag{13.55b}$$

For a unit ramp input,

$$R(z) = \frac{Tz}{(z-1)^2}$$

Following the procedure for the step input, we get

$$e_{ss}^{*} = \frac{1}{K_v} \tag{13.56a}$$

where the *velocity-error constant*

$$K_v = \frac{1}{T}\lim_{z \to 1} [(z-1)D(z)G_{h0}G(z)] \tag{13.56b}$$

For a unit parabolic input,

$$R(z) = \frac{T^2 z(z+1)}{2(z-1)^3}$$

and

$$e_{ss}^{*} = \frac{1}{K_a} \tag{13.57a}$$

where the *acceleration-error constant*

$$K_a = \frac{1}{T^2}\lim_{z \to 1} [(z-1)^2 D(z)G_{h0}G(z)] \tag{13.57b}$$

We can always express the forward path transfer function $D(z)G_{h0}G(z)$ as

$$D(z)G_{h0}G(z) = \frac{K\prod_i (z-z_i)}{(z-1)^N \prod_j (z-p_j)};\ p_j \neq 1,\ z_i \neq 1 \tag{13.58}$$

$D(z)G_{h0}G(z)$ in Eqn. (13.58) involves the term $(z-1)^N$ in the denominator. As $z \to 1$, this term dominates in determining the steady-state error. Digital control systems are, therefore, classified in accordance with the number of poles at $z = 1$ in the forward-path transfer function. From Eqns (13.55)–(13.57) we see that ***a system with $N = 0$ (called a type-0 system) has a finite nonzero position error, infinite velocity and acceleration errors at steady-state. A system with $N = 1$ (called a type-1 system) has zero position error, a finite nonzero velocity error and infinite acceleration error at steady state. A system with $N = 2$ (called a type-2 system) has zero position and velocity errors and a finite nonzero acceleration error at steady state.***

Lead Compensator

A simple lead compensator model in the s-plane is described by a transfer function,

$$D(s) = \frac{1+s\tau}{1+\alpha s\tau};\ \alpha < 1,\ \tau > 0 \tag{13.59a}$$

The pole and zero of $D(s)$ always lie on the real axis in the left half of s-plane; the zero is always to the right of the pole. On the z-plane, the pole and zero of $D(z)$ will always lie on the real axis inside the unit circle; with the zero always to the right of the pole (by s-plane to z-plane mapping). A typical pole-zero configuration of a lead compensator

$$D(z) = K_{c1}\frac{z-\alpha_1}{z-\alpha_2} \tag{13.59b}$$

is shown in Fig. 13.30a.

Lag Compensator

A simple lag compensator model in the s-plane is described by a transfer function,

$$D(s) = \frac{1+s\tau}{1+\beta s\tau};\ \beta > 1,\ \tau > 0 \tag{13.60a}$$

The pole and zero of $D(s)$ always lie on the real axis in the left half of s-plane; the pole is always to the right of the zero. On the z-plane, the pole and the zero of $D(z)$ will always lie on the real axis inside

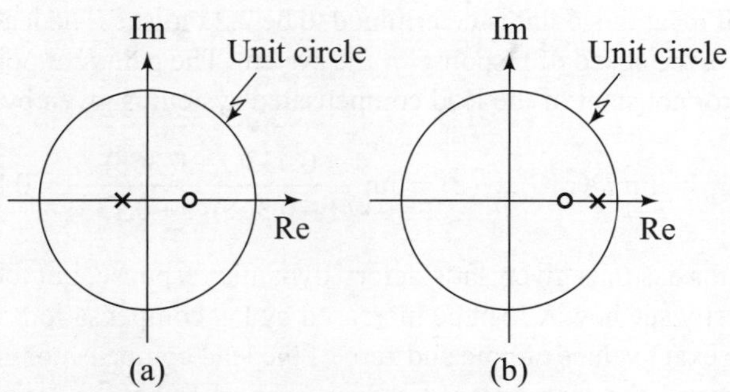

Fig. 13.30 *Pole-zero configurations of compensators*

the unit circle; with pole always to the right of the zero (by s-plane to z-plane mapping). A typical pole-zero configuration of a lag compensator

$$D(z) = K_{c2}\frac{z-\beta_1}{z-\beta_2} \tag{13.60b}$$

is shown in Fig. 13.30b.

Example 13.9 To study the effect of a lead compensator on the root loci, we consider a unity feedback sampled-data system with open-loop transfer function,

$$G_{h0}G(z) = \frac{K(z+0.368)}{(z-0.368)(z-0.135)}; T = 1 \text{ sec} \tag{13.61}$$

The root locus plot of the uncompensated system is shown in Fig. 13.31a. The plot intersects the $\zeta = 0.5$ locus at point P. The value of gain K at this point is determined to be 0.3823.

Constant-ω_d locus passing through point P is a radial line at an angle of 82° with the real axis (Fig. 13.31a). Therefore,

$$\omega_d T = \omega_n T\sqrt{1-\zeta^2} = \frac{82\pi}{180}$$

This gives $\omega_n = 1.65$ rad/sec

Since $G_{h0}G(z)$ given by Eqn. (13.61) is a type-0 system, we will consider position error constant K_p to study steady-state accuracy. For $K = 0.3823$,

$$K_p = \lim_{z\to 1} G_{h0}G(z) = \frac{0.3823(1+0.368)}{(1-0.368)(1-0.135)} = 0.957$$

We now cancel the pole of $G_{h0}G(z)$ at $z = 0.135$ by the zero of the lead compensator and add a pole at $z = -0.135$, i.e., we select

$$D(z) = \frac{z-0.135}{z+0.135}$$

Figure 13.31b shows the root locus plot of lead compensated system. The modified locus has moved to the left, toward the more stable part of the plane. The intersection of the locus with the $\zeta = 0.5$ line is at point Q. The value of ω_n at this point is determined to be 2.2 rad/sec. The lead compensator has, thus, increased ω_n and, hence, the speed of response of the system. The gain K at point Q is determined to be 0.433. The position error constant of the lead compensated system is given by

$$K_p = \lim_{z\to 1} D(z)G_{h0}G(z) = \lim_{z\to 1}\frac{0.433(z+0.368)}{(z-0.368)(z+0.135)} = 0.826$$

The lead compensator has thus given satisfactory dynamic response, but the position-error constant is too low. We will shortly see how K_p can be increased by lag compensation.

The selection of the exact values of pole and zero of the lead compensator is done by experience and by trial and error. In general, the zero is placed in the neighborhood of the desired dominant closed-loop poles, and the pole is located at a reasonable distance to the left of the zero location.

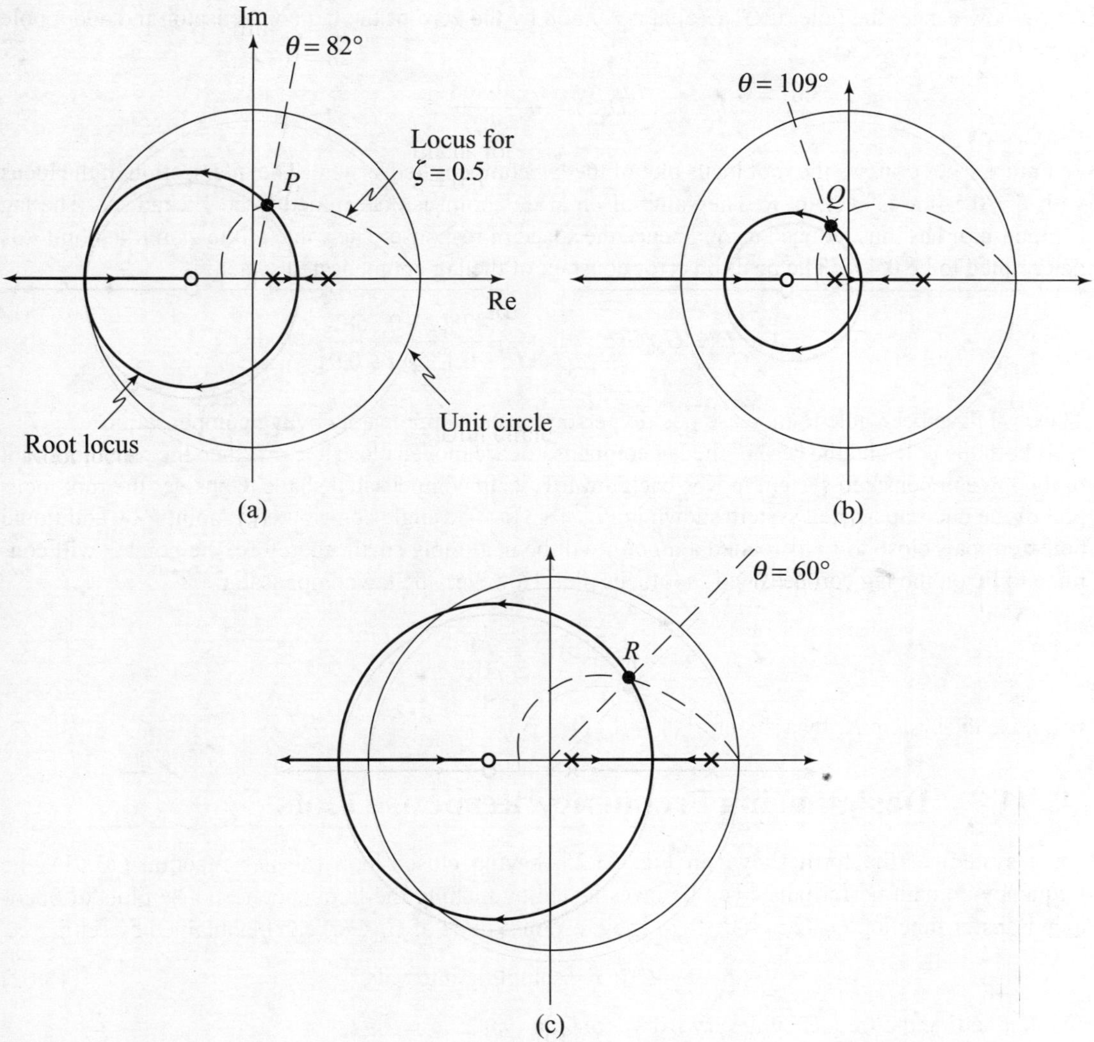

Fig 13.31 *Root locus plot for (a) Uncompensated; (b) Lead compensated; and (c) Lag compensated system*

Example 13.10 To study the effect of lag compensator on the root loci, we reconsider the system described by Eqn. (13.61):

$$G_{h0}G(z) = \frac{K(z+0.368)}{(z-0.368)(z-0.135)}; T = 1 \text{ sec}$$

The root locus plot of the uncompensated system is shown in Fig. 13.31a. At point P, $\zeta = 0.5$, $\omega_n = 1.65$ and $K = 0.3823$ ($K_p = 0.957$).

We now cancel the pole of $G_{h0}G(z)$ at $z = 0.368$ by the zero of the lag compensator and add a pole at $z = 0.9$, i.e., we select

$$D(z) = \frac{z-0.368}{z-0.9}$$

Figure 13.31c shows the root locus plot of the lag compensated system. The intersection of the locus with $\zeta = 0.5$ line is at point R. The value of ω_n at this point is determined to be 1.2 rad/sec. The lag compensator has thus reduced ω_n and hence the speed of response. The value of the gain K at point R is determined to be 0.478. The position error constant of the lag compensated system is

$$K_p = \lim_{z\to 1} D(z)G_{h0}G(z) = \lim_{z\to 1} \frac{0.478(z+0.368)}{(z-0.135)(z-0.9)} = 7.56$$

Thus, we have been able to increase position error constant appreciably by lag compensation.

If both the pole and the zero of the lag compensator are moved close to $z = 1$, then the root locus plot of the lag compensated system moves back towards its uncompensated shape. Consider the root locus plot of the uncompensated system shown in Fig. 13.31a. The angle contributed at point P by additional pole-zero pair close to $z = 1$ (called a *dipole*) will be negligibly small; therefore, the point P will continue to lie on the lag compensated root locus plot. However, the lag compensator

$$D(z) = \frac{z-\beta_1}{z-\beta_2}$$

will raise the system K_p by a factor of $(1 - \beta_1)/(1 - \beta_2)$.

13.11.2 Design using Frequency-Response Plots

For a system of the form shown in Fig. 13.25, having closed-loop transfer function (13.51), the frequency-domain performance can be investigated by making the frequency response plots of open-loop transfer function $G_{h0}G(z)$. The frequency response plots of $G_{h0}G(z)$ can be obtained by setting

$$z = e^{j\omega T};\ T = \text{sampling interval} \tag{13.62}$$

and then letting the frequency ω vary from $-\omega_s/2$ to $+\omega_s/2$; $\omega_s = 2\pi/T$.

Since the frequency appears in the form $e^{j\omega T}$, the discrete-time transfer functions are typically not rational functions and the simplicity of Bode's design techniques (refer to Chapter 12) is altogether lost in the z-plane. Computer assistance is normally required to make the frequency response plots.

The following sequential steps are quite helpful for designing digital compensators:

1. Design an analog compensator $D(s)$ for the given continuous-time plant model using the procedure detailed in Chapter 12.
2. Discretize the analog compensator using bilinear transformation; this will yield $D(z)$. Note that bilinear transformation will result in rational transfer function $D(z)$.
3. Analyze in time domain the sampled-data feedback system composed of discretized plant model and the compensator $D(z)$ obtained from step (2).
4. Obtain frequency response plots of discrete-time open-loop transfer function and fine-tune the digital compensator of step (2) using frequency-domain design preocedures.

Refer to the companion book [155] for more details.

13.12 REVIEW EXAMPLES

Review Example 13.1 Consider the sampled-data system shown in Fig. 13.32a. From the block diagram we obtain (refer to Eqn. (13.43))

$$\frac{Y(z)}{R(z)} = \frac{G_{h0}G(z)}{1+G_{h0}G(z)} \tag{13.63}$$

Figure 13.32b gives the z-domain equivalent of Fig. 13.32a. The forward path transfer function,

$$G_{h0}G(z) = \mathscr{Z}[G_{h0}(s)G(s)]$$

$$= (1-z^{-1})\mathscr{Z}\left[\frac{G(s)}{s}\right] = (1-z^{-1})\mathscr{Z}\left[\frac{1}{s^2(s+1)}\right]$$

$$= (1-z^{-1})\mathscr{Z}\left[\frac{1}{s^2} - \frac{1}{s} + \frac{1}{s+1}\right] = (1-z^{-1})\left[\frac{Tz}{(z-1)^2} - \frac{z}{z-1} + \frac{z}{z-e^{-T}}\right]$$

$$= \frac{z(T-1+e^{-T}) + (1-e^{-T} - Te^{-T})}{(z-1)(z-e^{-T})}$$

When $T = 1$, we have

$$G_{h0}G(z) = \frac{ze^{-1} + 1 - 2e^{-1}}{(z-1)(z-e^{-1})} = \frac{0.3679z + 0.2642}{z^2 - 1.3679z + 0.3679}$$

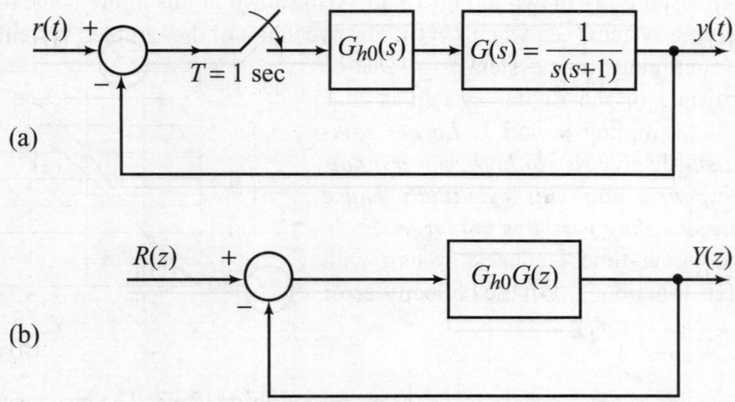

Fig. 13.32 *A closed-loop sampled-data system*

Substituting in Eqn. (13.63), we obtain

$$\frac{Y(z)}{R(z)} = \frac{0.3679z + 0.2642}{z^2 - z + 0.6321}$$

For a unit-step input,

$$R(z) = \frac{z}{z-1}$$

and, therefore,

$$Y(z) = \frac{z(0.3679z + 0.2642)}{(z-1)(z^2 - z + 0.6321)} = \frac{0.3679z^2 + 0.2642z}{z^3 - 2z^2 + 1.6321z - 0.6321}$$

$$= \frac{0.3679z^{-1} + 0.2642z^{-2}}{1 - 2z^{-1} + 1.6321z^{-2} - 0.6321z^{-3}}$$

By long-division process, we get

$$Y(z) = 0.3679z^{-1} + z^{-2} + 1.3996z^{-3} + 1.3996z^{-4} + 1.1469z^{-5} + 0.8944z^{-6} + 0.8015z^{-7} + \cdots$$

Therefore, the sequence ($k = 1, 2, ...$)

$$y(kT) = \{0.3679, 1, 1.3996, 1.3996, 1.1469, 0.8944, 0.8015, ...\}$$

Note that the final value of $y(kT)$ is (refer to Eqn. (13.35)),

$$\lim_{k\to\infty} y(kT) = \lim_{z\to 1} (z-1)Y(z) = \frac{0.3679 + 0.2642}{0.6321} = 1$$

The unit-step response is shown in Fig. 13.33. Also shown in this figure is the unit-step response of the continuous-time system (i.e., when $T = 0$). The overshoot of the sampled system is 45%, in contrast to 17% for the continuous-time system.

The performance of the digital system is thus dependent on the sampling period T. ***Larger sampling periods usually give rise to higher overshoots in the step response, and may eventually cause instability if the sampling period is too large.***

For the continuous-time feedback system with process transfer function $G(s)$, the velocity-error constant

$$K_v = \lim_{s\to 0} sG(s) = \lim_{s\to 0} s\left[\frac{1}{s(s+1)}\right] = 1$$

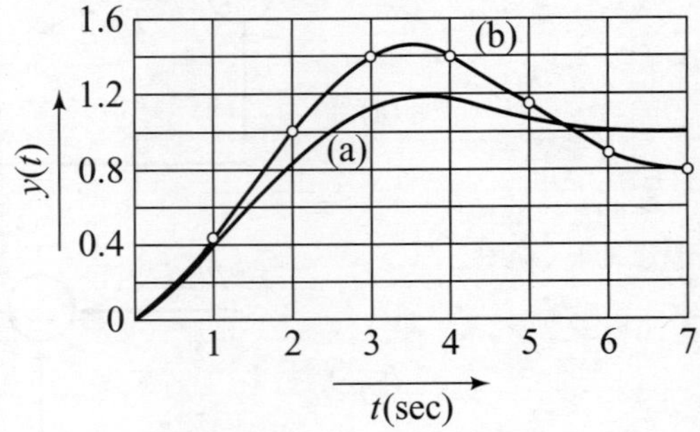

Fig. 13.33 *The response of a second-order system: (a) analog; (b) sampled*

When the sampler and the ZOH are introduced in the feedback loop (Fig. 13.32a), the velocity-error constant becomes

$$K_v = \frac{1}{T}\lim_{z\to1}[(z-1)G_{h0}G(z)]$$

$$= \frac{1}{T}\lim_{z\to1}\left\{(z-1)\left[\frac{z(T-1+e^{-T})+(1-e^{-T}-Te^{-T})}{(z-1)(z-e^{-T})}\right]\right\}$$

$$=1$$

Thus for a type-1 system, the system with sampler and zero-order hold has exactly the same steady-state error as the continuous-time system with the same process transfer function. This, in fact, is true for type-0 and type-2 systems also.

Review Example 13.2 Consider the block diagram of a digital control system shown in Fig. 13.34; $r(t)$ stands for reference input and $w(t)$ stands for disturbance.

For $r = 0$, the block diagram reduces to the one shown in Fig. 13.35. From this figure, it immediately follows that (note that functional equivalence will be lost if we introduce a sampler to discretize disturbance input)

$$Y(s) = G(s)W(s) - G_{h0}(s)G(s)U^*(s)$$
$$\triangleq WG(s) - G_{h0}G(s)U^*(s)$$

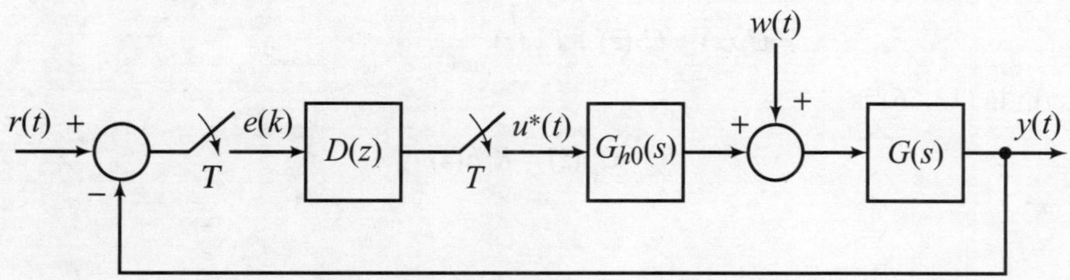

Fig. 13.34

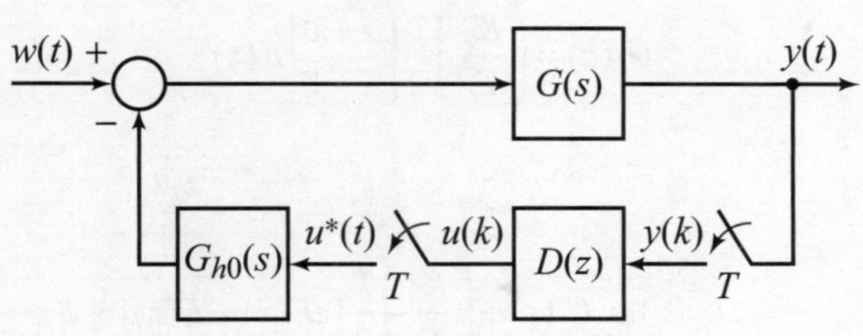

Fig. 13.35

Therefore,

$$Y(z) = \mathscr{Z}[WG(s)] - \mathscr{Z}[G_{h0}G(s)]U(z)$$
$$= WG(z) - G_{h0}G(z)U(z)$$

Also,

$$U(z) = D(z)Y(z)$$

This gives

$$Y(z) = WG(z) - D(z)G_{h0}G(z)Y(z)$$

or

$$Y(z) = \frac{WG(z)}{1 + D(z)G_{h0}G(z)} \tag{13.64}$$

Review Example 13.3 Let us reconsider the problem of discretization of a PID controller discussed earlier in Example 13.1. Equation (13.8) is equivalent to

$$U(s) = K_c\left[1 + \frac{1}{T_I s} + T_D s\right]E(s) = D(s)E(s) \tag{13.65}$$

Discretization of $D(s)$ using bilinear mapping (13.50) gives

$$U(z) = K_c\left[1 + \left(\frac{1}{T_I}\right)\frac{T}{2}\left(\frac{z+1}{z-1}\right) + (T_D)\frac{2}{T}\left(\frac{z-1}{z+1}\right)\right]E(z) \tag{13.66}$$

$$= U_P(z) + U_I(z) + U_D(z)$$

The first term in (13.66) is

$$U_P(z) = K_c E(z)$$

This gives

$$u_P(k) = K_c e(k) \tag{13.67a}$$

The second term in (13.66) is

$$U_I(z) = \left(\frac{K_c}{T_I}\right)\frac{T}{2}\left[\frac{z+1}{z-1}\right]E(z)$$

or

$$zU_I(z) = U_I(z) + \left(\frac{K_c}{T_I}\right)\frac{T}{2}[zE(z) + E(z)]$$

Therefore,

$$U_I(z) = z^{-1}U_I(z) + \left(\frac{K_c}{T_I}\right)\frac{T}{2}[E(z) + z^{-1}E(z)]$$

This gives

$$u_I(k) = u_I(k-1) + \left(\frac{K_c}{T_I}\right)\frac{T}{2}[e(k) + e(k-1)] \tag{13.67b}$$

The third term in (13.66) is

$$U_D(z) = (K_cT_D)\frac{2}{T}\left[\frac{z-1}{z+1}\right]E(z)$$

or

$$z\,U_D(z) = -U_D(z) + (K_cT_D)\frac{2}{T}[zE(z) - E(z)]$$

Therefore,

$$U_D(z) = -z^{-1}U_D(z) + (K_cT_D)\frac{2}{T}[E(z) - z^{-1}E(z)$$

This gives

$$u_D(k) = (K_cT_D)\frac{2}{T}[e(k) - e(k-1)] - u_D(k-1) \tag{13.67c}$$

From (13.67a), (13.67b) and (13.67c), we obtain

$$u(k) = u_P(k) + u_I(k) + u_D(k) \tag{13.68}$$

CONCLUDING REMARKS

In this chapter, we covered design and analysis of digital control systems using transform domain techniques. State space techniques will be covered in Chapters 14 and 15.

What we have learnt?

- The *z*-transform replaces the Laplace transform as the transform of choice for sampled-data systems. It provides direct parallels to the *s*-plane analysis of transients, steady-state errors, stability, block diagram reduction, etc.
- We can map points on the *s*-plane into points on the *z*-plane and attach familiar transient response characteristics like peak overshoot, settling time, etc., to the *z*-plane points. The unit circle in *z*-plane becomes the boundary of stability, replacing the imaginary axis of the *s*-plane.
- A digital system with high sampling rate approximates an analog system. Using this fact, we can design an analog controller and then translate it into a digital one, i.e., performing discretization *after* design.

- Desired performance with lower sampling rates can generally be achieved by performing discretization *before* design; analog plants are transformed into sampled-data plants and then the design is carried out.
- The rules for sketching the z-plane root locus are same as those for s-plane root locus. The concepts of root locus design discussed in Chapter 9 are easily carried into the z-plane.
- The simplicity of Bode plots for s-domain transfer functions gets lost in z-domain. Computer assistance is normally required for constructing Bode plots for z-domain transfer functions.

Review Questions

13.1 Describe important advantages offered by the use of digital computers as compensator device in a control system. What are the main problems associated with implementation of digital control?

13.2 Sketch a schematic diagram of a speed control system using the following units: motor, tacho-generator, A/D convertor, digital computer, D/A convertor, and power amplifier. Explain the function of each section of the diagram. How do we select the sampling rate?

13.3 Explain with the aid of appropriate diagrams the construction and operation of a shaft encoder. Give a schematic diagram showing how the shaft encoder would be embodied in a digital position control system.

13.4 Given the tuning rules for analog controllers, how do we obtain tuning rules for digital controllers?

13.5 Viewing sampler as an 'impulse modulator', derive suitable mathematical models of A/D and D/A convertors.

13.6 State and prove the final value theorem of the z-transform. What is the condition under which the theorem is valid?

13.7 The sampling theorem states that a continuous-time signal whose frequency spectrum is bounded by upper limit ω_m, can be completely reconstructed from its samples when the sampling frequency $\omega_s > 2\omega_m$. Explain the theorem using $z = e^{sT}$ mapping.

13.8 (a) Given a desired second-order characteristic equation on the s-plane, how can one find the associated desired characteristic equation on the z-plane?

(b) Given a second-order characteristic equation on the z-plane, how can one determine the associated percent overshoot, settling time, and peak time?

13.9 Define the regions of stability, marginal stability, and instability on the s-plane. How are these regions translated to z-plane by the mapping: $z = e^{sT}$?

13.10 Explain bilinear mapping for conversion of analog compensator $D(s)$ to an equivalent digital compensator $D(z)$.

13.11 Using appropriate examples, show that sampling has detrimental effect on the transient response of a closed-loop system, but the steady-state response remains unaffected.

Self-Appraisal Quiz

A set of questions with multiple answer choices is given below. These questions have been designed to capture important aspects of the subject covered in the chapter. For each of these questions, identify the best of the given answer choices. Compare your answers with the Master Key given at the end of the session. This quiz should serve the purpose of a self-appraisal test for readers.

13.1 A mathematical model of zero-order hold is

(A) e^{-sT} (B) $1 - e^{-sT}$

(C) $\dfrac{1-e^{-sT}}{s}$ (D) None of the answers in (A), (B) or (C) is correct

13.2 z-transform of sampled ramp function $r(k) = kT\mu(k)$ is ($\mu(\cdot)$ is a step function)

(A) $\dfrac{z}{z-1}$ (B) $\dfrac{Tz}{(z-1)^2}$

(C) $\dfrac{z}{z-e^{-T}}$ (D) None of the answers in (A), (B), or (C) is correct

13.3 z-transform of $\mu(k-2)$ is ($\mu(\cdot)$ is a step function)

(A) z^{-2} (B) z^2

(C) $\dfrac{z^{-1}}{z-1}$ (D) None of the answers in (A), (B), or (C) is correct

13.4 The inverse z-transform of $1/(z-0.2)$ is ($\mu(\cdot)$ is a step function)

(A) $(0.2)^k \mu(k)$ (B) $(0.2)^{-k} \mu(k)$

(C) $(0.2)^{k-1} \mu(k-1)$ (D) $(0.2)^{-k+1} \mu(k-1)$

13.5 A dynamic system is represented by the difference equation,

$$y(k) - \frac{3}{4}y(k-1) + \frac{1}{8}y(k-2) = r(k-2)$$

The transfer function of the system is

(A) $\dfrac{1}{z^2 + \frac{3}{4}z + \frac{1}{8}}$ (B) $\dfrac{1}{\left(z-\frac{1}{4}\right)\left(z-\frac{1}{2}\right)}$

(C) $\dfrac{1}{\left(z+\frac{1}{4}\right)\left(z+\frac{1}{2}\right)}$ (D) None of the answers in (A), (B), or (C) is correct

13.6 A closed-loop sampled-data system has the transfer function,

$$\frac{Y(z)}{R(z)} = \frac{0.3679z + 0.2642}{z^2 - z + 0.6321}$$

The steady-state value y_{ss} of the output $y(k)$ in response to unit-step input $\mu(k)$, is

(A) $\dfrac{0.2642}{0.6321}$ (B) $\dfrac{0.3679}{0.6321}$

(C) 1 (D) None of the answers in (A), (B) or (C) is correct.

13.7 z-transform of $\left(\dfrac{1-e^{-sT}}{s}\right)G(s)$ is

(A) $(1-z^{-1})\,Z(G(s))$ (B) $(1-z^{-1})Z\left(\dfrac{G(s)}{s}\right)$

(C) $(z-1)Z(G(s))$ (D) $(z-1)Z\left(\dfrac{G(s)}{s}\right)$

13.8 The bilinear mapping is given by

(A) $s=\dfrac{2}{T}\left(\dfrac{1-z^{-1}}{1+z^{-1}}\right)$ (B) $s=\dfrac{2}{T}\left(\dfrac{1+z^{-1}}{1-z^{-1}}\right)$

(C) $s=\dfrac{2}{T}\left(\dfrac{z-1}{z+1}\right)$ (D) $s=\dfrac{2}{T}\left(\dfrac{z+1}{z-1}\right)$

13.9 A unity-feedback sampled-data system has open-loop transfer function

$$G_{h0}G(z)=\frac{0.478(z+0.368)}{(z-0.135)(z-0.9)}$$

The steady-state error to unit step input is

(A) zero (B) infinity

(C) $\dfrac{1}{7.56}$ (D) None of the answers in (A), (B), or (C) is correct

13.10 The pole-zero configuration of Fig. Q13.10 represents a

(A) lead compensator (B) lag compensator
(C) Both of the two (D) None of the two

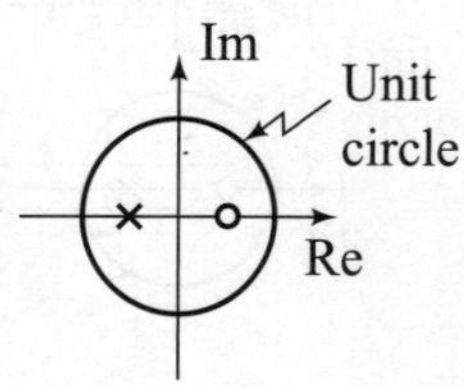

Fig. Q13.10

13.11 An open-loop unity feedback sampled-data system is given by the transfer function

$$G_{h0}G(z)=\frac{0.5}{z^2(z-0.5)}$$

The system is

(A) type-0 (B) type-1
(C) type-2

13.12 Consider the sampled-data system shown in Fig. Q13.12. Steady-state error for unit-ramp input is

(A) $\dfrac{1}{KT}$ (B) $\dfrac{T}{K}$

(C) $\dfrac{1}{K}$ (D) None of the answers in (A), (B), or (C) is correct

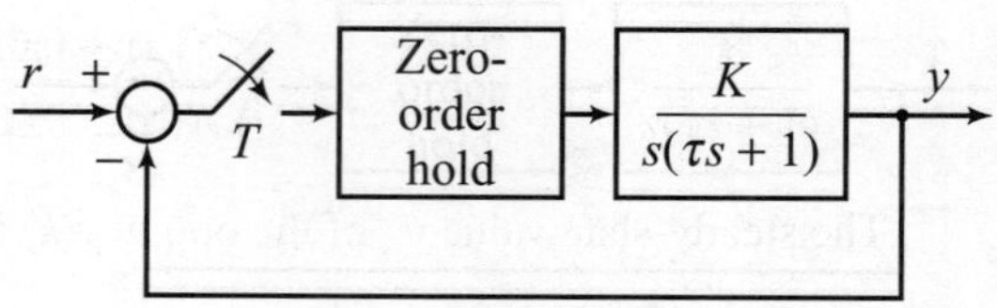

Fig. Q13.12

13.13 In a digital control system, selection of a large sampling period

(A) increases the stability margin
(B) decreases the stability margin
(C) has no effect on stability
(D) has an effect on stability that depends on plant parameters

13.14 In a digital control scheme, selection of a large sampling interval

(A) improves the steady-state performance
(B) deteriorates the steady-state performance
(C) has no effect on steady-state performance
(D) has an effect on steady-state performance that depends on plant parameters

13.15 In a sampled-data control system, delay introduced by sampling and reconstruction process is approximately equal to

(A) sampling interval
(B) twice the sampling interval
(C) half the sampling interval
(D) None of the answers in (A), (B), or (C) is correct

13.16 For a compensator

$$\frac{U(s)}{E(s)} = \frac{1}{s+a},$$

the algorithm (with sampling period T), based on trapezoidal rule for integration, is given by

(A) $u(k+1) = (1-aT)u(k) + Te(k)$
(B) $u(k+1) = \frac{1}{1+aT}u(k) + \frac{T}{1+aT}e(k+1)$
(C) None of the two

13.17 A PI controller given by

$$u(t) = K_c\left[e(t) + \frac{1}{T_I}\int e(\tau)d\tau\right]$$

has been approximated by the following algorithm (with sampling period T):

$$u(k) = u_P(k) + u_I(k)$$
$$u_P(k) = K_c e(k)$$
$$u_I(k) = u_I(k-1) + \left(\frac{K_c}{T_I}\right)\frac{T}{2}[e(k) + e(k-1)]$$

The discretization is based on

(A) rectangular rule for integration
(B) trapezoidal rule for integration
(C) None of the two

MASTER KEY

13.1 (C)	13.2 (B)	13.3 (C)	13.4 (C)	13.5 (B)
13.6 (C)	13.7 (B)	13.8 (A)	13.9 (D)	13.10 (A)
13.11 (A)	13.12 (C)	13.13 (B)	13.14 (C)	13.15 (C)
13.16 (C)	13.17 (B)			

Problems

13.1 A PID controller is described by the following relation between input $e(t)$ and output $y(t)$:

$$u(t) = K_c\left[e(t) + \frac{1}{T_I}\int_0^t e(t)dt + T_D\frac{de(t)}{dt}\right]$$

Using the trapezoidal rule for integration, obtain the difference equation model for the PID algorithm.

13.2 Consider the analog control system shown in Fig. P13.2. By root-locus analysis, it is found that the closed-loop natural frequency $\omega_n \cong 0.6$ rad/sec, and damping ratio $\zeta = 0.5$.

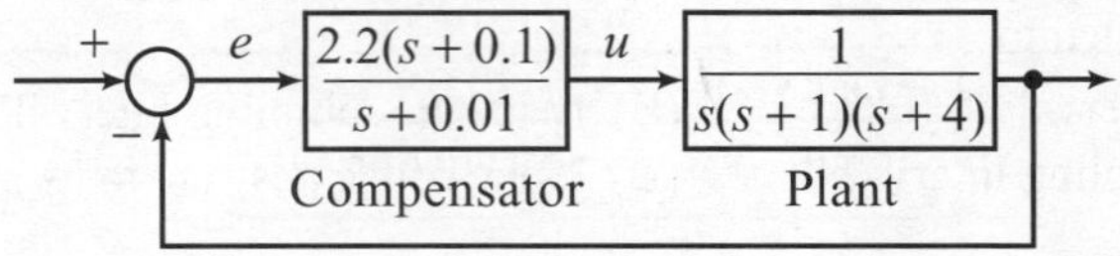

Fig. P13.2

We wish to replace the analog compensator by a digital one. Select a suitable sampling interval and obtain an equivalent digital control algorithm using bilinear mapping. Give a block diagram of the digital control system.

MATLAB Exercise Use Simulink/Control System Toolbox to obtain the step responses of the analog and digital implementations. Are there significant differences?

13.3 Figure P13.3 shows a minor-loop feedback compensation scheme. We wish to realize the control action by a digital controller with a uniform sampling interval of T sec. Obtain digital control algorithm using trapezoidal rule for integration. Give a block diagram of the digital control system.

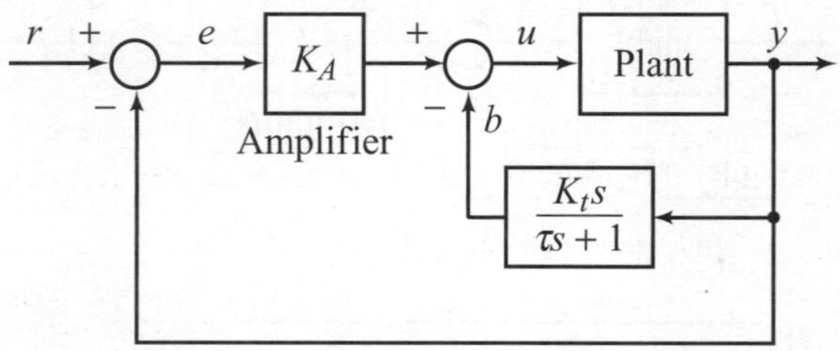

Fig. P13.3

13.4 The most widely used approach for tuning digital PID controllers, is to apply rules given in Table 7.2 developed for analog PID controllers with a simple correction to account for the effect of A/D sampling; the D/A reconstructed signal is effectively delayed by approximately one-half of the sampling interval T. The correction for sampling is then simply to add $T/2$ to the dead-time τ_D. The tuning formulas of Table 7.2 can directly be used for digital PID controllers with τ_D replaced by corrected dead time.

A process has a gain of 2, a time constant of 10 min, and a dead time of 2 min.

(a) Calculate the QDR tuning parameters for a PID controller.

(b) Readjust the tuning parameters calculated in (a) if the PID control is to be carried out with a processing period of 8 sec on a computer control installation.

13.5 Table 7.1 can be directly used for tuning of digital PID controllers; correction for sampling is not required if the test has been conducted on a computer-controlled loop.

In a low-pass process, the following critical quantities were determined by an oscillation test conducted on a computer-controlled loop with sample time $T = 4$ sec.

Ultimate gain, $K_{cu} = 5$

Ultimate period, $T_u = 34$ sec

Determine tuning parameters for a digital PI controller.

13.6 Consider the discrete-time system shown in Fig. P13.6

(a) Obtain the difference equation model and therefrom the transfer function model of the system.

(b) Find the impulse response of the system.

(c) Find the response of the system to unit step input $\mu(k)$.

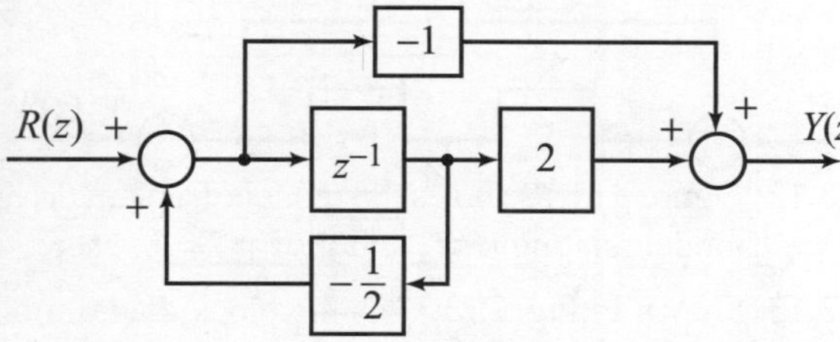

Fig. P13.6

MATLAB Exercise Using Control System Toolbox, plot the impulse and step responses.

13.7 Solve the following difference equation using z-transform approach:

$2y(k) - 2y(k-1) + y(k-2) = r(k)$; $y(k) = 0$ for $k < 0$, and

$$r(k) = \begin{cases} 1; & k = 0, 1, 2, \ldots \\ 0; & k < 0 \end{cases}$$

13.8 Solve the following difference equation using z-transforms:

$$y(k) - 3y(k-1) + 2y(k-2) = r(k)$$

where
$$r(k) = \begin{cases} 1 \text{ for } k = 0, 1 \\ 0 \text{ for } k \geq 2 \end{cases}; y(-2) = y(-1) = 0$$

Will the final value theorem give the correct value of $y(k)$ as $k \to \infty$? Why?

13.9 For the transfer function models and inputs given below, find the response $y(k)$ as a function of k:

(a) $$G(z) = \frac{Y(z)}{R(z)} = \frac{2z - 3}{(z - 0.5)(z + 0.3)}; \; r(k) = \begin{cases} 1; & k = 1 \\ 0; & k = 0, 2, 3, 4, \ldots \end{cases}$$

(b) $$G(z) = \frac{Y(z)}{R(z)} = \frac{-6z + 1}{\left(z - \frac{1}{2} + j\frac{1}{4}\right)\left(z - \frac{1}{2} - j\frac{1}{4}\right)}; \; r(k) = \begin{cases} 0; & k < 0 \\ 1; & k = 0, 1, 2, 3, \ldots \end{cases}$$

13.10 For the transfer function models and inputs given below, find the response $y(k)$ as a function of k:

(a) $$G(z) = \frac{Y(z)}{R(z)} = \frac{1}{(z - 0.5)(z + 0.3)}; \; r(k) = \begin{cases} 1; & k \text{ even} \\ 0; & k \text{ odd} \end{cases}$$

(b) $$G(z) = \frac{Y(z)}{R(z)} = \frac{1}{(z - 0.5)^2 (z - 0.1)}; \; r(k) = \begin{cases} 0; & k < 0 \\ 1; & k = 0, 1, 2, 3, \ldots \end{cases}$$

13.11 Figure P13.11 shows a discrete-time system. Determine the transfer function of this system assuming that the samplers operate synchronously at intervals of T sec. Also find the unit-step response of the system.

MATLAB Exercise Use Control System Toolbox to obtain the step response.

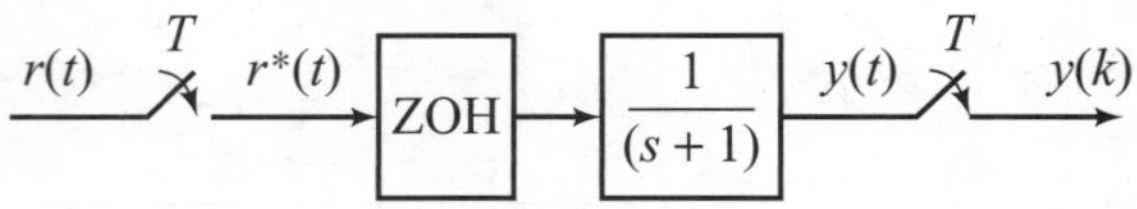

Fig. P13.11

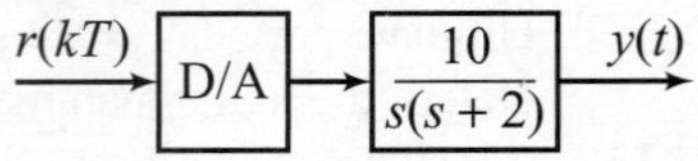

Fig. P13.12

13.12 Figure P13.12 shows the model of a plant driven by a D/A converter. Derive the transfer function model relating $r(kT)$ and $y(kT)$; $T = 0.4$ sec.

13.13 Shown in Fig. P13.13 is the block diagram of the servo control system for one of the joints of a robot. With $D(z) = 1$, find the transfer function model of the closed-loop system. Sampling period $T = 0.25$ sec.

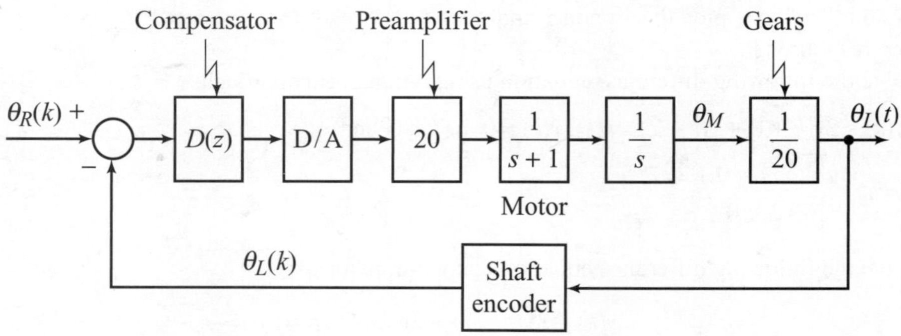

Fig. P13.13

13.14 For the system shown in Fig. P13.14, the computer solves the difference equation $u(k) = u(k - 1) + 0.5\, e(k)$, where $e(k)$ is the filter input and $u(k)$ is the filter output. If the sampling rate $f_s = 5$ Hz, find $Y(z)/R(z)$.

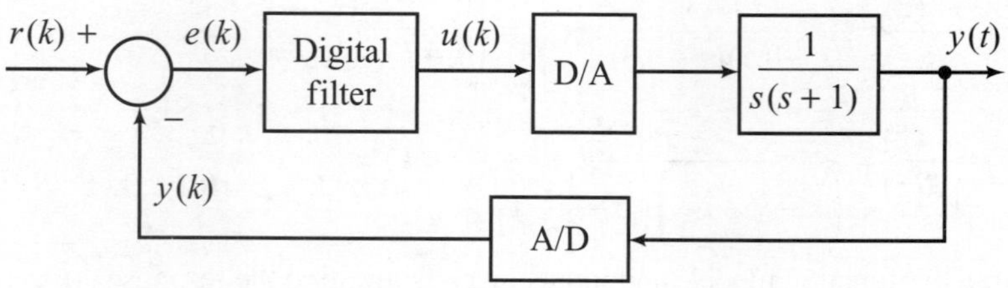

Fig. P13.14

13.15 For the sampled-data system shown in Fig. P13.15, find the output $y(k)$ for $r(t)$ = unit step.

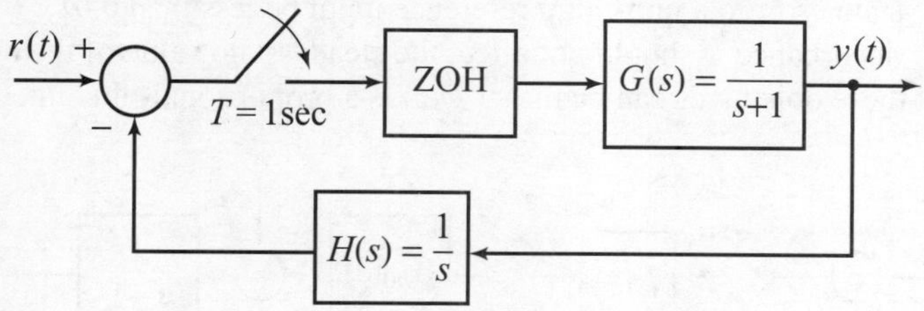

Fig. P13.15

13.16 For the sampled-data system of Fig. P13.16, find the response $y(kT)$; $k = 0, 1, 2, \ldots$, to a unit-step input $r(t)$.

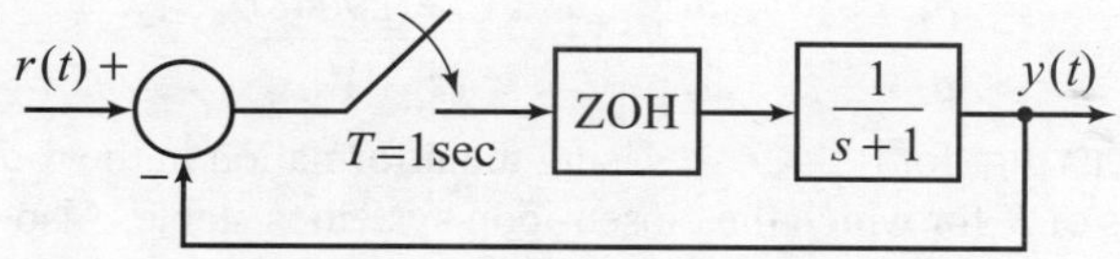

Fig. P13.16

13.17 For the system shown in Fig. P13.17, find
(a) position-error constant, K_p,
(b) velocity-error constant, K_v, and
(c) acceleration-error constant, K_a.
Express the results in terms of K_1, K_2, J, and T.

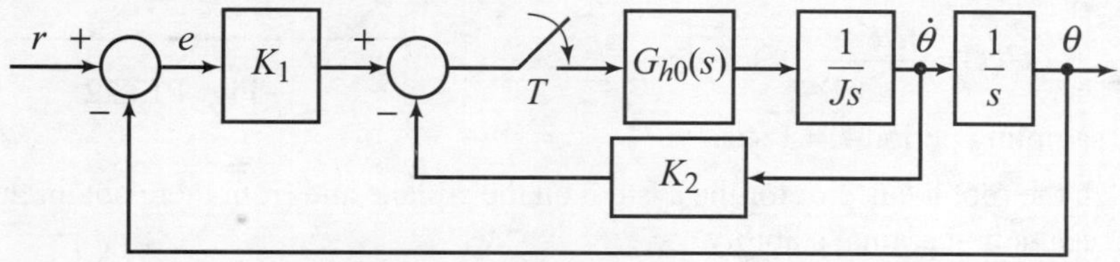

Fig. P13.17

13.18 A unity feedback system is characterized by the open-loop transfer function,

$$G_{h0}G(z) = \frac{0.2385(z + 0.8760)}{(z - 1)(z - 0.2644)}$$

The sampling period $T = 0.2$ sec.
Determine steady-state errors for unit-step, unit-ramp, and unit-acceleration inputs.

13.19 Predict the nature of the transient response of a discrete-time system whose characteristic equation is given by

$$z^2 - 1.9z + 0.9307 = 0$$

The sampling interval $T = 0.02$ sec.

13.20 Consider the system of Fig. P13.20. The design specifications for the system require that
(a) the steady-state error to a unit-ramp reference input be less than 0.01,
(b) a constant disturbance w should not affect the steady-state value of the output.
Show that these objectives can be met if $D(z)$ is a proportional-plus-integral compensator.

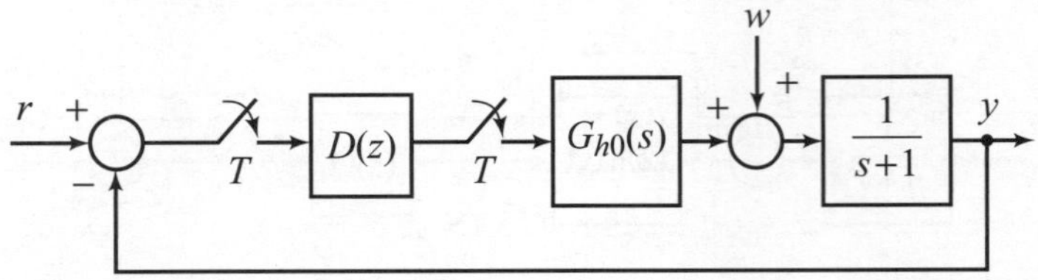

Fig. P13.20

13.21 Given

$$G_{h0}G(z) = \frac{K(z-0.9048)}{(z-1)^2}$$

Sketch the root locus plot for $0 \le K < \infty$. Using the information in the root locus plot, determine the range of values of K for which the closed-loop system is stable. Also determine the value of K for which the system closed-loop poles are real and multiple.

MATLAB Exercise Obtain root locus plot using Control System Toolbox. Verify your solution obtained by hand calculation.

13.22 A sampled-data feedback control system is shown in Fig. P13.22. The controlled process of the system is described by the transfer function,

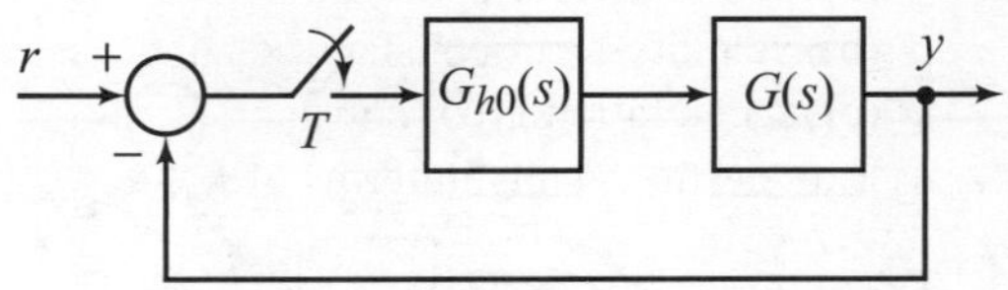

Fig. P13.22

$$G(s) = \frac{K}{s(s+1)}; 0 \le K < \infty$$

The sampling period $T = 1$ sec.

(a) Sketch the root locus plot for the system on the z-plane and from there obtain the value of K that results in marginal stability.
(b) Repeat part (a) for (i) $T = 2$ sec, (ii) $T = 4$ sec, and compare the stability properties of the system with different values of sampling interval.

MATLAB Exercise (a) Use Control System Toolbox to draw root locus of the system (i) on s-plane (sampler and ZOH not present), and (ii) on z-plane ($T = 1$sec). Determine the range of K for stability in each case.
(b) Use Simulink/Control System Toolbox to obtain the step response of both the continuous-time and discrete-time systems with K chosen to yield damping ratio $\zeta = 0.5$.

13.23 Consider the system shown in Fig. P13.23. Using root-locus analysis, find the range of $K > 0$ for which the system is stable.

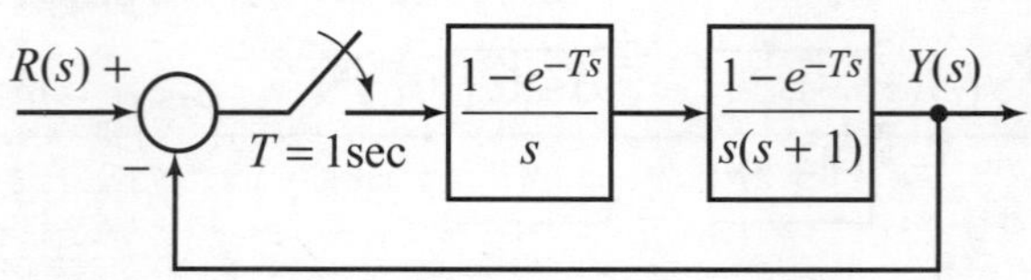

Fig. P13.23

13.24 The digital process of a unity feedback system is described by the transfer function,

$$G_{h0}G(z) = \frac{K(z+0.717)}{(z-1)(z-0.368)}; T = 1\,\text{sec}$$

Sketch the root locus plot for $0 \le K < \infty$ and from there obtain the following information:

(a) The value of K that results in marginal stability. Also find the frequency of oscillations.

(b) The value of K that results in $\zeta = 1$. What are the time constants of the closed-loop poles?

(c) The value of K that results in $\zeta = 0.5$. Also find the natural frequency ω_n for this value of K. You may use the following table to construct a constant ζ locus on the z-plane corresponding to $\zeta = 0.5$.

Re	0.891	0.64	0.389	0.169	0	−0.113	−0.174	−0.188	−0.163
Im	0.157	0.37	0.463	0.464	0.404	0.310	0.207	0.068	0

13.25 The characteristic equation of a feedback control system is

$$z^2 + 0.2A\,z - 0.1\,A = 0$$

Sketch the root loci for $0 \le A < \infty$, and therefrom obtain the range of parameter A for which the system is stable.

13.26 Consider the system shown in Fig. P13.26 with $G(s) = \dfrac{1}{s(s+1)}$ and $T = 0.2$ sec.

(a) Design a lead compensator so that the dominant closed-loop poles of the system will have $\zeta = 0.5$ and $\omega_n = 4.5$.

(b) Obtain the velocity-error constant K_v of the lead compensated system.

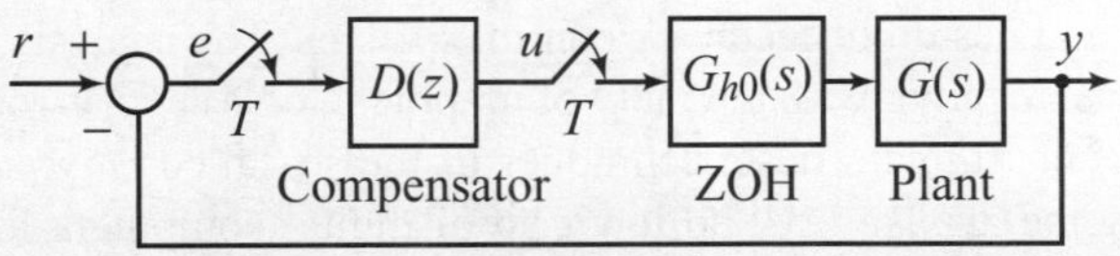

Fig. P13.26

13.27 Consider the system shown in Fig. P13.26 with

$$G(s) = \frac{1}{(s+1)(s+2)}; T = 1\,\text{sec}$$

Design a compensator $D(z)$ that meets the following specifications on system performance:

(a) $\zeta = 0.5$

(b) $\omega_n = 1.5$, and

(c) $K_p \ge 7.5$

Use root locus method.

MATLAB Exercise Use LTI Viewer to determine the peak time and settling time of the closed-loop step response.

CONTROL SYSTEM ANALYSIS USING STATE VARIABLE METHODS

14

Preview

In this chapter, the coverage of state space techniques is to be regarded as an introduction to the subject; a springboard to advanced studies, and an alternative approach to frequency domain techniques (For more details, refer to [155]). State space approach is applicable to a wide range of systems: multi-input, multi-output systems, time-varying systems, nonlinear systems. We will limit our discussion is this chapter to single-input, single-output, time-invariant, linear systems. The fact that we have a state variable model for the plant will lead us to a very different kind of control design, called state-feedback control. This topic will unfold in the next chapter.

For readers with no background of state variable methods, a natural way of introducing the subject is to give the treatment for continuous-time systems, followed by a brief presentation for discrete-time systems. This approach has been followed in the present and the next chapter.

14.1 INTRODUCTION

In the preceding chapters, we showed that the root-locus and the frequency-response methods are quite powerful for the analysis and design of feedback control systems. The analysis and design are carried out using transfer functions, together with a variety of graphical tools such as root-locus plots, Nyquist plots, Bode plots, Nichols chart, etc. These techniques of the so-called *classical control theory* have been greatly enhanced by the availability and low cost of digital computers for system analysis and simulation. The graphical tools can now be more easily used with computer graphics.

The classical design methods suffer from certain limitations due to the fact that the transfer function model is applicable only to linear time-invariant systems, and there too it is generally restricted to SISO systems, as the classical design approach becomes highly cumbersome for use in MIMO systems. Another limitation of the transfer function technique is that it reveals only the system output for a given input and provides no information about the internal behavior of the system. There may be situations where the output of a system is stable and yet some of the system elements may have a tendency to exceed their specified ratings. In addition to this, it may sometimes be necessary and advantageous to provide a feedback proportional to the internal variables of a system, rather than the output alone, for the purpose of stabilizing and improving the performance of a system.

The limitations of classical methods based on transfer function models, have led to the development of a state space approach of analysis and design. It is a direct time-domain approach which provides a basis for *modern control theory*. It is a powerful technique for the analysis and design of linear and nonlinear, time-invariant or time-varying MIMO systems. The organization of the state-space approach is such that it is easily amenable to solution through digital computers.

It will be incorrect to conclude from the foregoing discussion that the state variable design methods can completely replace the classical design methods. In fact, the classical control theory comprising a large body of use-tested knowledge is going strong. State variable design methods prove their mettle in applications which are intractable by classical methods.

The state variable formulation contributes to the application areas of classical control theory in a different way. It is the most efficient form of system representation from the standpoint of computer simulation. To compute the response of $G(s)$ to an input $R(s)$ requires expansion of $\{G(s)R(s)\}$ into partial fractions; which, in turn, requires computation of all the poles of $\{G(s)R(s)\}$, or all the roots of a polynomial. The roots of a polynomial are very sensitive to their coefficients. Furthermore, to develop a computer program to carry out partial fraction expansion is not simple. On the other hand, the response of state variable equations is easy to program. Its computation does not require computation of roots or eigenvalues; therefore, it is less sensitive to parameter variations. For these reasons, it is desirable to compute the response of $G(s)$ through state variable equations.

Many CAD packages handling both the classical and the modern tools of control system design use state variable formulation. It is, therefore, helpful for the control engineer to be familiar with state variable methods of system representation and analysis. This chapter introduces the main concepts of state variable analysis. This material may be used as a brief terminating study or as an introduction to a more in-depth discussion of state variable methods in Chapter 15.

We have been mostly concerned with SISO systems in the text so far. In this chapter our emphasis will also be on SISO systems. However, many of the methods based on state variable concepts are applicable to both SISO and MIMO systems with almost equal convenience; the only difference being the additional computational effort for MIMO systems which is taken care of by CAD packages.

14.2 MATRICES

This section is intended to be a concise summary of facts about matrices[1] that the reader will need to know in reading the present chapter. Having them all at hand will minimize the need to consult a book on matrix theory. It also serves to define the notation and terminology which are, regrettably, not entirely standard.

No attempt has been made at proving every statement made in this section. The interested reader is urged to consult a suitable book (for example [27, 28]) for details of proofs.

Basic definitions and algebraic operations associated with matrices are given below.

Matrix

The matrix

$$\mathbf{A} = \begin{bmatrix} a_{11} & a_{12} & \cdots & a_{1m} \\ a_{21} & a_{22} & \cdots & a_{2m} \\ \vdots & \vdots & & \vdots \\ a_{n1} & a_{n2} & \cdots & a_{nm} \end{bmatrix} = [a_{ij}] \tag{14.1}$$

is a rectangular array of nm elements. It has n rows and m columns. a_{ij} denotes (i, j)th element, i.e., the element located in ith row and jth column. $\mathbf{A}$ is said to be a *rectangular matrix* of order $n \times m$.

When $m = n$, i.e., the number of columns is equal to that of rows, the matrix is said to be a *square matrix* of order n.

An $n \times 1$ matrix, i.e., a matrix having only one column is called a *column matrix*. A $1 \times n$ matrix, i.e., a matrix having only one row is called a *row matrix*.

[1] We will use upper case bold letters to represent matrices and lower case bold letters to represent vectors.

Diagonal matrix

A diagonal matrix is a square matrix whose elements off the *principal diagonal* are all zeros ($a_{ij} = 0$ for $i \neq j$). The following matrix is a diagonal matrix:

$$\mathbf{\Lambda} = \begin{bmatrix} a_{11} & 0 & \cdots & 0 \\ 0 & a_{22} & \cdots & 0 \\ \vdots & \vdots & & \vdots \\ 0 & 0 & \cdots & a_{nn} \end{bmatrix} \tag{14.2}$$

Unit (identity) matrix

A unit matrix **I** is a diagonal matrix whose diagonal elements are all equal to unity ($a_{ii} = 1$, $a_{ij} = 0$ for $i \neq j$).

$$\mathbf{I} = \begin{bmatrix} 1 & 0 & \cdots & 0 \\ 0 & 1 & \cdots & 0 \\ \vdots & \vdots & & \vdots \\ 0 & 0 & \cdots & 1 \end{bmatrix}$$

Whenever necessary, an $n \times n$ unit matrix will be denoted by $\mathbf{I}_n$.

Null (zero) matrix

A null matrix **0** is a matrix whose elements are all equal to zero.

$$\mathbf{0} = \begin{bmatrix} 0 & 0 & \cdots & 0 \\ 0 & 0 & \cdots & 0 \\ \vdots & \vdots & & \vdots \\ 0 & 0 & \cdots & 0 \end{bmatrix}$$

Whenever necessary, the dimensions of the null matrix will be indicated by two subscripts: $\mathbf{0}_{nm}$.

Matrix transpose

If the rows and columns of an $n \times m$ matrix **A** are interchanged, the resulting $m \times n$ matrix, denoted as $\mathbf{A}^T$, is called the *transpose* of the matrix **A**. Namely, if **A** is given by Eqn. (14.1), then

$$\mathbf{A}^T = \begin{bmatrix} a_{11} & a_{21} & \cdots & a_{n1} \\ a_{12} & a_{22} & \cdots & a_{n2} \\ \vdots & \vdots & & \vdots \\ a_{1m} & a_{2m} & \cdots & a_{nm} \end{bmatrix}$$

Some properties of the matrix transpose are

1. $(\mathbf{A}^T)^T = \mathbf{A}$
2. $(k\mathbf{A})^T = k\mathbf{A}^T$, where k is a scalar
3. $(\mathbf{A} + \mathbf{B})^T = \mathbf{A}^T + \mathbf{B}^T$
4. $(\mathbf{AB})^T = \mathbf{B}^T\mathbf{A}^T$

Determinant of a matrix

Determinants are defined for square matrices only. The determinant of the $n \times n$ matrix **A**, written as $|\mathbf{A}|$ or *det* **A**, is a scalar-valued function of **A**. It is found through the use of minors and cofactors.

The *minor* m_{ij} of the element a_{ij} is the determinant of a matrix of order $(n-1)\times(n-1)$ obtained from **A** by removing the row and the column containing a_{ij}.

The *cofactor* c_{ij} of the element a_{ij} is defined by the equation,

$$c_{ij} = (-1)^{i+j}\, m_{ij}$$

Determinants can be evaluated by the method of *Laplace expansion*. If **A** is an $n \times n$ matrix, any arbitrary row k can be selected and $|\mathbf{A}|$ is then given by

$$|\mathbf{A}| = \sum_{j=1}^{n} a_{kj} c_{kj}$$

Similarly, Laplace expansion can be carried out with respect to any arbitrary column l, to obtain

$$|\mathbf{A}| = \sum_{i=1}^{n} a_{il} c_{il}$$

Laplace expansion reduces the evaluation of an $n \times n$ determinant down to the evaluation of a string of $(n-1)\times(n-1)$ determinants, namely the cofactors.

Some properties of determinants are

1. $det\ \mathbf{AB} = (det\ \mathbf{A})(det\ \mathbf{B})$
2. $det\ \mathbf{A}^T = det\ \mathbf{A}$
3. $det\ k\mathbf{A} = k^n det\ \mathbf{A}$; **A** is $n \times n$ matrix and k is scalar
4. The determinant of any diagonal matrix is the product of its diagonal elements.

Singular matrix

A square matrix is called singular if the associated determinant is zero.

Nonsingular matrix

A square matrix is called nonsingular if the associated determinant is nonzero.

Adjoint matrix

The adjoint matrix of a square matrix **A** is found by replacing each element a_{ij} of matrix **A** by its cofactor c_{ij} and then transposing.

$$adj\ \mathbf{A} = \mathbf{A}^+$$

$$= \begin{bmatrix} c_{11} & c_{21} & \cdots & c_{n1} \\ c_{12} & c_{22} & \cdots & c_{n2} \\ \vdots & \vdots & & \vdots \\ c_{1n} & c_{2n} & \cdots & c_{nn} \end{bmatrix} = [c_{ji}]$$

Note that

$$\mathbf{A}(adj\ \mathbf{A}) = (adj\ \mathbf{A})\mathbf{A} = |\mathbf{A}|\ \mathbf{I} \tag{14.3}$$

Matrix inverse

The inverse of a square matrix is written as $\mathbf{A}^{-1}$, and is defined by the relation,

$$\mathbf{A}^{-1}\mathbf{A} = \mathbf{A}\mathbf{A}^{-1} = \mathbf{I}$$

From Eqn. (14.3) and the definition of the inverse matrix, we have

$$\mathbf{A}^{-1} = \frac{adj\ \mathbf{A}}{|\mathbf{A}|} \tag{14.4}$$

Some properties of matrix inverse are

1. $(\mathbf{A}^{-1})^{-1} = \mathbf{A}$
2. $(\mathbf{A}^T)^{-1} = (\mathbf{A}^{-1})^T$
3. $(\mathbf{AB})^{-1} = \mathbf{B}^{-1}\mathbf{A}^{-1}$
4. $det\ \mathbf{A}^{-1} = \dfrac{1}{det\ \mathrm{A}}$
5. $det\ \mathbf{P}^{-1}\mathbf{AP} = det\ \mathbf{A}$
6. Inverse of diagonal matrix given by Eqn. (14.2) is

$$\mathbf{\Lambda}^{-1} = \begin{bmatrix} 1/a_{11} & 0 & \cdots & 0 \\ 0 & 1/a_{22} & \cdots & 0 \\ \vdots & \vdots & & \vdots \\ 0 & 0 & \cdots & 1/a_{nn} \end{bmatrix}$$

Rank of a matrix

The rank $\rho(\mathbf{A})$ of a matrix **A** is the dimension of the largest array in **A** with a nonzero determinant. Some properties of rank are

1. $\rho(\mathbf{A}^T) = \rho(\mathbf{A})$
2. The rank of a rectangular matrix cannot exceed the lesser of the number of rows or the number of columns. A matrix whose rank is equal to the lesser of the number of rows and number of columns is said to be of *full rank.*

$$\rho(\mathbf{A}) \le \min(n, m);\ \mathbf{A} \text{ is } n \times m \text{ matrix}$$

3. The rank of a product of two matrices cannot exceed the rank of the either:

$$\rho(\mathbf{AB}) \le \min[\rho(\mathbf{A}), \rho(\mathbf{B})]$$

Partitioned matrix

A matrix can be partitioned into submatrices. Broken lines are used to show the partitioning when the elements of the submatrices are explicitly shown. For example,

$$\mathbf{A} = \left[\begin{array}{cc:c} a_{11} & a_{12} & a_{13} \\ a_{21} & a_{22} & a_{23} \\ \hdashline a_{31} & a_{32} & a_{33} \end{array}\right]$$

The broken lines indicating the partitioning are sometimes omitted when the context makes it clear that partitioned matrices are being considered.

We will be frequently using the following forms of partitioning:

1. Matrix **A** partitioned into its columns:

$$\mathbf{A} = [\mathbf{a}_1\ \mathbf{a}_2 \ldots \mathbf{a}_m] \tag{14.5a}$$

where

$$\mathbf{a}_i = \begin{bmatrix} a_{1i} \\ a_{2i} \\ \vdots \\ a_{ni} \end{bmatrix} = i\text{th column in } \mathbf{A}$$

2. Matrix **A** partitioned into its rows:

$$\mathbf{A} = \begin{bmatrix} \alpha_1 \\ \alpha_2 \\ \vdots \\ \alpha_n \end{bmatrix} \tag{14.5b}$$

where

$$\alpha_i = [\alpha_{i1}\ \alpha_{i2} \ldots \alpha_{im}] = i\text{th row in } \mathbf{A}$$

3. A *block diagonal matrix* is a square matrix that can be partitioned so that the nonzero elements are contained only in square submatrices along the main diagonal,

$$\mathbf{A} = \begin{bmatrix} \mathbf{A}_1 & \mathbf{0} & \cdots & \mathbf{0} \\ \mathbf{0} & \mathbf{A}_2 & \cdots & \mathbf{0} \\ \vdots & \vdots & & \vdots \\ \mathbf{0} & \mathbf{0} & \cdots & \mathbf{A}_m \end{bmatrix} \tag{14.6a}$$

$$= \text{diag}\,[\mathbf{A}_1\ \mathbf{A}_2 \ldots \mathbf{A}_m]$$

For this case
(a) $|\mathbf{A}| = |\mathbf{A}_1|\,|\mathbf{A}_2| \ldots |\mathbf{A}_m|$
(b) $\mathbf{A}^{-1} = \text{diag}\left[\mathbf{A}_1^{-1}\ \mathbf{A}_2^{-1} \cdots \mathbf{A}_m^{-1}\right]$, provided that $\mathbf{A}^{-1}$ exists.

4. A square matrix which has all of its elements *below* (*above*) the main diagonal equal to zero, is called an *upper triangular* (*lower triangular*) matrix. The determinant of any triangular matrix is the product of its diagonal elements.

The determinant of *block triangular* matrices is the product of determinants of square submatrices along the main diagonal. For example, for

$$\mathbf{A} = \begin{bmatrix} \mathbf{A}_{11} & \mathbf{A}_{12} \\ \mathbf{0} & \mathbf{A}_{22} \end{bmatrix}; |\mathbf{A}| = |\mathbf{A}_{11}||\mathbf{A}_{22}| \tag{14.6b}$$

Quadratic forms and definiteness of matrices

An expression such as

$$V(x_1, x_2, \ldots, x_n) = \sum_{i=1}^{n}\sum_{j=1}^{n} q_{ij}x_i x_j$$

involving terms of second degree in x_i and x_j is known as the *quadratic form* of *n* variables. When x_i, x_j and q_{ij} are all real, the value of *V* is real, and the quadratic form can be expressed in the vector-matrix notation as [155]

$$V(\mathbf{x}) = \mathbf{x}^T\mathbf{Q}\mathbf{x} \tag{14.7a}$$

where

$$\mathbf{x} = \begin{bmatrix} x_1 \\ x_2 \\ \vdots \\ x_n \end{bmatrix}; \ (n \times 1) \text{ vector of variables}$$

$$\mathbf{Q} = \begin{bmatrix} q_{11} & q_{12} & \cdots & q_{1n} \\ q_{21} & q_{22} & \cdots & q_{2n} \\ \vdots & \vdots & & \vdots \\ q_{n1} & q_{n2} & \cdots & q_{nn} \end{bmatrix}; (n \times n) \text{ constant } \textit{symmetric matrix } (\mathbf{Q} = \mathbf{Q}^T).$$

If for all $\mathbf{x} \neq \mathbf{0}$,

1. $V(\mathbf{x}) = \mathbf{x}^T\mathbf{Q}\mathbf{x} \geq 0$, then $V(\mathbf{x})$ is called a *positive semidefinite function* and $\mathbf{Q}$ is called a *positive semidefinite matrix*;
2. $V(\mathbf{x}) = \mathbf{x}^T\mathbf{Q}\mathbf{x} > 0$, then $V(\mathbf{x})$ is called a *positive definite function* and $\mathbf{Q}$ is called a *positive definite matrix*.

The Sylvester's criterion states that the necessary and sufficient conditions for $\mathbf{Q}$ to be positive definite are that all the successive *principal minors* of $\mathbf{Q}$ be positive, i.e.,

$$q_{11} > 0; \begin{vmatrix} q_{11} & q_{12} \\ q_{21} & q_{22} \end{vmatrix} > 0; \begin{vmatrix} q_{11} & q_{12} & q_{13} \\ q_{21} & q_{22} & q_{23} \\ q_{31} & q_{32} & q_{33} \end{vmatrix} > 0; \cdots ; |\mathbf{Q}| > 0 \tag{14.7b}$$

The necessary and sufficient conditions for $\mathbf{Q}$ to be positive semidefinite are that $\mathbf{Q}$ is singular and all the other principal minors of $\mathbf{Q}$ are non-negative.

14.3 STATE VARIABLE REPRESENTATION

In this book, we will be mainly concerned with linear, time-invariant, lumped systems. Application of physical laws to such a system, as we shall see in this chapter, results in a set of ordinary differential equations which directly provide us with the rates of change of significant variables characterizing the dynamics of the system. These differential equations when rearranged in the form of a set of first-order differential equations are called *state equations*. Laplace transformation of these differential equations with zero initial conditions gives us *transfer function* representation. To obtain a transfer function model, one first writes the state equations of the system which 'naturally' arise from the application of physical laws. State variable modeling is thus a 'natural' step required for transfer function modeling.

State variable analysis and design methods use the matrix theory and are, to some extent, different from those based on transfer function models.

14.3.1 State Variable Modeling

Consider the resistance-inductance-capacitance (RLC) network shown in Fig. 14.1. The input is a voltage source. The desired output information is usually the voltages and currents associated with various elements of the network. This information at time t can be obtained if the voltage across the capacitor and the current through the inductor of the network at that time are known, in addition to the values of the input. The voltage

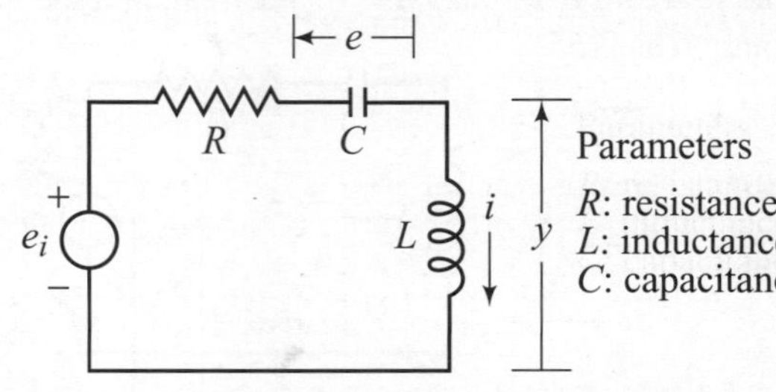

Fig. 14.1 *An RLC network*

$e(t)$ across the capacitor and the current $i(t)$ through the inductor thus constitute a set of *characterizing variables* of the network. The selection of characterizing variables is linked with the energy concept. At time t, energy stored in the capacitor is $\frac{1}{2}Ce^2(t)$ and energy stored in the inductor is $\frac{1}{2}Li^2(t)$. Dynamical changes in characterizing variables are caused by the redistribution of energy within the network. The number of independent energy storage possibilities thus equals the number of characterizing variables of the network. The values of the characterizing variables at time t describe the *state* of the network at that time; these variables are therefore called *state variables* of the network.

Network analysis usually requires setting up of dynamical equations (using Kirchhoff's voltage and current laws) in terms of rates of change of capacitor voltages and inductor currents. The solution of these equations describes the state of the network at time t. The desired output information is then obtained from the state using an algebraic relation.

Let us set up dynamical equations for the RLC network of Fig. 14.1, by using Kirchhoff's voltage law. The governing equations of the system are

$$R\,i(t)+e(t)+L\frac{di(t)}{dt}=e_i(t);C\frac{de(t)}{dt}=i(t)$$

Rearrangement of these equations gives the rates of change of capacitor voltage and inductor current.

$$\left.\begin{aligned}\frac{de(t)}{dt}&=\frac{1}{C}i(t)\\ \frac{di(t)}{dt}&=-\frac{1}{L}e(t)-\frac{R}{L}i(t)+\frac{1}{L}e_i(t)\end{aligned}\right\}\qquad(14.8a)$$

The solution of these equations for the given input $e_i(t)$ applied at $t = 0^+$, and the given initial state $\{e(0), i(0)\}$ yields the state $\{e(t), i(t)\}$, $t > 0$. If $y(t)$ shown in Fig. 14.1 is the desired output information, we have the following algebraic relation to obtain $y(t)$:

$$y(t) = -\,e(t) - R\,i(t) + e_i(t)\qquad(14.8b)$$

Equations (14.8a) give the rates of change of state variables (capacitor voltage $e(t)$, and inductor current $i(t)$) in terms of the state variables and the input. These equations are called the *state equations*. Equation (14.8b) is an instantaneous relation, reading output $y(t)$ from the state variables $\{e(t), i(t)\}$ and the input $e_i(t)$. This equation is called the *output equation*.

It may be noted that the state of a system is not uniquely specified. For the network of Fig. 14.1, the charge $q(t)$ deposited on the capacitor is given by

$$q(t) = C\,e(t)$$

Therefore, the variables $q(t)$ and $i(t)$ also qualify as state variables for this network.

In terms of these state variables, the state and output equations are given below:

$$\left.\begin{aligned}\frac{dq(t)}{dt}&=i(t)\\ \frac{di(t)}{dt}&=-\frac{1}{LC}q(t)-\frac{R}{L}i(t)+\frac{1}{L}e_i(t)\end{aligned}\right\}\qquad(14.9a)$$

$$y(t)=-\frac{1}{C}q(t)-Ri(t)+e_i(t)\qquad(14.9b)$$

In mass-spring-damper systems, we are usually interested in the displacements, velocities, and forces associated with various elements of the systems. Normally, an independent set of displacements and velocities of the masses, and displacements of the springs constitutes a set of state variables of the system. The values of these variables at time t describe the status of energy storage elements at that time, and also ensure that the desired output information can be 'read' from the state variables.

For the mass-spring-damper system of Fig. 14.2a, the displacement $x(t)$ and velocity $v(t)$ are state variables. A systematic way of setting up dynamical equations for mass-spring-damper systems is to draw a free-body diagram. In a free-body diagram, each mass is isolated from the rest of the system; forces acting on each free-body are due to the rest of the system, including external forces. The free-body mass moves under the action of the resultant force. By applying Newton's law of motion to each free-body, we get the force equations describing the dynamics of the system.

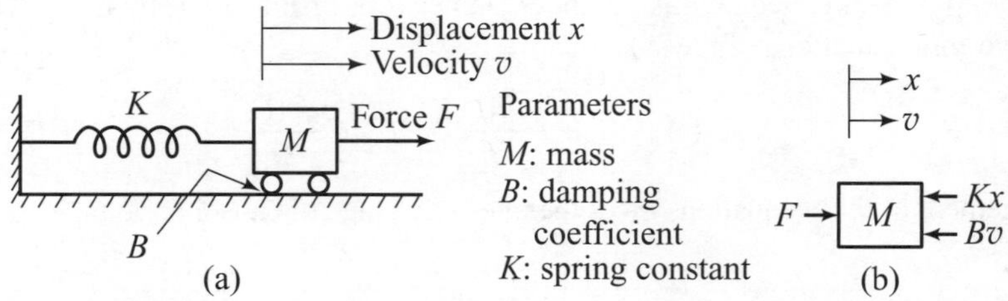

Fig. 14.2 *(a) A mass-spring-damper system (b) Free-body diagram*

Figure 14.2b shows the free-body diagram of the system of Fig. 14.2a. Applying Newton's law of motion, the force equation can be written as

$$F(t) - B\,v(t) - K\,x(t) = M\frac{dv(t)}{dt}$$

Since $$v(t) \triangleq \frac{dx(t)}{dt}, \text{ we have}$$

$$M\frac{d^2x(t)}{dt^2} + B\frac{dx(t)}{dt} + K\,x(t) = F(t)$$

Rearrangement of these equations gives the rates of change of displacement $x(t)$ and velocity $v(t)$:

$$\left.\begin{aligned}\frac{dx(t)}{dt} &= v(t)\\ \frac{dv(t)}{dt} &= -\frac{K}{M}x(t) - \frac{B}{M}v(t) + \frac{1}{M}F(t)\end{aligned}\right\} \tag{14.10a}$$

The solution of these equations for given input (force $F(t)$) applied at $t = 0^+$ and given initial state $\{x(0), v(0)\}$ yields the state $\{x(t), v(t)\}$, $t > 0$. If displacement of the mass is the desired output $y(t)$, we have the output equation,

$$y(t) = x(t) \tag{14.10b}$$

Consider now the general SISO system shown in the block diagram of Fig. 14.3a. We will denote the system *state* by n *state variables* $\{x_1(t), x_2(t), ..., x_n(t)\}$. In compact notation, this information may be represented by the *state vector*

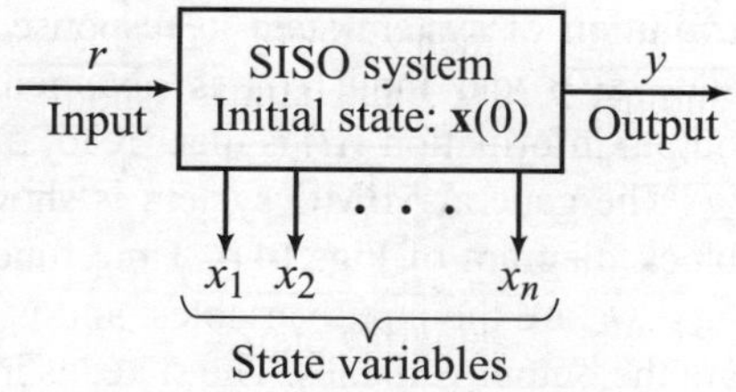

Fig. 14.3a *A SISO system*

$$\mathbf{x}(t) \triangleq \begin{bmatrix} x_1(t) \\ x_2(t) \\ \vdots \\ x_n(t) \end{bmatrix}$$

The dimension of the state vector defines the *order of the system.*

The set of equations that describe the relationship among input, output, and the state variables, is called the *state variable model.* For nth-order linear time-invariant system, the state variable model takes the form (refer to Eqns (14.8), (14.9), (14.10)):

$$\left.\begin{aligned} \dot{x}_1(t) &\triangleq \frac{dx_1(t)}{dt} = a_{11}x_1(t) + a_{12}x_2(t) + \cdots + a_{1n}x_n(t) + b_1 r(t) \\ \dot{x}_2(t) &\triangleq \frac{dx_2(t)}{dt} = a_{21}x_1(t) + a_{22}x_2(t) + \cdots + a_{2n}x_n(t) + b_2 r(t) \\ &\vdots \\ \dot{x}_n(t) &\triangleq \frac{dx_n(t)}{dt} = a_{n1}x_1(t) + a_{n2}x_2(t) + \cdots + a_{nn}x_n(t) + b_n r(t) \end{aligned}\right\}$$

$$y(t) = c_1 x_1(t) + c_2 x_2(t) + \cdots + c_n x_n(t) + d\, r(t)$$

or in vector-matrix form

$$\dot{\mathbf{x}}(t) = \mathbf{A}\mathbf{x}(t) + \mathbf{b}\, r(t) : \textit{state equation} \tag{14.11a}$$

$$y(t) = \mathbf{c}\mathbf{x}(t) + d\, r(t) : \textit{output equation} \tag{14.11b}$$

with

$$\mathbf{x}(t) = \begin{bmatrix} x_1(t) \\ x_2(t) \\ \vdots \\ x_n(t) \end{bmatrix}; \mathbf{A} = \begin{bmatrix} a_{11} & a_{12} & \cdots & a_{1n} \\ a_{21} & a_{22} & \cdots & a_{2n} \\ \vdots & \vdots & & \vdots \\ a_{n1} & a_{n2} & \cdots & a_{nn} \end{bmatrix}; \mathbf{b} = \begin{bmatrix} b_1 \\ b_2 \\ \vdots \\ b_n \end{bmatrix}; \mathbf{c} = [c_1\ c_2 \cdots c_n]$$

where $\mathbf{x}(t)$, $r(t)$ and $y(t)$ are the state, the input and the output of a system, respectively. $\mathbf{A}$ is $n \times n$ constant matrix; $\mathbf{b}$ is $n \times 1$ constant column vector; $\mathbf{c}$ is $1 \times n$ constant row vector; and d is a constant scalar representing direct coupling between input and output.

Equations (14.11a) are n first-order differential equations expressing time derivatives of each of the components of the state vector $\mathbf{x}(t)$ as a linear combination of system states and input. The solution of these equations for given input $r(t)$ applied at $t = 0^+$, and given initial state $\mathbf{x}(0)$ yields the system state $\mathbf{x}(t)$, $t > 0$. The output $y(t)$ (Eqn. (14.11b)) is a linear combination of system states and input.

Equations (14.11a) and (14.11b) together provide the input-output-state description of a dynamic system, represented diagrammatically in Fig. 14.3b. Note that in this model structure, the dynamic evolution of system state in response to initial state $\mathbf{x}(0)$ and input $r(t)$ is obtained; desired output information $y(t)$ is then 'read' from $\mathbf{x}(t)$.

The general MIMO system is shown in the block diagram of Fig. 14.4. Time functions, r_1, r_2, ..., r_p are the input variables, and y_1, y_2, ..., y_q are the output variables. The state, as in the case of SISO systems, is represented by the variables x_1, x_2, ..., x_n. For notational economy, the different variables may be represented by the input vector $\mathbf{r}(t)$, the output vector $\mathbf{y}(t)$, and the state vector $\mathbf{x}(t)$, where

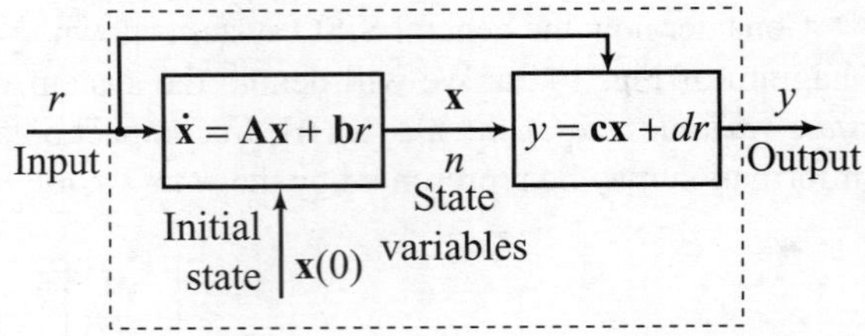

Fig. 14.3b *Input-output-state description*

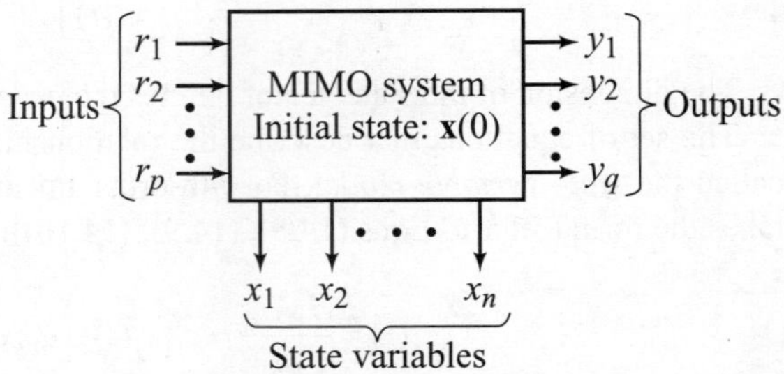

Fig. 14.4 *A MIMO system*

$$\mathbf{r}(t) = \begin{bmatrix} r_1(t) \\ r_2(t) \\ \vdots \\ r_p(t) \end{bmatrix}; \mathbf{y}(t) = \begin{bmatrix} y_1(t) \\ y_2(t) \\ \vdots \\ y_q(t) \end{bmatrix}; \mathbf{x}(t) = \begin{bmatrix} x_1(t) \\ x_2(t) \\ \vdots \\ x_n(t) \end{bmatrix}$$

For MIMO systems, the derivative of each state variable becomes a linear combination of system states $x_1, x_2, \ldots, x_n$, and the inputs $r_1, r_2, \ldots, r_p$. Also, each output variable at time t is a linear combination of the states and the inputs at time t. The state variable model for MIMO systems takes the following form:

$$\dot{\mathbf{x}}(t) = \mathbf{Ax}(t) + \mathbf{Br}(t) : \textit{state equation} \tag{14.12a}$$

$$\mathbf{y}(t) = \mathbf{Cx}(t) + \mathbf{Dr}(t) : \textit{output equation} \tag{14.12b}$$

with

$$\mathbf{A} = \begin{bmatrix} a_{11} & a_{12} & \cdots & a_{1n} \\ a_{21} & a_{22} & \cdots & a_{2n} \\ \vdots & \vdots & & \vdots \\ a_{n1} & a_{n2} & \cdots & a_{nn} \end{bmatrix}; \mathbf{B} = \begin{bmatrix} b_{11} & b_{12} & \cdots & b_{1p} \\ b_{21} & b_{22} & \cdots & b_{2p} \\ \vdots & \vdots & & \vdots \\ b_{n1} & b_{n2} & \cdots & b_{np} \end{bmatrix}; \mathbf{C} = \begin{bmatrix} c_{11} & c_{12} & \cdots & c_{1n} \\ c_{21} & c_{22} & \cdots & c_{2n} \\ \vdots & \vdots & & \vdots \\ c_{q1} & c_{q2} & \cdots & c_{qn} \end{bmatrix}; \mathbf{D} = \begin{bmatrix} d_{11} & d_{12} & \cdots & d_{1p} \\ d_{21} & d_{22} & \cdots & d_{2p} \\ \vdots & \vdots & & \vdots \\ d_{q1} & d_{q2} & \cdots & d_{qp} \end{bmatrix}$$

A, **B**, **C** and **D** are respectively. $n \times n$, $n \times p$, $q \times n$ and $q \times p$ constant matrices.

We will be mostly concerned with SISO system configurations of the form shown in the block diagram of Fig. 14.5. The plant in the figure is a physical process, characterized by the state variables $x_1, x_2, \ldots, x_n$; the output variable y, and the input variable u. Application of physical laws to this process results in a state variable model of the form (refer to Eqns (14.11)):

$$\dot{\mathbf{x}}(t) = \mathbf{Ax}(t) + \mathbf{b}u(t); \ \mathbf{x}(t_0) \triangleq \mathbf{x}^0 \tag{14.13a}$$

$$y(t) = \mathbf{cx}(t) + du(t) \tag{14.13b}$$

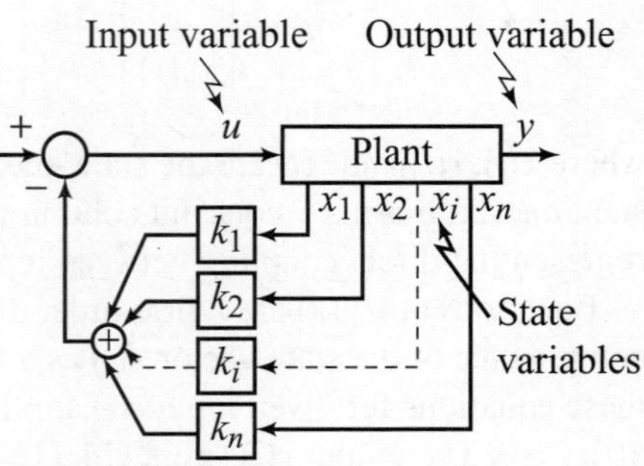

Fig. 14.5 *A state-feedback control configuration*

Example 14.1 Two very usual applications of dc motors are in speed and position control systems. Figure 14.6 gives the basic block diagram of a speed control system. A separately excited dc motor drives the load. A dc tachogenerator is attached to the motor shaft; speed signal is fedback and the error signal is used to control the armature voltage of the motor.

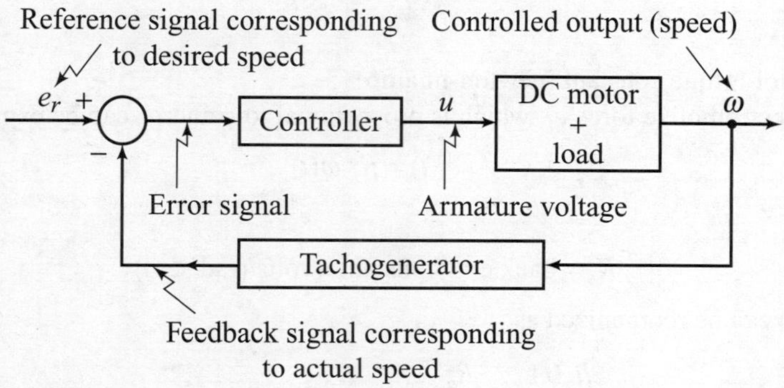

Fig. 14.6 *Basic block diagram of a closed-loop speed control system*

In the following, we derive the plant model for the speed control system. A separately excited dc motor with armature voltage control is shown in Fig. 14.7.

The loop equation for the armature circuit is

$$u(t) = L_a \frac{d i_a(t)}{dt} + R_a\, i_a(t) + e_b(t) \tag{14.14a}$$

where

L_a = inductance of armature winding (henrys)
R_a = resistance of armature winding (ohms)
i_a = armature current (amperes)
e_b = back emf (volts)
u = applied armature voltage (volts)

The torque balance equation is

$$T_M(t) = J \frac{d\omega(t)}{dt} + B\,\omega(t) \tag{14.14b}$$

where

T_M = torque developed by the motor (newton-m)

J = equivalent moment of inertia of motor and load referred to motor shaft (kg-m^2)

B = equivalent viscous friction coefficient of motor and load referred to motor shaft(newton-m/(rad/sec))

ω = angular velocity of motor shaft (rad/sec)

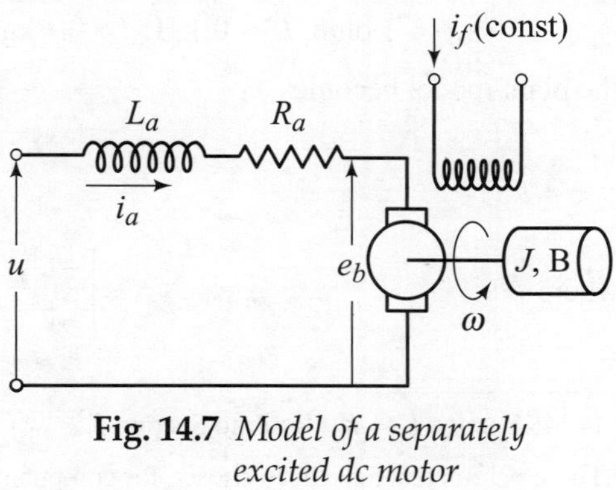

Fig. 14.7 *Model of a separately excited dc motor*

In servo applications, the dc motors are generally used in the linear range of the magnetization curve. Therefore, the air gap flux ϕ is proportional to the field current. For the armature controlled motor, the field current i_f is held constant. Therefore, the torque T_M developed by the motor, which is proportional to the product of the armature current and the air gap flux, can be expressed as

$$T_M(t) = K_T\, i_a(t) \tag{14.14c}$$

where

K_T = motor torque constant (newton-m/amp)

The counter electromotive force e_b, which is proportional to ϕ and ω, can be expressed as

$$e_b(t) = K_b\, \omega(t) \tag{14.14d}$$

where

K_b = back emf constant[2] (volts/(rad/sec))

Equations (14.14) can be reorganized as

$$\begin{aligned} \frac{di_a(t)}{dt} &= -\frac{R_a}{L_a} i_a(t) - \frac{K_b}{L_a}\omega(t) + \frac{1}{L_a}u(t) \\ \frac{d\omega(t)}{dt} &= \frac{K_T}{J} i_a(t) - \frac{B}{J}\omega(t) \end{aligned} \tag{14.15}$$

$x_1(t) = \omega(t)$, and $x_2(t) = i_a(t)$ is the obvious choice for state variables. The output variable is $y(t) = \omega(t)$.

The plant model of the speed control system, organized into the vector-matrix notation, is given below:

$$\begin{bmatrix} \dot{x}_1(t) \\ \dot{x}_2(t) \end{bmatrix} = \begin{bmatrix} -\dfrac{B}{J} & \dfrac{K_T}{J} \\ -\dfrac{K_b}{L_a} & -\dfrac{R_a}{L_a} \end{bmatrix} \begin{bmatrix} x_1(t) \\ x_2(t) \end{bmatrix} + \begin{bmatrix} 0 \\ \dfrac{1}{L_a} \end{bmatrix} u(t)$$

$$y(t) = x_1(t)$$

Let us assign numerical values to the system parameters.

For the parameters[3]

$$R_a = 1 \text{ ohm}, L_a = 0.1 \text{ H}, J = 0.1 \text{ kg-m}^2, B = 0.1 \text{ newton-m/(rad/sec)}, K_b = K_T = 0.1 \tag{14.16}$$

the plant model becomes

$$\begin{aligned} \dot{\mathbf{x}}(t) &= \mathbf{A}\mathbf{x}(t) + \mathbf{b}u(t) \\ y(t) &= \mathbf{c}\mathbf{x}(t) \end{aligned} \tag{14.17}$$

where $$\mathbf{A} = \begin{bmatrix} -1 & 1 \\ -1 & -10 \end{bmatrix}; \mathbf{b} = \begin{bmatrix} 0 \\ 10 \end{bmatrix}; \mathbf{c} = [1 \quad 0]$$

[2] In MKS units, $K_b = K_T$. Refer to Section 3.2.

[3] These parameters have been chosen for computational convenience.

Example 14.2 Figure 14.8 gives the basic block diagram of a position control system. The controlled variable is now the angular position $\theta(t)$ of the motor shaft:

$$\frac{d\theta(t)}{dt} = \omega(t) \tag{14.18}$$

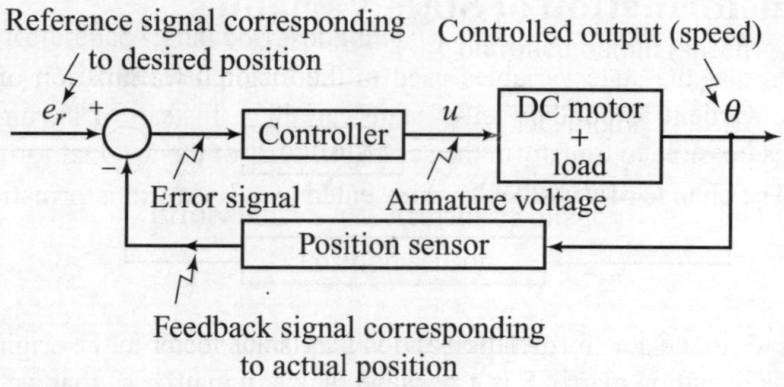

Fig. 14.8 *Basic block diagram of a closed-loop position control system*

We make the following choice for state and output variables:

$$x_1(t) = \theta(t),\ x_2(t) = \omega(t),\ x_3(t) = i_a(t),\ y(t) = \theta(t)$$

For this choice, we obtain the following plant model from Eqns (14.15) and (14.18).

$$\begin{bmatrix} \dot{x}_1(t) \\ \dot{x}_2(t) \\ \dot{x}_3(t) \end{bmatrix} = \begin{bmatrix} 0 & 1 & 0 \\ 0 & -\dfrac{B}{J} & \dfrac{K_T}{J} \\ 0 & -\dfrac{K_b}{L_a} & -\dfrac{R_a}{L_a} \end{bmatrix} \begin{bmatrix} x_1(t) \\ x_2(t) \\ x_3(t) \end{bmatrix} + \begin{bmatrix} 0 \\ 0 \\ \dfrac{1}{L_a} \end{bmatrix} u(t)$$

$$y(t) = x_1(t)$$

For the system parameters given by (14.16), the plant model for position control system becomes

$$\begin{aligned} \dot{\mathbf{x}}(t) &= \mathbf{A}\mathbf{x}(t) + \mathbf{b}\mathbf{u}(t) \\ y(t) &= \mathbf{c}\mathbf{x}(t) \end{aligned} \tag{14.19}$$

where

$$\mathbf{A} = \begin{bmatrix} 0 & 1 & 0 \\ 0 & -1 & 1 \\ 0 & -1 & -10 \end{bmatrix};\ \mathbf{b} = \begin{bmatrix} 0 \\ 0 \\ 10 \end{bmatrix};\ \mathbf{c} = \begin{bmatrix} 1 & 0 & 0 \end{bmatrix}$$

In Examples 14.1 and 14.2 discussed above, the selected state variables are the physical quantities of the systems, which can be measured. It may sometimes be necessary and advantageous to provide a feedback proportional to the state variables of a system, rather than the output alone, for the purpose of stabilizing and improving the performance of a system (refer to Chapter 15). The implementation of design with state variable feedback becomes straightforward if the state variables are available for feedback.

The choice of physical variables of a system as state variables therefore helps in the implementation of design. Another advantage of selecting physical variables for state variable formulation is that the solution of state equation gives time variation of variables which have direct relevance to the physical system.

14.3.2 Transformation of State Variables

It frequently happens that the state variables used in the original formulation of the dynamics of a system are not as convenient as another set of state variables. Instead of having to reformulate the system dynamics, it is possible to transform the set $\{\mathbf{A}, \mathbf{b}, \mathbf{c}, d\}$ of the original formulation (14.13) to a new set $\{\bar{\mathbf{A}}, \bar{\mathbf{b}}, \bar{\mathbf{c}}, \bar{d}\}$. The change of variables is represented by a linear transformation,

$$\mathbf{x} = \mathbf{P}\bar{\mathbf{x}} \tag{14.20a}$$

where $\bar{\mathbf{x}}$ is a state vector in the new formulation and $\mathbf{x}$ is the state vector in the original formulation. It is assumed that the transformation matrix $\mathbf{P}$ is a nonsingular $n \times n$ matrix, so that we can always write

$$\bar{\mathbf{x}} = \mathbf{P}^{-1}\mathbf{x} \tag{14.20b}$$

We assume, moreover, that $\mathbf{P}$ is a constant matrix.

The original dynamics are expressed by

$$\dot{\mathbf{x}}(t) = \mathbf{A}\mathbf{x}(t) + \mathbf{b}u(t); \mathbf{x}(t_0) \triangleq \mathbf{x}^0 \tag{14.21a}$$

and the output by

$$y(t) = \mathbf{c}\mathbf{x}(t) + du(t) \tag{14.21b}$$

Substitution of $\mathbf{x}$ as given by Eqn. (14.20a) into these equations, gives

$$\mathbf{P}\,\dot{\bar{\mathbf{x}}}(t) = \mathbf{A}\mathbf{P}\bar{\mathbf{x}}(t) + \mathbf{b}u(t)$$
$$y(t) = \mathbf{c}\mathbf{P}\,\bar{\mathbf{x}}(t) + du(t)$$

or

$$\dot{\bar{\mathbf{x}}}(t) = \bar{\mathbf{A}}\bar{\mathbf{x}}(t) + \bar{\mathbf{b}}\,u(t); \bar{\mathbf{x}}(t_0) = \mathbf{P}^{-1}\mathbf{x}(t_0) \tag{14.22a}$$

$$y(t) = \bar{\mathbf{c}}\,\bar{\mathbf{x}}(t) + \bar{d}\,u(t) \tag{14.22b}$$

with

$$\bar{\mathbf{A}} = \mathbf{P}^{-1}\mathbf{A}\mathbf{P}, \bar{\mathbf{b}} = \mathbf{P}^{-1}\mathbf{b}, \bar{\mathbf{c}} = \mathbf{c}\mathbf{P},\ \bar{d} = d$$

In the next section, we will prove that both the linear systems (14.21) and (14.22) have identical output responses for the same input. The linear system (14.22) is said to be *equivalent* to the linear system (14.21), and $\mathbf{P}$ is called an *equivalence* or *similarity transformation.*

It is obvious that there exist an infinite number of equivalent systems since the transformation matrix $\mathbf{P}$ can be arbitrarily chosen. Some transformations have been extensively used for the purposes of analysis and design. Five of such special (*canonical*) transformations will be introduced in this chapter.

Example 14.3 Example 14.1 revisited

For the system of Fig. 14.7, we have taken angular velocity $\omega(t)$ and armature current $i_a(t)$ as state variables:

$$\mathbf{x} = \begin{bmatrix} x_1 \\ x_2 \end{bmatrix} = \begin{bmatrix} \omega \\ i_a \end{bmatrix}$$

We now define new state variables as

$$\bar{x}_1 = \omega, \ \bar{x}_2 = -\omega + i_a$$

or
$$\bar{\mathbf{x}} = \begin{bmatrix} \bar{x}_1 \\ \bar{x}_2 \end{bmatrix} = \begin{bmatrix} x_1 \\ -x_1 + x_2 \end{bmatrix} = \begin{bmatrix} 1 & 0 \\ -1 & 1 \end{bmatrix} \begin{bmatrix} x_1 \\ x_2 \end{bmatrix}$$

We can express velocity $x_1(t)$ and armature current $x_2(t)$ in terms of the variables $\bar{x}_1(t)$ and $\bar{x}_2(t)$:

$$\mathbf{x} = \mathbf{P}\bar{\mathbf{x}} \tag{14.23}$$

with
$$\mathbf{P} = \begin{bmatrix} 1 & 0 \\ 1 & 1 \end{bmatrix}$$

Using Eqns (14.22) and (14.17), we obtain the following state variable model for the system of Fig. 14.7 in terms of the transformed state vector $\bar{\mathbf{x}}(t)$:

$$\begin{aligned} \dot{\bar{\mathbf{x}}}(t) &= \bar{\mathbf{A}}\,\bar{\mathbf{x}}(t) + \bar{\mathbf{b}}\,u(t) \\ y(t) &= \bar{\mathbf{c}}\,\bar{\mathbf{x}}(t) \end{aligned} \tag{14.24}$$

where

$$\bar{\mathbf{A}} = \mathbf{P}^{-1}\mathbf{A}\,\mathbf{P} = \begin{bmatrix} 1 & 0 \\ -1 & 1 \end{bmatrix} \begin{bmatrix} -1 & 1 \\ -1 & -10 \end{bmatrix} \begin{bmatrix} 1 & 0 \\ 1 & 1 \end{bmatrix} = \begin{bmatrix} 0 & 1 \\ -11 & -11 \end{bmatrix}$$

$$\bar{\mathbf{b}} = \mathbf{P}^{-1}\mathbf{b} = \begin{bmatrix} 1 & 0 \\ -1 & 1 \end{bmatrix} \begin{bmatrix} 0 \\ 10 \end{bmatrix} = \begin{bmatrix} 0 \\ 10 \end{bmatrix}$$

$$\bar{\mathbf{c}} = \mathbf{c}\mathbf{P} = \begin{bmatrix} 1 & 0 \end{bmatrix} \begin{bmatrix} 1 & 0 \\ 1 & 1 \end{bmatrix} = \begin{bmatrix} 1 & 0 \end{bmatrix}$$

$$\bar{x}_1(t_0) = x_1(t_0); \ \bar{x}_2(t_0) = -x_1(t_0) + x_2(t_0) \tag{14.25}$$

Equations (14.24) give an alternative state variable model of the system previously represented by Eqns (14.17). $\bar{\mathbf{x}}(t)$ and $\mathbf{x}(t)$ both qualify to be state vectors of the given system (the two vectors individually characterize the system completely at time t), and the output $y(t)$, as we shall see shortly, is uniquely determined from either of the models (14.17) and (14.24). State variable model (14.24) is thus *equivalent* to the model (14.17), and the matrix **P** given by Eqn. (14.23) is an equivalence or similarity transformation.

The state variable model given by Eqn. (14.24) is in a *canonical* (special) form.

14.4 CONVERSION OF STATE VARIABLE MODELS TO TRANSFER FUNCTIONS

We shall derive the transfer function of a SISO system from the Laplace-transformed version of the state and output equations. Refer to Section 14.2 for the matrix operations used in the derivation.

Consider the state variable model (Eqns (14.13)):

$$\dot{\mathbf{x}}(t) = \mathbf{A}\mathbf{x}(t) + \mathbf{b}u(t); \mathbf{x}(t_0) \triangleq \mathbf{x}^0$$
$$y(t) = \mathbf{c}\mathbf{x}(t) + du(t) \tag{14.26}$$

Taking the Laplace transform of Eqns (14.26), we obtain

$$s\mathbf{X}(s) - \mathbf{x}^0 = \mathbf{A}\mathbf{X}(s) + \mathbf{b}U(s)$$
$$Y(s) = \mathbf{c}\mathbf{X}(s) + dU(s)$$

where

$$\mathbf{X}(s) \triangleq \mathscr{L}[\mathbf{x}(t)]; U(s) \triangleq \mathscr{L}[u(t)]; Y(s) \triangleq \mathscr{L}[y(t)]$$

Manipulation of these equations gives

$$(s\mathbf{I} - \mathbf{A})\mathbf{X}(s) = \mathbf{x}^0 + \mathbf{b}U(s); \ \mathbf{I} \text{ is } n \times n \text{ identity matrix}$$

or

$$\mathbf{X}(s) = (s\mathbf{I} - \mathbf{A})^{-1}\mathbf{x}^0 + (s\mathbf{I} - \mathbf{A})^{-1}\mathbf{b}U(s) \tag{14.27a}$$

$$Y(s) = \mathbf{c}(s\mathbf{I} - \mathbf{A})^{-1}\mathbf{x}^0 + [\mathbf{c}(s\mathbf{I} - \mathbf{A})^{-1}\mathbf{b} + d]U(s) \tag{14.27b}$$

Equations (14.27) are algebraic equations. If $\mathbf{x}^0$ and $U(s)$ are known, $\mathbf{X}(s)$ and $Y(s)$ can be computed from these equations.

In the case of a zero initial state (i.e., $\mathbf{x}^0 = \mathbf{0}$), the input–output behavior of the system (14.26) is determined entirely by the transfer function,

$$\frac{Y(s)}{U(s)} = G(s) = \mathbf{c}(s\mathbf{I} - \mathbf{A})^{-1}\mathbf{b} + d \tag{14.28}$$

We can express the inverse of the matrix $(s\mathbf{I} - \mathbf{A})$ as

$$(s\mathbf{I} - \mathbf{A})^{-1} = \frac{(s\mathbf{I} - \mathbf{A})^{+}}{|s\mathbf{I} - \mathbf{A}|} \tag{14.29}$$

where

$$|s\mathbf{I} - \mathbf{A}| = \text{determinant of the matrix } (s\mathbf{I} - \mathbf{A})$$
$$(s\mathbf{I} - \mathbf{A})^{+} = \text{adjoint of the matrix } (s\mathbf{I} - \mathbf{A})$$

Using Eqn. (14.29), the transfer function $G(s)$ given by Eqn. (14.28) can be written as

$$G(s) = \frac{\mathbf{c}(s\mathbf{I} - \mathbf{A})^{+}\mathbf{b}}{|s\mathbf{I} - \mathbf{A}|} + d \tag{14.30}$$

14.4.1 Eigenvalues of a Matrix

For a general nth-order matrix

$$\mathbf{A} = \begin{bmatrix} a_{11} & a_{12} & \cdots & a_{1n} \\ a_{21} & a_{22} & \cdots & a_{2n} \\ \vdots & \vdots & & \vdots \\ a_{n1} & a_{n2} & \cdots & a_{nn} \end{bmatrix},$$

the matrix $(s\mathbf{I} - \mathbf{A})$ has the following appearance:

$$(s\mathbf{I} - \mathbf{A}) = \begin{bmatrix} s - a_{11} & -a_{12} & \cdots & -a_{1n} \\ -a_{21} & s - a_{22} & \cdots & -a_{2n} \\ \vdots & \vdots & & \vdots \\ -a_{n1} & -a_{n2} & \cdots & s - a_{nn} \end{bmatrix}$$

If we imagine calculating *det* $(s\mathbf{I} - \mathbf{A})$, we see that one of the terms will be the product of diagonal elements of $(s\mathbf{I} - \mathbf{A})$:

$$(s - a_{11})(s - a_{22}) \cdots (s - a_{nn}) = s^n + \alpha_1' s^{n-1} + \cdots + \alpha_n',$$

a polynomial of degree n with the leading coefficient of unity. There will be other terms coming from the off-diagonal elements of $(s\mathbf{I} - \mathbf{A})$, but none will have a degree as high as n. Thus, $|s\mathbf{I} - \mathbf{A}|$ will be of the following form:

$$|s\mathbf{I} - \mathbf{A}| = \Delta(s) = s^n + \alpha_1 s^{n-1} + \cdots + \alpha_n \tag{14.31}$$

where α_i are constant scalars.

This is known as the *characteristic polynomial* of the matrix **A**. It plays a vital role in the dynamic behavior of the system. The roots of this polynomial are called the *characteristic roots* or *eigenvalues* of matrix **A**. These roots, as we shall see in Section 14.6, determine the essential features of the unforced dynamic behavior of the system (14.26).

The adjoint of an $n \times n$ matrix is itself an $n \times n$ matrix whose elements are the cofactors of the original matrix. Each cofactor is obtained by computing the determinant of the matrix that remains when a row and a column of the original matrix are deleted. It, thus, follows that each element in $(s\mathbf{I} - \mathbf{A})^+$ is a polynomial in s of maximum degree $(n - 1)$. Adjoint of $(s\mathbf{I} - \mathbf{A})$ can, therefore, be expressed as

$$(s\mathbf{I} - \mathbf{A})^+ = \mathbf{Q}_1 s^{n-1} + \mathbf{Q}_2 s^{n-2} + \cdots + \mathbf{Q}_{n-1} s + \mathbf{Q}_n \tag{14.32}$$

where $\mathbf{Q}_i$ are constant $n \times n$ matrices. We can express transfer function $G(s)$ given by Eqn. (14.30) in the following form:

$$G(s) = \frac{\mathbf{c}[\mathbf{Q}_1 s^{n-1} + \mathbf{Q}_2 s^{n-2} + \cdots + \mathbf{Q}_{n-1} s + \mathbf{Q}_n]\,\mathbf{b}}{s^n + \alpha_1 s^{n-1} + \cdots + \alpha_{n-1} s + \alpha_n} + d \tag{14.33}$$

$G(s)$ is thus a rational function of s. When $d = 0$, the degree of numerator polynomial of $G(s)$ is strictly less than the degree of the denominator polynomial and, therefore, the resulting transfer function is a *strictly proper transfer function*. When $d \neq 0$, the degree of numerator polynomial of $G(s)$ will be equal to the degree of the denominator polynomial, giving a *proper transfer function*. Further,

$$d = \lim_{s\to\infty} [G(s)] \tag{14.34}$$

From Eqns (14.31) and (14.33) we observe that the characteristic polynomial of matrix **A** of the system (14.26) is same as the denominator polynomial of the corresponding transfer function $G(s)$. If there are no cancelations between the numerator and denominator polynomials of $G(s)$ in Eqn. (14.33), the *eigenvalues* of matrix **A** are same as the *poles* of $G(s)$. We will take up in Section 14.9, this aspect of the correspondence between state variable models and transfer functions. It will be proved that for a completely controllable and completely observable state variable model, the eigenvalues of matrix **A** are same as the poles of the corresponding transfer function.

14.4.2 Invariance Property

It is recalled that the state variable model for a system is not unique, but depends on the choice of a set of state variables. A transformation

$$\mathbf{x}(t) = \mathbf{P}\bar{\mathbf{x}}(t);\ \mathbf{P} \text{ is a nonsingular matrix} \tag{14.35}$$

results in the following alternative state variable model (refer to Eqns (14.22)) for the system (14.26):

$$\dot{\bar{\mathbf{x}}}(t) = \bar{\mathbf{A}}\bar{\mathbf{x}}(t) + \bar{\mathbf{b}}u(t);\ \bar{\mathbf{x}}(t_0) = \mathbf{P}^{-1}\mathbf{x}(t_0) \tag{14.36a}$$

$$y(t) = \bar{\mathbf{c}}\bar{\mathbf{x}}(t) + du(t) \tag{14.36b}$$

where $\bar{\mathbf{A}} = \mathbf{P}^{-1}\mathbf{A}\mathbf{P}, \bar{\mathbf{b}} = \mathbf{P}^{-1}\mathbf{b}, \bar{\mathbf{c}} = \mathbf{c}\mathbf{P}$

The definition of new set of internal state variables should evidently not affect the eigenvalues or input-output behavior. This may be verified by evaluating the characteristic polynomial and the transfer function of the transformed system.

1.
$$|s\mathbf{I} - \bar{\mathbf{A}}| = |s\mathbf{I} - \mathbf{P}^{-1}\mathbf{A}\mathbf{P}| = |s\mathbf{P}^{-1}\mathbf{P} - \mathbf{P}^{-1}\mathbf{A}\mathbf{P}|$$
$$= |\mathbf{P}^{-1}(s\mathbf{I} - \mathbf{A})\mathbf{P}| = |\mathbf{P}^{-1}|\,|s\mathbf{I} - \mathbf{A}|\,|\mathbf{P}| = |s\mathbf{I} - \mathbf{A}| \tag{14.37}$$

2. System output in response to input $u(t)$ is given by the transfer function,

$$\begin{aligned}\bar{G}(s) &= \bar{\mathbf{c}}(s\mathbf{I} - \bar{\mathbf{A}})^{-1}\bar{\mathbf{b}} + d\\ &= \mathbf{c}\mathbf{P}(s\mathbf{I} - \mathbf{P}^{-1}\mathbf{A}\mathbf{P})^{-1}\mathbf{P}^{-1}\mathbf{b} + d\\ &= \mathbf{c}\mathbf{P}(s\mathbf{P}^{-1}\mathbf{P} - \mathbf{P}^{-1}\mathbf{A}\mathbf{P})^{-1}\mathbf{P}^{-1}\mathbf{b} + d\\ &= \mathbf{c}\mathbf{P}[\mathbf{P}^{-1}(s\mathbf{I} - \mathbf{A})\mathbf{P}]^{-1}\mathbf{P}^{-1}\mathbf{b} + d\\ &= \mathbf{c}\mathbf{P}\mathbf{P}^{-1}(s\mathbf{I} - \mathbf{A})^{-1}\mathbf{P}\mathbf{P}^{-1}\mathbf{b} + d\\ &= \mathbf{c}(s\mathbf{I} - \mathbf{A})^{-1}\mathbf{b} + d = G(s)\end{aligned} \tag{14.38}$$

3. System output in response to initial state $\bar{\mathbf{x}}(t_0)$ is given by (refer to Eqn. (14.27b))

$$\begin{aligned}\bar{\mathbf{c}}(s\mathbf{I} - \bar{\mathbf{A}})^{-1}\bar{\mathbf{x}}(t_0) &= \mathbf{c}\mathbf{P}(s\mathbf{I} - \mathbf{P}^{-1}\mathbf{A}\mathbf{P})^{-1}\mathbf{P}^{-1}\mathbf{x}(t_0)\\ &= \mathbf{c}(s\mathbf{I} - \mathbf{A})^{-1}\mathbf{x}(t_0)\end{aligned} \tag{14.39}$$

The input-output behavior of the system (14.26) is thus *invariant* under the transformation (14.35).

Example 14.4 Consider the position control system of Example 14.2. The plant model of the system is reproduced below:

$$\begin{aligned}\dot{\mathbf{x}}(t) &= \mathbf{A}\mathbf{x}(t) + \mathbf{b}u(t)\\ y(t) &= \mathbf{c}\mathbf{x}(t)\end{aligned} \tag{14.40}$$

with

$$\mathbf{A} = \begin{bmatrix} 0 & 1 & 0 \\ 0 & -1 & 1 \\ 0 & -1 & -10 \end{bmatrix}; \mathbf{b} = \begin{bmatrix} 0 \\ 0 \\ 10 \end{bmatrix}; \mathbf{c} = [1 \;\; 0 \;\; 0]$$

The characteristic polynomial of matrix **A** is

$$|s\mathbf{I} - \mathbf{A}| = \begin{vmatrix} s & -1 & 0 \\ 0 & s+1 & -1 \\ 0 & 1 & s+10 \end{vmatrix} = s(s^2 + 11s + 11)$$

The transfer function

$$G(s) = \frac{Y(s)}{U(s)} = \frac{\mathbf{c}(s\mathbf{I} - \mathbf{A})^+\mathbf{b}}{|s\mathbf{I} - \mathbf{A}|}$$

$$= \frac{[1 \;\; 0 \;\; 0]\begin{bmatrix} s^2 + 11s + 11 & s+10 & 1 \\ 0 & s(s+10) & s \\ 0 & -s & s(s+1) \end{bmatrix}\begin{bmatrix} 0 \\ 0 \\ 10 \end{bmatrix}}{s(s^2 + 11s + 11)}$$

$$= \frac{10}{s(s^2 + 11s + 11)} \tag{14.41}$$

14.5 CONVERSION OF TRANSFER FUNCTIONS TO CANONICAL STATE VARIABLE MODELS

In the last section, we studied the problem—finding the transfer function from the state variable model of a system. The converse problem—finding a state variable model from the transfer function of a system, is the subject of discussion in this section. This problem is quite important because of the following reasons:

1. Quite often the system dynamics is determined experimentally using standard test signals like a step, impulse, or sinusoidal signal. A transfer function is conveniently fitted to the experimental data in some best possible manner.
   There are, however, many design techniques developed exclusively for state variable models. In order to apply these techniques, experimentally obtained transfer function descriptions must be realized into state variable models.
2. Realization of transfer functions into state variable models is needed even if the control system design is based on frequency-domain design methods. In these cases, the need arises for the purpose of transient response simulation. Many algorithms and numerical integration computer programs designed for solution of systems of first-order equations are available, but there is not much software for the numerical inversion of Laplace transforms. Thus, if a reliable method is needed for calculating

the transient response of a system, one may be better off converting the transfer function of the system to state variable description and numerically integrating the resulting differential equations rather than attempting to compute the inverse Laplace transform by numerical methods.

We shall discuss here the problem of realization of transfer function into state variable models. Note the use of the term 'realization'. A state variable model that has a prescribed rational function $G(s)$ as its transfer function, is the *realization* of $G(s)$. The term 'realization' is justified by the fact that by using the *state diagram,* a pictorial representation of the state variable model, the system with the transfer function $G(s)$ can be built in the real world by an op amp circuit. We shall see shortly that a state diagram is essentially an analog-computer program for the given system, which is characterized by three basic operations: (i) multiplication of a variable by a constant, (ii) addition of several variables, and (iii) integration of variables.

The following three problems are involved in the realization of a given transfer function into state variable models:

1. Is it possible at all to obtain state variable description from the given transfer function?
2. If yes, is the state variable description unique for a given transfer function?
3. How do we obtain the state variable description from the given transfer function?

The answer to the first problem has been given in the last section. A rational function $G(s)$ is realizable by a finite dimensional linear time-invariant state model, if and only if, $G(s)$ is a proper rational function. A proper rational function will have state model of the form:

$$\begin{aligned}\dot{\mathbf{x}}(t) &= \mathbf{A}\mathbf{x}(t) + \mathbf{b}u(t)\\ y(t) &= \mathbf{c}\mathbf{x}(t) + du(t)\end{aligned} \tag{14.42}$$

where **A**, **b**, **c** and d are constant matrices of appropriate dimensions. A strictly proper rational function will have state model of the form,

$$\begin{aligned}\dot{\mathbf{x}}(t) &= \mathbf{A}\mathbf{x}(t) + \mathbf{b}u(t)\\ y(t) &= \mathbf{c}\mathbf{x}(t)\end{aligned} \tag{14.43}$$

Let us now turn to the second problem. In the last section, we saw that there are innumerable systems that have the same transfer function. Hence, the representation of a transfer function in state variable form is obviously not unique. However, all these representations will be equivalent.

In the remaining part of this section, we deal with the third problem. We shall develop three standard, or 'canonical' representations of transfer functions.

A linear time-invariant SISO system is described by transfer function of the form,

$$G(s) = \frac{\beta_0 s^m + \beta_1 s^{m-1} + \cdots + \beta_m}{s^n + \alpha_1 s^{n-1} + \cdots + \alpha_n}; m \le n$$

where the coefficients α_i and β_i are real constant scalars. Note that there is no loss in generality to assume the coefficient of s^n to be unity.

In the following, we derive results for $m = n$; these results may be used for the case $m < n$ by setting appropriate β_i coefficients equal to zero. Therefore, our problem is to obtain a state variable model corresponding to the transfer function,

$$G(s) = \frac{\beta_0 s^n + \beta_1 s^{n-1} + \cdots + \beta_n}{s^n + \alpha_1 s^{n-1} + \cdots + \alpha_n} \tag{14.44}$$

14.5.1 First Companion Form

Our development starts with a transfer function of the form,

$$\frac{Z(s)}{U(s)} = \frac{1}{s^n + \alpha_1 s^{n-1} + \cdots + \alpha_n} \tag{14.45}$$

which can be written as

$$(s^n + \alpha_1 s^{n-1} + \cdots + \alpha_n) Z(s) = U(s)$$

The corresponding differential equation is

$$p^n z(t) + \alpha_1 p^{n-1} z(t) + \cdots + \alpha_n z(t) = u(t)$$

where

$$p^k z(t) \triangleq \frac{d^k z(t)}{dt^k}$$

Solving for highest derivative of $z(t)$, we obtain

$$p^n z(t) = -\alpha_1 p^{n-1} z(t) - \alpha_2 p^{n-2} z(t) - \cdots - \alpha_n z(t) + u(t) \tag{14.46}$$

Now consider a chain of n integrators as shown in Fig. 14.9. Suppose that the output of the last integrator is $z(t)$. Then the output of the just previous integrator is $pz = dz/dt$, and so forth. The output from the first integrator is $p^{n-1}z(t)$, and the input to this integrator is thus $p^n z(t)$. This leaves only the problem of obtaining $p^n z(t)$ for use as input to the first integrator. In fact, this is already specified by Eqn. (14.46). Realization of this equation is shown in Fig. 14.9.

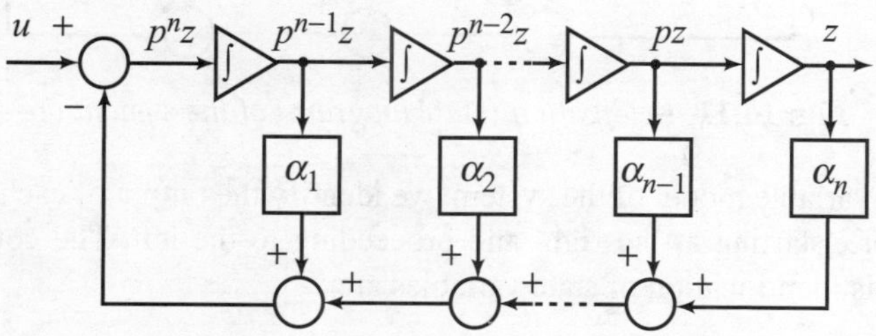

Fig. 14.9 *Realization of the system (14.46)*

Having developed a realization of the simple transfer function (14.45), we are now in a position to consider the more general transfer function (14.44). We decompose this transfer function into two parts, as shown in Fig. 14.10. The output $Y(s)$ can be written as

$$Y(s) = (\beta_0 s^n + \beta_1 s^{n-1} + \cdots + \beta_n) Z(s) \tag{14.47a}$$

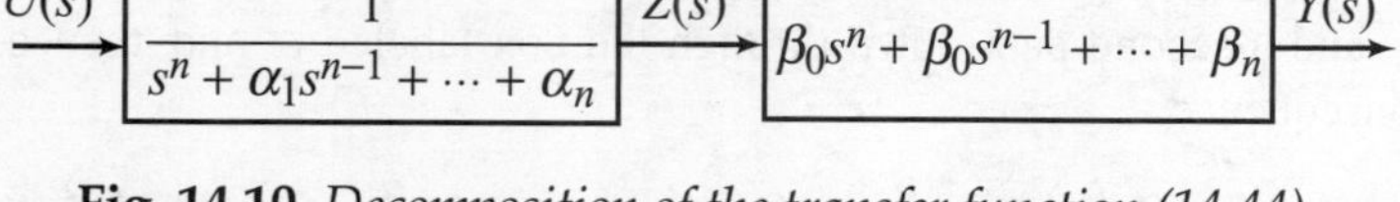

Fig. 14.10 *Decomposition of the transfer function (14.44)*

where $Z(s)$ is given by

$$\frac{Z(s)}{U(s)} = \frac{1}{s^n + \alpha_1 s^{n-1} + \cdots + \alpha_n} \tag{14.47b}$$

A realization of the transfer function (14.47b) has already been developed. Figure 14.9 shows this realization. The output of the last integrator is $z(t)$ and the inputs to the integrators in the chain from the last to the first are the n successive derivatives of $z(t)$.

Realization of the transfer function (14.47a) is now straightforward. The output

$$y(t) = \beta_0 p^n z(t) + \beta_1 p^{n-1} z(t) + \cdots + \beta_n z(t),$$

is nothing but the sum of the scaled versions of the inputs to the n integrators. Figure 14.11 shows complete realization of the transfer function (14.44). All that remains to be done is to write the corresponding differential equations.

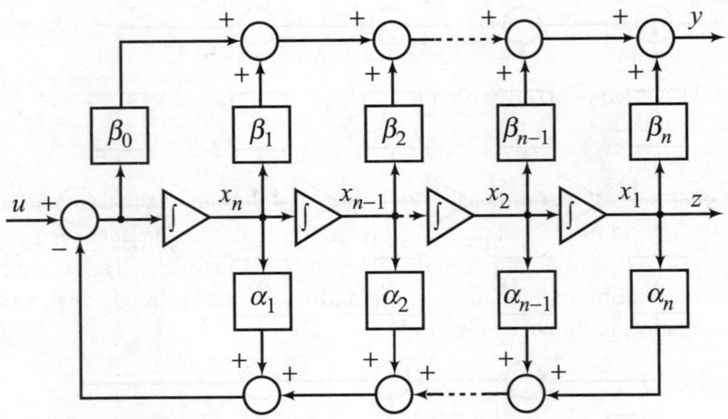

Fig. 14.11 *Realization (state diagram) of the system (14.44)*

To get one state variable model of the system, we identify the output of each integrator in Fig. 14.11 with a state variable, starting at the right and proceeding to the left. The corresponding differential equations, using this identification of state variables are

$$\begin{aligned} \dot{x}_1 &= x_2 \\ \dot{x}_2 &= x_3 \\ &\vdots \\ \dot{x}_{n-1} &= x_n \\ \dot{x}_n &= -\alpha_n x_1 - \alpha_{n-1} x_2 - \cdots - \alpha_1 x_n + u \end{aligned} \tag{14.48a}$$

The output equation is found by careful examination of the block diagram of Fig. 14.11. Note that there are two paths from the output of each integrator to the system output—one path upward through the box labeled β_i, and a second path down through the box labeled α_i and, thence, through the box labeled β_0. As a consequence,

$$y = (\beta_n - \alpha_n \beta_0) x_1 + (\beta_{n-1} - \alpha_{n-1}\beta_0) x_2 + \cdots + (\beta_1 - \alpha_1 \beta_0) x_n + \beta_0 u \tag{14.48b}$$

The state and output Eqns (14.48), organized in vector-matrix form, are given below:

$$\begin{aligned}\dot{\mathbf{x}}(t) &= \mathbf{A}\mathbf{x}(t) + \mathbf{b}u(t)\\ y(t) &= \mathbf{c}\mathbf{x}(t) + du(t)\end{aligned} \tag{14.49}$$

with

$$\mathbf{A} = \begin{bmatrix} 0 & 1 & 0 & \cdots & 0 \\ 0 & 0 & 1 & \cdots & 0 \\ \vdots & \vdots & \vdots & & \vdots \\ 0 & 0 & 0 & \cdots & 1 \\ -\alpha_n & -\alpha_{n-1} & -\alpha_{n-2} & \cdots & -\alpha_1 \end{bmatrix}; \mathbf{b} = \begin{bmatrix} 0 \\ 0 \\ \vdots \\ 0 \\ 1 \end{bmatrix}$$

$$\mathbf{c} = [\beta_n - \alpha_n\beta_0,\ \beta_{n-1} - \alpha_{n-1}\beta_0, ..., \beta_1 - \alpha_1\beta_0];\ d = \beta_0$$

If the direct path through β_0 is absent (refer to Fig. 14.11), then the scalar d is zero and the row matrix **c** contains only the β_i coefficients.

The matrix **A** in Eqns (14.49) has a very special structure: the coefficients of the denominator of the transfer function preceded by minus signs form a string along the bottom row of the matrix. The rest of the matrix is zero except for the 'superdiagonal' terms which are all unity. In matrix theory, a matrix with this structure is said to be in *companion form*. For this reason, we identify the realization (14.49) as *companion-form realization* of the transfer function (14.44). We call this the *first companion form*: another companion form follows.

14.5.2 Second Companion Form

In the first companion form, the coefficients of the denominator of the transfer function appear in one of the rows of the **A** matrix. There is another companion form in which the coefficients appear in a column of the **A** matrix. This can be obtained by writing Eqn. (14.44) as

$$(s^n + \alpha_1 s^{n-1} + \cdots + \alpha_n)\ Y(s) = (\beta_0 s^n + \beta_1 s^{n-1} + \cdots + \beta_n)\, U(s)$$

or

$$s^n[Y(s) - \beta_0 U(s)] + s^{n-1}[\alpha_1 Y(s) - \beta_1 U(s)] + \cdots + [\alpha_n Y(s) - \beta_n U(s)] = 0$$

On dividing by s^n and solving for $Y(s)$, we obtain

$$Y(s) = \beta_0 U(s) + \frac{1}{s}[\beta_1 U(s) - \alpha_1 Y(s)] + \cdots + \frac{1}{s^n}[\beta_n U(s) - \alpha_n Y(s)] \tag{14.50}$$

Note that $1/s^n$ is the transfer function of a chain of n integrators. Realization of $\frac{1}{s^n}[\beta_n U(s) - \alpha_n Y(s)]$ requires a chain of n integrators with an input $[\beta_n u - \alpha_n y]$ to the first integrator in the chain from left to right. Realization of $\frac{1}{s^{n-1}}[\beta_{n-1}U(s) - \alpha_{n-1}Y(s)]$ requires a chain of $(n-1)$ integrators with input $[\beta_{n-1} u - \alpha_{n-1} y]$ to the first integrator in the chain from left to right, and so forth. This immediately leads to the structure shown in Fig. 14.12. The signal y is fed back to each of the integrators in the chain and the signal u is fed forward. Thus the signal $[\beta_n u - \alpha_n y]$ passes through n integrators; the signal $[\beta_{n-1} u - \alpha_{n-1} y]$ passes through $(n-1)$ integrators and so forth to complete the realization of Eqn. (14.50). The structure retains the ladder-like shape of the first companion form, but the feedback paths are in different directions.

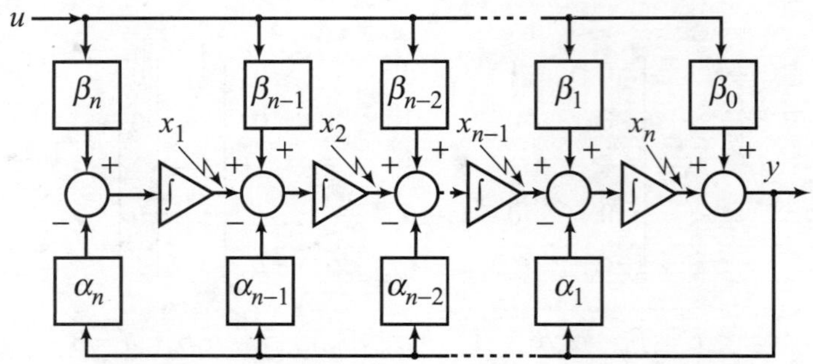

Fig. 14.12 *Realization (state diagram) of Eqn. (14.50)*

We can now write differential equations for the realization given by Fig. 14.12. To get one state variable model, we identify the output of each integrator in Fig. 14.12 with a state variable starting at the left and proceeding to the right. The corresponding differential equations are

$$\begin{aligned}\dot{x}_n &= x_{n-1} - \alpha_1(x_n + \beta_0 u) + \beta_1 u\\ \dot{x}_{n-1} &= x_{n-2} - \alpha_2(x_n + \beta_0 u) + \beta_2 u\\ &\vdots\\ \dot{x}_2 &= x_1 - \alpha_{n-1}(x_n + \beta_0 u) + \beta_{n-1} u\\ \dot{x}_1 &= -\alpha_n(x_n + \beta_0 u) + \beta_n u\end{aligned}$$

and the output equation is

$$y = x_n + \beta_0 u$$

The state and output equations organized in vector-matrix form are given below:

$$\begin{aligned}\dot{\mathbf{x}}(t) &= \mathbf{A}\mathbf{x}(t) + \mathbf{b}u(t)\\ y(t) &= \mathbf{c}\mathbf{x}(t) + du(t)\end{aligned} \tag{14.51}$$

with

$$\mathbf{A} = \begin{bmatrix} 0 & 0 & \cdots & 0 & -\alpha_n\\ 1 & 0 & \cdots & 0 & -\alpha_{n-1}\\ 0 & 1 & \cdots & 0 & -\alpha_{n-2}\\ \vdots & \vdots & & \vdots & \vdots\\ 0 & 0 & \cdots & 1 & -\alpha_1 \end{bmatrix}; \quad \mathbf{b} = \begin{bmatrix} \beta_n - \alpha_n\beta_0\\ \beta_{n-1} - \alpha_{n-1}\beta_0\\ \\ \vdots\\ \beta_1 - \alpha_1\beta_0 \end{bmatrix}$$

$$\mathbf{c} = [0 \;\; 0 \;\; \cdots \;\; 0 \;\; 1]; \; d = \beta_0$$

Compare **A**, **b**, and **c** matrices of the second companion form with that of the first. We observe that **A**, **b**, and **c** matrices of one companion form correspond to the transpose of the **A**, **c**, and **b** matrices, respectively, of the other.

There are many benefits derived from the companion forms of state variable models. One obvious benefit is that both the companion forms lend themselves easily to simple analog computer models. Both the companion forms also play an important role in pole-placement design through state feedback (refer to Chapter 15).

14.5.3 Jordan Canonical Form

In the two canonical forms (14.49) and (14.51), the coefficients of the denominator of the transfer function appear in one of the rows or columns of matrix **A**. In another of the canonical forms, the poles of the transfer function form a string along the main diagonal of the matrix. This canonical form follows directly from the partial fraction expansion of the transfer function.

The general transfer function under consideration is (refer to Eqn. (14.44))

$$G(s)=\frac{\beta_0 s^n+\beta_1 s^{n-1}+\cdots+\beta_n}{s^n+\alpha_1 s^{n-1}+\cdots+\alpha_n}$$

By long division, $G(s)$ can be written as

$$G(s)=\beta_0+\frac{\beta_1' s^{n-1}+\beta_2' s^{n-2}+\cdots+\beta_n'}{s^n+\alpha_1 s^{n-1}+\cdots+\alpha_n}=\beta_0+G'(s)$$

The results are simplest when the poles of the transfer function are all distinct. The partial fraction expansion of the transfer function then has the form,

$$G(s)=\frac{Y(s)}{U(s)}=\beta_0+\frac{r_1}{s-\lambda_1}+\frac{r_2}{s-\lambda_2}+\cdots+\frac{r_n}{s-\lambda_n} \tag{14.52}$$

The coefficients r_i $(i = 1, 2, ..., n)$ are the residues of the transfer function $G'(s)$ at the corresponding poles at $s = \lambda_i$ $(i = 1, 2, ..., n)$. In the form of Eqn. (14.52), the transfer function consists of a direct path with gain β_0, and n first-order transfer functions in parallel. A block diagram representation of Eqn. (14.52) is shown in Fig. 14.13. The gains corresponding to the residues have been placed at the outputs of the integrators. This is quite arbitrary. They could have been located on the input side, or indeed split between the input and the output.

Identifying the outputs of the integrators with the state variables results in the following state and output equations:

$$\begin{aligned}\dot{\mathbf{x}}(t)&=\Lambda\mathbf{x}(t)+\mathbf{b}u(t)\\ y(t)&=\mathbf{c}\mathbf{x}(t)+du(t)\end{aligned} \tag{14.53}$$

with

$$\Lambda=\begin{bmatrix}\lambda_1 & 0 & \cdots & 0\\ 0 & \lambda_2 & \cdots & 0\\ \vdots & \vdots & & \vdots\\ 0 & 0 & \cdots & \lambda_n\end{bmatrix};\mathbf{b}=\begin{bmatrix}1\\1\\ \vdots\\1\end{bmatrix}$$

$$\mathbf{c}=\begin{bmatrix}r_1 & r_2 & \cdots & r_n\end{bmatrix};\ d=\beta_0$$

It is observed that for this canonical state variable model, the matrix $\mathbf{\Lambda}$ is a diagonal matrix with the poles of $G(s)$ as its diagonal elements. The unique *decoupled* nature of the canonical model is obvious from Eqns (14.53); the n first-order differential equations are independent of each other:

$$\dot{x}_i(t) = \lambda_i x_i(t) + u(t);\ i = 1, 2, ..., n \tag{14.54}$$

This decoupling feature greatly helps in system analysis (refer to [155]).

The block diagram representation of Fig. 14.13 can be turned into hardware only if all the poles at $s = \lambda_1, \lambda_2, ..., \lambda_n$ are real. If they are complex, the feedback gains and the gains corresponding to the residues are complex. In this case, the representation must be considered as being purely conceptual; valid for theoretical studies, but not physically realizable. A realizable representation can be obtained by introducing an equivalence transformation.

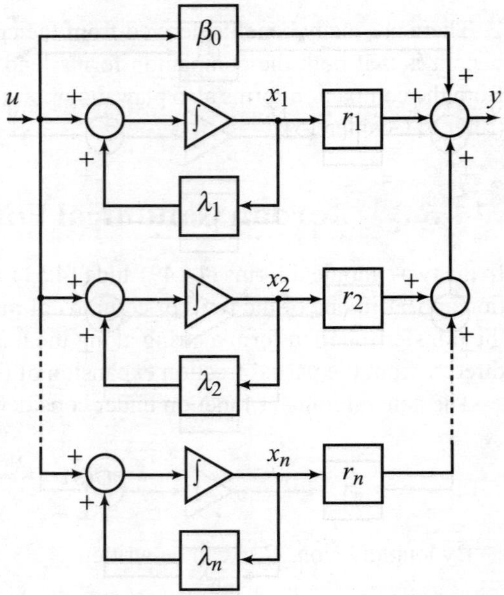

Fig. 14.13 *Realization (state diagram) of G(s) in Eqn. (14.52)*

Suppose that $s = \sigma + j\omega$, $s = \sigma - j\omega$ and $s = \lambda$ are the three poles of a transfer function. The residues at the pair of complex conjugate poles must be themselves complex conjugates. Partial fraction expansion of the transfer function with a pair of complex conjugate poles and a real pole has the form,

$$G(s) = d + \frac{p + jq}{s - (\sigma + j\omega)} + \frac{p - jq}{s - (\sigma - j\omega)} + \frac{r}{s - \lambda}$$

A state variable model for this transfer function is given below (refer to Eqns 14.53)):

$$\begin{aligned}\dot{\mathbf{x}} &= \mathbf{\Lambda x} + \mathbf{b}u \\ y &= \mathbf{cx} + du\end{aligned} \tag{14.55}$$

with $$\mathbf{\Lambda} = \begin{bmatrix} \sigma + j\omega & 0 & 0 \\ 0 & \sigma - j\omega & 0 \\ 0 & 0 & \lambda \end{bmatrix}; \mathbf{b} = \begin{bmatrix} 1 \\ 1 \\ 1 \end{bmatrix}$$

$$\mathbf{c} = [p + jq \quad p - jq \quad r]$$

Introducing an equivalence transformation

$$\mathbf{x} = \mathbf{P}\bar{\mathbf{x}}$$

with $$\mathbf{P} = \begin{bmatrix} 1/2 & -j1/2 & 0 \\ 1/2 & j1/2 & 0 \\ 0 & 0 & 1 \end{bmatrix}$$

we obtain (refer to Eqns 14.22))

$$\dot{\bar{\mathbf{x}}}(t) = \bar{\mathbf{A}}\,\bar{\mathbf{x}}(t) + \bar{\mathbf{b}}u(t)$$
$$y(t) = \bar{\mathbf{c}}\,\bar{\mathbf{x}}(t) + du(t) \tag{14.56}$$

where

$$\bar{\mathbf{A}} = \mathbf{P}^{-1}\mathbf{\Lambda}\mathbf{P} = \begin{bmatrix} 1 & 1 & 0 \\ j & -j & 0 \\ 0 & 0 & 1 \end{bmatrix}\begin{bmatrix} \sigma + j\omega & 0 & 0 \\ 0 & \sigma - j\omega & 0 \\ 0 & 0 & \lambda \end{bmatrix}\begin{bmatrix} 1/2 & -j1/2 & 0 \\ 1/2 & j1/2 & 0 \\ 0 & 0 & 1 \end{bmatrix} = \begin{bmatrix} \sigma & \omega & 0 \\ -\omega & \sigma & 0 \\ 0 & 0 & \lambda \end{bmatrix}$$

$$\bar{\mathbf{b}} = \mathbf{P}^{-1}\mathbf{b} = \begin{bmatrix} 2 \\ 0 \\ 1 \end{bmatrix}$$

$$\bar{\mathbf{c}} = \mathbf{c}\mathbf{P} = \begin{bmatrix} p & q & r \end{bmatrix}$$

When the transfer function $G(s)$ has repeated poles, the partial fraction expansion will not be as simple as Eqn. (14.52). Assume that $G(s)$ has m distinct poles at $s = \lambda_1, \lambda_2, \ldots, \lambda_m$ of multiplicity $n_1, n_2, \ldots, n_m$ respectively; $n = n_1 + n_2 + \cdots + n_m$. That is, $G(s)$ is of the form,

$$G(s) = \beta_0 + \frac{\beta_1' s^{n-1} + \beta_2' s^{n-2} + \cdots + \beta_n'}{(s-\lambda_1)^{n_1}(s-\lambda_2)^{n_2}\cdots(s-\lambda_m)^{n_m}} \tag{14.57}$$

The partial fraction expansion of $G(s)$ is of the form,

$$G(s) = \beta_0 + H_1(s) + \cdots + H_m(s) = \frac{Y(s)}{U(s)} \tag{14.58}$$

where

$$H_i(s) = \frac{r_{i1}}{(s-\lambda_i)^{n_i}} + \frac{r_{i2}}{(s-\lambda_i)^{n_i-1}} + \cdots + \frac{r_{in_i}}{(s-\lambda_i)} = \frac{Y_i(s)}{U(s)}$$

The first term in $H_i(s)$ can be synthesized as a chain of n_i identical, first-order systems, each having transfer function $1/(s - \lambda_i)$. The second term can be synthesized by a chain of $(n_i - 1)$ first-order systems, and so forth. The entire $H_i(s)$ can be synthesized by the system having the block diagram shown in Fig. 14.14.

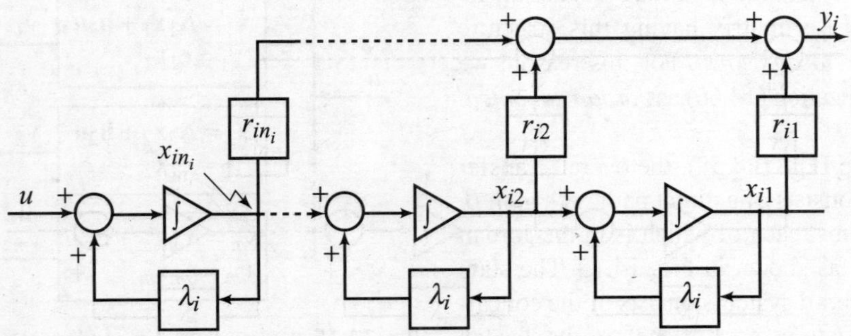

Fig. 14.14 *Realization (state diagram) of $H_i(s)$ in Eqn. (14.58)*

We can now write differential equations for the realization of $H_i(s)$ given by Fig. 14.14. To get one state variable formulation, we identify the output of each integrator with a state variable starting at the right and proceeding to the left. The corresponding differential equations are

$$\begin{aligned}\dot{x}_{i1} &= \lambda_i x_{i1} + x_{i2}\\ \dot{x}_{i2} &= \lambda_i x_{i2} + x_{i3}\\ &\vdots\\ \dot{x}_{in_i} &= \lambda_i x_{in_i} + u\end{aligned} \tag{14.59a}$$

and the output is given by

$$y_i = r_{i1}x_{i1} + r_{i2}x_{i2} + \cdots + r_{in_i}x_{in_i} \tag{14.59b}$$

If the state vector for the subsystem is defined by

$$\mathbf{x}_i = [x_{i1} \quad x_{i2} \cdots x_{in_i}]^T$$

then Eqns (14.59) can be written in the standard form,

$$\begin{aligned}\dot{\mathbf{x}}_i &= \mathbf{\Lambda}_i\mathbf{x}_i + \mathbf{b}_i u\\ y_i &= \mathbf{c}_i\mathbf{x}_i\end{aligned} \tag{14.60}$$

where
$$\mathbf{\Lambda}_i = \begin{bmatrix}\lambda_i & 1 & 0 & \cdots & 0 & 0\\ 0 & \lambda_i & 1 & \cdots & 0 & 0\\ \vdots & \vdots & \vdots & & \vdots & \vdots\\ 0 & 0 & 0 & \cdots & \lambda_i & 1\\ 0 & 0 & 0 & \cdots & 0 & \lambda_i\end{bmatrix}; \mathbf{b}_i = \begin{bmatrix}0\\0\\ \vdots\\0\\1\end{bmatrix}$$

$$\mathbf{c}_i = \begin{bmatrix} r_{i1} & r_{i2} & \cdots & r_{in_i}\end{bmatrix}$$

Note that matrix $\mathbf{\Lambda}_i$ has two diagonals—the principal diagonal has the corresponding characteristic root (pole) and the superdiagonal has all 1s. In matrix theory, a matrix having this structure is said to be in *Jordan form*. For this reason, we identify the realization (14.60) as *Jordan canonical form*.

According to Eqn. (14.58), the overall transfer function $G(s)$ consists of a direct path with gain β_0 and m subsystems, each of which is in the Jordan canonical form as shown in Fig. 14.15. The state vector of the overall system consists of the concatenation of the state vectors of each of the Jordan blocks:

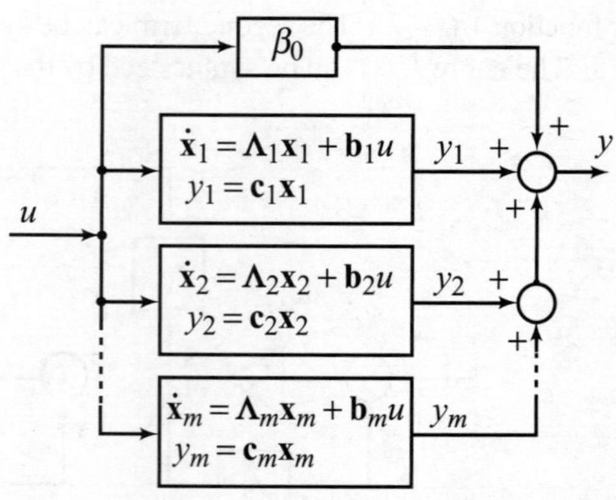

Fig. 14.15 *Subsystems in Jordan canonical form combined into overall system*

$$\mathbf{x} = \begin{bmatrix} \mathbf{x}_1 \\ \mathbf{x}_2 \\ \vdots \\ \mathbf{x}_m \end{bmatrix} \tag{14.61a}$$

Since there is no coupling between any of the subsystems, the $\mathbf{\Lambda}$ matrix of the overall system is 'block diagonal':

$$\mathbf{\Lambda} = \begin{bmatrix} \mathbf{\Lambda}_1 & \mathbf{0} & \cdots & \mathbf{0} \\ \mathbf{0} & \mathbf{\Lambda}_2 & \cdots & \mathbf{0} \\ \vdots & \vdots & & \vdots \\ \mathbf{0} & \mathbf{0} & \cdots & \mathbf{\Lambda}_m \end{bmatrix} \tag{14.61b}$$

where each of the submatrices $\mathbf{\Lambda}_i$ is in the Jordan canonical form (14.60). The **b** and **c** matrices of the overall system are the concatenations of the $\mathbf{b}_i$ and $\mathbf{c}_i$ matrices respectively of each of the subsystems:

$$\mathbf{b} = \begin{bmatrix} \mathbf{b}_1 \\ \mathbf{b}_2 \\ \vdots \\ \mathbf{b}_m \end{bmatrix}; \; \mathbf{c} = \begin{bmatrix} \mathbf{c}_1 & \mathbf{c}_2 & \cdots \mathbf{c}_m \end{bmatrix}; \; d = \beta_0 \tag{14.61c}$$

The state variable model (14.53) derived for the case of distinct poles, is a special case of Jordan canonical form (14.61) where each Jordan block is of 1×1 dimension.

Example 14.5 In the following, we obtain three different realizations for the transfer function,

$$G(s) = \frac{s+3}{s^3 + 9s^2 + 24s + 20} = \frac{Y(s)}{U(s)}$$

First companion form Note that the given $G(s)$ is a strictly proper fraction; the realization will therefore be of the form (14.43), i.e., the parameter d in the realization $\{\mathbf{A}, \mathbf{b}, \mathbf{c}, d\}$ is zero.

The state variable formulation in the first companion form can be written just by inspection of the given transfer function. Referring to Eqns (14.49), we obtain

$$\begin{bmatrix} \dot{x}_1 \\ \dot{x}_2 \\ \dot{x}_3 \end{bmatrix} = \begin{bmatrix} 0 & 1 & 0 \\ 0 & 0 & 1 \\ -20 & -24 & -9 \end{bmatrix} \begin{bmatrix} x_1 \\ x_2 \\ x_3 \end{bmatrix} + \begin{bmatrix} 0 \\ 0 \\ 1 \end{bmatrix} u$$

$$y = \begin{bmatrix} 3 & 1 & 0 \end{bmatrix} \begin{bmatrix} x_1 \\ x_2 \\ x_3 \end{bmatrix}$$

Figure 14.16a shows the state diagram in signal flow graph form.

Second companion form Referring to Eqns (14.51), we obtain

$$\begin{bmatrix} \dot{x}_1 \\ \dot{x}_2 \\ \dot{x}_3 \end{bmatrix} = \begin{bmatrix} 0 & 0 & -20 \\ 1 & 0 & -24 \\ 0 & 1 & -9 \end{bmatrix} \begin{bmatrix} x_1 \\ x_2 \\ x_3 \end{bmatrix} + \begin{bmatrix} 3 \\ 1 \\ 0 \end{bmatrix} u$$

$$y = x_3$$

Figure 14.16b shows the state diagram.

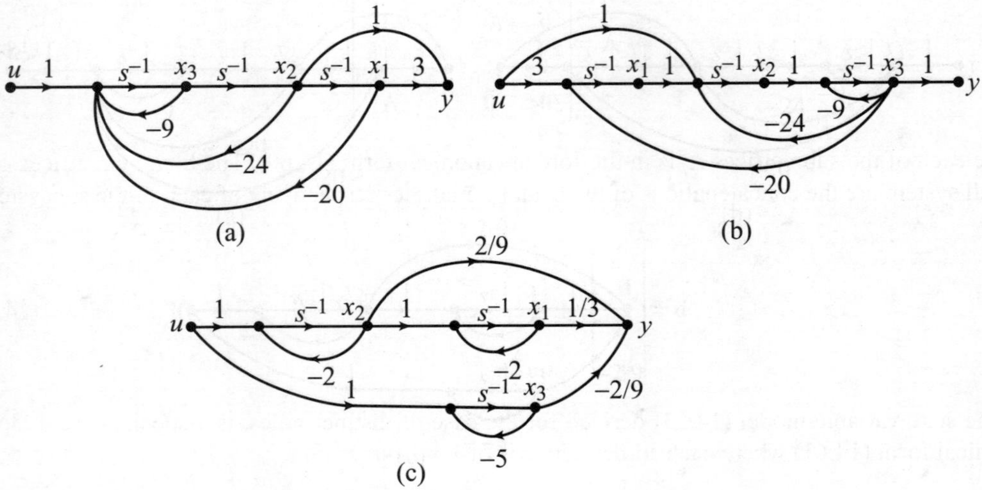

Fig. 14.16 *Three realizations of given G(s)*

Jordan canonical form The given transfer function $G(s)$ in the factored form,

$$G(s) = \frac{s+3}{(s+2)^2(s+5)}$$

Using partial fraction expansion, we obtain

$$G(s) = \frac{1/3}{(s+2)^2} + \frac{2/9}{s+2} + \frac{-2/9}{s+5}$$

A matrix of the state variable model in Jordan canonical form will be block diagonal; consisting of two Jordan blocks (refer to Eqns (14.60)):

$$\mathbf{\Lambda}_1 = \begin{bmatrix} -2 & 1 \\ 0 & -2 \end{bmatrix}; \mathbf{\Lambda}_2 = [-5]$$

The corresponding $\mathbf{b}_i$ and $\mathbf{c}_i$ vectors are (refer to Eqns (14.60)):

$$\mathbf{b}_1 = \begin{bmatrix} 0 \\ 1 \end{bmatrix}; \mathbf{c}_1 = \begin{bmatrix} \frac{1}{3} & \frac{2}{9} \end{bmatrix}; \mathbf{b}_2 = [1]; \mathbf{c}_2 = \begin{bmatrix} -\frac{2}{9} \end{bmatrix}$$

The state variable model of the given $G(s)$ in Jordan canonical form is therefore given by (refer to Eqns (14.61))

$$\begin{bmatrix} \dot{x}_1 \\ \dot{x}_2 \\ \dot{x}_3 \end{bmatrix} = \begin{bmatrix} -2 & 1 & 0 \\ 0 & -2 & 0 \\ 0 & 0 & -5 \end{bmatrix} \begin{bmatrix} x_1 \\ x_2 \\ x_3 \end{bmatrix} + \begin{bmatrix} 0 \\ 1 \\ 1 \end{bmatrix} u$$

$$y = \begin{bmatrix} \frac{1}{3} & \frac{2}{9} & -\frac{2}{9} \end{bmatrix} \begin{bmatrix} x_1 \\ x_2 \\ x_3 \end{bmatrix}$$

Figure 14.16c shows the state diagram. We note that Jordan canonical state variables are not completely decoupled. The decoupling is blockwise; state variables of one block are independent of state variables of all other blocks. However, the state variables of one block among themselves are coupled; the coupling is unique and simple.

14.6 SOLUTION OF STATE EQUATIONS

In this section, we investigate the solution of the state equation,

$$\dot{\mathbf{x}}(t) = \mathbf{A}\mathbf{x}(t) + \mathbf{b}u(t); \mathbf{x}(t_0) \triangleq \mathbf{x}^0 \tag{14.62}$$

where $\mathbf{x}$ is $n \times 1$ state vector, u is a scalar input, $\mathbf{A}$ is $n \times n$ constant matrix, and $\mathbf{b}$ is $n \times 1$ constant vector.

14.6.1 Matrix Exponential

Functions of square matrices arise in connection with the solution of vector differential equations. Of immediate interest to us are matrix infinite series.

Consider the infinite series in a scalar variable x:

$$f(x) = \alpha_0 + \alpha_1 x + \alpha_2 x^2 + \cdots = \sum_{i=0}^{\infty} \alpha_i x^i \tag{14.63a}$$

with the radius of convergence r.

We can define infinite series in a matrix variable $\mathbf{A}$ as

$$f(\mathbf{A}) = \alpha_0 \mathbf{I} + \alpha_1 \mathbf{A} + \alpha_2 \mathbf{A}^2 + \cdots = \sum_{i=0}^{\infty} \alpha_i \mathbf{A}^i \tag{14.63b}$$

An important relation between the scalar power series (14.63a) and the matrix power series (14.63b) is that if the absolute values of eigenvalues of $\mathbf{A}$ are smaller than r, then the matrix power series (14.63b) converges (for proof, refer to Lefschetz [33]).

Consider, in particular, the scalar power series

$$f(x) = 1 + x + \frac{1}{2!}x^2 + \cdots + \frac{1}{k!}x^k + \cdots = \sum_{i=0}^{\infty} \frac{1}{i!}x^i \tag{14.64a}$$

It is well known that this power series converges on to the exponential e^x for a finite x, so that

$$f(x) = e^x \tag{14.64b}$$

It follows from this result that the matrix power series

$$f(\mathbf{A}) = \mathbf{I} + \mathbf{A} + \frac{1}{2!}\mathbf{A}^2 + \cdots + \frac{1}{k!}\mathbf{A}^k + \cdots = \sum_{i=0}^{\infty} \frac{1}{i!}\mathbf{A}^i$$

converges for all **A**. By analogy with the power series in Eqns (14.63) for the ordinary exponential function, we adopt the following nomenclature:

If **A** is an $n \times n$ matrix, the *matrix exponential* of **A** is

$$e^{\mathbf{A}} \triangleq \mathbf{I} + \mathbf{A} + \frac{1}{2!}\mathbf{A}^2 + \cdots + \frac{1}{k!}\mathbf{A}^k + \cdots = \sum_{i=0}^{\infty} \frac{1}{i!}\mathbf{A}^i$$

The following matrix exponential will appear in the solution of state equations:

$$e^{\mathbf{A}t} = \mathbf{I} + \mathbf{A}t + \frac{1}{2!}\mathbf{A}^2t^2 + \cdots + \frac{1}{k!}\mathbf{A}^kt^k + \cdots = \sum_{i=0}^{\infty} \frac{1}{k!}\mathbf{A}^it^i \tag{14.65}$$

It converges for all **A** and all finite t.

In the following we examine some of the properties of the matrix exponential.

1. $$e^{\mathbf{A}0} = \mathbf{I} \tag{14.66}$$

This is easily verified by setting $t = 0$ in Eqn. (14.65).

2. $$e^{\mathbf{A}(t+\tau)} = e^{\mathbf{A}t}e^{\mathbf{A}\tau} = e^{\mathbf{A}\tau}e^{\mathbf{A}t} \tag{14.67}$$

This is easily verified by multiplying out the first few terms for $e^{\mathbf{A}t}$ and $e^{\mathbf{A}\tau}$.

3. $$(e^{\mathbf{A}t})^{-1} = e^{-\mathbf{A}t} \tag{14.68}$$

Setting $\tau = -t$ in Eqn. (14.67), we obtain

$$e^{\mathbf{A}t}e^{-\mathbf{A}t} = e^{\mathbf{A}0} = \mathbf{I}$$

Thus, the inverse of $e^{\mathbf{A}t}$ is $e^{-\mathbf{A}t}$.

Since the inverse of $e^{\mathbf{A}t}$ always exists, the matrix exponential is nonsingular for all finite values of t.

4. $$\frac{d}{dt}e^{\mathbf{A}t} = \mathbf{A}e^{\mathbf{A}t} = e^{\mathbf{A}t}\mathbf{A} \tag{14.69}$$

Term-by-term differentiation of Eqn. (14.65) gives

$$\frac{d}{dt}e^{\mathbf{A}t} = \mathbf{A} + \mathbf{A}^2t + \frac{1}{2!}\mathbf{A}^3t^2 + \cdots + \frac{1}{(k-1)!}\mathbf{A}^kt^{k-1} + \cdots$$

$$= \mathbf{A}\left[\mathbf{I} + \mathbf{A}t + \frac{1}{2!}\mathbf{A}^2t^2 + \cdots + \frac{1}{(k-1)!}\mathbf{A}^{k-1}t^{k-1} + \cdots\right] = \mathbf{A}e^{\mathbf{A}t}$$

$$= \left[\mathbf{I} + \mathbf{A}t + \frac{1}{2!}\mathbf{A}^2t^2 + \cdots + \frac{1}{(k-1)!}\mathbf{A}^{k-1}t^{k-1} + \cdots\right] = \mathbf{A} = e^{\mathbf{A}t}\mathbf{A}$$

14.6.2 Solution of Homogeneous State Equation

The simplest form of the general differential Eqn. (14.62) is the homogeneous, i.e., unforced equation

$$\dot{\mathbf{x}}(t) = \mathbf{A}\mathbf{x}(t); \mathbf{x}(t_0) \triangleq \mathbf{x}^0 \tag{14.70}$$

We assume a solution $\mathbf{x}(t)$ of the form

$$\mathbf{x}(t) = e^{\mathbf{A}t}\mathbf{k} \tag{14.71}$$

where $e^{\mathbf{A}t}$ is the matrix exponential function defined in Eqn. (14.65) and $\mathbf{k}$ is a suitably chosen constant vector.

The assumed solution is, in fact, the true solution since it satisfies the differential Eqn. (14.70) as is seen below.

$$\dot{\mathbf{x}}(t) = \frac{d}{dt}[e^{\mathbf{A}t}\mathbf{k}] = \frac{d}{dt}[e^{\mathbf{A}t}]\mathbf{k}$$

Using property (14.69) of the matrix exponential, we obtain

$$\dot{\mathbf{x}}(t) = \mathbf{A}e^{\mathbf{A}t}\mathbf{k} = \mathbf{A}\mathbf{x}(t)$$

To evaluate the constant $\mathbf{k}$ in terms of the known initial state $\mathbf{x}(t_0)$, we substitute $t = t_0$ in Eqn. (14.71):

$$\mathbf{x}(t_0) = e^{\mathbf{A}t_0}\mathbf{k}$$

Using property (14.68) of the matrix exponential, we obtain

$$\mathbf{k} = (e^{\mathbf{A}t_0})^{-1}\mathbf{x}(t_0) = e^{-\mathbf{A}t_0}\mathbf{x}(t_0)$$

Thus, the general solution to Eqn. (14.70) for the state $\mathbf{x}(t)$ at time t, given the state $\mathbf{x}(t_0)$ at time t_0, is

$$\mathbf{x}(t) = e^{\mathbf{A}t}e^{-\mathbf{A}t_0}\mathbf{x}(t_0) = e^{\mathbf{A}(t-t_0)}\mathbf{x}(t_0) \tag{14.72a}$$

We have used the property (14.67) of the matrix exponential to express the solution in this form.

If the initial time $t_0 = 0$, i.e., the initial state $\mathbf{x}^0$ is known at $t = 0$, we have from Eqn. (14.72a):

$$\mathbf{x}(t) = e^{\mathbf{A}t}\mathbf{x}(0) \tag{14.72b}$$

From Eqn. (14.72b) it is observed that the initial state $\mathbf{x}(0) \triangleq \mathbf{x}^0$ at $t = 0$ is driven to a state $\mathbf{x}(t)$ at time t. This transition in state is carried out by the matrix exponential $e^{\mathbf{A}t}$. Due to this property, $e^{\mathbf{A}t}$ is known as the *state transition matrix*, and is denoted by $\boldsymbol{\phi}(t)$.

Properties of state transition matrix

Properties of the matrix exponential, given earlier in Eqns (14.66) – (14.69), are restated below in terms of state transition matrix $\boldsymbol{\phi}(t)$.

1. $$\frac{d}{dt}\boldsymbol{\phi}(t) = \mathbf{A}\boldsymbol{\phi}(t); \boldsymbol{\phi}(0) = \mathbf{I}$$
2. $$\boldsymbol{\phi}(t_2 - t_1)\,\boldsymbol{\phi}(t_1 - t_0) = \boldsymbol{\phi}(t_2 - t_0) \text{ for any } t_0, t_1, t_2$$
   This property of the state transition matrix is important since it implies that a state transition process can be divided into a number of sequential transitions. The transition form t_0 to t_2:

$$\mathbf{x}(t_2) = \boldsymbol{\phi}(t_2 - t_0)\,\mathbf{x}(t_0);$$

is equal to the transition from t_0 to t_1 and then from t_1 to t_2:

$$\mathbf{x}(t_1) = \boldsymbol{\phi}(t_1 - t_0)\mathbf{x}(t_0)$$
$$\mathbf{x}(t_2) = \boldsymbol{\phi}(t_2 - t_1)\mathbf{x}(t_1)$$

3. $\boldsymbol{\phi}^{-1}(t) = \boldsymbol{\phi}(-t)$
4. $\boldsymbol{\phi}(t)$ is a nonsingular matrix for all finite t.

Evaluation of state transition matrix

Taking the Laplace transform on both sides of Eqn. (14.70) yields

$$s\mathbf{X}(s) - \mathbf{x}^0 = \mathbf{A}\mathbf{X}(s)$$

where

$$\mathbf{X}(s) \triangleq \mathscr{L}[\mathbf{x}(t)]$$
$$\mathbf{x}^0 \triangleq \mathbf{x}(0)$$

Solving for $\mathbf{X}(s)$, we get

$$\mathbf{X}(s) = (s\mathbf{I} - \mathbf{A})^{-1}\mathbf{x}^0$$

The state vector $\mathbf{x}(t)$ can be obtained by inverse transforming $\mathbf{X}(s)$:

$$\mathbf{x}(t) = \mathscr{L}^{-1}[s\mathbf{I} - \mathbf{A})^{-1}]\mathbf{x}^0$$

Comparing this equation with Eqn. (14.72b), we get

$$e^{\mathbf{A}t} = \boldsymbol{\phi}(t) = \mathscr{L}^{-1}[s\mathbf{I} - \mathbf{A})^{-1}] \qquad (14.73)$$

The matrix $(s\mathbf{I} - \mathbf{A})^{-1} = \boldsymbol{\Phi}(s)$ is known in mathematical literature as the resolvent of $\mathbf{A}$. The entries of the resolvent matrix $\boldsymbol{\Phi}(s)$ are rational functions of s.

Example 14.6 Consider the system

$$\dot{\mathbf{x}} = \begin{bmatrix} 0 & 0 & -2 \\ 0 & 1 & 0 \\ 1 & 0 & 3 \end{bmatrix}\mathbf{x};\ \mathbf{x}(0) = \begin{bmatrix} 0 \\ 1 \\ 0 \end{bmatrix}$$

The resolvent matrix

$$\boldsymbol{\Phi}(s) = (s\mathbf{I} - \mathbf{A})^{-1} = \begin{bmatrix} s & 0 & 2 \\ 0 & s-1 & 0 \\ -1 & 0 & s-3 \end{bmatrix}^{-1} = \frac{(s\mathbf{I} - \mathbf{A})^{+}}{|s\mathbf{I} - \mathbf{A}|}$$

$$|s\mathbf{I} - \mathbf{A}| = (s-1)^2(s-2)$$

$$(s\mathbf{I} - \mathbf{A})^{+} = \begin{bmatrix} (s-1)(s-3) & 0 & -2(s-1) \\ 0 & (s-1)(s-2) & 0 \\ (s-1) & 0 & s(s-1) \end{bmatrix}$$

$$e^{\mathbf{A}t} = \mathscr{L}^{-1}[\mathbf{\Phi}(s)] = \mathscr{L}^{-1}[s\mathbf{I}-\mathbf{A})^{-1}]$$

$$= \mathscr{L}^{-1}\begin{bmatrix} \dfrac{(s-3)}{(s-1)(s-2)} & 0 & \dfrac{-2}{(s-1)(s-2)} \\ 0 & \dfrac{1}{(s-1)} & 0 \\ \dfrac{1}{(s-1)(s-2)} & 0 & \dfrac{s}{(s-1)(s-2)} \end{bmatrix} = \begin{bmatrix} 2e^t - e^{2t} & 0 & 2e^t - 2e^{2t} \\ 0 & e^t & 0 \\ -e^t + e^{2t} & 0 & 2e^{2t} - e^t \end{bmatrix}$$

Consequently, the free response of the system is

$$\mathbf{x}(t) = e^{\mathbf{A}t}\mathbf{x}(0) = \begin{bmatrix} 0 \\ e^t \\ 0 \end{bmatrix}$$

Note that $\mathbf{x}(t)$ could be more easily computed by taking the inverse Laplace transform of

$$\mathbf{X}(s) = [(s\mathbf{I} - \mathbf{A})^{-1}\mathbf{x}(0)]$$

14.6.3 Solution of Nonhomogeneous State Equation

When an input $u(t)$ is present, the complete solution $\mathbf{x}(t)$ is obtained from the nonhomogeneous Eqn. (14.62).

By writing Eqn. (14.62) as

$$\dot{\mathbf{x}}(t) - \mathbf{A}\mathbf{x}(t) = \mathbf{b}u(t)$$

and premultiplying both sides of this equation by $e^{-\mathbf{A}t}$, we obtain

$$e^{-\mathbf{A}t}[\dot{\mathbf{x}}(t) - \mathbf{A}\mathbf{x}(t)] = e^{-\mathbf{A}t}\mathbf{b}u(t) \tag{14.74}$$

By applying the rule for the derivative of the product of two matrices, we can write (refer to Eqn. (14.69))

$$\begin{aligned}\frac{d}{dt}[e^{-\mathbf{A}t}\mathbf{x}(t)] &= e^{-\mathbf{A}t}\frac{d}{dt}(\mathbf{x}(t)) + \frac{d}{dt}(e^{-\mathbf{A}t})\mathbf{x}(t) \\ &= e^{-\mathbf{A}t}\dot{\mathbf{x}}(t) - e^{-\mathbf{A}t}\mathbf{A}\mathbf{x}(t) \\ &= e^{-\mathbf{A}t}[\dot{\mathbf{x}}(t) - \mathbf{A}\mathbf{x}(t)]\end{aligned}$$

Use of this equality in Eqn. (14.74) gives

$$\frac{d}{dt}[e^{-\mathbf{A}t}\mathbf{x}(t)] = e^{-\mathbf{A}t}\mathbf{b}u(t)$$

Integrating both sides with respect to t between the limits 0 and t, we get

$$e^{-\mathbf{A}t}\mathbf{x}(t)\Big|_0^t = \int_0^t e^{-\mathbf{A}\tau}\mathbf{b}u(\tau)d\tau$$

or
$$e^{-\mathbf{A}t}\mathbf{x}(t)-\mathbf{x}(0)=\int_0^t e^{-\mathbf{A}\tau}\mathbf{b}u(\tau)d\tau$$

Now premultiplying both sides by $e^{\mathbf{A}t}$, we have

$$\mathbf{x}(t)=e^{\mathbf{A}t}\mathbf{x}(0)+\int_0^t e^{\mathbf{A}(t-\tau)}\mathbf{b}u(\tau)d\tau \tag{14.75}$$

If the initial state is known at $t = t_0$, rather than $t = 0$, Eqn. (14.75) becomes

$$\mathbf{x}(t)=e^{\mathbf{A}(t-t_0)}\mathbf{x}(t_0)+\int_{t_0}^t e^{\mathbf{A}(t-\tau)}\mathbf{b}u(\tau)d\tau \tag{14.76}$$

Equation (14.76) can also be written as

$$\mathbf{x}(t)=\boldsymbol{\phi}(t-t_0)\,\mathbf{x}(t_0)+\int_{t_0}^t \boldsymbol{\phi}(t-\tau)\mathbf{b}u(\tau)d\tau \tag{14.77}$$

where
$$\boldsymbol{\phi}(t)=e^{\mathbf{A}t}$$

Equation (14.77) is the solution of Eqn. (14.62). This equation is called the *state transition equation*. It describes the change of state relative to the initial conditions $\mathbf{x}(t_0)$ and the input $u(t)$.

Example 14.7 For the speed control system of Fig. 14.6, the following plant model was derived in Example 14.1 (refer to Eqns (14.17)):

$$\dot{\mathbf{x}}=\mathbf{A}\mathbf{x}+\mathbf{b}u$$
$$y=\mathbf{c}\mathbf{x}$$

with

$$\mathbf{A}=\begin{bmatrix}-1 & 1\\ -1 & -10\end{bmatrix};\mathbf{b}=\begin{bmatrix}0\\10\end{bmatrix};\mathbf{c}=[1\;\;0]$$

State variables x_1 and x_2 are the physical variables of the system:

$x_1(t)=\omega(t)$, angular velocity of the motor shaft; $x_2(t)=i_a(t)$, armature current

The output

$$y(t)=x_1(t)=\omega(t)$$

In the following, we evaluate the response of this system to a unit step input under zero initial conditions.

$$(s\mathbf{I}-\mathbf{A})^{-1}=\begin{bmatrix}s+1 & -1\\ 1 & s+10\end{bmatrix}^{-1}=\frac{1}{s^2+11s+11}\begin{bmatrix}s+10 & 1\\ -1 & s+1\end{bmatrix}$$

$$=\begin{bmatrix}\dfrac{s+10}{(s+a_1)(s+a_2)} & \dfrac{1}{(s+a_1)(s+a_2)}\\ \dfrac{-1}{(s+a_1)(s+a_2)} & \dfrac{s+1}{(s+a_1)(s+a_2)}\end{bmatrix};\; a_1=1.1125,\; a_2=9.8875$$

$$e^{\mathbf{A}t} = \mathscr{L}^{-1}[s\mathbf{I} - \mathbf{A})^{-1}]$$

$$= \begin{bmatrix} 1.0128e^{-a_1t} - 0.0128e^{-a_2t} & 0.114e^{-a_1t} - 0.114e^{-a_2t} \\ -0.114e^{-a_1t} + 0.114e^{-a_2t} & -0.0128e^{-a_1t} + 1.0128e^{-a_2t} \end{bmatrix}$$

$$u(t) = 1,\ t \geq 0; \mathbf{x}(0) = \mathbf{0}$$

Therefore,

$$\mathbf{x}(t) = \int_0^t e^{\mathbf{A}(t-\tau)}\mathbf{b}d\tau = \int_0^t \begin{bmatrix} 1.14(-e^{-a_1(t-\tau)} - e^{-a_2(t-\tau)} \\ 1.14(-0.1123e^{-a_1(t-\tau)} + 8.8842e^{-a_2(t-\tau)}) \end{bmatrix} d\tau$$

$$= \begin{bmatrix} 0.9094 - 1.0247e^{-a_1t} + 0.1153e^{-a_2t} \\ -0.0132 + 0.1151e^{-a_1t} - 0.1019e^{-a_2t} \end{bmatrix}$$

The output

$$y(t) = \omega(t) = 0.9094 - 1.0247\, e^{-1.1125t} + 0.1153e^{-9.8875t};\ t \geq 0.$$

14.7 CONCEPTS OF CONTROLLABILITY AND OBSERVABILITY

Controllability and observability are properties which describe structural features of a dynamic system. These properties play an important role in modern control system design theory; the conditions on controllability and observability often govern the control solution.

To motivate the investigation of controllability and observability properties, we first consider the problem of the stabilization of an inverted pendulum on a motor-driven cart; the definitions are presented thereafter.

Example 14.8 Figure 14.17 shows an inverted pendulum with its pivot mounted on a cart. The cart is driven by an electric motor. The motor drives a pair of wheels of the cart; the whole cart and the pendulum become the 'load' on the motor. The motor at time t exerts a torque *T(t)* on the wheels. The linear force applied to the cart is *u(t)*; *T(t)* = *Ru(t)*, where *R* is the radius of the wheels.

The pendulum is obviously unstable. It can, however, be kept upright by applying a proper control force *u*(*t*). This somewhat artificial system example represents a dynamic model of a space booster on take off—the booster is balanced on top of the rocket engine thrust vector.

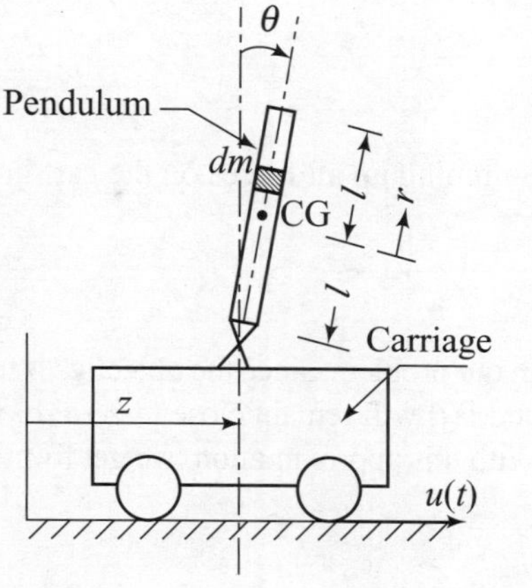

Fig. 14.17 *Inverted pendulum system*

From inspection of Fig. 14.17, we construct the differential equations describing the dynamics of the inverted pendulum and the cart. The horizontal displacement of the pivot on the cart with respect to the fixed nonrotating coordinate frame, is $z(t)$, while the rotational angle of the pendulum is $\theta(t)$. The parameters of the system are as follows.

M = the mass of the cart
L = the length of the pendulum = $2l$
m = the mass of the pendulum
J = the moment of inertia of the pendulum with respect to center of gravity (CG)

$$J = \int_{-l}^{l} r^2 dm = \int_{-l}^{l} r^2(\rho A dr) = \rho A\left[\frac{r^3}{3}\right]_{-l}^{l} = \rho A\left(\frac{2l^3}{3}\right)$$

$$= \rho A(2l)\left(\frac{l^2}{3}\right) = \frac{ml^2}{3}$$

A = area of cross section; and ρ = density.

The horizontal and vertical positions of the CG of the pendulum are given by $(z + l\sin\theta)$ and $(l\cos\theta)$, respectively.

The forces exerted on the pendulum are the force mg on the center of gravity, a horizontal reaction force H and a vertical reaction force V (Fig. 14.18). H is the horizontal reaction force that the cart exerts on the pendulum, whereas $-H$ is the force exerted by the pendulum on the cart. Similar convention applies to forces V and $-V$. Taking moments around CG of the pendulum, we get

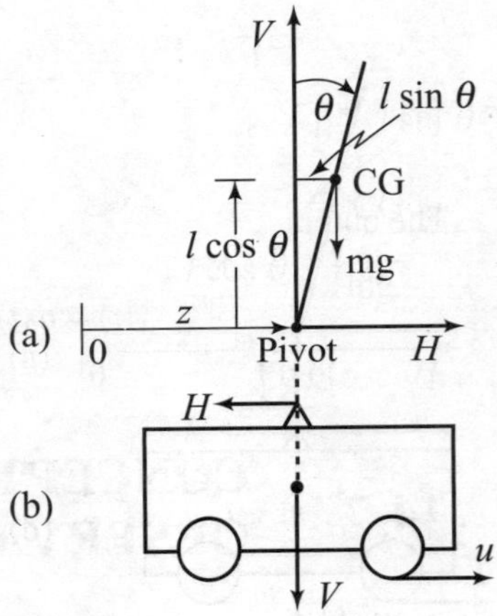

Fig. 14.18

$$J\frac{d^2\theta(t)}{dt^2} = V(t)\, l\sin\theta(t) - H(t)\, l\cos\theta(t) \quad (14.78a)$$

Summing up all forces on the pendulum in vertical and horizontal directions, we obtain

$$m\frac{d^2}{dt^2}(l\cos\theta(t)) = V(t) - mg \quad (14.78b)$$

$$m\frac{d^2}{dt^2}(z(t) + l\sin\theta(t)) = H(t) \quad (14.78c)$$

Summing up all forces on the cart in the horizontal direction, we get

$$M\frac{d^2z(t)}{dt^2} = u(t) - H(t) \quad (14.78d)$$

In our problem, since the objective is to keep the pendulum upright, it seems reasonable to assume that $\dot{\theta}(t)$ and $\theta(t)$ will remain close to zero. In view of this, we can set with sufficient accuracy $\sin\theta \cong \theta$; $\cos\theta \cong 1$. With this approximation, we get from Eqns (14.78),

$$ml\,\ddot{\theta}(t) + (m+M)\,\ddot{z}(t) = u(t)$$

$$(J + ml^2)\ddot{\theta}(t) + ml\,\ddot{z}(t) - mgl\,\theta(t) = 0$$

These equations may be rearranged as

$$\ddot{\theta}(t) = \frac{ml(M+m)g}{\Delta}\theta(t) - \frac{ml}{\Delta}u(t) \tag{14.79a}$$

$$\ddot{z}(t) = -\frac{m^2l^2 g}{\Delta}\theta(t) + \frac{(J+ml^2)}{\Delta}u(t) \tag{14.79b}$$

where

$$\Delta = (M+m)J + Mm\,l^2$$

Suppose that the system parameters are $M = 1$ kg, $m = 0.15$ kg, and $l = 0.5$ m. Recall that $g = 9.81$ m/sec^2.
For these parameters, we have from Eqns (14.79),

$$\ddot{\theta}(t) = 16.3106\,\theta(t) - 1.4458\,u(t) \tag{14.80a}$$

$$\ddot{z}(t) = -1.0637\,\theta(t) + 0.9639\,u(t) \tag{14.80b}$$

Choosing the states $x_1 = \theta$, $x_2 = \dot{\theta}$, $x_3 = z$, and $x_4 = \dot{z}$, we obtain the following state model for the inverted pendulum on moving cart:

$$\dot{\mathbf{x}} = \mathbf{A}\mathbf{x} + \mathbf{b}u \tag{14.81}$$

with $$\mathbf{A} = \begin{bmatrix} 0 & 1 & 0 & 0 \\ 16.3106 & 0 & 0 & 0 \\ 0 & 0 & 0 & 1 \\ -1.0637 & 0 & 0 & 0 \end{bmatrix};\ \mathbf{b} = \begin{bmatrix} 0 \\ -1.4458 \\ 0 \\ 0.9639 \end{bmatrix}$$

The plant (14.81) is said to be *completely controllable* if every state $\mathbf{x}(t_0)$ can be affected or controlled to reach a desired state in finite time by some unconstrained control $u(t)$. Shortly we will see that the plant (14.81) satisfies this condition and therefore, a solution exists to the following control problem:

Move the cart from one location to another without causing the pendulum to fall.

The solution to this control problem is not unique. We normally look for a feedback control scheme so that the destabilizing effects of disturbance forces (due to wind, for example) are filtered out. Figure 14.19a shows a state-feedback control scheme for stabilizing the inverted pendulum. The closed-loop system is formed by feeding back the state variables through a real constant matrix **k**;

$$u(t) = -\mathbf{k}\mathbf{x}(t)$$

The closed-loop system is thus described by

$$\dot{\mathbf{x}}(t) = (\mathbf{A} - \mathbf{b}\mathbf{k})\,\mathbf{x}(t)$$

The design objective in this case is to find the feedback matrix **k** such that the closed-loop system is stable. The existence of solution to this design problem is directly based on the controllability property of the plant (14.81).

Implementation of the state-feedback control solution requires access to all the state variables of the plant model. In many control situations of interest, it is possible to install sensors to measure all the state variables. This may not be possible or practical in some cases. For example, if the plant model includes nonphysical state variables, measurement of these variables using physical sensors is not possible. Accuracy requirements or cost considerations may prohibit the use of sensors for some physical variables also.

The input and the output of a system are always physical quantities, and are normally easily accessible to measurement. We therefore need a subsystem that performs the estimation of state variables based on the information received from the input $u(t)$ and the output $y(t)$. This subsystem is called an *observer* whose design is based on *observability property* of the controlled system.

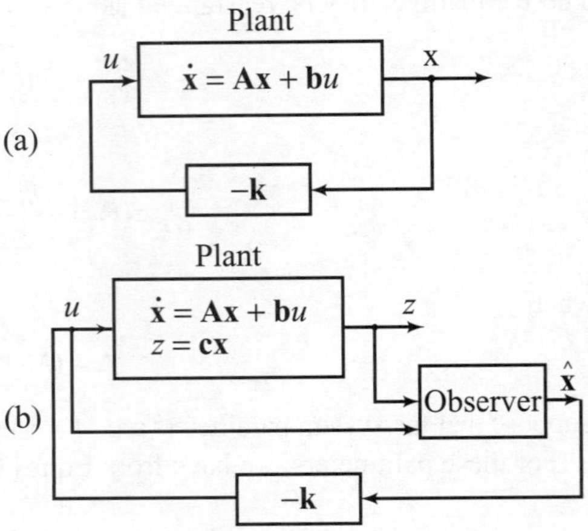

Fig. 14.19 *Control system with state feedback*

The plant (14.81) is said to be *completely observable* if all the state variables in $\mathbf{x}(t)$ can be observed from the measurements of the output $y(t) = \theta(t)$ and the input $u(t)$. Shortly we will see that the plant (14.81) does not satisfy this condition, and therefore a solution to the observer-design problem does not exist when the inputs to the observer subsystem are $u(t)$ and $\theta(t)$.

Cart position $z(t)$ is easily accessible to measurement and, as we shall see, the observability condition is satisfied with this choice of input information to the observer subsystem. Figure 14.19b shows the block diagram of the closed-loop system with an observer that estimates the state vector from measurements of $u(t)$ and $z(t)$. The observed or estimated state vector, designated as $\hat{\mathbf{x}}$, is then used to generate the control u through the feedback matrix $\mathbf{k}$.

A study of controllability and observability properties, presented in this section, provides a basis for the state-feedback design problems (Chapter 15). Further, these properties establish the conditions for complete equivalence between the state variable and transfer function representations (Section 14.9).

14.7.1 Definitions of Controllability and Observability

In this section, we study the controllability and observability of linear time-invariant systems described by state variable model of the following form:

$$\dot{\mathbf{x}}(t) = \mathbf{A}\mathbf{x}(t) + \mathbf{b}u(t) \tag{14.82a}$$

$$y(t) = \mathbf{c}\mathbf{x}(t) + du(t) \tag{14.82b}$$

where $\mathbf{A}$, $\mathbf{b}$, $\mathbf{c}$ and d are respectively $n \times n$, $n \times 1$, $1 \times n$ and 1×1 matrices, $\mathbf{x}(t)$ is $n \times 1$ state vector, $y(t)$ and $u(t)$ are, respectively, output and input variables.

Controllability

For the linear system given by (14.82), if there exists an input $u_{[0,\,t_1]}$ which transfers the initial state $\mathbf{x}(0) \triangleq \mathbf{x}^0$ to the state $\mathbf{x}^1$ in a finite time t_1, the state $\mathbf{x}^0$ is said to be controllable. If all initial states are controllable, the system is said to be *completely controllable*, or simply *controllable*. Otherwise, the system is said to be *uncontrollable*.

From Eqn. (14.75), the solution of Eqn. (14.82a) is

$$\mathbf{x}(t) = e^{\mathbf{A}t}\mathbf{x}^0 + \int_0^t e^{\mathbf{A}(t-\tau)}\mathbf{b}u(\tau)\, d\tau$$

To study the controllability property, we may assume without loss of generality that $\mathbf{x}^1 \equiv \mathbf{0}$. Therefore, if the system (14.82) is controllable, there exists an input $u_{[0,\, t_1]}$ such that

$$-\mathbf{x}^0 = \int_0^{t_1} e^{-\mathbf{A}\tau}\mathbf{b}u(\tau)\, d\tau \tag{14.83}$$

From this equation, we observe that complete controllability of a system depends on **A** and **b**, and is independent of output matrix **c**. The controllability of the system (14.82) is frequently referred to as the controllability of the pair $\{\mathbf{A}, \mathbf{b}\}$.

It may be noted that according to the definition of controllability, there is no constraint imposed on the input or on the trajectory that the state should follow. Further, the system is said to be uncontrollable although it may be 'controllable in part'.

From the definition of controllability, we observe that by complete controllability of a plant, we mean that we can make the plant do whatever we please. Perhaps this definition is too restrictive in the sense that we are asking too much of the plant. But if we are able to show that system equations satisfy this definition, certainly there can be no intrinsic limitation on the design of the control system for the plant. However, if the system turns out to be uncontrollable, it does not necessarily mean that the plant can never be operated in a satisfactory manner. Provided that a control system will maintain the important variables in an acceptable region, the fact that the plant is not completely controllable is immaterial.

Another important point which the reader must bear in mind is that almost all physical systems are nonlinear in nature to a certain extent and a linear model is obtained after making certain approximations. Small perturbations of the elements of **A** and **b** may cause an uncontrollable system to become controllable.

A common source of uncontrollable state variable models arises when redundant state variables are defined. No one would intentionally use more state variables than the minimum number needed to characterize the behavior of the dynamic system. In a complex system with unfamiliar physics, one may be tempted to write down differential equations for everything in sight and in doing so, may write down more equations than are necessary. This will invariably result in an uncontrollable model for the system.

Observability

For the linear system given by Eqns (14.82), if the knowledge of the output y and the input u over a finite interval of time $[0, t_1]$ suffices to determine the state $\mathbf{x}(0) \triangleq \mathbf{x}^0$, the state $\mathbf{x}^0$ is said to be observable. If all initial states are observable, the system is said to be *completely observable*, or simply *observable*. Otherwise, the system is said to be *unobservable*.

The output of the system (14.82) is given by

$$y(t) = \mathbf{c}\, e^{\mathbf{A}t}\mathbf{x}^0 + \mathbf{c}\int_0^t e^{\mathbf{A}(t-\tau)}\mathbf{b}u(\tau)d\tau + du(t)$$

The output and the input can be measured and used, so that the following signal $\eta(t)$ can be obtained from u and y.

$$\eta(t) \triangleq y(t) - \mathbf{c}\int_0^t e^{\mathbf{A}(t-\tau)}\mathbf{b}\,u(\tau)d\tau - d\,u(t) = \mathbf{c}\,e^{\mathbf{A}t}\mathbf{x}^0 \tag{14.84}$$

Premultiplying by $e^{\mathbf{A}^T t}\mathbf{c}^T$ and integrating from 0 to t_1, gives

$$\left\{\int_0^{t_1} e^{\mathbf{A}^T t}\mathbf{c}^T\mathbf{c}e^{\mathbf{A}t}dt\right\}\mathbf{x}^0 = \int_0^{t_1} e^{\mathbf{A}^T t}\mathbf{c}^T\eta(t)dt \tag{14.85}$$

When the signal $\eta(t)$ is available over a time interval $[0, t_1]$, and the system (14.82) is observable, then the initial state $\mathbf{x}^0$ can be uniquely determined from Eqn. (14.85).

From Eqn. (14.85) we see that complete observability of a system depends on **A** and **c**, and is independent of **b**. The observability of the system (14.82) is frequently referred to as the observability of the pair {**A**, **c**}.

Note that the system is said to be unobservable, although it may be 'observable in part'. For plants that are not completely observable, one may examine feedback control schemes which do not require complete state feedback.

14.7.2 Controllability Test

It is difficult to guess whether a system is controllable or not from the defining Eqn. (14.83). Some simple mathematical tests which answer the question of controllability have been developed. The following theorem gives a simple controllability test[4].

Theorem 14.1

The necessary and sufficient condition for the system (14.82) to be completely controllable is that the *$n \times n$ controllability matrix*

$$\mathbf{U} \triangleq [\mathbf{b}\quad \mathbf{A}\mathbf{b}\quad \mathbf{A}^2\mathbf{b}\cdots\mathbf{A}^{n-1}\mathbf{b}] \tag{14.86}$$

has rank equal to n, i.e., $\rho(\mathbf{U}) = n$.

Example 14.9 Recall the inverted pendulum of Example 14.8, shown in Fig. 14.17, in which the object is to apply a force *u(t)* so that the pendulum remains balanced in the vertical position. We found the linearized equations governing the system to be:

$$\dot{\mathbf{x}} = \mathbf{A}\mathbf{x} + \mathbf{b}u$$

where

$$\mathbf{x} = [\theta\ \ \dot{\theta}\ \ z\ \ \dot{z}]^T$$

$$\mathbf{A} = \begin{bmatrix} 0 & 1 & 0 & 0 \\ 16.3106 & 0 & 0 & 0 \\ 0 & 0 & 0 & 1 \\ -1.0637 & 0 & 0 & 0 \end{bmatrix};\ \mathbf{b} = \begin{bmatrix} 0 \\ -1.4458 \\ 0 \\ 0.9639 \end{bmatrix}$$

[4] For proof, refer to Chapter 5 of reference [155].

$z(t)$ = horizontal displacement of the pivot on the cart
$\theta(t)$ = rotational angle of the pendulum.

To check the controllability of this system, we compute the controllability matrix **U**:

$$\mathbf{U} = [\mathbf{b}\ \ \mathbf{Ab}\ \ \mathbf{A}^2\mathbf{b}\ \ \mathbf{A}^3\mathbf{b}] = \begin{bmatrix} 0 & -1.4458 & 0 & -23.5816 \\ -1.4458 & 0 & -23.5816 & 0 \\ 0 & 0.9639 & 0 & 1.5379 \\ 0.9639 & 0 & 1.5379 & 0 \end{bmatrix}$$

Since $|\mathbf{U}| = 420.4851$, **U** has full rank, and by Theorem 14.1, the system is completely controllable. Thus, if the angle θ departs from equilibrium by a small amount, a control always exists which will drive it back to zero.[5] Moreover, a control also exists which will drive both θ and z, as well as their derivatives, to zero.

Example 14.10 Consider the electrical network shown in Fig. 14.20. Differential equations governing the dynamics of this network can be obtained by various standard methods. By use of nodal analysis, for example, we get

$$C_1 \frac{de_1}{dt} + \frac{e_1 - e_2}{R_3} + \frac{e_1 - e_0}{R_1} = 0$$

$$C_2 \frac{de_2}{dt} + \frac{e_2 - e_1}{R_3} + \frac{e_2 - e_0}{R_2} = 0$$

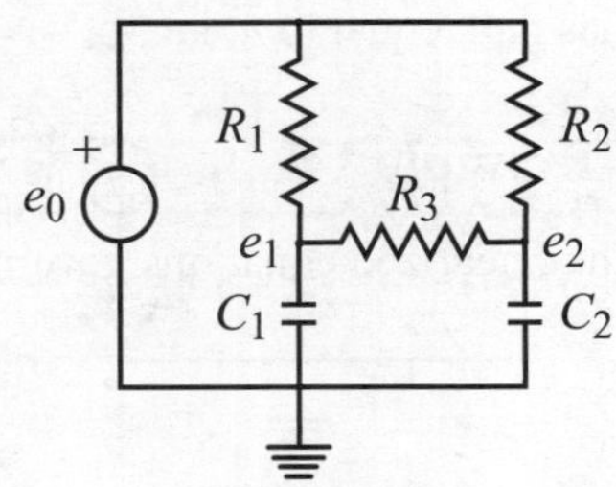

Fig. 14.20

The appropriate state variables for the network are the capacitor voltages e_1 and e_2. Thus, the state equations of the network are

$$\dot{\mathbf{x}} = \mathbf{Ax} + \mathbf{b}e_0$$

where $$\mathbf{x} = [e_1\ e_2]^T$$

$$\mathbf{A} = \begin{bmatrix} -\left(\frac{1}{R_1} + \frac{1}{R_3}\right)\frac{1}{C_1} & \frac{1}{R_3C_1} \\ \frac{1}{R_3C_2} & -\left(\frac{1}{R_2} + \frac{1}{R_3}\right)\frac{1}{C_2} \end{bmatrix}; \mathbf{b} = \begin{bmatrix} \frac{1}{R_1C_1} \\ \frac{1}{R_2C_2} \end{bmatrix}$$

The controllability matrix of the system is

$$\mathbf{U} = [\mathbf{b}\ \ \mathbf{Ab}] = \begin{bmatrix} \frac{1}{R_1C_1} & -\frac{1}{(R_1C_1)^2} + \frac{1}{R_3C_1}\left(\frac{1}{R_2C_2} - \frac{1}{R_1C_1}\right) \\ \frac{1}{R_2C_2} & -\frac{1}{(R_2C_2)^2} + \frac{1}{R_3C_2}\left(\frac{1}{R_1C_1} - \frac{1}{R_2C_2}\right) \end{bmatrix}$$

[5] This justifies the assumption that $\theta(t) \cong 0$, provided we choose an appropriate control strategy.

We see that under the condition

$$R_1C_1 = R_2C_2,$$

$\rho(\mathbf{U}) = 1$ and the system becomes 'uncontrollable'. This condition is the one required to balance the bridge, and in this case, the voltage across the terminals of R_3 cannot be influenced by the input e_0.

14.7.3 Observability Test

The following theorem gives a simple observability test[6].

Theorem 14.2

The necessary and sufficient condition for the system (14.82) to be completely observable is that the *$n \times n$ observability matrix*

$$\mathbf{V} \triangleq \begin{bmatrix} \mathbf{c} \\ \mathbf{cA} \\ \vdots \\ \mathbf{cA}^{n-1} \end{bmatrix} \tag{14.87}$$

has rank equal to n, i.e., $\rho(\mathbf{V}) = n$.

Example 14.11 We now return to the inverted pendulum of Example 14.9. Assuming that the only output variable for measurement is $\theta(t)$, the position of the pendulum, then the linearized equations governing the system are

$$\dot{\mathbf{x}} = \mathbf{Ax} + \mathbf{b}u$$
$$y = \mathbf{cx}$$

where
$$\mathbf{A} = \begin{bmatrix} 0 & 1 & 0 & 0 \\ 16.3106 & 0 & 0 & 0 \\ 0 & 0 & 0 & 1 \\ -1.0637 & 0 & 0 & 0 \end{bmatrix}; \mathbf{b} = \begin{bmatrix} 0 \\ -1.4458 \\ 0 \\ 0.9639 \end{bmatrix}$$
$$\mathbf{c} = [1\ 0\ 0\ 0]$$

The observability matrix

$$\mathbf{V} = \begin{bmatrix} \mathbf{c} \\ \mathbf{cA} \\ \mathbf{cA}^2 \\ \mathbf{cA}^3 \end{bmatrix} = \begin{bmatrix} 1 & 0 & 0 & 0 \\ 0 & 1 & 0 & 0 \\ 16.3106 & 0 & 0 & 0 \\ 0 & 16.3106 & 0 & 0 \end{bmatrix}$$

$|\mathbf{V}| = 0$, and therefore by Theorem 14.2, the system is not completely observable.

Consider now the displacement $z(t)$ of the cart as the output variable. Then

$$\mathbf{c} = [0\ 0\ 1\ 0]$$

[6] For proof, refer to Chapter 5 of reference [155].

and the observability matrix

$$\mathbf{V} = \begin{bmatrix} 0 & 0 & 1 & 0 \\ 0 & 0 & 0 & 1 \\ -1.0637 & 0 & 0 & 0 \\ 0 & -1.0637 & 0 & 0 \end{bmatrix}$$

$|\mathbf{V}| = 1.1315 \neq 0$; the system is therefore completely observable. The values of $\dot{z}(t)$, $\theta(t)$ and $\dot{\theta}(t)$ can all be determined by observing $z(t)$ over an arbitrary time interval.

14.7.4 Invariance Property

It is recalled that the state variable model for a system is not unique, but depends on the choice of a set of state variables. A transformation

$$\mathbf{x}(t) = \mathbf{P}\,\bar{\mathbf{x}}(t);\ \mathbf{P} \text{ is a nonsingular constant matrix,}$$

results in the following alternative state variable model (refer to Eqns (14.22)) for the system (14.82):

$$\dot{\bar{\mathbf{x}}}(t) = \bar{\mathbf{A}}\,\bar{\mathbf{x}}(t) + \bar{\mathbf{b}}\,u(t);\ \bar{\mathbf{x}}(t_0) = \mathbf{P}^{-1}\mathbf{x}(t_0)$$
$$y(t) = \bar{\mathbf{c}}\,\bar{\mathbf{x}}(t) + du(t)$$

where

$$\bar{\mathbf{A}} = \mathbf{P}^{-1}\mathbf{A}\mathbf{P},\ \bar{\mathbf{b}} = \mathbf{P}^{-1}\mathbf{b},\ \bar{\mathbf{c}} = \mathbf{c}\mathbf{P}$$

The definition of a new set of internal state variables should evidently not affect the controllability and observability properties. This may be verified by evaluating the controllability and observability matrices of the transformed system.

I.

$$\bar{\mathbf{U}} = [\bar{\mathbf{b}}\ \ \bar{\mathbf{A}}\bar{\mathbf{b}}\ \cdots\ (\bar{\mathbf{A}})^{n-1}\bar{\mathbf{b}}]$$
$$\bar{\mathbf{b}} = \mathbf{P}^{-1}\mathbf{b}$$
$$\bar{\mathbf{A}}\bar{\mathbf{b}} = \mathbf{P}^{-1}\mathbf{A}\mathbf{P}\mathbf{P}^{-1}\mathbf{b} = \mathbf{P}^{-1}\mathbf{A}\mathbf{b}$$
$$(\bar{\mathbf{A}})^2\bar{\mathbf{b}} = (\bar{\mathbf{A}})(\bar{\mathbf{A}}\bar{\mathbf{b}}) = \mathbf{P}^{-1}\mathbf{A}\mathbf{P}\mathbf{P}^{-1}\mathbf{A}\mathbf{b} = \mathbf{P}^{-1}\mathbf{A}^2\mathbf{b}$$
$$\vdots$$
$$(\bar{\mathbf{A}})^{n-1}\bar{\mathbf{b}} = \mathbf{P}^{-1}\mathbf{A}^{n-1}\mathbf{b}$$

Therefore,

$$\bar{\mathbf{U}} = [\mathbf{P}^{-1}\mathbf{b}\ \ \mathbf{P}^{-1}\mathbf{A}\mathbf{b} \cdots \mathbf{P}^{-1}\mathbf{A}^{n-1}\mathbf{b}] = \mathbf{P}^{-1}\mathbf{U}$$

where

$$\mathbf{U} = [\mathbf{b}\ \ \mathbf{A}\mathbf{b} \cdots \mathbf{A}^{n-1}\mathbf{b}]$$

Since $\mathbf{P}^{-1}$ is nonsingular,

$$\rho(\bar{\mathbf{U}}) = \rho(\mathbf{U})$$

II. A similar relationship can be shown for the observability matrices.

14.8 STABILITY OF LINEAR SYSTEMS

Controllability and observability, introduced in the previous section, are two important qualitative properties of linear systems. Another important qualitative property is *stability*.

Consider an *n*-dimensional linear time-invariant system described by a state variable model of the form (refer to Eqns (14.11))

$$\dot{\mathbf{x}}(t) = \mathbf{A}\mathbf{x}(t) + \mathbf{b}r(t);\ \ \mathbf{x}(t=0) \triangleq \mathbf{x}^0 \tag{14.88a}$$

$$y(t) = \mathbf{c}\mathbf{x}(t) + dr(t) \tag{14.88b}$$

where $\mathbf{x}$ is the $n\times 1$ state vector, r is the scalar input, y is the scalar output, $\mathbf{A}$ is $n\times n$ real constant matrix, $\mathbf{b}$ and $\mathbf{c}$ are, respectively, $n\times 1$ and $1\times n$ real constant vectors, and d is a constant scalar.

The system (14.88) has two sources of excitation:

(i) the initial state $\mathbf{x}^0$ representing initial energy storage; and
(ii) external input $r(t)$; $t \geq 0$

The response of the linear system (14.88) to these two sources of excitation can always be decomposed as the *zero-state response* (i.e., response with $\mathbf{x}^0 = \mathbf{0}$), and zero-input response (i.e., response with $r(t) = 0$; $t \geq 0$). It is customary to study the stability of these two responses separately.

In the stability study of linear systems, we are generally concerned with the following questions:

1. If the system is relaxed ($\mathbf{x}^0 = \mathbf{0}$), will the bounded input $r(t)$; $t \geq 0$, produce a bounded output $y(t)$ for all t?
2. For the system with $r(t) = 0$; $t \geq 0$, we obtain from Eqn. (14.88a),

$$\dot{\mathbf{x}}(t) = \mathbf{A}\mathbf{x}(t); \mathbf{x}(t=0) = \mathbf{x}^0 \tag{14.89}$$

At $\mathbf{x}(t) = \mathbf{0}, \dot{\mathbf{x}}(t) = \mathbf{0}$ and the system is in steady state. Therefore, a relaxed system ($\mathbf{x}^0 = \mathbf{0}$) with input $r(t) = 0$; $t \geq 0$, will continue to be in the zero state for all time. This condition can be viewed as the *equilibrium state* of the system: the system is said to be in the equilibrium state $\mathbf{x}^e = \mathbf{0}$ when both the internal energy storage and the external input are zero.

Taking changes in the initial state as the stimulus and the system state $\mathbf{x}(t)$ as the response, we pose the following question:

If the system with zero input ($r(t) = 0$; $t \geq 0$) is *perturbed* from its equilibrium state $\mathbf{x}^e = \mathbf{0}$ at $t = 0$, will the system state $\mathbf{x}(t)$ return to $\mathbf{x}^e$, remain 'close' to $\mathbf{x}^e$, or diverge from $\mathbf{x}^e$?

The first notion of stability is concerned with the boundedness of the output, and is called *bounded-input, bounded-output (BIBO) stability*. This concept of stability was discussed in Chapter 5. There, the conditions of BIBO stability were given in terms of the pole locations of the input-output transfer functions.

The second notion of stability is concerned with the 'boundedness' of the state of an *unforced system* in response to arbitrary initial state, and is called *zero-input stability*. The goal of this section is to study this concept of stability of dynamical systems described in terms of state variables. State variable model (14.89) is the most appropriate model for the study of dynamic evolution of the state $\mathbf{x}(t)$ in response to the initial state $\mathbf{x}^0$.

We may classify stability as follows:

1. *Unstable*: There is at least one finite initial state $\mathbf{x}^0$ such that $\mathbf{x}(t)$ grows thereafter without bound at $t \rightarrow \infty$.
2. *Asymptotically stable*: Every initial state $\mathbf{x}^0$ excites a bounded response, which, in addition, approaches $\mathbf{0}$ as $t \rightarrow \infty$.
3. *Marginally stable*: For all initial states $\mathbf{x}^0$, $\mathbf{x}(t)$ remains thereafter within finite bounds for $t > 0$.

Marginal stability is useful only in the design of oscillators. Other than oscillators, every linear system is designed to be asymptotically stable.

The solution of (14.89) is given by (refer to Section 14.6.2) the inverse Laplace transformation of

$$\mathbf{X}(s) = (s\mathbf{I} - \mathbf{A})^{-1}\mathbf{x}^0 \tag{14.90a}$$

$$(s\mathbf{I} - \mathbf{A})^{-1} = \frac{(s\mathbf{I} - \mathbf{A})^{+}}{|s\mathbf{I} - \mathbf{A}|} = \frac{adj(s\mathbf{I} - \mathbf{A})}{det(s\mathbf{I} - \mathbf{A})} \tag{14.90b}$$

Note that for an $n \times n$ matrix $\mathbf{A}$, $|s\mathbf{I} - \mathbf{A}|$ is an nth-order polynomial in s. Also, each element of the adjoint matrix $(s\mathbf{I} - \mathbf{A})^{+}$ is a polynomial in s of order less than or equal to $(n - 1)$. Therefore, each element of $(s\mathbf{I} - \mathbf{A})^{-1}$ is a strictly proper rational function, and can be expanded in partial fraction expansion.

The nature (modes) of response $\mathbf{x}(t) = \mathscr{L}^{-1}[\mathbf{X}(s)]$, depends on the roots of the characteristic polynomial $|s\mathbf{I} - \mathbf{A}|$ (refer to Section 5.3). The roots of $|s\mathbf{I} - \mathbf{A}|$ are the *eigenvalues* of matrix $\mathbf{A}$ (Section 14.4.1).

The conditions for zero-input stability given in terms of eigenvalues of matrix $\mathbf{A}$, are as follows:

1. If all the eigenvalues of matrix $\mathbf{A}$ have negative real parts, the system is asymptotically stable.
2. If any eigenvalue of matrix $\mathbf{A}$ has positive real part, or if there is a repeated eigenvalue with zero real part, the system is unstable.
3. If condition (i) is satisfied except for the presence of one or more nonrepeated eigenvalues with zero real part, the system is marginally stable.

These conditions can easily be established from inverse Laplace transformation of Eqns (14.90).

In the next section, we will see that under mildly restrictive conditions (namely, the system must be both controllable and observable), the eigenvalues of matrix $\mathbf{A}$ are same as the poles of the corresponding transfer function, and asymptotic stability ensures BIBO stability and *vice versa*.

We must emphasize here that these concepts of stability are applicable only to linear systems. A more general concept of stability, applicable to both nonlinear and linear systems, follows from *Lyapunov stability theory*. In Chapter 16, we will introduce Lyapunov stability theory; and there an alternative method of determining asymptotic stability of linear systems using *Lyapunov equation*, will also be given.

14.9 EQUIVALENCE BETWEEN TRANSFER FUNCTION AND STATE VARIABLE REPRESENTATIONS

In frequency-domain analysis, it is tacitly assumed that the dynamic properties of a system are completely determined by the transfer function of the system. That this is not always the case is illustrated by the following examples:

Example 14.12 Consider the system

$$\dot{\mathbf{x}} = \mathbf{A}\mathbf{x} + \mathbf{b}u \qquad (14.91)$$
$$y = \mathbf{c}\mathbf{x}$$

with

$$\mathbf{A} = \begin{bmatrix} -2 & 1 \\ 1 & -2 \end{bmatrix}; \mathbf{b} = \begin{bmatrix} 1 \\ 1 \end{bmatrix}; \mathbf{c} = [0 \quad 1]$$

The controllability matrix

$$\mathbf{U} = [\mathbf{b} \quad \mathbf{A}\mathbf{b}] = \begin{bmatrix} 1 & -1 \\ 1 & -1 \end{bmatrix}$$

Since $\rho(\mathbf{U}) = 1$, the second-order system (14.91) is not completely controllable.
The eigenvalues of matrix **A** are the roots of the characteristic equation

$$|s\mathbf{I} - \mathbf{A}| = \begin{vmatrix} s+2 & -1 \\ -1 & s+2 \end{vmatrix} = 0$$

The eigenvalues are obtained as –1, –3. The *modes* of the transient response are therefore e^{-t} and e^{-3t}.
The transfer function of the system (14.91) is calculated as

$$G(s) = \mathbf{c}(s\mathbf{I} - \mathbf{A})^{-1}\mathbf{b} = [0 \quad 1]\begin{bmatrix} s+2 & -1 \\ -1 & s+2 \end{bmatrix}^{-1}\begin{bmatrix} 1 \\ 1 \end{bmatrix}$$

$$= [0 \quad 1]\begin{bmatrix} \dfrac{s+2}{(s+1)(s+3)} & \dfrac{1}{(s+1)(s+3)} \\ \dfrac{1}{(s+1)(s+3)} & \dfrac{s+2}{(s+1)(s+3)} \end{bmatrix}\begin{bmatrix} 1 \\ 1 \end{bmatrix} = \frac{1}{s+1}$$

We find that because of pole-zero cancelation, both the eigenvalues of matrix **A** do not appear as poles in $G(s)$. The dynamic mode e^{-3t} of the system (14.91) does not show up in input-output characterization given by the transfer function $G(s)$. Note that the system under consideration is not a completely controllable system.

Example 14.13 Consider the system

$$\dot{\mathbf{x}} = \mathbf{A}\mathbf{x} + \mathbf{b}u \qquad (14.92)$$
$$y = \mathbf{c}\mathbf{x}$$

with

$$\mathbf{A} = \begin{bmatrix} -2 & 1 \\ 1 & -2 \end{bmatrix}; \mathbf{b} = \begin{bmatrix} 1 \\ 0 \end{bmatrix}; \mathbf{c} = [1 \quad -1]$$

The observability matrix

$$\mathbf{V} = \begin{bmatrix} \mathbf{c} \\ \mathbf{c}\mathbf{A} \end{bmatrix} = \begin{bmatrix} 1 & -1 \\ -3 & 3 \end{bmatrix}$$

Since $\rho(\mathbf{V}) = 1$, the second-order system (14.92) is not completely observable.

The eigenvalues of matrix **A** are $-1, -3$. The transfer function of the system (14.92) is calculated as

$$G(s) = \mathbf{c}(s\mathbf{I} - \mathbf{A})^{-1}\mathbf{b}$$

$$= \begin{bmatrix} 1 & -1 \end{bmatrix} \begin{bmatrix} \dfrac{s+2}{(s+1)(s+3)} & \dfrac{1}{(s+1)(s+3)} \\ \dfrac{1}{(s+1)(s+3)} & \dfrac{s+2}{(s+1)(s+3)} \end{bmatrix} \begin{bmatrix} 1 \\ 0 \end{bmatrix} = \frac{1}{s+3}$$

The dynamic mode e^{-t} of the system (14.92) does not show up in the input-output characterization given by the transfer function $G(s)$. Note that the system under consideration is not a completely observable system.

In the following, we give two specific state transformations to reveal the underlying structure imposed upon a system by its controllability and observability properties (for proof, refer [105]). These results are then used to establish equivalence between transfer function and state variable representations.

Theorem 14.3

Consider an nth-order system

$$\begin{aligned} \dot{\mathbf{x}} &= \mathbf{A}\mathbf{x} + \mathbf{b}u \\ y &= \mathbf{c}\mathbf{x} \end{aligned} \tag{14.93a}$$

Assume that

$$\rho(\mathbf{U}) = \rho[\mathbf{b}\ \mathbf{A}\mathbf{b}\ \cdots\ \mathbf{A}^{n-1}\mathbf{b}] = m < n$$

There exists an equivalence transformation

$$\mathbf{x} = \mathbf{P}\bar{\mathbf{x}} \tag{14.93b}$$

that transforms the system (14.93a) to the following form:

$$\begin{aligned} \begin{bmatrix} \dot{\bar{\mathbf{x}}}_1 \\ \dot{\bar{\mathbf{x}}}_2 \end{bmatrix} &= \begin{bmatrix} \bar{\mathbf{A}}_c & \bar{\mathbf{A}}_{12} \\ \mathbf{0} & \bar{\mathbf{A}}_{22} \end{bmatrix} \begin{bmatrix} \bar{\mathbf{x}}_1 \\ \bar{\mathbf{x}}_2 \end{bmatrix} + \begin{bmatrix} \bar{\mathbf{b}}_c \\ \mathbf{0} \end{bmatrix} u \\ &= \bar{\mathbf{A}}\bar{\mathbf{x}} + \bar{\mathbf{b}}u \end{aligned} \tag{14.93c}$$

$$y = \begin{bmatrix} \bar{\mathbf{c}}_1 & \bar{\mathbf{c}}_2 \end{bmatrix} \begin{bmatrix} \bar{\mathbf{x}}_1 \\ \bar{\mathbf{x}}_2 \end{bmatrix} = \bar{\mathbf{c}}\bar{\mathbf{x}}$$

where the m-dimensional subsystem

$$\dot{\bar{\mathbf{x}}}_1 = \bar{\mathbf{A}}_c\bar{\mathbf{x}}_1 + \bar{\mathbf{b}}_c u + \bar{\mathbf{A}}_{12}\bar{\mathbf{x}}_2$$

is controllable from u (the additional driving term $\bar{\mathbf{A}}_{12}\bar{\mathbf{x}}_2$ has no effect on controllability), and the $(n-m)$ dimensional subsystem

$$\dot{\bar{\mathbf{x}}}_2 = \bar{\mathbf{A}}_{22}\bar{\mathbf{x}}_2$$

is not affected by the input, and is, therefore, entirely uncontrollable.

This theorem shows that any system which is not completely controllable can be decomposed into controllable and uncontrollable subsystems shown in Fig. 14.21. The state model (14.93c) is said to be in *controllability canonical form*.

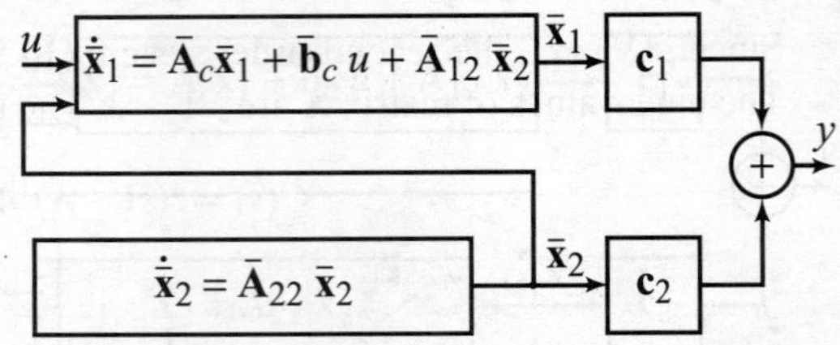

Fig. 14.21 *The controllability canonical form of a state variable model*

In Section 14.4, it was shown that the characteristic equations and transfer functions of equivalent systems are identical. Thus, the set of eigenvalues of matrix **A** of system (14.93a) is same as the set of eigenvalues of matrix **A** of system (14.93c), which is a union of the subsets of eigenvalues of matrices $\bar{\mathbf{A}}_c$ and $\bar{\mathbf{A}}_{22}$. Also the transfer function of system (14.93a) must be the same as that of (14.93c). The transfer function of (14.93a) is calculated from Eqn. (14.93c) as[7]

$$G(s) = \begin{bmatrix} \bar{\mathbf{c}}_1 & \bar{\mathbf{c}}_2 \end{bmatrix} \begin{bmatrix} s\mathbf{I} - \bar{\mathbf{A}}_c & -\bar{\mathbf{A}}_{12} \\ \mathbf{0} & s\mathbf{I} - \bar{\mathbf{A}}_{22} \end{bmatrix}^{-1} \begin{bmatrix} -\bar{\mathbf{b}}_c \\ \mathbf{0} \end{bmatrix}$$

$$= \begin{bmatrix} \bar{\mathbf{c}}_1 & \bar{\mathbf{c}}_2 \end{bmatrix} \begin{bmatrix} (s\mathbf{I} - \bar{\mathbf{A}}_c)^{-1} & (s\mathbf{I} - \bar{\mathbf{A}}_c)^{-1}\bar{\mathbf{A}}_{12}(s\mathbf{I} - \bar{\mathbf{A}}_{22})^{-1} \\ \mathbf{0} & (s\mathbf{I} - \bar{\mathbf{A}}_{22})^{-1} \end{bmatrix} \begin{bmatrix} \bar{\mathbf{b}}_c \\ \mathbf{0} \end{bmatrix}$$

$$= \bar{\mathbf{c}}_1 (s\mathbf{I} - \bar{\mathbf{A}}_c)^{-1} \bar{\mathbf{b}}_c$$

Therefore, the input–output relationship for the system is dependent only on the controllable part of the system. We will refer to the eigenvalues of $\bar{\mathbf{A}}_c$ as *controllable poles* and the eigenvalues of $\bar{\mathbf{A}}_{22}$ as *uncontrollable poles*.

Only the controllable poles appear in the transfer function model; the uncontrollable poles are canceled by the zeros.

Theorem 14.4

Consider the *n*th order system

$$\begin{aligned} \dot{\mathbf{x}} &= \mathbf{A}\mathbf{x} + \mathbf{b}u \\ y &= \mathbf{c}\mathbf{x} \end{aligned} \qquad (14.94a)$$

Assume that

$$\rho(\mathbf{V}) = \rho \begin{bmatrix} \mathbf{c} \\ \mathbf{c}\mathbf{A} \\ \vdots \\ \mathbf{c}\mathbf{A}^{n-1} \end{bmatrix} = l < n$$

There exits an equivalence transformation

$$\bar{\mathbf{x}} = \mathbf{Q}\mathbf{x} \qquad (14.94b)$$

---

7 $\begin{bmatrix} \mathbf{A}_1 & \mathbf{A}_2 \\ \mathbf{0} & \mathbf{A}_3 \end{bmatrix} \begin{bmatrix} \mathbf{B}_1 & \mathbf{B}_2 \\ \mathbf{B}_3 & \mathbf{B}_4 \end{bmatrix} = \begin{bmatrix} \mathbf{I} & \mathbf{0} \\ \mathbf{0} & \mathbf{I} \end{bmatrix}$

gives $\begin{bmatrix} \mathbf{B}_1 & \mathbf{B}_2 \\ \mathbf{B}_3 & \mathbf{B}_4 \end{bmatrix} = \begin{bmatrix} \mathbf{A}_1^{-1} & -\mathbf{A}_1^{-1}\mathbf{A}_2\mathbf{A}_3^{-1} \\ 0 & \mathbf{A}_3^{-1} \end{bmatrix}$

that transforms the system (14.94a) to the following form:

$$\begin{bmatrix}\dot{\bar{\mathbf{x}}}_1\\ \dot{\bar{\mathbf{x}}}_2\end{bmatrix} = \begin{bmatrix}\bar{\mathbf{A}}_0 & \mathbf{0}\\ \bar{\mathbf{A}}_{21} & \bar{\mathbf{A}}_{22}\end{bmatrix}\begin{bmatrix}\bar{\mathbf{x}}_1\\ \bar{\mathbf{x}}_2\end{bmatrix} + \begin{bmatrix}\bar{\mathbf{b}}_1\\ \bar{\mathbf{b}}_2\end{bmatrix} u = \bar{\mathbf{A}}\,\bar{\mathbf{x}} + \bar{\mathbf{b}}\,u \tag{14.94c}$$

$$y = \begin{bmatrix}\bar{\mathbf{c}}_0 & \mathbf{0}\end{bmatrix}\begin{bmatrix}\bar{\mathbf{x}}_1\\ \bar{\mathbf{x}}_2\end{bmatrix} = \bar{\mathbf{c}}\,\bar{\mathbf{x}}$$

where the l-dimensional subsystem

$$\dot{\bar{\mathbf{x}}}_1 = \bar{\mathbf{A}}_0\bar{\mathbf{x}}_1 + \bar{\mathbf{b}}_1\,u$$
$$y = \bar{\mathbf{c}}_0\bar{\mathbf{x}}_1$$

is observable from y, and the $(n - l)$-dimensional subsystem

$$\dot{\bar{\mathbf{x}}}_2 = \bar{\mathbf{A}}_{22}\bar{\mathbf{x}}_2 + \bar{\mathbf{b}}_2 u + \bar{\mathbf{A}}_{21}\bar{\mathbf{x}}_1$$

has no effect upon the output y, and is therefore entirely unobservable, i.e., nothing about $\bar{\mathbf{x}}$ can be inferred from output measurement.

This theorem shows that any system which is not completely observable can be decomposed into the observable and unobservable subsystems shown in Fig. 14.22. The state model (14.94c) is said to be in *observability canonical form*.

Since systems (14.94a) and (14.94c) are equivalent, the set of eigenvalues of matrix $\mathbf{A}$ of system (14.94a) is same as the set of eigenvalues of matrix $\bar{\mathbf{A}}$ of system (14.94c), which is a union of the subsets of eigenvalues of matrices $\bar{\mathbf{A}}_0$ and $\bar{\mathbf{A}}_{22}$. The transfer function of the system (14.94a) may be calculated from (14.94c) as follows:

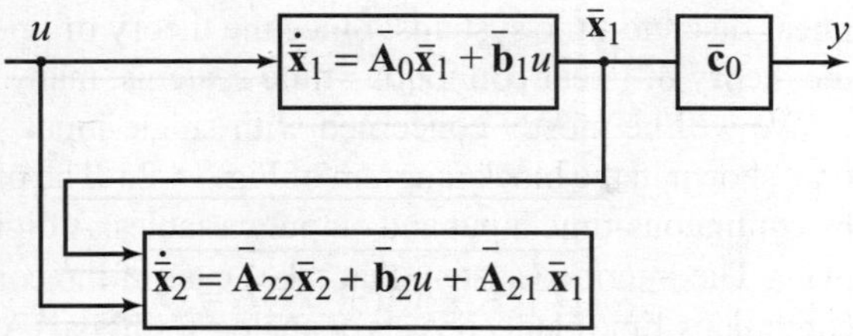

Fig. 14.22 *The observability canonical form of a state variable model*

$$G(s) = \begin{bmatrix}\bar{\mathbf{c}}_0 & \mathbf{0}\end{bmatrix}\begin{bmatrix}s\mathbf{I} - \bar{\mathbf{A}}_0 & \mathbf{0}\\ -\bar{\mathbf{A}}_{21} & s\mathbf{I} - \bar{\mathbf{A}}_{22}\end{bmatrix}^{-1}\begin{bmatrix}\bar{\mathbf{b}}_1\\ \bar{\mathbf{b}}_2\end{bmatrix}$$

$$= \begin{bmatrix}\bar{\mathbf{c}}_0 & \mathbf{0}\end{bmatrix}\begin{bmatrix}(s\mathbf{I} - \bar{\mathbf{A}}_0)^{-1} & \mathbf{0}\\ (s\mathbf{I} - \bar{\mathbf{A}}_{22})^{-1}\bar{\mathbf{A}}_{21}(s\mathbf{I} - \bar{\mathbf{A}}_0)^{-1} & (s\mathbf{I} - \bar{\mathbf{A}}_{22})^{-1}\end{bmatrix}\begin{bmatrix}\bar{\mathbf{b}}_1\\ \bar{\mathbf{b}}_2\end{bmatrix}$$

$$= \bar{\mathbf{c}}_0(s\mathbf{I} - \bar{\mathbf{A}}_0)^{-1}\bar{\mathbf{b}}_1 \tag{14.95}$$

which shows that the unobservable part of the system does not affect the input–output relationship. We will refer to the eigenvalues of $\bar{\mathbf{A}}$ as *observable poles* and the eigenvalues of $\bar{\mathbf{A}}_{22}$ as unobservable poles.

We now examine the use of state variable and transfer function models of a system to study its dynamic properties.

We know that a system is asymptotically stable if all the eigenvalues of the characteristic matrix $\mathbf{A}$ of its state variable model are in the left half of complex plane. Also we know that a system is (bounded-input bounded-output) BIBO stable if all the poles of its transfer function model are in the left half of complex plane. ***Since, in general, the poles of the transfer function model of a system are a subset of the eigenvalues of the characteristic matrix A of the system, asymptotic stability always implies BIBO stability.***

The reverse, however, may not always be true because the eigenvalues of the uncontrollable and/or unobservable part of the system are hidden from the BIBO stability analysis. These may lead to instability of a BIBO stable system. When a state variable model is both controllable and observable, all the eigenvalues of characteristic matrix **A** appear as poles in the corresponding transfer function. Therefore, ***BIBO stability implies asymptotic stability only for completely controllable and completely observable system.***

To conclude, we may say that the transfer function model of a system represents its complete dynamics only if the system is both controllable and observable.

14.10 STATE VARIABLE ANALYSIS OF DIGITAL CONTROL SYSTEMS

So far in this chapter, we treated the analysis of linear continuous-time system using state variable methods in considerable detail. In this section, we give a condensed review of the same methods for linear discrete-time systems. Since the theory of linear discrete-time systems—very closely—parallels the theory of linear continuous-time systems, many of the results are similar.

We will be mostly concerned with single-input, single-output (SISO) system configurations of the type shown in the block diagram of Fig. 14.23. The plant in the figure, is a physical process characterized by continuous-time input and output variables. A digital computer is used to control the continuous-time plant. The interface system that takes care of the communication between the digital computer and the continuous-time plant consists of analog-to-digital (A/D) converter and digital-to-analog (D/A) converter. In order to analyse such a system, it is often convenient to represent the continuous-time plant, together with the D/A converter and the A/D converter by an equivalent discrete-time system.

State variable model of a SISO discrete-time system consists of a set of first-order difference equations relating state variables $x_1(k), x_2(k), ..., x_n(k)$ of the discrete-time system to the input $u(k)$; the output $y(k)$ is algebraically related to the state variables and the input. Assuming that the input is switched on to the system at $k = 0$ ($u(k) = 0$ for $k < 0$), then the initial state is given by

$$\mathbf{x}(0) \triangleq \mathbf{x}^0; \text{ a specified } n \times 1 \text{ vector}$$

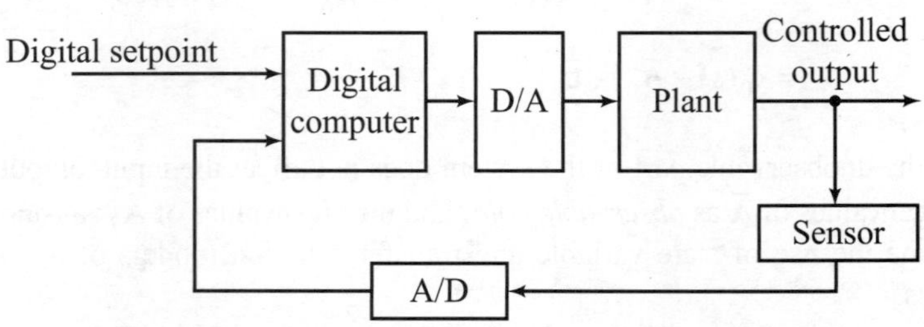

Fig. 14.23 *Basic structure of digital control systems*

The dynamics of a linear time-invariant system are described by equations of the form,

$$\mathbf{x}(k+1) = \mathbf{F}\mathbf{x}(k) + \mathbf{g}u(k); \mathbf{x}(0) \triangleq \mathbf{x}^0 \tag{14.96a}$$

$$y(k) = \mathbf{c}\mathbf{x}(k) + du(k) \tag{14.96b}$$

where

$$\mathbf{x}(k) = \begin{bmatrix} x_1(k) \\ x_2(k) \\ \vdots \\ x_n(k) \end{bmatrix} = n \times 1 \text{ state vector of } n\text{th-order system; } u(k) = \text{system input; } y(k) = \text{defined output.}$$

$$\mathbf{F} = \begin{bmatrix} f_{11} & f_{12} & \cdots & f_{1n} \\ f_{21} & f_{22} & \cdots & f_{2n} \\ \vdots & \vdots & & \vdots \\ f_{n1} & f_{n2} & \cdots & f_{nn} \end{bmatrix} = n \times n \text{ constant matrix; } \mathbf{g} = \begin{bmatrix} g_1 \\ g_2 \\ \vdots \\ g_n \end{bmatrix} = n \times 1 \text{ constant column matrix}$$

$\mathbf{c} = [c_1 \; c_2 \; \cdots \; c_n] = 1 \times n$ constant row matrix

d = scalar, representing direct coupling between input and output

Equation (14.96a) is called the *state equation* of the system, Eqn. (14.96b) is called the *output equation*; the two equations together give the *state variable model* of the system.

14.10.1 Conversion of State Variable Models to Transfer Functions

In the study of linear time-invariant discrete-time equations, we may also apply the z-transform techniques. Taking the z-transform of Eqns (14.96), we obtain

$$z\mathbf{X}(z) - z\mathbf{x}^0 = \mathbf{F}\mathbf{X}(z) + \mathbf{g}U(z)$$

$$Y(z) = \mathbf{c}\mathbf{X}(z) + dU(z)$$

where $\mathbf{X}(z) \triangleq \mathscr{Z}[\mathbf{x}(k)]; U(z) \triangleq \mathscr{Z}[u(k)]; Y(z) \triangleq \mathscr{Z}[y(k)]$

Manipulation of these equations, gives

$$(z\mathbf{I} - \mathbf{F})\,\mathbf{X}(z) = z\mathbf{x}^0 + \mathbf{g}U(z); \; \mathbf{I} \text{ is } n \times n \text{ identity matrix}$$

or

$$\mathbf{X}(z) = (z\mathbf{I} - \mathbf{F})^{-1} z\mathbf{x}^0 + (z\mathbf{I} - \mathbf{F})^{-1}\mathbf{g}U(z) \tag{14.97a}$$

$$Y(z) = \mathbf{c}(z\mathbf{I} - \mathbf{F})^{-1} z\mathbf{x}^0 + [\mathbf{c}(z\mathbf{I} - \mathbf{F})^{-1}\mathbf{g} + d]\ U(z) \tag{14.97b}$$

Equations (14.97) are algebraic equations. If $\mathbf{x}^0$ and $U(z)$ are known, $\mathbf{X}(z)$ can be computed from the equations.

In the case of zero initial state (i.e., $\mathbf{x}^0 = \mathbf{0}$), the input-output behavior of the system (14.96) is determined entirely by the transfer function,

$$\frac{Y(z)}{U(z)} = G(z) = \mathbf{c}(z\mathbf{I} - \mathbf{F})^{-1}\mathbf{g} + d \tag{14.98a}$$

$$= \mathbf{c}\frac{(z\mathbf{I} - \mathbf{F})^{+}\mathbf{g}}{|z\mathbf{I} - \mathbf{F}|} + d \tag{14.98b}$$

where $(z\mathbf{I} - \mathbf{F})^{+}$ = adjoint of the matrix $(z\mathbf{I} - \mathbf{F})$; $|z\mathbf{I} - \mathbf{F}|$ = determinant of the matrix $(z\mathbf{I} - \mathbf{F})$
$|\lambda\mathbf{I} - \mathbf{F}|$ is the *characteristic polynomial* of matrix **F**. The roots of this polynomial are the *characteristic roots* or *eigenvalues* of matrix **F**.

From Eqn. (14.98b), we observe that the characteristic polynomial of matrix **F** of the system (14.96), is same as the *denominator* polynomial of the corresponding transfer function $G(z)$. If there are no cancelations between the numerator and denominator polynomials of $G(z)$ in Eqn. (14.98b), the eigenvalues of matrix **F** are same as the poles of $G(z)$.

14.10.2 State Description of Sampled Continuous-time plants

Simple models of the interface actions of D/A and A/D converters (refer to Fig. 14.23) have been developed in Chapter 13. A brief review is in order here.

A simple model of A/D converter is shown in Fig. 14.24. A continuous-time function $f(t)$, $t \geq 0$, is the input, and the sequence of real numbers $f(k)$, $k = 0, 1, 2, \ldots$, is the output; the following relation holds between input and output:

$$f(k) = f(t = kT);\ T \text{ is the time interval between samples} \tag{14.99a}$$

A simple model of D/A converter is shown in Fig. 14.25. A sequence of numbers $f(k)$, $k = 0, 1, 2, \ldots$, is the input, and the continuous-time function $f^{+}(t)$, $t \geq 0$, is the output; the following relation holds between input and output:

$$f^{+}(t) = f(k);\ kT \leq t < (k+1)T \tag{14.99b}$$

Figure 14.26 illustrates a typical example of an interconnection of discrete-time and continuous-time systems. In order to analyse such a system, it is often convenient to represent the continuous-time system together with the zero-order hold (ZOH) and the sampler, by an *equivalent discreter-time system.*

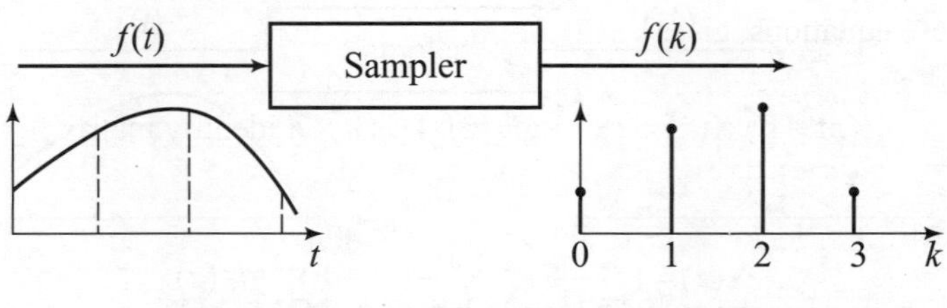

Fig. 14.24 *A model of an A/D converter*

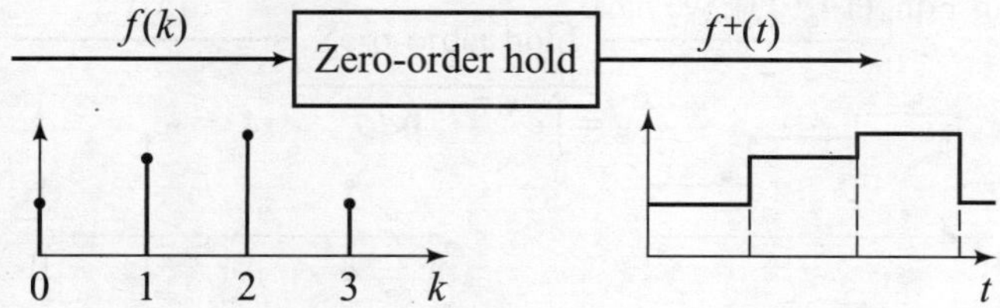

Fig. 14.25 *A model of a D/A converter*

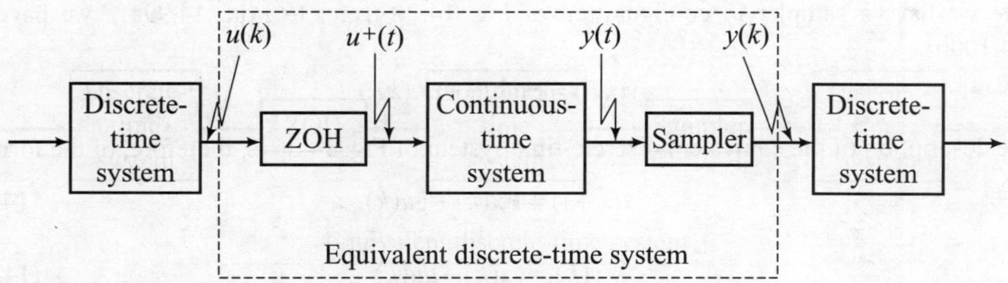

Fig. 14.26 *Interconnection of discrete-time and continuous-time systems*

We assume that the continuous-time system of Fig. 14.26 is a linear system with state variable model

$$\dot{\mathbf{x}}(t) = \mathbf{A}\mathbf{x}(t) + \mathbf{b}u^+(t) \tag{14.100a}$$

$$y(t) = \mathbf{c}\mathbf{x}(t) + du^+(t) \tag{14.100b}$$

where $\mathbf{x}(t)$ is $n \times 1$ state vector, $u^+(t)$ is scalar input, $y(t)$ is scalar output; $\mathbf{A}$, $\mathbf{b}$, $\mathbf{c}$ and d are, respectively, $n \times n$, $n \times 1$, $1 \times n$, and 1×1 real constant matrices.

The solution of Eqn. (14.100a) with t_0 as initial time is

$$\mathbf{x}(t) = e^{\mathbf{A}(t-t_0)}\mathbf{x}(t_0) + \int_{t_0}^{t} e^{\mathbf{A}(t-\tau)}\mathbf{b}u^+(\tau)d\tau \tag{14.101}$$

Since we use a ZOH (refer to Eqn. 14.99b)),

$$u^+(t) = u(kT); kT \le t < (k+1)T; k = 0,1,2,...$$

Then from Eqn. (14.101) we can write,

$$\mathbf{x}(t) = e^{\mathbf{A}(t-kT)}\mathbf{x}(kT) + \left[\int_{kT}^{t} e^{\mathbf{A}(t-\tau)}\mathbf{b}\, d\tau\right]u(kT); kT \le t < (k+1)T \tag{14.102}$$

In response to the input $u(kT)$, the state settles to the value $\mathbf{x}((k+1)T)$ prior to the application of the input $u((k+1)T)$, where

$$\mathbf{x}((k+1)T) = e^{\mathbf{A}T}\mathbf{x}(kT) + \left[\int_{kT}^{(k+1)T} e^{\mathbf{A}[(k+1)T-\tau]}\mathbf{b}\, d\tau\right]u(kT)$$

$$= \mathbf{F}\mathbf{x}(kT) + \mathbf{g}u(kT) \tag{14.103}$$

Letting $\sigma = (\tau - kT)$ in Eqn. (14.103), we have

$$\mathbf{g} = \int_0^T e^{\mathbf{A}(T-\sigma)}\mathbf{b}d\sigma$$

With $\theta = T - \sigma$, we get

$$\mathbf{g} = \int_0^T e^{\mathbf{A}\theta}\mathbf{b}d\theta$$

Since we have a sampler in configuration of Fig. 14.26 (refer to Eqn. 14.99a)), we have from Eqn. 14.100b),

$$y(kT) = \mathbf{c}\mathbf{x}(kT) + du(kT)$$

State description of the equivalent discrete-time system of Fig. 14.26 is, therefore, of the form,

$$\mathbf{x}(k+1) = \mathbf{F}\mathbf{x}(k) + \mathbf{g}u(k) \tag{14.104a}$$

$$y(k) = \mathbf{c}\mathbf{x}(k) + du(k) \tag{14.104b}$$

where

$$\mathbf{F} = e^{\mathbf{A}T}; \mathbf{g} = \int_0^T e^{\mathbf{A}\theta}\mathbf{b}d\theta \tag{14.104c}$$

Example 14.14 Consider a unity feedback sampled-data system shown in Fig. 14.27. We are interested in discrete-time state model for the feedback system. The continuous-time state model of the plant is given:

$$\dot{\mathbf{x}}(t) = \mathbf{A}\mathbf{x}(t) + \mathbf{b}u(t) \tag{14.105a}$$

$$y(t) = \mathbf{c}\mathbf{x}(\mathrm{t}) \tag{14.105b}$$

with

$$\mathbf{A} = \begin{bmatrix} 0 & 1 \\ 0 & -5 \end{bmatrix}; \mathbf{b} = \begin{bmatrix} 0 \\ 1 \end{bmatrix}; \mathbf{c} = \begin{bmatrix} 1 & 0 \end{bmatrix}$$

The state transition matrix

$$e^{\mathbf{A}t} = \mathscr{L}^{-1}[s\mathbf{I} - \mathbf{A})^{-1}] = \mathscr{L}^{-1}\begin{bmatrix} \frac{1}{s} & \frac{1}{s(s+5)} \\ 0 & \frac{1}{s+5} \end{bmatrix} = \begin{bmatrix} 1 & \frac{1}{5}(1-e^{-5t}) \\ 0 & e^{-5t} \end{bmatrix}$$

Discretizing time axis into intervals of width $T = 0.1$ sec, we get the following discrete-time model for the plant:

$$\mathbf{x}(k+1) = \mathbf{F}\mathbf{x}(k) + \mathbf{g}u(k) \tag{14.106}$$

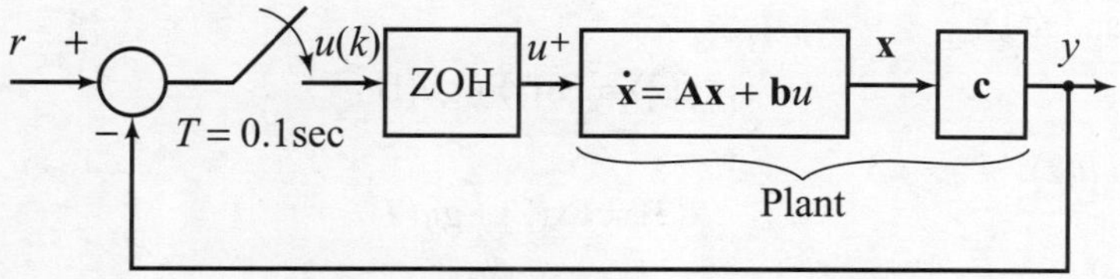

Fig. 14.27 *A sampled-data system*

where

$$\mathbf{F} = e^{\mathbf{A}T} = \begin{bmatrix} 1 & \frac{1}{5}(1-e^{-5T}) \\ 0 & e^{-5T} \end{bmatrix} = \begin{bmatrix} 1 & 0.0787 \\ 0 & 0.6065 \end{bmatrix}$$

$$\mathbf{g} = \int_0^T e^{\mathbf{A}\theta}\mathbf{b}d\theta = \begin{bmatrix} \int_0^T \frac{1}{5}(1-e^{-5\theta})d\theta \\ \int_0^T e^{-5\theta}d\theta \end{bmatrix} = \begin{bmatrix} \frac{1}{5}(T - \frac{1}{5} + \frac{1}{5}e^{-5T}) \\ \frac{1}{5}(1-e^{-5T}) \end{bmatrix} = \begin{bmatrix} 0.0043 \\ 0.0787 \end{bmatrix}$$

The plant input is derived from reference input and the feedback (Fig. 14.27):

$$u(k) = r(k) - y(k)$$

From Eqn. (14.105b), we obtain

$$y(k) = \mathbf{c}\,\mathbf{x}(k) = x_1(k)$$

Therefore,

$$u(k) = r(k) - x_1(k)$$

Substituting in Eqn. (14.106), we have

$$\begin{bmatrix} x_1(k+1) \\ x_2(k+1) \end{bmatrix} = \begin{bmatrix} 1 & 0.0787 \\ 0 & 0.6065 \end{bmatrix}\begin{bmatrix} x_1(k) \\ x_2(k) \end{bmatrix} + \begin{bmatrix} 0.0043 \\ 0.0787 \end{bmatrix}(r(k) - x_1(k))$$

$$= \begin{bmatrix} 0.9957 & 0.0787 \\ -0.0787 & 0.6065 \end{bmatrix}\begin{bmatrix} x_1(k) \\ x_2(k) \end{bmatrix} + \begin{bmatrix} 0.0043 \\ 0.0797 \end{bmatrix} r(k)$$

This is the state model of the sampled-data feedback system of Fig. 14.27.

14.10.3 Properties of Discrete-Time State Variable Models

Solution by Recursion

The solution of Eqn. (14.104a) for any positive integer k may be obtained directly by recursion as follows. From $\mathbf{x}(0)$ and $u(0)$, $\mathbf{x}(1)$ can be calculated:

$$\mathbf{x}(1) = \mathbf{F}\mathbf{x}(0) + \mathbf{g}u(0)$$

Then using $\mathbf{x}(1)$ and $u(1)$:

$$\mathbf{x}(2) = \mathbf{F}\mathbf{x}(1) + \mathbf{g}u(1)$$

From $\mathbf{x}(2)$ and $u(2)$:

$$\mathbf{x}(3) = \mathbf{F}\mathbf{x}(2) + \mathbf{g}u(2)$$

$$\vdots$$

From $\mathbf{x}(k-1)$ and $u(k-1)$:

$$\mathbf{x}(k) = \mathbf{F}\mathbf{x}(k-1) + \mathbf{g}u(k-1)$$

Controllability

For the linear system given by Eqns (14.96), if there exists an input $u(k); k \in [0, N-1]$ with N a finite positive integer, which transfers the initial state $\mathbf{x}(0) \triangleq \mathbf{x}^0$ to the state $\mathbf{x}^1$ at $k = N$, the state $\mathbf{x}^0$ is said to be controllable. If all initial states are controllable, the system is said to be *completely controllable* or simply *controllable*. Otherwise, the system is said to be *uncontrollable*.
The following theorem gives a simple controllability test.

Theorem 14.5
The necessary and sufficient condition for the system (14.96) to be completely controllable is that the *$n \times n$ controllability matrix,*

$$\mathbf{U} \triangleq [\mathbf{g} \quad \mathbf{F}\mathbf{g} \quad \mathbf{F}^2\mathbf{g} \quad \cdots \quad \mathbf{F}^{n-1}\mathbf{g}] \tag{14.107}$$

has rank equal to n, i.e., $\rho(\mathbf{U}) = n$.

Observability

For the linear system given by Eqns (14.96), if the knowledge of the input $u(k); k \in [0, N-1]$ and the output $y(k); k \in [0, N-1]$ with N a finite positive integer, suffices to determine the state $\mathbf{x}(0) \triangleq \mathbf{x}^0$, the state $\mathbf{x}^0$ is said to be observable. If all initial conditions are observable, the system is said to be *completely observable*, or simply *observable*. Otherwise, the system is said to be *unobservable*.
The following theorem gives a simple observability test.

Theorem 14.6
The necessary and sufficient condition for the system (14.96) to be completely observable is that the *$n \times n$ observability matrix*

$$\mathbf{V} = \begin{bmatrix} \mathbf{c} \\ \mathbf{c}\mathbf{F} \\ \mathbf{c}\mathbf{F}^2 \\ \vdots \\ \mathbf{c}\mathbf{F}^{n-1} \end{bmatrix} \tag{14.108}$$

has rank equal to n, i.e., $\rho(\mathbf{V}) = n$.

Stability

The conditions for zero-input stability given in terms of eigenvalues of matrix $\mathbf{F}$, are as follows:

1. If all the eigenvalues of matrix $\mathbf{F}$ have magnitudes less than unity, the system is asymptotically stable.

2. If any eigenvalue of matrix **F** has magnitude greater than unity, or if there is a repeated eigenvalue with unity magnitude, the system is unstable.
3. If condition (1) is satisfied except for the presence of one or more non-repeated eigenvalues with unity magnitude, the system is marginally stable.

Under mildly restrictive conditions (namely, the system must be both controllable and observable), the eigenvalues of matrix **F** are same as the poles of the corresponding transfer function, and asymptotic stability ensures BIBO stability and *vice versa*.

14.11 REVIEW EXAMPLES

Review Example 14.1 A feedback system has a closed-loop transfer function,

$$\frac{Y(s)}{R(s)} = \frac{10(s+4)}{s(s+1)(s+3)}$$

Construct three different state models for this system:

(a) One where the system matrix **A** is a diagonal matrix.
(b) One where **A** is in first companion form.
(c) One where **A** is in second companion form.

Solution

(a) The given transfer function can be expressed as

$$\frac{Y(s)}{R(s)} = \frac{10(s+4)}{s(s+1)(s+3)} = \frac{40/3}{s} + \frac{-15}{s+1} + \frac{5/3}{s+3}$$

Therefore,

$$Y(s) = \frac{40/3}{s}R(s) + \frac{-15}{s+1}R(s) + \frac{5/3}{s+3}R(s)$$

Let

$$X_1(s) = \frac{40}{3}R(s)\text{; this gives } \dot{x}_1 = \frac{40}{3}r$$

$$X_2(s) = \frac{-15}{s+1}R(s)\text{; this gives } \dot{x}_2 + x_2 = -15r$$

$$X_3(s) = \frac{5/3}{s+3}R(s)\text{; this gives } \dot{x}_3 + 3x_3 = \frac{5}{3}r$$

In terms of x_1, x_2 and x_3, the output $y(t)$ is given by

$$y(t) = x_1(t) + x_2(t) + x_3(t)$$

A state variable formulation for the given transfer function is defined by the following matrices:

$$\mathbf{\Lambda} = \begin{bmatrix} 0 & 0 & 0 \\ 0 & -1 & 0 \\ 0 & 0 & -3 \end{bmatrix}; \mathbf{b} = \begin{bmatrix} 40/3 \\ -15 \\ 5/3 \end{bmatrix}; \mathbf{c} = [1 \;\; 1 \;\; 1];\ d = 0$$

Note that the coefficient matrix $\mathbf{\Lambda}$ is diagonal, and the state model is in Jordan canonical form. We now construct two state models for the given transfer function in companion form. To do this, we express the transfer function as

$$\frac{Y(s)}{R(s)} = \frac{10(s+4)}{s(s+1)(s+3)} = \frac{10s+40}{s^3+4s^2+3s} = \frac{\beta_0 s^3+\beta_1 s^2+\beta_2 s+\beta_3}{s^3+\alpha_1 s^2+\alpha_2 s+\alpha_3}$$

$$\beta_0 = \beta_1 = 0,\ \beta_2 = 10,\ \beta_3 = 40,\ \alpha_1 = 4,\ \alpha_2 = 3,\ \alpha_3 = 0$$

(b) With reference to Eqns (14.49), we obtain the following state model in the first companion form:

$$\mathbf{A} = \begin{bmatrix} 0 & 1 & 0 \\ 0 & 0 & 1 \\ 0 & -3 & -4 \end{bmatrix}; \mathbf{b} = \begin{bmatrix} 0 \\ 0 \\ 1 \end{bmatrix}; \mathbf{c} = [40 \;\; 10 \;\; 0];\ d = 0$$

(c) With reference to Eqns (14.51), the state model in second companion form becomes

$$\mathbf{A} = \begin{bmatrix} 0 & 0 & 0 \\ 1 & 0 & -3 \\ 0 & 1 & -4 \end{bmatrix}; \mathbf{b} = \begin{bmatrix} 40 \\ 10 \\ 0 \end{bmatrix}; \mathbf{c} = [0 \;\; 0 \;\; 1]; d = 0$$

Review Example 14.2 A linear time-invariant system is characterized by the homogeneous state equation,

$$\begin{bmatrix} \dot{x}_1 \\ \dot{x}_2 \end{bmatrix} = \begin{bmatrix} 0 & 1 \\ 0 & -2 \end{bmatrix} \begin{bmatrix} x_1 \\ x_2 \end{bmatrix}$$

(a) Compute the solution of the homogeneous equation assuming the initial state vector,

$$\mathbf{x}(0) = \begin{bmatrix} 1 \\ 0 \end{bmatrix}$$

(b) Consider now that the system has a forcing function and is represented by the following nonhomogeneous state equation:

$$\begin{bmatrix} \dot{x}_1 \\ \dot{x}_2 \end{bmatrix} = \begin{bmatrix} 0 & 1 \\ 0 & -2 \end{bmatrix} \begin{bmatrix} x_1 \\ x_2 \end{bmatrix} + \begin{bmatrix} 0 \\ 1 \end{bmatrix} u$$

where u is a unit step input.

Compute the solution of this equation assuming initial conditions of part (a).

Solution

(a) Since

$$(s\mathbf{I}-\mathbf{A})=\begin{bmatrix} s & -1 \\ 0 & s+2 \end{bmatrix}$$

we obtain

$$(s\mathbf{I}-\mathbf{A})^{-1}=\begin{bmatrix} \dfrac{1}{s} & \dfrac{1}{s(s+2)} \\ 0 & \dfrac{1}{s+2} \end{bmatrix}$$

Hence,
$$e^{\mathbf{A}t}=\mathscr{L}^{-1}[(s\mathbf{I}-\mathbf{A})^{-1}]=\begin{bmatrix} 1 & \dfrac{1}{2}(1-e^{-2t}) \\ 0 & e^{-2t} \end{bmatrix}$$

$$\mathbf{x}(t)=e^{\mathbf{A}t}\mathbf{x}(0)=\begin{bmatrix} 1 \\ 0 \end{bmatrix}$$

(b) $\mathbf{x}(t)=e^{\mathbf{A}t}\mathbf{x}(0)+\int_0^t e^{\mathbf{A}(t-\tau)}\mathbf{b}u(\tau)d\tau$

Now

$$\int_0^t e^{\mathbf{A}(t-\tau)}\mathbf{b}u(\tau)d\tau=\begin{bmatrix} \dfrac{1}{2}\displaystyle\int_0^t [1-e^{-2(t-\tau)}]d\tau \\ \displaystyle\int_0^t e^{-2(t-\tau)}d\tau \end{bmatrix}=\begin{bmatrix} -\dfrac{1}{4}+\dfrac{1}{2}t+\dfrac{1}{4}e^{-2t} \\ \dfrac{1}{2}(1-e^{-2t}) \end{bmatrix}$$

Therefore,

$$x_1(t)=\frac{3}{4}+\frac{1}{2}t+\frac{1}{4}e^{-2t}$$

$$x_2(t)=\frac{1}{2}(1-e^{-2t})$$

Review Example 14.3 The motion of a satellite in the equatorial (r, θ) plane is given by the state equation [122]:

$$\begin{bmatrix} \dot{x}_1 \\ \dot{x}_2 \\ \dot{x}_3 \\ \dot{x}_4 \end{bmatrix}=\begin{bmatrix} 0 & 1 & 0 & 0 \\ 3\omega^2 & 0 & 0 & 2\omega \\ 0 & 0 & 0 & 1 \\ 0 & -2\omega & 0 & 0 \end{bmatrix}\begin{bmatrix} x_1 \\ x_2 \\ x_3 \\ x_4 \end{bmatrix}+\begin{bmatrix} 0 \\ 1 \\ 0 \\ 0 \end{bmatrix}u_1+\begin{bmatrix} 0 \\ 0 \\ 0 \\ 1 \end{bmatrix}u_2$$

where ω is the angular frequency of the satellite in circular, equatorial orbit, $x_1(t)$ and $x_3(t)$ are, respectively, the deviations in position variables $r(t)$ and $\theta(t)$ of the satellite, and $x_2(t)$ and $x_4(t)$ are, respectively, the deviations in velocity variables $\dot{r}(t)$ and $\dot{\theta}(t)$. The inputs $u_1(t)$ and $u_2(t)$ are the thrusts u_r and u_θ in the radial and tangential directions respectively, applied by small rocket engines or gas jets ($\mathbf{u} = \mathbf{0}$ when $\mathbf{x} = \mathbf{0}$). Sensors have been installed for measuring $r(t)$ and $\theta(t)$.

(a) Suppose that the tangential thruster becomes inoperable. Determine the controllability of the system with the radial thruster alone.
(b) Suppose that the radial thruster becomes inoperable. Determine the controllability of the system with the tangential thruster alone.
(c) Suppose that the tangential measuring device becomes inoperable. Determine the observability of the system from radial position measurement ($x_1 = r$) alone.
(d) Suppose that the radial measurements are lost. Determine the observability of the system from tangential position measurement ($x_3 = \theta$) alone.

Solution

(a) With $u_2 = 0$,

$$\mathbf{b} = \begin{bmatrix} 0 \\ 1 \\ 0 \\ 0 \end{bmatrix}$$

The controllability matrix

$$\mathbf{U} = \begin{bmatrix} \mathbf{b} & \mathbf{Ab} & \mathbf{A}^2\mathbf{b} & \mathbf{A}^3\mathbf{b} \end{bmatrix} = \begin{bmatrix} 0 & 1 & 0 & -\omega^2 \\ 1 & 0 & -\omega^2 & 0 \\ 0 & 0 & -2\omega & 0 \\ 0 & -2\omega & 0 & 2\omega^3 \end{bmatrix}$$

$$|\mathbf{U}| = -\begin{vmatrix} 1 & 0 & -\omega^2 \\ 0 & -2\omega & 0 \\ -2\omega & 0 & 2\omega^2 \end{vmatrix} = -[-2\omega(2\omega^3 - 2\omega^3)] = 0$$

Therefore, $\rho(\mathbf{U}) < 4$, and the system is not completely controllable with u_1 alone.

(b) With $u_1 = 0$,

$$\mathbf{b} = \begin{bmatrix} 0 \\ 0 \\ 0 \\ 1 \end{bmatrix}$$

The controllability matrix

$$\mathbf{U} = \begin{bmatrix} 0 & 0 & 2\omega & 0 \\ 0 & 2\omega & 0 & -2\omega^3 \\ 0 & 1 & 0 & -4\omega^2 \\ 1 & 0 & -4\omega^2 & 0 \end{bmatrix}$$

$$|\mathbf{U}| = 2\omega \begin{vmatrix} 0 & 2\omega & -2\omega^3 \\ 0 & 1 & -4\omega^2 \\ 1 & 0 & 0 \end{vmatrix} = -12\omega^4 \neq 0$$

Therefore, $\rho(\mathbf{U}) = 4$, and the system is completely controllable with u_2 alone.

(c) With

$$y = x_1,$$

$$\mathbf{c} = [1 \ 0 \ 0 \ 0]$$

The observability matrix

$$\mathbf{V} = \begin{bmatrix} \mathbf{c} \\ \mathbf{cA} \\ \mathbf{cA}^2 \\ \mathbf{cA}^3 \end{bmatrix} = \begin{bmatrix} 1 & 0 & 0 & 0 \\ 0 & 1 & 0 & 0 \\ 3\omega^2 & 0 & 0 & 2\omega \\ 0 & -\omega^2 & 0 & 0 \end{bmatrix}$$

$$|\mathbf{V}| = 0$$

Therefore, $\rho(\mathbf{V}) < 4$, and the system is not completely observable from $y = x_1$ alone.

(d) With $y = x_3$,

$$\mathbf{c} = [0 \ 0 \ 1 \ 0]$$

The observability matrix

$$\mathbf{V} = \begin{bmatrix} 0 & 0 & 1 & 0 \\ 0 & 0 & 0 & 1 \\ 0 & -2\omega & 0 & 0 \\ -6\omega^3 & 0 & 0 & -4\omega^2 \end{bmatrix}$$

$$|\mathbf{V}| = -12\omega^4 \neq 0$$

Therefore, $\rho(\mathbf{V}) = 4$, and the system is completly observable from $y = x_3$ alone.

CONCLUDING REMARKS

This chapter focused on the state variable techniques for analysis of linear control systems. Control design using state variable feedback will be covered in the next chapter. We will expand on the application of state variable representation to nonlinear systems in Chapter 16.

What we have learnt?

- The application of physical laws to dynamical systems, under the assumption of linearity and time-invariance, results in a set of ordinary linear differential equations with real coefficients. These differential equations when rearranged in the form of a set of first-order differential equations, result in state variable model.
- Laplace transformation of the first-order differential equations (state equations) with zero initial conditions gives us transfer function. Thus, to obtain transfer function model, one first writes the state equations, which 'naturally' arise from the application of physical laws.

- Dynamic evolution of a system to the two sources of excitation: external input and internal energy (initial conditions), is given by the state variable model, while transfer function assumes zero initial conditions.
- Eigenvalues of state model matrix **A** are important indicators of stability of a system. These eigenvalues take the same values as the poles of the corresponding transfer function if the system is both controllable and observable.
- Why do we use Laplace transformation from naturally arising state equations to transfer function, when state variable model appears to be more powerful? Transformation is beneficial only if it helps in our understanding of the characteristics of signals and systems, and understanding in transformed domain leads to better designs. The transfer function model, through better understanding, has given birth to the most important design tools—the root locus, and the Bode plots. We should not jump to the conclusions of merits of state variable model over transfer function model and *vice versa*; both the tools are very important in the toolkit of control engineers.
- State variable analysis of linear discrete-time systems very closely parallels the analysis of continuous-time systems.

Review Questions

14.1 Given a single-input single-output state variable model,

$$\dot{\mathbf{x}} = \mathbf{A}\mathbf{x} + \mathbf{b}u$$
$$y = \mathbf{c}\mathbf{x}$$

Prove that the eigenvalues of matrix **A** are invariant under state transformation $\mathbf{x} = \mathbf{P}\bar{\mathbf{x}}; \mathbf{P}$ is a constant nonsingular matrix.

14.2 Given a single-input single-output state variable model,

$$\dot{\mathbf{x}} = \mathbf{A}\mathbf{x} + \mathbf{b}u$$
$$y = \mathbf{c}\mathbf{x}$$

(a) Prove that $\dfrac{Y(s)}{U(s)} = G(s) = \mathbf{c}(s\mathbf{I} - \mathbf{A})^{-1}\mathbf{b}$.

(b) Prove that $G(s)$ is invariant under state transformation $\mathbf{x} = \mathbf{P}\bar{\mathbf{x}}; \mathbf{P}$ is a constant nonsingular matrix.

14.3 Is the state variable description for a given transfer function,

$$G(s) = \frac{\beta_0 s^n + \beta_1 s^{n-1} + \cdots + \beta_{n-1}s + \beta_n}{s^n + \alpha_1 s^{n-1} + \cdots + \alpha_{n-1}s + \alpha_n}$$

unique? If not, explain how three different state variable descriptions for the given transfer function can be obtained.

14.4 **(a)** Write two $n \times n$ general matrices, one having companion form and the other having Jordan form structure.

(b) An nth-order state variable model in Jordan canonical form always yields n decoupled first-order differential equations. Is the statement true? Justify your answer.

14.5 Given a single-input state variable equation,

$$\dot{\mathbf{x}} = \mathbf{A}\mathbf{x} + \mathbf{b}u$$

Prove that

$$\mathbf{x}(t) = e^{\mathbf{A}t}\mathbf{x}(0) + \int_0^t e^{\mathbf{A}(t-\tau)}\mathbf{b}\,u(\tau)\,d\tau$$

14.6 List the properties of the state transition matrix

$$\phi(t) = e^{\mathbf{A}t}$$

of the system

$$\dot{\mathbf{x}} = \mathbf{A}\mathbf{x};\, \mathbf{x}(0) \triangleq \mathbf{x}^0$$

Prove that

$$e^{\mathbf{A}t} = \mathscr{L}^{-1}[(s\mathbf{I} - \mathbf{A})^{-1}]$$

Why is the matrix exponential $e^{\mathbf{A}t}$ called the state transition matrix?

14.7 Given a single-input state equation,

$$\dot{\mathbf{x}} = \mathbf{A}\mathbf{x} + \mathbf{b}u$$

For discrete-time solution, the equation can be discretized to the following form:

$$\mathbf{x}(kT + T) = \mathbf{F}\mathbf{x}(kT) + \mathbf{g}u(kT);\ T = \text{sampling interval}$$

Find the matrices **F** and **g**.

14.8 A system is described by the single-input single-output state variable model,

$$\dot{\mathbf{x}} = \mathbf{A}\mathbf{x} + \mathbf{b}u$$

$$y = \mathbf{c}\mathbf{x}$$

What is the motivation behind the concept of controllability of the system? Give a precise definition of controllability and describe a controllability test in terms of matrices **A** and **b**.

Prove that controllability property is invariant under state transformation $\mathbf{x} = \mathbf{P}\bar{\mathbf{x}}; \mathbf{P}$ is a constant nonsingular matrix.

14.9 A system is described by the single-input single-output state variable model,

$$\dot{\mathbf{x}} = \mathbf{A}\mathbf{x} + \mathbf{b}u$$

$$y = \mathbf{c}\mathbf{x}$$

What is the motivation behind the concept of observability? Give a precise definition of observability and describe an observability test in terms of matrices **A** and **c**.

Prove that observability property is invariant under state transformation $\mathbf{x} = \mathbf{P}\bar{\mathbf{x}}; \mathbf{P}$ is a constant nonsingular matrix.

14.10 (a) Show that if a continuous-time linear time-invariant system is asymptotically stable, it is also BIBO stable.

(b) Show that a BIBO stable continuous-time linear time-invariant system is asymptotically stable only if the system is completely controllable and completely observable.

14.11 Prove that the transfer function model of the single-input single-output system,

$$\dot{\mathbf{x}} = \mathbf{A}\mathbf{x} + \mathbf{b}u$$

$$y = \mathbf{c}\mathbf{x}$$

is a complete characterization of the system only if the system is both controllable and observable.

Self-Appraisal Quiz

A set of questions with multiple choices is given below. These questions have been designed to capture important aspects of the subject covered in the chapter. For each of these questions, identify the correct answer out of the given choices. Compare your answers with the Master Key given at the end of the session. This quiz should serve the purpose of a self-appraisal test for readers.

14.1 Given

$$\mathbf{A} = \begin{bmatrix} 1 & 0 & -1 \\ 2 & 1 & -1 \\ 2 & 3 & 2 \end{bmatrix}$$

The top row of $\mathbf{A}^{-1}$ is

(A) [5 6 4] (B) [5 −3 1] (C) [2 0 −1] (D) [2 −1 1/2]

14.2 Given a state variable model

$$\dot{\mathbf{x}} = \mathbf{A}\mathbf{x} + \mathbf{b}u$$
$$y = \mathbf{c}\mathbf{x} + du$$

Under the transformation $\mathbf{x} = \mathbf{P}\bar{\mathbf{x}}$; $\mathbf{P}$ a constant nonsingular matrix, the model becomes

$$\dot{\bar{\mathbf{x}}} = \bar{\mathbf{A}}\bar{\mathbf{x}} + \bar{\mathbf{b}}u$$
$$y = \bar{\mathbf{c}}\bar{\mathbf{x}} + du$$

(A) $\bar{\mathbf{A}} = \mathbf{P}\mathbf{A}\mathbf{P}^{-1}; \bar{\mathbf{b}} = \mathbf{P}^{-1}\mathbf{b}; \bar{\mathbf{c}} = \mathbf{c}\mathbf{P}$

(B) $\bar{\mathbf{A}} = \mathbf{P}^{-1}\mathbf{A}\mathbf{P}; \bar{\mathbf{b}} = \mathbf{P}^{-1}\mathbf{b}; \bar{\mathbf{c}} = \mathbf{c}\mathbf{P}$

(C) $\bar{\mathbf{A}} = \mathbf{P}^{-1}\mathbf{A}\mathbf{P}; \bar{\mathbf{b}} = \mathbf{P}\mathbf{b}; \bar{\mathbf{c}} = \mathbf{c}\mathbf{P}$

(D) $\bar{\mathbf{A}} = \mathbf{P}^{-1}\mathbf{A}\mathbf{P}; \bar{\mathbf{b}} = \mathbf{P}\mathbf{b}; \bar{\mathbf{c}} = \mathbf{c}\mathbf{P}^{-1}$

14.3 A state variable formulation of a system is given by the equations,

$$\begin{bmatrix} \dot{x}_1 \\ \dot{x}_2 \end{bmatrix} = \begin{bmatrix} -1 & 0 \\ 0 & -3 \end{bmatrix} \begin{bmatrix} x_1 \\ x_2 \end{bmatrix} + \begin{bmatrix} 1 \\ 1 \end{bmatrix} u$$

$$y = [1 \ \ 0] \begin{bmatrix} x_1 \\ x_2 \end{bmatrix}$$

The transfer function of the system is

(A) $\dfrac{1}{(s+1)(s+3)}$ (B) $\dfrac{1}{s+1}$

(C) $\dfrac{1}{s+3}$ (D) None of the answers in (A), (B), and (C) is correct

14.4 A state variable formulation of a system is given by the equations,

$$\begin{bmatrix} \dot{x}_1 \\ \dot{x}_2 \end{bmatrix} = \begin{bmatrix} -1 & 0 \\ 0 & -3 \end{bmatrix} \begin{bmatrix} x_1 \\ x_2 \end{bmatrix} + \begin{bmatrix} 1 \\ 1 \end{bmatrix} u; x_1(0) = x_2(0) = 0$$

$$y = [1 \ 0] \begin{bmatrix} x_1 \\ x_2 \end{bmatrix}$$

The response $y(t)$ to unit-step input is

(A) $1 + e^{-t}$ (B) $\frac{1}{3}(1 - e^{-3t})$

(C) $1 - e^{-t}$ (D) None of the answers in (A), (B), and (C) is correct

14.5 The eignevalues of the matrix

$$\mathbf{A} = \begin{bmatrix} 2 & 1 \\ 4 & 5 \end{bmatrix}$$

can be found by solving

(A) $\lambda^2 - 7\lambda + 6 = 0$ (B) $\lambda^2 - 7\lambda + 10 = 0$ (C) $\lambda^2 - 5\lambda + 4 = 0$ (D) $\lambda^2 - 2\lambda + 4 = 0$

14.6 The eigenvalues of the matrix

$$\mathbf{A} = \begin{bmatrix} 0 & 1 & 0 \\ 0 & 0 & 1 \\ 0 & -3 & -4 \end{bmatrix}$$

are

(A) 0, –1, –3 (B) 0, –3, –4

(C) 0, 0, –4 (D) None of the answers in (A), (B), and (C) is correct

14.7 Given the system

$$\dot{\mathbf{x}} = \begin{bmatrix} 0 & 0 & -20 \\ 1 & 0 & -24 \\ 0 & 1 & -9 \end{bmatrix} \mathbf{x} + \begin{bmatrix} 3 \\ 1 \\ 0 \end{bmatrix} u$$

$$y = [0 \ 0 \ 1]\mathbf{x}$$

The characteristic equation of the system is

(A) $s^3 + 20s^2 + 24s + 9 = 0$ (B) $s^3 + 9s^2 + 24s + 20 = 0$

(C) $s^3 + 24s^2 + 9s + 20 = 0$ (D) None of the answers in (A), (B), and (C) is correct

14.8 A state variable model of a system is given by

$$\begin{bmatrix} \dot{x}_1 \\ \dot{x}_2 \end{bmatrix} = \begin{bmatrix} 1 & 1 \\ -2 & -1 \end{bmatrix} \begin{bmatrix} x_1 \\ x_2 \end{bmatrix} + \begin{bmatrix} 0 \\ 1 \end{bmatrix} u$$

$$y = [1 \ 0] \begin{bmatrix} x_1 \\ x_2 \end{bmatrix}$$

The system is
(A) controllable and observable (B) controllable but unobservable
(C) observable but uncontrollable (D) uncontrollable and unobservable

14.9 The transfer function

$$G(s) = \mathbf{c}(s\mathbf{I} - \mathbf{A})^{-1}\mathbf{b}$$

of the system

$$\dot{\mathbf{x}} = \mathbf{A}\mathbf{x} + \mathbf{b}u$$
$$y = \mathbf{c}\mathbf{x} + du$$

has pole-zero cancelation. The system
(A) is uncontrollable and unobservable (B) is observable but uncontrollable
(C) is controllable but unobservable (D) may be any one of (A), (B), and (C)

14.10 The transfer function

$$G(s) = \mathbf{c}(s\mathbf{I} - \mathbf{A})^{-1}\mathbf{b}$$

of the system

$$\dot{\mathbf{x}} = \mathbf{A}\mathbf{x} + \mathbf{b}u$$
$$y = \mathbf{c}\mathbf{x} + du$$

has no pole-zero cancelation. The system
(A) is controllable and observable (B) is observable but uncontrollable
(C) is controllable but unobservable (D) may be any one of (A), (B), and (C)

14.11 Consider the system

$$\mathbf{A} = \begin{bmatrix} 0 & 1 & 0 \\ 0 & 0 & 1 \\ -6 & -11 & -6 \end{bmatrix}; \mathbf{b} = \begin{bmatrix} 0 \\ 0 \\ 1 \end{bmatrix}; \mathbf{c} = [4 \;\; 5 \;\; 1]$$

The transfer function of the system has pole-zero cancelation. The system is
(A) controllable and observable (B) uncontrollable and unobservable
(C) controllable but unobservable (D) observable but uncontrollable

14.12 Consider the system

$$\mathbf{A} = \begin{bmatrix} 0 & -2 \\ 1 & -3 \end{bmatrix}; \mathbf{b} = \begin{bmatrix} 1 \\ 1 \end{bmatrix}; \mathbf{c} = [0 \;\; 1]$$

The transfer function of the system has pole-zero cancelation. The system is
(A) controllable and observable (B) uncontrollable and unobservable
(C) controllable but unobservable (D) observable but uncontrollable

14.13 In a control system design exercise for a marine engine, an engineer is using the state-space framework. The system model is given by

$$\dot{\mathbf{x}} = \begin{bmatrix} 13/6 & 4/3 \\ -4/3 & -7/6 \end{bmatrix}\mathbf{x} + \mathbf{b}u;\; \mathbf{b} = \begin{bmatrix} 0 \\ 0.5 \end{bmatrix}$$

The engineer decides to experiment with a different method of actuation for the system and this leads to the different input matrix

$$\mathbf{b}_1 = \begin{bmatrix} -1/6 \\ 1/3 \end{bmatrix}$$

(A) Both the systems are controllable
(B) System with input matrix **b** is controllable, and it becomes uncontrollable when **b** is changed to $\mathbf{b}_1$
(C) System with input matrix **b** is uncontrollable and it becomes controllable when **b** is changed to $\mathbf{b}_1$
(D) Both the systems are uncontrollable

14.14 Given the state variable model $\{\mathbf{A}, \mathbf{b}, \mathbf{c}\}$ of a single-input, single-output system. The asymptotic stability is determined from

(A) **A**, **b** and **c** matrices (B) **A** and **b** matrices
(C) **A** and **c** matrices (D) **A** matrix

14.15 For all finite values of t, the matrix exponential $e^{\mathbf{A}t}$ is nonsingular for

(A) singular **A** (B) nonsingular **A**
(C) all **A** (D) Nothing can be said, in general, about nonsingularity of $e^{\mathbf{A}t}$ for a given **A**

14.16 Consider the state variable model

$$\dot{\mathbf{x}} = \begin{bmatrix} 0 & 1 \\ 0 & 0 \end{bmatrix}\mathbf{x} + \begin{bmatrix} 0 \\ 1 \end{bmatrix}u$$

Discrete-time description of the system when input is applied through a zero-order hold (sampling interval T) is of the form : $\mathbf{x}(k+1) = \mathbf{F}\mathbf{x}(k) + \mathbf{g}\,u(k)$. The matrix **F** is equal to

(A) $\begin{bmatrix} 0 & 1 \\ 0 & 0 \end{bmatrix}$ (B) $\begin{bmatrix} 0 & T \\ 0 & 0 \end{bmatrix}$

(C) $\begin{bmatrix} 1 & T \\ 0 & 1 \end{bmatrix}$ (D) None of the answers in (A), (B) and (C) is correct

14.17 Consider the state variable model,

$$\dot{\mathbf{x}} = \begin{bmatrix} 0 & 1 \\ 0 & -5 \end{bmatrix}\mathbf{x} + \begin{bmatrix} 0 \\ 1 \end{bmatrix}u$$

For a sampling interval $T = 0.1$ sec, the discrete-time system is given by

$$\mathbf{x}(k+1) = \begin{bmatrix} 1 & \frac{1}{5}(1-e^{-5T}) \\ 0 & e^{-5T} \end{bmatrix}\mathbf{x}(k) + \mathbf{g}u(k)$$

The matrix **g** is given by

(A) $\begin{bmatrix} 0.0043 \\ 0.0787 \end{bmatrix}$ (B) $\begin{bmatrix} 0.043 \\ 0.787 \end{bmatrix}$

(C) $\begin{bmatrix} 0.43 \\ 7.87 \end{bmatrix}$ (D) None of the answers in (A), (B) and (C) is correct

MASTER KEY

14.1 (B)	14.2 (B)	14.3 (B)	14.4 (C)	14.5 (A)
14.6 (A)	14.7 (B)	14.8 (A)	14.9 (D)	14.10 (A)
14.11 (C)	14.12 (D)	14.13 (B)	14.14 (D)	14.15 (C)
14.16 (C)	14.17 (A)			

Problems

14.1 Figure P14.1 shows a control scheme for controlling the azimuth angle of a rotating antenna. The plant consists of an armature-controlled dc motor with dc generator used as an amplifier. The parameters of the plant are given below:

Motor torque constant, $K_T = 1.2$ newton-m/amp
Motor back emf constant, $K_b = 1.2$ V/(rad/sec)
Generator gain constant, $K_g = 100$ V/amp
Motor to load gear ratio, $n = (\dot{\theta}_L / \dot{\theta}_M) = 1/2$

$R_f = 21\Omega$, $L_f = 5$H, $R_g = 9\Omega$, $L_g = 0.06$H, $R_a = 10\Omega$, $L_a = 0.04$H, $J = 1.6$ newton-m/(rad/sec^2), $B = 0.04$ newton-m/(rad/sec), motor inertia and friction are negligible.

Taking physically meaningful and measurable variables as state variables, derive a state model for the system.

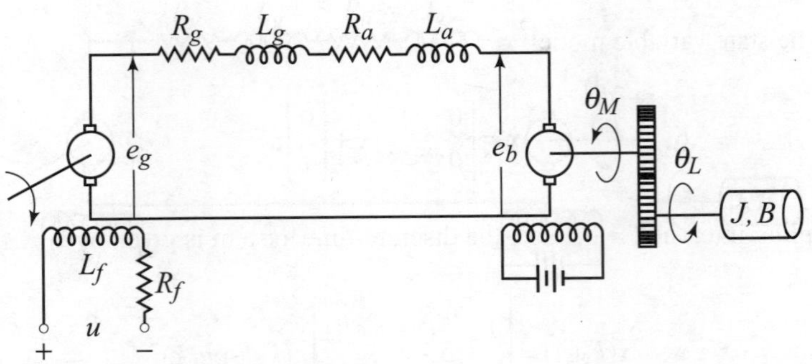

Fig. P14.1

14.2 Figure P14.2 shows a position control system with state variable feedback. The plant consists of a field-controlled dc motor with a dc amplifier. The parameters of the plant are given below:

Amplifier gain, $K_A = 50$ volt/volt
Motor field resistance, $R_f = 99\ \Omega$
Motor field inductance, $L_f = 20$ H
Motor torque constant, $K_T = 10$ newton-m/amp
Moment of inertia of load, $J = 0.5$ newton-m/(rad/sec^2)
Coefficient of viscous friction of load, $B = 0.5$ newton-m/(rad/sec)
Motor inertia and friction are negligible.

Taking $x_1 = \theta, x_2 = \dot{\theta}$, and $x_3 = i_f$ as the state variables, $u = e_f$ as the input, and $y = \theta$ as the output, derive a state variable model for the plant.

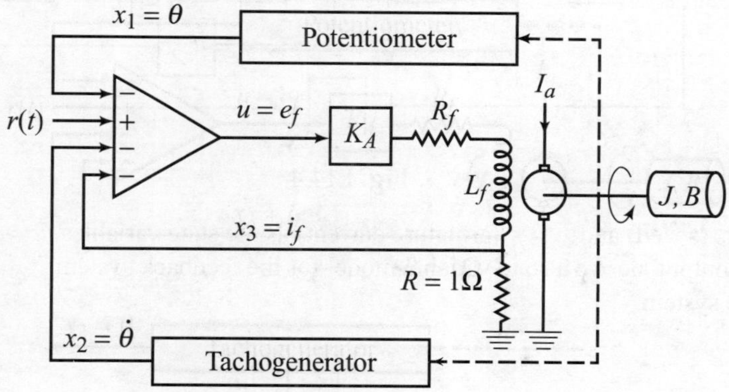

Fig. P14.2

14.3 Figure P14.3 shows the block diagram of a motor-driven single-link robot manipulator with position and velocity feedback. The drive motor is an armature controlled dc motor; e_a is armature voltage, i_a is armature current, θ_M is the motor shaft position and $\dot{\theta}_M$ is motor shaft velocity. θ_L is the position of the robot arm.

Taking θ_M, $\dot{\theta}_M$, and i_a as state variables, derive a state model for the feedback system.

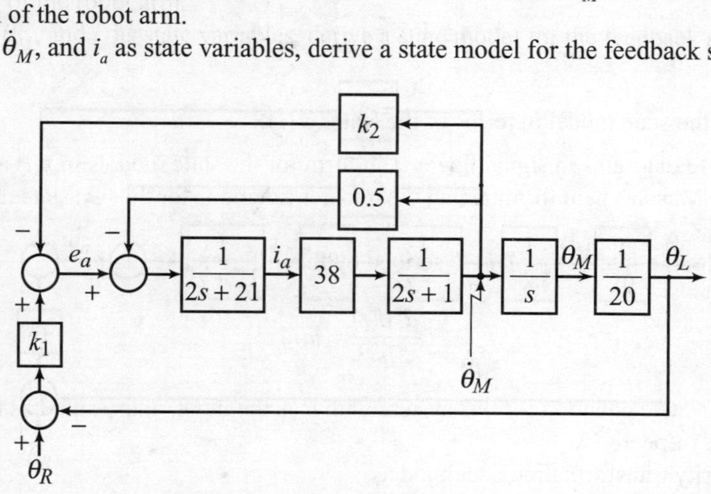

Fig. P14.3

14.4 Figure P14.4 shows the block diagram of a speed control system with state variable feedback. The drive motor is an armature-controlled dc motor with armature resistance R_a, armature inductance L_a, motor-torque constant K_T, inertia referred to motor shaft J, viscous-friction coefficient referred to motor shaft B, back-emf constant K_b, and tachogenerator constant K_t. The applied armature voltage is controlled by a three-phase full converter. We have assumed a linear relationship between the control voltage e_c and the armature voltage e_a. e_r is the reference voltage corresponding to the desired speed.

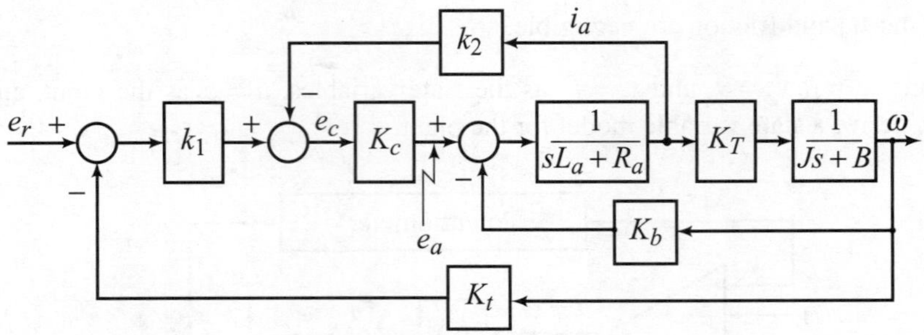

Fig. P14.4

Taking $x_1 = \omega$ (speed) and $x_2 = i_a$ (armature current) as the state variables, $u = e_r$ as the input, and $y = \omega$ as the output, derive a state variable model for the feedback system.

14.5 Consider the system,

$$\dot{\mathbf{x}} = \begin{bmatrix} -3 & 1 \\ -2 & 0 \end{bmatrix} \mathbf{x} + \begin{bmatrix} 0 \\ 1 \end{bmatrix} u$$

$$y = [1 \;\; 0]\,\mathbf{x}$$

A similarity transformation is defined by

$$\mathbf{x} = \mathbf{P}\,\bar{\mathbf{x}} = \begin{bmatrix} 2 & -1 \\ -1 & 1 \end{bmatrix} \bar{\mathbf{x}}$$

(a) Express the state model in terms of the states $\bar{\mathbf{x}}(t)$.

(b) Draw state diagrams in signal flow graph form for the state models in $\mathbf{x}(t)$ and) $\bar{\mathbf{x}}(t)$.

(c) Show by Mason's gain formula that the transfer functions for the two state diagrams in (b) are equal.

14.6 Consider a double-integrator plant described by the differential equation,

$$\frac{d^2\theta(t)}{dt^2} = u(t)$$

(a) Develop a state equation for this system with u as the input, and θ and $\dot{\theta}$ as the state variables x_1 and x_2, respectively.

(b) A similarity transformation is defined as

$$\mathbf{x} = \mathbf{P}\overline{\mathbf{x}} = \begin{bmatrix} 1 & 0 \\ 1 & 1 \end{bmatrix} \overline{\mathbf{x}}$$

Express the state equation in terms of the states $\overline{\mathbf{x}}(t)$.

(c) Show that the eigenvalues of the system matrices of the two state equations in (a) and (b) are equal.

14.7 A system is described by the state equation,

$$\dot{\mathbf{x}} = \begin{bmatrix} 0 & 1 & 0 \\ 0 & 0 & 1 \\ -1 & 0 & -3 \end{bmatrix} \mathbf{x} + \begin{bmatrix} 0 \\ 0 \\ 1 \end{bmatrix} u; \mathbf{x}(0) = \mathbf{x}^0$$

Using the Laplace transform technique, transform the state equation into a set of linear algebraic equations in the form,

$$\mathbf{X}(s) = \mathbf{G}(s)\mathbf{x}^0 + \mathbf{H}(s)\, U(s)$$

MATLAB Exercise Using Control System Toolbox, plot the response of the unforced ($u = 0$) system to the initial condition $\mathbf{x}(0) = \begin{bmatrix} 0 & -1 & 1 \end{bmatrix}^T$, for $0 \le t \le 10$.

14.8 Give a block diagram for the programming of the system of Problem 14.7 on an analog computer.

14.9 The state diagram of a linear system is shown in Fig. P14.9. Assign the state variables and write the dynamic equations of the system.

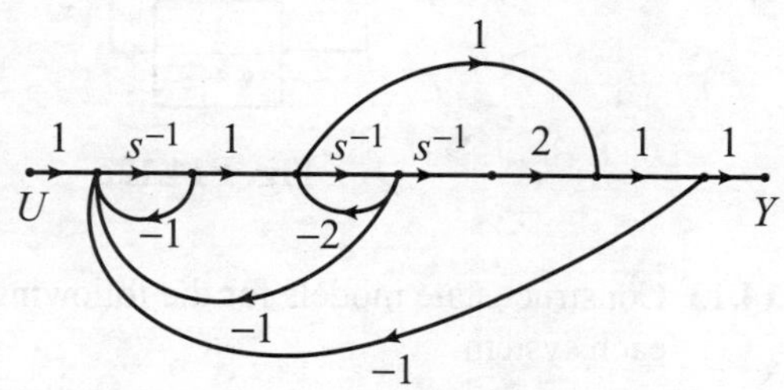

Fig. P 14.9

14.10 Derive transfer functions corresponding to the following state models:

(a) $\dot{\mathbf{x}} = \begin{bmatrix} 0 & 1 \\ -2 & -3 \end{bmatrix} \mathbf{x} + \begin{bmatrix} 1 \\ 0 \end{bmatrix} u; y = [1 \;\; 0]\,\mathbf{x}$

(b) $\dot{\mathbf{x}} = \begin{bmatrix} -3 & 1 \\ -2 & 0 \end{bmatrix} \mathbf{x} + \begin{bmatrix} 0 \\ 1 \end{bmatrix} u; y = [1 \;\; 0]\,\mathbf{x}$

14.11 Figure P14.11 shows the block diagram of a control system with state variable feedback and integral control. The plant model is

$$\begin{bmatrix} \dot{x}_1 \\ \dot{x}_2 \end{bmatrix} = \begin{bmatrix} -3 & 2 \\ 4 & -5 \end{bmatrix} \begin{bmatrix} x_1 \\ x_2 \end{bmatrix} + \begin{bmatrix} 1 \\ 0 \end{bmatrix} u$$

$$y = [0 \;\; 1]\mathbf{x}$$

(a) Derive a state model of the feedback system.

(b) Derive the transfer function $Y(s)/R(s)$.

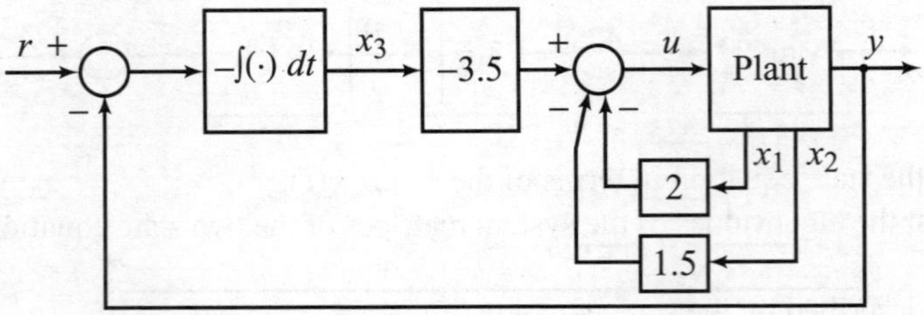

Fig. P14.11

MATLAB Exercise (i) Use Control System Toolbox to obtain closed-loop transfer function from the state model and verify your result in part (b). (ii) Using the state model, obtain the unit step response of the feedback system, with zero initial conditions.

14.12 Construct state models for the systems of Fig. P14.12a and Fig. P14.12b, taking outputs of simple lag blocks as state variables.

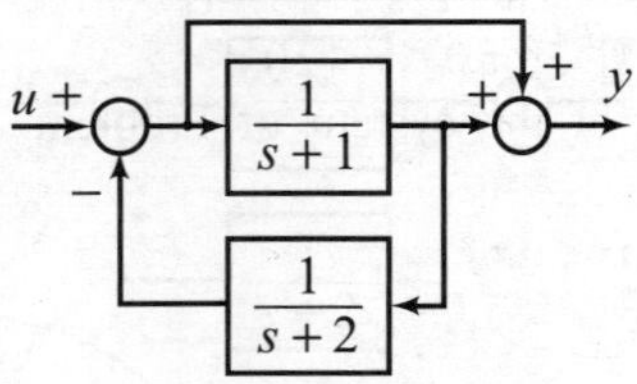

Fig. P14.12a

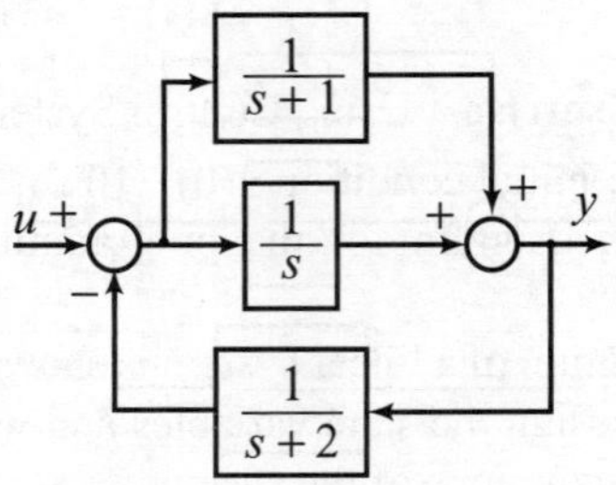

Fig. P14.12b

14.13 Construct state models for the following transfer functions. Obtain different canonical form for each system.

(i) $\dfrac{s+3}{s^2+3s+2}$ **(ii)** $\dfrac{5}{(s+1)^2(s+2)}$ **(iii)** $\dfrac{s^3+8s^2+17s+8}{(s+1)(s+2)(s+3)}$

Give block diagrams for the analog computer simulation of these transfer functions.

14.14 Construct state models for the following differential equations. Obtain a different canonical form for each system.

(i) $\dddot{y}+3\ddot{y}+2\dot{y}=\dot{u}+u$ **(ii)** $\dddot{y}+6\ddot{y}+11\dot{y}+6y=u$

(iii) $\dddot{y}+6\ddot{y}+11\dot{y}+6y=\dddot{u}+8\ddot{u}+17\dot{u}+8u$

14.15 Derive two state models for the system with transfer function,

$$\frac{Y(s)}{U(s)}=\frac{50(1+s/5)}{s(1+s/2)(1+s/50)}$$

(a) One for which the system matrix is a companion matrix.

(b) One for which the system matrix is diagonal.

14.16 **(a)** Obtain state variable model in Jordan canonical form for the system with transfer function

$$\frac{Y(s)}{U(s)} = \frac{2s^2+6s+5}{(s+1)^2\,(s+2)}$$

(b) Find the response $y(t)$ to a unit-step input using the state variable model in (a)

(c) Give a block diagram for analog computer simulation of the transfer function.

14.17 Given the system,

$$\dot{\mathbf{x}} = \begin{bmatrix} -2 & 1 \\ 1 & -2 \end{bmatrix}\mathbf{x} + \begin{bmatrix} 1 \\ 1 \end{bmatrix}u$$

(a) Obtain a state diagram in signal flow graph form.

(b) From the signal flow graph, determine the state equation in the form,

$$\mathbf{X}(s) = \mathbf{G}(s)\mathbf{x}(0) + \mathbf{H}(s)U(s)$$

(c) Using inverse Laplace transformation, obtain the

(i) zero-input response to initial condition

$$\mathbf{x}(0) = \begin{bmatrix} x_1^0 & x_2^0 \end{bmatrix}^T; \text{ and}$$

(ii) zero-state response to unit-step input.

14.18 Consider the system,

$$\dot{\mathbf{x}} = \begin{bmatrix} 0 & 1 \\ -2 & -3 \end{bmatrix}\mathbf{x} + \begin{bmatrix} 0 \\ 1 \end{bmatrix}u; \mathbf{x}(0) = \begin{bmatrix} 1 \\ 1 \end{bmatrix}$$

$$y = [1 \;\; 0]\mathbf{x}$$

(a) Determine the stability of the system.

(b) Find the output response of the system to unit-step input.

14.19 Figure P14.19 shows the block diagram of a control system with state variable feedback and feedforward control. The plant model is

$$\dot{\mathbf{x}} = \begin{bmatrix} -3 & 2 \\ 4 & -5 \end{bmatrix}\mathbf{x} + \begin{bmatrix} 1 \\ 0 \end{bmatrix}u$$

$$y = [0 \;\; 1]\mathbf{x}$$

(a) Derive a state model for the feedback system.

(b) Find the output $y(t)$ of the feedback system to a unit-step input $r(t)$; the initial state is assumed to be zero.

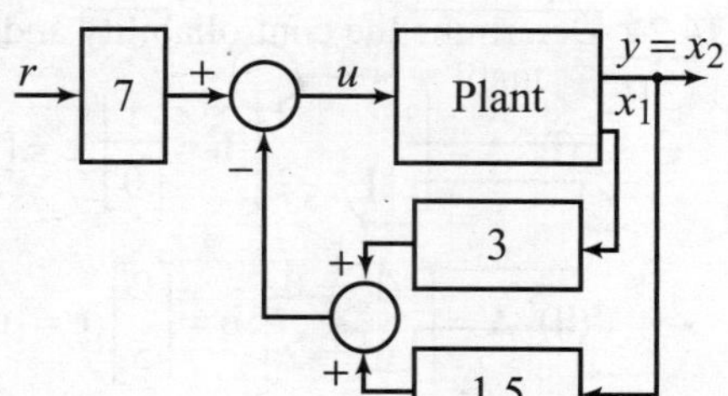

Fig. P14.19

14.20 Consider the state equation,

$$\dot{\mathbf{x}} = \begin{bmatrix} 0 & 1 \\ -1 & -2 \end{bmatrix}\mathbf{x}$$

Find a set of states $x_1(1)$ and $x_2(1)$ such that $x_1(2) = 2$.

14.21 The following facts are known about the linear system:

$$\dot{\mathbf{x}}(t) = \mathbf{A}\mathbf{x}(t)$$

If $\mathbf{x}(0) = \begin{bmatrix} 1 \\ -2 \end{bmatrix}$, then $\mathbf{x}(t) = \begin{bmatrix} e^{-2t} \\ -2e^{-2t} \end{bmatrix}$

If $\mathbf{x}(0) = \begin{bmatrix} 1 \\ -1 \end{bmatrix}$, then $\mathbf{x}(t) = \begin{bmatrix} e^{-t} \\ -e^{-t} \end{bmatrix}$

Find $e^{\mathbf{A}t}$ and hence $\mathbf{A}$.

14.22 Show that the pair $\{\mathbf{A}, \mathbf{c}\}$ is completely observable for all values of α_i's.

$$\mathbf{A} = \begin{bmatrix} 0 & 0 & 0 & \cdots & 0 & -\alpha_n \\ 1 & 0 & 0 & \cdots & 0 & -\alpha_{n-1} \\ 0 & 1 & 0 & \cdots & 0 & -\alpha_{n-2} \\ \vdots & \vdots & \vdots & & \vdots & \vdots \\ 0 & 0 & 0 & \cdots & 0 & -\alpha_2 \\ 0 & 0 & 0 & \cdots & 1 & -\alpha_1 \end{bmatrix}$$

$$\mathbf{c} = [0 \;\; 0 \;\; \cdots 0 \;\; 1]$$

14.23 Show that the pair $\{\mathbf{A}, \mathbf{b}\}$ is completely controllable for all values of α_i's.

$$\mathbf{A} = \begin{bmatrix} 0 & 1 & 0 & \cdots & 0 \\ 0 & 0 & 1 & \cdots & 0 \\ \vdots & \vdots & \vdots & & \vdots \\ 0 & 0 & 0 & \cdots & 1 \\ -\alpha_n & -\alpha_{n-1} & -\alpha_{n-2} & \cdots & -\alpha_1 \end{bmatrix}; \mathbf{b} = \begin{bmatrix} 0 \\ 0 \\ \vdots \\ 0 \\ 1 \end{bmatrix}$$

14.24 Determine the controllability and observability properties of the following systems:

(i) $\mathbf{A} = \begin{bmatrix} -2 & 1 \\ 1 & -2 \end{bmatrix}; \mathbf{b} = \begin{bmatrix} 1 \\ 0 \end{bmatrix}; \mathbf{c} = [1 \;\; -1]$

(ii) $\mathbf{A} = \begin{bmatrix} -1 & 0 \\ 0 & -2 \end{bmatrix}; \mathbf{b} = \begin{bmatrix} 2 \\ 5 \end{bmatrix}; \mathbf{c} = [0 \;\; 1]$

(iii) $\mathbf{A} = \begin{bmatrix} 0 & 1 & 0 \\ 0 & 0 & 1 \\ 0 & -2 & -3 \end{bmatrix}; \mathbf{b} = \begin{bmatrix} 0 \\ 0 \\ 1 \end{bmatrix}; \mathbf{c} = [10 \;\; 0 \;\; 0]$

(iii) $\mathbf{A} = \begin{bmatrix} 0 & 0 & 0 \\ 1 & 0 & -3 \\ 0 & 1 & -4 \end{bmatrix}; \mathbf{b} = \begin{bmatrix} 40 \\ 10 \\ 0 \end{bmatrix}; \mathbf{c} = [0 \;\; 0 \;\; 1]$

MATLAB Exercise Verify your results using Control System Toolbox.

14.25 The following models realize the transfer function $G(s) = \dfrac{1}{s+1}$.

(i) $\mathbf{A} = \begin{bmatrix} -2 & 1 \\ 1 & -2 \end{bmatrix}$; $\mathbf{b} = \begin{bmatrix} 1 \\ 1 \end{bmatrix}$; $\mathbf{c} = [0 \;\; 1]$

(ii) $\mathbf{A} = \begin{bmatrix} -1 & 0 \\ 0 & -3 \end{bmatrix}$; $\mathbf{b} = \begin{bmatrix} 1 \\ 1 \end{bmatrix}$; $\mathbf{c} = [1 \;\; 0]$

(iii) $\mathbf{A} = \begin{bmatrix} -2 & 0 \\ 0 & -1 \end{bmatrix}$; $\mathbf{b} = \begin{bmatrix} 0 \\ 1 \end{bmatrix}$; $\mathbf{c} = [0 \;\; 1]$

Investigate the controllability and observability properties of these models. Find a state variable model for the given transfer function which is both controllable and observable.

14.26 Consider the systems,

(i) $\mathbf{A} = \begin{bmatrix} 0 & -2 \\ 1 & -3 \end{bmatrix}$; $\mathbf{b} = \begin{bmatrix} 1 \\ 1 \end{bmatrix}$; $\mathbf{c} = [0 \;\; 1]$

(ii) $\mathbf{A} = \begin{bmatrix} 0 & 1 & 0 \\ 0 & 0 & 1 \\ -6 & -11 & -6 \end{bmatrix}$; $\mathbf{b} = \begin{bmatrix} 0 \\ 0 \\ 1 \end{bmatrix}$; $\mathbf{c} = [4 \;\; 5 \;\; 1]$

Determine the transfer function in each case. What can we say about controllability and observability properties without making any further calculations?

14.27 Consider the system,

$$\dot{\mathbf{x}} = \begin{bmatrix} 1 & 1 & 0 \\ 0 & -2 & 1 \\ 0 & 0 & -1 \end{bmatrix}\mathbf{x} + \begin{bmatrix} 0 \\ 1 \\ -2 \end{bmatrix}u; \; y = [1 \;\; 0 \;\; 0]\mathbf{x}$$

(a) Find the eigenvalues of **A** and from there determine the stability of the system.

(b) Find the transfer function model and from there determine the stability of the system.

(c) Are the two results same? If not, why?

MATLAB Exercise Verify your results using Control System Toolbox.

14.28 Given a transfer function,

$$G(s) = \frac{10}{s(s+1)} = \frac{Y(s)}{U(s)}$$

Construct three different state models for this system:

(a) One which is both controllable and observable.

(b) One which is controllable but not observable.

(c) One which is observable but not controllable.

14.29 Obtain the transfer function description for the following system:

$$\begin{bmatrix} x_1(k+1) \\ x_2(k+1) \end{bmatrix} = \begin{bmatrix} 2 & -5 \\ \frac{1}{2} & -1 \end{bmatrix} \begin{bmatrix} x_1(k) \\ x_2(k) \end{bmatrix} + \begin{bmatrix} 1 \\ 0 \end{bmatrix} u(k);\; y(k) = 2x_1(k)$$

14.30 The block diagram of a sampled-data system is shown in Fig. P14.30. Obtain the discrete-time state model of the system.

Given $\mathbf{A} = \begin{bmatrix} 0 & 1 \\ -2 & -3 \end{bmatrix}; \mathbf{b} = \begin{bmatrix} 0 \\ 1 \end{bmatrix}; \mathbf{c} = [1 \;\; 0]$

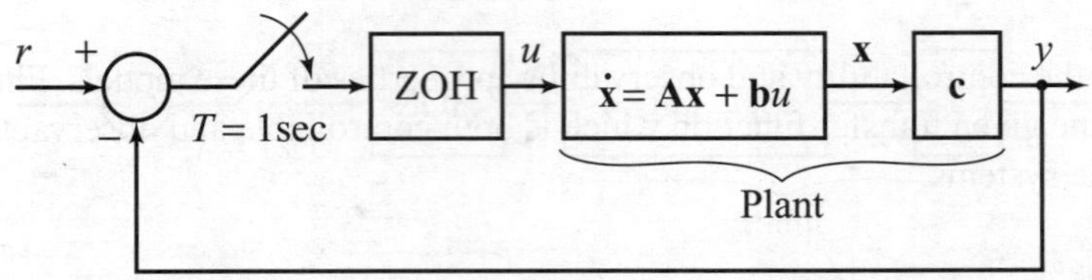

Fig. P14.30

14.31 Consider a continuous-time system,

$$\dot{\mathbf{x}}(t) = \begin{bmatrix} -2 & 2 \\ 1 & -3 \end{bmatrix} \mathbf{x}(t) + \begin{bmatrix} -1 \\ 5 \end{bmatrix} u(t)$$

$$y(t) = \begin{bmatrix} 2 & -4 \end{bmatrix} \mathbf{x}(t) + 6u(t)$$

Insert sample-and-hold devices and determine the vector difference state model for digital simulation of the continuous-time system when the computation interval $T = 0.2$.

MATLAB Exercise Verify your results using Control System Toolbox.

14.32 The plant of a unity-feedback continuous-time control system is described by the equations,

$$\dot{\mathbf{x}} = \begin{bmatrix} 0 & 1 \\ 0 & -2 \end{bmatrix} \mathbf{x} + \begin{bmatrix} 0 \\ 2 \end{bmatrix} u$$

$$y = x_1$$

(a) Show that the continuous-time closed-loop system is stable.

(b) A sampler and zero-order hold are now introduced in the forward loop. Show that the stable linear continuous-time system becomes unstable upon the introduction of a sampler and a zero-order hold with sampling period $T = 3$ sec. Determine the stability with $T = 0.4$ sec.

14.33 Investigate the controllability and observability of the following system:

$$\mathbf{x}(k+1) = \begin{bmatrix} -1 & 1 \\ 0 & -1 \end{bmatrix} \mathbf{x}(k) + \begin{bmatrix} 0 \\ 1 \end{bmatrix} u(k)$$

$$y(k) = [1 \;\; 1]\, \mathbf{x}(k)$$

MATLAB Exercise Verify your results using Control System Toolbox.

14.34 Consider the state variable model,

$$\mathbf{x}(k+1) = \mathbf{F}\mathbf{x}(k) + \mathbf{g}\,r(k)$$
$$y(k) = \mathbf{c}\mathbf{x}(k)$$

with
$$\mathbf{F} = \begin{bmatrix} 0 & 1 \\ -\frac{1}{8} & \frac{3}{4} \end{bmatrix}; \mathbf{g} = \begin{bmatrix} 0 \\ 1 \end{bmatrix}; \mathbf{c} = \begin{bmatrix} -\frac{1}{2} & 1 \end{bmatrix}$$

(a) Find the eigenvalues of matrix **F**.

(b) Find the transfer function $G(z) = Y(z)/R(z)$ and determine the poles of the transfer function.

(c) Comment upon the controllability and observability properties of the given system.

MATLAB Exercise Verify your results using Control System Toolbox.

CONTROL SYSTEM DESIGN USING STATE VARIABLE METHODS

15

Preview

In addition to the transform design techniques of root locus (Chapter 9) and Bode plots (Chapter 12), there is another major method of designing feedback control systems: the state-space method. The transform design is based on input-output transfer function model; output feedback and the use of a compensator is the design structure. The state-space method, on the other hand, is based on state variable model; direct feedback of state vector is the design structure. We will see in this chapter, the merits and limitations of state-space method in comparison with transform design methods. The approach of presentation is similar to that followed in the previous chapter—state-feedback design is first introduced for continuous-time systems, followed by a brief parallel presentation for discrete-time systems.

This chapter should be considered only an introduction to state-space design; we introduce the pole-placement design and optimal control design techniques, and apply them to single-input linear systems. For more details, refer [155].

15.1 INTRODUCTION

State variable methods have provided control theory with a deep and satisfying mathematical basis. With state variable models, we use an internal description of the system dynamics as opposed to a straightforward input-output transfer function model. We will see in this chapter that we gain from the knowledge of these internal states, and benefit from the use of this information in control system design.

The very fact that we describe a system using a state variable model leads directly to new terms and analysis/design methods. In the previous chapter on state variable analysis, we used system eigenvalues when we discussed system stability. These eigenvalues were found to be the same as system poles for systems that are both controllable and observable. When we looked at the initial condition and forced system responses, we found that these naturally occurred in the state variable framework.

When we discuss the control of a plant or process using state variable model, we will find many new fundamental insights about control theory. We will learn how state variable models lead to new methods for the design of compensators. These developments are relatively new; whereas PID control dates from 1940 onwards, state variable methods emerged from around 1960 and underwent many refinements before reaching the highly developed level available today. In this chapter, we introduce state variable control methods.

Before we begin our introduction to state variable control methods, a recapitulation of (Eqns (14.13)) the general state variable system structure will be in order here. Consider the general state variable system diagram shown in Fig. 15.1. The link between the real world and this rather abstract representation of a system is through the system input, output, and states.

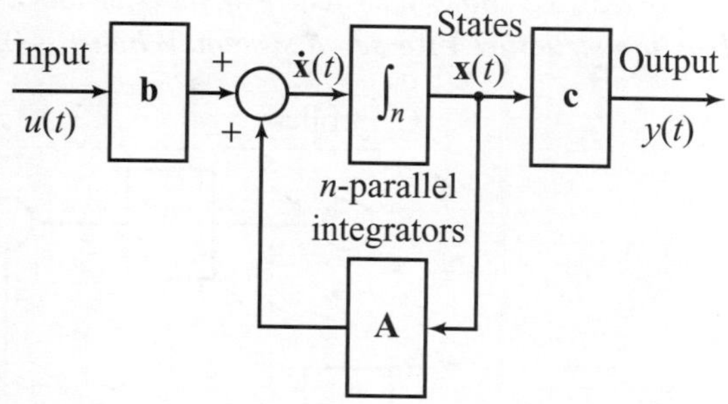

Fig. 15.1 *General state variable system structure*

The *system states* are the important set of system variables. It is the $n \times 1$ state vector $\mathbf{x}$. The $n \times n$ system matrix $\mathbf{A}$ governs how the system internal dynamics will operate. This means that $\mathbf{x}(t)$ is the key *information-carrying vector* in a state variable system model.

The *system inputs* are the control inputs to the system. They are able to influence and shape the dynamical behavior of the system. In the system structure of Fig. 15.1, we have assumed that there is only one control input $u(t)$ available to us to manipulate. The input is connected to the system state through the $n \times 1$ input matrix $\mathbf{b}$.

The *system outputs* are generally those variables we can measure and wish to control. In the structure of Fig. 15.1, we have assumed that there is only one output $y(t)$ we can measure and wish to control. The state is connected to the output via the $1 \times n$ output matrix $\mathbf{c}$. Thus, $\mathbf{c}$ is *information-reducing matrix* in the sense that it specifies just how much we are able to view of the whole system state vector $\mathbf{x}(t)$.

In the structure of Fig. 15.1, we have assumed that there is no direct coupling between output y and input u (i.e., $d = 0$ in Eqns (14.13)). Direct coupling is rare in control systems.

The state variable model for the SISO system of Fig. 15.1 takes the form,

$$\dot{\mathbf{x}}(t) = \mathbf{A}\mathbf{x}(t) + \mathbf{b}u(t) \tag{15.1a}$$

$$y(t) = \mathbf{c}\mathbf{x}(t) \tag{15.1b}$$

15.2 STATE VARIABLE FEEDBACK STRUCTURE

Output feedback and the use of a compensator has been the common structure applied in this book so far. The compensators were used in the context of input-output transfer function model. The solution of the design problem was based on the choice of the parameters of the compensator to move poles of the closed-loop system to new locations to maintain stability of the closed-loop, and achieve some desired performance measures on speed of response and steady-state accuracy. The performance measures on speed of response were translated to specifying two dominant closed-loop poles. The locations of the other $(n - 2)$ poles of the nth-order closed-loop system were not specified; however, it was required to push these poles to the regions in the complex plane from where the contribution of these poles to the dynamics of the closed-loop system is insignificant.

One obvious question may be raised here. Why did we not specify all the n closed-loop poles to shape the dynamics of the closed-loop system? The answer to this question is straightforward: with output feedback and the selected compensator structures (PID, lag/lead), a full pole-placement solution cannot be obtained, and we settle for the partial control over the closed-loop pole locations.

An important constraint here is that the output y is only a partial view of the state vector $\mathbf{x}$; the state passes through the output matrix $\mathbf{c}$ to give the output variable y. We now investigate what happens if we can remove this constraint and have direct access to all the state variables, i.e., instead of using output in a feedback control law, we investigate the outcome of using the state vector *directly* in the feedback control law. We assume that we have a set of measurements available that allow a *full* view of the state vector.

Figure 15.2 shows a structure of state-feedback control for tracking the constant reference signal r. It is the structure of a servo system. When $r = 0$, the structure corresponds to a regulator system.

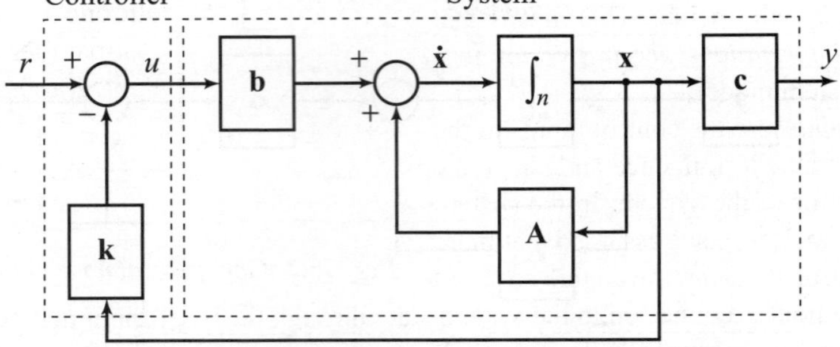

Fig. 15.2 *State-feedback control structure*

The state-feedback law is of the form,

$$u(t) = r - \mathbf{kx}(t) \tag{15.2}$$

The gain $\mathbf{k}$ is a $1 \times n$ row matrix with n tunable design parameters $k_1, k_2, \ldots, k_n$. We anticipate that we should be able to achieve more with state feedback. Closed-loop system analysis for the general state-feedback system of Fig. 15.2 is as follows:

Direct substitution of the control law (15.2) into the system Eqn. (15.1a) gives

$$\begin{aligned}\dot{\mathbf{x}}(t) &= \mathbf{Ax}(t) + \mathbf{b}[r - \mathbf{kx}(t)] \\ &= (\mathbf{A} - \mathbf{bk})\mathbf{x}(t) + \mathbf{b}r\end{aligned} \tag{15.3}$$

where $(\mathbf{A} - \mathbf{bk})$ is the closed-loop system matrix.

We assume that a state-feedback gain $\mathbf{k}$ can be found to stabilize the closed-loop, so that all the eigenvalues of $(\mathbf{A} - \mathbf{bk})$ have a strictly negative real part. As we shall see shortly, this is possible if the system (15.1) is controllable.

We have earlier observed in Chapter 2 that the transient response of a stable system vanishes as $t \to \infty$, and the steady-state response has the same nature as the input itself except for a modification in magnitude caused by the system's behavior to the specified input. Therefore, with the state-feedback gain $\mathbf{k}$ that stabilizes the closed-loop system of Fig. 15.2, the constant reference signal r causes the output y to reach constant level as $t \to \infty$. We denote this steady level as y_s. This output is connected to the state $\mathbf{x}_s$ via the output matrix $\mathbf{c}$.

Equation (15.3), as $t \to \infty$, becomes

$$\dot{\mathbf{x}}_s = (\mathbf{A} - \mathbf{bk})\mathbf{x}_s + \mathbf{b}r = \mathbf{0}$$

which gives

$$\mathbf{x}_s = -(\mathbf{A} - \mathbf{bk})^{-1}\mathbf{b}r$$

and

$$y_s = \mathbf{c}\left[-(\mathbf{A} - \mathbf{bk})^{-1}\right]\mathbf{b}r \tag{15.4}$$

Thus, ***with the stabilizing feedback matrix* k*, the state-feedback control structure of Fig. 15.2 results in a finite steady-state error.***

We have two design issues in hand:

1. Find the gain matrix **k** so that the eigenvalues of (**A** – **bk**) are at the desired locations in the complex plane. Iterate on the desired locations to obtain acceptable transient as well as steady-state response.
2. Another design issue arises when finite steady-state error resulting from solution to the first design problem is not acceptable, and it is desired to reduce the error to zero. This requires modifying the control structure of Fig. 15.2.

We first take up design problem 1. Consideration of design problem 2 will follow subsequently.

15.3 POLE-PLACEMENT DESIGN USING STATE FEEDBACK

Consider that the process/plant is described by the following state equation:

$$\dot{\mathbf{x}}(t) = \mathbf{A}\mathbf{x}(t) + \mathbf{b}u(t) \tag{15.5}$$

where $\mathbf{x}(t)$ is the $n \times 1$ state vector and $u(t)$ is the scalar control. The state-feedback control law is

$$u(t) = -\mathbf{k}\mathbf{x}(t) + r \tag{15.6}$$

where $\mathbf{k}$ is the $1 \times n$ feedback matrix with constant gain elements.

By substituting Eqn. (15.6) into Eqn. (15.5), the closed-loop system is represented by the state equation,

$$\dot{\mathbf{x}}(t) = (\mathbf{A} - \mathbf{b}\mathbf{k})\mathbf{x}(t) + \mathbf{b}r \tag{15.7}$$

It will be shown in the following that ***if the pair {A, b} is completely controllable, then a matrix k exists that can give an arbitrary set of eigenvalues of* (A – bk),** i.e., the roots of the characteristic equation (poles of the closed-loop system)

$$|s\mathbf{I} - (\mathbf{A} - \mathbf{b}\mathbf{k})| = 0 \tag{15.8}$$

can be arbitrarily placed.

To show that this is true, we refer to the findings in Chapter 14 that a state equation in first companion form (refer to Eqns (14.49)) is always completely controllable (refer to Problem 14.23). For this reason, the companion-form realization given by Eqns (14.49) is referred to as *controllable canonical form*[1]. It implies that if a system is completely controllable, it can always be represented in the controllable canonical form.

We assume that there exists a transformation

$$\bar{\mathbf{x}} = \mathbf{P}\mathbf{x} \tag{15.9}$$

$$= \begin{bmatrix} p_{11} & p_{12} & \cdots & p_{1n} \\ p_{21} & p_{22} & \cdots & p_{2n} \\ \vdots & \vdots & & \vdots \\ p_{n1} & p_{n2} & \cdots & p_{nn} \end{bmatrix} \mathbf{x} = \begin{bmatrix} \mathbf{p}_1 \\ \mathbf{p}_2 \\ \vdots \\ \mathbf{p}_n \end{bmatrix} \mathbf{x}$$

$$\mathbf{p}_i = \begin{bmatrix} p_{i1} & p_{i2} & \cdots & p_{in} \end{bmatrix};\ i = 1, 2, \ldots, n$$

[1] A state model in second companion form (refer to Eqns (14.51)) is always completely observable (refer to Problem 14.22). For this reason, the companion form realization given by Eqns (14.51) is referred to as *observable canonical form*.

which transforms system (15.5) to the following controllable canonical model:

$$\dot{\bar{\mathbf{x}}} = \bar{\mathbf{A}}\bar{\mathbf{x}} + \bar{\mathbf{b}}\,u \tag{15.10}$$

where

$$\bar{\mathbf{A}} = \mathbf{PAP}^{-1} = \begin{bmatrix} 0 & 1 & 0 & \dots & 0 \\ 0 & 0 & 1 & \dots & 0 \\ \vdots & \vdots & \vdots & & \vdots \\ 0 & 0 & 0 & & 1 \\ -\alpha_n & -\alpha_{n-1} & -\alpha_{n-2} & \dots & -\alpha_1 \end{bmatrix}$$

$$\bar{\mathbf{b}} = \mathbf{Pb} = \begin{bmatrix} 0 \\ 0 \\ \vdots \\ 0 \\ 1 \end{bmatrix}$$

$$|s\mathbf{I} - \bar{\mathbf{A}}| = s^n + \alpha_1 s^{n-1} + \cdots + \alpha_{n-1} s + \alpha_n = |s\mathbf{I} - \mathbf{A}|$$

(Characteristic polynomial is invariant under equivalence transformation)
The first equation in the set (15.9) is given by

$$\bar{x}_1 = p_{11}x_1 + p_{12}x_2 + \cdots + p_{1n}x_n = \mathbf{p}_1\mathbf{x}$$

Taking the derivative on both sides of this equation, we get

$$\dot{\bar{x}}_1 = \mathbf{p}_1\dot{\mathbf{x}} = \mathbf{p}_1\mathbf{Ax} + \mathbf{p}_1\mathbf{b}u$$

But $\dot{\bar{x}}_1$ $(= \bar{x}_2)$ is a function of $\mathbf{x}$ only as per the canonical model (15.10).
Therefore,

$$\mathbf{p}_1\mathbf{b} = 0 \text{ and } \bar{x}_2 = \mathbf{p}_1\mathbf{Ax}$$

Taking derivative on both sides once again, we get

$$\mathbf{p}_1\mathbf{Ab} = 0 \text{ and } \bar{x}_3 = \mathbf{p}_1\mathbf{A}^2\mathbf{x}$$

Continuing the process, we obtain

$$\mathbf{p}_1\mathbf{A}^{n-2}\mathbf{b} = 0 \text{ and } \bar{x}_n = \mathbf{p}_1\mathbf{A}^{n-1}\mathbf{x}$$

Taking derivative once again, we obtain

$$\mathbf{p}_1\,\mathbf{A}^{n-1}\mathbf{b} = 1$$

Thus,

$$\bar{\mathbf{x}} = \mathbf{Px} = \begin{bmatrix} \mathbf{p}_1 \\ \mathbf{p}_1\mathbf{A} \\ \vdots \\ \mathbf{p}_1\mathbf{A}^{n-1} \end{bmatrix}\mathbf{x}$$

where $\mathbf{p}_1$ must satisfy the conditions

$$\mathbf{p}_1\mathbf{b} = \mathbf{p}_1\mathbf{A}\mathbf{b} = \cdots = \mathbf{p}_1\mathbf{A}^{n-2}\mathbf{b} = 0, \mathbf{p}_1\mathbf{A}^{n-1}\mathbf{b} = 1$$

From Eqn. (15.10), we have

$$\mathbf{P}\mathbf{b} = \begin{bmatrix} 0 \\ 0 \\ \vdots \\ 0 \\ 1 \end{bmatrix} = \begin{bmatrix} \mathbf{p}_1\mathbf{b} \\ \mathbf{p}_1\mathbf{A}\mathbf{b} \\ \vdots \\ \mathbf{p}_1\mathbf{A}^{n-2}\mathbf{b} \\ \mathbf{p}_1\mathbf{A}^{n-1}\mathbf{b} \end{bmatrix}$$

or

$$\mathbf{p}_1 = \begin{bmatrix} \mathbf{b} & \mathbf{A}\mathbf{b} & \cdots & \mathbf{A}^{n-2}\mathbf{b} & \mathbf{A}^{n-1}\mathbf{b} \end{bmatrix} = \begin{bmatrix} 0 & 0 & \cdots & 0 & 1 \end{bmatrix}$$

This gives

$$\mathbf{p}_1 = \begin{bmatrix} 0 & 0 & \cdots & 0 & 1 \end{bmatrix}\mathbf{U}^{-1}$$

where

$$\mathbf{U} = \begin{bmatrix} \mathbf{b} & \mathbf{A}\mathbf{b} & \cdots & \mathbf{A}^{n-1}\mathbf{b} \end{bmatrix}$$

is the controllability matrix, which is nonsingular because of the assumption of controllability of the system (15.5).

Therefore, the controllable state model (15.5) can be transformed to the canonical form (15.10) by the transformation,

$$\bar{\mathbf{x}} = \mathbf{P}\mathbf{x} \tag{15.11}$$

where $\mathbf{P} = \begin{bmatrix} \mathbf{p}_1 \\ \mathbf{p}_1\mathbf{A} \\ \vdots \\ \mathbf{p}_1\mathbf{A}^{n-1} \end{bmatrix}; \mathbf{p}_1 = \begin{bmatrix} 0 & 0 \cdots & 0 & 1 \end{bmatrix}\mathbf{U}^{-1}$

Under the equivalence transformation (15.11), the state-feedback control law (15.6) becomes

$$u = -\mathbf{k}\mathbf{x} + r = -\bar{\mathbf{k}}\,\bar{\mathbf{x}} + r \tag{15.12a}$$

where $\bar{\mathbf{k}} = \mathbf{k}\mathbf{P}^{-1} = \begin{bmatrix} \bar{k}_1 & \bar{k}_2 \cdots & \bar{k}_n \end{bmatrix}$ (15.12b)

With this control law, system (15.10) becomes

$$\dot{\bar{\mathbf{x}}} = (\bar{\mathbf{A}} - \bar{\mathbf{b}}\bar{\mathbf{k}})\bar{\mathbf{x}} + \bar{\mathbf{b}}r$$

$$= \begin{bmatrix} 0 & 1 & 0 & \cdots & 0 & 0 \\ 0 & 0 & 1 & \cdots & 0 & 0 \\ \vdots & \vdots & \vdots & & \vdots & \vdots \\ 0 & 0 & 0 & \cdots & 0 & 1 \\ \vdots & \vdots & \vdots & & \vdots & \vdots \\ -\alpha_n - \bar{k}_1 & -\alpha_{n-1} - \bar{k}_2 & -\alpha_{n-2} - \bar{k}_3 & \cdots & -\alpha_2 - \bar{k}_{n-1} & -\alpha_1 - \bar{k}_n \end{bmatrix} \bar{\mathbf{x}} + \begin{bmatrix} 0 \\ 0 \\ \vdots \\ 0 \\ 1 \end{bmatrix} r \tag{15.13}$$

$$\left| s\mathbf{I} - (\bar{\mathbf{A}} - \bar{\mathbf{b}}\bar{\mathbf{k}}) \right| = s^n + (\alpha_1 + \bar{k}_n))s^{n-1} + (\alpha_2 + \bar{k}_{n-1})s^{n-2} + \cdots + (\alpha_{n-1} + \bar{k}_2)s + (\alpha_n + \bar{k}_1) \tag{15.14}$$

Since the coefficients $\bar{k}_i$ are arbitrarily chosen real numbers, the coefficients of the characteristic polynomial of (**A** – **bk**) can be given any desired values. Hence, the closed-loop poles can be placed at any desired locations in the complex plane (subject to conjugate pairing: coefficients of a characteristic polynomial will be real only if the complex poles are present in conjugate pairs).

Assume that the desired characteristic polynomial of (**A** – **bk**), and hence $(\bar{\mathbf{A}} - \bar{\mathbf{b}}\bar{\mathbf{k}})$, is

$$s^n + a_1 s^{n-1} + \cdots + a_n$$

From Eqn. (15.14), it is obvious that this requirement is met if $\bar{\mathbf{k}}$ is chosen as

$$\bar{\mathbf{k}} = [a_n - \alpha_n \quad a_{n-1} - \alpha_{n-1} \quad \cdots \quad a_1 - \alpha_1]$$

Transforming this feedback controller to the original coordinates, we obtain

$$\mathbf{k} = \bar{\mathbf{k}}\mathbf{P} = [a_n - \alpha_n \quad a_{n-1} - \alpha_{n-1} \quad \cdots \quad a_1 - \alpha_1]\mathbf{P} \tag{15.15}$$

This proves that if (15.5) is controllable, the closed-loop poles can be arbitrarily assigned

We can find the parameters of the feedback matrix **k** by matching the coefficients of $|s\mathbf{I} - (\mathbf{A} - \mathbf{bk})|$ with the coefficients of the desired characteristic polynomial of the closed-loop system. This technique, in general, leads to difficult calculations of the feedback gains, especially for higher-order systems not represented in controllable canonical form. For higher-order systems represented in controllable canonical form, the feedback gains **k** can be easily determined by matching the coefficients. However, the physics of the problem generally requires feedback from state variables that are not canonical. This requires transforming the system to controllable canonical form, designing the feedback gains, and transforming the designed system back to the original state variable representation. This technique, derived earlier in this section, is summarized below.

Characteristic polynomial of the plant:

$$|s\mathbf{I} - \mathbf{A}| = s^n + \alpha_1 s^{n-1} + \cdots + \alpha_{n-1}s + \alpha_n \tag{15.16a}$$

Desired characteristic polynomial of the feedback system:

$$\Delta(s) = s^n + a_1 s^{n-1} + \cdots + a_{n-1}s + a_n \tag{15.16b}$$

The feedback matrix

$$\mathbf{k} = [a_n - \alpha_n \quad a_{n-1} - \alpha_{n-1} \quad \cdots \quad a_1 - \alpha_1]\mathbf{P} \tag{15.16c}$$

$$\mathbf{P} = \begin{bmatrix} \mathbf{p}_1 \\ \mathbf{p}_1\mathbf{A} \\ \vdots \\ \mathbf{p}_1\mathbf{A}^{n-1} \end{bmatrix}; \quad \begin{matrix} \mathbf{p}_1 = [0 \;\; 0 \;\; \cdots \;\; 0 \;\; 1]\mathbf{U}^{-1} \\ \mathbf{U} = \begin{bmatrix} \mathbf{b} & \mathbf{Ab} & \cdots & \mathbf{A}^{n-1}\mathbf{b} \end{bmatrix} \end{matrix} \tag{15.16d}$$

Example 15.1 Consider the problem of designing an attitude control system for a rigid satellite. The dynamics of the system and the objectives of control were described earlier in Example 4.2. Figure 4.7 depicts the control schematic for one axis of the full three-axis attitude control system. The control signal comes from the reaction jets that produce a torque $T(t)$ (= Fd) about the mass center.

The satellite is assumed to be in frictionless environment. If $T(t)$ is the system input and $\theta(t)$ is the system output, we have

$$T(t) = J\frac{d^2\theta(t)}{dt^2}$$

where J is the moment of inertia of the satellite. Normalizing, we define

$$u = T(t)/J$$

and obtain

$$\ddot{\theta} = u \text{ or } \frac{\theta(s)}{U(s)} = \frac{1}{s^2}$$

This is a reasonably accurate model of a rigid satellite in a frictionless environment, and is useful in examples because of its simplicity.

Choosing $x_1 = \theta$ and $x_2 = \dot{\theta}$ as state variables, we obtain the following state equation for the system:

$$\dot{\mathbf{x}} = \mathbf{Ax} + \mathbf{b}u = \begin{bmatrix} 0 & 1 \\ 0 & 0 \end{bmatrix}\mathbf{x} + \begin{bmatrix} 0 \\ 1 \end{bmatrix}u \tag{15.17}$$

It is desired to maintain the attitude of the satellite at $\theta = 0$ (i.e., $r = 0$). This is a regulator problem. To stabilize the system, the input signal is chosen to be of the form,

$$u(t) = -k_1x_1(t) - k_2x_2(t) = -\mathbf{kx}(t)$$

The state equation for the closed-loop system (Fig. 15.3), then becomes

$$\dot{\mathbf{x}} = (\mathbf{A} - \mathbf{bk})\,\mathbf{x} = \left(\begin{bmatrix} 0 & 1 \\ 0 & 0 \end{bmatrix} - \begin{bmatrix} 0 \\ 1 \end{bmatrix}[k_1 \quad k_2]\right)\mathbf{x} = \begin{bmatrix} 0 & 1 \\ -k_1 & -k_2 \end{bmatrix}\mathbf{x}$$

The characteristic equation of the closed-loop system is

$$|s\mathbf{I} - (\mathbf{A} - \mathbf{bk})| = s^2 + k_2s + k_1 = 0 \tag{15.18a}$$

Suppose that the design specifications for this system require $\zeta = 0.707$ with a settling time of 1 sec $\left(\frac{4}{\zeta\omega_n} = 1 \text{ or } \zeta\omega_n = 4\right)$. The closed-loop pole locations needed are at $s = -4 \pm j4$. The desired characteristic equation is

$$(s+4+j4)(s+4-j4) = s^2 + 8s + 32 \quad (15.18b)$$

Equating the coefficients with like powers of s in Eqns (15.18a) and (15.18b) yields

$$k_1 = 32,\ k_2 = 8$$

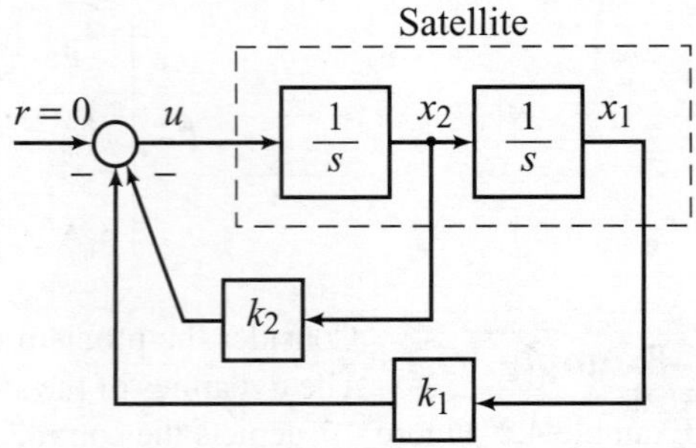

Fig. 15.3 *State diagram for the satellite control system*

Example 15.2 The plant of a servo system is described by the equations,

$$\dot{\mathbf{x}} = \mathbf{A}\mathbf{x} + \mathbf{b}u$$
$$y = \mathbf{c}\mathbf{x} \quad (15.19)$$

where

$$\mathbf{A} = \begin{bmatrix} -3 & 2 \\ 4 & -5 \end{bmatrix};\ \mathbf{b} = \begin{bmatrix} 0 \\ 1 \end{bmatrix};\ \mathbf{c} = \begin{bmatrix} 0 & 1 \end{bmatrix}$$

The characteristic polynomial of this system is

$$|s\mathbf{I} - \mathbf{A}| = s^2 + 8s + 7 = s^2 + \alpha_1 s + \alpha_2$$

The desired characteristic polynomial is

$$(s+4)(s+7) = s^2 + 11s + 28 = s^2 + a_1 s + a_2$$

A control law of the form

$$u = -\mathbf{k}\mathbf{x} + r \quad (15.20a)$$

is proposed; r is a unit-step reference input. The gain matrix (refer to Eqns (15.16))

$$\mathbf{k} = [a_2 - \alpha_2 \quad a_1 - \alpha_1]\mathbf{P} = [21 \quad 3]\mathbf{P}$$

where

$$\mathbf{P} = \begin{bmatrix} \mathbf{p}_1 \\ \mathbf{p}_1\mathbf{A} \end{bmatrix}$$

$$\mathbf{P}_1 = [0 \quad 1][\mathbf{b} \quad \mathbf{A}\mathbf{b}]^{-1} = [0 \quad 1]\begin{bmatrix} 1 & \frac{3}{4} \\ 0 & \frac{1}{4} \end{bmatrix} = \begin{bmatrix} 0 & \frac{1}{4} \end{bmatrix};\mathbf{p}_1\mathbf{A} = \begin{bmatrix} 1 & \frac{-5}{4} \end{bmatrix}$$

This gives

$$\mathbf{k} = [3 \quad 1.5] \tag{15.20b}$$

Substituting the control law (15.20) into (15.19), we get the state equations of the feedback system:

$$\dot{\mathbf{x}} = (\mathbf{A} - \mathbf{bk})\mathbf{x} + \mathbf{b}r = \begin{bmatrix} -6 & 0.5 \\ 4 & -5 \end{bmatrix}\mathbf{x} + \begin{bmatrix} 1 \\ 0 \end{bmatrix} r \tag{15.21}$$

At steady state, $\dot{\mathbf{x}} = \mathbf{0}$. This gives (for $r = 1$)

$$4x_{1s} - 5x_{2s} = 0$$

and

$$-6x_{1s} + 0.5x_{2s} = 1$$

From these equations, we get

$$x_{2s} = y_s = 1/7$$

Steady-state error $= r - y_s = 6/7$

15.4 STATE FEEDBACK WITH INTEGRAL CONTROL

This section is concerned with steady-state error design via integral control. Recall that state-feedback control structure of Fig. 15.2 results in a finite steady-state error to step commands. We modify this control structure to the one shown in Fig. 15.4. A feedback path from the output has been added to form the error e, which is fed forward to the controlled plant via an integrator. The integrator increases the system type and reduces the finite steady-state error to zero. We will now derive the form of the state equations for the system of Fig. 15.4, and then use that form to design a controller. Thus, we will be able to design a system for zero steady-state error to step input, as well as obtain the desired transient response.

An additional state variable x_{n+1} has been added at the output of the left-most integrator. The error e is the derivative of this variable. Now from Fig.15.4,

$$\dot{x}_{n+1}(t) = r - \mathbf{cx}(t) = e(t) \tag{15.22a}$$

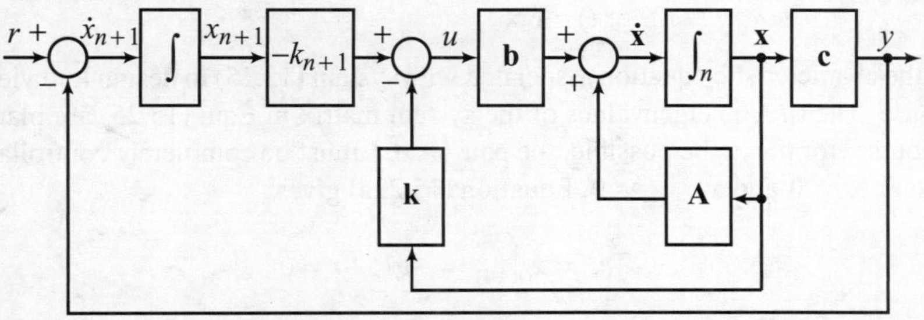

Fig. 15.4 *State feedback with integral control*

Writing the state equations from Fig. 15.4, we have

$$\dot{\mathbf{x}}(t) = \mathbf{A}\mathbf{x}(t) + \mathbf{b}u(t)$$
$$\dot{x}_{n+1}(t) = -\mathbf{c}\mathbf{x}(t) + r \tag{15.22b}$$
$$y(t) = \mathbf{c}\mathbf{x}(t)$$

These equations can be written as augmented vectors and matrices. Hence,

$$\begin{bmatrix} \dot{\mathbf{x}} \\ \dot{x}_{n+1} \end{bmatrix} = \begin{bmatrix} \mathbf{A} & \mathbf{0} \\ -\mathbf{c} & 0 \end{bmatrix} \begin{bmatrix} \mathbf{x} \\ x_{n+1} \end{bmatrix} + \begin{bmatrix} \mathbf{b} \\ 0 \end{bmatrix} u + \begin{bmatrix} \mathbf{0} \\ 1 \end{bmatrix} r$$

(15.23a)

or

$$y = \begin{bmatrix} \mathbf{c} & 0 \end{bmatrix} \begin{bmatrix} \mathbf{x} \\ x_{n+1} \end{bmatrix}$$

$$\dot{\bar{\mathbf{x}}}(t) = \bar{\mathbf{A}}\bar{\mathbf{x}}(t) + \bar{\mathbf{b}}u(t) + \begin{bmatrix} 0 \\ 1 \end{bmatrix} r$$
$$y(t) = \bar{\mathbf{c}}\,\bar{\mathbf{x}}(t) \tag{15.23b}$$

where $\bar{\mathbf{x}}(t)$ is the $(n+1)\times 1$ state vector, $u(t)$ and $y(t)$ are the scalar control signal and output, respectively; r is the scalar constant reference input. The coefficient matrices are represented by $(n+1)\times(n+1)$ matrix, $\bar{\mathbf{A}}$, $(n+1)\times 1$ matrix $\bar{\mathbf{b}}$, and $1\times(n+1)$ matrix $\bar{\mathbf{c}}$.

The control signal $u(t)$ is related to the state variables through state and integral feedback:

$$u(t) = -\mathbf{k}\mathbf{x}(t) - k_{n+1}x_{n+1} = -\bar{\mathbf{k}}\bar{\mathbf{x}}(t) \tag{15.24a}$$

where

$$\mathbf{k} = \begin{bmatrix} k_1 & k_2 & \cdots & k_n \end{bmatrix} \tag{15.24b}$$

is the state-feedback gain vector, and k_{n+1} is the scalar integral-feedback gain; and

$$\bar{\mathbf{k}} = \begin{bmatrix} k_1 & k_2 & \cdots & k_n & k_{n+1} \end{bmatrix} \tag{15.24c}$$

Substituting Eqn. (15.24a) into Eqn. (15.23b), the $n + 1$ state equations of the feedback system are written as

$$\dot{\bar{\mathbf{x}}}(t) = (\bar{\mathbf{A}} - \bar{\mathbf{b}}\bar{\mathbf{k}})\bar{\mathbf{x}}(t) + \begin{bmatrix} \mathbf{0} \\ 1 \end{bmatrix} r$$
$$y(t) = \bar{\mathbf{c}}\,\bar{\mathbf{x}} \tag{15.25}$$

We can use the characteristic equation associated with system (15.25) to design $\bar{\mathbf{k}}$ to yield the desired transient response. The $(n + 1)$ eigenvalues of the system matrix in Eqn. (15.25) are placed at desired locations. Of course, for this to be possible, the pair $\{\bar{\mathbf{A}}, \bar{\mathbf{b}}\}$ must be completely controllable.

At steady-state, $\dot{\mathbf{x}}_s = \mathbf{0}$ and $\dot{x}_{(n+1)s} = 0$. Equation (15.22a) gives

$$e_{ss} = \dot{x}_{(n+1)s} = -\mathbf{c}\mathbf{x}_s + r = 0$$

ensuring zero steady-state error to step input r.

Example 15.3 Suppose the system is given by

$$\frac{Y(s)}{U(s)} = \frac{1}{s+3}$$

with a constant reference command signal. We wish to have integral control with closed-loop poles at $\omega_n = 5$ and $\zeta = 0.5$, which is equivalent to asking for a desired characteristic equation,

$$s^2 + 5s + 25 = 0$$

The plant model is

$$\dot{x}_1 = -3x_1 + u$$

$$y = x_1$$

Augmenting the plant state x_1 with the integral state x_2 defined by the equation,

$$x_2(t) = \int_0^t (r - y(t))dt$$

we obtain

$$\begin{bmatrix} \dot{x}_1 \\ \dot{x}_2 \end{bmatrix} = \begin{bmatrix} -3 & 0 \\ -1 & 0 \end{bmatrix} \begin{bmatrix} x_1 \\ x_2 \end{bmatrix} + \begin{bmatrix} 1 \\ 0 \end{bmatrix} u + \begin{bmatrix} 0 \\ 1 \end{bmatrix} r$$

We can find **k** from

$$det\left(s\mathbf{I} - \begin{bmatrix} -3 & 0 \\ -1 & 0 \end{bmatrix} + \begin{bmatrix} 1 \\ 0 \end{bmatrix} \mathbf{k} \right) = s^2 + 5s + 25$$

or

$$s^2 + (3 + k_1)s - k_2 = s^2 + 5s + 25$$

Therefore,

$$\mathbf{k} = [2 \quad -25] = [k_1 \quad k_2]$$

The control

$$u = -k_1x_1 - k_2x_2 = -2x_1 + 25\int_0^t (r - y(t))dt$$

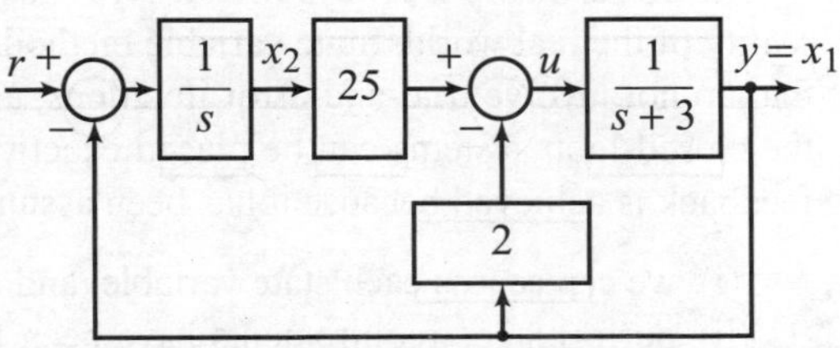

Fig. 15.5 *Integral control example*

The control configuration is shown in Fig. 15.5. This system will behave according to the desired closed-loop roots ($\omega_n = 5$, $\zeta = 0.5$) and will exhibit the characteristics of integral control: zero steady-state error to a step r.

15.5 CRITIQUE OF POLE-PLACEMENT STATE FEEDBACK CONTROL

Chapters 1–12 of this book were devoted to the classical methods of control design using output feedback. From Chapters 9 and 12, we have observed that the output-feedback control law is restricted in design achievements; by trial-and-error we place the two dominant poles of the closed-loop system at the desired locations, and all other poles are made insignificant by pole-zero cancelation and/or by shifting them to the region of insignificance on the complex plane. Rigorous trial-and-error is involved in the design procedure and many a times we terminate trials when dominant poles *close to* the desired locations (not *at* the desired locations) have been achieved.

The state-feedback control law is able to give total control over the system dynamics; all the poles of the closed-loop system can be placed exactly at the desired locations. Also there is no trial-and-error involved; state variable methods provide control theory with deep and satisfying mathematical design tools.

The important step in the pole-placement design approach using state-feedback is to choose the locations of the desired closed-loop poles. The design procedure, however, gives no guidance as to where in the left half of the complex plane, the closed-loop poles should be placed. It appears that we can choose the magnitude of the real part of the closed-loop poles arbitrarily large, making the system respond arbitrarily fast. However, to increase the rate at which the plant responds, the input signal to the plant must become larger, requiring large values of gains. As the magnitudes of the signals in a system increase, the likelihood of the system entering nonlinear regions of operation increases. For very large signals, this nonlinear operation will occur for almost every physical system. Hence, the linear model that is used in design no longer accurately models the physical system.

Thus, the selection of desired closed-loop poles requires a proper balance of bandwidth, overshoot, control effort, and other design requirements. If the system is of second-order, then the system dynamics (response characteristics) can be precisely correlated to the locations of the desired closed-loop poles. For higher-order systems, the locations of the closed-loop poles and the response characteristics are not easily correlated. Hence, in determining the state-feedback gain matrix **k** for a given system, it is desirable to examine by computer simulations, the response characteristics for several different matrices **k** (based on several different characteristic equations), and choose the one that gives the best overall performance. In fact, if critically examined, we observe that the trial-and-error involved in classical design procedures is also involved in state-variable methods, though in a different form.

Therefore, theory is always what it says it is: *just theory*, which has to be matched against the practical reality of the real world. State variable methods provide control theory with mathematical design tools that do not involve trial-and-error iterations, and that guarantee the design properties: all the poles of the closed-loop systems can be placed exactly at the desired locations. The design flexibility of state feedback is achieved because it has been assumed that

(i) we can access each state variable; and
(ii) performance requirements have been translated into desired locations for all the closed-loop poles.

The assumption of accessibility of each state variable, in practice, means that we must be able to measure every state variable, or we must have a more complicated control system when we include the model of the estimator which provides us with the state information. *Observer-based state-feedback controllers* incorporate model of the state estimator and are used when all the states are not able to be measured and we have a good model of the process.

Most individual control systems will have many unmeasurable system states and, further, industrial process models are not usually so well known or accurate. Adequate stability margins are required to protect against the model uncertainties. In some situations we may design a controller that works perfectly under computer simulations but turns out to be unstable in practice. Therefore, any design must be tested thoroughly to prevent disastrous results.

The assumption of availability of performance specifications in the form of exact locations for desired poles of the closed-loop system does not match against the practical reality. ***With the freedom***

of choice rendered by state feedback (of relocating all the closed-loop poles at desired locations, in contrast with classical design whereby the designer may only hope to achieve a pair of complex conjugate poles that are dominant) comes with the responsibility of selecting closed-loop poles judiciously. An obvious temptation of pushing the poles too far into the left hand of the complex plane to get a faster system, is costlier because a faster system may require accurate sensors and a larger or stronger actuator (such as a motor) to perform the task. Also the system bandwidth increases, and the system becomes sensitive to noise. Further, the signals in the system may become very large, with the result that the system may become nonlinear. No magic choice of closed-loop poles is possible; trial-and-error involved in classical design procedure gets shifted to trial-and-error in selection of closed-loop poles when state-feedback freedom is utilized.

Another approach is based in *linear quadratic optimal control*. This approach will determine the desired closed-loop poles such that it balances between the acceptable response and the amount of control energy required. The linear quadratic optimal control approach, as we shall see in this chapter, also involves trial-and-error but the number of trials are drastically reduced. The task of the designer shifts from selection of closed-loop poles to one of specifying various parameters in the performance index.

The classical output feedback procedures have more tolerance on uncertainity of the process models. State variable methods require 'good' models. In spite of this, we can say that state variable methods provide control theory with a deep mathematical basis. Of course, there is a gap between theory and practice, but it is the interplay between them that makes the practical outcomes so interesting and useful. We need the precision of mathematical control tools to illuminate control application problems in industry.

In the following, we present an introduction to observer-based state-feedback control design. Formulation and solution of linear quadratic optimal control will also be given; however, for want of space we refer the reader to the companion book [155] for the complete view of optimal control problem.

15.6 OBSERVER-BASED STATE FEEDBACK CONTROL

The control law designed in Section 15.3 assumed that all the state variables are available for feedback. In most cases, not all the state variables are measured. The cost of the required sensors may be prohibitive, or it may not be physically possible to measure all the state variables.

In this section, we demonstrate how to reconstruct all the state variables of a system from a few measurements. If the estimate of the state is denoted by $\hat{\mathbf{x}}$, it would be convenient if we could replace the true value in the control law given in Eqn.(15.6) with the estimate so that the control law becomes

$$u(t) = -\mathbf{k}\hat{\mathbf{x}}(t) + r \tag{15.26}$$

This is indeed possible, as we shall see in this section. So the construction of a state estimate is a key part of state feedback control design.

A device (or a computer program) that estimates the state variables is called a *state observer*, or simply an *observer*. If the state observer estimates all the state variables of the system, regardless of whether some state variables are available for direct measurement, it is called a *full-order state observer*. However, if accurate measurements of certain states are possible, we may estimate only the remaining states, and the accurately measured signals are then used directly for feedback. The resulting observer is called a *reduced-order state observer*. In this chapter, we discuss full-order observers. Refer to the companion book [155] for reduced-order observers.

15.6.1 Full-Order State Observer

Consider a process described by the state equation,

$$\dot{\mathbf{x}}(t) = \mathbf{A}\mathbf{x}(t) + \mathbf{b}u(t) \tag{15.27a}$$

where **A** and **b** are, respectively, $n \times n$ and $n \times 1$ real constant matrices. The measurement $y(t)$ is related to the state by the equation,

$$y(t) = \mathbf{c}\mathbf{x}(t) \tag{15.27b}$$

where **c** is $1 \times n$ real constant matrix. Without loss of generality, the direct transmission part has been assumed to be zero.

One method of estimating all the state variables that we may consider is to construct a model of the plant dynamics:

$$\dot{\hat{\mathbf{x}}}(t) = \mathbf{A}\hat{\mathbf{x}}(t) + \mathbf{b}u(t) \tag{15.28}$$

where $\hat{\mathbf{x}}$ is the estimate of the actual state **x**. We know **A**, **b** and $u(t)$, and hence, this estimator is satisfactory if we can obtain the correct initial condition $\mathbf{x}(0)$ and set $\hat{\mathbf{x}}(0)$ equal to it. Figure 15.6 depicts this 'open-loop' estimator. However, it is precisely the lack of information on $\mathbf{x}(0)$ that requires the construction of an estimator. If $\hat{\mathbf{x}}(0) \neq \mathbf{x}(0)$, the estimated state $\hat{\mathbf{x}}(t)$ obtained from the open-loop scheme of Fig. 15.6 would have a continually growing error, or an error that goes to zero too slowly to be of any use. Furthermore, small errors in our knowledge of the system (**A**, **b**), and the disturbances that enter the system but not the model would also cause the estimate to slowly diverge from the true state.

In order to speed up the estimation process and provide a useful state estimate, we feedback the difference between the measured and the estimated outputs and correct the model continuously with this error signal. This scheme, commonly known as 'Luenberger state observer', is shown in Fig. 15.7, and the equation for it is

$$\dot{\hat{\mathbf{x}}}(t) = \mathbf{A}\hat{\mathbf{x}}(t) + \mathbf{b}u(t) + \mathbf{m}(y(t) - \hat{y}(t)) \tag{15.29}$$

where **m** is an $n \times 1$ real constant gain matrix.

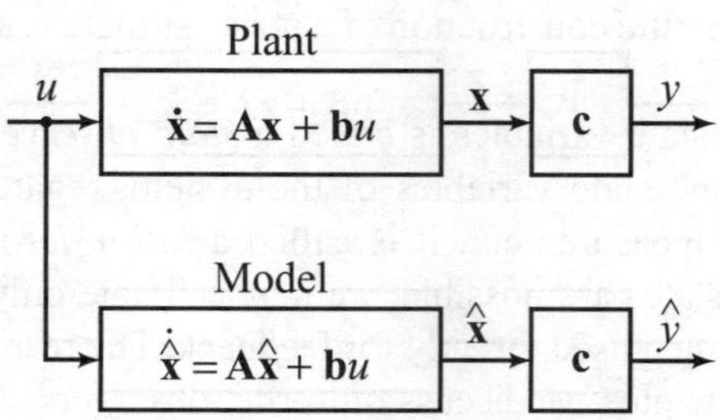

Fig. 15.6 *Open-loop estimator*

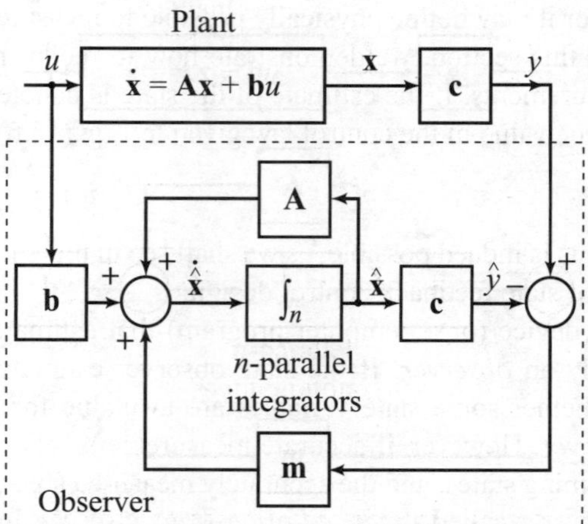

Fig. 15.7 *Luenberger state observer*

The state error vector

$$\tilde{\mathbf{x}}(t) = \mathbf{x}(t) - \hat{\mathbf{x}}(t) \tag{15.30}$$

Differentiating both sides, we get

$$\dot{\tilde{\mathbf{x}}}(t) = \dot{\mathbf{x}}(t) - \dot{\hat{\mathbf{x}}}(t)$$

Substituting for $\dot{\mathbf{x}}(t)$ and $\dot{\hat{\mathbf{x}}}(t)$ from Eqns (15.27) and (15.29) respectively, we get

$$\begin{aligned}\dot{\tilde{\mathbf{x}}}(t) &= \mathbf{A}\mathbf{x}(t) + \mathbf{b}u(t) - \mathbf{A}\hat{\mathbf{x}}(t) - \mathbf{b}u(t) - \mathbf{mc}(\mathbf{x}(t) - \hat{\mathbf{x}}(t)) \\ &= (\mathbf{A} - \mathbf{mc})\tilde{\mathbf{x}}(t)\end{aligned} \tag{15.31}$$

The characteristic equation of the error is given by

$$|s\mathbf{I} - (\mathbf{A} - \mathbf{mc})| = 0 \tag{15.32a}$$

If **m** can (we hope) be chosen so that (**A** – **mc**) has stable and reasonably fast roots, $\tilde{\mathbf{x}}(t)$ will decay to zero irrespective of $\tilde{\mathbf{x}}(0)$. This means that $\hat{\mathbf{x}}(t)$ will converge to $\mathbf{x}(t)$ regardless of the value of $\hat{\mathbf{x}}(0)$, and furthermore, the dynamics of the error can be chosen to be faster than the open-loop dynamics. Note that Eqn. (15.31) is independent of applied control. This is a consequence of assuming **A**, **b** and **c** to be identical in the plant and the observer. Therefore, the estimation error $\tilde{\mathbf{x}}$ converges to zero and remains there, independent of any known forcing function $u(t)$ on the plant and its effect on the state $\mathbf{x}(t)$. If we do not have a very accurate model of the plant (**A**, **b**, **c**), the dynamics of the error are no longer governed by Eqn. (15.31). However, **m** can typically be chosen so that the error system is stable and the error is acceptably small, even with small modeling errors and disturbance inputs.

The selection of **m** can be approached in exactly the same fashion as the selection of **k** is in the control law design. If we specify the desired location of the observer error roots as

$$s = \lambda_1, \lambda_2, \ldots, \lambda_n,$$

the desired observer characteristic equation is

$$(s - \lambda_1)(s - \lambda_2)\cdots(s - \lambda_n) = 0 \tag{15.32b}$$

and one can solve for **m** by comparing coefficients in Eqns (15.32a) and (15.32b). However, as we shall see shortly, this can be done only if the system (15.27) is completely observable.

We can directly use the equations of the control-law design for computing the observer gain matrix **m** if we examine the resemblance between the estimation and control problems. In fact, the two problems are mathematically equivalent. This property is called *duality*. The design of a full-order observer requires the determination of the gain matrix **m** such that (**A** – **mc**) has desired eigenvalues λ_i; $i = 1, 2, \ldots, n$. This is mathematically equivalent to designing a full state-feedback controller for the 'transposed auxiliary system',

$$\dot{\zeta}(t) = \mathbf{A}^T\zeta(t) + \mathbf{c}^T\eta(t) \tag{15.33a}$$

with feedback

$$\eta(t) = \mathbf{m}^T\zeta(t) \tag{15.33b}$$

so that the closed-loop auxiliary system

$$\dot{\zeta}(t) = (\mathbf{A}^T - \mathbf{c}^T\mathbf{m}^T)\zeta(t) \tag{15.34}$$

has eigenvalues λ_i; $i = 1, 2, \ldots, n$.

Since

$$det\ \mathbf{W} = det\ \mathbf{W}^T,$$

one obtains

$$det\ [s\mathbf{I} - (\mathbf{A}^T - \mathbf{c}^T\mathbf{m}^T)] = det\ [s\mathbf{I} - (\mathbf{A} - \mathbf{mc})]$$

i.e., the eigenvalues of $(\mathbf{A}^T - \mathbf{c}^T\mathbf{m}^T)$ are same as the eigenvalues of $(\mathbf{A} - \mathbf{mc})$.

By comparing the characteristic equation of the closed-loop system (15.7) and that of the auxiliary system (15.34), we obtain the duality relations given in Table 15.1 between the control and estimation problems. The control-design relations given by Eqns (15.16) become the observer-design relations if the substitutions of Table 15.1 are made.

Table 15.1 *Duality between control and estimation*

Control	*Estimation*
$\mathbf{A}$	$\mathbf{A}^T$
$\mathbf{b}$	$\mathbf{c}^T$
$\mathbf{k}$	$\mathbf{m}^T$

A necessary and sufficient condition for determination of the observer gain matrix $\mathbf{m}$ for the desired eigenvalues of $(\mathbf{A} - \mathbf{mc})$ is that the auxiliary system (15.33) be completely controllable. The controllability condition for this system is that the rank of

$$[\mathbf{c}^T \quad \mathbf{A}^T\mathbf{c}^T \cdots (\mathbf{A}^T)^{n-1} \quad \mathbf{c}^T]$$

is n. This is the condition for complete observability of the original system defined by Eqns (15.27). This means that a necessary and sufficient condition for estimation of the state of the system defined by Eqns (15.27) is that the system be completely observable.

Example 15.4 We will consider the satellite control system of Example 15.1. The state equation of the plant is

$$\dot{\mathbf{x}} = \mathbf{A}\mathbf{x} + \mathbf{b}u$$

with

$$\mathbf{A} = \begin{bmatrix} 0 & 1 \\ 0 & 0 \end{bmatrix}; \mathbf{b} = \begin{bmatrix} 0 \\ 1 \end{bmatrix}$$

$$x_1 = \theta, \text{ the orientation of the satellite; } x_2 = \dot{\theta}$$

We assume that the orientation θ can be accurately measured from the antenna signal; therefore,

$$y = \mathbf{cx}(t)$$

with

$$\mathbf{c} = [1 \quad 0]$$

Let us design a state observer for the system. We choose the observer to be critically damped with a settling time of 0.4 sec $\left(\dfrac{4}{\zeta\omega_n} = 0.4;\ \zeta\omega_n = 10\right)$. To satisfy these specifications, the observer poles will be placed at $s = -10, -10$.

The transposed auxiliary system is given by

$$\dot{\zeta} = \mathbf{A}^T \zeta + \mathbf{c}^T \eta$$

$$\eta = -\mathbf{m}^T \zeta$$

The desired characteristic equation of the closed-loop auxiliary system is

$$s^2 + a_1 s + a_2 = (s + 10)(s + 10) = s^2 + 20s + 100$$

To apply pole-placement relations given by Eqns (15.16), we compute

$$\mathbf{U}^{-1} = \begin{bmatrix} \mathbf{c}^T & \mathbf{A}^T\mathbf{c}^T \end{bmatrix}^{-1} = \begin{bmatrix} 1 & 0 \\ 0 & 1 \end{bmatrix}$$

$$\mathbf{p}_1 = \begin{bmatrix} 0 & 1 \end{bmatrix} \mathbf{U}^{-1} = \begin{bmatrix} 0 & 1 \end{bmatrix}$$

$$\mathbf{p}_1 \mathbf{A}^T = \begin{bmatrix} 0 & 1 \end{bmatrix} \begin{bmatrix} 0 & 0 \\ 1 & 0 \end{bmatrix} = \begin{bmatrix} 1 & 0 \end{bmatrix}$$

$$\mathbf{P} = \begin{bmatrix} 0 & 1 \\ 1 & 0 \end{bmatrix}$$

$$\left| s\mathbf{I} - \mathbf{A}^T \right| = s^2 = s^2 + \alpha_1 s + \alpha_2$$

$$\Delta(s) = s^2 + 20s + 100$$

$$\mathbf{m}^T = \begin{bmatrix} a_2 - \alpha_2 & a_1 - \alpha_1 \end{bmatrix} \mathbf{P} = \begin{bmatrix} 100 & 20 \end{bmatrix} \begin{bmatrix} 0 & 1 \\ 1 & 0 \end{bmatrix} = \begin{bmatrix} 20 & 100 \end{bmatrix}$$

Therefore,

$$\mathbf{m} = \begin{bmatrix} 20 \\ 100 \end{bmatrix}$$

15.6.2 Effects of the Addition of an Observer on a Closed-Loop System

In the pole-placement design process, we assumed that the actual state $\mathbf{x}(t)$ was available for feedback. In practice, however, the actual state $\mathbf{x}(t)$ may not be measurable; so we will need to design an observer and use the observed state $\hat{\mathbf{x}}(t)$ for feedback. The design process therefore becomes a two-stage process; the first stage being the determination of the feedback gain matrix $\mathbf{k}$ to yield the desired characteristic equation of the closed-loop system, and the second stage being the determination of the observer gain matrix $\mathbf{m}$ to yield the desired observer characteristic equation.

Let us now investigate the effects of the use of the observed state $\hat{\mathbf{x}}(t)$, rather than the actual state $\mathbf{x}(t)$, on the characteristic equation of the closed-loop system.

Consider the completely controllable and completely observable system defined by the equations,

$$\dot{\mathbf{x}} = \mathbf{A}\mathbf{x} + \mathbf{b}u$$
$$y = \mathbf{c}\mathbf{x} \tag{15.35}$$

Suppose we have designed a state-feedback control law

$$u = -\mathbf{k}\mathbf{x} + r \tag{15.36}$$

using the method of Section 15.3. And also suppose we have designed a full-order observer

$$\dot{\hat{\mathbf{x}}} = \mathbf{A}\hat{\mathbf{x}} + \mathbf{b}u + \mathbf{m}(y - \mathbf{c}\hat{\mathbf{x}}) \tag{15.37}$$

using the method described earlier in this section.

For the state-feedback control based on the observed state $\hat{\mathbf{x}}$,

$$u = -\mathbf{k}\hat{\mathbf{x}} + r \tag{15.38}$$

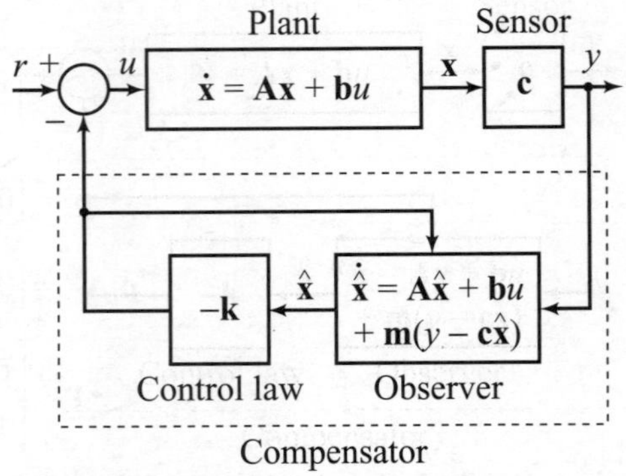

Fig. 15.8 *Combined state-feedback control and state estimation.*

The control system based on combining the state-feedback control law and state observer, has the configuration shown in Fig. 15.8. Note that the number of state variables in the compensator is equal to the order of the embedded observer and hence is equal to the order of the plant. Thus, the order of the overall closed-loop system, when a full-order observer is used in the compensator, is $2n$ for a plant of order n. We are interested in the dynamic behavior of the $2n$th-order system comprising the plant and the compensator. With the control law (15.38) used, the plant dynamics become

$$\dot{\mathbf{x}} = \mathbf{A}\mathbf{x} - \mathbf{b}\mathbf{k}\hat{\mathbf{x}} + \mathbf{b}r = (\mathbf{A} - \mathbf{b}\mathbf{k})\mathbf{x} + \mathbf{b}\mathbf{k}(\mathbf{x} - \hat{\mathbf{x}}) + \mathbf{b}r \tag{15.39}$$

The difference between the actual state $\mathbf{x}$ and observed state $\hat{\mathbf{x}}$ has been defined as the error $\tilde{\mathbf{x}}$:

$$\tilde{\mathbf{x}} = \mathbf{x} - \hat{\mathbf{x}}$$

Substitution of the error vector into Eqn. (15.39) gives

$$\dot{\mathbf{x}} = (\mathbf{A} - \mathbf{b}\mathbf{k})\mathbf{x} + \mathbf{b}\mathbf{k}\,\tilde{\mathbf{x}} + \mathbf{b}r \tag{15.40}$$

Note that the observer error was given by Eqn. (15.31), repeated here:

$$\dot{\tilde{\mathbf{x}}} = (\mathbf{A} - \mathbf{m}\mathbf{c})\tilde{\mathbf{x}} \tag{15.41}$$

Combining Eqns (15.40) and (15.41), we obtain

$$\begin{bmatrix} \dot{\mathbf{x}} \\ \dot{\tilde{\mathbf{x}}} \end{bmatrix} = \begin{bmatrix} \mathbf{A} - \mathbf{b}\mathbf{k} & \mathbf{b}\mathbf{k} \\ \mathbf{0} & \mathbf{A} - \mathbf{m}\mathbf{c} \end{bmatrix} \begin{bmatrix} \mathbf{x} \\ \tilde{\mathbf{x}} \end{bmatrix} \tag{15.42}$$

Equation (15.42) describes the dynamics of the $2n$-dimensional system of Fig. 15.8. The characteristic equation for the system is

$$|s\mathbf{I} - (\mathbf{A} - \mathbf{b}\mathbf{k})|\,|s\mathbf{I} - (\mathbf{A} - \mathbf{m}\mathbf{c})| = 0$$

In other words, ***the closed-loop poles of the observed state-feedback control system consist of the poles due to the pole-placement design alone and the poles due to the observer design alone.***

This means that the pole-placement design and the observer design are independent of each other. They can be designed separately and combined to form the observed state feedback control system. If the order of the plant is n, then the observer is also of nth order (if full-order state observer is used) and the resulting characteristic equation for the entire closed-loop system becomes of order $2n$.

The poles of the observer are usually chosen so that the observer response is much faster than the system response. A rule of thumb is to choose an observer response atleast two to five times faster than the system response. This is to ensure a faster decay of estimation errors compared with the desired dynamics, thus causing the closed-loop poles generated by state feedback to determine the total response.

15.7 DIGITAL CONTROL DESIGN USING STATE FEEDBACK

Most of the industrial processes to be controlled are continuous-time systems, and today most of the complex industrial processes are under computer control. The interface of continuous-time plants with digital computers was discussed at length in Chapter 13. In that chapter, we also learnt that there are two approaches to carrying out digital control design. The first approach uses the classical methods—Bode plots, root locus plots, etc., to design an analog controller and then transforms the analog controller into a digital one. This approach is based on the fact that a digital system with high sampling rate approximates to an analog system. However, the use of high sampling rates wastes computer power and can lead to the problem of arithmetic precision, etc. One is therefore driven to find methods of design which take into account the sampling process. The alternative approach is to first transform analog plants into sampled-data plants and then carry out the design using digital techniques. The first approach performs discretization *after* design and the second approach performs discretization *before* design. The sampling rates are generally lower in the second approach than for discretized analog design.

These observations of Chapter 13 directly apply to control design using state variable feedback. So far in this chapter, we have focused on the continuous-time plants and design of analog controllers. The analog controller can be digitally implemented using A/D, D/A, and computer in the feedback loop. We now present techniques for direct digital design based on sampled-data plant models.

Fortunately, the analysis and design method using state feedback, involved in the second approach—direct digital design, parallels the first approach—analog design and digital implementation. Because of this strong analogy, our discussion here on second approach will be brief. For more details, refer [155].

We have seen in Chapter 14 that linear time-invariant SISO sampled-data systems can be represented by a first-order matrix difference equation of the form (refer to Eqns (14.104)

$$\mathbf{x}(k+1) = \mathbf{F}\mathbf{x}(k) + \mathbf{g}u(k) \tag{15.43a}$$

The $n \times 1$ vector $\mathbf{x}(k)$ is the system state, and the $n \times n$ matrix $\mathbf{F}$ is the system matrix that governs how the system internal dynamics will operate. The scalar input $u(k)$ is connected to the system state through $n \times 1$ input matrix $\mathbf{g}$.

The output equation can be expressed as

$$y(k) = \mathbf{c}\mathbf{x}(k) \tag{15.43b}$$

The state $\mathbf{x}(k)$ is connected to the scalar output $y(k)$ through the $1 \times n$ output matrix $\mathbf{c}$.

Much of the algebra for state-space design of sampled-data systems is the same as for the continuous-time case discussed in earlier sections of this chapter. Figure 15.9 shows a structure of state-feedback control for tracking the constant signal r.

The state feedback control law is

$$u(k) = r - \mathbf{kx}; \quad \mathbf{k} = \begin{bmatrix} k_1 & k_2 & \cdots & k_n \end{bmatrix} \tag{15.44a}$$

By substituting Eqn. (15.44a) into Eqn. (15.43a), the closed-loop system is represented by the state equation,

$$\mathbf{x}(k+1) = (\mathbf{F} - \mathbf{gk})\mathbf{x}(k) + \mathbf{g}\,r \tag{15.44b}$$

From Section 14.11.3, it follows that the closed-loop system (15.44b) is stable if all the eigenvalues of the matrix $(\mathbf{F}-\mathbf{gk})$, given by the characteristic equation,

$$\left|z\mathbf{I} - (\mathbf{F} - \mathbf{gk})\right| = 0 \tag{15.44c}$$

have magnitudes less than uiny.

It will be shortly shown that a matrix $\mathbf{k}$ exists that can give an arbitrary selection of the eigenvalues of $(\mathbf{F} - \mathbf{gk})$, i.e.,

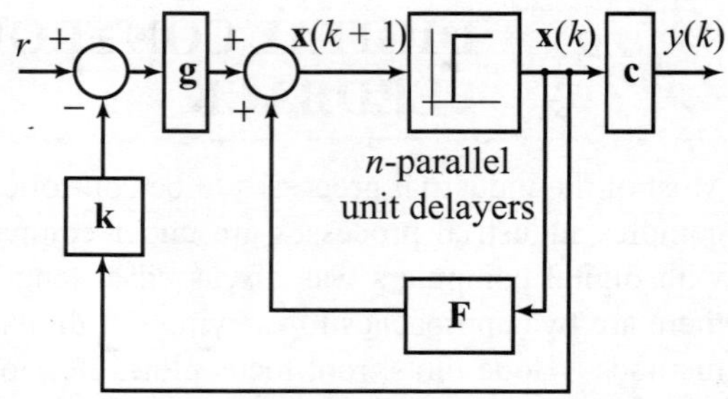

Fig. 15.9 *State-feedback control structure for sampled-data systems*

$$\left|z\mathbf{I} - (\mathbf{F} - \mathbf{gk})\right| = \text{desired characteristic polynomial } \Delta(z) \tag{15.44d}$$

provided the system is controllable.

The system is controllable if the controllability matrix (refer to Eqn. (14.107))

$$\mathbf{U} = \begin{bmatrix} \mathbf{g} & \mathbf{Fg} & \cdots & \mathbf{F}^{n-1}\mathbf{g} \end{bmatrix} \tag{15.45}$$

is of rank n.

Given the desired characteristic polynomial

$$\begin{aligned} \Delta(z) &= (z - \lambda_1)(z - \lambda_2) \cdots (z - \lambda_n) \\ &= z^n + a_1 z^{n-1} + \cdots + a_{n-1} z + a_n \end{aligned} \tag{15.46a}$$

the required elements of $\mathbf{k}$ are obtained from the following equations (refer to Eqn. (15.16))

$$\left|z\mathbf{I} - \mathbf{F}\right| = z^n + \alpha_1 z^{n-1} + \cdots + \alpha_{n-1} z + \alpha_n \tag{15.46b}$$

$$\mathbf{k} = \begin{bmatrix} a_n - \alpha_n & a_{n-1} - \alpha_{n-1} & \cdots & a_1 - \alpha_1 \end{bmatrix}\mathbf{P} \tag{15.46c}$$

$$\mathbf{P} = \begin{bmatrix} \mathbf{p}_1 \\ \mathbf{p}_1\mathbf{F} \\ \vdots \\ \mathbf{p}_1\mathbf{F}^{n-1} \end{bmatrix}; \mathbf{p}_1 = \begin{bmatrix} 0 & 0 & \cdots & 0 & 1 \end{bmatrix}\mathbf{U}^{-1} \tag{15.46d}$$

Direct substitution of control law (15.44a) into the system equation (15.43a) gives

$$\mathbf{x}(k+1) = (\mathbf{F} - \mathbf{gk})\,\mathbf{x}(k) + \mathbf{g}\,r \tag{15.47a}$$

As $k \to \infty$, $\mathbf{x}(k+1) - \mathbf{x}(k) \to \mathbf{0}$; $\mathbf{x}(k+1) = \mathbf{x}(k) = \mathbf{x}_s$. At steady state, Eqn. (15.47a) becomes

$$\mathbf{x}_s = (\mathbf{F} - \mathbf{gk})\mathbf{x}_s + \mathbf{g}\,r$$

which gives

$$\mathbf{x}_s = -(\mathbf{F} - \mathbf{gk} - \mathbf{I})^{-1}\mathbf{g}\,r \tag{15.47b}$$

or

$$y_s = \mathbf{c}[-(\mathbf{F} - \mathbf{gk} - \mathbf{I})^{-1}]\mathbf{g}\,r \tag{15.47c}$$

A discrete-time state observer has the form,

$$\hat{\mathbf{x}}(k+1) = \mathbf{F}\hat{\mathbf{x}}(k) + \mathbf{g}u(k) + \mathbf{m}(y(k) - \mathbf{c}\hat{\mathbf{x}}(k)) \tag{15.48a}$$

where $\hat{\mathbf{x}}$ is the state estimate, and

$$\mathbf{m} = \begin{bmatrix} m_1 & m_2 & \cdots & m_n \end{bmatrix}^T \tag{15.48b}$$

are the observer gains.

The error equation,

$$\tilde{\mathbf{x}}(k+1) = (\mathbf{F} - \mathbf{mc})\tilde{\mathbf{x}}(k);\ \tilde{\mathbf{x}} = \mathbf{x} - \hat{\mathbf{x}} \tag{15.48c}$$

can be given any arbitrary dynamics, provided that the system (15.43) is observable, which requires that the observability matrix (refer to Eqn. (14.108))

$$\mathbf{V} = \begin{bmatrix} \mathbf{c} \\ \mathbf{cF} \\ \vdots \\ \mathbf{cF}^{n-1} \end{bmatrix} \tag{15.49}$$

be of rank n.

To introduce integrator in the control scheme shown in Fig. 15.9, we augment the plant state vector $\mathbf{x}$ with the 'integral state' x_{n+1} that integrates the difference between the output $y(k)$ and the constant reference input r (Fig. 15.10). The 'integral state' x_{n+1} is defined by

$$x_{n+1}(k) = x_{n+1}(k-1) + r - y(k)$$

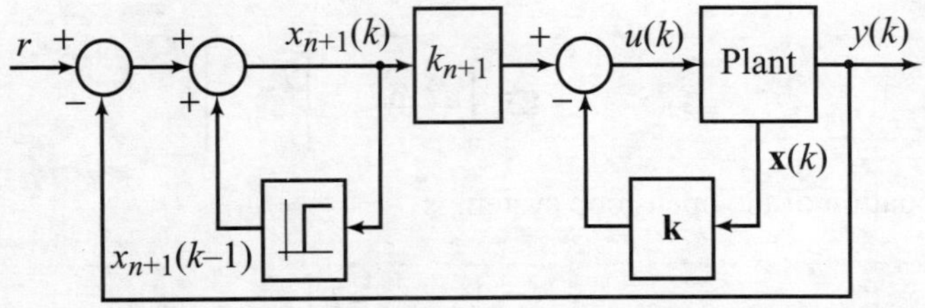

Fig. 15.10 *State feedback with integral control for sampled-data systems*

This equation may be rewritten as

$$\begin{aligned} x_{n+1}(k+1) &= x_{n+1}(k) + r - y(k+1) \\ &= x_{n+1}(k) - \mathbf{c}[\mathbf{F}\mathbf{x}(k) + \mathbf{g}u(k)] + r \\ &= -\mathbf{c}\mathbf{F}\mathbf{x}(k) + x_{n+1}(k) - \mathbf{c}\mathbf{g}u(k) + r \end{aligned} \tag{15.50}$$

From Eqns (15.43) and (15.50), we obtain

$$\begin{bmatrix} \mathbf{x}(k+1) \\ x_{n+1}(k+1) \end{bmatrix} = \begin{bmatrix} \mathbf{F} & \mathbf{0} \\ -\mathbf{c}\mathbf{F} & 1 \end{bmatrix} \begin{bmatrix} \mathbf{x}(k) \\ x_{n+1}(k) \end{bmatrix} + \begin{bmatrix} \mathbf{g} \\ -\mathbf{c}\mathbf{g} \end{bmatrix} u(k) + \begin{bmatrix} \mathbf{0} \\ 1 \end{bmatrix} r \tag{15.51a}$$

or

$$\bar{\mathbf{x}}(k+1) = \bar{\mathbf{F}}\bar{\mathbf{x}}(k) + \bar{\mathbf{g}}u(k) + \begin{bmatrix} \mathbf{0} \\ 1 \end{bmatrix} r \tag{15.51b}$$

The control

$$u(k) = -\mathbf{k}\mathbf{x}(k) - k_{n+1}x_{n+1}(k) = -\bar{\mathbf{k}}\bar{\mathbf{x}}(k) \tag{15.52}$$

may be obtained from the equations,

$$|z\mathbf{I} - (\bar{\mathbf{F}} - \bar{\mathbf{g}}\bar{\mathbf{k}})| = \text{desired } (n+1)^{\text{th}} \text{-order characteristic polynomial} \tag{15.53}$$

Example 15.5 Consider the problem of attitude control of a rigid satellite. A state variable model of the plant is (refer to Example (15.1))

$$\dot{\mathbf{x}} = \mathbf{A}\mathbf{x} + \mathbf{b}u = \begin{bmatrix} 0 & 1 \\ 0 & 0 \end{bmatrix} \mathbf{x} + \begin{bmatrix} 0 \\ 1 \end{bmatrix} u$$

where $x_1 = \theta$ is the attitude angle and u is the system input.

The discrete-time description of the plant (assuming that the input u is applied through a zero-order hold (ZOH)) is given below (refer to Eqn. (14.104)):

$$\mathbf{x}(k+1) = \mathbf{F}\mathbf{x}(k) + \mathbf{g}u(k) \tag{15.54}$$

where

$$\mathbf{F} = e^{\mathbf{A}T} = \begin{bmatrix} 1 & T \\ 0 & 1 \end{bmatrix}$$

$$\mathbf{g} = \int_0^T e^{\mathbf{A}\tau}\mathbf{b}\,d\tau = \begin{bmatrix} T^2/2 \\ T \end{bmatrix}$$

The characteristic equation of the open-loop system is

$$|z\mathbf{I} - \mathbf{F}| = \begin{vmatrix} z-1 & -T \\ 0 & z-1 \end{vmatrix} = (z-1)^2 = 0$$

It is desired to maintain the attitude of the satellite at $\theta = 0$ (i.e., $r = 0$). This is a regulator problem. With the control law

$$u(k) = -\mathbf{k}\mathbf{x}(k) = -\begin{bmatrix} k_1 & k_2 \end{bmatrix}\mathbf{x}(k)$$

the closed-loop system becomes

$$\mathbf{x}(k+1) = (\mathbf{F} - \mathbf{g}\mathbf{k})\mathbf{x}(k)$$

The characteristic equation of the closed-loop system is

$$|z\mathbf{I} - (\mathbf{F} - \mathbf{g}\mathbf{k})| = z^2 + (Tk_2 + (T^2/2)k_1 - 2)z + (T^2/2)k_1 - Tk_2 + 1 = 0 \qquad (15.55a)$$

We assume that $T = 0.1$ sec, and the desired characteristic roots of the closed-loop system are $z_{1.2} = 0.875 \angle \pm 17.9°$.

Note that these roots correspond to $\zeta = 0.5$, and $\omega_n = 3.6$ (refer to Eqn. (15.47b)):

$$z_{1,2} = e^{-\zeta\omega_n T \pm j\omega_n T\sqrt{1-\zeta^2}}$$

The desired characteristic equation is then (approximately)

$$z^2 - 1.6z + 0.70 = 0 \qquad (15.55b)$$

Matching coefficients in Eqns (15.55a) and (15.55b), we obtain

$$k_1 = 10, \; k_2 = 3.5$$

Example 15.6 Consider the problem of digital control of a plant described by the transfer function,

$$G(s) = \frac{1}{s+3}$$

Discretization of the plant model gives (refer to Chapter 13)

$$G_{h0}G(z) = \frac{Y(z)}{U(z)} = \mathscr{Z}\left[\left(\frac{1-e^{-sT}}{s}\right)\left(\frac{1}{s+3}\right)\right]$$

$$= (1 - z^{-1})\,\mathscr{Z}\left[\frac{1}{s(s+3)}\right] = \frac{1}{3}\left(\frac{1-e^{-3T}}{z-e^{-3T}}\right)$$

For a sampling interval $T = 0.1$ sec,

$$G_{h0}G(z) = \frac{0.0864}{z - 0.741}$$

The difference equation model of the plant is

$$y(k+1) = 0.741y(k) + 0.0864u(k)$$

The plant has a constant reference command signal. We wish to design a PI control algorithm that results in system response characteristics: $\zeta = 0.5$, $\omega_n = 5$. This is equivalent to asking for the closed-loop poles at

$$z_{1,2} = e^{-\zeta\omega_n T} e^{\pm j\omega_n T\sqrt{1-\zeta^2}} = 0.7788\angle \pm 24.82° = 0.7068 \pm j0.3269$$

The desired characteristic equation is, therefore,

$$\begin{aligned}&(z - 0.7068 - j0.3269)(z - 0.7068 + j0.3269)\\&= z^2 - 1.4136z + 0.6065 = z^2 + a_1 z + a_2 = 0\end{aligned}$$

Augmenting the plant state $x_1(k) = y(k)$ with the 'integral state' $x_2(k)$ defined by

$$x_2(k) = x_2(k-1) + r - y(k),$$

we obtain

$$\begin{bmatrix} x_1(k+1) \\ x_2(k+1) \end{bmatrix} = \begin{bmatrix} 0.741 & 0 \\ -0.741 & 1 \end{bmatrix} + \begin{bmatrix} 0.0864 \\ -0.0864 \end{bmatrix} u(k) + \begin{bmatrix} 0 \\ 1 \end{bmatrix} r$$

or

$$\bar{\mathbf{x}}(k+1) = \bar{\mathbf{F}}\bar{\mathbf{x}}(k) + \bar{\mathbf{g}}u(k) + \begin{bmatrix} 0 \\ 1 \end{bmatrix} r$$

Following the pole-placement relations given by Eqns (15.46), we get

$$\mathbf{U}^{-1} = \begin{bmatrix} \bar{\mathbf{g}} & \bar{\mathbf{F}}\bar{\mathbf{g}} \end{bmatrix}^{-1} = \begin{bmatrix} 0.0864 & 0.064 \\ -0.0864 & -0.15 \end{bmatrix}^{-1} = \begin{bmatrix} 20.1873 & 8.6133 \\ -11.628 & -11.628 \end{bmatrix}$$

$$\begin{aligned}\mathbf{p}_1 &= \begin{bmatrix} 0 & 1 \end{bmatrix}\mathbf{U}^{-1} = \begin{bmatrix} -11.628 & -11.628 \end{bmatrix}\\ \mathbf{p}_1\bar{\mathbf{F}} &= \begin{bmatrix} 0 & -11.628 \end{bmatrix}\end{aligned}$$

$$\mathbf{P} = \begin{bmatrix} \mathbf{p}_1 \\ \mathbf{p}_1\bar{\mathbf{F}} \end{bmatrix}$$

$$|z\mathbf{I} - \bar{\mathbf{F}}| = z^2 - 1.741z + 0.741 = z^2 + \alpha_1 z + \alpha_2$$

$$\begin{aligned}\mathbf{k} &= \begin{bmatrix} a_2 - \alpha_2 & a_1 - \alpha_1 \end{bmatrix}\mathbf{P} = \begin{bmatrix} -0.1345 & 0.328 \end{bmatrix}\mathbf{P}\\ &= \begin{bmatrix} 1.564 & -2.25 \end{bmatrix}\end{aligned}$$

15.8 OPTIMAL CONTROL SYSTEMS

In the control design procedures discussed so far, compensators were designed to satisfy specified requirements for steady-state error, transient response, stability margins, or closed-loop pole locations. We have observed that various trade-offs have to be made because of the limitations of the design techniques. None of the techniques result in best possible performance.

In this section, we introduce optimization-based techniques that result in an optimal action. The word "optimal" intuitively means doing a job in the best of all possible ways. Before a more definitive meaning can be attached, the "job" must be defined, a mathematical scale must be established for handling what "best" means, and the "possible" alternatives must be spelled out. Without these qualifiers, a claim that a system is optimal is really meaningless.

The task to be performed is to control the dynamics of a controllable system described by the state equations,

$$\dot{\mathbf{x}}(t) = \mathbf{A}\mathbf{x}(t) + \mathbf{b}u(t) \tag{15.56a}$$

It is assumed that all states **x** are available as output measurements. The states are to be regulated to the steady-state solution. In *regulator problems* we have considered earlier (refer to Example 15.1), the control law is of the form,

$$u(t) = -\,\mathbf{k}\mathbf{x}(t) \tag{15.56b}$$

The feedback system obeys the equation,

$$\dot{\mathbf{x}}(t) = (\mathbf{A} - \mathbf{b}\mathbf{k})\mathbf{x}(t) \tag{15.56c}$$

$\dot{\mathbf{x}}(t) = \mathbf{0}$ in steady-state provided that the feedback system is stable. This means that as $t \to \infty$, the state $\mathbf{x}(t)$ of the regulator (15.56c) must approach the origin of the state space. Therefore, at any time t, the state $\mathbf{x}(t)$ represents a deviation of the state dynamics from this desired value, i.e., $\mathbf{x}(t)$ is an error vector. It is desired that the state $\mathbf{x}(t)$, for all possible initial states, asymptotically decays to zero as $t \to \infty$.

The dynamics (speed of response, overshoot, etc.) of the decay, as we have seen earlier in this chapter, can be manipulated by locating the eigenvalues of $(\mathbf{A} - \mathbf{b}\mathbf{k})$ at appropriate locations in the complex plane. Here in this section, a somewhat different approach to design is taken. The performance of the system (the decay of state $\mathbf{x}(t)$ to zero as $t \to \infty$) is measured with a single scalar quantity—the *performance index*. The free parameters **k** in the state-feedback configuration that optimize the performance index are determined.

Selection of an appropriate performance index is as much a part of the design process as the optimization of the index. We know that the performance of a control system can be adequately specified in terms of settling time, peak overshoot, and steady-state error. An index which is a function of these performance parameters may be selected and optimization (minimization) of this index should lead to an optimal control strategy. Such an index looks quite reasonable; however, it is not trackable analytically. A compromise must be made between specifying a performance index which includes all the desired system characteristics, and a performance index which can be minimized with a reasonable amount of computation.

In the following, we present a performance index which includes the desired system characteristics and in addition has good mathematical trackability. This index involves integrating square of system error, and is therefore called *quadratic performance index*:

$$J = \int_0^\infty \left[\mathbf{x}^T(t)\mathbf{x}(t)\right] dt$$

$$= \int_0^\infty \left[x_1^2(t) + x_2^2(t) + \cdots + x_n^2(t)\right] dt$$

where $\mathbf{x}(t)$, a deviation of the system state from the origin of the state space (equilibrium point), is the system error at time t.

In practical systems, the regulation of all the states of the system is not equally important. To be more general,

$$J = \int_0^\infty \left[\mathbf{x}^T(t)\mathbf{Q}\mathbf{x}(t) \right] dt \tag{15.57a}$$

with $\mathbf{Q}$, a $n \times n$ real, symmetric, positive definite constant matrix (refer to Chapter 14 for matrix definitions), can be used as a performance measure (A positive semidefinite matrix $\mathbf{Q}$ may also be selected for some control problems (reference [155])).

The simplest form of $\mathbf{Q}$ one can use is a diagonal matrix:

$$\mathbf{Q} = \begin{bmatrix} q_1 & 0 & \cdots & 0 \\ 0 & q_2 & \cdots & 0 \\ \vdots & \vdots & & \vdots \\ 0 & 0 & \cdots & q_n \end{bmatrix} \tag{15.57b}$$

The ith entry of $\mathbf{Q}$ represents the weight the designer places on the constraint on the ith variable. The larger the value of q_i relative to the other values of q, the more control effort is spent to regulate $x_i(t)$.

The design of systems carried out to meet time response specifications through minimization of performance index (15.57a), may be unsatisfactory because it may lead to excessively large control signals. A more realistic solution to the problem is reached if the performance specifications also account for the constraint: $|u(t)| \le M$, for some constant M.

In the pole-placement design method, the control effort constraints cannot be embedded in the design procedure. Trial-and-error on the selection of poles of the closed-loop system is done and design for each selection is carried out; the design that meets the control effort constraints becomes the outcome of the design procedure. The optimal control techniques, however, allow for the inclusion of control effort constraints in the performance index in the following way:

$$J = \int_0^\infty \left[\mathbf{x}^T(t)\mathbf{Q}\mathbf{x}(t) + Ru^2(t) \right] dt \tag{15.57c}$$

where R, a positive constant, is called the weighting factor. By adjusting R, we can weigh the relative importance of regulation of $x_i(t)$ and the control effort. By increasing R, i.e., by giving sufficient weight to control effort, the amplitude of the control signal which minimizes the overall performance index may be kept within practical bounds, although at the expense of poor regulation of $\mathbf{x}(t)$ to the origin.

***Compared to state feedback design procedure based on pole placement, in the quadratic optimal control design procedure, the trial-and-error in the selection of the desired closed-loop poles reduces to the trial-and-error in the selection of* $\mathbf{Q}$ *and* R.** For each selection of $\{\mathbf{Q}, R\}$, state feedback matrix $\mathbf{k}$ that minimizes the performance index is calculated; the eigenvalues of the matrix $(\mathbf{A} - \mathbf{bk})$ are the corresponding closed-loop poles. In practice, a faster convergence to acceptable performance is achieved this way. This is followed by observer-based state feedback design according to the procedure of Section 15.6, if the assumption of accessibility of each state variable is not satisfied.

For a class of optimal control problems, we have defined the "job" to be performed, and also specified a mathematical scale for handling what "best" means. ***The solution to the optimization problem thus formulated, is the best with respect to the specified mathematical criterion, but, unfortunately, not***

necessarily the best with respect to all alternatives of design discussed so far in the book. The reason is that whether the mathematical performance criterion truly represents the performance requirements imposed on a physical system, is not guaranteed. Selection of **Q** and R is a trial-and-error exercise. Therefore, the optimal control design method gives an additional tool, though an important one, in the toolkit of the control engineers.

15.8.1 Servo Problem/Shifted Regulator Problem

The command-following control structure of Fig.15.2 controls the plant

$$\dot{\mathbf{x}}(t) = \mathbf{A}\,\mathbf{x}(t) + \mathbf{b}\,u(t) \tag{15.58a}$$

with the control law

$$u(t) = -\mathbf{k}\,\mathbf{x}(t) + r \tag{15.58b}$$

The feedback system obeys the equation

$$\dot{\mathbf{x}}(t) = (\mathbf{A} - \mathbf{bk})\mathbf{x}(t) + \mathbf{b}r \tag{15.58c}$$

For a constant r, $\dot{\mathbf{x}}(t) = \mathbf{0}$ in steady-state provided that the system is stable. This means that steady-state solution $\mathbf{x}_s$ must satisfy the equation,

$$\mathbf{x}_s = -(\mathbf{A} - \mathbf{bk})^{-1}\mathbf{b}r \tag{15.59}$$

Therefore, at any time t, $(\mathbf{x}(t) - \mathbf{x}_s)$ represents a deviation of the state dynamics from the desired value, i.e., $\mathbf{x}(t) - \mathbf{x}_s = \tilde{\mathbf{x}}$ is an error vector.

We can formulate this servo problem as a shifted regulator problem by shifting the origin of the state space to the steady-state point $\mathbf{x}_s$. The resulting regulator system has state $\tilde{\mathbf{x}}(t)$ and it is desired that $\tilde{\mathbf{x}}(t)$ decays to the origin of the state space asymptotically.

15.8.2 Solution of Regulator Problem

For want of space, the solution of the linear quadratic optimal control problem is not being developed in this chapter; the detailed account of the subject is given in the companion book [155].

A method suitable for computer calculation is stated without proof in the following paragraph. Consider the uncompensated single-input controllable system,

$$\dot{\mathbf{x}} = \mathbf{Ax} + \mathbf{b}u, \tag{15.60a}$$

with feedback

$$u = -\mathbf{kx}$$

The performance index is

$$J = \frac{1}{2}\int_0^\infty (\mathbf{x}^T\mathbf{Q}\mathbf{x} + Ru^2)dt; \ \mathbf{Q} \text{ is positive definite} \tag{15.61}$$

The index is minimized when

$$\mathbf{k} = R^{-1}\mathbf{b}^T\mathbf{P} \tag{15.62}$$

The $(n \times n)$ symmetric, positive definite matrix $\mathbf{P}$ is determined from the solution of the equation,

$$\mathbf{A}^T\mathbf{P} + \mathbf{PA} - \mathbf{Pb}R^{-1}\mathbf{b}^T\mathbf{P} + \mathbf{Q} = \mathbf{0} \tag{15.63}$$

Equation (15.63) can be easily programmed for a computer or solved using MATLAB (refer to MATLAB Module 15 in the Appendix). Equation (15.63) is often called the *Riccati equation.*

15.9 OPTIMAL DIGITAL CONTROL

In the following, we give the formulation and solution (without proof) for the design of optimal controllers for discrete-time systems. Our coverage is very brief because of the strong analogy between the discrete-time and continuous-time cases.

Consider the discretized model of the given controllable plant:

$$\mathbf{x}(k+1) = \mathbf{F}\,\mathbf{x}(k) + \mathbf{g}\,u(k) \tag{15. 64a}$$

with feedback

$$u(k) = -\,\mathbf{k}\,\mathbf{x}(k) \tag{15.64b}$$

The performance index is

$$J = \frac{1}{2}\sum_{k=0}^{\infty}[\mathbf{x}^T(k)\mathbf{Q}\mathbf{x}(k) + Ru^2(k)];\mathbf{Q} \text{ is positive definite} \tag{15.65}$$

The index is minimized when [155]

$$\mathbf{k} = (R + \mathbf{g}^T\mathbf{P}\mathbf{g})^{-1}\mathbf{g}^T\mathbf{P}\mathbf{F} \tag{15.66}$$

The $(n \times n)$ symmetric positive definite matrix $\mathbf{P}$ is determined from the solution of the equation,

$$\mathbf{F}^T\mathbf{P}\mathbf{F} - \mathbf{P} + \mathbf{Q} - \mathbf{F}^T\mathbf{P}\mathbf{g}(R + \mathbf{g}^T\mathbf{P}\mathbf{g})^{-1}\mathbf{g}^T\mathbf{P}\mathbf{F} = \mathbf{0} \tag{15.67}$$

This equation is often called the *discrete Riccati equation.*

Example 15.7 Consider feedback system with the plant

$$\dot{\mathbf{x}}(t) = \mathbf{A}\mathbf{x}(t) + \mathbf{b}u(t)$$

$$\mathbf{A} = \begin{bmatrix} 0 & 1 \\ 0 & 0 \end{bmatrix}; \quad \mathbf{b} = \begin{bmatrix} 0 \\ 1 \end{bmatrix}$$

and the control signal

$$u(t) = -\,\mathbf{k}\,\mathbf{x}(t)$$

Determine the optimal gain matrix $\mathbf{k}$ such that the following performance index is minimized:

$$J = \frac{1}{2}\int_0^{\infty}(\mathbf{x}^T\mathbf{Q}\mathbf{x} + u^2)\,dt\ ; \quad \mathbf{Q} = \begin{bmatrix} 1 & 0 \\ 0 & 1 \end{bmatrix}$$

Solution: Let us solve equation (15.63). Note that $R = 1$, and solution **P** is required to be symmetric positive definite, Equation (15.63) can be written as

$$\begin{bmatrix} 0 & 0 \\ 1 & 0 \end{bmatrix} \begin{bmatrix} p_{11} & p_{12} \\ p_{12} & p_{22} \end{bmatrix} + \begin{bmatrix} p_{11} & p_{12} \\ p_{12} & p_{22} \end{bmatrix} \begin{bmatrix} 0 & 1 \\ 0 & 0 \end{bmatrix}$$

$$- \begin{bmatrix} p_{11} & p_{12} \\ p_{12} & p_{22} \end{bmatrix} \begin{bmatrix} 0 \\ 1 \end{bmatrix} [1] \begin{bmatrix} 0 & 1 \end{bmatrix} \begin{bmatrix} p_{11} & p_{12} \\ p_{12} & p_{22} \end{bmatrix} + \begin{bmatrix} 1 & 0 \\ 0 & 1 \end{bmatrix} = \begin{bmatrix} 0 & 0 \\ 0 & 0 \end{bmatrix}$$

Simplifying we get

$$\begin{bmatrix} 0 & 0 \\ p_{11} & p_{12} \end{bmatrix} + \begin{bmatrix} 0 & p_{11} \\ 0 & p_{12} \end{bmatrix} - \begin{bmatrix} p_{12}^2 & p_{12}\,p_{22} \\ p_{12}\,p_{22} & p_{22}^2 \end{bmatrix} + \begin{bmatrix} 1 & 0 \\ 0 & 1 \end{bmatrix} = \begin{bmatrix} 0 & 0 \\ 0 & 0 \end{bmatrix}$$

from which we obtain the following three equations:

$$1 - p_{12}^2 = 0$$

$$p_{11} - p_{12}\,p_{22} = 0$$

$$1 + 2p_{12} - p_{22}^2 = 0$$

Solving for p_{11}, p_{12}, and p_{22}, requiring **P** to be positive definite, we get

$$\mathbf{P} = \begin{bmatrix} p_{11} & p_{12} \\ p_{12} & p_{22} \end{bmatrix} = \begin{bmatrix} \sqrt{3} & 1 \\ 1 & \sqrt{3} \end{bmatrix}$$

The optimal gain matrix

$$\mathbf{k} = R^{-1}\mathbf{b}^T\mathbf{P} = [1] \begin{bmatrix} 0 & 1 \end{bmatrix} \begin{bmatrix} \sqrt{3} & 1 \\ 1 & \sqrt{3} \end{bmatrix} = \begin{bmatrix} 1 & \sqrt{3} \end{bmatrix}$$

15.10 REVIEW EXAMPLES

Review Example 15.1 In the following example, we consider the problem of design of a servo system when the plant has integrating property. General configuration of such a system is shown in Fig. 15.11. Here we assumed that $y = x_1$ (by a proper choice of a set of state variables, it is possible to choose the output equal to one of the state variables; refer to Chapter 14).

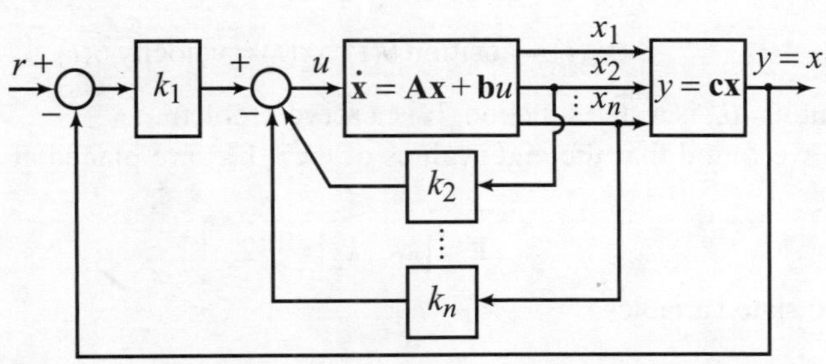

Fig. 15.11 *Design of a type-1 servo system when the plant has an integrator*

The state feedback control law for the configuration of Fig. 15.11 is

$$u = k_1(r - x_1) - k_2x_2 - \cdots - k_nx_n$$
$$= -\mathbf{kx} + k_1r;\ \mathbf{k} = [k_1\ \ k_2\ \ \cdots\ \ k_n] \quad (15.68)$$

For a constant reference input r, the desired steady state is

$$\mathbf{x}_s = [r\ \ 0\ \ 0 \cdots 0]^T \quad (15.69)$$

which is a non-null state.

As the plant has integrating property, the steady-state value u_s of the input u must be zero (otherwise the output cannot stay constant).

The state error vector

$$\tilde{\mathbf{x}}(t) = \mathbf{x}(t) - \mathbf{x}_s \quad (15.70a)$$

Differentiating both sides, we get

$$\dot{\tilde{\mathbf{x}}}(t) = \dot{\mathbf{x}}(t) - \dot{\mathbf{x}}_s = \mathbf{Ax}(t) + \mathbf{b}u(t) - \mathbf{Ax}_s - \mathbf{b}u_s$$

Since $u_s = 0$, we have

$$\dot{\tilde{\mathbf{x}}}(t) = \mathbf{A}\tilde{\mathbf{x}}(t) + \mathbf{b}u(t) \quad (15.70b)$$

In terms of the state error vector $\tilde{\mathbf{x}}$, the state-feedback control law (15.68) becomes

$$u(t) = -\mathbf{k}\tilde{\mathbf{x}}(t) \quad (15.71)$$

The dynamics of $\tilde{\mathbf{x}}$ is thus given by

$$\dot{\tilde{\mathbf{x}}}(t) = (\mathbf{A} - \mathbf{bk})\tilde{\mathbf{x}}(t) \quad (15.72)$$

We shall design the servo system such that the eigenvalues of ($\mathbf{A} - \mathbf{bk}$) are located at desired positions. The designed system will be an asymptotically stable system: as $t \to \infty$, $\tilde{\mathbf{x}} \to \mathbf{0}$, ($\mathbf{x} \to [r\ 0\ \cdots\ 0]$), $y = x_1 \to r$ and $u \to 0$.

As an example, consider the attitude control system for a rigid satellite. The plant equations are (refer to Example 15.1)

$$\dot{\mathbf{x}} = \mathbf{Ax} + \mathbf{b}u$$
$$y = \mathbf{cx}$$

where

$$\mathbf{A} = \begin{bmatrix} 0 & 1 \\ 0 & 0 \end{bmatrix};\ \mathbf{b} = \begin{bmatrix} 0 \\ 1 \end{bmatrix};\ \mathbf{c} = [1\ \ 0]$$

$$x_1(t) = \text{position } \theta(t);\ x_2(t) = \text{velocity } \dot{\theta}(t)$$

The reference input $r = \theta_r$, is a step function. It is a servo problem.

In Example 15.1, we found that the eigenvalues of ($\mathbf{A} - \mathbf{bk}$) are placed at the desired locations $-4 \pm j4$ when

$$\mathbf{k} = [k_1\ \ k_2] = [32\ \ 8]$$

In terms of the shifted state variables,

$$\tilde{x}_1 = x_1 - \theta_r,\ \tilde{x}_2 = x_2$$

the control law is

$$u = -k_1\tilde{x}_1 - k_2\tilde{x}_2$$

In terms of original state variables,

$$u = -k_1 x_1 - k_2 x_2 + k_1\theta_r = -\mathbf{kx} + k_1\theta_r$$

Figure 15.12 shows a configuration for attitude control of the satellite.

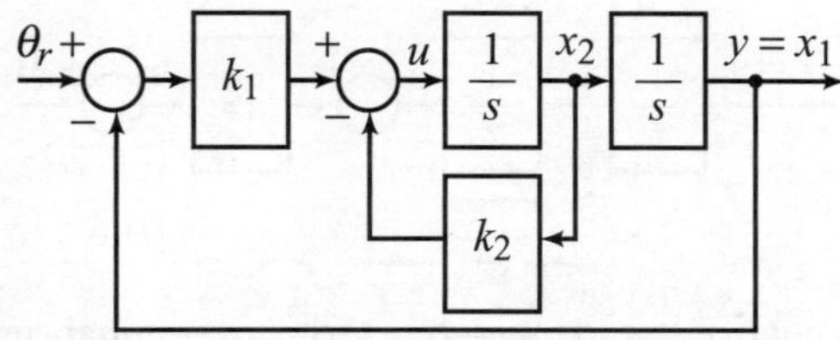

Fig. 15.12 *Attitude control of a satellite*

Review Example 15.2

The system considered in this example is the attitude control system for a rigid satellite. The plant equations are (refer to Example 15.5)

$$\mathbf{x}(k+1) = \mathbf{Fx}(k) + \mathbf{g}u(k)$$

where

$$\mathbf{F} = \begin{bmatrix} 1 & T \\ 0 & 1 \end{bmatrix}; \mathbf{g} = \begin{bmatrix} T^2/2 \\ T \end{bmatrix}; T = 0.1\,\text{sec}$$

$x_1(k)$ = position state θ ; $x_2(k)$ = velocity state ω

The reference input $r = \theta_r$, a step function. The desired steady state

$$\mathbf{x}_s = [\theta_r \quad 0]^T$$

which is a non-null state. It is a servo problem.

As the plant has integrating property, the steady-state value u_s of the input must be zero (otherwise the output cannot stay constant). For this case, the shifted regulator problem may be formulated as follows:

$$\tilde{x}_1 = x_1 - \theta_r$$

$$\tilde{x} = x_2$$

Shifted state variables satisfy the equations

$$\tilde{\mathbf{x}}(k+1) = \mathbf{F}\tilde{\mathbf{x}}(k) + \mathbf{g}u(k)$$

The state-feedback control

$$u(k) = -\mathbf{k}\tilde{\mathbf{x}}(k)$$

results in the dynamics of $\tilde{\mathbf{x}}$ given by

$$\tilde{\mathbf{x}}(k+1) = (\mathbf{F} - \mathbf{gk})\tilde{\mathbf{x}}(k)$$

We now determine the gain matrix **k** such that the desired characteristic equation is

$$z^2 = 0 = z^2 + a_1 z + a_2$$

$$\mathbf{U}^{-1} = [\mathbf{g} \quad \mathbf{Fg}]^{-1} = \begin{bmatrix} 0.005 & 0.015 \\ 0.1 & 0.1 \end{bmatrix}^{-1} = \begin{bmatrix} -100 & 15 \\ 100 & -5 \end{bmatrix}$$

$$\mathbf{p}_1 = [0 \quad 1]\mathbf{U}^{-1} = [100 \quad -5]$$

$$\mathbf{p}_1\mathbf{F} = [100 \quad 5]$$

$$\mathbf{P} = \begin{bmatrix} 100 & -5 \\ 100 & 5 \end{bmatrix}$$

$$|z\mathbf{I}-\mathbf{F}| = z^2 - 2z + 1 = z^2 + \alpha_1 z + \alpha_2$$
$$\mathbf{k} = [a_2 - \alpha_2 \quad a_1 - \alpha_1]\mathbf{P} = [-1 \quad 2]\mathbf{P}$$
$$= [100 \quad 15]$$

The control law expressed in terms of original state variables is given as

$$u(k) = -k_1\tilde{x}_1(k) - k_2\tilde{x}_2(k) = -100(x_1(k) - \theta_r) - 15x_2(k)$$

Review Example 15.3 Reconsider the problem of attitude control of a satellite. For implementation of the design of the previous review example, we require the states $x_1(k)$ and $x_2(k)$ to be measurable. Assuming that the output $y(k) = x_1(k)$ is the only state variable that can be measured, we design a state observer for the system. It is desired that the error vector exhibit response given by the characteristic equation $z^2 = 0 = z^2 + a_1 z + a_2$.

The measurement equation is

$$y(k) = \mathbf{c}\mathbf{x}(k) = [1 \quad 0]\mathbf{x}(k)$$

The state observer for the system is given as

$$\hat{\mathbf{x}}(k+1) = \mathbf{F}\hat{\mathbf{x}}(k) + \mathbf{g}u(k) + \mathbf{m}(y(k) - \mathbf{c}\hat{\mathbf{x}}(k))$$

The gains **m** may be calculated by solving the state regulator design problem for the 'transposed auxiliary system'

$$\zeta(k+1) = \mathbf{F}^T\zeta(k) + \mathbf{c}^T\eta(k)$$

$$\eta(k) = -\mathbf{m}^T\zeta(k)$$

$$\mathbf{p}_1 = [0 \quad 1]\mathbf{U}^{-1} = [0 \quad 10]$$

$$\mathbf{p}_1\mathbf{F}^T = [1 \quad 10]$$

$$\mathbf{P} = \begin{bmatrix} 0 & 10 \\ 1 & 10 \end{bmatrix}$$

$$|z\mathbf{I} - \mathbf{F}^T| = z^2 - 2z + 1 = z^2 + \alpha_1 z + \alpha_2$$

$$\mathbf{m}^T = [a_2 - \alpha_2 \quad a_1 - \alpha_1]\mathbf{P} = [-1 \quad 2]\mathbf{P} = [2 \quad 10]$$

CONCLUDING REMARKS

State space techniques provide many new fundamental insights about control theory, that lead us to new methods for design of compensators. In this chapter, we have briefly introduced the pole placement and optimal control methods.

What we have learnt?

- One of the limitations of transform domain design is that after designing the location of the dominant second-order pair of poles, we keep our fingers crossed, hoping that the higher-order poles do not affect the second-order approximation. What we would like to be able to do is specify *all* closed-loop poles of the higher-order system. Transform design methods do not allow us to do this because they do not allow for a sufficient number of unknown parameters to place all of the closed-loop poles *uniquely*. State space methods solve this problem by introducing additional parameters through state variable feedback.
- There is an advantage in placing all the closed-loop poles uniquely as long as we know where to place the higher-order poles. Correspondence between the desired performance and the location of higher-order poles is, unfortunately, not so obvious. This, in fact, is a limitation of the design method.
- Alternative approach is to use optimal control with state feedback wherein the desired performance is translated into a performance index.
- If the state variables are not available for feedback, an observer is designed to emulate the plant and provide estimated state variables. The response of the observer is designed to be faster than that of the controller; so the estimated state variables effectively appear instantaneously at the controller.

Review Questions

15.1 Briefly describe the configuration of a state feedback control system. What is the advantage that state variable methods have over the root-locus techniques in the placement of closed-loop poles for transient response design? What are the associated problems to be tackled to utilize this advantage?

15.2 Under what conditions can the state feedback matrix be selected so that the roots of the characteristic equation of the closed-loop system are arbitrarily placed? What is the advantage of using the linear quadratic optimal control in designing the state feedback matrix?

15.3 A high-order plant is controllable but state variables have not been selected in controllable canonical form. Derive an algorithm for state feedback design of such a system.

15.4 Pole-placement design of a servo system using state feedback leads to an unacceptable steady-state error. Describe a configuration employing state feedback with integral control for improving the steady-state performance.

15.5 What is an observer? Under what conditions would you use an observer in your state feedback design of a control system? Briefly describe the configuration of an observer.

15.6 An observer-based state feedback control system has a characteristic equation of order $2n$. Show that the placement of $2n$ roots of this equation can be done in a two-stage process: the first stage being the determination of the feedback gain matrix to yield the desired n^{th}-order characteristic equation of the closed-loop system, and the second stage being determination of the observer gain matrix to yield the desired n^{th}-order observer characteristic equation.

15.7 Briefly describe the configuration of a sampled-data system employing state feedback. Modify this configuration by introducing 'integral state' to improve steady-state performance.

Self-Appraisal Quiz

A set of questions with multiple answer choices is given below. These questions have been designed to capture the important aspects of the subject covered in the chapter. For each of these questions, identify the correct answer from the given choices. Compare your answers with the Master Key given at the end of the session. This quiz should serve the purpose of a self-appraisal test for readers.

15.1 A process furnace unit has a dynamical state variable model

$$\dot{x}(t) = -10x(t) + 0.2u(t)$$
$$y(t) = 2550x(t)$$

where $u(t)$ represents the fuel flow in m^3/min, $x(t)$ represents the thermocouple output in volts, and $y(t)$ represents the measured temperature in ^{0}C/volt.
A steady operating condition arises when the fuel flow input is 10 m^3/min.
The furnace operating temperature at steady-state is

(A) 510°C (B) 255°C
(C) 51°C (D) None of the answers in (A), (B), and (C) is correct

15.2 Arbitrary pole placement by state feedback is possible for the systems

$$\dot{\mathbf{x}} = \mathbf{A}\mathbf{x} + \mathbf{b}u; y = \mathbf{c}\mathbf{x}$$

(i) $\mathbf{A} = \begin{bmatrix} 13/6 & 4/3 \\ -4/3 & -7/6 \end{bmatrix}; \mathbf{b} = \begin{bmatrix} 0 \\ 0.5 \end{bmatrix}; \mathbf{c} = [1 \ \ 1]$

(ii) $\mathbf{A} = \begin{bmatrix} 13/6 & 4/3 \\ -4/3 & -7/6 \end{bmatrix}; \mathbf{b} = \begin{bmatrix} -1/6 \\ 1/3 \end{bmatrix}; \mathbf{c} = [1 \ \ 1]$

(A) Both (i) and (ii) are true (B) (i) is true, but (ii) is false
(C) (i) is false, but (ii) is true (D) Both (i) and (ii) are false

15.3 A pendulum with frequency ω_0, is described by the state space model

$$\begin{bmatrix} \dot{x}_1 \\ \dot{x}_2 \end{bmatrix} = \begin{bmatrix} 0 & 1 \\ -\omega_0^2 & 0 \end{bmatrix} \begin{bmatrix} x_1 \\ x_2 \end{bmatrix} + \begin{bmatrix} 0 \\ 1 \end{bmatrix} u$$

The control law is $u = -3\omega_0^2 x_1 - 4\omega_0 x_2$. The characteristic equation of the closed-loop system is

(A) $\lambda^2 + \omega_0^2 = 0$ (B) $(\lambda + \omega_0)^2 = 0$
(C) $\lambda^2 + 4\omega_0\lambda + 4\omega_0^2 = 0$ (D) None of the answers in (A), (B) and (C) is correct

15.4 For a system with state variable description:

$$\dot{\mathbf{x}} = \begin{bmatrix} 0 & 1 \\ 0 & 0 \end{bmatrix} \mathbf{x} + \begin{bmatrix} 0 \\ 1 \end{bmatrix} u$$

the gains k_1 and k_2 $(u = -k_1 x_1 - k_2 x_2)$ that yield $\omega_n = 1$ rad/sec and $\zeta = 0.7$, are

(A) $k_1 = 1, k_2 = 1$ (B) $k_1 = 1, k_2 = \sqrt{2}$
(C) $k_1 = \sqrt{2}, k_2 = 1$ (C) None of the answers is (A), (B), and (C) is correct

15.5 Given a system:

$$\dot{\mathbf{x}} = \mathbf{A}\mathbf{x} + \mathbf{b}u;\ y = \mathbf{c}\mathbf{x}$$

$$\mathbf{A} = \begin{bmatrix} 0 & 1 \\ 0 & 0 \end{bmatrix}; \mathbf{b} = \begin{bmatrix} 0 \\ 1 \end{bmatrix}; \mathbf{c} = [1 \quad 0]$$

The estimator $\qquad \dot{\hat{\mathbf{x}}} = \mathbf{A}\hat{\mathbf{x}} + \mathbf{b}u + \mathbf{m}\,(y - \mathbf{c}\hat{\mathbf{x}})$

places the estimator-error roots at $\omega_n = 5$ rad/sec and $\zeta = 0.5$. The gain matrix **m** is

(A) $[25 \quad 5]$ (B) $[5 \quad 25]$ (C) $\begin{bmatrix} 25 \\ 5 \end{bmatrix}$ (D) $\begin{bmatrix} 5 \\ 25 \end{bmatrix}$

15.6 (i) The condition under which a state-feedback matrix can be selected so that roots of the closed-loop characteristic polynomial are arbitrary is that the system is controllable.
(ii) In an observer-based state-feedback control system, the poles of the observer are usually chosen so that the system response is much faster than the observer response.
(A) Both the statements are true
(B) Statement (i) is true but statement (ii) is false
(C) State (i) is false but statement (ii) is true
(D) Both the statements are false

15.7 Given a model of a pendulum with frequency ω_0:

$$\dot{\mathbf{x}} = \mathbf{A}\mathbf{x} + \mathbf{b}u;\ y = \mathbf{c}\mathbf{x}$$

$$\mathbf{A} = \begin{bmatrix} 0 & 1 \\ -\omega_0^2 & 0 \end{bmatrix}; \mathbf{b} = \begin{bmatrix} 0 \\ 1 \end{bmatrix}; \mathbf{c} = [1 \quad 0]$$

The gain matrix of the estimator

$$\dot{\hat{\mathbf{x}}} = \mathbf{A}\hat{\mathbf{x}} + \mathbf{b}u + \mathbf{m}(y - \mathbf{c}\hat{\mathbf{x}})$$

is $\mathbf{m} = \begin{bmatrix} 20\omega_0 & 99\omega_0^2 \end{bmatrix}^T$. The estimator-error poles are located at

(A) $+j\omega_0, -j\omega_0$ (B) $-\omega_0, -\omega_0$ (C) $+j10\omega_0, -j10\omega_0$ (D) $-10\omega_0, -10\omega_0$

15.8 For the discrete-time system;

$$\mathbf{x}(k+1) = \begin{bmatrix} 1 & 0.1 \\ 0 & 1 \end{bmatrix}\mathbf{x}(k) + \begin{bmatrix} 0.005 \\ 0.1 \end{bmatrix}u(k)$$

the gain matrix **k** ($u = -\mathbf{k}\mathbf{x}$) that results in characteristic polynomial: $z^2 = 0$, is

(A) $[100 \quad 15]$ (B) $\begin{bmatrix} 100 \\ 15 \end{bmatrix}$ (C) $[10 \quad 3.5]$ (D) $\begin{bmatrix} 10 \\ 3.5 \end{bmatrix}$

15.9 For specified values: $\zeta = 0.5$, $\omega_n = 3.6$, $T = 0.1$ sec, the desired characteristic polynomial of a discrete-time system is (approximately)
(A) $z^2 = 0$ (B) $z^2 - 1.6z + 0.7 = 0$
(C) $z^2 + 1.6z + 0.7 = 0$ (D) None of the answers in (A), (B), and (C) is correct

15.10 For a system $\dot{\mathbf{x}} = \mathbf{A}\mathbf{x} + \mathbf{b}u$, with performance index $J = \int_0^\infty (\mathbf{x}^T\mathbf{Q}\mathbf{x} + u^2)dt$, sufficient condition for existence of a solution $u = -\mathbf{k}\mathbf{x}$ is

(A) (**A**, **b**) is controllable
(B) **Q** is positive definite
(C) Both of these conditions
(D) None of these conditions

MASTER KEY

15.1 (A)	15.4 (B)	15.7 (D)	15.10 (C)
15.2 (B)	15.5 (D)	15.8 (A)	
15.3 (C)	15.6 (B)	15.9 (B)	

Problems

15.1 A regulator system has the plant

$$\dot{\mathbf{x}} = \begin{bmatrix} 0 & 1 & 0 \\ 0 & 0 & 1 \\ -6 & -11 & -6 \end{bmatrix}\mathbf{x} + \begin{bmatrix} 0 \\ 0 \\ 1 \end{bmatrix}u$$

$$y = [1 \quad 0 \quad 0]\mathbf{x}$$

(a) Design a state-feedback controller which will place the closed-loop poles at $-2 \pm j3.464$, -5. Give a block diagram of the control configuration.

(b) Design a full-order state observer; the observer-error poles are required to be located at $-2 \pm j3.464$, -5. Give all the relevant observer equations and a block diagram description of the observer structure.

MATLAB Exercise (a) Using Control System Toolbox, verify your design of state-feedback controller. Using the state-feedback gain matrix thus obtained, examine the performance of the system by computer simulation. Assume arbitrary initial conditions.

(b) Verify your design of full-order state observer using Control System Toolbox. Simulate the observer-based state-feedback system for initial conditions of part (a) for the original system, and zero initial conditions for observer system.

15.2 A regulator system has the plant

$$\dot{\mathbf{x}} = \mathbf{A}\mathbf{x} + \mathbf{b}u$$

$$y = \mathbf{c}\mathbf{x}$$

with

$$\mathbf{A} = \begin{bmatrix} 0 & 0 & -6 \\ 1 & 0 & -11 \\ 0 & 1 & -6 \end{bmatrix}; \mathbf{b} = \begin{bmatrix} 1 \\ 0 \\ 0 \end{bmatrix}; \mathbf{c} = [0 \quad 0 \quad 1]$$

(a) Compute **k** so that the control law $u = -\mathbf{k}\mathbf{x}$ places the closed-loop poles at $-2 \pm j3.464$, -5. Give the state variable model of the closed-loop system.

(b) For the estimation of the state vector **x**, we use an observer defined by

$$\dot{\hat{\mathbf{x}}} = (\mathbf{A} - \mathbf{mc})\hat{\mathbf{x}} + \mathbf{b}u + \mathbf{m}y$$

Compute **m** so that the eigenvalues of (**A** – **mc**) are located at $-2 \pm j3.464, -5$.

MATLAB Exercise Simulate your observer-based feedback system using Simulink. Assume that the initial conditions for the original system are $\mathbf{x}(0) = [-1 \quad -10 \quad -3]^T$; and the observer has zero initial conditions.

15.3 Consider the system with the transfer function,

$$\frac{Y(s)}{U(s)} = \frac{9}{s^2 - 9}$$

(a) Find (**A**, **b**, **c**) for this system in observable canonical form.
(b) Compute **k** so that the control law $u = -\mathbf{kx}$ places the closed-loop poles at $-3 \pm j3$.
(c) Design a full-order observer such that the observer-error poles are located at $-6 \pm j6$. Give all the relevant observer equations.

15.4 The equation of motion of an undamped oscillator with frequency ω_0 is

$$\ddot{y} + \omega_0^2 y = u$$

(a) Write the equations of motion in the state variable form with $x_1 = y$ and $x_2 = \dot{y}$ as the state variables.
(b) Find k_1 and k_2 such that $u = -k_1x_1 - k_2x_2$ gives closed-loop characteristic roots with $\omega_n = 2\omega_0$ and $\zeta = 1$.
(c) Design a second-order observer that estimates x_1 and x_2, given measurements of x_1. Pick the characteristic roots of the state-error equation with $\omega_n = 10\omega_0$ and $\zeta = 1$. Give a block diagram of the observer-based state-feedback control system.

15.5 A regulator system has the plant

$$\dot{\mathbf{x}} = \begin{bmatrix} 0 & 1 \\ 20.6 & 0 \end{bmatrix} \mathbf{x} + \begin{bmatrix} 0 \\ 1 \end{bmatrix} u$$

$$y = [1 \quad 0]\,\mathbf{x}$$

(a) Design a control law $u = -\mathbf{kx}$ so that the closed-loop system has eigenvalues at $-1.8 \pm j2.4$.
(b) Design a full-order state observer to estimate the state vector. The observer matrix is required to have eigenvalues at $-8, -8$.

15.6 A regulator system has the double integrator plant

$$\frac{Y(s)}{U(s)} = \frac{1}{s^2}$$

(a) Taking $x_1 = y$ and $x_2 = \dot{y}$ as state variables, obtain the state variable model of the plant.
(b) Compute **k** such that $u = -\mathbf{kx}$ gives closed-loop characteristic roots with $\omega_n = 1$, $\zeta = \sqrt{2}/2$.
(c) Design a full-order observer that estimates x_1 and x_2, given measurements of x_1. Pick the characteristic roots of the state-error equation with $\omega_n = 5$, $\zeta = 0.5$.

15.7 A servo system has the plant with integrating property, described by the equation,

$$\dot{\mathbf{x}} = \mathbf{A}\mathbf{x} + \mathbf{b}u$$
$$y = \mathbf{c}\mathbf{x}$$

where

$$\mathbf{A} = \begin{bmatrix} 0 & 1 & 0 \\ 0 & -1 & 1 \\ 0 & 0 & -2 \end{bmatrix}; \mathbf{b} = \begin{bmatrix} 0 \\ 0 \\ 1 \end{bmatrix}; \mathbf{c} = \begin{bmatrix} 1 & 0 & 0 \end{bmatrix}$$

(a) If $u = -\mathbf{kx} + Nr$, compute **k** and N so that the closed-loop poles are located at $-1 \pm j1, -2$; and $y(\infty) = r$, a constant reference input.

(b) For the estimation of the state vector **x**, we use a full-order observer

$$\dot{\hat{\mathbf{x}}} = (\mathbf{A} - \mathbf{mc})\hat{\mathbf{x}} + \mathbf{b}u + \mathbf{m}y$$

Compute **m** so that observer-error poles are located at $-2 \pm j2, -4$.

(c) Replace the control law in (a) by $u = -\mathbf{k}\hat{\mathbf{x}} + Nr$, and give a block diagram of the observer-based servo system.

15.8 A plant is described by the equation,

$$\dot{\mathbf{x}} = \begin{bmatrix} -1 & 0 \\ 0 & -2 \end{bmatrix}\mathbf{x} + \begin{bmatrix} 1 \\ 1 \end{bmatrix}u$$
$$y = \begin{bmatrix} 1 & 3 \end{bmatrix}\mathbf{x}$$

Add to the plant equations an integrator $\dot{x}_3 = r - y$(r is a constant reference input) and select gains **k**, k_3 so that if $u = -\mathbf{kx} - k_3x_3$, the closed-loop poles are at $-2, -1 \pm j\sqrt{3}$. Give a block diagram of the control configuration.

15.9 The plant of a servo system is described by the equations,

$$\dot{\mathbf{x}} = \mathbf{A}\mathbf{x} + \mathbf{b}u$$
$$y = \mathbf{c}\mathbf{x}$$

where

$$\mathbf{A} = \begin{bmatrix} -3 & 2 \\ 4 & -5 \end{bmatrix}; \mathbf{b} = \begin{bmatrix} 1 \\ 0 \end{bmatrix}; \mathbf{c} = \begin{bmatrix} 0 & 1 \end{bmatrix}$$

A control law of the form $u = -\mathbf{kx} + r$ is proposed; r is a constant reference input.

(a) Compute **k** so that the eigenvalues of $(\mathbf{A} - \mathbf{bk})$ are $-4, -7$.

(b) Show that the steady-state error to a constant input r is nonzero.

(c) Add to the plant equation, an integrator equation ($x_3(t)$ being the state of the integrator):

$$\dot{x}_3(t) = r - y(t)$$

and select gains k_1, k_2 and k_3 so that the control law $u = -k_1x_1(t) - k_2x_2(t) - k_3x_3(t)$ places the eigenvalues of closed-loop system matrix at $-1, -2, -7$.

(d) Draw a block diagram of the control scheme employing integral control and show that the steady-state error to constant reference input is zero.

15.10 A discrete-time regulator system has the plant

$$\mathbf{x}(k+1) = \begin{bmatrix} 0 & 1 & 0 \\ 0 & 0 & 1 \\ -4 & -2 & -1 \end{bmatrix}; \mathbf{x}(k) = \begin{bmatrix} 0 \\ 0 \\ 1 \end{bmatrix} u(k)$$

Design a state-feedback controller which will place the closed-loop poles at $-\frac{1}{2} \pm j\,\frac{1}{2}$, 0. Give a block diagram of the control configuration.

MATLAB Exercise Verify your result using Control System Toolbox.

15.11 Consider the system defined by

$$\mathbf{x}(k+1) = \begin{bmatrix} 0 & 1 & 0 \\ 0 & 0 & 1 \\ -0.5 & -0.2 & 1.1 \end{bmatrix} \mathbf{x}(k) + \begin{bmatrix} 0 \\ 0 \\ 1 \end{bmatrix} u(k)$$

Determine the state-feedback gain matrix **k** such that when the control signal is given by $u(k) = -\mathbf{kx}(k)$, the closed-loop system will have all poles at $z = 0$. Give the state variable model of the closed-loop system.

MATLAB Exercise Verify your result using Control System Toolbox

15.12 A discrete-time regulator system has the plant

$$\mathbf{x}(k+1) = \begin{bmatrix} 2 & -1 \\ -1 & 1 \end{bmatrix} \mathbf{x}(k) + \begin{bmatrix} 4 \\ 3 \end{bmatrix} u(k)$$

$$y(k) = [1\ 1]\,\mathbf{x}(k)$$

(a) Design a state-feedback control algorithm $u(k) = -\mathbf{kx}(k)$ which places the closed-loop characteristic roots at $\pm j\frac{1}{2}$.

(b) Design a state observer which has both the characteristic roots at $z = 0$. Give the relevant observer equations.

MATLAB Exercise (i) Using the state-feedback gain matrix **k** obtained in part (a), examine the performance of the system by computer simulation. Assume arbitrary initial conditions.

(ii) Simulate the observer-based state-feedback system for initial conditions of part (a) for the original system, and zero initial conditions for observer system.

15.13 A double integrator plant is to be controlled by a digital computer employing state feedback. Figure P15.13 shows a model of the control scheme. Both the state variables x_1 and x_2 are assumed to be measurable.

(a) Obtain the discrete-time state variable model of the plant.

(b) Compute k_1 and k_2 so that the response $y(t)$ of the closed-loop system has the parameters: $\zeta = 0.5$, $\omega_n = 4$.

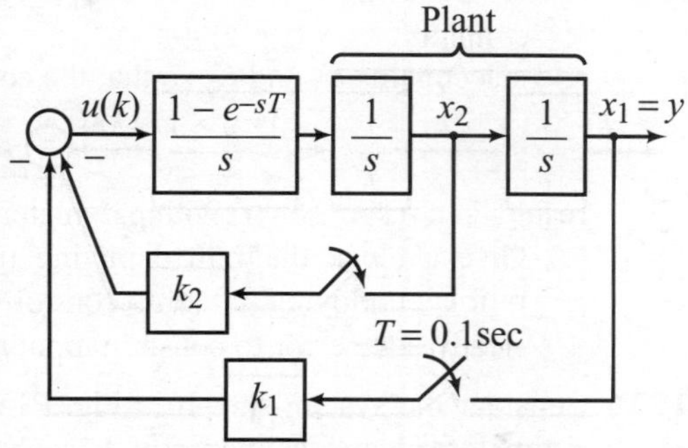

Fig. P15.13

(c) Assume now that only x_1 is measurable. Design a state observer to estimate the state vector **x**; both the characteristic roots of the observer are required to be at $z = 0$.

15.14 Figure P15.14 shows the block diagram of a digital positioning system. The plant is a dc motor driving inertia load. Both the position θ and velocity $\dot{\theta}$ are measurable.

(a) Obtain matrices (**F**, **g**, **c**) of the discrete-time state variable model of the plant.
(b) Compute k_1 and k_2 so that the closed-loop system has all characteristic roots at $z = 0$.
(c) Assume now that the position θ is measured by a shaft encoder and a second-order state observer is used to estimate the state vector **x** from plant input u and measurements of θ. Design a state observer with characteristic roots at $z = 0$. Give a block diagram of the observer-based digital positioning system.

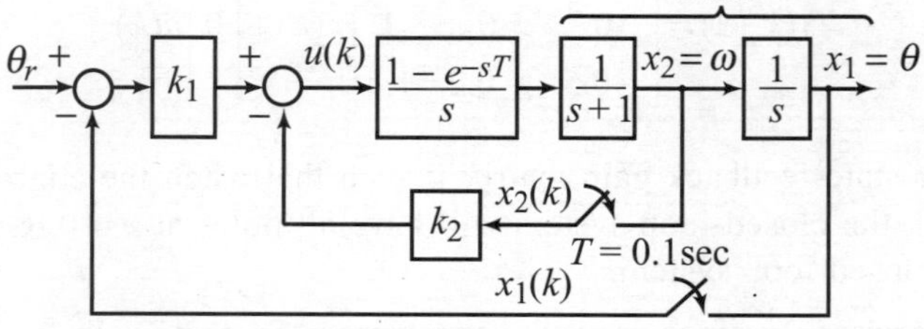

Fig. P15.14

15.15 A continuous-time plant described by the equation,

$$\dot{y} = -y + u$$

is to be controlled by a digital computer; y is the output, and u is the input. Sampling interval $T = 1$ sec.

(a) Obtain a discrete-time state variable model of the plant.
(b) Compute K so that the control law

$$u(k) = -Ky(k) + r$$

results in a response $y(t)$ with time constant 0.5 sec.

(c) Show that the steady-state error to a constant reference input r is non-zero for the above choice of the control scheme.
(d) Add to the plant equation, an integrator equation ($x_2(k)$ being the integral state; $x_1(k) = y(k)$)

$$x_2(k) = x_2(k-1) + r - x_1(k)$$

and select gains K_1 and K_2 so that the control law

$$u = -K_1x_1(k) - K_2x_2(k)$$

results in a response $y(t)$ with parameters: $\zeta = 0.5$, $\omega_n = 4$.

(e) Give a block diagram depicting the control configuration employing integral control and show that the steady-state error to constant input r is zero.

15.16 Consider the system shown in Fig. P15.16. Determine the optimal feedback gain matrix **k** such that the following performance index is minimized:

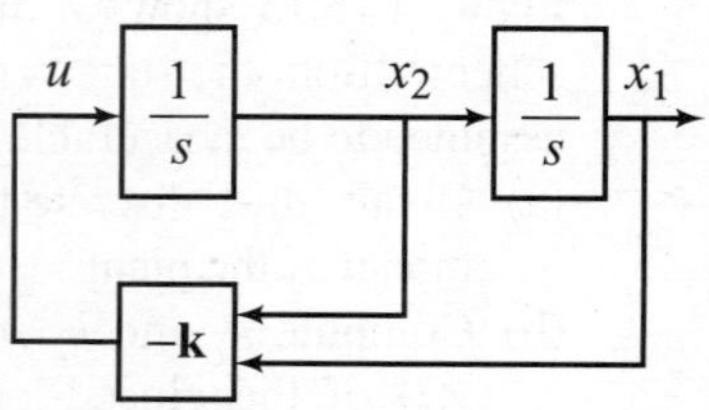

Fig. P15.16

$$J = \frac{1}{2}\int_0^\infty (\mathbf{x}^T\mathbf{Q}\mathbf{x} + 2u^2)\, dt \;;\; \mathbf{Q} = \begin{bmatrix} 2 & 0 \\ 0 & 2 \end{bmatrix}$$

MATLAB Exercise Verify your result using Control System Toolbox.

15.17 Consider the plant

$$\dot{\mathbf{x}} = \mathbf{A}\mathbf{x} + \mathbf{b}u$$

$$\mathbf{A} = \begin{bmatrix} 0 & 1 \\ 0 & -1 \end{bmatrix} ; \mathbf{b} = \begin{bmatrix} 0 \\ 1 \end{bmatrix}$$

with the performance index

$$J = \int_0^\infty (x_1^2 + x_2^2 + u^2)\, dt$$

Choose a control law $u = -\mathbf{k}\mathbf{x}$ that minimizes J.

MATLAB Exercise Verify your result using Control System Toolbox.

15.18 Consider the system

$$x(k+1) = 0.368x(k) + 0.632u(k)$$

Find the control sequence

$$u(k) = -K\,x(k)$$

that minimizes the performance index

$$J = \sum_{k=0}^{\infty} [x^2(k) + u^2(k)]$$

MATLAB Exercise Verify your result using Control System Toolbox.

NONLINEAR SYSTEMS ANALYSIS

16

Preview

All systems are nonlinear especially if large signals are considered. On the other hand, almost all physical systems can be well approximated by linear models if the signals are small. For this reason, the analysis and design methods presented thus far in this book have considered only the enormously powerful techniques available for linear models. However, when the required operation range is large or when the system includes nonlinearities whose discontinuous nature does not allow linear approximation, then the nonlinear effects must be taken into account to explain the behavior of the system. In this chapter, a few of the important tools available for this purpose will be described. For more details, refer to [155].

16.1 INTRODUCTION

As we have tried to make clear in the previous chapters, every real control system is nonlinear, and the linear analysis and design methods we have described use linear approximations to the real models. The linear control design relies on the key assumption of a small range of operation for the linear model to be valid. In practice, because of unforeseen reasons, when the operation range exceeds and invalidates the key assumption of design, the linear controller is likely to perform very poor or to be unstable, because the nonlinearities in the system cannot be properly compensated for. Also, in designing linear controllers it is usually necessary to assume that the parameters of the system model are reasonably well known. However, as we have seen in earlier chapters, many control problems involve uncertainties in the model parameters. A linear controller based on inaccurate values of the model parameters may exhibit significant performance degradation or even instablility in practice. Thus, the design based on approximation of the real models requires rigorous analysis before it is put to practice. The analysis must be carried out to determine tolerance bounds on uncertainties in models used for design, i.e., the analysis must determine a family of models for which the design gives an acceptable performance. The analysis must also consider the situations wherein the linearity range is violated in practice.

Perhaps the single most valuable asset in the field of engineering is the simulation tool—constructing a model of the proposed or actual system and using a numerical solution of the model to reveal the behavior of the system. Simulation is the only general method of analysis applicable to finding effects of uncertainties and/or nonlinearities on the performance of a control system. Of course, simulation finds *specific* solutions; that is, solutions to the system response with specific inputs, initial conditions; and parametric conditions. Important properties such as stability and conditional stability are not *proven* with simulations. When the complexity of a system precludes the use of any analytical approach, simulation will be the only way to obtain necessary information for design purposes. Throughout the book so for, we have been dependent on simulations for the analysis of the effects of uncertainties and/or nonlinearities.

This chapter also does not provide a magic solution to the analysis problem. In fact, no universal analytical technique exists that can cater to our demand on analysis of the effects of uncertainties and/or nonlinearities. Our focus in this chapter is only on some important categories of nonlinear systems for which significant analysis (and design) can be done.

Describing Function Analysis

For the so-called *separable* systems, which comprise a linear part defined by its transfer function, and a nonlinear part defined by a time-independent relationship between its input and output variables, the *describing function method* is most practically useful for analysis. It is an approximate method but experience with real systems and computer simulation results shows adequate accuracy in many cases. Basically, the method is an approximate extension of frequency response methods (including Nyquist stability criterion) to nonlinear systems.

In terms of mathematical properties, nonlinearities may be categorized as *continuous* and *discontinuous*. Because discontinuous nonlinearities cannot be locally approximated by linear functions, they are also called "hard" nonlinearities. Hard nonlinearities are commonly found in control systems, both in small range operation and large range operation. Whether a system in small range operation should be regarded as nonlinear or linear depends on the magnitude of the hard nonlinearities and on the extent of their effects on the system performance.

The continuous or so-called "soft" nonlinearities are present in every control system, though not visible because these are not separable. Throughout the book we have neglected these nonlinearities in our derivations of transfer function and state variable models. For example, we have assumed linear restoring force of a spring, a constant damping coefficient independent of the position of the mass, etc. In practice, none of these assumptions is true for a large range operation. Also there are situations, not covered in this book, wherein the linearity assumption gives too small range of operation to be useful; linear design methods cannot be applied for such systems. If the controlled systems are not too complex and the performance requirements are not too stringent, the linearity assumptions made in this book give satisfactory results in practice.

For the category of separable systems, as we shall see later in this chapter, the predictions of describing function analysis usually are a good approximation to actual behavior when the linear part of the system provides a sufficiently strong filtering effect. Filtering characteristics of the linear part of a system improve as the order of the system goes up. The 'lowpass filtering' requirement is never completely satisfied; for this reason the describing function method is mainly used for stability analysis and is not directly applied to the optimization of system design.

Phase Plane Analysis

Another practically useful method for nonlinear system analysis is the *phase-plane method*. While phase-plane analysis does not suffer from any approximations and hence can be used for stability analysis as well as optimization of system design, its main limitation is that it is applicable to systems which can be well approximated by second-order dynamics. Its basic idea is to solve second-order differential equation and graphically display the result as a family of system motion trajectories on a two-dimensional plane, called the *phase-plane*, which allows us to visually observe the motion patterns of the system. The method is equally applicable to both hard and soft nonlinearities.

Lyapunov Stability Analysis

The most fundamental analysis tool is the concept of a Lyapunov function and its use in nonlinear stability analysis. The power of the method comes from its generality; it is applicable to all kinds of control systems; systems with hard or soft nonlinearities, and of second-order or higher-order. The limitation of the method lies in the fact that it is often difficult to find a Lyapunov function for a given system.

The aim of this chapter is to introduce the two classical, yet practically important tools—the describing function method and the phase-plane method—for a class of nonlinear systems. The two methods are complementary to a large extent, each being available for the study of the systems which are most likely to be beyond the scope of the other. The phase-plane analysis applies primarily to systems described by second-order differential equations. Systems of order higher than the second are likely to be well filtered and tractable by the describing function method.

A brief introduction to Lyapunov stability analysis will also be given in this chapter.

16.1.1 Unique Characteristics of Nonlinear Systems

Several inherent properties of linear systems, which greatly simplify the solution for this class of systems, are not valid for nonlinear systems. Nonlinear systems, in fact, display some unusual characteristics peculiar to them; these phenomena do not occur in linear systems.

As a minimum, it is important to be aware of the main characteristics of nonlinear behavior, only to permit recognition if these are encountered experimentally or in system simulations.

The previous chapters have been predominantly concerned with the study of linear time-invariant control systems. We have observed that these systems have quite simple properties, such as

(i) a linear system $\dot{\mathbf{x}} = \mathbf{A}\mathbf{x}$, with $\mathbf{x}$ being the vector of states and $\mathbf{A}$ being the system matrix, has a *unique equilibrium point* (if $\mathbf{A}$ is nonsingular; normally true for feedback system matrices);

(ii) the equilibrium point is stable if all eigenvalues of $\mathbf{A}$ have negative real parts, *regardless of initial conditions*;

(iii) the transient response is composed of the natural modes of the system, and the general solution can be solved analytically;

(iv) in the presence of the external input $u(t)$, the system response has a number of interesting properties: (1) it satisfies the *principle of superposition*, (2) the asymptotic stability of the system implies bounded-input bounded-output stability, and (3) a sinusoidal input leads to a sinusoidal output of the same frequency.

The behavior of nonlinear systems, however, is much more complex. Due to the lack of superposition property, nonlinear systems respond to external inputs and initial conditions quite differently from linear systems. Some common nonlinear system properties are as follows [126]:

1. Nonlinear systems frequently have more than one equilibrium point. For a linear system, stability is seen by noting that for *any* initial condition, the motion of a stable system always converges to the equilibrium point. However, a nonlinear system may converge to an equilibrium point starting with one set of the initial conditions, and may go to infinity starting with another set of initial conditions. This means that the stability of nonlinear systems may depend on initial conditions. In the presence of a bounded external input, unlike linear systems, stability of nonlinear systems may also be dependent on the input value.
2. Nonlinear systems can display oscillations of fixed amplitude and fixed period without external excitation. These oscillations are called *limit cycles*. Consider the well-known *Van der Pol's differential equation:*

$$M\ddot{y} + B(y^2 - 1)\,\dot{y} + Ky = 0;\ M > 0, B > 0, K > 0 \tag{16.1}$$

which describes physical situations in many nonlinear systems. It can be regarded as describing a mass-spring-damper system with a position-dependent damping coefficient $B(y^2 - 1)$. For large values of y, the damping coefficient is positive and the damper removes energy from the system. This implies that the system motion has a convergent tendency. However, for small values of y, the damping coefficient is negative and the damper adds energy into the system. This suggests that the system motion has a divergent tendency. Therefore, because the nonlinear damping varies with y, the system motion can neither grow unboundedly nor decay to zero. Instead, it displays a sustained oscillation independent of initial conditions, as illustrated in Fig. 16.1.

Of course, sustained oscillations can also be found in linear systems, e.g., in the case of marginally stable linear systems. However, the oscillation of a marginally stable linear system has its amplitude determined by its initial conditions, and such a system is very sensitive to changes in system parameters (a slight change in parameters is capable of leading either to stable convergence or to instability). In nonlinear systems, on the other hand, the amplitude of sustained oscillations is independent of the initial conditions, and limit cycles are not easily affected by parameter changes.

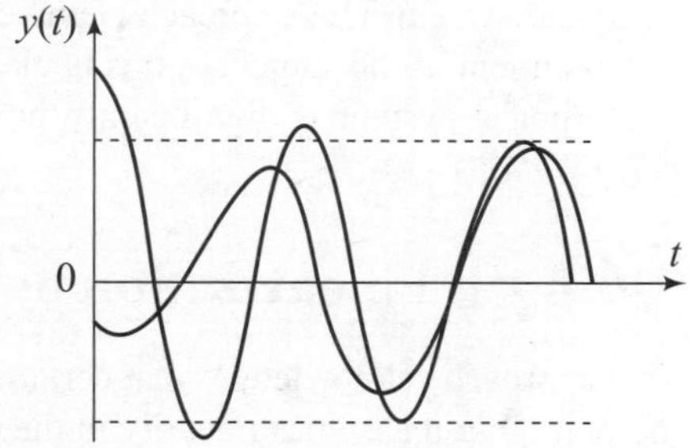

Fig. 16.1 *Responses of the Van der Pol oscillator*

3. A nonlinear system with a periodic input may exhibit a periodic output whose frequency is either a subharmonic or a harmonic of the input frequency. For example, an input of frequency of 10 Hz may result in an output of 5 Hz for the subharmonic case or 30 Hz for the harmonic case.
4. A nonlinear system can display *jump resonance*, a form of hysteresis, in its frequency response. Consider a mass-spring-damper system:

$$M\ddot{y} + B\dot{y} + K_1 y + K_2 y^3 = F\cos\omega t; M > 0, B > 0, K_1 > 0, K_2 > 0 \quad (16.2)$$

Note that the restoring force of the spring is assumed to be nonlinear. If in an experiment, the frequency ω is varied and the input amplitude F is held constant, frequency-response curve of the form shown in Fig. 16.2 may be obtained. As the frequency ω is increased, the response y follows the curve through the points A, B and C. At point C, a small change in frequency results in a discontinuous jump to point D. The response then follows the curve to point E upon further increase in frequency. As the frequency is decreased from point E, the response follows the curve through points D and F. At point F, a small change in frequency results in a discontinuous jump to point B. The response follows the curve to point A for further decrease in frequency. Observe from this description that the response never actually follows the segment CF. This portion of the curve represents a condition of unstable equilibrium.

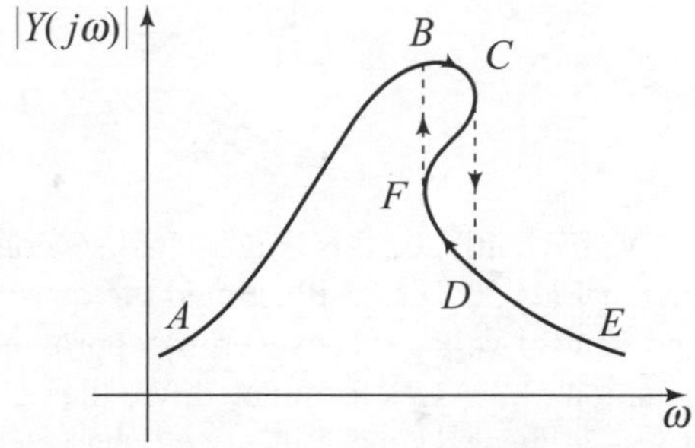

Fig. 16.2 *Frequency response of a system with jump resonance*

16.2 LINEARIZING APPROXIMATIONS

In systems where the deviation from linearity is not too great, linear approximations may permit the application of linear system theory to nonlinear systems. Every real-world system is nonlinear to some extent; however, if the system is only mildly nonlinear, the simplest approach might be to ignore the nonlinearity in designing a controller (i.e., to omit the nonlinearity in the design mode), but to include its effect in evaluating the system performance through simulations. The inherent robustness of the control law designed for the approximating nonlinear system is relied upon to carry it over to the nonlinear system. This approach is universally used by engineers. In fact, all through Chapters 1 to 15, our treatment of the subject in this book was based on this approach, without explicitly highlighting it.

When a system is significantly nonlinear, it is traditionally dealt with by the *theory of small perturbations*.

16.2.1 Linearization by Small-Signal Analysis

This approach acknowledges that certain system characteristics change from operating point to operating point, but it assumes linearity in the neighborhood of a specific operating point.

Many real-world nonlinearities involve a 'smooth' curvilinear relation between an independent variable x and a dependent variable y:

$$y = f(x) \tag{16.3}$$

A linear approximation to the curve, accurate in the neighborhood of selected operating point x_0, is the tangent line to the curve at that point. This approximation is given conveniently by the first two terms of the *Taylor series* expansion of $f(x)$:

$$y = f(x_0) + \left[\left. \frac{df}{dx} \right|_{x=x_0} \right] (x - x_0) + \left[\left. \frac{d^2 f}{dx^2} \right|_{x=x_0} \right] \frac{(x - x_0)^2}{2!} + \cdots \tag{16.4}$$

The linear approximation is

$$y \cong f(x_0) + \left[\left. \frac{df}{dx} \right|_{x=x_0} \right] (x - x_0) \tag{16.5}$$

We have utilized this approach at several places in the book. For example, for the two-phase servomotor, Figs 3.24–3.25 illustrated the quasilinear characteristics relating developed torque with speed and control voltage. However, by approximating these curves by straight lines, the linear differential equation (3.33) was formulated. We then obtained the transfer function of the ac motor, assuming that it was a linear device. Linearized characteristic curves of a hydraulic actuator (Fig. 3.37), linearized flow-pressure relationship for orifice flow (Fig. 3.47), and linearized heating-rate vs input-voltage relationship for a heater (Fig. 3.50) are other examples.

A nonlinear differential equation is an equation for which the derivatives of the state have a nonlinear relationship to the state itself and/or the control. In other words, the differential equation *cannot* be written in the form (refer to Eqn. (14.11a))

$$\dot{\mathbf{x}}(t) = \mathbf{A}\mathbf{x}(t) + \mathbf{b}u(t)$$

but must be left in the form

$$\dot{\mathbf{x}}(t) = \mathbf{f}(\mathbf{x}(t),\ u(t)) \tag{16.6}$$

For small signal linearization, we first determine equilibrium point ($\mathbf{x}_0$, u_0), such that

$$\dot{\mathbf{x}}_0 = \mathbf{0} = \mathbf{f}(\mathbf{x}_0, u_0) \tag{16.7}$$

and we let $\mathbf{x}(t) = \mathbf{x}_0 + \delta\mathbf{x}(t)$ and $u(t) = u_0 + \delta u(t)$. We then expand the nonlinear equation in terms of perturbations from the equilibrium point, which yields

$$\begin{aligned}\frac{d}{dt}\big(\mathbf{x}_0 + \delta\mathbf{x}(t)\big) = \dot{\mathbf{x}}_0 + \delta\dot{\mathbf{x}}(t) &= \mathbf{f}(\mathbf{x}_0 + \delta\mathbf{x}(t), u_0 + \delta u(t)) \\ &\cong \mathbf{f}(\mathbf{x}_0, u_0) + \mathbf{A}\delta\mathbf{x}(t) + \mathbf{b}\delta u(t)\end{aligned} \tag{16.8a}$$

where **A** and **b** are the best linear fits to the nonlinear function $\mathbf{f}(\mathbf{x},u)$ at $\mathbf{x}_0$ and u_0, computed as

$$\mathbf{A} = \left[\left.\frac{\partial \mathbf{f}}{\partial \mathbf{x}}\right|_{\mathbf{x}_0, u_0}\right]; \quad \text{and} \quad \mathbf{b} = \left[\left.\frac{\partial \mathbf{f}}{\partial u}\right|_{\mathbf{x}_0, u_0}\right] \tag{16.8b}$$

The *Jacobian matrices*

$$\frac{\partial \mathbf{f}}{\partial \mathbf{x}} = \begin{bmatrix} \frac{\partial f_1}{\partial x_1} & \frac{\partial f_1}{\partial x_2} & \cdots & \frac{\partial f_1}{\partial x_n} \\ \frac{\partial f_2}{\partial x_1} & \frac{\partial f_2}{\partial x_2} & \cdots & \frac{\partial f_2}{\partial x_n} \\ \vdots & \vdots & \vdots & \vdots \\ \frac{\partial f_n}{\partial x_1} & \frac{\partial f_n}{\partial x_2} & \cdots & \frac{\partial f_n}{\partial x_n} \end{bmatrix}; \quad \frac{\partial \mathbf{f}}{\partial u} = \begin{bmatrix} \frac{\partial f_1}{\partial u} \\ \frac{\partial f_2}{\partial u} \\ \vdots \\ \frac{\partial f_n}{\partial u} \end{bmatrix} \tag{16.8c}$$

Subtracting out the equilibrium solution (16.7), this reduces to

$$\delta\dot{\mathbf{x}}(t) = \mathbf{A}\delta\mathbf{x}(t) + \mathbf{b}\delta u(t) \tag{16.9}$$

which is a linear differential equation approximating the dynamics of motion about the equilibrium point. Normally, the δ function is dropped and it is understood that **x** and u refer to the deviation from the equilibrium.

In developing the models discussed so far in this book, we have encountered nonlinear differential equations on several occasions. For example, for the inverted pendulum system in Example 14.8, the steps followed essentially involved finding **A** and **b** in order to lineariaze the differential equations to the form of Eqn. (16.9). Another illustration is given below.

Example 16.1 The Dutch mathematician Balthasar Van der Pol, in studying a certain class of oscillators, found that their action can be represented by the equations (refer to Eqn. (16.1).

$$\begin{aligned}\dot{x}_1 &= x_2 \\ \dot{x}_2 &= \mu(1 - x_1^2)x_2 - x_1; \mu > 0\end{aligned} \tag{16.10}$$

We can rewrite the equations of motion in the vector-matrix form, with $\mathbf{x} = [x_1\ x_2]^T$, as

$$\dot{\mathbf{x}} = \begin{bmatrix} x_2 \\ \mu(1 - x_1^2)x_2 - x_1 \end{bmatrix} = \begin{bmatrix} f_1(\mathbf{x}) \\ f_2(\mathbf{x}) \end{bmatrix} = \mathbf{f}(\mathbf{x}) \tag{16.11}$$

$\mathbf{x}_0 = \mathbf{0}$ is the equilibrium state.
The state-space matrix of the linearized model is given by

$$\mathbf{A} = \begin{bmatrix} \dfrac{\partial f_1}{\partial x_1} & \dfrac{\partial f_1}{\partial x_2} \\ \dfrac{\partial f_2}{\partial x_1} & \dfrac{\partial f_2}{\partial x_2} \end{bmatrix}\Bigg|_{\mathbf{x}_0} = \begin{bmatrix} 0 & 1 \\ -1 & \mu \end{bmatrix} \tag{16.12}$$

The linearized system has eigenvalues of $\dfrac{\mu}{2} \pm \sqrt{\dfrac{\mu^2}{4} - 1}$. Since it represents an unstable system ($\mu > 0$), all solutions of the linearized system leave the equilibrium point as t increases.

16.2.2 Piecewise-Linear Approximations

Approximating any nonlinearity by means of a piecewise-linear segmentation is a very useful tool for analysis. Each segment leads to a relatively simple linear differential equation which can be solved by conventional linear techniques. Then these segmented solutions are pieced together to obtain the solution for nonlinear analysis. This technique has the advantage of yielding an exact solution for nonlinearities that are piecewise linear or can be approximated by piecewise-linear segments.

In the next section, we will see that common nonlinearities found in control systems can be approximated by piecewise-linear segments. We will exploit this property in Section 16.9 for constructing phase portraits for *phase-plane analysis* of nonlinear systems.

16.2.3 Harmonic Linearization

The frequency response method is a powerful tool for the analysis and design of linear control systems. It is based on describing a linear system by a complex-valued function, the frequency response, instead of a differential equation. Frequency domain analysis cannot be directly applied to nonlinear systems because frequency response functions cannot be defined for nonlinear systems. However, it is natural to try to adapt familiar tools for new problems.

For some nonlinear systems and under certain conditions, an extended version of the frequency response method of linear control can be used to *approximately* analyze nonlinear control systems. This extended version is called the *describing function method*. Though it is only an approximation method, the desirable properties it inherits from the frequency response method of linear control, and the shortage of other systematic tools for nonlinear system analysis, make it an indispensable component of the bag of tools of practicing control engineers.

A describing function, as we shall see in Section 16.4, is defined as *the ratio of the fundamental component of the output of a nonlinear device to the amplitude of a sinusoidal input signal.* All harmonics and subharmonics in the output of the nonlinear element when input is sinusoidal, are neglected.

This assumption is reasonable, because the harmonic terms are often small when compared with the fundamental term. In addition, a feedback control system usually provides additional attenuation of the harmonic terms because of its inherent filtering action. The describing function method can, thus, be viewed as "harmonic linearization" of a nonlinear element.

16.3 COMMON NONLINEARITIES IN CONTROL SYSTEMS

In this section, we take a closer look at the nonlinearities found in control systems. Consider the typical block diagram of closed-loop system shown in Fig. 16.3. It is composed of four parts: a plant to be controlled, sensors for measurements, actuators for control action, and a control law usually implemented on a computer. Nonlinearities may occur in any part of the system.

Saturation

Saturation is probably the most commonly encountered nonlinearity in control systems. It is often associated with amplifiers and actuators. In transistor amplifiers, the output varies linearly with the input only for small amplitude limits. When the input amplitude gets out of the linear range of the amplifier, the output changes very little and stays close to its maximum value. Figure 16.4a shows a linear-segmented approximation of saturation nonlinearity.

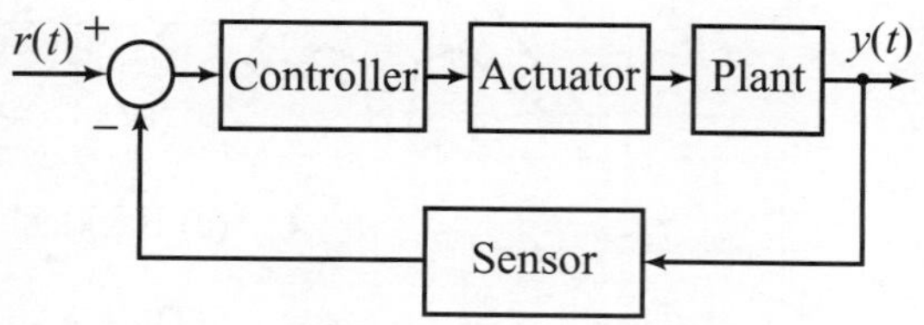

Fig. 16.3 *General block diagram of a control system*

Most actuators display saturation characteristics. For example, the output torque of a servo motor cannot increase infinitely and tends to saturate due to the properties of the magnetic material. Similarly, valve-controlled hydraulic actuators are saturated by the maximum flow rate.

Deadzone

A deadzone nonlinearly may occur in sensors, amplifiers and actuators. In a dc motor, we assume that any voltage applied to the armature windings will cause the armature to rotate if the field current is maintained constant. In reality, due to static friction at the motor shaft, rotation will occur only if the torque provided by the motor is sufficiently large. This corresponds to a so-called deadzone for small voltage signals. Similar deadzone phenomena occur in valve-controlled pneumatic and hydraulic actuators. Figure 16.4b shows linear-segmented approximation of deadzone nonlinearity.

Backlash

A backlash nonlinearity commonly occurs in mechanical components of control systems. In gear trains, small gaps exist between a pair of mating gears (refer to Fig. 16.4c). As a result, when the driving gear rotates a smaller angle than the gap H, the driven gear does not move at all, which corresponds to the deadzone (OA segment in Fig. 16.4c); after contact has been established between the two gears, the driven gear follows the rotation of the driving gear in a linear fashion (AB segment). When the driving gear rotates in the reverse direction by a distance of $2H$, the driven gear again does not move, corresponding to the segment BC in Fig. 16.4c. After the contact between the two gears is re-established, the driven gear linearly follows the rotation of the driving gear in the reverse direction (CD segment).

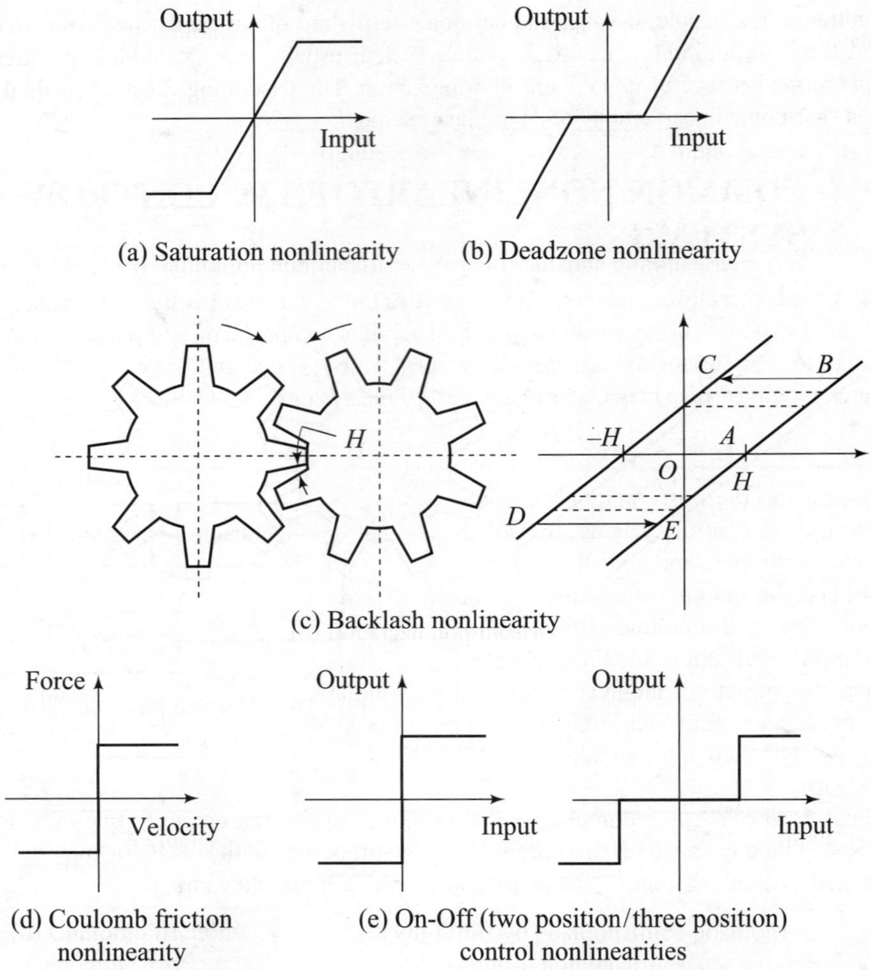

Fig. 16.4 *Common nonlinearities in control systems*

Therefore, if the driving gear is in periodic motion, the driven gear will move in the fashion represented by the closed path *EBCD*.

A critical feature of backlash, a form of hysteresis, is its multivalued nature. Corresponding to each input, two output values are possible. Which one of the two occurs depends on the history of the input.

Coulomb Friction

In any system where there is a relative motion between contacting surfaces, there are several types of friction: all of them nonlinear except the viscous components. Coulomb friction is, in essence, a drag (reaction) force which opposes motion, but is essentially constant in magnitude regardless of velocity (Fig. 16.4d). The common example is an electric motor in which we find Coulomb friction drag due to the rubbing contact between the brushes and the commutator.

On-Off Nonlinearity

In this book, we have primarily covered three *modes of control*:

(i) proportional control;
(ii) integral control; and
(iii) derivative control.

Another important mode of feedback control is the *on-off control*. This class of controllers have only two fixed states rather than a continuous output. In its wider application, the states of an *on-off controller* may not be simply on and off however, but could represent any two values of a control variable. Oscillatory behavior is a typical response characteristic of a system under *two-position control,* also called *bang-bang control*. The oscillatory behavior may be avoided using a *three-position control* (*on-off controller with a deadzone*).

The on-off mode of control results in a *variable structure system* whose structure changes in accordance with the current value of its state. A variable structure system can be viewed as a system composed of independent structures together with a switching logic between each of the structures. With appropriate switching logic, a variable structure system can exploit the desirable properties of each of the structures the system is composed of. Even more, a variable structure system may have a property that is not a property of any of its structures. The variable structure sliding mode control law is usually implemented on a computer.

In this chapter, due to space considerations, we cannot introduce the reader to variable structure sliding mode control systems. This, however, is a highly important and widely practiced robust control strategy. The companion book [155] covers this subject.

We accommodate in this chapter the analysis of nonlinear systems with on-off controllers in the feedback loop. Figure 16.4e shows the characteristics of these controllers.

We may classify the nonlinearities as *inherent* and *intentional*. Inherent nonlinearities naturally come with the system's hardware (saturation, deadzone, backlash, coulomb friction). Usually such nonlinearities have undesirable effects, and control systems have to properly compensate for them. Intentional nonlinearities, on the other hand, are artificially introduced by the designer. Nonlinear control laws, such as bang-bang optimal control laws [155] and adaptive control laws [155], are typical examples of intentional nonlinearities.

16.4 DESCRIBING FUNCTION FUNDAMENTALS

Of all the analytical methods developed over the years for nonlinear systems, the describing function method is generally agreed upon as being the most practically useful. It is an approximate method but experience with real systems and computer simulation results shows adequate accuracy in many cases. The method predicts whether limit cycle oscillations will exist or not, and gives numerical estimates of oscillation frequency and amplitude when limit cycles are predicted. Basically the method is an approximate extension of frequency-response methods (including Nyquist stability criterion) to nonlinear systems.

To discuss the basic concept underlying the describing function analysis, let us consider the block diagram of a nonlinear system shown in Fig. 16.5, where the blocks $G_1(s)$ and $G_2(s)$ represent the linear elements, while the block N represents the nonlinear element.

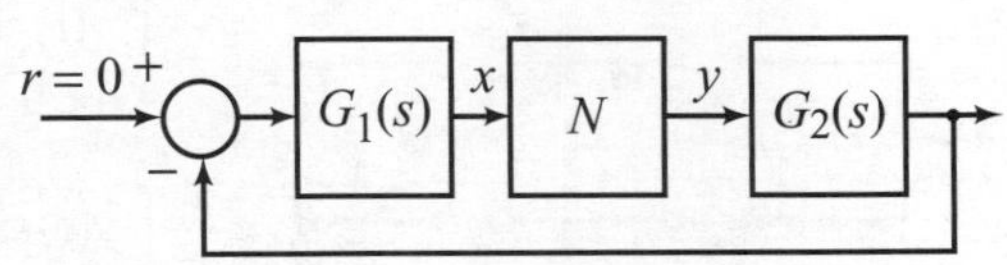

Fig. 16.5 *A nonlinear system*

The describing function method provides a "linear approximation" to the nonlinear element based on the assumption that the input to the nonlinear element is a sinusoid of known, constant amplitude. ***The fundamental harmonic of the element's output is compared with the input sinusoid to determine the steady-state amplitude and phase relation. This relation is the describing function for the nonlinear element.*** The method can thus be viewed as "harmonic linearization" of a nonlinear element.

The describing function method is based on the Fourier series. A review of the Fourier series will be in order here.

16.4.1 Fourier Series

We begin with the definition of a periodic signal. A signal $y(t)$ is said to be periodic with the period T if $y(t + T) = y(t)$ for every value of t. The smallest positive value of T for which $y(t + T) = y(t)$, is called fundamental period of $y(t)$. We denote the fundamental period as T_0. Obviously, $2T_0$ is also a period of $y(t)$, and so is any integer multiple of T_0.

A periodic signal $y(t)$ may be represented by the series [31]:

$$y(t) = \frac{a_0}{2} + \sum_{n=1}^{\infty} [a_n \cos n\omega_0 t + b_n \sin n\omega_0 t] \tag{16.13a}$$

$$= \frac{a_0}{2} + \sum_{n=1}^{\infty} Y_n \sin(n\omega_0 t + \phi_n) \tag{16.13b}$$

where

$$a_n = \frac{2}{T_0}\int_0^{T_0} y(t) \cos n\omega_0 t \, dt; n = 0,1,2\ldots \tag{16.13c}$$

$$b_n = \frac{2}{T_0}\int_0^{T_0} y(t) \sin n\omega_0 t \, dt; n = 1,2\ldots \tag{16.13d}$$

$$Y_n = \sqrt{a_n^2 + b_n^2} \tag{16.13e}$$

$$\phi_n = \tan^{-1}\left(\frac{a_n}{b_n}\right) \tag{16.13f}$$

In Eqn. (16.13b), the term for $n = 1$ is called *fundamental* or *first-harmonic* and always has the same frequency as the repetition rate of the original periodic waveform; whereas n = 2, 3, ..., give second, third, and so forth harmonic frequencies as integer multiples of the fundamental frequency.

Introducing a change of variable, $\psi = \omega_0 t$, we obtain the following alternative equations for the coefficients of Fourier series:

$$a_n = \frac{1}{\pi}\int_0^{2\pi} y(t) \cos n\omega_0 t \, d(\omega_0 t); n = 0,1,2,\ldots \tag{16.14a}$$

$$b_n = \frac{1}{\pi}\int_0^{2\pi} y(t) \sin n\omega_0 t \, d(\omega_0 t); n = 1,2,\ldots \tag{16.14b}$$

Certain simplifications are possible when $y(t)$ has a symmetry of one type or another.

(i) Even symmetry: $y(t) = y(-t)$ results in

$$b_n = 0; n = 1, 2, \dots \tag{16.14c}$$

(ii) Odd symmetry: $y(t) = -y(-t)$ results in

$$a_n = 0; n = 0, 1, 2, \dots \tag{16.14d}$$

(iii) Odd half-wave symmetry: $y(t \pm T_0/2) = -y(t)$ results in

$$a_n = b_n = 0; n = 0, 2, 4, \dots \tag{16.14e}$$

16.4.2 Describing Function for the Nonlinear Element

Let us assume that input x to the nonlinearity in Fig. 16.5 is sinusoidal, i.e.,

$$x = X \sin \omega t$$

With such an input, the output y of the nonlinear element will in general be a nonsinusoidal periodic function which may be expressed in terms of Fourier series as follows (refer to Eqns (16.14)–(16.14)):

$$y = Y_0 + A_1 \cos \omega t + B_1 \sin \omega t + A_2 \cos 2\omega t + B_2 \sin 2\omega t + \cdots$$

The nonlinear characteristics listed in the previous section are all odd-symmetrical/odd half-wave symmetrical; the mean value Y_0 for all such cases is zero and therefore the output

$$y = A_1 \cos \omega t + B_1 \sin \omega t + A_2 \cos 2\omega t + B_2 \sin 2\omega t + \cdots$$

In the absence of an external input (i.e., $r = 0$ in Fig. 16.5), the output y of the nonlinear element N is fedback to its input through the linear elements $G_2(s)$ and $G_1(s)$ in tandem. ***If $G_2(s)G_1(s)$ has lowpass characteristics (this is usually the case in control systems), it can be assumed to a good degree of approximation that all the higher harmonics of y are filtered out in the process, and the input x to the nonlinear element N is mainly contributed by the fundamental component (first harmonic) of y, i.e., x remains sinusoidal.*** Under such conditions, the second and higher harmonics of y can be thrown away for the purpose of analysis and the fundamental component of y, i.e.,

$$y_1 = A_1 \cos \omega t + B_1 \sin \omega t$$

need only be considered.

The above procedure heuristically linearizes the nonlinearity since for a sinusoidal input, only a sinusoidal output of the same frequency is now assumed to be produced. This type of linearization, called the *first-harmonic approximation*, is valid for large signals as well, so long as the filtering condition is satisfied.

We can write $y_1(t)$ in the form,

$$y_1(t) = A_1 \sin(\omega t + 90°) + B_1 \sin \omega t = Y_1 \sin(\omega t + \phi_1) \tag{16.15a}$$

where, by using phasors

$$Y_1 \angle \phi_1 = B_1 + jA_1 = \sqrt{B_1^2 + A_1^2} \angle \tan^{-1}(A_1/B_1) \tag{16.15b}$$

The coefficients A_1 and B_1 of the Fourier series are given by (refer to Eqns (16.13))

$$A_1 = \frac{1}{\pi}\int_0^{2\pi} y\cos\omega t\, d(\omega t) \tag{16.15c}$$

$$B_1 = \frac{1}{\pi}\int_0^{2\pi} y\sin\omega t\, d(\omega t) \tag{16.15d}$$

As we shall see shortly, the amplitude Y_1 and the phase shift ϕ_1 are both functions of X, but independent of ω. We may combine the amplitude ratio and the phase shift in a *complex equivalent gain* $N(X)$ such that

$$N(X) = \frac{Y_1(X)}{X}\angle\phi_1(X) = \frac{B_1 + jA_1}{X} \tag{16.16}$$

Under first-harmonic approximation, the nonlinear element is completely characterized by the function $N(X)$; this function is usually referred to as the *describing function* of the nonlinearity.

The describing function differs from a linear system transfer function in that its numerical value will vary with input amplitude X. Also, it does not depend on frequency ω (there are, however, a few situations in which the describing function for the nonlinearity is a function of both the input amplitude X and the frequency ω (refer to [128–129]). When embedded in an otherwise linear system (Fig. 16.6), the describing function can be combined with the 'ordinary' sinusoidal transfer function of the rest of the system to obtain the complete open-loop function; however, we will get a different open-loop function for every different amplitude X. We can check all of these open-loop functions for closed-loop stability using Nyquist stability criterion.

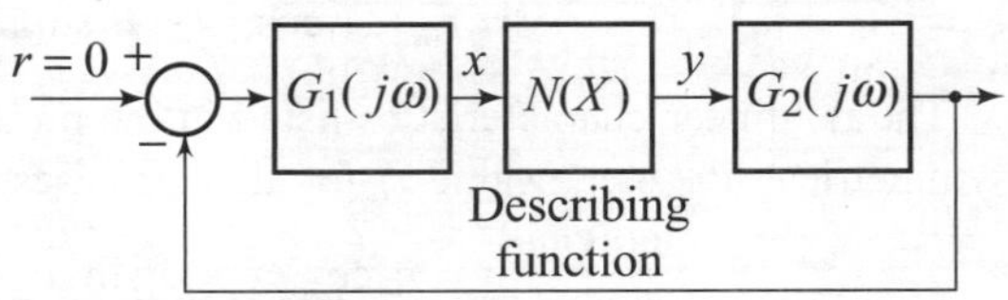

Fig. 16.6 *Nonlinear system with nonlinearity replaced by describing function*

It is important to remind ourselves here that the simplicity in analysis of nonlinear systems using describing functions has been achieved at the cost of certain limitations, the foremost being the assumption that in traversing the path through the linear parts of the system from nonlinearity output back to nonlinearity input, the higher harmonics will have been effectively lowpass filtered, relative to the first harmonic. When the linear part of the system does indeed provide a sufficiently strong filtering effect, then the predictions of describing function analysis usually are a good approximation to actual behavior. ***Filtering characteristics of the linear part of the system improve as the order of the system goes up.***

The 'lowpass filtering' requirement is never completely satisfied; for this reason the describing function method is mainly used for stability analysis and is not directly applied to the optimization of system design. Usually the describing function analysis will correctly predict the existence and characteristics of limit cycles. However, false indications cannot be ruled out; therefore the results must be verified by simulation. Simulation, in fact, is an almost indispensable tool for analysis and design of nonlinear systems; describing function and other analytical methods provide the background for intelligent planning of the simulations.

We will limit our discussion to separable nonlinear systems with reference input $r = 0$, and with symmetrical nonlinearities (listed in Section 16.3) in the loop. Refer to [128–129] for situations wherein dissymmetrical nonlinearities are present, and/or the reference input is nonzero.

16.5 DESCRIBING FUNCTIONS OF COMMON NONLINEARITIES

Before coming to the stability study by the describing function method, it is worthwhile to derive the describing functions of some common nonlinearities. Our first example is an on–off controller with a deadzone as in Fig. 16.7. If X is less than deadzone Δ, then the controller produces no output; the first harmonic component of the Fourier series is of course zero, and the describing function is also zero. If $X > \Delta$, the controller produces the 'square wave' output y. One cycle of this periodic function of period 2π is described as follows:

$$y = \begin{cases} 0; & 0 \le \omega t < \alpha \\ M; & \alpha \le \omega t < (\pi - \alpha) \\ 0; & (\pi - \alpha) \le \omega t < (\pi + \alpha) \\ -M; & (\pi + \alpha) \le \omega t < (2\pi - \alpha) \\ 0; & (2\pi - \alpha) \le \omega t \le 2\pi \end{cases} \tag{16.17}$$

where $X \sin\alpha = \Delta$; or $\alpha = \sin^{-1}(\Delta/X)$

This periodic function has odd symmetry:

$$y(\omega t) = -y(-\omega t)$$

Therefore, the fundamental component of y is given by (refer to Eqn. (16.14d))

$$y_1 = B_1 \sin \omega t$$

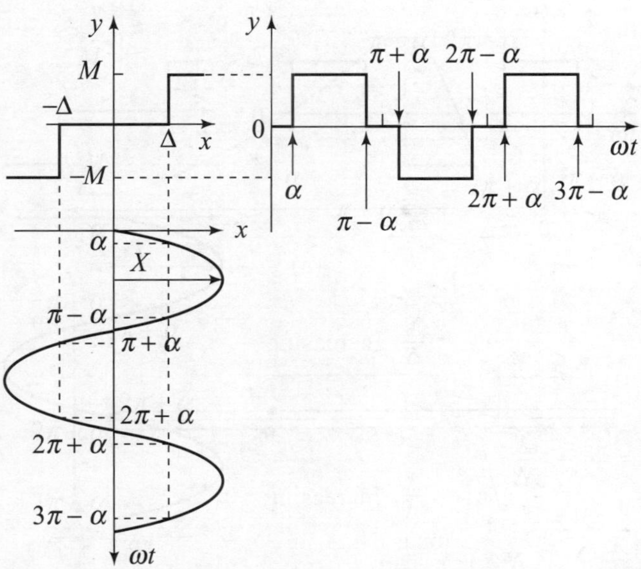

Fig. 16.7 *Fourier-series analysis of an on–off controller with deadzone*

where $$B_1 = \frac{1}{\pi}\int_0^{2\pi} y \sin \omega t \, d(\omega t)$$

Due to the symmetry of y (refer to Fig. 16.7), the coefficient B_1 can be calculated as follows:

$$B_1 = \frac{4}{\pi}\int_0^{\frac{\pi}{2}} y \sin \omega t \, d(\omega t) = \frac{4M}{\pi}\int_\alpha^{\frac{\pi}{2}} \sin \omega t \, d(\omega t) = \frac{4M}{\pi}\cos \alpha \quad (16.18)$$

Since A_1 (the Fourier series cosine coefficient) is zero, the first harmonic component of y is exactly in phase with $X \sin \omega t$, and the describing function $N(X)$ is given by (refer to Eqns (16.16)–(16.18))

$$N(X) = \begin{cases} 0 & ; X < \Delta \\ \dfrac{4M}{\pi X}\sqrt{1 - \left(\dfrac{\Delta}{X}\right)^2} & ; X \geq \Delta \end{cases} \quad (16.19)$$

For a given controller, M and Δ are fixed and the describing function is a function of input amplitude X, which is graphed in Fig. 16.8a together with peak location and value found by standard calculus maximization procedure. Note that for a given X, $N(X)$ is just a pure real positive number, and thus plays the role of a steady-state gain in a block diagram of the form shown in Fig. 16.6. However, this gain term is unusual in that it changes when X changes.

A describing function $N(X)$ may be equivalently represented by a plot of

$$-\frac{1}{N(X)} = \left|-\frac{1}{N(X)}\right| \angle(-1/N(X)) \quad (16.20)$$

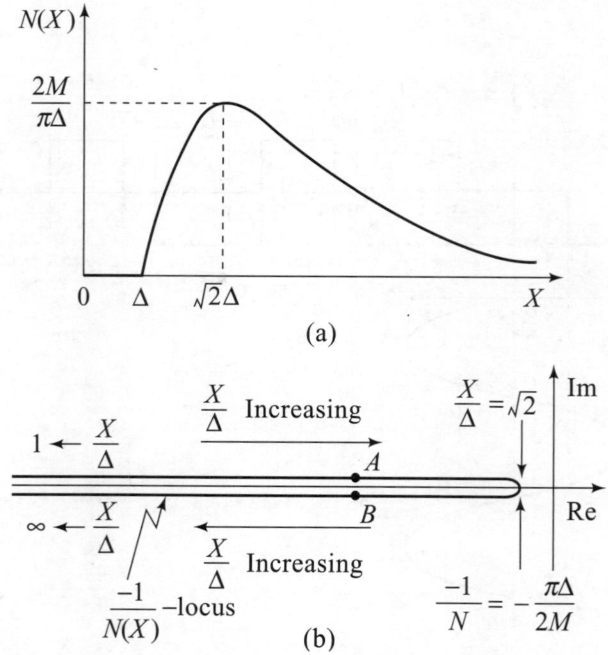

Fig. 16.8 *Describing function of an on-off controller with deadzone*

as a function of X on the polar plane. We will use this form of representation in the next section for stability analysis.

Rearrangement of Eqn. (16.19) gives

$$-\frac{1}{N(X)} = -\frac{\pi\Delta}{4M}\frac{(X/\Delta)^2}{\sqrt{(X/\Delta)^2-1}} \tag{16.21}$$

Figure 16.8b gives the representation on the polar plane of the describing function for an on–off controller with deadzone. It may be noted that though the points A and B lie at the same place on the negative real axis, they belong to different values of X/Δ.

We choose as another example the backlash since its behavior brings out certain features not encountered in our earlier example. The characteristics of backlash nonlinearity and its response to sinusoidal input are shown in Fig. 16.9. The output y is again a periodic function of period 2π; one cycle of this function is described as follows:

$$y = \begin{cases} x-H; & 0 \le \omega t < \pi/2 \\ X-H; & \pi/2 \le \omega t < (\pi-\beta) \\ x+H; & (\pi-\beta) \le \omega t < 3\pi/2 \\ -X+H; & 3\pi/2 \le \omega t < (2\pi-\beta) \\ x-H; & (2\pi-\beta) \le \omega t \le 2\pi \end{cases} \tag{16.22}$$

where $X\sin\beta = X-2H$; or $\beta = \sin^{-1}\left(1-\frac{2H}{X}\right)$.

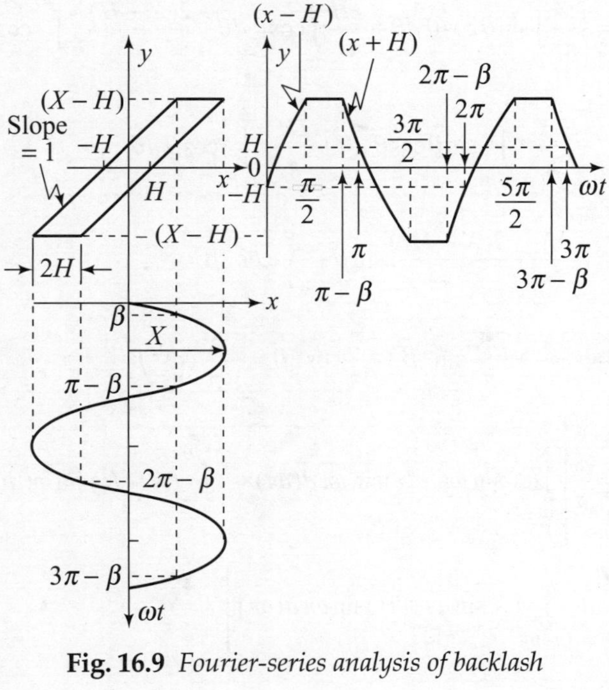

Fig. 16.9 *Fourier-series analysis of backlash*

The periodic function does not possess odd symmetry:

$$y(\omega t) \neq -y(-\omega t),$$

but possesses odd half-wave symmetry:

$$y(\omega t \pm \pi) = -y(\omega t)$$

Therefore, the fundamental component of y is given by (refer to Eqn. (16.14e))

$$y_1 = A_1 \cos \omega t + B_1 \sin \omega t$$

where

$$A_1 = \frac{1}{\pi}\int_0^{2\pi} y \cos \omega t \, d(\omega t)$$

$$B_1 = \frac{1}{\pi}\int_0^{2\pi} y \sin \omega t \, d(\omega t)$$

Due to the symmetry of y, only the positive half wave needs to be considered (Fig.16.9):

$$A_1 = \frac{2}{\pi}\left[\int_0^{\frac{\pi}{2}} (X \sin \omega t - H)\cos \omega t \, d(\omega t) + \int_{\frac{\pi}{2}}^{(\pi-\beta)} (X - H)\cos \omega t \, d(\omega t) + \int_{(\pi-\beta)}^{\pi} (X \sin \omega t + H)\cos \omega t \, d(\omega t)\right]$$

$$= \frac{2X}{\pi}\int_0^{\frac{\pi}{2}} \sin\theta \cos\theta \, d\theta - \frac{2H}{\pi}\int_0^{\frac{\pi}{2}} \cos\theta \, d\theta + \frac{2(X-H)}{\pi}\int_{\frac{\pi}{2}}^{(\pi-\beta)} \cos\theta \, d\theta + \frac{2X}{\pi}\int_{(\pi-\beta)}^{\pi} \sin\theta\cos\theta \, d\theta + \frac{2H}{\pi}\int_{(\pi-\beta)}^{\pi} \cos\theta \, d\theta$$

$$= -\frac{3X}{2\pi} + \frac{2(X-2H)}{\pi}\sin\beta + \frac{X}{2\pi}\cos 2\beta$$

$$= -\frac{3X}{2\pi} + \frac{2X}{\pi}\sin^2\beta + \frac{X}{2\pi}\cos 2\beta = -\frac{X}{\pi}\cos^2\beta \qquad (16.23a)$$

$$B_1 = \frac{2}{\pi}\left[\int_0^{\frac{\pi}{2}} (X \sin \omega t - H)\sin \omega t \, d(\omega t) + \int_{\frac{\pi}{2}}^{(\pi-\beta)} (X - H)\sin \omega t \, d(\omega t) + \int_{(\pi-\beta)}^{\pi} (X \sin \omega t + H)\sin \omega t \, d(\omega t)\right]$$

$$= \frac{2X}{\pi}\int_{0}^{\frac{\pi}{2}} \sin^2\theta\, d\theta - \frac{2H}{\pi}\int_{0}^{\frac{\pi}{2}} \sin\theta\, d\theta + \frac{2(X-H)}{\pi}\int_{\frac{\pi}{2}}^{(\pi-\beta)} \sin\theta\, d\theta$$

$$+ \frac{2X}{\pi}\int_{(\pi-\beta)}^{\pi} \sin^2\theta\, d\theta + \frac{2H}{\pi}\int_{(\pi-\beta)}^{\pi} \sin\theta\, d\theta$$

$$= \frac{X}{\pi}\left[\frac{\pi}{2}+\beta\right] + \frac{2(X-2H)}{\pi}\cos\beta - \frac{X}{2\pi}\sin 2\beta$$

$$= \frac{X}{\pi}\left[\frac{\pi}{2}+\beta\right] + \frac{2X}{\pi}\sin\beta\cos\beta - \frac{X}{2\pi}\sin 2\beta = \frac{X}{\pi}\left[\frac{\pi}{2}+\beta+\frac{1}{2}\sin 2\beta\right] \tag{16.23b}$$

It is clear that the fundamental component of y will have a phase shift with respect to $X \sin \omega t$ (a feature not present in our earlier example). The describing function $N(X)$ is given by (refer to Eqns (16.16), (16.22), (16.23))

$$N(X) = \frac{1}{X}(B_1 + jA_1) = \frac{1}{\pi}\left[\frac{\pi}{2}+\beta+\frac{1}{2}\sin 2\beta - j\cos^2\beta\right]$$
$$\beta = \sin^{-1}\left(1-\frac{2H}{X}\right) \tag{16.24}$$

Note that $N(X)$ is a function of the non dimensional ratio H/X; we can thus tabulate or plot a single graph of $N(X)$ that will be usable for any numerical value of H (see Table 16.1, and Fig. 16.10).

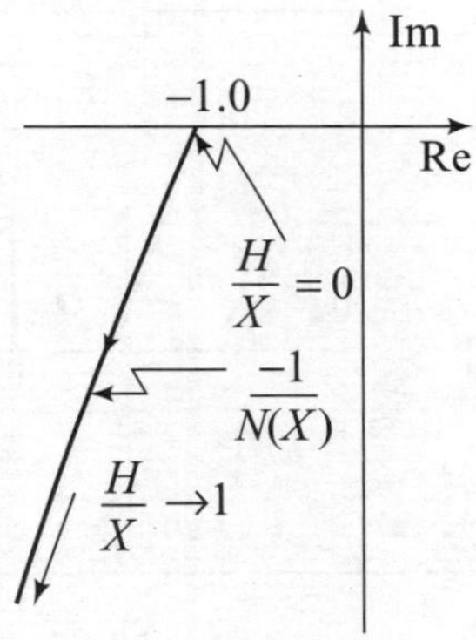

Fig. 16.10 *Describing function for backlash*

Table 16.1 *Describing function for backlash*

H/X	*|–1/N(X)|*	*(N(X))*
0.000	1.000	180.0
0.050	1.017	183.5
0.125	1.066	188.5
0.200	1.134	193.4
0.300	1.259	199.7
0.400	1.435	206.0
0.500	1.687	212.5
0.600	2.072	219.3
0.700	2.720	226.7
0.800	4.024	235.1
0.850	5.330	239.9
0.900	7.946	245.6
0.925	10.560	248.9
0.950	15.800	252.8
0.975	31.500	257.9

We have so far given illustrative derivations of describing functions for on–off controller with deadzone, and backlash. By similar procedures, the describing functions of other common nonlinearities can be derived; some of these are tabulated in Table 16.2.

Table 16.2 *Describing functions of common nonlinearities*

Nonlinearity	*Describing function (input = X sin ωt)*
1. 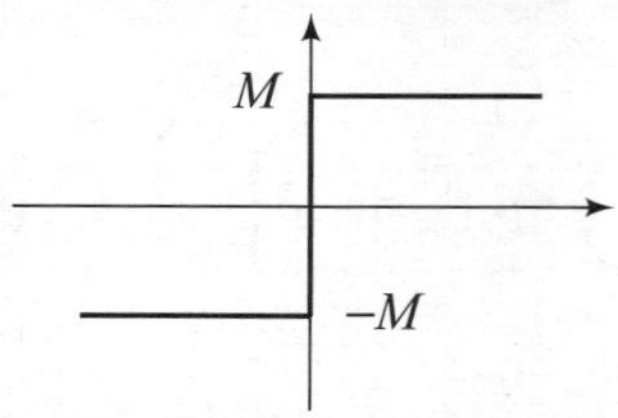	$N(X) = \frac{4M}{\pi X}$
2. 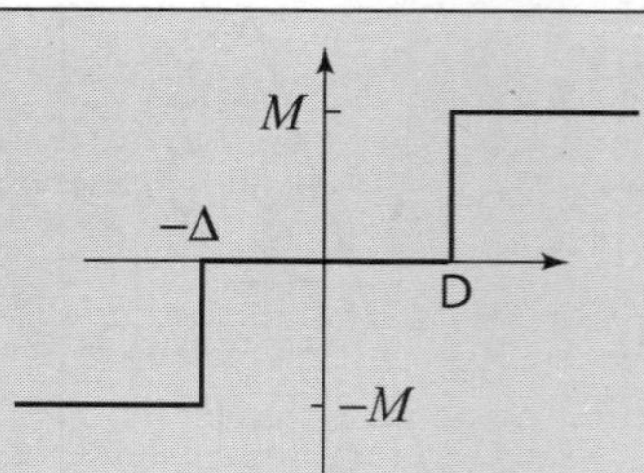	$N(X) = \begin{cases} 0 & ; X < \Delta \\ \frac{4M}{\pi X}\sqrt{1-\left(\frac{\Delta}{X}\right)^2} & ; X \geq \Delta \end{cases}$
3. 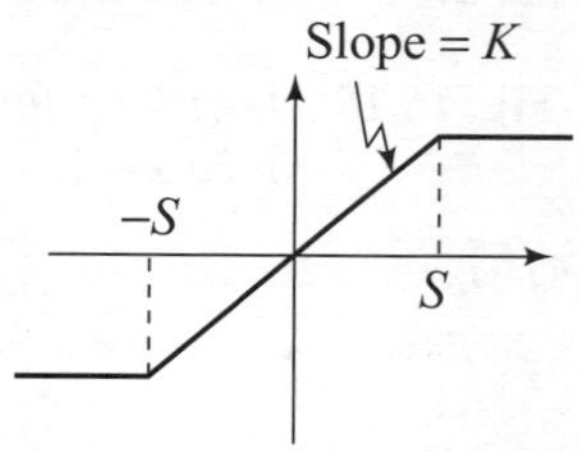	$N(X) = \begin{cases} K & ; X < S \\ \frac{2K}{\pi}\left[\sin^{-1}\frac{S}{X} + \frac{S}{X}\sqrt{1-\left(\frac{S}{X}\right)^2}\right] & ; X \geq S \end{cases}$
4. 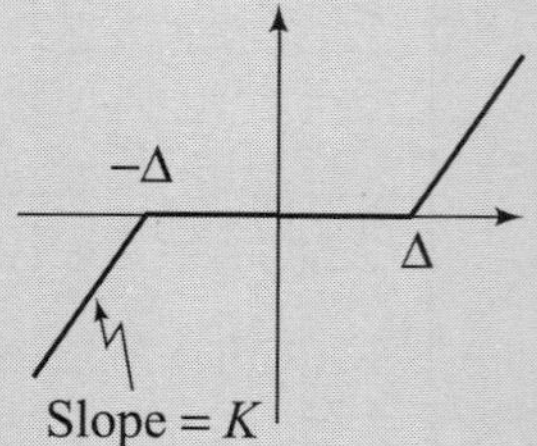	$N(X) = \begin{cases} 0 & ; X < \Delta \\ \frac{2K}{\pi}\left[\frac{\pi}{2} - \sin^{-1}\frac{\Delta}{X} - \frac{\Delta}{X}\sqrt{1-\left(\frac{\Delta}{X}\right)^2}\right] & ; X \geq \Delta \end{cases}$
5. 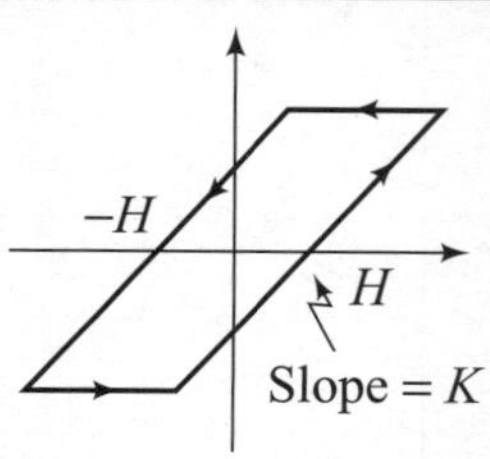	$N(X) = \frac{1}{\pi}\left[\frac{\pi}{2} + \beta + \frac{1}{2}\sin 2\beta - j\cos^2\beta\right]$ $\beta = \sin^{-1}\left(1 - \frac{2H}{X}\right)$

16.6 STABILITY ANALYSIS BY THE DESCRIBING FUNCTION METHOD

Consider the linear system of Fig. 16.11a. Application of Nyquist stability criterion (refer to Chapter 10) to this system involves the following steps:

1. Define the Nyquist contour in the s-plane that encloses the entire right-hand side (unstable region) of the s-plane (Fig. 16.11b).
2. Sketch the Nyquist plot, which is the locus of $KG(s)H(s)$ when s takes on values along the Nyquist contour (Fig. 16.11c).
3. The characteristic equation of the system is

$$1 + KG(s)H(s) = 0$$

or

$$KG(s)H(s) = -1 \tag{16.25}$$

The stability of the closed-loop system is determined by investigating the behavior of the Nyquist plot of $KG(s)H(s)$ with respect to the *critical point* $(-1 + j0)$ in the $KG(s)H(s)$-plane.

For the predominant case of systems wherein open-loop transfer function $KG(s)H(s)$ has no poles in the right half of the of s-plane, the Nyquist stability criterion is stated below:

If the Nyquist plot of the open-loop transfer function $KG(s)H(s)$ corresponding to the Nyquist contour in the s-plane does not encircle the critical point $(-1 + j0)$, the closed-loop system is stable.

4. The characteristic Eqn. (16.25) may be rearranged as follows:

$$G(s)H(s) = -1/K \tag{16.26}$$

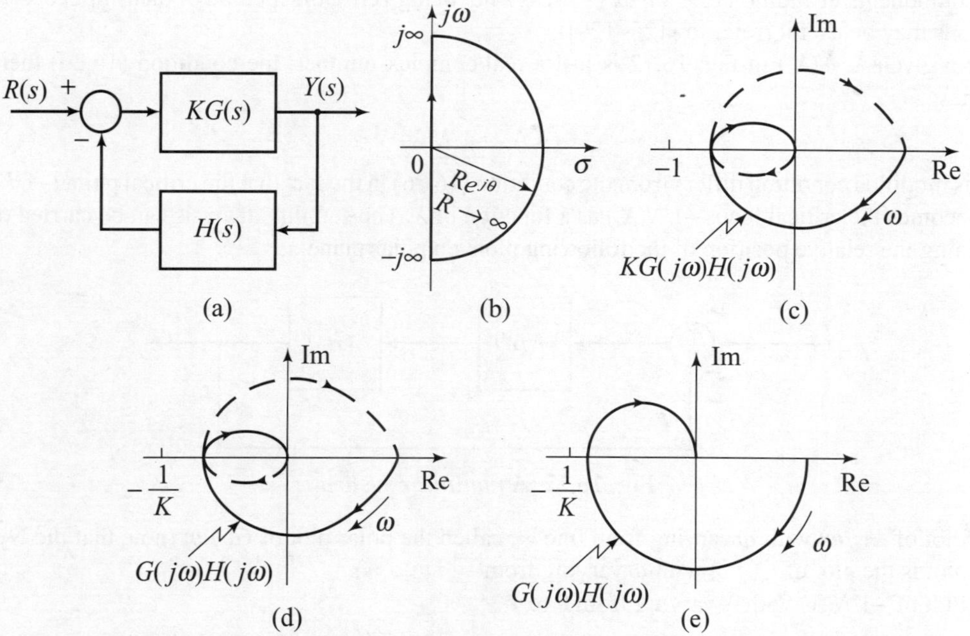

Fig. 16.11 *Application of the Nyquist stability criterion*

For the linear system with open-loop transfer function $KG(s)H(s)$, we can count the number of encirclements of $(-1/K + j0)$ point if the Nyquist plot of $G(s)H(s)$ is constructed (Fig. 16.11d).

5. When the Nyquist plot of $G(s)H(s)$ passes through $(-1/K + j0)$ point, the number of encirclements is indeterminate. This corresponds to the condition where $1 + KG(s)H(s)$ has zeros on the imaginary axis (i.e., the closed-loop system has poles on the imaginary axis). The gain corresponding to this situation will yield oscillatory behavior (we have assumed that the zeros are nonrepeated).
6. The most commonly occurring situation in control system design is that the system becomes unstable if the gain increases past a certain critical value. Stability condition for such systems becomes

$$|G(j\omega)H(j\omega)| < 1/K \text{ at } G(j\omega)H(j\omega) = -180°$$

The stability may therefore be examined from polar plot (plot of $G(j\omega)H(j\omega)$ on polar plane with ω varying from 0 to ∞) only (Fig. 16.11e).

Consider now a nonlinear system of Fig. 16.12; $N(X)$ is the describing function of the nonlinear element and $G(s)$ is the transfer function of the linear part of the system. $G(s)$ is assumed to have no poles in the right half s-plane.

The validity of the block diagram shown in Fig. 16.12 is based on the assumption that the input to the nonlinearity is a pure sinusoid $x = X \sin\omega t$. This necessarily requires that r is zero, since nonzero values of the system input usually result in the nonlinearity input signal containing components in addition to the assumed sine wave. So the describing function approach is applicable when the input r is zero and the system is excited by some initial conditions. For different values of the initial conditions, a signal of the form $x = X \sin\omega t$ will be generated at the input of an odd-symmetrical/odd half-wave symmetrical nonlinearity, with X varying from 0 to ∞. This is true only if the linear part of the system possesses the required lowpass characteristics (for situations where the nonlinearity input signal contains components in addition to $X \sin \omega t$ (such as r not being zero), the method of dual-input describing functions may be useful (refer to [128–129]).

For a given X, $N(X)$ in Fig. 16.12 is just a real/complex number; the condition (16.26) therefore becomes

$$G(s) = -1/N(X) \tag{16.27}$$

This modified condition differs from the condition (16.26) in the fact that the critical point $(-1/K + j0)$ now becomes the critical locus $-1/N(X)$ as a function of X. The stability analysis can be carried out by examining the relative position of the following plots on polar plane.

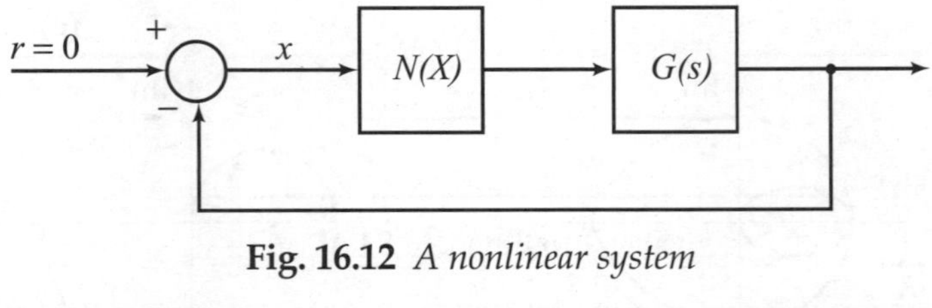

Fig. 16.12 *A nonlinear system*

1. Plot of $G(j\omega)$ with ω varying from 0 to ∞, called the polar plot of $G(j\omega)$ (note that the Nyquist plot is the plot of $G(j\omega)$ with ω varying from $-\infty$ to $+\infty$).
2. Plot of $-1/N(X)$ with X varying from 0 to ∞.

When the critical points of $-1/N(X)$ lie to the left of the polar plot of $G(j\omega)$ (or are not encircled by the Nyquist plot of $G(j\omega)$), the closed-loop system is stable; any disturbances which appear in the

system will tend to die out. Conversely, if any part of the $-1/N(X)$ locus lies to the right of the polar plot of $G(j\omega)$ (or is enclosed by the Nyquist plot of $G(j\omega)$), it implies that any disturbances which are characterized by the values of X corresponding to the enclosed critical points will provide unstable operations. The intersection of $G(j\omega)$ and $-1/N(X)$ loci corresponds to the possibility of a periodic oscillation (limit cycle) characterized by the value of X on the $-1/N(X)$ locus and the value of ω on the $G(j\omega)$ locus.

Figure 16.13a shows a $G(j\omega)$ plot superimposed on a $-1/N(X)$ locus. The values of X for which the $-1/N(X)$ locus lies in the region to the right of an observer traversing the polar plot of $G(j\omega)$ in the direction of increasing ω, correspond to unstable conditions. Similarly, the values of X for which the $-1/N(X)$ locus lies in the region to the left of an observer traversing the polar plot of $G(j\omega)$ in the direction of increasing ω, correspond to the stable conditions. The locus of $-1/N(X)$ and the polar plot of $G(j\omega)$ intersect at the point $A(\omega = \omega_2, X = X_2)$ which corresponds to the condition of limit cycle. The system is unstable for $X < X_2$ and is stable for $X > X_2$. The stability of the limit cycle can be judged by the perturbation technique described below.

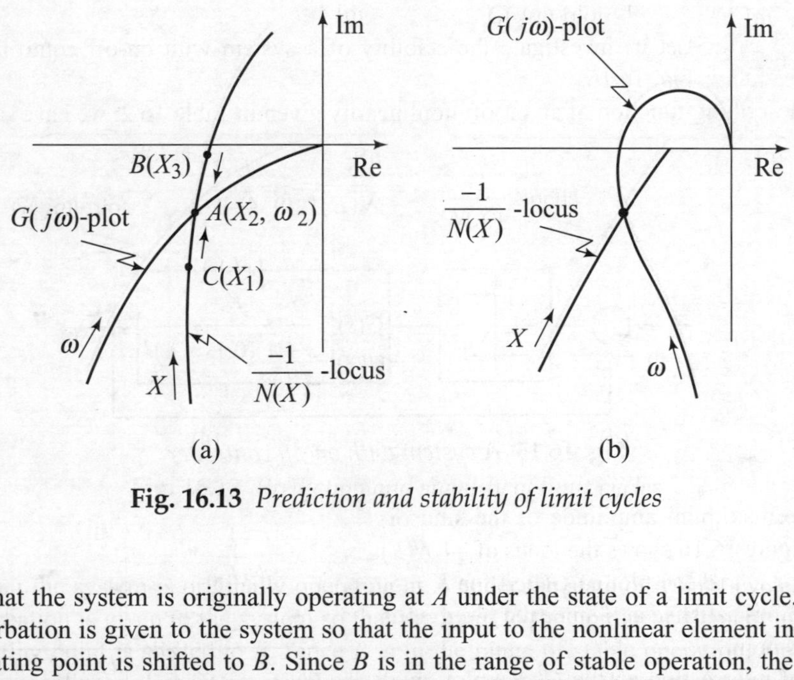

Fig. 16.13 *Prediction and stability of limit cycles*

Suppose that the system is originally operating at A under the state of a limit cycle. Assume that a slight perturbation is given to the system so that the input to the nonlinear element increases to X_3, i.e., the operating point is shifted to B. Since B is in the range of stable operation, the amplitude of the input to the nonlinear element progressively decreases and hence the operating point moves back towards A. Similarly, a perturbation which decreases the amplitude of input to the nonlinearity shifts the operating point to C which lies in the range of unstable operation. The input amplitude now progressively increases and the operating point again returns to A. Therefore, the system has a stable limit cycle at A.

Figure 16.13b shows the case of an unstable limit cycle. For systems having $G(j\omega)$ plots and $-1/N(X)$ loci as shown in Figs 16.14a and 16.14b, there are two limit cycles; one stable and the other unstable.

Describing function method usually gives sufficiently accurate information about stability and limit cycles. This analysis is invariably followed by a simulation study.

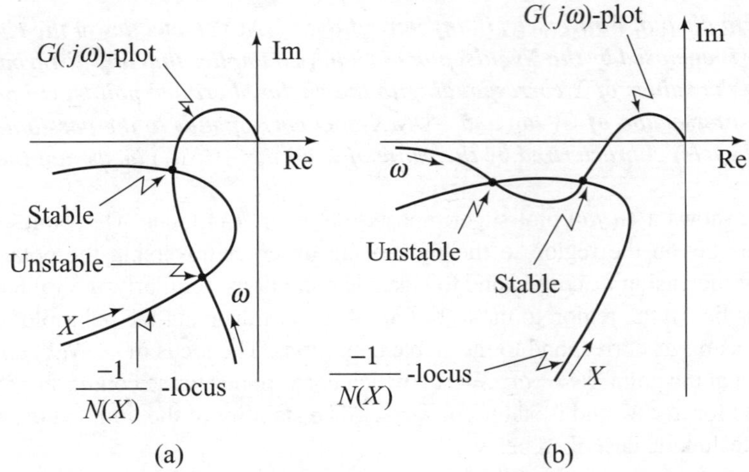

Fig. 16.14 *Prediction and stability of limit cycles*

Example 16.2 Let us investigate the stability of a system with on-off controller, shown in Fig. 16.15.

Using the describing function of an on-off nonlinearity given in Table 16.2, we have

$$-\frac{1}{N(E)} = -\frac{\pi E}{4} \tag{16.28}$$

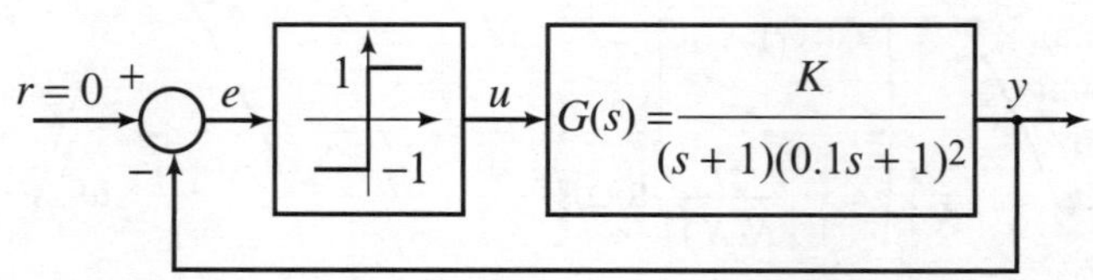

Fig. 16.15 *A system with on-off controller*

where E is the maximum amplitude of the sinusoidal signal e. Figure 16.16 shows the locus of $-1/N(E)$ as a function of E and the plot of $G(j\omega)$ for $K = 5$. Equation (16.27) is satisfied at A since the two graphs intersect at this point.

The point of intersection on the $G(j\omega)$ plot gives a numerical value ω_1 for the frequency of the limit cycle; whereas the same point on the $-1/N(E)$ locus gives us the predicted amplitude E_1 of the oscillation. As an observer traverses the $G(j\omega)$ plot in the direction of increasing ω, the portion O-A of the $-1/N(E)$ locus lies to its right and the portion A-C lies to its left. Using the arguments presented previously, we can conclude that the limit cycle is a stable one.

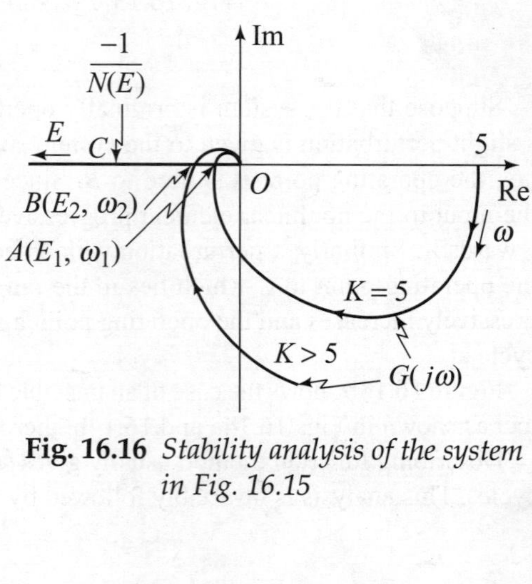

Fig. 16.16 *Stability analysis of the system in Fig. 16.15*

Since $-1/N(E)$ is a negative real number, it is clear that intersection occurs at $-180°$ phase angle. The frequency ω_1 that gives $\angle G(j\omega_1) = -180°$ is 10.95 rad/sec. Furthermore, at point A

$$|G(j\omega_1)| = \left| -\frac{1}{N(E_1)} \right|$$

At $\omega_1 = 10.95$, $|G(j\omega_1)| = 0.206$ and therefore (refer to Eqn. (16.28)).

$$\left| -\frac{1}{N(E_1)} \right| = \frac{\pi E_1}{4} = 0.206$$

This gives $E_1 = 0.262$.

The describing function analysis thus predicts a limit cycle (sustained oscillation)

$$y(t) = -e(t) = -0.262 \sin 10.95t$$

For $K > 5$, the intersection point shifts to B (Fig. 16.16) resulting in a limit cycle of amplitude $E_2 > E_1$ and frequency $\omega_2 = \omega_1$. It should be observed that the system has a limit cycle for all positive values of gain K.

To gain some further insight into on–off control behavior and describing function analysis, let us modify the system of Fig. 16.15 by letting the linear portion be of second-order with

$$G(s) = \frac{5}{(s+1)(0.1s+1)}$$

Figure 16.17 shows the plot of $G(j\omega)$ superimposed on the locus of $-1/N(E)$. The intersection of the two graphs is now impossible since the phase angle of neither $G(j\omega)$ nor $-1/N(E)$ can be more lagging than $-180°$. Describing function analysis thus seems to predict no limit cycling, whereas the fact that the control signal u must be either +1.0 or –1.0 dictates that the system oscillate. One possible interpretation to this analysis would be that the second-order linear system provides less of the lowpass filtering assumed in the describing function method than did the third-order system and thus the approximation has become inaccurate to the point of predicting no limit cycle when actually one occurs. Another interpretation would be that the curves actually do 'intersect' at the origin, predicting a limit cycle of infinite frequency and infinitesimal amplitude. This latter interpretation, even though it predicts a physically impossible result, agrees with the rigorous mathematical solution of the differential equations: for some nonzero initial value of y, we find that $y(t)$ oscillates about zero with ever-decreasing amplitude and ever-increasing frequency. We will examine this solution on the phase plane in a later section.

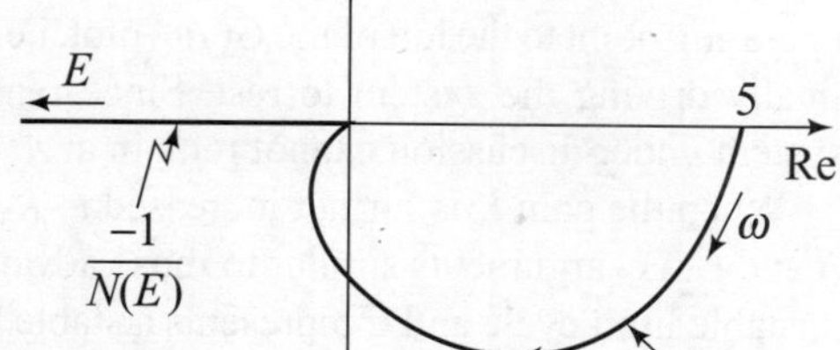

Fig. 16.17 *Describing function analysis of a system with second-order plant and on–off controller*

Let us now modify the system of Fig. 16.15 by giving the controller a deadzone Δ, as shown in Fig. 16.18. The $-1/N(E)$ locus for this type of controller is given by Fig. 16.8b. Plots of $G(j\omega)$ for different values of K, superimposed on $-1/N(E)$ locus, are shown in Fig. 16.19. From this figure, we observe that for $K = K_1$, the $G(j\omega)$ plot crosses the negative real axis at a point to the right of $-\pi\Delta/2$ such that no intersection

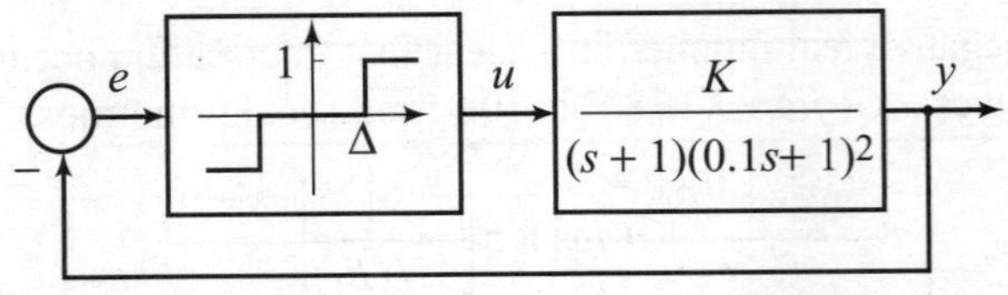

Fig. 16.18 *A system controlled by on–off controller with deadzone*

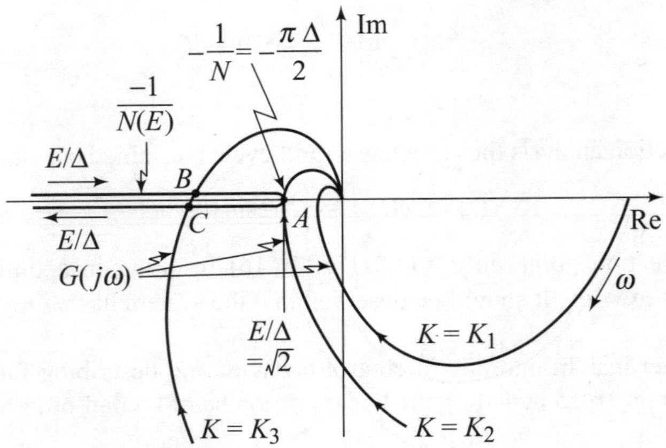

Fig. 16.19 *Stability analysis of the system of Fig. 16.18*

takes place between the graphs of $G(j\omega)$ and $-1/N(E)$, and therefore no limit cycle results. With such a gain, the $-1/N(E)$ locus lies entirely to the left of the $G(j\omega)$ plot; the system is therefore stable, i.e., it has effective positive damping.

If the gain K is now increased to a value K_2 such that the $G(j\omega)$ plot intersects the $-1/N(E)$ locus at the point A (i.e., on the negative real axis at $-\pi\Delta/2$), then there exists a limit cycle. Now suppose that the system is operating at the point A. Any increase in the amplitude of E takes the operating point to the left so that it is not enclosed by the $G(j\omega)$ plot, which means that the system has positive damping. This reduces E till the operating point comes back to A. Any decrease in the amplitude of E again takes the operating point to the left of the $G(j\omega)$ plot, i.e., the system has positive damping which further reduces E, finally driving the system to rest. Since random disturbances are always present in any system, the system under discussion cannot remain at A. Therefore, the limit cycle represented by A is unstable.

When the gain K is further increased to K_3, the graphs of $G(j\omega)$ and $-1/N(E)$ intersect at two points B and C. By arguments similar to those advanced earlier, it can be shown that the point B represents an unstable limit cycle and C represents a stable limit cycle. It may be noted that though the points B and C lie at the same place on the negative real axis, they belong to different values of E/Δ.

It is also clear that limit cycling is predicted only for deadzone Δ smaller than the value given by

$$\frac{\pi\Delta}{2} = |G(j\omega_1)|$$

where ω_1 is the frequency at which the plot $G(j\omega)$ intersects the negative real axis. A deadzone in on–off controllers appears to be a desirable feature to avoid limit cycling. However, as we shall see later in this chapter, a large value of Δ would cause the steady-state performance of the system to deteriorate.

Example 16.3 Figure 16.20a shows a block diagram for a servo system consisting of an amplifier, a motor, a gear train, and a load (gear 2 shown in the diagram includes the load element). It is assumed that the inertia of the gears and load element is negligible compared with that of the motor, and backlash exists between gear 1 and gear 2. The gear ratio between gear 1 and gear 2 is unity.

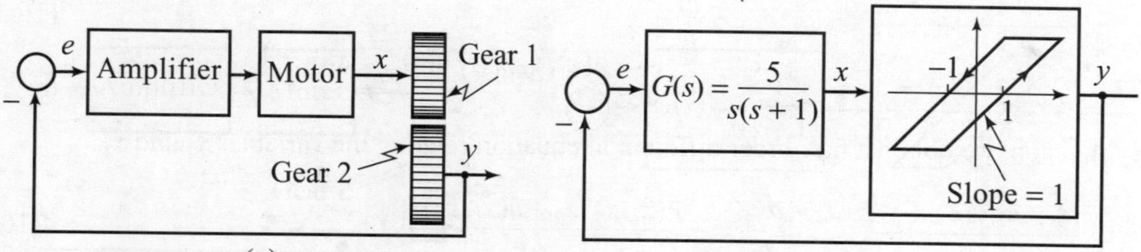

Fig. 16.20 *A servo system with backlash in gears*

The transfer function of the amplifier-motor combination is given by $5/s(s + 1)$ and the backlash amplitude is given as unity ($H = 1$).

From the problem statement, the block diagram of the system may be redrawn as shown in Fig. 16.20b. Let us investigate the stability of this system. The $-1/N(X)$ locus for the backlash nonlinearity is given by Fig. 16.10 (Table 16.1). Plot of $G(j\omega)$ superimposed on $-1/N(X)$ locus is shown in Fig. 16.21. As seen from this figure, there are two intersections of the two loci. Applying the stability test for the limit cycle reveals that point A corresponds to a stable limit cycle and point B corresponds to an unstable limit cycle. The stable limit cycle has a frequency of 1.6 rad/sec and an amplitude of 2 (the unstable limit cycle cannot physically occur). To avoid limit-cycle behavior, the gain of the amplifier must be decreased sufficiently so that the entire $G(j\omega)$ plot lies to the left of $-1/N(X)$ locus.

Note that checking for an intersection must be done graphically/numerically since no analytical solution for limit-cycle amplitude or frequency is possible. A computer program that tabulates $G(j\omega)$ and $-1/N(X)$ is useful in searching for intersections and is not difficult to write. Once the general region of an intersection is found, we can use smaller increments of H/X and ω to pinpoint the intersection as accurately as we wish.

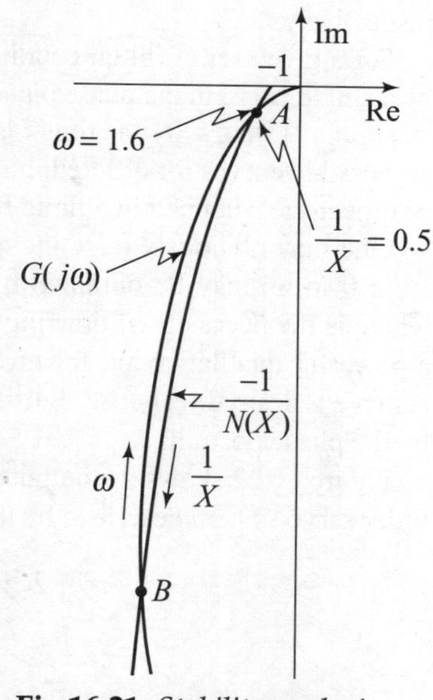

Fig 16.21 *Stability analysis of the system of Fig. 16.20*

16.7 CONCEPTS OF PHASE PLANE ANALYSIS

The free motion of any second-order nonlinear system can always be described by an equation of the form,

$$\ddot{y} + g(y, \dot{y})\dot{y} + h(y, \dot{y})y = 0 \tag{16.29}$$

The state of the system at any moment can be represented by a point of coordinates $(y, \dot{y})$ in a system of rectangular coordinates. Such a coordinate plane is called a 'phase plane'.

In terms of the state variables

$$x_1 = y, x_2 = \dot{y}, \tag{16.30a}$$

second-order system (16.29) is equivalent to the following canonical set of state equations:

$$\begin{aligned}\dot{x}_1 &= \frac{dx_1}{dt} = x_2 \\ \dot{x}_2 &= \frac{dx_2}{dt} = -g(x_1,x_2)x_2 - h(x_1,x_2)x_1\end{aligned} \tag{16.30b}$$

By division, we obtain a first-order differential equation relating the variables x_1 and x_2:

$$\frac{dx_2}{dx_1} = -\frac{g(x_1,x_2)x_2 + h(x_1,x_2)x_1}{x_2} \tag{16.31}$$

Thus, we have eliminated the independent variable t from the set of first-order differential equations given by (16.30b). In Eqn. (16.31), we consider x_1 and x_2 as independent and dependent variables respectively.

For a given set of initial conditions $\{x_1(0), x_2(0)\}$, the solution to Eqn. (16.31) may be represented by a single curve in the phase plane for which the coordinates are x_1 and x_2. The curve traced out by the state point $\{x_1(t), x_2(t)\}$ as time t is varied from 0 to ∞ is called the *phase trajectory*, and the family of all possible curves for different initial conditions is called the *phase portrait*. Normally, a finite number of trajectories, defined in a finite region, is considered a portrait.

One may obviously raise the question that when time solutions $x_1(t)$ and $x_2(t)$ as time t is varied from 0 to ∞, may be obtained by direct integration of Eqns (16.30b) analytically or numerically, where is the necessity of drawing phase portraits? In fact, as we shall see, the phase portraits provide a powerful qualitative aid for investigating system behavior and the design of system parameters to achieve a desired response. Furthermore, the existence of limit cycles is sharply brought into focus by the phase portrait.

Figure 16.22 shows the output response and the corresponding phase trajectory for a linear second-order servo system described by the differential equation,

$$\ddot{y} + 2\zeta\dot{y} + y = 0; \quad y(0) = y^0, \quad \dot{y}(0) = 0, \quad 0 < \zeta < 1$$

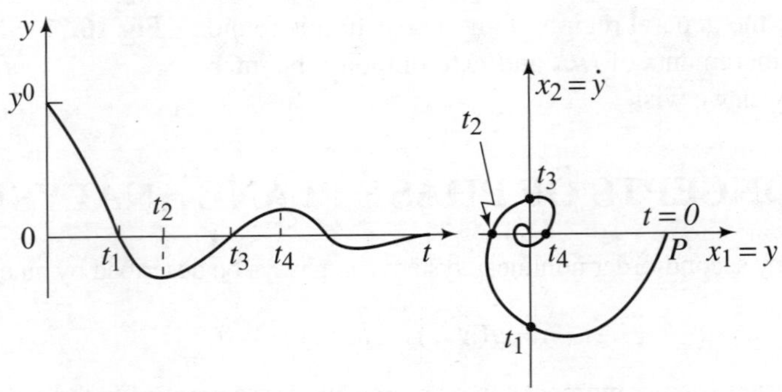

Fig. 16.22 *A second-order linear system on the phase plane*

In terms of the state variables $x_1 = y$ and $x_2 = \dot{y}$, the system model is given by the equations

$$\dot{x}_1 = x_2; \quad \dot{x}_2 = -2\zeta x_2 - x_1; \quad x_1(0) = y^0, \quad x_2(0) = 0$$

The origin of the phase plane ($x_1 = 0$, $x_2 = 0$) is the *equilibrium point* of the system since at this point the derivatives $\dot{x}_1$ and $\dot{x}_2$ are zero (the system continues to lie at the equilibrium point unless otherwise disturbed). The nature of the transient can be readily inferred from the phase trajectory of Fig. 16.22; starting from the point *P*, i.e., with initial deviation but no initial velocity, the system returns to rest, i.e., to the origin, with damped oscillatory behavior.

Consider now the well-known *Van der Pol's differential equation* (refer to Eqn. (16.1))

$$\ddot{y} - \mu(1 - y^2)\dot{y} + y = 0$$

which describes physical situations in many nonlinear systems. In terms of the state variables $x_1 = y$ and $x_2 = \dot{y}$, we obtain

$$\dot{x}_1 = x_2; \dot{x}_2 = \mu(1 - x_1^2)x_2 - x_1$$

where μ is a 'small' number (refer to Example 16.7).

Origin of the phase plane is the equilibrium point of the system. Figure 16.23 shows phase portraits for (i) $\mu > 0$, (ii) $\mu < 0$. In the case of $\mu > 0$, we observe that for large values of $x_1(0)$, the system response is damped and the amplitude of $x_1(t) = y(t)$ decreases till the system state enters the limit cycle as shown by the outer trajectory. On the other hand, if initially $x_1(0)$ is small, the damping is negative, hence the amplitude of $x_1(t) = y(t)$ increases till the system state enters the limit cycle as shown by the inner trajectory. The limit cycle is a stable one, since the paths in its neighborhood converge toward the limit cycle. Figure 16.23 shows an unstable limit cycle for $\mu < 0$.

The *phase plane* for second-order systems is indeed a special case of *phase space* or *state space* defined for *n*th-order systems. Much work has been done to extend this approach of analysis to third-order systems. Though a phase trajectory for a third-order system can be graphically visualized through its projections on two planes, say (x_1, x_2) and (x_2, x_3) planes, this complicacy causes the technique to lose its major power of quick graphical visualization of the total system response. The phase trajectories are therefore generally restricted to second-order systems only.

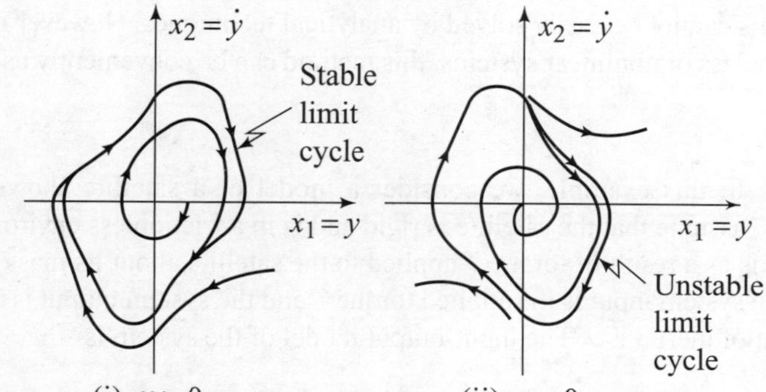

Fig. 16.23 *A second-order nonlinear system on the phase plane*

For time-invariant systems, the entire phase plane is covered with trajectories with one and only one curve passing through each point of the plane except for certain critical points through which either infinite number or none of the trajectories pass. Such points (called *singular points*) are discussed later in Section 16.9.

If the parameters of a system vary with time, or if a time-varying driving function is imposed, two or more trajectories may pass through a single point in a phase plane. In such cases, the phase portrait becomes complex and more difficult to work with and interpret. Therefore, the use of phase-plane analysis is restricted to second-order systems with constant parameters and constant or zero input. However, it may be mentioned that investigators have made fruitful use of the phase-plane method in investigating second-order time-invariant systems under simple time-varying inputs such as ramp. Some simple time-varying systems have also been analysed by this method. Our discussion will be limited to second-order time-invariant systems with constant or zero input.

From the above discussion, we observe that the phase-plane analysis applies primarily to systems described by second-order differential equations. In the case of feedback control systems, systems of order higher than the second are likely to be well filtered and tractable by the describing-function method discussed earlier in this chapter. The two methods of the phase plane and of the describing function are therefore complementary to a large extent, each being available for the study of the systems which are most likely to be beyond the scope of the other.

16.8 CONSTRUCTION OF PHASE PORTRAITS

Today, phase portraits are routinely computer-generated. However (as for example, in the case of root locus for linear systems), it is still practically useful to learn how to roughly sketch phase portraits or quickly verify the plausibility of computer outputs.

For some special nonlinear systems, particularly piecewise linear systems (whose phase portraits can be constructed by piecing together the phase portraits of the related linear systems), phase portraits can be constructed analytically. Analytical methods are useful for systems modeled by differential equations that can be easily solved. If the system of differential equations cannot be solved analytically, we can use graphical methods. A number of graphical methods for constructing phase-plane trajectories are now available; we will describe in this section the method of isoclines.

16.8.1 Analytical Method

Most nonlinear systems cannot be easily solved by analytical techniques. However for piecewise linear systems, an important class of nonlinear systems, this method can be conveniently used, as the following examples show.

Example 16.4 In this example, we consider a model of a satellite shown in Fig. 15.3. We assume that the satellite is rigid and is in a frictionless environment. It can rotate about the reference axis as a result of torque T applied to the satellite about its mass center by firing the thrusters ($T = Fd$). The system input is the applied torque T and the system output is the attitude angle θ. The satellite's moment of inertia is J. The input-output model of the system is

$$J\frac{d^2\theta}{dt^2} = T \tag{16.32a}$$

We assume that when the thrusters fire, the thrust is constant; that is, $T = A$, a constant greater than or less than zero. In terms of output variable $y(= \theta)$, we obtain the equation,

$$J\ddot{y} = A \tag{16.32b}$$

In terms of the state variables $x_1 = y$ and $x_2 = \dot{y}$, the state equations become

$$\dot{x}_1 = x_2; \dot{x}_2 = \frac{A}{J} \tag{16.32c}$$

Elimination of t by division yields the equation of the trajectories:

$$\frac{dx_2}{dx_1} = \frac{A}{J\,x_2} \tag{16.33}$$

or

$$J x_2\, dx_2 = A dx_1$$

This equation is easily integrated; the general solution is

$$x_1(t) = \frac{J x_2^2(t)}{2A} + C \tag{16.34a}$$

where C is a constant of integration and is determined by initial conditions, i.e.,

$$C = x_1(0) - \frac{J x_2^2(0)}{2A} \tag{16.34b}$$

For an initial state point $(x_1(0), x_2(0))$, the trajectory is a parabola passing through the point $x_1 = C$ on the x_1-axis where C is defined by Eqn. (16.34b).

A family of parabolas in the (x_1, x_2) plane is shown in Fig. 16.24a for $A > 0$. As time t increases, each trajectory is described in the clockwise direction as indicated by the arrows. The direction of the phase trajectories is dictated by the relationship $\dot{x}_1 = x_2$: x_1 increases with time in the upper half of the phase plane and the state point therefore moves from left to right (→); in the lower half of the phase plane, x_1 decreases with time and the state point must therefore move from right to left (←).

The time interval between two points of a trajectory is given by $\Delta t = \Delta x_1/x_{2av}$. The trajectories may be provided with a time scale by means of this equation. This operation is, however, often unnecessary since the phase portrait is mainly used to display the general features of the system transients.

The phase portrait for $A < 0$ is shown in Fig. 16.24b. In the special case of $A = 0$ (no driving torque), the integration of the trajectory equation (16.33) gives $x_2(t) = x_2(0)$. The trajectories are, therefore, straight lines parallel to x_1-axis.

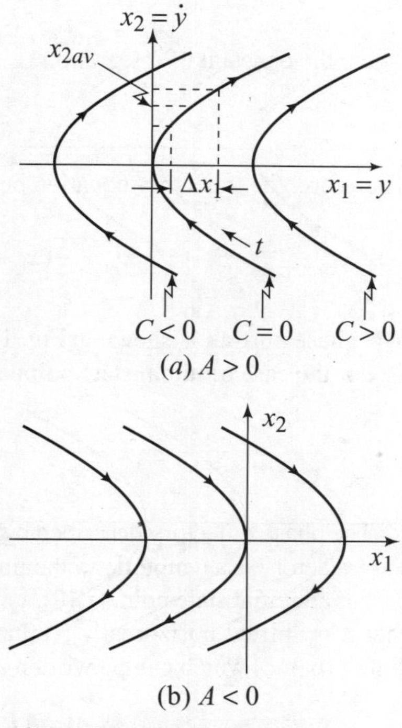

Fig. 16.24 *Phase portraits for system (16.32)*

Example 16.5 Consider now the equation,

$$J\ddot{\theta}+B\ddot{\theta}=T \tag{16.35a}$$

corresponding to a torque T driving a load comprising inertia J and viscous friction B. For a constant torque, the equation may be expressed as

$$\tau\ddot{y}+\dot{y}=A \tag{16.35b}$$

where $y = \theta$ is the system output and A represents normalized torque; a constant greater than or less than zero.

The equivalent system is

$$\dot{x}_1 = x_2;\ \tau\dot{x}_2 = A - x_2 \tag{16.35c}$$

Let us take a new variable z such that

$$A - x_2 = z;\ dx_2 = -dz$$

Eliminating the time variable by division, we obtain

$$\frac{1}{\tau}\frac{dx_1}{dz} = -\frac{A-z}{z} = 1 - \frac{A}{z}$$

This first-order equation is readily integrated.

$$\frac{1}{\tau}x_1 = z - A\ln z + C$$

or

$$\frac{1}{\tau}x_1(t) = A - x_2(t) - A\ln(A - x_2(t)) + C \tag{16.36a}$$

where the constant of integration C is determined by the initial conditions, i.e.,

$$C = \frac{1}{\tau}x_1(0) - A + x_2(0) + A\ln(A - x_2(0)) \tag{16.36b}$$

Therefore, the trajectory equation becomes

$$\frac{1}{\tau}(x_1 - x_1(0)) = -(x_2 - x_2(0)) - A\ln\left(\frac{A-x_2}{A-x_2(0)}\right) \tag{16.36c}$$

The phase portrait is shown in Fig. 16.25a for $A > 0$.

For the case of initial state point at the origin ($x_1(0) = x_2(0) = 0$), Eqn. (16.36c) reads

$$\frac{1}{\tau}x_1 = -x_2 - A\ln\left(\frac{A-x_2}{A}\right) \tag{16.36d}$$

The phase trajectory described by this equation is shown in Fig. 16.25a as the curve Γ_0. It is seen that the trajectory is asymptotic to the line $x_2 = A$, which is the final velocity.

For an initial state point ($x_1(0), x_2(0)$), the trajectory will have the same shape as the curve Γ_0, except that it is shifted horizontally so that it passes through the point ($x_1(0)$, $x_2(0)$). This is obvious from Eqn. (16.36c) which can be written as

$$\frac{1}{\tau}(x_1 - K) = -x_2 - A\ln\left(\frac{A-x_2}{A}\right)$$

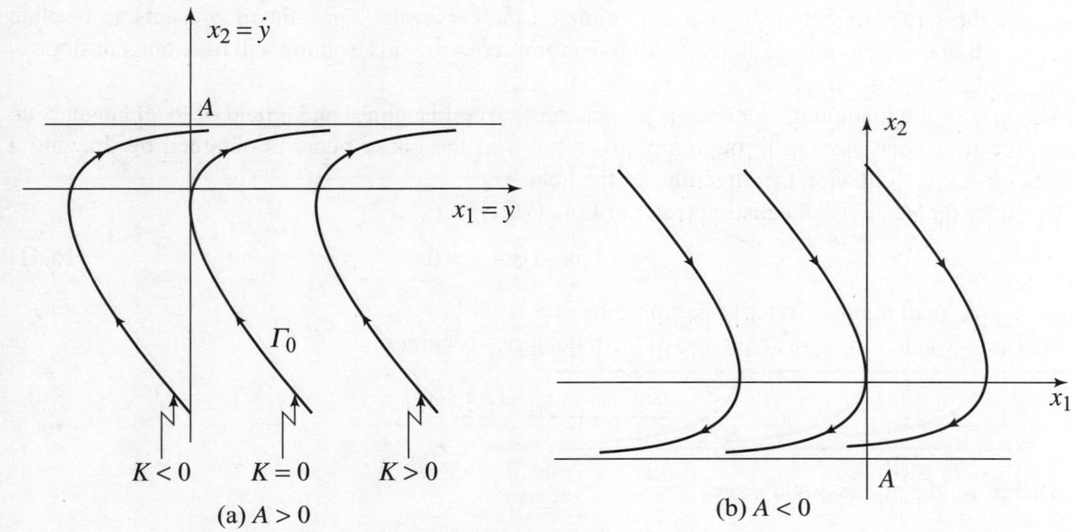

Fig. 16.25 *Phase portraits for system (16.35)*

where $$K = x_1(0) + \tau x_2(0) + \tau A \ln\left(\frac{A - x_2(0)}{A}\right)$$

For an initial state point $(x_1(0), x_2(0))$, the trajectory is Γ_0 shifted horizontally by K units. The phase portrait for $A < 0$ is shown in Fig. 16.25b. In the special case of $A = 0$, the phase portrait consists of a family of straight lines of slope $-1/\tau$.

16.8.2 The Method of Isoclines

Consider a time-invariant second-order system described by equations of the form (refer to Eqns (16.30b))

$$\dot{x}_1 = x_2$$
$$\dot{x}_2 = f(x_1, x_2) \tag{16.37}$$

The equation of the trajectories is

$$\frac{dx_2}{dx_1} = \frac{f(x_1, x_2)}{x_2} \tag{16.38}$$

At a point (x_1^*, x_2^*) in the phase plane, the slope m^* of the tangent to the trajectory can be determined from

$$\frac{f(x_1^*, x_2^*)}{x_2^*} = m^* \tag{16.39}$$

An *isocline* is defined to be the locus of the points corresponding to a given constant slope m of the trajectories on the phase plane. All trajectories passing through the points on the curve

$$f(x_1, x_2) = m x_2 \tag{16.40}$$

will have the same tangent slope m at the points on the curve; the curve thus represents an isocline corresponding to trajectories of slope m. All trajectories crossing this isocline will have tangent slope m at the points on the isocline.

The idea of the method of isoclines is to construct several isoclines and a field of local tangents m. Then, the trajectory passing through any given point in the phase plane is obtained by drawing a continuous curve following the directions of the field.

Consider the Van der Pol equation (refer to Eqn. (16.1))

$$\ddot{y} + \mu(y^2 - 1)\dot{y} + y = 0 \tag{16.41}$$

where μ is a small number (refer to Example 16.7).

With $x_1 = y$ and $x_2 = \dot{y}$, the equation of the trajectories becomes

$$\frac{dx_2}{dx_1} = \frac{-\mu(x_1^2 - 1)x_2 - x_1}{x_2}$$

Therefore, the points on the curve

$$\frac{-\mu(x_1^2 - 1)x_2 - x_1}{x_2} = m$$

all have the same slope m. The isocline equation becomes

$$x_2 = \frac{x_1}{(\mu - \mu x_1^2) - m}$$

By taking m of different values, different isoclines can be obtained. Short line segments are drawn on the isoclines to generate a field of tangent directions. A trajectory starting at any point can be constructed by drawing short lines from one isocline to another at average slope corresponding to the two adjoining isoclines as shown in Fig. 16.26.

Of course, the construction is much simpler if isoclines are straight lines.

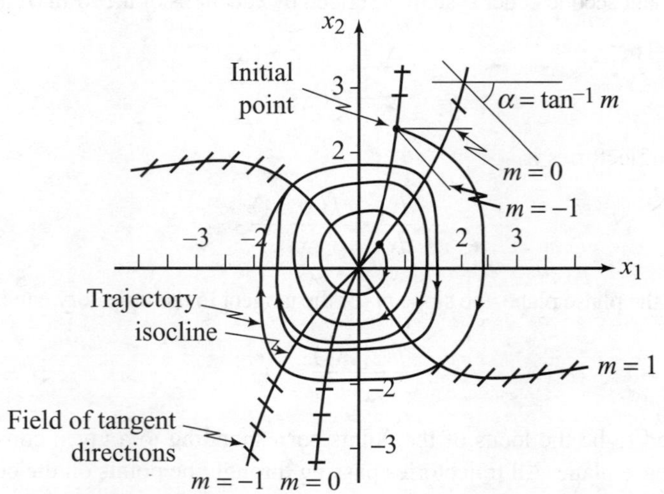

Fig. 16.26 *Phase portrait of the Van der Pol equation*

Example 16.6 The satellite in Example 16.4 is now placed in a feedback configuration in order to maintain the attitude θ at 0°. This feedback control system, called an attitude control system, is shown in Fig. 16.27. When θ is other than 0°, the appropriate thruster will fire to force θ towards 0°. When $\theta = x_1$ is greater than 0°, u (torque T) = $-U$, and the trajectories of Fig. 16.24b (corresponding to $A < 0$) apply. When $\theta = x_1$ is less than 0°, u (torque T) = U, and the trajectories of Fig. 16.24a (corresponding to $A > 0$) apply. Note that the switching of u occurs at $x_1 = 0$. Thus the line $x_1 = 0$ (the x_2-axis) is called the *switching line*. From this discussion we see that Fig. 16.28a illustrates a typical trajectory for the system corresponding to the initial condition (x_1^0, x_2^0). The system response is thus a periodic motion. Figure 16.28b shows many closed curves for different initial conditions.

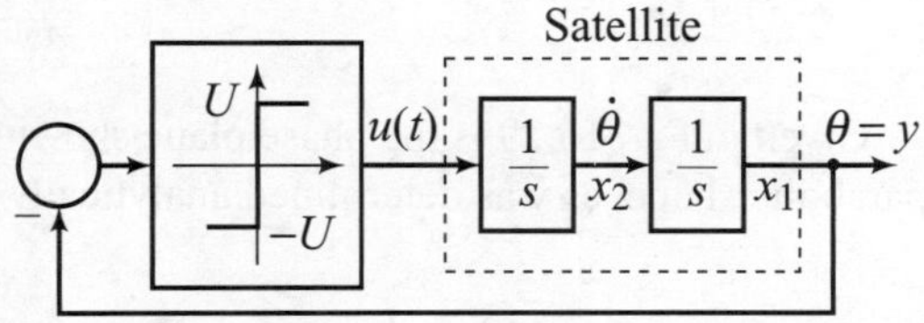

Fig. 16.27 *A satellite-attitude control system*

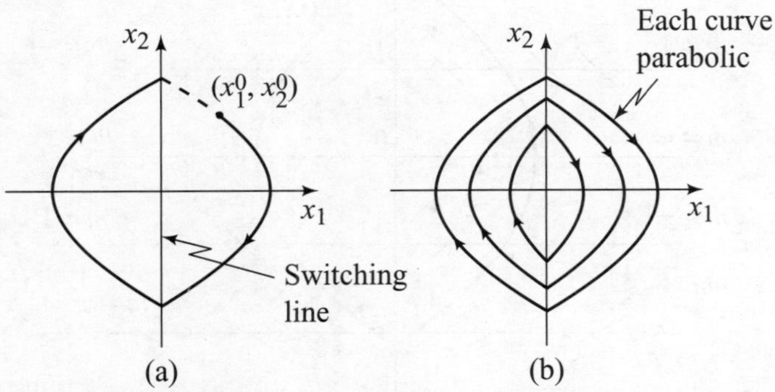

Fig. 16.28 *Typical trajectories for the system of Fig.16.27*

By controlling the switching line in the phase plane, we can control the performance of the attitude control system. This simple control strategy leads to a robust nonlinear control structure: the *variable structure sliding mode control*. Refer to the companion book [155] for details.

In the following we obtain the phase portrait of the closed-loop system of Fig. 16.27 using the method of isoclines. The purpose here is to illustrate the method of isoclines.

The state equations are

$$\dot{x}_1 = x_2$$
$$\dot{x}_2 = -U \operatorname{sgn} x_1$$

where

$$\operatorname{sgn} x_1 = \begin{bmatrix} 1, x_1 > 0 \\ -1, x_1 < 0 \end{bmatrix}$$

Then

$$m = \frac{dx_2}{dx_1} = \frac{-U \operatorname{sgn} x_1}{x_2}$$

Suppose that U is normalized to a value of unity for convenience. Then

$$x_2 = -\frac{1}{m} \operatorname{sgn} x_1$$

For $x_1 > 0$, sgn $x_1 = 1$, and the isocline equation is

$$x_2 = -\frac{1}{m};\ x_1 > 0$$

For $x_1 < 0$, sgn $x_1 = -1$, and the isocline equation is

$$x_2 = \frac{1}{m};\ x_1 < 0$$

Given in Fig. 16.29 is the phase plane showing the isoclines and a typical phase trajectory. Note the parabolic shape, as was determined analytically earlier in this example.

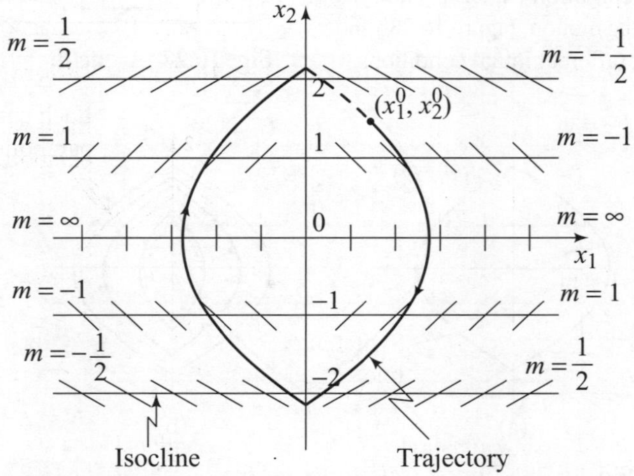

Fig.16.29 *The isoclines and a typical trajectory for the system of Fig. 16.27*

16.9 SYSTEM ANALYSIS ON THE PHASE PLANE

In the phase plane analysis of nonlinear systems, two points should be kept in mind:

1. Phase plane analysis of nonlinear systems is related to that of linear systems because the local behavior of a nonlinear system can be approximated by a linear system behavior.
2. Still, nonlinear systems can display much more complicated patterns on the phase plane, such as multiple equilibrium points, and limit cycles.

Consider a time-invariant second-order system described by equations of the form (refer to Eqns (16.37))

$$\dot{x}_1 = x_2;\ \dot{x}_2 = f(x_1, x_2) \tag{16.42}$$

Elimination of independent variable t gives the equation of the trajectories of phase plane (refer to Eqn. (16.38)):

$$\frac{dx_2}{dx_1} = \frac{f(x_1, x_2)}{x_2} \tag{16.43}$$

In this equation, x_1 and x_2 are independent and dependent variables, respectively. Integration of the equation, analytically, graphically or numerically, for various initial conditions yields a family of phase trajectories which displays the general features of the system transients.

16.9.1 Singular Points

Every point (x_1, x_2) of the phase plane has associated with it the slope of the trajectory which passes through that point. The slope m at the point (x_1, x_2) is given by the equation,

$$m = \frac{dx_2}{dx_1} = \frac{f(x_1, x_2)}{x_2}$$

With the function $f(x_1, x_2)$ assumed to be single valued, there is usually a definite value for this slope at any given point in phase plane. This implies that the phase trajectories will not intersect. The only exceptions are the *singular points* at which the trajectory slope is indeterminate:

$$\frac{dx_2}{dx_1} = \frac{0}{0} = \frac{f(x_1, x_2)}{x_2} \tag{16.44a}$$

Many trajectories may intersect at such points. This indeterminacy of the slope accounts for the adjective 'singular'.

Singular points are very important features on the phase plane. Examination of the singular points can reveal a great deal of information about the properties of a system. In fact, the stability of linear systems is uniquely characterized by the nature of their singular points. For nonlinear systems, besides singular points, there may be more complex features such as limit cycles.

We need to know the following:

1. Where will the singular points be and how many will be there?
2. What is the behavior of trajectories (i.e., the system) in the vicinity of a singular point?

The first question is answered by our definition of the singular point. There will be singular points at all the points of the phase plane for which the slope of the trajectory is undefined. These points are given by the solution of the equations,

$$x_2 = 0;\ f(x_1, x_2) = 0 \tag{16.44b}$$

Singular points of the nonlinear system (16.42), thus, lie on the x_1-axis of the phase plane.

Since at singular points on the phase plane, $\dot{x}_1 = \dot{x}_2 = 0$, these points in fact correspond to the *equilibrium states* of the nonlinear system. We know a nonlinear system often has multiple equilibrium states.

To determine the behavior of the trajectories in the vicinity of a singular point (equilibrium state of the nonlinear system), we first linearize the nonlinear equations at the singular point and then determine the nature of phase trajectories around the singular point by linear system analysis. If the singular point of interest is not at the origin, by defining the difference between the original state and the singular point as a new set of state variables, one can always shift the singular point to the origin. Therefore, without loss of generality, we can simply consider Eqns (16.42) with a singular point at **0**. Using Taylor series expansion, Eqns (16.42) can be rewritten as

$$\dot{x}_1 = x_2$$
$$\dot{x}_2 = ax_1 + bx_2 + g_2(x_1, x_2)$$

where $g_2(\cdot)$ contains higher-order terms.

In the vicinity of the origin, the higher-order terms can be neglected and, therefore, the nonlinear system trajectories essentially satisfy the linearized equations,

$$\dot{x}_1 = x_2$$
$$\dot{x}_2 = ax_1 + bx_2$$

Transforming these equations into a scalar second-order equation, we get

$$\ddot{x}_1 = a x_1 + b\dot{x}_1$$

Therefore, we will simply consider the second-order linear system described by

$$\ddot{y} + 2\zeta\omega_n \dot{y} + \omega_n^2 y = 0 \tag{16.45a}$$

The characteristic roots of this equation are assumed to be λ_1 and λ_2:

$$\ddot{y} + 2\zeta\omega_n \dot{y} + \omega_n^2 y = (s - \lambda_1)(s - \lambda_2) = 0 \tag{16.45b}$$

The corresponding canonical state model is

$$\dot{x}_1 = x_2;\ \dot{x}_2 = -2\zeta\omega_n x_2 - \omega_n^2 x_1 \tag{16.45c}$$

and the differential equation of the trajectories is

$$\frac{dx_2}{dx_1} = \frac{-2\zeta\omega_n x_2 - \omega_n^2 x_1}{x_2} \tag{16.45d}$$

By inspection of this equation it is easily seen that at $x_1 = x_2 = 0$, the slope dx_2/dx_1 is indeterminate:

$$\frac{dx_2}{dx_1} = \frac{0}{0}$$

In the following, we will discuss the behavior of the trajectories in the vicinity of this point with undefined slope (the *singular point*).

According to the values of λ_1 and λ_2, one is led to distinguish between the six types of singular points shown in Fig. 16.30. Let us examine each of these cases in detail.

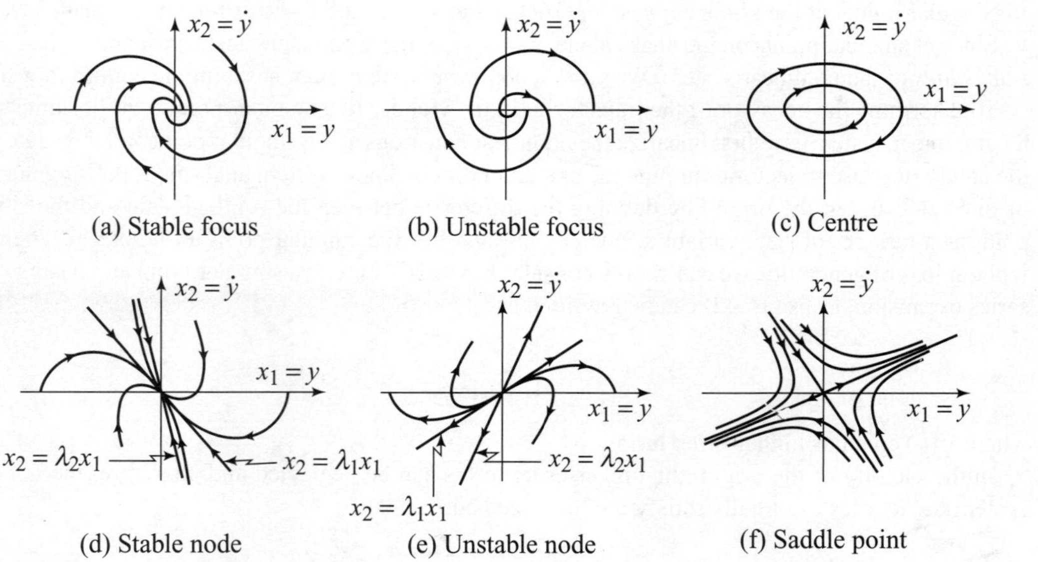

Fig. 16.30 *Phase portraits for system (16.45)*

1. ***Stable system with complex roots***

$$\lambda_1 = -\alpha + j\beta,\ \lambda_2 = -\alpha - j\beta;\ \alpha > 0,\ \beta > 0$$

The response $$y(t) = C_1 e^{-\alpha t}\sin(\beta t + C_2) \tag{16.46}$$
where the constants C_1 and C_2 are determined by the initial conditions.

Using Eqn. (16.46), we can construct a phase portrait on the (x_1, x_2)-plane with $x_1 = y$ and $x_2 = \dot{y}$. A typical phase trajectory is shown in Fig. 16.30a which is a logarithmic spiral into the singular point. This type of singular point is called a *stable focus*.

2. ***Unstable system with complex roots***

$$\lambda_1 = \alpha + j\beta,\ \lambda_2 = \alpha - j\beta;\ \alpha > 0,\ \beta > 0$$

The response $$y(t) = C_1 e^{\alpha t}\sin(\beta t + C_2) \tag{16.47}$$
The transient is an exponentially increasing sinusoid; the phase trajectory on the $(x_1 = y, x_2 = \dot{y})$-plane is a logarithmic spiral expanding out of the singular point (Fig. 16.30b). This type of singular point is called an *unstable focus*.

3. ***Marginally stable system with complex roots***

$$\lambda_1 = j\beta,\ \lambda_2 = -j\beta;\ \beta > 0$$

The response $$y(t) = C_1 \sin(\beta t + C_2) \tag{16.48}$$
The phase trajectories are closed curves (elliptical) concentric with the singular point (Fig. 16.30c). This type of singular point is called a *center*, or a *vortex*.

4. ***Stable system with real roots*** Assume that λ_1 and λ_2 are two real, distinct roots in the left half of the s-plane; λ_1 is the root with the smaller modulus. The response

$$y(t) = C_1 e^{\lambda_1 t} + C_2 e^{\lambda_2 t} \tag{16.49a}$$

It is an overdamped system. The phase portrait in the vicinity of the singular point on the $(x_1 = y, x_2 = \dot{y})$-plane is shown in Fig. 16.30d. Such a singular point is called a *stable node*.

The phase portrait has two straightline trajectories defined by the equations,

$$x_2(t) = \lambda_1 x_1(t);\quad x_2(t) = \lambda_2 x_1(t) \tag{16.49b}$$

It can easily be verified that these trajectories satisfy the differential equation of the given system.

The transient term $e^{\lambda_2 t}$ decays faster than the term $e^{\lambda_1 t}$. Therefore, as t increases indefinitely, $x_1 \to C_1 e^{\lambda_1 t} \to 0$, and $x_2 \to \lambda_1 C_1 e^{\lambda_1 t} \to 0$, so that all the trajectories are tangential at the origin to the straightline trajectory $x_2(t) = \lambda_1 x_1(t)$. The other straightline trajectory, $x_2(t) = \lambda_2 x_1(t)$, is described only if the initial conditions are such that $x_2(0) = \lambda_2 x_1(0)$.

For stable systems with repeated real roots, the two straightline trajectories coalesce into a single trajectory, again with the slope determined by the root value.

5. ***Unstable system with positive real roots*** Assume that λ_1 and λ_2 are two real distinct roots in the right half of the s-plane; λ_1 is the smaller root. The phase portrait in the vicinity of the singular point on the $(x_1 = y, x_2 = \dot{y})$-plane is shown in Fig. 16.30e. Such a singular point is called an *unstable node*.

All trajectories emerge from the singular point and go to infinity. The trajectories are tangential at the origin to the straightline trajectory, $x_2(t) = \lambda_1 x_1(t)$.

6. ***Unstable system with one negative real root and one positive real root*** The phase portrait in the vicinity of the singular point on the $(x_1 = y, x_2 = \dot{y})$-plane is shown in Fig. 16.30f. Such a singular point is called a *saddle*.

There are two straightline trajectories with slopes defined by the root values. The straightline due to the negative root provides a trajectory that *enters* the singular point, while the straight-line trajectory due to the positive root *leaves* the singular point. All other trajectories approach the singular point adjacent to the incoming straight line, then curve away and leave the vicinity of the singular point, eventually approaching the second straight line asymptotically.

Example 16.7 Let us reconsider the Van der Pol equation (16.10). Its linearized version is given in (16.12). The eigenvalues of matrix **A** are

$$\lambda = \frac{\mu}{2} \pm \sqrt{\frac{\mu^2}{4} - 1}\ ;\ \mu > 0$$

For $\frac{\mu^2}{4} < 1$, the origin is an unstable focus, and for $\frac{\mu^2}{4} \geq 1$, the origin is an unstable node.

Example 16.8 Consider the nonlinear system shown in Fig. 16.31. The nonlinear element is an on–off controller with deadzone whose characteristics are shown in the figure.

The differential equation governing the dynamics of the system is given by

$$\ddot{y} + \dot{y} = u; \quad \text{or} \quad \ddot{e} + \dot{e} = -\phi(e) \tag{16.50}$$

where $e = r - y$; r is constant, and

$$\phi(e) = \begin{cases} +1; & e > 1 \\ 0; & -1 < e < 1 \\ -1; & e < -1 \end{cases}$$

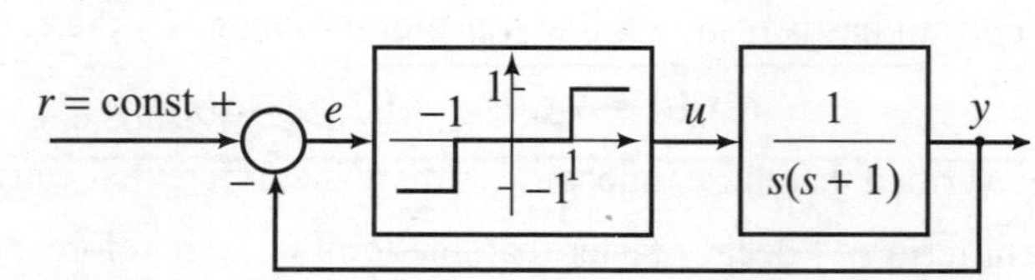

Fig. 16.31 *A system controlled by on–off controller with deadzone*

Choosing the state variables $x_1 = e$ and $x_2 = \dot{e}$, we obtain the following first-order equations:

$$\dot{x}_1 = x_2;\ \ \dot{x}_2 = -x_2 - \phi(x_1)$$

These equations are same as Eqns (16.35c) with $\tau = 1$ and $A = -\phi(x_1)$.

The phase plane may be divided into three regions:

1. *Region* I (defined by $x_1 > 1$): The trajectories in this region are given by the equation (refer to Eqn. (16.36c): $\tau = 1, A = -1$)

$$x_1 - x_1(0) = -(x_2 - x_2(0)) + \ln\left(\frac{1 + x_2}{1 + x_2(0)}\right) \tag{16.51a}$$

The trajectories are asymptotic to the ordinate –1.

2. *Region* II (defined by $-1 < x_1 < 1$): The trajectories in this region are given by the equation (refer to Eqn. (16.36c): $\tau = 1, A = 0$)

$$x_1 - x_1(0) = -(x_2 - x_2(0)) \tag{16.51b}$$

The trajectories are straightlines of slope –1.

3. *Region* III (defined by $x_1 < -1$): The trajectories in this region are given by the equation (refer to Eqn. (16.36c): $\tau = 1, A = 1$)

$$x_1 - x_1(0) = -(x_2 - x_2(0)) - \ln\left(\frac{1 - x_2}{1 - x_2(0)}\right) \tag{16.51c}$$

The trajectories are asymptotic to the ordinate +1.

For a step input $r = 3$ and zero initial conditions, the initial point of the phase trajectory is located at P in Fig. 16.32. The figure also shows a phase trajectory, constructed using Eqns (16.51).

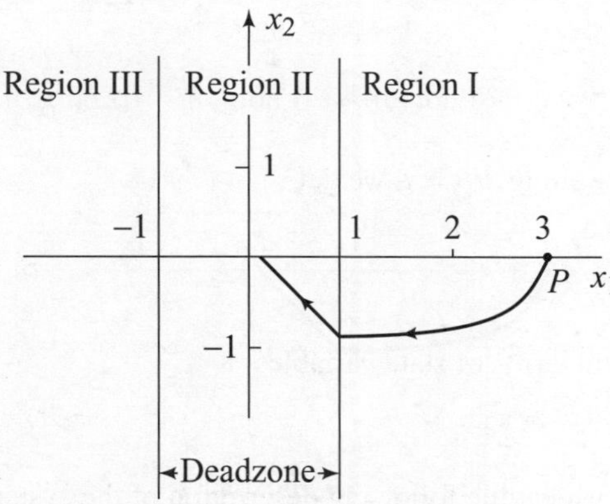

Fig. 16.32 *A typical trajectory for the system in Fig. 16.31*

It is important to note that a small deadzone region is not always undesirable in on–off controllers. Let us investigate the behavior of the system of Fig. 16.31 using on–off with no deadzone as a controller. For such a controller, the width of region II (corresponding to deadzone) in the phase plane reduces to zero. The phase trajectory of such a system with $r = 3$ is shown in Fig. 16.33: $e(t)$ oscillates about the origin with ever-decreasing amplitude and ever-increasing frequency.

Comparison of Figs 16.32 and 16.33 reveals that deadzone in on–off controller characteristic helps to reduce system oscillations, thereby reducing settling time. However the on–off controller with deadzone drives the system to a point within the deadzone width. A large deadzone would of course cause the steady-state performance of the system to deteriorate.

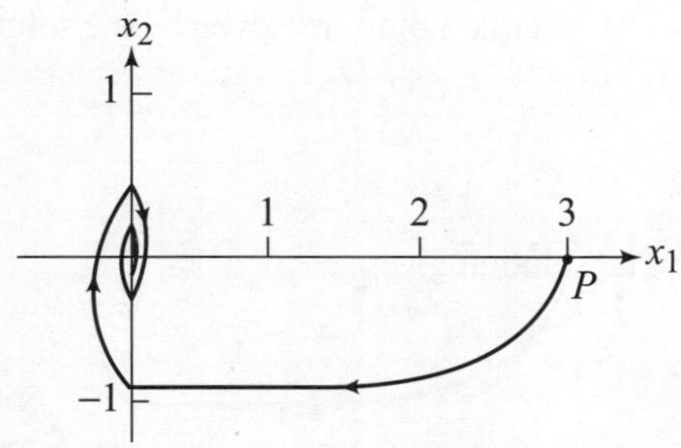

Fig. 16.33 *Phase trajectory for the system in Fig. 16.31 when the deadzone is absent*

Example 16.9 Let us investigate the performance of a second-order position control system with Coulomb friction. Figure 16.34 is a model for a motor position servo with Coulomb friction on the motor shaft. The dynamics of the system is described by the following differential equation:

$$K e - T_c \operatorname{sgn}(\dot{y}) = J\ddot{y} + B\dot{y}$$

where T_c is the Coulomb frictional torque.

For constant input r, $\dot{y} = -\dot{e}$ and $\ddot{y} = -\ddot{e}$. Therefore,

$$J\ddot{e} + B\dot{e} + T_c \operatorname{sgn}(\dot{e}) + Ke = 0$$

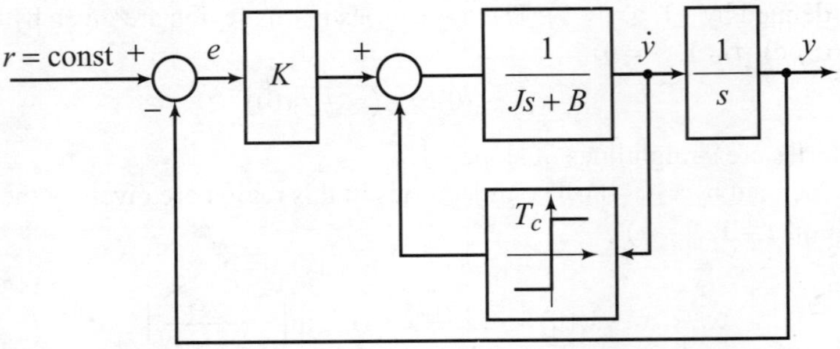

Fig. 16.34 *A motor position servo with Coulomb friction on the motor shaft*

or

$$\frac{J}{B}\ddot{e}+\dot{e}+\frac{T_c}{B}\text{sgn}(\dot{e})+\frac{K}{B}e=0 \tag{16.52}$$

Letting $J/B = \tau$, we get

$$\tau\ddot{e}+\dot{e}+\frac{T_c}{B}\text{sgn}(\dot{e})+\frac{K}{B}e=0 \tag{16.53}$$

In terms of state variables

$$x_1 = e; \quad x_2 = \dot{e}$$

we get the following description of the system:

$$\dot{x}_1 = x_2 \quad \tau\dot{x}_2 = -\frac{K}{B}x_1 - x_2 - \frac{T_c}{B}\text{sgn}(x_2) \tag{16.54}$$

The singular points are given by the solution of the equations (refer to Eqns (16.34b))

$$0 = x_2; \quad 0 = -\frac{K}{B}x_1 - x_2 - \frac{T_c}{B}\text{sgn}(x_2)$$

The solution gives

$$x_1 = -\frac{T_c}{K}\text{sgn}(x_2)$$

Thus, there are two singular points. Their location can be interpreted physically—they are at a value of $e = x_1$, such that $|Ke| = |T_c|$, i.e., the drive torque is exactly equal to the Coulomb-friction torque. We note that both the singular points are on the x_1-axis ($x_2 \equiv 0$) and that the singular point given by $x_1 = T_c/K$ is related to the lower-half phase plane (x_2 negative) and the singular point given by $x_1 = -T_c/K$ is related to the upper-half phase plane (x_2 positive).

Let us now investigate the stability of the singular points. For $\dot{e} > 0$, Eqn. (16.53) may be expressed as

$$\tau\frac{d^2}{dt^2}\left(e+\frac{T_c}{K}\right)+\frac{d}{dt}\left(e+\frac{T_c}{K}\right)+\frac{K}{B}\left(e+\frac{T_c}{K}\right)=0 \tag{16.55}$$

This is a linear second-order system with the singular point at $(-T_c/K, 0)$ on the $(e, \dot{e})$-plane. The characteristic equation of this system is given by

$$\lambda^2 + \frac{1}{\tau}\lambda + \frac{K}{\tau B} = 0$$

Let us assume the following parameter values for the system under consideration:

$$(K/B) = 5, \tau = 4 \tag{16.56}$$

With these parameters, the roots of the characteristic equation are complex conjugate with negative real parts; the singular point is therefore a stable focus (refer to Fig. 16.30a).

Let us now investigate the system behavior when large inputs are applied. Phase trajectories may be obtained by solving the following second-order differential equations for given initial state points (refer to Eqns (16.45b)–(16.46)).

Region I (defined by $x_2 > 0$):

$$4\ddot{z} + \dot{z} + 5z = 0; \quad z = x_1 + \frac{T_c}{K}; \quad x_2 = \dot{x}_1$$

Region II (defined by $x_2 < 0$):

$$4\ddot{z} + \dot{z} + 5z = 0; \quad z = x_1 - \frac{T_c}{K}; x_2 = \dot{x}_1$$

Figure 16.35 shows a few phase trajectories. It is observed that for small as well as large inputs, the resulting trajectories terminate on a line along the x_1-axis from $-T_c/K$ to $+T_c/K$, i.e., the line joining the singular points. Therefore, the system with Coulomb friction is stable; however, there is a possibility of large steady-state error.

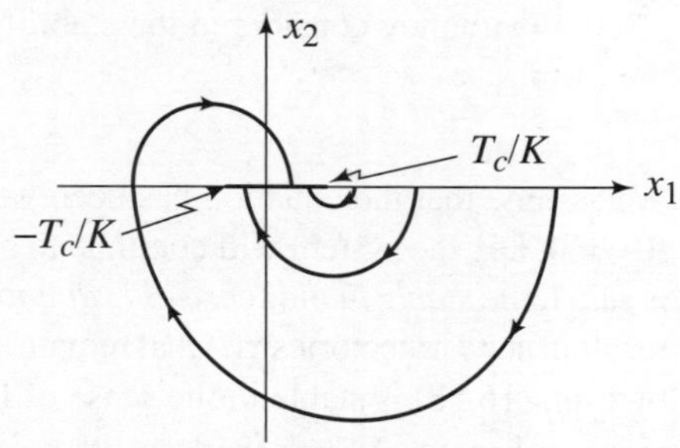

Fig. 16.35 *Phase portrait for the system in Fig. 16.34*

16.9.2 Limit Cycles

In the phase portrait of the nonlinear Van der Pol equation, shown in Fig. 16.26, one observes that there is a closed-curve in the phase portrait. Trajectories inside the curve and those outside the curve all tend to this curve, while a motion started on this curve will stay on it forever. This curve is an instance of the so-called "limit cycle" phenomenon. Limit cycles are unique features of nonlinear systems.

On the phase plane, *a limit cycle is defined as an isolated closed curve*. The trajectory has to be both closed, indicating the periodic nature of motion, and isolated, indicating the limiting nature of the cycle (with neighboring trajectories converging to or diverging from it). Thus, while there are many closed curves in Fig. 16.28b (in the satellite system in Example 16.5), these are not limit cycles because they are not isolated.

A limit cycle is stable if all trajectories in the vicinity of the limit cycle converge to it as $t \to \infty$ (Fig. 16.23a). A limit cycle is unstable if all trajectories in the vicinity of the limit cycle diverge from it as $t \to \infty$ (Fig. 16.23b).

16.10 LYAPUNOV STABILITY

It should be obvious by now that stability plays a major role in control systems design. We have introduced earlier in Chapters 5 and 14, the concept of stability based on the dynamic evolution of the system state in response to arbitrary initial state, representing initial energy storage. State variable model

$$\dot{\mathbf{x}}(t) = \mathbf{A}\mathbf{x}(t); \mathbf{x}(t=0) \triangleq \mathbf{x}^0 \tag{16.57}$$

is most appropriate to study dynamic evolution of the state $\mathbf{x}(t)$ in response to the initial state $\mathbf{x}^0$ with zero external input. At the origin of the state space, $\dot{\mathbf{x}}(t) = \mathbf{0}$ for all t; the origin is thus the *equilibrium point* of the system, and $\mathbf{x}^e = \mathbf{0}$ is the *equilibrium state*. This system is *marginally stable* if for all possible initial states $\mathbf{x}^0$, $\mathbf{x}(t)$ remains thereafter within finite bounds for $t > 0$. This is true if none of the eigenvalues of $\mathbf{A}$ are in the right half of the complex plane and eigenvalues on the imaginary axis, if any, are simple (A multiple eigenvalue on the imaginary axis would have a response that grows in time and could not be stable). Furthermore, the system is *asymptotically stable* if for all possible initial states $\mathbf{x}^0$, $\mathbf{x}(t)$ eventually decays to zero as t approaches infinity. This is true if all the eigenvalues of $\mathbf{A}$ are inside the left half of the complex plane.

AM Lyapunov considered the stability of general nonlinear systems described by state equation of the form,

$$\dot{\mathbf{x}}(t) = \mathbf{f}(\mathbf{x}(t)); \mathbf{x}(0) \triangleq \mathbf{x}^0 \tag{16.58}$$

We assume that the equation has been written so that $\mathbf{x} = \mathbf{0}$ is an *equilibrium point*, which is to say that $\mathbf{f}(\mathbf{0}) = \mathbf{0}$, i.e., the system will continue to be in *equilibrium state* $\mathbf{x}^e = \mathbf{0}$ for all time. This equilibrium point is said to be *stable in the sense of Lyapunov* if we are able to select a bound on initial condition $\mathbf{x}^0$ that will result in state trajectories $\mathbf{x}(t)$ that remain within a chosen finite limit. More formally, the system described by Eqn. (16.58) is stable in the sense of Lyapunov at $\mathbf{x} = \mathbf{0}$ if, for every real number $\varepsilon > 0$ there exists a real number $\delta > 0$, such that[1] $\|\mathbf{x}(0)\| < \delta$ results in $\|\mathbf{x}(t)\| < \varepsilon$ for all $t \geq 0$. The system is asymptotically stable at $\mathbf{x} = \mathbf{0}$ if it is stable in the sense of Lyapunov and in addition, the state $\mathbf{x}(t)$ approaches zero as time t approaches infinity. Responses that are stable in the sense of Lyapunov, and asymptotically stable are shown in Figs 16.36.

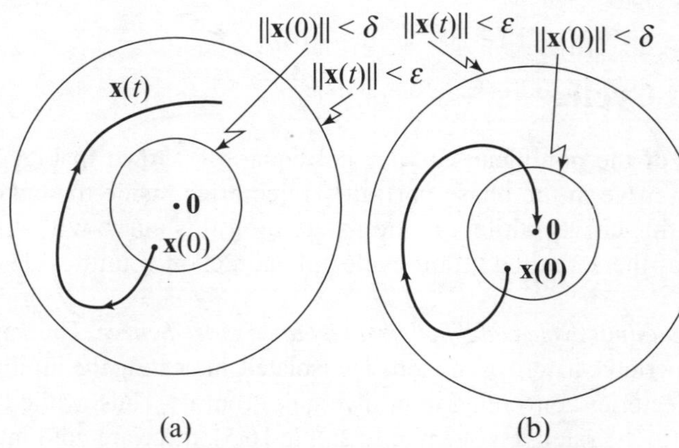

Fig. 16.36 *Stability definitions*

[1] $\|\mathbf{x}\|$ = *a norm of vector* $\mathbf{x} = \left(\sum_{i=1}^{n} x_i^2\right)^{1/2} = \left(\mathbf{x}^T\mathbf{x}\right)^{1/2}$ is a measure of the length of vector $\mathbf{x}$.

No new results are obtained by the use of Lyapunov's method for the stability analysis of linear time-invariant systems. Simple and powerful methods discussed in earlier chapters are adequate for such systems. However, Lyapunov functions supply certain performance indices and synthesis data of linear time-invariant systems [155].

The Lyapunov's method of stability analysis, in principle, is the most general method for determination of stability of nonlinear systems. The major drawback which seriously limits its use in practice is the difficulty often associated with the construction of the Lyapunov function required by the method. Unfortunately, there is no universal method for selecting the Lyapunov function which is unique for a given nonlinear system. Several techniques have been devised for the systematic construction of Lyapunov functions; each is applicable to a particular class of systems. If a Lyapunov function cannot be found, it in no way implies that the system is unstable. It only means that our attempt in trying to establish the stability of an equilibrium state has failed. Therefore, faced with specific systems, one has to use experience, intuition, and physical insights to search for an appropriate Lyapunov function. An elegant and powerful Lyapunov analysis may be possible for complex systems if engineering insight and physical properties are properly exploited. Inspite of these limitations, Lyapunov's method is the most powerful technique available today for the stability analysis of nonlinear systems. For linear time-invariant systems, the quadratic function (refer to Eqn. (14.7a)) is adequate for demonstrating Lyapunov stability.

Since our brief presentation is meant as a first exposure of the student to Lyapunov stability, we restrict our choice of Lyapunov functions to quadratic form in our examples on stability analysis of both linear and nonlinear systems. For further details on the subject, refer to the companion book [155].

16.10.1 Lyapunov Stability Theorem

To prove stability results for nonlinear systems, Lyapunov introduced a function that has many of the properties of energy. For the system (16.58), sufficient conditions of stability are as follows.

Suppose that there exists a scalar function $V(\mathbf{x})$ which satisfies the following properties:

1. $V(\mathbf{x}) > 0;\ \mathbf{x} \neq \mathbf{0}$
2. $V(\mathbf{0}) = 0$ (16.59)
3. $V(\mathbf{x})$ is continuous and has continuous partial derivatives with respect to all components of $\mathbf{x}$.
4. $\dot{V}(\mathbf{x}) \leq 0$ along trajectories of Eqn. (16.58)

We call $V(\mathbf{x})$ having these properties a *Lyapunov function* for the system. Properties 1 and 2 mean that, like energy, $V(\mathbf{x})>0$ if any state is different from zero, but $V(\mathbf{x}) = 0$ when the state is zero. Property 3 ensures that $V(\mathbf{x})$ is a smooth function and generally has the shape of a bowl near the equilibrium.

A visual analysis may be obtained by considering the surface

$$V(x_1, x_2) = \frac{1}{2}p_1x_1^2 + \frac{1}{2}p_2x_2^2;\ p_1 > 0, p_2 > 0 \quad (16.60)$$

This is a paraboloid (a solid generated by rotation of parabola about its axis of symmetry) surface as shown in Fig. 16.37. The value $V(x_1, x_2) = k_i$ (a constant) is represented by the intersection of $V(x_1, x_2)$ surface and

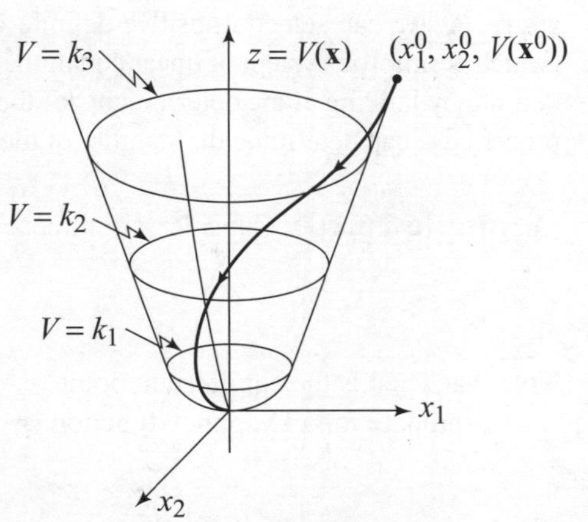

Fig. 16.37 *Constant-V curves*

the plane $z = k_i$. This intersection results in a closed curve; an oval. If one plots a trajectory from the point $(x_1^0, x_2^0, V(x^0))$, the trajectory crosses the ovals $V(x_1, x_2) = k_i$ for successively smaller values of $V(x_1, x_2)$ and moves towards the point corresponding to $V(x_1, x_2) = 0$, which is the equilibrium point. Figure 16.37 shows a typical trajectory.

Property 4 guarantees that any trajectory moves so as never to climb higher on the bowl than where it started out. If property 4 was made stronger so that $\dot{V}(\mathbf{x}) < 0$ for $\mathbf{x} \neq \mathbf{0}$, then the trajectory must be drawn to the origin (The trajectory in Fig. 16.37, in fact, corresponds to this case, i.e., $\dot{V}(\mathbf{x}) < 0$).

The *Lyapunov stability theorem states that, given the system of equations* $\dot{\mathbf{x}} = \mathbf{f}(\mathbf{x})$ *with* $\mathbf{f}(\mathbf{0}) = \mathbf{0}$, *if there exists a Lyapunov function for this equation, then the origin is stable in the sense of Lyapunov; in addition, if* $\dot{V}(\mathbf{x}) < 0$ for $\mathbf{x} \neq \mathbf{0}$, *then the stability is asymptotic.*

16.10.2 Lyapunov Function for Linear Systems

For linear time-invariant systems, the quadratic function is adequate for demonstrating Lyapunov stability. Consider the function,

$$V(\mathbf{x}) = \mathbf{x}^T\mathbf{P}\mathbf{x} \tag{16.61}$$

where $\mathbf{P}$ is a symmetric positive definite matrix. The quadratic function (16.61) satisfies properties 1, 2 and 3 of a Lyapunov function. We need to consider property 4, the derivative condition. To compute the derivative of $V(\mathbf{x})$, we use the chain rule to get

$$\begin{aligned} \dot{V}(\mathbf{x}) &= \frac{d}{dt}(\mathbf{x}^T\mathbf{P}\mathbf{x}) \\ &= \dot{\mathbf{x}}^T\mathbf{P}\mathbf{x} + \mathbf{x}^T\mathbf{P}\dot{\mathbf{x}} \end{aligned} \tag{16.62}$$

Using Eqns (16.57) and (16.62), we get

$$\dot{V}(\mathbf{x}) = \mathbf{x}^T\mathbf{A}^T\mathbf{P}\mathbf{x} + \mathbf{x}^T\mathbf{P}\mathbf{A}\mathbf{x} = \mathbf{x}^T(\mathbf{A}^T\mathbf{P} + \mathbf{P}\mathbf{A})\mathbf{x} = -\mathbf{x}^T\mathbf{Q}\mathbf{x}$$

where $\mathbf{Q}$ is defined by

$$\mathbf{A}^T\mathbf{P} + \mathbf{P}\mathbf{A} = -\mathbf{Q} \tag{16.63}$$

Lyapunov showed that for any positive definite $\mathbf{Q}$, the solution $\mathbf{P}$ of the *Lyapunov Eqn.* (16.63) is positive definite if and only if all the eigenvalues of $\mathbf{A}$ have negative real parts. In other words, given the system matrix $\mathbf{A}$, we can select a positive definite $\mathbf{Q}$ (such as identify matrix $\mathbf{I}$), solve the Lyapunov equation (which is simply a system of linear equations in $n(n-1)/2$ unknowns), and test to see whether $\mathbf{P}$ is positive definite by looking at the determinants of the n principal minors (refer to inequalities (14.7b)). From this process, we can determine the stability of the system (16.57) without finding the eigenvalues of matrix $\mathbf{A}$.

Example 16.10 Consider a nonlinear system governed by the equations,

$$\begin{aligned} \dot{x}_1 &= -x_1 + 2x_1^2 x_2 \\ \dot{x}_2 &= -x_2 \end{aligned}$$

Note that $\mathbf{x} = \mathbf{0}$ is the equilibrium point.

A candidate for a Lyapunov function is

$$V = p_{11}x_1^2 + p_{22}x_2^2; p_{11} > 0, p_{22} > 0$$

which is a positive definite function.

Then

$$\frac{dV}{dt} = 2p_{11}x_1\dot{x}_1 + 2p_{22}x_2\dot{x}_2 = 2p_{11}x_1(-x_1 + 2x_1^2x_2) + 2p_{22}x_2(-x_2)$$

$$= -2p_{11}x_1^2(1 - 2x_1x_2) - 2p_{22}x_2^2$$

dV/dt is negative definite if

$$1 - 2x_1x_2 > 0 \tag{16.64}$$

Therefore, for asymptotic stability we require that the condition (16.64) is satisfied. The region of state space where this condition is not satisfied is *possibly* the region of instability. Let us concentrate on the region of state space where this condition is satisfied. The limiting condition for such a region is

$$1 - 2x_1x_2 = 0$$

The dividing lines lie in the first and the third quadrants and are rectangular hyperbolas as shown in Fig. 16.38. In the second and the fourth quadrants, the inequality is satisfied for all values of x_1 and x_2. Figure 16.38 shows the regions of stability and possible instability. Since the choice of the Lyapunov function is not unique, it may be possible to choose another Lyapunov function for the system under consideration which yields a larger region of stability.

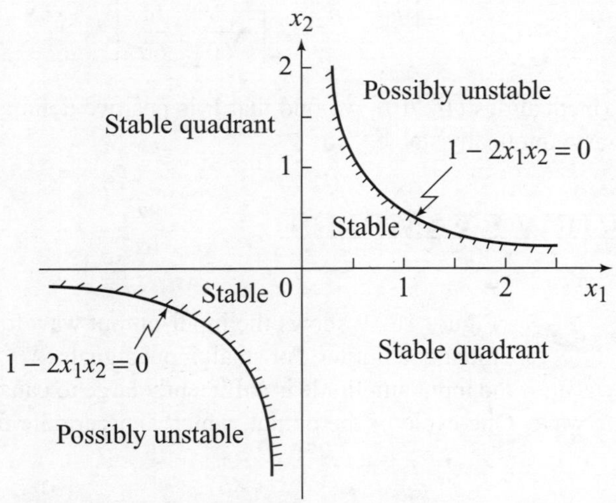

Fig. 16.38 *Stability regions for the nonlinear system of Example 16.10*

Example 16.11 Let us determine the stability of the system described by the following equation:

$$\dot{\mathbf{x}} = \mathbf{A}\mathbf{x}$$

with

$$\mathbf{A} = \begin{bmatrix} -1 & -2 \\ 1 & -4 \end{bmatrix}$$

We will first solve Eqn. (16.63) for **P** for an arbitrary choice of real symmetric positive definite matrix **Q**. We may choose **Q** = **I**, the identity matrix. Equation (16.63) then becomes

$$\mathbf{A}^T\mathbf{P} + \mathbf{PA} = -\mathbf{I}$$

or

$$\begin{bmatrix} -1 & 1 \\ -2 & -4 \end{bmatrix}\begin{bmatrix} p_{11} & p_{12} \\ p_{12} & p_{22} \end{bmatrix} + \begin{bmatrix} p_{11} & p_{12} \\ p_{12} & p_{22} \end{bmatrix}\begin{bmatrix} -1 & -2 \\ 1 & -4 \end{bmatrix} = \begin{bmatrix} -1 & 0 \\ 0 & -1 \end{bmatrix} \quad (16.65)$$

Note that we have taken $p_{12} = p_{21}$. This is because the solution matrix **P** is known to be a positive definite real symmetric matrix for a stable system.

From Eqn. (16.65), we get

$$-2p_{11} + 2p_{12} = -1$$

$$-2p_{11} - 5p_{12} + p_{22} = 0$$

$$-4p_{12} - 8p_{22} = -1$$

Solving for p_{ij}'s, we obtain

$$\mathbf{P} = \begin{bmatrix} p_{11} & p_{12} \\ p_{12} & p_{22} \end{bmatrix} = \begin{bmatrix} \frac{23}{60} & -\frac{7}{60} \\ -\frac{7}{60} & \frac{11}{60} \end{bmatrix}$$

Using Sylvester's test (inequalities (14.7b)), we find that **P** is positive definite. Therefore, the system under consideration is asymptotically stable.

16.11 REVIEW EXAMPLES

Review Example 16.1 Figure 16.39 shows the input-output waveforms of a saturating element or a limiter. For small input signals ($X < S$), the output is proportional to the input. However, if the input amplitude is sufficiently large to cause saturation ($X > S$), the output is a clipped sine wave. One cycle of the output, which is a periodic function of period 2π, is described as follows:

$$y = \begin{cases} Kx; & 0 \le \omega t < \alpha \\ KS; & \alpha \le \omega t < (\pi - \alpha) \\ Kx; & (\pi - \alpha) \le \omega t < (\pi + \alpha) \\ -KS; & (\pi + \alpha) \le \omega t < (2\pi - \alpha) \\ Kx; & (2\pi - \alpha) \le \omega t \le 2\pi \end{cases}$$

where $\alpha = \sin^{-1}(S/X)$

This periodic function has odd symmetry

$$y(\omega t) = -y(-\omega t)$$

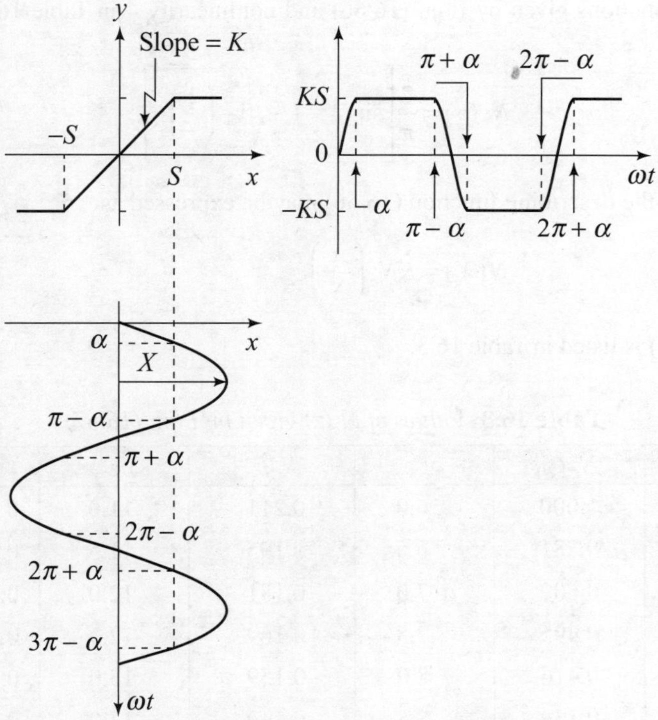

Fig. 16.39 *Fourier-series analysis of a saturating element*

Therefore, the fundamental component of y is given by (refer to Eqn. (16.14d))

$$y_1 = B_1 \sin \omega t$$

where

$$B_1 = \frac{1}{\pi}\int_0^{2\pi} y \sin \omega t \; d(\omega t)$$

Due to symmetry of y (refer to Fig. 16.39), the coefficient B_1 can be calculated as follows:

$$B_1 = \frac{4}{\pi}\int_0^{\frac{\pi}{2}} y \sin\theta \, d\theta = \frac{4K}{\pi}\left[\int_0^{\alpha} X \sin^2\theta \, d\theta + \int_{\alpha}^{\frac{\pi}{2}} S \sin\theta \, d\theta\right] = \frac{4K}{\pi}\left[\frac{X}{2}(\alpha - \sin\alpha\cos\alpha) + S\cos\alpha\right]$$

$$\frac{B_1}{X} = \frac{2K}{\pi}\left[\sin^{-1}\frac{S}{X} - \frac{S}{X}\sqrt{1-\left(\frac{S}{X}\right)^2} + \frac{2S}{X}\sqrt{1-\left(\frac{S}{X}\right)^2}\right]$$

Therefore,

$$N(X) = \begin{cases} \dfrac{2K}{\pi}\left[\sin^{-1}\dfrac{S}{X} + \dfrac{S}{X}\sqrt{1-\left(\dfrac{S}{X}\right)^2}\right]; & X \ge S \\ K; & X < S \end{cases} \qquad (16.66)$$

The describing functions given by Eqn. (16.66) and nonlinearity 4 in Table 16.2, have a common term of the form,

$$N_c(z) = \frac{2}{\pi}\left[\sin^{-1}\frac{1}{z} + \frac{1}{z}\sqrt{1-\left(\frac{1}{z}\right)^2}\right] \tag{16.67}$$

In terms of $N_c(z)$, the describing function (16.66) may be expressed as

$$N(X) = KN_c\left(\frac{X}{S}\right) \tag{16.68}$$

The function $N_c(z)$ is listed in Table 16.3.

Table 16.3 *Values of $N_c(z)$ given by Eqn. (16.67)*

z	$N_c(z)$	z	$N_c(z)$	z	$N_c(z)$
1.0	1.000	6.0	0.211	11.0	0.116
1.5	0.781	6.5	0.195	11.5	0.111
2.0	0.609	7.0	0.181	12.0	0.106
2.5	0.495	7.5	0.169	12.5	0.102
3.0	0.416	8.0	0.159	13.0	0.0978
3.5	0.359	8.5	0.149	14.0	0.0909
4.0	0.315	9.0	0.141	15.0	0.0848
4.5	0.281	9.5	0.134	19.0	0.0670
5.0	0.253	10.0	0.127	25.0	0.0509
5.5	0.230	10.5	0.121	30.0	0.0424
				50.0	0.0255
				100.0	0.0127

Review Example 16.2 Consider the nonlinear system of Fig. 16.40a, with a saturating amplifier having gain K in its linear region. Determine the largest value of gain K for the system to stay stable. What would be the frequency, amplitude and nature of the limit cycle for a gain $K = 3$?

Solution: It is convenient to regard the amplifier to have unit gain and the gain K to be attached to the linear part. From Eqns (16.66)–(16.67), we obtain for $S = 1$ and $K = 1$,

$$N(E) = N_c(E);$$

the function $N_c(E)$ is listed in Table 16.3.

The locus of $-1/N(E)$ thus starts from $(-1 + j0)$ and travels along the negative real axis for increasing E, as shown in Fig. 16.40b. Now, for the equation,

$$KG(j\omega) = -1/N(X)$$

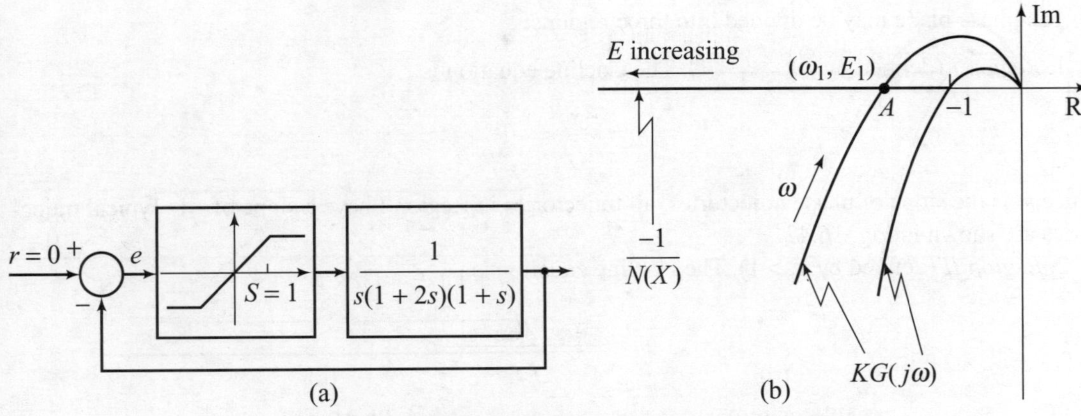

Fig. 16.40 *Nonlinear system with saturating amplifier*

to be satisfied, $G(j\omega)$ must have an angle of $-180°$:

$$\angle G(j\omega) = -90° - \tan^{-1}2\omega - \tan^{-1}\omega = -180°$$

This gives

$$\frac{2\omega + \omega}{1 - 2\omega^2} = \tan 90° = \infty \quad \text{or} \quad \omega = 1/\sqrt{2} \text{ rad/sec.}$$

The largest value of K for stability is obtained when $KG(j\omega)$ passes through $(-1 + j0)$, i.e.,

$$\left|KG(j\omega)\right|_{\omega=1/\sqrt{2}} = 1 \quad \text{or} \quad \frac{K}{\left(\frac{1}{\sqrt{2}}\right)(\sqrt{3})\left(\frac{\sqrt{3}}{\sqrt{2}}\right)} = 1 \quad \text{or} \quad K = 3/2$$

For $K = 3$, $KG(j\omega)$ plot intersects $-1/N(X)$ locus resulting in a limit cycle at (ω_1, E_1) where $\omega_1 = 1/\sqrt{2}$, while E_1 is obtained from the relation,

$$\left|-1/N(E_1)\right| = \left|3G(j\omega_1)\right| = 2 \quad \text{or} \quad \left|N(E_1)\right| = 0.5$$

From Table 16.3, we obtain

$$E_1 \cong 2.5$$

Applying the stability test for the limit cycle reveals that point A in Fig. 16.40b corresponds to a stable limit cycle.

Review Example 16.3 Consider the servomechanism of Fig. 16.41, having a deadzone nonlinearity in the feedback loop. With $x_1 = y$ and $x_2 = \dot{y}$, we get

$$\dot{x}_1 = x_2$$

$$\dot{x}_2 = \begin{cases} -x_2 & ; \quad -1 < x_1 < 1 \\ -x_2 - 2(x_1 - 1) & ; \quad x_1 > 1 \\ -x_2 - 2(x_1 + 1) & ; \quad x_1 < -1 \end{cases}$$

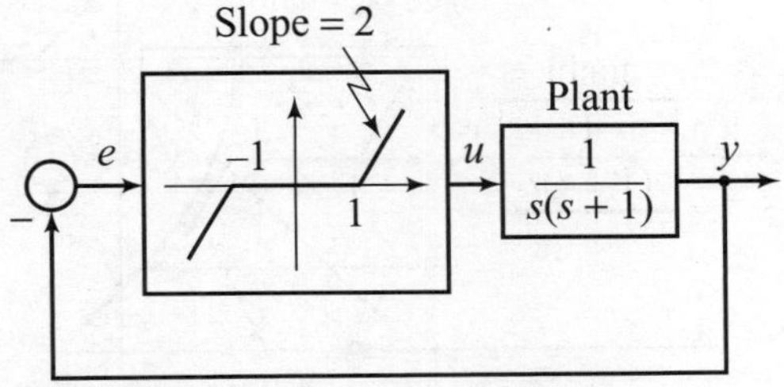

Fig. 16.41 *A system with deadzone nonlinearity*

The phase plane may be divided into three regions:

1. *Region I* (defined by $-1 < x_1 < 1$): The isocline equation is

$$m = \frac{-x_2}{x_2} = -1$$

Since m is the slope of phase trajectories, all trajectories in Region I have a slope of –1. Typical trajectories are shown in Fig. 16.42.

2. *Region II* (defined by $x_1 > 1$): The isocline equation is

$$m = \frac{-x_2 - 2(x_1 - 1)}{x_2}$$

or

$$x_2 = \frac{-2(x_1 - 1)}{m + 1}$$

The isoclines are straightlines intersecting the x_1-axis at $x_1 = 1$, with slope equal to $-2/(m + 1)$. Some of these isoclines are shown in Fig. 16.42.

3. *Region III* (defined by $x_1 < -1$): The isocline equation is

$$x_2 = \frac{-2(x_1 + 1)}{m + 1}$$

These isoclines are straight lines intersecting the x_1-axis at $x_1 = -1$ with slope equal to $-2/(m + 1)$. Some of these isoclines are shown in Fig. 16.42.

Some typical phase trajectories are also shown in Fig. 16.42. Note that for the given system, we have an *equilibrium zone*, $-1 \le x_1 \le 1$ with $x_2 = 0$. The system comes to rest at any point within this zone.

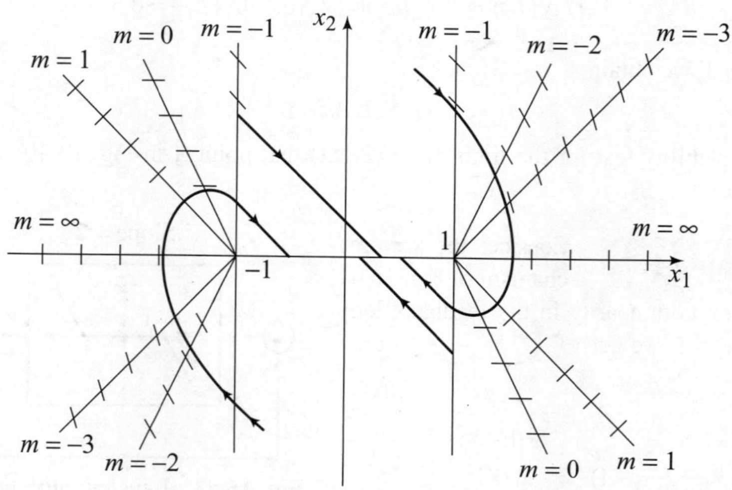

Fig. 16.42 *Isoclines and typical trajectories for the system of Fig. 16.41*

CONCLUDING REMARKS

Because every nonlinear system is in many ways unique, a vast number of approaches are used for nonlinear control systems. In this chapter, we have covered the techniques that have application to relatively larger classes of nonlinear systems.

What we have learnt?

- To apply describing function method, a sinusoid is applied to the nonlinear part of the system and the first harmonic of the periodic response is computed. The ratio of the output to the input is taken as if it were a linear but variable frequency response. Thus, the Nyquist plot is the natural domain in which to consider the system behavior.
- The predictions of the describing function analysis usually are a good approximation to actual behavior when the linear part of the system provides a sufficiently strong filtering effect. The filtering characteristics of the linear part improve as the order of the system goes up.
- Phase-plane analysis is a graphical method used to study second-order dynamic systems. It allows visual examination of the global behavior of the system in time domain. The phenomena of limit cycles is clearly seen.
- The limitation of phase plane analysis is its applicability to second-order systems (though with no approximation); and that of describing function analysis is its applicability to higher-order systems with approximation; the two techniques complement each other.
- The Lyapunov's method of stability analysis is the most general method for determination of stability of nonlinear systems. The major limitation of its use is the difficulty associated with the construction of the Lyapunov function required by the method.

Review Questions

1. Describe the following common nonlinear system behaviors:
   (a) Limit cycles
   (b) Jump resonance
2. Phase plane analysis applies primarily to second-order systems while the describing function method gives satisfactory results for higher-order systems. Why?
3. Phase portraits are today routinely computer generated. It is still practically useful to learn how to roughly sketch phase portraits. Describe the isoclines method of construction of phase portraits. Illustrate through an example.
4. What are singular points on phase plane? Describe the behavior of trajectories in the vicinity of commonly found singular points in phase portraits.
5. State and explain Lyapunov stability theorem. How do we generate Lyapunov functions for nonlinear systems?
6. State and derive Lyapunov equation for stability analysis of linear systems.

Self-Appraisal Quiz

A set of questions with multiple answer choices is given below. These questions have been designed to capture important aspects of the subject covered in the chapter. For each of these questions, identify the

Correct answer of the given choices. Compare your answers with the Master Key given at the end of the session. This quiz should serve the purpose of a self-appraisal test for readers.

16.1 A linear approximation at the operating point $(x^* = 0.5, u^* = 1.5)$ to $y(x,u) = 2x^2 + xu + u^2$, is

(A) $\delta y = 3.5\,\delta x + 3.5\,\delta u$ (B) $\delta y = 4\,\delta x + 2\,\delta u$

(C) $\delta y = 2\,\delta x + 3\,\delta u$ (D) $\delta y = 4.75\,\delta x + 3.75\,\delta u$

16.2 The voltage buildup of a simple electronic oscillator is given by the equation,

$$\ddot{v}(t) - (1 - v^2(t))\dot{v}(t) + v(t) = 0$$

The equilibrium point $(v, \dot{v})$ of this system is

(A) (0,0) (B) (1,0) (C) (−1,0)

16.3 Among the three types of friction described, which type is governed by a linear mathematical model?

(A) Stiction (B) Coulomb friction (C) Viscous friction

16.4 Stability of nonlinear systems depends on

(A) system parameters (B) initial conditions

(C) nature of the input signal (D) all of these in (A), (B), and (C)

16.5 The describing function analysis is more reliable when the linear part of the nonlinear loop is of

(A) first-order (B) second-order

(C) higher than second-order (D) any order

16.6 The phase-plane analysis is more suitable for nonlinear systems of

(A) second-order (B) higher than second-order

(C) any order

16.7 Output of a nonlinear system excited by a sinusoidal signal of frequency ω is of

(A) frequency ω (B) frequencies higher than ω

(C) frequencies less than ω (D) None of the answers in (A), (B) and (C) is correct.

16.8 Consider the nonlinear system:

$$\dot{x}_1 = x_2; \quad \dot{x}_2 = -x_1 - x_1^2 x_2$$

The linearized system about the origin indicates the singular point as

(A) vortex (B) stable focus (C) stable node

16.9 Consider the nonlinear system:

$$\dot{x}_1 = x_2; \dot{x}_2 = -(x_1^2 - 1)x_2 - x_1$$

The linearized system about the origin indicates the singularity as

(A) unstable focus (B) unstable node (C) center

16.10 Consider the following statements:

(i) Lyapunov function for a linear system is unique.

(ii) Exceeding the stability limits based on a particular Lyapunov function will always cause the equilibrium point of the nonlinear system to be unstable.

(A) Both the statements are true (B) Statement (i) is true, but statement (ii) is false

(C) Statement (i) is false, but statement (ii) is true (D) Both the statements are false

MASTER KEY

16.1 (A) 16.2 (A) 16.3 (C) 16.4 (D) 16.5 (C)
16.6 (A) 16.7 (D) 16.8 (A) 16.9 (A) 16.10 (D)

Problems

16.1 For a sinusoidal input $x = X \sin \omega t$, find the output waveforms for each of the nonlinearities listed in Table 16.2. By Fourier-series analysis of the output waveforms, derive the describing function for each entry of the table.

16.2 Consider the system shown in Fig. P16.2. Using the describing-function analysis, show that a stable limit cycle exists for all values of $K > 0$. Find the amplitude and frequency of the limit cycle when $K = 4$, and plot $y(t)$ versus t.

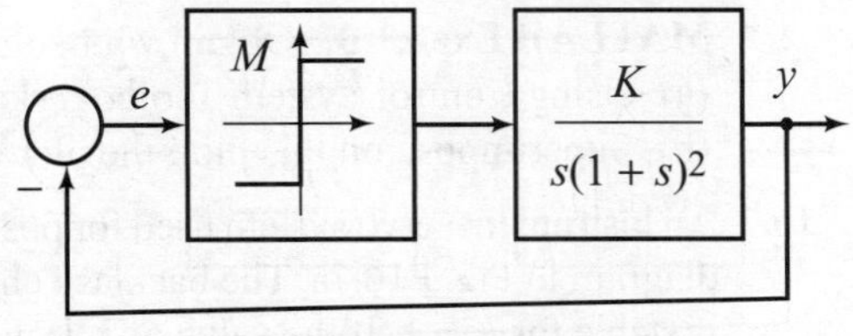

Fig. P16.2

16.3 Consider the system shown in Fig. P16.3. Use the describing function technique to investigate the possibility of limit cycles in this system. If a stable limit cycle is predicted, determine its amplitude and frequency.

MATLAB Exercise Set up a Simulink block diagram as per the given feedback structure. Simulate the system for various initial conditions. If the Simulink response shows a limit cycle, determine its amplitude and frequency.

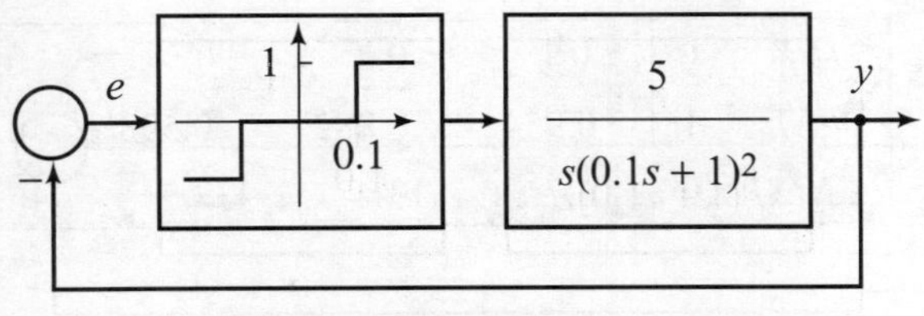

Fig. P16.3

16.4 Using the describing-function analysis, prove that no limit cycle exists in the system shown in Fig. P16.4. Find the range of values of the deadzone of the on-off controller for which limit cycling will be predicted.

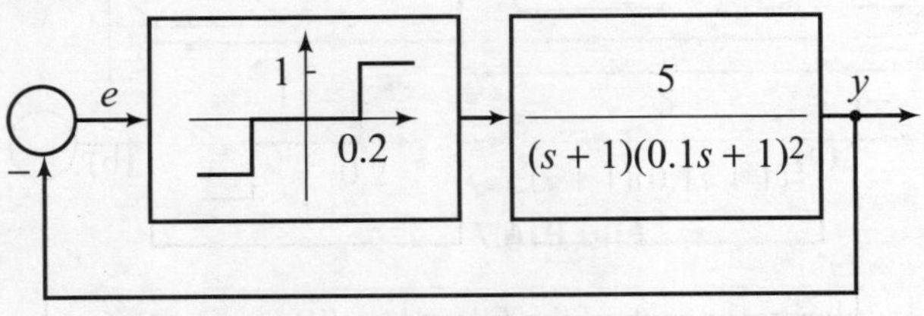

Fig. P16.4

16.5 Consider the system shown in Fig. P16.5. Using the describing-function technique, show that a stable limit cycle cannot exist in this system for any $K > 0$.

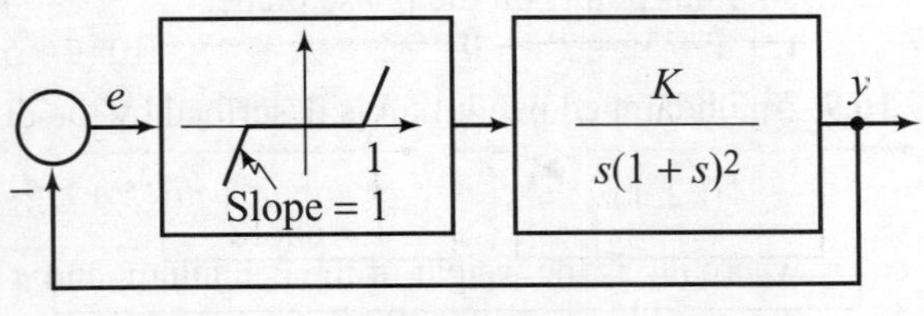

Fig. P16.5

16.6 Consider the system shown in Fig. P16.6. Using the describing-function analysis, investigate the possibility of a limit cycle in the system. If a limit cycle is predicted, determine its amplitude and frequency, and investigate its stability.

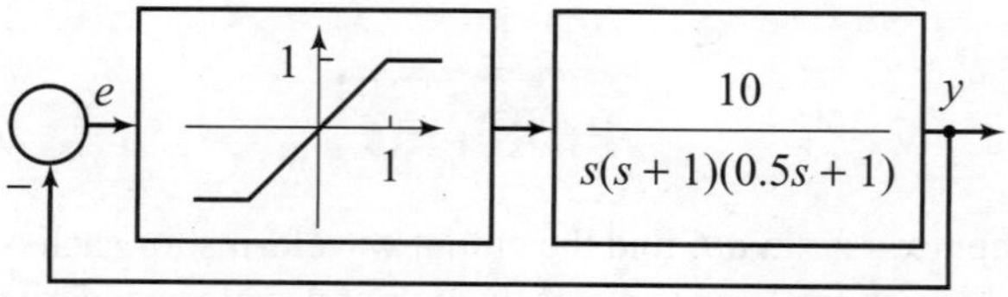

Fig. P16.6

MATLAB Exercise Obtain your solution using MATLAB-generated plot:
(a) Using Control System Toolbox, sketch the Nyquist plot for the linear transfer function.
(b) Superimpose on this plot, the plot of describing function of saturation nonlinearity.

16.7 An instrument servo system used for positioning a load may be adequately represented by the block diagram in Fig. P16.7a. The backlash characteristic is shown in Fig. P16.7b. Show that the system is stable for $K = 1$. If the value of K is now raised to 2, show that limit cycles exist. Investigate the stability of these limit cycles. Determine the amplitude and frequency of the stable limit cycle.

Given:

H/X	0	0.1	0.2	0.5	0.7	0.8	0.9	0.95	1.0
$\lvert N(X)\rvert$	1	0.954	0.882	0.593	0.367	0.249	0.126	0.063	0
$\angle N(X)$	0	– 6.7°	– 13.4°	– 32.5°	–46.6°	– 55.1°	– 65.6°	– 72.8°	– 90°

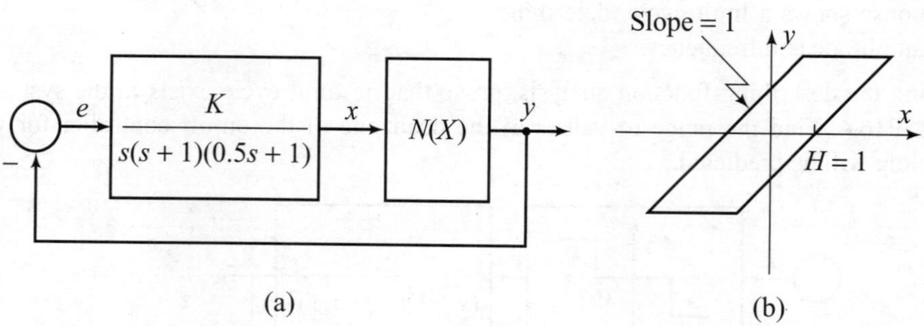

Fig. P16.7

16.8 Determine the kind of singularity for each of the following differential equations. Also locate the singular points on the phase plane.
(a) $\ddot{y} + 3\dot{y} + 2y = 0$ (b) $\ddot{y} + 5\dot{y} + 6y = 6$ (c) $\ddot{y} - 8\dot{y} + 17y = 34$

16.9 An undamped pendulum is described by the differential equation,

$$m\ell^2\ddot{\theta} + mg\ell \sin\theta = 0$$

where mg is the weight of the pendulum, and ℓ is its length. Show that this nonlinear system has two equilibrium points: $\theta = 0$, $\theta = \pi$. Develop linear state models using Taylor's series approximation around the two equilibrium points, and therefrom identify the kind of singularity at each point.

16.10 A linear second-order servo is described by the equation,

$$\ddot{y} + 2\zeta\omega_n\dot{y} + \omega_n^2 y = \omega_n^2$$

where $\omega_n = 1, y(0) = 2.0, \dot{y}(0) = 0$

Determine the singular points when (i) $\zeta = 0$, (ii) $\zeta = 0.15$. Construct the phase trajectory in each case.

16.11 Consider the block diagram of a system, shown in Fig. P16.11.

(a) Derive state variable model with $x_1 = e$ and $x_2 = \dot{e}$.

(b) Write equations of the isoclines on the x_1 versus x_2 plane.

(c) Given: $x_1(0) = 1, x_2(0) = 0$, obtain a trajectory on the x_1 versus x_2 plane. Show singular points (if any) and some isoclines.

MATLAB Exercise Use Simulink to obtain the trajectory on the x_1 versus x_2 plane.

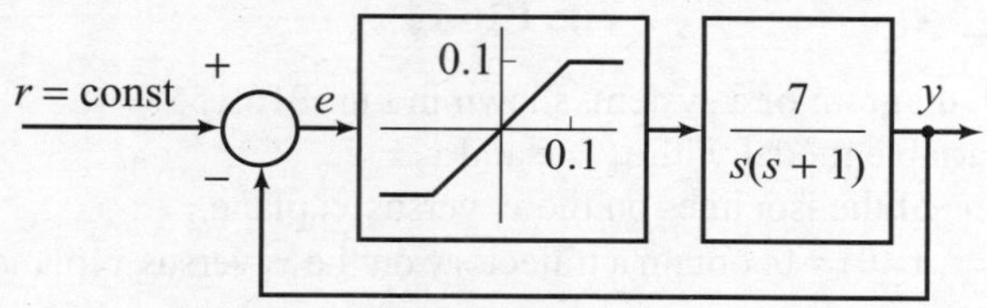

Fig. P16.11

16.12 Consider the block diagram of a system, shown in Fig. P16.12.

(a) Derive state variable model with $x_1 = e$ and $x_2 = \dot{e}$.

(b) Write equations of the isoclines on the x_1 versus x_2 plane.

(c) Given: $x_1(0) = 1, x_2(0) = 0$, obtain a trajectory on the x_1 versus x_2 plane. Show singular points (if any) and some isoclines.

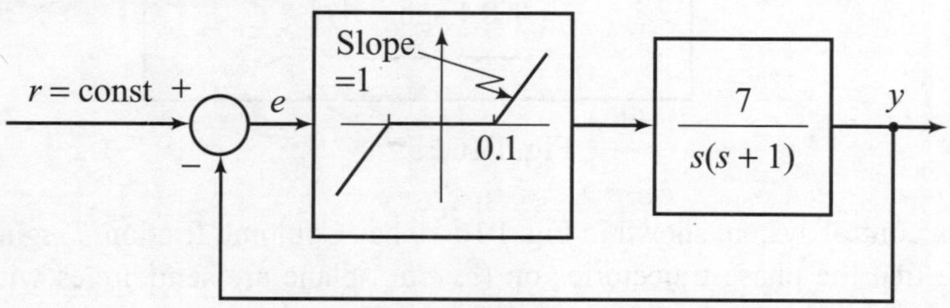

Fig. P16.12

16.13 Consider the block diagram of a system, shown in Fig. P16.13.

(a) Derive state variable model with $x_1 = e$ and $x_2 = \dot{e}$.

(b) Write equations of the isoclines on the x_1 versus x_2 plane.

(c) Given: $x_1(0) = 1, x_2(0) = 0$, obtain a trajectory on the x_1 versus x_2 plane. Show singular points (if any) and some isoclines.

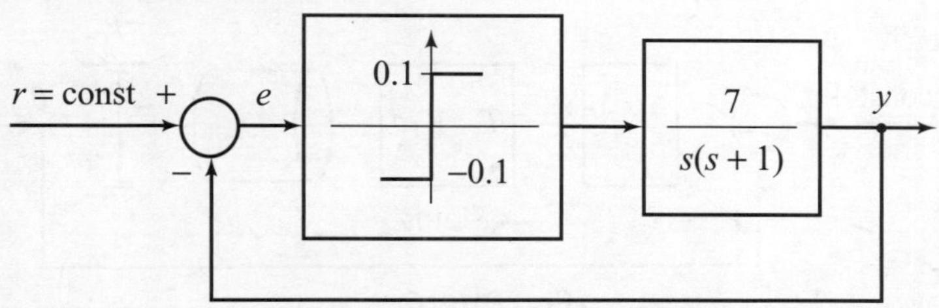

Fig. P16.13

16.14 Consider the block diagram of a system, shown in Fig. P16.14.

(a) Derive state variable model with $x_1 = e$ and $x_2 = \dot{e}$.

(b) Write equations of the isoclines on the x_1 versus x_2 plane.

(c) Given: $x_1(0) = 1, x_2(0) = 0$, obtain a trajectory on the x_1 versus x_2 plane. Show singular points (if any) and some isoclines.

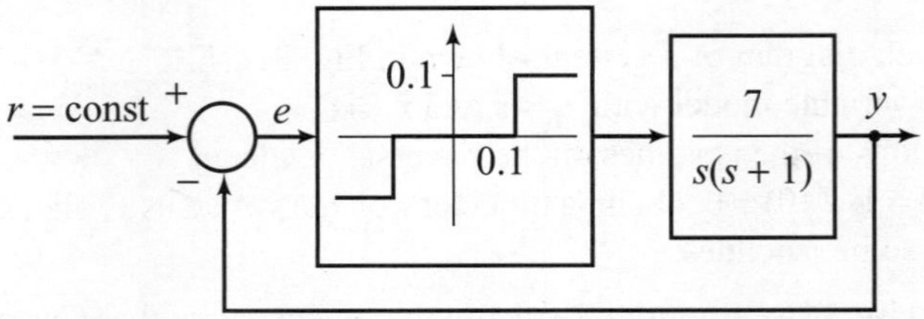

Fig. P16.14

16.15 Consider the block diagram of a system, shown in Fig. P16.15.

(a) Derive state variable model with $x_1 = e$ and $x_2 = \dot{e}$.

(b) Write equations of the isoclines on the x_1 versus x_2 plane.

(c) Given: $x_1(0) = 1, x_2(0) = 0$, obtain a trajectory on the x_1 versus x_2 plane. Show singular points (if any) and some isoclines.

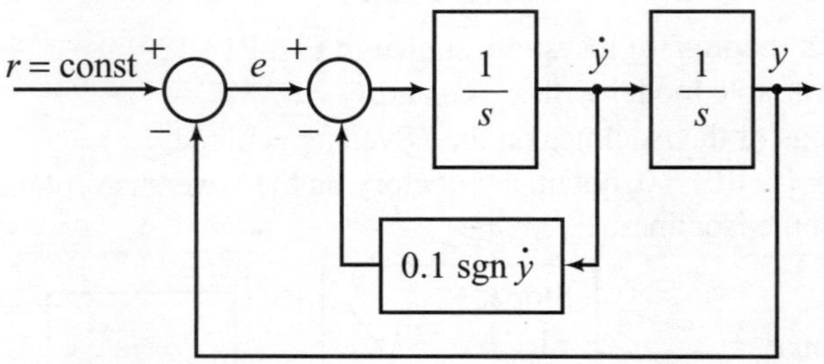

Fig. P16.15

16.16 The position control system shown in Fig. P16.16 has Coulomb friction $T_c \operatorname{sgn}(\dot{\theta})$ at the output shaft. Prove that the phase trajectories on $(e, \dot{e}/\omega_n)$-plane are semicircles with the centre on horizontal axis at $+T_c/K$ for $\dot{e} < 0$ and $-T_c/K$ for $\dot{e} > 0$.

Examine the phase trajectory corresponding to θ_R = unit step, $\dot{\theta}(0) = \theta(0) = 0$; and find the value of the steady-state error. What is the largest possible steady-state error which the system in Fig. P16.16 can possess?

Given:

$$\omega_n = \sqrt{K/J} = 1.2 \text{ rad/sec}; \quad T_c/K = 0.3 \text{ rad where } \quad K = K_A K_1.$$

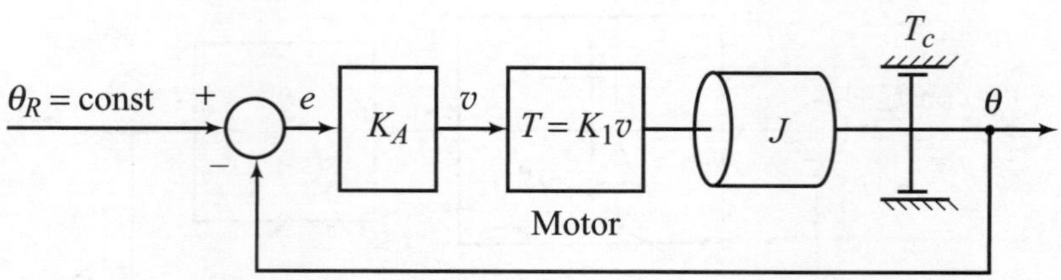

Fig. P16.16

16.17 Consider the nonlinear system with deadzone shown in Fig. P16.17. Using the method of isoclines, sketch some typical phase trajectories with and without deadzone, and comment upon the effect of deadzone on transient and steady-state behavior of the system.

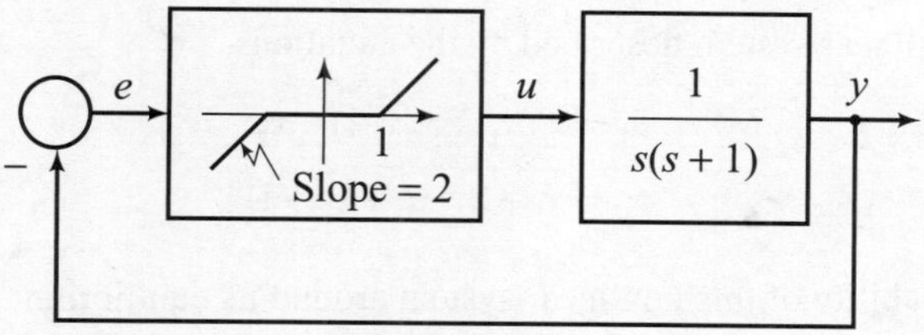

Fig. P16.17

16.18 Consider the system shown in Fig. P16.18 in which the nonlinear element is a power amplifier with gain equal to 1.0, which saturates for error magnitudes greater than 0.4. Given the initial condition: $e(0) = 1.6$, $\dot{e}(0) = 0$, plot phase trajectories with and without saturation, and comment upon the effect of saturation on the transient behavior of the system.

MATLAB Exercise Using Simulink, obtain the system response for given initial conditions (a) with saturation, and (b) without saturation; and comment upon the effect of saturation on the transient behavior of the system.

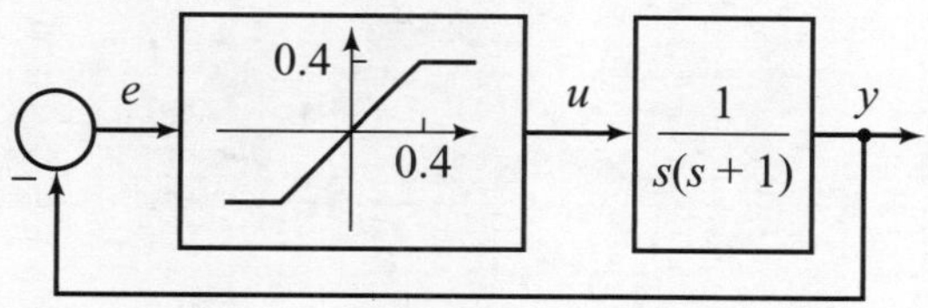

Fig. P16.18

16.19 Consider the linear autonomous system,

$$\dot{\mathbf{x}} = \begin{bmatrix} 0 & 1 \\ -1 & -1 \end{bmatrix} \mathbf{x}$$

Using Lyapunov analysis, determine the stability of the equilibrium state.

16.20 Using Lyapunov analysis, determine the stability of the equilibrium state of the system,

$$\dot{\mathbf{x}} = \mathbf{A}\mathbf{x}$$

with
$$\mathbf{A} = \begin{bmatrix} 0 & 1 \\ -1 & 1 \end{bmatrix}$$

16.21 Consider the nonlinear system described by the equations,

$$\dot{x}_1 = x_2$$
$$\dot{x}_2 = -(1 - |x_1|)x_2 - x_1$$

Find the region in the state plane for which the equilibrium state of the system is asymptotically stable.

16.22 Check the stability of the equilibrium state of the system described by

$$\dot{x}_1 = x_2$$
$$\dot{x}_2 = -x_1 - x_1^2 x_2$$

16.23 Consider the nonlinear system described by the equations,

$$\dot{x}_1 = x_1(x_1^2 + x_2^2 - 1) - x_2$$
$$\dot{x}_2 = x_1 + x_2(x_1^2 + x_2^2 - 1)$$

Investigate the stability of this nonlinear system around its equilibrium point at the origin.

16.24 Consider a nonlinear system described by the differential equation,

$$\ddot{x} + [K_1 + K_2(\dot{x})^2]\,\dot{x} + x = 0$$

Check the stability of the equilibrium state of this system when (i) $K_1 > 0$ and $K_2 > 0$ (ii) $K_1 < 0$ and $K_2 < 0$ (iii) $K_1 > 0$ and $K_2 < 0$.

REFERENCES

APPLICATIONS

1. Siouris, GM; *Missile Guidance and Control Systems*: New York: Springer-Verlag, 2004.
2. Lawrence, A; *Modern Inertial Technology: Navigation, Guidance, and Control;* 2nd Edition, New York: Springer-Verlag, 1998.
3. Leonhard, W; *Control of Electric Drives*; 3rd Edition, New York: Springer-Verlag, 2001.
4. Bose, BK; *Modern Power Electronics and AC Drives*: Englewood Cliffs, NJ: Prentice-Hall, 2001.
5. Sen, PC; *Thyristor DC Drives*; 2nd Edition, New York: Wiley-Interscience, 1991.
6. Seborg, DE, TF Edgar, and DA Mellichamp; *Process Dynamics and Control*; 2nd Edition, New York: John Wiley, 2004.
7. Lee, J.,W. Cho, and T.F. Edgar, *An Improved Technique for PID Controller Tuning from Closed-Loop Tests*, *AIChE J.*, Vol. 36, No. 12, pp. 1891–1895, 1990.
8. Ziegler, J.G., and N.B. Nichols, *Optimum Settings for Automatic Controllers*, *Trans. ASME*, Vol. 64, pp. 759–768, 1942.
9. Perry, R.H., and D. Green (ed.), *Perry's Chemical Engineering Handbook*, 6th Edition, New York: McGraw-Hill, 1997.
10. Shinskey, FG; *Process Control Systems*; 4th Edition, New York: McGraw-Hill, 1996.
11. Astrom, KJ, and T Hagglund; *PID Controllers: Theory, Design, and Tuning*; 2nd Edition, Seattle, WA: International Society for Measurement and Control, 1995.
12. Corripio, AB; *Tuning of Industrial Control Systems*; Research Tringle Park, North Carolina: Instrument Society of America, 1990.
13. Stephanopoulos, G; *Chemical Process Control—An Introduction to Theory and Practice*; Englewood Cliffs, NJ: Prentice-Hall, 1984.
14. Steven, B, and F Lewis; *Aircraft Control and Simulation*; New York: Wiley Interscience, 2003.
15. Nelson, RC; *Flight Stability and Automatic Control*, 2nd Edition, New York: McGraw-Hill, 1997.
16. Etkin, B, and LD Reid; *Dynamics of Flight: Stability and Control*; 3rd Edition, New York: John Wiley, 1996.
17. Craig, JJ; *Introduction to Robotics: Mechanics and Control*; 3rd Edition, Englewood Cliffs, NJ: Prentice-Hall, 2003.
18. Koivo, AJ; *Fundamentals for Control of Robotics Manipulators*; New York; John Wiley, 1989.
19. Asda, H, and K Youcef-Touni, *Direct Drive Robots*: *Theory and Practice*, Cambridge, Massachusetts: The MIT Press, 1987.
20. Valentino, JV, and J Goldenberg; *Introduction to Computer Numerical Control* (CNC); 3rd Edition, Englewood Cliffs, NJ: Prentice-Hall, 2002.
21. Groover, MK; *Automation, Production Systems, and Computer-Integrated Manufacturing*; 2nd Edition, Englewood Cliffs, NJ: Prentice-Hall, 2000.
22. Olsson, G, and G Piani; *Computer Systems for Automation and Control*; London: Prentice-Hall International, 1992.
23. Hughes, TA; *Programmable Controllers*; 4th Edition, North Carolina: The Instrumentation, Systems, and Automation Society, 2004.

24. Petruzella, FD; *Programmable Logic Controllers*; 3rd Edition, New York: McGraw-Hill, 2004.
25. Bolton, W; *Programmable Logic Controllers*; 3rd Edition, Burlington, MA: Newnes Publications, 2003.
26. Beards, CF; *Vibration Analysis and Control System Dynamics*; 2nd Edition, Englewood Cliffs, NJ: Prentice-Hall, 1995.

MATHEMATICAL BACKGROUND

27. Noble, B; and JW Daniel, *Applied Linear Algebra*, 3rd Edition, Englewood Cliffs, NJ: Prentice-Hall 1988.
28. Lancaster, P, and M Tismenetsky, *The Theory of Matrices*, 2nd Edition, Orlando, Florida: Academic Press, 1985.
29. Nash, SG, and A Sofer, *Linear and Nonlinear Programming,* Singapore: McGraw-Hill, 1996.
30. Oppenheim, AV, RW Shafer, and JR Buck; *Discrete-Time Signal Processing*; 2nd Edition, Englewood Cliffs, NJ: Prentice-Hall, 1999.
31. Oppenheim, AV, AS Willsky and S Hamid Nawab; *Signals and Systems*; 2nd Edition, Upper Saddle River, NJ: Prentice-Hall, 1997.
32. Brown, JW, and RV Churchill; *Complex Variables and Applications*; 7th Edition, New York: McGraw-Hill, 2003.
33. Lefschetz, S, *Differential Equations: Geometric Theory*; 2nd Edition, New York: Dover Publications, 1977.

DYNAMICAL SYSTEMS AND MODELING

34. Palm, WJ, III; *System Dynamics*; New York: McGraw-Hill, 2004.
35. Zak, SH; *Systems and Control*; New York: Oxford University Press, 2003.
36. Ogata, K; *System Dynamics*; 4th Edition, Englewood Cliffs, NJ: Prentice-Hall, 2003.
37. Clark, RN; *Control System Dynamics*; New York: Cambridge University Press, 1996.
38. Belanger, PR, *Control Engineering*: *A Modern Approach*; Orlando, Florida: Saunders College Publishing, 1995.
39. Moschytz, GS, *Linear Integrated Networks: Fundamentals*, New York: Van Nostrand Reinhold, 1974.
40. Schwarz, RJ, and B Friedland, *Linear Systems*, New York: McGraw-Hill, 1965.
41. Mason, SJ, "Feedback Theory: Further Properties of Signal Flow Graphs", *Proc. IRE*, Vol. 44, No. 7, pp. 920–926, July 1956.
42. Mason, SJ, "Feedback Theory: Some Properties of Signal Flow Graphs", *Proc. IRE*, Vol. 41, No. 9, pp. 1144–1156, Sept. 1953.

INDUSTRIAL CONTROL DEVICES

43. Necsulescu, D; *Mechatronics*; Singapore: Pearson Education, 2002.
44. Gupta, S; *Elements of Control Systems*; Upper Saddle River, N.J: Pearson Education, 2002.
45. Kilian,CT, *Modern Control Technology: Components and Systems*; 2nd Edition, Singapore: Delmar Publishers, 2001.
46. Morris, NM; *Control Engineering*; 4th Edition, London: McGraw-Hill, 1991.
47. De Silva, CW, *Control Sensors and Actuators*, Englewood Cliffs, NJ: Prentice-Hall, 1989.
48. Anderson, WR, *Controlling Electrohydraulic Systems*, New York: Marcel Dekker, 1988.
49. Parr, EA, *Industrial Control Handbook*, Vol.1–Vol.3, Oxford: BSP Professional Books, 1987.
50. Schuler, CA, and WL McNamee, *Industrial Electronics and Robotics*, New York: McGraw-Hill, 1986.
51. Kenjo, T, and S Nagamori, *Permanent-magnet and Brushless DC Motors*, Oxford: Clarendon Press, 1985.
52. Kenjo, T *Stepping Motors and their Microprocessor Controls*, Oxford: Clarendon Press, 1984.
53. Ahrendt, WR, and CJ Savant Jr, *Servomechanism Practice*, 2nd Edition, New York: McGraw-Hill, 1960.
54. Kuo, S, and WS Gan; *Digital Signal Processors: Architectures, Implementations, and Applications*; Upper Saddle River, NJ: Pearson Education, 2005.

55. Pack, D, and S Barrett; *68HCI2 Microcontroller: Theory and Applications*; Upper Saddle River, NJ: Prentice-Hall, 2002.
56. Wolf, W; *Computers as Components: Principles of Embedded Computing System Design*; San Franscisco: Morgan Kaufmann, 2001.
57. Mackenzie, I; *The 8051 Microcontroller*; Upper Saddle River, NJ: Prentice-Hall, 1998.

FEEDBACK CONTROL THEORY

58. Franklin, GF, JD Powell, and A Emami-Naeini; *Feedback Control of Dynamical Systems*; 5th Edition, Upper Saddle River, NJ: Pearson Education, 2005.
59. Dorf, RC, and RH Brishop; *Modern Control Systems*, 10th Edition, Upper Saddle River, NJ: Pearson Education, 2004.
60. Nise, NS; *Control Systems Engineering;* 4th Edition, Danvers, MA: John Wiley, 2003.
61. Kuo, BC, and F. Golnaraghi: *Automatic Control systems*; 8th Edition, Danvers, MA: John Wiley, 2003.
62. D'Azzo, JJ, CH Houpis, and SN Sheldon; *Linear Control System Analysis and Design with MATLAB*, 5th Edition, New York: Marcel Dakker, 2003.
63. Wilkie, J, M Johnson, and R Ketabi; *Control Engineering: An Introductory Course*; New York: Palgrave, 2003.
64. Dorsey, J; *Continuous and Discrete Control Systems*; New York: McGraw-Hill, 2002.
65. Stefani, RT, B Shahian, CJ Savant, and GH Hostetter; *Design of Feedback Control Systems*; 4th Edition, New York: Oxford University Press, 2002.
66. Ogata, K; *Modern Control Engineering*; 4th Edition, Englewood Cliffs, NJ: Prentice-Hall, 2001.
67. Goodwin, GC, SF Graebe, and ME Salgado; *Control System Design*; Upper Saddle River, NJ: Pearson Education, 2001.
68. Phillips, CL, and RD Harbor; *Feedback Control Systems*; 4th Edition, Englewood Cliffs, NJ: Prentice-Hall, 1999.
69. Shinners, SM; *Modern Control System Theory and Design*; 2nd Edition, New York: John Wiley, 1998.
70. Raven, FH; *Automatic Control Engineering*; 5th Edition, New York: McGraw-Hill, 1995.
71. Wolovich, WA; *Automatic Control Systems: Basic Analysis and Design*; Orlando, Florida: Saunders College Publishing, 1994.
72. Doebelin, EO, *Control System Principles and Design*, New York: John Wiley, 1986.
73. Truxal, JG, *Automatic Feedback Control System Synthesis*, New York: McGraw-Hill, 1955.

ROBUST CONTROL

74. Dullerud, GE, and F Paganini; *A Course in Robust Control Theory*; New York: Springer-Verlag, 2000.
75. Zhou, K, and JC Doyle; *Essentials of Robust Control*; Englewood Cliffs, NJ: Prentice-Hall, 1997.
76. Saberi, A, BM Chen, and P Sannuti, *Loop Transfer Recovery—Analysis and Design*, London: Springer-Verlag, 1993.
77. Doyle, JC, BA Francis, and AR Tannenbaum, *Feedback Control Theory*, New York: MacMillan, 1992.
78. Francis, BA, *A Course in Control Theory*, *Lecture Notes in Control and Information Sciences*, No. 88, Berlin: Springer-Verlag, 1987.
79. Rosenwasser, E, and R Yusupov; *Sensitivity of Automatic Control Systems*; Boca Raton, FL: CRC Press, 2004.

DIGITAL CONTROL

80. Franklin, GF, JG Powell, and ML Workman; *Digital Control of Dynamic Systems*; 3rd Edition, San Diego, CA: Addison-Wesley, 1997.

81. Astrom, KJ, and B Wittenmark; *Computer-Controlled Systems*; 3rd Edition, Englewood Cliffs, NJ: Prentice-Hall, 1996.
82. Santina, MS, and AR Stubberud; *Sample-Rate Selection*, Boca Raton, FL: CRC Press, 1996.
83. Santina, MS, and AR Stubberud; *Quantization Effects*; Boca Raton, FL: CRC Press, 1996.
84. Ogata, K; *Discrete-Time Control Systems*; 2nd Edition, Upper Saddle River, NJ: Pearson Education, 1995.
85. Santina, MS, AR Stubberud, and GH Hostetter; *Digital Control System Design*; 2nd Edition, Stamford, CT: International Thomson Publishing, 1994.
86. Phillips, CL, and HT Nagle, Jr; *Digital Control System Analysis and Design*; 3rd Edition, Upper Saddle River, NJ: Pearson Education, 1994.
87. Kuo, BC; *Digital Control Systems*; 2nd Edition, Orlando, Florida: Saunders College Publishing, 1992.
88. Houpis, CH, and GB Lemont; *Digital Control Systems: Theory, Hardware, and Software*; 2nd Edition, New York: McGraw-Hill, 1992.
89. Hristu-Varsakelis, D, and WS Levine (eds); *Handbook of Networked and Embedded Control Systems*; Boston: Birkhauser Publishers, 2005.
90. Ibrahim, D; *Microcontroller Based Temperature Monitoring and Control*; Woburn, MA: Newnes Publications, 2002.
91. Chidambaram, M; *Computer Control of Processes*; New Delhi: Narosa Publishers, 2002.
92. Ozkul, T; *Data Acquisition and Process Control using Personal Computers*; New York: Marcel Dekkar, 1996.
93. Rigby, WH, and T Delby; *Computer Interfacing: A Practical Approach to Data Acquisition and Control*; Englewood Cliffs, NJ: Prentice-Hall, 1995.
94. Tooly, M; *PC-Based Instrumentation and Control*; 2nd Edition, Oxford: Newnes Publications, 1995.
95. Gupta, S, and JP Gupta; *PC Interfacing for Data Acquisition and Process Control*, 2nd Edition, Research Triangle Park, North Carolina: Instrument Society of America, 1994.
96. Gopal, M, *Digital Control Engineering*; New Delhi: New Age International, 1988.
97. Rao, MVC, and AK Subramanium "Elimination of Singular Cases in Jury's Test" *IEEE Trans. Automatic Control*, Vol. AC-21, pp.114–115, February 1976.
98. Jury, EI, and J Blanchard, "A Stability Test for Linear Discrete-time Systems in Table Form", *Proc. IRE*, Vol. 49, pp. 1947–1948, 1961.

STATE SPACE AND LINEAR SYSTEMS

99. Tewari, A; *Modern Control Design with MATLAB and Simulink*; Singapore: John Wiley, 2003.
100. Chen, C-T; *Linear System Theory and Design*; 3rd Edition, New York: Oxford University Press, 1998.
101. DeRusso, PM, RJ Roy, CM Close, and AA Desrochers; *State Variables for Engineers;* 2nd Edition, New York: John Wiley, 1998.
102. Szidarovszky, F, and AT Bahill; *Linear System Theory*; 2nd Edition, Boca Raton, FL: CRC Press, 1998.
103. Antsaklis, PJ, and AN Michel; *Linear Systems*; New York: McGraw-Hill, 1997.
104. Rugh, WJ; *Linear System Theory*; 2nd Edition, Upper Saddle River, NJ: Prentice-Hall, 1995.
105. Gopal, M; *Modern Control System Theory*; 2nd Edition, New Delhi: New Age International, 1993.
106. Brogan, WL; *Modern Control Theory;* 3rd Edition, Englewood Cliffs, NJ: Prentice-Hall, 1991.
107. Friedland, B, *Control System Design*: *An Introduction to State-Space Methods*, New York: McGraw-Hill, 1986.
108. Zadeh, LA, and CA Desoer, *Linear System Theory: A State Space Approach*, New York: McGraw-Hill, 1963.

MULTIVARIABLE AND OPTIMAL CONTROL

109. Bertsekas, DP; *Dynamic Programming and Optical Control;* 3rd Edition, Belmont: Athena Scientific, 2005.

110. Naidu, DS, *Optimal Control Systems;* Boca Raton, FL: CRC Press, 2003.
111. Camacho, EF, and C. Bordons; *Model Predictive Control*; 2nd Edition, London: Springer-Verlag, 2003.
112. Bryson, AE; *Applied Linear Optimal Control: Examples and Algorithms*; Cambridge, UK: Cambridge University Press, 2002.
113. Locatelli, A; *Optimal Control: An Introduction*; Basel, Switzerland: Birkhauser Verlag, 2001.
114. Chen, T, and B Francis; *Optimal Sampled-Data Control Systems*; 2nd Edition, London: Springer, 1996.
115. Lewis, FL, and VL Syrmos; *Optimal Control*; 2nd Edition, New York: John Wiley, 1995.
116. Zhou, K, JC Doyle, and K Glover; *Robust and Optimal Control*; Upper Saddle River, NJ: Prentice-Hall, 1996.
117. Anderson, BDO, and JB Moore, *Optimal Control: Linear Quadratic Methods*, Englewood Cliffs, NJ: Prentice-Hall, 1990.
118. Grimble, MJ, and MA Johnson, *Optimal Control and Stochastic Estimation: Theory and Applications*, Vol. I, Chichester: John Wiley, 1988.
119. Albertos, P, and A Sala; *Multivariable Control Systems: An Engineering Approach*; Springer, 2004.
120. Clarke, DW, C Mothadi, and PS Tuffs, "Generalized Predictive Control–Part I. The Basic Algorithm", *Automatica*, Vol. 23, No. 2, pp. 137–148, 1987.
121. Clarke, DW, C Mothadi, and PS Tuffs, "Generalized Predictive Control–Part II. Extensions and Interpretations", *Automatica,* Vol. 23, No. 2, pp. 149–160, 1987.
122. Fortman, TE, and KL Hitz, *An Introduction to Linear Control Systems,* New York: Marcel Dekker, 1977.
123. Kautsky, J, NK, Nichols, and P Dooren, "Robust Pole Assignment by Linear State Feedback", *Int. J. Control*, Vol. 41, No. 5, pp.1129–1155, 1985.
124. Doyle, JC, and G Stein, "Robustness with Observers", *IEEE Trans. Automatic Control*, Vol. AC-24, No. 4, pp. 607–611, Aug.1979.

NONLINEAR CONTROL SYSTEMS

125. Khalil, HK.; *Nonlinear Systems*; 3rd Edition, Upper Saddle River, NJ: Prentice-Hall, 2001.
126. Slotine, JJE, and W Li, *Applied Nonlinear Control*, Englewood Cliffs, NJ: Prentice-Hall, 1991.
127. Itkis, U, *Control Systems of Variable Structure*, New York:John Wiley, 1976.
128. Atherton, DP, *Nonlinear Control Engineering*, London: Van Nostrand Reinhold Company, 1975.
129. Minorsky N, *Theory of Nonlinear Control System,* New York: McGraw-Hill, 1969.

SYSTEM IDENTIFICATION AND ADAPTIVE CONTROL

130. Narendra, KS, and AM Annaswamy; *Stable Adaptive Systems*; New York: Dover Publications, 2005.
131. Ljung, L; *System Identification: Theory for the User*; 2nd Edition, Englewood Cliffs, NJ: Prentice-Hall, 1998.
132. Astrom, KJ, and B Wittenmark; *Adaptive Control*; 2nd Edition, Reading, MA: Prentice-Hall, 1994.
133. Landau, ID, *System Identification and Control Design*, Englewood Cliffs, NJ: Prentice-Hall, 1990.
134. Sastry, S, and M Bodson, *Adaptive Control: Stability, Convergence and Robustness*, Englewood Cliffs, NJ: Prentice-Hall, 1989.
135. Grimble, MJ, and MA Johnson, *Optimal Control and Stochastic Examination: Theory and Applications*, Vol. II, Chichester: John Wiley, 1988.
136. Harris, CJ, and SA Billings (eds); *Self-Tuning and Adaptive Control: Theory and Applications;* 2nd Edition, London: Peter Peregrinus, 1985.

INTELLIGENT CONTROL

137. Negnevitsky, M; *Artificial Intelligence: A Guide to Intelligent Systems*; 2nd Edition, Essex, UK: Pearson Education, 2005.

138. Kecman, V; *Learning and Soft Computing;* Cambridge, MA: The MIT Press, 2001.
139. Norgaard, M, O Ravn, NK Poulsen, and LK Hansen, *Neural Networks for Modeling and Control of Dynamic Systems*, London: Springer-Verlag, 2000.
140. Lewis, FL, S Jagannathan, and A Yesildirek, *Neural Network Control of Robot Manipulators and Nonlinear Systems*, London: Taylor & Francis, 1999.
141. Haykin, S, *Neural Networks: A Comprehensive Foundation*, 2nd Edition, Prentice-Hall, 1998.
142. Jang, JSR, C-T Sun, and E Mizutani; *Neuro-Fuzzy and Soft Computing: A Computational Approach to Learning and Machine Intelligence*; Upper Saddle River, NJ: Pearson Education, 1997.
143. Lin C-T, and CS George Lee, *Neural Fuzzy Systems*, Upper Saddle River, NJ: Prentice-Hall, 1996.
144. Ying, Hao, *Fuzzy Control and Modeling: Analytical Foundations and Applications*, New York: IEEE Press, 2000.
145. Passino, KM, and S Yurkovich, *Fuzzy Control*, California: Addison-Wesley, 1998.
146. Goldberg, DE, *Genetic Algorithms in Search Optimization, and Machine Learning*, Reading, Massachusetts: Addison-Wesley, 1989.
147. Werbos, P, *A Handbook of Learning and Approximate Dynamic Programming*, New York: Wiley-IEEE Press, August 2004.
148. Sutton, RS, and AG Barto, *Reinforcement Learning: An Introduction,* Cambridge, Mass.: MIT Press, 1998.
149. Mitchell, TM, *Machine Learning*, Singapore: McGraw-Hill, 1997.
150. Bertsekas, DP, and JN Tsitsiklis, *Neuro-Dynamic Programming*, Belmont, Mass: Athena Scientific, 1996.

TECHNICAL COMPUTING

151. NATIONAL PROGRAMME ON TECHNOLOGY ENHANCEMENT LEARNING
Course: Electrical Engineering (Control Engineering)
Faculty Coordinator: Prof. M Gopal
Web Content: MATLAB Modules for Control System Principles and Design
www.nptel.iitm.ac.in
152. MATLAB & Simulink Software
www.mathworks.com
The MathWorks, Inc.
3 Apple Hill Drive
Natick, MA 01760–2098, USA
153. MATHEMATICA Software
www.wolfram.com
Wolfram Research, Inc.
100 Trade Center Drive
Champaign, IL 61820–7237, USA
154. MAPLESOFT Software
www.maplesoft.com
Maplesoft
615 Kumpf Drive
Waterloo, Ontario
Canada N2V1K8

COMPANION BOOK

155. Gopal, M, *Digital Control and State Variable Methods: Conventional and Intelligent Control Systems,* 4th Edition, New Delhi: Tata McGraw-Hill, 2012.

ANSWERS TO PROBLEMS

2.1 (a) $y(t) = \frac{1}{2}e^{-t} - \frac{1}{3}e^{-2t} - \frac{1}{6}e^{-5t}$ (b) $y(t) = \frac{1}{2}e^{-t} + \frac{3}{2}e^{-2t} - e^{-5t}$

2.2 $y(t) = [1 - 1.193e^{-17.25t} \sin(26.5t + 56.94°)]\mu(t)$

2.3 $y(t) = \left(2e^{-t}\cos 2t - \frac{1}{2}e^{-t}\sin 2t\right)\mu(t)$

2.4 (a) (i) $y(t) = \frac{1}{\tau}e^{-t/\tau}; y_{ss} = 0$ (ii) $y(t) = 1 - e^{-t/\tau}; y_{ss} = 1$

(iii) $y(t) = t - \tau + \tau e^{-t/\tau}; y_{ss} = t - \tau$

(iv) $y(t) = \frac{KA\omega\tau}{\tau^2\omega^2 + 1}e^{-t/\tau} + \frac{KA}{\sqrt{\tau^2\omega^2 + 1}}\sin(\omega t - \tan^{-1}\omega\tau)$

$y_{ss} = \frac{KA}{\sqrt{\tau^2\omega^2 + 1}}\sin(\omega t - \tan^{-1}\omega\tau)$

(b) (i) $y(t) = 1 - \frac{e^{-\zeta\omega_n t}}{\sqrt{1-\zeta^2}}\sin\left(\omega_d t + \tan^{-1}\frac{\sqrt{1-\zeta^2}}{\zeta}\right)$

$\omega_d = \omega_n\sqrt{1-\zeta^2}$

$y_{ss} = 1$

(ii) $y(t) = t - \frac{2\zeta}{\omega_n} + \frac{e^{-\zeta\omega_n t}}{\omega_d}\sin\left(\omega_d t + 2\tan^{-1}\frac{\sqrt{1-\zeta^2}}{\zeta}\right); y_{ss} = t - \frac{2\zeta}{\omega_n}$

2.6 (a) $\frac{s^2 + s + 1}{(s+1)(s^2 + 2s + 2)}$; DC gain $= \frac{1}{2}; y_{ss} = 0.5$ (b) $\frac{1}{4} + \frac{1}{2}t - \frac{1}{4}e^{-2t}; y_{ss} = \frac{1}{4} + \frac{1}{2}t$

2.7 Steady-state error = 1.5

2.8 $\zeta = 0.5; \omega_n = 10; \omega_d = 5\sqrt{3}; y(t) = 0.01\left[1 - \frac{2}{\sqrt{3}}e^{-5t}\sin\left(5\sqrt{3}\,t + \tan^{-1}\sqrt{3}\right)\right]$

2.9 0.632 sin (2t – 108.4°)

2.10 $\frac{E_0(j\omega)}{E_i(j\omega)} = \frac{1}{j\omega(RC) + 1}$; Bandwidth = 497 Hz

2.11 $\dfrac{E_0(j\omega)}{E_i(j\omega)} = \dfrac{R}{R-\omega^2 RLC + j\omega L}$; Bandwidth = 742 rad/sec

2.12 $G(j\omega) = \dfrac{Y(j\omega)}{F(j\omega)} = \dfrac{1/K}{1-(\omega/\omega_n)^2}$; $\omega_n = \sqrt{K/M}$

2.13 $\dfrac{X(s)}{Y(s)} = \dfrac{0.1667}{(0.0033s+1)(0.0217s+1)}$; $x(\text{peak}) = 1.2113\,\text{cm}$

2.14 (a) Peak amplitude of the force transmitted to the foundation

$$= \frac{A\sqrt{1+\left(\dfrac{B\omega}{K}\right)^2}}{\sqrt{\left(1-\dfrac{M\omega^2}{K}\right)^2+\left(\dfrac{B\omega}{K}\right)^2}}$$

(b) Peak amplitude of machine vibration

$$= \frac{A\sqrt{1+\left(\dfrac{B\gamma}{K}\right)^2}}{\sqrt{\left(1-\dfrac{M\gamma^2}{K}\right)^2+\left(\dfrac{B\gamma}{K}\right)^2}}$$

2.15 $M_1\ddot{y}_1 + K_1(y_1-y_0) + B_1(\dot{y}_1-\dot{y}_0) + K_2(y_1-y_2) + B_2(\dot{y}_1-\dot{y}_2) = 0$
$M_2\ddot{y}_2 + K_2(y_2-y_1) + B_2(\dot{y}_2-\dot{y}_1) = 0$

2.16 $\dfrac{Y(s)}{R(s)} = \dfrac{G_1(G_2G_3+G_4)}{1+(G_2G_3+G_4)(G_1+H_2)+G_1H_1G_2}$

2.17 $\dfrac{Y(s)}{R(s)} = G_4 + \dfrac{G_1G_2G_3}{1+G_2G_3H_2+G_2H_1(1-G_1)}$

2.18 $\dfrac{Y(s)}{R(s)} = \dfrac{(1+G_4H_2)(G_2+G_3)G_1}{1+G_1H_1H_2(G_2+G_3)}$

2.19 $M(s) = \dfrac{G_1G_2G_3}{1+G_2H_3+G_3H_2+G_1G_2G_3H_1}$; $M_w(s) = \dfrac{G_3(1+H_3G_2)}{1+H_3G_2+G_3(G_1H_1G_2+H_2)}$

2.20 $\dfrac{G_1G_2G_3G_4G_5+G_1G_6G_4G_5+G_1G_2G_7(1+G_4H_1)}{1+G_4H_1+G_2G_7H_2+G_6G_4G_5H_2+G_2G_3G_4G_5H_2+G_4H_1G_2G_7H_2}$

3.1 (a) $\dfrac{\theta_L(s)}{\theta_R(s)} = \dfrac{G_1G_2G_3}{1+G_1G_2G_3}$; $\dfrac{\theta_L(s)}{nT_w(s)} = -\dfrac{NG_3}{1+G_1G_2G_3}$; $G = K_pK_AR_f/R$;

$G_2 = \dfrac{K_T/R_a}{Js+B+K_TK_b/R_a}$; $G_3 = n/s$

(b) $\dfrac{\theta_2(s)}{\theta_r(s)} = \dfrac{K_s K_A K_e K_v G_p(s)}{1+K_s K_A K_e K_v G_p(s)}; \dfrac{\theta_2(s)}{\theta_i(s)} = \dfrac{N(s)}{1+K_s K_A K_e K_v G_p(s)}$

3.2 $\dfrac{\omega(s)}{E_r(s)} = \dfrac{50}{s+10.375}; \omega(t) = 481.8(1-e^{-10.375t})$

3.3 $\dfrac{\theta_L(s)}{\theta_R(s)} = \dfrac{1}{s(0.1s+1)(0.2s+1)+1}$

3.4 $\dfrac{\theta_L(s)}{E_c(s)} = \dfrac{1}{s(0.32s+1)}$

3.5 (i) $\dfrac{\theta_M(s)}{E_a(s)} = \dfrac{16.7}{s(s^2+100s+19.2)}$

3.6 $\dfrac{\theta_M(s)}{\theta_R(s)} = \dfrac{13.33}{s^2+3.55s+13.33}$

3.7 (a) $\dfrac{\omega(s)}{E_r(s)} = \dfrac{K}{\dfrac{1}{\omega_n^2}s^2 + \dfrac{2\zeta}{\omega_n}s+1}; K = 17.2, \zeta = 1.066, \omega_n = 61.2$

3.8 (b) $\dfrac{\theta_L(s)}{\theta_R(s)} = \dfrac{K_P K_A K_T G(s)}{1+K_P K_A K_T G(s)}; \dfrac{\theta_L(s)}{-T_w(s)} = \dfrac{N(s)}{1+K_P K_A K_T G(s)}$

$$G(s) = \frac{Bs+K}{s^2\left[J_M J_L s^2 + (J_M+J_L)Bs + (J_M+J_L)K\right]};$$

$$N(s) = \frac{J_M s^2 + Bs + K}{s^2\left[J_M J_L s^2 + (J_M+J_L)Bs + (J_M+J_L)K\right]}$$

3.9 $\dfrac{Y(s)}{R(s)} = \dfrac{2}{s^2+0.02s+2}$

3.10 $\dfrac{Y(s)}{E_r(s)} = \dfrac{G(s)}{1+G(s)H(s)}; G(s) = \dfrac{(K_A K_v K_1 A/K_2)e^{-s\tau_D}}{s(Ms+B+A^2/K_2)}; \tau_D = d/v; H(s) = K_s$

3.11 $G(s) = \dfrac{R/\rho g}{\tau s+1}; \tau = AR/\rho g,; \dfrac{H(s)}{H_r(s)} = \dfrac{G(s)K_v K_e K_a K_s}{1+G(s)K_v K_e K_a K_s}$

3.12 $G(s) = \dfrac{R_t}{\tau s+1}; \tau = R_t C_t; \dfrac{\theta(s)}{E_r(s)} = \dfrac{G(s)K_h R_f R_1}{RR_1 + G(s)K_h K_t R_f R_2}$

3.13 $G(s) = \dfrac{K_g}{\dfrac{1}{\omega_n^2}s^2 + \dfrac{2\zeta}{\omega_n}s+1}; K_g = \dfrac{1}{Q\rho c}, \omega_n = \sqrt{\dfrac{Q}{R_1 C_1 V}}; 2\zeta\omega_n = \dfrac{R_1 C_1 Q\rho c + V\rho c + C_1}{VR_1 C_1 \rho c}$

$$\frac{\theta(s)}{\theta_r(s)} = \frac{G(s)K_hK_AK_s}{1+G(s)K_hK_AK_s}; \frac{\theta(s)}{\theta_i(s)} = \frac{G(s)[(\tau s+1)/K_g]}{1+G(s)K_hK_AK_s}; \tau = R_1C_1$$

4.1 $\left|S_G^M\right| = 0.0289; \left|S_H^M\right| = 1.02$

4.2 $\frac{Y(s)}{R(s)} = M(s) = \frac{5K_1}{s^2(s+1+5K_1)+5K_1+5}; S_{K_1}^M = 0.5$

4.3 $y(t)\,|_{\text{open-loop}} = 20(1-e^{-t}); y(t)\,|_{\text{closed-loop}} = \frac{20}{21}(1-e^{-21t})$

$y'(t)\,|_{\text{open-loop}} = 50(1-e^{-0.4t}); y'(t)\,|_{\text{closed-loop}} = \frac{20}{20.4}(1-e^{-20.4t})$

4.4 $e_{ss}\,|_{\text{open-loop}} = 0; e_{ss}\,|_{\text{closed-loop}} = 0.0099$

$e'_{ss}\,|_{\text{open-loop}} = -0.1; e'_{ss}\,|_{\text{closed-loop}} = 0.009$

4.5 $K = 3.9$

4.6 Case I: $y_{ss} = 0$; Case II: $y_{ss} = 1$

4.7 (a) $y_0(t) = 50(1-e^{-0.1t}); y_{0ss} = 50$ (b) $y_{0ss} = 45; y_{css} = \frac{4995}{101}$

$y_c(t) = \frac{5000}{101}(1-e^{\frac{-101}{10}t}); y_{css} = \frac{5000}{101}$

4.8 (a) $e_r = 1.5$ volts (b) $S_{K_e}^M = \frac{(s+1)(5s+1)}{(s+1)(5s+1)+9.75}; M(s) = \frac{V(s)}{E_r(s)}$ (c) 25%

4.9 (a) $\omega_0(t) = \frac{125}{13}(1-e^{-130t}); \omega_{0ss} = 9.61$ rad / sec (b) $\omega_0(t) = 9.61(1-e^{-5t})$

(c) $\frac{\omega_0(s)}{\omega_r(s)} = M(s)$ with feedback; $\frac{\omega_0(s)}{\omega_r(s)} = G(s)$ without feedback

$\left|S_{K_A}^M\right| = \frac{2s+10}{2s+10+K_AK_g} = S_{\omega_g}^M; S_{K_A}^G = 1 = S_{\omega_g}^G$

4.10 (b) $\left|S_{K_g}^M\right| = 4\times10^{-4}$ (c) $\theta_{Lss} = 0.1$ rad

4.11 (c) $S_R^M = \frac{1}{1+KR}; M(s) = \frac{H(s)}{H_r(s)}$ (d) $e_{ss} = \frac{1}{1+KR}$

4.12 (b) Open-loop case: (i) $S_K^G = 1, G(s) = \frac{\theta(s)}{E_r(s)}$ (ii) $\theta_{ss} = 1$ (iii) $\tau = R_1C$

Closed-loop case: (i) $S_K^M = \frac{R_1Cs+1}{R_1Cs+1+KK_tR_1}, M(s) = \frac{\theta(s)}{E_r(s)}$

(ii) $\theta_{ss} = \dfrac{1}{1+KK_t R_1}$ (iii) $\tau = \dfrac{R_1 C}{1+KK_t R_1}$

(c) $\theta_{ss}|_{\text{open-loop}} = \dfrac{R_1}{R_1+R_2}; R_2 = \dfrac{1}{UA}; \theta_{ss}|_{\text{closed-loop}} = \dfrac{R_1}{R_1+R_2+KK_t R_1 R_2}$

4.13 (a) $\tau|_{\text{without current feedback}} = \dfrac{L_f}{R_f+1}; \tau|_{\text{with current feedback}} = \dfrac{L_f}{R_f+K_A+1}$

4.14 $|S_K^M| = 1.41$ for Fig. P 4.14a; $|S_K^M| = 1$ for Fig. P4.14b

---

5.1 (a) All roots in the left half of the plane
(b) Two roots in the right half of the plane, and the rest in the left half of the plane
(c) Two roots in the right half of the plane, and the rest in the left half of the plane
(d) Two roots on the imaginary axis, and the rest in the left half of the plane
(e) One root in the right half of the plane, and the rest in the left half of the plane
(f) Two roots in the right half of the plane, and the rest in the left half of the plane

5.2 (a) $(s+j3)(s-j3)(s+1+j1)(s+1-j1)=0$ (b) $(s+2)(s+1)(s-1)(s+j5)(s-j5)=0$

(c) $(s+1)(s+2)(s+j1)(s-j1)\left(s+j\sqrt{2}\right)\left(s-j\sqrt{2}\right)=0$

5.3 The largest time constant is equal to 1.0 sec.

5.4 (a) Unstable for all values of K (b) Unstable for all values of K
(c) $K < 14/9$ (d) $K > 0.528$

5.5 $K > 0.528$

5.6 $K < 43.3$

5.7 (i) $K < 45$ (ii) $K < 7.5$ (iii) $K < 10.1$

5.8 (a) (i) $K < 70$ (ii) $K = 70$ (iii) $\Delta(s) = (s+j9.54)(s-j9.54)(s+10)$
(b) (i) $K > 7.5$ (ii) $K = 7.5$ (iii) $\Delta(s) = (s+j1.223)(s-j1.223)(s+5)$
(c) (i) $K < 28.1$ (ii) $K = 28.1$ (iii) $\Delta(s) = (s+j2.75)(s-j2.75)(s+5.7)(s+1.3)$

5.9 (a) System stable for $K < 666.25$. Oscillations of frequency 4.06 rad/sec when $K = 666.25$.
(b) System stable for $23.3 < K < 35.7$. Oscillations of frequency 1.56 rad/sec when $K = 23.3$, and 2.56 rad/sec when $K = 35.7$.

5.10 $K = 2$; $a = 0.75$

5.11 (a) $K < 15$ (b) 7.07 rad/sec for $K = 15$
(c) Two characteristic roots have time constants larger than 1 sec.

5.12 (a) Unstable for all K_c (b) For stability, $T_D > \tau$

5.13 (a) $K < 30$ (b) $\tau < 0.2$ (c) $K < \left(\dfrac{2}{\tau}+10\right)$

5.14 $K_c < (30 - 25/T_I)$

5.15 $K < (7.5 + 15K_t)$

6.1 (b) (i) ω_n = 10 rad/sec; ζ = 0.5; ω_d = 8.66 rad/sec
(ii) t_r = 0.242 sec; t_p = 0.363 sec; M_p = 16.32%; t_s = 0.8 sec
(iii) $K_p = \infty$; K_v = 10; K_a = 0 (iv) e_{ss} = 0; e_{ss} = 0.1; $e_{ss} = \infty$

6.2 (a) K_A = 2.5 (b) e_{ss} = 0.4

6.3 (a) (i) e_{ss} = 0 (ii) e_{ss} = 2 (b) The system of Fig P6.3b is unstable

6.4 e_{ss} = 7/3

6.5 (a) K_A < 23.84 (b) K_A = 11.25

6.6 (i) M_p = 10.68%; t_s (2% tolerance) = 0.4 sec
(ii) 3.69×10^{-3}

6.7 $K = 100, p = 12, e_{ss} = 0.12$

6.8 M = 79.71 kg; B = 182.47 newtons/(m/sec); K = 300 newtons/m

6.9 J = 2.5 kg-m^2; B = 40 newtons-m/(rad/sec); K_T = 1000 newtons-m/rad

6.10 $\omega(t) = 1.5\pi(1 - e^{-0.5t})$ rad/sec; e_{ss} = 5 rpm

6.11 $\theta(t) = \dfrac{\pi}{3}\left[1 - 1.1547e^{-2t}\sin\left(3.464t + \dfrac{\pi}{3}\right)\right]$ rad; M_p = 16.3%

6.12 (b) K_A = 904.5; M_p = 0.85%; t_s = 0.0796 sec (c) e_{ss} = 0; (d) M_p = 0.772%; t_s = 0.08 sec

6.13 K_A = 2.67; fraction of the tachogenerator voltage fedback = 0.024

6.14 (b) K_A = 1010; ζ = 0.355; M_p = 30.33% (c) e_{ss} = 0.01, ζ = 0.6; M_p = 9.5%

6.15 (a) $e_{ss} = 0.01/K_A$ (b) $y_{ss} = 0.1/K_A$ (c) Minimum $e_{ss} = 7.81 \times 10^{-3}$; Minimum y_{ss} = 0.078

6.16 (a) System-type number = 2 (b) K_P > 10; $K_D > 0.1K_P$

6.17 M_p = 20%; t_s = 2 sec.

7.1 (a) $G(s) = \dfrac{0.5}{10s+1}; \tau_{CL} = 10/1.5; e_{ss} = 1/1.5$
(b) $K_P = 7/1.5; e_{ss} = 0.3$

7.2 (i) e_{ss} = 1/3 (ii) e_{ss} = 0 (iii) e_{ss} = 1/3

7.3 (i) Oscillatory (ii) stable (iii) unstable

7.4 (a) K_c = 4; T_D = 0.5

7.5 (b) $0 < K_I < 3$

7.6 (a) K_c = 10; K_D = 0.09 (b) ζ is not an accurate estimate of M_p.

7.7 (a) K_I = 10 (b) K_c = 0.41
(c) $-1 \pm j\,11.14$; ζ = 0.09 (d) ζ approximately represents M_p.

7.8 (a) T_D = 0.216 (c) T_I > 0.0198

7.9 (a) ζ = 0.158 (b) K_t = 0.216; e_{ss} = 0.316 (c) K_t = 0.9; K_A = 10

7.10 (a) $K_1 = 735.39; K_2 = 0.31$ (b) $K_p = \infty; K_v = 114.9; K_a = 0$

7.11 K = 8; K_t = 0.79

7.16 (i) K_c = 0.54; T_I = 3.75 min (ii) K_c = 0.9; T_I = 2.81 min; T_D = 0.45 min

7.17 K = 1; τ = 1.875; τ_D = 1.375 (a) K_c = 1.36 (b) K_c = 1.23; T_I = 4.58 min

7.18 $K_P = \dfrac{R_4}{R_3}; K_D = R_1C$

7.19 $K_P = \frac{R_4}{R_3}; K_I = \frac{1}{R_1C}$

7.20 $K_P = \frac{R_6R_4}{R_5R_3} + \frac{R_1C_1}{R_2C_2}; K_I = \frac{R_4}{R_3R_2C_2}; K_D = \frac{R_6R_1C_1}{R_5}$

7.21 (a) $K_2 = \frac{R_4}{R_3}; K_1 = RC; \tau = R_1C$

(b) $K_c = K_2; \tau = \tau_1 + \frac{K_1}{K_2}; \alpha = \frac{\tau_1}{\tau_1 + \frac{K_1}{K_2}}$

7.22 (a) $K_2 = \frac{R_4}{R_3}; K_1 = R_1RC; \tau = R_1C$

(b) $K_c = \frac{K_2}{1+\frac{K_1}{K_2}}; \tau = \frac{\tau_1}{1+\frac{K_1}{K_2}}; \beta = 1+\frac{K_1}{K_2}$

8.2 (i) Two roots at $s = -1.1835$; Two roots at $s = -2.8165$
(ii) Three roots at $s = -2$ (iii) Three roots at $s = -3$
(iv) Two roots at $s = -1$; Two roots at $s = -1 + j1.22$; Two roots at $s = -1 - j1.22$
(v) Four roots at $s = -1$

8.3 $K = 1.04$; Closed-loop transfer function $= \frac{1.04}{(s+0.33+j0.58)(s+0.33-j0.58)(s+2.33)}$

8.4 $K = 130$; real part of complex pair away from $j\omega$-axis is approximately four times as large as that of the pair near $j\omega$-axis.

8.5 $s_d = -2 + j3.4; K = 65$; Third pole at $s = -4$

8.6 $K = 1.128$; oscillations of $\sqrt{5}$ rad/sec when $K = 30$

8.8 $\zeta = 0.82$; $K = 2.1$

8.10 There exists a circular root locus with center at $\left(\frac{1}{2}, 0\right)$ and the radius equal to $\sqrt{5}/2$.

8.11 $K = 5$; $M_p = 4.3\%$, $t_s = \frac{4}{3}$ sec, $e_{ss} = 11.11\%$
8.12 $K = 14$
8.13 $K = 1.475$
8.14 (a) $K = 10$ (b) $\zeta = 0.4$
8.15 $K = 10.154$; $s_d = -0.75 + j1.3$; Third pole at $s = -4.5$
8.16 $K_t = 0.3427$
8.17 $K_t = 0.449$
8.18 $\alpha = 2.16$; closed-loop poles: $-0.63 \pm j1.09$, -2.75

8.19 $\alpha = 2.8$; closed-loop poles: $-1 \pm j1.732$, -7

8.20 There exists a circular root locus with center at (2, 0) and the radius equal to $\sqrt{12}$.

9.1 Uncompensated system: $K = 7.0$, $K_v = 1.4$, $t_s = 11.6$ sec
Compensated system: $K = 60$, $K_v = 12.0$, $t_s = 12.7$ sec

9.2 $\alpha = 5.3$; $K = 23.2$; third pole at $s = -3.1$

9.3 (c) System is stable for $K > 19.2$; steady-state error to unit-ramp input $= -16/K$.

9.4 $D(s)$ improves transient performance considerably; no effect on steady-state performance

9.5 (c) $K_P = 8$; $K_D = 2$

9.9 $-0.33 \pm j0.58$; $K_v = 0.53$

9.11 $K = 820$; $K_v = 4.1$

9.14 (a) $D(s) = 3 + 2s$; $K_p = 3$

9.15 (a) $K = 56$ (b) $t_s = 3.6$ sec; $K_v = 2.07$

10.2 (a) Unstable (b) Stable

10.3 (i) Yes (ii) No (iii) Yes

10.4 (a) Two (b) Two (c) Stable
(d) Stable (e) Nyquist criterion not applicable (f) Stable

10.5 (i) Stable (ii) Unstable

10.6 (a) Stable for $K < 5/4$ (b) Stable for all K
(c) Unstable for all K (d) Stable for $K < 2.65$

10.7 $K > 0.75$

10.8 $K > 6$

10.9 $K > 2$

10.10 (a) $K = 0.385$ (b) Unstable for all values of K

10.11 $\tau_D = 1.2$ sec

10.12 $\omega_g = 6.2$ rad/sec; $\omega_\phi = 15.9$ rad/sec; $\Phi M = 31.7°$; $GM = 5.5$

10.13 $GM = 0.7$; $\Phi M = 10°$

10.14 (a) 16 rad/sec (b) 10.2 rad/sec (c) 10 rad/sec (d) 26 rad/sec

10.15 (a) $\omega_g = 7.86$ rad/sec; ω_ϕ: phase never reaches – 180°; $\Phi M = 51.8°$; $GM = \infty$
(b) $\omega_g = 6.8$ rad/sec; $\omega_\phi = 10$ rad/sec; $\Phi M = 21.4°$; $GM = 6$ dB
(c) $\omega_g = 9$ rad/sec; ω_ϕ: Phase never reaches – 180°; $\Phi M = 73.4°$; $GM = \infty$
(d) $\omega_g = 9$ rad/sec; $\omega_\phi = 13.65$ rad/sec; $\Phi M = 22°$; $GM = 4.2$ dB
(e) ω_g: gain never reaches 0 dB; $\omega_\phi = 7$ rad/sec; $\Phi M = \infty$; $GM = 20$ dB
(f) $\omega_g = 3.16$ rad/sec; ω_ϕ: phase never reaches – 180°; $\Phi M = -33°$: $GM = -\infty$

10.16 (i) Stable system with $\Phi M = 21°$ and $GM = 8$ dB
(ii) Unstable system with $\Phi M = -30°$, and $GM = -12$ dB

10.17 (a) $K < 400$ (b) $K < 55$ (c) $K < 1.67$

10.18 (a) $GM = 12$ dB; $\Phi M = 33°$ (b) $GM = 2.5$ dB; $\Phi M = 18°$ (c) $\tau_D = 0.078$ sec

10.19 (a) $\dfrac{4(1+s/2)}{s(1+s/10)}$ (b) $\dfrac{5s}{(1+s/2)(1+/10)(1+s/30)}$ (c) $\dfrac{0.355(1+s)(1+0.05s)}{s(1+0.025s)}$

(d) $\dfrac{250}{s(1+0.4s)(1+0.025s)}$ (e) $\dfrac{79.6s^2}{(1+2s)(1+s)(1+0.2s)}$

10.20 (a) $\dfrac{31.623s}{(1+s)\left(\frac{1}{5}s+1\right)\left(\frac{1}{20}s+1\right)}$ (b) $\omega_{g1} = 0.0316$ rad/sec (c) 56.234 rad/sec

10.21 $GM = -2.5$ dB

10.22 $G(s) = \dfrac{5\left(1+\frac{1}{10}s\right)}{s\left(1+\frac{1}{2}s\right)\left[1+\frac{0.6}{50}s+\frac{1}{2500}s^2\right]}$

10.23 $G(s) = \dfrac{0.1}{\left(\frac{1}{3}s+1\right)^2}$

11.4 (a) $K = 475$; $a = 26.2$ (b) $t_s = 0.305$ sec; $\omega_b = 25.1$ rad/sec

11.5 $M_r = 1.11$; $\omega_r = 13.25$ rad/sec; $\omega_b = 24.6$ rad/sec

11.6 $\zeta = 0.01\,\phi_m; K = \dfrac{1}{K_t^2}\left[2\times10^{-4}\phi_m^2\,\tau - K_t \pm 2\times10^{-2}\phi_m\sqrt{\tau}\sqrt{10^{-4}\phi_m^2\tau - K_t}\right]$

11.7 $K = \dfrac{1}{G_m}\left(\dfrac{1}{\tau_1}+\dfrac{1}{\tau_2}\right)$

11.8 (a) $\omega_g = 1$ rad/sec; $\Phi M = 59.2°$ (b) $M_p = 9.48\%$; $t_s = 4.76$ sec

11.9 $\Phi M = 50°$; $GM = 24$ dB; $M_p \cong 18\%$

11.10 (a) $\Phi M = 12°$ (b) Reduce the gain by a factor of 3.5 (c) $M_p \cong 18\%$

11.11 $K = 1.826$; $GM = \infty$

11.12 (a) $GM = 3.88$ dB; $\Phi M = 10°$ (b) Gain should be changed by a factor of 0.156
(c) Gain should be changed by a factor of 0.147

11.13 $K = 582.51$; $t_p = 0.19$ sec

11.14 (a) $GM = 5$ dB, $\Phi M = 30°$, $\omega_g = 0.83$ rad/sec (b) Gain should be reduced by 5 dB
(c) Gain should be increased by 4.8 dB (d) Gain should be reduced by 3.5 dB

11.15 $K = 37.67$

11.16 (a) $M_p = 33\%$ (b) $M_p = 7.3\%$

11.17 (a) $\Phi M = 40°$ (b) $\Phi M = -3°$; Reduce the gain by a factor of 1.78
(c) Increase in steady-state error as the loop gain is reduced

11.18 (a) $M_r = 1.8$; $\omega_b = 3.61$ rad/sec (b) $M_p = 38.6\%$; $t_s = 5.58$ sec

11.19 (b) $M_r = 1.4$; $\omega_r = 4$ rad/sec (c) $M_p = 26.43\%$; $t_s = 2.14$ sec

11.20 (a) $\omega_g = 8.3$ rad/sec, $\omega_\phi = 14.14$ rad/sec
$GM = 8.2$ dB, $\Phi M = 27.7°$

(b) $M_r = 6.7$ dB, $\omega_r = 9$ rad/sec,
$\omega_b = 13.6$ rad/sec

11.21 $GM = 7$ dB; $\Phi M = 17°$; $M_r = 10$ dB; $\omega_r = 2.75$ rad/sec; $\omega_b = 4.2$ rad/sec

11.22 (a) $\Phi M = 12°$ (b) Gain should be reduced to the value 4/3.5
(c) 0.911 rad/sec (d) $M_p = 18\%$; $t_s = 11.93$ sec

11.23 (a) $M_r = 1.4$; $\omega_r = 6.9$ rad/sec (b) $\zeta = 0.39$; $\omega_n = 8.27$ rad/sec

11.24 (a) $K = 0.54$ (b) $\omega_b = 9.2$ rad/sec

11.25 (a) $K = 0.5$ (b) $K = 0.446$ (c) $K = 0.63$, $\omega_r = 0.5$ rad/sec, $\omega_b = 1$ rad/sec
(d) $K = 1.35$

11.26 $GM = 16.5$ dB; $\Phi M = 59°$
Gain should be increased by a factor of 1.75. The gain compensated system has $GM = 11.6$ dB, $\Phi M = 42.5°$

11.27 K should be reduced by a factor of 3.5.; $\Phi M = 42.5°$, $M_r = 1.4$ gives $\zeta = 0.387$, $\Phi M = 42.5°$ gives $\zeta = 0.394$

12.1 (a) $GM = 6$ dB; $\Phi M = 17°$ (b) $K_c = 1.2$

12.2 (a) $\omega_g = 4.08$ rad/sec; $\Phi M = 3.9°$, $GM = 1.6$ dB
(b) $\omega_g = 5$ rad/sec; $\Phi M = 37.6°$, $GM = 18$ dB

12.3 (a) $K = 12$ (b) $\Phi M = 15°$; $\omega_b = 5.5$ rad/sec; $M_r = 12$ dB; $\omega_r = 3.5$ rad/sec

12.16 (a) $\Phi M = 0.63°$ (b) $\Phi M = 9.47°$; no effect on steady-state performance

12.17 $K_v = 4.8$

13.2 $T = 0.5$ sec; $\dfrac{U(z)}{E(z)} = \dfrac{2.2[4.1 - 3.9z^{-1}]}{4.01 - 3.99z^{-1}}$

13.3 (b) $u(k) = \dfrac{1 - \dfrac{aT}{2}}{1 + \dfrac{aT}{2}} u(k-1) + \dfrac{K\left(1 + \dfrac{bT}{2}\right)}{1 + \dfrac{aT}{2}} e(k) - \dfrac{K\left(1 - \dfrac{bT}{2}\right)}{1 + \dfrac{aT}{2}} e(k-1)$

13.4 (a) $K_c = 3.75$; $T_I = 5$ min; $T_D = 0.8$ min
(b) $K_c = 3.63$; $T_I = 5.17$ min; $T_D = 0.83$ min

13.5 $K_c = 2.25$; $T_I = 28.33$ sec

13.6 (a) $y(k+1) + \dfrac{1}{2} y(k) = -r(k+1) + 2r(k)$

$$\frac{Y(z)}{R(z)} = \frac{-z+2}{z + \dfrac{1}{2}}$$

(b) $y(k) = \begin{cases} -1 & ; k = 0 \\ \dfrac{5}{2}\left(-\dfrac{1}{2}\right)^{k-1} & ; k \ge 1 \end{cases}$

(c) $y(k) = \dfrac{2}{3} - \dfrac{5}{3}\left(-\dfrac{1}{2}\right)^k ; k \ge 0$

13.7 $y(k) = 1 + \dfrac{1}{2}\left(\dfrac{1}{\sqrt{2}}\right)^k \left[\sin\dfrac{k\pi}{4} - \cos\dfrac{k\pi}{4}\right]; k \ge 0$

13.8 $y(k) = 3(2)^k - 2; k \geq 0$

13.9 (a) $y(k) = -40\delta(k) + 20\delta(k-1) - 10(0.5)^k + 50(-0.3)^k; k \geq 0$

(b) $y(k) = -16 + (0.56)^k[7.94\sin(0.468k) + 16\cos(0.468k)]; k \geq 0$

13.10 (a) $y(k) = -0.833(0.5)^k - 0.41(-0.3)^k + 0.476(-1)^k + 0.769; k \geq 0$

(b) $y(k) = -10k(0.5)^k + 2.5(0.5)^k - 6.94(0.1)^k + 4.44 : k \geq 0$

13.11 $G(z) = \dfrac{1-e^{-T}}{z-e^{-T}}; y(k) = 1 - e^{-kT}; k \geq 0$

13.12 $\dfrac{Y(s)}{R(z)} = \dfrac{10}{16}\left[\dfrac{z+0.76}{(z-1)(z-0.46)}\right]$

13.13 $G_{h0}G(z) = 0.0288\left[\dfrac{z+0.92}{(z-1)(z-0.7788)}\right]; \dfrac{\theta_L(z)}{\theta_R(z)} = \dfrac{G_{h0}G(z)}{1+G_{h0}G(z)}$

13.14 $\dfrac{Y(z)}{R(z)} = 0.0095\left[\dfrac{z(z+0.92)}{z^3 - 2.81z^2 + 2.65z - 0.819}\right]$

13.15 $y(k) = 1.02(0.795)^k \sin(0.89k); k \geq 0$

13.16 $y(k) = 0.5[1 - (-0.264)^k]\mu(k)$

13.17 $K_p = \infty; K_v = K_1/K_2; K_a = 0.$

13.18 0, 1/3.041, ∞

13.19 Underdamped response with $\zeta = 0.199$ and $\omega_n = 8.93$.

13.21 $0 < K < 2.1$; $K = 0.38$

13.22 (a) $K = 2.3925$ (b) (i) $K = 1.4557$; (ii) $K = 0.9653$

13.23 $0 < K < 0.785$

13.24 (a) $K = 0.88$; 1.33 rad/sec (b) $K = 0.072$; $\tau = 2.3$ sec (c) $K = 0.18$; $\omega_n = 0.644$ rad/sec

13.25 $0 < A < 3.33$

13.26 (a) $D_1(z) = 13.934\left(\dfrac{z-0.8187}{z-0.1595}\right)$ (b) $K_v = 3$

13.27 $D(z) = 1.91\left(\dfrac{z-0.84}{z-0.98}\right)$

14.1 $x_1 = \theta_M, x_2 = \dot{\theta}_M, x_3 =$ motor armature current i_a, $x_4 =$ generator filed current i_f; $y = \theta_L$

$$\mathbf{A} = \begin{bmatrix} 0 & 1 & 0 & 0 \\ 0 & -0.025 & 3 & 0 \\ 0 & -12 & -190 & 1000 \\ 0 & 0 & 0 & -4.2 \end{bmatrix}; \mathbf{b} = \begin{bmatrix} 0 \\ 0 \\ 0 \\ 0.2 \end{bmatrix}; \mathbf{c} = [0.5 \ \ 0 \ \ 0 \ \ 0]$$

14.2 $\mathbf{A} = \begin{bmatrix} 0 & 1 & 0 \\ 0 & -1 & 20 \\ 0 & 0 & -5 \end{bmatrix}; \mathbf{b} = \begin{bmatrix} 0 \\ 0 \\ 2.5 \end{bmatrix}; \mathbf{c} = [1 \ 0 \ 0]$

14.3 $x_1 = \theta_M, x_2 = \dot{\theta}_M, x_3 = i_a$

$$\mathbf{A} = \begin{bmatrix} 0 & 1 & 0 \\ 0 & -0.5 & 19 \\ \dfrac{-k_1}{40} & \dfrac{-(k_2+0.5)}{2} & \dfrac{-21}{2} \end{bmatrix}; \mathbf{b} = \begin{bmatrix} 0 \\ 0 \\ \dfrac{k_1}{2} \end{bmatrix}; \mathbf{c} = \begin{bmatrix} \dfrac{1}{20} & 0 & 0 \end{bmatrix}$$

14.4 $\mathbf{A} = \begin{bmatrix} \dfrac{-B}{J} & \dfrac{K_T}{J} \\ \dfrac{-(k_1 K_t K_c + K_b)}{L_a} & \dfrac{-(R_a + k_2 K_c)}{L_a} \end{bmatrix}; \mathbf{b} = \begin{bmatrix} \dfrac{k_1^0 K_c}{L_a} \end{bmatrix}; \mathbf{c} = [1 \ \ 0]$

14.5 (a) $\bar{\mathbf{A}} = \begin{bmatrix} -11 & 6 \\ -15 & 8 \end{bmatrix}; \bar{\mathbf{b}} = \begin{bmatrix} 1 \\ 2 \end{bmatrix}; \bar{\mathbf{c}} = [2 \ \ -1]$

(b) $= \dfrac{1}{s^2 + 3s + 2}$

14.6 (a) $\mathbf{A} = \begin{bmatrix} 0 & 1 \\ 0 & 0 \end{bmatrix}; \mathbf{b} = \begin{bmatrix} 0 \\ 1 \end{bmatrix}$ (b) $\bar{\mathbf{A}} = \begin{bmatrix} 1 & 1 \\ -1 & -1 \end{bmatrix}; \bar{\mathbf{b}} = \begin{bmatrix} 0 \\ 1 \end{bmatrix}$ (c) $|\lambda \mathbf{I} - \mathbf{A}| = |\lambda \mathbf{I} - \bar{\mathbf{A}}| = \lambda^2$

14.7 $\mathbf{G}(s) = \dfrac{1}{\Delta} \begin{bmatrix} s(s+3) & s+3 & 1 \\ -1 & s(s+3) & s \\ -s & -1 & s^2 \end{bmatrix}; \mathbf{H}(s) = \dfrac{1}{\Delta} \begin{bmatrix} 1 \\ s \\ s^2 \end{bmatrix}; \Delta = s^3 + 3s^2 + 1$

14.9 x_1, x_2, x_3: outputs of integrators starting at the right and proceeding to the left.

$$\mathbf{A} = \begin{bmatrix} 0 & 1 & 0 \\ 0 & -2 & 1 \\ -2 & 1 & -2 \end{bmatrix}; \mathbf{b} = \begin{bmatrix} 0 \\ 0 \\ 1 \end{bmatrix}; \mathbf{c} = [2 \ \ -2 \ \ 1]$$

14.10 (a) $G(s) = \dfrac{s+3}{(s+1)(s+2)}$ (b) $G(s) = \dfrac{1}{(s+1)(s+2)}$

14.11 (a) $\mathbf{A} = \begin{bmatrix} -5 & 0.5 & -3.5 \\ 4 & -5 & 0 \\ 0 & 1 & 0 \end{bmatrix}; \mathbf{b} = \begin{bmatrix} 0 \\ 0 \\ -1 \end{bmatrix}; \mathbf{c} = [0 \ 1 \ 0]$

(b) $G(s) = \dfrac{14}{(s+1)(s+2)(s+7)}$

14.12 (a) x_1 = output of lag $1/(s+2)$; x_2 = output of lag $1/(s+1)$

$$\mathbf{A}=\begin{bmatrix}-2 & 1\\ -1 & -1\end{bmatrix};\mathbf{b}=\begin{bmatrix}0\\ 1\end{bmatrix};\mathbf{c}=[-1 \quad 1];d=1$$

(b) x_1 = output of lag $1/(s+2)$; x_2 = output of lag $1/s$; x_3 = output of lag $1/(s+1)$

$$\mathbf{A}=\begin{bmatrix}-2 & 1 & 1\\ -1 & 0 & 0\\ -1 & 0 & -1\end{bmatrix};\mathbf{b}=\begin{bmatrix}0\\ 1\\ 1\end{bmatrix};\mathbf{c}=[0 \quad 1 \quad 1]$$

14.13 (i) $$\mathbf{\Lambda}=\begin{bmatrix}-1 & 0\\ 0 & -2\end{bmatrix};\mathbf{b}=\begin{bmatrix}1\\ 1\end{bmatrix};\mathbf{c}=[2 \quad -1]$$

(ii) $$\mathbf{A}=\begin{bmatrix}0 & 0 & -2\\ 1 & 0 & -5\\ 0 & 1 & -4\end{bmatrix};\mathbf{b}=\begin{bmatrix}5\\ 0\\ 0\end{bmatrix};\mathbf{c}=[0 \quad 0 \quad 1]$$

(iii) $$\mathbf{A}=\begin{bmatrix}0 & 1 & 0\\ 0 & 0 & 1\\ -6 & -11 & -6\end{bmatrix};\mathbf{b}=\begin{bmatrix}0\\ 0\\ 1\end{bmatrix};\mathbf{c}=[2 \quad 6 \quad 2]$$

14.14 (i) $$\mathbf{A}=\begin{bmatrix}0 & 0 & 0\\ 1 & 0 & -2\\ 0 & 1 & -3\end{bmatrix};\mathbf{b}=\begin{bmatrix}1\\ 1\\ 0\end{bmatrix};\mathbf{c}=[0 \quad 0 \quad 1]$$

(ii) $$\mathbf{A}=\begin{bmatrix}0 & 1 & 0\\ 0 & 0 & 1\\ -6 & -11 & -6\end{bmatrix};\mathbf{b}=\begin{bmatrix}0\\ 0\\ 1\end{bmatrix};\mathbf{c}=[1 \quad 0 \quad 0]$$

(iii) $$\mathbf{\Lambda}=\begin{bmatrix}-1 & 0 & 0\\ 0 & -2 & 0\\ 0 & 0 & -3\end{bmatrix};\mathbf{b}=\begin{bmatrix}1\\ 1\\ 1\end{bmatrix};\mathbf{c}=[-1 \quad 2 \quad 1];d=1$$

14.15 (a) $$\mathbf{A}=\begin{bmatrix}0 & 1 & 0\\ 0 & 0 & 1\\ 0 & -100 & -52\end{bmatrix};\mathbf{b}=\begin{bmatrix}0\\ 0\\ 1\end{bmatrix};\mathbf{c}=[5000 \quad 1000 \quad 0]$$

(b) $$\mathbf{\Lambda}=\begin{bmatrix}0 & 0 & 0\\ 0 & -2 & 0\\ 0 & 0 & -50\end{bmatrix};\mathbf{b}=\begin{bmatrix}50\\ -31.25\\ -18.75\end{bmatrix};\mathbf{c}=[1 \quad 1 \quad 1]$$

14.16 (a) $$\mathbf{\Lambda}=\begin{bmatrix}-1 & 1 & 0\\ 0 & -1 & 0\\ 0 & 0 & -2\end{bmatrix};\mathbf{b}=\begin{bmatrix}0\\ 1\\ 1\end{bmatrix};\mathbf{c}=[1 \quad 1 \quad 1]$$

(b) $y(t)=2.5-2e^{-t}-0.5e^{-2t}-te^{-t}$

14.17 (b) $\frac{X_1(s)}{x_1^0} = G_{11}(s) = \frac{1/2}{s+1} + \frac{1/2}{s+3}; \frac{X_1(s)}{x_2^0} = G_{12}(s) = \frac{1/2}{s+1} + \frac{-1/2}{s+3}$

$\frac{X_2(s)}{x_1^0} = G_{21}(s) = \frac{1/2}{s+1} + \frac{-1/2}{s+3}; \frac{X_2(s)}{x_2^0} = G_{22}(s) = \frac{1/2}{s+1} + \frac{1/2}{s+3}$

$\frac{X_1(s)}{U(s)} = H_1(s) = \frac{1}{s+1}; \frac{X_2(s)}{U(s)} = H_2(s) = \frac{1}{s+1}$

(c) (i) $\mathbf{x}(t) = \frac{1}{2}\begin{bmatrix} e^{-t}(x_1^0 + x_2^0) + e^{-3t}(x_1^0 - x_2^0) \\ e^{-t}(x_1^0 + x_2^0) + e^{-3t}(-x_1^0 + x_2^0) \end{bmatrix}$

(ii) $\mathbf{x}(t) = \begin{bmatrix} 1-e^{-t} \\ 1-e^{-t} \end{bmatrix}$

14.18 (a) Asymptotically stable (b) $y(t) = \frac{1}{2} + 2e^{-t} - \frac{3}{2}e^{-2t}$

14.19 (a) $\mathbf{A} = \begin{bmatrix} -6 & 0.5 \\ 4 & -5 \end{bmatrix}; \mathbf{b} = \begin{bmatrix} 7 \\ 0 \end{bmatrix}; \mathbf{c} = [0 \;\; 1]$ (b) $y(t) = \frac{28}{3}\left[\frac{1}{4}(1-e^{-4t}) - \frac{1}{7}(1-e^{-7t})\right]$

14.20 $\begin{bmatrix} x_1(1) \\ x_2(1) \end{bmatrix} = \begin{bmatrix} 2.7183 - k \\ 2k \end{bmatrix}$ for any $k \neq 0$

14.21 $e^{\mathbf{A}t} = \begin{bmatrix} 2e^{-t} - e^{-2t} & e^{-t} - e^{-2t} \\ 2e^{-2t} - 2e^{-t} & 2e^{-2t} - e^{-t} \end{bmatrix}; \mathbf{A} = \begin{bmatrix} 0 & 1 \\ -2 & -3 \end{bmatrix}$

14.24 (i) Controllable but not observable (ii) Controllable but not observable
(iii) Both controllable and observable (iv) Both controllable and observable

14.25 (i) Observable but not controllable (ii) Controllable but not observable
(iii) Neither controllable nor observable
$A = -1; b = 1; c = 1$

14.26 (i) $G(s) = \frac{1}{s+2}$ (ii) $G(s) = \frac{s+4}{(s+2)(s+3)}$

The state model is not controllable The state model is not observable

14.27 (a) $\lambda_1 = 1; \lambda_2 = -2; \lambda_3 = -1$; unstable

(b) $G(s) = \frac{1}{(s+1)(s+2)}$; stable

14.28 (a) $\mathbf{A} = \begin{bmatrix} 0 & 1 \\ 0 & -1 \end{bmatrix}; \mathbf{b} = \begin{bmatrix} 0 \\ 1 \end{bmatrix}; \mathbf{c} = [10 \;\; 0]$

(b) $\mathbf{A} = \begin{bmatrix} 0 & 1 & 0 \\ 0 & 0 & 1 \\ 0 & -2 & -3 \end{bmatrix}; \mathbf{b} = \begin{bmatrix} 0 \\ 0 \\ 1 \end{bmatrix}; \mathbf{c} = [20 \;\; 10 \;\; 0]$

(c) $\mathbf{A}=\begin{bmatrix}0&0&0\\1&0&-2\\0&1&-3\end{bmatrix};\mathbf{b}=\begin{bmatrix}20\\10\\1\end{bmatrix};\mathbf{c}=\begin{bmatrix}0&0&1\end{bmatrix}$

14.29 $G(z)=\dfrac{2z+2}{z^2-z+\dfrac{1}{2}}$

14.30 $\mathbf{x}(k+1)=\begin{bmatrix}0.4&0.233\\-0.698&-0.0972\end{bmatrix}\mathbf{x}(k)+\begin{bmatrix}0.2\\0.233\end{bmatrix}r(k);\ y(k)=[1\ \ 0]\,\mathbf{x}(k)$

14.31 $\mathbf{F}=\begin{bmatrix}0.696&0.246\\0.123&0.572\end{bmatrix};\mathbf{g}=\begin{bmatrix}-0.021\\0.747\end{bmatrix};\mathbf{c}=[2\ \ -4];d=6$

14.32 Stable with $T=0.4$ sec.

14.33 Both controllable and observable

14.34 (a) $\lambda_1=\dfrac{1}{4},\lambda_2=\dfrac{1}{2}$ (b) $G(z)=\dfrac{1}{z-\dfrac{1}{4}}$ (c) Controllable but not observable

---

15.1 (a) $k_1=74,k_2=25,k_3=3$ (b) $\dot{\hat{\mathbf{x}}}=(\mathbf{A}-\mathbf{mc})\hat{\mathbf{x}}+\mathbf{b}u+\mathbf{m}y$; $\mathbf{m}^T=\begin{bmatrix}3&7&-1\end{bmatrix}$

15.2 (a) $\mathbf{k}=\begin{bmatrix}3&7&-1\end{bmatrix}$; $\dot{\mathbf{x}}=(\mathbf{A}-\mathbf{bk})\mathbf{x}$ (b) $\mathbf{m}^T=\begin{bmatrix}74&25&3\end{bmatrix}$

15.3 (a) $\mathbf{A}=\begin{bmatrix}0&9\\1&0\end{bmatrix};\mathbf{b}=\begin{bmatrix}9\\0\end{bmatrix};\mathbf{c}=\begin{bmatrix}0&1\end{bmatrix}$ (b) $\mathbf{k}=\begin{bmatrix}\frac{2}{3}&3\end{bmatrix}$ (c) $\dot{\hat{\mathbf{x}}}=(\mathbf{A}-\mathbf{mc})\hat{\mathbf{x}}+\mathbf{b}\mathrm{u}+\mathbf{m}y$; $\mathbf{m}^T=\begin{bmatrix}81&12\end{bmatrix}$

15.4 (a) $\mathbf{A}=\begin{bmatrix}0&1\\-\omega_0^2&0\end{bmatrix};\mathbf{b}=\begin{bmatrix}0\\1\end{bmatrix};\mathbf{c}=\begin{bmatrix}1&0\end{bmatrix}$ (b) $k_1=3\omega_0^2;k_2=4\omega_0$ (c) $\dot{\hat{\mathbf{x}}}=(\mathbf{A}-\mathbf{mc})\hat{\mathbf{x}}+\mathbf{b}u+\mathbf{m}y$; $\mathbf{m}^T=\begin{bmatrix}20\omega_0&99\omega_0^2\end{bmatrix}$

15.5 (a) $\mathbf{k}=\begin{bmatrix}29.6&3.6\end{bmatrix}$ (b) $\dot{\hat{\mathbf{x}}}=(\mathbf{A}-\mathbf{mc})\hat{\mathbf{x}}+\mathbf{b}u+\mathbf{m}y$; $\mathbf{m}^T=\begin{bmatrix}16&84.6\end{bmatrix}$

15.6 (a) $\mathbf{A}=\begin{bmatrix}0&1\\0&0\end{bmatrix};\mathbf{b}=\begin{bmatrix}0\\1\end{bmatrix};\mathbf{c}=\begin{bmatrix}1&0\end{bmatrix}$ (b) $\mathbf{k}=\begin{bmatrix}1&\sqrt{2}\end{bmatrix}$ (c) $\dot{\hat{\mathbf{x}}}=(\mathbf{A}-\mathbf{mc})\hat{\mathbf{x}}+\mathbf{b}u+\mathbf{m}y$; $\mathbf{m}^T=\begin{bmatrix}5&25\end{bmatrix}$

15.7 (a) $k_1=4;k_2=3;k_3=1;N=k_1$ (b) $\mathbf{m}^T=\begin{bmatrix}5&7&8\end{bmatrix}$

15.8 (a) $\mathbf{k}=\begin{bmatrix}-1.4&2.4&-1.6\end{bmatrix}$

15.9 (a) $\mathbf{k}=\begin{bmatrix}3&1.5\end{bmatrix}$ (b) For a unit-step input, the steady-state error in the output is 6/7.
(c) $k_1=2;k_2=1.5;k_3=-3.5$

15.10 $k_1=-4;k_2=-3/2;k_3=0$

15.11 $\mathbf{k} = \begin{bmatrix} -0.5 & -0.2 & 1.1 \end{bmatrix}$

$\mathbf{x}(k+1) = (\mathbf{F} - \mathbf{gk})\mathbf{x}(k)$

15.12 (a) $\mathbf{k} = \begin{bmatrix} \frac{111}{76} & -\frac{18}{19} \end{bmatrix}$ (b) $\hat{\mathbf{x}}(k+1) = (\mathbf{F} - \mathbf{mc})\,\hat{\mathbf{x}}(k) + \mathbf{g}u(k) + \mathbf{m}y(k)$

$\mathbf{m}^T = \begin{bmatrix} 8 & -5 \end{bmatrix}$

15.13 (a) $\mathbf{F} = \begin{bmatrix} 1 & 0.1 \\ 0 & 1 \end{bmatrix}; \mathbf{g} = \begin{bmatrix} 0.005 \\ 0.1 \end{bmatrix}$ (b) $k_1 = 13; k_2 = 3.95$

(c) $\hat{\mathbf{x}}(k+1) = (\mathbf{F} - \mathbf{mc})\,\hat{\mathbf{x}}(k) + \mathbf{g}u(k) + \mathbf{m}y(k)$

$\mathbf{m}^T = \begin{bmatrix} 2 & 10 \end{bmatrix}$

15.14 (a) $\mathbf{F} = \begin{bmatrix} 1 & 0.0952 \\ 0 & 0.905 \end{bmatrix}; \mathbf{g} = \begin{bmatrix} 0.00484 \\ 0.0952 \end{bmatrix}; \mathbf{c} = [1 \quad 0]$ (b) $k_1 = 105.1; k_2 = 14.625$

(c) $\hat{\mathbf{x}}(k-1) = (\mathbf{F} - \mathbf{mc})\hat{\mathbf{x}}(k) + \mathbf{g}u(k) + \mathbf{m}y(k)$

$\mathbf{m}^T = [1.9 \quad 8.6]$

15.15 (a) $y(k+1) = 0.368\, y(k) + 0.632\, u(k)$ (b) $K = 0.3687$ (c) $K_1 = 0.553; K_2 = -2.013$

15.16 $\mathbf{k} = \begin{bmatrix} 1 & \sqrt{3} \end{bmatrix}$

15.17 $\mathbf{k} = [1 \quad 1]$

15.18 $K = 0.178$

16.2 $\dfrac{8M}{\pi}$; 1 rad/sec; $y(t) = \dfrac{-8M}{\pi}\sin t$

16.3 0.3; 10 rad/sec

16.4 $\Delta < 0.131$

16.6 4.25; $\sqrt{2}$ rad/sec, stable limit cycle

16.7 3.75; 1 rad/sec

16.8 (a) Stable node; (0, 0) point in $(y, \dot{y})$-plane (b) Stable node; (1, 0) point in $(y, \dot{y})$-plane

(c) Unstable focus; (2, 0) point in $(y, \dot{y})$-plane

16.9 For $\theta = 0$, the singular point is Center for $\theta = \pi$, it is Saddle.

16.10 (i) Singularity (1, 0) in $(y, \dot{y})$-plane is a Center

(ii) Singularity (1, 0) in $(y, \dot{y})$-plane is a stable Focus.

16.11 (a) For $-0.1 < x_1 < 0.1, \ddot{x}_1 + \dot{x}_1 + 7x_1 = 0$

For $x_1 > 0.1, \ddot{x}_1 + \dot{x}_1 + 0.7 = 0$

For $x_1 < -0.1, \ddot{x}_1 + \dot{x}_1 - 0.7 = 0$

(b) Isocline equations:

$x_2 = \dfrac{-7x_1}{m+1}; -0.1 < x_1 < 0.1$

$x_2 = \dfrac{-0.7}{m+1}; x_1 > 0.1$

$x_2 = \dfrac{0.7}{m+1}; x_1 < -0.1$

(c) A singular point at the origin

16.12 (a) For $-0.1 < x_1 < 0.1, \ddot{x}_1 + \dot{x}_1 = 0$

For $x_1 < -0.1, \ddot{x}_1 + \dot{x}_1 + 7(x_1 + 0.1) = 0$

For $x_1 > 0.1, \ddot{x}_1 + \dot{x}_1 + 7(x_1 - 0.1) = 0$

(b) Isocline equations:

$m = -1;\ -0.1 < x_1 < 0.1$

$$x_2 = \frac{-7x_1 - 0.7}{m+1}; x_1 < -0.1$$

$$x_2 = \frac{-7x_1 - 0.7}{m+1}; x_1 < -0.1$$

$$x_2 = \frac{-7x_1 + 0.7}{m+1}; x_1 > 0.1$$

(c) Singular points at $(x_1 = \pm 0.1, x_2 = 0)$

16.13 (a) $\ddot{x}_1 + \dot{x}_1 + 0.7 \operatorname{sgn} x_1 = 0$

(b) Isocline equations:

$$x_2 = \frac{-0.7}{m+1}; x_1 > 0$$

$$x_2 = \frac{0.7}{m+1}; x_1 < 0$$

(c) No singular points

16.14 (a) For $-0.1 < x_1 < 0.1, \ddot{x}_1 + \dot{x}_1 = 0$

For $x_1 > 0.1, \ddot{x}_1 + \dot{x}_1 + 0.7 = 0$

For $x_1 < -0.1, \ddot{x}_1 + \dot{x}_1 - 0.7 = 0$

(b) Isocline equations:

$m = -1; -0.1 < x_1 < 0.1$

$$x_2 = \frac{-0.7}{m+1}; x_1 > 0.1$$

$$x_2 = \frac{0.7}{m+1}; x_1 < -0.1$$

(c) No singular points

16.15 (a) $\ddot{x}_1 + 0.1 \operatorname{sgn} \dot{x}_1 + x_1 = 0$

(b) Isocline equation:

$$x_2 = \frac{-x_1 - 0.1 \operatorname{sgn} x_2}{m}$$

(c) Singualar points at $(x_1 = \mp 0.1, x_2 = 0)$

16.16 Steady-state error to unit-step input = − 0.2 rad
Maximum steady-state error = ± 0.3 rad.

16.17 Deadzone helps to reduce system oscillations, and introduces steady-state error

16.18 Saturation has a slowing down effect on the transient.

16.19 Asymptotically stable

16.20 Unstable

16.21 $|x_1| < 1$; origin is the equilibrium state

16.22 Asymptotically stable; origin is the equilibrium state.

16.23 Origin is asymptotically stable for $x_1^2 + x_2^2 < 1$

16.24 (i) Globally asymptotically stable (ii) Unstable
(iii) Asymptotically stable for $0 < x_2^2 < K_1 / K_2$

INDEX

MATLAB MODULES (www.mhhe.com/gopal/cs4e)

To realize the objective of preparing the student for early productivity upon entering industrial practice, the use of Computer-Aided-Design (CAD) software must be encouraged while teaching control engineering. With this goal in mind, the book has been designed to be MATLAB/ Simulink-friendly. The choice of MATLAB/Simulink software is guided by the consideration that today it is the most widely used CAD software in universities. However, it is not freely distributed software. Alternative software package SCILAB/SCICOS, which is freely distributed, could be used to provide the CAD platform.

The appendix in the book comprises of a set of sixteen MATLAB modules that provide an introduction to the MATLAB and Simulink environment, and give an extensive MATLAB and Simulink support to each chapter of the book. Practice problems with help script files and Simulink files are also included.

Module 1: Window Environment and Base Program

The opening chapter of the book doesn't require the use of MATLAB. A tutorial on MATLAB window environment and some of its basic commands presented in this module, will be helpful for subsequent MATLAB modules.

Module 2: MATLAB Support for Chapter 2

In Module 1, we have presented a tutorial on MATLAB window environment and some of its basic commands. In the present module, our objective is to take a preliminary look at MATLAB Control System Toolbox, which expands MATLAB base program to include control-system specific com¬mands. Transfer function tools are highlighted in this module.

Module 3: MATLAB/Simulink Support for Chapter 3

In Chapter 3, we have developed models for industrial control systems in diverse fields of applications. In the present module, the objective is to create a simulation environment provided by MATLAB's Simulink Toolkit. The Simulink model mimics the block diagram of a feedback control system and is used to evaluate the response of the controlled variable to any test input. It also provides the response of any internal variable of the control system (output variable of a subsystem block).

Module 4: MATLAB/Simulink Support for Chapter 4

In this module, the MATLAB tools described in Modules 1–3, are put to use for the study of feedback control system characteristics.

Module 5: MATLAB Support for Chapter 5

In this module, we reiterate the fact emphasized in the book: the real power of Routh's test lies in determining analytical solutions to stability regions in parametric space. When it comes to qualitative analysis using numerical techniques, root locus analysis is the most powerful tool. For such a numerical study, the reader is advised to refer to Module 8.

Module 6: MATLAB Support for Chapter 6

MATLAB has a powerful GUI tool, called LTI Viewer, for obtaining both the time and frequency response characteristics of a given linear time-invariant system. In this module, we give the basic features of this GUI tool and use it for time response analysis. Later in other modules, this tool will be exploited for frequency response analysis.

Module 7: MATLAB/Simulink Support for Chapter 7

In this module, the MATLAB tools described in earlier modules, are put to use for the study of PID control characteristics.

Module 8: MATLAB Support for Chapter 8

In this module, we demonstrate the use of powerful MATLAB commands for root locus analysis. For a given set of open-loop poles and zeros, MATLAB immediately plots root loci for the gain parameter varying from zero to infinity.

Module 9: MATLAB Support for Chapter 9

In this module, we carry out root locus design using MATLAB dialogs. The MATLAB SISO Design Tool—a powerful GUI tool developed on the dialog pattern, will also be described here.

Module 10: MATLAB Support for Chapter 10

In this module, we demonstrate the use of powerful MATLAB commands for sketching Bode and Nyquist plots, and there from relative stability analysis.

Module 11: MATLAB Support for Chapter 11

The basic features of the MATLAB GUI tool—LTI Viewer, and its use for time response analysis have already been given in Module 6. In the present module, we demonstrate the use of LTI Viewer for frequency response analysis.

Module 12: MATLAB Support for Chapter 12

In this module, we carry out frequency response design using MATLAB dialogs. The MATLAB SISO Design tool, described earlier in Module 9, is also exploited for frequency response design.

Module 13: MATLAB/Simulink Support for Chapter 13

Many of the functions for continuous-time systems covered in Modules 1–12, have equivalent coun¬terparts for sampled-data systems. A ready reference source along with application to sampled-data systems is available in the present module.

Module 14: MATLAB Support for Chapter 14

State variable formulation is an efficient form of representation from the standpoint of computer simulation. MATLAB commands for state space analysis are illustrated in this module.

Module 15: MATLAB Support for Chapter 15

Important MATLAB commands for design of state-feedback control by pole-placement, optimal control design, and design of state observers are illustrated in this module.

Module 16: MATLAB/Simulink Support for Chapter 16

In this module, the emphasis is on simulation tool for stability analysis of nonlinear systems.